How Things Work

To my children, Elana and Aaron, who have waited so patiently for me to finish this book and who are always so much fun, particularly in my class on snow days.

To my wife, Karen, whose support, encouragement, and friendship have sustained me and who has quietly done far more than her share for such a long time.

To my parents, Frances and Daniel, whose confidence and interest in whatever I do have never wavered.

And to the students of the University of Virginia, whose excitement and enthusiasm make teaching worthwhile.

How Things Work

The Physics of Everyday Life

Louis A. Bloomfield
The University of Virginia

John Wiley & Sons, Inc.
New York Chichester Brisbane Toronto Singapore

ACQUISITIONS EDITORS	Stuart Johnson, Clifford Mills
DEVELOPMENTAL EDITOR	Barbara Heaney
MARKETING MANAGER	Ethan Goodman
SENIOR PRODUCTION EDITOR	Katharine Rubin
DESIGNER	Madelyn Lesure
MANUFACTURING MANAGER	Mark Cirillo
PHOTO EDITOR	Hilary Newman
PHOTO RESEARCHER	Ramon Rivera-Moret
ILLUSTRATION EDITOR	Edward Starr
COVER PHOTOS	Top photo: © 1992 David Madison Center photo: John Lamb/Tony Stone Worldwide Bottom photo: © H. Martin/Sharpshooters

This book was set in 10 pt Palatino, and printed and bound by Courier/Stoughton. The cover was printed by Phoenix Color.

Recognizing the important of preserving what has been written, it is a policy of John Wiley & Sons, Inc. to have books of enduring value published in the United States printed on acid-free paper, and we exert our best efforts to that end.

The paper in this book was manufactured by a mill whose forest management programs include sustained yield harvesting of its timberlands. Sustained yield harvesting principles ensure that the number of trees cut each year does not exceed the amount of new growth.

Library of Congress Cataloging in Publication Data:

Bloomfield, Louis.
How things work : the physics of everyday life / Louis A. Bloomfield.
p. cm.
Includes index.
ISBN 0-471-59473-3 (pbk. : alk. paper)
1. Physics. I. Title.
QC21.2.B59 1996
530—dc20 96-14288
CIP

Printed in the United States of America

10 9 8 7 6 5 4

CONTENTS

PREFACE

This book offers a non-conventional view of physics and science that starts with whole objects and looks inside them to see what makes them work. It is written for liberal arts students who are seeking a connection between science and the world they live in. While these students may be reluctant to study science merely as an intellectual exercise, they show remarkable enthusiasm for it when it appears in a useful context. Many of the almost 3,000 students I have taught during the past five years have been surprised at their own interest in the course, looking forward to classes, asking questions, experimenting on their own, and explaining to friends and family how things they use, and care about, work.

The idea for this course occurred to me while working with a committee to assess the undergraduate curriculum in our College of Arts and Sciences. While the committee was eager to have all of our students study science, course requirements seemed the only way to achieve that aim. But there is another way—bring science to the students rather than the reverse. This book, like the course on which it's based, has always been for non-scientists and is written with their interests in mind. Nonetheless, it has attracted students from the sciences, engineering, architecture, and other technical fields who are eager to put familiar scientific concepts into context in their world.

While this book starts with objects and looks within them for scientific principles, most physics texts instead choose to develop the principles of physics first and to delay the search for real-life examples of these principles until later. That teaching method, starting with principles and then finding examples, provides few conceptual footholds for the students as they struggle to understand the principles. After all, the comforts of experience and intuition lie in the examples, not the principles. While a methodical and logical development of scientific principles can be very satisfying to the seasoned physicist, it can appear alien to an individual who isn't even familiar with the language being used.

No matter how basic its mathematics, a course following the "physics for physicists" paradigm can easily become scary and dull for people inexperienced at thinking scientifically. That traditional course structure also neglects the pattern in which much of science developed: people came to understand many of the natural laws by watching and studying the objects around them.

This book attempts to convey an understanding and appreciation for the concepts and principles of physics and science by finding them within specific objects of everyday experience. Many of these objects are interesting in their own rights and their appeal helps to motivate studies of the scientific principles that govern our universe. Because its structure is defined by real-life examples, this book necessarily discusses various physical concepts as they're needed. To show the universality of the natural laws, this book is explicit about reusing principles discussed in earlier chapters. The frequent revisiting of concepts is both a way to show how physical laws connect all objects and a useful technique for learning the laws themselves.

The Goals of this Book

As they read this book, students should:

1. Begin to see science in everyday life. This goal is the central focus of the book. Science is all around us; we only need to keep our eyes open to see it. While this book looks at many objects as it surveys the world, it doesn't examine every object that exists nor does it explore every concept in those it considers. Instead, the students must learn to generalize the ideas discussed in this book so that when they look at something new, they'll be able to figure out how it works. They're surrounded by things that can be understood at various levels and it's within their reaches to see science in each of these things. That doesn't mean that they should look at an oil painting and see only preferential absorption of incident light by organic and inorganic molecules. Rather, they should realize that there is a beauty to science that can supplement aesthetic beauty. They can learn to look at a glorious red sunset and appreciate both its appearance *and* why it exists.

2. Learn that science isn't frightening. The increased technological complexity of our environment has done more than just stop people from looking for science in the world around them. It has instilled within them a significant fear of science. A generation or two ago, many students grew up tinkering with gadgets, taking clocks apart, building amateur radios, and "souping" up their automobiles. But complexity and miniaturization have ended much of that activity. Inside a modern clock, there's just a battery, a small circuit board, and some wires. Amateur radio has been consumed by CB radios and cellular telephones that are prewired and virtually unrepairable when they break. And modern cars are filled with sensors, computers, and actuators that make it nearly impossible for tinkerers to perform useful modifications.

To make matters worse, a significant fraction of the time once spent studying physical science in schools may now be devoted to the study of computers. Computer science is an interesting and worthy field and is certain to have a tremendous impact on citizens of the twenty-first century. However, physical science and computer science are sufficiently different that they need to be studied separately. One is not a substitute for the other.

Thus one of the goals of this book is to remind students that science can be understood and that it need not be so scary. However, because science requires thought, students who are afraid to think will continue to fear it. Physics, in particular, is an ordered system built on rules, not individual facts. Since straight memorization won't help students see that order, they must learn the rules and begin to think about them.

3. Learn to think logically in order to solve problems. Because the universe obeys a system of well-defined rules, it permits a logical understanding of its behaviors. Like mathematics and computer science, physics offers a field of study where logic reigns supreme. Having learned two or three simple rules, students can combine them logically to obtain more complicated rules and be certain that those new rules are true. So the study of physical systems is a good place to practice logical thinking.

The applications of logic to everyday situations don't stop with the physical sciences. In virtually every field of human activity, logic provides a useful way to solve problems. Faced with new situations, new problems, students can often combine knowledge and logic to find new solutions.

As the world becomes more technically complicated and jobs become more specialized, people find themselves more inclined to throw up their hands and take every unfamiliar problem to a specialist. However, the majority of problems students will encounter are fairly straightforward so that, with a little courage and a sense of adventure, they can solve those problems themselves. A student who pays the plumber every time a washer wears out in a sink or throws away the espresso maker when its heating element fails had better plan on earning a lot of extra money.

4. Develop and expand their physical intuitions. The natural laws that this book examines are the essential tools of physicists but are of limited direct use to the average person. Most people know that when they slam on their brakes at 60 miles per hour, they may be thrown through the windshield if they're not held in by a seat belt. They don't have to consider velocity, acceleration, and inertia to determine that they should brake gradually—they already have physical intuition that tells them the consequences of such actions. Within the range of everyday experience, this physical intuition may be all most people ever need. But they regularly come to the edge of the domain in which they have good intuition and may prefer not to cross into new territory. When the faucet starts to leak, most people call the plumber because they lack the physical intuition necessary to do the job themselves.

If they don't have to work with some object or physical process and aren't interested in experimenting with it, students will never develop any physical intuition about it. Those of them who never tinker with cars may not know how hard they can turn a wrench before they'll shear the head off of a bolt or that shock-absorbers damp out oscillations of the car body following a sudden impulse from the road. Those of them who never cook may not know which pots transmit heat best or how hard it is to get rid of the excess salt once they've added too much. Each of these observations represents physical intuition and could be generalized and distilled into basic physical laws. For most students, the physical intuition will be far more useful and memorable than the fundamental laws behind it. This book is intended to broaden their physical intuition to situations that they regularly avoid and ones they've never even encountered. That is, after all, one of the purposes of reading books and studying other peoples' scholarship: to learn from other peoples' experiences.

5. Learn how things work. This book is about objects from everyday life. Each of its sections deals with a specific object and with one or more of the mechanisms that permit that object to work. Some of these mechanisms are obvious and intuitive, such as wheels on a bicycle. In those cases, this book supplements intuition with systematic understanding. Other mechanisms are more foreign to the students' experiences; for example, the operation of a magnetron in a microwave oven. In such cases, this book attempts to convey a basic understanding.

As this book explores each object, it exposes bits and pieces of the overall physical laws that govern the universe. After all, that's how the laws of physics have generally been found. While each of the objects in this book may only touch on one or two physical concepts, as a whole, the material in this book presents many or most of the laws of physics. The students should begin to see the similarities between objects, recurring themes that have been reused by nature or by humans. The book makes an effort to remind them of these connections and is ordered so that later objects enhance their understandings of concepts encountered earlier.

6. Begin to understand that the universe is predictable rather than magical. One of the most fundamental principles of science, not described by equations or complex diagrams, is the notion that every effect has its causes. Things don't just occur willy-nilly. Whatever happens, students can look backward in time to find its real causes. They can also predict the future based on insight acquired from the past and a knowledge of the present. Thus students who don't change the oil in their cars can anticipate engine trouble before 100,000 miles and those that throw rocks straight up in the air can plan on wearing helmets. What distinguishes the physical sciences and mathematics from other fields is that there are often absolute answers, free from inconsistency, contradiction, or paradox.

Once students begin to understand how the physical laws govern the universe, they can start to appreciate how orderly it is. From a practical point of view, they can replace a sense of magic at seeing a certain behavior with a sense of structure and understanding. They can begin to think "physically," at least on occasion. Of course, they may still feel the magic aesthetically, but to actually believe in magic is not likely to help them function in the real world.

7. Obtain a perspective on the history of science and technology. None of the objects that this book examines appeared suddenly and spontaneously in the workshop of a single individual who was oblivious to what had been done before. These objects were developed in the context of history by people who were fairly aware of what they were doing and usually familiar with any similar objects that already existed.

Both science and technology have evolved slowly and steadily since the dawn of humankind. There have been slow times, and there have been times of fairly rapid change. Only rare geniuses have succeeded in making the huge leaps of understanding that push science ahead dramatically in a single step. Nearly everything else is discovered or developed when related activities make their discoveries or developments inevitable and timely. To establish that historical context, this book describes some of the history behind the objects it discusses.

Artwork

Because this book is about real things, its illustrations and photographs are about real things, too. Whenever possible, artwork is built around familiar objects so that the concepts the artwork is meant to convey become associated with objects the students already know. Many students are visual learners—if they see it, they can learn it. By superimposing the abstract concepts of physics onto simple realistic artwork, this book attempts to connect physics with everyday life. This idea is particularly evident at the opening of each section, where the object examined in that section appears in a carefully rendered drawing. This drawing provides the students with something concrete to keep in mind as they encounter the more abstract physical concepts that appear in that section. By lowering the boundaries between what the students see in the book and what they see in their environment, the artwork of this book makes science a part of the students' world.

Features

This book contains 51 sections, each of which discusses how a particular thing works. The sections are grouped together in 19 chapters according to major physical themes they develop. In addition to the discussion itself, the sections and chapters include a number of features intended to strengthen the educational value of this book. Among these features are:

- **Chapter introductions, experiments, and itineraries.** Each of the 19 chapters begins with a brief introduction to the principal physical theme underlying that chapter. It then presents an experiment that students can do with household items to observe first hand some of the issues associated with that physical theme. Lastly, it presents a general itinerary for the chapter, identifying some of the physical issues that will come up as the objects in the chapter are discussed.

- **Section introductions, questions, and experiments.** Each of the 51 sections explains how something works.

Often that something is a specific object, but it can be more general. A section begins by introducing the object and then asks a number of questions about it; questions that might occur to students as they think about the object and that are answered by the section. Lastly, it suggests some simple experiments that the students can do to observe some of the physical concepts that are involved the object.

• **Check your understandings and check your figures.** Sections are divided into a number of brief segments, each of which ends with a "Check Your Understanding" question. These questions apply the physics of the preceding segment to new situations and are answered and explained in the summary material at the end of the section. Segments that introduce important equations also end with a "Check Your Figures" question. These questions show how the equations can be applied and are also answered and explained in the summary material.

• **Section summary, laws and equations, physics concepts, and glossary.** Each section ends with a brief summary of how the object works. This summary is followed by a restatement of the important laws and equations in the section, an outline of the principal physical concepts discussed in the section, and a glossary of key physics terms. These key physics terms are also marked in bold when they first appear and are defined in the text itself.

• **Section review questions, exercises, and problems.** Following the summary material at the end of each section is a collection of questions dealing with the physics concepts in that section. Review questions revisit material from the section itself to help students assess their grasp of the key physics concepts in that section. Exercises ask the students to apply those concepts to new situations. Problems ask the students to apply the equations in that section and to obtain quantitative results.

• **Chapter epilogue and explanation of experiment.** Each chapter ends with an epilogue that reminds the students of how the objects they studied in that chapter fit within the chapter's physical theme. Following the epilogue is a brief explanation of the experiment suggested at the beginning of the chapter, using physical concepts that were developed in the intervening sections.

• **Chapter cases.** The final items in each chapter are cases, extended exercises that apply the physical concepts from that and previous chapters to new circumstances. Each case asks at least four questions and leads the student through an examination of the physical principles involved in a new object or situation. In addition to tying the chapters together, these cases introduce many interesting objects, worthy of sections of their own. By studying these cases, students learn to explain how new objects work.

• **Three-way approach to the equations of physics.** The laws and equations of physics are the groundwork upon which everything else is built. But because each student responds differently to the equations, this book presents them carefully and in context. Rather than making one size fit all, these equations are presented in three different forms. The first is a word equation, identifying each physical quantity by name to avoid any ambiguities. The second is a symbolic equation, using the standard format and notation. The third is a sentence that conveys the meaning of the equation in simple terms and often by example. Each student is likely to find one of these three forms more comfortable, meaningful, and memorable than the others.

• **Historical, technical, and biographical asides.** To show how issues discussed in this book fit into the real world of people, places, and things, a number of brief asides have been placed in the margins of the text. An appropriate time at which to read a particular aside is marked in the text by a ❐.

Organization

The 51 sections that make up this book are ordered so that they follow a familiar path through physics: mechanics, heat, resonance and waves, electricity and magnetism, light and optics, solid state, and modern physics. Because there are too many topics here to cover in a single semester, the book is designed to allow short cuts through the material. In general, the final sections in each chapter and the final chapters in each of the main groups mentioned above can be omitted without serious impact on the material that follows. The only exception to that rule is the first chapter, which should be covered in its entirety as the introduction to any course taught from this book. The book also divides neatly in half so that it can be used for two independent one-semester courses—the first covering Chapters 1-10 and the second covering Chapters 1 and 11-19. A detailed guide to short cuts appears in the instructor's manual.

Supplements

The instructor's manual for this book contains (1) organizational ideas for designing a course using this book, (2) additional information about the object in each section, including aspects not covered in the section, anticipated student questions with answers, and a list of outside sources, (3) demonstration ideas for each section, and (4) supplementary sections on objects not included in the text.

Acknowledgments

There are a great many people who have contributed to this book over the years and to whom I am enormously grateful. First among these are my editors Barbara Heaney, Cliff Mills, and Stuart Johnson, whose steadfast belief in the concept of this book have carried it steadily forward along the difficult path to completion. I am also grateful to Hilary Newman and Ramon Rivera-Monet for finding such interesting photographs and to Ed Starr for coordinating the illustrations program so effectively. Maddy Lesure has done a wonderful job designing this book so that its central features, the objects themselves, shine though. Thanks to Katy Rubin for coordinating the gradual transition from development to production so masterfully and to Ethan Goodman and Catherine Faduska for all their help in learning what instructors and students want from this book. Thanks also to Ishaya Monokoff for his help with illustrations and to Cathy Donovan and Monica Stipanov for helping to keep everything flowing smoothly.

It has been a great pleasure to have my friend Steve Farmer help me with writing style for this book and I am delighted to have had my graduate student Jeff Emmert critique the problems. Thanks also to my family, Karen, Elana, and Aaron Bloomfield, for helping me with everything from questions about weight to photographs of whirling wine glasses.

I have also enjoyed invaluable assistance from a number of colleagues here and elsewhere. I am particularly grateful to those who have already used the manuscript in their classes, including Gordon Donaldson at the University of Strathclyde, William McNairy at Virginia Military Institute, John Shearin at Nansemond-Suffolk Academy, Larry Hunter at Amherst College, Robert Welsh at the College of William and Mary, and Bascom Deaver at the University of Virginia. Their enthusiasm for the concept of this course and their constructive comments about the manuscript have been extremely helpful to me along the way.

Because this book is about real objects, the course I teach from it is enriched by countless demonstrations. I could not have done those demonstrations, nor taken many of the photographs for this book, without the vast experiences and enormous enthusiasms of John Malone and Betsy Noble. Together with several talented student assistants, they make as fine a lecture-demonstration group as one could ever want.

I am also extremely grateful to the students of the University of Virginia for being so eager and interested, and also so tolerant of the evolving nature of the manuscript that has served as their text until now. Many of these students have contributed directly or indirectly to the book itself, particularly Dina Alvarez, Jennifer Annis, Faith Andrews Bedford, Sarah Friend, Darius Johnson, Evan Macbeth, and Haley Whitlock.

Throughout the time I have been working on this book, I have enjoyed the support and insights of many people at the University of Virginia. Among them are Marva Barnett, Alan Dorsey, Paul Fishbane, Michael Fowler, Tom Gallagher, Suzie Garrett, Julia Hsu, Dan Larson, Jim McDaniel, Jude Reagan, Steve Schnatterly, Michael Smith, and Steve Thornton.

Lastly, this book has benefited more than most from the constructive criticism of a number of talented reviewers. In addition to getting a better sense of how to present the material in this book, I have learned a great deal of physics from their reviews. My sincere thanks to all of these fine people:

Charles Ardary
Edmonds Community College

Ali Badakhshan
University of Northern Iowa

Keith Bonin
Wake Forest University

Edward R. Borchardt
Mankato State University

Arthur J. Braundmeier
Southern Illinois University

David Buckley
East Stroudsburg University

James J. Donaghy
Washington & Lee University

Gordon Donaldson
University of Strathclyde, Scotland

Bob Hallock
University Massachusetts-Amherst

Glenn M. Julian
Miami University

Mary Lu Larsen
Towson State University

David Markowitz
University of Connecticut

David C. McKenna
Rensselear Polytechnic Institute

William McNairy
Virginia Military Institute

Adam Niculescu
Virginia Commonwealth University

John C. Raich
Colorado State University

Paul Schuyler
Spokane Community College

Romeo A. Segnan
American University

Stanley J. Sobolewski
Indiana University of Pennsylvania

Robert Tremblay
Southern Connecticut State University

Tim Usher
California State University-San Bernardino

David Wagner
Edinboro University of Pennsylvania

John Yelton
University of Florida

I have always felt that the real test of this book, and of any course taught from it, is its impact on students' lives long after final grades have been recorded. It is my sincere hope that many of these students will find themselves looking at objects in the world around them years later with understanding and insight that they would not have had were it not for their encounter with this book.

Louis A. Bloomfield
Charlottesville, Virginia
lab3e@virginia.edu

TO THE STUDENT:

For many students, physics has been their downfall in high school or college. Up until this point, a student who has had a physics class probably recalls memorizing constants and equations in order to study for a test. I know that that is how I thought of physics until January of 1994. It was then, in my second semester at the University of Virginia, that I took a class entitled *How Things Work* (Physics 106) with Professor Louis Bloomfield. In that class, I learned for the first time that physics could make sense, that it was so much more than equations and variables. Physics is not math, although it makes use of math. Physics is an art of looking at the world, and things in it, explaining the processes by which it works. And that is the entire point of the book you hold here, to explain how the things around us work.

How Things Work approaches physics backwards to help the student better understand physics. The basic premise of this book is to show the student how common items around us function and only then to discuss the physics that underlies them. It is not a matter of "Which Law of Motion is in effect here?" but rather, "Why is it harder to start a ball rolling than to keep it rolling?" Since we were children, we have been asking questions like "why does a bird fly?" It is these why questions that *How Things Work* answers.

Yes, *How Things Work* does make use of principles, laws and math, but it does so in an easily understandable way. Professor Bloomfield discusses the physical principles in terms of physical objects. When he discusses Newton's second law of motion, he does so by talking about a falling ball. By the time he tells you that the law is in effect, you will have realized the principle for yourself, without ever hearing of Newton. The approach used in this book is common sense physics. By the time a principle, law or equation is introduced, the reader will have understood it to be common sense.

This *How Things Work* text is a final product development of this "backwards" approach to physics. There is far more here than simply a unique teaching method. There are also many aids throughout the book that will help you to study and understand. The chapter summaries will help you to review the physics in each chapter. If you find something unclear upon your review, it is a simple matter to go back into the chapter for the practical explanation of that piece of physics. The start of every chapter asks a series of questions, "why" this and "why" that. Every single why question will be answered in the chapter, and the questions will help you to better understand the physics. *How Things Work* is a book written for students, not for physicists. This is a book written for **you**.

How Things Work essays to make physics common sense, but it cannot do so without your cooperation as the student. This will involve work. When I took the course, I did well, even though I had had difficulties with physics in high school. A friend of mine got a 5 on the physics AP test in high school, and had real difficulty with *How Things Work*. The difference? Work. I did the homework; he did not. It sounds silly and obvious, but doing the homework and reading will help you significantly in understanding this material, and through it, the world around you.

How Things Work is real physics, but it successfully disguises itself as common sense to the point where anyone, physics major or not, can do physics without realizing it. *How Things Work* is interesting. It will allow you to understand and explain the functioning of the human-made world around you. Finally, *How Things Work* is useful. I successfully fixed a photocopier at an office where I worked thanks to this class. It was not a major repair, just a minor problem, but I knew where to look, and I knew how the photocopier was supposed to work. I sincerely hope that you will enjoy this book and class, and carry what you learn here with you through your life. I know that I do, and it has served me well.

Go Hoos!

Evan Macbeth

Evan Macbeth
The University of Virginia '97

CHAPTER 1

THE LAWS OF MOTION

This book aims to broaden your perspectives on familiar objects and situations by helping you understand the physical processes that make them work. While science is part of our daily existence—not some special activity we do only occasionally, if at all—most of us ignore it or take it for granted. In this book we'll counter that tendency by seeking out science in the world around us, in the objects we encounter every day. We'll see that seemingly magical objects and effects are really very straightforward once we know a few of the physical concepts that make them possible. In short, we'll learn about *physics*—the study of the material world and the rules that govern its behavior.

To help us get started, this first chapter will do two main things: introduce some of the language of physics, which we'll be using throughout the book, and present the basic laws of motion, upon which everything else will rest. In later chapters, we'll explore objects that are interesting and important, both in their own right and because of the scientific issues they raise. Most of these objects, as we'll see, involve many different aspects of physics, and thus they bring variety to each section and chapter. But this first chapter is special, because it must provide an orderly introduction to the discipline of physics itself.

Experiment: Pulling a Tablecloth from a Table

One famous effect is the feat of removing a tablecloth from a table without breaking the dishes on top. The person performing this stunt pulls the tablecloth out from under the place settings with a quick snap of the wrist. With a little luck—not to mention a smooth, slippery tablecloth and very rapid hand move-

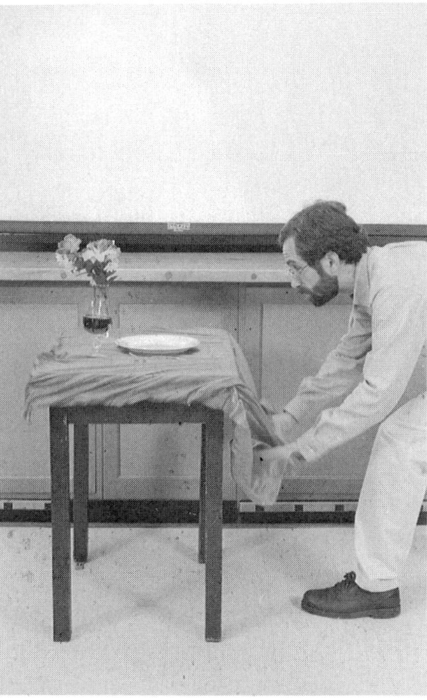

ment—the tablecloth slides off the table abruptly, leaving behind the objects that were resting on it.

With a little practice, you too can do this stunt. Choose an unhemmed tablecloth, one with no projections that might catch on the dishes; try also to find a soft, slippery material, such as silk, so that you can pull the cloth downward and over the edge of the table. When you finally get up the nerve to try this stunt—with unbreakable dishes, of course—make sure you pull as abruptly as possible, keeping the time you spend moving the cloth out from under the dishes to an absolute minimum. It helps to hold the cloth with your palms downward and to let the cloth hang loosely between each hand and the table so that you can get your hands moving before the cloth snaps taut and begins to slide off the table. Don't make the mistake of starting slowly, or the dishes will end up all over the floor.

Before you perform this stunt, try to **predict** what will happen when you pull the cloth. How far will the dishes move—or will they not move at all? How important is the speed with which you pull the cloth? How does the weight of each dish affect its movement or lack of movement? Is the surface texture of each dish important? How would rubbing each dish with wax paper alter the results? Write down your predictions.

Now give the tablecloth a pull and **observe** what happens. Hopefully, the table will remain set. If it doesn't, try again, but this time vary the tablecloth's speed or the types of dishes or the way you pull the cloth. See if you can **measure** the effects of these variations on the dishes. Does everything work as you expected? Do the results **verify** your predictions, or were those predictions wrong?

If you don't have a suitable tablecloth, or any dishes you care to risk, there are many similar experiments you can try. Put several coins on a sheet of paper and whisk that sheet out from under them. Or stack several books on a table and use a stiff ruler to knock out the bottom one. Most impressive of all is to balance a short eraser-less pencil on top of a wooden needlepoint ring that is itself balanced on the open mouth of a glass bottle. If you yank the ring away quickly enough, the pencil will be left behind and will drop right into the bottle.

We'll return to the tablecloth stunt at the end of this chapter. In the meantime, we'll explore some of the physics concepts that help explain why your stunt worked—or, if it didn't, why you now have a floor full of dishes.

Chapter Itinerary

To examine these concepts, we'll look carefully at four kinds of everyday objects: (1) *falling balls*, (2) *ramps*, (3) *seesaws*, and (4) *wheels*. In *falling balls*, we'll see how objects move and how that movement can be influenced by gravity. In *ramps*, we'll explore mechanical advantage and how gradual inclines make it possible to lift heavy objects without pushing very hard. In *seesaws*, we'll look at rotation and how two children manage to rock a seesaw back and forth. Finally, in *wheels*, we'll examine how friction affects motion and why wheels make a vehicle more mobile.

These objects may seem mundane, but understanding them in terms of basic physical laws requires considerable thought. This introductory chapter will be like climbing up the edge of a high plateau: the ascent won't be easy, and our destination will be hidden from view. But once we arrive at the top, with the language and basic concepts of physics in place, we'll be able to explain a broad variety of objects with only a small amount of additional effort. And so we begin the ascent.

Section 1.1

Falling Balls

We've all dropped balls from our hands or seen them arc gracefully through the air after being thrown. These motions seem like simplicity itself, and in fact they're so predictable that they're governed by a few universal rules. These basic laws of motion were formulated by Sir Isaac Newton three centuries ago, but they remain the foundations of science today. Indeed, Newton's laws are so important that we must learn them before we can understand how most objects work. Like Newton, who reportedly began his investigations after seeing an apple fall from a tree, we'll start simply and find the laws of motion in falling objects.

Questions To Think About: *What do we mean by "movement"? What makes a ball move and, once it's moving, what keeps it moving? What does it take to stop a moving ball? What do we mean by "falling," and why do balls fall?*

Experiments To Do: *A few seconds with a baseball will help you see some of the behaviors that we'll be exploring. Toss the ball into the air to various heights, catching it in your hand as it returns. Work with a classmate to help you time the flight of the ball. As you toss the ball higher, how much more time does it spend in the air? How does it feel coming back to your hands? Is there any difference in the impact it makes? Which takes the ball longer: rising from your hand to its peak height or returning from its peak height back to your hand?*

Now drop two different balls—a baseball, say, and a golf ball. If you drop them simultaneously, without pushing either one up or down, does one ball strike the ground first, or do they arrive together? Now throw one ball horizontally while dropping the second. If they both leave your hands at the same time and the first one's initial motion is truly horizontal, which one reaches the ground first?

The Motion of An Isolated Ball

Fig. 1.1.1 - A ball isolated in space. If it's stationary, it will tend to remain stationary; if it's moving, it will tend to continue moving.

Before we can understand the motion of a falling apple, ball, or anything else, we must understand how it would move if it weren't falling. Because falling is a result of the pull of gravity, an influence that we'll examine later in this section, we must begin by eliminating gravity. Fortunately, the pull of gravity gradually diminishes as you move farther away from the earth, so that in deep, empty space its effects are negligible. Let's imagine taking a ball to this empty space, where it essentially will be free of outside influences (Fig. 1.1.1). Once we know how the ball moves in isolation, we can begin to see how it responds to outside influences in general and to the pull of gravity in particular.

You, the observer, are located near the ball and are also free of outside influences. You look over at the ball and watch it carefully. What does that ball do?

The correct answer to this apparently simple question eluded people for thousands of years; even Aristotle, perhaps the most learned philosopher of the classical world, was mistaken about it (see ❒). What makes the question so tricky is that no objects on earth are truly isolated from each other; instead they all push on, rub against, or interact with one another in some way. As a result, it took the remarkable Italian astronomer, mathematician, and physicist Galileo Galilei many years of careful observation and logical analysis to answer this question ❒. The solution he came up with, like the question itself, sounds simple: if the ball is stationary, it will remain stationary; if it's moving in some particular direction, it will continue moving in that direction at a steady pace, following a straightline path. This property of steady motion in the absence of any outside influence is called **inertia.**

INERTIA:
A body in motion tends to remain in motion; a body at rest tends to remain at rest.

❒ Aristotle (Greek philosopher, 384–322 BC) theorized that objects' velocities were proportional to the forces exerted on them. While this theory correctly predicted the behavior of a sliding object, it incorrectly predicted that heavier objects should fall faster than lighter objects. Nonetheless, Aristotle's theory was retained for a long time, in part because finding the simpler and more complete theory was hard and in part because the scientific method of relating theory and observation took time to develop.

Although the motion of an isolated ball is fairly simple, that of a falling ball is complicated. If we want to describe a falling ball accurately, we'll need to understand several physical quantities and their relationships to one another: position, speed, velocity, mass, acceleration, and force. For simplicity's sake, let's start by studying these physical quantities in an isolated ball.

Begin by describing this ball's location. At any particular moment, the ball is located at a **position**—that is, at a specific point. Whenever you note an object's position, it's always as a **distance** and **direction** from some reference point: how many kilometers west of Cleveland or how many centimeters above your head. In this case, you can report the ball's position as its distance away from you and the direction from you to the ball.

Position is an example of a **vector quantity**. A vector quantity consists of both a **magnitude** and a direction; the magnitude tells you how much of the quantity there is, while the direction tells you which way the quantity is pointing. Vector quantities are common in nature. When you encounter one, pay attention to the direction part; if you're looking for buried treasure 30 paces from the old tree but forget that it's due east of the tree, you'll have a lot of digging ahead of you.

❒ While a professor in Pisa, Galileo Galilei (Italian scientist, 1564–1642) was obliged to teach the natural philosophy of Aristotle. Troubled with the conflict between Aristotle's theory and observations of the world around him, Galileo devised experiments that measured the speeds at which objects fall and determined that all falling objects fall at the same rate.

If a ball is moving, then its position is changing. In other words, it has a **velocity**. Velocity measures how quickly the ball's position changes; it's our second vector quantity and consists of the **speed** with which the ball is moving and the direction in which it's heading. The ball's speed is the distance it travels in a certain amount of time:

$$\text{speed} = \frac{\text{distance}}{\text{time}}.$$

You can report the ball's velocity as its speed and direction of travel as observed from our vantage point.

But an isolated ball has a particularly simple velocity. Since the ball travels at a steady pace, along a straightline path, its velocity is constant and never changes. If the ball is heading to your right at a speed of 10 meters-per-second, it continues to travel with that velocity forever. A speed of 10 meters-per-second means that, if the ball is allowed to travel for 1 second at its present speed, it will cover a distance of 10 meters. Since its velocity is constant, the ball will travel 100 meters in 10 seconds, 1000 meters in 100 seconds, and so on. Furthermore, the path it takes is a straight line.

Thanks to your imaginary, isolated ball, we can now restate the previous description of inertia in terms of velocity: an object that's not subject to any outside influences moves at a constant velocity, covering equal distances in equal times along a straight line path. This statement is frequently identified as **Newton's first law of motion**, after its discoverer, the English mathematician and physicist Sir Isaac Newton ❐. The outside influences referred to in this law are called **forces**, a technical term for pushes and pulls.

❐ In 1664, while Sir Isaac Newton (English scientist and mathematician, 1642–1727) was a student at Cambridge University, the university was forced to close for 18 months because of the plague. Newton retreated to the country and discovered the laws of motion and gravitation and invented the mathematical basis of calculus. These discoveries, along with his observation that celestial objects such as the moon obey the same simple physical laws as terrestrial objects such as an apple (a new idea for the time), are recorded in his *Philosophiae Naturalis Principia Mathematica*, first published in 1687. This book is perhaps the most important and influential scientific and mathematical work of all time.

NEWTON'S FIRST LAW OF MOTION:
An object that is not subject to any outside forces moves at a constant velocity, covering equal distances in equal times along a straightline path.

CHECK YOUR UNDERSTANDING #1: A Puck on Ice

Why does a moving hockey puck continue to slide across an ice rink even though no one is pushing on it?

How Forces Affect Balls

Why doesn't the isolated ball change speed or direction? Because it has **mass**. Mass is the measure of an object's inertia, its resistance to changes in its velocity. Almost everything in the universe has mass. Because the ball has mass, its velocity will change only if something pushes on it—that is, only if it experiences a force. The ball will keep moving steadily in a straight path until something pushes on it with a force to stop it or to send it in another direction. Force is our third vector quantity, having both a magnitude and a direction. A push to the right is different from a push to the left.

When something pushes on the ball, its velocity changes; in other words, it accelerates. **Acceleration**, our fourth vector quantity, measures how quickly the ball's velocity changes. *Any* change in the ball's velocity is an acceleration, whether the ball is speeding up, slowing down, or even turning. If either its speed or direction of travel is changing, it's accelerating!

Like any vector quantity, acceleration has a magnitude and a direction. To see how these two parts of acceleration work, imagine that you're driving a car that's stopped at a red light. When the light turns green and you push down on the gas pedal, your car begins to accelerate—its speed increases and you cover ground more and more quickly. The magnitude of this acceleration depends on how far you depress the pedal. If you're not in a hurry and you only give the car a little gas, the car's velocity changes slowly because the magnitude of its acceleration is small. But if you're late for a final exam and you press the accelerator of your Ferrari to the floor, the magnitude of its acceleration is enormous. The car's velocity changes rapidly and you feel pressed back into your seat. That sensation of being thrown backward is an indication that you're accelerating forward. As we'll discuss in Section 2.3, you can *feel* when you're accelerating.

But acceleration has more than just a *magnitude*. When you start out at the light, you also select a *direction* for your acceleration. This acceleration is in the same direction as the force causing it. If you select a forward gear, the force on the car will be in the forward direction and the car will accelerate forward. If you select a reverse gear, well... The driver behind you realizes the importance of direction in the definitions of force and acceleration.

Once the car is going fast enough, you ease off the gas pedal and the car maintains a constant speed. However that doesn't mean that the car isn't accelerating. Anything that changes its direction of travel makes the car accelerate—a turn left or right or a change in the slope of the roadway.

Finally you come to a red light and step on the brake pedal. The car begins to accelerate again. This time it's accelerating backward, in the direction opposite its forward velocity. While we often call this process *deceleration*, it's just a special type of acceleration. The car's forward velocity gradually diminishes until it comes to rest at the light.

To help you recognize acceleration, here are some accelerating objects:

1. A runner who's leaping forward at the start of a race—the runner's velocity is changing from zero to forward so the runner is accelerating *forward*.
2. A bicycle that's stopping at a crosswalk—its velocity is changing from forward to zero so it's accelerating *backward* (it's decelerating).
3. An elevator that's just starting upward from the 1st floor to the 5th floor—its velocity is changing from zero to upward so it's accelerating *upward*.
4. An elevator that's stopping at the 5th floor after coming from the 1st floor—its velocity is changing from upward to zero so it's accelerating *downward*.
5. A car that's turning left at an intersection—its velocity is changing from forward to leftward so it's accelerating toward *the inside of the turn*.
6. An airplane that's just beginning its descent—its velocity is changing from level forward to descending forward so it's accelerating mostly *downward*.
7. Children riding a carousel around in a circle—while their speeds are constant, their directions of travel are always changing. We'll discuss the directions in which they're accelerating in Section 2.3.

Here are some objects that are *not* accelerating:

1. A parked car—its velocity is always zero.
2. A car traveling straight forward on a level road at a steady speed—no change in its speed or direction of travel.
3. A bicycle that's climbing up a smooth, straight hill at a steady speed—no change in its speed or direction of travel.
4. An elevator that's moving straight upward at a steady pace, half way between the 1st floor and the 5th floor—no change in its speed or direction of travel.

Now that we've learned what acceleration is, let's see how a ball accelerates in response to a particular force. The ball's acceleration depends on its mass; the more massive the ball, the less it accelerates. For example, it's much easier to change the velocity of a golf ball than that of a bowling ball. A particular ball's acceleration also depends on the force exerted on it—the stronger the force, the more the ball accelerates. The harder you push on a bowling ball, the faster its velocity changes.

There is a simple relationship between a ball's acceleration, its mass, and the force exerted on it. The force exerted on the ball equals the product of the ball's mass times its acceleration. The acceleration, as we've seen, is in the same direction as the force. This relationship was deduced by Newton from his observations of motion and is referred to as **Newton's second law of motion**.

NEWTON'S SECOND LAW OF MOTION:
The force exerted on an object is equal to the product of that object's mass times its acceleration. The acceleration is in the same direction as the force.

This law can be written in a word equation:

$$\text{force} = \text{mass} \cdot \text{acceleration}, \tag{1.1.1}$$

in symbols:

$$\mathbf{F} = m \cdot \mathbf{a},$$

and in everyday language:

Throwing a baseball is much easier than throwing a bowling ball.

Remember that in Eq. 1.1.1 the direction of the acceleration is the same as the direction of the force.

Because this is an equation, the left side of Eq. 1.1.1 must be equal to the right side. The force on an object must therefore be equal to the product of the object's mass times its acceleration. If we choose a specific object, perhaps a bowling ball, we're also choosing the mass, since an object's mass can't change unless the object changes (for example, if the bowling ball is nicked). Equation 1.1.1 indicates that an increase in the force on that bowling ball is accompanied by a similar increase in its acceleration. That way, as the left side of the equation increases, the right side increases to keep the two sides equal. The harder we push on the bowling ball, the more rapidly its velocity changes. (For another useful form of Eq. 1.1.1, see ❐.)

❐ Equation 1.1.1 can be rearranged algebraically as:

$$\text{acceleration} = \frac{\text{force}}{\text{mass}}.$$

This form makes it clear that an object's acceleration depends both on the force causing that acceleration and on the object's mass. Increasing the force increases the acceleration while replacing the object with one that has a larger mass decreases the acceleration.

We can also compare the effects of a specific force on two different masses. Since the left side of the equation doesn't change, the right side must also remain constant; thus the equation indicates that a decrease in mass must be accompanied by a corresponding increase in acceleration. Sure enough, the velocity of a billiard ball changes more rapidly than the velocity of a bowling ball when the two are subjected to identical forces (see Fig. 1.1.2).

So far we've explored five principles:

1. The location of an object is its position.
2. The object's velocity measures how quickly its position changes.
3. The object's acceleration measures how quickly its velocity changes.
4. In order for the object to accelerate, something must exert a force on it.
5. The more mass the object has, the more force will have to be exerted on it in order for it to have a certain acceleration.

We've also encountered five important physical quantities—mass, force, acceleration, velocity, and position—as well as some of the rules that relate them to one another. Much of the groundwork of physics rests on these five quantities and on their interrelationships.

As we move through this book, we'll use these concepts to explain many everyday experiences. However, we'll move toward the real world somewhat gradually at first, since the real world has a number of complications that would only distract us early on. For the remainder of this section and the two that follow, for example, we'll neglect the ever-present influences of friction and air resistance on the motions of objects in our surroundings. These familiar forces, while important to everyday life, can be ignored without serious consequences as we explore a few more basic physical laws. As long as a falling ball is dense, like a baseball and not like a beach ball, the air usually will have only subtle effects on its motion.

Fig. 1.1.2 - A baseball accelerates easily because of its small mass. A bowling ball has a large mass and is harder to accelerate.

CHECK YOUR UNDERSTANDING #2: Hard to Stop

It's much easier to stop a bicycle traveling toward you at 5 kilometers-per-hour than an automobile traveling toward you at the same velocity. What accounts for this difference?

CHECK YOUR FIGURES #1: At the Bowling Alley

Bowling balls come in various masses. Suppose that you try bowling two different balls, one with twice the mass of the other. If you push on them with equal forces, which one will accelerate faster and how much faster?

Measure for Measure: The Importance of Units

If you went to the grocery store and asked for "6 sugar," the clerk wouldn't know how much sugar you wanted. The number *6* wouldn't be enough information; you'd need to specify the units you were referring to—cups or pounds or cubes or tons. This need to specify units applies to almost all physical quantities—velocity, force, mass, and so on—and it has led our society to develop units that everyone agrees on and can understand, also known as **standard units**.

For example, if a car has a velocity of 60 miles-per-hour toward the east, the units "miles-per-hour" are essential to give meaning to the number *60*. This same velocity can be specified using different units—for example, as 96.5 kilometers-per-hour to the east. In fact, we can use many different units to specify velocity: feet-per-second, yards-per-day, and inches-per-century, to name only a few. We may use any units that are convenient, and we can always find a simple relationship to convert one unit into another. For example, to convert miles-per-hour into kilometers-per-hour, we'd simply multiply by 1.609.

The value of standard units comes from a general agreement about what they mean. Everyone agrees how far a mile is and how long an hour is. When combined, these two basic units form a composite unit for measuring speed, the mile-per-hour, that everyone also can agree on.

Many of the common units in the United States come from the old **English System of Units**, which most of the world has abandoned in favor of the **SI Units** (Systéme Internationale d'Unités). The continued use of English units in the United States often makes life difficult. If you have to triple a cake recipe that calls for ¾ of a cup of milk, you must work hard to calculate that you need 2 and ¼ cups. Then you go to buy 2 and ¼ cups of milk, which is slightly more than half a quart, but end up buying two pints instead. You now have 14 ounces of milk more than you need. But is that 14 fluid ounces or 14 ounces of weight? And so it goes.

The SI system has two important characteristics that distinguish it from the English system and make it easier to use:

1. Different units for the same physical quantity are related by factors of 10.
2. Most of the units are constructed out of a few basic units: the meter, the kilogram, and the second.

Let's take the first characteristic first: Different units for the same physical quantity are related by factors of 10. When measuring volume, 1000 milliliters is exactly 1 liter and 1000 liters is exactly 1 meter^3. When measuring mass, 1000 grams is exactly 1 kilogram and 1000 kilograms is exactly 1 metric ton. Because of this consistent relationship, enlarging a recipe that's based on the SI system is as simple as multiplying a few numbers. You never have to think about teaspoons or tablespoons; instead, if you want to triple a recipe that calls for 250 milliliters of sugar, you just multiply the recipe by 3 to obtain 750 milliliters of sugar. (See Appendix B for more conversion factors.)

SI units remain somewhat mysterious to many U.S. residents, even though some of the basic units are slowly appearing on our grocery shelves and highways. As a result, while the SI system really is more sensible than the old English system, developing a feel for some SI units is still difficult. How many of us know our heights in meters (the SI unit of length) or our masses in kilograms (the SI unit of mass)? If your car is traveling 200 kilometers-per-hour and you pass a police car, are you in trouble? Yes, because 200 kilometers-per-hour is about 125 miles-per-hour. Actually, the hour is not an SI unit—the SI unit of time is the second—but it remains customary for describing long periods of time. Thus, the kilometer-per-hour is a unit that is half SI (the kilometer part) and half customary (the hour part).

The second characteristic of the SI system is its relatively small number of basic units. So far, we've noted the SI units of mass (the **kilogram**, abbreviated kg), length (the **meter**, abbreviated m), and time (the **second**, abbreviated s). One kilogram is about the mass of a liter of water or soda; one meter is about the length of a long stride; one second is about the time it takes to say "one banana." From these three basic units, we can create several others, such as the SI units of velocity (the **meter-per-second**, abbreviated m/s) and acceleration (the **meter-per-second^2**, abbreviated m/s^2). One meter-per-second is a healthy walking speed; one meter-per-second^2 is about the acceleration of an elevator after the door closes and it begins to head upward. This conviction that many units are best constructed out of other, more basic units dramatically simplifies the SI system; the English system doesn't usually suffer from such sensibility.

The SI unit of force is also constructed out of the basic units of mass, length, and time. If we choose a 1 kilogram object and ask just how much force is needed to make that object accelerate at 1 meter-per-second^2, we define a specific amount of force. Since 1 kilogram is the SI unit of mass and 1 meter-per-second^2 is the SI unit of acceleration, it's only reasonable to let the force that causes this acceleration be the SI unit of force: the *kilogram-meter-per-second*^{2}. Since this composite unit is pretty unwieldy but very important, it has been given its own name: the **newton** (abbreviated N)—after, of course, Sir Isaac, whose second law defines the relationship among mass, length, and time that the unit expresses. One newton is about the weight of 10 U.S. quarter dollars; if you hold 10 quarters steady in your hand, you'll feel a downward force of about 1 newton.

Because of our slow but steady move to the SI system, we'll be using SI units for most of the book. But whenever we might benefit from an intuitive feel for a physical quantity, we'll examine it in English and customary units as well. A bullet train traveling "67 meters-per-second" won't mean much to most of us, while one moving "150 miles-per-hour" (150 mph) or "241 kilometers-per-hour" (241 km/h) might elicit our well-deserved respect.

CHECK YOUR UNDERSTANDING #3: Going for a Walk
If you're walking at a pace of 1 meter-per-second, how many miles will you travel in an hour?

Weight and Gravity

As we've seen, 10 quarters weigh about 1 N (1 newton). But what is weight? Evidently it's a force, since the newton is a unit of force. But to understand what weight really is—and, in particular, where it comes from—we need to look at **gravity**. (That's not as inconvenient as it at first may seem; since our overall goal in this section is to understand falling balls, gravity is on our agenda anyway.)

Gravity is a physical phenomenon that produces an attractive force between every pair of objects in the universe. In our daily lives, however, the only object large enough and near enough to have an obvious direct effect on us is our planet, the earth. While the moon and the sun are also important, their weak gravitational attractions cause subtler effects, such as the ocean tides.

On the earth's surface, gravity creates a downward force on any object, attracting that object directly toward the center of the earth. We call this downward gravitational force the object's **weight** (Fig. 1.1.3). For example, a baseball has a weight of about 1.4 N. An automobile has a weight of about 10,000 N. A fly has a weight of about 0.001 N.

One remarkable characteristic of gravity is that the force it exerts on an object is exactly proportional to the object's mass. If one ball has twice the mass of another ball, it also has twice the weight. An object's weight is also proportional to the local strength of gravity, which is measured by a quantity called the **acceleration due to gravity**. We can express these observations in a word equation:

$$\text{weight} = \text{mass} \cdot \text{acceleration due to gravity}, \tag{1.1.2}$$

in symbols:

$$\mathbf{w} = m \cdot \mathbf{g},$$

and in everyday language:

You can lose weight either by reducing your mass or by going someplace, like a small planet, where the gravity is weaker.

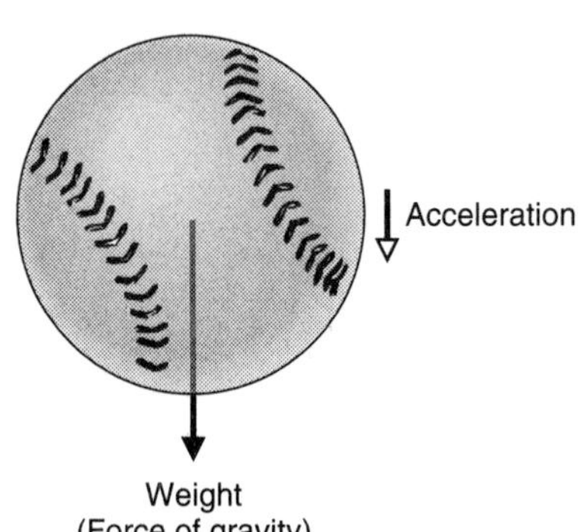

Fig. 1.1.3 - A ball experiencing the force of gravity. It accelerates downward.

The acceleration due to gravity is determined by various properties of the earth and doesn't depend on the object being considered. At the surface of the earth, the acceleration due to gravity is about 9.8 m/s^2 (9.8 meters-per-second^2, or 32 feet-per-second^2) towards the center of the earth. While this quantity doesn't appear to involve force, its units can be rewritten in terms of newtons as 9.8 newtons-per-kilogram; thus a ball (or any other object, for that matter) weighs 9.8 newtons for each kilogram of its mass.

If the only force on a ball is its weight, the ball accelerates downward; in other words, it falls. While a ball moving through the earth's atmosphere encounters additional forces due to air resistance, we'll ignore those forces for the time being. Doing so will cost us only a little in terms of accuracy—the effects of air resistance are negligible as long as the ball is dense and its speed relatively small—and it will allow us to focus exclusively on the effects of gravity.

How much does the falling ball accelerate? According to Eq. 1.1.1, the force on the ball is equal to the ball's mass times its acceleration. According to Eq. 1.1.2, the ball's weight is equal to the ball's mass times the acceleration due to gravity. But for a falling ball, the force on the ball is the ball's weight! We can put the two equations together by equating the force on the ball with ball's weight:

$$m \cdot \mathbf{a} = \mathbf{F} = \mathbf{w} = m \cdot \mathbf{g}$$

$$\mathbf{a} = \mathbf{g}.$$

As you can see, the ball's acceleration is equal to the acceleration due to gravity!

Thus a ball, or any other object falling near the earth's surface, experiences a downward acceleration of 9.8 m/s^2, regardless of its mass. This downward acceleration is substantially more than that of an elevator starting its descent. When you drop a ball, it picks up speed very quickly in the downward direction.

Because all falling objects at the earth's surface accelerate downward at exactly the same rate, a billiard ball and a bowling ball dropped simultaneously from the same height will reach the ground together. (Remember that we're not considering air resistance yet.) Although the bowling ball weighs more than the billiard ball, it also has more mass; so while the bowling ball experiences a larger downward force, its larger mass ensures that its downward acceleration is equal to that of the lighter and less massive billiard ball.

CHECK YOUR UNDERSTANDING #4: Weight and Mass

Out in deep space, far from any celestial object that exerts significant gravity, would an astronaut weigh anything? Would that astronaut have a mass?

CHECK YOUR FIGURES #2: Weighing In on the Moon

You're in your spacecraft on the surface of the moon. Before getting into your suit, you weigh yourself and find that your moon weight is almost exactly 1/6th your earth weight. What is the moon's acceleration due to gravity?

The Velocity of a Falling Ball

We're now ready to examine the motion of a falling ball near the earth's surface. A falling ball is one that has only the force of gravity acting on it; and gravity, as we've seen, causes any falling object to accelerate downward at a constant rate. But we're usually less interested in the falling object's acceleration than we are in its position and velocity. Where will the object be in 3 seconds, and what will its velocity be then? When you're trying to summon up the courage to jump off the

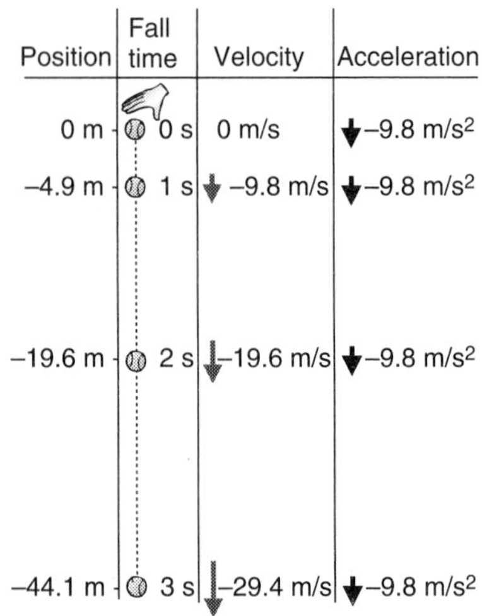

Fig. 1.1.4 - The moment you let go of a ball that was resting in your hand, it begins to fall. Its weight causes it to accelerate downward. After 1 second, it has fallen 4.9 m and has a velocity of 9.8 m/s downward. After 2 seconds, it has fallen 19.6 m and has a velocity of 19.6 m/s downward, and so on. As the ball continues to accelerate downward, its velocity continues to increase downward. Negative values for the position and velocity are meant to indicate downward movement, caused by a negative or downward acceleration.

high dive, you want to know how long it'll take you to reach the water and how fast you'll be going when you hit.

The first step in answering these questions is to look at how a ball's velocity is related to the time you've been watching it fall. To do that, you'll need to know the ball's *initial velocity*, that is its speed and direction at the moment you start watching it. If you drop the ball from rest, its initial velocity is zero.

You can then describe the ball's present velocity in terms of its initial velocity, its acceleration, and the time that has passed since you started watching it. Because a constant acceleration causes the ball's velocity to change by the same amount each second, the ball's present velocity differs from its initial velocity by the product of the acceleration times the time over which you've been watching it. We can relate these quantities as a word equation:

$$\text{present velocity} = \text{initial velocity} + \text{acceleration} \cdot \text{time}, \tag{1.1.3}$$

in symbols:

$$\mathbf{v} = \mathbf{v}_0 + \mathbf{a} \cdot t,$$

and in everyday language:

> *A stone dropped from rest descends faster with each passing second, but you can give it a head start by throwing it downward instead of just letting go.*

For a ball falling from rest, the initial velocity is zero, the acceleration is downward at $9.8\ \text{m/s}^2$, and the time you've been watching it is simply the time since it started to drop (Fig. 1.1.4). After one second, the ball has a downward velocity of 9.8 m/s (9.8 meters-per-second, or 32 feet-per-second). After two seconds, the ball has a downward velocity of 19.6 m/s. After three seconds, its downward velocity is 29.4 m/s, and so on. We often put a negative sign in front of the acceleration to indicate the direction: by convention, we say that a negative sign means "down."

CHECK YOUR UNDERSTANDING #5: Half a Fall

You drop a marble from rest and after 1 s, its velocity is 9.8 m/s in the downward direction. What was its velocity after only 0.5 s of falling?

CHECK YOUR FIGURES #3: The High Dive

If it takes you about 1.4 s to reach the water from the 10-meter diving platform, how fast are you going just before you enter the water?

The Position of a Falling Ball

The ball's velocity continues to increase as it falls, but where exactly is the ball located? To answer that question, you need to know the ball's *initial position*, that is where it was when you started watching it fall. If you drop the ball from rest, the initial position is your hand and you can define that spot as 0.

You can then describe the ball's present position in terms of its initial position, its initial velocity, its acceleration, and the time that has passed since you started watching it. However, because the ball's velocity is changing, you can't simply multiply its present velocity by the time that it's been falling to determine how much the ball's present position differs from its initial position. Instead, you need to use the ball's average velocity during the whole period you've been watching it. Since the ball's velocity has been changing smoothly from its initial velocity to its present velocity, the ball's average velocity is exactly half way in between the two:

$$\text{average velocity} = \text{initial velocity} + \tfrac{1}{2} \cdot \text{acceleration} \cdot \text{time}.$$

The ball's present position differs from its initial position by the product of this average velocity times the time over which you've been watching it. We can relate these quantities as a word equation:

$$\text{present position} = \text{initial position} + \text{initial velocity} \cdot \text{time} + \tfrac{1}{2} \cdot \text{acceleration} \cdot \text{time}^2, \quad (1.1.4)$$

in symbols:

$$\mathbf{x} = \mathbf{x}_0 + \mathbf{v}_0 \cdot t + \tfrac{1}{2} \cdot \mathbf{a} \cdot t^2,$$

and in everyday language:

The longer a stone has been falling, the more its height diminishes with each passing second. However, it won't overtake a stone that was dropped next to it at an earlier time or dropped from beneath it at the same time.

For a ball falling from rest, the initial velocity is zero, the acceleration is downward at 9.8 m/s^2, and the time you've been watching it is simply the time since it started to drop (Fig. 1.1.4). After one second, the ball has fallen 4.9 m (16 feet). After two seconds, the ball has fallen downward a total of 19.6 m. After three seconds, the ball has fallen a total of 44.1 m, and so on.

Eqs. 1.1.3 and 1.1.4 depend on the definition of acceleration as the measure of how quickly *velocity* changes and the definition of velocity as the measure of how quickly *position* changes. Because the acceleration of a falling ball doesn't change with time, the two equations can be derived using algebra. But in more complicated situations, where an object's acceleration changes with time, predicting its position and velocity usually requires the use of *calculus*. Calculus is the mathematics of change, invented by Newton to address just these sorts of problems.

We've been discussing what happens to a falling ball, but we could've chosen another object instead. Everything falls the same way; heavy or light, large or small, all objects take the same amount of time to fall some distance near the earth's surface, as long they're dense enough to overcome air resistance. If there were no air, this statement would be exactly true for any object; a feather and a lead brick would plummet downward together if you dropped them simultaneously.

Now that we've explored acceleration due to gravity, we can see why a ball dropped from a tall ladder is more dangerous than the same ball dropped from a short stool. The farther the ball has to fall, the longer it takes to reach the ground and the more time it has to accelerate. During its long fall from the tall ladder, the ball acquires a large downward velocity and becomes very hard to stop. If you try to catch it, you'll have to exert a very large upward force on it to accelerate it upward and bring it to rest quickly. Exerting that large upward force may hurt your hand.

The same notion holds if you're the falling object. If you leap off a tall ladder, a substantial amount of time will pass before you reach the ground. By the time you hit the ground, you will have acquired considerable downward velocity. The ground will then accelerate you upward and bring you to rest with a very large and unpleasant upward force. (For an interesting application of long falls, see ❐.)

❐ In 1782, William Watts, a plumber from Bristol, England, patented a technique for forming perfectly spherical, seamless lead shot for use in guns. His idea was to pour molten lead through a sieve suspended high above a pool of water. The lead droplets cool in the air as they fall, solidifying into perfect spheres before reaching the water. Shot towers based on this idea soon appeared throughout Europe and eventually in the United States. Nowadays, iron shot has all but replaced environmentally dangerous lead shot. Iron shot is cast, rather than dropped, because the longer cooling time needed to solidify molten iron would require impractically tall shot towers.

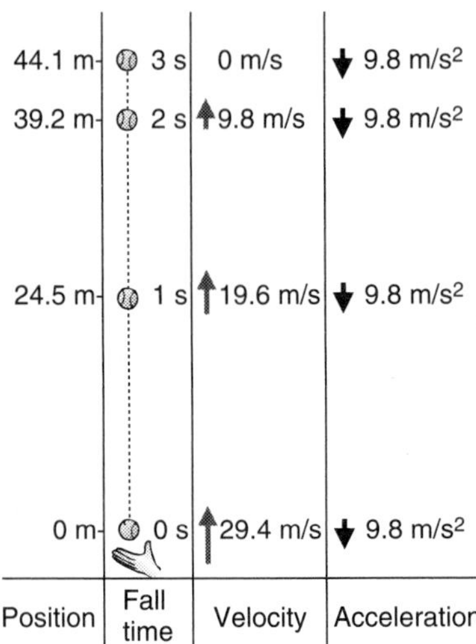

Fig. 1.1.5 - The moment you let go of a ball thrown straight upward, it begins to accelerate downward at 9.8 m/s^2. The ball rises but its upward velocity diminishes steadily until it momentarily comes to a stop. It then descends with its downward velocity increasing steadily. In this example, the ball rises for 3 s and comes to rest. It then descends for 3 s before returning to your hand in a very symmetrical flight.

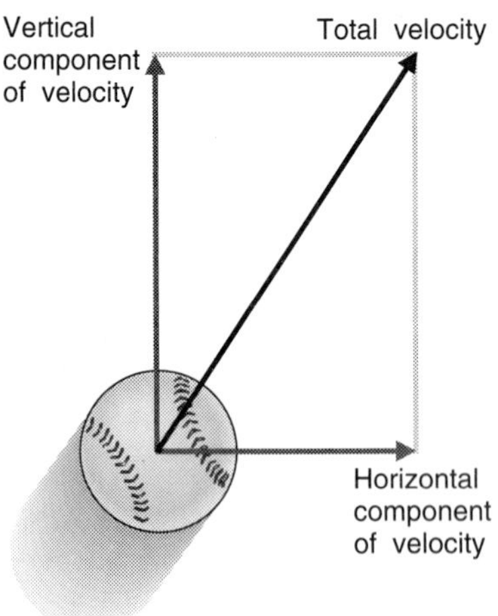

Fig. 1.1.6 - Even if the ball has a velocity that is neither purely vertical nor purely horizontal, its velocity may nonetheless be viewed as having a vertical component and a horizontal component. Part of its total velocity acts to move this ball upward and part of its total velocity acts to move this ball in a horizontal direction.

CHECK YOUR UNDERSTANDING #6: Half a Fall Again

You drop a marble from rest and after 1 s, it has fallen downward a distance of 4.9 m. How far had it fallen after only 0.5 s?

CHECK YOUR FIGURES #4: Extreme Physics

You're planning to construct a bungee-jumping amusement at the local shopping center. If you want your customers to have a 5-second free-fall experience, how tall will you need to build the tower from which they'll jump? (Don't worry about the extra height needed to stop people after the bungee pulls taut.)

How a Thrown Ball Moves: Projectile Motion

If the only force acting on an object is its weight, then the object is falling. So far, we've explored this principle only as it pertains to balls dropped from rest. However a thrown ball is falling, too; once it leaves your hand, it's subject only to the downward force of gravity and it falls. It may seem odd but even though it's initially traveling upward, the tossed ball is accelerating downward at 9.8 m/s^2. As a result, the tossed ball's upward velocity diminishes, it stops rising, its velocity becomes downward, and it eventually returns to the ground.

Eq. 1.1.3 still describes how the ball's velocity depends on the fall time, but now the initial velocity is not zero and points in the upward direction. If you toss a ball straight up in the air, it leaves your hand with a large upward velocity (Fig. 1.1.5). As soon as you let go of it, it begins to accelerate downward. If the ball's initial upward velocity is 29.4 m/s, then after one second its upward velocity is 19.6 m/s. After another second, its upward velocity is only 9.8 m/s. After a third second, the ball momentarily comes to a complete stop with a velocity of zero. It then begins to descend, falling just as it did when you dropped it from rest.

The ball reaches its peak height at the instant it stops rising. Its flight before and after that peak is symmetrical. It travels upward quickly at first, since it has a large upward velocity. As its upward velocity diminishes, it travels more and more slowly until it comes to a stop. It then begins to descend, slowly at first and then faster and faster as it continues its constant downward acceleration. The time the ball takes to rise from the ground to its peak height is exactly equal to the time it takes to descend back down from that peak to the ground. Eq. 1.1.4 indicates how the position of the ball depends on the fall time, with the initial velocity being the upward velocity of the ball as it leaves your hand.

The larger the initial velocity of the ball, the higher it rises before its upward velocity is reduced to zero and it reaches its peak height. It descends for the same amount of time it took to rise to its peak. The higher the ball goes before it begins to descend, the longer it takes to return to the ground and the faster it's traveling when it arrives. That's why catching a high fly ball with your bare hands stings so much: the ball is traveling very, very fast when it hits your hands, and a large force is required to bring the ball to rest.

What happens if you don't toss the ball exactly straight up? Suppose you throw the ball upward at some angle. The ball still rises to a peak height and then begins to descend; but as it rises and descends, it also travels away from you horizontally so that it strikes the ground at some distance from your feet. How much does this horizontal travel complicate the motion of a falling ball?

The answer is not very much. One of the beautiful simplifications of physics is that we can often treat an object's vertical motion independently of its horizontal motion. This technique involves separating the vector quantities—

acceleration, velocity, and position—into **components**, those portions of the quantities that lie along a specific direction. For example, the vertical component of an object's position is that object's altitude.

If you know a ball's altitude, you know only part of its position; you still need to know where it is to your left or right and to your front or back. In fact, you can completely specify its position (or any other vector quantity) in terms of three components along three directions that are at right angles to one another. This means that you can completely specify the ball's position by its vertical altitude, its horizontal distance to your left or right, and its horizontal distance in front or in back of you. For example, the ball might be 10 m above you, 3 m to your left, and 2 m behind you. These three distances indicate precisely where the ball is located.

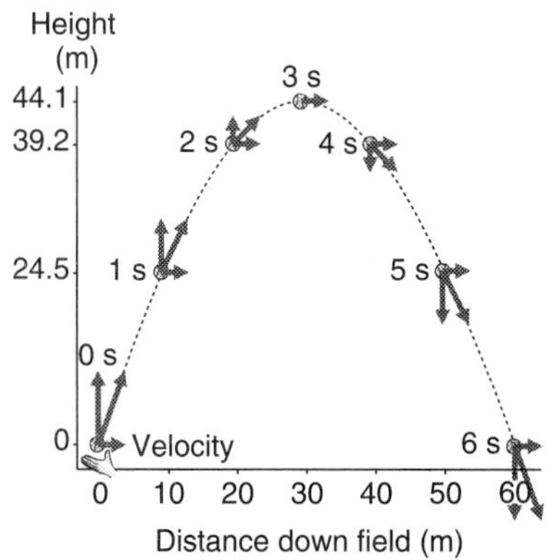

Fig. 1.1.7 - If you throw a ball upward, at an angle, part of the initial velocity will be in the vertical direction and part will be in the horizontal direction. The vertical and horizontal motions will take place independently of one another. The ball will rise and fall just as it did in Figs. 1.1.4 and 1.1.5; at the same time, however, it will move downfield. Because there is no horizontal acceleration (gravity acts only in the vertical direction), the horizontal velocity remains constant, as indicated by the velocity arrows (shaded arrows). In this example, the ball travels 10 m downfield each second and strikes the ground 60 m from your feet after a 6 s flight through the air.

Up until now, we've actually been examining only the vertical components of position, velocity, and acceleration. But now we're going to let the ball move horizontally so that we also have to consider what happens to the two horizontal components of each vector quantity. Keeping track of all three components of each quantity is difficult. Since we're interested in the motion of a tossed ball, we can eliminate the left-right component by throwing the ball forward; that way, the ball will move only in a vertical plane that extends directly in front of us. We can then specify the ball's position as its altitude and its horizontal distance in front of us.

Because the falling ball's acceleration is constant and doesn't depend on where the ball is near the earth's surface, the ball's horizontal motion is independent of its vertical motion. We already know about the falling ball's vertical motion, but what is its horizontal motion? It appears we need some new relationships to describe how the horizontal components of position, velocity, and acceleration change with time. Fortunately, we can reuse Eqs. 1.1.3 and 1.1.4.

While Eqs. 1.1.3 and 1.1.4 were introduced to describe a ball's vertical motion, they're actually more general. They relate three vector quantities—position, velocity, and constant acceleration—to one another. They describe how an object's position and velocity change with time when the object undergoes constant acceleration, regardless of the direction of that constant acceleration.

These equations also apply to the components of position, velocity, and constant acceleration. If you add the words "vertical component of" in front of each vector quantity in Eqs. 1.1.3 and 1.1.4, the equations correctly describe the object's vertical motion. The same can be done with the object's horizontal motion. Our previous look at vertical motion implicitly inserted the words "vertical component of" into the equations. Now we'll try inserting the words "horizontal component of" to understand the tossed ball's horizontal motion.

As soon as the ball leaves your hand, the motion can be broken into two parts: a vertical motion and a horizontal motion (Figs. 1.1.6, 1.1.7, and 1.1.8). Part of the ball's initial velocity is in the upward direction, and that vertical velocity component determines the object's ascent and descent. Part of the ball's initial velocity is in the downfield direction, and that horizontal velocity component determines the ball's drift downfield.

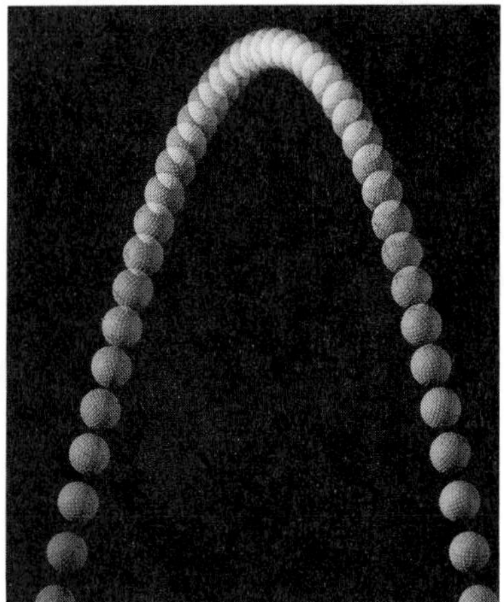

Fig. 1.1.8 - This ball drifts steadily to the right after being thrown because gravity only affects the ball's vertical component of velocity.

There is no horizontal component of acceleration because gravity acts only in the downward vertical direction and no other force is acting. The horizontal velocity component remains constant from the moment the ball leaves your hand to the moment it strikes the ground a few seconds later. The ball travels downfield at a steady rate during its flight. The upward vertical component of the ball's initial velocity determines how high the ball goes and how long the ball stays aloft before striking the ground; the horizontal component of the initial velocity determines how quickly the ball travels downfield during its time aloft.

When the ball hits the ground it still has its original horizontal velocity component, but its vertical velocity component is now in the downward direc-

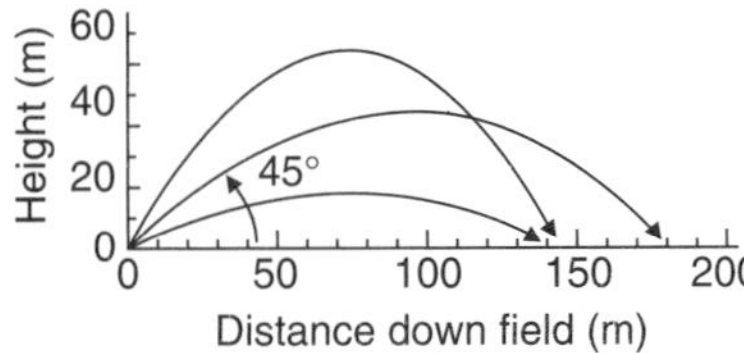

Fig. 1.1.9 - If you want the ball to hit the ground as far from your feet as possible, given a certain initial velocity, throw the ball at 45° from the horizontal or vertical directions. That way, the ball will have equal initial vertical and horizontal components of velocity. The ball will stay aloft for a relatively long time and will make good use of that flight time to travel downfield.

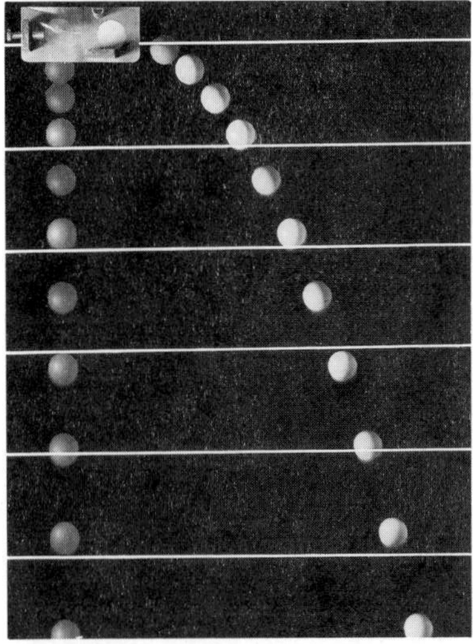

Fig. 1.1.10 - When these two balls were dropped, they accelerated downward at the constant rate of 9.8 m/s^2 and their velocities increased steadily in the downward direction. Even though one of them was initially moving to the right, they descended together.

tion. The total velocity of the ball is composed of these two components. The ball starts with its velocity up and forward and ends with its velocity down and forward.

If you want a ball or shot-put to hit the ground as far from your feet as possible, you should keep it aloft for a long time *and* give it a sizable horizontal component of velocity; in other words, you must achieve a good balance between the vertical and horizontal velocity components (Fig. 1.1.9). These components of velocity together determine the ball's flight path, its **trajectory**. If you throw the ball straight up, it will stay aloft for a long time but will not travel downfield at all (and you will need to wear a helmet). If you throw the ball directly downfield, it will have a large initial horizontal velocity component but will hit the ground almost immediately.

Neglecting air resistance and the altitude difference between your throwing arm and the ground that the ball will eventually hit, your best choice is to throw the ball at an angle of 45° above horizontal. At that angle, the initial upward velocity component will be the same as the initial downfield velocity component. The ball will stay aloft for a reasonably long time and will make good use of that time to move downfield. Other angles won't make such good use of the initial speed to move the ball downfield. (A discussion of how to determine the horizontal and vertical components of a velocity appears in Appendix A.)

These same ideas apply to two baseballs, one dropped from a cliff and the other thrown horizontally from that same cliff. If both leave your hands at the same time, they will both hit the ground below at the same time (Fig. 1.1.10). The fact that the second ball has an initial horizontal velocity doesn't affect the time it takes to descend to the ground, because the horizontal and vertical motions are independent. Of course, the ball thrown horizontally will strike the ground far from the base of the cliff, while the dropped ball will land directly below your hand.

CHECK YOUR UNDERSTANDING #7: Aim High

Why must a sharpshooter or an archer aim somewhat above her target? Why can't she simply aim directly at the bull's-eye to hit it?

Summary

How Falling Balls Work: Any ball that is subject only to the force of gravity is a falling ball. Gravity affects only the ball's vertical motion, causing the ball's vertical component of velocity to increase steadily in the downward direction. Since the ball's horizontal motion is unaffected by gravity, the ball moves downfield at a steady pace.

A falling ball that is initially rising soon stops rising and begins to descend. The larger its initial upward component of velocity, the longer the ball rises and the greater its peak height. When the ball then begins to descend, the peak height determines how long it takes for the ball to reach the ground.

When you throw a ball, the vertical component of initial velocity determines how long the ball remains aloft. The horizontal component of initial velocity determines how quickly the ball moves downfield. A thrower intuitively chooses an initial speed and direction for a ball so that it moves just the right distance downfield by the time it descends to the proper height.

Important Laws and Equations

1. Newton's First Law of Motion: An object that is free from all outside forces travels at a constant velocity, covering equal distances in equal times along a straight line path.

2. Newton's Second Law of Motion: An object's acceleration is equal to the force exerted on that object divided by the object's mass, or

$$\text{force} = \text{mass} \cdot \text{acceleration}. \quad (1.1.1)$$

3. Relationship Between Mass and Weight: An object's weight is equal to its mass times the acceleration due to gravity, or

$$\text{weight} = \text{mass} \cdot \text{acceleration due to gravity}. \quad (1.1.2)$$

4. The Velocity of an Object Experiencing Constant Acceleration: The object's present velocity differs from its initial velocity by the product of its acceleration times the time since it was at that initial velocity, or

$$\text{present velocity} = \text{initial velocity} + \text{acceleration} \cdot \text{time}. \quad (1.1.3)$$

5. The Position of an Object Experiencing Constant Acceleration: The object's present position differs from its initial position by the product of its average velocity since it was at that initial position times the time since it was at that initial position, or

$$\text{present position} = \text{initial position} + \text{initial velocity} \cdot \text{time} + \tfrac{1}{2} \cdot \text{acceleration} \cdot \text{time}^2. \quad (1.1.4)$$

The Physics of Falling Balls

1. An object not subject to any outside influences (i.e., experiencing no forces) travels along at a constant velocity. That means it travels at a steady pace in a straight line path. An object at rest has a constant velocity of zero.

2. For an object's velocity to change, it must accelerate. This acceleration must be caused by the exertion of a force on the object. The more force exerted on an object, the more it accelerates.

3. Mass is the measure of an object's inertia, its resistance to acceleration. The more mass an object has, the less it accelerates in response to a specific force.

4. Acceleration is a measure of how quickly an object's *velocity* changes. Velocity is a measure of how quickly an object's *position* changes.

5. An object that is maintaining a constant speed but that is changing its direction of motion is accelerating. Sometimes an accelerating object only changes the direction of its velocity and not the magnitude of that velocity.

6. Gravity produces an attractive force between any pair of objects. Near the earth's surface, gravity attracts all objects downward toward the earth's center by an amount proportional to the object's mass. This downward gravitational force is called the object's weight.

7. Because an object's weight is exactly proportional to its mass, all objects experience exactly the same downward acceleration due to gravity in the absence of air resistance.

8. As an object falls, its velocity increases steadily in the downward direction. An object that is initially moving upward slows to a stop and then begins to descend.

9. An object's motion can often be separated into components along three directions that are at right angles to one another. When this separation is possible, as it is for a falling object, the object's vertical motion can be handled separately from the two horizontal motions (left-right and front-back).

10. Gravity acts only in the downward vertical direction. A falling object experiences no horizontal components of force or acceleration. If a falling object initially has some horizontal component of velocity, that horizontal component will remain constant and the object will move downfield at a steady rate.

Check Your Understanding - Answers

1. The puck travels at constant velocity across the ice because it has inertia.

Why: A hockey puck resting on the surface of wet ice is almost completely free of horizontal influences. If someone pushes on the puck, so that it begins to travel with some horizontal velocity across the ice, inertia will ensure that the puck continues to slide.

2. The automobile has a much greater mass than the bicycle.

Why: To stop a moving vehicle, you must exert a force on it in the direction opposite its velocity. The vehicle will then accelerate in the direction opposite its velocity (decelerate in this case) and will eventually come to rest. If the vehicle is heading toward you, you must push it away from you. The more mass the vehicle has, the less it will accelerate in response to a certain force and the longer you will have to push on it to stop it completely. While it's easy to stop a bicycle by hand, stopping even a slowly moving automobile by hand requires a large force exerted for a substantial amount of time.

3. About 2.24 miles.

Why: There are many different units in this example, so we must do some converting. First, an hour is 3600 seconds, so that in an hour of walking at 1 meter-per-second, you will have walked 3600 meters. Second, a mile is about 1609 meters so that each time you travel 1609 meters, you have traveled a mile. By walking 3600 meters, you have completed 2 miles and are about a quarter of the way into your third mile.

4. The astronaut would have zero weight but would still have a normal mass.

Why: Weight is a measure of the force exerted on the astronaut by gravity. Far from the earth or any other large object, the astronaut would experience virtually no gravitational force and would have zero weight. But mass is a measure of inertia and doesn't depend at all on gravity. Even in deep space, it would be much harder to accelerate a school bus than to accelerate a baseball because the school bus has more mass than the baseball.

5. 4.9 m/s in the downward direction.

Why: A freely falling object accelerates downward at a steady rate. Its velocity changes by 9.8 m/s in the downward direction each and every second. In half a second, the marble's velocity changes by only half that amount or 4.9 m/s.

6. About 1.2 m.

Why: While a freely falling object's velocity changes steadily in the downward direction, its height is more complicated. When you drop the marble from rest, it starts its descent slowly but picks up speed and covers the downward distance faster and faster. In the first 0.5 s, it travels only a quarter of the distance it travels in the first 1 s, or about 1.2 m.

7. The bullet or arrow will fall in flight so she must compensate for its loss of height.

Why: To hit the bull's-eye, the sharpshooter or archer must aim above the bull's-eye because the projectile will fall in flight. The longer the bullet or arrow is in flight, the more it will fall and the higher she must aim. As the distance to the target increases, the flight time increases and her aim must move upward.

Check Your Figures - Answers

1. The less massive ball will accelerate twice as rapidly.

Why: You can rearrange Eq. 1.1.1 to show that an object's acceleration is inversely proportional to its mass:

$$\text{acceleration} = \frac{\text{force}}{\text{mass}}.$$

If you push on both bowling balls with equal forces, then their accelerations will only depend on their masses. Doubling the mass on the right side of this equation halves acceleration on the left side. That means that the more massive ball will only accelerate half as fast as the other ball.

2. About 1.6 m/s^2.

Why: You can rearrange Eq. 1.1.2 to show that the acceleration due to gravity is proportional to an object's weight:

$$\text{acceleration due to gravity} = \frac{\text{weight}}{\text{mass}}.$$

Your mass doesn't change in going to the moon, so any change in your weight must be due to a change in the acceleration due to gravity. Since your moon weight is $1/6^{\text{th}}$ of your earth weight, the moon's acceleration due to gravity must be $1/6^{\text{th}}$ that of the earth or about 1.6 m/s^2.

3. About 14 m/s (50 km/h or 31 mph).

Why: The downward acceleration of gravity is 9.8 m/s^2. You fall for 1.4 s, during which time your velocity increases steadily in the downward direction. Since you start with zero velocity, Eq. 1.1.3 gives a final velocity of:

$$\text{final velocity} = 9.8\ \text{m/s}^2 \cdot 1.4\ \text{s} = 13.72\ \text{m/s}.$$

Since the time of the fall is only given to two digits of accuracy (1.4 s could really be 1.403 s or 1.385 s), we shouldn't claim that our calculated final velocity is accurate to 4 digits. We should round the value to 14 m/s.

4. About 122 meters (402 feet or a 40 story building).

Why: As they fall, the jumpers will travel downward at ever increasing speeds. Since the jumpers start from rest and fall *downward* for 5 seconds, we can use Eq. 1.1.4 to determine how far they fall:

$$\begin{aligned}\text{final height} &= \text{initial height} - \tfrac{1}{2} \cdot 9.8\ \text{m/s}^2 \cdot (5\ \text{s})^2 \\ &= \text{initial height} - 122.5\ \text{m}.\end{aligned}$$

The downward acceleration is indicated here by the negative change in height. At the end of 5 seconds, the jumpers will have fallen more than 122 m and will be traveling downward at about 50 m/s. The tower will need additional height to slow the jumpers down and begin bouncing them back upward. Clearly, a 5 second free-fall is pretty unrealistic. Try for a 2 or 3 second free-fall instead.

Glossary

acceleration A vector quantity that measures how quickly an object's velocity is changing—the greater the acceleration, the more the object's velocity changes each second. It consists of both the amount of acceleration and the direction in which the object is accelerating. This direction is identical to the direction of the force causing the acceleration.

The SI unit of acceleration is the meter-per-second^2.

acceleration due to gravity A physical constant that specifies how quickly a freely falling object accelerates and also relates an object's weight to its mass. At the earth's surface, the acceleration due to gravity is 9.8 m/s^2 (or 9.8 N/kg).

components The portions of a vector quantity that lie along particular directions.

direction The line or course on which something is moving, is aimed to move, or along which something is pointing or facing.

distance The length between two positions in space.

English System of Units An assortment of antiquated units that were used throughout the English colonies and remain in common use in the United States today. Units in this system include feet, ounces, and miles-per-hour.

force An influence that if exerted on a free body results chiefly in an acceleration of the body and sometimes in deformation and other effects. A force is a vector quantity, consisting of both the amount of force and its direction. The SI unit of force is the newton.

gravity The gravitational attraction of the mass of the earth, the moon, or a planet for bodies at or relatively near its surface. All objects exert gravitational forces on all other objects.

inertia A property of matter by which it remains at rest or in uniform motion in the same straight line unless acted upon by some outside force.

kilogram (kg) The SI unit of mass. (The standard kilogram is a platinum-iridium cylinder kept at the International Bureau of Weights and Measures near Paris.) A liter of water has a mass of about 1 kilogram.

magnitude The amount of some physical quantity.

mass The property of a body that is a measure of its inertia or resistance to acceleration, that is commonly taken as a measure of the amount of material it contains, and that causes it to have weight in a gravitational field. The SI unit of mass is the kilogram.

meter (m) The SI unit of length. (1 meter is formally defined as the distance light travels through empty space in 1/299,792,458th of a second.) 1 meter is about the length of a long stride or about 3.28 feet.

meter-per-second (m/s) The SI unit of velocity or speed. 1 meter-per-second is a typical walking pace or about 2.2 mph.

meter-per-second^2 (m/s^2) The SI unit of acceleration. 1 meter-per-second^2 is about the acceleration of an elevator as it first begins to move upward.

newton (N) The SI unit of force (synonymous with the kilogram-meter-per-second^2). Ten United States quarters have a weight equal to about 1 newton. The common English unit of force, the pound, is about 4.45 newtons.

Newton's first law of motion An object that is free from all outside forces travels at a constant velocity, covering equal distances in equal times along a straight line path.

Newton's second law of motion An object's acceleration is equal to the force exerted on that object divided by the object's mass. This equality can be manipulated algebraically to state that the force on the object is equal to the product of the object's mass times its acceleration (Eq. 1.1.1).

position A vector quantity that specifies the location of an object relative to some reference point. It consists of both the length and the direction from the reference point to the object.

second (s or sec) The SI unit of time. (1 second is formally defined as the duration of 9,192,631,770 periods of the radiation corresponding to the transition between two hyperfine levels of the ground state of the cesium-133 atom.)

SI Units A system of units (Systéme International d'Unités) that carefully defines related units according to powers of 10. SI units are now used almost exclusively throughout most of the world, with the notable exception of the United States.

speed A measure of the distance an object travels in a certain amount of time. The SI unit of speed is the meter-per-second.

trajectory The path taken by an object as it moves.

vector quantity A quantity, characterizing some aspect of a physical system, that consists of both a magnitude and a direction in space.

velocity A vector quantity that measures how quickly an object's position is changing—the greater the velocity, the farther the object travels each second. It consists of both the object's speed and the direction in which the object is traveling. The SI unit of velocity is the meter-per-second.

weight (near the earth's surface) The downward force exerted on an object due to its gravitational interaction with the earth. An object's weight is equal to the product of that object's mass times the acceleration due to gravity. The direction of the weight is always toward the center of the earth.

Review Questions

1. If nothing is pushing on a moving object, why doesn't it come to a stop?

2. One car is going north at 88 km/h (55 mph) while another car is going east at 88 km/h. Compare their velocities.

3. When your car loses traction on the ice, it travels in a straight line. Why?

4. Which of the following is accelerating: a rocket traveling straight up at 10,000 km/h, a ball hitting a wall, a person running around a circular track at 3 m/s, a hockey puck experiencing a net force of 5 N, and a television traveling along a conveyor belt at a constant velocity?

5. In what direction are these items accelerating: a bus starting forward from a bus stop, a taxi stopping to pick up a passenger, a croquet ball being struck by a mallet, a bean-bag hitting a wall, a bicycle turning right at an intersection, a diver who has just run off a diving platform, a wagon experiencing a net force toward the right, a pilot performing a loop-the-loop in an airplane, and a ball swinging around in a circle at the end of a string?

6. What is the difference between 3 m and 3 km?

7. If you were to go far away from the earth, into empty space, your weight would diminish but your mass would remain unchanged. Why?

8. If an object accelerates steadily, what happens to its velocity?

9. An accelerating object doesn't always travel the same distance each second. When will the distance an object travels each second increase with the passage of time?

10. When will the distance an accelerating object travels each second decrease with the passage of time?

11. Can you think of a case in which an accelerating object continues to travel the same distance each second?

12. After you throw a ball upward, when is its acceleration most rapid: on its way up, at the top, or on its way down?

13. Why doesn't an object's downfield motion affect its rise and fall in height?

Exercises

1. A dolphin can leap several meters above the ocean's surface. Why doesn't gravity stop the dolphin from leaving the water?

2. Some shoes have lights that flash with every step. These lights are triggered by a drop of liquid that moves inside the shoe. Which way does the liquid travel within the shoe when the shoe suddenly stops moving forward?

3. The back of your car seat has a headrest to protect your neck during a collision. What type of collision causes your head to press against the headrest?

4. As you jump across a small stream, does a horizontal force keep you moving forward? If so, what is that force?

5. A blacksmith usually hammers hot metal on the surface of a massive steel anvil. Why is this more effective than hammering the hot metal on the surface of a thin steel plate?

6. One type of home coffee grinder has a small blade that rotates very rapidly and cuts the coffee beans into powder. Nothing prevents the coffee beans from moving so why don't they get out of the way when the blade begins to push on them?

7. If you pull slowly on the top sheet of pad of paper, the whole pad will move. But if you yank suddenly on that sheet, it will tear away from the pad. What causes these different behaviors?

8. Why is it so difficult to start moving forward or come to a stop when you are wearing roller skates on your feet?

9. A yellow car is heading east at 100 km/h while a red car is heading north at 100 km/h. Do they have the same velocities?

10. Why is your velocity continuously changing as you ride on a carousel?

11. You have been walking north and suddenly come to a stop. What was the direction of your acceleration as you stopped?

12. A ball falls from rest for 5 seconds. Neglecting air resistance, during which of the 5 seconds does the ball's speed increase most?

13. A diver leaps from a 50 m cliff into the water below. The cliff is not perfectly vertical so the diver must travel forward several meters in order to avoid the rocks beneath him. In fact, he leaps directly forward rather than upward. Explain why a forward leap allows him to miss the rocks.

14. The kicker in a sporting event isn't always concerned with how far downfield the ball travels. Sometimes the ball's flight time is more important. If he wants to keep the ball in the air as long as possible, which way should he kick it?

15. The heads of different golf clubs are angled to give the golf ball different initial velocities. The golf ball's speed remains almost constant, but the angle changes with the different clubs. Neglecting any air effects, how does changing the initial angle of the ball affect the distance the ball travels?

16. An acorn falls from a branch located 9.8 m above the ground. After 1 second of falling, the acorn's velocity will be 9.8 m/s downward. Why hasn't the acorn hit the ground?

17. Two children of equal size and weight run along a horizontal platform side by side and dive into a swimming pool at exactly the same instant. The only difference between the two children's dives is that one child jumps upward off the platform while the other child simply runs forward off the end. Which child reaches the water first?

18. A sprinter who is running a 200 m race travels the second 100 m in much less time than the first 100 m. Why?

19. When you apply the brakes on your bicycle, which way do you accelerate?

20. An unseatbelted driver can be injured by the steering wheel during a head-on collision. Why does the driver hit the steering wheel when the car suddenly comes to a stop?

21. If you drop a ball from a height of 4.9 m, it will hit the ground 1 s later. If you fire a bullet exactly horizontally from a height of 4.9 m, it will also hit the ground 1 s later. Explain.

22. In the movies, rooftop chases often involve death defying leaps from one building to another. If the two rooftops are at the same height, why must the leaper jump upward in order to cross the gap successfully?

23. Why do loose objects on the dashboard slide to the right when the car turns suddenly to the left?

24. If you drive fast down an icy road and slam on the brakes, your car will begin to slide. If the road is straight, your car will stay on it but if the road curves, your car may end up in a ditch. Why does the road's shape determine whether you stay on it or not?

Problems

1. If your car has a mass of 800 kg, how much force is required to accelerate it forward at 4 m/s^2?

2. If your car can accelerate at a steady rate of 4 m/s^2, how soon does it reach 88.5 km/h (55.0 mph or 24.6 m/s) when it starts from rest?

3. On Mars, the acceleration due to gravity is 3.71 m/s^2. What would a rock's velocity be 3 s after you dropped it on Mars?

4. How far would a rock fall in 3 s if you dropped it on Mars? (See Problem **3**.)

5. How would your Mars weight compare to your earth weight? (See Problem **3**.)

6. A basketball player can leap upward 0.5 m. What is his initial velocity at the start of the leap?

7. How long does the basketball player in Problem **6** remain in the air?

8. A sprinter can reach a speed of 10 m/s in 1 s. If the sprinter's acceleration is constant during that time, what is the sprinter's acceleration?

9. If a sprinter's mass is 60 kg, how much forward force must be exerted on the sprinter to make the sprinter accelerate at 0.8 m/s^2?

10. How much does a 60 kg person weigh on earth?

11. If you jump upward with a speed of 2 m/s, how long will it take before you stop rising?

12. How high will you be when you stop rising in Problem **11**?

13. How much force must a locomotive exert on a 12,000 kg box car to make it accelerate forward at 0.4 m/s^2?

14. How long will it take the box car in Problem **13** to reach its cruising speed of 100 km/h (62 mph or 28 m/s)?

15. The county fair has a game in which you can send a 2 kg metal block up a track by hitting a button with a huge hammer. If you hit the button hard enough, the block will hit a bell at the top of the track and you'll win a prize. How much does the block weigh?

16. When you pound the button in Problem **15**, the block accelerates upward at 1000 m/s^2. How much force must the button exert on the block to make it accelerate that quickly? (You can neglect gravity while making this calculation.)

17. If the block in Problems **15** & **16** starts up the track at 19.6 m/s, how long does it take before it stops rising?

18. If the bell in Problems **15**, **16**, & **17** is 20.0 m above the ground, will the block reach the bell before it stops rising?

19. During the summer, you're working at the local airport. You have several luggage carts to move. Some are full and others are empty, but you can't see inside them. The empty ones weigh 1000 N while the full ones weigh 10,000 N. What are the masses of these two types of carts?

20. The carts in Problem **19** roll on wheels, so you can ignore friction. If you exert a horizontal force of 200 N on one of the empty carts, how quickly will it accelerate?

21. How quickly will one of the full carts in Problems **19** & **20** accelerate if you exert a horizontal force of 500 N on it?

22. Use your answers to Problems **19**, **20**, & **21** to explain why you'd have no trouble identifying the full carts, even without looking inside them.

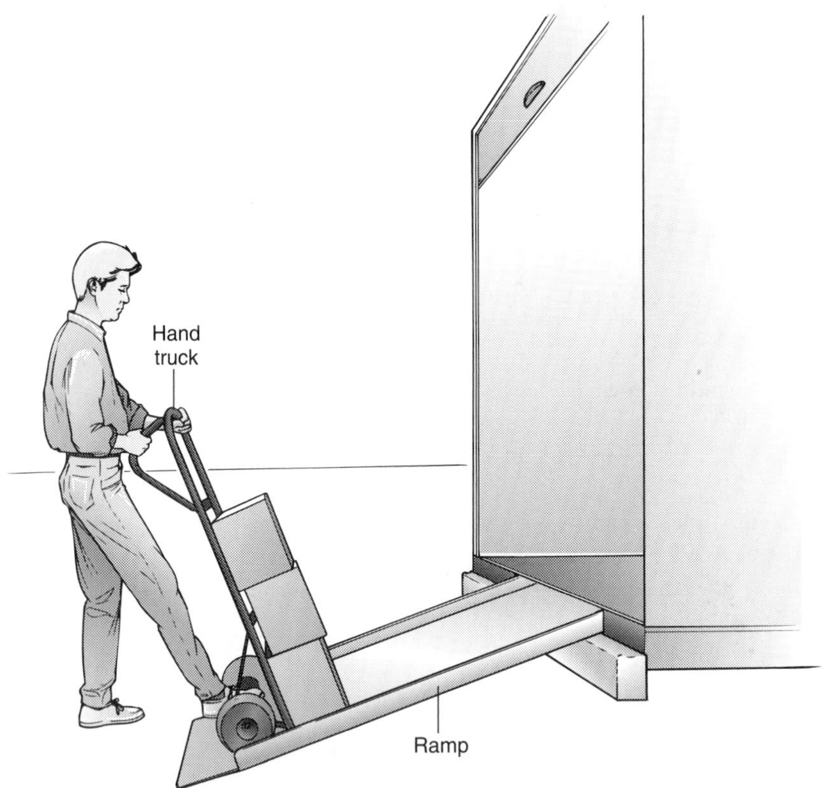

Section 1.2

Ramps

In the previous section, we looked at what happens to an object experiencing only a single force: the downward force of gravity. But what happens to objects that experience two or more forces at the same time? Imagine, for example, an object resting on the floor. That object experiences both the downward force of gravity and an upward force from the floor. If the floor is level, the object doesn't accelerate; but if the floor is tilted, so that it forms a ramp, the object accelerates downhill. In this section, we'll examine the motion of objects traveling along ramps. Also called inclined planes, ramps appear in many devices as tools to ease the movements of objects.

Questions to Think About: *How does a ramp make it possible for one person to lift a very heavy object? Why is it so much easier to lower a heavy object than to raise that same object? What is the difference between a heavy object resting on the ground and that same object high up in the air? Why is it scarier to ski or sled down a steep hill than it is to slide down a more gradual slope? Why is it more difficult to ride a bicycle up a steep hill than a gradual hill?*

Experiments to Do: *Place a book on a smooth, flat table or board. Hold the book steady for a second and then let go of it without pushing on it. What happens to the book?*

Now, equip yourself with a pencil. Have a friend tilt the surface of the table or board slightly so that the book begins to slide downhill. Can you stop the book by pushing on it with the pencil? Place the book back on the table and have your friend tilt the table more sharply. Does the tilt of the table affect your ability to stop the book from sliding? Now try to push the book uphill with the pencil (a) when the table is slightly tilted and (b) when it is more sharply tilted. Which task requires more force? Why?

Evidently a gentle push is all that may be needed to raise a relatively heavy object if you use a ramp to help you. To understand why this feat is possible, we need to explore a handful of physical concepts and a few basic laws of motion.

A Piano on the Sidewalk

Imagine that you have a friend who's a talented but undiscovered pianist. She's renting a new apartment, and because she can't afford professional movers (Fig. 1.2.1), she's asked you to help her move her baby grand. Fortunately, her new apartment is only on the second floor. But the two of you still face a difficult challenge: how do you get that heavy piano up there? More important, how do you keep it from falling on you during the move?

Fig. 1.2.1 - A ramp would make this move much easier.

The problem is that you can't push upward hard enough to lift the entire piano at once. One solution to this problem, of course, would be to break the piano into pieces and carry them up one by one. But this method has some obvious drawbacks: it would ruin the piano and disappoint your friend. A better solution would be to find something else to help you push upward, and one of your best choices would be the simple machine known as a ramp.

Throughout the ages, **ramps**, also known as inclined planes, have made tasks like piano-moving possible. Because ramps can exert the enormous upward forces needed to lift stone and steel, they've been essential building equipment since the days of the pyramids. To see how ramps provide these lifting forces, we'll continue to explore the example of the piano, looking first at the nonfrictional forces that the piano experiences when it touches a surface. For the time being, we'll continue to ignore friction and air resistance; they'd needlessly complicate our discussion. Besides, as long as the piano is on wheels, friction is negligible.

CHECK YOUR UNDERSTANDING #1: Brick Work
Which requires larger forces: lifting a pile of bricks one at a time or lifting them all together?

An Egg Hits the Floor: Newton's Third Law

If it hurts to drop a bowling ball on your foot, as we noted in the previous section, then it hurts even more to drop a piano on your foot. The reason either mishap hurts is that the falling object exerts a large downward force on your foot as it hits, and this force may damage your foot.

Now consider what happens when you drop an egg on the floor. By the time the egg hits, it too has acquired enough downward velocity to cause some damage. Dropped from sufficient height, that egg might make quite a crater. But the egg itself is also in trouble. When the egg arrives at the floor, the floor begins to exert an upward force on it to change its velocity and prevent it from passing through the floor. This force on the egg, localized in a small spot on the egg's surface, is what causes the egg to break.

The bowling ball, the piano, and the egg illustrate several important attributes of falling objects and floors. When a falling object reaches the floor, it exerts a downward force on the floor; that's why a falling ball or a dropped piano hurts your foot. But the floor also exerts an upward force on the object; that's why the falling egg breaks. These downward and upward forces are exactly equal in magnitude but exactly opposite in direction.

This observation, that two things exert equal but opposite forces on one another, isn't unique to floors and falling objects; in fact, it's always true. If you push on any object with a certain amount of force, that object will push on you with exactly the same amount of force in exactly the opposite direction. This rule—often expressed as "for every action, there is an equal but opposite reaction"—is known as **Newton's third law of motion**, the last of his three laws.

> **NEWTON'S THIRD LAW OF MOTION:**
> For every force that one object exerts on a second object, there is an equal but oppositely directed force that the second object exerts on the first object.

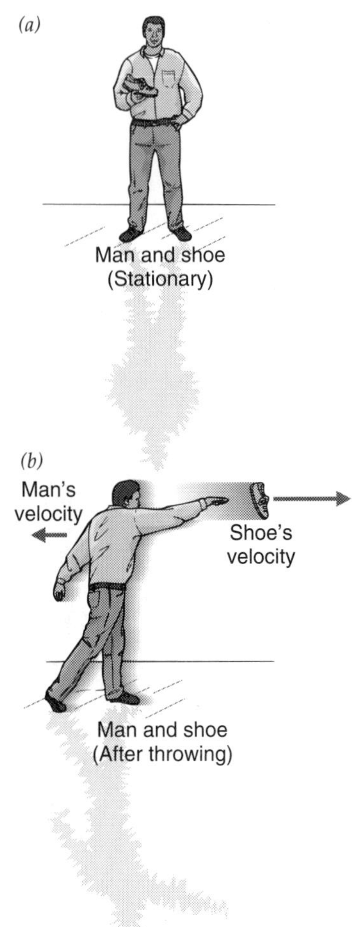

Fig. 1.2.2 - (*a*) With nothing to push him horizontally, this man is stuck on the slippery ice. (*b*) But when he pushes on his shoe, the shoe pushes back with an equal but oppositely directed force. The shoe accelerates to the right, the man to the left, and both slide off the ice in opposite directions.

One interesting but perhaps unlikely application of Newton's third law is in escaping from a slippery frozen pond. Suppose that you're stuck in the middle of a frozen pond, and that the ice is so slick that you can't obtain any horizontal force to help you off the pond (Fig. 1.2.2). Without a horizontal force, you won't accelerate sideways and your velocity will remain constant at zero.

To escape from the ice, you can take off your boot or any other object that you have with you and throw that object toward the farthest shore with all your might. By throwing the boot, you exert a force on it, and the boot accelerates and slides away from you in the direction you've thrown it. But because of Newton's third law, the boot also exerts a force on you. That force is equal to the force you exert on the boot but in exactly the opposite direction, and it makes you accelerate so that you also slide off the ice. Once you reach the shore, you can walk around the pond to retrieve the boot.

> **SUMMARY OF NEWTON'S LAWS OF MOTION:**
> 1. An object that is not subject to any outside forces moves at a constant velocity, covering equal distances in equal times along a straight-line path.
> 2. The force exerted on an object is equal to the product of that object's mass times its acceleration. The acceleration is in the same direction as the force.
> 3. For every force that one object exerts on a second object, there is an equal but oppositely directed force that the second object exerts on the first object.

> **CHECK YOUR UNDERSTANDING #2: Swinging**
>
> You are pushing a child on a playground swing. If you exert a 50 N force on him as he is swinging away from you, how much force will he exert back on you?

Adding Up the Forces

So far we've examined two sources of force: gravity and the floor. Objects near the earth experience gravity all the time, but they only feel a force from the floor when they touch its surface. Since a floor doesn't let most objects pass through it, it exerts upward **support forces** on them to keep them away. An upward support force can stop a falling object by accelerating that object upward, or it can hold up an object that's resting on the floor's surface. The piano in Fig. 1.2.3 is being held up by a support force from the sidewalk.

A support force is always directed exactly away from the surface that creates it. In the absence of friction, for example, a level floor doesn't push you to the left or to the right; it pushes you exactly upward. Forces that are directed exactly away from surfaces are called **normal forces**, since the term **normal** is used by mathematicians to describe something that points exactly away from a surface at right angles to that surface.

Let's return to the motionless piano in Fig. 1.2.3. You know from experience that even a motionless object has a weight; you feel "heavy" even when you're standing still, and you feel the weight of a motionless person sitting on your shoulders. So why isn't the piano in Fig. 1.2.3 accelerating downward? The reason is that the sidewalk is exerting just enough upward support force on the piano to cancel out its downward weight. The balance is perfect: the forces are of exactly identical magnitudes and in exactly opposite directions. The piano doesn't accelerate and remains motionless.

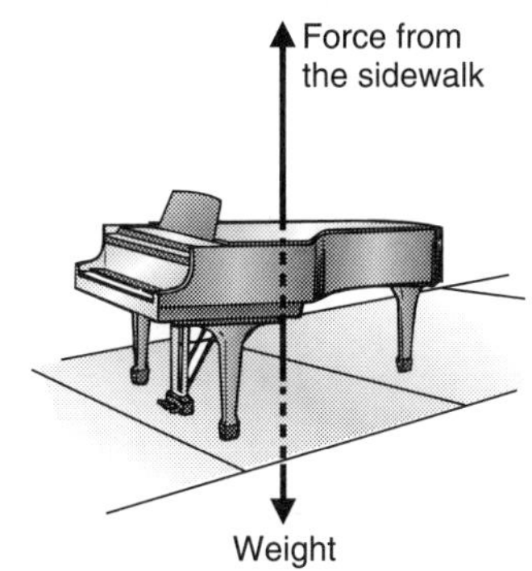

Fig. 1.2.3 - A piano resting on the sidewalk. The sidewalk exerts an upward force that exactly balances the piano's downward weight. The net force on the piano is zero, so the piano doesn't accelerate.

This perfect balance is reached because the support force from the sidewalk adjusts automatically to the piano's presence. If the sidewalk weren't exerting enough upward force on the piano, the piano would accelerate downward and penetrate the sidewalk's surface; if the sidewalk were exerting too much upward force on the piano, the piano would accelerate upward and move away from the sidewalk. A balance is quickly reached where the piano neither penetrates nor moves away from the sidewalk's surface.

Another way to state that the upward support force on an object exactly cancels the object's downward weight is to say that the **net force** on the object is zero, meaning that the sum of all the forces on the object is zero. Objects often experience more than one force at a time, and it's the net force, together with the object's mass, that determines how it accelerates. When several forces are exerted in the same direction, they add up, assisting one another so that the object accelerates in that direction (Fig. 1.2.4*a*). When several forces are exerted in opposite directions, they oppose and at least partially cancel one another (Fig. 1.2.4*b*).

Sometimes forces are exerted in two or more different directions so that the net force is pointing in yet another direction (Fig. 1.2.4*c*). For example, if you're standing on a frozen pond and someone pushes you north while some else pushes you east, the net force will point somewhere toward the north-east and you'll accelerate in that direction. The precise angle of the net force and your subsequent acceleration depend on exactly how hard each person pushes. For most of the following discussion, we'll only need a rough estimate of the net force's magnitude and direction, and we'll obtain that estimate using common sense.

CHECK YOUR UNDERSTANDING #3: Riding the Elevator
As you ride upward in an elevator at a constant velocity, what two forces act on you and what is the net force on you?

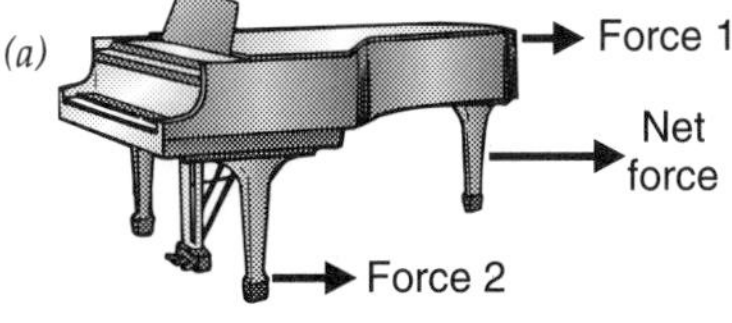

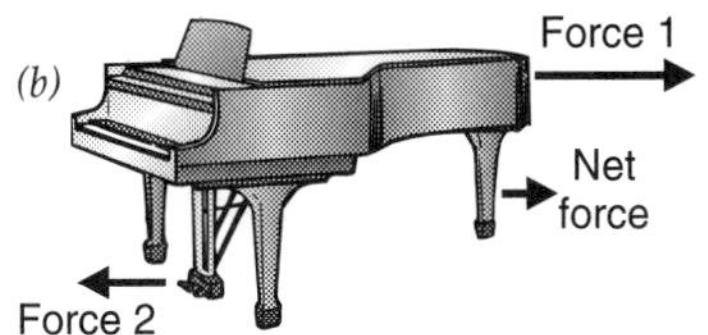

Fig. 1.2.4 - When several forces act simultaneously on an object, the object responds to the sum of the forces. This sum is called the net force, and it has both a magnitude and a direction. Here, as elsewhere in this book, the length of each force arrow indicates its magnitude.

Lifting the Piano: Energy and Work

As you approach the task of moving your friend's piano, you might begin to worry about safety. There is clearly a difference between the piano resting on the sidewalk and the piano suspended on a board just outside the second floor apartment. After all, which one would you rather be sitting beneath? The elevated piano has something that the piano on the sidewalk doesn't have: the ability to produce motion and structural rearrangement (i.e., damage) in itself and the things beneath it. This capacity to make things happen is called *energy*, and the process of making them happen is called *work*.

Energy and work are both important physical *quantities*, meaning that both are measurable. For example, you can measure the amount of energy in the suspended piano, and you can measure the amount of work the piano does when the board breaks and it falls to the sidewalk. As you may suspect, the physical definitions of energy and work are somewhat different from those of common English. Physical **energy** isn't the exuberance of a 5 year old at the amusement park or the contents of a large cup of coffee; instead, it's defined as the capacity to do work. Similarly, physical **work** doesn't refer to activities at the office or in the yard; instead, it refers to the process of transferring energy.

Energy is what's transferred, and work does the transferring. The most important characteristic of energy is that it's **conserved**. In physics, a conserved quantity is one that can't be created or destroyed but can be transferred between objects or converted from one form to another. Conserved quantities are very special in physics; there are only a few of them, and energy is one. An object that has energy can't simply make that energy disappear; it can only get rid of energy by giving it to another object, and it makes this transfer by doing work on that object.

This situation is analogous to money and spending: money is what is transferred, and spending does the transferring. Sensible, law-abiding citizens don't create or destroy money; they simply transfer it among themselves through spending. Just as the most interesting aspect of money is spending it, so the most interesting aspect of energy is doing work with it. We can define money as the capacity to spend, just as we define energy as the capacity to do work.

So far we've been using a circular definition: work is the transfer of energy and energy is the capacity to do work. But what is involved in doing work on an object? You do work on an object by exerting a force on it as it moves in the direction of that force. As you throw a ball, exerting a forward force on the ball and moving the ball forward, you do work on the ball; as you lift a rock, pushing upward and moving the rock upward, you do work on the rock. In both cases, you transfer energy from yourself to an object by doing work on it.

Sometimes transferring energy to an object produces an obvious change in that object. In these situations, the added energy is easy to observe. When you throw a ball, it picks up speed and its energy increases, so that it can then do work on the objects it hits. The moving ball has energy of motion or **kinetic energy**. When you lift a rock, its distance from the earth increases and so does its energy, so that it can then do work on the objects beneath it. Energy stored in the forces between or within objects is called **potential energy**, and since the rock's energy is stored in the gravitational forces between the earth and the rock, its energy is called **gravitational potential energy**.

If you spent the day holding up a heavy anvil it might seem as though you were doing an awful lot of work on it. However, the anvil's energy isn't changing; you're not transferring any energy to it because you're not doing any work on it. To do work on an object, you have to push on it while covering some distance in the direction you're pushing. The amount of work you do on the object depends on how hard you push and how much distance the object covers. We can express this relationship as a word equation:

$$\text{work} = \text{force} \cdot \text{distance}, \tag{1.2.1}$$

in symbols:

$$W = \mathbf{F} \cdot \mathbf{d},$$

and in everyday language:

If you're not pushing or it's not moving, then you're not working.

Note that the force and the distance traveled must be *in the same direction*.

Sometimes the force and distance traveled aren't in the same direction or are in the opposite directions. In these cases, the work you do on an object is the product of how far it travels times the component of your force along its direction of motion. If you push more or less in the direction of its motion, you do work on the object. If you push more or less in the direction opposite its motion, the amount of work you do on the object is negative, and the object is doing work on you. If you push at right angles to its motion or it doesn't move, no work is done at all.

The fact that when you do negative work on an object, it does work on you is the reason why energy is conserved. When you lift an anvil, you push it up as it moves up and do work on it. At the same time, it pushes your hand down but your hand moves up, so it does negative work on your hand. Overall, the anvil's energy increases by exactly the same amount that your energy decreases! You are transferring energy to the anvil. When you lower the anvil, the process is reversed and it transfers energy to you.

Let's return again to the piano on the sidewalk. Pretend that you're strong enough to carry it up a ladder all by yourself. You do work on the piano as you lift it because you exert an upward force on it and it moves upward. You're transferring energy to the piano, energy that originates in the food you eat, and the piano's gravitational potential energy increases. If you subsequently lower the piano, it does work on you. The piano transfers energy to you and its gravitational potential energy decreases. While you receive energy from the piano, your body can't turn that energy back into food energy. Instead, you become hotter, as we'll discuss in Section 1.4. But despite your body's inability to reuse work done on it, it's usually easier to have work done on you than to do work on something else. That's why it's harder to lift most objects than to lower them.

CHECK YOUR UNDERSTANDING #4: Pitching
When you throw a baseball horizontally, you're not pushing against gravity. Are you doing any work on the baseball?

CHECK YOUR FIGURES #1: Light Work, Heavy Work
You are moving books to a new shelf, 1.2 m above the old shelf. The books weigh 10 N each and you have 10 of them to move. How much work must you do on them as you move them? Does it matter how many you move at once?

Gravitational Potential Energy

Just how much work do you do in carrying the piano straight up to the apartment? You must exert an upward force on the piano equal to its weight and push it upward from the sidewalk to the second floor. Actually, to get the piano moving, you must push a little harder at the beginning; the net force on the piano then points upward and the piano accelerates upward. Once the piano is rising, however, you only have to support its weight so that the net force on it is zero. The piano then continues to move upward at constant velocity. Because you're pushing upward and the piano is moving upward, the work you do on the piano is the product of the piano's weight times the distance you lift it.

The piano's gravitational potential energy increases as it rises. The amount of that increase is equal to the work you do on the piano in lifting it. If we agree that the piano has zero gravitational potential energy when it rests on the sidewalk, then the suspended piano's gravitational potential energy is simply its weight times its height above the sidewalk. Since the piano's weight is equal to its mass times the acceleration due to gravity, its gravitational potential energy is its mass times the acceleration due to gravity times its height above the sidewalk.

These ideas are not limited to pianos. You can determine the gravitational potential energy of any object by multiplying its mass times the acceleration due to gravity times its height above the level at which its gravitational potential energy is zero. This relationship can be expressed as a word equation:

$$\text{gravitational potential energy} = \text{mass} \cdot \text{acceleration due to gravity} \cdot \text{height}, \qquad (1.2.2)$$

in symbols:

$$U = m \cdot g \cdot h,$$

and in common language:

The bigger they are, the harder they fall.

Of course, if you know the object's weight, you can use it in place of the object's mass times the acceleration due to gravity.

So what is the piano's gravitational potential energy when it reaches the second floor? If it weighs 2000 newtons (450 pounds) and the second floor is 5 meters above the sidewalk, you will have done 10,000 newton-meters of work in lifting it up there, and the piano's gravitational potential energy will thus be 10,000 newton-meters. The **newton-meter** is the SI unit of energy and work; it's so important that it has its own name, the **joule** (abbreviated J). At the second floor, the piano's gravitational potential energy is 10,000 J (10,000 joules).

A few everyday examples should give you a feeling for how much energy a joule is. Lifting a liter bottle of water 10 centimeters (4 inches) upward requires about 1 J of work. A 100-watt light bulb needs 100 J every second to operate. Your body is able to extract about 2,000,000 J from a slice of cherry pie. When you're bicycling or rowing hard, your body can do about 1000 J of work each second. A typical flashlight battery has about 10,000 J of stored energy.

CHECK YOUR UNDERSTANDING #5: Mountain Biking

Bicycling to the top of a mountain is much harder than rolling back down to the bottom. At which place do you have the most gravitational potential energy?

CHECK YOUR FIGURES #2: Watch Out Below

If you carry a United States penny (0.0025 kg) to the top of the Empire State Building (449 m), how much gravitational potential energy will it have?

Lifting the Piano with a Ramp

Unfortunately, you probably can't carry a grand piano up a ladder by yourself. You're going to need a ramp. What happens when you place the piano on a ramp? The ramp exerts a support force on the piano to prevent the piano from passing into its surface. However, that support force doesn't point directly upward (Fig. 1.2.5). Except for friction, which we're ignoring for now, the ramp doesn't resist motion along its surface, and it pushes the piano directly away from (or normal to) its surface. Since the ramp isn't exactly horizontal, this support force isn't exactly vertical. Because the piano's weight is directed straight down and the ramp's support force isn't directed straight up, the two forces can't balance one another. As a result, there is a non-zero net force on the piano.

Most of the piano's downward weight is balanced by the nearly upward support force of the ramp's surface. The components of force normal to the ramp's surface cancel one another, but the components of force along the ramp's surface don't cancel. Because the small net force that remains points down the ramp, the piano accelerates down the ramp. But because the net force is much smaller than the piano's weight, the acceleration down the ramp is slower than it would be if the piano were falling freely. This effect is familiar to anyone who has bicycled downhill or watched a glass slip slowly off a tilted table. Gravity still accelerates these objects, but more slowly than falling and in the direction of the downward slope.

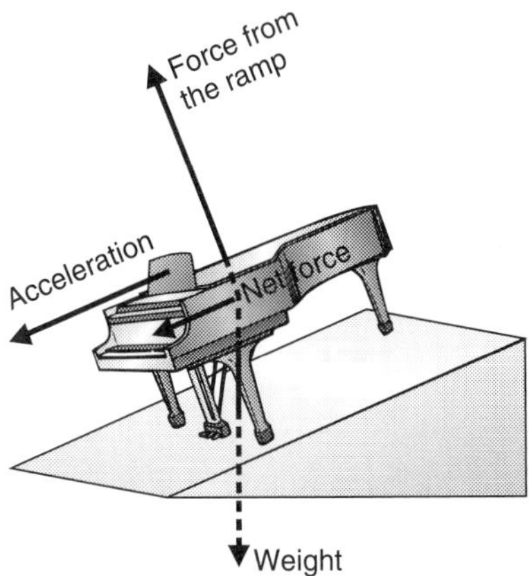

Fig. 1.2.5 - A piano sliding on a ramp while experiencing a force due to gravity. It accelerates down the ramp more slowly than it would if it were falling freely.

Herein lies the beauty of the ramp. By putting the piano on a ramp, you let the ramp supply most of the force needed to keep the piano from accelerating downward. The piano only experiences a small residual net force which pushes it downhill along the ramp. If you now push uphill on the piano with a force that exactly balances the downhill force, the new net force on the piano is zero and the piano stops accelerating. If you push uphill a little harder, the piano will accelerate up the ramp!

How does the ramp change the job of moving the piano? Suppose that you build a 50 m ramp that extends from the sidewalk to the apartment's balcony, 5 m above the pavement. This ramp is sloped so that traveling 10 m uphill along its surface actually lifts the piano only 1 m upward (Fig. 1.2.6). You can push the 2000 N piano up this 10 to 1 grade at constant velocity with a force of only 200 N (45 pounds). Most people can push that hard, so the moving job is now realistic. To reach the apartment, you must push the piano 50 m along this ramp with a force of 200 N so that you will do a total of 10,000 J of work.

By pushing the piano up a ramp, you've used physical principles to help you perform a task that would otherwise be nearly impossible. But you didn't get something for nothing. The ramp is much longer than the ladder, and you have had to push the piano for a longer distance in order to raise it to the second floor. Of course, you have had to push with less force.

Remarkably, the amount of work you do in either case is 10,000 J. In carrying the piano up the ladder, the force you exert is large but the distance the piano travels in the direction of that force is small. In pushing the piano up the ramp, the force is small but the distance is large. Either way, the final result is the same: the piano ends up on the second floor with an additional 10,000 J of gravitational potential energy and you have done 10,000 J of work. Expressed graphically in an equation, this relationship would appear:

$$\text{work} = \text{large force} \cdot \text{small distance}$$
$$= \text{small force} \cdot \text{large distance}.$$

(a)

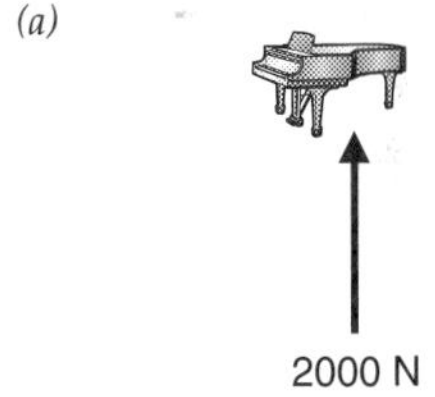

(b)

Fig. 1.2.6 - To lift a piano weighing 2000 N, you can either (*a*) push it straight up or (*b*) push it along a ramp. To keep the piano moving at a constant velocity, you must make sure it experiences a net force of zero. If you lift it straight up, you must exert an upward lifting force of 2000 N to balance the piano's downward weight. If you push it up the ramp shown in (*b*), you will only have to push the piano uphill with a force of 200 N in order to give the piano a net force of zero.

In the absence of friction, the amount of work you do on the piano to get it to the second floor doesn't depend on how you raise it. By doing work on the piano, you're increasing its energy, transferring energy to it from yourself. The amount of energy you transfer to the piano depends only on how its situation changes, not on how you achieve this change. No matter how you move that piano up to the second floor, you'll have to do 10,000 J of work on it and its energy will have to increase by 10,000 J. Even if you disassemble the piano into parts, carry them individually up the stairs, and reconstruct the piano in your friend's living room, you will have done 10,000 J of work lifting the piano.

Unless you're an experienced piano tuner, you'll probably be better off sticking with the ramp. It offers an easy method for one person to lift a baby grand piano. The ramp provides **mechanical advantage**, the process whereby a mechanical device redistributes the amounts of force and distance that go into performing a specific amount of mechanical work. In moving the piano with the help of the ramp, you've performed a task that would normally require a large force over a small distance by supplying a small force over a large distance. You might wonder whether the ramp itself does any work on the piano; it doesn't. Although the ramp exerts a support force on the piano and the piano moves along the ramp's surface, this force and the distance traveled are at right angles to one another. The ramp does no work on the piano.

Mechanical advantage occurs in many situations involving ramps. For example, it appears when you ride a bicycle up a hill. Climbing a gradual hill that is 500 m (1640 feet) high takes far less uphill force than climbing a steep hill of the same height. Since your peddling ultimately provides the uphill force, it's much easier to climb the gradual hill than the steep one. Of course, you must travel a longer distance along the road as you climb the gradual hill than you do on the steep hill, so the work you do is the same in either case. (If you use a mountain bike to peddle up one of these hills, you will also obtain mechanic advantage from gears and levers, as we'll discuss later on in this book.)

Ramps and inclined planes show up in many devices, where they usually increase the forces available at the expense of distances and allow us to perform tasks that would otherwise be difficult. They also change the character of certain activities. Skiing wouldn't be very much fun if the only slopes available were horizontal or vertical. By choosing ski slopes of various grades, you can select the net forces that set you in motion. Gentle slopes leave only small net forces and small accelerations; steep slopes produce large net forces and large accelerations. The steeper the slope, the more rapid the downhill acceleration and the larger the maximum downhill velocity. (The existence of a maximum downhill velocity is due to friction and air resistance, which we're ignoring for the moment but will return to later in this book.)

Finally, our observation about mechanical advantage is this: mechanical advantage allows you to do the same work, but you must make a trade-off: you must choose whether you want a large force or a large distance. The product of the two parts, force times distance, remains the same.

Fig. 1.2.7 - An access ramp allows this woman to raise herself to the height of the door using modest forces exerted over a long distance.

CHECK YOUR UNDERSTANDING #6: Access Ramps

Ramps for handicap entrances to buildings are often quite long and may even involve several sharp turns (Fig. 1.2.7). A shorter, straighter ramp would seem much more convenient. What consideration leads the engineers designing these ramps to make them so long?

Summary

How Ramps Work: An object at rest on the floor experiences two forces: its weight and a support force from the floor. The floor exerts an upward support force on the object that exactly balances the object's downward weight. The net force on the object is zero. If the floor is tilted to form a ramp, the support force is no longer directly upward and the net force on the object can't be zero. The net force is downhill, along the ramp, and is equal to the weight of the object multiplied by the ratio of the rise over the distance traveled along the ramp's surface. If the ramp is designed so that the object rises 1 m in height as it moves 10 m along the ramp, then this ratio is 1 m divided by 10 m or 0.10. The net force downhill along the ramp will thus be only 10% of the object's weight.

To stop an object from accelerating down a ramp, you must exert enough force up the ramp to exactly balance the force it experiences down the ramp. If you exert more force up the ramp than it experiences down the ramp, the object will begin to accelerate up the ramp. It takes less force to push an object up a ramp than to lift it directly upward, but you must push that object a long distance along the ramp. Overall, the work you do in raising the object remains the same as if you simply lifted it straight up.

The ramp gives you mechanical advantage, allowing you to do work that would require an unrealistically large amount of force for a short distance by instead exerting a much smaller force for a much longer distance.

Important Laws and Equations

1. Newton's Third Law: For every force that one object exerts on a second object, there is an equal but oppositely directed force that the second object exerts on the first object.

2. The Definition of Work: The work done on an object is equal to the product of the force exerted on that object times the distance that object travels in the direction of the force, or

$$\text{work} = \text{force} \cdot \text{distance}\,. \tag{1.2.1}$$

3. Gravitational Potential Energy: An object's gravitational potential energy is its mass times the acceleration due to gravity times its height above a zero level, or

$$\text{gravitational potential energy} = \text{mass} \cdot \text{acceleration due to gravity} \cdot \text{height}\,. \tag{1.2.2}$$

The Physics of Ramps

1. For every force that one object exerts on a second object, there is an equal but oppositely directed force that the second object exerts on the first object.

2. When two objects push each other away, they both accelerate but in opposite directions.

3. Stopping a moving object requires a force in the direction opposite that object's velocity. The faster you wish to stop the object, the more force you must exert on it.

4. When several forces act simultaneously on an object, they can be considered together as exerting a single net force on that object. The object accelerates as though experiencing only this single net force. The net force is the sum of all the exerted forces, taking into account the directions of those forces.

5. Two objects that touch one another exert support forces on one another. These support forces prevent the objects from passing into one another. Support forces are exerted normal to the surfaces.

6. The work you do on an object is the product of the force you exert on the object times the distance the object travels in the direction of that force. In doing work on that object, you are transferring energy from yourself to that object.

7. Raising an object a certain height takes a certain amount of work, no matter how you choose to perform that work. This work becomes gravitational potential energy in that object.

8. Energy is the capacity to do work. An object that is high in the air, or traveling at high speed has more energy than that same object resting motionless on the ground.

Check Your Understanding - Answers

1. Lifting the bricks all together.

Why: To lift a brick, you must exert an upward force on it that is greater than its downward weight so that the brick will begin to accelerate upward. If you must lift several bricks at once, you will have to accelerate all of them upward together. Each one will require a large upward force and the overall upward force you exert will be much larger than for a single brick.

2. 50 N.

Why: If you exert a 50 N force on any object you encounter, whether it's moving or stationary, it will exert a 50 N force back on you. If you push on a friend's hands with 50 N of force, it doesn't matter whether your friend is stationary or moving or wearing roller-skates or even sound asleep: she will push back with 50 N of force. She has no choice in the matter. Similarly, if someone pushes on you, you will feel yourself pushing back. That's how Newton's third law works.

3. The two forces are the downward force of gravity (your weight) and an upward support force from the floor. They cancel, so that the net force on you is zero.

Why: Whenever anything is moving with constant velocity, it's not accelerating and thus has zero net force on it. Although the elevator is moving upward, the fact that you are not accelerating means that the car must exert an upward support force on you that exactly balances your weight. You experience zero net force.

4. Yes.

Why: Any time you exert a force on an object and the object moves in the direction of that force, you are doing work on the object. Since gravity doesn't affect horizontal motion, the work you do on the baseball as you throw it ends up in the baseball as kinetic energy (energy of motion). As anyone who has been hit by a pitch can attest, a moving baseball has more energy than a stationary baseball.

5. At the top.

Why: Bicycling up the mountain is hard because you must do work against the force of gravity. You are storing work as gravitational potential energy, which increases all the way to the top. Gravity then does work on you as you roll back down and your gravitational potential energy decreases.

6. The engineers must limit the amount of force needed to propel a wheelchair up the ramp at constant velocity. The steeper the ramp, the more force required.

Why: A person traveling in a wheelchair on a level surface experiences little horizontal force and can move at constant velocity with very little effort. But climbing a ramp at constant velocity requires a substantial uphill force equal in magnitude to the downhill force from gravity. The steeper the ramp, the more uphill force is needed to maintain constant velocity. A 12 to 1 grade (12 meters of ramp surface for each meter of height rise) is the accepted limit to how steep such a long ramp can be.

Check Your Figures - Answers

1. It takes 120 J, no matter how many you lift at once.

Why: To keep each book from accelerating downward, you must support its weight with an upward force of 10 N. You must then move it upward 1.2 m. The work you do pushing upward on the book as it moves upward is given by Eq. 1.2.1:

$$\text{work} = \text{force} \cdot \text{distance} = 10\ \text{N} \cdot 1.2\ \text{m} = 12\ \text{J}.$$

It takes 12 J of work to lift each book, whether you lift it together with other books or all by itself. The total work you must do on all 10 books is 120 J.

2. About 11 J.

Why: The penny's gravitational potential energy is given by Eq. 1.2.2:

$$\begin{aligned}\text{gravitational potential energy} &= 0.0025\ \text{kg} \cdot 9.8\ \text{N/kg} \cdot 449\ \text{m}\\ &= 11\ \text{N} \cdot \text{m} = 11\ \text{J}\end{aligned}$$

This 11 J increase in energy would be quite evident if you were to drop the penny. The penny would accelerate to very high speed (up to 338 km/h or 210 mph) and do lots of damage when it hit the ground.

Glossary

conserved quantity A physical quantity, such as energy, that is neither created nor destroyed within an isolated system when that system undergoes changes. A conserved quantity may pass among the objects within an isolated system but its total amount remains constant.

energy The capacity to do work. Each object has a precise quantity of energy, which determines exactly how much work that object could do in an ideal situation. The SI unit of energy is the joule.

gravitational potential energy Potential energy stored in the gravitational forces between objects.

joule (J) The SI unit of energy and work (synonymous with the newton-meter). Lifting a liter of water upward 10 centimeters near the earth's surface requires about 1 joule of work.

kinetic energy The form of energy contained in an object's translational and rotational motion.

mechanical advantage The process whereby a mechanical device redistributes the amounts of force and distance that go into performing a particular amount of mechanical work.

net force The sum of all forces acting on an object, considering both the magnitude of each individual force and its direction. The magnitude of the net force is often less than the sum of the magnitudes of the individual forces, since they often oppose one another in direction.

Newton's third law of motion For every force that one object exerts on a second object, a force of equal magnitude but opposite direction is exerted by the second object on the first object.

newton-meter (N·m) The SI unit of energy and work (synonymous with the joule).

normal Directed exactly away from (perpendicular to) a surface. A line that is normal to a surface meets that surface at right angles.

normal force The force exerted by a surface on an object pressing against it. This force is normal to the surface and acts to prevent the object from entering that surface. "Normal force" and "support force" can be used interchangeably.

potential energy The stored form of energy that can produce motion. Potential energy is stored in the forces between or within objects.

ramp An inclined plane that allows work to be done over a longer distance, thereby requiring less force.

support force A force that is exerted when two objects come into contact. Each object exerts a force on the other object to keep the two from passing through one another. Support forces are always normal, or perpendicular, to the surfaces of objects. "Support force" and "normal force" can be used interchangeably.

work The mechanical means of transferring energy. Work is defined as the force exerted on an object times the distance that object travels in the direction of the force. A large force exerted for a short distance or a small force exerted for a long distance can perform the same amount of work. The SI unit of work is the joule.

Review Questions

1. When two pool balls collide, what forces do they experience? How do the various forces compare in magnitude and direction?

2. If you push on the following objects with a leftward force of 25 N, which ones will push on you with a rightward force of 25 N: a wall, a bowling ball, a Ping-Pong ball, a stationary truck, a truck heading toward the left, a truck heading toward the right, a friend sitting on a park bench, a friend sitting on an office chair with wheels, a sleeping lion, a wide-awake tiger, and a partridge in a pear tree?

3. Two children are struggling over a blanket. They both pull hard but it doesn't move. What forces are acting on the blanket and why doesn't it accelerate?

4. What must you do to a bowling ball in order to transfer some of your energy to it?

5. Why is it easier to roll a bowling ball up a ramp than to lift it straight upward?

6. Which requires a larger force: rolling a cart up a ramp to a height of 1 m or lifting it straight up to a height of 1 m? Which involves traveling the greatest distance? Which involves doing the most work?

Exercises

1. A heavy barbell accidentally falls on the floor. It breaks through and falls to the floor below. What forces do the floor and the barbell experience during this event?

2. During the accident in Exercise **1**, does either of the two object do (positive) work on the other object?

3. We said early in this section that an egg dropped to the floor is more damaged than the floor is upon collision. However, that was assuming the egg was dropped from your hand. What might happen if the egg were dropped from a high altitude, say 10 stories up, and hit a car below? In that case, does either of the two objects do (positive) work on the other? Why or why not?

4. Why does less snow and other debris accumulate on a steep roof than on a shallow roof?

5. Which does more work in lifting a grain of rice over its head: an ant or a person? Use this result to explain how insects can perform seemingly incredible feats of lifting and jumping.

6. You roll a marble down a playground slide that starts level, then curves downward, and finally curves very gradually upward so that it's level again at the end. Where along its travel is the marble undergoing the largest acceleration?

7. Where does the marble in Exercise **6** have its greatest speed?

8. When you fly a kite, there is a time when you must do (positive) work on the kite. Is that time when you let the kite out or when you pull it in?

9. Your suitcase weighs 50 N. As you ride up an escalator toward the second floor, carrying that suitcase, you are traveling at a constant velocity. How much upward force must you exert on the suitcase to keep it moving with you?

10. Comic book superheros often catch a falling person only a hairsbreadth from the ground. Why would this rescue actually be just as fatal for the victim as hitting the ground itself?

11. What is the net force on (a) the first car, (b) the middle car, and (c) the last car of a metro train traveling at constant velocity?

12. A speedboat is pulling a water-skier with a rope, exerting a large forward force on her. The skier is traveling forward in a straight line at constant speed. What is the net force she experiences?

13. Why does a ball lose altitude more slowly when it rolls down a ramp than when it falls directly downward?

14. A car passes by, heading to your left, and you reach out and push it toward the left with a force of 50 N. Does this moving car push on you and, if so, with what force?

15. Which is larger: the force the earth exerts on you or the force you exert on the earth?

16. Why does a plate shatter when you drop it on a hard kitchen floor?

17. The steel ball in a pinball game rolls around a flat, tilted surface. If you flick the ball straight uphill, it gradually slows to a stop and then begins to roll downhill. Which way is the ball accelerating as it rolls uphill? downhill?

18. When you kick a soccer ball, which pushes on the other harder: your foot or the soccer ball?

19. The earth exerts a downward force of 850 N on a veteran astronaut as he works outside the space shuttle. What force (if any) does the astronaut exert on the earth?

20. You accidentally miss the doorway and run into the wall. You suddenly experience a backward force that is several times larger than your weight. What's the origin of this force?

21. The wall dented slightly when you hit it in Exercise **20**. Did you do work on the wall?

22. You're cutting wood with a hand saw. You have to push the saw away from you as it moves away from you and pull the saw toward you as it moves toward you. When are you doing work on the saw?

23. When a professional boxer takes a jab at his opponent, he pushes his hand forward. His hand accelerates forward until it reaches his opponent's chest, at which point the opponent recoils backward. Describe how energy is being transferred in this situation.

24. Are you doing work while kneading bread? If so, when?

25. Two teams are having a tug-of-war with a sturdy rope. It has been an even match so far, with neither team moving. What is the net force on the left team?

26. The left team in Exercise **25** pulls extra hard and the two teams begin to accelerate together toward the left. What is the net force on the left team now?

27. The two teams in Exercises **25** & **26** are now moving steadily toward the left in a straight line. What is the net force on the left team?

28. When you're roller skating on level pavement, you can maintain your speed for a long time. But as soon as you start up a gradual hill, you begin to slow down. What slows you?

29. As you slow down on the hill in Exercise **28**, what happens to your kinetic energy, your energy of motion?

30. You come to a complete stop on the hill in Exercises **28** & **29** and begin to roll backward, down the hill. What force is causing you to accelerate?

31. Why does stamping your feet clean the snow off of them?

32. Why does tapping your toothbrush on the sink dry it off?

Problems

1. The builders of the pyramids used a long ramp to lift 20,000 kg (22 ton) blocks. If a block rose 1 m in height while traveling 20 m along the ramp's surface, how much uphill force was needed to push it up the ramp at constant velocity?

2. How much work was done in raising one of the blocks in Problem **1** to a height of 50 m?

3. What is the gravitational potential energy of one of the blocks in Problem **1** if it's now 75 m above the ground?

4. As water descends from the top of a tall hydroelectric dam, its gravitational potential energy is converted to electric energy. How much gravitational potential energy is released when 1000 kg of water descends 200 m to the generators?

5. While throwing a football, you exert a forward force of 50 N on it and push it forward a distance of 1.2 m. How much work do you do on the football?

6. If you did the same work on the football in Problem **5** while throwing it straight up, how high would it go before all of the work you did on it became gravitational potential energy? (The football's mass is 0.450 kg).

7. The tire of your bicycle needs air so you attach a bicycle pump to it and begin to push down on the pump's handle. If you exert a downward force of 25 N on the handle and the handle moves downward 0.5 m, how much work do you do?

8. You're using a wedge to split a log. You are hitting the wedge with a large hammer to drive it into the log. It takes a force of 2000 N to push the wedge into the wood. If the wedge moves 0.2 m into the log, how much work have you done on the wedge?

9. The wedge in Problem **8** acts like a ramp, slowly splitting the wood apart as it enters the log. The work you do on the wedge, pushing it into the log, is the work it does on the wood, separating its two halves. If the two halves of the log only separate by a distance of 0.05 m while the wedge travels 0.2 m into the log, how much force is the wedge exerting on the two halves of the log to rip them apart?

10. You're sanding a table. You must exert a force of 30 N on the sandpaper to keep it moving steadily across the table's surface. You slide the paper back and forth for 20 minutes, during which time you move it 1000 m. How much work have you done?

11. You're climbing a 380 m tall mountain on the moon, where the acceleration due to gravity is 1.6 m/s^2. If your mass is 60 kg, how much does your gravitation potential energy increase as the result of your upward climb?

12. Compare the work you did lifting yourself up the moon mountain in Problem **11** with the work you would do climbing a similar mountain on earth. Use this result to explain why the Apollo astronauts were able to climb up and down moon hills with ease, even while carrying massive objects.

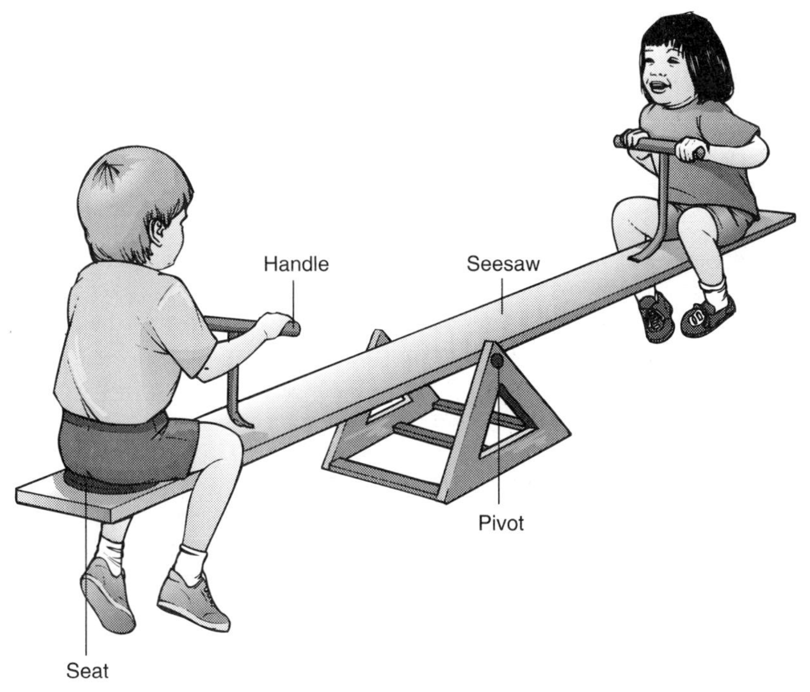

Section 1.3

Seesaws

The ramp that we examined in Section 1.2 is only one tool that provides mechanical advantage. In this section, we'll look at another such device: the type of lever known as a seesaw. As we discuss seesaws, we'll revisit many of the laws of motion that we encountered in the previous two sections. However, we'll see these laws in a new context: rotational motion.

Questions to Think About: *A playground seesaw only balances when the children riding it are properly situated. What do we mean by a balanced seesaw? Why does it matter just where the children sit on the board? What are they doing to make the balanced seesaw rock back and forth? Who is doing work on whom as the seesaw rocks back and forth? Along the same lines, how does a pry-bar—another kind of lever—help you pull out nails from a board or lift a very heavy object? Why does a hockey stick help a hockey player flick the puck with exceptionally high speed?*

Experiments to Do: *To get a feel for how levers work, find a rigid ruler with a hole in its center—the kind that can be clipped into a three-ring notebook. If you support the ruler by putting the tip of an upright pencil into the central hole, you'll find that the ruler balances; that is, it either remains stationary, at whatever orientation you choose, or rotates steadily about the central hole. (Eventually, the ruler comes to rest because of friction, a detail that we'll continue to ignore for now.) Now, push on one end of the ruler: what happens? Try pushing the ruler's end toward its central hole; what happens then? What is the most effective way to make the ruler spin?*

Now lay the pencil on a table and place the ruler flat on top of it so that the pencil and the ruler are at right angles, or perpendicular, to each other. If you center the ruler on the pencil, the ruler will balance. Load the two ends of the ruler with coins or other small weights, trying as you do to keep the ruler balanced. Try placing the coins at different positions relative to the pencil. Is there any way you can balance a light weight on one end with a heavy weight on the other end?

The Seesaw

Any child who has played on a seesaw with friends of different sizes knows that the toy works best for two children of roughly the same weight; evenly matched riders balance each other, and this balance allows them to rock back and forth easily. In contrast, when a light child tries to play seesaw with a heavy child, the heavy child's side of the seesaw drops rapidly and hits the ground with a thud, and the light child is tossed into the air.

There are several solutions to the heavy child/light child problem. Of course, two light children could try to balance one heavy child. But most children eventually figure out that if the heavy child sits closer to the seesaw's pivot, the board will balance (Fig. 1.3.1). The children can then make the board tip back and forth easily, just as it does when two evenly matched children ride it at its ends. This is a pretty useful trick, and we'll explore it in detail by the end of this section. First, though, we'll need to look carefully at the nature of rotational motion.

(a)

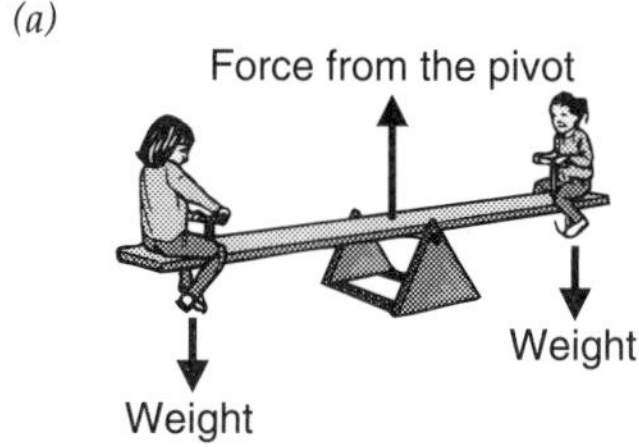

(b)

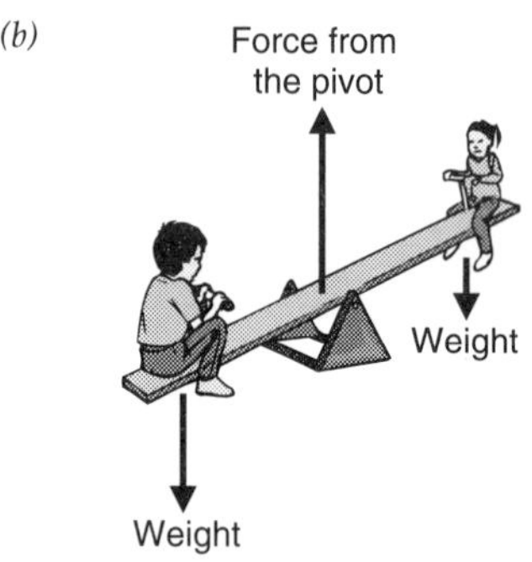

(c)

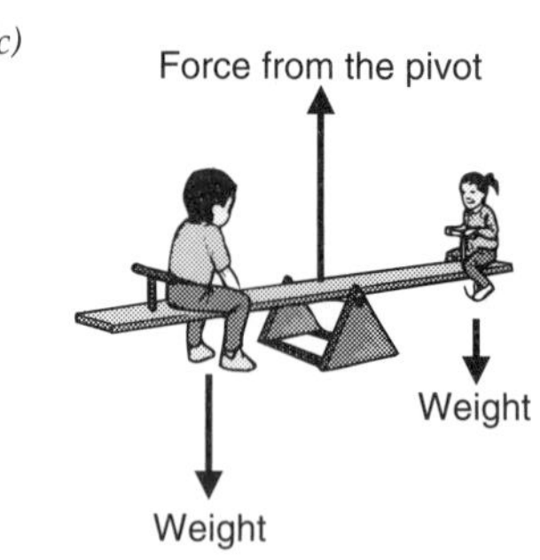

Fig. 1.3.1 - (*a*) When two children of equal weight sit at opposite ends of a seesaw, it balances. (*b*) When their weights are not equal, the heavy child descends. (*c*) But if the heavy child moves closer to the pivot, the seesaw can balance.

If, to keep things simple, we ignore the mass and weight of the board itself, three forces are acting on the seesaw shown in Fig. 1.3.1: two downward forces (the weights of the two children) and one upward force (the support force of the central pivot). Seeing those three forces, we may immediately think about net forces and begin to look for some overall acceleration of this toy and its riders. But we know that the seesaw remains where it is in the playground and isn't likely to head off for Kalamazoo or the center of the earth anytime soon. Because the seesaw's fixed pivot always provides just enough upward or sideways force to keep the board from accelerating as a whole in any particular direction, the board always experiences zero net force and never leaves the playground. Such overall movement of an object from one place to another is called **translational motion**. While the board never experiences this kind of motion, it can turn around the pivot, and thus it experiences a different kind of motion. Motion around a fixed point (which prevents rolling) is called **rotational motion**.

Rotational motion is what makes a seesaw interesting. The whole point of a seesaw is that it can rotate so that one child rises and the other descends. (You may not think of going up and down as rotating, but if the ground weren't there, the seesaw would be able to rotate in a big circle.) What causes the seesaw to rotate, and what observations can we make about the process of rotation?

To answer those questions, we'll need to examine several new physical quantities associated with rotation and explore the laws of rotational motion that relate them to one another. We'll do these things both by studying the workings of seesaws and other rotating objects and by looking for analogies between translational motion and rotational motion.

Imagine holding onto the seesaw in Fig. 1.3.1*a* to keep it level for a moment while the left child climbs off the board. Now imagine letting go of the board. As soon as you let go, the board will begin to rotate, and the child on the right will descend toward the ground. The board's motion will be fairly slow at first, but it will move more and more quickly until that child strikes the ground with a teeth-rattling thump.

If we focus only on the rotation itself, we might describe the motion of the seesaw board in the following way:

> "The board starts out not rotating at all. When we release the board, it begins to rotate clockwise. The board's rate of rotation increases

steadily in the clockwise direction until the moment the board strikes the ground."

This description sounds a lot like the description of a falling ball released from rest:

> "The ball starts out not moving at all. When we release the ball, it begins to move downward. The ball's rate of translation increases steadily in the downward direction until the moment the ball strikes the ground."

The statement about the seesaw involves rotational motion, while the statement about the ball involves translational motion. The similarities between these two descriptions are not coincidental; they're similar because the concepts and laws of rotational motion have many analogies in the concepts and laws of translational motion. The familiarity that we've acquired with translational motion will help us examine rotational motion.

CHECK YOUR UNDERSTANDING #1: Wheel of Fortune Cookies
The guests at a large table in a Chinese restaurant use a revolving tray, a lazy Susan, to share the food dishes. How does the lazy Susan's motion differ from that of the passing dessert cart?

The Motion of An Isolated Seesaw

Early in this chapter, we looked at the concept of translational inertia, which holds that a body in motion tends to stay in motion and a body at rest tends to stay at rest. This concept led us to Newton's first law of translational motion. Inserting the word "translational" here is a useful revision, because we're about to encounter analogous concepts associated with rotational motion. First we'll continue our examination of the rotational motion of an isolated seesaw board; once we've seen how such an isolated seesaw rotates, we can begin to explore how it responds to outside influences such as the pivot or a handful of young riders. Because of the similarities between rotational and translational motion, this section will closely parallel our earlier examination of falling balls.

Imagine an unoccupied seesaw somewhere out in space with nothing pushing on it or twisting it (Fig. 1.3.2). Such an isolated seesaw is free to turn in any direction, even completely upside down. You, the observer, are located near the seesaw and have nothing pushing on you or twisting you. When you look over at the seesaw, what does it do?

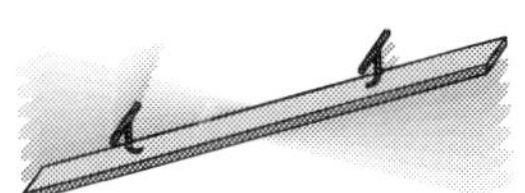

Fig. 1.3.2 - A seesaw that is all alone. Since nothing twists it, the seesaw rotates steadily about a fixed line in space.

If the seesaw is stationary, then it will remain stationary. However, if it's rotating, it will continue rotating at a steady pace, about a fixed line in space. What keeps the seesaw rotating? Its **rotational inertia**. A body that's rotating tends to remain rotating; a body that's not rotating tends to remain not rotating. Once again, that's how our universe works.

To describe the seesaw's rotational inertia and rotational motion more accurately, we'll need to understand several physical quantities associated with rotational motion. The first is the seesaw's orientation. At any particular moment, the seesaw is oriented in a certain way—that is, it has a particular **angular position**. Angular position describes the seesaw's orientation relative to some reference orientation; it can be specified by determining how far the seesaw has rotated away from its reference orientation and the pivot line about which that rotation has occurred. The seesaw's angular position points along the pivot line and has a magnitude equal to the amount of rotation. For example, to describe

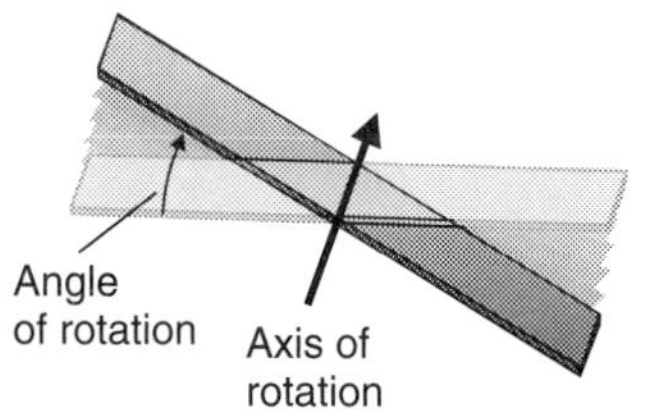

Fig. 1.3.3 - You can specify this seesaw's angular position, relative to its horizontal reference orientation, as the axis about which it was rotated to reach its new orientation and the angle through which it was rotated.

the seesaw's current angular position, we could define the reference orientation as horizontal, from left to right, and then figure out what line we would have to pivot the seesaw about and how far we would have to turn it to reach its current orientation (Fig. 1.3.3).

If the seesaw is rotating, then its angular position is changing; in other words, it has an angular velocity. **Angular velocity** is a vector quantity that measures how quickly the seesaw's angular position changes; it consists of the angular speed at which the seesaw is rotating and the axis or line about which that rotation proceeds. The seesaw's **angular speed** is the angular displacement it undergoes in a certain amount of time:

$$\text{angular speed} = \frac{\text{angular displacement}}{\text{time}},$$

where **angular displacement** is the difference between the angular position at the end of that period and the angular position at the beginning of that period.

The seesaw's **axis of rotation** is the line in space about which the seesaw is rotating. But we've neglected one subtle issue about this axis of rotation. Imagine that the seesaw is directly in front of you, turning once per second about a horizontal line pointing straight forward (its axis of rotation) (Fig. 1.3.4). While you know the seesaw's angular speed (1 rotation per second) and the line about which it's rotating (a forward horizontal line), you don't yet know which way the board is turning, clockwise or counter-clockwise. The direction part of angular velocity, therefore, must specify more than just the line about which the seesaw rotates; it must also specify whether that rotation is clockwise or counter-clockwise, a characteristic called the **sense of rotation**.

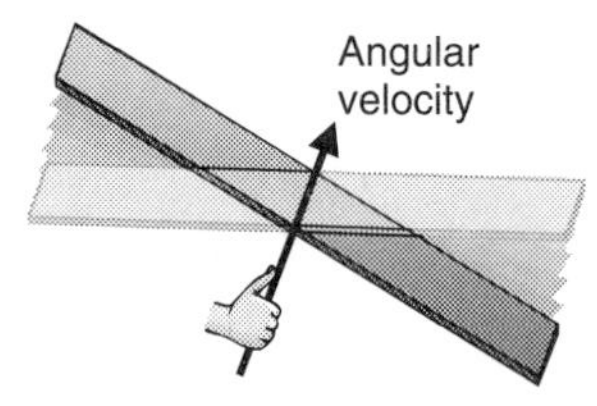

Fig. 1.3.4 - This seesaw is spinning about the rotation axis shown. The direction of the seesaw's angular velocity is defined by the right-hand rule.

To specify this sense of rotation, we take advantage of the fact that any line has two directions to it. Once we have identified the line about which the seesaw is rotating, we can look down that line at the seesaw from either direction. From one direction, the seesaw appears to be rotating clockwise; from the other direction, it appears to be rotating counter-clockwise. By convention, we choose the direction in which the seesaw appears to be rotating clockwise and say that the seesaw's axis of rotation points away from our eye toward the seesaw. This convention is called the **right-hand rule**, because if the four fingers of your right hand are curling around to point in the sense of the seesaw's rotation, then your thumb is pointing along the seesaw's axis of rotation (Fig. 1.3.4).

Remembering this convention isn't as important as understanding why we must specify the direction about which rotation occurs when describing a rotating object's angular velocity. Just as translational velocity consists of a translational speed and a direction in which the translational motion occurs, so angular velocity consists of a rotational speed and a direction about which the rotational motion occurs.

Now that we've developed these concepts of angular position and angular velocity, we can describe precisely the rotational motion of an isolated seesaw. The isolated seesaw has a particularly simple angular velocity because of its isolation and its rotational inertia. Since the isolated seesaw rotates at a steady pace, about a fixed axis of rotation, its angular velocity is constant and never changes. The seesaw just keeps on turning and turning, always at the same angular speed, always about the same axis of rotation.

As you might suspect, these observations about the rotational motion of an isolated seesaw are general. Taken together, they are **Newton's first law of rotational motion**, which states that a rigid object that is not wobbling and is not subject to any outside influences rotates at a constant angular velocity, turning equal amounts in equal times about a fixed axis of rotation. The outside influences referred to in this law are called **torques**, a technical term for twists and spins. When you twist off the lid of a jar or spin a top with your fingers, you're exerting a torque. The word "rigid" appears in this law because this law doesn't apply to an object that can change its shape as it rotates.

The phrase "not wobbling" appears in the law because the complicated motions of wobbling objects are governed by a more general principle that we'll learn in the next section. As long as an object is symmetric about its axis of rotation—like a ball or a seesaw board turning end over end—it won't wobble and will obey Newton's first law of rotational motion.

NEWTON'S FIRST LAW OF ROTATIONAL MOTION:
A rigid object that is not wobbling and is not subject to any outside torques rotates at a constant angular velocity, experiencing equal angular displacements in equal times as it turns about a fixed axis of rotation.

CHECK YOUR UNDERSTANDING #2: Going for a Spin

A rubber basketball floats in a swimming pool. It experiences zero torque, no matter which end of it is up. If you spin the basketball and then let go, how will it move?

The Seesaw's Center of Mass

Although you've never seen an isolated seesaw, you have seen other objects that are effectively isolated and nearly free from torques: a baton thrown overhead by a baton twirler, for example, or a football spiraling through the air off the fingers of a quarterback. These motions, however, are complicated because they rotate and translate at the same time. The spinning baton travels up and down and the turning football arcs through the air. How can we distinguish their translational motions from their rotational motions?

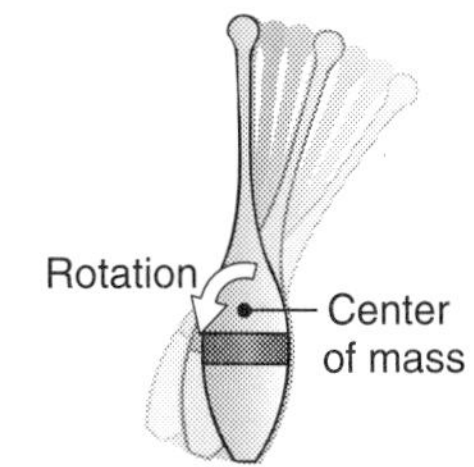

Fig. 1.3.5 - This club spins about its center of mass, which remains stationary.

Once again, we can make use of a wonderful simplification of physics. We can always find a particular point in or near an isolated object about which it naturally spins. The axis of rotation always passes right through this special point so that, as the object rotates, this point doesn't move unless the object has an overall translational velocity. We call this point the object's **center of mass**, and it's the single point about which all that object's mass is evenly balanced. The center of mass of a typical ball is at its geometrical center, while the center of mass of a less symmetrical object depends on how the mass of that object is distributed. You can begin to find a small object's center of mass by spinning it on a smooth table and looking for the fixed point about which it spins (Fig. 1.3.5).

This center of mass allows us to separate the object's translational motion from its rotational motion. As the football arcs through space, its center of mass follows the simple path we discussed in the section on falling balls. This translational motion isn't affected by any spin the ball might have. At the same time, the football's rotational motion about its center of mass is that of an isolated object—if it's not wobbling, it rotates with a constant angular velocity. The same separation is possible for a juggler's club (Fig. 1.3.6).

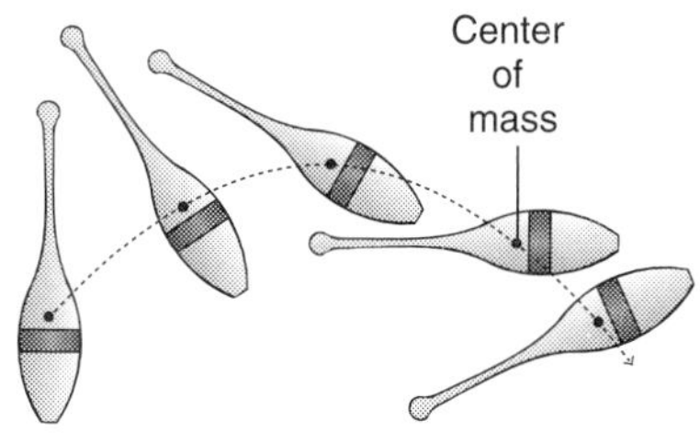

Fig. 1.3.6 - A juggler's club that is traveling through space rotates about its center of mass as its center of mass travels in the simple arc associated with a falling object.

While the seesaw is hardly an isolated object, we'll examine many effectively isolated objects in the course of this book, and it's worth remembering that their translational and rotational motions can be separated if we pay attention to their centers of mass. Even in the case of the seesaw, the pivot is strategically located at or very near the board's center of mass. As a result, the pivot prevents any translational motion of the seesaw while permitting nearly free rotational motion of the board about its center of mass, at least in one direction.

CHECK YOUR UNDERSTANDING #3: Tracking the High Dive

When a diver does a rigid, open somersault off a high diving board, his motion appears quite complicated. Can this motion be described simply?

How the Seesaw Responds to Torques

Why doesn't the isolated seesaw board change its rotational speed or axis of rotation? Because it has a moment of inertia. **Moment of inertia** is the measure of an object's *rotational* inertia, its resistance to changes in its *angular* velocity. It's analogous to mass, which is the measure of an object's *translational* inertia, its resistance to changes in its *translational* velocity. An object's moment of inertia depends both on its mass and on how that mass is distributed within the object. Because the seesaw has a moment of inertia, its angular velocity will change only if something twists or spins it. In other words, it must experience a torque.

Torque—our second important vector quantity of rotational motion—has both a magnitude and a direction. The more torque you exert on the seesaw, the more rapidly its angular velocity changes. Depending on the direction of the torque, you can make the seesaw turn more rapidly or less rapidly or even rotate about a different axis. But how do you determine the direction of a particular torque? One way is to imagine exerting this torque on a stationary ball floating in water (Fig. 1.3.7*a*,*b*). The ball will begin to rotate, acquiring a non-zero angular velocity (Fig. 1.3.7*c*). The direction of this angular velocity is that of the torque. The SI unit of torque is the **newton-meter** (abbreviated N·m).

(*a*)

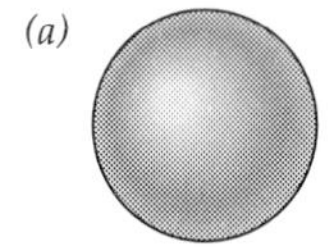

(*b*)

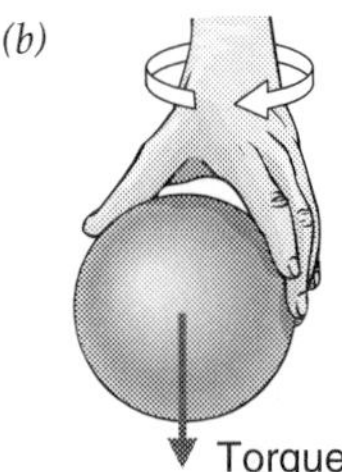

(*c*)

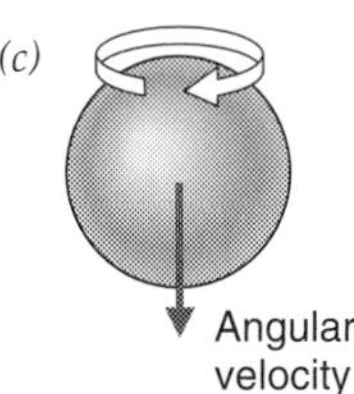

Fig. 1.3.7 - If you start with a ball that's not spinning (*a*), and twist it with a torque (*b*), the ball will acquire an angular velocity (*c*) that's in the same direction as that torque.

The larger an object's moment of inertia, the more slowly its angular velocity changes in response to a specific torque (Fig. 1.3.8). You can easily spin a basketball with the tips of your fingers, but it's much harder to spin a bowling ball. The bowling ball's larger moment of inertia comes about primarily because it has a greater mass than the basketball. But moment of inertia also depends on the shape of an object and how far its mass is from the axis of rotation. An object that has most of its mass located near the axis of rotation will have a smaller moment of inertia than an object of the same mass that has most of its mass located far from that axis. Thus a ball of pizza dough has a smaller moment of inertia than the finished pizza. The bigger the pizza gets, the harder it is to spin.

Fig. 1.3.8 - Spinning a merry-go-round is hard work because of its large moment of inertia. Despite the large torque exerted by this boy, the merry-go-round's angular velocity increases slowly.

Because an object's moment of inertia depends on how far its mass is from the axis of rotation, changes in the axis of rotation are likely to change its moment of inertia. For example, less torque is required to spin a tennis racket about its handle (Fig. 1.3.9*a*) than to flip the racket head-over-handle (Fig. 1.3.9*b*). When you spin the tennis racket about its handle, the axis of rotation runs right through the handle so that most of the racket's mass is fairly close to the axis and the moment of inertia is small. When you flip the tennis racket head-over-handle, the axis of rotation runs across the handle so that both the head and the handle are far away from the axis and the moment of inertia is large. The tennis racket's moment of inertia becomes even larger when you hold it in your hand and make it rotate about your shoulder rather than its center of mass (Fig. 1.3.9*c*).

When something exerts a torque on the seesaw, its angular velocity changes; in other words, it undergoes angular acceleration, our third important vector quantity of rotational motion. **Angular acceleration** measures how quickly the seesaw's *angular* velocity changes. It's analogous to acceleration, which measures how quickly an object's *translational* velocity changes. Just as with acceleration, angular acceleration involves both a magnitude and a direction. An object undergoes angular acceleration when its angular speed increases or decreases or when its angular velocity changes directions.

There is a simple relationship between the seesaw's angular acceleration, its moment of inertia, and the torque exerted on it. The torque exerted on the seesaw equals the product of the seesaw's moment of inertia times its angular acceleration. This relationship between torque, moment of inertia, and angular acceleration is **Newton's second law of rotational motion** and can be written in a word equation:

$$\text{torque} = \text{moment of inertia} \cdot \text{angular acceleration}, \tag{1.3.1}$$

in symbols:

$$\tau = I \cdot \alpha,$$

and in everyday language:

Spinning a marble is much easier than spinning a merry-go-round.

It resembles Newton's second law of translational motion (force=mass·acceleration), except that torque has replaced force, moment of inertia has replaced mass, and angular acceleration has replaced (translational) acceleration. However this new law doesn't apply to wobbling objects because they're being affected by more than one moment of inertia simultaneously (see the discussion of tennis rackets above) and follow a much more complicated law. (For another useful form of Eq. 1.3.1, see ❐.)

NEWTON'S SECOND LAW OF ROTATIONAL MOTION:
The torque exerted on an object that is not wobbling is equal to the product of that object's moment of inertia times its angular acceleration. The angular acceleration points in the same direction as the torque.

Let's take a moment to see how Eq. 1.3.1 works. As an equation, its left side must equal its right side. Any change in the torque you exert on a rigid object must be accompanied by a proportional change in its angular acceleration. As a result, the harder you twist or spin a seesaw, the more rapidly its angular velocity changes.

We can also compare the effects of a specific torque on two objects with different moments of inertia. In this case, since the left side of the equation doesn't change, the right side must remain constant. If we replace the seesaw with a baton, which has a much smaller moment of inertia, then the angular acceleration must increase to keep the right side of the equation constant. The angular velocity of a baton thus changes more rapidly than the angular velocity of a seesaw when the two are subjected to identical torques.

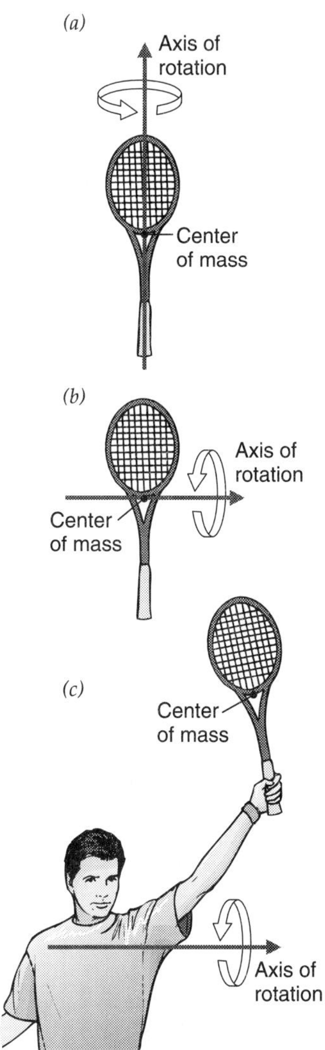

Fig. 1.3.9 - A tennis racket's moment of inertia depends on the axis about which it rotates. Its moment of inertia is small (*a*) when it rotates about its handle and large (*b*) when it rotates head-over-handle. (*c*) If you make it rotate about your shoulder, its moment of inertia becomes even larger.

❐ Equation 1.3.1 can be rearranged algebraically as:

$$\text{angular acceleration} = \frac{\text{torque}}{\text{moment of inertia}}.$$

This form shows that an object's angular acceleration depends both on the torque causing that angular acceleration and on the object's moment of inertia. Increasing the torque increases the angular acceleration while increasing the object's moment of inertia by reshaping it decreases the angular acceleration.

In sum:

1. The orientation of an object is its angular position.
2. The object's angular velocity measures how quickly its angular position changes.
3. The object's angular acceleration measures how quickly its angular velocity changes.
4. In order for the object to undergo angular acceleration, something must exert a torque on it.
5. The larger the object's moment of inertia, the more torque will have to be exerted on it in order for it to undergo a certain angular acceleration.

This summary of the physical quantities of rotational motion is analogous to the one for translational motion on p. 7. Take a moment to compare the two.

CHECK YOUR UNDERSTANDING #4: The Merry-Go-Round

The merry-go-round is a popular playground toy (Fig. 1.3.8). It's already hard to spin empty but when there are lots of children on it, it's even harder to start or stop. Why is it so difficult to change a full merry-go-round's angular velocity?

CHECK YOUR FIGURES #1: Hard to Turn

Automobile tires are normally hollow and filled with air. If they were made of solid rubber, their moments of inertia would be about 10 times as large. How much more torque would an automobile have to exert on a solid tire to make it undergo the same angular acceleration as a hollow tire?

Forces, Torques, and Seesaws

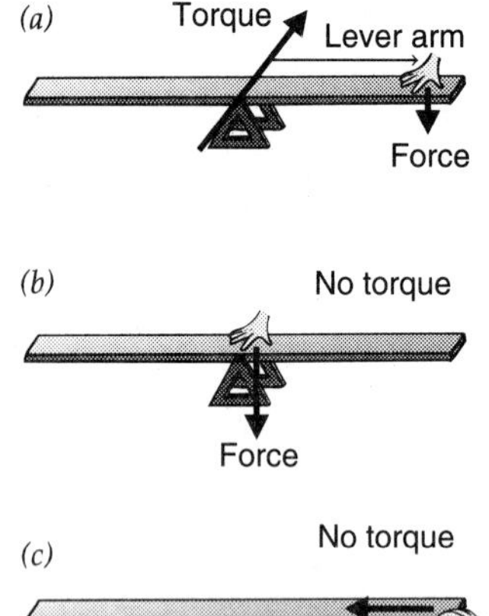

Fig. 1.3.10 - (*a*) When you push down on the seesaw, far from its pivot, you exert a torque on it. But when you (*b*) exert your force at the pivot or (*c*) exert your force toward the pivot, you exert no torque.

We're now ready to mount the seesaw on its pivot. Although the pivot limits the seesaw's motion, it still permits the board to rotate freely in one direction. Since the pivot passes directly through the seesaw's center of mass, it prevents gravity from exerting a torque on the seesaw. With gravity out of the picture, the unoccupied seesaw is essentially isolated and obeys Newton's first law of rotational motion, if only in one direction. That is, it rotates with constant angular velocity about its pivot.

The unoccupied seesaw is balanced, meaning that it has zero torque on it. As a result, it experiences no angular acceleration. You might think that a balanced seesaw always remains horizontal, but that isn't necessarily so. What is certain is that its angular velocity is constant. If it's rotating, then it continues to rotate steadily about the pivot; if it's stationary, then it remains stationary at its current tilt angle, whether horizontal or not.

To change the seesaw's angular velocity, you must exert a torque on it. But how do you actually exert a torque? You put your hand on one end of the seesaw and push that end down (Fig. 1.3.10*a*). The seesaw begins to rotate and your end soon hits the ground. You have exerted a torque on the board.

You pushed down on the board, exerting a force on its surface. But the board rotated, indicating that it experienced a torque about the pivot. Somehow your force has produced a torque about the pivot. Forces and torques must therefore be related; a force can produce a torque and a torque can produce a force. To help us explore the relationship between force and torque, we can begin by thinking of all the ways not to exert a torque on a seesaw.

What happens if you push on the seesaw board right where the pivot passes through it (Fig. 1.3.10*b*)? Nothing—no angular acceleration. If you move a little away from the pivot, you can get the board rotating, but doing so requires a large force. You do much better to push on the end of the board, where even a small force can start the seesaw rotating. The distance from the pivot to the place where you push on the board is called the **lever arm**; in general, the longer the lever arm, the less force it takes to cause a particular angular acceleration. We can make our first observation about producing a torque with a force: you obtain more torque by exerting that force farther from the pivot or axis of rotation. In other words, the torque is proportional to the lever arm.

Another ineffective way to start the seesaw board rotating is to push its end directly toward or away from the pivot (Fig. 1.3.10*c*). A force exerted toward or away from the axis of rotation doesn't produce any torque about that axis. At least a component of the force you exert must be at right angles to the lever arm, which is actually a vector pointing along the board's surface from the pivot to the place where you push on the board. Our second observation about producing a torque with a force is that you must exert at least a component of that force at right angles to the lever arm. Only that component of force contributes to the torque. To produce the most torque, push at a right angle to the lever arm.

We can summarize these two observations as follows: the torque produced by a force is equal to the product of the lever arm times that force, where we include only the component of the force that is at right angles to the lever arm. This relationship can be written as a word equation:

$$\text{torque} = \text{lever arm} \cdot \text{force}, \tag{1.3.2}$$

in symbols:

$$\tau = \mathbf{r} \times \mathbf{F},$$

and in everyday language:

When twisting an unyielding object, it helps to use a long stick.

The directions of the force and lever arm also determine the direction of the torque. The three directions follow a right-hand rule (Fig. 1.3.11). If you extend the index finger of your right hand in the direction of the lever arm and bend the middle figure of that hand in the direction of the force, then your thumb will point in the direction of the torque. Thus in Fig. 1.3.11*a*, the lever arm points to the right, the force points downward, and the resulting torque points into the page so that the seesaw undergoes angular acceleration in the clockwise direction. In Fig. 1.3.11*b*, the lever arm has reversed directions and so has the torque.

What happens if you and a friend push down simultaneously on both seats at once? Then you produce two torques on the seesaw about its pivot, and these torques have opposite directions. The seesaw responds to the **net torque** it experiences, the sum of all the individual torques on the seesaw. Since your two torques oppose one another, they at least partially cancel. If you carefully exert identical downward forces at identical distances from the pivot, the magnitudes of the two torques will be exactly equal and they will cancel perfectly. The seesaw will experience zero net torque and it will be balanced.

This observation explains the need for careful seating of the children on the seesaw. Each child's weight exerts a downward force; by properly distributing those weights on both sides of the pivot, the torques that the weights produce can be made to cancel perfectly. The net torque on the seesaw about its pivot is then zero and the seesaw balances.

(a)

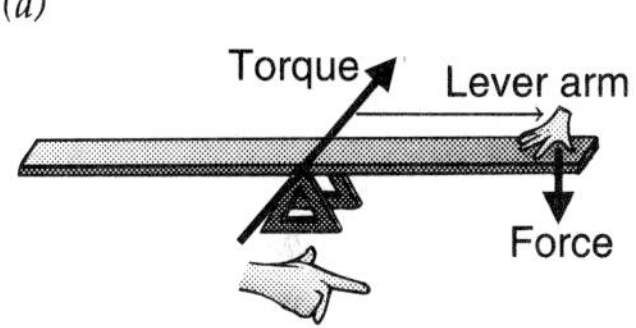

(b)

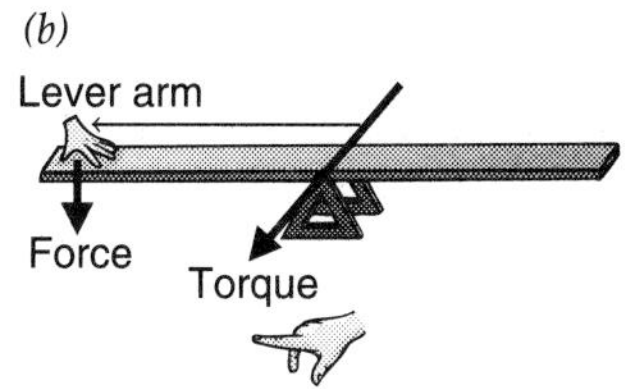

Fig. 1.3.11 - The torque on a seesaw obeys a right-hand rule: if your index finger points along the lever arm and your middle finger points along the force, your thumb points along the torque.

CHECK YOUR UNDERSTANDING #5: Cutting Up Cardboard

When you cut cardboard with a pair of scissors, it's best to move the cardboard as close as possible to the scissors' pivot. Explain.

CHECK YOUR FIGURES #2: A Few Loose Screws

You're trying to remove some rusty screws from your refrigerator, using an adjustable wrench with a 0.2 meter (20 centimeter) handle. Although you push as hard as you can on the handle, you can't produce enough torque to loosen one of the screws. You have a 1.0 meter (100 cm) pipe that you can slip over the handle of the wrench to make the wrench effectively 1 meter long. How much more torque will you then be able to exert on the screw?

Net Torque and Mechanical Advantage

(a)

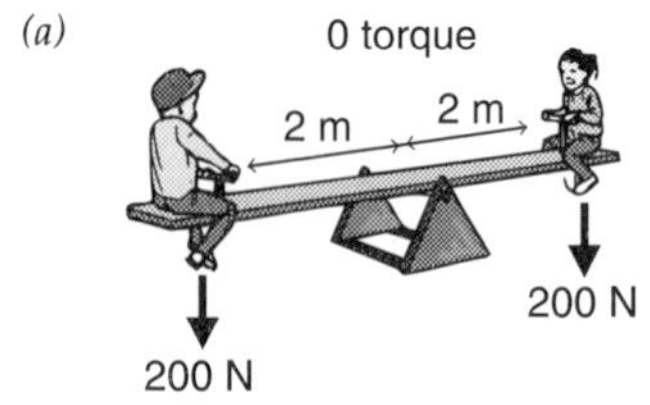

(b)

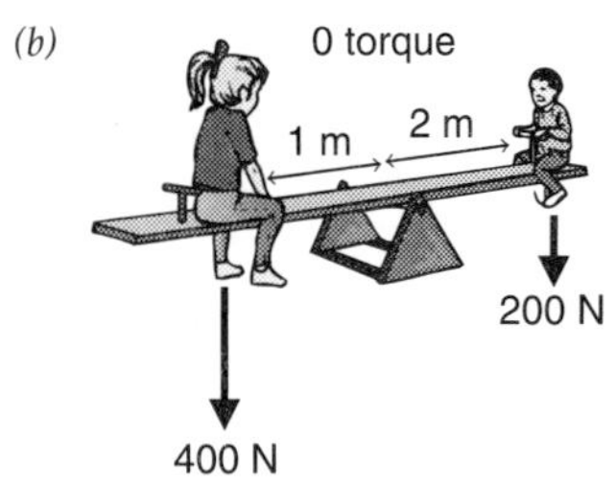

Fig. 1.3.12 - (*a*) When two children of equal weight sit at equal distances from the pivot, they produce equal but oppositely directed torques about the pivot. These torques cancel so that the seesaw experiences zero net torque. (*b*) When one child weighs twice as much as the other, the seesaw balances when the heavy child sits at half the distance from the pivot.

The amount of torque that a child's weight produces on a seesaw depends on that child's distance from the pivot. If the child sits on the pivot, the lever arm is zero, and she produces no torque about the pivot; but if she sits at the extreme end of the board, the lever arm is long and she produces a large torque. She can adjust her torque by moving along the board because the seesaw provides her with mechanical advantage. As we saw in Section 1.2, mechanical advantage appears when a device redistributes the amounts of force and distance used to produce a particular amount of work. The seesaw allows a small force exerted at the end of the rotating board to do the same work as a large force exerted near the board's pivot.

To see how mechanical advantage appears in a seesaw, think of what happens when two children sit on its ends. If two 5 year olds, each weighing 200 N, sit at opposite ends of the seesaw, 2 m from the pivot (Fig. 1.3.12*a*), each one exerts a torque of 400 N·m on the seesaw about its pivot (200 N·2 m=400 N·m). But because these torques are oppositely directed, they cancel. The net torque on the seesaw is zero and the seesaw balances.

If you replace one of the 5 year olds with a 400 N teenager, the teenager must sit at half the distance from the pivot (Fig. 1.3.12*b*). Doubling the force while halving the lever arm leaves the torque unchanged at 400 N·m. The two children again produce equal but oppositely directed torques about the pivot, so that the net torque on the seesaw is zero and the seesaw balances. This effect explains how a small child at the end of the seesaw can balance a large child nearer the pivot.

With the seesaw balanced, nothing is accelerating. Each child experiences zero net force; the seesaw pushes up on the small child with a force of 200 N and on the large child with a force of 400 N. Ultimately, it's the small child's 200 N weight that gives rise to the 400 N supporting force experienced by the large child. The seesaw's mechanical advantage allows the small child to support and lift the much heavier child. This effect, where a small force on one part of a rotating system produces a large force elsewhere in that system, is an example of the mechanical advantage associated with levers.

CHECK YOUR UNDERSTANDING #6: Pulling Nails

Some hammers have a special claw designed to remove nails from wood. When you slide the claw under the nail's head and rotate the hammer by pulling on its handle, the claw pulls the nail out of the wood. The hammer's head contacts the wood to form a pivot that's about 10 times closer to the nail than to the handle. The torque you exert on the hammer twists it in one direction, while the torque that the nail exerts on the hammer twists it in the opposite direction. The hammer isn't undergoing any significant angular acceleration, so the torques must be balanced. If you're exerting a force of 100 N on the hammer's handle, how much force is the nail exerting on the hammer's claw?

Riding a Seesaw

Each seesaw in Fig. 1.3.12 is balanced, meaning that the net torque on it is zero. Although each child's weight exerts a torque on the board, the two torques cancel one another. Since the seesaw experiences zero net torque and no angular acceleration, it continues rotating at constant angular velocity.

However, something seems to be missing from this description of a balanced seesaw. As it presently stands, a balanced seesaw should either remain motionless forever or else rotate endlessly in the same direction. Children are unlikely to wait motionless forever, and endless rotation implies that the children will be upside-down periodically. We've obviously neglected a few details.

First, what do the children do when the board is motionless? To start the board moving, they have to unbalance the seesaw. One of the children must change the torque she exerts on it. She can either change the downward force she exerts on the board or change the distance between that force and the pivot. Actually, children change both the force and the lever arm frequently without even thinking about it. If a child leans inward, toward the pivot, the lever arm decreases and the child exerts less torque on the board; as a result, the board begins to rotate and the child rises. If the child pushes on the ground with his feet, the ground exerts an upward force on him, reducing the force and torque he exerts on the board; again, the board begins to rotate and the child rises.

So either by leaning or by pushing on the ground, the children can start an initially motionless, balanced seesaw rotating. Similar actions prevent the seesaw from rotating endlessly in the same direction. When one end of the seesaw touches the ground, the ground exerts a strong, upward support force on that end. Because this force is exerted far from the pivot and almost at right angles to the board, this force produces a torque on the board. That new torque causes a large angular acceleration and abruptly stops the board from rotating. The angular acceleration is so large that it's uncomfortable for the riders. Most children push on the ground with their feet to cushion the impact. Once the seesaw has stopped turning, the child on the ground can then push down with her feet to start the seesaw rotating in the opposite direction. That child begins to rise and the other child descends. When the other end of the seesaw reaches the ground, this cycle begins again.

As they play on a seesaw, the two children frequently change the torques they exert on the board so that the seesaw tips back and forth. During the moments when a child is pushing on the ground or leaning inward or outward to get a stationary seesaw moving, the seesaw is no longer balanced. A balanced seesaw has zero angular acceleration; it's only by unbalancing the seesaw that the children can change the angular velocity of the seesaw.

CHECK YOUR UNDERSTANDING #7: Rocking the Boat

Loading a large container ship requires some care in balancing the cargo and fastening it down firmly. The effective pivot about which the ship can rotate in the water is located roughly along the centerline of the ship, between its bow to its stern. Why is improperly fastened-down cargo so dangerous on such a ship, possibly causing it to risk capsizing during a storm

Up and Down: Seesaws and Work

We've identified a balanced seesaw as one that experiences zero net torque and therefore has no angular acceleration. It can still move, but it moves with a constant angular velocity. What about work? If the seesaw in Fig. 1.3.12*a* is rotating clockwise, who's doing work on whom?

The girl is exerting a downward force on the board, and her end of the board is moving downward; she is doing work on the board. The boy is exerting a downward force on the board, but his end of the board is moving upward; the board is doing work on him. The two weights are equal and the distances the children travel are equal, so the work the girl does on the board is equal to the work the board does on the boy. Overall, the girl is simply doing work on the boy. The girl is transferring some of her gravitational potential energy to the boy.

Even when two children of different weights ride a balanced seesaw, all of the work done on the board by the descending child is needed to lift the ascending child. That doesn't mean that the two children exert the same forces on the board or that the distances they travel are the same. However, the product of force times distance must be the same. For example, as the boy in Fig 1.3.12*b* goes down, the teenager moves up only half as far. Although the teen exerts twice the force on the board, she moves only half the distance. Thus the work done by the boy as he descends is exactly the work done on the teen as she rises.

CHECK YOUR UNDERSTANDING #8: Seesaw Acrobatics

In a favorite circus act, several heavy people jump from a modest height onto one end of a seesaw and toss a single light person on the other end of the seesaw high into the air, usually to end up landing on their shoulders. What provides the work needed to lift the single light person so high above the ground?

Levers

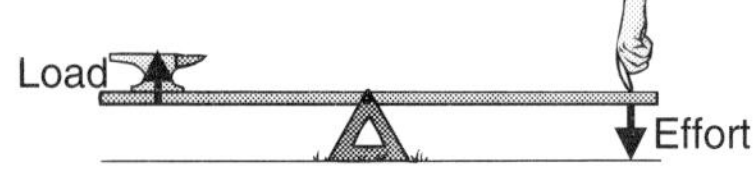

Fig. 1.3.13 - This lever pivots about its middle so that as you push the right end down (the effort), the left end of the lever pushes up (the load). The two forces are equal in magnitude.

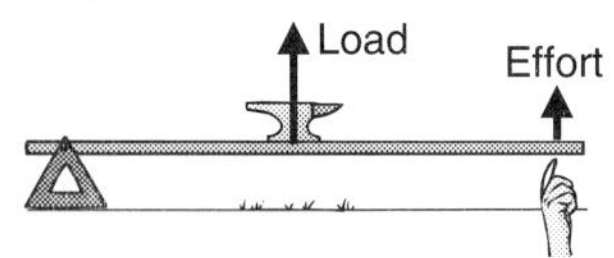

Fig. 1.3.14 - This lever pivots about its left end so that as you push the right end up (the effort), the middle of the lever also pushes up (the load). The load is twice as large as the effort and can be used to lift a heavy object a short distance.

A light child can use a seesaw to lift a much heavier child, albeit a reduced distance. In Fig. 1.3.12*b*, the light child is half the weight of the heavy child, but we could imagine more extreme differences in weight. If the heavy child were very close to the pivot, the light child would be able to lift him by exerting a very small force on the opposite end of the seesaw.

That's how a pry-bar or crowbar works. A pry-bar is designed so that the pivot is very close to one end of the bar. A modest downward force on the long end of the pry-bar exerts an enormous upward force on the short end of the pry-bar as it rotates about its pivot. You obtain mechanical advantage, meaning that you convert a small force into a large force, but again you don't get something for nothing. While it takes only a small force to lift a heavy object, the long end of the pry-bar moves much farther than the short end. You're still doing the same amount of work in lifting the heavy object, but you've traded distance for force. You exert less force but move farther in the direction of that force in order to lift the heavy object.

Obviously, a seesaw and a pry-bar aren't the only levers we use in everyday life. There are countless examples, many of which will appear later in this book. Three possible levers appear in Figs. 1.3.13-1.3.15. In each case, you exert a force (the **effort**) on the lever and produce a torque on the lever about its pivot. The lever then exerts a force (the **load**) on the object you're trying to move.

The lever shown in Fig. 1.3.15 may seem useless, since the load is actually less than the effort. However, the object that the load is pushing on moves both farther and faster than your hand does as the lever rotates. That can be useful in some situations. No matter how hard you try, you simply can't throw a baseball faster than about 160 km/h (100 mph) because you can't move your hand faster than that speed. But if you use a lever, you can easily get the lever's tip to move considerably faster. If you attach a hockey puck to the tip of a lever, it can acquire a large speed, too. That's how a hockey stick works.

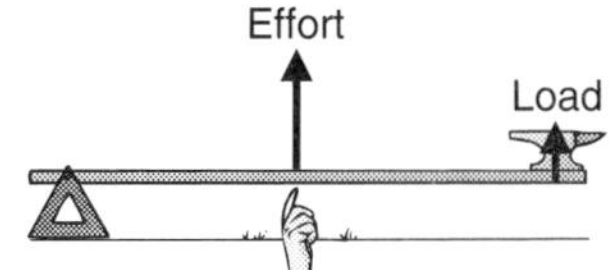

Fig. 1.3.15 - This lever pivots about its left end so that as you push the middle up (the effort), the right end of the lever also pushes up (the load). The load is half as large as the effort and can be used to lift a light object a long distance.

CHECK YOUR UNDERSTANDING #9: Hook, Line, and Sinker

How does a long fishing rod help you to cast the hook, line, and sinker farther than you could throw them by hand?

Summary

How Seesaws Work: A seesaw is a rotating toy that works best when it's almost perfectly balanced, meaning that it experiences zero net torque. The seesaw's pivot usually passes through the board's center of mass so that the board balances when it's not occupied. The riders arrange themselves so that the torques they exert on the seesaw cancel one another completely. The board then experiences zero net torque and zero angular acceleration, and it rotates with constant angular velocity. It either remains motionless or turns steadily in one direction or the other.

To make the seesaw tip back and forth, the riders subtly adjust the torques they exert on the seesaw. They do this either by leaning, thus varying their distances from the pivot, or by pushing against the ground with their feet, thus varying the forces they exert on the board. In either case, they unbalance the seesaw and it experiences both a net torque and an angular acceleration. By rhythmically changing the net torque on the seesaw, the riders cause it to rotate back and forth.

Important Laws and Equations

1. Newton's First Law of Rotational Motion: A rigid object that is not wobbling and is not subject to any outside torques rotates at a constant angular velocity, experiencing equal angular displacements in equal times as it turns about a fixed axis of rotation.

2. Newton's Second Law of Rotational Motion: The torque exerted on an object is equal to the product of that object's moment of inertia times its angular acceleration, or

$$\text{torque} = \text{moment of inertia} \cdot \text{angular acceleration}. \quad (1.3.1)$$

The angular acceleration points in the same direction as the torque. This law doesn't apply to objects that are wobbling.

3. Relationship Between Force and Torque: The torque produced by a force is equal to the product of the lever arm times that force, or

$$\text{torque} = \text{lever arm} \cdot \text{force}, \quad (1.3.2)$$

where we include only the component of the force that's at right angles to the lever arm.

The Physics of Seesaws

1. An object that is not wobbling and that is not subject to any outside influences (i.e., experiencing no torques) stays stationary or rotates with a constant angular velocity—it rotates steadily about a fixed axis. It has rotational inertia.

2. Angular velocity has a direction as well as a rate of rotation. The angular velocity points along the axis of rotation, in the direction established by the right-hand rule.

3. For a rigid object's angular velocity to change, it must undergo angular acceleration. This angular acceleration must be caused by the exertion of a torque on the object. The more torque exerted on an object, the greater its angular acceleration.

4. The object's rotational inertia is measured by its moment of inertia. The greater an object's moment of inertia, the smaller its angular acceleration in response to a specific torque.

5. An object's moment of inertia is determined both by the object's mass and by its shape. The farther an object's mass is from the axis of rotation, the larger its moment of inertia. An object's moment of inertia may change if you move the axis of rotation or change its direction.

6. An isolated object naturally rotates about its center of mass. An isolated object's axis of rotation always passes through its center of mass. A rigid object's motion can often be viewed most simply as translational motion of its center of mass and rotational motion about its center of mass.

7. Angular acceleration determines how quickly an object's angular velocity changes. Angular velocity determines how quickly an object's angular position changes.

Check Your Understanding - Answers

1. The lazy Susan undergoes rotational motion while the dessert cart undergoes translational motion.

Why: The lazy Susan has a fixed pivot at its center. This pivot never goes anywhere, regardless of how much you turn the lazy Susan. The tray simply rotates about this fixed pivot. In contrast, the dessert cart moves about the room and has no fixed point. The server can rotate the dessert cart when necessary, but its principal motion is translational.

2. It will continue to spin at a steady pace about a fixed rotational axis (although friction with the water will gradually slow the ball's rotation).

Why: Because the basketball is free of torques, the outside influences that affect rotational motion, it has a constant angular velocity. If you spin the basketball, it will continue to spin about whatever axis you chose. If you don't spin the basketball, its angular velocity will be zero and it will be stationary.

3. Yes. His center of mass falls smoothly, obeying the rules governing falling objects. As he falls, his body rotates at constant angular velocity about his center of mass.

Why: Like a thrown football or tossed baton, the diver is a rigid, rotating object. His motion can be separated into translational motion of his center of mass (it falls) and rotational motion about his center of mass (he rotates about it at constant angular velocity). While the diver may never think of his motion in these terms, he is aware intuitively of the need to handle both his rotational and translational motions carefully. Hitting the water with his chest because he did not rotate properly isn't much more fun than hitting the board because he didn't handle translation properly.

4. The full merry-go-round has a huge moment of inertia.

Why: Starting or stopping a merry-go-round involves angular acceleration. As the pusher, you exert a torque on the merry-go-round and it undergoes angular acceleration. But this angular acceleration depends on the merry-go-round's moment of inertia, which in turn depends on how much mass it has and how far that mass is from the axis of rotation. With many children adding to the merry-go-round's moment of inertia, its angular acceleration tends to be small.

5. The closer the cardboard is to the pivot, the more force it must exert on the scissors to produce enough torque to keep the scissors from rotating closed. When the cardboard is unable to produce enough torque, the scissors cut through it.

Why: When you place paper close to the pivot of a pair of scissors, you are requiring that paper to exert enormous forces on the scissors to keep them from rotating closed. Rotations are started and stopped by torques and forces exerted close to the pivot exert relatively small torques.

6. About 1000 N.

Why: Since the nail is 10 times closer to the pivot, the nail must exert 10 times the force on the hammer to create the same magnitude of torque as you do pulling on the handle. As the nail pulls on the hammer, the hammer pulls on the nail. Although the wood exerts frictional forces on the nail to keep it from moving, the extracting force overwhelms this friction and the nail slides slowly out of the wood.

7. A substantial shift in the cargo's position during a storm can unbalance the ship, creating a net torque on the ship, and cause it to begin rotating about the effective pivot. The boat may then capsize.

Why: Although most boats can compensate for some amount of cargo imbalance, shifting cargo can easily flip even a fairly stable boat. It happens frequently in real life, often with fatal consequences. Some boats, particularly canoes and racing shells, are notoriously sensitive to unbalanced loading and are easily flipped by careless or moving occupants.

8. The heavy people do a large amount of work on the seesaw, which does a large amount of work on the light person.

Why: When they land on one end of the seesaw, the heavy people exert a large downward force on that end and it moves downward. Since the force and motion are in the same direction, they do work on the seesaw. The other end of the seesaw exerts a huge upward force on the light person as that person moves upward. The seesaw does work on that person, tossing them into the air.

9. The fishing rod acts as a lever, giving the hook, line, and sinker large velocities during the cast so that they travel long distances before falling into the water. When you hold the rod's handle in your hands, you create a pivot near the end of the handle and exert the effort only a short distance above that pivot. A relatively small movement of your hands twists the rod about the pivot and produces a large swing of the fishing rod's tip.

Why: The sinker, sitting at the tip of the fishing rod, travels very fast during the cast, much faster than your hand can move. The distance the hook and sinker will travel depends on this initial velocity and the angle of the cast. With a little practice, a person who is fishing learns to release the line at the moment the sinker reaches both maximum speed and an optimal angle with respect to the water.

Check Your Figures - Answers

1. About 10 times as much torque.

Why: To keep the angular acceleration in Eq. 1.3.1 unchanged while increasing the moment of inertia by a factor of 10, the torque must also increase by a factor of 10. Solid tires are extremely hard to spin or to stop from spinning, which is why automobiles use hollow tires.

2. 5 times as much torque as before.

Why: The pipe increases the wrench's lever arm by a factor of five, from 0.2 meter to 1.0 meter. According to Eq. 1.3.2, the same force exerted 5 times as far from the pivot will produce 5 times as much torque about that pivot. Extending the handle of a lever-like tool is a common technique to increase the available torque, although it can be hazardous for both the tool and its user. Some tools that are designed for such extreme use come with removable handle extensions.

Glossary

angular acceleration A vector quantity that measures how quickly an object's angular velocity is changing—the greater the angular acceleration, the more the object's angular velocity changes each second. It consists of both the amount of angular acceleration and the direction about which the angular acceleration occurs. This direction is identical to the direction of the torque causing the angular acceleration.

angular displacement The change in angular position of a rotating body.

angular position A quantity that describes an object's orientation relative to some reference orientation.

angular speed A measure of the amount an object rotates in a certain amount of time.

angular velocity A vector quantity that measures how rapidly an object's angular position is changing—the greater the angular velocity, the farther the object turns each second. It consists of both the object's angular speed and the direction about which the object is rotating. This direction points along the axis of rotation in the direction established by the right-hand rule.

axis of rotation The straight line in space about which an object or group of objects rotates. More specifically, the axis of rotation points in a particular direction along that line to reflect the sense of rotation according to the right-hand rule.

center of mass The unique point about which all of the object's mass is balanced. The center of mass is the natural pivot point for a free object. In the absence of outside forces or torques, an object's center of mass travels at constant velocity while the object rotates at constant angular velocity about this center of mass.

effort The force exerted on a lever to cause it to rotate.

lever arm The directed distance from the pivot or axis of rotation to the point at which the force is exerted.

load The force that a lever exerts on the object it's lifting or moving.

moment of inertia The property of a body that is a measure of its rotational inertia. An object's moment of inertia is determined by its mass and how far that mass is from the axis of rotation. The SI unit of moment of inertia is the kilogram-meter^2.

net torque The sum of all torques acting on an object, considering both the magnitude of each individual torque and its direction. The magnitude of the net torque is less than the sum of the magnitudes of the individual torques, since they often oppose one another in direction.

Newton's first law of rotational motion An object that is not wobbling and is free from all outside torques rotates with constant angular velocity, spinning steadily about a fixed axis.

Newton's second law of rotational motion An object's angular acceleration is equal to the torque exerted on that object divided by the object's moment of inertia. This equality can be manipulated algebraically to state that the torque on the object is equal to the product of the object's moment of inertia times its angular acceleration (Eq. 1.3.1). The law doesn't apply to wobbling objects.

newton-meter (N·m) The SI unit of torque, exerted by a 1 newton force located 1 meter from the axis of rotation. 1 newton-meter is about the torque exerted on your shoulder by the weight of a baseball held in your outstretched arm.

right-hand rule The convention whereby the specific direction of an object's angular velocity is established. According to this rule, if the fingers of your right hand are curled to point in the direction of the object's rotation, your thumb will point in the direction of the angular velocity.

rotational inertia A property of matter by which it remains at rest or in steady rotation about the same rotational axis unless acted upon by some outside torque.

rotational motion Motion in which an object rotates about an axis. The orientation of an object undergoing only rotational motion will change but its position will remain unchanged.

sense of rotation The specific direction in which an object is rotating, either clockwise or counter-clockwise.

torque An influence that if exerted on a free body results chiefly in an angular acceleration of the body. A torque is a vector quantity, consisting of both the amount of torque and its direction. The SI unit of torque is the newton-meter.

translational motion Motion in which an object moves as a whole along a straight or curved line.

Review Questions

1. People seated in the dining car of a train are moving. So are people in the revolving restaurant of a skyscraper. How do their motions differ?

2. Compare your tendency to keep rolling forward on roller skates, with your tendency to keep spinning around in a swivel chair. How are they the same or different?

3. Just before it hits the pins at the end of the bowling lane, a bowling ball rolls steadily forward on the wooden surface and doesn't slip. Describe this rolling motion in terms of translation and rotation.

4. If your swivel chair is well oiled, you can continue to spin for a long time. What keeps you spinning?

5. As you spin in a swivel chair, what about you is changing? what is staying the same?

6. You can spin a swivel chair either clockwise or counterclockwise. Identify the direction of the angular velocity in both cases.

7. Why must you touch the floor with your foot to get yourself started spinning in a swivel chair?

8. Which of the following is undergoing angular acceleration: a diver who is just starting a forward somersault from a standing position, a toy top spinning steadily on its point, a windmill turning steadily in the breeze, a fan that's starting up, a fan running normally, a fan that's stopping, and a spinning fan that's being redirected to blow air in a different direction?

9. Why is the steering wheel of a bus so much larger than that of a car?

10. If you put a spoon on the table, bowl side up, and push down on the end of the bowl, the handle will rise up in the air. Describe the lifting of the handle in terms of the work you do on the end of the bowl.

Exercises

1. Why can't you open a door by pushing on its hinged side?

2. Why can't you open a door by pushing its doorknob directly toward or away from its hinges?

3. One way to crack open a walnut is to put it in the hinged side of a door and then begin to close the door. Why does a small force on the door produce a large force on the shell?

4. The chairs in an auditorium aren't all facing the same direction. How could you describe their angular positions in terms of a reference orientation and a rotation?

5. A mechanic balances the wheels of your car to make sure that their centers of mass are located exactly on their rotational axes. Neglecting friction and air resistance, how would an improperly balanced wheel behave if it were rotating all by itself?

6. The vast majority of wood screws have "right-handed" threads, meaning that each screw moves in the direction of its angular velocity as defined by the "right-hand rule." For example, if a screw's angular velocity points into the wood, the screw will move into the wood. If you are using such a screw to hold down a loose floorboard, in which direction should you turn the screwdriver (a tool for exerting torques on screws) so that the screw will move into the floor?

7. Tightrope walkers often use long poles for balance. Although the poles don't weight much, they can exert substantial torques on the walkers to keep them from tipping and falling off the ropes. Why are the poles so long?

8. An object's center of mass isn't always inside the object, as you can see by spinning it. Where is the center of mass of a boomerang or a horseshoe?

9. Some racing cars are designed so that their massive engines are near their geometrical centers. Why does this design make it easier for these cars to turn quickly?

10. It's much easier to carry a weight in your hand when your arm is at your side than it is when you arm is pointing straight out in front of you. Use the concept of torque to explain this effect.

11. A beach umbrella is attached to a long pole that sticks into the sand. Why does the umbrella tip over so easily in the wind if you don't push the pole deep into the sand?

12. You can exert much more crushing force on food with the teeth at the back of your mouth than you can with the teeth at the front of your mouth. Why?

13. You can make a shelf by nailing a thin board so that it extends outward at right angles to the wall. However, this shelf won't be able to support much weight unless you brace it from above or below with something that can exert an upward force on the free edge of the board. Why is the board itself unable to support much weight and why does a brace help so much?

14. How does the string of a yo-yo get the yo-yo spinning?

15. A common pair of pliers has a place for cutting wires, bolts, or nails. Why is it so important that this cutter be located very near the pliers' pivot?

16. How does a bottle opener use mechanical advantage to pry the top off a soda bottle?

17. A jar opening tool grabs onto a jar's lid and then provides a long handle for you to turn. Why does this handle's length help you to open the jar?

18. When you climb out on a thin tree limb, there's a chance that the limb will break off near the trunk. Why is this disaster most likely to occur when you're as far out on the limb as possible?

19. How does a crowbar make it easier to lift the edge of a heavy box a few centimeters off the ground?

20. The basket of a wheelbarrow is located in between its wheel and its handles. How does this arrangement make it relatively easy for you to lift a heavy load in the basket?

21. You can do push-ups with either your toes or your knees acting as the pivot about which your body rotates. When you pivot about your knees, your feet actually help you to lift your head and chest. Explain.

22. When an airplane starts its propellers, they spin slowly at first and gradually pick up speed. Why does it take so long for them to reach their full rotational speed?

23. A gristmill is powered by falling water, which pours into buckets on the outer edge of a giant wheel. The weight of the water turns the wheel. Why is it important that those buckets be on the wheel's outer edge?

24. Why is it hard to start the wheel of a roulette table spinning and what keeps it spinning once it's started?

25. The wheel of fortune game uses a large spinning disk. This disk is slowed to a stop by a flexible strip that pushes against nails projecting from the wheel's outer edge. How do the tiny forces that the strip exerts on the nails stop the massive wheel?

26. If the wheel in Exercise **25** had its nails nearer its center, how would the strip's ability to slow its rotation be affected?

27. If you lift the front wheel of your bicycle off the ground, it will probably begin turning until the wheel's air inlet nipple is at the bottom of the wheel. Use the concept of balance to explain why the nipple's weight causes the wheel to turn.

28. An automobile wheel has an air inlet nipple, just like the bicycle in Exercise **27**. Why is a small weight usually added to the wheel on the opposite side of the wheel from the nipple?

29. Long ago, cars had a crank in front to start their engines. How did exerting a force on the end of that crank produce a torque that turned the engine?

30. The pedals of a bicycle are attached to its crank. Suppose you stop pedaling your bicycle while the arms of the crank are horizontal. Then one pedal will be in front of the crank's center while the other pedal will be behind that center. If you then push downward on both pedals at once, with the same force on each, the crank won't undergo any angular acceleration. Explain.

Problems

1. When you ride a bicycle, your foot pushes down on a pedal that's 17.5 cm (0.175 m) from the axis of rotation. Your force produces a torque on the crank attached to the pedal. Suppose that you weigh 700 N. If you put all your weight on the pedal while it's directly in front of the crank's axis of rotation, what torque do you exert on the crank?

2. If you repeat Problem **1**, but this time put all your weight on the pedal while it's directly above the crank's axis of rotation, how much torque do you exert on the crank?

3. Toe clips allow you to pull upward on the pedals of a bicycle. If you arrange the bicycle as in Problem **1**, but this time pull up on the pedal with a force of 700 N, what torque do you exert on the crank?

4. Tractor tires are often filled with water to give them better traction. This water may increase a tire's mass and moment of inertia by a factor of 10. If you exert the same torque on two tires, one empty and one filled with water, which one will undergo the larger angular acceleration and by how much will their angular accelerations differ?

5. When you start your computer, the hard disk begins to spin. It takes 6 seconds of constant angular acceleration to reach full speed, at which time the computer can begin to access it. If you wanted the disk drive to reach full speed in only 2 seconds, how much more torque would the disk drive's motor have to exert on it during the starting process?

6. An electric saw uses a circular spinning blade to slice through wood. When you start the saw, the motor needs 2 seconds of constant angular acceleration to bring the blade to its full angular velocity. If you change the blade so that the rotating portion of the saw now has 3 times its original moment of inertia, how long will the motor need to bring the blade to its full angular velocity?

7. When the saw in Problem **6** slices wood, the wood exerts a 100 N force on the blade, 0.125 m from the blade's axis of rotation. If that force is at right angles to the lever arm, how much torque does the wood exert on the blade? Does this torque make the blade turn faster or slower?

8. An antique carousel that's powered by a large electric motor undergoes constant angular acceleration from rest to full rotational speed in 5 seconds. When the ride ends, a brake causes it to decelerate steadily from full rotational speed to rest in 10 seconds. Compare the torque that starts the carousel to the torque that stops it.

9. When you push down on the handle of a doll-like wooden nutcracker, its jaw pivots upward and cracks a nut. If the point at which you push down on the handle is 5 times as far from the pivot as the point at which the jaw pushes on the nut, how much force will the jaw exert on the nut if you exert a force of 20 N on the handle? (Assume all forces are at right angles to the lever arms involved.)

10. You're drilling a hole in a piece of wood. The drill's motor is exerting a downward torque of 2 N·m on the drill and the wood is exerting an upward torque of 2 N·m on the drill. What is the drill's angular acceleration? How does the drill's angular velocity change with time?

Section 1.4

Wheels

We're now almost finished with our first look at the basic laws of motion. We've explored translational motion, what initiates it, and how it proceeds; we've also examined rotational motion at some length. But we still have one more law of rotational motion to cover—Newton's third law—and we'll use another everyday object, the wheel, to explain it. Like ramps and levers, wheels are simple tools that make our lives easier. But the wheel's main purpose isn't mechanical advantage, it's overcoming friction. Up until now, we've ignored friction, looking at the laws of motion as they apply only in idealized situations. But our real world does have friction, and an object in motion tends to slow down and stop because of it. One of our first tasks in this section will therefore be to understand friction—though, for the time being, we'll continue to neglect air resistance.

Questions to Think About: *If objects in motion tend to stay in motion, why is it so hard to drag a heavy box across the floor? If objects should accelerate downhill on a ramp, why doesn't a plate slide off a slightly tilted table? Why does an egg roll off that table? What makes the wheels of a cart turn as you pull the cart forward? Why do the wheels of a moving cart continue to spin for a while if you lift the cart off the ground? How does turning the wheels of an automobile propel the car forward? Why does a parked car roll forward after being struck by a moving car?*

Experiments to Do: *To observe the importance of wheels in eliminating friction, try sliding a book along a flat table. Give the book a push and see how quickly it slows down and stops. Which way is friction pushing on the book? Does the force that friction exerts on the book depend on how fast the book is moving? Let the book come to a stop. Is friction still pushing on the book when isn't moving? If you push gently on the stationary book, what force does friction exert on it?*

Lay down three or four round pencils, parallel to one another and a few inches apart. Rest the book on top of the pencils and give the book a push in the direction that the pencils can roll. Describe how the book now moves. What do you think has caused the difference?

Friction isn't the only force that can stop the book from moving. If you slide one book into another, the collision will slow the first book and cause the second to begin moving. To illustrate this effect, place a coin on a smooth table and flick a second, identical coin so that it slides along the table and strikes the stationary coin squarely. What happens? Now line up several identical coins so that they touch and slide another coin into one end of this line. How does the collision affect the coin that was originally moving? How does it affect the line of coins? What is transferred among the coins by the collision?

Moving a File Cabinet: Friction

When we imagined moving your friend's piano into a new apartment back in Section 1.2, we neglected a familiar force: friction. Luckily for us, your friend's piano had wheels on its legs, and wheels facilitate motion by reducing the effects of friction. We'll focus on wheels in this section. But first, to help us understand the relationship between wheels and friction, we'll look at another item that needs to be moved: a file cabinet.

The file cabinet is resting on a smooth and level hardwood floor; it's full of sheet music and weighs about 1000 N (225 pounds). Despite its large mass, you know that it should accelerate in response to a horizontal force, so you give it a gentle push toward the door. Nothing happens. Something else must be pushing on the file cabinet in just the right way to cancel your force and keep it from accelerating. Undaunted, you push harder and harder until finally, with a tremendous shove, you manage to get the file cabinet sliding across the floor. But the cabinet moves slowly even though you continue to push on it with all your might. Something else is pushing on the file cabinet, stopping it from moving.

That something else is **friction**, a force that opposes the **relative motion** of two surfaces in contact with one another. Two surfaces that are in relative motion are traveling with different velocities so that a person standing still on one surface would observe that the other surface is moving. In opposing relative motion, friction exerts forces on both surfaces in directions that tend to bring them to a single velocity.

For example, when the file cabinet slides by itself toward the left, the floor exerts on it a rightward frictional force (Fig. 1.4.1). The frictional force exerted on the file cabinet, *toward the right*, is in the direction opposite the file cabinet's velocity, *toward the left*. The file cabinet's acceleration is in the direction opposite its velocity, so the file cabinet slows down. It slides more and more slowly and eventually comes to a stop.

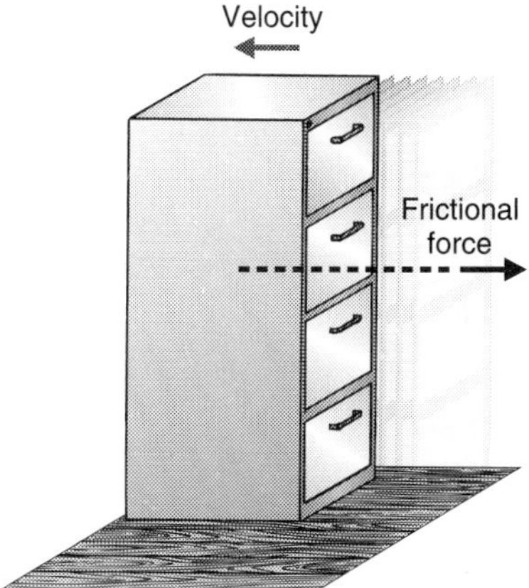

Fig. 1.4.1 - A file cabinet sliding to the left across the floor. The file cabinet experiences a frictional force toward the right that gradually brings it to a stop.

According to Newton's third law of motion, an equal but oppositely directed force must be exerted by the file cabinet on the floor. Forces always appear in matched pairs. Sure enough, the file cabinet does exert a leftward frictional force on the floor. However, the floor is rigidly attached to the earth, so it accelerates very little. The file cabinet does almost all the accelerating, and soon the two objects are traveling at the same velocity.

Frictional forces always oppose relative motion, but they vary in strength according to how tightly the two surfaces are pressed against one another, how slick the surfaces are, and whether or not the surfaces are actually moving relative to one another. The harder you press two surfaces together, the larger the frictional forces they experience. For example, an empty file cabinet slides more easily than a full one. Roughening the surfaces generally increases friction, while smoothing or lubricating the surfaces generally reduces it. Riding a toboggan down the driveway is much more interesting when the driveway is covered with snow or ice than it is when the driveway is bare asphalt.

CHECK YOUR UNDERSTANDING #1: The One That Got Away

Your table at the restaurant isn't level and your water glass begins to slide slowly downhill toward the edge. Which way is friction exerting a force on it?

A Microscopic View of Friction

Fig. 1.4.2 - The surface of a metal plate is rough on the microscopic scale. This cross section of a metal surface shows the peaks and valleys that are responsible for friction.

As the file cabinet slides by itself across the floor, it experiences a horizontal frictional force that gradually brings it to a stop. But where does this frictional force come from? The obvious forces on the file cabinet are both vertical, not horizontal: the cabinet's weight is downward and the support force from the floor is upward. How can the floor exert a horizontal force on the file cabinet?

The answer lies in the fact that neither the bottom of the file cabinet nor the floor is perfectly smooth. They both have microscopic hills and valleys of various sizes (Fig. 1.4.2). Because of this structure, the file cabinet is actually supported by thousands of tiny contact points, where the file cabinet directly touches the floor (Fig. 1.4.3). As the file cabinet slides across the floor, the microscopic projections on the bottom of the file cabinet pass through similar projections on the floor. Each time two microscopic projections collide, they experience forces. These tiny forces oppose the relative motion and give rise to the overall frictional forces experienced by the file cabinet and floor. Because even an apparently smooth surface still has some microscopic surface structure, all surfaces experience friction as they rub across one another.

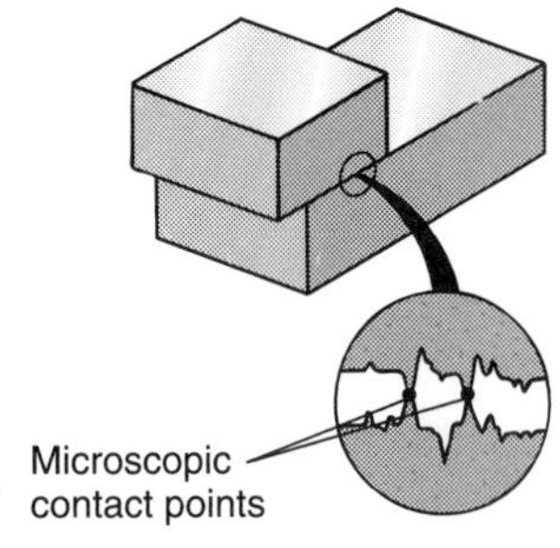

Fig. 1.4.3 - Two surfaces that are pressed against one another actually touch only at specific contact points. When the surfaces slide across one another, these contact points collide, producing sliding friction and wear.

The more these microscopic projections collide, the more friction there is. To increase the rate at which projections collide, you can increase their size and number or move the surfaces closer together. Rough surfaces have more and larger microscopic projections and create more friction. If you put sandpaper on the bottom of the file cabinet, it will experience larger frictional forces as it slides across the floor. On the other hand, if you put a particularly smooth surface on the bottom of the file cabinet, it will not experience much friction at all. Many non-stick pots and pans have surfaces that are unusually smooth, particularly at the microscopic level, and exert modest frictional forces on objects that slide across them. They also offer little surface structure for a fried egg to stick to, which is why they are called "non-stick."

To move the surfaces closer together, you must push them together with greater force. You could add some more sheet music to the file cabinet; the cabinet's increased weight would then squeeze it more tightly against the floor, the number of contact points between the file cabinet and floor would multiply, and the frictional force would consequently increase. Doubling the file cabinet's weight would roughly double the number of contact points and make it about twice as hard to slide the file cabinet across the floor. A useful rule of thumb is that the frictional forces between two surfaces are proportional to the forces pressing those two surfaces together.

Friction also causes wear when the colliding contact points break one another off (Fig. 1.4.4). Each time a wear chip is split away from a surface, the surface loses some of its mass. With time, friction can remove large amounts of material. Thus even seemingly indestructible stone steps are gradually worn away by foot traffic. The best way to reduce wear between two surfaces (other than to insert a lubricant between them) is to polish them so that they are extremely smooth. The smooth surfaces will still touch at contact points and experience friction as they slide across one another, but their contact points will be broad and round and will rarely break one another off during a collision.

CHECK YOUR UNDERSTANDING #2: Weight and Friction

How much harder is it to slide a stack of two identical books across a table than it is to slide just one of those books?

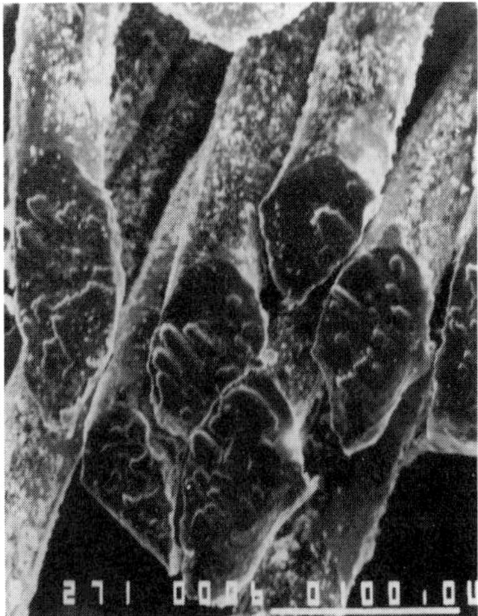

Fig. 1.4.4 - The ends of these wire brush bristles have been worn away by sliding friction. This microscope image shows wear chips that have been torn free from the bristles.

Static and Sliding Friction

There are really two kinds of friction: sliding and static. When two surfaces are moving across one another, they experience **sliding friction**, which acts to stop the surfaces from sliding across one another. But even when two surfaces have the same velocity, they may still experience **static friction**, which acts to prevent two surfaces from starting to slide across one another.

When you try to slide the file cabinet across the floor, you find it particularly hard at the beginning. Because the contact points between the cabinet and floor have settled into one another and are acting to keep the cabinet from moving, a small push does nothing; you need a mighty shove to get the cabinet going. Static friction is exerting a force in the direction opposite to your push, and this frictional force is always exactly equal to the force of your push. The net force on the file cabinet is therefore zero, and it doesn't accelerate.

However, the force that static friction can exert is limited. To get the file cabinet moving, all you have to do is exert more horizontal force on it than static friction can exert. If you push on it hard enough, the net force on the file cabinet will no longer be zero and it will accelerate.

Once the file cabinet is moving, static friction is replaced by sliding friction. Because sliding friction acts to bring the file cabinet back to rest, you must push on the cabinet to keep it moving. When the file cabinet is sliding across the floor, however, the contact points between the surfaces no longer have time to settle into one another, and they consequently experience weaker horizontal forces. Thus the force of sliding friction is generally weaker than that of static friction, which explains why it's easier to keep the file cabinet moving than it is to get it started.

CHECK YOUR UNDERSTANDING #3: Skidding to a Stop

Antilock brakes keep an automobile's wheels from locking and skidding during a sudden stop. Apart from issues of steering, what is the advantage of preventing the wheels from skidding (sliding) on the pavement?

Work, Energy, and Power

There is another difference between static and sliding friction: sliding friction wastes energy. It can't make that energy disappear altogether, because energy, as we've seen, is a conserved quantity and can't be created or destroyed. But energy can be transferred between objects or converted from one form to another. What sliding friction does is convert useful, **ordered energy**—energy that can easily be used to do work—into relatively useless, disordered energy. This disordered energy is called **thermal energy** and is the energy we associate with temperature. It's sometimes called *internal energy* or *heat*. Sliding friction makes things hotter by turning work into thermal energy.

To put this effect into perspective, let's take another look at energy. As we saw in Section 1.2, energy is the capacity to do work, and it's transferred between objects by doing that work. Energy can also change forms, appearing as either kinetic energy in the motions of objects or as potential energy in the forces between or within those objects. With a little practice, you can "watch" energy flow through a system just as an accountant watches money flow through an economy.

The most obvious form of energy is kinetic energy, the energy of motion. It's easy to see when kinetic energy is transferred into or out of an object. As kinetic energy leaves an object, the object slows down; thus moving water slows down as it turns a gristmill, and a bowling ball slows down as it knocks over bowling pins. Conversely, as kinetic energy enters an object, the object speeds up. A baseball moves faster as you do work on it during a pitch; you're transferring energy from your body into the baseball, where the energy becomes kinetic energy in the baseball's motion.

Potential energy usually isn't as visible as kinetic energy. It can take many different forms, some of which appear in the following list (Table 1.4.1). In each case nothing is moving; but because the objects still have a great potential to do work, they contain potential energy.

Table 1.4.1 - Several forms of potential energy with examples of those potential energies.

Form of Potential Energy	*Example*
Gravitational potential energy	A bowling ball at the top of a hill
Elastic potential energy	A wound clock spring
Electrostatic potential energy	A cloud in a thunderstorm
Chemical potential energy	A firecracker
Nuclear potential energy	Uranium

We measure energy in many common units: joules (J), calories, food Calories (also called kilocalories), and kilowatt-hours, to name only a few. All of these units measure the same thing, and they differ from one another only by numerical conversion factors, some of which can be found in Appendix B. For example, 1 food Calorie is equal to 1000 calories or 4,184 J. Thus, a jelly donut with about 250 food Calories contains about 1,000,000 J of energy. Since a joule is identical to a newton-meter, 1,000,000 J is the energy you'd use to lift your friend's file cabinet into the second-floor apartment 200 times (1000 N times 5 m upward is 5,000 J of work per trip). No wonder eating donuts is hard on your physique!

Of course, you can eventually use up the energy in a jelly donut; it just takes time. You can only do so much work each second. The measure of how quickly you do work is **power**—the amount of work you do in a certain amount of time, or

$$\text{power} = \frac{\text{work}}{\text{time}}.$$

The SI unit of power is the **joule-per-second**, also called the **watt** (abbreviated W). Other units of power include Calories-per-hour and horsepower; like the units for energy, these units differ only by numerical factors, which are again listed in Appendix B. For example, 1 horsepower is equal to 745.7 W. Since a 1-horsepower motor does 745.7 J of work each second, and since it takes 5,000 J of work to move the file cabinet to the second floor, that motor has enough power to do the job in about 6.7 s.

CHECK YOUR UNDERSTANDING #4: Apple Overtures
Trace the flow of energy as an archer shoots an apple off the head of her assistant with an arrow.

Friction and Thermal Energy

So far we've looked at kinetic energy and a variety of different potential energies. But what about the thermal energy produced by sliding friction? Is thermal energy a new kind of potential energy or an alternative to kinetic energy? In truth, it's neither. Thermal energy is actually a mixture of the same kinetic and potential energies that we've seen before. But unlike the kinetic energy in a moving ball or the potential energy in an elevated piano, the kinetic and potential energies in thermal energy are disordered at the atomic and molecular level. Thermal energy makes every microscopic particle in an object jiggle about independently; at any moment, each particle has its own tiny supply of potential and kinetic energies, and this dispersed energy is collectively referred to as thermal energy.

As you push the file cabinet across the floor, you do work on the file cabinet, but it doesn't pick up speed. Instead, sliding friction converts your work into thermal energy, so that the cabinet becomes hotter as the energy you transfer to it is dispersed among its atoms and molecules. But while sliding friction is an easy way to turn work into thermal energy, there's no easy way to turn thermal energy back into work. Disorder makes anything harder to use; while it's easy to disorder something, it's hard to undo that disordering. When you drop your favorite coffee mug on the floor and it shatters into a thousand pieces, the cup is still all there, but it's disordered and thus much less useful. Just as dropping the pieces on the floor a second time isn't likely to reassemble your cup, energy converted into thermal energy can't easily be reassembled into useful, ordered energy.

Unlike sliding friction, static friction doesn't convert work into thermal energy. Since two surfaces experiencing static friction don't move relative to one another, there is no distance traveled and thus no work done. You can push against the stationary file cabinet all day without doing any work on it. Even if you lift the file cabinet upward with your hands (no easy task), static friction between your hands and the file cabinet's sides merely assists you in doing work on the file cabinet itself. As you lift the file cabinet upward, all of your work goes into increasing the file cabinet's gravitational potential energy.

In contrast, sliding friction converts at least some work into thermal energy. Since two surfaces sliding across one another experience frictional forces that oppose their relative motion, sliding friction does negative work on the surfaces; it extracts energy from a sliding object and converts it into thermal energy. Thus, when you do work on the file cabinet by pushing it across the floor, sliding friction does negative work on it. The file cabinet's kinetic energy doesn't change very much, but its thermal energy continues to increase.

CHECK YOUR UNDERSTANDING #5: Burning Rubber

If you push too hard on your car's accelerator pedal when the traffic light turns green, your wheels will skid and you'll leave a black trail of rubber behind. Such a "jack-rabbit start" can cause as much wear on your tires as 50 km (31 miles) of normal driving. Why is skidding so much more damaging to the tires than normal driving?

Wheels

You've wrestled your friend's file cabinet out the door of the old apartment and are now dragging it along the sidewalk. You're doing work against sliding friction the whole way, producing large amounts of thermal energy in both the bot-

Fig. 1.4.5 - (*a*) A file cabinet that's supported on turning rollers experiences only static friction. (*b*) Since the top surface of a roller moves forward with the file cabinet, while its bottom surface stays behind with the sidewalk, the roller's center of mass moves only half as fast as the file cabinet. As a result, the rollers are soon left behind.

Fig. 1.4.6 - As this stage coach rolls forward, sliding friction between its axles and hubs converts some of its kinetic energy into thermal energy. To reduce this wasted energy, the coach has narrow axles that are lubricated with axle grease.

tom of the cabinet and the surface of the sidewalk. You're also damaging both objects, since sliding friction is causing wear on their surfaces. The four-drawer file cabinet may be down to three drawers by the time you arrive at the new apartment.

Fortunately, you can move one object across another object without sliding, and thus without sliding friction, if you use a mechanical system. The classic example is a roller (Fig. 1.4.5). If you place the file cabinet on rollers, you can push it along without any sliding friction; the rollers rotate as the file cabinet moves so that their surfaces never slide across the bottom of the cabinet or the top of the sidewalk. To see how the rollers work, make a fist with one hand and roll it across the palm of your other hand. The skin of one hand doesn't slide across the skin of the other hand; since the motion is silent and doesn't convert work into thermal energy, your skin remains cool. Now slide your two open palms across one another; this time, because there is sliding friction, the motion makes noise and warms your skin.

Although the rollers don't experience sliding friction, they do experience static friction. The top of each roller is touching the bottom of the cabinet, and the two surfaces move along together because of static friction; they grip one another tightly until the roller's rotation pulls them apart. A similar process takes place between the rollers and the top of the sidewalk; static friction exerts torques on the rollers and hence is what makes them rotate in the first place. Again, you can illustrate this behavior with your hands. Try to drag your fist across your open palm. Just before your fist begins to slide, you'll feel a torque on it. Static friction between the skin of your two hands, acting to prevent sliding, causes your fist to begin rotating just like a roller.

Once you get the file cabinet moving on rollers, you can keep it rolling along the level sidewalk indefinitely. Without any sliding friction, the cabinet doesn't lose kinetic energy, so it continues at constant velocity without your having to push it. However, the rollers move out from under the file cabinet as it travels, and you frequently have to move a roller from the back of the cabinet to the front. In fact, you need at least three rollers to ensure that the file cabinet never falls to the ground when a roller pops out the back. Although the rollers have eliminated sliding friction, they've created another headache—one that makes the prospect of traveling cross-country in a roller-supported, horse-drawn vehicle very unappealing. Is there another device that can reduce sliding friction without requiring constant attention?

One alternative would be a four-wheeled cart. The simplest cart rests on fixed poles or axles that pass through central holes or hubs in the four wheels (Fig. 1.4.6). The ground exerts upward support forces on the wheels, the wheels exert upward support forces on the axles, and the axles support the cart and its contents. As the cart moves forward, its wheels turn so that their bottom surfaces don't slide or skid across the ground; instead, each wheel lowers a portion of its surface onto the sidewalk, leaves it there briefly to experience static friction, and then raises it back off the sidewalk, with a new portion of wheel surface taking its place. Thus there is only static friction between the cart's wheels and the ground.

Because static friction produces no thermal energy, all of the work you do in pushing the cart forward should become kinetic energy in the cart's forward motion and in its wheels' rotations. When you stop pushing on the cart, it coasts along indefinitely at a constant speed, as long as no outside force does work on it and no sliding friction turns its kinetic energy into thermal energy. When the cart descends a hill, for example, gravity does work on it and it accelerates; when the cart climbs a hill, gravity does negative work on it and it decelerates.

Unfortunately, we've neglected the relative motion between each wheel's hub and the fixed axle passing through it. As each wheel turns, the inside of its hub rubs across the outside of the fixed axle (Fig. 1.4.7). Since the turning wheel does work against sliding friction, some of its kinetic energy is converted into thermal energy. This sliding friction also exerts a torque on the wheel which slows its rotation. However, this torque is small because it comes from a frictional force located very close to the axis of rotation. Static friction between the wheel and the ground occurs far from the axis of rotation and easily exerts enough torque on the wheel to keep it rotating without sliding along the ground.

The smallness of the hub and axle also reduces the amount of kinetic energy lost as thermal energy. Although the force of sliding friction between the hub and axle may be large, the distance the two slide across one another each second is small. As a result, only a small amount of work is done against sliding friction each second and relatively little of the cart's kinetic energy is converted into thermal energy each second. You don't have to supply very much power (work per second) to keep the cart moving. You can further reduce the production of thermal energy by lubricating each wheel hub with grease or oil and by lightening the cart's load, so as to reduce the force of sliding friction.

This type of wheel turns well enough to have been standard on carts and wagons of all kinds for centuries (Fig. 1.4.8). Sliding friction eventually wears away the hub and axle, but not before the wheel has given months or years of good service.

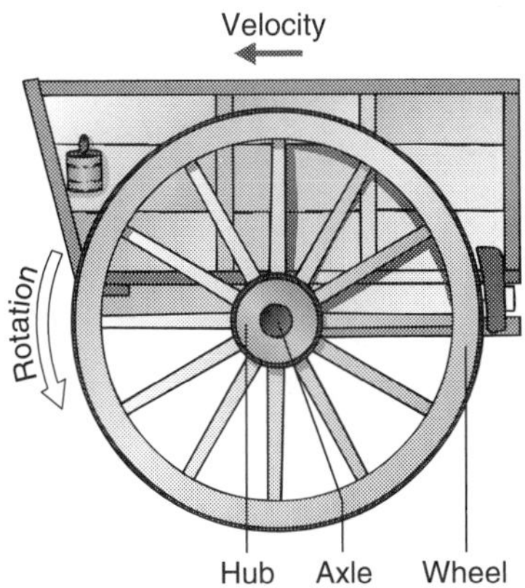

Fig. 1.4.7 - A wheel rotates as it travels so that no sliding friction occurs between it and the ground. The load is imparted to the wheel through an axle that enters the central hole of the wheel, the hub. As the wheel turns, the hub rubs against the axle, causing some sliding friction and wear.

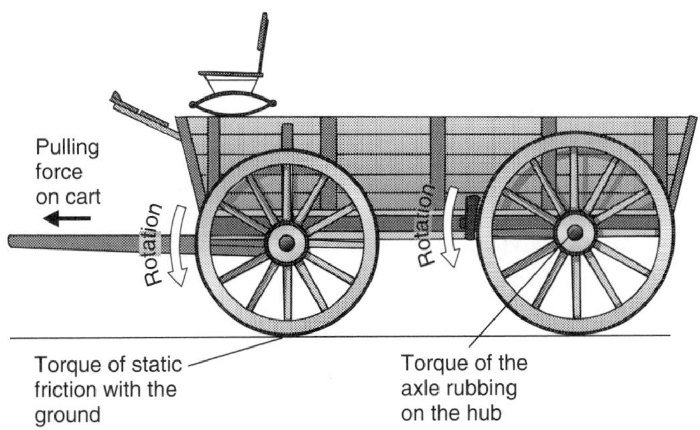

Fig. 1.4.8 - A cart often makes use of four wheels, attached to fixed axles. As the cart starts forward, each wheel experiences two torques: a large one due to static friction between the wheel and the ground and a small one due to sliding friction between the hub and axle. Static friction with the ground produces the larger torque and the wheel rotates.

Lubrication isn't the only way to handle the sliding friction problem between the hub and axle. You can also use a *bearing*, a device specifically designed to minimize the sliding friction between a hub and an axle. One of the best bearings is a roller bearing (Fig. 1.4.9), which introduces rollers between the hub and axle in much the same way that we introduced rollers between the file cabinet and the sidewalk. Unlike our crude imaginary rollers, however, the roller bearing arranges those rollers so that they recirculate automatically. The axle is supported by the rollers located below it. and these rollers are supported by the wheel hub. As the wheel turns, static friction makes the rollers rotate; they travel along the surface of the fixed axle, moving up, over, and down the other side. In this way each roller circles endlessly around the axle, returning under it periodically to provide support.

Freely turning automobile wheels operate in this manner. Each freely turning wheel is suspended on a fixed axle by two separate roller bearings. Large rollers in these bearings easily support the heavy load of the automobile. Because no surfaces slide against one another in the bearing, there is essentially no sliding

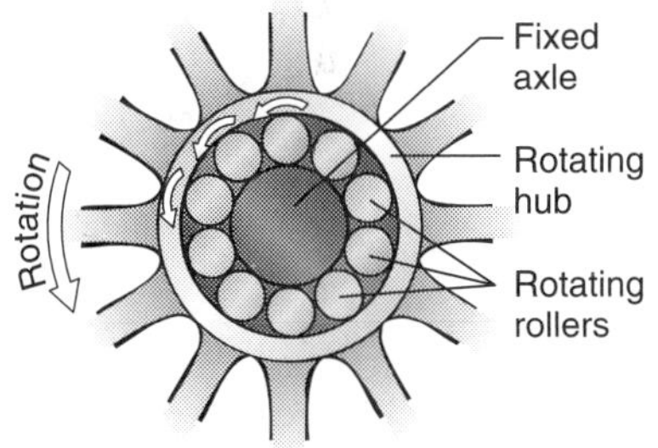

Fig. 1.4.9 - In a roller bearing, the hub of the wheel doesn't touch the axle directly. Instead, the two are separated by a set of rollers that turn with the hub. The bottom few rollers bear most of the load since the hub pushes up on them and they push up on the axle. As the wheel turns, the rollers recirculate, traveling up to the right and over the top of the axle before returning down to the left to bear the load once again. The rollers, wheel, and axle experience only static friction, not sliding friction.

friction, and the bearings last for tens of thousands of miles on the road without wearing out. In a real roller bearing, some additional components usually keep the individual rollers from slipping out, touching one another, and getting clogged with dust and dirt. Grease is added to reduce accidental sliding friction.

In a bicycle, the cylindrical rollers are replaced with balls to form ball bearings. While these balls serve exactly the same purpose as rollers, they're lighter and easier to work with; as a result, they're perfect for applications that don't require heavy loads, and they appear in many household machines with rotating parts. Still, ball bearings have some drawbacks. If they're overloaded, for example, the balls inside them may be crushed, since only a small region of each ball's surface exerts the forces that support the axle inside the hub. Once the balls have been damaged in this way, the bearing will begin to grind itself up. To reduce the chances of damage, the balls used in ball bearings are manufactured with truly remarkable precision into nearly perfect spheres of uniform size.

CHECK YOUR UNDERSTANDING #6: Jewel Movements

Many precision mechanical watches and clocks proudly proclaim that they have "jewel movements." Gears in these timepieces turn on axles that are pointed at either end and are supported at those ends by very hard, polished gemstones. What is the advantage of having needle-like ends on an axle and supporting those needles with smooth, hard jewels?

Powering Wheels

So far, we've discussed only the kind of wheel that rotates freely on an axle, with or without bearings. Such a wheel begins to turn as the vehicle to which it's attached accelerates from a stop. When the vehicle first begins to move—say, towards your left—static friction with the ground exerts a rightward force on the bottom of the wheel (Fig. 1.4.10*a*). This frictional force produces a torque on the wheel so that it begins to rotate counter-clockwise. The wheel turns (rotational motion) as the direct result of the movement of the cart (translational motion).

But not all wheels turn freely. A typical car has two freely turning wheels and two that are powered by the car's engine. Instead of spinning freely on fixed axles, these powered wheels are usually attached directly to axles that are themselves suspended in bearings. The car's engine exerts torques on the axles and powered wheels. If there are no other torques on the powered wheels, they undergo angular acceleration.

(a)

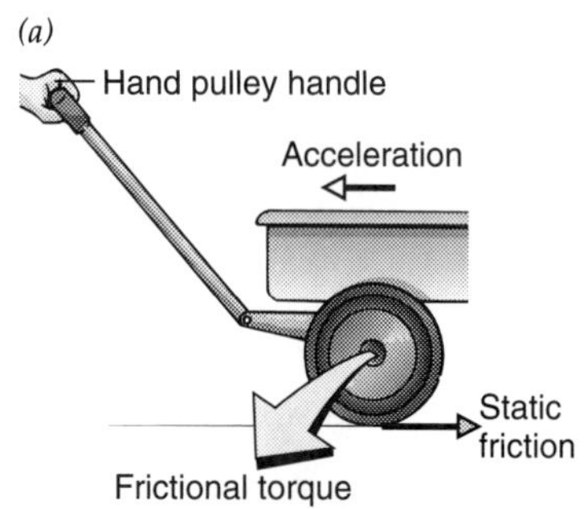

(b)

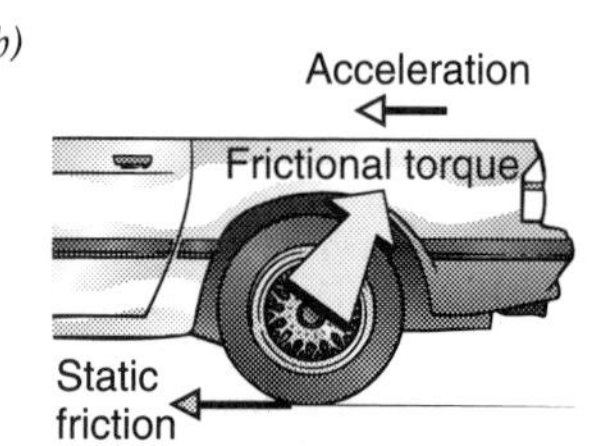

Fig. 1.4.10 - (*a*) When you pull a wagon forward, static friction from the ground exerts torques on its wheels and those wheels begin to rotate. (*b*) When the engine of a car turns its powered wheels, static friction from the ground exerts torques on the wheels to slow their rotation. This same static friction also pushes the car forward.

Like all wheels that are in contact with the ground, a powered wheel experiences static friction (Fig. 1.4.10*b*). This static friction can exert a torque on the wheel and prevent the wheel's outer surface from sliding or skidding across the ground. Since the engine's torque is always acting to make the wheel skid, the frictional torque always opposes the engine's torque. If the engine's torque is too large or static friction too weak, then the wheel will skid, its surface sliding across the pavement so that sliding friction produces thermal energy. Anyone who has driven a powerful, manual-transmission car has probably experienced such skidding when shifting gears incorrectly or releasing the clutch too abruptly.

However, if static friction is able to keep the powered wheels from skidding, the whole car will accelerate. It accelerates because the forces of static friction exerted on the wheels by the ground are the only horizontal forces experienced by the car. While the engine and frictional torques compete with one another over each wheel's angular acceleration, the whole car responds to whatever horizontal forces are exerted on it by its surroundings. The same static frictional forces exerted by the ground on the wheels to keep them from skidding also indirectly push the entire car, and the car accelerates as a result.

The direction of the car's acceleration depends on the direction of the engine's torque. With the car in a forward gear, static friction with the ground pushes the car forward and causes it to accelerate in the forward direction. In a reverse gear, the engine's torque is reversed and static friction with the ground pushes the car backward, causing it to accelerate in the backward direction. The car moves (translational motion) as the direct result of the turning wheels (rotational motion). Thus we have found simple mechanisms for converting rotational motion into translational motion and vice versa.

CHECK YOUR UNDERSTANDING #7: Front-Wheel Drive

Why is a front-wheel drive car, with its powered wheels located directly under the heavy engine, less prone to skidding and better able to get good traction than a rear-wheel drive car?

Coasting Forward: Linear Momentum

It's now clear that you should use wheels to help you move the file cabinet. Lacking a cart and unable to get the cabinet into the trunk of your Jaguar, you place it on your friend's office chair, which has a wheeled base that allows both the chair and the cabinet to roll freely down the sidewalk. The chair also swivels so that you can rotate the file cabinet at will. With the chair thus free to move or turn, you quickly become aware of the file cabinet's translational and rotational inertias; its translational inertia makes it hard either to start or stop, and its rotational inertia makes it difficult either to spin or stop spinning. Let's take another look at these two types of inertia and how they affect the file cabinet on wheels.

As we've seen, energy is a conserved quantity. But it's not the only conserved quantity in nature. Translational inertia and rotational inertia give rise to two others: linear momentum and angular momentum.

Linear momentum (often simply called "**momentum**") is the measure of an object's translational motion—its tendency to continue moving in a particular direction. Roughly speaking, an object's momentum indicates which way the object is heading and just how difficult it was to get the object moving with its current velocity. The linear momentum of a simple object, such as the file cabinet, is its mass times its velocity and can be written as a word equation:

$$\text{linear momentum} = \text{mass} \cdot \text{velocity}, \tag{1.4.1}$$

in symbols:

$$\mathbf{p} = m \cdot \mathbf{v},$$

and in everyday language:

Be careful trying to stop a fast-moving truck,

where the direction of momentum is the same as the direction of velocity. Linear momentum is a vector quantity, having both a magnitude and a direction. As we might expect, the faster an object is moving or the more mass it has, the more momentum it has in the direction of its motion.

Because linear momentum is a conserved quantity, it can't be created or destroyed; instead, it can only be transferred between objects. For the stationary file cabinet to begin moving forward, you must transfer momentum to it. If you run toward the file cabinet and crash into it, you'll give it some of your forward momentum; it will begin moving forward, and you'll slow down. If the moving cabinet then collides with a passerby, it will transfer to him some of its forward momentum; the cabinet will slow down and the passerby will suddenly begin heading forward. The passerby may then transfer this momentum to the earth, during a painful encounter with a lamppost, and you may find yourself in need of a lawyer.

Momentum is transferred to an object by giving it an **impulse**—that is, a force exerted on it for a certain amount of time. When you push the file cabinet forward for a few seconds, you give it an impulse and transfer momentum to it. This impulse is the change in the file cabinet's momentum and is equal to the product of the force exerted on the file cabinet times the length of time that force is exerted. This relationship can be written as a word equation:

$$\text{impulse} = \text{force} \cdot \text{time}, \tag{1.4.2}$$

in symbols:

$$\Delta\mathbf{p} = \mathbf{F} \cdot t,$$

and in everyday language:

The harder and longer you push a bobsled forward at the start of a race, the more momentum it will have when it starts down the hill.

The more force or the longer that force is exerted, the larger the impulse and the more the file cabinet's momentum changes.

You can also use different forces exerted for different amounts of time to transfer the same momentum to an object:

$$\begin{aligned} \text{impulse} &= \textbf{large force} \cdot \text{short time} \\ &= \text{small force} \cdot \textbf{long time}. \end{aligned} \tag{1.4.3}$$

Thus you can get the file cabinet moving with a certain forward momentum either by pushing it forward gently with a small force of long duration or by shoving it forward with a large, brief force. You can also stop it slowly or let it crash into a wall. While the latter process will extract all of its forward momentum in a very short period of time, it will involve very large forces that are likely to bend or break the cabinet. Sudden transfers of momentum are often quite destructive.

Why should momentum be a conserved quantity? It's conserved because of Newton's third law of motion. When one object exerts a force on a second object for a certain amount of time, the second object exerts an equal but oppositely directed force on the first object for exactly the same amount of time. Because of the equal but oppositely directed nature of the two forces, objects that push on one another receive impulses that are equal in amount but opposite in direction. The momentum gained by the second object is exactly equal to the momentum lost by the first object. Momentum is transferred from the first object to the second object.

When two automobiles collide with one another they transfer momentum. As they collide, each car exerts a force on the other and they accelerate in opposite directions. If a moving car strikes a parked car, the parked car accelerates forward while the moving car accelerates backward; the moving car transfers forward momentum to the parked car. The more mass a vehicle has, the less its velocity changes as a consequence of the momentum transfer. That's why the city bus barely slows at all when it crashes into your friend's file cabinet, while the cabinet accelerates dramatically. The bus has so much forward momentum that transferring a small fraction of it to the file cabinet causes a large change in the file cabinet's velocity. The same concept explains why bugs do virtually all the accelerating during their encounters with car windshields, even though the momentum transfer is equal.

CHECK YOUR UNDERSTANDING #8: Bowling Them Over

When a bean-bag hits the wall, it transfers all of its forward momentum to the wall and comes to a stop. When a rubber ball hits the wall, it transfers all of its forward momentum, comes to a stop, and then rebounds. During the rebound it transfers still more forward momentum to the wall. If you wanted to knock over a weighted bowling pin at the county fair, which would be the most effective projectile: the rubber ball or the bean-bag, assuming they have identical masses and you throw them with identical velocities?

CHECK YOUR FIGURES #1: Follow That Train!

The bad guys are getting away in a four-car train and you're trying to catch them. The train has a mass of 20,000 kg and it's rolling forward at 22 m/s (80 km/h or 50 mph). What is the train's momentum?

CHECK YOUR FIGURES #2: Stop That Train!

The engine of the train you're chasing (see CYF #1) has broken down, but it's still rolling forward. To stop it, you grab onto the last car and begin to drag your heels on the ground. The backward force on the train is 200 N. How long will it take you to stop the train?

Spinning in Circles: Angular Momentum

Angular momentum is the measure of an object's rotational motion—its tendency to continue spinning about a particular axis. Simply put, an object's angular momentum indicates the direction of its rotation and just how difficult it was to get it spinning. The angular momentum of a simple object, such as the file cabinet, is its moment of inertia times its angular velocity and can be written as a word equation:

$$\text{angular momentum} = \text{moment of inertia} \cdot \text{angular velocity}, \qquad (1.4.4)$$

in symbols:

$$\mathbf{L} = I \cdot \omega,$$

and in everyday language:

Be careful trying to stop a spinning carousel,

where the direction of angular momentum is the same as the direction of angular velocity. The faster an object is spinning or the larger its moment of inertia, the more angular momentum it has in the direction of its angular velocity.

Angular momentum is another conserved quantity and hence can only be transferred between objects. For the stationary file cabinet (which survived its collision with the bus) to begin rotating, you must transfer angular momentum to it; in other words, you must spin it, thereby transferring angular momentum from yourself to the file cabinet. The file cabinet will then continue to spin until it transfers the angular momentum elsewhere. If your friend were to grab onto the spinning file cabinet, she would also begin to spin, because the file cabinet would give her some of its angular momentum.

Angular momentum is transferred to an object by giving it an angular impulse. An **angular impulse** is a torque exerted on an object for a certain amount of time. When you spin the file cabinet for a few seconds, you give it an angular impulse and transfer angular momentum to it. This angular impulse is the change in the file cabinet's angular momentum and is equal to the product of the torque exerted on the file cabinet times the length of time that torque is exerted. This relationship can be written as a word equation:

$$\text{angular impulse} = \text{torque} \cdot \text{time}, \tag{1.4.5}$$

in symbols:

$$\Delta \mathbf{L} = \tau \cdot t,$$

and in everyday language:

To get a merry-go-round spinning rapidly, you must twist it hard and for a long time.

The more torque or the longer that torque is exerted, the larger the angular impulse and the more the file cabinet's angular momentum changes.

You can also use different torques exerted for different amounts of time to transfer the same angular momentum to an object:

$$\begin{aligned} \text{angular impulse} &= \Large\text{large torque} \normalsize\cdot \scriptsize\text{short time} \\ &= \scriptsize\text{small torque} \normalsize\cdot \Large\text{long time}. \end{aligned} \tag{1.4.6}$$

Thus you can get the file cabinet rotating with a certain angular momentum either by spinning it gently with a small torque of long duration or by twisting it hard with a large, brief torque. As with linear momentum, sudden transfers of angular momentum can break things, so you will probably want to start or stop the file cabinet's rotation gradually.

Why should angular momentum be a conserved quantity? Like linear momentum, angular momentum is conserved because of Newton's third law of motion. In this case, we are referring to **Newton's third law of rotational motion**: If one object exerts a torque on a second object, then the second object will exert an equal but oppositely directed torque on the first object. For example, when you use a screwdriver to exert a torque on a screw, the screw exerts an equal but oppositely directed torque back on the screwdriver and you. Because of the equal but oppositely directed nature of the two torques, objects that exert torques on one another receive angular impulses that are equal in amount but opposite in direction. The angular momentum gained by the second object is exactly equal to the angular momentum lost by the first object. Angular momentum is transferred from the first object to the second object.

NEWTON'S THIRD LAW OF ROTATIONAL MOTION:
For every torque that one object exerts on a second object, there is an equal but oppositely directed torque that the second object exerts on the first object.

Since an object's angular momentum depends on its moment of inertia, two different objects may end up rotating at different angular velocities even though you give them identical angular momenta. A small toy top spins much faster than a large merry-go-round when both have the same angular momentum. The same sort of behavior occurs with linear momentum, where a small hockey puck moves much faster across the ice than a large manhole cover when both have the same linear momentum. But while an object can't change its mass, it can sometimes change its moment of inertia. If it does change its moment of inertia, its angular *momentum* won't change, but its angular *velocity* will!

To see this change in angular velocity, imagine sitting on your friend's swivel chair. If you have such a chair, give it a try. Get yourself spinning by pushing on the floor with your foot. You will exert a torque on the earth, transferring angular momentum from yourself to the earth; it will spin one way while you spin the other. If the chair has good bearings, you will spin relatively freely. Now bring your legs and arms in close to the axis of rotation; you'll begin to spin faster because you've reduced your moment of inertia, and you'll no longer be a freely turning rigid object covered by Newton's first law of rotational motion. Angular momentum, however, is conserved, and you can't transfer it anywhere while you're isolated on the swivel chair. Because the chair prevents you from exerting torques on the ground, your angular momentum can't change. Since your moment of inertia has become smaller, your angular velocity must increase. And so it does. This effect of changing moment of inertia explains how an ice skater can achieve an enormous angular velocity by pulling herself into a thin, spinning object on ice (Fig. 1.4.11).

Fig. 1.4.11 - When a spinning ice skater pulls in her arms, she reduces her moment of inertia. She begins to spin more rapidly so as to maintain the same angular momentum.

CHECK YOUR UNDERSTANDING #9: Spinning the Merry-Go-Round

A person who is initially motionless starts a merry-go-round spinning and then returns to being motionless. If angular momentum is truly conserved, where did the angular momentum, that the spinning merry-go-round now, has come from?

CHECK YOUR FIGURES #3: Want to Go for a Spin?

Some satellites are set spinning to make them more stable. Suppose that the astronauts launching a particular satellite decide to increase its angular velocity by a factor of 5. How will that change affect the satellite's angular momentum?

CHECK YOUR FIGURES #4: Spin Away!

How much longer will it take the astronauts launching the satellite in CYF #4 to bring it to the faster angular velocity, if they use the initially planned torque?

The Conserved Quantities and Potential Energy

As you wheel the file cabinet across town to the new apartment, the file cabinet's motion is governed in large part by the three conserved quantities: energy, linear momentum, and angular momentum (Table 1.4.2). You're responsible for transfers of those three quantities to and from the file cabinet. As you roll the cabinet back and forth on the sidewalk, push it up and down hills, and turn it around corners, you control its energy, linear momentum, and angular momentum. Sometimes you do work on the file cabinet; sometimes it does work on you. You regularly give it impulses and angular impulses to change its speed, direction, and angular velocity.

When you come to a low point in the sidewalk, you discover an interesting effect. The file cabinet rolls back and forth about that low point, all by itself, so that it is always accelerating toward the depression. We've seen this tendency to accelerate downhill before with ramps, but now we can look at it in terms of energy. The file cabinet always accelerates in the direction that reduces its total po-

Conserved Quantity	*Transfer Mechanism*
Energy	Work
Linear Momentum	Impulse
Angular Momentum	Angular Impulse

Table 1.4.2 - The three conserved quantities of motion and the mechanism by which they are transferred between objects.

tential energy as quickly as possible. Since its only potential energy is gravitational potential energy, it accelerates in such a way as to reduce its gravitational potential energy as quickly as possible: down the steepest route to the bottom of the depression.

This behavior of accelerating in the direction that reduces total potential energy as quickly as possible is universal. Potential energy and forces are related to one another, so this rule is really just a way to determine the direction of the net force on an object or its parts. An object accelerates in the direction of the net force on it, which is also the direction that will reduce its total potential energy as quickly as possible. This rule is a useful way to determine how motion will proceed: which way a spring will leap, a chair will tip, or a ball will roll. We'll use it frequently in this book.

CHECK YOUR UNDERSTANDING #10: Heading Down
When you pull a child back on a playground swing and let go, which way does that child accelerate?

Kinetic Energy

Before leaving wheels and the laws of motion, there is one more question to answer: how do you determine an object's kinetic energy? One way is to calculate just how much work you'd have to do on it to bring it from rest to its current speed. The result of such a calculation is that a moving object's kinetic energy is equal to ½ of its mass times the square of its speed. This relationship can be written as a word equation:

$$\text{kinetic energy} = \tfrac{1}{2} \cdot \text{mass} \cdot \text{speed}^2, \tag{1.4.7}$$

in symbols:

$$K = \tfrac{1}{2} \cdot m \cdot v^2,$$

and in everyday language:

Racing around at twice the speed takes more than twice the energy.

Thus a car traveling 100 km/h (62 mph) has 4 times the kinetic energy of a car traveling 50 km/h (31 mph). This huge increase in energy with a modest increase in speed explains why high-speed crashes are far more deadly than those at lower speeds. At highway speeds, a car has so much kinetic energy to transfer during a crash that extensive damage occurs.

A rotating object, such as a wheel or toy top, also has kinetic energy. Like the kinetic energy of translational motion, the kinetic energy of rotational motion depends on an object's inertia and speed. But for a spinning object, it's the rotational inertia and rotational speed that matter. Its kinetic energy is equal to ½ of its moment of inertia times the square of its angular speed. This relationship can be written as a word equation:

$$\text{kinetic energy} = \tfrac{1}{2} \cdot \text{moment of inertia} \cdot \text{angular speed}^2, \tag{1.4.8}$$

in symbols:

$$K = \tfrac{1}{2} \cdot I \cdot \omega^2,$$

and in everyday language:

It takes a very energetic person to spin their wheels twice as fast.

If the file cabinet is rotating and translating at the same time—as it did after it was hit by the bus—its total kinetic energy is the sum of the translation and rotational parts. The translational kinetic energy depends on the speed of the file cabinet's center of mass and the rotational kinetic energy depends on the angular speed at which the file cabinet spins about its center of mass.

CHECK YOUR UNDERSTANDING #11: Throwing a Fastball

A typical grade-school pitcher can throw a baseball at 80 km/h (50 mph), but only a few professional athletes have the extraordinary strength needed to throw a baseball at twice that speed. Why is it so much harder to throw the baseball only twice as fast?

CHECK YOUR FIGURES #5: Blowing In The Wind

The air in a hurricane travels at 200 km/h (124 mph). How much more kinetic energy does 1 kg of this air have than 1 kg of air moving at only 20 km/h?

CHECK YOUR FIGURES #6: Playing Around at the Playground

When children climb onto a playground merry-go-round, they increase its rotational inertia. If the children triple the merry-go-round's moment of inertia, how will they alter the kinetic energy it has when it spins at a certain angular speed?

Summary

How Wheels Work: Wheels facilitate motion by eliminating or reducing sliding friction between an object and a surface. The wheels convey the support forces needed to hold the object up but allow the object to move without sliding. As a cart with freely turning wheels moves along the surface, static friction between each wheel and the surface exerts a torque on that wheel and causes it to turn. However, rubbing may occur between the wheel's hub and the axle, where sliding friction can waste energy and cause wear. To eliminate this sliding friction, roller or ball bearings are often used.

The torque that causes a powered wheel on a vehicle to turn comes from an engine by way of an axle. In this case, static friction between the outside of the wheel and the ground exerts a torque on the wheel to oppose the torque from the engine. This static frictional force also contributes to the net force on the vehicle and causes it to accelerate.

Once supported on wheels and bearings, objects can move freely and can retain linear momentum, angular momentum, and energy for long periods of time. By eliminating sliding friction, wheels can also keep objects from converting their ordered energies, either kinetic or potential, into thermal energy. Wheels allow vehicles to hold onto these conserved quantities for extended periods and make transportation far more practical.

Important Laws and Equations

1. Linear Momentum: An object's linear momentum is its mass times its velocity, or

$$\text{linear momentum} = \text{mass} \cdot \text{velocity}. \qquad (1.4.1)$$

2. The Definition of Impulse: The impulse given to an object is equal to the product of the force exerted on that object times the length of time that force is exerted, or

$$\text{impulse} = \text{force} \cdot \text{time}. \qquad (1.4.2)$$

3. Angular Momentum: An object's angular momentum is its moment of inertia times its angular velocity, or

$$\text{angular momentum} = \text{moment of inertia} \cdot \text{angular velocity}. \qquad (1.4.4)$$

4. The Definition of Angular Impulse: The angular impulse given to an object is equal to the product of the torque exerted on that object times the length of time that torque is exerted, or

$$\text{angular impulse} = \text{torque} \cdot \text{time}. \qquad (1.4.5)$$

5. Newton's Third Law of Rotational Motion: For every torque that one object exerts on a second object, there is an equal but oppositely directed torque that the second object exerts on the first object.

6. Kinetic Energy: An object's translational kinetic energy is ½ of its mass times the square of its speed, or

$$\text{kinetic energy} = \tfrac{1}{2} \cdot \text{mass} \cdot \text{speed}^2 . \quad (1.4.7)$$

An object's rotational kinetic energy is ½ of its moment of inertia times the square of its angular speed, or

$$\text{kinetic energy} = \tfrac{1}{2} \cdot \text{moment of inertia} \cdot \text{angular speed}^2 . \quad (1.4.8)$$

The Physics of Wheels

1. Frictional forces oppose relative motion between two solid surfaces. There are two types of friction: sliding friction between surfaces in relative motion and static friction between surfaces at rest relative to one another.

2. The amount of friction between two surfaces depends on the characteristics of the surfaces and is roughly proportional to the support force exerted by one surface on the other.

3. No work can be done against static friction because the surfaces travel no distance across one another. Work done against sliding friction produces thermal energy in the objects involved. Thermal energy is a disordered form of energy.

4. Momentum is a conserved vector quantity that measures an object's translational motion. It can't be created or destroyed but can be transferred between objects by impulses.

5. Angular momentum is a conserved vector quantity that measures an object's rotational motion. It can't be created or destroyed but can be transferred between objects by angular impulses.

6. Power measures how quickly work is done on an object; the amount of work done per unit of time.

7. An object accelerates in the direction of the net force on it, which is also the direction that reduces its total potential energy as quickly as possible.

Check Your Understanding - Answers

1. Friction is pushing the glass uphill.

Why: The glass is sliding downhill across the top of the stationary table. Since friction always opposes relative motion, it pushes the glass uphill, in the direction opposite its motion.

2. About twice as hard.

Why: The frictional forces between the table and books are roughly proportional to the force pressing them together. The book's weight is what pushes them together. When you effectively double the book's weight, by stacking a second book on top of it, you double the frictional forces between the table and the books.

3. If the wheels continue to turn, they experience static friction with the pavement. If they lock and begin to skid, they experience sliding friction. The maximum force of static friction is larger than the force of sliding friction, so the car will decelerate faster if the wheels don't skid.

Why: For a rapid stop, the car needs the maximum possible force in the direction opposite its velocity. The most effective way to obtain that stopping force from the road is with static friction between the turning wheels and the pavement. Sliding friction, the result of skidding tires, is much less effective at stopping the car, wears out the tires, and diminishes the driver's ability to steer the vehicle.

4. The archer does work on the string and bow as she draws the arrow back. (Chemical energy from her body is transferred to the bow, where it is stored as elastic potential energy.) As she releases the arrow, the string and bow do work on the arrow. (Elastic potential energy is transferred to the arrow, where it becomes kinetic energy.) Finally, the arrow does work on the apple, knocking it off the head of the assistant. (Kinetic energy is transferred from the arrow to the apple.)

Why: Because energy is conserved, we could in principle follow it back to the origins of the universe. Whatever energy we see around us now was somewhere in our universe yesterday, last week, and a million years ago, although its form may have changed. It will still be in our universe next year, too, but maybe not in as useful a form.

5. Normal driving involves mostly static friction because the surfaces of the tires don't slide across the pavement. Skidding involves sliding friction as the tire surfaces move independently of the pavement. Because it involves sliding friction, skidding creates thermal energy and damages the tires.

Why: The expression "burn rubber" is an appropriate name for skidding during a "jack-rabbit start." Substantial thermal energy is produced and a trail of hot rubber is left on the pavement behind the car. At drag races, the frictional heating that results from skidding at the start can be so severe that the tires actually catch on fire.

6. Because all the supporting forces are very close to the axis of rotation, the jewels exert almost zero torque on the axle. The axle turns remarkably freely.

Why: Mechanical timepieces need almost ideal motion to keep accurate time. One of the best ways to allow a rotating object free movement is to support it exactly on the axis of rotation, where the support can't exert torque on the object.

7. The forces of friction on the powered wheels are larger in the front-wheel drive car because the weight of the engine pushes the powered wheels firmly against the pavement.

Why: The frictional forces between the tires and the pavement are roughly proportional to the forces pressing the tires against the pavement. Since a front-wheel drive car has its heavy engine resting directly above its powered wheels, there is lots of static friction between those wheels and the pavement. The car can pull itself along successfully, even on fairly slippery roads. Rear-wheel drive cars or pickup trucks have better traction when their trunks or truck beds are loaded with heavy sand bags or even the engine.

8. The bouncy rubber ball would be more effective.

Why: Either projectile will transfer all of its momentum to the bowling pin while coming to a stop. But then the bouncy rubber ball will bounce back and continue to exert a force on the bowling pin. The impulse (force·time) delivered by the rubber ball will be greater than that delivered by the bean bag because the ball will exert its forward force for a longer time (during stopping *and* rebounding).

9. It came from the entire earth.

Why: Because the person stood on the earth as he started the merry-go-round spinning, he transferred angular momentum from the earth to the merry-go-round. The merry-go-round spins in one direction and the earth's rotation changes ever so slightly in the other direction. Because the earth is so enormous and has such an enormous moment of inertia, its slight change in rotation is undetectable.

10. The child accelerates forward, in the direction that will reduce their potential energy as quickly as possible.

Why: The child has only one form of potential energy: gravitational potential energy. This gravitational potential energy is lowest when the child is directly below the swing's supporting bar. The child accelerates forward because that will put the child below the support as quickly as possible.

11. Doubling the speed of the baseball requires quadrupling the energy transferred to it by the pitcher.

Why: To throw a 160 km/h (100 mph) fast ball, a major league pitcher must put 4 times as much kinetic energy into both the ball and his arm as when pitching an 80 km/h slow ball. He also pitches the fast ball in half the time needed to pitch the slow ball. Overall, he must do 4 times as much work throwing a fast ball and he must do that work in ½ the time. That means that he produces 8 times as much power while throwing a fast ball as when throwing a slow ball. No wonder amateurs have trouble duplicating that feat.

Check Your Figures - Answers

1. 440,000 kg·m/s.

Why: You can use Eq. 1.4.1 to calculate the train's momentum from its mass and velocity:

$$\begin{aligned}\text{linear momentum} &= 20{,}000\text{ kg}\cdot 22\text{ m/s}\\ &= 440{,}000\text{ kg}\cdot\text{m/s}.\end{aligned}$$

That momentum is in the forward direction.

2. 2,200 s.

Why: To stop the train, you must give it a backward impulse that completely cancels its forward momentum. Since its forward momentum is 440,000 kg·m/s, the backward impulse must be 440,000 kg·m/s. Since 200 N can also be written as 200 kg·m/s^2, we can use Eq. 1.4.2 to find the time:

$$\begin{aligned}\text{time} &= \frac{440{,}000\text{ kg}\cdot\text{m/s}}{200\text{ kg}\cdot\text{m/s}^2}\\ &= 2{,}200\text{ s}.\end{aligned}$$

3. The angular momentum will increase by a factor of 5.

Why: Because the satellite's angular momentum is proportional to its angular velocity, spinning it 5 times faster will increase its angular momentum by that same factor.

4. It will take them 5 times as long.

Why: To reach the new, faster angular velocity, the astronauts will need an angular impulse that's 5 times as large as originally planned. Since they will be using the same torque, they will have to exert that torque for 5 times as long.

5. 100 times as much kinetic energy.

Why: Because kinetic energy is proportional to speed squared, the kilogram of air in the hurricane moves 10 times as fast but has 100 times as much kinetic energy as the slower moving air. This enormous increase in energy is what makes a hurricane's wind dangerous. The air's terrific speed also brings large quantities of it to you quickly, so that the wind power arriving each second is overwhelming.

6. The children will triple its kinetic energy.

Why: The kinetic energy of a spinning object is proportional to its moment of inertia. Since the children triple the merry-go-round's moment of inertia, they triple its kinetic energy.

Glossary

angular impulse The mechanical means for transferring angular momentum. One object gives an angular impulse to a second object by exerting a certain torque on the second object for a certain amount of time. In return, the second object gives an equal but oppositely directed angular impulse to the first object.

angular momentum A conserved vector quantity that measures an object's rotational motion. It is the product of that object's moment of inertia times its angular velocity.

friction The force that resists relative motion between two surfaces in contact. Frictional forces are exerted parallel to the surfaces in the directions opposing their relative motion.

impulse The mechanical means for transferring momentum. One object gives an impulse to a second object by exerting a certain force on the second object for a certain amount of time. In return, the second object gives an equal but oppositely directed impulse to the first object.

joule-per-second The SI unit of power (synonymous to the watt).

linear momentum A conserved vector quantity that measures an object's motion. It is the product of that object's mass times its velocity.

momentum Linear momentum.

Newton's third law of rotational motion For every torque that one object exerts on a second object, there is an equal but oppositely directed torque that the second object exerts on the first object.

ordered energy Energy that can easily be used to do work.

power The measure of how quickly work is done on an object. The SI unit of power is the watt.

relative motion The movement of one object from the perspective of another object. Two objects that are moving relative to one another have different velocities.

sliding friction The forces that resist relative motion as two touching surfaces slide across one another.

static friction The forces that resist relative motion as outside forces try to make two touching surfaces begin to slide across one another.

thermal energy A disordered form of energy contained in the kinetic and potential energies of the individual atoms and molecules that make up a substance. Because of its random distribution, this disordered energy can't be converted directly into useful work. Other names for thermal energy include internal energy and heat.

watt (W) The standard SI unit of power, equal to the transfer of 1 joule-per-second. One watt is the power used by the bulb of a typical flashlight.

Review Questions

1. What's the difference between static and sliding friction?

2. How do the frictional forces between your shoes and the floor depend on your weight?

3. Why does scuffing your shoes on the ground wear away their soles?

4. Where is energy located in the following items: a speeding locomotive, a spinning cement mixer, a stretched spring, and a candle? How can that energy be transformed into other forms?

5. Why do your hands get hot when you rub them together?

6. If you roll a pencil between your hands, they don't get hot. What has the pencil done to the friction between them?

7. If you spin a basketball on your hand, what stops it?

8. When an 18-wheel truck collides with a sports car, which object experiences the largest change in velocity? in momentum?

9. Describe the process of spinning a toy top in terms of angular momentum and angular impulse. Where does the top's angular momentum come from?

10. Follow the flow of energy that occurs when children roll a snowball down a hill and into a collection of garbage cans.

Exercises

1. A perfectly balanced bicycle wheel experiences no torque due to gravity. However, if you lift the bicycle off the ground and spin the wheel, it will continue to spin for a very long time. What keeps this balanced wheel spinning? Is some torque keeping it going?

2. Why can't an acrobat stop himself from spinning while he is in midair?

3. When you spin a toy top with your fingers, you increase its angular momentum. How are you transferring angular momentum to the top?

4. Falling into a leaf pile is much more comfortable than falling onto the bare ground. In both cases you come to a complete stop, so why does the leaf pile feel so much better?

5. It's easier to injure your knees and legs while hiking downhill than while hiking uphill. Use the concept of energy to explain this observation.

6. A horse does work on a cart it's pulling along a straight, level road at a constant speed. The horse is transferring energy to the cart, so why doesn't the cart go faster and faster? Where is the energy going?

7. Skiers often stop by turning their skis sideways and skidding them across the snow. How does this trick remove energy from a skier and what happens to that energy?

8. If you give a horseshoe a twist as you throw it, it will continue to rotate until it hits something. Why?

9. Explain why a rolling pin flattens a pie crust without encountering very much sliding friction as it moves.

10. If you sit in a good swivel chair with your feet off the floor, the chair will turn slightly as you move about but will immediate stop moving when you do. Why can't you make the chair spin without touching something?

11. A yo-yo is a spool-shaped toy that spins on a string. In a sophisticated yo-yo, the end of the string forms a loop around the yo-yo's central rod so that the yo-yo can spin almost freely at the end of the string. Why does the yo-yo spin longest if the central rod is very thin and very slippery?

12. Firefighters slide down a pole to get to their trucks quickly. What happens to their gravitational potential energy and how does it depend on the slipperiness of the pole?

13. In countless movie and television scenes, the hero

punches a brawny villain who doesn't even flinch at the impact. Why is the immovable villain a Hollywood fantasy?

14. Professional sprinters wear spikes on their shoes to prevent them from sliding on the track at the start of a race. Why is energy wasted whenever a sprinter's foot slides backward along the track?

15. While a gymnast is in the air during a leap, which of the following quantities must remain constant for her: velocity, momentum, angular velocity, or angular momentum?

16. As you begin pedaling your bicycle and it accelerates forward, what is exerting the forward force that the bicycle needs to accelerate?

17. A newspaper sliding along the sidewalk quickly comes to rest. Why doesn't its inertia keep it moving forward indefinitely?

18. If you are pulling a sled along a level field at constant velocity, how does the force you are exerting on the sled compare to the force of sliding friction on its runners?

19. When you begin to walk forward, what exerts the force that allows you to accelerate?

20. A toy top spins for a very long time on its sharp point. Why does it take so long for friction to slow the top's rotation?

21. Why does putting sand in the trunk of a car help to keep the rear wheels from skidding on an icy road?

22. When you're driving on a level road and there's ice on the pavement, you hardly notice that ice while you're heading straight at a constant speed. Why is it that you only notice how slippery the road is when you try to turn left or right, or to speed up or slow down?

23. When a star runs out of nuclear fuel, gravity may crush it into a neutron star about 20 km (12 miles) in diameter. While the star may have taken a year or so to rotate once before its collapse, the neutron star rotates several times a second. Explain this dramatic increase in angular velocity.

24. The earth spins slightly faster in July and August than it does in April, shortening the day by about 0.0012 seconds. This seasonal effect is caused by the melting of snow from the vast mountains of the Northern Hemisphere. Why?

25. If you jump onto a spinning merry-go-round, you will begin to rotate with it. Describe this process in terms of the merry-go-round's angular momentum and your angular momentum (treat them separately—don't include your angular momentum in that of the merry-go-round).

26. When you first let go of a bowling ball, it's not rotating. But as it slides down the alley, it begins to rotate. Use the concept of energy to explain why the ball's forward speed decreases as it begins to spin.

27. While a coin tips over easily when you stand it on edge, a coin that's spinning on its edge stays up for a remarkably long time. What conserved quantity helps to keep it up?

28. Describe the process of writing with chalk on a blackboard in terms of friction and wear.

29. The binding that holds a ski boot to a ski is designed to prevent the boot from moving relative to the ski. However, when the force on the boot exceed a certain limit, the boot begins to move relative to the ski. How is this behavior similar to the behavior of friction?

Problems

1. Some special vehicles have spinning disks (flywheels) to store energy while they roll downhill. They use that stored energy to lift themselves uphill later on. Their flywheels have relatively small moments of inertia but spin at enormous angular speeds. How would a flywheel's kinetic energy change if its moment of inertia were five times larger but its angular speed were five times smaller?

2. What is the momentum of a fly if it's traveling 1 m/s and has a mass of 0.0001 kg?

3. If the fly in Problem **2** hits you and transfers all of its momentum to you, how much does your velocity change as a result?

4. If the fly in Problems **2** & **3** transfers its momentum to you in a period of 0.01 s, what average force does it exert on you?

5. Your car is broken, so you're pushing it. If your car has a mass of 800 kg, how much momentum does it have when it's moving forward at 3 m/s (11 km/h)?

6. You begin pushing the car forward from rest (see Problem 5). Neglecting friction, how long will it take you to push your car up to a speed of 3 m/s on a level surface if you exert a constant force of 200 N on it?

7. When your car is moving at 3 m/s (see Problems **5** & **6**), how much translational kinetic energy does it have?

8. No one is driving your car (see Problems **5**, **6**, & **7**) and it crashes into a parked car at 3 m/s. Your car comes to a stop in just 0.1 s. What force did the parked car exert on it to stop it that quickly?

9. You're at the rollerskating rink with a friend who weighs twice as much as you do. The two of you stand motionless in the middle of the rink so that your combined momentum is zero. You then push on one another and begin to roll apart. If your momentum is then 450 kg·m/s to the left, what is your friend's momentum?

10. A 6 kg bowling ball is rolling down the alley at a speed of 5 m/s. What is the ball's momentum?

11. To bowl the bowling ball in Problem **10**, you push it forward for 0.25 s. How much forward force must you exert on the ball to bring it up to speed that quickly?

12. How much translational energy does the bowling ball in Problem **10** have?

13. You usually give your closet door a gentle push and it swings closed gently in 5 seconds. But today you're in a rush and exert 3 times the normal torque on it. If you push on it for the usual time with this increased torque, how will its angular momentum differ from the usual value?

14. How long will it take the closet door in Problem **13** to swing closed after your hurried push?

15. Compare the door's energy (see Problems **13** & **14**) as it swings closed on a usual day with its energy after your hurried push.

Epilogue for Chapter 1

In this chapter we focused on four everyday objects and examined the basic physical laws that govern their behaviors. In *falling balls*, we explored the concept of inertia, noting that a force is what makes an object accelerate. In *seesaws* we examined rotational inertia and saw how a torque causes angular acceleration. We also took a first look at three important forces: gravitational forces in *falling balls*, support forces in *ramps*, and frictional forces in *wheels*.

In *ramps*, we saw how the work performed in changing an object's altitude doesn't depend on the method used to move that object, since work is actually the mechanical means for transferring energy from one object to another. Energy, we also noted, is one of three conserved physical quantities that govern the motion of objects in our universe. In *wheels*, we examined the other two conserved physical quantities: momentum and angular momentum. These conserved quantities, as we noted, can't be created or destroyed, but they can be transferred within objects or between them. Following the flow of energy, momentum, and angular momentum among objects often helps in understanding how they work.

In *ramps*, we explored gravitational potential energy (potential energy associated with gravity). In *wheels*, we looked at several other forms of energy: kinetic energy (energy associated with motion) and thermal energy (energy associated with heat and temperature). Regardless of the object we were examining, energy played an important role. For example, as a ball falls, it converts energy from one form to another; on the way up, kinetic energy becomes gravitational potential energy and on the way down, gravitational potential energy becomes kinetic energy.

Explanation: Pulling a Tablecloth from a Table

The dishes remain in place because of their inertia. As we've seen, an object in motion tends to remain in motion, while an object at rest tends to remain at rest. Before you pull on the tablecloth, the dishes sit motionless on its surface, and they tend to remain that way. By sliding the tablecloth off the table as quickly and smoothly as possible, you ensure that whatever force the tablecloth exerts on a dish occurs only for a very short time. As a result, the tablecloth transfers very little momentum to the dish.

Friction plays two roles in this stunt. Friction between the tablecloth and the dishes accelerates the dishes toward the edge of the table, where they might fall. That's why it's important to use a smooth, slippery tablecloth; since such a cloth experiences little sliding friction as it slides under the dishes, it reduces the unwanted acceleration. On the other hand, friction between the dishes and the bare table decelerates the dishes so that they come to rest before falling on the floor. The table itself should not be too slippery, or the dishes will slide off of it after the tablecloth is gone.

Cases

1. You're riding on a playground swing. You're traveling back and forth once every few seconds.

a. At what point(s) in your motion is your velocity zero?

b. At what point(s) in your motion is your gravitational potential energy at its maximum?

c. At what point(s) in your motion is your kinetic energy at its maximum?

d. As you reach the bottom of a swing, when the swing's ropes are exactly vertical, are you accelerating?

e. At the moment described in **d**, is the force that the ropes exert on you more, less, or equal to your weight?

2. You're juggling 3 baseballs so that one of them is always in the air.

a. As you catch one of the falling balls, which way does the ball accelerate?
b. As you toss the ball back into the air, which way does the ball accelerate?
c. In both **a** and **b**, how does the upward force that you exert on the ball compare with the ball's weight?
d. Even though one of the balls is always in the air, the average upward force you must exert on the balls is equal to their combined weight. Explain.

3. A 100,000 kg train is traveling north along a straight, level track at a constant speed of 30 m/s (108 km/h or 67 mph).

a. As it moves forward at constant velocity, what is the net force on the train?
b. What is the train's momentum?
c. To stop the train, how much momentum must you transfer to it?
***d.** If you exert a 500 N southward force on the coasting train, how long will it take to stop the train?

4. In bowling, you roll a heavy plastic ball down a smooth wooden lane and try to knock over 10 upright pins.

a. When you first let go of the ball, it is not rotating and skids along the wooden surface. What is happening to its kinetic energy as it skids?
b. The ball gradually starts to rotate. What provides the torque that causes it to spin?
c. When the ball strikes the pins, they go flying while the ball continues forward. Why doesn't the ball simply bounce off of the pins?
d. Bowling balls come in a variety of different weights. How does the ball's mass affect its behavior when it reaches the pins?

5. You have just started to coast down a hill on your bicycle. The hill has a very steady slope so that you descend 1 m for every 5 m of travel along the hill.

***a.** Explain why the amount of force accelerating you and the bicycle downhill is only a fifth of your combined weights.
***b.** How much does your velocity increase during the 1st second of coasting?
***c.** How much does your velocity increase during the 2nd second of coasting?
d. Why do you travel farther during the 2nd second than during the 1st second of coasting?

6. Two cars are driving at 88 km/h (55 mph) toward an intersection. A 1500 kg sedan is heading east and a 700 kg subcompact is heading north. The subcompact has a green light and drives into the intersection. The tired driver of the sedan runs the red light and crashes directly into the driver's side of the subcompact. The subcompact experiences a sudden, strong force to its right (toward the east).

a. Which car experiences the largest force due to the collision?
b. Which car experiences the largest acceleration due to the collision?
c. Which car is most likely to be thrown off the roadway because of the collision?
d. If the force on the subcompact is directly toward that car's right (directly toward the east), what happens to the north/south component of the subcompact's velocity?

7. Diving boards and platforms offer a nearly ideal opportunity in which to experience the various laws of motion. When you jump off the high diving board, you are the falling object and, if you can keep your presence of mind as you fall, you can learn something about physics. Imagine yourself diving off a platform 10 m above the water below.

a. If you walk very slowly off the platform, so that you fall directly downward, roughly how long will it take for you to reach the water? Look at Fig. 1.1.4 and make a reasonable estimate.
b. In the situation described in **a**, about how fast will you be traveling downward when you reach the water? Estimate your velocity from Fig. 1.1.4.
c. If you jump upward as you leave the platform, so that you begin with a modest upward velocity, will your downward velocity when you hit the water be more or less than in **b**?
d. You leave the platform simultaneously with a friend. She walks slowly off the platform and you jump to give yourself a modest initial upward velocity. Who will reach the water first?
e. You leave the platform simultaneously with a friend. He walks slowly off the platform and you run off the platform, so that your initial velocity is in the horizontal direction. Who will reach the water first?
f. In the situation described in **e**, who hits the water with the largest speed or are their speeds equal?

8. A high-jumper and a long-jumper are both human falling objects, but they have slightly different goals. The high-jumper wants to travel over a bar without touching it and the long-jumper wants to travel as great a distance as possible without touching the ground.

a. The high jumper approaches the bar at a moderate forward velocity. She then jumps directly upward (she obtains a purely upward force from the ground). Why is her initial horizontal velocity important?
b. The long jumper approaches the jumping pit at peak sprint speed. He then jumps directly upward (he obtains a purely upward force from the ground). Why doesn't he use all his strength to push himself forward instead of upward?
c. In a standing long-jump, the jumper leaps from a stationary position. Since the jumper has only so much strength and energy, she can only give herself a certain initial velocity. What direction should the jumper's initial velocity have so that she travels the greatest distance?
d. The earth's gravity is not perfectly uniform. Gravity is effectively 1% weaker at the equator than it is at the earth's poles. Explain why both a high-jumper and a long-jumper benefit from weaker gravity.

9. Imagine that you're standing on a slick ice-skating rink. No matter how hard you try, you can't move closer to the railing. You're initially at rest in the middle of the rink and you have no initial velocity.

a. To escape from the ice, you throw your boot toward one side of the rink. Which ends up moving faster: the boot or you, or do you end up with equal but oppositely directed velocities? Explain.
b. Once you begin to slide across the ice, why do you continue to slide (assuming there is no air resistance or friction)?
c. If you throw the boot forward with your right hand, while it's extended well out to your right, you'll experience a torque. How will that torque affect your subsequent motion?

d. Suppose that you didn't have a boot to throw away. Your lungs are full of air. If you pushed that air out of your mouth, what would happen? In which direction should you push it and in which direction would you then move?

10. Cable cars have traversed the hills of San Francisco for generations. As they ascend the hills, the cable cars are pulled up by moving, underground cables. As they descend the hills, the cable cars are held back by those same cables.

a. As the cable pulls the car uphill, which does (positive) work on which? As the cable lowers the car downhill, which does (positive) work on which?
***b.** On a gentle hill, the car rises 1 m in height as it travels 10 m along the road. On a second, steeper hill, the car rises 1 m in height as it travels 5 m along the road. If the cable must exert 3,000 N of force on the car to pull it up the gentle hill at constant velocity, how much force will be needed to pull the car up the steeper hill at constant velocity?
***c.** The car now begins to descend the gentle hill. How much force is needed to keep the car from accelerating downhill and in which direction does the cable exert its force on the car?
***d.** What is the weight of the cable car in newtons?
***e.** What is the mass of the cable car in kilograms?

11. An escalator is essentially a moving staircase. The individual steps are supported by metal tracks that run on either side of the escalator. These steps follow one another in a complete loop, driven by an electric motor. When you step onto an escalator at the ground floor, it soon begins to carry you upward and forward at a constant velocity toward the second floor.

a. While you're moving toward the second floor at a constant velocity, what is the net force exerted on you by all outside forces (specify the amount and the direction of the net force)?
b. You know that gravity gives you a weight in the downward direction. What force does the escalator exert on you as you move toward the second floor at a constant velocity (specify the amount and the direction of the force)?
c. Is the escalator doing work on you as you move toward the second floor?
d. As you first step onto the escalator, you begin to accelerate toward the second floor. Is the net force exerted on you by all outside forces the same as in **a**?
e. If the rapidly moving escalator suddenly stopped moving, you would be thrown forward and might even fall over. What causes you to be thrown forward?
f. You have more energy when you reach the second floor than you had on the first floor. Why aren't you moving faster as a result?

12. Imagine that you're sledding alone down a steep hill on a toboggan and that you left the top of the hill at the same time as an identical toboggan, loaded with six people.

a. Neglecting the effects of air resistance and friction, which toboggan will reach the bottom of the hill first?
b. During the descent, your toboggan brushes up against the six person toboggan. Which toboggan will experience the largest change in velocity as the result of the impact?
c. If you take a steeper route down the hill, how will that affect the speed of your descent? Explain.
d. Before each downhill run, you must pull the toboggan back to the top of the hill. Explain how the toboggan's gravitational potential energy changes on the way up the hill and on the way down it.
e. When are you doing (positive) work on the toboggan?
f. When is gravity doing (positive) work on the toboggan?

13. Three children are at a playground, playing with a seesaw. Two of the children are 3 years old and each has a mass of 15 kg; the third child is 7 years old and has a mass of 30 kg. The seesaw is made of a high-tech low-mass plastic, so that its mass can be ignored throughout this problem. The seesaw's pivot is in the middle of the board. The two 15 kg children sit together at the east end of the seesaw and the 30 kg child sits at the west end.

a. The seesaw balances, meaning that it experiences no net torques when the children and the seesaw are not touching the ground. Describe the nature of any possible motion of the seesaw during the periods when the children and the seesaw are not touching the ground.
b. What happens to the seesaw if the 30 kg child at the west end pushes downward on the ground?
c. One of the 15 kg children loses interest (as 3 year olds often do) and jumps off the board while the seesaw is exactly horizontal. What happens to the seesaw and the other 2 children?
d. The two remaining children decide to continue playing on the seesaw. If the 15 kg child stays at the east end of the seesaw, exactly where should the 30 kg child sit in order for the seesaw to balance?
e. The other 15 kg child returns and sits on the middle of the board, right above the pivot. The child has essentially no effect on the seesaw. Why not?

14. Revolving doors are popular in northern hotels and office buildings as a way to prevent cold outside air from blowing directly into the lobby. Most revolving doors have 4 panels arranged in a cross, as viewed from above. You step in between two panels of the door and push on the panel in front of you. The revolving door begins to rotate and once you reach the inside of the building, you step out into the lobby.

a. It is much easier to make the revolving door rotate by pushing on the panel far from the central pivot than it is by pushing near the pivot. Why?
b. As you push on the door, it begins to turn more and more quickly. What is your pushing doing to the door?
c. One of the dangers of revolving doors is being hit by the panel behind you as you step out of the door. The door tends to keep on turning after you stop pushing, and it can bump you if you're not careful. Why does it keep on turning after you stop pushing?
d. What eventually stops the revolving door when no one uses it for a minute or two?
e. The people who built the revolving door wanted it to be easy to start and stop. They had to be most careful to minimize the mass of which edge of each door panel: the upper edge, lower edge, inside edge (nearest the pivot), or outside edge (farthest from the pivot)?

15. A popular exercise machine in a modern weight room simulates the act of climbing stairs. You stand on two pedals and move your feet up and down. The pedals move with your feet and hidden machinery inside the device makes a whirring sound. It takes a lot of force to push each pedal downward but that pedal rises upward easily as you lift your foot. As long as you keep moving your feet up and down at the right pace, you remain stationary. If you slow your pace, you begin to sink downward. If you speed up, you begin to rise upward.

a. When you're moving your feet up and down at the right pace, so that you remain stationary, how much total downward force are you exerting on the two pedals?
b. As you push your foot and the pedal downward, you exert a large downward force on the pedal. Between your foot and the pedal, which is doing (positive) work on which?
c. As you lift your foot upward, the pedal exerts a small upward force on your foot. Between your foot and the pedal, which is doing (positive) work on which?
d. On the average, what is doing (positive) work on what (you and the machine)?
e. Clearly, energy is being transferred out of you and into the machine. The machine converts that energy into a form that can be carried away in the air. What is that final form of energy?

16. You're a pilot for the Navy. For your airplane to be able to lift itself off the ground, it must be traveling forward with a speed of 208 km/h (130 mph). At this take-off speed your airplane will have about 50,000,000 N·m (or 50,000,000 J) of kinetic energy.

***a.** During takeoff, your airplane's jet engine exerts a force of 250,000 N in the backward direction on the air leaving the engine. What force does that same air exert on the airplane (specify the amount and the direction of force)?
***b.** The force exerted by the air on the airplane causes it to accelerate down the runway. How long must the runway be for the airplane to reach its take-off speed?
***c.** An aircraft carrier runway is only about 100 m long. As your answer to **b** indicates, this distance is not enough for the airplane to reach take-off speed on its own. The aircraft carrier must assist the airplane by exerting an extra force on it. The aircraft carrier uses a steam-powered catapult to help push the airplane forward. How much additional force must the catapult exert on the airplane to bring the airplane to take-off speed at the end of the 100 m runway?
***d.** During an aircraft carrier take-off, the airplane and the catapult exert forces on one another. Which of these two objects does (positive) work on the other and which object transfers some of its energy to the other?
***e.** During an aircraft carrier landing, the airplane hooks onto a cable that slows the airplane to a stop. The airplane and cable exert forces on one another. Which of these two object does (positive) work on the other and which object transfers some of its energy to the other?

17. You're seated in a crowded bus that has stopped at a bus stop. Two people board the bus, one wearing rubber-soled shoes and one wearing inline skates (roller skates — i.e., no friction). They both have to stand in the middle of the aisle and neither one holds onto anything.

a. The bus driver is new and the bus lurches forward from the bus stop. Which way does the person wearing inline skates move in relationship to the bus?
b. The bus is accelerating forward as it pulls away from the bus stop. Why doesn't the person wearing inline skates accelerate with the bus?
c. The person wearing rubber-soled shoes remains in place as the bus starts moving. What provides the force that causes that person to accelerate with the bus?
d. Once the bus has reached a constant velocity (it's traveling along a straight road at a steady pace), the person wearing inline skates is able to stand comfortably in the aisle without rolling anywhere. What is the net force on the person wearing skates?
e. The driver accidentally runs into a curb and the bus stops so abruptly that everything slides forward, including the person wearing rubber-soled shoes. In stopping quickly, the bus experiences an enormous backward acceleration. Why doesn't the person wearing rubber-soled shoes accelerate backward with the bus?

18. You're about to go on a bicycle trip through the mountains. Being ambitious, you decide to take two children along. The children sit in a trailer that you pull with your bicycle.

a. As you wait to begin your trip, you and your bicycle are motionless. What is the net force on your body?
b. You begin to bicycle into the mountains. You soon find yourself ascending a steep grade. The road rises smoothly uphill and you're traveling up it at a steady pace in a straight line. You're traveling at a constant velocity up the hill. What is the net force on your body?
***c.** The road rises 1 m upward for every 10 m you travel along its surface. If the children and their trailer weigh 400 N, how much uphill force must you exert on the trailer to keep it moving uphill at a constant velocity?
***d.** How much work must you do on the trailer and children as you pull them to the top of a 500 m tall hill? Does the amount of work you do on them depend on whether you take the long, gradually sloping road or the short, steep road? (Answer both questions.)
***e.** How much work must you do on the trailer and children as you pull them back down the 500 m tall hill at constant velocity?

19. You have recently taken up track and field as a way to keep in shape. You soon begin to notice how simple physical laws appear in many of the events.

a. You notice that great sprinters have extremely strong legs. Why is it so important that a sprinter be able to push back hard on the starting blocks at the beginning of a race?
b. You find that throwing a heavy metal shot is far more difficult than throwing a baseball. Weight isn't the whole problem. Even if you try to throw the shot horizontally or downward, so that weight is not an issue, you have great difficulty getting the shot to move quickly. Why?
c. As you land on the soft foam pad beneath the pole vault, you realize that its job is to bring you to rest by accelerating you upward gradually with only modest support forces. If there were no pad there, only concrete, what would the acceleration and support forces be like during your landing?
d. You cross the finish line at the end of a race. The net force on your body points in what direction as you slow down?
e. In the long jump, you run rapidly down a path and then leap into the air. You find that the best distance comes from pushing yourself upward rather than forward during the leap. Why is it so important to have a large upward component of velocity at the start of the leap?

20. You're piloting a very large, very massive oil tanker. The ship is so massive that it is barely affected by air and water resistance. If you turn off all the engines, it travels at nearly constant velocity through the water.

a. The ship is at rest in the water when you first turn the engines on full forward. The ship begins to push the water toward the east. Which way does the water push on the ship and which way does the ship accelerate?
b. After 1 minute of acceleration, the ship has reached a speed of 1 km/h. If the force that the water exerts on the ship

remains constant, how fast will the ship be traveling after another minute?

c. The ship has reached its cruising speed of 20 km/h and you have steered it so that it is traveling directly northward. Its velocity is thus constant at 20 km/h to the north. If you cause the ship to accelerate directly toward the west for a few minutes, in which direction will its new velocity point? (to the south? to the south-west? ...?)

d. A small sailboat cuts directly in front of your ship as you move forward at cruising speed. You immediately turn the engines on full reverse so that you now accelerate backward at the same rate you accelerated forward before. How long will it take your ship to stop?

e. Unfortunately, you collide with the sailboat. Each vessel exerts a force of 50,000 N on the other vessel. Which vessel experiences the largest change in velocity and why?

21. Even though it has nothing preventing it from turning, the giant door to the local museum is surprisingly hard to open. You're inside the museum, trying to open the door so you can leave.

a. As you push gently on the door handle, the door begins to open. It turns more and more quickly as time passes. What kind of influence are you exerting on the door that causes it to begin turning?

b. If you try to open the door by pushing on the hinged side of the door, it won't open. Why does it matter where on the door you push?

c. If you stop pushing on the door after it has begun to turn rapidly, it will continue turning until it crashes into a doorstop. What keeps it turning?

d. How does the torque that the doorstop exerts on the door as the door crashes into it compare to the torque you exerted on the door to start it opening?

e. The museum has decided to make the door easier to open and close by reducing its mass. To have the greatest effect, from which part of the door should they remove the most mass: the hinged edge, the top edge, the doorknob edge, the bottom edge, or the middle of the door? Why?

22. Local fairs and amusement parks usually offer games in which you can win a large prize by performing a seemingly easy task. In many cases, these tasks are made surprisingly difficult by simple physical principles and few people receive prizes. Here are several of those games.

a. A pitching game requires that you knock over three milk bottles with a baseball. The bottles are filled with sand. Why does filling the bottles with sand make them so hard to knock over with the baseball?

b. A tossing game requires that you throw a coin forward and have it come to a stop on a smooth glass plate. Why doesn't the coin stop when it hits the smooth glass plate?

c. Another game requires that you knock over a narrow wooden peg with a ball hanging from a string. The string is suspended from a point directly above the peg. To win, you must swing the ball past the peg and have the ball knock over the peg on its return swing. This feat can't be done. The ball keeps circling around the peg at a relatively constant distance. The ball can't stop this circling to hit the peg and knock it over. Why does the ball keep circling the peg?

d. Still another tossing game requires that you throw a basketball into a shallow basket that is tipped toward you. Whenever you throw the ball into the basket, it bounces back out of the basket and falls onto the floor. What conserved physical quantity is the basketball unable to get rid of in time to remain in the basket?

23. Sledding makes use of the nearly frictionless nature of snow. Let's look at how sledding works, assuming that the snow is truly frictionless and there is no air resistance.

***a.** You're pulling a child up a hill on a sled, traveling at a constant velocity. The child and sled weigh 400 N and the smooth hill rises 1 m for every 5 m you travel along its surface. What is the net force on the child and sled?

***b.** How much uphill force are you exerting on the child and sled?

***c.** How much work must you do to pull the child and sled to the top of the 50 m high hill?

***d.** If the hill had been less steep but still 50 m high, would you have had to do more work, less work, or the same amount of work to pull the child and sled to the top?

***e.** How much has the child and sled's gravitational potential energy increased in going from the bottom of the hill to the top?

***f.** You release the child and sled and they slide down the hill, faster and faster. By the time they reach the bottom, how much kinetic energy do they have?

g. The sled continues on along the flat region at the bottom of the hill and onto some bare ground. The sled slows down abruptly and the child falls off the front. Why?

24. You are riding your bicycle on a north-south road and have just stopped at an intersection. Use that intersection as the reference point for position.

***a.** You leave the intersection with a northward velocity of 3 m/s. After 60 s, how far are you from the intersection?

***b.** What is your new position?

***c.** You abruptly change your velocity to 3 m/s toward the south. After another 60 s, what is your position?

***d.** Later in the day, you stop at the same intersection. This time you leave the intersection with an acceleration of 1 m/s^2 toward the north. After 10 s, what is your velocity?

***e.** What is your position?

***f.** You abruptly change your acceleration to 1 m/s^2 toward the south. After another 10 s, what is your velocity?

***g.** What is your position?

h. A trip during which you reverse your velocity is evidently quite different from one in which you reverse your acceleration. Explain this difference.

25. Advanced skiers turn by sliding the backs of their skis across the snow. Since the fronts of their skis don't move much, the skis end up pointed in a new direction.

a. The amount of sideways force that a skier must exert on the skis to slide them sideways is proportional to how hard the skis press down on the snow beneath them. Why?

b. To make it easier to slide the skis sideways, the skier "unweights"—reduces the force pressing their skis downward against the snow. The skier does this by jumping upward. How can the downward force that the skier exerts on the snow be less than the skier's weight?

c. When during the jump is it easiest to slide the skis sideways?

d. Rather than jumping, some racers simply pull their legs upward suddenly. Why does this action reduce the force pressing their skis against the snow?

e. Less skilled skiers sometime turn without unweighting—they push their skis sideways so hard that the skis slide anyway. This technique is exhausting. Why does it require so much work?

CHAPTER 2

SIMPLE MECHANICAL OBJECTS

Now that we've surveyed the laws of motion, we can begin using those laws to explain the behavior of everyday objects. But while we can already understand some of the central features at work in a toy wagon, a weight machine, or a ski lift, we're still missing a number of mechanical concepts that are important in the world around us. In this chapter, we'll look at some of those additional concepts.

One of the most important will be acceleration. If we treat acceleration passively, it can be fairly uninteresting: we push on the cart and the cart accelerates. But if we think of the concept more actively—for example, if we envision ourselves on a roller-coaster as it plummets down that first big hill—then acceleration becomes much more intriguing. In fact, we might even need to hold on to our hats.

Experiment: Swinging Water Overhead

To examine some of the novel effects of acceleration, try experimenting with a bucket of water. If you're careful, you can swing this bucket over your head and upside-down without spilling a drop. In the process, you'll be demonstrating a number of important physical concepts.

To do this experiment, you'll need a bucket with a handle. (You might substitute some equivalent container; even a plastic cup will do in a pinch.) Fill the bucket part way full of water and then hold it by the handle so that it hangs down by your side.

Now swing the bucket backward about an eighth of a turn and bring it

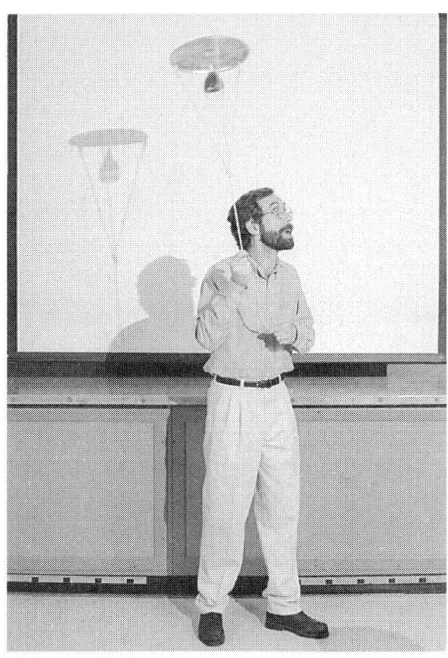

forward rapidly. In one smooth, fluid motion, swing it forward, up, and over your head. Continue this smooth motion all the way around behind you and then bring the bucket forward again. You'll need to swing the bucket quickly and smoothly to avoid getting wet.

You'll look like a windmill, turning the bucket around and around. As the bucket travels over your head, you'll notice that it's upside down—and yet the water stays in the bucket. Why doesn't the water fall out?

You can carry this experiment a step further by swinging the bucket at various speeds—that is, if you don't mind getting wet. First, try to **predict** what will happen if you change speeds. What will happen if you swing the bucket less rapidly? more rapidly? Now do the experiments and **observe** what happens. Did the experiments **verify** your predictions?

Take your experiments further still. As you swing the bucket over your head, try to **measure** how strongly the bucket pulls on your hand. Is the pull stronger or weaker when you swing less rapidly? more rapidly? Is there any relationship between the upward pull you feel from the inverted bucket and the water's tendency to remain inside?

You might vary this experiment in several ways. Try swinging a plastic cup held in your fingers, or placing a full wine glass in the bucket and swinging the two objects together. In the latter case, you'll find that the wine will stay in the glass and the glass at the bottom of the bucket, even when the bucket is upside-down.

By the way, the hardest part of all these tricks is stopping. To avoid a catastrophe, you'll need to do the same thing you did to get started, only in reverse. Come to a smooth, gradual stop about an eighth of a turn in front of you, and then let the bucket loosely drop back to your side. If you stop moving the bucket too abruptly, the water, wine, or glass will spill or smash. Why do you suppose stopping the bucket abruptly spills its contents? We'll return to this question, and the others raised by the swinging bucket, at the end of the chapter.

Chapter Itinerary

In the meantime, we'll examine three types of everyday objects: (1) *spring scales*, (2) *bouncing balls*, and (3) *centrifuges and roller coasters*. In *spring scales*, we'll review the relationship between mass and weight and explore how the distortion of a spring can be used to measure an object's weight. In *bouncing balls*, we'll study how balls store and return energy and how their bouncing depends both on their own characteristics and on those of the objects they hit. And in *centrifuges and roller coasters*, we'll look at how acceleration gives rise to gravity-like apparent forces that can wring water out of our clothes or make us scream with delight at the amusement park.

As we examine these objects, keep in mind that they illustrate concepts that can help you explain other phenomena as well. Almost any solid object, from a mattress to a diving board to a tire, behaves like a spring scale's spring when you push on it. Bouncing balls offer a view of collisions that help you comprehend what happens when two cars crash or when a hammer hits a nail. And the sensations associated with acceleration that you experience on a roller coaster are also present when you ride in airplanes, on subways, or on swing sets. When it comes to the physics of everyday objects, there really is nothing new under the sun.

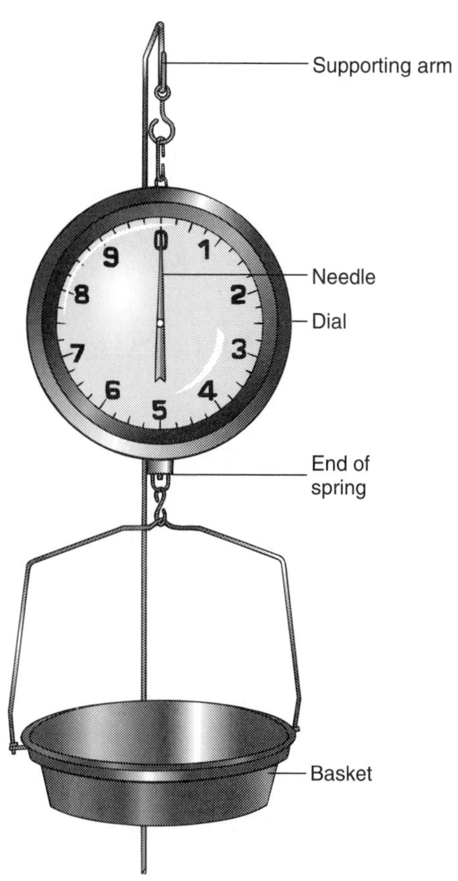

Section 2.1

Spring Scales

How much of you is there? From day to day, depending on how much you eat, the amount of you stays approximately the same. But how can you tell how much that is? The best measure of quantity is mass: kilograms of gold, beef, grain, or you. Mass is the measure of an object's inertia and, as we saw in Section 1.1, doesn't depend on the object's environment or on gravity. A kilogram box of cookies always has a mass of one kilogram, no matter where in the universe you take it.

But mass is difficult to measure directly. Moreover, the very concept of mass is only about 300 years old. Consequently, people began quantifying the material in an object by measuring its weight. Spring scales eventually became one of the simplest and most practical tools for accomplishing this task, and they are still found in bathrooms and grocery stores today. They really do contain springs, although these are normally hidden from view.

Questions to Think About: *How is your weight related to your mass? If the earth's gravity became twice as strong, how would your mass be affected? What about your weight? Does jumping up and down change either your mass or your weight? If you stand on a strong spring, how does your weight affect the shape of the spring? Why should there be a relationship between your weight and how much the spring bends?*

Experiments to Do: *Find a hanging spring scale of the type used in the produce section of a grocery store and watch the scale's basket and weight indicator as you put objects in the basket. What happens to the basket as you fill it up? Can you find a relationship between the basket's height and the weight reported by the scale? If you drop something into the basket, instead of placing it gently in the basket, how does the scale respond?*

Why does the weight indicator bounce back and forth rhythmically? What happens to the gravitational potential energy of the dropped item?

Now find a spring bathroom scale—the short, flat kind with a rotating dial. Stand on it. Why does it only read your correct weight when you are standing still? If you jump upward, how does the scale's reading change? What about if you let yourself drop? Bounce up and down gently. How does the scale's average reading compare with your normal weight? Does bouncing really change your weight? You can also change the scale's reading by pushing on the floor, wall, or other nearby objects. Which way must you push to increase the scale's reading? To decrease it? When you change the reading in these ways, are you actually changing your weight?

Why You Must Stand Still on a Scale

Whenever you stand on a scale in your bathroom or place a melon on a scale at the grocery store, you are measuring weight. An object's weight is the force exerted on it by gravity, usually the earth's gravity. (Refer to Section 1.1 for a review.) When you stand on a bathroom scale, the scale measures just how much upward force it must exert on you in order to keep you from moving downwards toward the earth's center. As in most scales you'll encounter, the bathroom scale uses a spring to provide this upward support. If you are stationary, you are not accelerating, so your downward weight and the upward force from the spring must cancel one another; that is, they must be equal in magnitude but opposite in-direction so that they sum to zero net force. Consequently, although the scale actually displays how much upward force it's exerting on you, that amount is also an accurate measure of your weight.

This subtle distinction in what is being reported is important. While an object's weight depends only on its gravitational environment, not on its motion, the weighing process itself is extremely sensitive to motion. If anything accelerates during the weighing process, the scale may not report the object's true weight. For example, if you jump up and down while you're standing on a scale, the scale's reading will vary wildly because you are accelerating. Since you're no longer experiencing zero net force, your downward weight and the upward force from the scale no longer cancel. If you want an accurate measurement of your weight, therefore, you have to stand still.

But even if you stand still, weighing is not a perfect way to quantify the amount of material in your body. While your mass is an intrinsic property of your body, your weight depends on your environment, on things around you. You've probably never noticed that your weight depends on your environment because you've most likely never lived anywhere but on the earth's surface. As a result, your weight remains pretty consistent, so long as you don't routinely eat a dozen jelly doughnuts for your snacks. If you move to the moon, however, you'll find that your weight is only about one-sixth what is on the earth's surface. You're still all there, but your weight has changed abruptly because the moon is much less massive than the earth and hence produces much smaller gravitational forces on objects at its surface. On a planet that's larger than the earth, the reverse would be true: on Jupiter, for example, your weight would be about two and a half times its earth value.

Even on the earth, gravity is not the same everywhere. The earth isn't exactly spherical; it bulges outward around the equator, lessening the weights of objects that are found there. As a result, your weight at the north or south pole would be about 0.5% higher than at the equator.

To make things even more complicated, the earth is rotating. As we'll see when we discuss a centrifuge, the earth's rotation does have a small effect on the reading of a scale. Near the equator, you and everything in your vicinity are accelerating as the earth turns. This acceleration again upsets the scale's ability to determine your weight, so it reads about 0.5% less than your true weight. This effect, together with the real difference in your weight due to the earth's non-spherical shape, will make the scale report that you're about 1% heavier at the north or south pole than at the equator.

CHECK YOUR UNDERSTANDING #1: Space Merchants
You're opening a company that will export gourmet food from the earth to the moon. You want the package labels to be accurate at either location. How should you label the amount of food in each package—by mass or by weight?

Stretching a Spring

So you know now that when you put a melon in the basket of a scale at the grocery store and read its weight from the scale's dial, the scale is actually reporting just how much upward force its spring is exerting on the melon. If you put the melon on a table instead, the table would also exert just enough upward force on the melon to stop it from accelerating and bring it to rest. But you'd have no simple way to determine just how much upward force the table was exerting on the melon. Therein lies the beauty, and the utility, of a spring: a simple relationship exists between its length and the forces it's exerting on its ends. The spring scale can therefore determine how much force it's exerting on the melon by measuring the length of its spring.

The simplest type of spring consists of a wire coil (Fig. 2.1.1). A coil spring pulls inward on its ends when it's stretched and pushes outward on them when it's compressed. If it's neither stretched nor compressed, a coil spring exerts no forces on its ends.

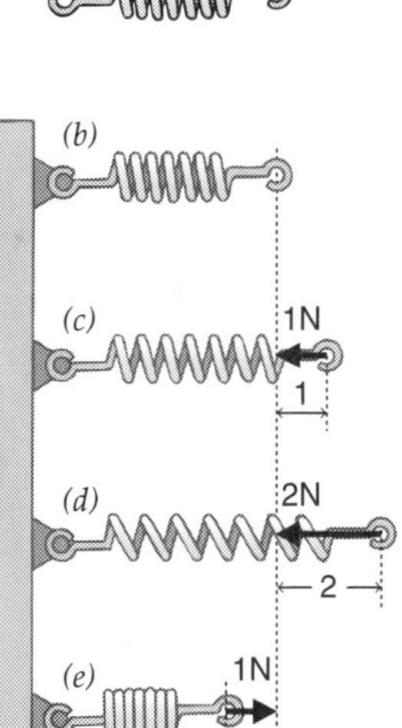

Fig. 2.1.1 - Five identical springs. The ends of spring (*a*) are free so that it can adopt its equilibrium length. The left ends of the other springs are fixed so that only their right ends can move. When the free end of a spring (*b*) is moved away from its original equilibrium position (*c*, *d*, and *e*), the spring exerts a restoring force on that end that is proportional to the distance between its new position and the original equilibrium position. The distances are in centimeters (cm).

The springs in Fig. 2.1.1 are supported by a table so that we can focus our attention on the forces that the springs themselves exert on their ends. The top spring (Fig. 2.1.1*a*) is neither stretched nor compressed and thus doesn't push or pull on its ends. If this spring is initially stationary and nothing outside is pushing or pulling on it, then it will remain stationary indefinitely because the net force on each of its ends is zero. In other words, it will be in **equilibrium**—experiencing zero net force. As the phrase "zero net force" suggests, equilibrium occurs whenever the forces acting on an object cancel one another perfectly so that the object doesn't accelerate. When you sit still in a chair, for example, you are in equilibrium.

When the spring in Fig. 2.1.1*a* is in equilibrium—that is, neither stretched nor compressed—it's at a special equilibrium length, its natural length when you leave it alone. No matter how you distort the spring, it tries to return to its equilibrium length. If you stretch it so that it's longer than its equilibrium length, it will pull inward on its end. If you compress it so that it's shorter than its equilibrium length, it will push outward on its ends.

Let's attach the left end of our spring to a post (Fig. 2.1.1*b*) and look at the behavior of its free right end. With nothing pushing or pulling on the spring, this free end will be in equilibrium at a particular location we can call its **equilibrium position**. If we now pull this free end to the right and let go, the stretched spring will pull it to the left; if we instead push the free end to the left and let go, the compressed spring will push it to the right. Because the spring automatically returns its free end toward the equilibrium position, the spring is in a **stable equilibrium**.

But what happens to the spring if we pull its free end to the right and don't let go? The spring now exerts a steady inward force on that end, trying to return it to its original equilibrium position. The more we stretch the spring, the more inward force it exerts on the end. This inward force is exactly proportional to how far we stretch the end away from its original equilibrium position. Since the end of the spring in Fig. 2.1.1*c* has been pulled 1 cm to the right of its original equilibrium position, the spring now pulls this end to the left with a force of 1 N; if the end is instead pulled 2 cm to the right, as it has been in Fig. 2.1.1*d*, the spring pulls it to the left with a force of 2 N. This proportionality continues to work even when you compress the spring: in Fig. 2.1.1*e*, the end has been pushed 1 cm to the left, and the spring is pushing it to the right with a force of 1 N.

In summary, the force exerted by a coil spring has two interesting properties. First, this force is always directed so as to return the spring to its equilibrium length. We call this kind of force a **restoring force** because it acts to restore the spring to its equilibrium length. Second, the spring's restoring force is proportional to how far it has been distorted (stretched or compressed) from its equilibrium length.

These two observations are expressed in **Hooke's law**, named after the Englishman Robert Hooke, who first demonstrated it in the late 17th century. This law can be written in a word equation:

$$\text{restoring force} = -\ \text{spring constant} \cdot \text{distortion}, \tag{2.1.1}$$

in symbols:

$$\mathbf{F} = -k \cdot \mathbf{x},$$

and in everyday language:

The more you compress a rubber eraser, the harder it pushes back.

Here the **spring constant**, k, is a measure of the spring's stiffness. The larger the spring constant—that is, the stiffer the spring—the larger the restoring forces the spring exerts. The negative signs in these equations indicate that a restoring force always points in the direction opposite the distortion.

HOOKE'S LAW:
The restoring force exerted by an elastic object is proportional to how far it has been distorted from its equilibrium shape.

Springs are distinguished by their stiffness, as measured by their spring constants. Some springs are very flexible and have small spring constants—for example, the one that pops the toast out of your toaster, which you can easily compress with your hand. Others, like the large springs that suspend an automobile chassis above the wheels, are very stiff and have large spring constants. But no matter the stiffness, all springs obey Hooke's law.

Hooke's law is remarkably general and isn't limited to the behavior of coil springs. Almost anything you push or pull on will pull or push back with a force that's proportional to how far you have distorted it away from its equilibrium length—or, in the case of a complicated object, its equilibrium shape. Equilibrium shape is the shape an object adopts when it's not subject to any outside forces. If you bend a tree, it will push back with a force proportional to how far it has been bent. If you pull on a rubber band, it will pull back with a force proportional to how far it has been stretched. If you squeeze a ball, it will push outward with a force proportional to how far it has been compressed. If a heavy truck bends a bridge downward, the bridge will push upward with a force proportional to how far it has been bent (Fig. 2.1.2).

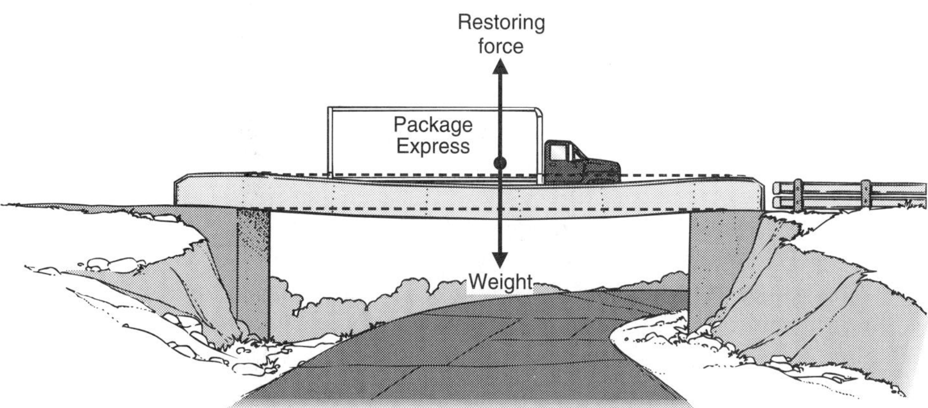

Fig. 2.1.2 - A steel bridge sags under the weight of a truck. The bridge bends downward until the upward restoring force it exerts on the truck exactly balances the truck's weight.

There is a limit to Hooke's law, however. If you distort an object too far, it will usually begin to exert less force than Hooke's law demands. This is because you will have exceeded the **elastic limit** of the object and will probably have permanently deformed it in the process. If you pull on a spring too hard, you'll stretch it forever; if you push on a tree too hard, you'll break it. But as long as you stay within the elastic limit, almost everything obeys Hooke's law: a rope, a ruler, an orange, a trampoline.

Distorting a spring requires work. When you stretch a spring with your hand, pulling its end outward, you transfer some of your energy to the spring. The spring stores this energy as **elastic potential energy**. (See Section 1.4 to review potential energy.) If you reverse the motion, the spring returns most of this energy to your hand, while a small amount is converted to thermal energy by frictional effects inside the spring itself. Work is also required to compress, bend, or twist a spring. In short, a spring that is distorted away from its equilibrium shape always contains elastic potential energy.

CHECK YOUR UNDERSTANDING #2: Going Down Anyone?

As you watch people walk off the diving board at a pool, you notice that it bends downward by an amount proportional to each diver's weight. Explain.

CHECK YOUR FIGURES #1: A Sinking Sensation

You're hosting a party in your 3rd floor apartment. When the first 10 guests begin standing in your living room, you notice that the floor has sagged 1 centimeter in the middle. How far will the floor sag when 20 guests are standing on it? when 100 guests are standing on it?

How a Hanging Grocery Scale Measures Weight

We're now ready to understand how spring scales work. Imagine a hanging spring scale of the kind used to weigh produce. Inside this scale is a coil spring, suspended from the ceiling by its upper end (Fig. 2.1.3). Hanging from its lower end is a basket. For the sake of simplicity, imagine that the scale's basket has little or no weight. With no force pulling down on it, the scale's spring adopts its equilibrium length and the basket, experiencing zero net force, is in a position of stable equilibrium. If you pull the basket downward and let go, the spring pulls

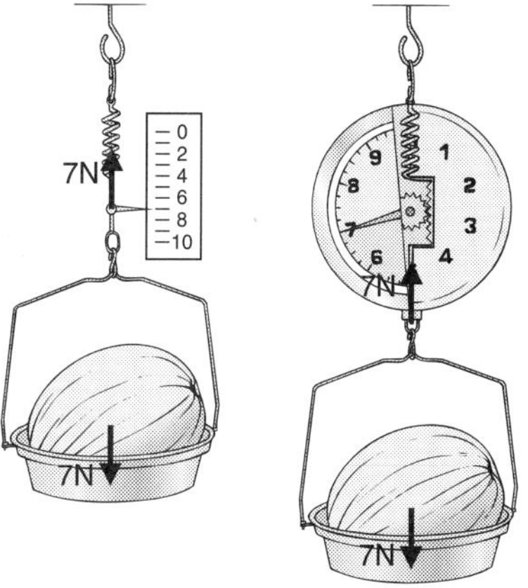

Fig. 2.1.3 - Two spring scales weighing melons. Each scale balances the melon's downward weight with the upward force of a spring. The heavier the melon, the more the spring will stretch before it exerts enough upward force to balance the melon's weight. The left scale has a pointer to indicate how far the spring has stretched and thus how much the melon weighs. The right scale has a rack and pinion gear that turns a needle on a dial. As the comb-like rack moves up and down, it turns the toothed pinion gear.

it upward. If you push the basket upward and let go, the spring pushes it downward.

When you place a melon in the basket, the melon is pulled downward by gravity, and its weight pulls downward on the basket. The spring is at its equilibrium length and is not exerting any force on the basket, so the basket accelerates downward. As the basket descends, the spring stretches and begins to exert an upward force on the basket. The more the spring stretches, the greater its upward force. Eventually, the spring is stretched just enough so that its upward force is equal in magnitude to the melon's weight.

The net force on the basket is now zero and the basket is in a new stable equilibrium position. Although the basket may bounce up and down briefly, it will eventually settle down to this new position—we'll discuss this settling process later on. If you then pull the basket downward and let go, the spring will still pull it upward; if you push the basket upward and let go, the spring and the melon's weight will still pull it downward. The spring will always return to its new equilibrium position.

How does it happen that the spring scale indicates the weight of the melon in its basket? Because of Hooke's law. Once the basket has adopted its new equilibrium position, where the melon's weight is exactly balanced by the upward force of the spring, the length of the stretched spring is an accurate measure of the melon's weight.

The scales in Fig. 2.1.3 differ only in the way they measure how far the spring has stretched beyond its equilibrium length. The left scale uses a pointer attached to the end of the spring, while the right scale uses a "rack and pinion" gear system that converts the small linear motion of the stretching spring into a much more visible rotary motion of the dial needle. The rack is the series of evenly spaced teeth attached to the lower end of the spring; the pinion is the toothed wheel attached to the dial needle. As the spring stretches, the rack moves downward, and its teeth cause the pinion to rotate. The farther the rack moves, the more the pinion turns, and the higher the weight reported by the needle.

Each of these spring scales reports a number for the weight of the melon you put in the basket. In order for that number to mean something, the scale has to be calibrated. **Calibration** is the process of comparing a local reference object to a generally accepted standard to ensure accuracy. To calibrate a spring scale, the device or its reference components must be compared against standard weights or masses. Someone must put a standard weight in the basket and measure just how far the spring stretches. Each spring is different, although spring manufacturers try to make all their springs as identical as possible.

Spring scales are popular because they are both simple and practical. Yet they have several inevitable shortcomings. As a spring ages, its equilibrium length may shift, and as a result an old spring scale may no longer read zero when its basket is empty. To correct for this drift, most spring scales have an adjustment that raises the upper end of the spring so that its lower end is back at the proper position to read zero. Harder to remedy is a second age-related defect: a spring's tendency to grow less stiff with time, which causes a scale to overestimate the weight in its basket. Because of this problem, grocery scales have to be checked periodically to ensure that they are still accurate; otherwise you would be getting less than you paid for.

We've already examined another difficulty with spring scales: their sensitivity to variations in gravity and acceleration, which cause your moon weight to be one-sixth of your earth weight or your weight at the equator to seem 1% less than your weight at the north pole. A final problem lies in the mechanical reporting mechanisms, such as the rack and pinion scheme in Fig. 2.1.2, which inevitably reduce these scales' accuracy. Because of friction in the mechanism, the rack and pinion exert small vertical forces on the spring and basket. Unavoidable and unpredictable, these forces move the basket's equilibrium position up or down slightly. As a result, each time you put a melon in the grocery store's hanging scale, it reports a slightly different weight. The melon's weight isn't really changing, just the scale's measurement of it.

CHECK YOUR UNDERSTANDING #3: Scaling Down

If you pull the basket of a hanging grocery store scale downward 1 centimeter, it reports a weight of 5 N (about 1.1 pounds) for the contents of its basket. If you pull the basket downward 3 centimeters, what weight will it report?

Other Kinds of Spring Scales

As we noted earlier, the most common type of bathroom scale is also a spring scale (Fig. 2.1.4). When you step on this kind of scale, you exert a force on several levers that make it insensitive to your exact position on it, and these levers exert a force on a spring hidden inside. The spring stretches until it exerts, through the levers, an upward force on you that is equal to your weight. As the spring stretches, a rack and pinion mechanism inside the scale turns a wheel with numbers printed on it. When the wheel stops moving, you can read one of these numbers through a window in the scale. Because which number you see depends on how far the spring has stretched, this number indicates your weight.

Fig. 2.1.4 - When you step on this bathroom scale, its surface moves downward slightly and compresses a stiff spring. The extent of this compression is proportional to your weight, which is reported by the dial on the left.

Suppose you want to weigh something that's too heavy for your bathroom scale. One solution to this problem would be to use two scales simultaneously. If you place them near one another on the floor and set the object across both of them at once, they will work together to support the object's weight. Each scale will report just how much upward force it's exerting on the object, so the sum of the two measurements will equal the total weight (Fig. 2.1.5).

The specific readings of the two scales will depend on the location of the object's center of gravity. Its **center of gravity** is the effective location of the object's weight and coincides with its center of mass. (Refer to Section 1.3 for a review of center of mass.) If the object's center of gravity is closer to one scale than the other, that scale will bear more of the weight and will report a higher value than the other scale. To keep from breaking either of the scales, you should position the object's center of gravity about half way in between them.

A more modern variation of the spring scale involves an electronic device called a load cell. Mechanically, a load cell is just another type of spring that exerts a restoring force proportional to how far it has been bent, compressed, or stretched. But the load cell responds in a unique way. Instead of turning a wheel or moving a needle, it changes its electric properties by an amount proportional to its distortion. Because no mechanical mechanism is involved, the actual amount of the distortion can be exquisitely small—so small, in fact, that it's usually hard to tell that anything in the scale is bending or compressing.

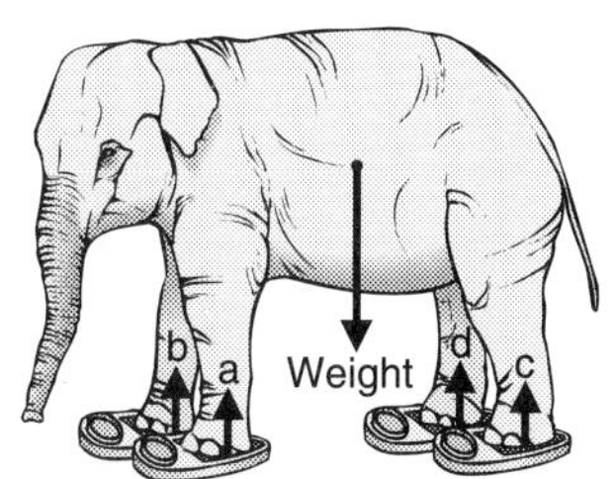

Fig. 2.1.5 - You can weigh a baby elephant by placing a spring scale under each of its four feet. These scales exert upward forces a, b, c, and d to support the elephant. The elephant's weight is equal to the sum of those four forces, as measured by the scales.

Load cell scales are gradually replacing conventional spring scales. They're already common in grocery stores and research laboratories, and they're even available as bathroom scales. Because load cell scales don't use mechanical mechanisms, they don't suffer from the inaccuracies caused by friction; consequently, they're often advertised as containing "no moving parts" and report consistent weights. Moreover, load cells age slowly, so scales made with these devices maintain their accuracy longer than scales made with metal springs.

CHECK YOUR UNDERSTANDING #4: Weighed Down

When you step on the surface of a spring bathroom scale, you can feel it move downward slightly. How is the distance that the scale's surface moves downward when you step on it related to the weight it reports?

Why the Basket Bounces: Vibrations

When you first drop a melon into the basket of a hanging spring scale, the basket bounces up and down for a second or two. This bouncing occurs because the scale can't transfer the melon's gravitational potential energy elsewhere. As the basket bounces up and down, the scale is converting that excess energy back and forth between potential and kinetic energies. Gradually this energy becomes thermal energy and the basket stops bouncing.

Let's watch this bouncing process more carefully. When you first put the melon in the basket, the spring is unable to support it, so the basket accelerates downward and descends, picking up speed as it does. By the time the basket reaches its new equilibrium position, it's descending rapidly and consequently overshoots the mark. The spring becomes too long, its upward force on the basket exceeds the melon's weight, and this net upward force causes the basket to accelerate upward, so that it gradually stops descending and begins to rise. Again the basket reaches a new equilibrium position, but it's now rising rapidly and again overshoots. When it does, the spring becomes too short to support the melon's weight, the net force on the basket becomes downward, and the basket accelerates downward. It gradually stops rising and begins to descend. This process repeats itself over and over again. As the basket bounces, the scale's reading fluctuates about the melon's actual weight.

The basket continues to bounce up and down until friction, air resistance, or your hand slows its motion. This bouncing is a **vibration**, a spontaneous repetitive and rhythmic motion about an equilibrium position. Vibrations are important in the mechanics of motion. You can easily see them in such diverse systems as playground swings, bouncing balls, and soup sloshing in a bowl. They are also essential to the operations of clocks and musical instruments, as we will see later in this book.

A vibrating system contains energy that is transferred back and forth from one form or location to another. The system vibrates until all of its excess energy is transferred elsewhere, typically as sound, or converted into thermal energy. In the case of the bouncing basket, that energy is transformed back and forth between potential and kinetic energies as the basket bounces. When the basket is far away from its new equilibrium position and moving slowly, most of its energy is in the melon's gravitational potential energy or the spring's elastic potential energy; when the basket is near its new equilibrium position and moving quickly, most of its energy is in the melon's kinetic energy. The vibration doesn't last long because the scale is specially designed to convert this excess energy into thermal energy. But when you stand on a bathroom scale and deliberately bounce up and down, you can maintain a similar vibration indefinitely.

CHECK YOUR UNDERSTANDING #5: Bowling with Marbles

If you roll a marble down the side of a round bowl, it will move back and forth about the center of the bowl a number of times before coming to a stop. What's happening to the marble's energy?

Weighing Astronauts

While spring scales work well as long as there is little or no acceleration, they become useless when everything is falling freely. In free fall, the only force acting on an object is gravity, so the object is accelerating downward at the full acceleration due to gravity. Imagine an astronaut orbiting the earth in a spaceship with its engines turned off. Although gravity still pulls on the astronaut so that she actually does have weight, both she and her ship are in free fall, and her acceleration makes her experience a sense of weightlessness. If she tries to substantiate this feeling by standing on a spring scale, the scale will report—erroneously—that she is indeed weightless. Since she and the spring scale are falling at the same rate, no forces are needed to keep them moving as a group, and the scale doesn't have to push on the astronaut to keep her from passing through its surface; as a result, the scale reads zero. For that reason it's hard to weigh the astronaut in space, even though she has a weight.

Recall that mass is the measure of an object's inertia, its resistance to acceleration. You can therefore measure an object's mass by exerting a force on it and seeing how quickly it accelerates. Remember, force equals mass times acceleration. This method is actually used to measure an astronaut's mass in space. The astronaut is given an impulse (a specific force exerted for a specific amount of time) and the resulting change in his velocity is used to determine his mass (Fig. 2.1.6). Of course, the astronaut has to be careful not to wiggle around, since accelerations will throw off the measurement. To help him keep still, the astronaut clings tightly to a rigid frame and the force is exerted on that frame. The acceleration of the frame and astronaut is measured and used to determine the mass of the frame and astronaut. Since the mass of the frame is already known, the astronaut's mass is easy to calculate.

Fig. 2.1.6 - Since a spaceship and its occupants are falling together, the astronauts can't measure their weights with a normal scale. Instead, this apparatus measures an astronaut's mass by pushing on him with a known force and recording his acceleration.

CHECK YOUR UNDERSTANDING #6: Avoid Heavy Lifting

You want to determine the mass and weight of a loaded railroad car. How can you make such measurements without lifting the car?

Summary

How Spring Scales Work: A spring scale measures an object's weight by supporting that object with a spring. When the object is at rest, the spring's upward force must exactly balance the object's downward weight; the scale then reports the upward force being exerted by its spring. As Hooke's law describes, the spring's distortion is proportional to the restoring force it's exerting, so the scale determines how hard its spring is pushing upward by measuring the spring's distortion. In a scale using a metal spring, the distortion measurement is done mechanically and is reported with a needle or dial. In a scale using a load cell as a spring, the distortion measurement is done electronically and is reported by a computer.

Spring scales will not function properly in a freely falling, accelerating environment, such as a spaceship. In space, astronauts have their masses measured rather than their weights. To determine an astronaut's mass, a machine measures the astronaut's acceleration in response to a known force.

Important Laws and Equations

1. Hooke's Law: The restoring force exerted by an elastic object is proportional to how far it is from its equilibrium shape, or

$$\text{restoring force} = -\text{ spring constant} \cdot \text{distortion}. \quad (2.1.1)$$

The Physics of Spring Scales

1. A coil spring exerts a restoring force on its ends that is proportional to the distance it is distorted from its equilibrium length. More generally, an elastic object exerts restoring forces that are proportional to how far it has been distorted from its equilibrium shape.

2. An object that is in equilibrium experiences zero net force and doesn't accelerate. If that equilibrium is stable, then the object will accelerate back toward its equilibrium position after a temporary disturbance.

3. It takes work to distort a spring from its equilibrium shape. Most of this energy is stored within the spring as elastic potential energy and most of that stored energy is released as kinetic energy when the spring returns to its equilibrium shape. Some of the work done in distorting a spring is converted to thermal energy.

4. A system that spontaneously undergoes a repetitive and rhythmic motion about a central point is vibrating. During a vibration, the system's excess energy moves back and forth between various forms or locations. The vibration continues until the system gets rid of its excess energy, either by transferring it elsewhere or converting it to thermal energy.

Check Your Understanding - Answers

1. You should sell by mass and label your foods in kilograms, not pounds.

Why: Packages that are labeled in pounds are specifying their contents by weight. Pounds are a unit of force, in this case the gravitational force the earth exerts on the package's contents. A 1 pound container of oatmeal will only weigh 1 pound at the surface of the earth. When it's exported to the moon, it will only weigh 1/6th of a pound and your company may be fined for selling underweight groceries. If you label the packages according to their masses, that labeling will remain correct no matter where you ship the packages. Mass is a measure of resistance to acceleration and depends only on the object, not on its environment.

2. The diving board is behaving as a spring, bending downward in proportion to the weight of each diver.

Why: The heavier the diver, the more the board bends downward before exerting enough upward force on the diver to balance the diver's weight.

3. 15 N (about 3.3 pounds).

Why: The scale's dial is simply reporting the position of its basket. The dial is calibrated so that a 1 centimeter drop in the basket indicates that the spring is exerting an upward force of 5 N on the basket. Since the spring's restoring force is described by Hooke's law, a 3 centimeter drop in the basket means that the spring is exerting an upward force of 15 N on the basket.

4. The distance the scale's surface moves downward is proportional to the weight it reports.

Why: The scale's spring is connected to its surface by levers so that as the surface moves downward, the spring distorts by a proportional amount. The spring's distortion is reported on the dial. Thus the dial's reading is proportional to the surface's downward movement.

5. The marble's energy transforms back and forth between gravitational potential energy and kinetic energy. However because these forms of energy are gradually converted into thermal energy, the marble settles to the bottom of the bowl.

Why: The marble has a stable equilibrium at the bottom of the bowl. When it's moved away from that equilibrium and released, it will vibrate about that equilibrium point. When it's close to the equilibrium point, it's moving quickly and most of its energy is kinetic. When it's far from the equilibrium point, it's moving slowly and most of its energy is gravitational potential energy. This motion gradually diminishes because friction and air resistance gradually convert the marble's energy into thermal energy.

6. You can give the railroad car a horizontal impulse (a specific force exerted for a specific amount of time) and measure its change in velocity. This change in velocity can be used to calculate the railroad car's mass and that mass can be used to calculate its weight.

Why: Even on earth, some objects are difficult to weigh. It may be easier to determine their masses by studying how they accelerate in response to forces. You can also distinguish a loaded container from an empty container by seeing which accelerates most easily. Once you know an object's mass, you can calculate its weight. At the earth's surface, each kilogram of mass weighs about 9.8 newtons.

Check Your Figures - Answers

1. 2 centimeters and 10 centimeters (assuming that the floor doesn't break).

Why: A floor, like most suspended surfaces, behaves like a spring. Your floor distorts 1 centimeter before it exerts an upward restoring force equal to the weight of 10 guests. It will thus distort 2 centimeters before supporting 20 guests and 10 centimeters before supporting 100 guests. However, 10 centimeters is a substantial distortion and will probably cause damage. Although the floor beams may not exceed their elastic limit, some rigid portions of the floor or the ceiling below will crack or crumble. If the beams break, the floor may collapse.

Glossary

calibration The process of comparing a local reference object to a generally accepted standard. For a spring scale to report accurate weights or masses, the device or its reference components must be compared against standard weights or masses.

center of gravity The unique point about which all of an object's weight is evenly distributed and therefore balanced. Because weight is proportional to mass, the center of mass is identical to the center of gravity for objects that are much smaller than the earth. For larger objects, the centers of mass and gravity differ slightly. An object suspended from its center of gravity will balance and will experience no net torque due to gravity. In many situations, you can accurately predict an object's behavior by assuming that all of the object's weight acts at its center of gravity.

elastic limit The most extreme distortion of an object from which it can return to its original size and shape without permanent deformation.

elastic potential energy The energy stored by the forces within a distorted elastic object.

equilibrium position The point at which an object experiences zero net force and doesn't accelerate.

equilibrium The state of an object in which zero net force (or zero net torque) acts on it. An object that is stationary or in uniform motion is in equilibrium.

free fall When an object is not constrained by any outside forces except gravity and accelerates downward at the acceleration due to gravity ($9.8\ \text{m/s}^2$ near the earth's surface).

Hooke's law The general law covering spring and elastic behavior. Hooke's law states that a spring exerts a restoring force that is proportional to the distance the spring is distorted from its equilibrium length.

restoring force A force that acts to return an object to its equilibrium shape. A restoring force is directed toward the position the object occupies when it's in its equilibrium shape.

spring constant A measure of the stiffness of an elastic object, the spring constant relates the object's distortion to the restoring force it exerts. The larger the spring constant, the stiffer the spring.

stable equilibrium A state of equilibrium to which an object will return if it's disturbed. At equilibrium, the object is free of net force or torque. However, if an object is moved away from that equilibrium state, the net force or torque that will then act on it will tend to return it to the equilibrium state.

vibration A spontaneous repetitive and rhythmic movement about an equilibrium position.

Review Questions

1. What must you do to exert a force larger than your weight on the floor? What about less than your weight?

2. Which of the following are in equilibrium: a falling apple, an apple resting on the floor, a truck traveling at constant velocity on a highway, a turning bicycle, a bird feeder hanging motionless from a tree, and a stationary trampoline?

3. Why is a marble in a round-bottomed bowl in a stable equilibrium?

4. When you compress a particular spring by 1 cm, it pushes outward on your hand with a force of 4 N. What force will it exert on your hand if you compress it by 5 cm? if you stretch it by 2 cm?

5. Identify at least 5 objects that behave as springs (that obey Hooke's law).

6. If your ruler or clock were improperly calibrated, how might that affect your life?

7. Why is it that every scale you encounter distorts slightly when you put something on it to be weighed?

8. Place a ruler flat on a table with one end extending over the edge. Hold the ruler's other end against the table. If you push down on the ruler's free end and then abruptly let go, the ruler will swing up and down rapidly. Describe the ruler's motion in terms of energy.

9. If you have two bowling ball that look identical but differ substantially in weight, how can you identify the heavy ball without lifting it off the floor?

10. If you taped a bathroom scale to your feet and jumped off the high diving board, what would the scale read as you fell?

Exercises

1. In what way does the string of a bow and arrow behave like a spring?

2. As you wind the mainspring of a mechanical watch or clock, why does the knob gets harder and harder to turn?

3. Curly hair behaves like a weak spring that can stretch under its own weight. Why is a hanging curl straighter at the top than at the bottom?

4. You take two boxes of cookies with you on your vacation to the moon. One of the boxes is labeled in terms of mass in kilograms, while the other is labeled in terms of weight in pounds. When you arrive at the moon, where gravity is much weaker than on earth, you find that one of the boxes is no longer accurately labeled. Which label is now wrong and what has caused the inaccuracy?

5. A dump truck supports its heavy load on springs that are themselves supported by the wheels. When the truck goes over a sudden bump, the wheels compress the springs rather than directly lifting the load upward. Why does this arrangement reduce the force that the truck exerts downward on the bump and prevent road damage?

6. Why are there no spring scales in which the basket moves upward as you fill it with objects?

7. Your pet pony weighs about 400 pounds, too much for a common bathroom scale. However, you do have two identical bathroom scales that can each weigh up to 300 pounds. How can you weigh the pony?

8. Small bridges on secondary roads have signs indicating their maximum capacities for each of several different types of trucks. The longer the type of truck, the more it's permitted to weigh. Why does a long truck exert a smaller fraction of its weight on a bridge at a given moment than a short truck?

9. If you pull down on the basket of a hanging grocery store scale so that it reads 15 N, how much downward force are you exerting on the basket?

10. When the Hubble Space Telescope was first built, the curvature of its mirror was adjusted with the help of a tool that hadn't been calibrated properly. As a result of this adjustment, the mirror was produced with the wrong curvature. What would happen if a scale manufacturing company adjusted its scales with the help of reference weights that hadn't been calibrated properly?

11. If you were to step on several different bathroom scales, one after the next, chances are that they would report slightly different weights. Explain this result in terms of calibration.

12. People often remark that a particular scale "reads heavy," meaning that it reports more than a person's real weight. What is wrong with the scale's spring?

13. Suppose that your scale "reads heavy" (see Exercise 12) but your neighbor's scale is accurate. How could you use your neighbor's scale to calibrate your scale, at least for weights in the vicinity of your own weight?

14. While you're weighing yourself on a bathroom scale, you reach out and push downward on a nearby table. Is the weight reported by the scale high, low, or correct?

15. There's a bathroom scale on your kitchen table and your friend climbs up to weigh himself on it. One of the table's legs is weak and you're afraid that he'll break it, so you hold up that corner of the table. The table remains level as you push upward on the corner with a force of 100 N. Is the weight reported by the scale high, low, or correct?

16. If you put your bathroom scale on a ramp and stand on it, will the weight it reports be high, low, or correct?

17. When you step on a scale, it reads your weight plus the weight of your clothes. Only your shoes are touching the scale, so how does the weight of the rest of your clothes contribute to the weight reported by the scale?

18. To weigh an infant you can step on a scale once with the infant and then again without the infant. Why is the difference between the scale's two readings equal to the weight of the infant?

19. If you stood on a scale in the bathroom of an airplane during a bumpy flight, the scale wouldn't give a consistent reading. Since your weight isn't changing, the scale can't be reporting your weight. What is the scale reporting?

20. When you lie on a spring mattress, it pushes most strongly on the parts of you that stick into it the farthest. Why doesn't it push up evenly on your entire body?

Problems

1. Your new designer chair has an S-shaped tubular metal frame that behaves just like a spring. When your friend, who weighs 600 N, sits on the chair, it bends downward 4 cm. What is the spring constant for this chair?

2. You have another friend who weighs 1000 N. When this friend sits on the chair from Problem **1**, how far does it bend?

3. You're squeezing a springy rubber ball in your hand. If you push inward on it with a force of 1 N, it dents inward 2 mm. How far must you dent it before it pushes outward with a force of 5 N?

4. When you stand on a particular trampoline, its springy surface shifts downward 0.12 m. If you bounce on it so that its surface shifts downward 0.30 m, how hard is it pushing up on you?

5. An elastic bungee cord behaves like a spring when you stretch it (though not when you compress it—it bends). If a force of 15 N stretches a particular bungee cord 10 cm, how far will a force of 60 N stretch that bungee cord?

Section 2.2

Bouncing Balls

If you visit a toy or sporting goods store, you'll find a lot of different balls—almost a unique ball for every sport or ball game. These balls differ in more than just size and weight. Some are very hard, others very soft; some are smooth, others rough or ridged.

In this section we'll focus primarily on another difference: the ability to bounce. A superball, for example, bounces extraordinarily well, while a foam rubber ball hardly bounces at all. Even balls that appear identical can be very different; a new tennis ball bounces much better than an old one. We'll begin this section by exploring these differences.

Questions to Think About: *Is it possible for a ball to bounce higher than the height from which it was dropped? Where does a ball's kinetic energy go as it bounces, and what happens to the energy that doesn't reappear after the bounce? What happens when a ball bounces off a moving object, such as a baseball bat? What role does the baseball bat's structure have in the bouncing process? Does it matter which part of a baseball bat hits the ball? When a pool or croquet ball strikes another identical ball, the first ball often stops and the other ball takes up the motion. How is the motion transferred? Why does a spinning Ping-Pong ball bounce in such a strange manner?*

Experiments to Do: *Drop a ball on a hard surface and watch it bounce. What happens to the ball's shape during the bounce? Hold the ball in your hands and push its surface inward with your fingers. What is the relationship between the force it exerts on your fingers and how far inward you dent it? Denting the ball takes work. Why? How does the ball's energy change as it dents? What happens to the ball's energy as its shape returns to normal (to equilibrium)?*

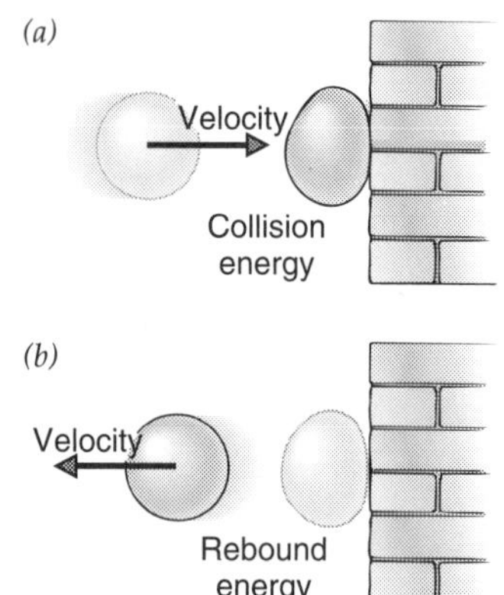

Fig. 2.2.1 - A bounce from a wall has two halves: (*a*) the collision and (*b*) the rebound. During the collision between the ball and the wall, some of their kinetic energy is transformed into other forms—an amount called the collision energy. During the rebound, some stored energy is released as kinetic energy; an amount called the rebound energy. The rebound energy is always less than the collision energy because some energy is lost as thermal energy. However, a lively ball wastes less energy than a dead one.

Drop the ball from various heights and see if you can find a simple relationship between the ball's initial height and the height to which it rebounds. Now drop the ball on a soft surface, such as a pillow. Why does that surface change the way the ball bounces? Drop the ball on a lively elastic surface, such as an air-filled plastic bag, and watch how well it bounces. Where is the ball's energy stored as it bounces from the springy surface? Now instead of a ball, drop a bag of beans, sand, or salt on the hard surface. Why doesn't the bag bounce? Where does the bag's energy go when it lands?

The Way the Ball Bounces: Balls as Springs

In many ways, balls are perfect objects. What would most sports be like without them? How would industrial machines function without ball bearings to keep them from grinding to a halt? Their simple shapes, uncomplicated motions, and ability to bounce make balls both fascinating and useful. Most balls are spherical, meaning that when no outside forces act on them they adopt spherical equilibrium shapes. But some balls, such as those used in U.S. football and rugby, have equilibrium shapes that are not spherical but oblong.

The term "equilibrium shape," of course, is one we've seen already: the previous section used it to describe springs. This is no coincidence, for a spherical ball behaves like a spherical spring. In fact, everything we associate with springs has some place in the behavior of balls. For example, when you push a ball's surface inward, it exerts an outward restoring force on you. When you do work on the ball as you distort its surface, it stores some of this work as elastic potential energy. When you let the ball return to its equilibrium shape, it releases its stored energy.

This spring-like behavior is evident when a ball collides with the floor or with a bat. The ball's surface distorts during the collision, giving it elastic potential energy that is released when the ball rebounds. If the ball is moving, then some of this stored energy comes from the ball's kinetic energy. If the ball hits a moving surface, then some of this stored energy comes from the surface's kinetic energy. Much of the stored energy reappears as kinetic energy in the ball and surface as the ball rebounds.

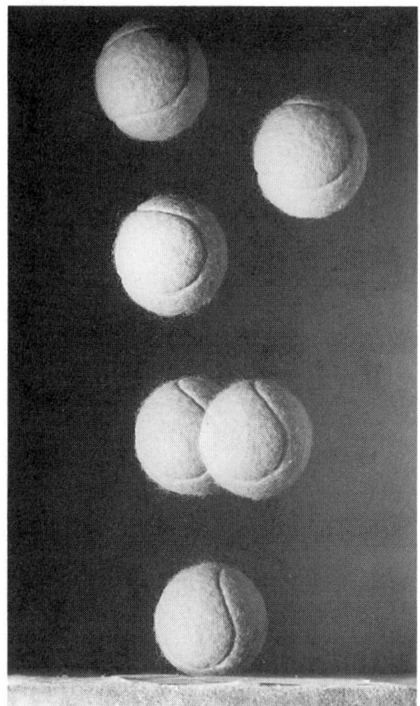

Fig. 2.2.2 - When a tennis ball hits the floor, it dents inward to store energy and then rebounds somewhat more slowly than it arrived. These images show the ball's position at 6 equally spaced times. Is the ball bouncing to the left or the right? How can you tell?

Some balls bounce better than others. We frequently call a very bouncy ball "lively" and a ball that doesn't bounce well "dead." One way to look at a ball's liveliness is to compare kinetic energies before and after the bounce. We can do that by dividing the bounce into two halves: the collision and the rebound (Fig. 2.2.1). During the collision, the ball and surface convert some of their overall kinetic energy into elastic potential energy and thermal energy. The amount of kinetic energy transformed at impact is called the **collision energy**. A lively ball does a good job of converting the collision energy into elastic potential energy, while a dead ball converts most of it into thermal energy (Fig. 2.2.2).

During the rebound, the ball and surface push away from one another, converting elastic potential energy back into kinetic energy. The total amount of kinetic energy released as the surface and ball push apart is the **rebound energy**. Collision energy that doesn't reappear as rebound energy has been transformed into thermal energy.

A particular ball is characterized by its **coefficient of restitution**: the ratio of its rebound speed to its collision speed when it bounces off a *hard, stationary* surface that *can't move*:

$$\text{coefficient of restitution} = \frac{\text{outgoing speed of ball}}{\text{incoming speed of ball}}. \tag{2.2.1}$$

Scientists have found that, for most balls, this speed ratio remains constant over a wide range of collision speeds (Table 2.2.1). A ball that rebounds with the same speed that it had when it collided with the stationary surface has a coefficient of restitution of 1.00. A superball is almost this lively, with a coefficient of restitution of about 0.90. Thus, when a superball traveling at 10 km/h collides with a concrete wall, it rebounds at about 9 km/h. In contrast, a foam rubber ball's coefficient of restitution is about 0.30, while that of a bean bag is almost zero.

Type of Ball	Coefficient of Restitution
Superball	0.90
Racket Ball	0.85
Golf Ball	0.82
Tennis Ball	0.75
Steel Ball Bearing	0.65
Baseball	0.55
Foam Rubber Ball	0.30
"Unhappy" Ball	0.10
Beanbag	0.05

Table 2.2.1 - Approximate coefficients of restitution for a variety of balls.

The coefficient of restitution is a speed ratio (rebound speed divided by collision speed), not an energy ratio (rebound energy divided by collision energy). But we can still use the speed ratio to determine what happens to the ball's kinetic energy. Remember, a ball's kinetic energy equals half its mass times its velocity squared, or $\frac{1}{2}mv^2$. Even if we don't know the masses of the balls, we know that the energy ratio is equal to the square of the speed ratio. The masses in this ratio would simply cancel out. Thus if a foam rubber ball rebounds at only 0.30 times its collision speed, it retains only 0.30^2—0.09 times, or 9%—of its original kinetic energy, and the remaining 91% is converted to thermal energy in the rubber and air that make up the ball. A superball, in contrast, retains about 81% of its original kinetic energy after a bounce.

Balls that rely on air pressure for their hardness lose much of their coefficient of restitution when they are underinflated. Compressed air stores energy much better than the rubber skin of a basketball, because rubber experiences a lot of internal friction when bent. A well-inflated basketball, which stores most of its energy in the compressed air inside it, has a high coefficient of restitution; an underinflated basketball does not, since most of the energy that goes into bending its surface is converted into thermal energy. Similarly, a tennis ball bounces best when new; after a while, the compressed air inside leaks out and the ball's coefficient of restitution drops.

Although rubber wastes energy when you bend it, it stores energy much more efficiently when you compress it or stretch it. When a solid rubber ball bounces, most of the collision energy is stored by compressing the ball's core rather than by bending its surface. That is why most solid rubber balls bounce well. Balls made of many hard woods, plastics, or metals also bounce well.

A ball's coefficient of restitution determines how high it bounces when you drop it on a hard floor (Fig. 2.2.3). If an ideal ball's coefficient of restitution were 1.00, it would rebound to its initial height. But a real ball wastes some of the collision energy as heat, so it rebounds to a lesser height. The lower its coefficient of restitution, the less kinetic energy the ball receives during the rebound and the weaker the bounce.

If the surface on which a ball bounces is not perfectly hard, that surface will contribute to the bouncing process. In other words, it will distort and store energy when the ball hits it and will return some of this stored energy to the ball as it rebounds. Overall, the collision energy is shared between the ball and the surface and each provides part of the rebound energy.

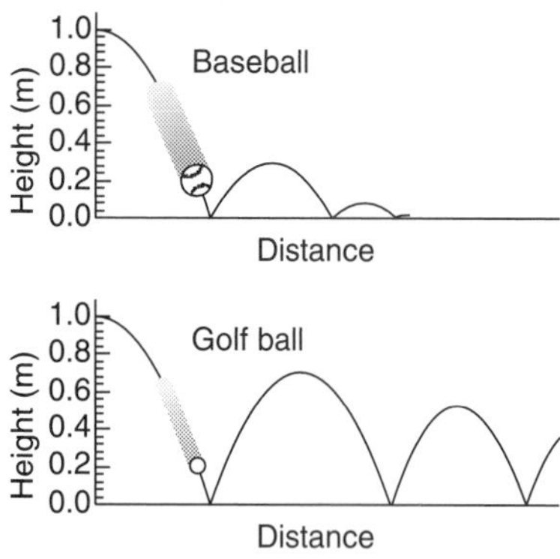

Fig. 2.2.3 - A baseball (above) wastes 70% of the collision energy as thermal energy and bounces weakly. In contrast, a golf ball (above and at right) wastes only 30% of the collision energy and bounces well.

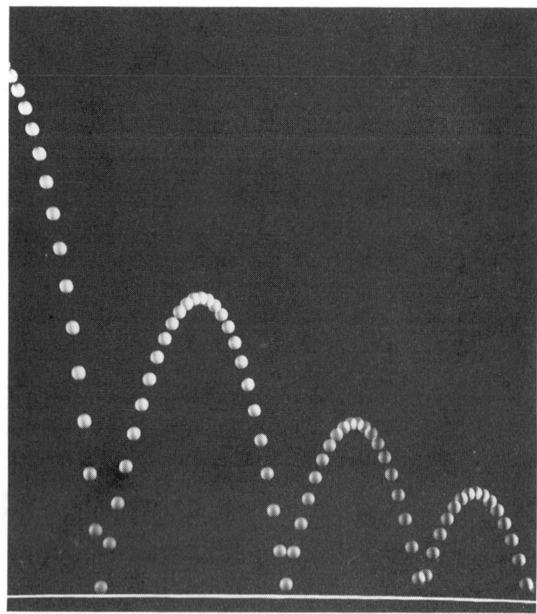

Just how the collision energy is distributed between the surface and ball depends on how stiff each one is. During the bounce, the surface and the ball both behave as springs, pushing on one another with equal but oppositely directed forces. Since the forces denting them inward are equal, the work done in distorting each object is proportional to how far inward it dents. Whichever object dents the furthest receives the most collision energy.

Fig. 2.2.4 - When a tennis ball bounces from a moving racket, both the ball and racket dent inward. The ball and racket share the collision energy almost evenly.

Since the ball usually distorts more than the surface it hits, most of the collision energy normally goes into the ball. As a result, you might expect the ball to provide most of the rebound energy, too. However, that is not always true. Some lively elastic surfaces store collision energy very efficiently and return almost all of it as rebound energy. Since a relatively dead ball wastes most of the collision energy it receives, a lively surface's contribution to the rebound energy can be very important to the bounce. A lively racket is critical to the game of tennis because the racket's strings provide much of the rebound energy as the ball bounces off the racket (Fig. 2.2.4). Trampolines and spring-boards are even more extreme examples, with surfaces so lively that they can even make people bounce. People, like bean bags, have coefficients of restitution near zero; when you land on a trampoline, it receives and stores most of the collision energy and then provides most of the rebound energy.

The stiffness of the ball and surface also determine how much force each object exerts on the other and thus how quickly the collision proceeds. Stiff objects resist denting much more strongly than soft objects. When both objects are very stiff, the forces involved are large and the acceleration is rapid. Thus a steel ball rebounds very quickly from a concrete floor because the two exert huge forces on one another. If the ball and/or surface are relatively soft, the forces are weaker and the acceleration is slower.

What if the surface a ball hits isn't very massive? In that case, the surface may do part or all of the "bouncing." During the bounce, the ball and the surface accelerate in opposite directions and share the rebound energy. Massive surfaces, such as floors and walls, accelerate little and receive almost none of the rebound energy. But when the surface a ball hits is not very massive, you may see it accelerate. When a ball hits a lamp on the coffee table, the ball will do most of the accelerating, but the lamp is likely to fall over, too.

Similarly, when a baseball strikes a baseball bat, the ball and bat accelerate in opposite directions. The more massive the bat, the less it accelerates and the smaller its share of the rebound energy. To ensure that most of the rebound energy went to the baseball, the legendary hitters of the first half of the 20th century used massive bats. Such bats are no longer in vogue because they are too hard to swing. But in the early days of baseball, when pitchers were less highly skilled, massive bats drove many long home runs.

CHECK YOUR UNDERSTANDING #1: A Game of Marbles

You are playing a game of marbles on a soft dirt field. The goal is to knock glass marbles out of a circle by hitting them with other marbles. You initially drop several marbles onto the ground inside the circle and they hardly bounce at all. What prevents them from bouncing well?

Moving Surfaces: Frames of Reference

The last paragraph describes the act of hitting a baseball with a moving baseball bat as though it were a case of "bouncing." That might sound a little strange. When a baseball hits a stationary bat, the ball will bounce. But if a moving bat hits a stationary baseball, is it proper to say that the ball bounces?

The answer is yes. In fact, which object is moving and which is stationary depends on your point of view—your **frame of reference**. A fly resting on the baseball will claim that the baseball is stationary and that it's struck by a moving bat. Another fly resting on the baseball bat will claim that the bat is stationary and that it's struck by a moving baseball. Which fly has the correct frame of reference?

Both frames of reference are equally valid. One of the remarkable discoveries of Galileo and Newton is that the laws of physics work perfectly in any inertial frame of reference. An **inertial frame of reference** is one that is not accelerating and is thus either stationary or moving at a constant velocity. As long as you view the world around you from an inertial frame of reference, everything you see will obey the laws of motion so that energy, momentum, and angular momentum will all be conserved.

For example, imagine a girl bouncing a basketball on the sidewalk . If you stand next to her on the sidewalk, adopting a particular inertial frame of reference, you will see that the ball obeys the laws of motion: it travels directly downward, bounces on the pavement, and returns toward her hand (Fig. 2.2.5*a*). Energy is always conserved, as are momentum and angular momentum.

Now imagine that you are driving eastward at 100 km/h and pass by that same girl. In your new inertial frame, the car is stationary and the landscape is moving westward. The girl, the sidewalk, and everything around her are moving at a constant westward velocity of 100 km/h. You see the basketball descend, bounce, and rise as before, but the ball now has a westward horizontal component to its velocity (Fig. 2.2.5*b*). Nonetheless, the laws of motion are obeyed. The ball still accelerates downward as it falls, stores collision energy as elastic potential energy, and rebounds toward her hand. Energy is still conserved, as are momentum and angular momentum.

Clearly, people in two different inertial frames of reference see the same situation somewhat differently. When they watch a bouncing ball, they do not agree on the velocities of the ball or the object it hits and, because kinetic energy depends on velocity, they do not agree on the kinetic energies of those two objects. However, they do agree on some special quantities: the collision and rebound energies and the difference between the velocities of the ball and object it hits — the *relative velocity* between these two.

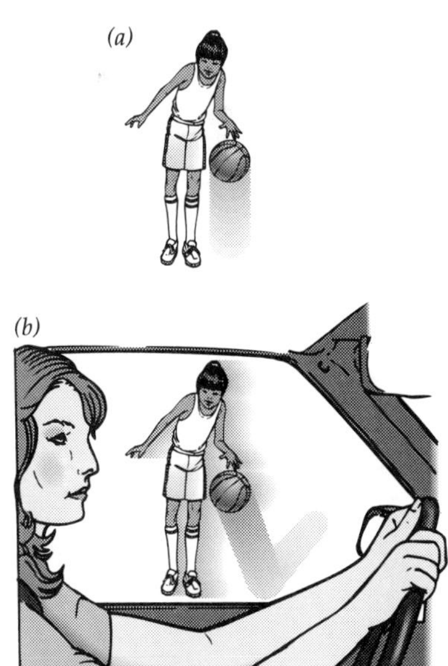

Fig. 2.2.5 - (*a*) If you stand still as you watch a girl bounce a basketball, the ball will travel directly downward and return to her hand. (*b*) But if you watch that same girl from a car heading eastward, the ball will have a westward component to its velocity as it bounces.

How does all this discussion about inertial frames of reference apply to balls and bats? We can now use inertial frames of reference to understand what happens when a baseball player hits a pitch into center field with a baseball bat. Imagine that the pitcher throws the ball toward home plate at 100 km/h (Fig. 2.2.6*a*). The baseball's coefficient of restitution is 0.50, so it rebounds at half its collision speed. The batter swings the bat toward the pitcher at 100 km/h and hits the ball back toward the pitcher at 200 km/h (Fig. 2.2.6*d*). How can the ball end up traveling faster than either the bat or the pitched ball?

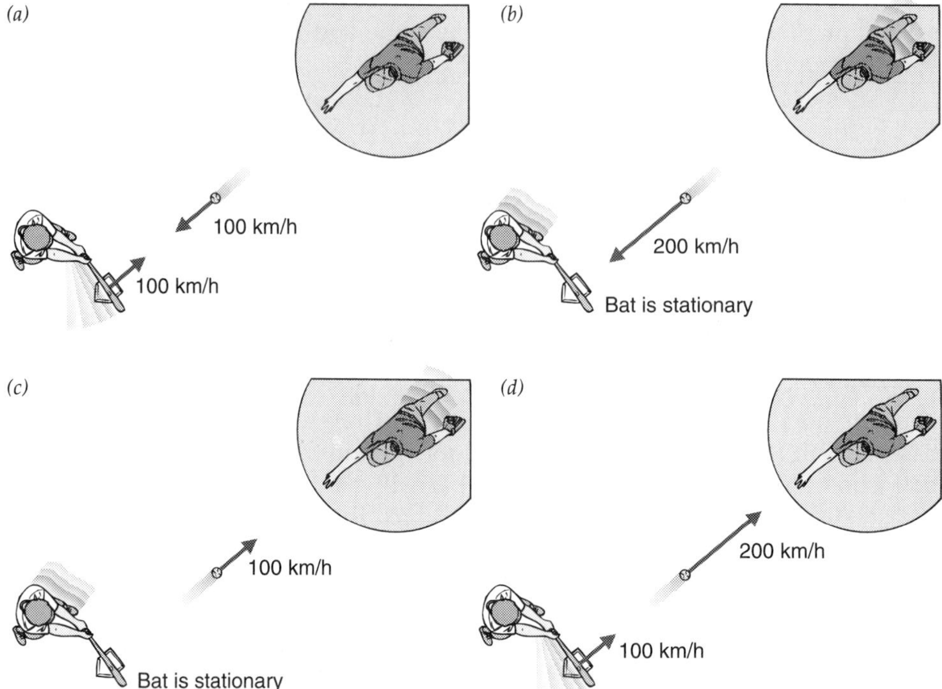

Fig. 2.2.6 - (*a*) From the fan's perspective, the pitcher sends the ball toward home plate at 100 km/h and the bat approaches it at 100 km/h. (*b*) From the bat's perspective, the ball is approaching it at 200 km/h. (*c*) From the bat's perspective, the ball rebounds from the bat at half its original speed, heading toward the outfield at 100 km/h. (*d*) From the fan's perspective, the ball is heading toward the outfield at 200 km/h.

For simplicity, imagine that the bat is extremely massive and perfectly hard, so that the ball does all the bouncing. We need two inertial frames of reference: the spectators' and the baseball bat's. Since all of the motion takes place along the line between the pitcher's mound and home plate, we can use these two landmarks to identify the two possible directions of motion: toward the pitcher's mound or toward home plate.

We begin with the spectators' inertial frame of reference shortly after the pitch, as the batter begins to swing the bat. This frame of reference is the most familiar to all of us; in it, the whole stadium is stationary and only the ball and bat are moving. The ball is moving toward home plate at 100 km/h, while the bat is moving toward the pitcher's mound at 100 km/h (Fig. 2.2.6*a*). The ball and bat are therefore approaching one another at 200 km/h, and the ball's velocity relative to that of the bat is 200 km/h toward home plate.

Now we switch to the baseball bat's inertial frame of reference. Although this may seem a strange perspective, it's useful to us now. In this frame of reference, only the bat is stationary. The whole stadium is moving 100 km/h toward home plate and the ball is moving toward home plate 100 km/h faster than the stadium. When we add these two velocities, we find that the ball is moving toward home plate and the stationary bat at 200 km/h (Fig. 2.2.6*b*). The ball's velocity relative to that of the bat is still 200 km/h toward home plate, even though we have changed reference frames.

The ball now bounces off the stationary bat at the collision speed of 200 km/h. Because it has a coefficient of restitution of 0.50, the ball rebounds from the bat at a speed of 100 km/h. The ball's new velocity relative to the bat is 100 km/h toward the pitcher (Fig. 2.2.6*c*).

We now return to the spectator's frame of reference. The ball's velocity relative to the bat is still 100 km/h toward the pitcher. However, in this frame of reference the bat itself is heading toward the pitcher at 100 km/h. If the bat is traveling toward the pitcher at 100 km/h and the ball is going 100 km/h faster than the bat in that same direction, then the ball must be traveling 200 km/h toward the pitcher (Fig. 2.2.6*d*). The ball sails past the pitcher and out into center field with a speed greater than either the ball or bat had initially.

So the baseball really does bounce off the moving bat, although this bounce is only obvious from the inertial reference frame of the bat itself. And the bat's motion affects the ball's bounce: the faster the bat is moving toward the pitcher, the faster the baseball approaches the bat's surface and the higher the ball's rebound speed after it bounces.

CHECK YOUR UNDERSTANDING #2: Marble Frames of Reference

Two of you flick your marbles into the circle simultaneously from opposite sides of the circle and they collide head on. Each marble was traveling forward at 1 m/s. From the inertial reference frame of your marble, what was the velocity of the other person's marble just before they hit?

Balls and Bats: Real Collisions

Of course, the real task of hitting a baseball is a bit more complicated. So far we've assumed that the bat is extremely massive and perfectly hard. In reality, neither of these assumptions is true. A real baseball bat has a limited mass, so it accelerates as it hits the ball; and its surface isn't perfectly hard, so it stores energy during the bounce. Both of these complications affect what happens when a batter hits a baseball.

For a batter who is trying to drive a baseball into the center field bleachers, the bat's mass is particularly important. It determines both how rapidly the bat can be swung and how much the bat accelerates as the ball bounces off it. Wooden baseball bats usually have masses of about 1 kg, which means they weigh between 31 and 36 ounces. These current bats are less massive and thus lighter than they were in the glory days of the New York Yankees, when Babe Ruth supposedly used a 56-ounce bat. We can explain this shift toward lighter bats in terms of simple physics.

The batter tries to swing the bat quickly so that the baseball approaches the bat's surface rapidly and rebounds at high speed. The batter also wants a bat that doesn't accelerate much as the ball bounces off it, so that the ball receives most of the rebound energy. Ideally, the batter would choose a bat that is easy to swing quickly and doesn't rebound when it hits the ball. Unfortunately, a bat can't have both of these properties at once, so the batter must choose a bat that affords the best compromise.

A heavy bat can't be swung as quickly as a light bat. Swinging a bat is a rotary motion and requires that the batter exert a huge torque to the bat's handle. A strong batter exerts a force of about 500 N on the handle, nearly a meter from the pivot point at the batter's shoulders. This force produces a torque of about

500 N·m (about 370 foot-pounds). The bat undergoes angular acceleration and rotates so that by the time the bat hits the ball, a few tenths of second into the swing, the batter has invested about 500 J (0.1 food Calories) of kinetic energy into the bat.

A heavy bat gives the batter a larger moment of inertia and a smaller angular acceleration. The swing is slow. When the heavy bat finally strikes the ball, it's not moving quite as fast as a light bat would be. For the fastest possible swing, the batter should opt for a light bat.

However, a light bat accelerates easily as the ball bounces from it and receives more of the rebound energy than a heavy bat would. So the batter is in a bind. A lighter bat can be swung more quickly, but it gives less of the rebound energy to the ball. What should the batter choose? Because these two effects of mass partially cancel one another, a professional ballplayer can use a pretty broad range of bat masses, between about 0.85 and 1.70 kg (weighing 30 and 60 ounces), to hit the ball almost the same distance. Within this range, the batter can choose according to taste.

For each batter, there is a specific optimum mass that he or she can swing to hit the ball the farthest. Modern players tend to opt for bats that are somewhat less than this optimum mass. They won't be able to hit the ball quite as far, but they will be able to swing faster and respond more quickly to a pitch. After all, the timing of the swing is extremely important. Hitting the ball as little as one hundredth of a second early or late in the swing will yield a foul ball. Some strong batters take a long, broad swing that is relatively insensitive to timing since the bat's orientation changes slowly. Less-strong batters often use a short, carefully timed swing to achieve similar effects.

For batters who have to make do with bats that are too massive for them to swing effectively, one solution is to "choke up." If they move their grip up the handle, away from the end of the bat, they reduce the bat's moment of inertia and won't need as much torque to swing it. The end of the bat will travel in a smaller arc, but the batter will be able to complete the arc more quickly. Overall, the struck ball will travel about the same distance, and the batter will be able to respond better to the pitch.

When the bat actually hits the ball, what happens to these two objects? The ball compresses dramatically (Fig. 2.2.7). A baseball is made of wool yarn wrapped on a small cork core and covered by cowhide. Since the rules require that it have a coefficient of restitution of between 0.514 and 0.578, it returns only about 30% of the energy that goes into it during the compression. The bat also compresses, but far less—about 50 times less—than the ball, so the ball receives 98% of the collision energy. Although the wooden bat is much more lively than the ball, it receives so little collision energy that it has little effect on the rebound.

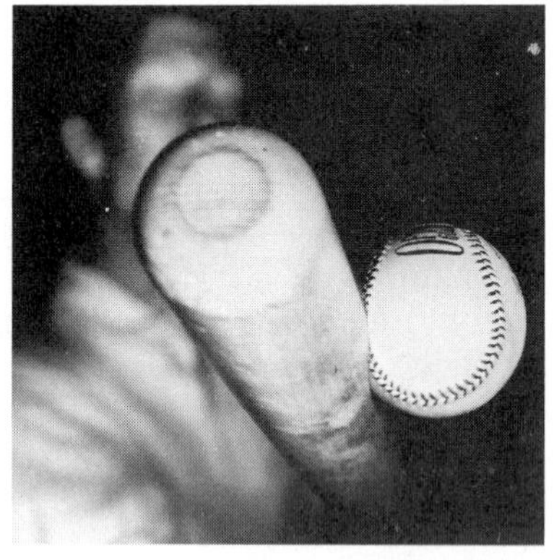

Fig. 2.2.7 - When a baseball bounces from a moving bat, the ball dents inward much more than the bat. The ball receives most of the collision energy.

We might wonder whether changing the construction of the bat itself can affect how far the ball travels. The answer is that it can. While "corking" a wooden bat to modify its mass distribution has little or no real effect on the bat's performance, switching to aluminum really does improve a bat. Legal in many non-professional leagues, aluminum bats have at least four technical advantages:

1. The same size bat can be made heavy or light just by changing the thickness of the aluminum.
2. Aluminum bats don't break.
3. The surface of an aluminum bat is more lively than that of a wooden bat.
4. Aluminum bats are stiffer along their length.

Despite its increased lengthwise stiffness, an aluminum bat's hollow surface is relatively flexible and lively. During the collision, the bat's surface may distort one tenth as much as the ball and receive about 10% of the collision energy. During the rebound, the aluminum bat returns most of this energy, while the ball itself wastes all but about 30% of its share. As a result, the bat makes a substantial contribution to the rebound energy and increases the distance the ball travels. In short, storing collision energy in the aluminum bat is much better than storing it in the ball. If the baseball were wound with rubber, like a golf ball, rather than wool yarn, it would store energy more efficiently and the choice of bat would make less of a difference.

An aluminum bat's increased lengthwise stiffness makes it harder to bend and reduces its tendency to vibrate. The result is a bigger "sweet spot," an ideal point on a bat with which to hit the ball. We'll explore the physics behind sweet spots in the next section.

CHECK YOUR UNDERSTANDING #3: Mass and Marbles

The marbles you're playing with are not all the same size and mass. You notice that larger marbles are particularly effective at knocking other marbles out of the circle. You decide to use a 10-cm-diameter glass ball as a marble, expecting to clean out the entire circle. But when you flick it with your thumb, your thumb merely bounces off. Why doesn't the glass ball move forward quickly?

The Sweet Spot: Acceleration and Vibration

It does matter where on a bat the ball strikes. The ideal location or "sweet spot" is located about 17.5 cm from the broad end and is special for two reasons: it's both the center of percussion and a vibrational node. Despite their coincidental overlap, these special features involve different mechanical issues.

To understand how the bat responds to hitting the ball, we'll find it useful to look at the bat's center of mass. As the section on seesaws explained, an object's center of mass is the single point, coincident with the object's center of gravity, about which all of that object's mass is evenly balanced. An object's motion can be separated conveniently into translational motion of its center of mass and rotational motion about its center of mass. A net force on the object causes its center of mass to accelerate. A net torque on the object causes it to undergo angular acceleration about its center of mass.

The object we wish to study is the bat itself, independent of the batter's hands. When the ball and bat collide, they exert huge forces on one another. These forces can exceed 20,000 times the weight of the ball and completely dominate the motion of the bat and ball during the collision. Thus we can forget that the batter is holding the bat and study what would happen if the bat were traveling freely through the air when it hits the ball.

The collision force on the bat produces a sudden acceleration of the bat's center of mass, slowing its forward motion. This acceleration consumes some of the rebound energy. But the bat can also rotate about its center of mass. Depending on where the ball hits, it will exert various torques on the bat about its center of mass and will almost certainly cause the bat to begin rotating one way or the other. So the collision will cause (1) the bat's center of mass to accelerate backward and (2) the bat to begin rotating about is center of mass. Each of these changes is likely to cause an abrupt acceleration of the bat's handle. The batter will feel the handle's acceleration as a sudden and unpleasant jerk.

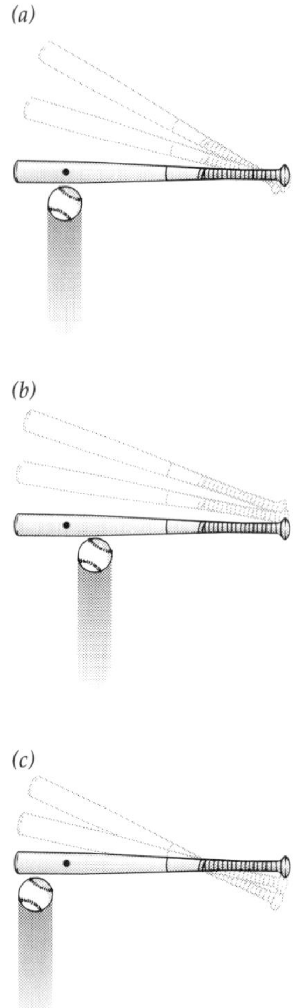

Fig. 2.2.8 - When the ball strikes the bat, the bat accelerates backward and begins to rotate about its center of mass. (*a*) If the ball hits the center of percussion, the handle will not accelerate. (*b*) If the ball hits far from the end, the handle will move away from the pitcher. (*c*) If the ball hits near the end of the bat, the handle will move toward the pitcher.

However, at a special point on the bat, a collision with a ball will not make the handle accelerate. Any collision will make the bat's center of mass accelerate away from the pitcher, taking the handle with it. But if the ball hits near the end of the bat, beyond its center of mass, the torque will cause the handle to rotate toward the pitcher. These two movements can cancel one another. If the ball hits at just the right point, the handle will have almost no acceleration (Fig. 2.2.8*a*). That special point of collision is called the **center of percussion**. It's one of the "sweet spots" that occur in baseball bats, tennis rackets, golf clubs, and other sporting equipment. If the ball hits closer to the bat's handle, the handle will suddenly jerk away from the pitcher (Fig. 2.2.8*b*). If the ball hits farther from the bat's handle, the handle will jerk toward the pitcher (Fig. 2.2.8*c*). But when the ball hits the center of percussion, the handle moves smoothly and is more comfortable to hold.

The other important sweet spot is a **vibrational node**: a point on the bat that doesn't move back and forth when the bat vibrates. A baseball bat is not perfectly rigid; like the bar of a xylophone, it can vibrate after being struck. A bat's normal and slowest way of vibrating, its **fundamental vibrational mode**, is for its two ends to move one way while the middle goes the other way (Fig. 2.2.9). Somewhere between the middle and each end must be a point on the bat that doesn't move. Such a point is called a vibrational node. Since the ends and the middle travel the farthest, they are called **vibrational antinodes**.

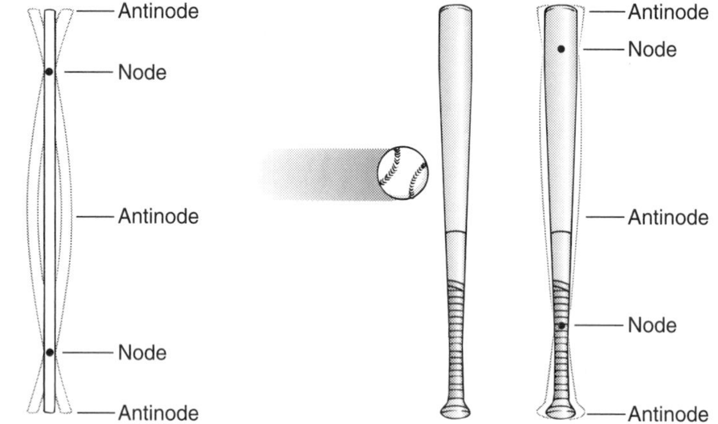

Fig. 2.2.9 - A free-standing rod can vibrate by having its middle and ends move back and forth in opposite directions. One point between the middle and each end doesn't move, however. This point is a vibrational node (no motion). The middle and ends are vibrational antinodes (maximum motion). A baseball bat has similar nodes, although their locations are slightly different due to the varying thickness of the bat. If the ball strikes the bat at a node, it will not vibrate. If the ball strikes the bat anywhere else, it will vibrate.

It may be harder to play music with a baseball bat than with a guitar, but some of the principles are the same. Just as you can't play a guitar by plucking it at its nodes, since doing so would mean plucking the very end of the string, you can't "play" a baseball bat by plucking it with a ball at its nodes. When the ball hits the bat at one of its vibrational nodes, the bat barely vibrates, and the sound of the collision is a crisp, clear "crack." When the ball misses the nodes, the bat vibrates loudly, the batter feels these vibrations in his or her hands, and the collision makes a dull "thunk." If the vibrations are severe enough, the bat can even shatter. Hitting the bat near the middle antinode is particularly likely to break the bat, which is why you shouldn't let the ball strike the bat too near its handle.

On a baseball bat, the center of percussion and the vibrational node almost coincide. Hitting the ball at this special spot, roughly 17.5 cm from its end, wastes very little energy in vibrations of the bat and transfers as much rebound energy as possible to the ball. Not only does hitting the ball at the sweet spot of the bat sound good, it makes the ball go farther. Outfielders often listen for the sound of the bat in order to guess how far the ball will travel.

CHECK YOUR UNDERSTANDING #4: Batting at Marbles

Your thumb is getting sore, so you begin flicking marbles with a wooden pencil. You notice that the sound the pencil emits depends on where the marble hits the pencil. How can the position of the impact affect the sound?

Balls Hitting Balls: Transferring Momentum

Baseball is played with a single ball, but games such as croquet and pool are played with several. One of the most interesting maneuvers in croquet is "sending" another player's ball. A player drives her ball into that of another player at very high speed. After the two balls hit, the other player's ball sails across the field while the first player's ball stops almost immediately. The same effect occurs on a pool table when one ball hits a second ball squarely. How does this trick work?

During the collision, the two balls push against one another. The moving ball pushes the stationary ball forward, causing it to accelerate forward. The stationary ball pushes the moving ball backward, causing it to accelerate backward. Newton's third law requires that the forces be equal in magnitude but oppositely directed so the moving ball slows down while the stationary ball starts to move forward. Overall, the collision transfers momentum and energy from one ball to the other.

But what causes this nearly perfect transfer of motion? Such a transfer has two requirements: the balls must have almost identical masses and their coefficients of restitution must be almost 1.00. When two ideal balls collide, they retain all of their kinetic energy and experience an **elastic collision**. However, real balls waste some of their kinetic energy as thermal energy and undergo an **inelastic collision**. But while kinetic energy can be converted to hidden thermal energy, momentum has nowhere to hide. As we saw in Section 1.4, momentum is a conserved quantity that is equal to an object's mass times its velocity. For the first ball to stop moving after the collision, it must transfer its energy and momentum to the second ball. If the two balls have different masses or if more than a little kinetic energy is wasted during the collision, then the second ball will end up traveling too slowly and will not carry away all of the first ball's momentum. Thus the equal masses and high coefficients of restitution are necessary for the moving ball to stop and the stationary ball to take over its motion.

Croquet, billiards, pool, and lawn bowling are each played with several balls of identical masses and very high coefficients of restitution. The balls are thus very effective at transferring momentum and kinetic energy from one to another. In pool, for example, when the cue ball strikes another ball directly, the cue ball almost stops (Fig. 2.2.10). This transfer of motion makes it possible to knock a ball into a pocket without having the cue ball follow it.

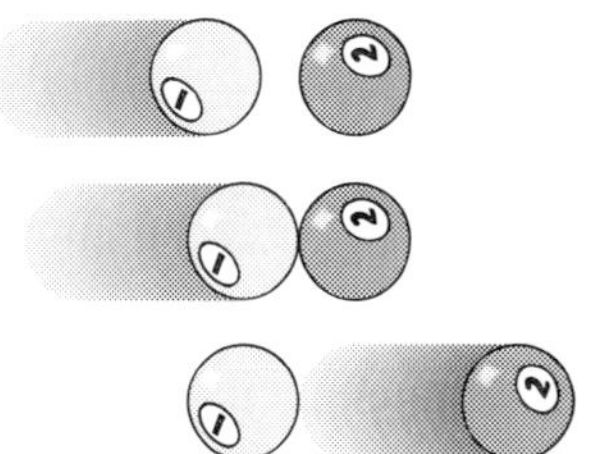

Fig. 2.2.10 - When one pool ball strikes a second, stationary pool ball directly, the transfer of energy and momentum is almost perfect. The second ball begins to move with the velocity of the first ball and the first ball stops.

CHECK YOUR UNDERSTANDING #5: Snagging an Aggie

You have your eye on a spectacular agate marble sitting near the middle of the circle. You have only one remaining marble which, while ugly, has the same size and mass as the pretty one you desire. If you hit the pretty marble head on with your ugly marble, will the pretty one leave the circle?

Hitting Surfaces: Rotation Affects Translation

When a cue ball stops following its collision with another ball, its movement becomes very sensitive to any rotational motion it might have. The collision between balls transfers linear momentum well, but very little angular momentum moves from one ball to the other. If the cue ball is rolling forward when it hits the second ball, its rotation will tend to carry it forward into the pocket. A good pool player compensates for this effect by hitting the cue ball below its center and giving it backspin. *Backspin*, where the top surface moves backward, away from the ball's direction of motion, reduces a ball's forward component of velocity whenever it touches a horizontal surface. Friction between the cue ball and the table slows the cue ball down after the collision. It can even make the cue ball roll backward if the backspin is fast enough.

Backspin is used in tennis, squash, Ping-Pong, basketball, and golf to slow the motion of a bouncing ball. As the ball bounces, it interacts with the surface through friction and slows down (Fig. 2.2.11*b*). In tennis, the backspin lob is hard to play since the ball bounces weakly and may even return toward the net. In basketball, backspin slows the ball when it hits the basket's rim or backboard and improves the chances that it will drop through the net. In golf, backspin helps to keep the golf ball from rolling off the green. Professional golfers use special golf balls that grip the club head during the swing and receive large amounts of backspin.

In contrast, a ball with *topspin*, where the top surface moves in the ball's direction of motion, accelerates forward when it bounces on a horizontal surface (Fig. 2.2.11*a*). Friction converts some of the ball's rotational energy into forward motion. Topspin, backspin, and even sidespin are useful in Ping-Pong, where much of the skill lies in hitting the ball so that it takes tricky bounces.

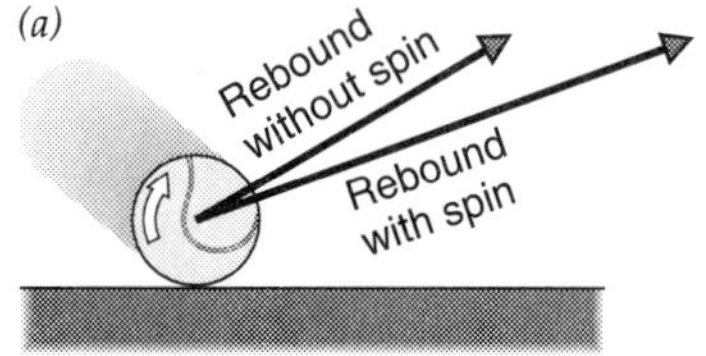

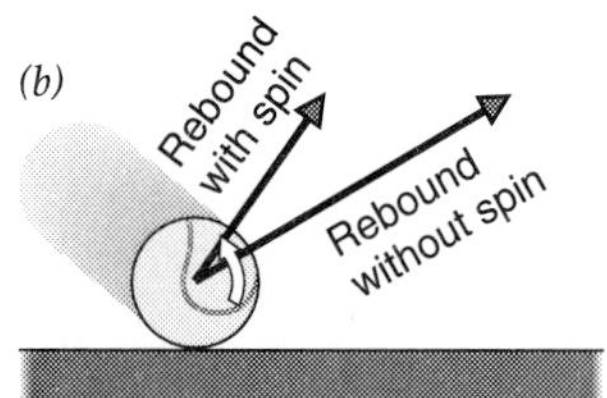

Fig. 2.2.11 - (*a*) When a ball with topspin bounces on a horizontal surface, the ball accelerates forward and rebounds more strongly than it would have without spin. (*b*) A ball with backspin will rebound weakly and can even reverse direction.

CHECK YOUR UNDERSTANDING #6: Losing Your Marbles

You are not satisfied with the idea of losing your ugly marble because you need it to continue playing. You refuse to risk losing the pretty one. How can you make the ugly marble follow the pretty one out of the circle after they collide?

Summary

How Bouncing Balls Work: A ball behaves like a spherical or oblong spring. It stores elastic potential energy as it distorts away from its equilibrium shape and releases some of that energy as it returns to its equilibrium shape. When a ball strikes a surface, some kinetic energy is removed from the ball and the surface and is either stored within those objects as elastic potential energy or lost as thermal energy. As the ball and surface rebound from one another, the stored energy becomes kinetic energy again. The kinetic energy returned, the rebound energy, is always less than the kinetic energy initially removed from the objects, the collision energy. The missing energy has been converted into thermal energy. The fraction of collision energy returned as rebound energy is determined by the ball and the surface from which it bounces.

The Physics of Bouncing Balls

1. An object's liveliness is usually specified by its coefficient of restitution, which is measured by bouncing the ball from a hard, immovable surface. It's the ratio of the object's rebound speed to its collision speed.

2. While people in different inertial reference frames will claim that the same objects have different velocities and kinetic energies, they will agree on the relative velocities of any two objects and on the collision and rebound energies when two objects collide. The laws of motion govern any situation, regardless of which inertial reference frame is used to observe that situation.

3. When an outside force acts on a free object, the object's center of mass accelerates in the direction of the force. If the force exerts a torque on the object about its center of mass, the object will also undergo angular acceleration.

4. When two objects collide, their stiffnesses determine the duration of the collision and the forces and accelerations involved. The stiffer the objects, the shorter the collision, and the larger the forces and accelerations.

5. An object that has a stable equilibrium shape exerts restoring forces when distorted. Most such objects vibrate when the forces causing the distortion are removed. The distortion stores energy in the object and it vibrates until it can transform that energy into another form or transfer it elsewhere. This energy is often transformed into sound or thermal energy.

6. An object struck at a vibrational node doesn't vibrate. When struck away from its vibrational nodes, particularly at a vibrational antinode, the object will vibrate strongly.

7. When two objects collide, they transfer both momentum and energy. If the two objects have identical masses, and coefficients of restitution near 1.00, this transfer can be almost complete.

8. A rotating object experiences frictional forces during a collision that affect its velocity after the rebound. Its velocity components along the surface it collided with may change as the result of these frictional forces. During bounces from a horizontal surface, the speed of an object with topspin will increase while that of an object with backspin will decrease.

Check Your Understanding - Answers

1. The soft dirt distorts more than a marble's glass surface and receives most of the collision energy. It converts most of that energy into thermal energy so the marble rebounds weakly.

Why: A marble has a very high coefficient of restitution and bounces well from a hard surface. However, the dirt is soft and receives virtually all of the collision energy when the marble hits it. The dirt distorts during the impact but stores little energy because it's not very elastic. The marble doesn't rebound much.

2. 2 m/s toward you.

Why: The velocities reported in the question are those observed by people sitting still with respect to the circle. From the inertial reference frame of your marble, the circle itself is heading in your direction at 1 m/s. Since the other person's marble is moving in your direction at 1 m/s faster than the circle, its total velocity according to your marble is 2 m/s in your direction.

3. The glass ball's mass is so much larger than that of your thumb that your thumb receives almost all the rebound energy. Your thumb bounces, not the glass ball.

Why: In any collision, it's the least massive object that experiences the greatest acceleration and that receives the largest share of the rebound energy. The effect is similar to what would happen if you swung a light aluminum baseball bat at a pitched bowling ball. The bat would rebound wildly but the bowling ball would continue to travel toward the catcher. This same effect is true in automobile collisions, where a massive sedan is much less disturbed than the tiny, subcompact with which it collides.

4. The pencil may vibrate after the marble hits it and the amount and type of vibration depend on where the marble hits the pencil.

Why: Like a baseball bat, a pencil's fundamental vibrational mode involves its middle moving in one direction while its ends move in the other. If the pencil hits the marble near one of its antinodes, either at its middle or at one of its ends, it will vibrate loudly in this fundamental mode. But if it hits the marble hear one of its nodes, between its ends and its middle, it will not vibrate in the fundamental mode and the sound it emits will be different.

5. Yes. Your marble will stop and the pretty one will roll out of the circle.

Why: Here is another case of equal mass objects having high coefficients of restitution. Like many very hard objects, glass marbles store energy well during compression and bounce nicely off equally hard surfaces. Your marble will exchange energy and momentum almost completely with the pretty marble. You will lose your marble, which will remain in the circle, but the pretty one will roll out.

6. Give the ugly marble a large amount of topspin so that it continues forward after the collision.

Why: If you give the marble topspin, it will experience frictional forces with the ground after the collision. The bottom surface of the marble will be turning toward you and the ground will exert a frictional force on it that is directed away from you. The marble will drive forward, just like a car. It will convert its rotational motion into translational motion.

Glossary

center of percussion The special spot on a bat or racket where a collision with a second object will not cause any acceleration of the bat's handle.

coefficient of restitution The measure of a ball's liveliness, determined by bouncing the ball from a rigid, immovable surface. It's the ratio of the ball's rebound speed to its collision speed.

collision energy The amount of kinetic energy removed from two objects as they collide.

elastic collision A collision between particles in which no kinetic energy is lost.

frame of reference A location from which motion is observed and measured. Such a frame of reference may be traveling at a constant velocity or it may be accelerating.

fundamental vibrational mode The slowest and often broadest vibration that an object can support.

inelastic collision A collision between particles in which at least some of the initial kinetic energy is converted into another form of energy.

inertial frame of reference A frame of reference that is not accelerating and is thus either stationary or traveling at constant velocity. The laws of motion accurately describe any situation that is observed from an inertial frame of reference.

rebound energy The amount of kinetic energy returned to two objects as they push apart following a collision.

vibrational antinode A region of a vibrating object that is experience maximal motion.

vibrational node A region of a vibrating object that is not moving at all.

Review Questions

1. What's the difference between the collision energy and the rebound energy when a ball hits a wall?

2. Since a ball's rebound energy is always less than its collision energy, what happens to the missing energy?

3. Which changes more during a bounce: a ball's speed or its kinetic energy? Why?

4. While a bean-bag won't bounce from a wall, it will bounce from a trampoline. Explain the difference.

5. When you're moving northward in a car at 80 km/h (50 mph), your frame of reference is different from when you're standing still. What velocities do the following objects have in those two reference frames: a tree, a person walking northward at 5 km/h, a bus moving northward at 80 km/h, and a truck moving southward at 80 km/h? (All the velocities listed are with respect to the ground itself.)

6. When a ball bounces off an object, how does that object's mass affect the ball's bounce? How does that object's motion affect the ball's bounce?

7. How does the impact between a ball and a bat affect the bat's velocity, angular velocity, and shape?

8. Identify at least five objects that vibrate when struck.

9. Why do two billiard balls exchange some momentum and kinetic energy whenever they collide?

10. If you want a ball to bounce forward when you drop it on the floor, how should that ball be spinning? What about if you want it to bounce backward?

Exercises

1. Why doesn't a marshmallow bounce well when you drop it on the floor?

2. The wind is blowing northward at 10 km/h, carrying the rain with it. When you stand still, the rain appears to be falling at an angle. If you begin to run northward at 10 km/h, why does the rain appear to be falling straight down?

3. Your car is on a crowded highway with everyone heading south at about 100 km/h (62 mph). The car ahead of you slows down slightly and your car bumps into it gently. Why is the impact so gentle?

4. Bumper cars are an amusement park ride in which people drive small electric vehicles around a rink and intentionally bump them into one another. All of the cars travel at about the same speed. Why are head-on collisions more jarring than other types of collisions?

5. A RIF baseball (Reduced Injury Factor) has the same coefficient of restitution as a normal baseball except that it deforms more severely during a collision. Why does this increased deformability lessen the forces exerted by the ball during a bounce and reduce the chances of it causing injury?

6. During rehabilitation after hand surgery, patients are often asked to squeeze and kneed putty to strengthen their muscles. How does the energy transfer in squeezing putty differ from that in squeezing a rubber ball?

7. Some athletic shoes have inflatable air pockets inside them. These air pockets act like springs that become stiffer as you pump up the air pressure. High pressure also makes you bounce back up off the floor sooner. Why does high pressure shorten the bounce time?

8. Padded soles in running shoes soften the blow of hitting the pavement. Why does padding reduce the forces involved in bringing your foot to rest?

9. What happens when you hit a baseball with a sturdy plastic bat that weighs less than the ball?

10. Steep mountain roads often have emergency ramps for trucks with failed brakes. Why are these ramps most effective when they are covered with deep, soft sand?

11. Why does a baseball bat slow down after it hits a baseball?

12. As you look out the window of a swiftly moving train you see a car moving in the same direction along a nearby highway. However, the car appears to be moving backward. Explain.

13. Line up several identical coins on a smooth table so that they are touching one another. Now slide another coin into the end of the line. This coin should come to a stop and a single coin should jump off the far end of the line to continue the motion. How is the motion transferred from the coin you pushed to the last one in the line?

14. Springs in the latch mechanism push outward on a car door and make it difficult to close the door slowly. It's much easier to get the door moving quickly while it's open so that it slams shut. Why?

15. If you drop a steel marble on a wooden floor, why does the floor receive most of the collision energy and contribute most of the rebound energy?

16. If you drop a ball onto the floor from a height of 1 m, it will rebound to a height of less than 1 m. Why can't it bounce higher than 1 m? Why doesn't it even reach 1 m?

17. Why is it so exhausting to run on soft sand?

18. The best running tracks have firm but elastic rubber surfaces. How does a lively surface assist a runner?

19. Why must the surface of a hammer be very hard and stiff for it to drive a nail into wood?

20. Why does it hurt less to land on a soft foam pad than on bare concrete after completing a high jump?

21. There have been baseball seasons in which so many home runs were hit that people began to suspect that something was wrong with the baseballs. What change in the baseballs would account for them traveling farther than normal?

22. The surface of a clay tennis court dents when the ball hits it and doesn't return to its original shape when the ball rebounds. Use the definition of work to show that the clay surface extracts energy from the ball (which is why clay courts are "slower" than asphalt courts).

23. An elastic ball that wastes 30% of the collision energy as heat when it bounces on a hard floor will rebound to 70% of the height from which it was dropped. Explain the 30% loss in height.

24. When a jar slips off the kitchen counter, you can prevent it from breaking on the floor by letting it hit your foot and allowing your foot to flex downward during the impact. Why is this process unlikely to break the jar?

25. When two trains are traveling side by side at break neck speed, it's still possible for people to jump from one train to the other. Explain why this can be done safely.

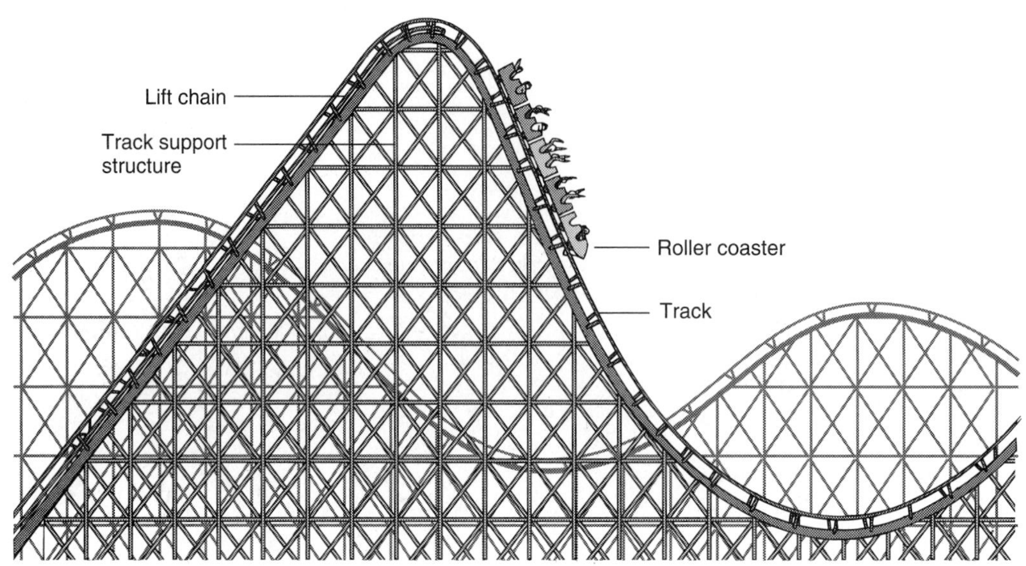

Section 2.3

Centrifuges and Roller Coasters

As your sports car leaps forward at a green light, you're pressed firmly back against your seat. It's as though gravity were somehow pulling you both down and backward at the same time. But it's not gravity that pulls back on you; it's your own inertia preventing you from accelerating forward with the car.

When this happens, you're experiencing the sensation of acceleration. We feel this sensation many times each day, whether through turning in an automobile or riding up several floors in a fast elevator. But nowhere is our sensation of acceleration more acute than at the amusement park. We accelerate up, down, and around on the carousel, back and forth in the bumper cars, and left and right in the scrambler. The ultimate ride, of course, is the roller coaster, which is one big, wild experience of acceleration. When you close your eyes on a straight stretch of highway, you can hardly tell the automobile is moving. But when you close your eyes on a roller coaster, you have no trouble feeling every last turn in the track. It's not the speed you feel, but the acceleration. What is often called motion sickness should really be called acceleration sickness.

Questions to Think About: *How does your body feel its own weight? When you swing a bucket full of water around in a circle, as we asked you to do in the chapter opening, why do you feel such a strong outward force from the bucket? Why can you swing that bucket completely over your head without spilling the water inside it? What keeps you from falling out of a roller coaster as it goes over the top of a loop-the-loop? Which car of a roller coaster should you sit in to experience the best ride?*

Experiments to Do: *To begin associating the familiar sensations of motion with the physics of acceleration, travel as a passenger in a vehicle that makes lots of turns and* *stops. Close your eyes and see whether you can tell which way the vehicle is turning and when it's starting or stopping. Which way do you feel pulled when the vehicle turns left? turns right? starts? stops? How is this sensation related to the direction of the vehicle's acceleration? Now find a time when the vehicle is traveling at constant velocity on a level path and see if you feel any sensations that tell you which way it's heading. Try to convince yourself that it's heading backward or sideways rather than forward.*

The Experience of Acceleration

Nothing is more central to the laws of motion than the relationship between force and acceleration. Up until now, we've looked at forces and noticed that they can produce accelerations; in this section we'll reverse that process, looking at accelerations and noticing that they require forces. For you to accelerate, something must push or pull on you. Just where and how that force is exerted on you determines what you feel when you accelerate.

The backward "force" you feel as your car accelerates forward is caused by your body's inertia, its tendency not to accelerate (Fig. 2.3.1). The car and your seat are accelerating forward, and since the seat acts to keep you from traveling through its surface, it exerts a forward support force on you that causes you to accelerate forward. But the seat can't exert a force uniformly throughout your body. Instead, it pushes only on your back, and your back then pushes on your bones, tissues, and internal organs to make them accelerate forward. Each piece of tissue or bone is responsible for the forward force needed to accelerate the tissue in front of it forward. A whole chain of forces, starting from your back and working forward toward your front, together cause your entire body to accelerate forward.

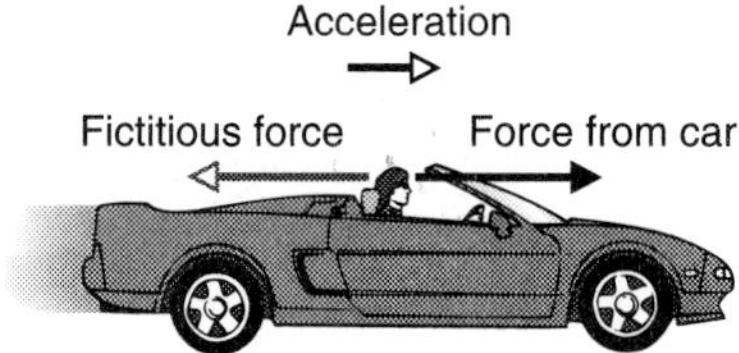

Fig. 2.3.1 - As you accelerate forward in a car, you feel a gravity-like fictitious force in the direction opposite to the acceleration. This fictitious force is really the mass of your body resisting acceleration.

Let's compare this situation with what happens when you're standing motionless on the floor. Since gravity exerts a downward force on you that's distributed uniformly throughout your body, each part of your body has its own independent weight; these individual weights, taken together, add up to your total weight. The floor, for its part, is exerting an upward support force on you that keeps you from accelerating downward through its surface. But the floor can't exert a force uniformly throughout your body. Instead, it pushes only on your feet, and your feet then push on your bones, tissues, and internal organs to keep them from accelerating downward. Each piece of tissue or bone is responsible for the upward force needed to keep the tissue above it from accelerating downward. A whole chain of forces, starting from your feet and working upward toward your head, together keep your entire body from accelerating downward.

As you probably noticed, the two previous paragraphs are very similar. But so are the sensations of gravity and acceleration. When the ground is preventing you from falling, you feel "heavy"; your body senses all the internal forces needed to support its pieces so that they don't accelerate, and you interpret these sensations as weight. When the car seat is causing you to accelerate forward, you also feel "heavy"; your body senses all the internal forces needed to accelerate its pieces forward, and you interpret these sensations as weight. This time the weight is experienced toward the back of the car.

The gravity-like "force" that you experience as you accelerate is truly indistinguishable from the force of gravity. And you're not the only one fooled by acceleration. Even the most sophisticated laboratory instruments can't determine directly whether they are experiencing gravity or are simply accelerating. However, despite the convincing sensations, the backward heavy feeling in your gut as you accelerate forward is not due to a real force. This experience of acceleration is called a **fictitious force**. It always points in the direction opposite the acceleration that causes it and its strength is proportional to that acceleration.

(a)

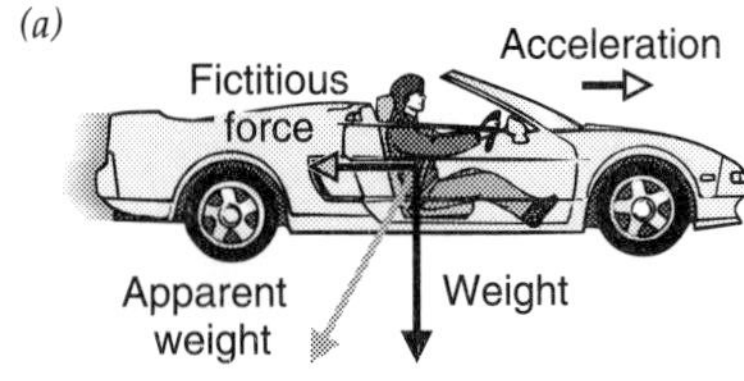

(b)

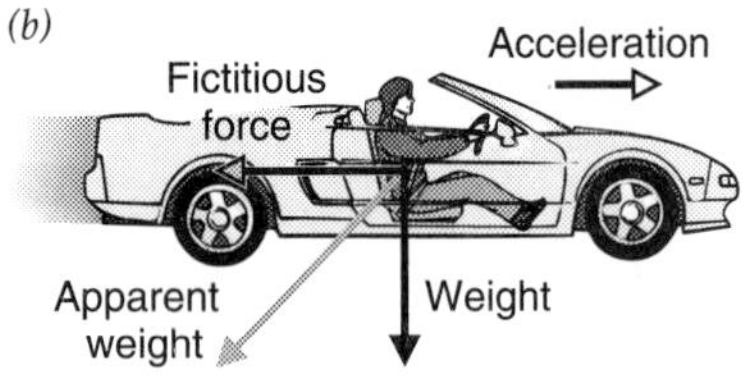

Fig. 2.3.2 - (*a*) When you accelerate forward gently, the backward fictitious force you feel is small and your apparent weight is mostly downward. (*b*) When you accelerate forward quickly, you experience a strong backward fictitious force and your apparent weight is backward and down.

If you accelerate forward quickly, the backward fictitious force you experience can be quite large. However, you don't experience this fictitious force all by itself; you also experience your downward weight, and together these effects feel like an especially strong gravitational force at an angle somewhere between straight down and the back of the car. We will call the combined effects of gravitational and fictitious forces **apparent weight**. The faster you accelerate, the stronger the backward fictitious force, and the more your apparent weight points toward the back of the car (Fig. 2.3.2).

CHECK YOUR UNDERSTANDING #1: The Feel of a Tight Turn

You're sitting in the passenger seat of a racing car that is moving rapidly along a level track. The track takes a sharp turn to the left and you find yourself thrown against the door to your right. What horizontal forces, real and fictitious, are acting on you?

Merry-Go-Rounds and Spin-Dryers

Merry-go-rounds and spin-dryers are two common examples of centrifuges. A **centrifuge** is a carefully balanced basket that spins about a central pivot. An object placed in the basket travels in a circle around that pivot. When left alone, no object ever travels in a circle. Newton's first law says that an object experiencing no net force will move in a straight line at constant speed. The only reason an object ever travels in a circle is because something is exerting a force on it. In that case, the object must be accelerating.

Which way is the object accelerating? Remarkably, the object is always accelerating toward the center of the circle. To see that it does, imagine looking down on a merry-go-round that is turning counter-clockwise at a steady pace (Fig. 2.3.3). At first, the boy riding the merry-go-round is directly east of its central pivot and is moving northward (Fig. 2.3.3*a*). If nothing were pulling on the boy, he would continue northward and fly off the merry-go-round. Instead, he follows a circular path by accelerating toward the pivot—that is, toward the west. As a result, his velocity turns toward the northwest and he heads in that direction. To keep from flying off the merry-go-round, he must continue to accelerate toward the pivot, which is now southwest of him (Fig. 2.3.3*b*). His velocity turns toward the west and he follows the circle in that direction. And so it goes (Fig. 2.3.3*c*).

The boy's body is always trying to go in a straight line, but the merry-go-round keeps pulling him inward so that he accelerates toward the central pivot. The boy is experiencing **uniform circular motion**. "Uniform" means that the boy is always moving at the same speed, although his direction keeps changing. "Circular" describes the path the boy follows as he moves, his trajectory.

Like any object undergoing uniform circular motion, the boy is always accelerating toward the center of the circle. An acceleration of this type, toward the center of a circle, is called a **centripetal acceleration** and is caused by a centrally directed force, a **centripetal force**. A centripetal force is not a new, independent type of force, like gravity, but the net result of whatever forces act on the object. Centripetal means "center-seeking" and a centripetal force pushes the object toward that center. The merry-go-round uses support forces and friction to exert a centripetal force on the boy and he experiences a centripetal acceleration. Amusement park rides often involve centripetal acceleration (Fig. 2.3.4).

The amount of acceleration the boy experiences depends on his speed and the radius of the merry-go-round. The faster the boy is moving and the smaller the radius of his circular trajectory, the more he accelerates. His acceleration is equal to the square of his speed divided by the radius of his circular trajectory. We can express this relationship as a word equation:

$$\text{acceleration} = \frac{\text{speed}^2}{\text{radius of circular trajectory}}, \tag{2.3.1}$$

in symbols:

$$a = \frac{v^2}{r},$$

and in everyday language:

Making a tight, high-speed turn involves lots of acceleration.

Since the boy is accelerating inward, toward the center of the circle, he feels a fictitious force outward, away from the center of the circle. This fictitious force is often called **"centrifugal force."** The quotation marks indicate that "centrifugal force" is not a force at all but an outward fictitious force due to inward acceleration. Fictitious forces such as "centrifugal force" don't contribute to the net force that acts on an object. Thus, if the centripetal force on an object is removed, the object will travel in a straight line with the velocity it had at the moment the centripetal force stopped. But despite its fictitious nature, "centrifugal force" creates a compelling sensation of gravity-like force. The boy on the merry-go-round feels as though gravity is pulling him outward as well as down and clings tightly to the merry-go-round to keep from falling off.

Gravitational force and "centrifugal force" can differ significantly in their amounts. While the gravitational force on a particular object has one set value on the earth's surface, the amount of "centrifugal force" that an object can experience has no limit. For an object going very rapidly around a small circle, the fictitious force can easily exceed the force of gravity.

Fictitious force is often measured relative to the earth's gravity. On the earth's surface, the acceleration due to gravity is $9.8\ \text{m/s}^2$ or $9.8\ \text{N/kg}$. Thus a 1 kg stone weighs 9.8 N. In a centrifuge or an airplane maneuvering sharply, that same stone may experience a fictitious force 5 times its weight, or 49 N. We describe this fictitious force as 5 gravities or 5 g's, for short.

Another example of a centrifuge is a clothes washer with a spin-dry cycle. When the basket is spinning to extract the water, the clothes undergo uniform circular motion. They move in a circular path, roughly 0.25 m in radius, at a speed of about 20 m/s. Equation 2.3.1 states that their acceleration is thus $(20\ \text{m/s})^2/0.25\ \text{m}$, or $1{,}600\ \text{m/s}^2$. Since the acceleration due to gravity is $9.8\ \text{m/s}^2$, the clothes experience about 163 g's. If the wet clothes have a mass of 5 kg, they really weigh 49 N (11 pounds) but their apparent weight in the spinning basket is

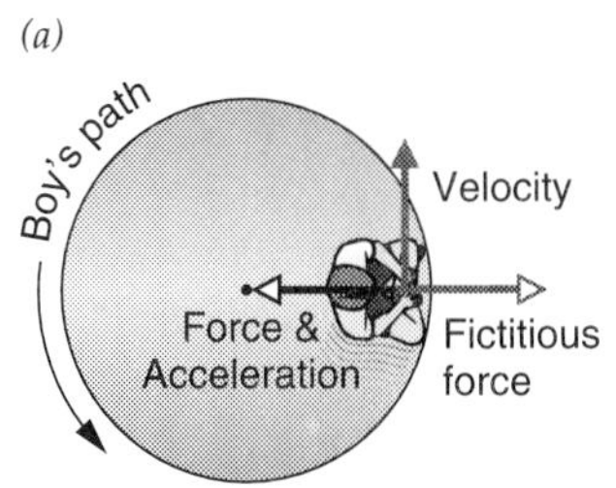

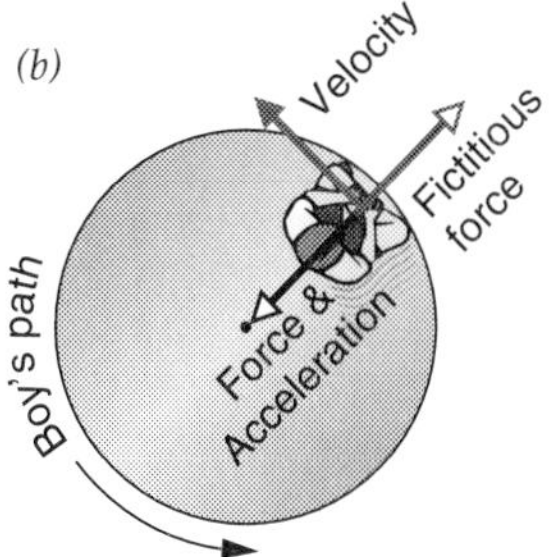

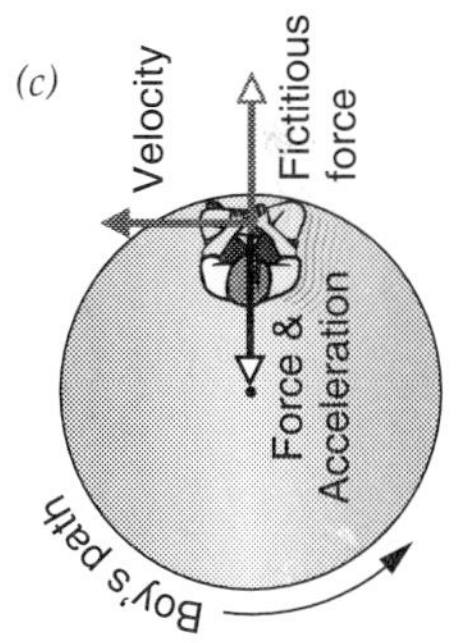

Fig. 2.3.3 - A boy riding on a turning merry-go-round is always accelerating toward the central pivot. His velocity vector shows that he is moving in a circle but his acceleration vector points toward the pivot. When he is heading north (*a*), he is accelerating toward the west. His velocity gradually changes direction until he is heading northwest (*b*), at which time he is accelerating toward the southwest. He turns further until he is heading west (*c*) and is then accelerating toward the south. (North is upward.)

Fig. 2.3.4 - The people on this ride travel in a circle, pulled inward by cords so that they accelerate toward the center of the circle.

❐ An amusement park rotor turns rapidly enough to press its occupants tightly against the outer wall of the enclosure. When the floor of the rotor is moved downward, the riders remain suspended by static friction with the wall.

about 163 times that much or almost 8000 N (1800 pounds).

The washer's basket exerts a centripetal force of 8000 N on the clothes to accelerate them around in a circle, and the clothes act to accelerate the water inside them. But the water, being fluid, is able to flow over the fabric, drip through the basket, and escape from the centrifuge. It leaves the centrifuge traveling with the velocity it had when the basket stopped pushing on it: it heads off tangent to the basket's surface (and begins to fall). It doesn't travel outward away from the basket's center because it wasn't heading that way when it left the basket!

The water doesn't accelerate with the clothes, which are spun dry. This same effect would happen if you placed a shallow pan of water on your lap in a stopped car and then accelerated forward quickly. The car would exert a force on the pan to accelerate it forward and the pan would try to accelerate the water. However, the water would just slosh out of the pan and be left behind (on you!) as the car picked up speed. You would now be ready to try the spin-dryer again.

In addition to merry-go-rounds and spin-dryers, centrifuges are found in water pumps, research labs, doctors' offices, amusement parks (see ❐), and even wineries. They can extract liquids from solids, as in a clothes washer or a winery, or they can separate heavy solids from light solids, as in research labs or doctors' offices. Labs use ultracentrifuges that produce as much as 100,000 g's to separate biological molecules (see ❐). Doctors' offices use less extreme centrifuges to separate components of a patient's blood to look for various abnormalities.

CHECK YOUR UNDERSTANDING #2: Banking on a Curve

Your racing car comes to a banked, left-hand turn, which it completes easily. Why is it essential that a racetrack turn be banked so that it slopes downhill toward the center of the turn?

CHECK YOUR FIGURES #1: Going for a Spin

Some children are riding on a playground merry-go-round with a radius of 1.5 m. The merry-go-round turns once every two seconds. How quickly are the children accelerating?

Roller Coasters

One of the most exciting variations on a centrifuge is the roller coaster. While roller coasters may present interesting visual effects, such as almost hitting an obstacle, and strange orientations, such as looking at the world upside down, the real thrill of roller coasters is in their accelerations. Plenty of other amusement park rides simply suspend you sideways or upside down, so that you feel normal gravity pulling at you from an unusual angle. These rides can be fairly interesting—but so can standing on your head, and you don't need to buy a ticket to do that. To experience real thrills, you need acceleration to give you the weightless feeling you experience as a roller coaster descends its first big hill or the several-g fictitious force you feel as you go around a sharp corner. Changes in the amount of "gravity" we feel are much more fun than changes in its direction.

❐ Fascinated by the large fictitious forces created by high-speed centrifuges, Jesse W. Beams (U.S. physicist, 1898–1977) developed a number of remarkable rotating devices. These included ultracentrifuges for biological research and gas centrifuges for separating the isotopes of uranium for nuclear power. During his studies, he built a tiny disk that spun over 1 million times per second and experienced fictitious forces of almost 1 billion g's.

Every time the roller coaster accelerates, you feel a fictitious force in the direction opposite the acceleration. That fictitious force is in addition to the real constant downward force of gravity. If you accelerate forward, you feel both your downward weight and a fictitious force backward. Your apparent weight is greater than your real weight and points down and toward the back of the car. If you accelerate upward, you feel both your downward weight and a fictitious force downward. Your apparent weight is greater than your real weight and points directly downward. But if you accelerate downward, you feel both your downward weight and a fictitious force upward. Now the real force of gravity and the fictitious force oppose one another and partially cancel. Your apparent weight is less than your real weight and points directly downward or perhaps, if the acceleration downward is fast enough, directly upward.

Consider that last possibility: if you accelerate downward at just the right rate, the upward fictitious force will exactly cancel your downward weight. You will feel perfectly weightless, as though gravity did not exist at all. The rate of downward acceleration that causes this perfect cancellation is the same as that of a *freely falling object*, an object that is not subject to any forces other than gravity itself. Your weight won't have changed, but the roller coaster will no longer be supporting you and you will be accelerating downward at $9.8\,\text{m/s}^2$. You will experience the same sensations as a skydiver who has just stepped out of an airplane.

Since freely falling objects are subject only to the force of gravity, they don't have to push on one another to keep their relative positions. As you fall, your hat and sunglasses will fall with you and won't require any support forces from your head. Even if your sunglasses come off, they will hover in front of you as the two of you accelerate downward together. Similarly, your internal organs don't need to support one another, and the absence of internal support forces gives rise to the exhilarating sensation of free fall.

A roller coaster is attached to a track, and its rate of downward acceleration can actually exceed that of a freely falling object. In that case, the track must assist gravity in pushing the roller coaster downward. As a rider, you will feel less than weightless. The upward fictitious force will be so large that your apparent weight will be in the upward direction, as though the world had turned upside down!

We're now prepared to look at a roller coaster and understand what you feel as you go over hills (Fig. 2.3.5) and loop-the-loops (Fig. 2.3.6). Fig. 2.3.7 shows a single-car roller coaster at various points along a simple track with one hill and one loop-the-loop. Weight, fictitious force, and apparent weight are all vector quantities, as are the car's velocity and acceleration. These quantities are indicated with arrows of varying lengths that show each vector's direction and

Fig. 2.3.5 - As this roller coaster plunges down the first hill, its last car is pulled over the edge by the cars in front of it and its riders feel almost weightless.

Fig. 2.3.6 - As it goes over the loop-the-loop, this roller coaster is accelerating rapidly toward the center of the circle. When it reaches the top of the loop, the track pushes the roller coaster downward and the riders feel pressed into their seats. If they close their eyes, they won't even be able tell that they're upside down.

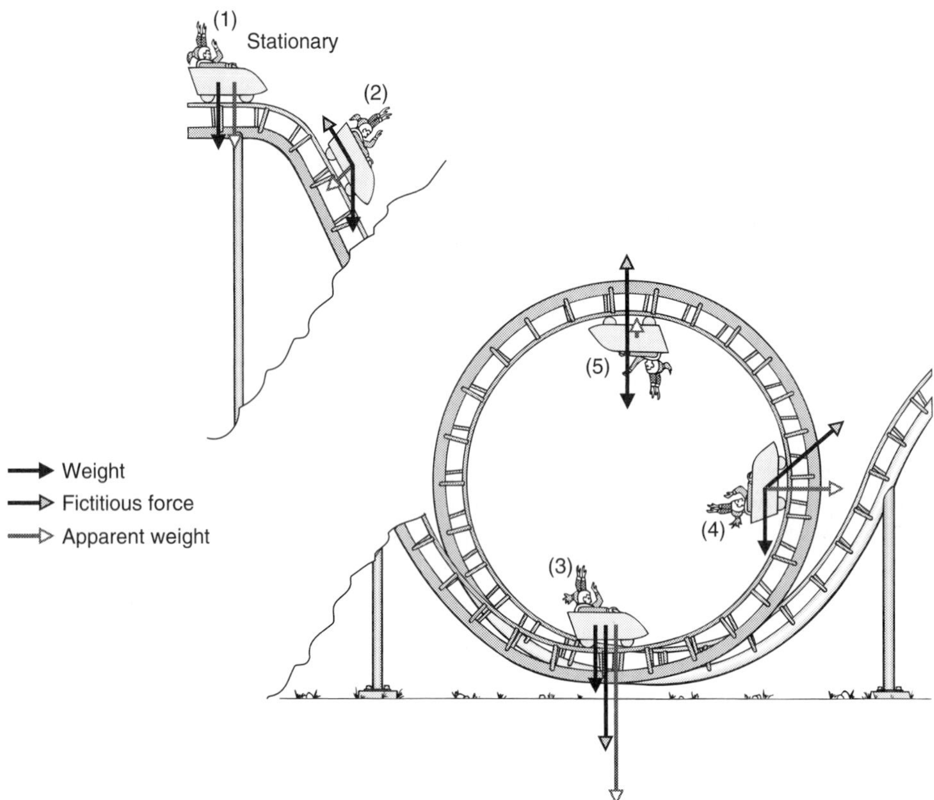

Fig. 2.3.7 - A single-car roller coaster going over the first hill and a loop-the-loop. At each point along the track, the car experiences its weight, a fictitious force due to its current acceleration, and an apparent weight that is the sum of those two. The apparent weight always points toward the track, which is why the car doesn't fall off it.

magnitude. The longer the arrow, the greater the magnitude of the quantity it represents.

At the top of the first hill of Fig. 2.3.7, the single-car roller coaster is almost stationary (1). You, the rider, feel only your weight, straight down—nothing exciting yet. But as soon as the car begins its descent, accelerating down the track, a fictitious force appears pointing up the track (2). The combination of this fictitious force and your weight gives you an apparent weight that is very small and points down and into the track. Most people find this sudden reduction in apparent weight terrifying. Our bodies are very sensitive to partial weightlessness and this falling sensation is half the fun of a roller coaster. An astronaut, weightless in space, has this disquieting feeling for days on end. No wonder astronauts have such frequent troubles with motion (or rather acceleration) sickness.

Fig. 2.3.8 - The roller coaster pushes these riders inward as it travels around the loop-the-loop. The riders feel a strong fictitious force outward and can hardly tell that they're turning upside down.

On earth, the weightless feeling can't last. It occurs only during downward acceleration and disappears as your car levels off near the bottom of the hill. By the time the car begins its rise into the loop-the-loop, it is traveling at maximum speed and has begun to accelerate upward (3). This upward acceleration creates a downward fictitious force so that your apparent weight is huge and downward. You feel pressed into your seat as you experience about 2 or 3 g's.

The trip through the loop-the-loop is almost uniform circular motion. In effect, you are taking a single turn around a vertical centrifuge. However, as the car rises into the loop-the-loop, some of its kinetic energy becomes gravitational potential energy and the car slows down. As the car descends out of the loop-the-loop, this gravitational potential energy returns to kinetic form and the car speeds up. As a result of these speed changes, your acceleration is not exactly inward toward the center of the loop-the-loop and the fictitious force you experience is not exactly outward. Still, an inward acceleration and an outward fictitious force are good approximations for what occurs.

Halfway up the loop-the-loop, your acceleration is roughly horizontal and inward, so the fictitious force you experience is roughly outward, into the track (4). Your apparent weight is still much more than your weight and is directed downward and outward. You feel pressed into your seat, and the car itself is pressed against the track (Fig. 2.3.8).

Finally you reach the top of the loop-the-loop (5). The car has slowed somewhat as the result of its climb against the force of gravity. But it's still accelerating toward the center of the circle, and you still experience a fictitious force outward and, in this case, upward. Your weight is downward but the upward fictitious force exceeds your weight. Your apparent weight is upward (Fig. 2.3.6)!

Not only does the car stay on its track, but you feel as though a weak gravity-like force is pulling you upward (away from the ground). You're pressed into your seat. If your hat were to come off, it would appear to "fall" upward (away from the ground) and might land in the bottom of the inverted car. In fact, you are accelerating almost directly downward at a rate that's faster than that of a freely falling object! The track is pushing you and the car downward, assisting gravity to make you accelerate downward so rapidly. A loose object such as your hat would simply fall in an arc, initially heading forward at the velocity it had when it left your head and then turning more and more downward as it fell. The car's rapidly increasing downward component of velocity would let it overtake the falling hat—your hat would be dropping but the car would be dropping even faster.

Many roller coaster tracks are designed so that their riders always have apparent weights toward their seats and the track, even when the cars go upside down. In principle, roller coasters that travel on such tracks don't need seat belts to prevent riders from falling out (although seat belts are comforting to the passengers and insurance companies). Some roller coaster tracks have special cars and seat belts that permit them to have apparent weights away from the track. These roller coasters can and do go upside down without enough downward acceleration to hold the riders in their seats. In such roller coasters, the riders feel like they are hanging and if one of them were to lose a hat, it would fall to the ground rather than into the car.

What about a roller coaster with more than one car? For the most part, the same rules apply. However, a new source of force now acts on each car: the other cars. The effects of other cars are most pronounced at the top of the very first hill. As the train approaches the descent, it's traveling slowly. It has just been raised to its present height by the lift chain and is beginning to roll forward. The first cars go slowly over the crest of the hill, dangling well over the edge before they begin to pick up much speed (Fig. 2.3.9*a*). They're pulling hard on the cars behind them and these cars pull back, slowing the descent. The first car travels rather slowly until it nears the bottom of the hill. By then, the track is beginning to turn upward and the first car's riders feel mostly upward acceleration and downward fictitious force. Riders in the first few cars of a roller coaster don't feel much weightlessness.

In contrast, the last car is pulled very quickly over the crest of the hill and is moving at high speed early in its descent. It undergoes a dramatic downward acceleration as it's pulled over the crest of the hill by the cars in front of it (Fig. 2.3.9*b*). As a result, its riders feel large upward fictitious forces and quite extreme weightlessness. In fact, the designers of the track must take care not to make the downward accelerations too rapid, or the roller coaster will flick the riders in the last car right out of their seats.

Obviously, it does matter where you sit on a roller coaster. The first seat offers the most exciting view, but less than spectacular fictitious forces. The last car

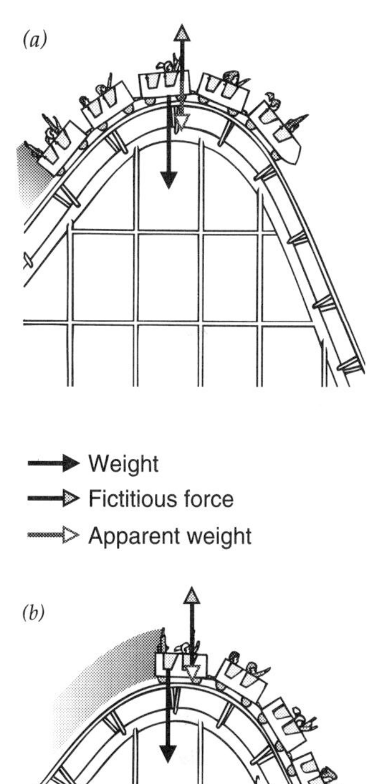

Fig. 2.3.9 - When a multi-car roller coaster descends the first hill, the ride experienced in the first cars is different from that in the last cars. (*a*) The first cars travel over the crest of the hill slowly and reach high speed only well down the hill. The cars behind them slow their descent. (*b*) The last cars are whipped over the top and are traveling very rapidly early on. The last cars accelerate downward dramatically as they go over the first hill and their riders experience a strong feeling of weightlessness.

almost always offers the best fictitious forces. Probably the dullest seat in the roller coaster is the second; it offers a view of the people in the front seat and a relatively tame ride.

CHECK YOUR UNDERSTANDING #3: Taking to the Air

Your racing car travels over a bump in the track and suddenly becomes airborne. What keeps you in the air? Has gravity disappeared?

Summary

How Centrifuges and Roller Coasters Work: A centrifuge uses rapid centripetal acceleration to produce large, outward fictitious forces on the objects it contains. An object in the centrifuge accelerates toward the center of the device at a rate that can be many times greater than the acceleration due to gravity. As a result, that object experiences an enormous gravity-like sensation in the outward direction. Movable objects try to follow straightline paths that take them out of the centrifuge: water is flung off clothes in a spin-dryer, chemicals separate in laboratories, and cars are flung off turns on highways.

A roller coaster also uses rapid acceleration to create unusual apparent weights for its riders. Each time the coaster accelerates on a hill or turn, the rider experiences a fictitious force in the direction opposite the acceleration. This fictitious force, combined with the rider's true weight, creates an apparent weight that varies dramatically in amount and direction throughout the ride. It's this fluctuating apparent weight, particularly its near approaches to zero, that make riding a roller coaster so exciting.

Important Laws and Equations

1. Acceleration of an Object in Uniform Circular Motion: An object in uniform circular motion has an acceleration equal to the square of its speed divided by the radius of its circular trajectory, or

$$\text{acceleration} = \frac{\text{speed}^2}{\text{radius of circular trajectory}}. \qquad (2.3.1)$$

The Physics of Centrifuges and Roller Coasters

1. Anything that accelerates experiences a fictitious force in the direction opposite the acceleration. While a fictitious force is not a real force, it feels exactly like the force of gravity. Both acceleration and the fictitious force it produces are vector quantities, pointing in opposite directions. The magnitude of the fictitious force is proportional to the magnitude of acceleration.

2. The gravitational and fictitious forces acting on a person combine to give the person an apparent weight. Since both forces are vectors, they can partially or completely cancel one another if they point in opposite directions. Such cancellation gives the person an apparent weight that is smaller in magnitude than their real weight, creating the disquieting sensation of partial or complete weightlessness. The apparent weight of a freely falling person is zero, complete weightlessness.

3. An object moving in a circular path at a constant speed is undergoing uniform circular motion. The object's acceleration is centripetal (toward the center) and is caused by a centripetal force. The fictitious force experienced by the object is away from the center and is called "centrifugal force." The "centrifugal force" increases as the object's speed increases and as the diameter of the circle decreases.

Check Your Understanding - Answers

1. The car seat and door are exerting a real leftward force on you, causing you to accelerate leftward with the car. You also experience a rightward fictitious force, as your inertia acts to keep you from accelerating leftward.

Why: As the car turns left, it accelerates toward the left. The car seat and right door together exert a real leftward force on you, to prevent the car from driving out from under you. Because it has mass, your body resists this leftward acceleration. You feel your body trying to go in a straight line, which would carry it out the right door of the car as the car accelerates toward the left.

2. So that the support force exerted by the racetrack on the car's wheels can provide at least some of the centripetal force needed to accelerate the car around the turn.

Why: As a car and driver travel around a circular turn, they are accelerating toward the center of the track and feel a huge "centrifugal force" outward. On a level track, the only centripetal force available is static friction between the ground and the car's tires. If static friction is unable to provide enough force, the car will skid off the track, following a straightline path. This type of accident is typical of a highway curve on an icy day and is why designers bank the curves. The banks are ramps sloping down toward the center so that the horizontal component of the support force exerted by the track on the car's wheels provides an additional, inward, centripetal force to help that car accelerate around the curve.

3. Gravity is still present but your inertia prevents you from following a rapid downturn in the surface you are traveling along.

Why: If the road you are traveling along suddenly turns downward, you must accelerate downward to stay in contact with its surface. The steeper and more abrupt the descent, the more downward acceleration you need. The only downward force you experience is your weight, which can cause a downward acceleration of no more than $9.8\ \text{m/s}^2$. If the surface drops out from under you faster than that, you will become airborne. You will then be falling freely and accelerating downward as fast as gravity will permit. Eventually, you will fall to the surface. In many sports, including skiing, motorcycle racing, and skateboarding, a person traveling along an uneven surface becomes airborne after passing over a bump.

Check Your Figures - Answers

1. About $15\ \text{m/s}^2$.

Why: The children are in uniform circular motion so their acceleration is given by Eq. 2.3.1. To use this equation, we need to know the children's speed and the radius of their circular trajectory. Since they complete one trip around the circumference of a 1.5 m radius circle every 2 seconds, their speed is

$$\begin{aligned}\text{children's speed} &= \frac{\text{distance}}{\text{time}}\\ &= \frac{2\pi\cdot\text{radius}}{\text{time}} = \frac{2\pi\cdot 1.5\ \text{m}}{2\ \text{s}} = 4.7\ \text{m/s}.\end{aligned}$$

We can now obtain their acceleration from Eq. 2.3.1:

$$\text{children's acceleration} = \frac{(4.7\ \text{m/s})^2}{1.5\ \text{m}} = 14.7\ \text{m/s}^2.$$

Since our measurements of the merry-go-round's radius and its turning time are only accurate to about 10%, our calculation of the children's acceleration is only accurate to about 10%. We report it as $15\ \text{m/s}^2$.

Glossary

apparent weight The sum of the gravitational and fictitious forces on an object. All three quantities are vectors, so that apparent weight can be quite large if the two forces point in the same direction or quite small if they point in opposite directions.

"centrifugal force" A fictitious force outward from the center of a circular trajectory, due to an object's centripetal acceleration.

centrifuge A carefully balanced basket or rotor that spins about a central pivot, used to separate substances of different densities, remove moisture, or simulate gravitational effects.

centripetal acceleration An acceleration that is always directed toward the center of a circular trajectory.

centripetal force A centrally directed force on an object. A centripetal force is not an independent force but rather the sum of other forces, such as gravity, acting on the object.

fictitious force Not a real force at all, but the object's mass resisting acceleration. An object undergoing acceleration experiences what appears to be a force in the direction exactly opposite the direction of acceleration. The amount of fictitious force is proportional to the amount of acceleration.

uniform circular motion Motion at a constant speed around a circular trajectory. An object undergoing uniform circular motion is accelerating toward the center of the circle and experiences an outward fictitious force, "centrifugal force."

Review Questions

1. If you close your eyes in a train or subway, can you tell which direction you're traveling?

2. If you close your eyes in a train or subway, can you tell when you're accelerating and in which direction?

3. Why is the pull that you feel when you're accelerating not really a force?

4. When you accelerate forward, why do you feel as though your weight is both downward and toward the back?

5. If you spin a ball on a string and suddenly let go, why will the ball travel in a straight line with the velocity it had at the moment you let go of the string, rather than heading outward, directly away from your hand?

6. When a bicyclist is moving steadily around a circular track, which way is that bicyclist accelerating?

7. What type of force acts on any object that is experiencing uniform circular motion?

8. Identify a case of uniform circular motion in which the centripetal force is a gravitational force. Find another case when it's the pull of a string. Find still another case when it's the support force from a curved surface.

9. If the strange feelings you experience while riding a roller coaster are caused by accelerations, how does the speed of the roller coaster along the track contribute to those feelings?

10. If all objects fall at the same rate near the earth's surface, how is it possible for you to swoop your hand downward to catch an object that's falling past you?

Exercises

1. You are traveling in a subway along a straight, level track at a constant velocity. If you close your eyes, you can't tell which way you're heading. Why not?

2. Some amusement park rides move you back and forth in a horizontal direction. Why is this motion so much more disturbing to your body than cruising at a high speed in a jet airplane?

3. How could you use a marble and a round-bottomed bowl to measure forward or backward acceleration?

4. Moving a can of spray paint rapidly in one direction will not mix it nearly as well as shaking it back and forth. Why is it so important to change directions as you mix the paint?

5. Railroad tracks must make only gradual curves to prevent trains from derailing at high speeds. Why is a train likely to derail if it encounters a sharp turn while it's traveling fast?

6. In some roller coasters, the cars travel through a smooth tube that bends left and right in a series of complicated turns. Why does the car always roll up the right-hand wall of the tube during a sharp left-hand turn?

7. When a moving hammer hits a nail, it exerts the enormous force needed to push the nail into wood. This force is far greater than the hammer's weight. How is it produced?

8. A hammer's weight is downward, so how can a hammer push a nail upward into the ceiling?

9. Could a hammer pound a nail into wood if gravity didn't exist?

10. Some amusement park rides swing you all the way upside down in a circle. Suppose that you're in one of these rides and that as the ride swings you over the top, your head is nearer to the ground than your feet are. If the ride goes over the top of the circle quickly, your hat stays on and you can hardly tell that you're upside down. But if the ride goes over the top slowly, your hat falls off and objects come out of your shirt pocket. What causes these different behaviors?

11. As you swing back and forth on a playground swing, your apparent weight changes. At what point do you feel the heaviest?

12. Why do skiers begin to skid across the snow when they try to turn too sharply?

13. Why does a baby's rattle only make noise when the baby moves it back and forth and not when the baby moves it steadily in one direction?

14. Some stores have coin-operated toy cars that jiggle back and forth on a fixed base. Why can't these cars give you the feeling of actually driving in a drag race?

15. A seeding machine slowly pours grass seeds onto the surface of a spinning disk. What causes the seeds to fly outward onto the ground as the disk spins? Which way do the seeds travel as they leave the disk? (Draw a picture.)

16. Police sometimes use metal battering rams to knock down doors. They hold the ram in their hands and swing it into a door from about 1 m away. How does the battering ram increase the amount of force the police can exert on the door?

17. A salad spinner is a rotating basket that dries salad after washing. How does the spinner extract the water?

18. Why do trains have brakes on each car, rather than just on the locomotive?

19. As a grand prix car wends its way around the curving streets of a European city, the driver must slow down to avoid sliding off the road. What limits the speed at which a level curve can be negotiated without sliding off the road?

20. In a seemingly impossible circus stunt, two motorcyclists drive around the inside of a spherical steel cage at highway speeds. Each motorcycle completes vertical loop-the-loops, carefully timed so that the two motorcycles don't collide. The hardest part of the stunt is starting and stopping. The motorcycles must be traveling full speed in order to complete the top half of a loop. Why?

21. If you swing a bucket of water over your head quickly enough, the water won't fall out. What keeps the water in the bucket?

22. People falling from a high diving board feel weightless. Has gravity stopped exerting a force on them? If not, why don't they feel it?

23. You board an elevator with a large briefcase in your hand. Why does that briefcase suddenly feel particularly heavy when the elevator begins to move upward?

24. As your car reaches the top in a smoothly turning Ferris wheel, which way are you accelerating?

25. When your car travels rapidly over a bump in the road, you suddenly feel weightless. Explain.

26. Astronauts learn to tolerate large g's by riding in a huge centrifuge. An astronaut's apparent weight depends on both the size of the centrifuge and on how quickly it turns. Explain.

27. Astronauts learn to tolerate weightlessness by riding in an airplane (nicknamed the "vomit comet") that follows an unusual trajectory. How does the pilot direct the plane in order to make its occupants feel weightless?

28. What keeps the ball pressed against the outside rim of a spinning roulette wheel? Why does the ball roll inward off the rim as the wheel slows down?

29. Slings have been used to fling stones for millennia. Suppose you are swinging a stone around in a circle at the end of cord and that it moves from right to left as it passes in front of you. If you suddenly let go of the cord as the stone passes directly in front of you, which way will the stone travel?

30. Why does swinging a wet towel around in a circle dry it?

31. Trucks have rubber anti-sail guards that extend downward behind their wheels. Each guard stops water and mud that leaves the wheel just above the road's surface from flying backward toward the car behind. Why would this water fly mostly backward (tangent to the wheel's surface) rather than downward (directly away from the center of the wheel)?

32. A rodeo rider must hold tight to a bucking bull to avoid being thrown off. The bull contorts its body so that its back accelerates downward faster than the acceleration due to gravity. Why does that movement tend to lose the rider?

33. Why does shaking your wet hands remove the water?

Problems

1. Engineers are trying to create artificial "gravity" in a ring-shaped space station by spinning it like a centrifuge. The ring is 100 m in radius. How quickly must the space station turn in order to give the astronauts inside it apparent weights equal to their real weights at the earth's surface?

2. A satellite is orbiting the earth just above its surface. The centripetal force making the satellite follow a circular trajectory is just its weight, so its centripetal acceleration is about 9.8 m/s^2 (the acceleration due to gravity near the earth's surface). If the earth's radius is about 6375 km, how fast must the satellite be moving? How long will it take for the satellite to complete one trip around the earth?

3. When you put water in a kitchen blender, it begins to travel in a 5 cm radius circle at a speed of 1 m/s. How quickly is the water accelerating?

4. In Problem **3**, how hard must the sides of the blender push inward on 0.001 kg of the spinning water?

5. To move ketchup to the mouth of a nearly empty bottle you swing the bottle in an arc with the mouth outward. The bottle moves at 5 m/s around an arc 0.5 m in radius. If the ketchup stays in the bottle, how rapidly is it accelerating inward?

6. Compare the ketchup's acceleration in Problem **5** with the acceleration due to gravity. Explain why the ketchup moves to the mouth of the bottle.

Epilogue for Chapter 2

In this chapter we looked at the physical concepts involved in three types of simple machines. In *spring scales,* we explored the relationship between the force acting on a spring and its distortion, and we examined how this distortion can be used to measure an object's weight. We also looked at oscillations—specifically, at how spring scales oscillate up and down while trying to eliminate their excess energies. Finally, we saw how acceleration complicates the measurement of weight but can be used to measure mass in a freely falling environment.

In *bouncing balls,* we examined the process of storing and releasing kinetic energy during a collision. As we saw, both the ball and the object it hits contribute to the bounce. We also studied the idea of relative velocity and noted how observers who are moving with respect to one another have different perceptions about the same events. Finally, we looked at the transfers of momentum and energy that can occur during collisions.

In *centrifuges and roller coasters,* we explored both the sensations we feel as we accelerate and why those sensations occur. We looked at uniform circular motion, noting that it involves a centrally directed acceleration caused by a centrally directed force. We also examined how that uniform circular motion gives rise to an outwardly directed apparent force. And we saw how rapid inward accelerations can fling water out of clothing in a spin-dryer and hold us in our seat during a loop-the-loop on a roller coaster.

Explanation: Swinging Water Overhead

As you swing the bucket over your head, you are pulling downward on it and causing it to accelerate downward very rapidly. The water remains in the inverted bucket because the bucket is accelerating downward faster than gravity alone can accelerate the water. Although the water is free to fall, the bucket overtakes the falling water. As a result, the water is pressed into the bottom of the bucket. The same effect occurs when you push a book downward rapidly with your open palm. Although the book is free to fall, it remains pressed into your palm because your hand is accelerating it downward faster than gravity. Finally, if you stop swinging the bucket too abruptly, the bucket's contents will not decelerate with it. Instead, they will spill or smash.

Cases

1. A spring bathroom scale is designed to report the amount of upward force it's exerting on the objects touching its surface.

a. When you first step on the scale, you usually have some downward velocity because you "land" on the scale. As it slows your motion to bring you to rest, does the scale report your correct weight, or does it report more than your weight or less than your weight?

b. If you stand motionless on one foot, rather than two feet, what fraction of your weight does the scale report?

c. If you jump upward, what does the scale report as you push yourself upward?

d. You stand motionless on two identical bathroom scales, one foot on the left scale and one foot on the right scale. What can you say about the weights that the two scales report?

e. You stand motionless on two identical bathroom scales stacked on top of each other. Each scale weighs 10 N. What weights do the two scales report?

2. You are doing arm exercises with a 10 kg dumbbell. You stand with your upper arm pointing downward and your elbow bent at a right angle so that your forearm points forward. The dumbbell is motionless in front of you.

a. What is the net torque on your lower arm?

b. Because the muscle that bends your elbow attaches to the bone very near the joint, the upward force that the muscle exerts on the bone is much more than the weight of the dumbbell. Why?

c. If you now extend your arm horizontally so that your elbow is straight, it becomes extremely difficult to support the dumbbell with the muscle that bends your elbow. Why?

***d.** How much does the barbell weigh?

3. A lazy Susan is a rotating tray that sits in the middle of a dinner table and makes it easy for everyone to help themselves to its contents. The flat circular tray turns on ball bearings when you push it to the right or left.

a. Why is it harder to spin the tray when it's heavily loaded with dishes?
b. Why does the tray continue to rotate (for a while) even after you stop pushing on it?
c. What eventually stops the tray from rotating after you stop pushing on it?
d. If you spin the tray too quickly, objects begin to fly off it. Why do they leave the tray and in what direction do they travel once they leave?

4. You are seated in a subway car, facing forward with your eyes closed. The only way that you can tell where you are going is by feeling the effects of motion on your body.

a. When the car is traveling at a steady speed on a straight section of track, what is the net force on your body?
b. Why is it very difficult to feel which way the car is going when it's traveling at a constant velocity?
c. Explain the sensations you experience as the car increases or decreases its forward speed.
d. How can you tell when the car turns to the left or the right?

5. A trampoline has an elastic surface, supported at the edges by springs or elastic bands so that it's normally flat. This surface stores energy during a bounce so that you can jump very high on it.

a. When you get on a particular trampoline and stand in its center, its surface distorts downward 10 cm. It's behaving like a spring that's distorted away from its equilibrium shape. The distorted trampoline surface is exerting a restoring force on you in which direction?
b. If someone with twice your weight climbed into the middle of the empty trampoline, how far downward would its surface distort?
c. You begin bouncing up and down on the trampoline. As you land on the trampoline during one of the bounces, its surface distorts downward 30 cm. Your weight hasn't changed so how can you make it distort so far downward?
d. While you are in the air above the trampoline, nothing is pushing on you except gravity. On both your way upward and your way downward (while in the air), do you feel weightless, your normal weight, or particularly heavy?
e. While you are touching the trampoline during a bounce, its surface is pushing upward on you. Near the bottom of the bounce, you distort the trampoline's surface downward more than 10 cm. On both your way downward and your way upward (while the surface is distorted downward more than 10 cm), do you feel weightless, your normal weight, or particularly heavy?

6. Bumper cars are a popular ride at many amusement parks. You drive about an enclosed area in a small, electrically powered vehicle. This vehicle has a large rubber bumper wrapped all the way around its exterior.

a. Half the fun of driving a bumper car is crashing into other people's cars. When you drive forward at high speed and slam into the car in front of you, you find yourself thrown forward in your car. Which way is your car accelerating?
b. When you are stopped and someone else slams into the front of your car, you find yourself thrown forward in your car. Which way is your car accelerating?
c. What would happen to the support forces between cars and to the accelerations those cars experience if the soft rubber bumpers were replaced by hard, steel bumpers?
d. Suppose that all of the cars are traveling at 10 km/h in various directions, as viewed by the people waiting in line for a turn. You crash first into a car that's heading in about the same direction as yours and feel a gentle thud. You then crash into a car that's heading toward you and experience a tremendous jolt. Explain briefly why these two collisions are different.
e. If you were to fill your car with metal bars so that its mass were 10 times that of any other bumper car, how would it affect the jolts you experience during crashes, as compared to riding in a normal car?

7. You want to weigh a horse with several bathroom scales. The bathroom scales can only handle about 1500 N (337 lb.) and the horse weighs about 4000 N (900 lb.).

a. Your friend suggests stacking four bathroom scales on top of one another and placing the horse on the upper scale. He claims that each scale will then read exactly a quarter of the horse's weight. Why isn't this right?
b. Another friend suggests placing one scale under each of the horse's four feet. She claims that each scale will then read exactly a quarter of the horse's weight. Although a much better suggestion, this isn't quite right, either. Why not?
c. You place one scale under each of the horse's four feet and correctly determine the horse's weight from the readings of the four scales. How?
d. If the horse lifts its foot off one of the scales and holds it in the air, what will happen to the readings of the other three scales?
e. If the horse now places that foot on the ground, what will happen to the readings of the other three scales?

8. Moving sidewalks are used in many airports to help people travel the long distances between gates. A moving sidewalk is essentially a conveyer belt for people. When you step on the sidewalk, its moving belt carries you along at a steady speed. Imagine a moving sidewalk that travels eastward at a speed of 1 m/s. You step on it and quickly reach that same eastward velocity. You are now looking at the world from a new inertial frame of reference.

a. As you stand on the moving sidewalk you pass a number of people who are not on it. One person is sitting in a chair but appears to be moving backward. Who is really doing the moving, or is that a matter of perspective?
***b.** From your perspective, what is the velocity of the sitting person?
***c.** To be at rest in your frame of reference, what must the sitting person do?
***d.** From the sitting person's perspective, what is your velocity?
***e.** An impatient person on the moving sidewalk walks past you at a speed of 1 m/s. From the sitting person's perspective, what is the impatient person's velocity?
***f.** If you turn around and begin walking toward the start of the sidewalk at 1 m/s, what is your velocity from the sitting person's perspective?

9. The body of a car is suspended above the four wheels on four huge coil springs. The road supports the wheels, the wheels support the springs, and the springs support the body.

***a.** When the manufacturer placed the body of the car on the springs, the springs were compressed downward 10 cm. The body weighs 10,000 N. When you and your friends climb into the car, the body descends an additional 2 cm. How much do you and your friends weigh?
***b.** While you and your friends are driving down a straight, level highway at a constant speed of 100 km/h, what is the net force on the car body and which way (if any) is it accelerating?
***c.** The car now passes over a very sudden rise in the pavement. The wheels move upward about 5 cm, instantly compressing the springs by that same amount. Which way does the car body now accelerate and how large is the acceleration compared to the acceleration due to gravity?
***d.** If the car body were attached directly to the wheels, with no springs in between, and it passed over the 5 cm rise in the pavement, which way would the car body accelerate and how large would that acceleration be relative to the acceleration due to gravity? (You must make some assumptions here; justify those assumptions.)
e. Use your answers to **c** and **d** to explain why cars have spring suspensions.

10. While you usually consider the force of the *earth's gravity* on you, you can also consider the force of *your gravity* on the earth. To do this, think of yourself as a very small planet. Since you're a planet, you can consider things as being "near your surface." But because your mass is relatively small, your gravity is extremely weak.

a. The only nearby object with a measurable weight near your surface is the earth. Explain why the earth has a substantial weight near you while other nearby objects do not.
b. What is the earth's weight on your surface?
c. Imagine placing a scale on your surface and then placing the earth on top of that scale. Will the scale accurately report the earth's weight near your surface? If not, what will it report?
d. If you suddenly push extra hard against the scale, it and the earth will begin to accelerate away from you, and you will accelerate away from them. As you push, will the scale read more, less, or the same as it did in **c**?
e. Which will accelerate more rapidly, the earth or you?
f. Suppose that you pushed hard enough that the earth and scale briefly left your surface completely (they stopped touching you). They will now begin to fall toward your surface. How will the acceleration due to gravity that they experience as they fall toward you compare to 9.8 m/s^2?
g. While the earth and scale are off your surface, will the scale read more, less, or the same as it did in **c**?

11. You're taking a step aerobics class. In front of you is a small platform that you step onto and off of during the course of the exercises. Most of the time, one foot remains stationary on the platform and you use it to lift your body up and down.

a. As you step up onto the platform, your leg does work on your body. What characteristics of you and the platform determine how much work your leg must do?
b. How much work is your leg doing on your body as it lowers you gently back down to the ground?
c. If you let yourself drop back down to the ground, rather than lowering yourself gently, you could injure the leg you land on. Why does wearing padded shoes reduce your risk of injury?
d. The platforms are all identical and are made of a sturdy plastic that acts like a stiff spring. When you step up onto your platform it distorts downward by about 4 mm. You're curious about the weight of the person to your right, so you watch the platform as that person steps onto it. It distorts downward by 8 mm. How much does that person weigh?

12. A foam rubber pillow is relatively springy.

a. When you drop your head onto the foam pillow, your head bounces. How is energy stored and returned during this collision between your head and the pillow?
b. Your head only bounces once or twice and then settles down. Explain this settling process.
c. Why is dropping your head onto a foam pillow so much more comfortable than dropping it onto a wooden pillow?
d. How is the distortion of the pillow related to the weight of your head?
e. As the pillow ages, your head settles deeper into it. What is happening to the relationship between the pillow's distortion and the restoring force it exerts on your head?

13. If you attach a string to a ball and begin swinging it around your head, the ball will travel in a circle. Let's suppose that the ball is travelling counter-clockwise, as viewed from overhead.

a. Although the ball is traveling with a constant speed, it is accelerating. Which way is it accelerating?
b. In which direction is the net force on the ball?
c. Which direction is the ball traveling at the moment it passes directly in front of you?
d. If you let go of the string at the moment the ball passes directly in front of you, which way will the ball travel? Why?
e. You feel an outward tug on the string, yet there is no outward force on the ball. The only way that the ball can tug on the string is if the string is tugging on the ball. What effect does the string's pull have on the ball if there isn't any outward force on the ball?

CHAPTER 3

FLUIDS

So far all of the everyday objects we've examined have been solids. But since gases and liquids are also important parts of the world around us—as the air we breathe, the water we swim in, and even the blood we pump through our veins—we'll now turn to objects that, unlike solids, do not have well-defined shapes. These objects are called fluids, and the study of their behavior and motion is a broad field, extending across the sciences and engineering. Fluid dynamics, or hydrodynamics as it's often called, is as important to an oil-well engineer as to an animal physiologist or a stellar astrophysicist. The tools used to analyze fluids are somewhat more complicated than for solids, because fluids themselves are more complicated: it's hard to exert a force directly on them, and, even if we could, they usually don't move as a single rigid object. In this chapter, we'll look at some of the concepts and tools needed to understand this complex behavior.

Experiment: A Cartesian Diver

One of these concepts is buoyancy. An object immersed in a fluid experiences an upward force from the fluid. This buoyant force is what lifts a helium balloon in the air and suspends a boat on the surface of water; it depends on the relative densities of the object and the fluid in which it's immersed, where density is the ratio of the object's or the fluid's mass to its volume. As we'll see in this chapter, an object that's more dense than the fluid around it sinks, while one that's less dense than the surrounding fluid floats.

To see the importance of density in determining whether an object floats or

sinks, try constructing a simple toy called a Cartesian Diver. This once-popular parlor gadget consists of a small air-filled vial floating in a sealed container of water. Normally, the air bubble inside the vial keeps it floating at the surface of the water, but whenever you squeeze the container, the vial sinks.

To make a Cartesian Diver, you'll need only a plastic soda bottle and a small vial that's open at one end. The vial can be made of almost anything—plastic, metal, or glass—as long as it's dense enough to sink in water when there's no air in it. Fill the plastic soda bottle full of water and float the vial in it upside-down; air trapped inside the vial should keep the vial afloat. Now slowly reduce the size of the air bubble inside the vial until the vial barely floats. You can make this adjustment by tipping the vial to let some of its air escape or by removing it from the bottle and pouring water into it. Once you have the vial floating only a few millimeters out of the water at the top of the soda bottle, cap the bottle and prepare to test your diver.

Before you squeeze the soda bottle, try to **predict** what will happen when you do. How will squeezing the bottle affect the air bubble and the vial's position in the water? Now squeeze the bottle hard and **observe** the results. Did they **verify** your predictions? **Measure** the air bubble's size as you squeeze the bottle. How does the bubble's size depend on how hard you squeeze the bottle? Why should the two be related? What is the relationship between the size of the air bubble and the diver's height in the water?

As you release the pressure on the bottle, the diver will float back up to the surface. Why does the sunken diver suddenly become buoyant again? By carefully squeezing the bottle, you can even make the diver hover in the middle of the bottle. Try making the diver hover while your eyes are closed. Why is achieving the hovering state so difficult? Why must you be watching the diver to make it hover?

Chapter Itinerary

We'll return to the diver at the end of the chapter. But first we'll examine two things from the world around us: (1) *balloons* and (2) *water distribution*. In *balloons*, we'll explore how the concepts of pressure and buoyancy help explain how the earth's atmosphere keeps hot air and helium balloons from falling to the ground. In *water distribution*, we'll see both how pressure propels water through plumbing and the ways in which water can contain energy.

The issues we'll be looking at crop up frequently in our everyday experiences. Pressure plays an important role in aerosol cans, steam engines, firecrackers, and even the weather; buoyancy supports ships on water and keeps oil above vinegar in a bottle of salad dressing. Just as important, these concepts will lay the groundwork for Chapter 4, where we'll examine objects in which motion affects the behavior of fluids.

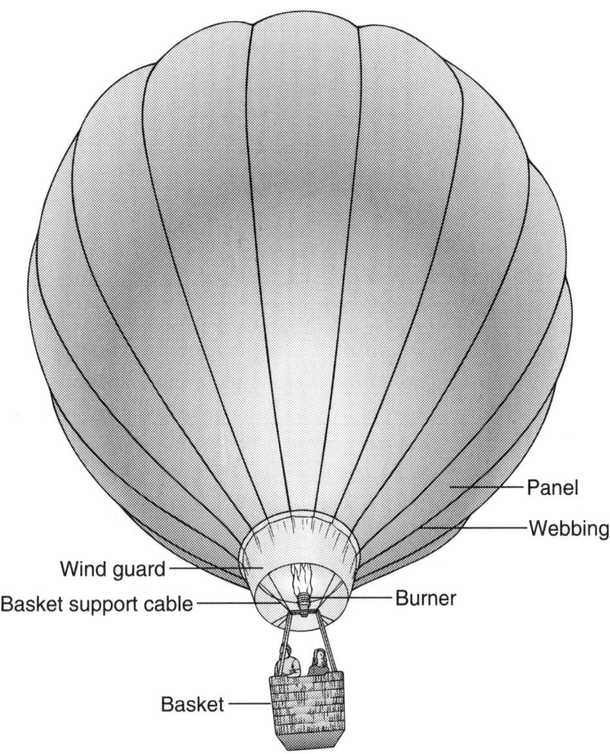

Section 3.1

Balloons

Because gravity gives every object near the earth's surface a weight proportional to its mass, objects fall when you drop them. Why then does a helium-filled balloon—which, after all, is just another object with a mass and a weight—sail upward into the sky when we drop it? Does the balloon have a negative mass and a negative weight, or are we forgetting something?

What we're forgetting is air—specifically, the layer of air that sits atop the earth's surface and is held in place by gravity. Since this air is difficult to see and moves out of our way so easily, we often forget that it's there. But it sometimes makes itself noticeable. When we ride a bicycle, we feel its forces; when we blow up a beach ball, we see that it takes up space. And when we drop a helium balloon, it's air that lifts the balloon upward.

Questions to Think About: *Since most objects fall to the ground through the atmosphere, why doesn't the atmosphere itself fall downward? Why is the air "thinner" in the mountains than at sea level? If we suck all of the air out of a plastic bag, what squeezes the bag into a thin sheet? Why does blowing air into the bag make it inflate? What happens to the bag's total mass when we fill it with air? with hot air? with helium? If we took a sturdy helium balloon to the moon, where there is no air, and then dropped it, which way would it move?*

Experiments to Do: *Pull on the string of a helium balloon to get a feel for how it behaves. If it pulls upward on your fingers, does that mean its weight (and mass) is negative? How would an object with a negative mass respond to a force? Pull on the balloon's string and convince yourself that the balloon's mass is positive. Do you think there are any objects with negative masses?*

Since the balloon's mass and weight are both positive, gravity must be pulling the balloon downward. But how can the balloon pull upward on your finger without accelerating downward? What other forces might be pushing the balloon upward? You can enhance these upward forces by partially submerging the balloon in a container of water. Where else do similar upward forces appear in everyday life?

Take the balloon for a ride in a car. Which way does the balloon move when you start suddenly? when you stop suddenly? Again, it seems as though the balloon's mass is negative. What is pushing on the balloon to make it move in this counter-intuitive way?

Air and Air Pressure

Hot-air and helium balloons are supported by the air around them. Although these balloons have positive masses and downward weights, the surrounding air pushes upward on them hard enough to balance their weights so that they float. Thus, if we want to understand balloons, we must start by understanding air.

Like the objects we've already studied, air has mass and weight. Unlike these objects, however, it has no fixed shape or size. If we collected 1 kg of air, we could shape it any way we liked. We could also make it occupy many different **volumes**. Since air is **compressible**—that is, since we can change the volume that a specific mass of it occupies—1 kg could fill a single scuba tank or a whole basketball arena.

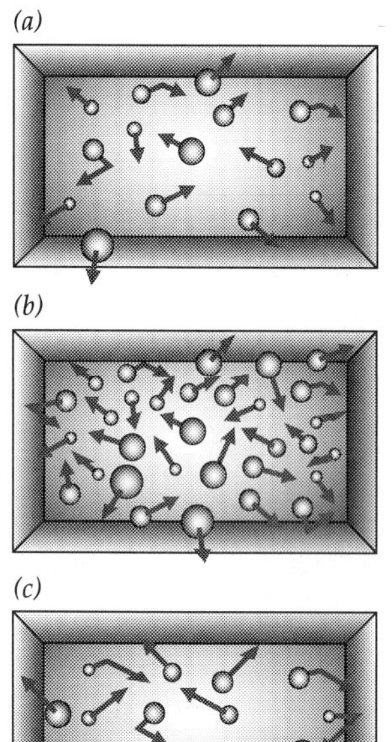

Fig. 3.1.1 - (*a*) As air molecules bounce off surfaces, they exert pressure on those surfaces; the amount of pressure depends on the air's temperature and on how densely its molecules are packed. (*b*) Packing the air molecules more densely increases the number of molecules that hit the surfaces each second. (*c*) Increasing the temperature of the air increases the speed of the molecules (shown by the arrows) so that they hit the surfaces harder and more frequently. Either change, in speed or collision frequency, increases the air's pressure.

This flexibility of size and shape originates in the microscopic nature of air. Air is a **gas**, a substance that consists of tiny, individual particles that travel around independently. These individual particles are atoms and molecules. An **atom** is the smallest portion of an element that retains all of the chemical characteristics of that element; a **molecule**, an assembly of two or more atoms, is the smallest portion of a chemical compound that retains all of the characteristics of that compound. The atoms in each molecule are held together by a **chemical bond**, a linkage formed by electromagnetic forces between the atoms.

The particles in air are extremely small, less than a millionth of a millimeter in diameter. Most are nitrogen molecules, which are formed by two nitrogen atoms, and oxygen molecules, which are formed by two oxygen atoms. But some are molecules of other compounds—carbon dioxide, water, methane, or hydrogen—while others are atoms of argon, neon, helium, krypton, or xenon. These atoms, which don't normally form molecules or make strong chemical bonds with other atoms, are called **inert gases** because of their chemical inactivity.

To see how these particles function, think of each one as a tiny baseball: each has a size, mass, and weight, and each falls because of gravity whenever it's free of other forces. But this comparison introduces an apparent problem. When you pour baseballs out of a box, they fall and quickly settle to the ground; but when you pour air molecules out of a cup, they don't seem to fall at all. If air molecules are like tiny baseballs, why don't they pile up on the earth's surface?

The answer has to do with the air's *internal kinetic energy*. This energy—the fraction of the air's sun-derived thermal energy that's contained in the motions of its molecules—keeps the air molecules moving, spinning, and away from the earth's surface. That fraction is large; since the weak forces between gas molecules allow only a small amount of the air's thermal energy to exist as *internal potential energy*, almost all of it instead takes the form of internal kinetic energy. This energy per molecule is measured as its **temperature**; the more internal kinetic energy per air molecule, the hotter that air is.

If you could observe the microscopic structure of air, you'd see countless individual molecules in frenetic **thermal motion** (Fig. 3.1.1*a*). At room temperature, these molecules travel at bullet-like speeds of roughly 500 m/s, but they collide so often that they make little progress in any particular direction. Between collisions, they travel in nearly straight-line paths because gravity doesn't have enough time to make them fall very far.

For the moment, let's ignore gravity and examine what would happen to 1 kg of air placed in a box far out in space. These air molecules whiz around inside the box, bouncing off its walls. As each molecule bounces, it exerts a force on the particular wall that it hits. Although the individual forces are tiny, the number of molecules is not, and together they produce a large average force. The size of this total force depends on the wall's **surface area**; the larger its surface area, the more average force it experiences. In order to characterize the air, however, we don't really need to know the wall's surface area; instead we can refer to the average force the air exerts on each unit of surface area, a quantity called **pressure**.

Pressure is measured in units of force-per-area. Since the SI unit of surface area is the **meter^2** (abbreviated m^2 and often referred to as the square meter), the SI unit of pressure is the **newton-per-meter^2**. This unit is also called the **pascal** (abbreviated Pa), after French mathematician and physicist Blaise Pascal. 1 Pa is a small pressure; in contrast, the air around you has a pressure of about 100,000 Pa, so that it exerts a force of about 100,000 N on a 1 m^2 surface. Since 100,000 N is about the weight of a city bus, air pressure can exert enormous forces on large surfaces.

Besides pushing on the outer walls of our hypothetical box, air also pushes on any object contained within that box. The molecules bounce off the object's surfaces, pushing them inward. As long as the object can withstand these compressional forces, the air won't greatly affect it, since the uniform air pressure ensures that the forces on all sides of the object cancel one another perfectly. A sheet of paper, for example, will experience zero net force, because the forces exerted on either side of it will cancel one another.

Air molecules also bounce off each other, so that air pressure exerts forces on air, too. A cube of air inserted into the box experiences all of the inward forces that a cube of metal would experience. The air around the cube pushes inward on it, and the cube pushes outward on the air around it. Since the net force on the cube of air is zero, the cube doesn't accelerate.

CHECK YOUR UNDERSTANDING #1: Getting a Grip on Suction

After you push a suction cup against a smooth wall, the elastic cup bends back and a small, empty space is created between the cup and the wall. What keeps the suction cup against the wall?

Pressure, Density, and Temperature

Since air pressure is produced by bouncing air molecules, it depends on two things: how often, and how hard, those molecules hit a particular region of surface. If the air molecules hit the surface more often or harder, the air pressure will increase.

To increase the rate at which air molecules hit a surface, we can pack them more tightly. If we add another 1 kg of air to our hypothetical box, we double the number of air molecules in the same volume, which doubles the rate at which they hit each surface and therefore doubles the pressure (Fig. 3.1.1*b*). Thus air's pressure is proportional to its **density**, to how much mass is contained in each unit of volume. Since the SI unit of volume is the **meter^3** (abbreviated m^3 and often referred to as the cubic-meter), the SI unit of density is the **kilogram-per-meter^3** (abbreviated kg/m^3). The air around you has a density of about

$1.25\ \mathrm{kg/m^3}$. Water, in contrast, is much denser, with a density of about $1000\ \mathrm{kg/m^3}$.

We can also increase the rate at which air molecules hit a surface by speeding them up (Fig. 3.1.1*c*). If we double the internal kinetic energy of the air in our box, we double the average kinetic energy of each molecule. Because a molecule's kinetic energy depends on the square of its speed, doubling its kinetic energy increases its speed by a factor of $\sqrt{2}$. As a result, each molecule hits the surface $\sqrt{2}$ times as often and exerts $\sqrt{2}$ times as much force when it hits. Overall, the pressure doubles. Thus air's pressure is proportional to the average kinetic energy of its molecules—to their average internal kinetic energies.

This average kinetic energy per molecule is measured by the air's temperature; the hotter the air, the larger the average kinetic energy per molecule and the higher the air's pressure. But the most convenient scale for relating the temperature of air to its pressure isn't the common **Celsius** (C) or **Fahrenheit** (F) scale; instead, it's a special **absolute temperature scale**. The SI scale of absolute temperature is the **Kelvin** scale (K). When the air's temperature is 0° K (–273.15° C or –459.67° F), it contains no internal kinetic energy at all and has no pressure; this temperature is called **absolute zero**. The Kelvin scale is identical to the Celsius scale, except that it's shifted so that 0° K is equal to –273.15° C. In addition to associating the zero of temperature with the zero of internal kinetic energy, the Kelvin scale avoids the need for negative temperatures. Room temperature is about 293° K.

Since air pressure is proportional to both the air's density and its absolute temperature, we can express the relationship among these things in the following way:

$$\text{pressure} \propto \text{density} \cdot \text{absolute temperature}. \tag{3.1.1}$$

This formula is useful, since it allows us to predict what will happen if we change the temperature or density of a specific gas, such as air. But it has its limitations; in particular, it doesn't work if we compare two different gases, such as air and helium, which differ in their chemical compositions. To make such a comparison, we'd need to add to the equation; we'll examine how later when we return to helium balloons.

Even in describing a specific gas, Eq. 3.1.1 has other shortcomings. The main problem is that real gas molecules are not completely independent of one another. If the temperature drops too low, the molecules begin to stick together to form a liquid, and Eq. 3.1.1 becomes invalid. Still, despite its limitations, this simple relationship between pressure, density, and temperature will prove useful in understanding how hot air balloons float: it will help us understand the basic structure of the earth's atmosphere, the origins of the upward force that keeps a hot air balloon aloft, and the reason why hot air rises.

CHECK YOUR UNDERSTANDING #2: Snacks that go Pop in the Night

If you remove a partially filled container of food from the refrigerator and allow it to warm to room temperature, the lid will often bow outward and may even pop off. What has happened?

The Earth's Atmosphere

Most of the mass of the earth's atmosphere is contained in a layer less than 6 km (4 miles) thick. Since the earth is 12,700 km in diameter, this layer is relatively very thin—so thin that, if the earth were the size of a basketball, it would be no thicker than a sheet of paper.

The atmosphere stays on the earth's surface because of gravity. Every air molecule, as we've seen, has a weight. Just as a baseball thrown upward eventually falls back to the ground, so the molecules of air keep returning toward the earth's surface. Although the molecules are moving too fast for gravity to affect their motions significantly over the short term, gravity works slowly to keep them relatively near the earth's surface. A molecule, like a rapidly moving baseball, may appear to travel in a straight line at first, but it will arc over and begin to fall downward eventually. Only the lightest and fastest moving particles in the atmosphere, hydrogen molecules and helium atoms, occasionally manage to escape from earth's gravity and drift off into space.

While gravity pulls the atmosphere downward, air pressure pushes the atmosphere upward. The air molecules try to fall to the earth's surface, but as they do, their density grows higher and higher. As more air molecules are compressed into the same volume, the air pressure increases. It's this air pressure that supports the atmosphere and prevents it from collapsing into a thin pile on the ground.

To understand how gravity and air pressure structure the atmosphere, think of a 1 m square column of the atmosphere as though it were a tall stack of air blocks (Fig 3.1.2). The 1 kg blocks support one another with air pressure to form a stack of about 10,000 blocks. The bottom block must support the weight of all the blocks above it and is tightly compressed, with a height of about 0.8 m, a density of about 1.25 kg/m^3, and a pressure of about 100,000 Pa. A block farther up in the stack has less weight to support and is less tightly compressed. The higher in the stack you look, the lower the density of the air and the less the air pressure. After all, blocks near the top of the stack have less weight above them to support.

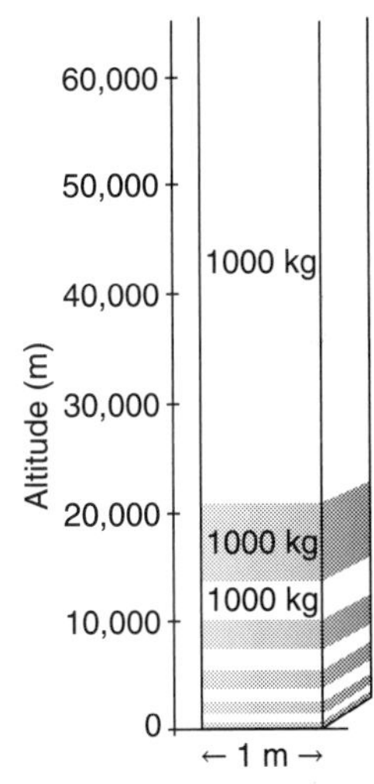

Fig. 3.1.2 - The air in a 1 m square column of atmosphere has a mass of about 10,000 kg. The bottom 1,000 kg is the most tightly compressed, because it supports the most weight above it. At higher altitudes, the air is less tightly compressed because it has less weight above it.

The atmosphere has essentially the same structure as this stack of blocks. The air near the ground supports the weight of several kilometers of air above it, giving it a density of about 1.25 kg/m^3 and a pressure of about 100,000 Pa; at higher altitudes, however, the air's density and pressure are reduced, since there is less atmosphere overhead and the air doesn't have to support as much weight. High-altitude air is thus "thinner" than low-altitude air. Whatever the altitude, the pressure of the surrounding air is referred to as **atmospheric pressure**.

With these large pressures and weights in the air around us, why are we so nearly oblivious to the presence of the atmosphere? One reason is that, although atmospheric pressure exerts large forces on everything the atmosphere touches, those forces nearly cancel one another. (Note the word "nearly": gravity spoils the perfect cancellation, as we'll soon see.) The atmospheric pressure around you now exerts an inward force of about 6000 N (about 1350 pounds) on each side of a sheet of typing paper, but the net force on the sheet of paper is almost zero. Our bodies are subjected to similar enormous inward forces that nearly cancel one another.

But why don't these huge forces cause us to compress? The reason is that we're built out of **liquids** and **solids**. In contrast to gases, which can change volume in response to pressure, liquids and solids are almost impossible to compress. If you stand on a coin, you exert an enormous pressure on its surface; but you probably won't observe any compression at all, and the coin's volume remains essentially unchanged. Because their volumes are nearly constant even when they're subjected to tremendous pressures, liquids and solids are said to be **incompressible**.

Even an object that contains hollow regions is virtually unaffected by air pressure as long as the gas pressure inside each hollow region is the same as the air pressure outside it. In such a situation, since the gas pushes out with the same force per area as the air pushes in, there is no net force on the object's shell. The skin of a lightly inflated beach ball, for example, experiences no net force because the air pressure outside the ball is equal to the air pressure inside. Although the forces on each side of the skin are very large, they cancel one another.

CHECK YOUR UNDERSTANDING #3: Mountain Travel is a Pain in the Ears

As you drive up and down in the mountains, you may feel a popping in your ears as air moves to equalize the pressures inside and outside your ear drum. What causes these pressure changes?

The Lifting Force on a Balloon: Buoyancy

So far we've examined air, air pressure, and the atmosphere. While it may seem that we've avoided dealing with balloons, these topics really are involved in keeping a hot-air or helium balloon aloft. As we've seen, the air in the earth's atmosphere is a **fluid**, a shapeless substance with mass and weight. This air has a pressure and exerts forces on the surfaces it touches; this atmospheric pressure is greatest near the ground and decreases with increasing altitude. Air pressure and its variation with altitude give rise to the principle of buoyancy, the effect that's responsible for air's ability to lift a hot-air or helium balloon.

The principle of buoyancy was first described more than two thousand years ago by the Greek mathematician Archimedes (287–222 BC). Archimedes realized that an object partially or wholly immersed in a fluid is acted upon by an upward **buoyant force** equal to the weight of the fluid it displaces. **Archimedes' principle** is actually very general and applies to objects floating or submerged in any fluid, including air, water, or oil. The buoyant force originates in the forces a fluid exerts on the surfaces of an object. We've seen that such forces can be quite large but tend to cancel one another. How then can pressure create a non-zero total force on an object, and why should that force be in the upward direction?

ARCHIMEDES' PRINCIPLE:
An object partially or wholly immersed in a fluid is acted upon by an upward buoyant force equal to the weight of the fluid it displaces.

Without gravity the forces would cancel each other perfectly, because the pressure of a stationary fluid would be uniform throughout. But gravity causes a stationary fluid's pressure to decrease with altitude. For example, when nothing is moving, the air pressure beneath an object is always higher than the air pressure above it. Thus air pushes upward on the object's bottom more strongly than it pushes downward on the object's top, and the object consequently experiences an upward force from the air; a buoyant force.

How large is the buoyant force on this object? It's equal in magnitude to the weight of the fluid the object displaces. To understand this simple result, imagine replacing the object with a similarly shaped portion of the fluid itself (Fig. 3.1.3*a*). Since the buoyant force is exerted by the surrounding fluid, not the object, it doesn't depend on what the object is made of. A balloon filled with helium will experience the same buoyant force as a similar balloon filled with water or lead or even air. So replacing the object with a similarly shaped portion of fluid will leave the buoyant force on it unchanged.

But a portion of fluid suspended in more of the same fluid doesn't accelerate anywhere; it just sits there, so the net force on it is zero. It has a downward weight, but that weight must be canceled by some upward force that can only come from the surrounding fluid. This upward force is the buoyant force, and it's

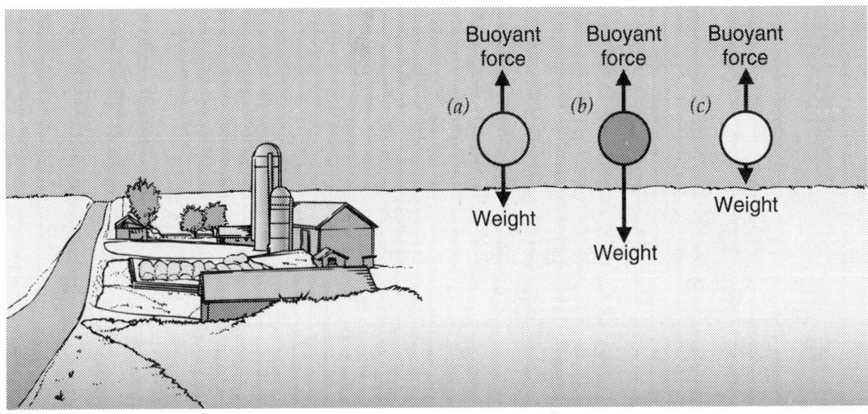

Fig. 3.1.3 - (*a*) A portion of air immersed in that same air experiences an upward buoyant force equal to its weight and doesn't accelerate. (*b*) An object that is heavier than the air it displaces sinks, while (*c*) another object that is lighter than the air it displaces floats.

always equal in magnitude to the weight of the object-shaped portion of fluid, the fluid displaced by the object.

This buoyant principle explains why one object floats upward in a fluid while another sinks downward. An object placed in a fluid experiences two forces: its downward weight and an upward buoyant force. If its weight is more than the buoyant force, the object will accelerate downward (Fig. 3.1.3*b*); if its weight is less than the buoyant force, it will accelerate upward (Fig. 3.1.3*c*). And if the two forces are equal, the object won't accelerate at all and will maintain a constant velocity.

Whether or not an object will float in a fluid can also be viewed in terms of density. An object that has an average density greater than that of the surrounding fluid sinks, while one that has a lower average density floats. A water-filled balloon, for example, will sink in air, because water and rubber are more dense than air. If you double the size of the balloon, you double both its weight and the buoyant force on it, so it still sinks. The total volume of an object is not as important as how its density compares to that of the surrounding fluid.

CHECK YOUR UNDERSTANDING #4: Why People Don't Float
If a person displaces 0.08 m^3 of air, what is the buoyant force he experiences?

Hot Air Balloons

Since air is very light, with a density of only 1.25 kg/m^3, few objects float in it. One of these rare objects is a balloon with absolutely nothing inside. Assuming that the balloon has a very thin outer shell or envelope, the whole object will weigh almost nothing and will have an average density of almost zero. It will experience a buoyant force far greater than its negligible weight, and it will float upward nicely.

Unfortunately, this empty balloon will be crushed by the air surrounding it. With an atmospheric pressure of about 100,000 Pa outside, each m^2 of surface area on the envelope will experience an inward force of 100,000 N. Because there is nothing inside to support the walls against this enormous force, the balloon will be flattened immediately. A thick, rigid envelope is needed to withstand the crushing pressure of the air outside, but then the average density of the balloon

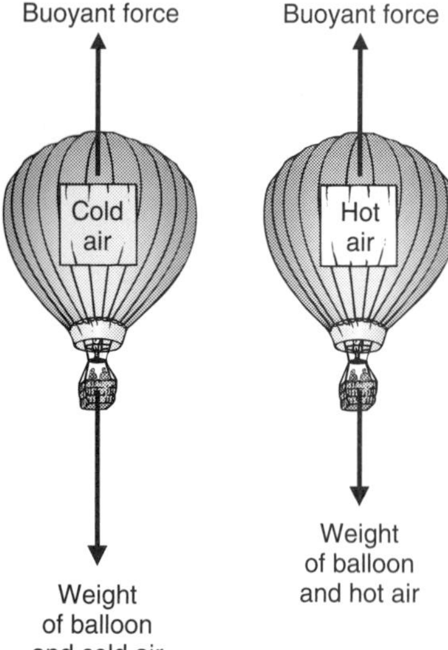

Fig. 3.1.4 - A balloon filled with hot air contains fewer air molecules and weighs less than a balloon filled with cold air. If the balloon's weight is small enough, the net force on the balloon will be in the upward direction and the balloon will accelerate upward.

would be large and it would weigh too much to float. So an empty balloon won't work.

What will work is a balloon filled with something that exerts an outward pressure on the envelope equal to the inward pressure of the surrounding air. Then each region of the envelope will experience zero net force and the balloon will not be crushed. We could fill the balloon with outside air, but that would make its average density too large. Instead, we need a gas that has the same pressure as the surrounding air but a smaller density.

One gas that has a low density at atmospheric pressure is hot air. Filling our balloon with hot air takes fewer molecules than filling it with cold air, since each hot air molecule is moving faster and contributes more to the overall pressure than does a cold air molecule. A hot-air balloon contains fewer molecules, has less mass, and weighs less than it would if it contained cold air. Now we have a practical balloon with an average density less than that of the surrounding air. The buoyant force it experiences is larger than its weight, and up it goes (Fig. 3.1.4).

Because the air pressure inside a hot-air balloon is the same as the air pressure outside the balloon, the air has no tendency to move in or out, and the balloon doesn't need to be sealed (Fig. 3.1.5). A large propane burner, located at the balloon's open end, heats the air that fills the envelope. The hotter the air in the envelope, the lower its density and the less the balloon weighs. The balloon's pilot controls the flame so that the balloon's weight is very nearly equal to the buoyant force on the balloon. If the pilot raises the air's temperature, molecules leave the envelope, the balloon's weight decreases, and the balloon rises. If the pilot allows the air to cool, molecules enter the envelope, the balloon's weight increases, and the balloon descends.

But even if the pilot heats the air very hot, the balloon will not rise upward forever. As the balloon ascends, the air becomes thinner and the pressure decreases both inside and outside the envelope. Although the balloon's weight decreases as the air thins out, the buoyant force on it decreases even more rapidly, and it becomes less effective at lifting its cargo. When the air becomes too thin to lift the balloon any higher, the balloon reaches a *flight ceiling* above which it can't rise, even if the pilot turns the flame on full blast. For each hot air temperature, then, there is a cruising altitude at which the balloon will hover. When the balloon reaches that altitude, it's in a stable equilibrium. If the balloon shifts downward for some reason, the net force on it will be upward; if it shifts upward, the net force on it will be downward.

Even when it's not enclosed in a balloon, hot air rises in cold air. A region of hotter air, surrounded by colder air, experiences an upward buoyant force greater than its weight and accelerates upward. This effect explains why hot air tends to build up near the ceiling of a room and why a ceiling fan helps to warm that room by pushing the hot air back down toward the floor.

Fig. 3.1.5 - The bottom of a hot air balloon is open so that heated air can flow in and cold air can flow out. The heated air displaces more than its weight in cold air and makes the balloon lighter.

CHECK YOUR UNDERSTANDING #5: Ballooning Weather
Can a hot air balloon lift more on a hot day or a cold day?

Helium Balloons

Although the molecules in hot and cold air are similar, there are fewer of them in each cubic-meter of hot air than in each cubic-meter of cold air. We call the number of molecules per unit of volume **particle density**, and hot air has a smaller particle density than cold air (Fig. 3.1.6). Because they contain similar molecules,

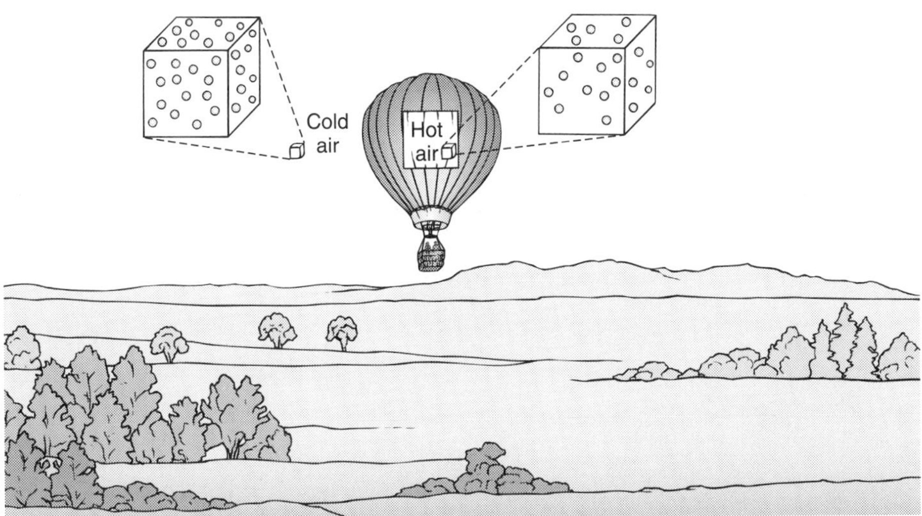

Fig. 3.1.6 - A cube of hot air contains fewer air particles than a similar cube of cold air. Since it weighs less than the cold air it displaces, the hot air inside the balloon experiences an upward buoyant force that is greater than its weight.

hot air also has a smaller density than cold air and is lifted upward by the buoyant force.

But there is another way to make one gas float in another: use a gas consisting of very light particles. Helium atoms, for example, are much lighter than air molecules. When they have equal pressures and temperatures, helium gas and air also have equal particle densities. Since each helium atom weighs on average 14% as much as the average air molecule, 1 m^3 of helium weighs only 14% as much as 1 m^3 of air. Thus a helium-filled balloon has only a fraction of the weight of the air it displaces, and the buoyant force carries it upward easily.

Why should air and helium have the same particle densities whenever their pressures and temperatures are equal? Because a gas particle's contribution to the pressure doesn't depend on its mass (or weight). At a particular temperature, each particle in a gas has the same average kinetic energy in its translational motion, regardless of its mass. Although a helium atom is much less massive than a typical air molecule, the average helium atom moves much faster and bounces more often. As a result, lighter but faster-moving helium atoms are just as effective at creating pressure as heavier but slower-moving air molecules.

Thus if you allow the helium atoms inside a balloon to spread out until the pressures and temperatures inside and outside the balloon are equal, the particle densities inside and outside the balloon will also be equal (Fig. 3.1.7). Since the helium atoms inside the balloon are lighter than the air molecules outside it, the balloon weighs less than the air it displaces, and it will be lifted upward by the buoyant force.

The pressure of a gas is proportional to the product of its particle density and its absolute temperature, as the following formula indicates:

$$\text{pressure} \propto \text{particle density} \cdot \text{absolute temperature}. \qquad (3.1.2)$$

This relationship holds, regardless of the gas's chemical composition. Our previous relationship, Eq. 3.1.1, worked only so long as the gas's composition didn't change, so that density and particle density remained proportional to one another. But now we have a relationship with a wider applicability.

Equation 3.1.2, with an associated constant of proportionality, is called the **ideal gas law**. This law relates pressure, particle density, and absolute temperature for a gas in which the particles are perfectly independent. It's also fairly ac-

❒ Helium gas is obtained as a by-product of natural gas production from underground reservoirs in the United States. While some of this gas is saved for industrial and commercial use, much is simply released into the atmosphere. The only other source of helium is the atmosphere, where helium is present at a level of 5 parts per million. Once the underground stores are consumed, helium will become a relatively rare and expensive gas.

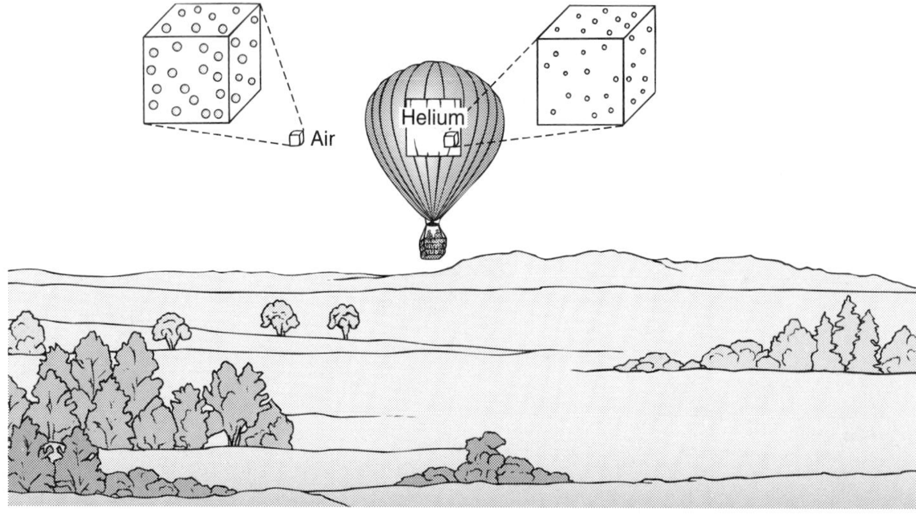

Fig. 3.1.7 - A cube of helium gas contains the same number of particles as a similar cube of air, but each helium particle weighs less than the average air particle. Since it weighs less than the air it displaces, the helium inside the balloon experiences an upward buoyant force that is greater than its weight.

Fig. 3.1.8 - Rigid airships were appropriately named because they were truly ships that floated through the air. Unfortunately, the hydrogen gas that filled most airships, making them light enough that air could lift them, is very flammable. The Hindenburg burned on May 6, 1937, while trying to land at Lakehurst, New Jersey. Because hydrogen is so buoyant in air, most of the combustion occurred above the airship and many passengers survived.

curate for real gases in which the particles do interact somewhat. All that remains to complete the ideal gas law is to replace the proportionality with an equality. The constant of proportionality is the **Boltzmann constant**, with a measured value of $1.381 \cdot 10^{-23}$ Pa·m^3/(particle·K). Using the Boltzmann constant, the ideal gas law can be written as a word equation:

$$\text{pressure} = \text{Boltzmann constant} \cdot \text{particle density} \cdot \text{absolute temperature}, \qquad (3.1.3)$$

in symbols:

$$p = k \cdot \rho_{particle} \cdot T,$$

and in everyday language:

Don't incinerate a spray can. A hot, dense gas tends to explode.

THE IDEAL GAS LAW:
The pressure of a gas is equal to the product of the Boltzmann constant times the particle density times the absolute temperature.

Helium is not the only "lighter-than-air" gas. Hydrogen gas, which is half as dense as helium, is also used to make balloons float. But don't expect hydrogen to lift twice as much weight as helium. A balloon's lifting capacity is the difference between the upward buoyant force it experiences and its downward weight. Although the gas in a hydrogen balloon weighs half that in a similar helium balloon, the balloons experience the same buoyant force. Thus the hydrogen balloon's lifting capacity is only slightly more than that of the helium balloon. Hydrogen's main advantage is that it's cheap and plentiful, while helium is scarce (see ❒). But because hydrogen is also dangerously flammable, it's avoided in situations where safety is important (Fig. 3.1.8). However, even helium-filled airships can have problems (see ❒).

❒ Even helium-filled airships were easily destroyed by bad weather. The Shenandoah, one of two U.S. airships based on the German designs, was destroyed by air turbulence on September 3, 1925, near Ava, Ohio. Crowds from a local fair immediately poured over the wreckage, collecting souvenirs.

CHECK YOUR UNDERSTANDING #6: What Not to Put in a Balloon

A carbon dioxide molecule is heavier than an average air molecule. If you pour carbon dioxide gas from a cup, which way does it flow in air, up or down?

CHECK YOUR FIGURES #1: Popped Out of the Refrigerator

When you take an air-filled plastic container out of the refrigerator, it warms from 2° C to 25° C. How much does the pressure of the air inside it change?

Elastic Balloons

The air molecules in a hot air balloon with an open bottom are constantly being exchanged with air molecules from outside the balloon; some air molecules migrate out the opening and others migrate in. This drift of molecules is called **diffusion**, and it results from the thermal motions of the individual molecules. Diffusion is what carries odors around a room, even when the air is still. The odor molecules are bouncing about rapidly and, although they collide frequently with air molecules, they slowly drift throughout the room. Diffusion is not important to a hot air balloon, because any cold air molecules that migrate into the balloon are quickly heated by the gas flame.

Fig. 3.1.9 - Helium balloons are sealed to keep helium from diffusing out and air from diffusing in.

But diffusion is a serious problem for a helium balloon. If a helium balloon is left open at the bottom, helium atoms will diffuse out and heavier air molecules will diffuse in; as a result, the balloon will slowly increase in weight until it can no longer float. To make matters worse, helium atoms move extremely rapidly, so that they diffuse very quickly. In fact, they move so quickly and are so tiny that they can actually diffuse right through certain solids—for example, the rubber envelope of a balloon.

To slow the outward diffusion of helium, helium balloons are usually sealed (Fig. 3.1.9). But a sealed balloon slowly deflates, since the helium atoms that diffuse out through the envelope's skin are not replaced by air molecules. The typical toy helium balloon, which has a rubber envelope filled with helium, takes about a week for most of its helium to diffuse out through the rubber.

The rubber balloon itself is an interesting elastic object. As it's inflated with helium, the rubber balloon is stretched away from its equilibrium shape, and it exerts restoring forces that try to return it to that equilibrium shape. As a result, each region of the balloon's surface experiences three forces: an inward force from the pressure of air outside, an outward force from the pressure of helium inside, and an inward force from the elastic skin of the balloon itself. Since each region of surface is stationary, it must be experiencing zero net force; the outward force must balance the two inward forces, so the pressure of the helium inside the balloon must be somewhat greater than the pressure of the outside air.

Sure enough, the pressure of helium inside the balloon is higher than that of outside air. This difference of pressure explains why a balloon "pops" when you stick a pin in it. The rubber tears, abruptly releasing the pressurized helium gas. The noise you hear is mostly this gas being released (Fig. 3.1.10).

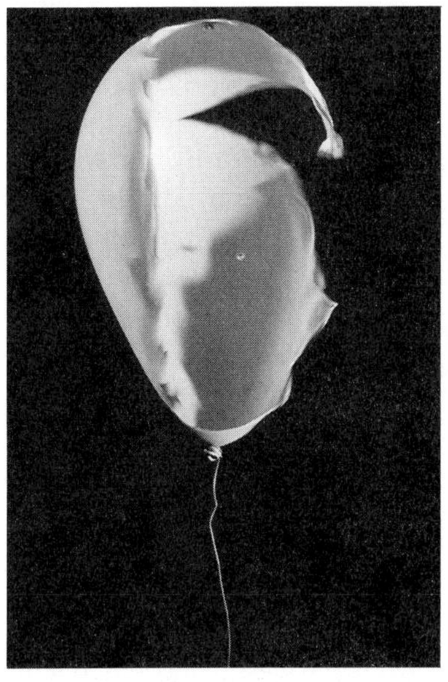

Fig. 3.1.10 - This balloon tears after being hit by a small pellet. The sudden release of high pressure gas creates a loud pop.

As a helium balloon rises upward in the atmosphere, the air pressure around it decreases. The pressure inside the balloon stretches the balloon's skin outward further until the skin exerts enough inward force to contain the helium. The balloon gets somewhat larger, so the pressure inside it diminishes. The higher the balloon travels, the less force the atmosphere exerts on the balloon's skin, and the more the skin itself must act to contain the helium. Eventually the skin can't withstand the tension and it tears, releasing the helium.

Special helium balloons are sometimes used to lift objects almost to the top of the atmosphere. These balloons are specially built so that the helium inside them can expand to much greater volume as the balloon rises upward. The balloon's envelope is so light that even a small amount of helium can lift the envelope. As the balloon rises, the atmospheric pressure around it decreases, and the helium inside expands to become less dense and to match the atmospheric pressure. The envelope slowly grows in size until it is fully inflated. By that time, the balloon is 50 km to 100 km (30 to 60 miles) above the ground.

CHECK YOUR UNDERSTANDING #7: Relieving the Pressure

When a firecracker burns, it fills a small paper shell with extremely hot, dense, and high-pressure gas. What causes the pop sound?

Summary

How Balloons Work: A hot-air balloon floats because its total weight (basket, envelope, and hot air) is less than that of the cooler air it displaces. By heating the air in the envelope with a propane flame, the pilot reduces its density. As the air warms up, fewer molecules are needed to fill the envelope, the extra molecules flow out of the envelope through the opening at its bottom, and the balloon becomes lighter. In flight, the pilot controls the balloon's weight by turning the burner on and off. When the burner is on, the air warms up, the balloon's weight decreases, and the balloon tends to accelerate upward; when the burner is off, the air cools down, the balloon's weight increases, and the balloon tends to accelerate downward. Because of the decrease in atmospheric pressure with altitude, the balloon usually adopts a cruising altitude that depends mostly on the temperature of the air in the envelope.

A helium balloon also floats because its total weight is less than the air it displaces. But its lower weight is due to the lightness of the individual helium atoms, each of which weighs much less than the air molecule it replaces. By filling a balloon with helium, the balloon's weight is dramatically reduced. Since the buoyant force exerted on the balloon by the surrounding air exceeds the balloon's weight, the balloon accelerates upward. To prevent helium from diffusing out of the envelope, the balloon is sealed at the bottom. The pilot controls the balloon's vertical motion by adding or releasing helium from the envelope. Adding helium, which is stored in tanks in the basket, increases the volume of the balloon without changing the balloon's total weight, so that the net force on the balloon becomes more upward; releasing helium decreases the volume of the balloon while only slightly decreasing its weight, which causes the net force on the balloon to become more downward.

Important Laws and Equations

1. Archimedes' Principle: An object partially or wholly immersed in a fluid is acted upon by an upward buoyant force equal to the weight of the fluid it displaces.

2. The Ideal Gas Law: The pressure of a gas is equal to the product of the Boltzmann constant times the particle density times the absolute temperature, or

$$\text{pressure} = \text{Boltzmann constant} \cdot \text{particle density} \cdot \text{absolute temperature}. \quad (3.1.3)$$

The Physics of Balloons

1. A gas is composed of countless tiny atoms and molecules that move about at great speed due to thermal energy in the gas. These particles bounce off surfaces that the gas touches and exert forces on those surfaces. The amount of force a gas exerts per unit of surface is the gas's pressure.

2. A gas's pressure is proportional to its density and its temperature, measured on a scale in which absolute zero is defined as 0°. Increasing the gas's density (compressing it) and/or increasing its temperature increases its pressure.

3. The earth's atmosphere is drawn to the earth's surface by gravity. It is supported by air pressure. Air near the ground is compressed by the weight of the air above it. The atmosphere's density and pressure are highest near sea level and decrease with increasing altitude.

4. An object immersed in a fluid experiences an upward buoyant force equal in amount to the weight of the fluid it displaces. The buoyant force is actually the result of differences in pressure between fluid near the bottom of the object and fluid near the top of the object.

5. Diffusion occurs because of the thermal motions of the atoms and molecules, which are present even in apparently stationary gas. Diffusion causes atoms and molecules to migrate slowly from one region of gas to another. Diffusion of very small particles can even occur through solids.

Check Your Understanding - Answers

1. Air pressure.

Why: Because the space between the suction cup and the wall is empty, the pressure there is zero. Pressure of the surrounding air exerts large inward forces on the outsides of both the cup and the wall, squeezing them together. As long as there is no air between them to push outward, the cup and wall remain tightly attached. Once air leaks into the suction cup, it's easily detached from the wall.

2. The pressure of the air trapped in the container increases as its temperature increases, causing the lid to bulge outward.

Why: Whenever a trapped quantity of gas changes temperature, it also changes volume or pressure or both. In this case, warming the air trapped in the container causes its pressure to increase. The unbalanced pressures inside and outside the container cause the lid to bow outward or even to pop off.

3. As you change altitude, the atmospheric pressure changes.

Why: The air inside your ears is trapped, so that its temperature, density, and pressure is normally constant. As you change altitudes, the pressure outside you ear changes and your ear drum experiences a net force. It bows inward or outward, muting the sounds you hear and causing some discomfort. The pressure imbalance is relieved during swallowing, when air can flow into or out of your ear drum.

4. About 1 N.

Why: Air exerts a buoyant force on him equal to the weight of the air he displaces. The density of air near sea level is about 1.25 kg/m^3, so 0.08 m^3 of air has a mass of about 0.1 kg (1.25 kg/m^3 times 0.08 m^3) and a weight of about 1 N. So the upward buoyant force on him is about 1 N. This buoyant force due to the air is real and reduces the weight you read when you stand on a scale by about 0.125%.

5. It can lift more on a cold day.

Why: On a cold day, the outside air is relatively dense and the buoyant force on a balloon is larger than it would be on a hot day. The hot air in the balloon will cool off more quickly on a cold day, but the balloon will be able to carry a heavier load. Airplanes also fly better on cold days.

6. It flows down.

Why: Carbon dioxide, found in carbonated beverages, dry ice, and fire extinguishers, is heavier than air because its molecules are heavier than air molecules. The carbon dioxide you pour from the cup has the same pressure and temperature as the air around it and thus the same particle density. But each carbon dioxide molecule weighs more so the carbon dioxide is the denser gas and flows down in air. This tendency to flow along the floor makes carbon dioxide very good at extinguishing low-lying flames by depriving them of oxygen.

7. When the shell can no longer withstand the enormous internal pressure, it tears and releases the high pressure gas. This sudden release of pressure creates the pop.

Why: A firecracker explodes just like a balloon. They both suddenly release the high pressure gas they contain, producing the sound we hear. However, the firecracker creates the high pressure gas very abruptly, using a chemical reaction. The higher the gas pressure and the more of it that is released when the shell tears, the louder the explosion.

Check Your Figures - Answers

1. It increases by 8.4%.

Why: To use Eq. 3.1.3 to determine the pressure change, we need temperatures measured on an absolute scale, such as the Kelvin scale. Since 0° C is about 273° K, 2° C is about 275° K and 25° C is about 298° K. We can write Eq. 3.1.3 twice, once for each temperature:

$$\text{pressure}_{298^\circ\text{ K}} = \text{Boltzmann constant} \cdot \text{particle density} \cdot 298^\circ\text{ K},$$

and:

$$\text{pressure}_{275^\circ\text{ K}} = \text{Boltzmann constant} \cdot \text{particle density} \cdot 275^\circ\text{ K}.$$

The particle density of the air in the container can't change as it warms up because its volume is fixed. Therefore, we can divide the upper equation by the lower one and cancel the Boltzmann constant and particle density on the right-hand side:

$$\frac{\text{pressure}_{298^\circ\text{ K}}}{\text{pressure}_{275^\circ\text{ K}}} = \frac{298^\circ\text{ K}}{275^\circ\text{ K}} = 1.084\ .$$

The pressure in the container thus increases by a factor of almost 1.084, or about 8.4%. This elevated pressure will cause the container to emit a "pop" sound when you open it.

Glossary

absolute temperature scale A scale for measuring temperature in which 0° corresponds to absolute zero.

absolute zero The temperature at which all thermal energy has been removed from an object or system of objects. Because it's impossible to find and remove all the thermal energy from an object, absolute zero can be approached but is not actually attainable.

Archimedes' principle The observation that an object partially or wholly immersed in a fluid is acted upon by an upward buoyant force equal to the weight of the fluid it's displacing.

atmospheric pressure The pressure of air in the earth's atmosphere. Atmospheric pressure reaches a maximum of about 100,000 Pa near sea level and diminishes with increasing altitude.

atom The smallest portion of a chemical element that retains the chemical properties of that element.

Boltzmann constant The constant of proportionality relating a gas's pressure to its particle density and temperature. It has a measured value of $1.381 \cdot 10^{-23}$ Pa·m^3/(particle·K).

buoyant force The upward force exerted by a fluid on an object immersed in that fluid. The buoyant force is actually caused by pressure from the fluid. That pressure is highest below the object so the force exerted upward on the object's bottom is greater than the force exerted downward on the object's top.

Celsius A temperature scale in which 0° is defined as the melting point of water and 100° is defined as the boiling point of water at sea level. Absolute zero is -273.15° C.

chemical bond The forces between two or more atoms that hold those atoms together to form a molecule. Chemical bonds are electromagnetic in origin, caused by the attractions and repulsions between charged particles in the atoms involved.

compressible A substance that changes density significantly as its pressure changes. A gas is compressible since its density is proportional to its pressure.

density The mass of an object divided by its volume. The SI unit of density is the kilogram-per-meter^3.

diffusion The gradual migration of atoms or molecules through a fluid, caused by their thermal motions. Diffusion also occurs in solids, although generally much more slowly than in fluids.

Fahrenheit A temperature scale in which 32° is defined as the melting point of water and 212° is defined as the boiling point of water at sea level. Absolute zero is -459.67° F.

fluid A substance that has mass but no fixed shape. Gases and liquids are both fluids. A fluid can flow to match its container. A gas expands in volume until it entirely fills its container. A liquid doesn't change volume easily and may only fill part of its container.

gas A form of matter consisting of tiny, individual particles (atoms or molecules) that travel around independently. A gas takes on the shape and volume of its container.

ideal gas law The law relating the pressure, temperature, and particle density of an ideal gas. An ideal gas is one that is composed of perfectly independent particles. The particles don't stick and bounce perfectly from one another.

incompressible A substance that doesn't change density significantly as its pressure changes. Liquids and solids are incompressible since their densities change very little as their pressures change dramatically.

inert gas A gas consisting of atoms that are chemically inactive, unable to bond permanently with other atoms or molecules. The common inert gases are helium, neon, argon, krypton, and xenon.

Kelvin The SI scale of absolute temperature scale, in which 0° is defined as absolute zero. The spacing between degrees is the same as that used in the Celsius scale.

kilogram-per-meter^3 (kg/m^3) The SI unit of density. 1 kilogram-per-meter^3 is about the density of air at 2000 m (about 1 mile) above sea level.

liquid A form of matter consisting of particles (atoms or molecules) that are touching one another but that are free to move relative to one another. A liquid has a fixed volume but takes the shape of its container.

meter^2 (m^2) The SI unit of area. 1 meter^2 is about twice the area of an opened newspaper.

meter^3 (m^3) The SI unit of volume. 1 meter^3 is about the volume of a 4-drawer file cabinet.

molecule A particle formed out of two or more atoms. A molecule is the smallest portion of a chemical compound that retains the chemical properties of that compound.

newton-per-meter^2 The SI unit of pressure (identical to the pascal).

particle density The number of particles in an object divided by its volume. The particle density of water is about $3.35 \cdot 10^{28}$ molecules-per-meter^3. The particle density of air at sea level is about $2.687 \cdot 10^{25}$ molecules-per-meter^3.

pascal (Pa) The SI unit of pressure (identical to the newton-per-meter^2). Atmospheric pressure at sea level is about 100,000 pascals. A 1-millimeter-high water droplet exerts a pressure of about 10 pascals on your hand.

pressure The average amount of force a fluid exerts on a certain region of surface area. Pressure is reported as the amount of force divided by the surface area over which that force is exerted.

solid A form of matter consisting of particles (atoms or molecules) that are touching one another and that are not free to move relative to one another. A solid has a fixed volume and shape.

surface area The extent of a two-dimensional surface bounded by a particular border. The SI unit of surface area is the meter^2.

temperature A measure of the average internal kinetic energy per particle in a material. In a gas, temperature measures the average kinetic energy of each atom or molecule.

thermal motion The random motions of individual particles in a material due to the internal or thermal energy of that material.

volume The extent of a three-dimensional region space bounded by a particular enclosure. The SI unit of volume is the meter^3.

Review Questions

1. Why does air take up space, rather than falling to the bottom of whatever it's in?

2. What else must you know before you can determine how much volume 1 kg of air occupies?

3. Where is the thermal energy in air?

4. How is pressure different from force?

5. What is the difference between density and particle density?

6. How does the Kelvin temperature scale differ from the Celsius temperature scale?

7. Why is the Kelvin temperature scale so useful for studying the behaviors of gases?

8. If the earth had twice as many air molecules as it does now, how would atmospheric pressure be affected?

9. What force supports a steel cruise ship in the middle of the ocean?

10. How can you determine whether an object is more or less dense than water?

11. If you compare hot air and cold air at atmospheric pressure, which has the greater density? Which has the greater particle density?

12. If you compare air and helium at atmospheric pressure and the same temperature, which has the greater density? Which has the greater particle density?

13. Why is the pressure inside a rubber balloon greater than the pressure outside that balloon?

Exercises

1. When a fish is floating motionless below the surface of a lake, what is the amount and direction of the force the water is exerting on it?

2. Why doesn't air fall into a well?

3. A helium-filled balloon floats in air. What will happen to an air-filled balloon in helium? Why?

4. A barometer is a device that measures air pressure which is often used to monitor the weather. How could you use a barometer to measure your altitude as you climbed in the mountains?

5. An automobile will float on water as long as it doesn't allow water to leak inside. In terms of density, why does admitting water cause the automobile to sink?

6. A log is much heavier than a stick, yet both of them float in water. Why doesn't the log's greater weight cause it to sink?

7. A car tire that is underinflated will flatten out so that more of its surface is in contact with the road. Why does the amount of tire surface on the road depend on the tire's air pressure?

8. Even when the air is not circulating in a room, the scent of an opened perfume bottle eventually reaches your nose. How do the perfume molecules move from the bottle to your nose through the still air?

9. Whales have large amounts of oil in their bodies to lower their average densities and make them buoyant even when they dive a kilometer below the surface. What would happen to a whale's buoyancy during a deep dive if it used only air to make itself buoyant? (Note: the pressure on a whale's body increases rapidly with its depth in water.)

10. If you seal a soft plastic bottle or juice container while hiking high in the mountains and then return to the valley, the container will be dented inward. What causes this compression?

11. You seal a container that is half full of hot food and put it in the refrigerator. Why is the container's lid bowed inward when you look at it later?

12. If you place a hot, wet cup upside down on a smooth counter for a few seconds, you may find it difficult to lift up again. What is holding that cup down on the counter?

13. Many grocery stores display frozen foods in bins that are open at the top. Why doesn't the warm room air enter the bins and melt the food?

14. Many jars have dimples in their lids that pop up when you open the jar. What holds the dimple down while the jar is sealed and why does it pop up when the jar is opened?

15. Some clear toys contain two colored liquids. No matter how you tilt one of those toys, one liquid remains above the other. What keeps the upper liquid above the lower liquid?

16. Why aren't there any thermometers that read temperatures down to –300° C?

17. Water settles to the bottom of a tank of gasoline. Which takes up more space: 1 kg of water or 1 kg of gasoline?

18. When the car you are riding in stops suddenly, heavy objects move toward the front of the car. Explain why a helium-filled balloon will move toward the rear of the car.

19. Oil and vinegar salad dressing settles with the oil floating on top of the water. Explain this phenomenon in terms of density.

20. You use your breath to inflate a large rubber tube and then ride down a snowy hill on it. After a few minutes in the snow the tube is underinflated. What happened to the air?

21. Use the concept of buoyancy to explain why the air over a fire rises.

22. Imagine filling a hollow metal cube, 1 m on a side, with air at the same density and temperature as the atmosphere. If you now sealed that cube and took it into empty space, how much outward force would each face of the cube experience?

23. The earth's crust is a thin layer that sits atop the molten mantle. How is the mantle supporting the crust?

24. When it rains, water molecules that were overhead fall to the surface of the earth. If nothing else changes, how does raining affect the air pressure around you?

25. A mylar balloon filled with helium gradually deflates in air because the tiny helium atoms can defuse out through the mylar but the larger air molecules can't defuse in. What will happen to a mylar balloon filled with air if you put it in a container filled with pure helium?

26. If you know the pressure inside a car's tires and how much of each tire's surface is pressing against the ground, you can make a pretty good estimate of the car's weight. Explain how you could make that estimate.

27. Some fish move extremely slowly and it's hard to tell whether they are even alive. However, if a fish is floating at a middle height in your aquarium and not at the top or bottom of the water, you can be pretty certain that it's alive. Why?

28. If you put a Ping-Pong ball in a container that's partially filled with sand and shake the container, the Ping-Pong ball will rise to the surface of the sand. What force lifts the Ping-Pong ball upward?

29. How can a boat float when the boat's anchor sinks?

30. A marshmallow is filled with air bubbles. Why does a marshmallow puff up when you toast it?

Problems

1. The particle density of standard atmospheric air at 273.15° K (0° C) is $2.687 \cdot 10^{25}$ particles/m^3. Using the ideal gas law, calculate the pressure of this air.

2. How much force is the air exerting on the front cover of this book?

3. If you fill a container with air at room temperature (300° K), seal the container, and then heat the container to 900° K, what will the pressure be inside the container?

4. An air compressor is a device that pumps air molecules into a tank. A particular air compressor adds air molecules to its tank until the particle density of the inside air is 30 times that of the outside air. If the temperature inside the tank is the same as that outside, how does the pressure inside the tank compare to the pressure outside?

5. If you seal a container of air at room temperature (20° C) and then put it in the refrigerator (2° C), how much will the pressure of the air in the container change?

6. If you submerge an 8 kg log in water and it displaces 10 kg of water, what will the net force on the log be the moment you let go of the log?

7. If your boat weighs 1200 N, how much water will it displace when it's floating motionless at the surface of a lake?

8. The density of gold is 19 times that of water. If you take a gold crown weighing 30 N and submerge it in water, what will the buoyant force on the crown be?

9. How much upward orce must you exert on the su b-merged crown in Problem **8** to keep it from accelerating?

10. How could you use your results from Problems **8** & **9** to determine whether the crown was actually gold rather than gold-plated copper? (The density of copper is 9 times that of water.)

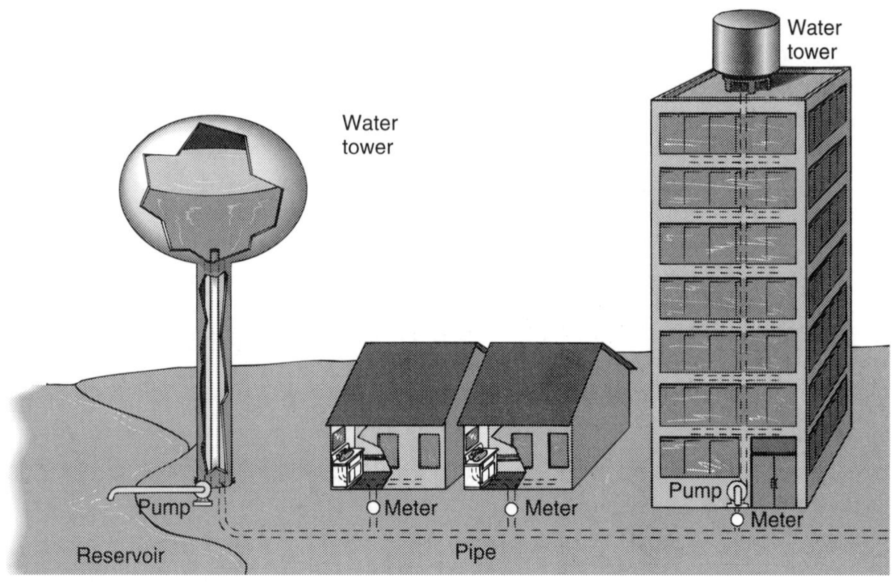

Section 3.2

Water Distribution

Now that we've explored the behavior of objects contained in fluids, let's turn to the behavior of fluids contained in objects. In this section, as we examine how plumbing distributes water, we'll see that pressure, density, and weight are just as important in plumbing as they are in ballooning. To keep things simple, we'll focus on the causes of water's motion through plumbing, leaving most of the complications associated with the motion itself for the next chapter. For example, we'll temporarily ignore drag and viscosity and the fascinating pressure changes that accompany fluid motion.

Questions to Think About: *If a water tower is only a storage device, why is it so tall? Why does a deep water well require a pump at the bottom? Why do skyscrapers have complicated plumbing systems that include reservoirs at various levels in the building? How does a siphon manage to transfer water from a high tank to a low tank? What causes water to flow up through a drinking straw to your mouth?*

Experiments to Do: *To see the effects of pressure and weight on water, try these simple experiments with a drinking straw. First, "suck" water up the straw from a glass to your mouth. Are you exerting an attractive force on the water, or is some other force pushing it upward toward your mouth? With the straw full of water, seal the top with your finger; keeping the seal tight, remove the straw from the glass. What happens to the water inside? What happens to the water when you release the seal? Now blow into one end of a straw full of water while sealing the far end with your finger. When you release the seal on the far end, what happens to the water? What forces are responsible for this effect?*

Water Pressure

All water distribution systems require two things: plumbing and water pressure. Plumbing, whether it's inside your home or under the city streets, is basically a complicated hollow object that contains water; water pressure is what induces water to begin moving through that object. Like everything else, a volume of water accelerates only in response to a net force. Without any forces, water that's initially motionless would remain motionless, and nothing would happen when you opened the faucet in your house. Because many of the forces that accelerate water result from differences in water pressure, we need to look carefully at how such pressure is created and controlled.

For the present, we'll ignore gravity. As we've seen with the atmosphere and the oceans, gravity creates **pressure gradients** in fluids—distributions of pressures that vary continuously with position. On earth, pressure decreases with altitude and increases with depth, creating a vertical pressure gradient that complicates plumbing in a hilly city or in a skyscraper. But if all of our plumbing is in a level region—for example, a single-story house in a very flat city—our job is much simpler. Without any significant changes in height, we can safely ignore gravity, since no water is supporting the weight of water above it and gravity's effects are minimal.

With gravity temporarily out of the picture, water accelerates only in response to unbalanced pressures. Just as unbalanced forces make a solid object accelerate, so unbalanced pressures make a fluid accelerate. If the water inside a pipe is exposed to a uniform pressure throughout, then each portion of the water will feel no net force and will not accelerate (Fig. 3.2.1). But if the pressure is out of balance, the water will accelerate toward the region of lower pressure.

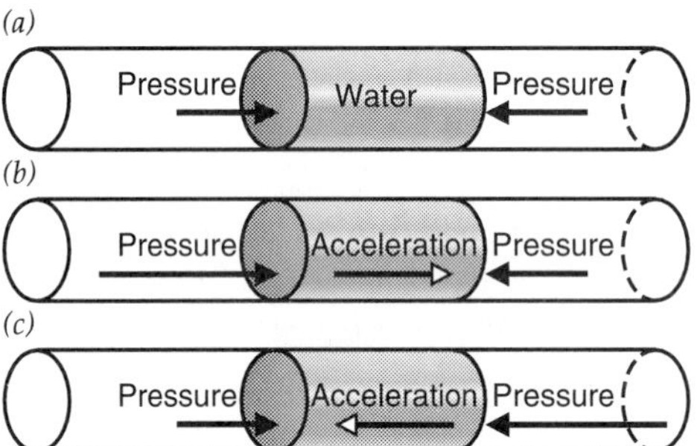

Fig. 3.2.1 - (*a*) If the water in a horizontal pipe is exposed to a uniform pressure, then it will not accelerate. (*b,c*) However, if the pressure along the pipe is not uniform, the imbalance will create a net force on each portion of the water and the water will accelerate toward the side with the lower pressure.

Whether the water is stationary or flowing, it will always accelerate away from higher pressure. This acceleration doesn't mean that the water will instantly begin moving toward lower pressure; because it has momentum, it changes velocity gradually. For example, water moving through a pipe will continue to move at constant velocity until it experiences a force. If the pressure in front of it becomes much higher than the pressure behind it, the water will accelerate backward; that is, it will decelerate.

The water flowing through plumbing in your home follows all the bends and curves because of imbalances in pressure. The water is always accelerating away from higher pressure. It starts at high pressure near the city pumping station, your well, or a water tower, and it flows toward low pressure at the open faucet in your house. Each change in its velocity during this trip is caused by unbalanced pressures.

How can you create these variations in pressure? One method is to exert pressure on the water from outside, with a device such as a pump. In this case, water passively experiences the pressure differences and responds to them by accelerating; these changes are **static** variations in pressure, ones that aren't created by water's motion. But the motion of water itself can also create variations in pressure. Called **dynamic** variations, these changes in pressure are complicated and fascinating, giving rise to such diverse effects as the lift on an airplane's wing, the curve of a curve ball, and the "hammer" of water in pipes. We'll examine these dynamic effects of fluid motion in the next chapter.

CHECK YOUR UNDERSTANDING #1: Under Pressure in the Garden

With the water faucet open and the nozzle at the end of your garden hose shut tightly, the hose is full of high-pressure water. Why doesn't this water accelerate?

Creating Water Pressure with Water Pumps

If you're going to deliver water to a house, even in a flat city, you'll need water pressure—the influence that accelerates water out of a shower head to produce the hard spray people expect. The simplest way to create this pressure is to squeeze water inside an enclosure.

When you grab a plastic bottle full of water and grip it tightly in your hands, you're creating pressure in the water inside. You push inward on the walls of the bottle, and the walls of the bottle push inward on the water; the water pushes outward on the walls of the bottle, and the walls of the bottle push outward on your hands. Newton's third law is satisfied because all forces appear in equal but oppositely directed pairs. None of the objects involved experiences any net force so nothing accelerates. But the outward force exerted by the water on the walls of the bottle stems from water pressure in the bottle. The harder you squeeze, the more water pressure is produced as the water tries to prevent you from compressing it.

If you remove the top from the bottle and then squeeze it, the water squirts out of the bottle. The water pressure inside the bottle still rises to oppose the inward forces on the water. This rise in water pressure creates a pressure imbalance between the higher pressure inside the bottle and the lower atmospheric pressure outside. Water near the mouth of the bottle accelerates toward the lower pressure outside and the bottle begins to empty. You've created a water **pump**.

Pumps appear throughout our world, taking many shapes and forms. In general, they're devices that move something from where it naturally tends to be to where it naturally tends not to be. They also normally go against the natural flow of whatever is being pumped; since water naturally flows downhill, it takes a pump to move it uphill, against its natural flow. There are pumps for just about any fluid, including water, air, and oil. Other devices "pump" solids, such as conveyer belts for gravel, corn, and coal. Still others, such as batteries and generators, pump electrons, the main microscopic participants in electricity. Some even pump thermal energy, such as refrigerators and air conditioners. In each case, the device is capable of pumping something against its natural flow. To operate against this flow, each pump requires ordered energy. Work is needed to pump water or gravel uphill or to generate electricity. Even pumping thermal energy from a cold object to a hot object requires work.

Your open plastic bottle is a primitive water pump. You do work on the sides of the bottle as you squirt water out of its mouth. This pump can only deliver a small amount of water before you have to refill it; a more practical alternative is the piston pump shown in Fig. 3.2.2. This pump also squeezes water in an enclosure, but it has two one-way valves that permit the enclosure to draw water in from a reservoir at low pressure and deliver water to an outlet at high pressure.

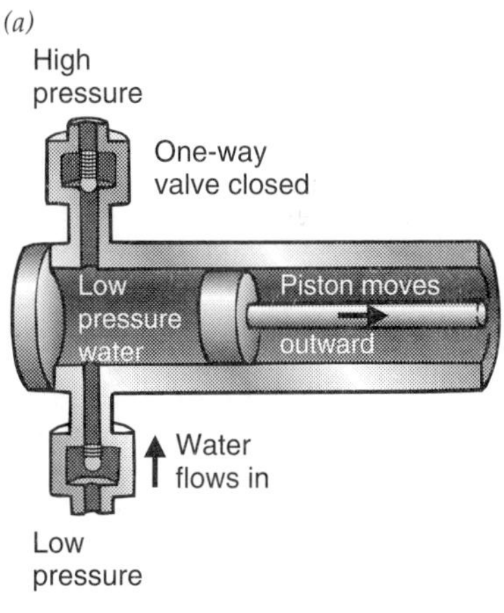

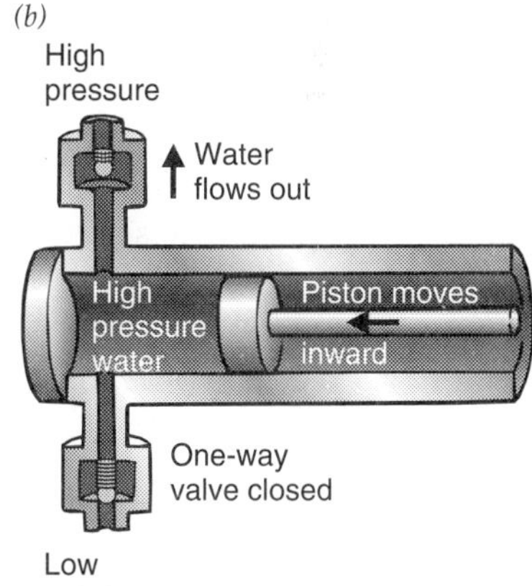

Fig. 3.2.2 - Water is pumped from a region of low pressure to a region of high pressure by a reciprocating piston pump. (*a*) As the piston is drawn outward, water flows into the cylinder from the low pressure region. (*b*) As the piston is pushed inward, the inlet one-way valve closes and water is driven out of the cylinder and into the high pressure region.

Fig. 3.2.3 - Raising the handle of a well pump draws water into its cylinder. Pushing the handle back down forces water out of the cylinder and up the pipe to the spigot. Several cycles of the pump are needed to fill the pipe with water. Once the pipe is full, water emerges from the spigot each time the pump handle is pushed down.

❒ A hand air pump draws air into its cylinder as its handle is raised. Pressurized air is driven out of the cylinder and into the air hose when the handle is pushed down. Pumps for high-pressure bicycle tires are narrow, so that a reasonable force on the pump's handle can produce a high pressure in the cylinder.

❒ A human heart contains two reciprocating pumps. The right side of the heart pressurizes blood flowing to the lungs, while the left side pressurizes blood flowing throughout the body. Its chambers contract to increase the pressure of the blood inside them. One-way valves in the heart permit low pressure blood to flow from veins into the auricles and high pressure blood to flow out of the ventricles into arteries.

The pump's enclosure is a hollow cylinder that's open at one end. A piston slides back and forth in that open end, making a water-tight seal. As you push the piston inward, water trapped in the cylinder experiences a rise in pressure. The piston spreads your force out over its surface area, creating a pressure on the water that is equal to your force divided by the piston's surface area; the water pushes back on the piston with an equal pressure, as required by Newton's third law. You can increase the water pressure either by pushing harder or by using a piston with a smaller surface area.

Two one-way valves control the flow of water into and out of the pump's cylinder. Each valve has a ball that can block the movement of fluid through a pipe. When the pressures on either side of this ball are different, the ball experiences a net force. When the pressure imbalance is in one direction, the ball moves to unblock the pipe; when the pressure imbalance is reversed, the ball moves to block the pipe.

As you push inward on the piston, the water pressure in the cylinder increases, and this high pressure affects the balls in both valves. The inlet valve closes to prevent the flow of water through the inlet, while the outlet valve opens the pipe to permit the high-pressure water to flow out of the cylinder through the outlet.

Once the cylinder is empty, it's time to refill the cylinder with water from the low pressure reservoir attached to the inlet valve. You now pull the piston out of the cylinder, creating a partial vacuum inside. Such a **partial vacuum** occurs whenever the pressure in a region falls below atmospheric pressure, and this one affects the balls in both valves. This time, the outlet valve closes to prevent the high pressure water from returning to the cylinder, and the inlet valve opens to permit water to flow into the cylinder through the inlet. Water flows into the inlet because it experiences a small pressure imbalance between the atmospheric pressure in the reservoir and the partial vacuum inside the cylinder.

You can repeat this cycle over and over, drawing in low pressure water from the reservoir and delivering high pressure water to the outlet pipe. Simple reciprocating pumps of this type are used in some hand-powered well pumps (Fig. 3.2.3) and in bicycle pumps (see ❒). Many other machines (including us, see ❒) also use reciprocating pumps, but with motors or engines pushing their pistons in and out. There are also rotary systems that rhythmically expand and contract the volume available to a liquid or gas and are also used in pumps.

CHECK YOUR UNDERSTANDING #2: Working a Water Pump
Which normally requires more work: pulling the piston of a water pump out of the cylinder or pushing it back in?

Moving Water: Pressure and Energy

The simple pump of Fig. 3.2.2 can draw low-pressure water from a pond and fill a hose with high pressure water. If the other end of the hose is open, the water will accelerate toward lower pressure at that end and will have considerable kinetic energy as it sprays out of the hose. Where does this kinetic energy come from?

The energy comes from you and the pump. As you push inward with the piston, pressurizing the water and squeezing it out through the outlet valve, you're doing work on the water because you exert an inward force on the water's surface and the water moves inward. The amount of work you do is equal to the product of the water pressure times the volume of water you pump. This simple relationship between work, pressure, and volume comes about because the inward *force* the piston exerts on the water is equal to the water pressure times the surface area of the piston, and because the inward *distance* the piston travels is equal to the volume of water being pumped divided by the surface area of the piston. *Force* times *distance* equals *work*.

How much work is involved in pumping water? Suppose you fill a backyard swimming pool with water from a nearby creek. You pump 50 m^3 (13,200 gallons) of water at a typical delivery pressure of 500,000 Pa (5 times atmospheric pressure or 5 atmospheres). The work you do pumping this water is 25,000,000 joules, the product of 50 m^3 times 500,000 Pa—or the energy needed to illuminate a 40 watt light bulb for a week. It's also the food energy in about 2 dozen doughnuts. You'll be pretty tired and hungry when you're done.

As you pump the water, it sprays out of the hose and into the swimming pool. The energy that makes the water accelerate out of the hose actually travels through the incompressible water directly from the pump to the end of the hose. The water never really stores any energy, but because the pump gives each m^3 of water a certain amount of energy as it leaves the hose, we can imagine that this energy is associated with the water and not with the pump. We create a useful fiction: pressure potential energy. Water that's under pressure and connected to a pump has a **pressure potential energy** equal to the product of the water's volume times its pressure. However, the pressure potential energy is really the energy the pump will provide when it delivers that water.

Because pressure potential energy actually comes from the pump, it vanishes as soon as you detach the water from the pump. You can't save a bottle of high-pressure water and expect it to retain this potential energy. The concept of pressure potential energy is only meaningful if the water is flowing freely so that water leaving the plumbing is immediately replaced by the pump; then whatever energy leaves the plumbing as kinetic energy in the water is put back into the plumbing by the pump. Actually, the details of the pump are not as important as the idea that any water moving through the plumbing is immediately replaced by more water with the same pressure. As long as the water is flowing steadily, you can safely use the concept of pressure potential energy, even if you don't know where the pump is or whether there even is one.

A situation in which fluid flows continuously and steadily, without starting or stopping or otherwise changing its characteristics at any particular location, is called **steady-state flow**. Pressure potential energy is most meaningful in **steady-state** flow. The water spraying steadily out of a hose, the wind blowing smoothly across your face, and the gentle current flowing in a quiet river are all cases of **steady-state** flow in fluids.

Without gravity, the energy in a certain volume of water in steady-state flow is equal to the sum of its pressure potential energy and its kinetic energy. We've already seen that the pressure potential energy is the product of the water's volume times its pressure. The water's kinetic energy is given by Eq. 1.4.7 as ½ the product of its mass times the square of its speed. The water's mass is its density times its volume. Together, this sum is:

$$\begin{aligned}\text{energy} &= \text{pressure potential energy} + \text{kinetic energy}\\ &= \text{pressure}\cdot\text{volume} + \tfrac{1}{2}\cdot\text{density}\cdot\text{volume}\cdot\text{speed}^2, \end{aligned} \tag{3.2.1}$$

If we divide both sides of this expression by the volume involved, we can obtain another useful form of this relationship:

$$\begin{aligned}\frac{\text{energy}}{\text{volume}} &= \frac{\text{pressure potential energy}}{\text{volume}} + \frac{\text{kinetic energy}}{\text{volume}}\\ &= \text{pressure} + \tfrac{1}{2}\cdot\text{density}\cdot\text{speed}^2. \end{aligned} \tag{3.2.2}$$

❐ As a professor in Basel, Daniel Bernoulli (Swiss mathematician, 1700–1782) taught not only physics, but also botany, anatomy, and physiology. He correctly proposed that the pressure a gas exerts on the walls of its container results from the countless impacts of tiny particles that make up the gas. He also derived an important relationship between the pressure, motion, and height of a fluid, Bernoulli's equation.

But each volume of water moves along with the flow and is soon replaced by a new volume of water. Because the flow is steady-state, the energy in the new volume of water must be exactly the same as in the volume that preceded it; thus the energy in each volume of water that flows along a particular path must be identical. The particular path that a volume of water takes is called a **streamline**, and the energy-per-volume of fluid along a streamline is constant:

$$\begin{aligned}\frac{\text{energy}}{\text{volume}} &= \frac{\text{pressure potential energy}}{\text{volume}} + \frac{\text{kinetic energy}}{\text{volume}} \\ &= \text{pressure} + \tfrac{1}{2}\cdot\text{density}\cdot\text{speed}^2 \\ &= \text{constant } (\textit{Along a Streamline}). \end{aligned} \quad (3.2.3)$$

Eq. 3.2.3 is called **Bernoulli's equation**, after Swiss mathematician Daniel Bernoulli ❐, whose work led to its development; although Swiss mathematician Leonhard Euler (1707–1783) actually completed it.

Because energy is conserved, an incompressible fluid such as water that's in steady-state flow can exchange pressure for speed or speed for pressure as it flows along a streamline. As water accelerates out of a hose, for example, its pressure drops but its speed increases, because it's converting pressure potential energy into kinetic energy. As the moving water sprays against the car you're washing and slows down, its pressure increases, because it's converting kinetic energy back into pressure potential energy. In both cases, the water's energy is conserved.

CHECK YOUR UNDERSTANDING #3: How Does Your Garden Grow?
Water in your garden hose has considerable pressure and arcs several meters through the air as you water your plants. What is the water pressure in the falling water at the end of the hose?

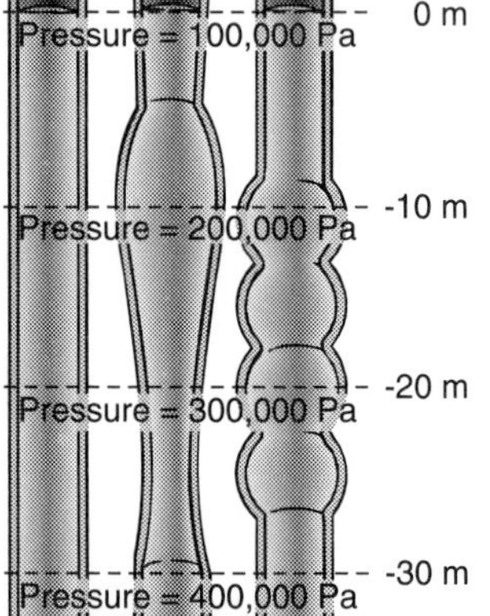

Fig. 3.2.4 - The pressure of stationary water in pipes increases with depth by about 10,000 Pa per meter of depth. *The shape of the pipes doesn't matter.* For plumbing that's open on top and connected near the bottom, as shown here, water will tend to flow until its height is uniform throughout the plumbing.

Gravity and Water Pressure

Gravity creates a pressure gradient in water: the deeper the water, the more weight there is overhead, and the greater the pressure. Since water is much denser than air, the water pressure increases very rapidly with depth. Only 10 m (33 feet) below the water's surface, the weight of the water overhead is equal to the weight of the air overhead, even though the atmosphere is several kilometers thick.

The pressure of stationary water increases by about 10,000 Pa per meter of depth, so the pressure gradient is 10,000 Pa/m. In a vertical pipe that's open at the top, the water's surface is at atmospheric pressure (about 100,000 Pa) and the pressure increases with depth at a rate of 10,000 Pa/m. This uniform pressure gradient creates an upward buoyant force on each volume of water in the pipe and prevents that water from falling.

The pressure gradient in water measures the increase in pressure with depth produced by the increase in weight over each unit of surface area. Since water's weight depends on its density times the acceleration due to gravity times its volume, water's pressure gradient is its density (1,000 kg/m^3) times the acceleration due to gravity (9.8 N/kg). This product gives a gradient of 9,800 Pa/m, or roughly 10,000 Pa/m.

If you know the pressure at the top of a vertical pipe, you can calculate the pressure at any depth you like. You simply multiply the pressure gradient by how far you are below the top of the pipe and add to that the pressure on top:

$$\text{pressure} = \text{density} \cdot \text{acceleration due to gravity} \cdot \text{depth} + \text{top pressure}. \quad (3.2.4)$$

This relationship, although obtained with water in mind, applies to any incompressible fluid regardless of its density. For example, since oil is less dense than water, the pressure in oil increases slightly less rapidly with depth, but the relationship still holds. The relationship also applies in situations where the acceleration due to gravity is different; while the water pressure on Jupiter would increase much more rapidly because the acceleration due to gravity there is larger, the equation would still allow us to calculate the pressure there at any depth. But Eq. 3.2.4 doesn't apply to compressible fluids such as air, because their densities change with pressure.

The shape of the pipe doesn't affect the relationship between pressure and depth. No matter how complicated the plumbing, the pressure of stationary water inside it increases with depth by 10,000 Pa/m (Fig 3.2.4). Whenever the plumbing is horizontal, the walls of the pipe provide the upward force that supports the water's weight, and the water pressure doesn't change along the pipe. Whenever the plumbing is vertical, water pressure provides the upward force and the water pressure increase as you move down the pipe. When the plumbing is in between horizontal and vertical, the walls and water pressure together provide the upward force.

Because the pressure depends only on height, water in a complicated system of plumbing, connected near the bottom and open to the air on top, will flow until it has filled that plumbing to a uniform height. If water in one region of the plumbing is not as high as it is elsewhere, the pressures in that region will be relatively low and water will accelerate toward it. This natural flow quickly equalizes the heights of water in the plumbing and is the origin of the expression "water seeks its level." This natural flow is often used in water delivery (see ❐).

❐ The Romans used gravity to convey water to Rome from sources up to 90 km away. A very gradual slope in the aqueducts kept the water moving in spite of frictional effects that opposed the water's progress. Poisoning from the lead pipes used in some of the aqueducts is blamed in part for the decay of the Roman Empire.

In general, water in a vertical pipe accelerates whenever the difference in pressures isn't what it should be due to gravity alone. The water maintains a constant velocity only if its pressure increases with depth by 10,000 Pa/m (Fig. 3.2.5*a*). If the pressure at the bottom of the pipe isn't high enough to support the water's weight, the water will accelerate downward; in other words, it will fall (Fig. 3.2.5*b*). If the pressure at the bottom of the pipe is too high, the water will accelerate upward (Fig. 3.2.5*c*).

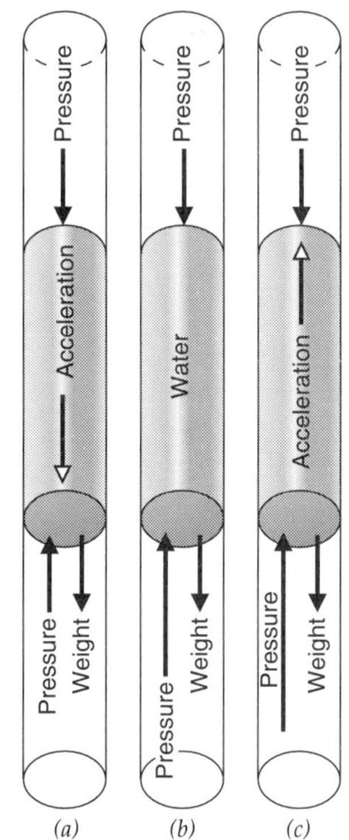

Fig. 3.2.5 - When a pipe is oriented vertically, gravity affects the motion of water in the pipe. (*a*) If the pressure in the water doesn't change with depth, the water will accelerate downward because of its weight. (*b*) If the water's pressure increases with depth by 10,000 Pa/m, the water won't accelerate. (*c*) If the water's pressure increases with depth by more than that amount, the water will accelerate upward.

The dependence of water pressure on depth has a number of important implications for water delivery. First, the water pressure at the bottom of a tall pipe is substantially higher than at the top of that same pipe. Consequently, if only a single pipe supplies water to a skyscraper, then the water pressure on the ground floor will be dangerously high when the pressure in the penthouse will be barely enough for a decent shower. Tall buildings must therefore handle water pressure very carefully; they can't supply water to every floor directly out of the same pipe.

Second, pressure in a city water main does more than simply accelerate water out of a shower head; it also supports water in the pipes of multistory buildings. Lifting water to the third floor against the downward force of gravity requires a large upward force, and that force is provided by water pressure. The higher you want to lift the water, the more water pressure you need at the bottom of the plumbing. Lifting the water also requires energy, and this energy is provided by the water pump. As you squeeze water out of the cylinder of a piston pump, you're providing the energy that lifts the water upward to the third floor.

Third, as water travels up and down the streets of a hilly city, its pressure varies with height. In the valleys the pressure can be very large, and at the tops of hills the pressure can be very small. Water mains in valleys must therefore be particularly strong to keep from bursting. The large pressure in a valley is very

Fig. 3.2.6 - The city of Los Angeles receives much of its water from Owens Valley, located 300 km to its north. The water negotiates the mountains and valleys in between, including the Sierra Nevada, driven by gravity alone. Giant pipes allow pressure to build during downhill stretches in order to push the water back uphill later on. Parts of the aqueduct, completed in 1913, support so much pressure that the steel pipe used in them has to be more than an inch thick.

useful, because it helps lift the water back uphill on the other side of the valley (Fig. 3.2.6). Nonetheless, a very hilly city must have pumping stations and other water pressure control systems located throughout in order to provide reasonable water pressures to all the buildings, regardless of their altitudes.

CHECK YOUR UNDERSTANDING #4: What's the Water Pressure?

If all of the water to a 400 m tall skyscraper were delivered from a single pipe, how much higher would the water pressure be on the ground floor than on the top floor?

Moving Water Again: Gravity

As we've seen, it takes pressure and energy to lift water to the third floor of a building. We can now expand our statement of energy conservation in fluids to include gravity. Either pressure potential energy or kinetic energy can be converted to gravitational potential energy. Whether you lift water up to the third floor with pressure in a pipe or spray it up there with kinetic energy through an open hose on the ground, it ends up with gravitational potential energy.

Water's gravitational potential energy is equal to its weight times its height (the force required to lift it times the distance it has been lifted). Its gravitational potential energy-per-volume is its weight-per-volume times its height. Since its weight-per-volume is its density times the acceleration due to gravity, water's gravitational potential energy-per-volume is its density times the acceleration due to gravity times its height.

If we include gravitational potential energy in Eq. 3.2.2 and recognize that, for fluid in steady-state flow along a streamline, the energy-per-volume is constant, we obtain a relationship that can be written as a word equation:

$$\begin{aligned}\frac{\text{energy}}{\text{volume}} &= \frac{\text{pressure potential energy}}{\text{volume}} + \frac{\text{kinetic energy}}{\text{volume}} \\ &\quad + \frac{\text{gravitational potential energy}}{\text{volume}} \\ &= \text{pressure} + \tfrac{1}{2}\cdot\text{density}\cdot\text{speed}^2 \\ &\quad + \text{density}\cdot\text{acceleration due to gravity}\cdot\text{height} \\ &= \text{constant } (\textit{Along a Streamline}),\end{aligned} \tag{3.2.5}$$

in symbols:

$$p + \tfrac{1}{2}\cdot\rho\cdot v^2 + \rho\cdot g\cdot h = \text{constant } (\textit{Along a Streamline}),$$

and in everyday language:

When a stream of water speeds up in a nozzle or flows uphill in a pipe, its pressure drops.

This is a revised version of Bernoulli's equation, one that includes gravity. It correctly describes steady-state flow in streamlines that change height.

BERNOULLI'S EQUATION:
For an incompressible fluid in steady-state flow, the sum of its pressure potential energy, its kinetic energy, and its gravitational potential energy is constant along a streamline. Eq. 3.2.5 expresses this law as a formula.

Bernoulli's equation (Eq. 3.2.5) only describes incompressible fluids because compressible fluids, such as gases, have another way to store energy. It takes work to compress a gas and increase its density, and the gas stores this work as thermal energy. Thermal energy is contained in the gas and doesn't vanish when the pump or source of pressure is removed. Popping a balloon, spraying paint from an aerosol container, and opening a recently shaken can of soda all show that compressed and pressurized gas contains additional energy.

This additional energy is often useful; for example, compressed air is frequently used to pressurize the water from private wells. But it prevents us from using Bernoulli's equation to describe airflow in situations where the air undergoes large changes in pressure. Fortunately, there are many important situations in which the pressure changes are so small that only a little energy goes into compressing the air. In these cases, air is essentially incompressible and Bernoulli's equation works pretty well. In the next chapter, we'll examine how Bernoulli's equation, when used with air, can help explain how airplanes fly.

CHECK YOUR UNDERSTANDING #5: Water Power

A hydroelectric power plant extracts energy from water that has descended from an elevated reservoir in a pipe. In the reservoir, this energy takes the form of gravitational potential energy. Just before the power plant, what form does the energy take?

CHECK YOUR FIGURES #1: Up or Out

If the water pressure at the entrance to a building is 1,000,000 Pa, how high can the water rise up inside the building and how fast will it flow out of a faucet right at the entrance? (A liter of water inside a pipe has a mass of 1 kg.)

Drinking Straws

While water pressure in an open-topped pipe never goes below atmospheric pressure, the water pressure in a pipe that's sealed at the top can go all the way to zero. With no air in the pipe, a **vacuum**—that is, a region of essentially empty space—would exist above the water's surface and the pressure there would be zero. While this situation may seem remote and unlikely, it's part of two common water delivery mechanisms: drinking straws and siphons.

Water in an open drinking straw inserted in a glass of water rises to the level of the water around it. But when you seal the straw's top with your mouth and remove air from within the straw, the water accelerates up the straw toward your mouth. By removing air from the straw, you reduce the pressure at the top of the water to less than atmospheric pressure; water then accelerates upward toward the lower pressure. Although you might imagine that you are somehow attracting the water, you are actually permitting water pressure in the glass to push water toward your mouth.

Water at the surface of the glass of water is at atmospheric pressure; it can support the weight of either the air overhead or a 10 m column of water, but not both. By removing the air from above the water, you permit the water pressure to support a column of water instead of a column of air. But the tallest column of water that atmospheric pressure can support is 10 m. If you try to "suck" water up an 11 m straw, it won't reach your mouth. Even if you remove all the air from the straw, a feat that you can't actually do with your mouth, atmospheric pressure will only lift the water 10 m up the straw.

Similarly, a water pump at the surface can't "suck" water out of a well if the water level is more than 10 m below the pump. That is why most well pumps are actually inserted into the well hole itself. Gravity-produced water pressure pushes water into the submerged pump, and the pump then provides the enormous pressure needed to lift the water up and out of the well.

CHECK YOUR UNDERSTANDING #6: Straw Games at the Restaurant

If you fill a drinking straw with water and seal the top with your finger, what will happen as you lift the straw out of the water glass?

Siphons

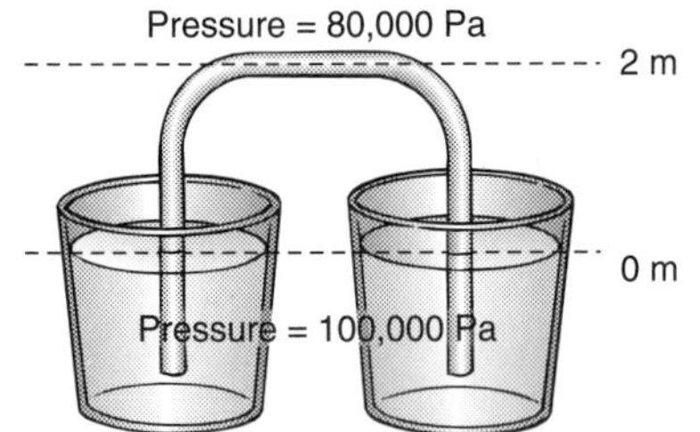

Fig. 3.2.7 - The two open containers of water are connected by a siphon. This U-shaped tube permits water to flow until its level is equal in both containers. The closed top of the siphon keeps air out and permits the water pressure in the siphon to drop below atmospheric pressure.

When water tries to fall out of a pipe that's sealed at the top, a region of low density and pressure is created above the water. This low pressure can be used to draw water up a second pipe. Figure 3.2.7 shows a U-shaped tube called a *siphon* that connects two containers of water. Because air can't directly enter the siphon, the pressure inside it can and does drop below atmospheric pressure. The siphon permits water to flow freely between the two containers, just as water would if the connection between containers were located below the water levels. Water "seeks its level" in the two open containers. If the level of water in one container is lower than that in the other container, water will flow through the siphon until the levels are equal.

By allowing water to seek its level, siphons help water flow downhill to reduce its potential energy. But while siphons are useful for moving water around between containers, they can't lift water higher than atmospheric pressure will push it. A siphon only works if its highest point is less than 10 m above the water level in the uppermost container. Above that height, an empty region will appear in the tube and the water won't flow between containers.

CHECK YOUR UNDERSTANDING #7: Down the Drain

You have a small swimming pool on the elevated deck behind your house. How can you drain this pool using only a hose, without making a hole in the bottom of the pool?

Water Towers

One way to maintain large water pressures in plumbing is to have a tall column of water connected to the pipes. That's why cities, communities, and even individual buildings have water towers (Fig. 3.2.8). A water tower is built at a relatively high site within the region it serves. A pump fills the water tower with water and then gravity maintains a constant high pressure throughout the plumbing that connects to it. The water is at atmospheric pressure at the top of the water tower, but the pressure is much higher at the bottom; at the base of a 50 m high water tower, for example, the pressure is about 600,000 Pa or 6 times atmospheric pressure.

Since the water leaving a water tower situated on a hill continues to flow downhill toward people's houses, the water pressure can become extremely high. It must be reduced by regulating devices near low-lying houses so as not to burst their pipes or hot water heaters. These regulating devices use friction-like effects to convert the water's pressure potential energy into thermal energy. We'll examine those effects in the next chapter.

❒ The streams of water from some toy water guns are accelerated by high-pressure air. A small hand-pump compresses air inside the water container, a process requiring work. When the water is released, the pressure imbalance causes it to accelerate and leave the gun as a narrow spray. The stored energy becomes kinetic energy in the water. As the water leaves the container, the air gradually expands and its stored energy decreases.

Using a water tower to maintain pressure in the water mains eliminates the need to run a water pump continuously. Since a real water pump, because of frictional effects, turns ordered energy into thermal energy even if there is no water flowing, the pump is most energy efficient when it moves as much water as possible. A city saves ordered energy and money by allowing the pump to fill the water tower at night, when electricity is relatively inexpensive and the pump can work at its full capacity and efficiency. The water level in the tower then drops slowly during the day as water is consumed by the houses. If the water level in the tower drops too far, reducing the pressure in the water mains, the pump turns on and refills the water tower.

In addition to providing a fairly steady pressure in the water mains, a water tower stores energy efficiently and can deliver that energy in a short time. When water is drawn out of the water tower, its gravitational potential energy at the top becomes pressure potential energy at the bottom. The water tower replaces a pump, supplying a steady flow of water at an almost constant high pressure. But unlike a pump, the water tower can supply this high pressure water at an enormous rate. As long as the water level doesn't drop too far, the high pressure water will keep flowing.

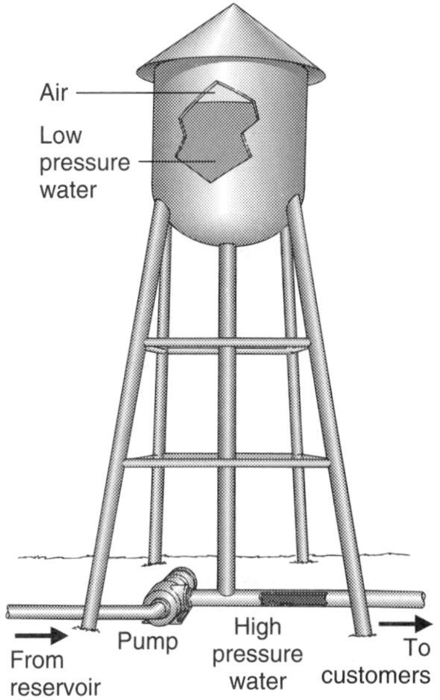

Fig. 3.2.8 - A water tower uses the weight of water to create a large water pressure near the ground. The higher the tower, the greater the pressure near the ground. The water tower is able to maintain the water pressure passively and doesn't require constant pumping. Even during periods of peak water consumption, it maintains a fairly steady pressure. When the water level in the water tower drops below a certain set point, a pump refills the tower.

If the water pressure is maintained by a pump, then the pump supplies the pressure potential energy. But the pump has a limited power capacity, typically rated in horsepower, and can transfer no more than a certain amount of energy to the water each second. If you draw water slowly from the pump, the pump will be able to deliver that water at high pressure. But if you open too many faucets at once, they will exceed the power capacity of the pump, and the pump will then deliver a large amount of water but at a lower pressure. Since people tend to use water in daily schedules, taking showers in the morning and washing dishes in the evening, a city served only by pumps can experience substantial drops in water pressure. Attaching a water tower to the water system nearly eliminates these pressure drops. (Compressed air can also maintain water pressure, see ❐.)

But skyscrapers present a serious problem for water delivery systems. It would take enormous water pressure to drive water all the way to the top of a skyscraper, and no city system can provide that much pressure. Instead, a skyscraper has pumps that lift the water up. It begins with water near the bottom of the building and pumps it to the top in several steps. The skyscraper actually has water towers inside it to maintain steady water pressures for its occupants (Fig. 3.2.9). These internal tanks are also essential for fire safety in the event of electric power loss. A fire truck would have difficulty pumping water to the top of the skyscraper, but it can make good use of water already stored at the top.

Fig. 3.2.9 - Many buildings in New York City have water towers on their roofs. These towers maintain water pressure in the plumbing and help in fire fighting.

CHECK YOUR UNDERSTANDING #8: Up, Up, and Away

A fire truck pumps water out of a hydrant and boosts its pressure so that it can spray high into the air. If the nozzle converts most of the water's pressure potential energy into kinetic energy, how much pressure is needed to spray the water 100 m into the air?

Summary

How Water Distribution Works: Water distribution begins with a reservoir of low-pressure water, such as a river, lake, or underground stream. A pump transfers water from this low-pressure source to high-pressure plumbing. In a city water system, the pump may lift water to the top of a water tower, where the energy given to the water is stored as gravitational potential energy. The water tower provides a steady pressure in the water system. A private water well may use an equivalent energy storage system (compressed air) to maintain a steady pressure in the local plumbing.

Along a level path, water accelerates toward regions of lower pressure, such as open hoses or shower heads. Pressure imbalances allow the water to negotiate bends in the pipes on route to its destination. During its travels, the water may rise or fall in height; as it does, its pressure changes, the pressure decreasing as the water moves upward and increasing as the water moves downward. In low-lying regions, where the water pressure may be too high to use directly, a pressure regulator may have to be added to the plumbing. In high-lying regions, where the water pressure may be too low to be practical, an additional pump may have to be employed to boost its pressure.

Water can be "sucked" (actually pushed) up a pipe or straw by reducing the pressure inside the pipe. The water is lifted by a pressure imbalance between the atmospheric pressure outside the pipe and the lower pressure inside. Atmospheric pressure at sea level can't lift the water more than about 10 m, even if there is no air pressure inside the pipe. In a pipe that is sealed from direct contact with the atmosphere, the water pressure can actually drop below atmospheric pressure so that the pipe behaves as a siphon.

Important Laws and Equations

1. Bernoulli's Equation: For an incompressible fluid in steady-state flow, the sum of its pressure potential energy, its kinetic energy, and its gravitational potential energy is constant along a streamline, or

$$\begin{aligned}\frac{\text{energy}}{\text{volume}} &= \frac{\text{pressure potential energy}}{\text{volume}} + \frac{\text{kinetic energy}}{\text{volume}} \\ &\quad + \frac{\text{gravitational potential energy}}{\text{volume}} \\ &= \text{pressure} + \tfrac{1}{2}\cdot\text{density}\cdot\text{speed}^2 \\ &\quad + \text{density}\cdot\text{acceleration due to gravity}\cdot\text{height} \\ &= \text{constant } (\textit{along a streamline}). \end{aligned} \tag{3.2.5}$$

The Physics of Water Distribution

1. Without drag or viscosity, fluids accelerate in response to imbalances in pressure. In the absence of other forces, each volume of fluid accelerates in the direction of lowest pressure.

2. A pump is a device that can move something against its natural direction of flow. To move against the natural flow, the pump requires an input of ordered energy.

3. A fluid pump moves fluid from a region of low pressure to a region of high pressure. The pump does work on the fluid in moving it, increasing the fluid's energy, and requires an input of ordered energy to function.

4. The work that a pump does on an incompressible fluid can be viewed as potential energy in the fluid, as long as the fluid is in a state of steady flow. An incompressible fluid can't store pressure energy internally because it can't be compressed. The pressure potential energy actually comes directly from the source of fluid pressure.

5. Pressure potential energy can make the fluid accelerate, becoming kinetic energy, or it can lift the fluid against gravity, becoming gravitational potential energy. In general, pressure potential energy, gravitational potential energy, and kinetic energy can be transformed into one another in a flowing incompressible fluid.

6. For an incompressible fluid in steady-state flow, the sum of pressure potential energy, kinetic energy, and gravitational potential energy in any volume of fluid is constant because energy is conserved.

7. A compressible fluid, such as a gas, stores work as thermal energy during compression and releases that energy during expansion. Conservation of energy is more complicated in a compressible fluid than in an incompressible fluid.

Check Your Understanding - Answers

1. The water pressure (force on a unit of surface area) inside the hose is uniform throughout, so the water experiences no net force and doesn't accelerate.

Why: In the absence of gravity, fluids accelerate only when they experience pressure imbalances. Since water throughout the hose is at the same pressure, there is no pressure imbalance and no acceleration. When you open the nozzle, the pressure at that end of the hose drops and the water accelerates toward it.

2. Pushing it back in usually requires more work.

Why: When you pull the piston out of the cylinder, you are moving air out of the way and creating a partial vacuum inside the cylinder. This action requires a modest amount of work because the air doesn't push terribly hard on the back of the piston and pressure from the water flowing into the cylinder assists you. But as you push the piston back into the cylinder, you are pressurizing the water. Depending on the pressure of water in the outlet hose, the water in the cylinder may exert a very large force on the piston. In that case, you must do a great deal of work on the water as you push the piston inward and drive the water out of the cylinder.

3. Atmospheric pressure.

Why: As the water accelerates out of the hose, its pressure drops. It is converting pressure potential energy into kinetic energy. The water pressure drops until it reaches the pressure of the surrounding air, atmospheric pressure.

4. About 4,000,000 Pa (40 atmospheres) higher.

Why: The weight of water inside the pipe would create an enormous excess pressure near the bottom of the building. Water spraying from an opened faucet on the first floor at this enormous pressure could accelerate to 319 km/h (200 mph), as it does in some high-pressure jet washing and cutting machines.

5. Pressure potential energy (and some kinetic energy).

Why: As the water descends inside the pipe, its gravitational potential energy is converted into pressure potential energy. The water reaching the power plant is under enormous pressure and it's this pressure that exerts the forces needed to turn the turbines that run the generators. Work is required to turn the turbines, so the water gives up much of its energy in the power plant. This energy leaves the power plant via the electric power lines.

6. The water will remain in the vertical straw even if the straw's bottom is open to the air.

Why: The water begins to descend but as it does, it creates an empty region near your finger. The pressure inside the straw near your finger drops below atmospheric pressure. The open bottom of the straw is still at atmospheric pressure so the water in the straw experiences a pressure imbalance, with the lower pressure on top. This pressure imbalance creates an upward force that prevents the water from descending further. The longer the straw, the larger the pressure imbalance needed to support the increasing weight of the water. If the straw were about 10 m long, the pressure near your finger would reach zero.

7. Use the hose as a siphon. Fill the hose with water, insert one end in the pool, and let the other end hang down below the deck. Water will flow out of the pool, through the hose, and onto the ground below.

Why: Water in each side of the siphon will try to fall, creating low pressure above it. Water in the tallest side of the hose will weigh the most and will create the lowest pressure above it. It will actually "suck" water up from the shortest side. As water descends off the side of the deck, it will create a low pressure in the hose that allows atmospheric pressure to push water into the hose from the pool. The flow will continue until air enters the hose and stops the siphon effect.

8. About 1,000,000 Pa, in addition to normal atmospheric pressure.

Why: Supporting the weight of a column of water requires 10,000 Pa per meter of height. Whether you lift the water inside a pipe or spray it up with a nozzle, you must exert 1,000,000 Pa of pressure on it to lift it 100 m upward. In the pipe, the pressure potential energy becomes gravitational potential energy directly. In the spray, it temporarily becomes kinetic energy first.

Check Your Figures - Answers

1. It can rise up about 100 m or emerge from the faucet at about 45 m/s (101 mph).

Why: As the water flows through the pipe (a streamline), its pressure potential energy can become gravitational potential energy or kinetic energy. At the start, the water's energy is all pressure potential energy so, from Eq. 3.2.5, the water's energy per volume is 1,000,000 Pa. If the water flows up the pipe, that energy will become gravitational potential energy. We can rearrange Eq. 3.2.5 to find the height it can reach:

$$\text{height} = \frac{\text{energy}}{\text{volume}} \cdot \frac{1}{\text{density} \cdot \text{acceleration due to gravity}}$$

$$\text{height} = 1{,}000{,}000\ \text{Pa} \cdot \frac{1}{1000\ \text{kg/m}^3 \cdot 9.8\,\text{m/s}^2} = 102\ \text{m}.$$

If the water flows out the faucet, that energy will become kinetic energy. We can also rearrange Eq. 3.2.5 to find the speed it will obtain:

$$\text{speed} = \sqrt{\frac{\text{energy}}{\text{volume}} \cdot \frac{2}{\text{density}}}$$

$$= \sqrt{\frac{2{,}000{,}000\ \text{Pa}}{1000\ \text{kg/m}^3}} = 45\ \text{m/s}.$$

Glossary

Bernoulli's equation An equation relating the total energy of an incompressible fluid in steady-state flow to the sum of its pressure potential energy, kinetic energy, and gravitational potential energy.

dynamic In motion or associated with motion. A dynamic system is a system in motion. Dynamic pressure in a fluid is associated with that fluid's motion.

partial vacuum A region of space containing a gas at a less than atmospheric pressure.

pressure gradient A distribution of pressures that varies continuously with position.

pressure potential energy The product of a fluid's volume times its pressure. However, this energy isn't really stored in the water. Instead, it's energy that's provided by a pump (or other source) when the fluid is delivered.

pump A device that can move a substance from a region where it naturally tends to be to a region where it naturally tends not to be. A pump can move its substance against the natural flow. A pump requires an input of ordered energy to operate when it moves its substance against the natural flow.

static Stationary or independent of motion. A static system is a system in which nothing is moving. Static pressure in a fluid is caused by effects other than the fluid's motion.

steady-state flow A situation in a fluid where the characteristics of the fluid at any fixed point in space don't change with time. The flow at that point neither starts nor stops nor changes its direction of travel during its period of steady-state flow.

streamline The path followed by a particular portion of a flowing fluid.

vacuum A region of essentially empty space.

Review Questions

1. What causes water in a horizontal pipe to accelerate?

2. When water in a horizontal pipe is all at the same pressure, it doesn't accelerate. But when water in a vertical pipe is all at the same pressure, it falls downward. Why is there a difference?

3. Can water in a horizontal pipe ever move toward a region of higher pressure? Can water in a horizontal pipe ever accelerate toward a region of higher pressure? What is the difference between those two questions?

4. Why does it take work to transfer water from a region of lower pressure to a region of higher pressure? What becomes of this work?

5. Think of three situations in which a fluid speeds up (usually as it passes through a narrowing) and its pressure potential energy becomes kinetic energy.

6. Think of three situations in which a fluid slows down (usually as it enters a large container or hits a surface) and its kinetic energy becomes pressure potential energy.

7. Why doesn't Bernoulli's equation allow you to relate the energy of air inside a passing airplane with the energy of air outside that airplane?

8. Why does the pressure decrease as water flows up a pipe to a third floor apartment? Why does the pressure increase as water flows down a pipe to the basement?

9. What causes water to rise up a drinking straw? Why can't you drink from a long straw if you're more than 10 m above the drink?

10. How does the height of the local water tower affect the water pressure in the adjacent houses?

11. Why must skyscrapers have their own water towers?

Exercises

1. There is some water in the bottom of your boat. This water is a meter or two below sea level. Why can't you use a siphon to transfer it to the surrounding sea?

2. When you open a large can of juice with a can punch, you do best to make two holes—one on each side of the top. If you try to pour from a can with only one hole, the juice will come out in spurts. What is the purpose of the second hole?

3. The attic bathroom of a house on a hill has low water pressure. What is the most likely reason for this problem?

4. Ice tea is often dispensed from a large jug with a faucet near the bottom. Why does the speed of tea flowing out of the faucet decrease as the jug empties?

5. A novelty drinking straw is a plastic tube that has been wound up into a complicated shape. Overall, the straw is 0.5 m tall. How many meters of tube can the straw have and it still be possible to drink through the straw?

6. How does pushing on the plunger of a syringe cause medicine to flow into a patient through a hollow hypodermic needle?

7. Wasp and hornet sprays proudly advertise just how far they can send insecticide. How does the pressure inside the spray can affect that distance and why is the direction of the spray important (vertical vs. horizontal)?

8. Why must the pressure inside a whistle teakettle exceed atmospheric pressure before the whistle can begin to make noise?

9. You can inflate a plastic bag by holding it up so that it catches the wind. Use Bernoulli's equation to explain this effect.

10. Waterproof watches have a maximum depth to which they can safely be taken while swimming. Why?

11. Why must tall dams be so much thicker at their bases than at their tops?

12. As the bowl of a toilet empties, the water in it flows through a short pipe that's shaped like an inverted "U". The water flows up one side of this pipe, over the top, and down the other side. How is it lifted up the first side of the pipe?

13. If you fill a balloon full of water, the water has pressure potential energy. This pressure potential energy becomes kinetic energy when you let the water spray out of the opening. But pressure potential energy isn't actually stored in water, so where is it contained in the water balloon?

14. If you stamp your foot on a foil ketchup packet, the ketchup will spray out of the packet's side at a terrific speed. Describe the origin of this speed in terms of kinetic energy and pressure potential energy.

15. Pressure potential energy isn't actually stored in the ketchup of Exercise **14**. What is really providing the energy that propels the ketchup out of the packet?

16. Each time you breathe in, air accelerates toward your mouth and lungs. How does the pressure in your lungs compare with that in the surrounding air as you breath in?

17. When you breathe out, air flows rapidly out of your mouth. Where is that air's kinetic energy coming from?

18. Some small animals, such as bats and opossums, hang upside down when they sleep. If you did that, the increased blood pressure in your head would give you a headache. Why don't small animals experience a similar large increase in blood pressure in their heads?

19. The jellied ink of a ball point pen used to be contained in a sealed tube, along with a certain amount of air. The only way out of that tube was around the writing ball. When one of those pens would warm up, typically inside a shirt pocket, the ink would flow out of the pen and make a mess. What caused the ink to begin moving?

20. A blacksmith must pump air into his fire to heat iron hot enough to shape. He uses a bellows—a mechanical pump resembling an accordion. When he lifts the top of the bellows, its volume increases and air enters through a one-way valve. When he pushes down on the top of the bellows, its volume decreases and the air blows out into the fire through a narrow tube. Show that the blacksmith is doing work on the bellows as he blows air out of it. What becomes of this work?

21. How is the amount of force that the blacksmith applies to the bellows in Exercise **20** related to the speed of the air as it enters the fire?

22. When you turn on a drinking fountain, a thin stream of water rises from its nozzle in a low arc so that you can drink. The drinking fountain contains a pressure regulating device that limits the water pressure in its chilled water storage tank to just a few thousand pascals above atmospheric pressure. Why is that low pressure important and what would happen if the pressure inside the tank were much higher?

23. When you stand in a pool with water up to your neck, you find that it's somewhat more difficult to breathe than when you're out of the water. Why?

24. When two approaching cars pass one another, the driver of each car observes that the air in the other person's car is moving extremely fast. Naively thinking that "high speed air is low pressure air," each driver concludes that the air in the other person's car must have a much lower pressure than the air in their own car. That turns out to be nonsense. The cars have their windows closed and the air inside each car is at atmospheric pressure. How can that air be at atmospheric pressure if it's moving so fast?

25. When water moves from a full bucket to a nearly empty bucket through a siphon, it rises above the buckets and moves through the siphon's pipe toward the empty bucket. The water thus has more gravitational potential energy and more kinetic energy as it moves through the siphon than it had when it was in the full bucket. Where does this extra energy come from?

26. When someone pulls a fire alarm in a skyscraper, pumps increase the water pressure in the section of the building nearest that alarm box. How does this pressure change assist firefighters who must battle the blaze?

Problems

1. Your town is installing a fountain in the main square. If the water is to rise 25 m (82 feet) above the fountain, how much pressure must the water have as it moves slowly toward the nozzle that sprays it up into the air?

2. Rather than putting a pump in the fountain (see Problem **2**), the town engineer puts a water storage tank in one of the nearby high-rise office buildings. How high up in that building should the tank be for its water to rise to 25 m when spraying out of the fountain? (Neglect friction.)

3. Firefighters are battling a fire in a tall apartment building. The water pressure in the adjacent fire hydrant is about 500,000 Pa above atmospheric pressure. The firefighters take their hose up the stairwell inside the building. What is the highest level at which the firefighters can expect water to flow out of their hose without additional pressure?

4. Firefighters on the ground in Problem **3** begin to spray water upward from their hoses. The water enters the hose traveling slowly at 500,000 Pa above atmospheric pressure. How fast will the water be traveling when it leaves the nozzle at atmospheric pressure?

5. How high will the water from Problem **4** rise if the firefighters send it straight up?

6. To boost the water pressure, the firefighters in Problem **3** send it sequentially through pumps in two fire engines. Each pump boosts the pressure by 500,000 Pa, for a total of 1,500,000 Pa above atmospheric pressure. How high will this water rise in hoses carried up the stairs inside the building?

7. To clean the outside of your house you rent a small high pressure water sprayer. The sprayer's pump delivers slow-moving water at a pressure of 10,000,000 Pa (about 150 atmospheres). How fast can this water move if all of its pressure potential energy becomes kinetic energy as it flows through the nozzle of the sprayer?

8. When the water from the sprayer in Problem **7** hits the side of your house, it slows to a stop. If it hasn't lost any energy since leaving the sprayer, what is the pressure of the water at the moment it stops completely?

9. To dive far below the surface of the water a submarine must be able to withstand enormous pressures. At a depth of 300 m, what pressure does water exert on the submarine's hull?

Epilogue for Chapter 3

In this chapter we investigated some of the basic concepts associated with fluids. In *balloons*, we explored the concept of pressure and the way that air pressure structures and supports the earth's atmosphere. We saw that increased air pressure beneath an object creates an upward buoyant force on that object, and that this buoyant force can support objects with densities less than that of the surrounding air. We also looked at two types of balloons that are less dense than air: hot air balloons and helium balloons.

In *water distribution*, we examined how water pressure causes water to accelerate through plumbing, from high pressure to low pressure; we then focused on ways to produce water pressure, either with pumps or with gravity. By studying the forms of energy in water, we were led to Bernoulli's equation, which describes the conversion of a fluid's energy between pressure potential energy, kinetic energy, and gravitational potential energy. Although the most dramatic applications of Bernoulli's equation are ahead of us, we've already used it to understand the changes in pressure and speed that accompany water's movement up and down in pipes and jets.

Explanation: A Cartesian Diver

The diver floats because its average density—that is, the mass of the vial and its contents divided by the volume of space those two components occupy—is less than the density of water. Since the upward buoyant force on the diver exceeds its downward weight, the diver floats upward toward the top of the bottle. When the diver begins to stick out of the water, it displaces less of that water and the buoyant force it experiences decreases. Eventually it experiences zero net force and floats without accelerating at the water's surface.

When you squeeze the soda bottle, you increase the pressure inside it. Because water is incompressible, its density doesn't change as the pressure goes up. However, the air bubble inside the vial is compressed and takes up less space inside the vial. Water flows into the vial and increases the average density of the vial and its contents. When the average density of the diver has increased until it exceeds the density of water, the diver sinks.

To keep the diver hovering in the water, you must adjust the water pressure until the diver's average density is exactly that of water. This adjustment is impossible to make without looking at the diver. Even the slightest over- or under-pressure will cause the diver to drift slowly down or up.

Cases

1. A large ship floats motionless at the surface of a still sea, supported by the buoyant force. Its average density, including the air it contains, is less than that of the sea water.

a. What is the net force on the ship?
b. The ship is displacing both water and air. If the ship were to move upward a few centimeters, what would happen to average density of the fluid the ship displaces?
c. If the ship were to move up a few centimeters, what would happen to the buoyant force on the ship?
d. If the ship were to move up a few centimeters, what would happen to the net force on the ship?
e. Show that the floating ship is in a stable equilibrium with respect to up and down motion.
f. Why does the ship settle deeper into the water when another passenger climbs aboard?
g. The ship's captain can estimate how much cargo it's carrying by how deep the ship rides in the water. How does that technique work?
h. Does the air exert a buoyant force on the ship? If so, why doesn't the ship float in air?

2. Like all bony fish, a bass has a gas-filled sac or air bladder that allows it to float motionless below the surface of the water.

a. What forces act on the motionless bass and what is the net force it experiences?

b. What is the bass's average density?
c. Salt water is more dense than fresh water. What are the relative sizes of the air bladders in salt and fresh water bass?
d. A shark is a cartilaginous fish that lacks an air bladder. If it can't move, it will sink. Why?

3. A professional house washer uses a powerful pump to create the very high pressure water needed to remove dirt and grime from a house's surface.

a. The same amount of water leaves the pump through the sprayer hose as enters it from a water faucet. In what way is the water's energy different in the sprayer hose than it was in the garden hose that delivered the water to the pump?
b. As the water flows out of the sprayer nozzle, what happens to its energy?
c. As the water flies through the air toward the house, it's at atmospheric pressure. Why then does the water pressure in the sprayer hose affect the water's ability to clean the house's surface?
d. Some gadget stores sell special "pressure-boosting" nozzles that are supposed to clean better than normal garden hose nozzles. These devices claim to increase the water pressure and are supposedly powered by the water itself. Why is this claim impossible?

4. A scuba diver swims below the surface of the water, breathing air from the steel tanks of an aqualung.

a. It's easiest for her to maintain a constant depth if the buoyant force on her and her equipment exactly balances their weight. Her wet suit floats so she wears a heavy weight belt to compensate. How should her average density compare with that of the water around her?
***b.** When she is 20 m below the surface of the water, what is the pressure pushing inward on her chest?
c. When she breathes, she expects air to flow into her lungs. She can change the pressure in her lungs slightly, using the muscles in her chest and diaphragm. How must the pressure in her mouth compare with that in her lungs in order for air to begin flowing into her lungs?
d. If instead of taking compressed air with her, she were to try to breathe surface air through a straw, the air entering her mouth would be at atmospheric pressure. What would happen when she tried to breathe?
e. The special pressure regulator of an aqualung delivers air to her mouth at exactly the same pressure as that of the surrounding water. She can breathe this air easily. However, the deeper she dives, the faster she consumes air molecules from the tanks and the sooner she must return to the surface. Why does she exhaust her tanks faster by going deeper?

5. Blimps are compact versions of the great airships of the early 20th century. They are filled with helium and are propelled forward by fans. But a blimp has more to it than meets the eye. Inside the cigar-shaped exterior skin are several flexible containers called ballonets. The blimp's buoyancy and orientation are controlled by pumping air into or out of these ballonets.

a. The exterior skin is rigid, so the blimp's total volume doesn't change. Why does pumping air into the ballonets cause the blimp to become less buoyant?
b. Why does pumping air from the forward ballonet to the rearward ballonet cause the blimp to tip its nose upward?
c. When the blimp flies into warmer air, what should it do with its ballonets in order to maintain a constant altitude?
d. The blimp is propelled forward by huge fans which push the air backward. Why does this procedure help the blimp to move forward?

6. A canoe is a particularly simple type of boat. Most modern canoes are made of aluminum or plastic. Suppose that you and a friend are paddling an aluminum canoe in a lake.

a. Aluminum is more dense than water, so how can an aluminum canoe weigh less than the water it displaces?
b. The canoe has gas-filled foam inserts in its ends to ensure that it will float even when filled with water. The foam is closed-pore and doesn't soak up water like a sponge. Why is it important that the foam not soak up water when the canoe tips over?
c. How does pulling the paddles backward through the water cause the boat to accelerate forward?
d. If both of you paddle the canoe on the right hand side, it will tend to turn toward the left. Like most effectively free objects, the canoe naturally pivots about its center of mass. What causes the canoe to begin turning toward the left? (Draw a picture.)

7. You are washing a car in front of your apartment. You have a garden hose attached to a water faucet and a bucket with a sponge in it.

a. When the water faucet is on, water flows through the hose and out the nozzle. What causes water to accelerate to high speed as it leaves the nozzle?
b. As it arcs through the air, the stream of water is at atmospheric pressure. But when you accidentally hit the bucket with the stream of water, the bucket accelerates away from you and falls off the car. Evidently, the water pressure increases when it touches the bucket. Why does the water pressure increase?
c. You decide that perhaps the water would clean more effectively if it were moving faster, so you let the water spray upward and fall back down on the car. But the falling water is traveling no faster than it was when it first left the nozzle. Why not?
d. The sponge sinks to the bottom of the bucket of water. What can you say about the sponge's average density?

8. Most sink drains have traps. A trap is an S-shaped pipe that carries water from the drain first downward, then upward, then downward to the sewer. Water fills the bottom of the trap and prevents unpleasant smells from working their way out of the drain.

a. As water pours down the drain, it begins to flow up the middle portion of the trap. What is lifting that water against the force of gravity?
b. If you drop a gold ring down the drain, it will usually remain in the first curve of the trap. Why won't the ring travel up the middle portion of the trap with the water?
c. If the trap fills up with debris, it may prevent the sink from draining. One way to clear the trap, without disassembling it, is to create a very high water pressure in the drain. A plumber's helper resembles a large suction cup and can be used to squeeze the water in the drain and raise its pressure. How does this high pressure unblock the trap?
d. Even without a trap, there is no overall movement of air in the drain pipe. What allows odor molecules to travel from the sewer to the drain?

9. A submarine hovers below the surface of the water, supported by the buoyant force. Although it also pushes on the water to propel itself forward and to adjust its height, the submarine's most important depth control is its average density.

a. If a submarine is hovering motionless, 50 m below the surface of the water, what is the net force it's experiencing?
b. What is the motionless submarine's average density?
c. A submarine has ballast tanks that control its buoyancy. To start descending, the submarine floods these tanks with outside sea water. How does this flooding affect the submarine's average density and the net force on the submarine?
d. To ascend, the submarine uses compressed air to purge the sea water from its ballast tanks. How does eliminating this sea water affect the submarine's average density and the net force on the submarine?
e. To control the submarine's tilt, water is pumped between tanks in the front and back of the vessel. Why doesn't this transfer of water affect the submarine's buoyancy?

10. When a patient receives an injection, fluid passes from a cylinder with a plunger in it, through a narrow, hollow needle, and into the patient's arm.

a. Before the nurse begins pushing the plunger into the cylinder, how do the pressures inside the cylinder and inside the needle compare to one another?
b. As the nurse squeezes fluid through the needle, how do the pressures inside the cylinder and inside the needle compare to one another?
c. Why will fluid travel more rapidly through the needle if the nurse squeezes harder on the plunger?
d. As the fluid flows out of the needle and slows down in the patient's tissue, what happens to its pressure?

11. When a person donates blood, the nurse inserts a narrow, hollow needle into the donor's vein. Blood passes through the needle, through a wider plastic tube, and collects in a plastic bag. The tube and bag are at the same height as the needle.

a. The donor's blood pressure is higher than atmospheric pressure. Why is that important to the flow of blood through the needle?
b. As blood flows through the needle and the tube, and into the bag, where is its pressure highest?
c. Where is its pressure lowest?
d. What factors determine how long it will take to collect a unit of blood?
e. Would lowering the collecting bag below the height of the needle change the time it would take to collect a unit of blood? Why?
f. How would raising the collecting bag above the height of the needle affect the blood collecting process?
g. Use your answer to **f** to explain how blood or other fluids can be administered to patients using gravity alone.

12. Some banks and businesses use pneumatic tubes to move documents or cash from one place to another. These items travel inside a cylindrical canister that fits snugly inside a tube and is propelled through that tube by air pressure. While friction is important, we'll neglect it for now.

a. To start the motionless cylinder moving forward through a level tube, how must the air pressure in front of the cylinder compare to the air pressure behind it?
b. To keep the cylinder moving forward through the level tube at constant velocity, how must the air pressure in front of the cylinder compare to the air pressure behind it?
c. The cylinder begins to rise up a vertical section of tube at constant velocity. How must the air pressure above the cylinder compare to the air pressure below it?
d. The cylinder reaches another level section and begins to slow down for arrival. How must the air pressure in front of the cylinder compare to the air pressure behind it as it decelerates?
e. How could you propel the cylinder through the tube if all you had was a device that produced low pressure (like a vacuum cleaner)?
f. How could you propel the cylinder through the tube if all you had was a device that produced high pressure (like a bicycle pump)?

13. Because the earth's crust floats on the surface of the molten mantle, supporting tall mountains is problematic.

a. How does the density of the earth's crust compare to the density of the mantle (more, less, the same)?
b. If you began to build a mountain by piling rocks over one region of the earth's crust, how would the crust respond?
c. What does your answer to **b** tell you about the thickness of the crust in the vicinity of a tall mountain?
d. The earth's crust is never more than about 80 km (50 miles) thick. Why does that small crust thickness impose a limit on the heights of the earth's mountains?

14. A traditional water cooler has a large bottle of water turned upside down so that its neck is submerged in a small chilled water reservoir at the top of the water cooler. The water level in the small reservoir remains just above the neck of the bottle. If you open the valve to let water out of the reservoir, bubbles of air rise up into the water bottle and the level of water in the bottle goes down.

a. Since there is no true seal between the neck of the bottle and the reservoir, what holds the water up inside the water bottle?
b. When you open the valve at the bottom of the water cooler, the water flows out into your glass. What provides the force needed to make the water accelerate?
c. Whenever the water level in the reservoir drops below the lip of the inverted water bottle, air bubbles enter the bottle and some water flows out. What force lifts the air bubbles upward inside the water bottle?
d. Once in a while the delivery person drops a water bottle and makes a tiny crack in its bottom (which is on top when the bottle is upside down on the water cooler). Although water can't pass through the crack, air can. What will happen if you put the cracked bottle on the water cooler, and why will this happen?

CHAPTER 4

FLUIDS AND MOTION

Fluids are fascinating when they move. Stationary water and air may be essential to life, but they are also fairly simple; only their pressures vary from place to place, and even these are determined primarily by gravity. But rushing rivers or gusts of wind, with their wonderful variety of simple and complicated behaviors, are much more interesting. And the motion of fluids isn't just interesting; it's also important, since our world is filled with objects and machines that work in whole or part because of the behaviors of moving fluids. In this chapter, we will look at several situations in which fluid motion contributes to the way things work.

Experiment: A Vortex Cannon

Fluids are real entities, with an existence that doesn't depend on the solids that move through them. This idea is easy to grasp in reference to water, since we can see that water doesn't wait for a boat to sail by before it moves in interesting ways. But air is harder to visualize in this way, since we seldom see it by itself, apart from its effects on buildings or airplanes or our skin.

To begin seeing air as a real entity, and to demonstrate the rich possibilities of motion available to air itself, build a vortex cannon—a device that sends rings of air sailing across a room. While a serious vortex cannon is best constructed from a large drum or crate, you can make a reasonably effective one from an empty cardboard cereal box.

Seal the rectangular box on all edges with tape, and then cut a circular hole 5 centimeters (2 inches) in diameter in the center of one face. To use your cannon, just tap hard on the other face of the box. Rings of air will leap out of the hole and sail across the room at about 5 m/s.

You can watch these rings move by looking for their effects on the objects they meet. They can easily blow out candles or rustle light window drapes across a small room. If you have a friend blow some rings at you, you'll feel exactly where they hit on your face or shirt.

To actually see these rings, fill the cereal box with smoke, mist, or dust. Try to **predict** what the rings will look like when they emerge from the hole. Will the smoke in each ring be stationary, or will it be swirling about the ring in some manner? How will the ring's size and speed of travel depend on the size of the hole it emerges from? Will it depend on how hard you tap the cereal box? Will the ring's size and speed change in flight?

Now tap the box and **observe** the smoke rings. What motions do you see? **Measure** the size and speed of the rings as they fly. Were your predictions verified? Change the hole size and see how this change affects the rings and their motion. As you can see, the air can execute some very complicated movements all by itself. In this chapter, we will examine how moving air, water, and other fluids affect our everyday life.

Chapter Itinerary

In particular, we'll explore (1) *water faucets,* (2) *vacuum cleaners,* (3) *balls and air,* and (4) *airplanes.* In *water faucets,* we'll look at how water's pressure and speed vary as it flows through a faucet into your sink. In *vacuum cleaners,* we'll see how moving air sweeps dust along with it and how passing that air through a narrow opening enhances its cleaning ability. In *balls and air,* we'll investigate the effects of air on the motions of balls and other flying toys. Lastly, in *airplanes,* we'll study the ways in which moving air supports an airplane in flight.

This chapter continues to develop the concept of energy conservation along a streamline that was introduced in Chapter 3. It also brings up several new types of forces that are present when fluids move past one another or past solid objects. These ideas are present not only in the topics of this chapter, but in many other commonplace activities, from washing windows with a hose to pumping water with a windmill.

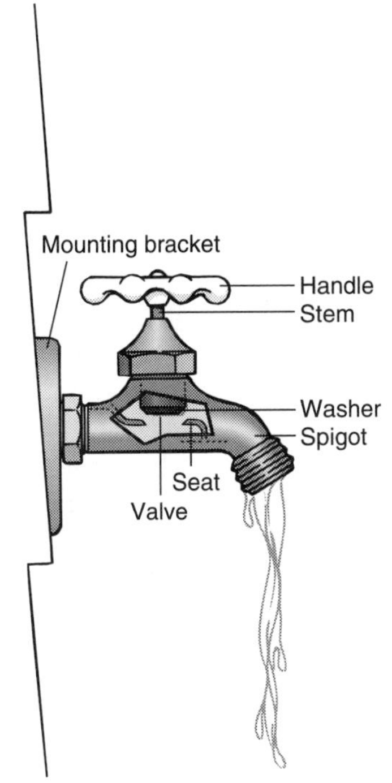

Section 4.1

Water Faucets

Water faucets are a common part of everyday life; without them, you wouldn't be able to control the flow of water into your home. Standard faucets have changed little since the advent of indoor plumbing more than a century ago. While most current models feature one or more recent innovations—for example, a device that directly mixes hot and cold water—they're all based on the same concept: they control the water by changing the sizes of the openings through which it flows.

Questions to Think About: *Where is the water when the faucet is closed, and why does it begin flowing from the outlet pipe when you open the faucet? What determines the rate at which water flows from the outlet? If honey ran through your water pipes instead of water, how would that affect the flow when you opened the faucet? Why does water sometimes make noise as it passes through the faucet? Why do the pipes sometimes clank when you abruptly close a faucet?*

Experiments to Do: *Find a faucet that opens and closes when you rotate its handle. Except in a few modern faucets, the handle does more than simply rotate as you turn it. How does the handle move when you turn the faucet on? when you turn it off? Look below the faucet and try to determine how the water enters and exits the faucet.*

With the water still on, use your fingers to stop it from leaving the outlet pipe. Can you block the flow of water completely? Now turn the handle to close the faucet. The water stops easily and the faucet doesn't leak.

Finally, listen for noises that accompany water flow through the faucet. When is the faucet quiet and when is it noisy? Record the kinds of sounds you hear when you stop the water flow slowly, at a moderate pace, and quickly. Is the sound different if the water is flowing at full force when you shut it off?

Water Flow from a Faucet: Effects of Viscosity

We brought water to your home in the previous chapter, but we haven't yet learned how to shut it off. Before your furniture starts floating around the room, it's time to consider water faucets. The operation of a water faucet is simple: as you turn its handle, you open a hole through which water can flow, and water flows through the outlet pipe and into the sink. The more you turn the handle, the larger the hole gets and the more water flows out of the faucet. But a hole is just a hole, so why should its size affect the flow of water? There are two parts to the answer.

1. *The speed with which water can flow through a hole is limited by the water's energy*: the sum of its pressure potential energy, its kinetic energy, and its gravitational potential energy. Water is an incompressible fluid and the flow here is steady-state, so Bernoulli's equation applies. As it passes through the hole, the water converts most of its pressure potential energy into kinetic energy. It moves through the narrow hole as quickly as it can, but its energy and peak speed are limited by the water pressure in the plumbing. Because of this limited speed, only so much water can pass through the hole each second. Think of a two-lane road with a fixed speed limit: it can only carry so much traffic. Widening the hole, however, also widens the flow of water and allows more water to pass—in the same way that widening a two-lane road to four lanes will roughly double its traffic capacity.

2. *The flow of water through a faucet is constrained by viscous forces.* **Viscous forces** appear whenever one layer of fluid tries to slide across another layer of fluid. These forces oppose this relative motion and produce friction-like effects within the fluid. You observe these effects when you pour honey out of a jar. Since honey has enormous difficulty sliding past itself—that is, since it's a "thick" or *viscous fluid*—the honey at the jar's surface is stuck there and remains stationary. But even honey that is far from the walls can't move easily; it experiences viscous forces as it tries to move relative to nearby honey. In this way, viscous forces try to keep all of the honey moving with the same velocity. Since the honey at the walls is fixed in place, viscous forces tend to prevent all of the honey from moving.

❒ Your car's engine is protected by motor oil with a carefully chosen viscosity. If that oil were too thin, it would flow out from between surfaces and wouldn't keep them from rubbing against one another. If that oil were too thick, the engine would waste power moving its parts through the oil. Years ago, you had to change your motor oil for the season. Thick 40 Weight motor oil was used in summer because hot weather made it thinner; thin 10 Weight oil was used in winter because cold weather made it thicker. But a modern, multigrade oil maintains a nearly constant viscosity over a wide range of temperatures and need not be changed with the seasons. This oil contains tiny molecular chains that ball up when cold but straighten out when hot. These chains thicken hot oil so that 10W-40 oil resembles 10 Weight oil in winter and 40 Weight oil in summer.

Water isn't as thick as honey, so it's less resistant to relative motion. The measure of this resistance to relative motion within a fluid is called **viscosity**, and water's viscosity is less than that of honey. But these two liquids are hardly extremes of viscosity. Superfluid liquid helium has no viscosity, while the viscosity of softened glass is extraordinarily high. Other fluids, including air, water, honey, and shampoo, fall somewhere in between (Table 4.1.1). However, not all liquids can be characterized by viscosity (see ❒).

Hot water is actually less viscous than cold water and thus flows more easily. In fact, most liquids become less viscous when hot. This decrease in viscosity with temperature reflects the molecular origins of viscous forces. The molecules in a liquid stick to one another, forming weak chemical bonds that require energy to break. In a hot liquid, the molecules have more thermal energy, so they

❒ Not all liquids can be characterized by viscosity. Egg white is a strange liquid. You can pour it part way out of a cup and then, with a flick of your wrist, have it come back into the cup. Other exotic liquids include Silly Putty and wet corn starch.

Table 4.1.1 - Approximate viscosities of a variety of fluids. The pascal-second (abbreviated Pa·s and synonymous with the kg/m·s) is the SI unit of viscosity. Only the superfluid portion of ultracold liquid helium exhibits zero viscosity.

Fluid	*Viscosity*
Helium (2° K)	0 Pa·s
Air (20° C)	0.0000183 Pa·s
Water (20° C)	0.00100 Pa·s
Olive Oil (20° C)	0.084 Pa·s
Shampoo (20° C)	100 Pa·s
Honey (20° C)	1000 Pa·s
Glass (540° C)	10^{12} Pa·s

break these bonds more easily in order to move past one another (see ❐).

When the faucet is open, water near the center of the opening accelerates toward low pressure and leaves the faucet at high speed. But water at the edges of the opening experiences friction with the walls and is essentially motionless. Since different portions of the water are moving at different velocities, viscous forces become important. The stationary water at the faucet walls exerts viscous forces on the moving water, converting some of that water's energy into thermal energy; thus these viscous forces act as an internal sliding friction within the water, making it hotter. Because of this wasted energy, the water flowing out of a faucet doesn't reach the speed that Bernoulli's equation predicts, and the faucet delivers less water per second than it would in the absence of viscosity.

CHECK YOUR UNDERSTANDING #1: Keeping Warm on a Windy Day

A loosely woven wool sweater has many tiny air passages between the wool fibers, yet it dramatically reduces the rate at which air flows through to your skin when you stand in a breeze. Why doesn't air flow easily through the gaps between the fibers?

Water Flow in Pipes

If viscosity slows the flow of water through faucets, it ought to slow the flow of water through pipes as well. And so it does. The stationary water at the pipe walls exerts viscous forces on the water moving through the pipe and slows that water down (Fig. 4.1.1).

These viscous forces impede water delivery. Water flowing through level pipes at a uniform pressure gradually comes to a stop as viscous forces turn its kinetic energy into thermal energy. These viscous forces are strongest when the layers of water are trying to slide past one another most rapidly, since high relative velocities between layers produce many collisions between fluid particles and cause the layers to push hard on one another. As a result, the faster the water moves through the pipes, the more kinetic energy it loses per meter.

To keep the water moving, a pressure imbalance must push it forward; in other words, the pressure behind the water must be greater than the pressure ahead of it. When the forward force created by the pressure imbalance exactly balances the backward viscous force, the water will stop accelerating and will flow through the pipe at a steady rate.

What determines just how much water a particular pipe will deliver per second?

1. The rate of delivery is inversely proportional to the fluid's viscosity. The more viscous the liquid, the more trouble it has flowing through the pipe.
2. The rate of delivery is inversely proportional to the length of the pipe. The longer the pipe, the more opportunity viscous forces have to stop the water from moving.
3. The rate of delivery is proportional to the pressure difference between the pipe's inlet and its outlet. The harder you pump on the water, the more effective the water is in overcoming viscous forces.
4. The rate of delivery also depends on the diameter of the pipe. Doubling its diameter provides four times as much space for the water to pass through and increases the speed of water near the pipe's center. Overall, the rate of delivery is proportional to the pipe's diameter to the fourth power.

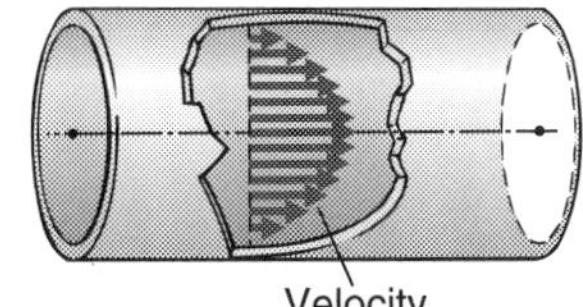

Fig. 4.1.1 - The speed of water flowing through a pipe is not constant across the pipe. The water near the walls is stationary, while the water at the center of the pipe moves the fastest. The differences in velocity are the results of viscous forces.

We can turn all these proportional relationships into an equation by adding a numerical constant ($\pi/128$). This relationship is called **Poiseuille's law** and can be written as a word equation:

$$\text{volume flow rate} = \frac{\pi \cdot \text{pressure difference} \cdot \text{pipe diameter}^4}{128 \cdot \text{pipe length} \cdot \text{fluid viscosity}}, \tag{4.1.1}$$

in symbols:

$$\frac{\Delta V}{\Delta t} = \frac{\pi \cdot \Delta p \cdot D^4}{128 \cdot L \cdot \eta},$$

and in everyday language:

It's hard to squeeze honey through a long, thin tube.

POISEUILLE'S LAW:
The volume of fluid flowing through a pipe each second is equal to ($\pi/128$) times the pressure difference (Δp) across that pipe times the pipe's diameter to the fourth power, divided by the pipe's length times the fluid's viscosity (η).

❐ To deliver large amounts of water at high pressure or velocity, fire hoses must have large diameters. When filled with high-pressure water, these wide hoses become stiff and heavy, making them difficult to handle. Chemical additives that decrease water's viscosity allow firefighters to use narrower, lighter, and more flexible hoses.

❐ Very large diameter pipes are required to transport crude oil across the Alaskan wilderness. The distances are long and the fluid is viscous, particularly during the winter.

It's hardly surprising that flow rate depends in this manner on the pressure difference, pipe length, and viscosity; we've all observed that low water pressure or a long hose lengthens the time needed to fill a bucket with water and that viscous syrups pour slowly from bottles. But the dependence of flow rate on the fourth power of diameter may come as a surprise. Even a small change in the diameter of a hose or pipe will significantly change the amount of water that it delivers. Over time, because house plumbing narrows due to the buildup of minerals inside its pipes, filling a bathtub takes longer and longer. Similarly, a modest narrowing in a person's artery substantially reduces the flow of blood through that artery. For two approaches to overcoming viscosity, see ❐s.

We can also look at viscous forces in terms of total energy. By opposing the flow of water through a pipe, viscous forces do negative work on it and reduce its "total energy" (the energy considered in Bernoulli's equation, which doesn't include thermal energy). Just how much total energy the water retains depends on how much water flows through the pipe. If you allow lots of water to leave the pipe, water will flow through it quickly and encounter large viscous forces; in the process, most of the water's total energy will be converted into thermal energy, and little will be left as it leaves the pipe. But if you partially block the pipe's opening, so that only a small amount of water can leave the pipe, water will flow through the pipe slowly and encounter small viscous forces. As a result, the water will keep most of its total energy and will still be at high pressure when it reaches your thumb; this high-pressure water will then accelerate to high speed as it squirts into the air. This is a time-honored trick used by anyone who has ever tried to water a garden with a water hose having very little pressure.

CHECK YOUR UNDERSTANDING #2: Air Ducts

The long air ducts used to ventilate homes and businesses usually have very large diameters. These ducts are often visible near the ceilings of modern warehouse-style stores and restaurants as pipes roughly 0.5 m across. Why must the ducts be so large in diameter?

CHECK YOUR FIGURES #1: Old Plumbing

When your friend's house was new, the kitchen faucet could deliver 0.50 liters per second (0.50 l/s). But mineral deposits have built up in the pipes over the years and reduced their effective diameters by 20%. How much water can the faucet deliver now?

Noise from Pipes and Faucets

As water flows through pipes and faucets, it often makes noise. The sound is usually a high-pitched hiss that is loudest when the water flows through a narrow opening. This sound is created when water swirls about erratically, a behavior called **turbulent flow**. The smoothly moving water in the rest of the pipe is in **laminar flow**, meaning that the velocity of the fluid at any specific point doesn't change in magnitude or direction. The water in laminar flow remains silent.

In laminar flow, adjacent regions of a fluid always remain near one another. If you place two drops of dye near one another in a smoothly flowing stream, they will remain close together indefinitely because the stream experiences laminar flow (Fig. 4.1.2). If you slowly pour several filaments of dye into that stream, they will form stripes in the water as it flows downstream. These stripes illustrate the presence of streamlines in the water.

But as the stream flows past rocks and obstacles, these streamlines follow tortuous paths and soon break up into the eddies and churning "white water" that make rafting exciting. The narrow stripes of dye are dispersed and the water becomes uniformly colored. The flow is now turbulent. In turbulent flow, adjacent regions of a fluid quickly become separated from one another as they move independently in unpredictable directions.

Whether a flow is laminar or turbulent depends on several characteristics of the fluid and its environment:

Fig. 4.1.2 - Water flows slowly past rocks in the stream on the left and its viscosity keeps it smooth and laminar. Water flows quickly past rocks in the stream on the right and its inertia separates it into swirling, splashing pockets of turbulence.

Fig. 4.1.3 - Honey's large viscosity keeps it flowing smoothly (laminar flow) when you pour it. Water's small viscosity allows it to splash about (turbulent flow).

1. The fluid's viscosity. Viscous forces tend to keep nearby regions of fluid moving together, so high viscosity favors laminar flow. For example, a thick or viscous syrup pours smoothly into a bowl, while thin water, with its lower viscosity, splashes about (Fig. 4.1.3).
2. The fluid's speed past a stationary obstacle. The faster the fluid is moving, the more quickly two nearby regions of fluid can become separated, and the harder it is for viscous forces to keep them together. A fast-flowing stream becomes turbulent more easily than a slowly flowing one.
3. The size of the obstacles the fluid encounters. The larger the obstacle, the more likely that it will cause turbulence because viscous forces will be unable to keep the fluid ordered over such a long distance.
4. The fluid's density. The denser the fluid, the less it responds to the viscous forces and the more likely it is to become turbulent.

Rather than keeping track of all four physical quantities independently, English mathematician and engineer Osborne Reynolds (1842–1912) found that they could be combined into a single number that permits a comparison of seemingly different flows. The **Reynolds number** is defined as

$$\text{Reynolds number} = \frac{\text{density} \cdot \text{obstacle length} \cdot \text{flow speed}}{\text{viscosity}}. \tag{4.1.2}$$

The units on the right side of Eq. 4.1.2 cancel one another so that the Reynolds number is dimensionless; that is, it's just a simple number, such as 10 or 25,000, with no dimensions or units. As the Reynolds number increases, the flow is likely to go from laminar to turbulent. At a low Reynolds number, viscous forces dominate the flow and keep it smooth and laminar. At a high Reynolds number, inertia dominates the flow and each portion of fluid moves according to its own momentum. These individual portions of fluid separate or collide frequently and the flow becomes swirling and turbulent.

In his experiments, Reynolds found that turbulence appears when the Reynolds number exceeds about 2300. If you move a 1 cm thick stick slowly through water (about 10 cm/s), the Reynolds number is about 1000 and the flow around the stick is laminar. If you move the stick faster (about 50 cm/s), the Reynolds number rises to about 5000 and the flow becomes turbulent. Try it, moving the stick either in a straight line or in a circle. A kitchen mixer makes good use of this same turbulence.

One of the most common features of turbulent flows is the **vortex**, a whirling region of fluid that moves in a circle above a central cavity. A vortex resembles a miniature tornado, with its cavity created by inertia as the fluid spins. These vortices are quite visible behind a canoe paddle or in a mixing bowl. Once an object moves fast enough through a fluid to create turbulence, these vortices begin to form. Each vortex builds up behind the object but is soon whisked away to form a regular pattern of *shed vortices* (Fig. 4.1.4).

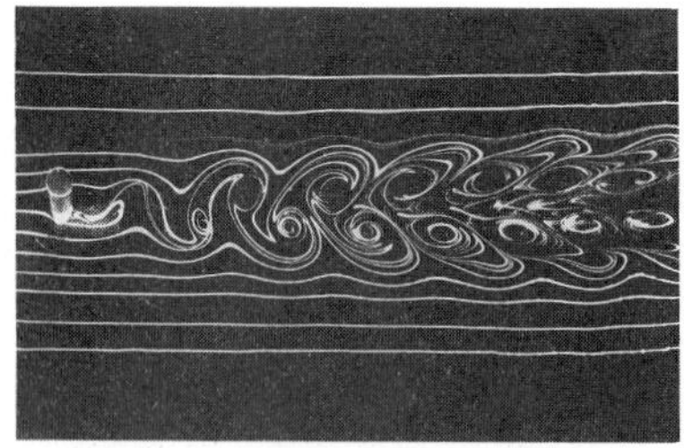

Fig. 4.1.4 - When water flows rapidly around a cylinder, its flow becomes turbulent. A pattern of swirling vortices forms after the cylinder.

CHECK YOUR UNDERSTANDING #3: Urban Wind Storms

On a windy day in a city with many tall buildings, leaves and papers can be seen swirling about in the air or on the sidewalks. What causes these whirling air currents?

CHECK YOUR FIGURES #2: Wind on the Open Road

Is the flow of air around a convertible laminar or turbulent as the convertible cruises down the highway?

Turbulent Water and Chaos

When the water passing through a faucet becomes turbulent, its flow begins to exhibit chaotic behavior or **chaos**—you can no longer predict exactly where any particular drop of water will go. The study of chaos is a relatively new field of science. Because a **chaotic system**—a system exhibiting chaos—is exquisitely sensitive to initial conditions, even the slightest change in those conditions may produce profound changes in its situation later on. For example, your motion on a crowded highway is a chaotic system. If the movement of just one other car prevents you from switching lanes as you pass through a complicated interchange, you could soon find yourself in an entirely different city.

The classic example of a chaotic system is turbulent flow in a fluid. The slightest change in almost any aspect of turbulent water flow through a faucet will dramatically alter its swirls and eddies. A grain of sand being carried along with the water will move unpredictably. Similarly, two grains of sand that enter the faucet right next to one another will quickly become separated; the longer you watch them, the more independent their motions will become. This divergence of particles or paths that are initially near each other is one of the hallmarks of chaos.

CHECK YOUR UNDERSTANDING #4: Weather Prediction?

The earth's atmosphere moves as a fluid experiencing turbulent flow. How does this turbulent flow make weather so difficult to predict?

Water Hammer

The other noise that comes from faucets is the "klunk" of **water hammer**, the impact that occurs when moving water is suddenly brought to a stop. When a faucet is opened wide, the water runs freely through the pipe leading to that faucet. This water has momentum; the longer the pipe and the faster the water is running, the more momentum the water has. When you suddenly close the faucet, all of that moving water has to stop. The faucet must exert a large force on the water and the water, in effect, bounces off the end of the pipe. Since the only way to exert a force on a fluid is through pressure, the pressure in front of the moving water surges to an enormous value. The column of water then experiences a pressure imbalance and accelerates backward, quickly coming to a stop.

The "klunk" occurs when this pressure surge makes the pipes move. The water suddenly transfers its momentum to the pipes and the faucet, just as a hammer does when it hits a nail. This phenomenon is more than just a nuisance; it can actually break a pipe. Water hammer can be reduced, however, by putting an air shock absorber in the pipe (Fig. 4.1.5). With such a device in place, the water compresses the air when it slams into the faucet and slowly comes to a stop; while the momentum transferred from the water to the pipe remains the same, the transfer occurs gently, over a longer period of time. A careful plumber may install sealed vertical tubes at water outlets for clothes washers and other machines that can stop the flow of water abruptly; these air pockets reduce water hammer as they fill with air that naturally comes out of solution in the water.

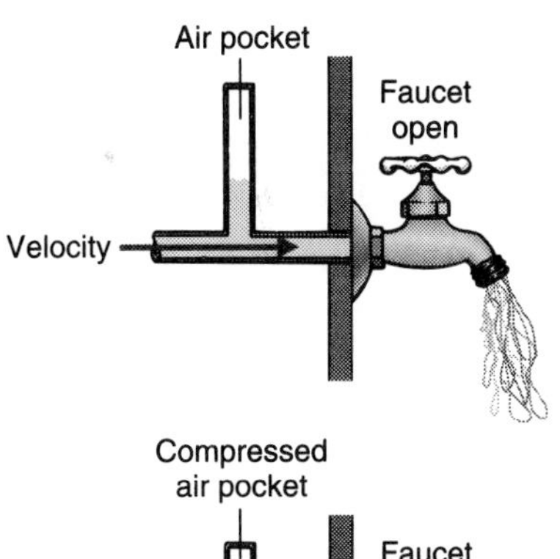

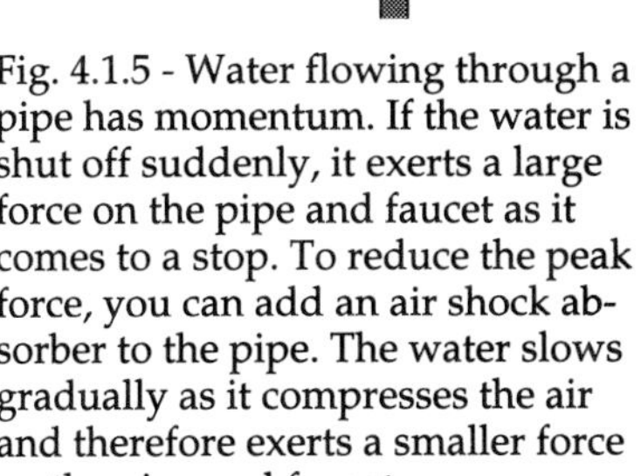

Fig. 4.1.5 - Water flowing through a pipe has momentum. If the water is shut off suddenly, it exerts a large force on the pipe and faucet as it comes to a stop. To reduce the peak force, you can add an air shock absorber to the pipe. The water slows gradually as it compresses the air and therefore exerts a smaller force on the pipe and faucet.

CHECK YOUR UNDERSTANDING #5: Flash Floods

A sudden heavy rain can send a flash flood coursing through a dry river bed. A car caught by this flash flood can be spun around, knocked over, or washed away. How can water exert such force?

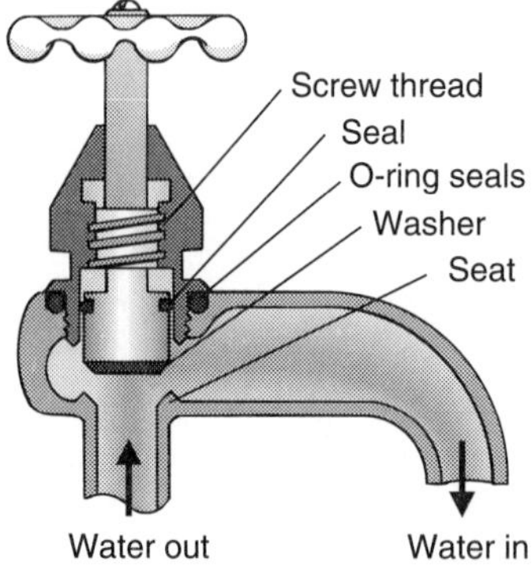

Fig. 4.1.6 - A water faucet controls the flow of water by partially or completely blocking the inlet pipe with an elastic washer. The position of the washer is set by a screw attached to the handle. Several other water seals prevent the faucet from leaking.

The Faucet's Simple Machine: The Screw

We're finally ready to look at the inside of a water faucet (Fig. 4.1.6). The water flow is controlled by a rubber washer. When the faucet is closed, the washer presses against the water inlet pipe, keeping the water from flowing; but when you open the faucet, the washer moves away from the pipe's opening, and water begins to flow through. This rubber washer sits at the end of a shaft that's raised or lowered by a screw as you turn the handle. But why does the faucet need this screw? Why not just push the washer down to block the flow of water and pull it up to unblock the flow of water?

The problem is that the water coming into the faucet is under pressure. When the washer is pressed against the inlet pipe so that no water flows, the pressure in the inlet pipe is city water pressure and the pressure in the outlet pipe is atmospheric pressure. There is a large pressure imbalance on the washer and thus a large upward force on it. As a result, you'd have to push down very hard on the shaft to keep the faucet closed, and it would fly upward when you let go. To make things easier, therefore, many faucets include a screw.

A **screw** is a rotating device that exhibits mechanical advantage. It's actually two ramps or inclined planes that are wrapped around cylinders so that they slide across one another as the cylinders rotate. When two ordinary ramps slide across one another, one goes up while the other goes down (Fig. 4.1.7*a*). However, the only parts of the ramps that are really important are their surfaces (Fig. 4.1.7*b*). If these surfaces are wrapped around cylinders, they become screws (Fig. 4.1.7*c*).

The cylindrical ramps themselves are called *screw threads*. In a screw, one of these screw threads attaches to an inner cylinder, while the other screw thread attaches to an outer cylinder. The threaded inner cylinder is similar to a bolt; the threaded outer cylinder is similar to a nut. When the inner cylinder of the screw turns relative to the outer cylinder, the two screw threads slide across one another. The inner cylinder moves up or down in the outer cylinder, depending on the direction of their relative rotation.

When two ramps slide across one another, modest horizontal forces on the ramps can produce very large vertical forces. The same holds for a screw, where a modest torque on the screw (the equivalent of the horizontal forces on the ramps) can produce very large vertical forces. Like the ramp, the screw exhibits mechanical advantage. In this case, the modest torque you exert as you rotate the handle several turns exerts a very large force as the shaft descends a very small distance. While the work you do in closing the faucet is the same whether you push the shaft down directly or use a screw, the screw spreads out the work and lets you do it with smaller forces or torques exerted over longer distances or angles. Here, the screw easily exerts the large force needed to keep the washer pressed against the inlet pipe.

(*a*)

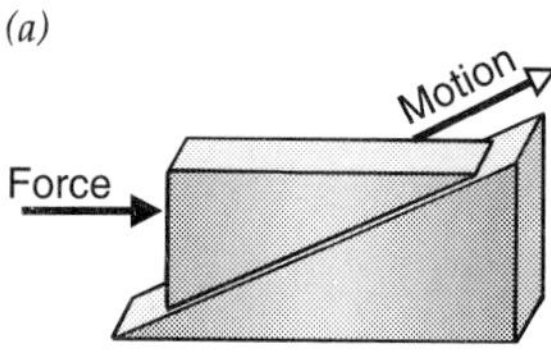

(*b*)

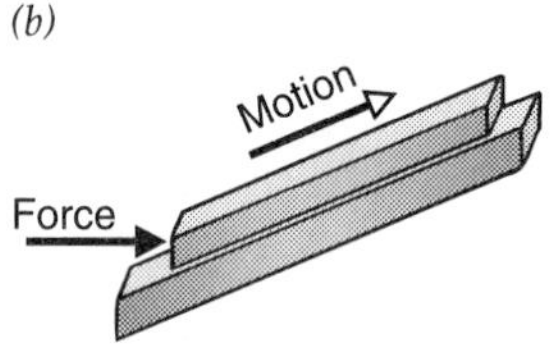

(*c*)

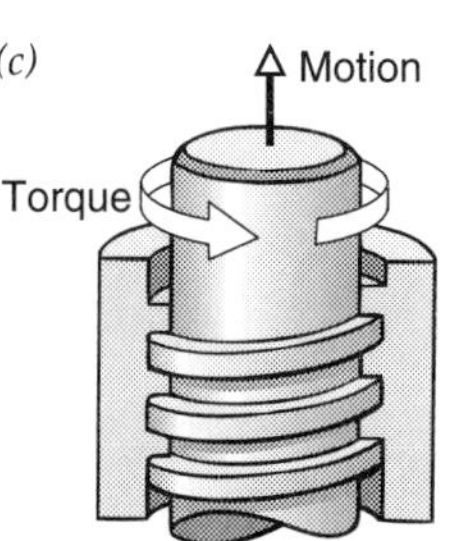

Fig. 4.1.7 - (*a*) When you push a movable ramp across a second, fixed ramp, the movable ramp slides upward relative to the other. (*b*) The same holds for two inclined surfaces. (*c*) When the inclined surfaces are wrapped around cylinders to form cylindrical ramps, a screw is created. As the movable cylinder rotates, it moves upward or downward relative to the outer, fixed cylinder.

CHECK YOUR UNDERSTANDING #6: A Turn of the Screw

A wood screw is able to penetrate hardwood with only a modest torque exerted on the screwdriver. Pushing the screw or a similar-sized nail directly into the same wood can't be done by hand. How does the screw make this penetration so nearly effortless?

Stopping Leaks: Water-tight Seals

A faucet isn't a solid object; it's built from several separate pieces, and there are joints between these pieces. Why doesn't the faucet leak through its joints?

The faucet doesn't leak because seals block the water's path through each of these joints (Fig. 4.1.6). The washer itself forms a seal when the faucet is closed, pressing down against the water inlet to stop the flow. But the faucet also contains at least two O-ring seals or their equivalent. O-rings are doughnut-shaped pieces of hard *elastomers*, rubber-like materials, that are confined in a groove and prevent liquid from flowing through joints between two pieces of metal (Fig. 4.1.8). Elastomers store energy when they are distorted and release that energy when they return to their equilibrium shapes. The faucet's O-rings keep water from flowing out beside the handle. One of these O-rings is attached to the faucet's shaft and maintains its seal even as the shaft rotates and moves up or down.

An O-ring is particularly effective at blocking the flow of fluid because it is **elastic**. In a seal, where it's distorted away from its equilibrium shape and experiences a strong restoring force, it pushes against the surfaces that confine it, filling all gaps and blocking the passage of fluid. As long as the O-ring remains compressed in the seal, it will always try to return to its equilibrium shape and keep the seal fluid-tight (see ❐).

As the shaft O-ring turns and slides inside the faucet, it experiences sliding friction. Although the water lubricates its motion, the O-ring experiences some wear when you turn the faucet on or off, and with enough wear it begins to leak and must be replaced. In order to reduce this wear and prolong the O-ring's useful life, the faucet manufacturer polishes the metal surfaces against which the O-ring slides.

The washer that controls the flow of water in a faucet eventually wears out, too. Each time it presses against the water inlet, the washer is distorted. Like any elastomer, it stores energy when it's distorted and releases that energy when it returns to its equilibrium shape. But sliding friction with the inlet pipe, long periods of distortion, and thermal damage from the hot water eventually change the washer's equilibrium shape forever. Since it no longer distorts from its new equilibrium shape when pressed against the inlet pipe, the washer fails to fill the gaps and the faucet leaks. To fix the faucet, you must unscrew the handle, shaft, and shaft holder from the faucet and replace the washer. During this repair, you can stop the water flow to the faucet using the cut-off valve (another faucet-like valve installed upstream of the faucet's water inlet and usually located underneath the sink or at the entrance to a house).

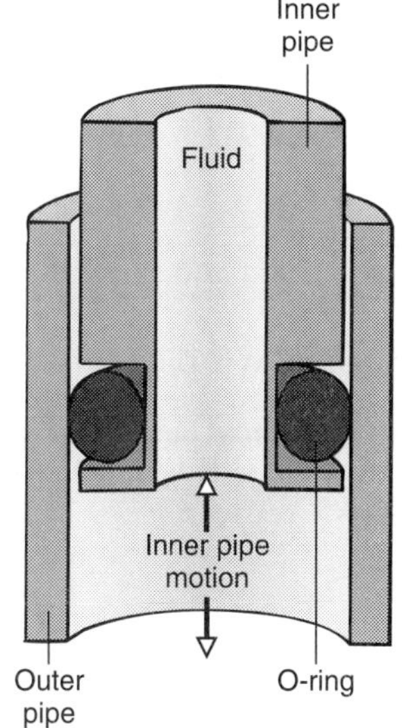

Fig. 4.1.8 - An O-ring seals an inner pipe to an outer pipe. The inner pipe has a groove cut in its outer edge to hold the O-ring. The motion of the inner pipe through the outer pipe causes the O-ring to twist, lubricating it with the fluid contained inside the pipes.

❐ O-ring seals were invented during World War II and revolutionized the hydraulic plumbing fittings used in aircraft. An O-ring seal between an inner and an outer pipe functions well even if the pipes slide through one another. As they slide, the O-ring twists slightly and is lubricated by the hydraulic fluid. Because of this lubricating effect, the O-ring receives little wear and lasts for years.

CHECK YOUR UNDERSTANDING #7: Steel Seals

Why can't a faucet washer be made of metal, which is more durable and heat resistant than rubber?

Summary

How Water Faucets Work: A faucet controls the rate at which water flows from a pipe by opening or closing a hole through which the water must flow. The water's speed through this hole is limited by its total energy and by viscous forces that oppose relative motion in the water. Because of this limited speed, the amount of water flowing through the faucet each second depends on the size of the hole. The hole size is adjusted by turning the faucet's handle, which operates a screw that raises or lowers a rubber washer. When the washer presses tightly against the water inlet pipe, no water can flow through the faucet, because the washer's elastic nature ensures a water-tight seal. But as the screw lifts the washer away from the water inlet pipe, water begins to flow. Because the screw produces mechanical advantage, a gentle torque on the handle can control the flow of very high pressure water.

Important Laws and Equations

1. Poiseuille's law: The volume of fluid flowing through a pipe each second is equal to ($\pi/128$) times the pressure difference across that pipe times the pipe's diameter to the fourth power, divided by the pipe's length times the fluid's viscosity, or

$$\text{volume flow rate} = \frac{\pi \cdot \text{pressure difference} \cdot \text{pipe diameter}^4}{128 \cdot \text{pipe length} \cdot \text{fluid viscosity}}. \quad (4.1.1)$$

The Physics of Water Faucets

1. The flow of a fluid through an opening or pipe is limited principally by its size and length, the pressure difference across it, and the viscosity of the fluid. Fluid that touches a solid object is stationary and tends to slow nearby fluid by way of viscous forces.

2. Fluid flowing at low Reynolds numbers tends to be laminar and quiet. The flow is dominated by the fluid's viscosity, which orders the flow.

3. Fluid flowing at high Reynolds numbers tends to be turbulent and noisy. The flow is dominated by the fluid's inertia, which tends to scramble the flow.

4. Fluids carry momentum. Stopping a moving fluid requires that the pressure in front of the flow be much larger than the pressure behind it.

5. A chaotic system is exquisitely sensitive to its initial situation. The slightest change in that initial situation will profoundly alter its future. Because of this sensitivity, it's impossible to make accurate predictions of a chaotic system's future.

Check Your Understanding - Answers

1. The air at the surfaces of the fibers is stationary and the air's viscosity slows the motion of air in the vicinity of the fibers.

Why: Although air trying to pass through the gaps may not directly touch the fibers, viscosity keeps all of the air moving together at the same velocity. As soon as some air is held up by a fiber, the air nearby is slowed by viscous forces. The fibers of a sweater are close enough together that viscous forces slow all the air trying to pass through the sweater. Imagine trying to pour honey through a sweater.

2. Air's viscosity slows its flow through ductwork. Moving large volumes of air rapidly without a large pressure difference between inlet and outlet requires a very large diameter pipe.

Why: The airflow through long ductwork is dominated by viscous forces. The volume of air moved through ductwork is often enormous and the pressure difference between the inlet and outlet is normally only a fraction of an atmosphere. To keep the air moving quickly through the ducts, their diameters must be large.

3. The air flowing through the buildings becomes turbulent and forms vortices that swirl the leaves and papers.

Why: Whether an object moves through a fluid or a fluid moves past an object, a Reynolds number is associated with the situation. When the Reynolds number becomes high enough that viscosity is unable to keep the fluid flowing in an orderly fashion, turbulence appears. Air is a fluid and when wind blows through in a big city, turbulence is commonplace. Whirling vortices are found everywhere.

4. The atmosphere is a chaotic system so that slight changes in the current situation will significantly affect the weather a day or a week from now.

Why: Surface winds are filled with vortices and eddies, common features in turbulent and chaotic flows. Tiny changes in the earth's surface can dramatically affect these turbulent flows and redirect the wind. It's frequently hypothesized that the turbulence caused by a single butterfly flapping its wings can redirect the flow of the wind and eventually affect the weather everywhere on earth. The butterfly changes the local winds, which in turn change other winds, and so on until the entire earth has been affected. Because of the atmosphere's chaotic nature, the weather is difficult to predict. Weather prediction may never be completely reliable nor will it extend much further into the future than it does today. You simply can't record the movements of every butterfly on earth.

5. The moving water carries considerable momentum and can transfer this momentum violently to a car caught in its path.

Why: Water hammer occurs whenever some obstacle suddenly blocks the path of rapidly moving water. The pressure that the water exerts on the obstacle surges upward, particularly if the water can't find a path around the obstacle. Waves crashing against cliffs also experience this effect. The pressure surge can squirt water high into the air through gaps or holes in the rocks.

6. The wood screw employs mechanical advantage to convert a modest torque on the screwdriver into an enormous forward force that drives the screw into the wood.

Why: As the wood screw turns, it forms a screw thread in the wood around it. The screw then uses this screw thread to push itself into the wood. As the screw turns, its thread slides across the thread of the wood. The screw moves forward slowly, exerting an enormous forward force on the wood.

7. Most metals don't distort significantly when pressure is exerted on them. They are not elastomers. As a result, they don't fill gaps easily and would leak.

Why: Rubber and other elastomeric materials fill gaps well because they can be distorted dramatically and still return to their original shape. Metals don't distort easily and when distorted, may not return to their original shapes. However, metal seals are used where high temperature or exceptional chemical purity is required. In such cases, the seals must be made very precisely so that the metal doesn't have to distort very much. Metal seals usually only work a few times before they must be replaced.

Check Your Figures - Answers

1. About 0.20 l/s.

Why: When the faucet is wide open, water flow is limited by the pipes. The diameters of the old pipes are 20% less than when they were new, or 0.80 times the diameter of the new pipes. We can write Poiseuille's law (Eq. 4.1.1) for the new pipes:

$$0.50\ \mathrm{l/s} = \frac{\pi \cdot \Delta p \cdot D^4}{128 \cdot L \cdot \eta},$$

and the old pipes:

$$\text{volume flow rate} = \frac{\pi \cdot \Delta p \cdot (0.80 \cdot D)^4}{128 \cdot L \cdot \eta},$$

and divide the second equation by the first:

$$\frac{\text{volume flow rate}}{0.50\ \mathrm{l/s}} = \frac{\dfrac{\pi \cdot \Delta p \cdot (0.80 \cdot D)^4}{128 \cdot L \cdot \eta}}{\dfrac{\pi \cdot \Delta p \cdot D^4}{128 \cdot L \cdot \eta}} = 0.80^4 = 0.4096.$$

We don't need to know the pressure difference (Δp), the pipe length (L), or water's precise viscosity (η) because they cancel in the last equation. If we multiply both sides of that equation by 0.5 l/s, we get the volume flow rate:

$$\text{volume flow rate} = 0.4096 \cdot 0.5\ \mathrm{l/s} = 0.2048\ \mathrm{l/s}.$$

Because our measured values are only accurate to two decimal digits (0.50 l/s is not as accurate a measurement as 0.5000 l/s), we can only report a two digit result: 0.20 l/s for the flow from the old pipes. It clearly doesn't take very much mineral accumulation to dramatically reduce the flow from a faucet.

2. Turbulent.

Why: To calculate the Reynolds number for the airflow around the car, you need air's viscosity from Table 4.1.1 (0.0000183 Pa·s or 0.0000183 kg/m·s), air's density from Section 3.1 (1.25 kg/m^3), the car's size (roughly 3 m), and its speed through the air (roughly 55 mph or 25 m/s). You can then calculate its approximate Reynolds number, using Eq. 4.1.2:

$$\text{Reynolds number} = \frac{1.25\ \mathrm{kg/m^3} \cdot 3\ \mathrm{m} \cdot 25\ \mathrm{m/s}}{0.0000183\ \mathrm{kg/m \cdot s}} = 5.1\ \text{million}.$$

The convertible's Reynolds number is far above the threshold for turbulence (2300) so the air swirls chaotically around the vehicle. That explains why your hair flies around wildly.

Glossary

chaos Unpredictable behavior in which minute changes in a system's initial arrangement lead to very different final arrangements. These differences grow more dramatic with each passing second.

chaotic system A dynamic system that is exquisitely sensitive to initial conditions. Minute changes in how you set up a chaotic system can lead to wildly different final configurations.

elasticity A property of a solid that allows it to be stretched or compressed by outside forces and then return to equilibrium shape once those forces are removed.

laminar flow Smooth, predictable fluid flow in which the velocity at any specific point doesn't change with time in either magnitude or direction and nearby portions of the fluid remain nearby as they travel along.

Poiseuille's law The volume of fluid flowing through a pipe each second is equal to (π/128) times the pressure difference across that pipe times the pipe's diameter to the fourth power, divided by the pipe's length times the fluid's viscosity.

Reynolds number A dimensionless number that characterizes fluid flow through a system. At low Reynolds numbers, a fluid's viscosity dominates the flow while at high Reynolds numbers, a fluid's inertia dominates.

screw A simple tool in which a cylindrical ramp is used to raise or lower an object.

turbulent flow Irregular, fluctuating, non-predictable fluid flow in which the velocity at any specific point changes with time in either magnitude or direction and nearby portions of the fluid become separated.

viscosity The measure of a fluid's resistance to relative motion within that fluid.

viscous forces The forces exerted within a fluid that oppose relative motion. Layers of fluid that are moving across one another exert viscous forces on each other.

vortex A whirling region of fluid that is moving in a circle above a central cavity.

water hammer The impact of a moving mass of water that is suddenly stopped.

Review Questions

1. What factors limit how quickly water can flow through a narrow opening?

2. List at least three situations in which viscosity affects the movements of familiar fluids.

3. Why does the water near the center of a pipe flow faster than the water near the wall of the pipe?

4. Why do the pressures at the start and end of a pipe affect the amount of water flowing through that pipe each second?

5. Why does water travel faster when it squirts out of a nozzle at the end of a garden hose than it does when it squirts out of the hose itself?

6. If water squirting out of an open hose travels relatively slowly, where has the rest of its energy gone?

7. How could you use a box of toothpicks to distinguish between water experiencing laminar flow and water experiencing turbulent flow?

8. Show that the numerator of Eq. 4.1.2 is closely related to momentum (see Eq. 1.4.1). Use this result to show that the Reynolds number increases as momentum increases and that it decreases as viscosity increases.

9. What does it mean that laminar flow is dominated by viscosity while turbulent flow is dominated by inertia?

10. How does water hammer demonstrate that fluids have momentum?

11. Why are elastic materials used in seals?

Exercises

1. A favorite college prank involves simultaneously flushing several toilets while someone is in the shower. The cold water pressure to the shower drops and the shower becomes very hot. Why does the cold water pressure suddenly drop?

2. Why is it so difficult to squeeze ketchup through a very small hole in its packet?

3. A baker is decorating a cake by squeezing frosting out of a sealed paper cone with the tip cut off. If the baker makes the hole at the tip of the cone too small, it's extremely difficult to get any frosting to flow out of it. Why?

4. Sometimes dried toothpaste partially blocks the outlet of a toothpaste tube. When this happens, it's not easy to squeeze toothpaste through the remaining open space. Why does a modest reduction in the opening size make so much difference?

5. Why does a relatively modest narrowing of the coronary arteries, the blood vessels supplying blood to the heart, cause a dramatic drop in the amount of blood flowing through them?

6. On hot days in the city people sometimes open up fire hydrants and play in the water. Why does this activity reduce the water pressure in nearby hydrants?

7. Why does the spray of water from a dormitory shower slow down when several showers are on at once? Why does the increased water usage reduce the speed of the shower spray?

8. Why is the wind stronger several meters above a flat field than it is just above the ground?

9. Why is it so difficult to get ketchup to flow out of a brand new glass bottle?

10. When you extrude cookie dough from a cookie press (a hollow cylinder with a plunger at one end and a nozzle at the other), you do lots of work on the plunger but the dough leaves the press with very little kinetic energy. Use the dough's viscosity to explain the apparent disappearance of energy. Where does your work end up?

11. If you drop a full can of applesauce and it strikes a cement floor squarely with its flat bottom, what happens to the pressures at the top and bottom of the can?

12. When you mix milk or sugar into your coffee, you should move the spoon quickly enough to produce turbulent flow around the spoon. Why does this turbulence aid mixing?

13. If you start two identical paper boats from the same point, you can make them follow the same path down a quiet stream. Why can't you do the same on a brook that contains eddies and vortices?

14. There is a story of a Dutch boy who prevented a flood by putting his finger in a hole in a stone dike. Why did the boy's finger work better than a rock at plugging this hole?

15. The Challenger space shuttle disaster occurred because the O-ring in a booster rocket became rigid in the cold January weather. Why did the O-ring's rigidity affect its ability to seal the booster?

16. An electric valve controls the water for the lawn sprinklers in your backyard. Why do the pipes in your home shake whenever this valve suddenly stops the water but not when this valve suddenly starts the water?

17. Though they feel very soft, water balloons can exert large forces on the objects they hit. How does the water pressure in the balloon change as it hits a solid object such as a window? Why can it break the window?

18. Raindrops make a familiar sound as they land on a tile roof. Each raindrop pushes downward on a tile as it hits and the moving tile emits sound. Explain how a soft, liquid raindrop can exert a downward force on the tile as it hits.

19. If you try to fill a bucket by holding it in a waterfall, you will find the bucket pushed downward with enormous force. How does the falling water exert such a huge downward force on the bucket?

20. Why are cylinders of elastic cork better for sealing wine bottles than cylinders of rigid oak?

21. The stiff plastic lids used on many beverage containers don't seal very well. Why would rubber lids form more reliable seals?

22. Why does hot maple syrup pour more easily than cold maple syrup?

23. Why is "molasses in January" slower than "molasses in July," at least in the northern hemisphere?

24. When you swing a stick slowly through the air, it's silent. But when you swing it quickly, you hear a "woosh" sound. What behavior of the air is creating that noise?

25. An ultrasonic cleaner uses high frequency sound to shake the cleaning solution around a piece of jewelry. The vigorous shaking creates empty cavities in the solution and these cavities close suddenly, moments after they're formed. When they close, the local pressure skyrockets and the solution pounds the dirt off the surface of the jewelry. Why does the pressure surge upward whenever one of these cavities closes?

Problems

1. About how fast can a small fish swim before experiencing turbulent flow around its body?

2. How much higher must your blood pressure get to compensate for a 5% narrowing in your blood vessels? (The pressure difference across your blood vessels is essentially equal to your blood pressure.)

3. If someone replaced the water in your home plumbing with olive oil, how much longer would it take you to fill a bathtub?

4. You are trying to paddle a canoe silently across a still lake and know that turbulence makes noise. How quickly can the canoe and the paddle travel through water without causing turbulence?

5. The pipes leading to the showers in your locker room are old and inadequate. While the city water pressure is 700,000 Pa, the pressure in the locker room when one shower is on is only 600,000 Pa. Use Eq. (4.1.1) to calculate the *approximate* pressure if three showers are on.

6. If the plumbing in your dorm carried honey instead of water, filling a cup to brush your teeth could take a while. If the faucet takes 5 s to fill a cup with water, how long would it take to fill your cup with honey, assuming all the pressures and pipes remain unchanged?

7. Instead of waiting for honey to flow through your dorm's plumbing (see Problem **6**), you might send it through pipes 20 times as large in diameter. How much faster would your cup fill through these larger pipes?

8. A fire truck carries hoses of two different diameters: 5 cm and 8 cm. The firefighters connect two hoses to the truck's pump, one of each diameter. These hoses are of equal lengths and are wide open at their ends. If the water pressure entering each hose is equal, how much more water will flow through the larger diameter hose?

9. How quickly would you have to move a 1 cm diameter stick through olive oil to reach a Reynolds number of 2000, so that you would begin to see turbulence around the stick? (Olive oil has a density of 918 kg/m^3.)

10. The effective obstacle length of a blimp is its width—the distance to which the air is separated as it flows around the blimp. How slowly would a 15-m-wide blimp have to move in order to keep the airflow around it laminar? (Air has a density of 1.25 kg/m^3).

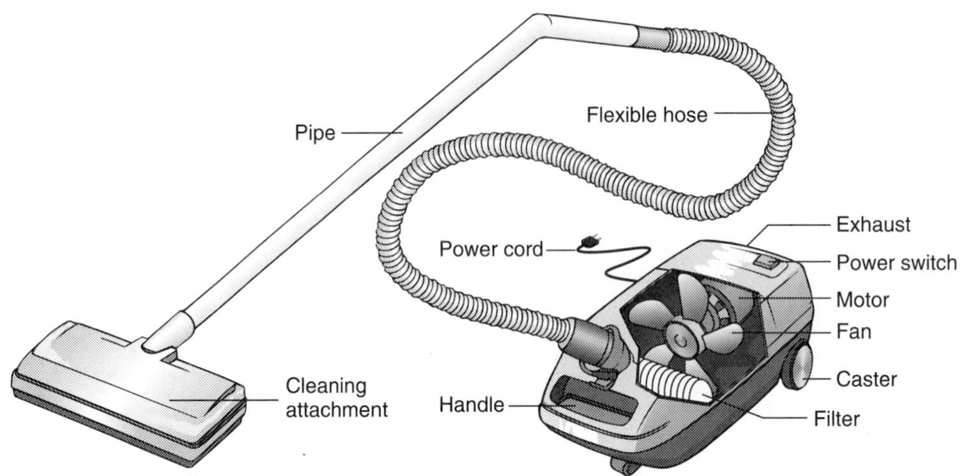

Section 4.2

Vacuum Cleaners

When we examined faucets in the previous section, we were looking at a tool that permits a fluid to flow out of a pipe. In this section, we'll look at the reverse, a device that draws fluid into a hose. The device is a vacuum cleaner, and the fluid that it draws inward is air. This moving air gathers dust and debris as it rushes into the vacuum cleaner, which is why vacuum cleaners are useful.

Questions to Think About: *Why does nature "abhor a vacuum"? If you remove some of the air from a region of space, how does the surrounding air respond? How does wind push on the objects it passes? Can you think of cases in which moving air actually picks up at least some of the objects it passes? How does the vacuum cleaner create the partial vacuum it uses to draw air into the hose? Why does the vacuum cleaner require electric power?*

Experiments to Do: *Find a vacuum cleaner with a hose and watch how the airflow draws in dust. As you shrink the diameter of the cleaning attachment or partially block the end of the hose, does the air entering the hose move faster or slower? Is a partially blocked hose more or less effective at trapping dust than an unblocked one? Try to vacuum up tiny objects and then larger ones; which are easier? Block the airflow into the hose completely and notice what happens to the motor's pitch. Why should the motor's rotation depend on the airflow?*

Even without a vacuum cleaner, you can try similar experiments with a drinking straw or a cardboard tube. You could pretend to be a vacuum cleaner by sucking dust into your mouth, but you'd do better to blow it around instead. Try blowing on different-sized objects with different-sized openings on the straw or tube. Is a narrow or a wide opening most effective? Which blow about more easily: small objects or large objects?

Air Flowing into the Vacuum Cleaner

Vacuum cleaners use swiftly moving air to sweep up dust. In this section, we'll examine how they create that swiftly moving air and why dust is so easily carried along by it. But let's forget dust for the moment and look at how air itself flows into a vacuum cleaner. In particular, let's try to understand the airflow in a canister vacuum cleaner with a long hose.

For simplicity, we'll reduce the machine to a hose and a fan (Fig. 4.2.1). The fan draws air through the hose. Outside the hose, the air is stationary at atmospheric pressure. When you turn the fan on, the pressure inside the hose drops and a partial vacuum is created (hence the name "vacuum cleaner"). Since air accelerates from high pressure to low pressure ("nature abhors a vacuum"), outside air rushes into the hose. Once steady-state flow is reached, a process that takes a second or two, the pressures and velocities of the air in the hose become steady, too. However, these pressures and velocities are different at different points along the hose. We can understand these differences using Bernoulli's equation.

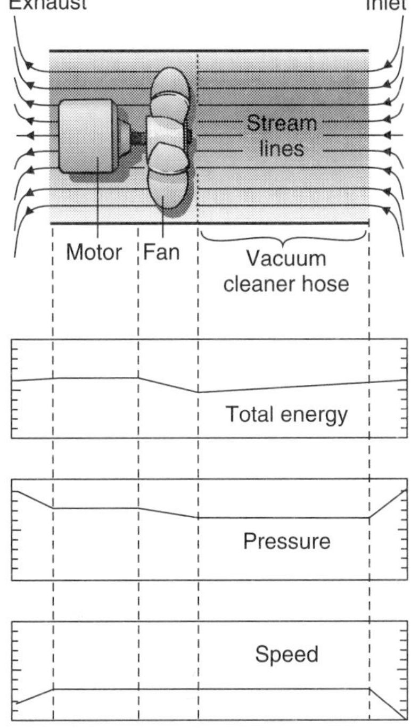

Fig. 4.2.1 - As air flows into the inlet of a vacuum cleaner hose, its pressure drops and its speed increases. The fan boosts the air's total energy, helping it overcome viscous losses of total energy so that it can return to the outside air through the outlet.

Strictly speaking, we shouldn't be using Bernoulli's equation; it applies only to incompressible fluids in perfect steady-state flow, and air certainly isn't incompressible. But if certain conditions are met—if the air's velocity is less than about 300 km/h, and if there are no pressure differences of more than one tenth of an atmosphere—then we can *consider* air to be incompressible, since its density will remain fairly constant. That will make things much easier for us, both now and when we discuss curve balls and airplanes. We can also ignore gravity. If the air were flowing up and down hundreds of meters, we would need to include it; but here it's inconsequential, since the whole vacuum cleaner is at one altitude.

With air behaving as an incompressible fluid and gravity out of the picture, Bernoulli's equation predicts a simple result. The sum of the air's pressure potential energy and kinetic energy should be constant along a streamline. Thus, if a portion of air speeds up as it moves along its streamline, its pressure must drop; if it slows down, its pressure must rise.

The pressure inside the vacuum cleaner hose is low, so air accelerates toward the inlet (Fig. 4.2.1) and its speed increases as its pressure drops. Inside the hose, the air has the same total energy that it had before it entered, but some of its pressure potential energy has been converted to kinetic energy.

We can see when such an energy conversion occurs if we look at the spacing of the streamlines. In Fig. 4.2.1, the streamlines bunch together as they enter the inlet. All of the air is being squeezed through a narrow channel. Since air takes up space, its speed must increase if all of it is to pass through that channel quickly enough. Consequently, its pressure must decrease.

This result is a general one. Whenever you see streamlines bunching together in this manner, you can be sure that the speed of the fluid is increasing and its pressure is decreasing. Conversely, if you see streamlines spreading apart, you know that the air is slowing down and its pressure is increasing.

Once inside the hose, the air continues at high speed and low pressure until it encounters the fan. The fan increases the air's total energy, using a technique we will look at later. Because the fan's inlet and outlet have the same diameter, the air's speed doesn't change as it moves through the fan; instead, its pressure and pressure potential energy increase. Low-pressure air enters the fan through the hose and higher-pressure air leaves the fan through the exhaust port. Thus the fan maintains a pressure imbalance inside the vacuum cleaner. The low pressure inside the hose is what got the air moving in the first place!

While the air leaving the fan may still be below atmospheric pressure, it's now traveling fast so that much of its total energy is in the form of kinetic energy.

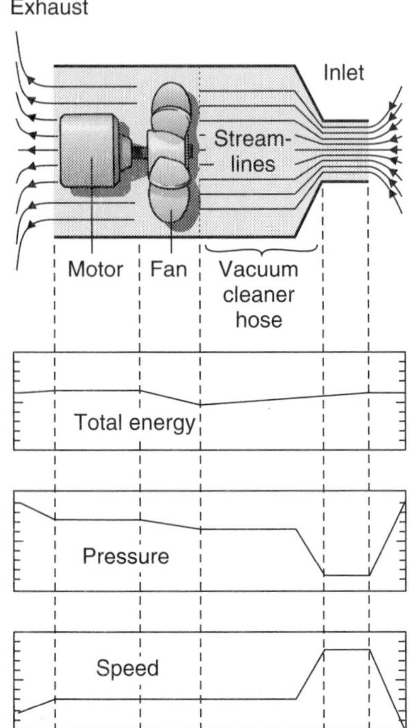

Fig. 4.2.2 - When you add a narrow cleaning attachment to the end of the vacuum cleaner hose, the air in the inlet reaches very high speed and relatively low pressure. The attachment's narrow channel is short to minimize the loss of total energy to viscous forces.

Consequently, as this air flows out of the exhaust port, its streamlines spread out (Fig. 4.2.1). This spreading indicates that the air is slowing down and that its pressure is rising to atmospheric pressure. The air has completed its trip through the vacuum cleaner.

But when you add a narrow cleaning attachment to the hose (Fig. 4.2.2), the airflow becomes more complicated. For the air to continue flowing through the fan at the same rate as before, it must rush rapidly through the narrow channel of the attachment. The streamlines bunch tightly together, indicating a dramatic rise in speed and drop in pressure. You see this same rise in speed when a gentle stream suddenly becomes a rushing torrent as it passes through a narrow gap.

The dramatic increase in speed and drop in pressure that occur when a steady flow of fluid passes through a narrow channel is called the **Venturi effect**, after the Italian physicist G. B. Venturi (1746–1822), who first noticed it. The Venturi effect is a special case of the **Bernoulli effect**, which recognizes that any increase in a fluid's velocity along a streamline is accompanied by a drop in pressure.

While the Bernoulli effect has many important applications in the world around us and will be used frequently in this chapter, it doesn't apply to every situation. You should remember that the Bernoulli effect only occurs *along a streamline*, in fluid that is in *steady-state flow*. Just because air is moving quickly doesn't mean that its pressure is low. The air in the cabin of a passing aircraft is really hustling along, relative to the earth, but its pressure is essentially atmospheric.

CHECK YOUR UNDERSTANDING #1: Spray Painting

In many paint and garden sprayers, air or water passes through a narrow channel on its way out of the nozzle. A thin tube joins the narrow channel at right angles and atmospheric pressure pushes paint or garden chemicals through that tube and into the channel. Evidently, the pressure inside the channel is less than atmospheric pressure. How can this be possible?

Dust and Drag Forces

Now that we understand speed and pressure in a vacuum cleaner, let's return to the dust. As air rushes toward the opening in the cleaning attachment, it carries dust with it. This phenomenon, in which a particle or portion of fluid is carried along in the flow of another fluid, is called **entrainment**. Dust entrainment is most effective in very high-speed air, which is why a narrow attachment cleans a carpet more thoroughly than a wide one.

Dust particles are entrained in air by **drag forces**. These friction-like forces appear whenever an object moves through a fluid and act to bring the object to rest relative to the fluid; they both slow an object moving through a stationary fluid and slow a fluid moving past a stationary object. In other words, drag forces, like all forces, appear in matched pairs: the fluid exerts a drag force on the object and the object exerts an equal but oppositely directed drag force on the fluid.

Since drag forces come in too many varieties to discuss all at once, we'll look at them one at a time throughout the following chapters. For now, we'll examine only **viscous drag**, the one that is active for a vacuum cleaner. Viscous drag—the sliding friction between a surface and a fluid as they move past one another—is what slows laminar water flow through a pipe. The pipe is held in place and can't move, so viscous drag slows the water down. But when air flows past a dust particle, the dust particle moves easily and viscous drag carries it along in the air.

The amount of viscous drag force that air exerts on a dust particle is proportional to the diameter of the particle and to the difference in velocities between the particle and the air. This relationship comes about because the amount of air the particle encounters via viscosity is proportional to its girth and to its velocity through the air. As always, the viscous drag force on the particle is directed so that it brings the particle to the same velocity as the air. Of course, the particle pushes back on the air with an equal but oppositely directed drag force.

Because dust particles are often just tiny rocks, we might expect viscous drag to affect rocks and dust particles equally. But that isn't so. At issue is how large the drag force is when compared to an object's weight and mass. A dust particle experiences less viscous drag than a rock because of their different diameters. While drag is proportional to diameter, the dust particle is much lighter and less massive than the rock—and mass and weight are proportional to diameter to the *third power*. Think of how many dust particles you can make by grinding up even a small rock, and consider how light each of those dust particles must be. With so little weight to hold them in place, it's not surprising that dust particles are easily blown about by viscous drag forces.

Dust particles are so incredibly light that they are easily borne aloft by air currents. Dust swept into the air by winds, emitted by industrial smokestacks, or blown upward in volcanic eruptions can remain in the atmosphere for days, weeks, or even years. Thus volcanic ash from the 1982 explosion of Mount St. Helens in Washington State was carried through the Midwest and even to the eastern United States.

Although dust tries to fall, viscous drag keeps it from descending rapidly. Like any object falling through the atmosphere, dust has a **terminal velocity**, the velocity at which the upward viscous drag force on it exactly cancels its downward weight. A falling object accelerates downward only up to its terminal velocity, at which point the net force on it is zero and it descends at constant velocity. While a person's terminal velocity may be 300 km/h without a parachute and 30 km/h with a parachute, a dust particle's terminal velocity may be only 0.0036 km/h (1 mm per second). Any upward air current will therefore lift it back into the sky. In a calm, sunlit room, you can often see dust particles drifting about with the air currents, prevented from falling by the viscous drag force.

This same viscous drag force is what lets air carry dust particles into the vacuum cleaner. The force reduces any difference in velocities between the air and the dust, so if the air rushes into the vacuum cleaner, the dust will, too. The faster the air moves, the larger the viscous drag force on a particle. This increase in force with air speed explains why a narrow attachment cleans better than the hose alone: the air speed is higher near the attachment and the drag force is larger.

Unfortunately, viscous drag also slows the air as it passes close to carpet fibers or the surface of the floor. It's hard to keep air moving quickly near those surfaces. To remove really ground-in dirt from a carpet or floor, you need a powerful fan and the high air speed that comes from making air pass through a narrow opening. The need for high air speed also explains why battery-powered or poor-quality vacuum cleaners don't clean well: their fans are too weak to move the air fast enough to remove the dirt effectively, particularly near surfaces. Beater brushes, which jostle the dirt away from surfaces, help to move the dirt into the fast-moving air stream so that it can be carried into the vacuum cleaner.

All this explains why vacuum cleaners can pick up small particles; but why should they have trouble collecting large objects? First, the viscous drag force on a marble is just not large enough to overcome its weight and lift it up into the vacuum cleaner; you'd probably have to sweep over it a couple of times unsuccessfully and then pick it up with your hand and drop it into the hose. Second, a marble is much larger than a dust particle, and the flow of air around it is characterized by a much larger Reynolds number. Viscous drag is the dominant drag force only when the flow is laminar. When vacuuming, the flow of air around a dust particle is laminar, but the flow of air around a marble includes turbulence. As a result, the marble experiences new drag forces that we will look at when we discuss balls and air. Nonetheless, these new drag forces aren't able to carry the marble into the vacuum cleaner.

CHECK YOUR UNDERSTANDING #2: Keeping Clouds in the Sky
Clouds consist of countless tiny water droplets. Water is more dense than air, so why don't these water droplets fall?

The Fan

We've looked at how a vacuum cleaner picks up dust; now it's time to look at the fan. Without the fan, viscous drag forces would quickly stop air from flowing through the cleaning hose by converting its total energy into thermal energy. This effect is most severe when you use a narrow cleaning attachment, because the high-speed air inside that attachment loses total energy particularly quickly.

To keep air moving through the hose, the fan pumps the air from the low-pressure region in the cleaning hose to the high-pressure region at the exhaust port, against its natural direction of flow. The fan does work on the air and replaces the energy that viscous drag has converted to thermal energy. Because the fan increases the air's total energy, the airflow through the fan is *not steady-state flow*; the air's pressure increases as it flows through the fan without a corresponding decrease in speed.

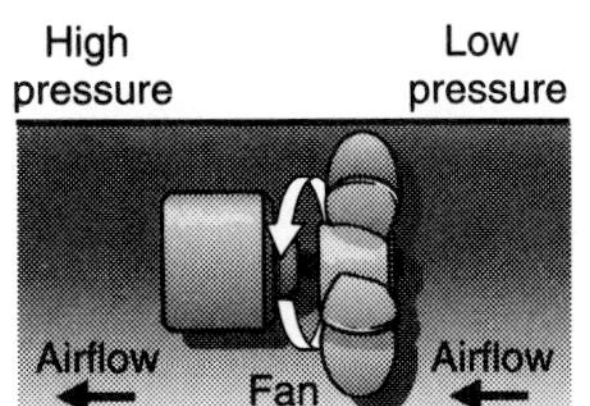

Fig. 4.2.3 - As a vacuum cleaner's fan rotates, its blades transfer air from the low-pressure side to the high-pressure side, doing work on the air in the process.

In its basic form, the fan is just a rotating array of ramp-like blades that push the air as they turn past (Fig. 4.2.3). These blades do work on the air, using energy provided by an electric motor, which of course gets its energy from the electric power company. What is interesting here is that work is being done by a rotational motion, not the translational motion that we normally associate with work. We have seen that work is done by exerting a force while traveling a distance in the direction of that force. But work can also be done by exerting a torque while rotating through an angle in the direction of that torque.

To see how you can do work with torques, think of a basketball. If you exert a torque on this basketball, it begins to rotate; you've done work on it and given it rotational kinetic energy. If you now exert a torque on it in the other direction, so that the torque and the direction of rotation are opposite one another, you do negative work on the basketball and it comes to rest.

The amount of work that you do in this manner is equal to the product of the torque you exert times the angle through which the object rotates. This relationship can be written as a word equation:

$$\text{work} = \text{torque} \cdot \text{angle of rotation}, \tag{4.2.1}$$

in symbols:

$$W = \tau \cdot \theta,$$

and in everyday language:

If you're not twisting it or it's not turning, then you're not working.

In this relationship, the angle of rotation is in the direction of the torque. If the object turns in the direction opposite your torque, it does work on you! To actually calculate the work you do on something when you twist it, you must measure the angle of rotation in the natural units of angle: **radians**. 1 radian is equal to $180/\pi$ degrees (about 57.3°). While degrees are probably more familiar to you than radians, using degrees in Eq. 4.2.1 will lead to incorrect results.

We will not use Eq. 4.2.1 often, but it will be important in devices such as bicycles and electric motors. For now, you only need to recognize that the motor does work on the fan via a torque and that the fan uses this work to pump air from low pressure to high pressure.

This concept helps explain the high-pitched whine that most vacuum cleaners emit when you block the airflow through their hoses. If no air can reach the low-pressure side of the fan, the fan can't pump any air to the high-pressure side. Since the fan moves no air, it does no work and spins idly. The motor is still turning the fan, but neither one is doing any work; the motor isn't exerting any torque on the fan, and the fan is turning freely. Since the fan offers no resistance to rotation, the motor and the fan turn faster and faster, so that the motor quickly exceeds its rated rotational speed and begins to whine. When you unblock the airflow, the fan and motor begin to do work on the air, the motor slows down, and the whine disappears.

CHECK YOUR UNDERSTANDING #3: Moving the Air
In terms of fluid flow, how does a household fan cause air to move?

CHECK YOUR FIGURES #1: Doing Work on Your Bicycle
You are using a wrench to tighten a bolt. To keep the bolt turning steadily, you must exert a torque of 10 N·m on it. By the time you have completed one full turn (360° or 2π radians), how much work have you done?

Filtering the Dust

A vacuum cleaner uses a fan to create a moving stream of air; this air entrains dust particles by way of viscous drag forces. But the vacuum cleaner doesn't make the air vanish; it returns the air to the room once that air has passed through the fan. What keeps the dust from returning to the room, too?

The answer, of course, is a filter, a device that blocks the dust particles while permitting air molecules to pass (Fig. 4.2.4). A typical filter is made of porous paper or cloth, with fibers that are loosely woven to create openings or pores large enough for air to pass but too small for dust to pass.

This simple filtration is complicated by viscous drag. The air that passes through the filter's pores loses some of its total energy trying to move past the stationary air at the surfaces of the pores. Just as the viscous drag force on a particle becomes larger as the particle's speed through the air increases, so the viscous drag force on the air becomes larger as the air's speed through a pore increases. The faster air moves through a pore, the more energy it loses during the trip. To minimize the energy lost to viscous drag, the vacuum cleaner must move the air slowly through the filter. It does this by using a very large filter so that the air has lots of surface area and many pores through which to flow.

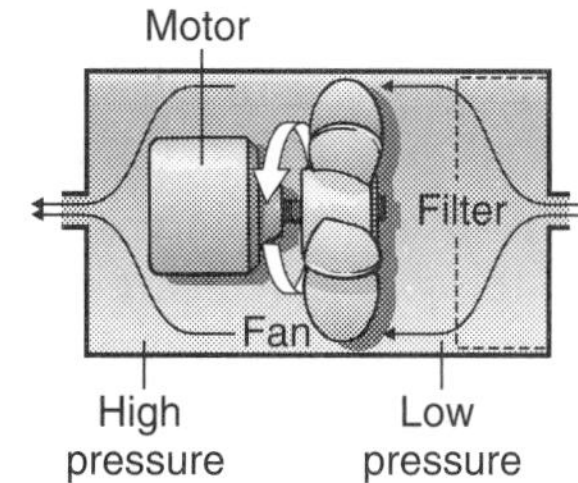

Fig. 4.2.4 - In a canister vacuum cleaner, air flowing through the hose first passes through the filter and then through the fan. The fan pumps air from the low-pressure region inside the vacuum cleaner to the atmospheric pressure outside.

When the filter is new, the air flows slowly through its pores and loses relatively little energy; the filter removes dust from the air without much effect on the air itself, and the vacuum cleaner works well. But as dust begins to plug the pores in the filter, the air must move more and more quickly through the pores that remain open. The air loses much of its energy as it struggles to pass through the filter. The filter thus slows the flow of air, and the vacuum cleaner doesn't clean well anymore.

CHECK YOUR UNDERSTANDING #4: The Beekeeper's Bind

Honey flows very slowly out of the honeycomb or through a cheesecloth filter. What slows it down?

Practical Vacuum Cleaners

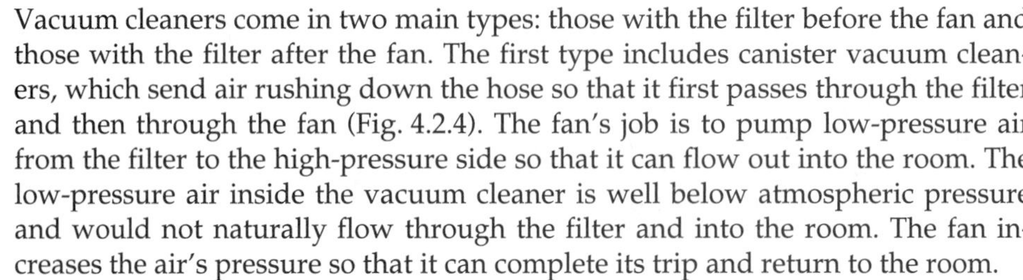

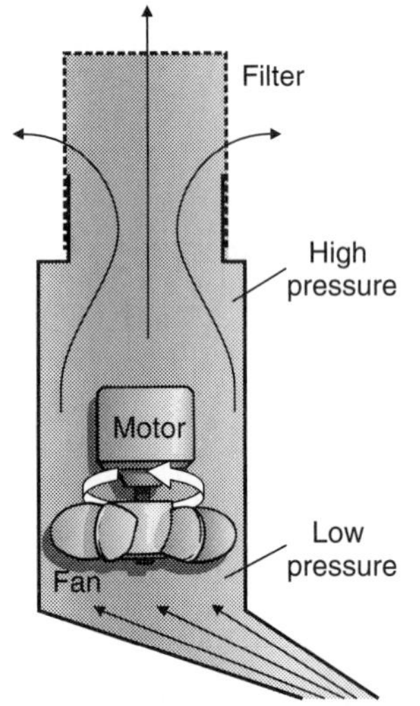

Fig. 4.2.5 - In an upright vacuum cleaner, air flowing up from the carpet is pumped by the fan and flows out into the room through the filter.

Vacuum cleaners come in two main types: those with the filter before the fan and those with the filter after the fan. The first type includes canister vacuum cleaners, which send air rushing down the hose so that it first passes through the filter and then through the fan (Fig. 4.2.4). The fan's job is to pump low-pressure air from the filter to the high-pressure side so that it can flow out into the room. The low-pressure air inside the vacuum cleaner is well below atmospheric pressure and would not naturally flow through the filter and into the room. The fan increases the air's pressure so that it can complete its trip and return to the room.

In contrast, upright vacuum cleaners place the filter after the fan, so that air flowing up from the carpet first passes through the fan, then past the motor, and then through the filter (Fig. 4.2.5). Perhaps the biggest difference between the canister and upright vacuum cleaners is in what happens when you vacuum up an object like a penny. In the canister vacuum, the penny stops in the filter and never gets into the fan; in the upright vacuum, the penny flies right through the fan. Since a penny is fairly heavy and the vacuum cleaner has trouble moving it with drag in the first place, the penny tends to rattle around in the fan, making a terrific racket.

CHECK YOUR UNDERSTANDING #5: When to Use a Mop

Why is it dangerous to vacuum up water with an upright vacuum cleaner?

Summary

How Vacuum Cleaners Work: Vacuum cleaners work by entraining dust and dirt in rapidly moving air. The air is accelerated by a difference in pressures. The air outside the vacuum cleaner is at atmospheric pressure and flows toward a partial vacuum created by a fan inside the vacuum cleaner. The air's speed increases as its pressure drops. Depending on the shape of the hose and the rate at which the fan can move air, the air speed near the end of the hose can become very large. Dust surrounded by this moving air experiences viscous drag forces that act to minimize the differences in their velocities. The dust quickly acquires the same velocity as the air and is carried along in it. Once inside the vacuum cleaner, dust is extracted from the air by a filter. The air, pumped to atmospheric pressure by the fan, returns to the room.

Important Laws and Equations

1. The Definition of Work in Rotational Motion: Work is the torque exerted on an object times the angle through which the object rotates in the direction of that torque, or

$$\text{work} = \text{torque} \cdot \text{angle of rotation}\,. \tag{4.2.1}$$

The angle is measured in radians.

The Physics of Vacuum Cleaners

1. Air is effectively an incompressible fluid as long as its speed is below about 300 km/h and its pressure is within about 10% of atmospheric. Its behavior is well characterized by Bernoulli's equation. In situations where its speed or pressure exceed these limits, its behavior may deviate noticeably from that predicted by Bernoulli's equation.

2. In steady-state flow, an effectively incompressible fluid that experiences an increase in speed also experiences a decrease in pressure. This drop in pressure is the Bernoulli effect.

3. In laminar fluid flow, the dominant drag force on an object is viscous drag. This drag tries to bring the object and the fluid to the same velocity.

4. Work is required to transfer a volume of fluid from low pressure to high pressure.

5. Work can be done by exerting a torque on an object and turning it through an angle in the direction of that torque. The work done is the product of the torque times the angle of rotation, measured in radians.

Check Your Understanding - Answers

1. As the air or water flows through the narrow channel, its speed increases and its pressure drops. Atmospheric pressure pushes liquid through the thin tube toward the region of low pressure in the channel.

Why: The pressure inside the narrow channel of the sprayer is much lower than atmospheric pressure. You could measure this low pressure through a tiny hole drilled in the side of the channel. Since the pressure in the channel is low, atmospheric pressure can push paint or chemicals through the hole and into the channel. The sprayer improves on this idea by adding a tube to allow atmospheric pressure to transfer the liquid from a container to the channel.

2. Viscous drag prevents the water droplets in clouds from falling at a perceptible rate.

Why: The tiny water droplets in clouds are so light that their motion is dominated by viscous drag. They move with the wind. For them to fall, they must grow much larger, so that their weight can begin to influence their motion. Once they grow to the size of an apple seed, they fall to the earth as rain, or perhaps snow or hail. In a violent thunderstorm, with rapid upward movements of air, drag forces can keep hail aloft until it becomes remarkably large.

3. It pumps the air, increasing the air's total energy. Since the air's pressure doesn't change in passing through the fan (it remains atmospheric), its speed must increase.

Why: A fan's main action is to increase the total energy of the air that passes through it. In a vacuum cleaner, the air can't increase its speed so that increase in energy takes the form of an increase in pressure. The vacuum cleaner's fan increases the air's pressure potential energy by pumping it from low pressure to high pressure. But in an open room fan, the air's pressure must remain roughly atmospheric. Instead, the air's speed increases. The room fan increases the air's kinetic energy by speeding it up. The room fan draws air from a wide region behind the fan and sweeps it into a narrow, fast moving beam in front of the fan.

4. Viscous drag.

Why: Honey is highly viscous and has great difficulty flowing through pores. The stationary honey on the surface of each pore exerts strong viscous forces on honey trying to pass through the pore. Gravity alone is often not enough to keep the honey moving at a reasonable speed so it's usually extracted with a centrifuge.

5. The water passes immediately through the fan and the electric motor on its way to the filter.

Why: The reason a shop vacuum adopts the canister approach of having the filter before the fan is so that water and shop debris don't go through the fan and motor. In an upright vacuum, anything you vacuum up goes right through the fan and past the motor. Water is not a good thing to put into a motor.

Check Your Figures - Answers

1. About 62.8 J of work.

Why: The work you do is equal to the force you exert on the bolt times the angle through which it rotates, as measured in radians (Eq. 4.2.1). Since the torque is 10 N·m and the angle is 2π or about 6.28 radians,

$$\text{work} = \text{torque} \cdot \text{angle of rotation} = 10\ \text{N} \cdot \text{m} \cdot 6.28 = 62.8\ \text{J}.$$

The unit of torque, the newton-meter, is the same as the unit of energy, the joule.

Glossary

Bernoulli effect The drop in pressure that occurs when the speed of an effectively incompressible fluid in steady-state flow increases as it moves along a streamline.

drag forces The friction-like forces exerted by a fluid and a solid on one another as the solid moves through the fluid. These forces act to reduce the relative velocity between the two.

entrainment The phenomenon in which a particle or portion of fluid is carried along in the flow of another fluid.

radians The natural unit in which angles are measured. 1 radian is $180/\pi$ degrees or approximately 57.3°.

terminal velocity The velocity at which an object moving through a fluid experiences enough drag force to balance the other forces on it and keep it from accelerating.

Venturi effect The increase in speed and drop in pressure that occur when an incompressible fluid in steady-state flow passes through a narrow channel.

viscous drag A drag force that results from viscous forces on a moving surface immersed in a fluid.

Review Questions

1. When air flows through a narrow portion of a hose, what happens to the air's speed and pressure?

2. When air flows through a wide portion of a hose, what happens to the air's speed and pressure?

3. How does a parachute slow a person's descent and why does the parachutist eventually stop accelerating downward?

4. What is the net force on a parachutist who is descending at terminal velocity?

5. To show that an object's volume and mass decrease more rapidly than its surface area as it shrinks, consider two buckets that are identical except that one is exactly half as tall and half as wide as the other. How much more water do you need to fill the larger bucket than to fill the smaller bucket? How much more paint do you need to coat the inside of the larger bucket than to coat the inside of the smaller bucket?

6. Why are small solid objects affected more strongly by viscous drag forces than are large solid objects?

7. If an object falls through air at its terminal velocity, how can a rising air current lift that object upward?

8. List three situations in which you do work on something by twisting it while it rotates in the direction of that twist.

9. List three situations in which something does work on you as you try to turn it in the direction opposite its rotation.

10. Why does the wind lose speed and pressure after it passes through the fabric of your shirt?

Exercises

1. Some disease-causing bacteria are carried by the air. Why don't these organisms fall to the ground as people and animals do?

2. To study single-celled organisms in water, biologists often add non-toxic chemicals that increase the water's viscosity. Why does this change slow the organisms' motions through the water solution?

3. Which falls faster: a large raindrop or a small raindrop?

4. An exhaust fan draws air out of a house. Fresh air will then flow into the house through opened windows. How does the number of open windows affect the air pressure in the house and the speed of the air as it flows through the windows?

5. Why do fuzzy coverings allow some seeds to be carried aloft by the wind?

6. A skydiver can control her terminal velocity by changing shape. Why does she fall faster when she's rolled into a ball than when she has her arms and legs outspread?

7. A pearl dropped into a jar of honey descends much more slowly than the same marble dropped into water. Explain this effect. (Neglect buoyancy.)

8. Most swimming pools and spas have porous filters such as packed sand and gravel that clean the water. Water pumps circulate water through these filters. Why is it important to clean the filters periodically?

9. Adding a wide diffuser to the end of a blow drier allows the airflow to broaden before it reaches your hair. What happens to the air's speed and pressure as it enters the diffuser?

10. Some vacuum systems are built right into the walls of a house. They use heavy, powerful motors to draw air through the long pipes that extend throughout the house. How do the long pipes affect such a system's performance? Why is a powerful motor important?

11. A shop vacuum cleaner is a canister vacuum cleaner that can pick up water. Its has a large bucket beneath the fan into which water falls after arriving through the narrow cleaning hose. Why does the bucket's large size keep water from flowing into the fan?

12. A leaf blower uses rapidly moving air to clear debris from sidewalks. How does the rushing air cause leaves to move?

13. How can a tornado pick up large objects such as cars?

14. As a large truck rolls down the highway, it leaves a trail of forward moving air. Explain how this air would affect a bicycle rider traveling behind that truck.

15. When you overtake and pass a truck on the highway, your car is drawn toward the truck. How does the air's passage between your car and the truck explain this attraction?

16. When wind blows through the canyons of skyscrapers in a big city, the doors in those buildings may spontaneously open outward. Why?

17. When the driver of a moving car opens the side window slightly, air is drawn out of the car. What creates the region of low pressure just outside that car window?

18. A sandblaster removes paint from a cement wall by pelting the paint with high-speed sand. The sandblaster is powered by compressed air. How can compressed air produce a stream of sand?

19. You can clean dust off a shelf by blowing on it hard. How does this work?

20. When a boatload of running shoes fell from a cargo ship in the Pacific Ocean in the 1980's, ocean currents carried the individual floating shoes along specific paths until they reached land. What was this motion an example of?

21. A sealed jar often has an indented lid. If the seal is broken, the lid will have popped up and you should beware of its contents. What causes the sealed jar's lid to remain indented?

22. When you open a sealed jar, the indented lid usually pops upward and you hear a hissing sound. Explain this effect in terms of pressure and force.

23. An aneurysm is a widening in a blood vessel. Aneurysms are particularly susceptible to bursting because the pressure inside them is unusually high. Explain.

24. Small insects often tumble from vast heights without injuring themselves. Why don't they hit the ground hard enough to cause damage?

25. If you were only 1 cm tall, you would be able to jump out of an airplane without a parachute and land without injury. Why?

26. If you drop a pearl into a bottle of thick shampoo, it falls slowly at a constant speed. Explain.

27. When a meteor enters the earth's atmosphere, it actually slows down. Gravity is pulling it downward, so why doesn't the meteor speed up?

28. A police forensics lab studies marks left on a bullet by its passage through a gun barrel. To catch the bullet without damaging it, they fire it into water. What kind of force slows the bullet as it moves through the water?

29. Builders often used corrugated plastic pipes to carry water away from a house. These accordion-like pipes alternately narrow and widen about every 5 cm, a feature that makes them much more flexible than straight plastic pipes. If water experienced laminar flow in one of these pipes, what would happen to its speed and pressure?

30. Why does cleaning the lint filter of a clothes dryer increase the volume of air that passes through the dryer?

31. A forced air heating system is almost identical to a giant canister vacuum cleaner, except that heat is added to the air after it passes through the fan. Both filter the air just before it enters the fan. What happens to the air pressures before and after the air filter if you forget to change it for a year or two?

32. An airbrush or paint sprayer produces a paint mist by sending pure air through a narrow channel. The channel is pierced at right angles by a tube that connects to the paint reservoir. Paint flows up the tube and into the airstream. What pushes the paint up the tube?

33. When you let the air out of an elastic rubber balloon, its narrow neck opens and closes rapidly, making a sputtering noise. Since the neck doesn't stretch, this effect isn't due to the rubber's elasticity. The neck opens whenever the pressure inside it rises above atmospheric pressure and it closes whenever that pressure falls below atmospheric pressure. Explain why the pressure in the neck rises above and falls below atmospheric pressure.

34. When you start a merry-go-round spinning, its kinetic energy increases. When you stop that merry-go-round, its kinetic energy decreases. Show that you do work on the merry-go-round as you start it spinning and that it does work on you as you stop it.

Problems

1. To open an old water faucet completely, you have to turn its handle three complete turns (3 times 360° or 1080°). How far is that in radians?

2. You are attaching bookshelves to a wall with wood screws. To keep one of the screws turning steadily, you must exert a torque of 1.5 N·m on it. If you must turn the screw 18 times to tighten it completely, how much work are you doing on that screw?

3. The fan in your room turns 20 times a second and consumes 80 W of electric power. If all of that electric power is used to do work on the fan blades, how much torque is the fan's motor exerting on those fan blades?

4. The ceiling fan in your favorite restaurant has three speeds: 1, 2, and 4 turns per second. At the higher speeds, the fan blades encounter more air and push on that air harder, so the torque that the motor must exert on the blades increases as the square of the fan's angular speed (i.e., speed^2). For example, the motor's torque at the medium speed is 4 times that at the slow speed. If the fan uses 2 J of energy to turn once at the lowest speed, how much energy does it use to turn once at the medium and high speeds?

5. How much power does the fan in Problem **4** use at each of its three speeds?

6. You're tightening bolts in your car with a torque wrench, a lever that grips the head of a bolt and allows you to exert a controlled torque on that bolt. It has a dial or needle that reports exactly how much torque you're exerting. In the present case, the manual tells you to tighten the bolt with a maximum of 12 N·m of torque. (If you exert more torque than that, you risk damaging the bolt or the car.) In the last full turn before reaching the specified torque, the average torque you exerted on the bolt was 10 N·m. How much work did you do on the bolt during that last turn?

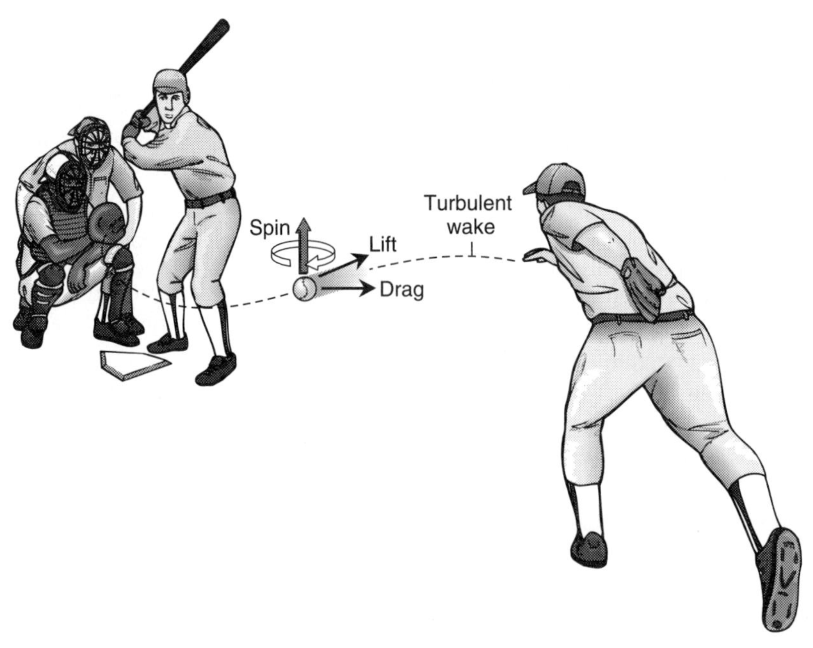

Section 4.3

Balls, Birdies, and Frisbees

Much of the subtlety and nuance in games such as baseball and Ping-Pong come from the way balls interact with air. If baseball were played on the moon, which has no air, the sinking fast ball would be the only interesting pitch; other thrown objects, such as flying disks, wouldn't work at all. In this section we will investigate how air affects the flight of balls and other related objects.

Questions to Think About: *Why can you a throw a real baseball so much farther than a hollow plastic one? Why does a long fly ball appear to drop straight down when you try to catch it in deep center field? What kind of force could make a curve ball curve? How can a well-hit golf ball hang in the air before falling to the green? How can a knuckle ball or spit ball jitter about in flight? Why does a shuttlecock always fly with its feathers behind? What keeps a Frisbee in the air?*

Experiments to Do: *To make air's effects most apparent, you need a ball that weighs little but has lots of surface area. A whiffle ball or hollow plastic ball will do nicely. See how far you can throw it. How does it stop? Does it slow down and lose height gradually, or does it stop rapidly and fall to the ground? Now make the ball spin as you throw it, by rolling it against your index and middle fingers as it leaves your hand. Does the ball curve in flight? Does a fast spin make the ball curve more or less? Which way is the ball spinning, and how is the spin related to the direction of its curve? Change the direction of spin by orienting your hand differently as you throw. Which way does the ball curve now?*

When a Ball Moves Slowly: Laminar Airflow

One of the first things you might notice if you joined a new baseball franchise on the moon would be that you could throw the ball farther than back at home. Part of this increased distance would be due to the weak gravity (the ball would take longer to fall), but part would stem from the absence of air resistance. Air acts to slow down anything that moves through it. We've seen that air is so effective at slowing dust that dust barely falls at all. Now we'll look at how air affects the motion of a larger object: a ball.

Our encounter with dust and viscous drag might seem to have taught us everything we need to know about **aerodynamics**, the study of the dynamic (moving) interactions of air with objects. But dust is a particularly simple case. Because a dust particle is so tiny and moves so slowly, air flows around it in an orderly fashion. The airstream spreads neatly as the dust particle drifts by and then comes back together once the dust particle has left. The airflow around the particle is laminar and the only force that the air exerts on the dust particle is viscous drag.

But there are other **aerodynamic forces**—that is, forces exerted on an object by the air moving past it. Some are *drag forces*, slowing the object's motion through the air, while others are *lift forces*, pushing the object to one side or the other. Since many of these new aerodynamic forces are caused by turbulent airflow, they are present only when the air's motion is dominated by inertia. And since viscosity dominates air's motion around small, slow-moving particles and keeps the airflow laminar, we need a large, fast-moving object to explore turbulent airflow. Let's examine the airflow around a ball. Except when it moves extremely slowly, the airflow around a ball is turbulent.

> Laminar airflow: viscosity dominates the airflow, keeping it smooth and orderly. Laminar airflow occurs around smaller objects that move slowly through the air.
>
> Turbulent airflow: inertia dominates the airflow, ripping it apart into swirling eddies. Turbulent airflow occurs around larger objects that move quickly through the air.

Before we explore turbulent airflow, however, we should look at the laminar airflow around a slow-moving ball (Fig. 4.3.1). Actually, the pattern of airflow will be the same whether the ball moves slowly through the air or the air flows slowly past the ball. To simplify our observations in this section, we'll study airflow from the reference frame of the ball, so that the ball appears stationary, with the air flowing past it. When the air moves slowly enough, it separates neatly around the front of the ball and comes back together behind it, producing a **wake**, an air trail left behind the ball, that is smooth and free of turbulence.

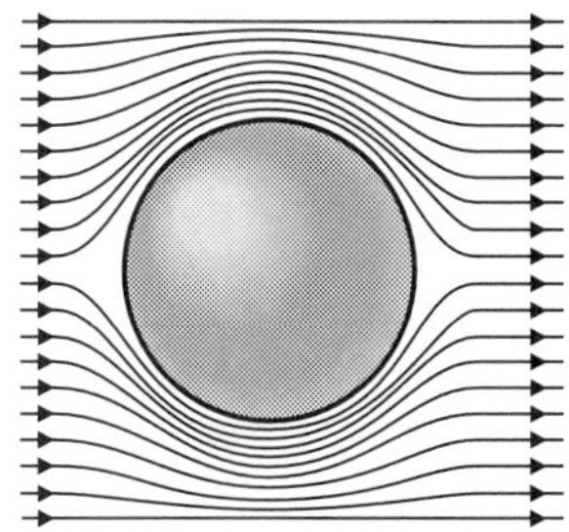

Fig. 4.3.1 - The airflow around a slowly moving ball is laminar. Air slows down in front of and behind the ball and its pressure there increases. Air speeds up at the sides of the ball and its pressure there decreases. However, the pressure forces on the ball balance one another perfectly and it experiences only viscous drag.

But the air's speed isn't uniform all the way around the ball. At the front, the air slows down and turns to move out of the ball's way. Since air in steady-state flow obeys Bernoulli's equation, this decrease in speed is accompanied by an increase in pressure. Thus the air pressure in front of the ball is actually greater than atmospheric pressure. The wide spacing of the streamlines in front of the ball indicates this drop in speed and increase in pressure.

Near the sides of the ball, the air speeds up as it rushes around the curved surfaces. Air following those curves must travel farther and faster than air flowing straight past the ball. According to Bernoulli's equation, this increase in speed is accompanied by a decrease in pressure. Around the sides of the ball, therefore, the air pressure is actually less than atmospheric pressure. The

❐ When the airflow around an object is laminar, the pressure forces on it cancel perfectly and it experiences no drag due to pressure imbalances—no *pressure drag*. The absence of pressure drag was a great puzzlement to early aerodynamicists, who knew that the airflow around dust is laminar and that it experiences a drag force. This mystery was named d'Alembert's paradox, after Jean le Rond d'Alembert (1717–1783), the French mathematician who first recognized it. D'Alembert and his contemporaries didn't know about the viscous drag force, which is what really slows dust's motion through the air.

bunching together of the streamlines on the ball's sides indicates this rise in speed and decrease in pressure.

The laminar airflow continues around to the back of the ball, where the air slows down and its pressure rises. The streamlines spread apart once again, indicating a drop in speed and increase in pressure. Just as at the front of the ball, the air pressure behind the ball is actually greater than atmospheric pressure.

It seems strange that the air pressure can be different at different points on the ball, but that is what happens in a flowing stream of air. It's particularly remarkable that low-pressure air at the sides of the ball is able to flow around to the back of the ball, where the pressure is higher! Your intuition may tell you that air should flow from high pressure to low pressure, not the reverse. But when you think that, you forget inertia. The low-pressure air flowing past the sides of the ball has lots of forward momentum that carries it around to the back of the ball. When the airflow moves from high pressure to low pressure, it speeds up. When the airflow moves from low pressure to high pressure, it slows down. Everything makes sense.

The airflow around the ball is symmetric, and the forces exerted by air pressure on the ball are also symmetric. These pressure forces cancel one another perfectly so that the ball experiences no net pressure force. In particular, the high pressure in front of the ball is balanced by the high pressure behind it. As a result of this symmetric arrangement, the only aerodynamic force acting on the ball is viscous drag (see ❐).

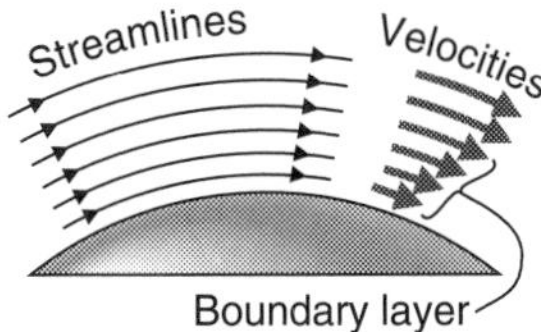

Fig. 4.3.2 - Airflow past a surface is laminar until the Reynolds number exceeds about 2,000. The main airstream then becomes turbulent, leaving only the thin layer of laminar flow near the surface that is shown here. This boundary layer is slowed by viscous drag and doesn't become turbulent until the Reynolds number exceeds about 100,000.

CHECK YOUR UNDERSTANDING #1: Smooth Flow in a Stream

When water in a stream flows slowly past a small rock, the water in front of the rock slows down and its increased pressure lifts the water level slightly. The water level behind the rock also rises slightly. Explain.

When a Ball Moves Fast: Turbulent Airflow

Not all objects experience laminar flow in air. Turbulence is everywhere, particularly in sports, bringing with it a new type of drag force. When the air flowing around a ball becomes turbulent, the air pressure around it is no longer symmetric and it experiences **pressure drag**, the *downstream* force exerted by unbalanced pressures in the moving air. This pressure imbalance produces a net force on the ball that slows its motion through the air.

A ball can experience turbulent airflow and pressure drag when its Reynolds number exceeds about 2000. The Reynolds number, introduced in Section 4.1, combines the ball's size and speed and air's density and viscosity to give an indication of whether the airflow is dominated by viscosity or inertia. At low Reynolds numbers, air's viscosity keeps it flowing smoothly and orderly over the ball's surface in laminar flow. But at high Reynolds numbers, those above about 2000, air's inertia prevents it from following the ball's curves so that it begins to swirl about. This turbulence, however, won't start until something triggers it. Something must cause the first swirling vortex in the air, and that something is viscosity.

❐ Among Ludwig Prandtl's (German engineer, 1875–1953) many pivotal contributions to aerodynamic theory is the concept of boundary layers in fluid motion. Prandtl was so engrossed in establishing Göttingen as the world's foremost aerodynamic research facility that he did not have time to court a wife. Deciding he should be married, Prandtl wrote his former advisor's wife, asking to marry one of her two daughters but not specifying which one. The family selected the eldest daughter and the wedding took place.

Even in a strong wind, air at the surface of the ball is stationary; it's held in place by viscous forces, which also slow down a thin layer of air near the ball's surface, the **boundary layer** (Fig. 4.3.2). This boundary layer, first recognized by Ludwig Prandtl ❐ with help from Gustave Eiffel (Fig. 4.3.3), experiences all the pressure changes of the freely flowing air farther from the surface, but it doesn't have as much speed or kinetic energy.

As air flows around toward the back of the ball, it slows down and its pressure increases. This situation, where air must flow from low pressure to high pressure, is called an **adverse pressure gradient**. The air outside the boundary layer has plenty of forward momentum and completes its trip through this adverse pressure gradient without difficulty. But the air in the boundary layer has been slowed by viscous forces and doesn't have as much kinetic energy as the freely flowing airstream.

If the Reynolds number is low (laminar flow), there is no boundary layer and the entire airstream helps push air near the ball's surface around to the back of the ball. But at high Reynolds numbers, viscous forces between the main airstream and the boundary layer are too weak to keep the boundary layer moving forward. As the boundary layer flows into the adverse pressure gradient behind the ball, it comes to a stop and reverses directions. This reversal is a disaster for laminar flow, which can't have parts of its stream heading backwards. The main airflow separates from the back of the ball, leaving a turbulent wake or air pocket behind the ball (Fig. 4.3.4).

Because of this turbulent wake, the air behind the ball no longer slows down and its pressure no longer rises. This swirling air upsets the perfect cancellation of pressure forces. Since the air in front of the ball exerts more pressure and force on the ball than the air behind it, the ball experiences pressure drag. In effect, the ball drags the air in the turbulent wake along with it.

Pressure drag slows the flight of almost any ball moving faster than a snail's pace. It also affects all large or fast-moving objects, including cars, bicycles, and blimps, since their Reynolds numbers easily exceed the value of 2000 that sets the upper limit for laminar flow. The pressure drag force is roughly proportional to the cross-sectional area of the turbulent air pocket and to the square of the object's speed through the air. For a ball moving at a modest speed, the air pocket is about as large around as the ball and the ball experiences a large pressure drag force.

Fig. 4.3.3 - Early experiments in aerodynamics were performed by Gustave Eiffel (French engineer, 1832–1923), who designed the tower that bears his name. In the 1890's, Eiffel dropped objects of various sizes and shapes from his tower and measured the drag that they experienced. His work was used by Prandtl to explain the reduction in drag that accompanies the appearance of turbulent boundary layers.

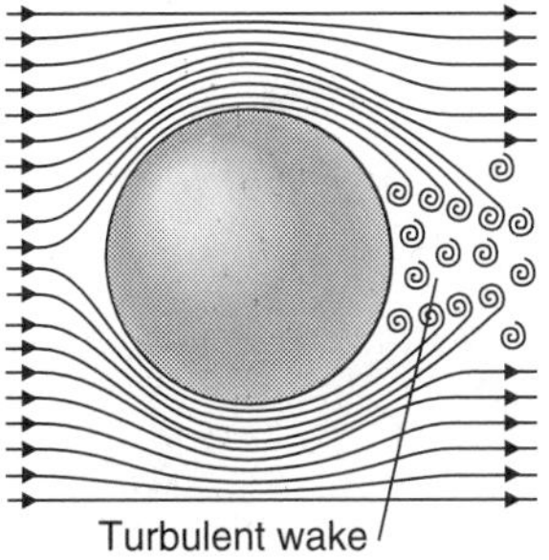

Fig. 4.3.4 - When a ball's speed gives it a Reynolds number between about 2,000 and 100,000, flow problems in the laminar boundary layer cause the main airflow to separate from the ball's surface, leaving a large turbulent wake. The pressure behind the ball remains low and the ball experiences a large pressure drag.

CHECK YOUR UNDERSTANDING #2: Leaving No Trace

When your canoe coasts extremely slowly across the water of a still lake, it leaves almost no trail in the water behind it. However, when you paddle it swiftly through the water, the canoe leaves a rippling wake. Explain this difference.

The Dimples on a Golf Ball

If this were the whole story, you would never hit a home run at a baseball game or a 250-yard drive on the golf course. At higher Reynolds numbers, however, something new happens. When the air moves fast enough over the ball's surface, the nature of the boundary layer changes. The air at the surface of the ball is still stationary, and its average speed still increases slowly as you look farther away from the surface. But the boundary layer air itself becomes turbulent (Fig. 4.3.5); it contains lots of tiny whirling eddies that effectively insulate the main airstream from the surface (Figure 4.3.6).

Because it's whirling about, air in a turbulent boundary layer has more kinetic energy than air in a laminar boundary layer. This extra kinetic energy allows the turbulent boundary layer to flow most of the way around the back of the ball before the rising pressure there stops its forward motion. The main airstream follows the turbulent boundary layer around the back of the ball, and its pressure rises as it slows down. Although the main airstream eventually separates from the back of the ball, the turbulent wake it creates is relatively small.

Since the air pocket behind the ball is smaller, the pressure drag is reduced from what it would be without the turbulent boundary layer. The effect of re-

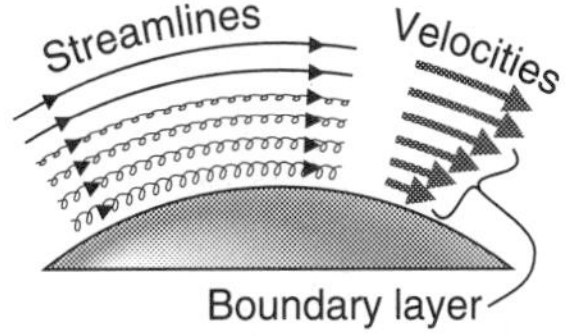

Fig. 4.3.5 - When the Reynolds number exceeds about 100,000, the boundary layer of air flowing past a surface becomes turbulent. This whirling gas effectively insulates the main airstream from the surface. Its relatively high downstream velocity and momentum can carry this turbulent boundary layer deep into a region of increasing pressure.

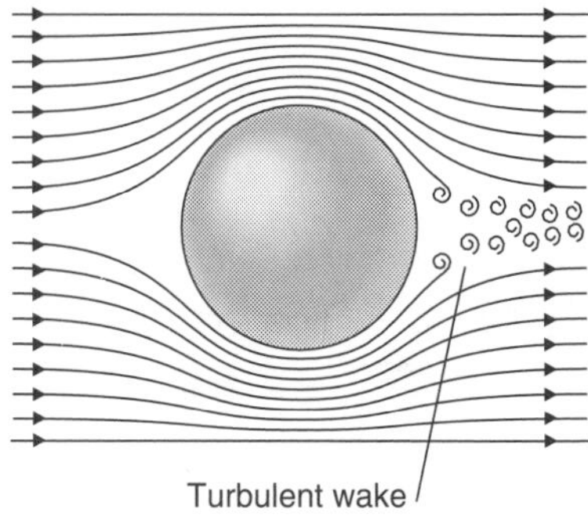

Fig. 4.3.6 - When a ball travels fast enough that its Reynolds number exceeds 100,000, its boundary layer becomes turbulent. This turbulent layer insulates the main airstream from the surface so that it continues most of the way around the ball, leaving only a small turbulent wake. The pressure behind the ball is relatively high and the ball experiences only a modest pressure drag.

placing the laminar boundary layer with a turbulent one is enormous; it's the difference between a golf drive of 70 yards and one of 250 yards! The effects of Reynolds number of the airflow around a ball are summarized in Table 4.3.1.

Table 4.3.1 - The effects of Reynolds number on the airflow around a ball or other object.

Reynolds Number	*Boundary Layer*	*Main Airstream*	*Type of Wake*	*Main Drag Force*
<2,000	Laminar	Laminar	Small Laminar	Viscous
2,000–100,000	Laminar	Turbulent	Large Turbulent	Pressure
>100,000	Turbulent	Turbulent	Small Turbulent	Pressure

Delaying the airflow separation behind the back of the ball is so important to distance and speed that the balls of many sports are designed to encourage a turbulent boundary layer (Fig. 4.3.7). Rather than waiting for the Reynolds number to exceed 100,000, the point near which the boundary layer spontaneously becomes turbulent, these balls "trip" the boundary layer deliberately (Fig. 4.3.8). They introduce some impediment to laminar flow, such as hair or surface irregularities, that causes the air near the ball's surface to tumble about and become turbulent. That's why a tennis ball has fuzz and a golf ball has dimples.

CHECK YOUR UNDERSTANDING #3: Designing a Great Sports Car

As an automobile designer, your job is to minimize the aerodynamic drag experienced by the car you are working on. Where should you try to locate the point at which the airflow separates from the car?

Fig. 4.3.7 - Early golf balls (*left*) were handmade of leather and stuffed with feathers. Golf became popular when cheap balls made of a hard rubber called gutta-percha became available. But new, smooth "gutties" didn't travel very far; they flew better when they were nicked and worn. Manufacturers soon began to produce balls with various patterns of grooves on them (*bottom* and *right*), and these balls traveled dramatically farther than smooth ones. Modern golf balls (*top*) have dimples instead of grooves.

How Drag Affects Sports

Drag has at least two important effects on objects in sports. First, it tends to slow down anything that moves relative to the air—a point that should be pretty obvious by now. Second, it also exerts torques on objects that are not symmetrical—a point that's new and hence less obvious. Let's begin by seeing how drag's slowing effects change the nature of sports.

While a large, dense ball such as a shot-put or bowling ball may have enough momentum to ignore the effects of drag, most other balls are decelerated by it. Drag forces increase dramatically with speed; as soon as a turbulent wake and pressure drag appear, the drag force increases as the square of an object's speed through air or water. As a result, baseball pitches slow significantly during their flight to home plate, and the faster they're thrown, the more speed they lose. A 90-mph fastball loses about 8 mph en route, while a 70-mph curve ball loses only about 6 mph.

A batted baseball fares slightly better because it travels fast enough for the boundary layer around it to become turbulent, an effect that appears at around 100 mph. While the resulting reduction in drag explains why it's possible to hit a home run, the presence of air drag still shortens the distance the ball travels by as much as 50%. Without air drag, a routine fly ball would become an out-of-the-park home run. To compensate for air drag, the angle at which the ball should be hit for maximum distance isn't the theoretical 45° above horizontal. Because of the ball's tendency to lose downfield velocity, it should be hit at a little lower angle, about 35° above horizontal (Fig. 4.3.9).

Since the ball loses much of its horizontal component of velocity during its trip to the outfield, a long fly ball tends to drop almost straight down as you catch it; gravity causes it to move downward, but drag almost stops its horizontal motion away from home plate. Drag also limits the speed of a falling baseball to about 160 km/h (100 mph). That is the baseball's terminal velocity, at which the upward drag force exactly balances its downward weight. Even if you drop a baseball from an airplane, its speed will be only about 160 km/h when it reaches the ground.

Many other ball sports would be very different without air drag, including tennis, golf, football, soccer, and basketball. But since drag's effects in these sports are similar to those we've already seen, let's move on to torques exerted by drag on non-symmetric objects.

The shuttlecock or "birdie" used in badminton has the peculiar property of always flying bumper first and feathers last (Fig. 4.3.10). This behavior is caused by air pressure, which exerts a torque on the birdie about its center of mass whenever it's not flying bumper first. Air pressure can exert such a torque because the point at which the total pressure forces on the birdie effectively act, the birdie's **center of pressure**, isn't located at the birdie's center of mass. Instead, since an object's center of pressure will shift toward any large surfaces that are pushed on by the passing air, the birdie's center of pressure is in its broad, air-catching feathers. Because the birdie's center of mass is in its heavy bumper, air pressure can exert a torque on the shuttlecock about its center of mass and cause it to undergoes angular acceleration. Hence the birdie rotates so that it flies bumper first.

With its feathers behind, the shuttlecock is in stable equilibrium. If it turns clockwise, air pressure exerts a torque on it that makes it accelerate counter-clockwise; if it turns counter-clockwise, air pressure turns it clockwise. This aerodynamic restoring force is strong and reliable. It flips the shuttlecock around quickly after each hit and then keeps the shuttlecock flying bumper forward until it's hit again.

Other sports, including darts and archery, use this aerodynamic torque to orient and stabilize a moving object. In some sports, however, this torque is a problem. A rifle bullet, for example, experiences tremendous pressure at its front as it sails through the air. Its center of pressure is in front of its center of mass, making it aerodynamically unstable. Without any compensating effect, it would tumble wildly in flight and wouldn't be very accurate. To keep this from happening, a properly made rifle has rifling grooves that spin the bullet as it travels through the gun barrel. The bullet's angular momentum keeps it from tumbling.

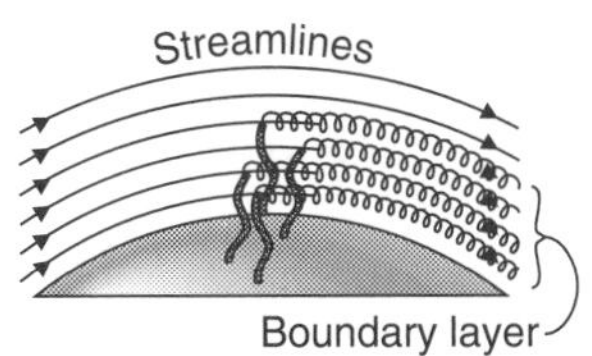

Fig. 4.3.8 - The boundary layer can be made turbulent at Reynolds numbers below 100,000 by "tripping" it with obstacles such as fuzz or dimples.

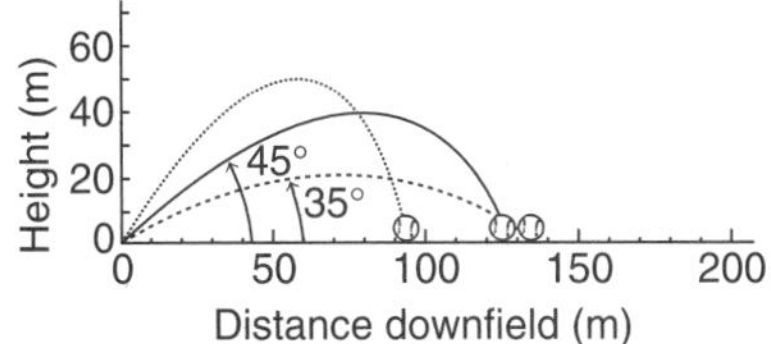

Fig. 4.3.9 - Air drag slows the flight of a batted ball so that the ideal angle at which to hit it isn't the theoretical 45° of Fig. 1.1.9. An angle of roughly 35° above horizontal will achieve the maximum distance.

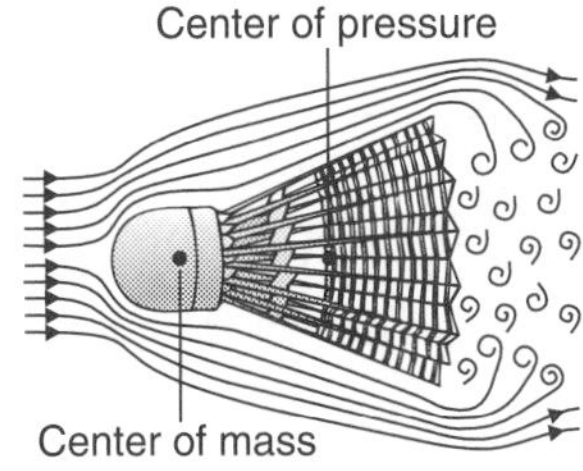

Fig. 4.3.10 - A badminton shuttlecock flies into the wind bumper first because the overall force exerted by air pressure is located at its center of pressure, far from its center of mass. If the feathers try to overtake the bumper, air resistance exerts a torque about the shuttlecock's center of mass and returns the feathers to the rear.

CHECK YOUR UNDERSTANDING #4: Going the Distance

Why must a cyclist continue to pedal her bicycle, even when she is traveling on a level road?

Curve Balls and Knuckle Balls

The drag forces on a ball oppose its motion through the air; these forces always push the ball downstream, parallel to the onrushing air (Fig. 4.3.11). But another type of aerodynamic force, **lift**, is exerted perpendicular to the airstream. When an airstream pushes a ball straight ahead of it, it's exerting a drag force on the ball; when that airstream pushes the ball to one side or the other, it's exerting a lift force on the ball. To experience drag, the ball only has to slow the airstream down; to experience lift, the ball must deflect the airstream to one side or the other. Although its name implies an upward force, lift can also act toward the side or even downward.

Curve balls and knuckle balls both use lift forces. In each of these famous baseball pitches, the ball deflects the airstream toward one side and the ball ac-

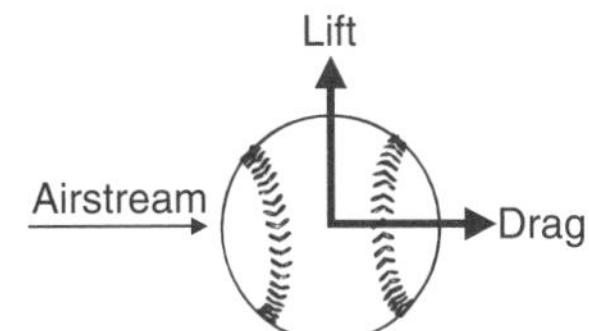

Fig. 4.3.11 - The two types of aerodynamic forces exerted on objects by air are drag and lift. Drag is exerted parallel to the onrushing airstream and slows the object's motion through the air. Lift is exerted perpendicular to that airstream so that it pushes the object to one side or the other. Lift is not necessarily in the upward direction.

celerates toward the other. Action and reaction; the air and the ball push off one another. Getting the air to push the ball sideways is no small trick. Explaining it isn't easy, either, but here we go.

A curve ball is thrown by making the ball spin rapidly about an axis perpendicular to its direction of motion. The choice of this axis determines which way the ball curves. In Fig. 4.3.12, the ball is spinning clockwise, as viewed from above. With this choice of rotation axis, the ball curves to the pitcher's right because the ball experiences two lift forces to the right. One is the **Magnus force**, named after the German physicist H. G. Magnus (1802–1870) who discovered it. The other is a force we will call the wake deflection force.

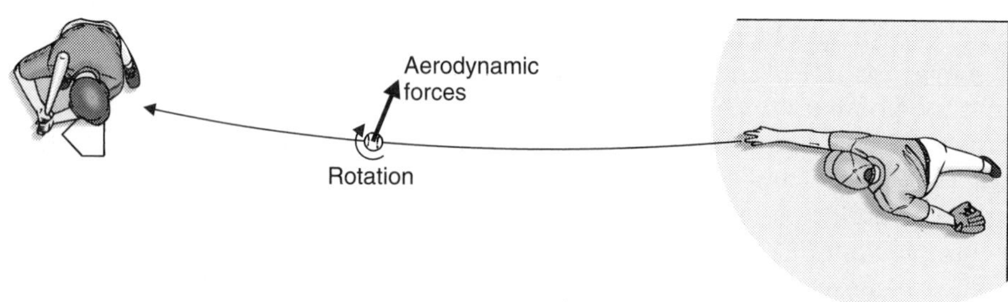

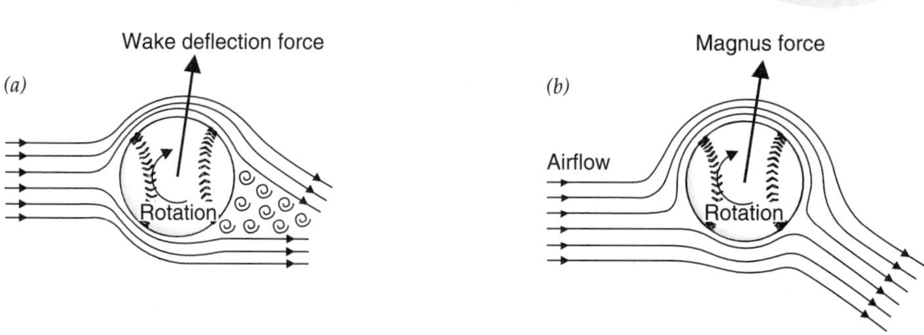

Fig. 4.3.12 - A rapidly rotating baseball experiences two lift forces that cause it to curve in flight. (*a*) The Magnus force occurs because air flows more rapidly over the ball in the direction of the ball's rotation and its pressure drops. (*b*) The wake deflection force occurs because air flowing around the ball in the direction of its rotation remains attached to the ball longer and the ball's wake is deflected.

The Magnus force occurs because the spinning ball carries some of the air in the airstream around with it (Fig. 4.3.12*a*). This extra air, flowing around the ball on the side turning back toward the pitcher, makes the air's speed on that side higher and its pressure lower. Because the pressure forces on the ball's sides no longer balance one another, the ball experiences a sideways force. This Magnus force points toward the low-pressure side and deflects the ball in that direction.

In laminar flow, the Magnus force is the only lift force acting on a spinning object. But a pitched baseball has a turbulent wake behind it and is also acted on by the **wake deflection force**. The ball's rotation pulls air around the ball on the side turning toward the pitcher and delays separation of the main airstream on that side (Fig. 4.3.12*b*). But on the side of the ball turning toward the batter, the ball's rotation slows air in the boundary layer and hastens separation of the main airstream. The overall wake of air behind the ball is thus deflected toward the side of the ball turning toward the batter. Since the ball has deflected the airstream one way, the airstream must push it the other way. The wake deflection force and the Magnus force both deflect the ball in the same direction.

Of these two forces, the wake deflection force is probably the most important for a curve ball (although the Magnus force is usually given all the credit). A skillful pitcher can make a baseball curve about 0.3 m (12 inches) during its flight from the mound to home plate—the more spin, the more curve. The pitcher counts on this change in direction to confuse the batter. The pitcher can also choose the *direction* of the curve by selecting the axis of the ball's rotation. The ball will always curve toward the side of the ball that is turning toward the pitcher. When thrown by a right-handed pitcher, a proper curve ball curves down and to the left, a slider curves horizontally to the left, and a screwball curves down and to the right.

Perhaps surprisingly, the most spectacular curves are thrown fairly slowly to give the ball time to accelerate to the side. As it curves, the ball is exchanging momentum with the air. The longer the Magnus and wake deflection forces can act, the larger the impulse on the ball and the more its momentum will change before it reaches the batter. Faster pitches curve relatively little and are called fastballs. Despite the name, most fastballs are thrown with some spin and experience lift forces in flight. Fastballs are usually thrown with backspin, the top of the ball turning toward the pitcher. That should make a fastball curve up; but while the baseball does experience an upward lift force, it's not as strong as gravity, and the ball still falls on its way to the plate. Still, the ball falls less rapidly than it would without spin and appears to "hop." A fastball thrown with relatively little spin falls naturally and is called a sinking fastball.

These two lift forces on spinning balls are critical in other sports as well. In golf, for example, backspin is the name of the game. A golf ball with backspin hangs a long time in the air and stops abruptly when it bounces on the green. Professional golfers use soft golf balls that grip the club heads and receive lots of backspin; while the ball leaves the driver only about 10° above horizontal, it remains aloft for a long time because of lift. For the less competent golfer, slices or hooks (curves toward or away from the golfer's dominant hand) are the results of various sidespins put on the ball by improper swings. A bad hook can curve the ball 15° in flight and send it into the woods. Novices, hoping to avoid these curves, use hard golf balls that don't grip the club heads and don't receive much spin of any kind.

In some cases, the special effect of a ball depends on its *lack* of spin. In baseball, for example, a knuckle ball is thrown by giving the ball almost no rotation at all. Because the knuckle ball is not turning quickly, the baseball's seams become very important. As air passes over a seam, its flow is disturbed. Because this flow disturbance is only on one side of the ball, the ball experiences a sideways aerodynamic force, a lift force.

The exact nature of the flow disturbance that creates the sideways force is not well understood. Whether the seams trip the boundary layer and make it turbulent or whether they cause the airstream to separate from the back of the ball, the ball still flutters around in a remarkably erratic manner. Releasing the ball without making it spin is difficult and requires great skill. Pitchers who are unable to throw a knuckle ball legally sometimes resort to putting a lubricant on their fingers to let the ball slip out of their hands without acquiring any spin. Like its legal relative the knuckle ball, the so-called "spit ball" dithers about and is hard to hit. The same is true for a scuffed ball.

CHECK YOUR UNDERSTANDING #5: Center Court

One of the most difficult and effective strokes in tennis is the topspin lob, in which the top of the ball spins away from the player who hit it. Which way is the lift force on this ball directed?

Flying Disks

Even if there were no such thing as lift, baseball and golf would still look familiar. But sports involving flying disks would not be the same at all. Frisbees and Aerobies are held aloft by aerodynamic lift and would fall like stones without it. In fact, flying disks are one of the best examples of aerodynamic lift, something we will find useful when discussing airplanes in the next section.

Fig. 4.3.13 - As this Frisbee flies forward, air flows faster over its top surface than under its open bottom. The resulting pressure imbalance lifts the Frisbee into the air.

A Frisbee's lift arises from its shape and orientation as it flies through the air (Fig. 4.3.13). The Frisbee's top surface bows upward so that air flowing over the top has to travel farther and faster than air flowing under the open bottom. This increased speed reduces the air pressure above the Frisbee. As a result, the Frisbee experiences a pressure force that pushes it upward and slightly toward the rear—a strong upward lift force and a weak rearward drag force. The lift force keeps the Frisbee aloft while the drag force gradually slows its forward motion.

Like an airplane wing, the Frisbee is an **airfoil**, an aerodynamically engineered surface that is designed to obtain particular lift and drag forces from the air flowing around it. But the lift force doesn't appear instantly when the Frisbee is thrown; it can only act when the airflow over the top of the Frisbee begins to travel faster than that below the Frisbee. Let's take a look at what must happen before the lift force can act.

The airstream first encounters an airfoil at its *leading edge* and leaves the airfoil from its *trailing edge*. At the start of the Frisbee's flight, the airstream splits apart near the bottom of the Frisbee's leading edge and begins to flow over and under the Frisbee (Fig. 4.3.14*a*). Initially, these two airflows travel at equal speeds, so they cover equal distances in equal times. Since air going up and over the curved top of the Frisbee has farther to go, air traveling straight underneath the Frisbee arrives at the trailing edge first and heads upward to meet the upper airflow. The airflows join together and the combined airstream leaves from the top of the trailing edge. Because the air speeds and pressures are the same above and below the Frisbee, the air exerts no lift on the Frisbee; the Frisbee doesn't deflect the airstream downward, and the airstream doesn't push the Frisbee upward.

However, air flowing up from behind the Frisbee must turn abruptly around the trailing edge. Because this flow pattern is unstable and can't continue indefinitely, it's soon blown away from the back of the Frisbee. The air in this portion of flow is rotating and becomes a vortex as it leaves the trailing edge of the Frisbee (Fig. 4.3.14*b*). The vortex is "shed" and left far behind in the air as the Frisbee flies downfield. The Frisbee now has a new pattern of airflow around it that gives it upward lift, perpendicular to the horizontal airstream (Fig. 4.3.14*c*).

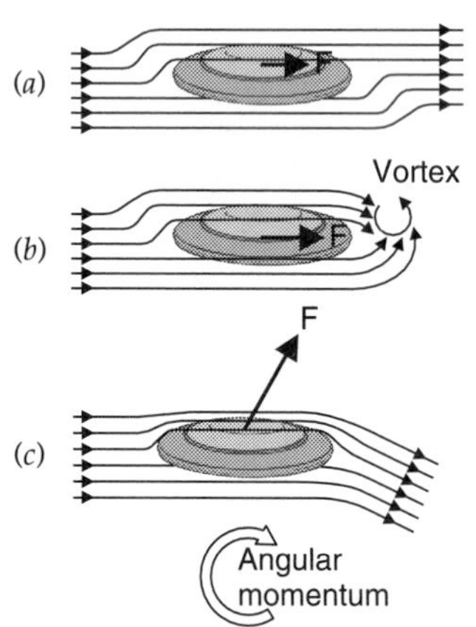

Fig. 4.3.14 - (*a*) When a Frisbee starts its flight, the airflow around it is symmetric and it experiences no lift, only drag. (*b*) The symmetric airflow is unstable and a vortex is soon shed from the Frisbee's trailing edge. (*c*) The Frisbee's wake is then deflected downward and the Frisbee experiences an upward lift force. The aerodynamic forces on the Frisbee are indicated by the arrow F.

When the Frisbee sheds its vortex, which it does only as it's being thrown, the air flowing over its upper surface begins to flow faster than the air under its lower surface. This difference in speeds occurs because the airflow around the Frisbee has acquired angular momentum. The whirling vortex took away angular momentum with it, so the air passing around the Frisbee must have received an equal amount of angular momentum in the opposite direction. Although the air flowing over the Frisbee is always heading toward its trailing edge, the fact that the top flow is faster than the bottom flow gives it angular momentum. In Fig. 4.3.14*c*, this angular momentum appears as a clockwise rotation because the top airstream is moving right faster than the bottom one. The shed vortex has left the scene with an equal amount of angular momentum in the opposite direction, appearing briefly as the counter-clockwise swirl in Fig. 4.3.14*b*. Overall, angular momentum is conserved.

If this lift force keeps a Frisbee from falling, what keeps it from tipping over or tumbling through the air? Its principal stabilizer is its rotation. A well-thrown Frisbee is spinning rapidly and has lots of angular momentum about a vertical axis. As long as the air exerts no torque on the Frisbee, it will keep turning about that same axis and will not tip. In fact, the air exerts remarkably little torque on a genuine Frisbee. The Frisbee's center of pressure, the point at which all the aerodynamic forces act, is extremely close to its center of mass. Without any lever arm between the center of pressure and the center of mass, air pressure can't exert a torque on the Frisbee.

Frisbee imitations don't always succeed in keeping their centers of pressure coincident with their centers of mass. Since these ersatz Frisbees catch the wind at the wrong points and have centers of pressure that aren't at their centers of mass, they tip over in flight as torques from the air slowly change the directions of their angular momenta.

An exception to this rule is the Aerobie, a flat, ring-shaped flying disk that flies even better than the Frisbee itself. Like the Frisbee, an Aerobie has its center of pressure aligned with its center of mass and doesn't tip over. In a Frisbee, this alignment is achieved with the help of turbulence in the bottom of the Frisbee. In an Aerobie, it's achieved with a tiny lip on the outside of the ring. The Frisbee, with its turbulence, experiences far more drag than the Aerobie and doesn't fly as far.

CHECK YOUR UNDERSTANDING #6: Flinging a Pie Dish

You can throw an upside-down pie dish like a Frisbee, but it soon tips to one side and falls over. What makes it tip?

Summary

How Balls, Birdies, and Frisbees Work: A ball traveling through the air experiences two major types of aerodynamic force: drag and lift. For a non-spinning ball traveling very slowly, the only drag force is viscous drag; it experiences no lift force. As the ball's speed through the air increases, a large turbulent wake appears behind the ball and the ball experiences pressure drag. At a still higher speed, the boundary layer of air near the ball's surface becomes turbulent and the size of the wake behind the ball shrinks. A ball with a turbulent boundary layer experiences less drag than one with a laminar boundary layer, which is why balls are designed to encourage turbulent boundary layers.

Rotating balls and flying disks experience lift forces. These forces occur both because of differences in air speeds and pressures on opposite sides of the object and because of deflections in the wakes the objects leave behind them. Lift forces can cause a ball to curve in flight or keep a flying disk from falling.

The Physics of Balls and Air

1. There are three ways in which air can flow around an object. At low Reynolds numbers, the airflow is laminar and only viscous drag is present. At medium Reynolds numbers, a large turbulent wake forms behind the object and pressure drag is present. At high Reynolds numbers, the boundary layer at the object's surface becomes turbulent, shrinking the turbulent wake and reducing the pressure drag.

2. A nonsymmetric object moving through a fluid tends to travel with its center of pressure located behind its center of mass. It is then in stable equilibrium.

3. A rotating object carries the airflow with it, creating a high-speed/low-pressure region on one side and a low-speed/high-pressure region on the other. This pressure imbalance gives rise to the Magnus force, which is directed toward the low-pressure side.

4. The turbulent wake of a rotating object is deflected to one side and the object experiences a deflecting force in the opposite direction. The direction of this wake deflection force is the same as that of the Magnus force.

5. An airfoil achieves lift by making the airstream over one surface move faster than the airstream under the opposite surface. It achieves this condition only after it has shed a vortex from its trailing edge. The airflow over the airfoil then contains angular momentum.

Check Your Understanding - Answers

1. The slow flow of water around the rock is laminar, so its pressure is highest at the front and back of the rock. The increased pressure behind the rock lifts the water level there.

Why: Laminar flow around an obstacle tends to create high pressure regions at the front and back and low pressure regions on the sides. In this case, the pressure differences are visible as changes in the water level. At the front and back of the rock, the relatively high pressures push the water level upward while at the sides of the rock, the water level is depressed.

2. The slow-moving canoe experiences laminar flow in the water while the fast moving canoe experiences turbulent flow.

Why: If the canoe's speed is less than about 1 cm/s, its Reynolds number will be less than 2000 and the water flow around it will be laminar. The water will pass smoothly around the canoe's sides and join back together behind it. But when the canoe is moving fast enough that its Reynolds number exceeds 2000, the water flow becomes turbulent and the canoe leaves a churning wake in the water behind it. This wake produces pressure drag on the canoe and extracts energy from it. Anyone who has paddled a canoe knows that overcoming this pressure drag can be exhausting.

3. As far back on the car as possible.

Why: As for all large, fast moving objects in air, pressure drag is the main source of air resistance. You want to minimize this drag by keeping the air flowing smoothly over the car until it leaves the rear around a small turbulent wake. The smaller the air pocket behind a car, the better. Aerodynamically designed production cars leave a turbulent wake that is only about a third as large in area as the thickest cross section of the car. While there is still room for improvement, these cars experience far less drag than the boxy cars of earlier times.

4. Air drag tends to slow her motion, an effect she must counter by pedaling.

Why: Bicycling would be effortless were it not for pressure drag. The human body is not very aerodynamic and the rush of air around the cyclist exerts far more pressure on her front than on her back. To keep from accelerating backward, she must cause the roadway to push her forward. She obtains this forward force by pedaling. Because pressure drag increases roughly as the square of her velocity through the air, she will have tremendous difficulty exceeding a speed of about 70 km/h on the level road. She can reduce the size of her wake and the associated pressure drag by wearing an aerodynamic helmet and leaning down low to the handlebars.

5. The lift force is directed downward, so that ball accelerates downward faster than it would by gravity alone.

Why: A ball with topspin falls faster than it would without a spin. In tennis, the topspin strokes appear to dive downward once they cross the net. Their downward curve means that they can travel very fast and still remain inside the court and are thus very hard to return.

6. The center of aerodynamic pressure on the dish is not at its center of mass.

Why: When thrown correctly, a pie dish obtains lift from the air but the overall aerodynamic force on the dish is located ahead of the dish's center of mass. As a result, air exerts a torque on the dish about its center of mass and changes the dish's axis of rotation. The dish tips over.

Glossary

adverse pressure gradient A region of fluid flow in which the fluid must flow toward higher pressure. The fluid's momentum and kinetic energy carry it through this situation, although the fluid does slow down.

aerodynamics The study of the dynamic (moving) interactions of air with objects.

aerodynamic forces The forces exerted on an object by the motion of the air surrounding it. The two types of aerodynamic forces are lift and drag.

airfoil An aerodynamically engineered surface, designed to obtain particular lift and drag forces from the air flowing around it.

boundary layer A thin region of fluid near a surface that, because of viscous drag, is not moving at the full speed of the surrounding airflow.

center of pressure The point on an object at which the total pressure forces effectively act. Pressure exerts no torque on the object about its center of pressure.

lift An aerodynamic force exerted at right angles to the airflow around an object.

Magnus force A lift force experienced by a spinning object as it moves through a fluid. The Magnus force points toward the side of the ball moving away from the onrushing airstream.

pressure drag The drag force that results from higher pressures at the front of an object than at its rear.

wake The trail left behind by an object as it moves through a fluid.

wake deflection force A lift force experienced by a spinning ball when it deflects its turbulent wake to one side. The wake deflection force points toward the side of the ball moving away from the onrushing airstream.

Review Questions

1. What is the difference between a drag force and a lift force?

2. Why is the drag force that a ball experiences always downwind, never upwind?

3. Explain why, as a ball's speed through the air increases, the air's inertia becomes more important and its viscosity becomes less important.

4. How can the low pressure air at the sides of a moving ball flow around the ball into the higher pressure behind it? Doesn't air always flow toward lower pressure?

5. Why does air in a ball's boundary layer have more trouble flowing into the higher pressure behind the ball than the air farther from the ball does?

6. A ball that leaves a small turbulent wake, and pulls only a narrow column of air along with it, experiences much less drag than one that leaves a large turbulent wake. Explain this effect in terms of a transfer of momentum from the ball to the air.

7. Why do modern golf balls have dimples on their surfaces?

8. If you hold a dart backward and throw it at a dart board fins first, it will turn around in flight and its point will still hit the board. Explain the forces and torques that make the dart turn around.

9. If a spinning ball curves toward the left as it flies away from you, which side of the ball is experiencing the lowest air pressure? In which direction is the lift force on the ball?

10. The air around the Frisbee in Fig. 4.3.14*c* has angular momentum because the airstream above it is traveling toward the right faster than the airstream below it is. Compare this situation to the motions of the upper and lower surfaces of a bowling ball rolling toward the right.

11. If you throw a Frisbee upside down, in which direction will the lift force it experiences be?

Exercises

1. Fish often swim just upstream or downstream from the posts supporting a bridge. How does the water's speed in these regions compare with its speed in the open stream and at the sides of the posts?

2. Why is a single isolated tree more vulnerable to wind damage than trees in a forest?

3. A water skier skims along the surface of a lake. What types of forces is the water exerting on the skier and what is the effect of these forces?

4. A weathervane is a pivoting device that turns to indicate the direction of the wind. What forces and torques are responsible for keeping it pointing in the right direction?

5. How does an arrow's heavy stone arrowhead help the arrow remain stable in flight?

6. Why do flyswatters have many holes in them?

7. How does running directly behind another runner reduce the wind resistance you experience?

8. If you ride your bicycle directly behind a large truck, you will find that you don't have to pedal very hard to keep moving forward. Why?

9. You have two golf balls that differ only in their surfaces. One has dimples on it while the other is smooth. If you drop these two balls simultaneously from a tall tower, which one will hit the ground first?

10. In 1971, astronaut Alan Shepard hit a golf ball on the moon. How did the absence of air affect the ball's flight?

11. To drive along a level road at constant velocity, your car's engine must be running and friction from the ground must be pushing your car forward. Since the net force on an object at constant velocity is zero, why do you need this forward force from the ground?

12. How would a Frisbee fly on the airless moon?

13. A bullet slows very quickly in water but a spear doesn't. What force acts to slow these two objects and why does the spear take longer to stop?

14. When a car is stopped, its flexible radio antenna points straight up. But when the car is moving rapidly down a highway, the antenna arcs toward the rear of the car. What force is bending the antenna?

15. You can buy special golf tees that wrap around behind the ball to prevent you from giving it any spin when you hit it. These tees are guaranteed to prevent hooks and slices (i.e., curved flights). But how do these tees affect the distance the ball travels? Why?

16. If you hang a tennis ball from a string, it will deflect downwind in a strong breeze. But if you wet the ball so that the fuzz on its surface lies flat, it will deflect even more than before. Why does smoothing the ball increase its deflection?

17. Explain why a parachute slows your descent when you leap out of an airplane.

18. How would putting dimples on the surface of a baseball bat increase the speed at which a player can swing it?

19. Some racing boats have fine dimples on the outsides of their hulls. How does this dimpling help them go faster?

20. Racing bicycles often have smooth disk-shaped covers over the spokes of their wheels. Why would these thin wire spokes be a problem for a fast-moving bicycle?

21. You sometimes see paper pressed tightly against the front of a moving car. What holds the paper in place?

22. A football experiences a smaller drag force when it travels point first than when it flies sideways through the air. Why does the drag force depend on the football's orientation?

23. A skillful volleyball player can serve the ball so that it barely spins at all. The ball dithers slightly from side to side as it flies over the net and is hard to return. What causes the ball to accelerate sideways?

24. Pedestrians on the surface of a wind-swept bridge don't feel the full intensity of the wind because near the bridge's surface, the air is moving relatively slowly. Explain this effect in terms of a boundary layer.

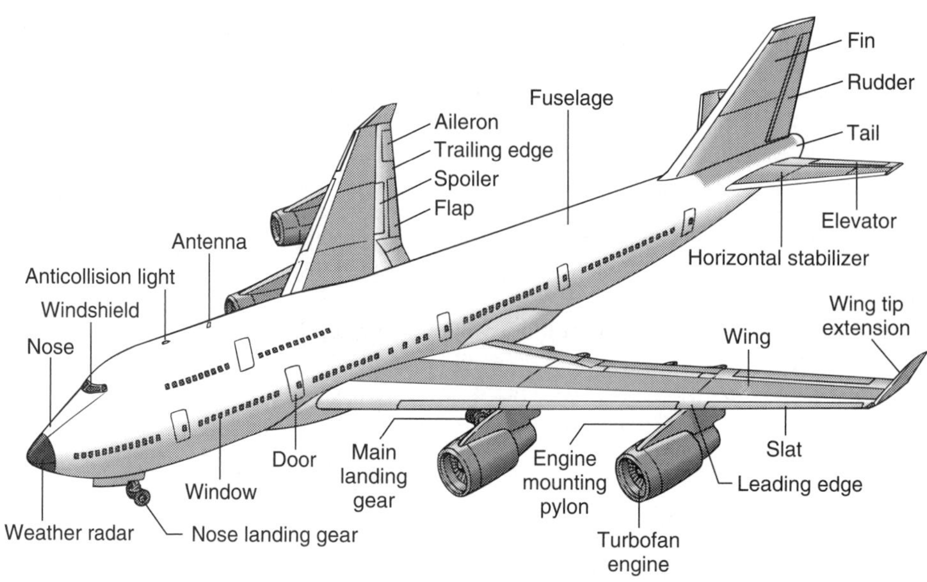

Section 4.4

Airplanes

We have now set the stage for that ultimate aerodynamic machine, the airplane. Free from contact with the ground, the airplane is affected only by aerodynamic forces and gravity, hopefully in that order. Despite their complex appearance, airplanes employ physical principles that we have already successfully examined. But while this section revisits many familiar concepts, it also explores lots of new territory. For example, you may have already figured out what type of aerodynamic force holds the airplane up, but what type of aerodynamic force keeps it moving forward?

Questions to Think About: *Why are the wings of small, propeller-driven aircraft relatively large and bowed in comparison to those of jets? Why does a commercial airplane extend slats and flaps from its wings during takeoffs and landings? How can an airplane change its speed in flight? Why does an airplane have a tail? How can some airplanes fly upside down? Why do most fast commercial aircraft employ jet engines and not propellers?*

Experiments to Do: *The best experiment for this section is to take a plane flight or at least visit the airport and watch the planes.*

As you sit in the plane during take off, feel the plane accelerating forward. If you're on a commercial jet, notice that the slats and flaps on the airplane's wings are extended during take off, making the wings wider and more curved. How could this increased width and curvature help the plane take off? The pilot holds the plane on the ground until it reaches the proper speed, then quickly tips it upward into the air. An invisible vortex of air peels away from the trailing edge of the wing and the plane lifts into the air.

Once airborne, the airplane retracts its landing gear, slats, and flaps. Watch the trailing edge of each wing as the plane turns or changes altitudes; you'll see various surfaces there move up or down. Similar motions occur on the tail. How do these surfaces control the plane's orientation?

Near its destination, the airplane prepares for landing. Again the slats and flaps are extended. Watch as spoilers on the tops of the wings pop up and down noisily. How do these surfaces affect the drag force on the plane? The plane's landing gear extends and it touches down on the runway. The propellers or jet engines abruptly begin to slow the airplane, assisted by the spoilers on the wings. Feel the plane accelerating backward. Another invisible vortex of air peels away from the trailing edge of the wing, rotating in the opposite direction from the first vortex, and the flight is over.

Airplane Wings

By now you may be wondering if the lift on an airplane wing is the same as that on a Frisbee. It is. The wing is an airfoil, shaped and oriented so that, during flight, air tends to travel faster over its top surface than under its bottom surface. Because of Bernoulli's effect, the pressure is lower above the wing than below it and the wing experiences an upward lift force. It's this lift force that supports the airplane in the sky.

To understand how the wing works, let's imagine that you are inside an airplane that has just begun to roll down the runway. From your perspective, air is beginning to flow past the airplane's wings. This moving air separates into two streams around each wing, one above the wing and one below it. Initially both airstreams travel at the same speed past the wing's surfaces.

Because of the wing's curved and tilted profile, the airstream flowing over it has farther to travel than the airstream flowing under it. The lower airstream reaches the wing's trailing edge first and finds the space above it empty—the upper airstream hasn't gotten there yet. To fill the void, the lower airstream flows up and around the trailing edge, heading backward to meet the upper airstream flowing over the wing.

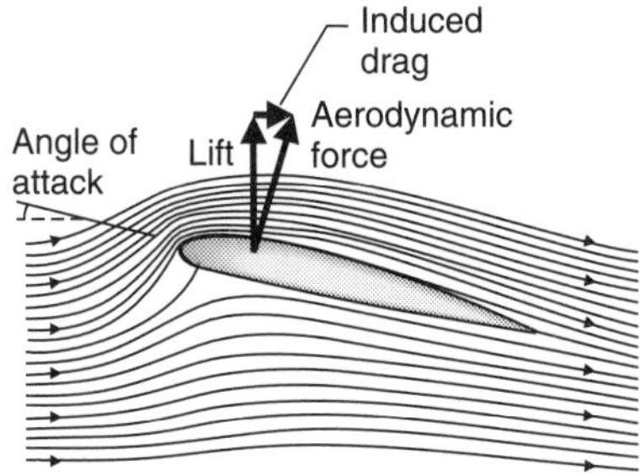

Fig. 4.4.1 - An airplane wing is an airfoil, shaped and oriented so that air flows faster over its top surface than under its bottom surface. The wing experiences a large aerodynamic force that points upward and slightly downstream. The upward component of this force is lift. The downstream component is induced drag.

But the lower airstream's inertia makes this initial flow pattern unstable. The faster the plane moves, the more that airstream's momentum must change as it turns around the trailing edge of the wing. The turning point drifts farther and farther behind the trailing edge until a vortex of air peels away from the wing. A new, stable flow pattern forms in which the airstream flowing over the wing travels faster than the airstream flowing under it. This flow pattern allows air particles that arrive together at the wing's leading edge to flow separately over and under the wing but still meet again at the wing's trailing edge. Because the pressure is now lower above the wing than below it, this new flow pattern produces lift. The air can now support your plane, and up you go.

Another way to think about lift is to follow the streamlines in Fig. 4.4.1. The closely spaced streamlines over the wing indicate high speed and low pressure, while the widely spaced streamlines under the wing indicate low speed and high pressure. But the streamlines also turn as they pass the wing. The airstream approaches the wing horizontally but is deflected downward as it leaves the wing. To produce this deflection, the wing must push the airstream downward. The airstream pushes back and produces lift on the wing. These streamlines can be made visible in a wind tunnel by blowing smoke streams past a wing (Fig. 4.4.2).

But the force on the wing is not directly upward; it tilts slightly downstream. The vertical component of this pressure force is lift, but the downstream horizontal component is a new type of drag force: induced drag. **Induced drag**, produced whenever an airfoil obtains lift by deflecting an airstream, is a consequence of energy conservation. The airplane's wing does work as it deflects the airstream downward: it pushes the airstream downward and the airstream moves downward. Thus the airstream extracts energy from the airplane. If nothing pushes the airplane forward, its kinetic energy will decrease and it will slow down. The force that slows it down is induced drag.

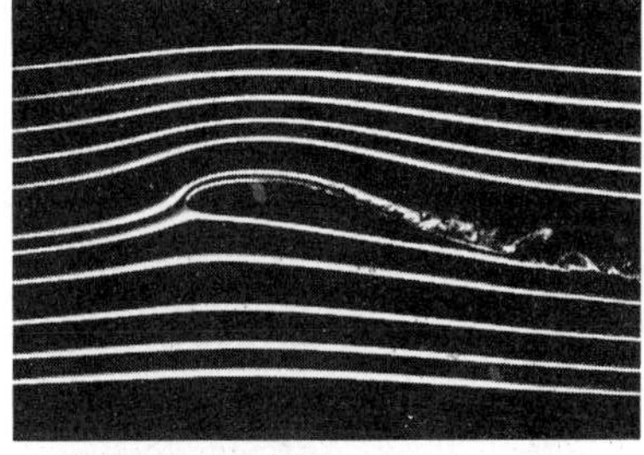
Fig. 4.4.2 - Smoke trails show the smooth flow of air past this wing. Air is traveling fastest over the wing, producing lift.

To minimize induced drag, a plane must do as little work as possible on the passing airstream. Since the downward force that the plane exerts on the airstream is roughly equal to the plane's weight, it can only reduce its work by pushing the air downward a smaller distance. This distance depends on how much air the airplane pushes on. By exerting the same downward force on more air, the plane can lessen that air's downward acceleration. The air then descends more slowly and travels less distance while the plane is pushing on it. Thus increasing wing size allows the airplane to deflect more air a smaller distance and experience less induced drag.

Aircraft builders design an airplane's wings to support it at its cruising speed with a minimum of *total* drag. While wings that are too small experience large induced drags or may not be able to lift the airplane at all, wings that are too large experience substantial viscous drags. Thus, choosing the right size for an airplane's wings is complicated. Moreover, the amount of lift experienced by a wing increases with the curvature of that wing's top surface and its *airspeed* (its speed relative to the air it encounters). Small propeller airplanes that move slowly through the air need relatively large, highly curved wings to support them, while commercial and military jets fly faster and can get by with relatively small, moderately curved wings.

To maintain a constant altitude, the lift force on the plane must exactly balance the plane's weight. However, to accelerate up or down, the lift must be made greater or less than the weight. How does the pilot adjust the lift to maneuver the plane? The lift can be varied by tilting the wings to change their *angle of attack*, the angle at which they approach the airstream. The larger the angle of attack, the greater the speed difference between air flowing over and under the wings and the greater the lift. Because the wings are rigidly attached to the plane, the pilot rotates the nose of the plane upward to accelerate upward, and downward to accelerate downward. While it may seem that the body of the plane must point roughly in the direction of acceleration, there's no reason why this must be so. If the wings could swivel without taking the body of the plane with them, the pilot would be able to maneuver the plane up and down while the passenger compartment remained perfectly level.

Since lift depends strongly on a wing's angle of attack, some planes can be flown upside-down. As long as the inverted wing is tilted properly, it obtains upward lift and supports the plane. But because its curvature is reversed, it must have a very large angle of attack in order to obtain lift. Nonetheless, stunt fliers regularly fly upside down at air shows. When they do, they are still supported by the Bernoulli effect, as air flows faster over the inverted wing than under it.

CHECK YOUR UNDERSTANDING #1: Blowing in the Wind

The sail of a small sailboat bows forward and outward so that wind traveling around the sail's outside surface travels farther and faster than wind traveling across its inside surface. How does this speed difference propel the sailboat across the water?

Lift has its Limits: Stalling a Wing

❐ Airplane designers can reduce the dangers of stalling by adding special boundary layer control devices to their aircraft. Narrow metal strips called *vortex generators,* that stick up from the surfaces of wings, add energy to the boundary layers over the wings and help keep the airstreams attached to the surfaces.

However, there is a limit to how much lift the pilot can obtain by increasing the wings' angle of attack, inverted or otherwise. Beyond a certain angle, the airflow over the top of the wing separates from its surface and the wing **stalls**. This separation occurs when air in the boundary layer on top of the wing is unable to flow into the adverse pressure gradient located after the wing's thickest point. The pressure is increasing as the upper airstream approaches the trailing edge of the wing, and air in the boundary layer finds itself turned around by the pressure imbalance. In other words, it just runs out of momentum and kinetic energy trying to flow into the high pressure.

During a stall, the airstream over the top of the wing breaks away from the surface, leaving a billowing storm of turbulence beneath it (Fig. 4.4.3). This airflow separation is an aerodynamic catastrophe for the airplane. The wing loses much of its lift because the pressure above the thickest part of the wing increases. Drag also increases because, when the wing is operating correctly, air above the wing's trailing edge slows down and its high pressure there exerts a modest forward force on the plane. During a stall, this high pressure recovery region is lost and the plane experiences pressure drag. These streamlines can be made visible in a wind tunnel by blowing smoke streams past a wing (Fig. 4.4.4).

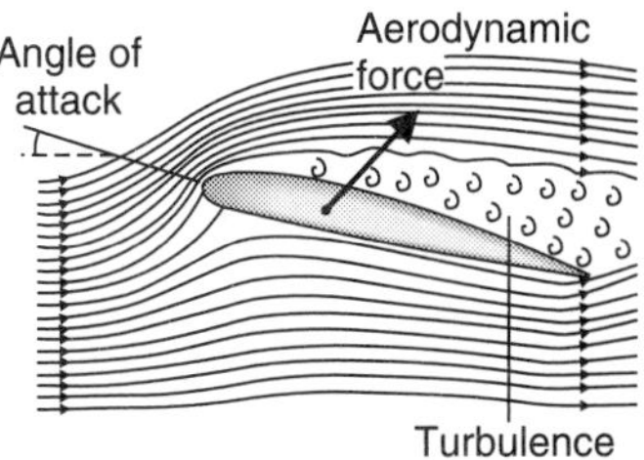

Fig. 4.4.3 - A wing stalls when the airstream over the top of the wing separates from its surface. A turbulent air pocket forms above the wing, making it much less efficient. The wing's lift decreases because the pressure above the thickest part of the wing becomes higher, and the drag increases because the pressure above the trailing edge becomes lower.

While a stalled wing is fine for a kite (which flies in exactly this manner), it's not good for an airplane. To avoid stalling, pilots keep the angle of attack within a safe range. But the possibility of stalling also limits the minimum speed at which the airplane will fly. As the airplane slows down, the pilot must increase its angle of attack to maintain lift. Below a certain speed, the airplane can't obtain enough lift without tilting its wings until they stall. It can no longer fly.

To keep from stalling, a plane must never fly slower than this minimum speed, particularly during landings and takeoffs. For a small, propeller-driven plane with highly curved wings, this minimum flight speed is so low that it's not a problem. For a commercial jet, however, the minimum airspeed is about 220 km/h (140 mph). Airplanes taking off or landing this fast would require very long runways so that they could build up or get rid of this speed. Instead, commercial jets have wings that can change shape during flight. Slats move forward and down from the leading edges of the wings, and flaps move back and down from the trailing edges (Fig. 4.4.5). With both slats and flaps extended, the wing becomes larger and more strongly curved, similar to the wings of a small propeller plane. Vanes near the flaps also emerge during landings to direct high-pressure air from beneath the wings onto the flaps. These jets of air keep the boundary layers moving downstream and help prevent stalling. The minimum safe airspeed for a commercial jet with slats and flaps extended is only about 150 km/h (95 mph), making it much easier for this plane to take off or land. (For another approach to stall prevention, see ❒ on previous page.

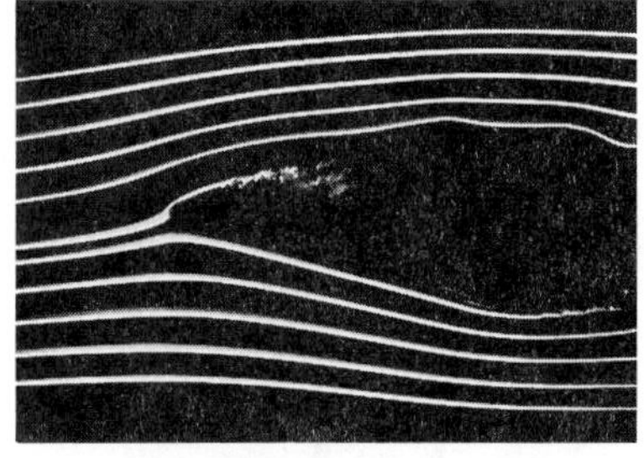

Fig. 4.4.4 - Smoke trails show that the air separates from the surface and becomes turbulent as it flows over this stalled wing. The wing is experiencing more drag than lift.

Once a commercial jet lands, flat panels on the top surfaces of its wings are tilted upward and cause the airflow to separate from the tops of the wings. The resulting turbulence created by these spoilers reduces the lift of the wings and increases their drag. On the ground, the airplane doesn't need lift from the wings, so the spoilers help to keep the plane from flying again. They also slow it down. The spoilers are sometimes used in flight to slow the plane and help it descend rapidly toward an airport.

One remaining issue with wings is that they have ends. What happens to the airflow at the end of a wing? Because the pressure is low above the wing and high beneath it, air tends to flow around the ends of the wing from bottom to top. That air is soon left behind by the airplane, but not before it has acquired lots of angular momentum and kinetic energy. As a result, the airplane creates a swirling vortex with the tip of each wing. These two vortices trail behind the plane for several kilometers, like invisible tornadoes. A wing-tip vortex from a jumbo jet is capable of flipping a small aircraft that flies through it and can even give passengers in a much larger plane an unexpected thrill. To avoid the hazards of flying through wing-tip vortices, air traffic controllers are careful to keep planes from flying through one another's wakes. This care explains why aircraft must wait about 90 s between takeoffs from a single runway; during that time, the wing-tip vortices from the previous plane will have decayed away. Some modern airplanes have wing tip extensions that reduce these vortices, both to save energy and to eliminate the hazard.

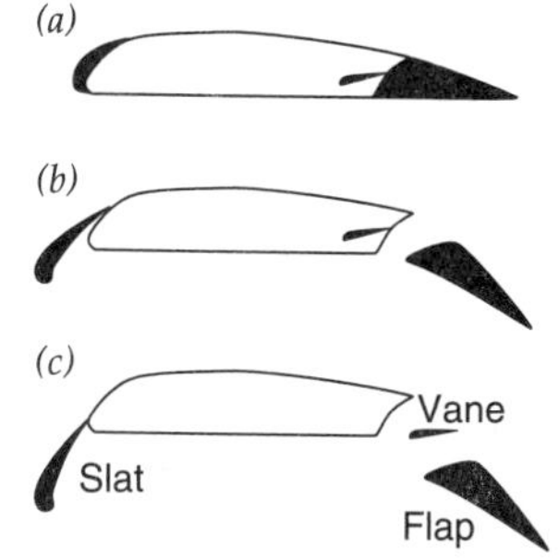

Fig. 4.4.5 - At cruising speed, an airplane's wings are moderately curved airfoils (*a*). But during takeoffs (*b*) and landings (*c*), slats are extended from the leading edges and flaps from the trailing edges. The airfoils become much more highly curved, generating more lift at low speeds. During landing, a vane is also extended for boundary layer control to prevent stalling.

CHECK YOUR UNDERSTANDING #2: Stunt Flying

A pilot normally tips the plane's nose upward in order to gain altitude. But if the pilot tries to make the plane rise too quickly, the plane will suddenly begin to drop. What is happening?

Propellers

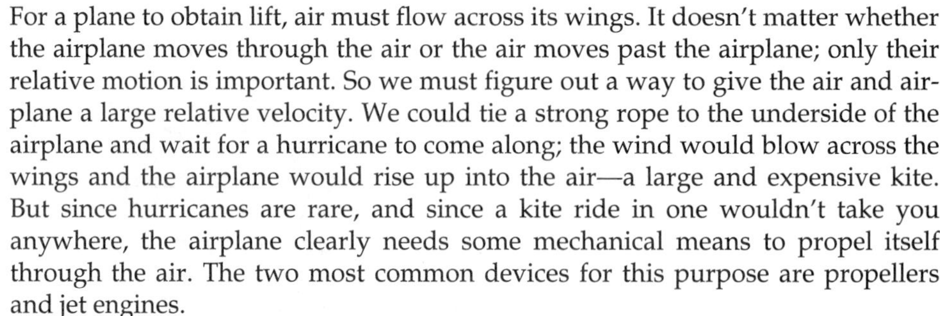

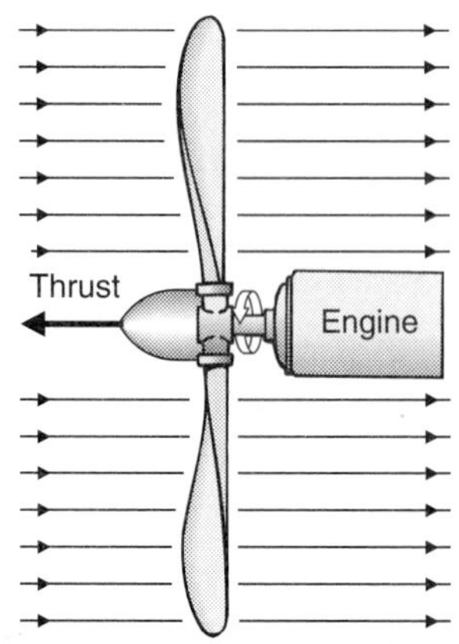

Fig. 4.4.6 - A propeller behaves like a rotating wing. As the propeller turns, its blades create lift in the forward direction. This lift pushes the engine and the aircraft forward through the air, so that it's called thrust.

For a plane to obtain lift, air must flow across its wings. It doesn't matter whether the airplane moves through the air or the air moves past the airplane; only their relative motion is important. So we must figure out a way to give the air and airplane a large relative velocity. We could tie a strong rope to the underside of the airplane and wait for a hurricane to come along; the wind would blow across the wings and the airplane would rise up into the air—a large and expensive kite. But since hurricanes are rare, and since a kite ride in one wouldn't take you anywhere, the airplane clearly needs some mechanical means to propel itself through the air. The two most common devices for this purpose are propellers and jet engines.

A propeller can be thought of as a rotating wing. Extending from a central hub are two or more blades that together form a symmetric fan (Fig. 4.4.6). These blades have airfoil cross sections and are designed to create a forward force when the propeller turns and the blades move through the air. While we've encountered this device before, as the fan in a vacuum cleaner, we're now prepared to see exactly how it works.

As a propeller blade slices through the air, air flows around it just as air flows around the wings of the plane. Because of the shape and orientation of the blade, the air must flow faster over the front surface of the blade than over the rear surface (Fig. 4.4.7). The pressure on the front of the propeller is less than on its rear, so the propeller experiences a forward lift force. Because this lift force propels the plane forward, it's called **thrust**.

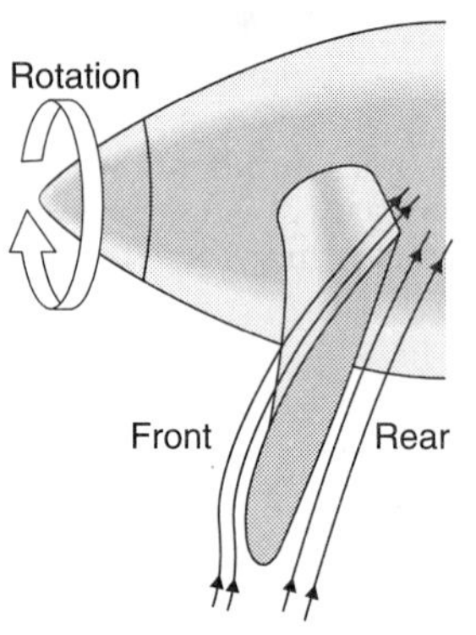

Fig. 4.4.7 - As the propeller blade rotates, the Bernoulli effect creates a low pressure flow in front of it (left) and a high pressure flow behind it (right). The blade experiences a lift force that pushes the propeller and plane forward (toward the left). Induced drag tends to slow the rotation of the propeller.

❐ One of the principal sources of noise in submarines is the turbulence created by their propellers. To reduce this turbulence, the propellers of modern nuclear submarines are designed to avoid water flow separation and stalling.

This rotating wing has all the features, good and bad, of an airplane wing. Its lift, or rather thrust, propels the plane forward. The amount of thrust increases with the propeller's size, front-edge curvature, angle of attack, and airspeed; in other words, the larger the propeller, the faster it turns, and the more its blades are angled into the wind, the more thrust it produces. Like a wing, a propeller stalls when the airflow separates from the front surfaces of its blades. It suddenly becomes more of an air-mixer than a propeller. This stalled-wing behavior was the standard operating condition for air and marine propellers (see ❐) before the work of Wilbur Wright in 1902 (see ❐). The Wrights were among the first people to study aerodynamics with a wind tunnel (Fig. 4.4.8) and their methodical and scientific approach to aeronautics allowed them to achieve the first powered flight (Fig. 4.4.9).

A propeller also experiences induced drag, which tends to slow its rotation. An **engine** must continuously do work on the propeller to keep it turning; while the propeller's thrust pushes the plane through the air, the air extracts energy from the propeller with induced drag, so something must do the work of turning the propeller. Propellers are driven by high-performance reciprocating engines, like those found in automobiles, and the turbojet engines that we'll discuss later.

Propellers aren't perfect; they have three serious limitations. First, since they exert a torque on the air that passes through them, they experience a reaction torque in return, and this large torque can tip the plane over. Many airplanes use pairs of propellers that turn in opposite directions so the net torque on the plane is zero. But some planes have only a single propeller. In these planes, the propeller is located in front of the wings, so that rotating air leaving the propeller can return some angular momentum to the plane as it passes over the wings.

Fig. 4.4.8 - The Wright Brothers were accomplished aerodynamicists, using this wind tunnel to study and perfect wings and propellers for their airplanes.

Fig. 4.4.9 - The era of powered flight began at 10:35AM on Dec. 17, 1903, when the Wright Flyer lifted Orville Wright into the air over Kitty Hawk, NC. His brother Wilbur stands beside him in this unique photograph of that first powered flight.

❐ In addition to achieving the first self-propelled flight of an airplane in 1903, Orville (1871–1948) and Wilbur (1867–1912) Wright (American aviators) were exceptionally accomplished aerodynamicists. In 1902, Wilbur was the first person to recognize that a propeller is actually a rotating wing. Propellers up until his time were little more than rotating paddles, more effective at stirring the air than propelling the plane. Wilbur's aerodynamically redesigned propeller made flight possible and dominated aircraft design for a decade.

A second problem with propellers is that their thrust diminishes as the plane's forward speed increases. When the airplane is stationary, the propeller blade turns past motionless air (Fig. 4.4.10*a*). But when the airplane is moving very fast, the air approaches that same propeller blade from the front of the plane (Fig. 4.4.10*b*). To retain lift or thrust at higher airspeeds, the propeller blade must swivel forward, increasing its pitch to meet this onrushing airstream. The blades themselves have a twisted shape to accommodate the variations in airspeed along their lengths, from hub to tip.

The third and most discouraging problem with propellers, especially in high-speed aircraft, is drag. To keep up with the onrushing air at high airspeeds, the propellers must turn at phenomenal rates. The tips of the blades travel so fast that they exceed the **speed of sound**—the fastest speed at which a fluid such as air can convey forces from one place to another. When the blade tip exceeds this speed, the air near the tip doesn't accelerate until the tip actually hits it. Instead of flowing smoothly around the tip, the air forms a **shock wave**—a narrow region of high pressure and temperature caused by the supersonic impact—and the propeller stalls. Because a propeller's thrust decreases and its drag increases when it stalls, propellers aren't useful on high-speed aircraft. However, swept-back propellers, which don't form shock waves, may someday change all that.

CHECK YOUR UNDERSTANDING #3: Circulating the Air

One difference between a fan and a propeller is what moves, the air or the object. Which side of each fan blade experiences the lowest air pressure, the inlet or the outlet side?

Jet Engines

As we have seen, propellers are well suited for slow-moving airplanes, but they don't work well at high speeds. That's why jet engines have replaced propellers on most long-distance commercial airplanes. While a propeller tries to operate directly in the stream of high-speed air approaching the plane, a jet engine first slows this air down to a manageable speed. To achieve this change in speeds, the jet engine makes wonderfully subtle use of Bernoulli's effect.

A turbojet engine is depicted in Fig. 4.4.11. During flight, air rushes into the engine's inlet at about 800 km/h (500 mph), the speed of the plane. Once inside the inlet, the air slows down and its pressure increases. Its total energy is unchanged. The air then passes through a series of fan-like compressor blades that

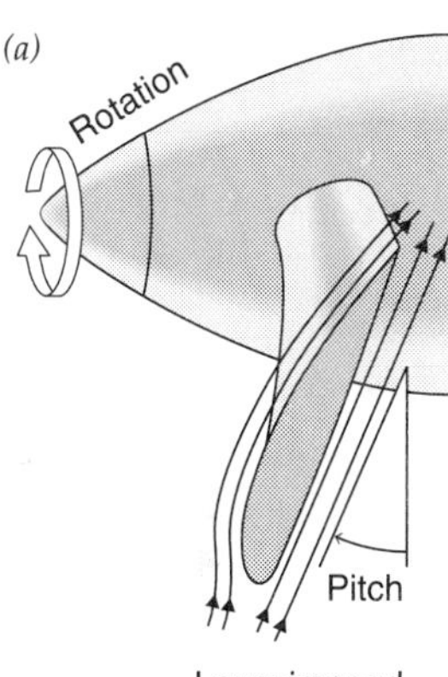

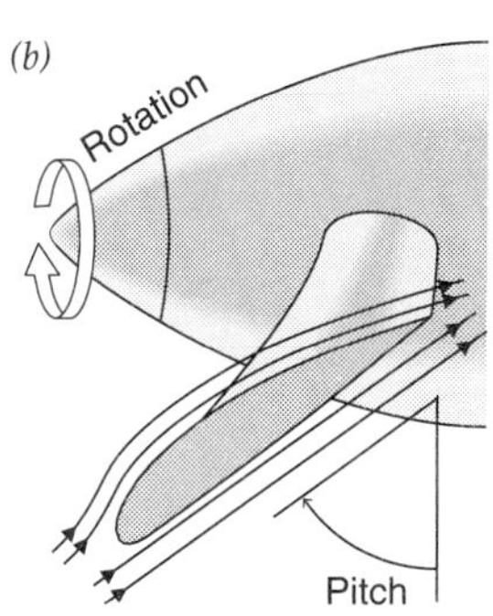

Fig. 4.4.10 - At low airspeeds (*a*), the propeller blade approaches nearly stationary air as it rotates. At high airspeeds (*b*), air rushes past the propeller, so the blade must swivel forward to meet it. The blade's angle of attack is called its pitch.

push it forward, doing work on it and increasing both its pressure and its total energy. By the time the air arrives at the combustion chamber, its pressure is far above atmospheric.

Now fuel is added to the air and the mixture is ignited. Since hot air is less dense than cold air, the hot exhaust gas takes up more space than it did before combustion. This hot exhaust gas pours out of the combustion chamber, traveling faster than when it entered. Its pressure is still very high and it accelerates out of the back of the engine, exchanging its pressure for speed. On its route out of the engine, the exhaust gas passes through a fan-like turbine. The turbine behaves like a windmill. The gas does work on this turbine and gives up a little of its energy in the process. The turbine drives the compressor for the incoming air.

Overall, the engine has slowed the air down, added energy to it, and then accelerated it back to high speed. Because the engine has added energy to the air, the air leaves the engine traveling faster than when it arrived. The jet engine has exerted a rearward force on the air to accelerate it rearward; the air has pushed back and produced a thrust that propels the airplane forward.

The turbojet isn't quite as efficient as it might be; it moves too little air and gives that air too much kinetic energy. To make a jet engine (or any action-reaction propulsion system) more efficient, it should move lots of air. That way, the air will leave the engine traveling relatively slowly and will thus take away relatively little energy.

The turbofan engine solves this problem by adding a huge fan to the front of a turbojet (Fig. 4.4.12), which provides the power needed to keep the fan turning. Air flowing into the turbofan engine's inlet slows down as it approaches the fan, and its pressure increases. The fan does work on the air passing through it so that the air leaves the fan at a higher pressure than when it entered. While about 5% of this air then enters the turbojet engine, the vast majority of it accelerates out the back of the fan duct and sails off behind the airplane. It converts its pressure energy into kinetic energy and leaves the engine traveling faster than when it arrived. The fan has pushed on the air and the air has pushed back on the fan, producing forward thrust.

Once again, the engine has slowed the air down, added energy to it, and returned it to high speed. Again, the air leaves the engine traveling faster than when it arrived. But the turbofan engine moves more air than a simple turbojet engine, giving that air less energy and using less fuel. The huge fan-like engines on many jumbo jets are turbofans. (For another type of jet engine, see ❐.)

❐ Ramjets are jet engines that have no moving parts. Air that approaches the engine at supersonic speeds interacts with carefully tapered surfaces so that its own forward momentum compresses it to high density. The engine then adds fuel to this pressurized air, ignites the mixture, and allows the hot burned gas to expand out of a nozzle. The engine pushes this exhaust backward and the exhaust propels the engine and airplane forward. Although the air enters the engine at supersonic speeds, it passes through the combustion chamber much more slowly. In a supersonic combustion ramjet or "scramjet," the fuel and air mixture flows through the combustion chamber at supersonic speeds. This motion makes it extremely difficult to keep the fuel burning because the flame tends to flow downstream and out of the engine. The flame can't advance through the mixture faster than the speed of sound, so it won't spread upstream fast enough to stay in the engine on its own.

CHECK YOUR UNDERSTANDING #4: Energy and a Jet Engine

A jet engine somehow slows the air down, adds energy to it at low speed, and then returns the air to high speed. Why doesn't slowing the air down waste lots of energy?

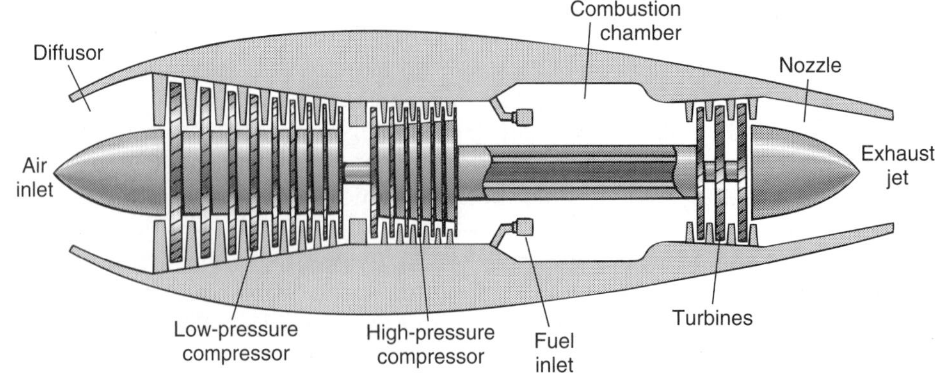

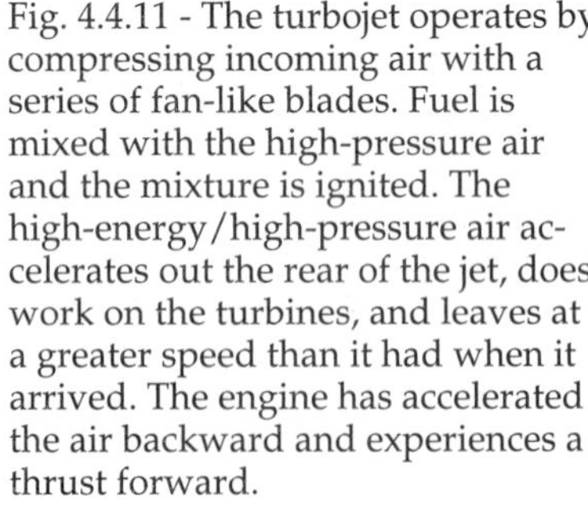

Fig. 4.4.11 - The turbojet operates by compressing incoming air with a series of fan-like blades. Fuel is mixed with the high-pressure air and the mixture is ignited. The high-energy/high-pressure air accelerates out the rear of the jet, does work on the turbines, and leaves at a greater speed than it had when it arrived. The engine has accelerated the air backward and experiences a thrust forward.

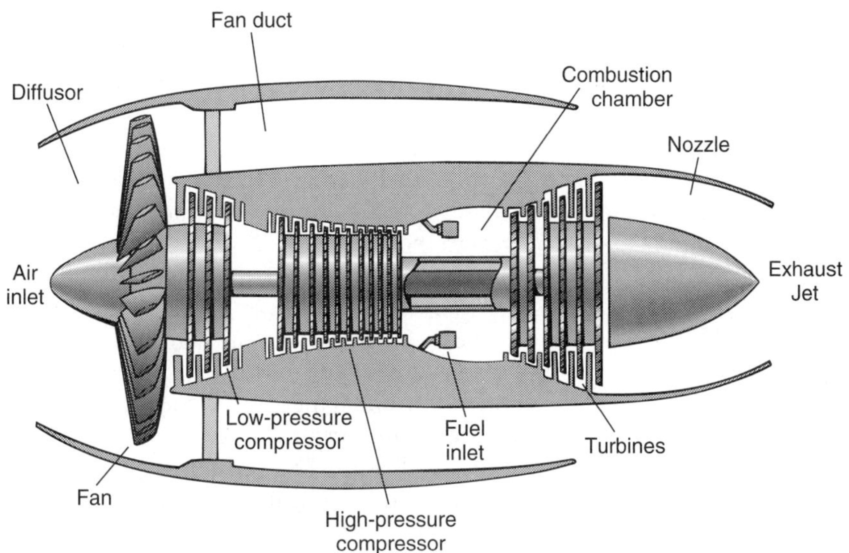

Fig. 4.4.12 - The turbofan engine adds a giant fan to the shaft of a normal turbojet engine. Most of the air passing through the fan bypasses the turbojet and returns directly to the airstream around the engine. Because the fan does work on this air, it leaves the engine at a higher speed than it had when it arrived. The air has transferred forward momentum to the engine and the plane.

Stability and Steering

For an airplane to fly from one place to another, it must remain stable in flight and the pilot must be able to steer it. Stability means that the plane will continue to fly nose forward rather than spinning around wildly. The airplane will be stable if its center of aerodynamic pressure is located behind its center of mass; that way, whenever the airplane's orientation is disturbed so that its tail begins to overtake its nose, the air will exert a torque on the airplane to return its tail to the back, where the tail belongs. We saw exactly the same principle of stability when we discussed badminton birdies.

On most airplanes, the tail is a large, three-finned object that moves the center of pressure toward the rear of the airplane. The tail ensures that the center of pressure is behind the airplane's center of mass and that the airplane always travels nose first. Because the passengers can move about and the cargo can be rearranged, the airplane must be designed carefully so that it never becomes unstable. On smaller airplanes, passengers are often asked to sit at the front of the airplane or at least not behind the wings. This seating balances the plane's weight over its wings and keeps its center of mass forward of its center of pressure.

On most airplanes, steering is performed by tilting one or more of five surfaces on the airplane's wings and tail. These movements change the airplane's orientation and the direction of the aerodynamic forces it experiences. The pilot controls the airplane's orientation by moving three types of control surfaces on the airplane: the two ailerons, the two elevators, and the rudder.

The ailerons are small horizontal panels located near the tips of the wings, on their trailing edges. By turning the steering yoke as though it were the steering wheel of a car, the pilot tilts the two ailerons in opposite directions. On one wing, the aileron tilts up so that it deflects the passing airstream upward. On the other wing, the aileron tilts down so that it deflects the airstream downward. The air pushes back and exerts a torque on the airplane, which rotates so that one wing becomes lower than the other.

The elevators are small horizontal panels located on the horizontal fins of the tail, at their trailing edges. By pushing or pulling on the steering yoke or stick, the pilot tilts the two elevators in the same direction. If the elevators tilt up, they deflect the airstream upward and the tail is pushed downward. The airplane rotates so that its nose becomes higher than its tail. The reverse happens when the elevators are tilted downward.

The rudder is a vertical panel on the vertical fin of the tail and is controlled by foot pedals. As the pilot moves the pedals, the rudder swivels to the right or to the left and begins to deflect the airstream in that direction. The airstream pushes back, exerting a torque on the airplane so that it rotates horizontally.

An airplane isn't steered like an automobile or a boat. To turn an airplane properly, the pilot must use the airplane's lift to accelerate it through the turn. Simply turning the rudder will make the airplane's body rotate, but the airplane will end up traveling sideways through the air.

Instead, the pilot turns the rudder *and* tilts one wing lower than the other. By dropping the inside wing lower than the outside wing, the pilot puts a component of the lift force into the direction of the turn. The airplane then accelerates around the turn and the passengers are virtually unaware of this change in direction. But tilting the wings produces an imbalance in the induced drag forces those wings experience—as the outside wing acts to obtain more lift, so that it can rise, it also experiences more induced drag. The rudder allows the pilot to keep the plane from twisting in response to these unbalanced drag forces on its wings. (For a discussion of steering in some modern aircraft, see ❐.)

❐ Planes that are very stable in flight are inherently hard to turn and maneuver. Their large tails also experience undesirable viscous drag forces. To increase their maneuverability, the Wright brothers' planes were designed to be unstable. They tended to turn around or tumble in flight and required great skill to fly. Modern fighter aircraft and recent commercial jets also obtain enhanced maneuverability and fuel efficiency by being unstable. These aircraft are flyable only with the aid of computers.

CHECK YOUR UNDERSTANDING #5: The Importance of Symmetry
Why must an airplane have wings on both sides of its body?

Supersonic Flight

In air, sound is a pressure disturbance that travels at a speed of about 1193 km/h (331 m/s or 741 mph) near sea level. It consists of pressure waves, patterns of compressions and rarefactions that propagate outward from their source at the speed of sound. Sound is created when something upsets air's normally uniform density. Just as upsetting the uniform surface of a still lake sends water waves rippling off in all directions, so upsetting the uniform density of air sends sound waves rippling off in all directions. When these sound waves reach our ears, we hear them. And just as the water waves take some time to reach the shore, so sound waves take some time to travel from their source to our ears.

When the sound source is stationary, the sound spreads out evenly in all directions. But when the sound source is moving, the pattern of sound is much more complicated. To understand this pattern, we can look at how ripples travel on the surface of a lake. Imagine a stone thrown into the lake. Ripples on the surface head out as nice, concentric circles in all directions (Fig. 4.4.13*a*). But if the source for the ripples is moving, the ripples bunch up in one direction and spread out in the other (Fig. 4.4.13*b*). If the source for the ripples is moving very quickly, the ripples will bunch up further in one direction until finally the source exceeds the speed of the ripples. Now the source actually outruns the ripples it generated earlier (Fig. 4.4.13*c*,*d*). The pattern of ripples looks more like a triangle than a circle. The triangle's forward edge is a shock wave. Power boats make shock waves like these whenever they move faster than about 15 km/h (9 mph).

The same kind of ripple behavior happens for sound waves in air. Because sound spreads in three dimensions, rather than the two dimensions available to surface waves on water, the ripples in air are spherical rather than circular. As long as an airplane moves slower than the speed of sound, it emits simple sound waves in all directions. But once it travels faster than sound (see ❐), its sound expands in a cone. This cone extends outward behind the airplane and moves along with it. The outer surface of the cone is a shock wave. When this shock wave passes over observers on the ground, they hear a "sonic boom."

❐ As they approached the speed of sound during dives, W.W.II aircraft encountered buffeting and stability problems that led people to believe that a plane could not fly faster than the speed of sound. This notion of a "sound barrier" was dispelled when, on October 14, 1947, Capt. Charles Yeager piloted his XS-1 rocket plane to 1.06 times the speed of sound. The flight was so uneventful that Yeager, who was flying with two broken ribs from a horse riding accident, could only tell that he had exceeded the speed of sound with the aid of instruments.

The edge of the cone, the shock wave, carries a great deal of energy. This energy is stored as a large difference in air pressure across the shock wave. The

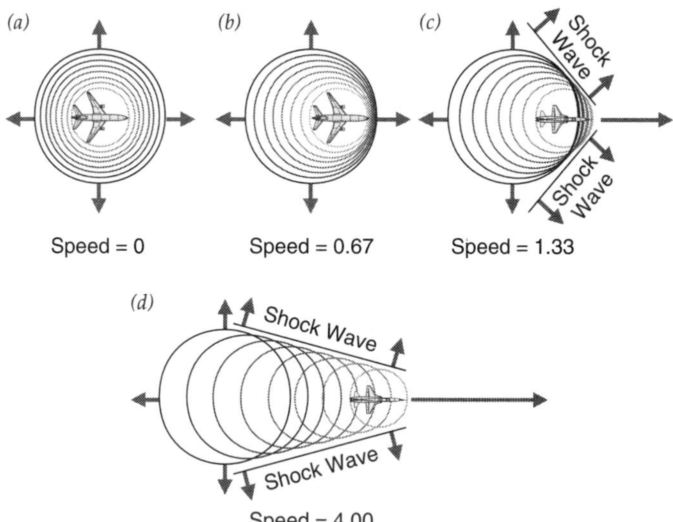

Fig. 4.4.13 - When a source of sound waves is not moving (*a*), the sound spreads out in a series of concentric circles. When the source begins to move, the circles of sound bunch up in the direction of the source's motion (*b*). Once the source travels faster than the speed of sound, the circle pattern appears as a triangle (*c*,*d*). The forward edge of the triangle is a shock wave. All of the speeds indicated in this figure are fractions of the speed of sound. Thus, a speed of 0.67 is 0.67 times the speed of sound or about 800 km/h (497 mph).

abrupt change in air pressure that occurs when the shock wave passes can hurt your ears or break your windows. It can also heat objects it touches, particularly on surfaces of a supersonic airplane that are constantly being hit by this shock wave. Supersonic airplanes must be designed so that they don't fly through too many of their own shock waves and so that surfaces that do fly through shock waves can withstand high temperatures. Since shock waves carry off energy from the airplane, it takes additional thrust to propel an airplane above the speed of sound. Supersonic airplanes are designed to minimize their energy loss to the formation of shock waves.

CHECK YOUR UNDERSTANDING #6: Wake-Up Time

You're sitting on a dock as a powerboat speeds by. A large ripple spreads out on both sides of the passing boat, forming a huge, moving triangle. As this ripple crosses your dock, waves splash you with water. What is this moving ripple?

Summary

How Airplanes Work: An airplane is supported in flight by air passing across its wings. Air traveling over the wings moves farther and faster than air traveling under the wings. As a result, the air pressure is lower above the wings and the wings experience an upward lift force. This lift force decreases at low airspeeds so that most commercial jet airplanes must alter the shapes of their wings during takeoffs and landings.

The airplane also experiences drag, which tends to slow it down. To keep the airplane moving forward, the plane employs propellers or jet engines. These devices push the air backward and the air reacts by pushing them forward. A propeller works directly in the oncoming air, increasing the air's energy and pushing it backward with rotating blades. A jet engine first slows the air, then increases its energy by burning fuel in it, and finally pushes it backward at high speed.

The plane is tilted and steered by deflecting the airstreams flowing across its wings and tail. The tail also ensures that the plane flies nose first and doesn't spin around in flight. The plane's tilt affects both the amount and direction of the lift. The pilot uses the plane's tilt to execute turns and to change the plane's altitude.

The Physics of Airplanes

1. An airfoil's lift increases with the size and curvature of its upper surface, its angle of attack, and its airspeed.

2. When an airfoil's angle of attack is too high, the airflow separates from its upper surface and the airfoil stalls. Its lift decreases and its drag increases.

3. An object that is moving faster than the speed of sound creates a shock wave. This shock wave forms a cone that trails behind the object. When this cone passes by an observer, that observer hears a sonic boom due to the large pressure difference across the shock wave.

Check Your Understanding - Answers

1. The air traveling around the outside of the sail has a lower pressure than that traveling across the inside of the sail. The sail experiences a lift force that pushes it and the boat across the water.

Why: Sails experience both lift and drag forces. The sail experiences an aerodynamic force that pushes it outward (lift) and slightly downwind (drag) just as an airplane wing experiences an aerodynamic force upward (lift) and slightly downstream (drag). The sailboat's keel, or centerboard, and its rudder provide additional forces so that the net force on the sailboat can be controlled and it can travel in a variety of directions.

2. The plane's wings are stalling.

Why: Tipping the plane's nose upward increases the wings' angle of attack. While this action increases lift up to a point, it can also cause the airflow to separate from the top surface of the wing. The sudden reduction in lift and increase in drag that accompanies stalling can cause the plane to drop. A stall during takeoff or landing is extremely dangerous.

3. The inlet side of each fan blade experiences the lowest air pressure.

Why: Air blowing toward you from a fan is like the air blown back by a propeller. The lowest pressures experienced by a propeller are on its forward surfaces. Similarly, the lowest pressures experienced by a fan are on its inlet side surfaces. This pressure imbalance pushes the fan away from you while the fan pushes the air toward you.

4. As the air is slowed down, its pressure increases. The air's total energy remains constant.

Why: The remarkable result of Bernoulli's effect is that you can slow the air down without squandering its kinetic energy. That energy becomes pressure potential energy and it passes through the jet engine in that form. As the air leaves the jet engine, its pressure potential energy becomes kinetic energy once again, so that no energy is wasted.

5. To position its center of aerodynamic pressure properly behind its center of mass.

Why: The only reason for having wings on both sides of the plane is to make it stable in flight. One wing could easily provide enough lift to support the plane, but that lift would be located far to one side of the airplane's center of mass. The plane would experience a terrible torque that would cause it to turn over. To avoid this torque, planes must have two balanced wings.

6. A shock wave.

Why: The powerboat is traveling faster than the speed of surface waves on water. As a result, it creates a shock wave that spreads out in a triangle behind it. This shock wave is analogous to the pressure shock wave that appears in air when an airplane flies faster than the speed of sound. In air, the shock wave is a cone rather than a triangle.

Glossary

engine A machine that converts chemical or internal energy into useful work.

induced drag The drag force that occurs when a wing deflects the stream of air passing across it to obtain lift.

shock wave A narrow region of high pressure and temperature that forms when the speed of an object through a medium exceeds the speed at which sound, waves, or other vibrations travel in that medium.

speed of sound The speed at which sound's compressions and rarefactions travel in a medium such as air or water.

stalling An aerodynamic problem that occurs when the airflow over a wing separates from the surface, leaving a pocket of turbulent air between it and the wing.

thrust A forward, propulsive force.

Review Questions

1. The streamlines below the wing of a flying airplane are more widely spaced than those above the wing. How are these spacings related to the pressure in those places?

2. Induced drag is an unavoidable consequence of energy conservation. Explain why from your perspective on the ground, an airplane flying past through calm air must transfer energy to the air in order to stay aloft.

3. Adding passengers to a plane increases its weight. Why does the plane then consume more fuel during its flight?

4. When a wing stalls, how do the speed and pressure of the air above it change?

5. How does stalling decrease lift and increase drag?

6. Describe the increased drag force on a stalled wing in terms of its turbulent wake and a transfer of momentum to the column of air that moves along with the wing.

7. Why can't the air in a spinning wing-tip vortex behind an airplane stop spinning all by itself?

8. Describe a plane's propulsion in terms of a momentum transfer between the plane and the air. Is energy exchanged during this transfer?

9. Why is a propeller a rotating wing?

10. How is air able to flow through a jet engine more slowly than it flows past the outside of that engine? What becomes of the air's kinetic energy when it slows down?

11. A plane with its "tail" in front tends to be unstable and therefore difficult to fly. Why does it tend to flip around?

12. Suppose that air were flowing past you and your friends faster than the speed of sound. If your friends were upwind of you, they wouldn't be able to hear you talking. Why not?

Exercises

1. An aerator mixes air with the water leaving a faucet. In the aerator, tiny streams of rapidly moving water flow around a spherical piece of plastic. Why does this arrangement cause air to accelerate toward the water and mix with it?

2. An airplane can fly upside down as long as its angle of attack is adjusted properly. While it's flying upside down, what is the relationship between the air pressure above and below its inverted wings?

3. Some military jets have afterburners which burn extra fuel in the tailpipes of their jet engines. How does this added energy affect the velocity of the burned gas once it has left the engine?

4. As your pet canary flies around the room, its weight is ultimately being supported by the floor. How does the air convey forces between the floor and the airborne canary?

5. Why does an airplane have a "flight ceiling," a maximum altitude above which it can't obtain enough lift to balance the downward force of gravity?

6. A hurricane or gale force wind can lift the roof off a house, even when the roof has no exposed eves. How can wind blowing across a roof produce an upward force on it?

7. If you let a stream of water from a faucet flow rapidly over the curved bottom of a spoon, the spoon will be drawn into the stream. Explain this effect.

8. Bicycle racers sometimes wear teardrop-shaped helmets that taper away behind their heads. Why does having this smooth taper behind them reduce the drag forces they experience relative to those they would experience with more ball-shaped helmets?

9. If you want the metal tubing in your bicycle to experience as little drag as possible while you're riding in a race, is cylindrical tubing the best shape? How should it be shaped?

10. The main rotating blades of a helicopter are essentially a propeller. How do these blades obtain the upward force that supports the helicopter against the force of gravity?

11. Some military planes can swivel their jet engines to draw in air from above and send their exhaust downward. As a result, these planes can take off vertically without using their wings. How do its jet engines provide the upward force needed to support the plane against the force of gravity?

12. A common misconception is that a plane emits a "sonic boom" at the moment it "breaks the sound barrier" (exceeds the speed of sound). At what times in its flight can a plane cause "sonic booms" and in what sense is the "sound barrier" a barrier?

13. Supersonic transport planes such as the Concorde are forbidden to fly at full speed over many populated areas because of their "sonic booms." Why can't they "break the sound barrier" over an unpopulated area and then continue their supersonic flights quietly anywhere they like?

14. If a meteor were to pass near you on its fiery supersonic trip through the earth's atmosphere, you would probably hear a "sonic boom." If your friends were far away and the meteor passed them first, would they hear the "sonic boom" earlier, later, or at the same time as you heard it?

15. The windmills that some farms use to pump water have broad, flat tails that keep their blades turned toward the wind. Draw a diagram to show how the airstream flowing past this tail always twists it downwind.

16. The tail of a classic diamond-shaped kite hangs from the bottom of the diamond and pulls it both downward and downwind. What produces the tail's downward force and what produces the downwind force?

17. A poorly designed household fan stalls, making it inefficient at moving air. Describe the airflow through the fan when its blades stall.

18. When a plane flies upside down, which way does it deflect the air that passes by its inverted wings?

19. If an airplane deflects the passing airstream downward, it's transferring downward momentum to the air. Where is the airplane getting this downward momentum from in order to pass it along to the air?

20. If you put your hand out the window of a moving car, so that your palm is pointing directly forward, the force on your hand is directly backward. Explain why the two halves of the airstream, passing over and under your hand, don't produce an overall up or down force on your hand.

21. If instead of holding your hand palm forward (see Exercise **20**), you tip your palm slightly downward, the force on your hand will be both backward *and* upward. How is the airstream exerting an upward force on your hand?

22. As the airstream in Exercise **21** leaves your hand, in which direction is it heading?

23. If an airplane experienced no drag forces at all, the only aerodynamic forces on it would be lift forces. As a result, the plane could sustain level flight indefinitely, even without engines. Explain why lift forces don't change the energy of a plane in level flight through stationary air.

24. From the perspective of someone standing on the ground on a calm day watching an airplane fly by, drag forces do negative work on the plane. Explain why this is true.

25. Show that drag forces discussed in Exercise **24** transfer forward momentum from the plane to the air.

26. The wings of most balsa wood gliders and paper airplanes are flat, rather than bowed outward. However the air flowing over their wings still flows faster than the air flowing under their wings. How can a flat wing make the streamlines spread apart beneath it and squeeze together above it?

27. When a paper airplane arcs upward after you throw it, the plane gradually loses speed. Eventually it stops rising and begins to plummet downward, usually nose first. Why does the lift force stop supporting the plane?

28. When a stunt plane flies in a loop-the-loop, it accelerates primarily toward the center of its circular path. What type of aerodynamic force causes this centripetal acceleration?

29. When a plane enters a steep dive, the air rushes toward it from below. If the pilot pulls up suddenly from such a dive, the wings may abruptly stall, even though the plane is oriented horizontally. Explain why the wings stall.

Epilogue for Chapter 4

In this chapter, we looked at four objects that use moving fluids to perform their tasks. In *water faucets*, we saw how water moves through pipes and openings. We looked at the effects of viscosity and the importance of Bernoulli's equation in describing the conversion of pressure potential energy into kinetic energy and vice versa. Two types of fluid flow appeared: laminar and turbulent. While laminar flow is smooth and predictable, we saw that turbulent flow involves unpredictability and chaos, a common feature of our complicated universe.

In *vacuum cleaners*, we saw how viscous drag forces permit air to exert forces on dust, and we looked at how those forces are increased by passing the air through a narrow opening. We considered how a fan can add energy to moving air so as to maintain flow against the slowing effects of viscous forces and drag forces.

In the section on *balls and airs*, we examined the ways in which moving air exerts forces on larger objects, both in the downstream direction as drag and in a perpendicular direction as lift. We learned about several different types of drag forces and how these can be controlled or reduced by choosing the shapes or motions of the balls. We also looked at airfoils, aerodynamic objects that are designed to obtain substantial lift as they pass through the air.

Finally, in *airplanes*, we explored the ultimate aerodynamic machine. We saw how it uses Bernoulli's effect to support itself on carefully designed wings. We also examined the limitations of those wings and what can go wrong if those limitations are exceeded. We studied propulsion in both propeller and jet aircraft and saw that both systems involve Bernoulli's equation and the forward force that comes from pushing the air backward.

Explanation: A Vortex Cannon

The vortex cannon creates ring vortices—tiny tornadoes that are bent into loops so that they have no beginnings or ends. Air swirls forward in the middle of each ring and backward on its outer edge. This circular tornado structure is created when air flows through the hole of the vortex cannon. Air flows forward in the middle of the hole while the hole's edges create the backward flow around the outside of the ring. After it leaves the cannon, each ring vortex crawls forward through the surrounding air until its kinetic energy has been exhausted and it slows to a stop. It finishes its existence swirling in place until viscous forces bring its moving air to rest.

Cases

1. A glider is a propulsionless airplane that is completely dependent on external energy to remain airborne. It must be pulled forward to obtain the lift it needs to start flying. Once aloft, it can use the air rising off hot spots on the ground to remain in the air almost indefinitely. But for the moment, let's suppose that the air is completely motionless.

a. One way to launch a glider is to pull it with a car, using a rope that can be detached from the glider when it's in the air. Just after the glider's wheels leave the runway, what four forces are acting on the glider?

b. When the glider releases the rope, one of the four forces disappears. With only three forces left, what happens to the glider's forward component of velocity?

c. If the glider begins to descend gradually, the air no longer approaches it horizontally. From the glider's perspective, the air is now blowing upward toward it at a shallow angle. How does this new wind direction affect the directions of the three forces on the glider, as viewed from the ground?

d. It's possible for the descending glider to maintain a constant forward component of velocity. Explain.

e. Now suppose that the air is rising quickly, perhaps heated by contact with a hot spot on the ground. In that case, the glider doesn't have to descend in order to maintain a constant forward component of velocity; it can glide straight and level forever. Why?

2. A helicopter suspends itself from a spinning rotor. Each blade in the rotor is actually a wing, with an airfoil shape that creates lift as it slices through the air. To hover, the helicopter's rotor creates just enough upward lift to exactly balance the helicopter's downward weight.

a. The rotor keeps turning at a constant rate. To control the rotor's lift, the helicopter adjusts the angles of attack of the blades as they turn. If a hovering helicopter increases the angles of attack of all of its blades simultaneously, what will happen to the forces on the helicopter and how will it move?
b. The helicopter can also adjust the angle of attack of a single blade. If it increases the angle of attack of whatever blade is currently on the pilot's right, the helicopter will experience a torque about its center of mass. Which way will the helicopter begin to tilt?
c. If the helicopter is tilted nose down and tail high, the helicopter will begin to accelerate forward. Why?
d. To turn the helicopter so that it faces a different direction, something must exert a torque on it about a vertical axis. That's one reason why the helicopter has a small rotor at the end of its tail boom. This rotor pushes air to the right or left, depending on the angles of attack of its blades. Which way should the tail rotor push the air to turn the helicopter to the right?
e. The rotor blades are actually flexible, but they extend almost straight outward when the rotor is spinning. Why?

3. You're watching a downhill ski race. In this race, the skiers are trying to descend a mountain as quickly as possible without crashing. The veteran skier standing nearby makes a number of observations about the racers that can actually be explained scientifically.

a. The veteran points out that one of the skiers is skidding around corners and that such skids slow her down. During a skid, the skis slip sideways across the snow rather than sliding straight forward through it. Why will skidding always slow the skier down?
b. The veteran notes that skiers slow down and lose time whenever they remain airborne too long after passing over a jump. Why should a skier regain contact with the slope as soon as possible after a jump?
c. A skier loses control and skis off the course into a net. The veteran remarks that it's much better to hit a net than a tree. In terms of force and momentum, why is it safer for the skier to crash into a net rather than into a tree?
d. The veteran is impressed by one skier's consistent compact form, a tight tucked shape. Why doesn't the skier want to push on the air during the descent?
e. The veteran remarks about the high air pressure that skiers feel in front of them as they travel swiftly downhill. What causes this high pressure?

4. A parachute slows your descent after you jump out of a plane. In the parachute's frame of reference, wind is blowing upward toward it and its large surface area creates a great deal of upward drag force. This drag opposes your weight and slows your fall.

a. As air rushes past the parachute, the pressures above and below the cloth become different. Why is the pressure below the parachute higher than atmospheric pressure?
b. Why is the pressure above the parachute roughly equal to atmospheric pressure?
c. How does air exert an upward force on the parachute?
d. Why does the amount of upward force exerted on the parachute depend on how fast you're falling?
e. If you open your parachute while you're falling very rapidly, the parachute slows your descent. Which way is the net force on you pointing, and which way do you accelerate?
f. A few seconds after your parachute opens, you reach terminal velocity. You are no longer accelerating. How does the parachute's drag force compare to your weight?
g. How heavy do you feel when you first jump out of the plane? when your opening parachute first begins to slow your descent? when you are descending at terminal velocity?

5. A ball point pen uses a tiny ball to apply ink to paper. As you draw the pen across the paper, the ball rotates and transfers a very viscous ink from a supply tube to the paper.

a. What force exerts the torque that makes the pen rotate?
b. What force carries ink around with the turning ball?
c. A ball point pen won't write upside down for long—the ink descends a small distance in the supply tube and the ball can't transfer any to the paper. But the ball prevents air from entering the supply tube, so the ink stops descending and won't leak from the pen. Why does the ink stop descending?
d. Occasionally air does get in between the ball and the ink and the pen stops working. How does shaking the pen drive the air past the ball so that the pen starts working again?

6. You and your best friend live on the 58th floors of two adjacent high-rise apartment buildings. You have windows that face one another across an open courtyard. One day, the city turns off all the water to your friend's building. You decide to help your friend obtain water. You immediately buy 500 m of garden hose, enough to reach from either apartment to the ground at least two times.

a. Your first thought is to run water from the gardener's faucet, which is at street level in the courtyard, up to your friend's apartment. Although water runs briefly into the hose, it never reaches your friend's apartment. Why not?
b. You decide to obtain water from a faucet in your apartment. You run the hose out your window, down to the courtyard, and up to your friend's apartment. To your surprise, the hose bursts in the courtyard when you turn on the water. Why does the hose burst?
c. You patch the hose and decide not to let it droop down to the courtyard. With the hose running almost directly across the gap between windows, you turn on the faucet. What happens this time?
d. Water is flowing rapidly through all 500 m of hose as your friend fills a bathtub. To stop the flow, your friend suddenly makes a kink in the end of the hose. The pressure in the hose increases and the hose bursts. What produced this pressure surge?
e. As water flows out of the narrow split in the hose, it accelerates to a very high speed. What happens to its pressure as its speed increases?

7. A kite is an airfoil that uses the wind to produce lift. Held in place by a string, a kite can remain aloft indefinitely. The diamond kite is a particularly simple kite.

a. As the wind passes under the lower surface of the diamond kite, it slows down. How does the air pressure there compare to atmospheric pressure?
b. As the wind passes over the upper surface of the diamond kite, it separates from the kite and creates a turbulent air pocket. How does the air pressure there compare to atmospheric pressure?
c. In what direction is the overall force that the air exerts on the kite?

d. If the upper surface of the kite were bowed outward so that the air didn't separate from it, the kite would experience more upward lift. Why?

e. Both the diamond kite's center of mass and center of aerodynamic pressure are roughly at its geometrical center. By itself, the kite is unstable. Why does the kite tend to turn around or flip upside down?

f. To make it stable, the diamond kite has a fabric tail attached to the bottom of the diamond. How does the tail's extra weight help to keep the kite from flipping upside-down?

g. How does the tail's extra drag help to keep the kite from flipping back to front?

8. The sail of a sailboat acts like a wing, creating both lift and drag. Because it deflects the wind horizontally, the lift is horizontal, too, and pulls the boat through the water.

a. Wind blows around both sides of the sail. It speeds up as it travels around the bowed outer surface of the sail. How does the pressure on that side of the sail compare with atmospheric pressure?

b. Air passing across the inside surface of the sail slows down. How does the pressure on that side of the sail compare with atmospheric pressure?

c. What is the direction of the overall force exerted on the sail by the wind?

d. A sailboat has a flat keel which projects into the water and prevents the boat from tilting or drifting sideways. The keel effectively produces a straight track through the water along which the boat can move. If the wind is blowing toward the boat from north to south and the boat's sail is experiencing an eastward lift force, show that the keel can be turned so that the net force on the boat is toward the northeast. (This is how sailboats manage to sail upwind!)

e. Without a keel, the only horizontal forces pushing the boat through the water come from the wind. These wind forces always push the boat at least partly downwind. Why can't forces from the wind push the boat partly upwind?

f. The wind and keel together can push the boat in an upwind direction. But the boat doesn't accelerate forever; it eventually reaches a maximum forward velocity. What other force acts on the boat to reduce the net force on the boat to zero so that it maintains a constant velocity?

g. As the boat moves forward, it's steered by its rudder. With the rudder turned so that it deflects water to one side as the boat travels forward, the boat begins to rotate. How do the water and rudder exert a torque on the sailboat?

h. If the sailors pull the free end of the sail in too close to the ship, the air passing around the sail's outside surface will break away from that surface and create a turbulent air pocket. The sail stalls and its cloth surface begins to flap. How does this stalling affect the sailboat's ability to travel upwind?

9. Centrifugal pumps are commonly used to move water through pipes. The pumping mechanism is a spinning rotor, usually turned by an electric motor. Water enters the rotor near its center and leaves from its outer edge. The rotor has vanes that help to push the water around in a circle. The pump has no valves or narrow passages, so it can be used to pump almost anything, even sewage.

a. If the outlet of the pump is sealed, the water will travel around in a circle with the rotor. Which way is the water accelerating as it turns?

b. For the water to accelerate in this direction, how must the pressures at the center and edge of the rotor compare to one another?

c. How do the energies of the water at the center and edge of the rotor compare to one another?

d. If the outlet of the pump is opened, water will begin to flow out of the pump. Why will additional water flow outward from the center of the rotor to replace it?

e. If the outlet of the centrifugal pump is opened to the air through a narrow nozzle, the water will leave it at high speed. Why must the water pass through a narrow nozzle before it can speed up?

f. How could a centrifugal pump be used to propel a boat?

10. Umbrellas are subject to many forces on a windy day.

a. When you hold the umbrella upright, its curved surface is above you like a dome. When the wind blows horizontally, why does the umbrella feel lighter than normal?

b. If you are walking into the wind and tip the umbrella forward so that it's in front of your face, why doesn't your face feel much wind anymore?

c. If you hold your hand just beyond the edge of the umbrella while it's still tipped into the wind, you'll feel that the airspeed there is even higher than that of the wind. Explain.

d. If you accidentally tip the umbrella backward, so that the inner surface of its dome is exposed to the wind, the wind will push the umbrella backward hard. Why is this force stronger than when the umbrella is tipped forward?

e. When the inner surface of the umbrella is exposed to a strong wind, the umbrella may turn inside out. Explain this effect in terms of a pressure imbalance.

11. To measure its airspeed, an airplane uses a device called a pitot tube. This rod-shaped instrument points forward into the onrushing airstream. It has two holes, each of which is connected to its own pressure gauge: one hole in its front to catch the onrushing air and one hole on its side to examine the moving air as it passes the tube.

a. When air enters the front hole of the pitot tube, it slows down. How is the pressure of the slowed air inside this hole related to the plane's airspeed?

b. When air passes by the side hole of the pitot tube, its speed doesn't change. The hole permits the pitot tube to measure the pressure of this passing air. Is that pressure related to the plane's airspeed? If so, what's the relationship?

c. How can the plane use pressure measurements made at these two holes to determine how fast the plane is moving?

d. An airplane's pitot tube is carefully suspended away from the plane's surface. Why shouldn't the tube be mounted right along the surface where it would be less vulnerable to damage?

e. The pitot tube is mounted in a region of the plane that isn't involved in lift. Why shouldn't the pitot tube be mounted on a lifting surface, such as the top of a wing?

CHAPTER 5

MECHANICAL OBJECTS AND FLUIDS

In the real world around us, few objects are based entirely on the mechanics of solids or on the dynamics of fluids. Most things involve at least a little of each. Nearly every solid object experiences air resistance sometime and nearly every fluid has to encounter a surface sooner or later. In this chapter, we will look at a few machines that can't be described adequately without an understanding of both areas. They have a little solid here and a little fluid there.

Experiment: High Flying Balls

Among the physical issues that we will discuss in this chapter are the importance of reaction effects in propelling rockets and the value of air pressure in giving shape and resilience to bicycle tires. Both of these issues appear in an interesting way in a simple experiment that involves two different sized balls, a basketball and a tennis ball.

If you can't find a basketball and a tennis ball, any two air-filled balls of very different sizes will do. Drop the balls separately and see how high they bounce. If they are properly inflated, they will bounce well because the air they contain stores and returns energy well during a collision. If they are not properly inflated, they will bounce poorly and waste energy bending rubber or leather. A poorly inflated bicycle tire also wastes energy.

But no matter how you inflate a single ball, it can't bounce higher than the point from which you dropped it. Such a rebound would give it more energy than it had originally. In fact, some of the ball's energy will be lost to thermal energy so that it will not even reach its original height.

But what will happen when you stack the smaller ball on top of the larger ball and drop the two balls together? Think about the sequence of events that will occur and try to **predict** what will happen. Now give it a try. Make sure that the small ball remains directly above the large ball as the two balls fall to the floor. **Observe** what happens and see whether the outcome **verifies** your prediction. Measure the heights of the rebounds and try to explain the effect. You could also try the same experiment with balls of various sizes to gather more insight into what is happening.

Chapter Itinerary

The objects in this chapter combine many of the concepts we have discussed in the previous four chapters. These objects are: (1) *rockets*, (2) *bicycles*, and (3) *elevators*. In *rockets*, we'll look at the principle of action and reaction and see how this basic idea makes it possible for vehicles to leave the earth's surface and travel toward the stars. In *bicycles*, we'll see how motion and clever design can make an apparently unstable object stable enough to ride without hands on the handlebars. In *elevators*, we'll discuss the physical mechanisms that propel elevator cars along their vertical tracks with remarkable speed and comfort.

This chapter revisits many ideas from the previous chapters but it also introduces a variety of new concepts we can use. In *rockets*, we'll see how gravity causes objects to orbit on another, giving rise to the intricate motion of the planets and stars. In *bicycles*, we'll discuss how gears and belts can yield mechanical advantage, forming the basis for thousands of machines from pocket watches to grain harvesters. And in *elevators*, we'll look at how hydraulic systems can obtain mechanical advantage with fluids, a concept used in most industrial lifting and moving equipment.

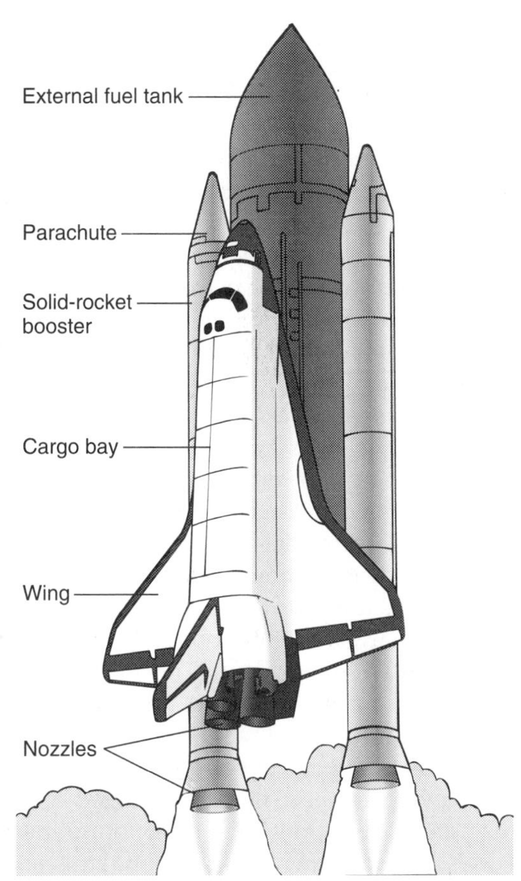

Section 5.1

Rockets

Despite the complexity of modern spacecraft, the rocket is one of the simplest of all machines. It makes use of the very basic principle that every action has a reaction. Each rocket is propelled forward by pushing material out its tail. Nonetheless, people have been developing better and better rockets for more than 700 years. Rockets are now used for such pursuits as space exploration, military weaponry, rescue operations, and amusement.

Questions to Think About: *What pushes a rocket forward? How can a rocket work in space, where there seems to be nothing to push against? What keeps a rocket pointing forward? Why do modern rockets have such fancy exhaust nozzles? What is the fastest speed that a rocket can achieve? Why do satellites travel endlessly around the earth?*

Experiments to do: *While an afternoon spent launching model rockets (available at hobby and toy stores) would be the best introduction to this section, a toy balloon will do just fine. In fact, the balloon illustrates well the need for stability in flight. Blow up the toy balloon and let it go. It will sail around the room. What pushes the balloon forward? Are the room's air and walls involved in the propulsion or is the balloon propelled by the very act of ejecting gas through its opening? How could you verify your answer to the previous question?*

The balloon careens wildly around the room when it's full. Why doesn't this "rocket" fly straight? Where does the force that propels the balloon forward act? Where does the air resistance force that slows the balloon down act? Can you explain why these two forces cause the balloon to tumble as it flies? Why does the balloon begin to fly straight just as it runs out of air? Can you attach something to the balloon to make it fly straight, even when relatively full?

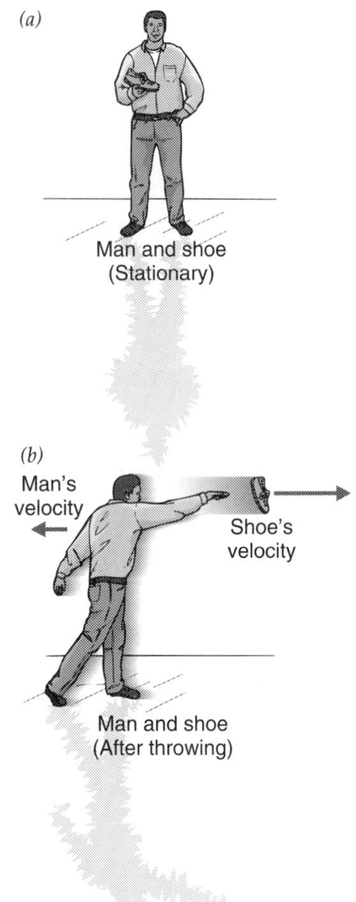

Fig. 5.1.1 - A man who is holding a shoe while standing still on ice has zero momentum. Once he has thrown the shoe to the right, the shoe has a momentum to the right and the man has a momentum to the left. The total momentum of the man and shoe is still zero. Because the man is much more massive than the shoe, the shoe moves much faster than the man.

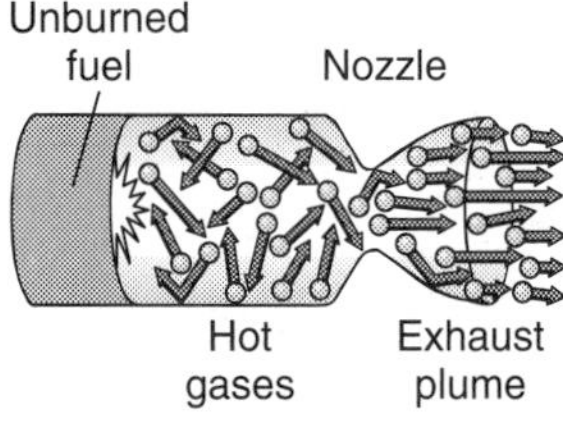

Fig. 5.1.2 - A molecular picture of what happens in a chemical rocket engine. The engine burns its fuel in a confined chamber and the exhaust gas flows out of a nozzle. The nozzle converts the random, thermal motions of the exhaust gas molecules into directed motion away from the rocket engine.

Rocket Propulsion

A rocket obtains its thrust, the force that accelerates it forward even against the pull of gravity, by pushing gas out its tail. The rocket pushes on the gas so the gas pushes back; action and reaction. The more gas it ejects and the faster that gas moves away from its tail, the more thrust the rocket experiences. But to understand just why the thrust depends on the amount of gas ejected and on its speed, we must look a little more carefully at how Newton's third law, the one describing action and reaction, applies to rockets.

We return to an old analogy, one that we first discussed back in Chapter 1. Imagine that you are sitting still in the middle of a frozen pond. It's a warm day and the ice is wet and remarkably slippery. Try as you like, you can't seem to get moving at all. How can you get off the ice?

You are stranded on the ice because you have a mass and no velocity. Your momentum is exactly zero. The only way to acquire some velocity is for you to be acted upon by a force. Since the ice is so slippery, you can't exert a horizontal force on it and it can't exert a horizontal force on you. You could wave your hands like a propeller and try to push the air in one direction, like an airplane. But this section is about rockets, so we will examine what would happen to you if you were to remove one of your shoes and throw it as hard as you could toward the far side of the pond (Fig. 5.1.1). As you throw the shoe, you exert a force on it with your hand. The shoe has no other horizontal force on it, so it accelerates horizontally. By the time the shoe leaves your hand, it has a substantial velocity and heads off rapidly across the ice.

What happens to you? You head off toward the near side of the pond! You are moving because the shoe exerted a force on your hand that was equal in magnitude but opposite in direction to the force your hand exerted on the shoe. For every action there is a reaction. Thus you pushed the shoe toward the far side of the pond while it pushed you toward the near side of the pond.

As you push on the shoe and accelerate it, you transfer momentum to it. The shoe pushes on you and transfers momentum (in the opposite direction) to you. Momentum isn't being created or destroyed, it's only being redistributed. Even after you let go of the shoe, your combined momentum remains at zero. The shoe has as much momentum in one direction as you have in the other.

Of course, you are much more massive than the shoe so that it ends up traveling faster than you do. Momentum is the product of mass times velocity. The more massive the object, the less velocity it needs for the same amount of momentum. The shoe must travel much faster than you do for its momentum to exactly cancel your momentum. Still, you have achieved what you set out to do: you slide slowly toward the near side of the pond.

Your final speed is limited because you can only transfer a small amount of momentum to the shoe and thus only received a small amount of momentum (in the opposite direction) in return. If you were able to throw the shoe at a greater speed or if you had a whole box of shoes to throw, you would be able to transfer more momentum and would end up moving faster.

Instead of throwing shoes, you would do better to throw very fast-moving gas molecules. Even at room temperature, the molecules in the air are traveling about 1,800 km/h. When exhaust gas molecules are heated to about 2800° C (5000° F), as they are in a liquid-fuel rocket engine, they move about 3 times that fast. If you could throw something in one direction at that kind of speed, you would receive quite a lot of momentum in the other direction.

That is what a conventional rocket engine does (Fig. 5.1.2). It uses a chemical reaction to create very hot exhaust gas from fuels contained entirely within the rocket itself. What began as potential energy in the stored chemical fuels becomes thermal energy in the hot, burned gas. This thermal energy is mostly kinetic energy, hidden in the random motion of the tiny molecules themselves. The rocket engine's nozzle converts much of this random motion into directed motion by only permitting the gas molecules to leave the engine from one side. Like an anxious crowd of people streaming out of a sports stadium, the molecules pour out of the nozzle at high speed and head mostly in one direction.

Fig. 5.1.3 - The space shuttle's nozzles are designed to push its rocket exhaust downward as long glowing plumes. The gas pushes back, lifting the shuttle upward and sending it off into space.

If you have ever watched the launch of a large rocket, you have probably noticed the huge, bell shaped nozzles through which the exhaust flows (Fig. 5.1.3). Why the special shape? A rocket nozzle's shape is determined by aerodynamics just as aerodynamics determines the shape of an airplane wing. The nozzle's purpose is to make sure that the rocket extracts as much energy and momentum as possible from the gas. While gas could leave the rocket through a straight section of pipe, it would escape at relatively high pressure. The gas would carry much of its thermal energy away with it and the momentum transfer between the rocket and gas would be less than ideal.

To avoid wasting energy, a properly built rocket has a nozzle so that the pressure of gas leaving the rocket is equal to or only slightly higher than the ambient pressure surrounding it. The shape that makes the most efficient use of the gas's energy is a converging-diverging nozzle, called a de Laval nozzle after its inventor Carl Gustaf de Laval ❒. The first half of the nozzle is converging; it narrows so that gas flowing through it must speed up. The gas experiences the Venturi effect and its increase in speed is accompanied by a drop in its pressure. The gas converts part of its pressure potential energy into kinetic energy.

At low speeds, we can pretend that the exhaust gas is an incompressible fluid. But by the time the gas reaches the narrowest part of the rocket nozzle, it's traveling through that channel at the speed of sound and has become highly compressible. To reduce its pressure further, the gas's density must decrease. The nozzle stops becoming narrower and instead flares outward to allow the gas itself to expand. The gas's density drops and so does its pressure. By the time the gas reaches the end of the nozzle, its pressure is very low and it has converted most of its internal energy into kinetic energy directed away from the nozzle.

The gas actually continues to burn even as it flows through the nozzle, so its kinetic energy continues to rise. With the help of the de Laval nozzle, gas molecules leave the rocket at between 10,000 km/h and 16,000 km/h. If you could throw your shoe that fast, you would move off the ice at about 40 km/h (25 mph). If you had a whole box of shoes to throw, you would be able to reach an enormous speed.

In directing the rush of exhaust out its nozzle, the rocket is exerting a substantial force on the gas. After all, the gas starts out stationary and ends up moving rapidly in one direction. The rocket nozzle exerts a force on the gas to give it this new momentum and the gas pushes back to complete the transfer. Overall, the rocket is pushed forward by its own exhaust. It doesn't need anything outside to push "against" and will operate perfectly well in empty space. Although the local air pressure affects the final shape of the plume leaving the nozzle, the rocket's thrust depends only slightly on the presence or absence of the atmosphere.

❒ Swedish inventor and engineer Carl Gustaf de Laval's (1845–1913) invention of the converging-diverging nozzle predates the modern development of rockets by several decades. He invented this nozzle as a way to make steam turbines more efficient and is credited with laying the foundation for all future turbine technology. De Laval is also known for his invention of the cream separator for milk.

The thrust exerted on the rocket by its exhaust plume can obviously cause it to accelerate, but is it enough to lift the rocket off the ground? The answer is clearly, yes. If the exhaust gas exerts enough thrust on the rocket, that upward force can exceed the rocket's downward weight so that the rocket feels an upward net force. The rocket begins to accelerate upward. For example, the Space Shuttle weighs about 20,000,000 N at launch but its thrust is about 30,000,000 N. The shuttle can not only support its own weight, it can also accelerate upward at about half the acceleration due to gravity! As the shuttle consumes its fuel, so that its weight and mass diminish, it can accelerate upward even more rapidly.

One important measure of a rocket engine's performance is its total impulse, the total amount of momentum it transfers to the rocket. An engine's impulse is its thrust times how long that thrust lasts. The more thrust an engine has and the longer it produces that thrust, the greater the engine's impulse and the more effective it is at pushing the rocket. Liquid fuel rockets pack more chemical potential energy per kilogram, burning hotter and producing higher exhaust velocities than solid fuel rockets (4,500 m/s versus 3,000 m/s). As a result, liquid fuels generate more thrust and higher impulses per kilogram than solid fuels.

Model rockets, available in hobby and toy stores, use solid fuel engines to loft small rockets as much as 800 m into the air. The engines are rated according to their impulses. A "C" engine supplies an impulse of 10 N·s (the newton-second, abbreviated N·s, is the SI unit of momentum or impulse and is synonymous with the kg·m/s). This engine as capable of giving its rocket an upward momentum of approximately 10 N·s. Although gravity and the effects of motion on the apparent exhaust velocity reduce the rocket's final speed, a typical model rocket can reach 500 km/h (311 mph).

CHECK YOUR UNDERSTANDING #1: A Rocket with a Head Start

When a missile is launched from beneath the wing of a fighter aircraft, what does it push against in order to accelerate forward?

A Rocket's Stability

Just because a rocket has enough thrust to overcome gravity and lift itself off the ground doesn't mean that the rocket will fly directly upward. Even sophisticated, modern rockets have fallen over during lift-off or flown in wild spirals in the sky. What makes a rocket stable on the ground and in flight?

On the launch pad, the rocket's stability depends on the behavior of its center of gravity. When the rocket's center of gravity moves upward, its gravitational potential energy increases. When the rocket's center of gravity moves downward, its gravitational potential energy decreases. Since an object accelerates in the direction that decreases its potential energy as rapidly as possible, the rocket will naturally try to lower its center of gravity. If there is a way for it to lower its center of gravity by tipping, the rocket will fall over. It will be stable only if its center of gravity rises no matter which way it tips.

These issues are illustrated in Fig. 5.1.4. The rocket resting on its nose is in **unstable equilibrium**. It's in equilibrium because it experiences zero net force when it's perfectly balanced. However, this equilibrium is unstable because the slightest disturbance causes the rocket to accelerate away from its equilibrium position. If it does begin to tip, the rocket's center of gravity descends so that it accelerates away from its upright equilibrium position. Gravity exerts a torque on the rocket about its nose and it rotates faster and faster until it crashes to the ground.

The rocket on the bottom of Fig. 5.1.4 is in stable equilibrium and doesn't tip over easily. If it does begin to tip, this rocket's center of gravity rises so that it accelerates back toward its upright equilibrium position. Gravity exerts a torque on the rocket about the pivot at its base and it rotates back toward the stable equilibrium. Its broad bottom establishes a **base of support** and its center of gravity lies above that base of support. With this arrangement, the rocket's center of gravity will always rise as the rocket tips and keeps the rocket from falling over easily.

But what happens once the rocket has left the ground? In flight, the rocket's natural pivot is its center of mass. The rocket will fly nose first only if it feels ei-

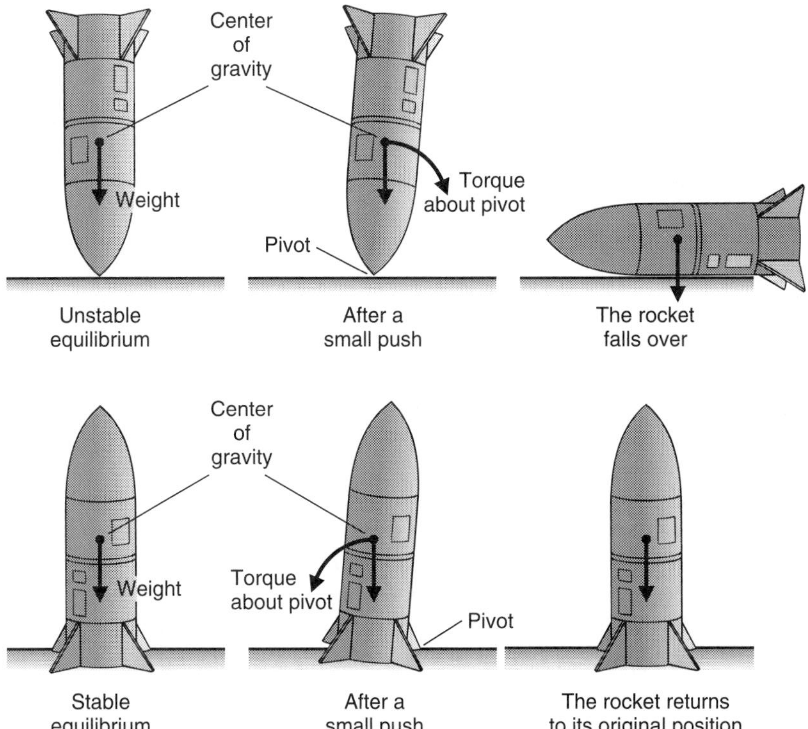

Fig. 5.1.4 - A rocket that is standing on its nose (top) is in an unstable equilibrium. Any disturbance causes it to fall. A rocket that is standing on its flat base (bottom) is in a stable equilibrium. When it's disturbed, it experiences a gravitational torque that returns it to its upright position.

ther no torques about that center of mass or torques that tend to return its nose to the front. If it experiences torques that tend to start it rotating, it's in trouble.

There are two sources of torque that the rocket designers must consider. First, there are torques due to the engine's thrust. The rear-mounted engine is pushing the rocket from behind, a situation that tends to cause trouble. After all, it's easier to steer when you are pulling a cart from in front than when you are pushing it from behind. To keep the rocket moving straight ahead, the engine must exert its thrust directly at the rocket's center of mass. If one of the engines is misaligned, it may exert a torque on the rocket and start it spinning. This unbalanced torque is a common cause of catastrophes in sophisticated rockets. Such rockets usually have steerable engines so that they can correct for misaligned torques on the rocket. But if one engine doesn't ignite or if the steering system fails, the rocket flies wildly.

The other source of torque on a rocket in the atmosphere is aerodynamic force. Once the rocket is moving quickly, air rushing past its surfaces exerts considerable force on the rocket. The effective location of this aerodynamic force is the rocket's center of pressure. If the center of pressure is ahead of the rocket's center of mass, aerodynamic torques will tend to turn the rocket around and it may tumble in flight. If the center of pressure is behind the rocket's center of mass, aerodynamic forces will help to keep the nose pointing forward, as they did for the badminton birdie we discussed in Section 4.3.

Primitive rockets rely exclusively on aerodynamic force to remain stable. They have fins or sticks attached to their tails so that their centers of pressure are naturally behind their centers of mass. Their nozzles are also carefully aligned so that the thrust exerts no torque about their centers of mass. These simple rockets fly straight but are difficult to steer. As we saw in the section on airplanes, a vehicle that is very stable is intrinsically hard to maneuver because it naturally tries to go straight.

Sophisticated rockets abandon fins and sticks and use their thrust to stabilize them in flight. These rockets sense their orientations and swivel their engine nozzles to make corrections. They may also have small rocket engines attached to their sides to push them around and keep them oriented properly. The Space Shuttle and most other modern launch vehicles have essentially no fins. Their stability and maneuverability are governed entirely by carefully controlled rocket engines.

Using rocket exhaust for steering is essential once a spacecraft leaves the atmosphere. Without air, there are no aerodynamic forces at all and the ship's orientation is handled exclusively with rocket engines. Special attitude control rockets are used to rotate the ship, emitting short bursts of gas to keep it pointed in the desired direction. The Space Shuttle's wings and tail are only effective during reentry, when it becomes a glider in the atmosphere. In orbit around the earth, these surfaces serve no purpose because there is no air to deflect.

Still, any self-respecting starship commander wants a vessel that looks as streamlined as possible, or so movies would have you believe. The uselessness of fins and wings doesn't prevent most movie spacecraft from having them. However, next time you see an intergalactic cruiser with wings and a tail, remember that it would work just fine if it were shaped like an oversized school bus.

CHECK YOUR UNDERSTANDING #2: Flying Backwards

If an orbiting Space Shuttle suddenly finds itself traveling tail-first, what will be the immediate consequences for its occupants?

History and Types of Rockets

Rockets date from 13th Century China, as a follow-up to the invention of gunpowder. Burning gunpowder sent hot exhaust gas out of a nozzle and propelled the rocket forward. To make these rockets stable in flight, a guide stick was attached to the engine. This long stick trailed behind the rocket and moved the center of aerodynamic pressure well back of the center of mass. Guide stick rockets flew straight once they were moving quickly but had to be carefully supported during launch. This type of rocket reached its peak of development in the 18th Century when the British colonel William Congreve designed a 28 kg guide stick rocket that could travel several kilometers. Congreve rockets were common weapons used, for example, in the bombardment of Fort McHenry during the War of 1812 and described by Frances Scott Key in his poem *The Star Spangled Banner*.

In the mid-18th Century, Englishman William Hale replaced the guide stick with vanes attached to the sides of the rocket. The Hale rocket was stable for the same reason an arrow is stable. Moreover, he canted the vanes so that air pressure caused the rocket to spin. This rotation made the rocket more accurate than the guide stick rocket. Any imperfections in the rocket engine would be averaged out as the rocket spun around and it would go quite straight in its path.

The value of liquid fuel in a rocket was first pointed out by the Russian schoolteacher Konstantin Tsiolkovsky in 1895 ❐. While solid fuel rockets are reliable and easy to make (Fig. 5.1.5), they don't contain as much chemical potential energy per kilogram as liquid fuels. Liquid fuel rockets can also vary their thrusts at will, while solid fuel rockets cannot. Tsiolkovsky advocated the use of liquid oxygen and liquid hydrogen as fuels, realizing that they contain more chemical energy per kilogram than any other substances.

❐ A self-taught mathematics teacher, Konstantin Tsiolkovsky (Russian schoolteacher, 1857–1935) was inspired by the writings of Jules Verne to consider the possibilities of space travel. He first identified the importance of using liquid fuels in rockets, of steering with the engine's exhaust, and of having multiple stages. In addition to his theoretical work on rockets, Tsiolkovsky wrote science fiction about traveling in space.

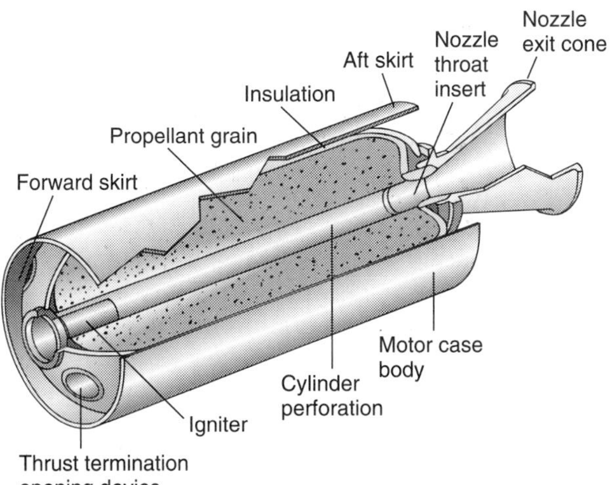

Fig. 5.1.5 - A solid fuel rocket engine is a canister filled with a propellant grain. The core of the propellant is hollow, so that the flame burns outward from the middle of the cylinder. Hot gas created in the combustion accelerates out of the nozzle and pushes the rocket engine forward.

Fig. 5.1.6 - On March 16, 1926, American physicist Robert Goddard (1882–1945) launched the world's first successful liquid fuel rocket. The rocket was powered by gasoline and liquid oxygen, pressure-fed from below to the engine near the top of this picture. Although it attained a height of only 12.5 m and landed 56 m from the launch pad, this was the first modern rocket. The top-mounted nozzle pulled the body of the rocket forward.

But in the early 1900's, few people understood what could be done with rockets and financial support for rocket development was difficult to obtain (see ❐). It was not until 1926 that Robert Goddard launched the first liquid fuel rocket (Fig. 5.1.6). Despite Goddard's successes, rocket development in the United States proceeded very slowly with little government interest. It was Germany that finally brought about the rapid development of liquid fuel rockets in the late 1930's and early 1940's. The German V2 rocket, developed under the direction of Wernher von Braun (1912–1977), was the first missile to travel faster than the speed of sound, covering approximately 200 km in 5 minutes. The V2 was a potent terror weapon and thousands were launched toward Antwerp and London. After the war, the Allies carted off every bit of V2 technology they could find, including von Braun himself.

Since then, liquid fuel rockets have advanced enormously. Engineers have developed pumps that can deliver hundreds or even thousands of kilograms of liquid fuels to the engines each second (Fig. 5.1.7). Nozzle cooling systems now keep the nozzles from burning up during launch so that they can be used for several flights. Hypergolic liquid fuels, chemicals that ignite spontaneously when mixed, make it possible to construct reliable engines that can be turned on and off thousands of times during a flight. And guidance systems and computers have allowed engineers to control rocket flights with enormous precision. In the past decades, rockets have left the earth far behind, taking objects toward other parts of the solar system or even the stars.

❐ On January 13, 1920, the New York Times ran an editorial attacking Robert Goddard for proposing that rockets could be used for travel in space. With modest financial support from the Smithsonian Institution, Goddard was pioneering the development of liquid fuel rockets. The editorial began:

"That Professor Goddard, with his 'chair' in Clark College and the countenancing of the Smithsonian Institution, does not know the relation of action to reaction, and of the need to have something better than a vacuum against which to react—to say that would be absurd. Of course he only seems to lack the knowledge ladled out daily in high schools."

CHECK YOUR UNDERSTANDING #3: Rocket Hood and His Merry Men

One of the earliest uses of rocket propulsion was in rocket-assisted fire arrows. Where should the rocket engines have been attached to the arrows?

The Ultimate Speed of A Rocket

It might seem that the ultimate speed of a rocket is limited by the exhaust velocity of the engine. Remarkably, there is no such limit. As long as the engine keeps pushing material backward, the rocket will continue to accelerate. However, to reach extremely high speed, the rocket must push a vast majority of its initial mass backward as exhaust. For example, if a rocket pushes 90% of its initial mass backward as exhaust, then the remaining portion (with only 10% of the initial mass) might be expected to head forward at about 9 times the speed of its exhaust gas. That analysis is oversimplified, since the rocket speeds up as its engine

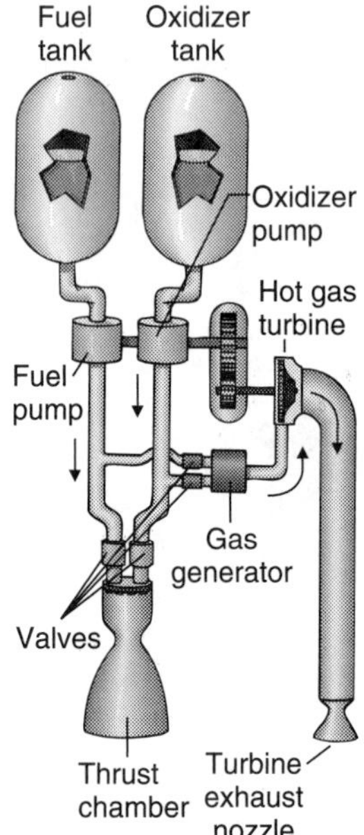

Fig. 5.1.7 - Liquid fuel rockets mix and ignite two liquids directly in the thrust chamber. Hot gas produced by combustion accelerates out the nozzle and pushes the rocket forward. The complex turbine-powered pumping scheme shown above is needed to deliver the huge quantities of fuel burned each second.

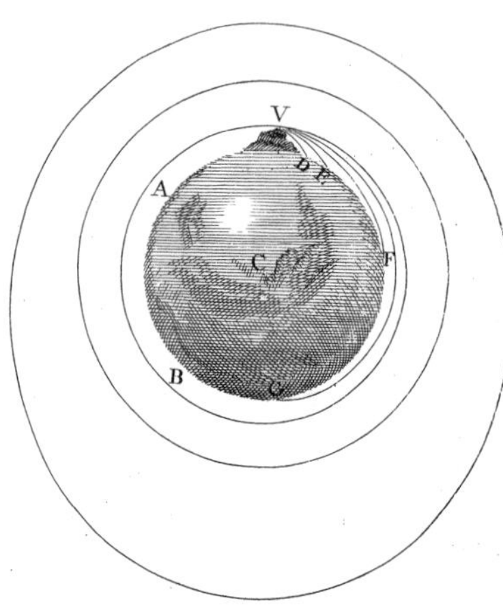

Fig. 5.1.8 - Newton's drawing of a cannonball fired horizontally from the top of a tall mountain. As the cannonball's speed increases, it travels farther from the mountain before hitting the earth. Eventually, the cannonball moves so quickly that the curved earth drops away beneath it and it never hits the earth at all. The cannonball then orbits the earth.

operates. Nonetheless, its implications are correct. Taking into account the rocket's changing speed and the effect of this motion on the exhaust gas, the rocket can actually reach 2.3 times the speed of its exhaust gas. If it can eject *more* than 90% of its initial mass as exhaust, it can go even faster!

But there is a problem with trying to burn up and eject a huge fraction of the rocket's original mass as exhaust. It's difficult to construct a rocket that is 99.99% fuel. Instead, space-bound rockets use several separate stages, each stage much smaller than the previous stage. Once the first stage has used up all of its fuel, the whole stage is discarded and a new, smaller rocket begins to operate. In this manner, the rocket behaves as though it's ejecting almost all of its mass as rocket exhaust. With the help of stages and lots of fuel, rockets can travel substantially faster than their exhaust velocities and reach earth orbit or the solar system beyond. (For recent developments in single-stage rockets, see ❐.)

CHECK YOUR UNDERSTANDING #4: Not Everything is Disposable

The Space Shuttle doesn't have the staged look of expendable rockets. How does it manage to eject most of its launch mass as exhaust?

Orbiting the Earth

Once it has reached a very high horizontal velocity, a spacecraft's engine stops firing and it begins to circle endlessly around the earth. With its engine off and no atmosphere pressing against it, the only significant force acting on the spacecraft is the earth's gravity. The spacecraft's weight causes it to accelerate toward the center of the earth so that, instead of traveling in a straight line at constant velocity, it travels in a huge elliptical loop around the earth.

The spacecraft is in orbit around the earth. An **orbit** is the path an object takes as it falls freely around a celestial object. The spacecraft is accelerating directly toward the earth's center at every moment but its huge horizontal speed prevents it from actually hitting the earth's surface. In effect, the spacecraft perpetually misses the earth as it falls.

Newton conceived of this possibility by imagining what would happen if a cannon were fired horizontally from a mountain top. The greater the cannonball's initial horizontal speed, the farther it would travel before hitting the ground (Fig. 5.1.8). At the enormous speed of 7.9 km/s (about 17,800 mph), the cannonball would travel beyond the horizon and would never actually hit the earth's surface at all. Instead, it would circle completely around the earth in an orbit. The orbiting cannonball would return to the cannon about 84 minutes after it was fired.

Of course, a real cannonball fired from a mountain top would pass through the earth's atmosphere. The ball would gradually lose its horizontal speed and would eventually hit the ground. But a spacecraft orbiting just above the atmosphere can continue to orbit indefinitely. It travels at a speed of 7.9 km/s and circles the earth once every 84 minutes.

However, as the spacecraft moves farther from the earth's surface, its **orbital period**, the time it takes to complete one orbit, increases significantly. First, the spacecraft must travel farther to complete the larger orbit so the trip takes longer. Second, the spacecraft must travel slower in order to follow a circular path around the earth because the earth's gravity becomes weaker with distance.

Chapter 1 noted that gravity attracts every object in the universe toward every other object in the universe (see ❐). In particular, objects are attracted toward the earth. Near the earth's surface, an object's weight is simply its mass times 9.8 N/kg, the acceleration due to gravity. But as the object's distance from the center of the earth increases, the acceleration due to gravity diminishes. Equation 1.1.2 is only an approximation, valid for objects near the earth's surface.

A more general formula relates the gravitational forces between two objects to their masses and the distance separating them. These forces are equal to the gravitational constant times the product of the two masses, divided by the square of the distance separating them. This relationship, discovered by Newton and called the **law of universal gravitation**, can be written as a word equation:

$$\text{force} = \frac{\text{gravitational constant} \cdot \text{mass}_1 \cdot \text{mass}_2}{(\text{distance between masses})^2}, \tag{5.1.1}$$

in symbols:

$$F = \frac{G \cdot m_1 \cdot m_2}{r^2},$$

and in common language:

The pull of gravity is strongest between massive objects but diminishes rapidly with distance.

where the force on mass_1 is directed toward mass_2 and the force on mass_2 is directed toward mass_1. Those two forces are equal in magnitude but oppositely directed. The **gravitational constant** is a fundamental constant of nature, with a measured value of $6.6720 \cdot 10^{-11}$ $\text{N} \cdot \text{m}^2/\text{kg}^2$ (see ❐).

THE LAW OF UNIVERSAL GRAVITATION:
Every object in the universe attracts every other object in the universe with a force equal to the gravitational constant times the product of the two masses, divided by the square of the distance separating the two objects.

This relationship describes any gravitational attraction, whether it's between two planets or between the earth and you. The effective location of an object's mass is its center of mass, so the distance used in Eq. 5.1.1 is the distance separating the two centers of mass. For a spacecraft orbiting the earth just above its atmosphere, that distance is roughly the earth's radius of 6378 km. But for a spacecraft far above the atmosphere, the distance is larger and the force of gravity is weaker. That spacecraft experiences a smaller acceleration due to gravity. To give it the additional time it needs for its path to bend around in a circle, the high altitude spacecraft must travel more slowly than the low altitude spacecraft. This reduced speed explains the long orbital periods of high altitude spacecraft.

The astronauts riding in an orbiting spaceship feel as though they are falling all the time. They actually are! Everything in the spaceship is in free fall, accelerating toward the center of the earth but missing it because of an enormous sideways velocity. Since everything moves and accelerates together, the astronauts are not pressed against the floor, walls, or ceiling of the spaceship. They are simply orbiting the earth inside a spaceship that is itself orbiting the earth. Their perceived weights are zero and they feel weightless. But before you sign up for astronaut training, think about what weightlessness feels like to your stomach. As an orbiting astronaut, you would feel like you had stepped off the highest diving board imaginable and were falling for days or weeks at a time.

To avoid trouble from the atmosphere, the Space Shuttle and most reconnaissance satellites orbit roughly 200 km above the earth's surface. At this altitude, the orbital period is about 90 minutes. If one of these spacecraft tried to move any faster, it would arc outward, away from the earth. If it tried to move any slower, it would crash into the surface.

❐ In 1989, the U.S. government began a program to develop a reusable rocket vehicle that could achieve earth orbit with only a single stage. Nothing but fuel would be jettisoned during launch so that the vehicle could travel to and from orbit repeatedly with only refueling, and minimal maintenance between flights. The challenges facing this program are formidable. Even with liquid hydrogen and oxygen as its fuels, almost 90% of this vehicle's launch weight must be fuel. Nonetheless, construction and testing of such Single Stage To Orbit (SSTO) vehicles is proceeding rapidly and a test vehicle, the DC-X1, has already shown the feasibility of the ideas. A full scale demonstration vehicle, the X-33, is expected to begin flight tests in 1999.

❐ English physicist Henry Cavendish (1731–1810) proved that terrestrial objects do exert gravitational forces on one another. His experiment, performed in 1798, measured the tiny forces that two metal spheres exert on one another, using a very sensitive torsion balance. Comparing the forces between the two spheres with those between the earth and those same spheres (their weights), Cavendish was able to deduce the mass of the earth.

For a spacecraft to take longer to complete its orbit, it must be farther from the earth. At 35,900 km (22,300) above the earth's surface, the orbital period reaches 24 hours. A satellite in such an orbit is said to be *geosynchronous*. If a geosynchronous satellite orbits the earth eastward around the equator, it's also *geostationary*. Geostationary satellites and the earth turn together in unison. A geostationary satellite always remains over the same spot on the earth's surface. Such a fixed orientation is very useful for communications and weather satellites.

Not all orbits are circular. Many spacecraft follow elliptical orbits so that their altitudes vary up and down once per orbit. At apogee, its greatest distance from the earth's center, a spacecraft travels relatively slowly because it has converted some of its kinetic energy into gravitational potential energy. At perigee, its smallest distance from the earth's center, the spacecraft travels relatively rapidly because it has converted some of its gravitational potential energy into kinetic energy. Of course, the perigee should not bring the spacecraft into the earth's atmosphere or it will crash.

The other type of orbit that a spacecraft can experience is a parabolic orbit. If the spacecraft is traveling too fast, the earth will be unable to bend its path into a closed loop and the spacecraft will coast off into interplanetary space. The spacecraft's path near the earth is then a parabola. The spacecraft only follows this parabolic path once and then drifts away from the earth forever.

A spacecraft usually enters a parabolic orbit by firing its rocket engine. It starts in an elliptical orbit around the earth and uses the rocket engine to increase its kinetic energy. This kinetic energy becomes gravitational potential energy as the spacecraft arcs away from the earth. But the earth's gravity becomes weaker with distance and the spacecraft's gravitational potential energy slowly approaches a maximum value even as its distance from the earth becomes infinite. If the spacecraft has more than enough kinetic energy to reach this maximum gravitational potential energy, it will be able to escape completely from the earth's gravity. The speed that a spacecraft needs in order to escape from the earth's gravity is called the **escape velocity**. This escape velocity depends on the spacecraft's altitude and is about 11.2 km/s near the earth's surface. A spacecraft traveling at more than the escape velocity follows a parabolic orbital path and heads off toward the other planets or beyond.

CHECK YOUR UNDERSTANDING #5: Speeding Up the Lunar Month

The moon orbits the earth every 27.3 days, at a distance of 384,400 km from the earth's center of mass. For the moon to orbit in less time, how would its distance from the earth have to change?

CHECK YOUR FIGURES #1: Attractive Cars

How much force does a 1000 kg automobile exert on an identical car located 10 m away?

Summary

How Rockets Work: A rocket obtains its thrust by ejecting gas from an engine in its tail. The rocket pushes on the gas and the gas pushes back. This gas usually comes from burning chemical fuels contained entirely inside the rocket itself and is released from the rocket's engine through a nozzle. A carefully designed nozzle permits the rocket to make efficient use of the energy stored in the fuel by ensuring that gas leaves the rocket at a very low pressure. The rocket's thrust is used to lift the rocket against the force of gravity and to accelerate it upward. On sophisticated rockets, the thrust is also used to steer the rocket by exerting torques about its center of mass. While aerodynamic forces play a role on rocket motion in the earth's atmosphere, they are not present in space. Once in space, with its engine inactive, the rocket will orbit the earth.

Important Laws and Equations

1. The Law of Universal Gravitation: Every object in the universe attracts every other object in the universe with a force equal to the gravitational constant times the product of the two masses, divided by the square of the distance separating the two objects, or

$$\text{force} = \frac{\text{gravitational constant} \cdot \text{mass}_1 \cdot \text{mass}_2}{(\text{distance between masses})^2}. \quad (5.1.1)$$

The Physics of Rockets

1. In certain situations, it's possible to convert the random, chaotic motion of internal energy in directed kinetic energy. For example, a nozzle can redirect the motion of molecules in a hot gas so that they flow together as a stream of gas.

2. When a force causes an object or fluid to accelerate in one direction, that object or fluid exerts a reaction force on whatever accelerated it. These equal but oppositely directed forces transfer momentum and can propel a rocket.

3. An object is always trying to lower its center of gravity, thus diminishing its gravitational potential energy. In the absence of other outside forces, the object will tend to move toward a situation in which its center of gravity is as low as possible.

4. An object resting on the ground is in stable equilibrium if its center of gravity rises when it's tipped in any direction. An object's center of gravity rises when tipped if that center of gravity lies above the object's base of support.

5. An object resting on the ground falls over if its center of gravity descends when it's tipped in one direction. An object's center of gravity descends when it's tipped in one direction if that center of gravity doesn't lie above the object's base of support.

6. Every object in the universe exerts an attractive force on every other object with forces described by the law of universal gravitation.

7. An object that experiences the gravitational attraction of a celestial object may orbit that object, following an elliptical path. The time taken to complete one full orbit increases with the distance between the two objects.

8. An object that is traveling fast enough past a celestial object will not follow a closed, elliptical path. Instead, it will follow a parabolic path and never return.

Check Your Understanding - Answers

1. It pushes against its own exhaust, as all rockets do.

Why: While it may appear that the plume of exhaust beneath a ground-launched rocket is what lifts that rocket upward, the ground itself doesn't contribute to the thrust. The very action of pushing the gas out the nozzle propels the rocket forward.

2. Nothing at all. The Shuttle frequently orbits tail-first.

Why: In orbit, with virtually no atmosphere to worry about, the Space Shuttle's orientation makes no difference at all. It can fly in any orientation it likes, including upside-down. Only during reentry must it be turned nose-first and right-side-up. Once it enters the atmosphere, it's a hypervelocity glider and it must live by the rules of aerodynamics.

3. Near the tips of the arrows.

Why: In flight, a rocket-assisted arrow is like a guide-stick rocket. Putting the engine near the front moves the arrow's center of mass forward while leaving the center of aerodynamic pressure far back near the feathers. Furthermore, putting the thrust up front allows the arrow to follow the engine. While the engine may not push the arrow in exactly the right direction, at least a badly made nozzle will not cause the arrow to tumble, as it would if it were pushing the arrow from behind.

4. It's actually staged subtly. Its two solid fuel boosters are effectively the first stage, the external liquid fuel tank is the second stage, and the orbiter itself is the third stage.

Why: Although the Space Shuttle is not stacked one stage on the next like a Saturn or Delta rocket, it doesn't travel from ground to space as a single object. It discards empty fuel containers as it accelerates upward. First to go are the two boosters, followed by the external fuel tank. The final mass of the orbiter itself is much less than what left the launch pad.

5. The distance between the earth and moon would have to decrease.

Why: The moon behaves just like a spacecraft orbiting the earth at a distance of 384,400 km. Such a spacecraft would also have an orbital period of 27.3 days. To reduce this orbital period, the moon would have to move closer to the earth so that the earth's gravity could bend its path more rapidly.

Check Your Figures - Answers

1. About $6.7 \cdot 10^{-7}$ N.

Why: We use Eq. 5.1.1 to obtain the force:

$$\text{force} = \frac{6.6720 \cdot 10^{-11}\ \text{N} \cdot \text{m}^2/\text{kg}^2 \cdot 1000\ \text{kg} \cdot 1000\ \text{kg}}{(10\ \text{m})^2}$$

$$\text{force} = 6.6720 \cdot 10^{-7}\ \text{N}.$$

This force is roughly equal to the weight of a grain of sand. No wonder it's hard to feel gravity from anything but the entire earth.

Glossary

base of support A surface outlined on the ground by the points at which an object is supported.

escape velocity The speed a spacecraft needs in order to follow a parabolic orbital path and escape forever from a particular celestial object.

gravitational constant The fundamental constant of nature that determines the gravitational forces two masses exert on one another. Its value is $6.6720 \cdot 10^{-11}\ \text{N} \cdot \text{m}^2/\text{kg}^2$

law of universal gravitation Every object in the universe attracts every other object in the universe with a force equal to the gravitational constant times the product of the two masses, divided by the square of the distance separating the two objects.

orbit The path an object takes as it falls freely around a celestial object.

orbital period The time required to complete one full orbit.

unstable equilibrium An equilibrium situation to which the object will not return if it's disturbed. At equilibrium, the object is free of net force or torque. However, if the object is moved away from that equilibrium situation, the net force or torque that will then act on it will tend to accelerate it further away from the equilibrium situation.

Review Questions

1. When two objects that are initially motionless push apart, how does their total momentum change? their individual momenta? their individual speeds?

2. Show that it takes energy for two objects to push apart from one another.

3. If the main purpose of rocket fuel is to provide something for the rocket to push against, why is it so important that the rocket fuel store large amounts of energy?

4. Why will a rocket that's resting on the ground tip over if its center of gravity isn't above its base of support?

5. Why is it important that a flying rocket push its exhaust almost directly away from its center of mass?

6. Why isn't a rocket's forward speed limited to the speed at which exhaust leaves its nozzles?

7. Astronauts in an earth-orbiting spaceship weigh almost as much as they do on the earth's surface. Why don't their weights vanish as soon as they leave the atmosphere?

8. If an orbiting spaceship is always accelerating toward the center of the earth, why doesn't it crash into the earth?

9. Why don't you notice the gravitational forces between you and the various objects in the room around you?

10. Why does the time it takes a satellite to orbit the earth increase as the radius of that satellite's orbit increases?

Exercises

1. An airboat travels through swamps, propelled by a large fan. How does the fan push the boat forward?

2. When a hummingbird hovers in front of a flower, what forces are acting on it and what is the net force it experiences?

3. Why does a round-bottomed bottle fall over while a flat-bottomed bottle tends to remain upright?

4. When a sharpshooter fires a pistol at a target, the gun recoils backward very suddenly, leaping away from the target. Explain this recoil effect in terms of the transfer of momentum.

5. When a cannon fires a shell, the cannon pushes the shell forward and the shell pushes the cannon backward. That's why the cannon suddenly jumps backward as it's fired. Why doesn't a rocket launcher experience this recoil effect when it launches a rocket?

6. A whirlybird sprinkler squirts water out of an S-shaped tube that pivots about its middle. Which way does the sprinkler tend to rotate and what exerts the torque that starts it spinning?

7. Some no-spill toddler training cups have a round weighted bottom. These cups always roll back upright when you tip them over. Explain this behavior in terms of the cup's center of gravity.

8. Face the wall with your toes touching it and try to rise up on your tiptoes. Why do you begin to fall over backward?

9. A catamaran is a boat with two widely separated parallel hulls. What makes this wide, low boat so hard to tip over?

10. You and a friend are each wearing roller skates. You stand facing each other on a smooth, level surface and begin tossing a heavy ball back and forth. Why do you drift apart?

11. Water effectively supports a boat at a certain point—the boat's center of buoyancy. As the boat rocks back and forth, its center of buoyancy doesn't move. Why is it important to load the boat so that its center of gravity is below its center of buoyancy?

12. Use your answer to Exercise **11** to explain why some large boats put heavy ballast deep in their holds to keep them from tipping over.

13. Use your answer to Exercise **11** to explain why it's dangerous to stand up in a canoe.

14. The time it takes a relatively small object to orbit a much larger object doesn't depend on the mass of the small object. Use an astronaut who is walking in space near the space shuttle to illustrate that point.

15. Suppose that you are in a small boat that's floating on a lake and you decide to take a swim. You and the boat are both motionless just before you dive northward into the water. If your momentum as you leave the boat is 200 kg·m/s toward the north, what is the boat's momentum at that same moment?

16. Suppose that the boat in Exercise **15** was heading southward at 3 km/h just before you dove into the water. In the process of diving forward, your momentum increased by 200 kg·m/s toward the north. How much did the boat's momentum change?

17. Which action will give you more momentum toward the north: throwing one shoe southward at 10 m/s or two shoes southward at 5 m/s?

18. Do both of the actions in Exercise **17** take the same amount of energy? If not, which one requires more energy?

19. The tops of some tables extend well beyond their legs. If you put too much weight near the edge of such a table, it will tip over. Why does the table only tip over when the weight is near its edge, rather than in its middle?

20. Why is it so difficult to balance a pencil on its point?

21. You are propelling yourself across the surface of a frozen lake by hitting tennis balls toward the southern shore. From your perspective, each ball you hit heads southward at 160 km/h. You have a huge bag of balls with you and are already approaching the northern shore at a speed of 160 km/h (100 mph). When you hit the next ball southward, will you still accelerate northward?

22. Can you use the tennis ball scheme of Exercise **21** to propel yourself to any speed or are you limited to the speed at which you can hit the tennis balls?

23. As the moon orbits the earth, which way is the moon accelerating?

24. Which object is exerting the stronger gravitational force on the other, the earth or the moon, or are the forces equal in magnitude?

25. To free an Apollo spacecraft from the earth's gravity took the efforts of a gigantic Saturn V rocket. Freeing a lunar module from the moon's gravity took only a small rocket in the lunar module's base. Why was it so much easier to escape from the moon's gravity than from that of the earth?

26. Spacecraft in low earth orbit take about 90 minutes to circle the earth. Why can't they be made to orbit the earth in half that amount of time?

Problems

1. If an 80 kg baseball pitcher wearing frictionless roller skates picks up a 0.145 kg baseball and pitches it toward the south at 42 m/s (153 km/h or 95 mph), how fast will he begin moving toward the north?

2. How high above the earth's surface would you have to be before your weight would be only half its current value?

3. As you walked on the moon, the earth's gravity would still pull on you weakly and you would still have an earth weight. How large would that earth weight be, compared to your earth weight on the earth's surface?

4. If you and a friend 10 m away each have masses of 70 kg, how much gravitational force are you exerting on your friend?

5. How quickly would your friend in Problem **4** accelerate in your direction if the only force on your friend were your gravity?

6. When stars and star fragments collide, their gravitational forces can release vast amounts of energy. Suppose that two neutron stars, only 20 km in diameter but with masses of 10^{30} kg each, have their centers of mass only 50 km apart. How much work will the force of gravity do on one of them if it moves 1 m closer to the other?

7. The gravity of a black hole is so strong that not even light can escape from within its surface or "event horizon." Even outside that surface, enormous energies are needed to escape. Suppose that you are 10 km away from the center of a black hole that has a mass of 10^{31} kg. If your mass is 70 kg, how much do you weigh?

8. In Problem **7**, how much work would something have to do on you to lift you 1 m farther away from the black hole?

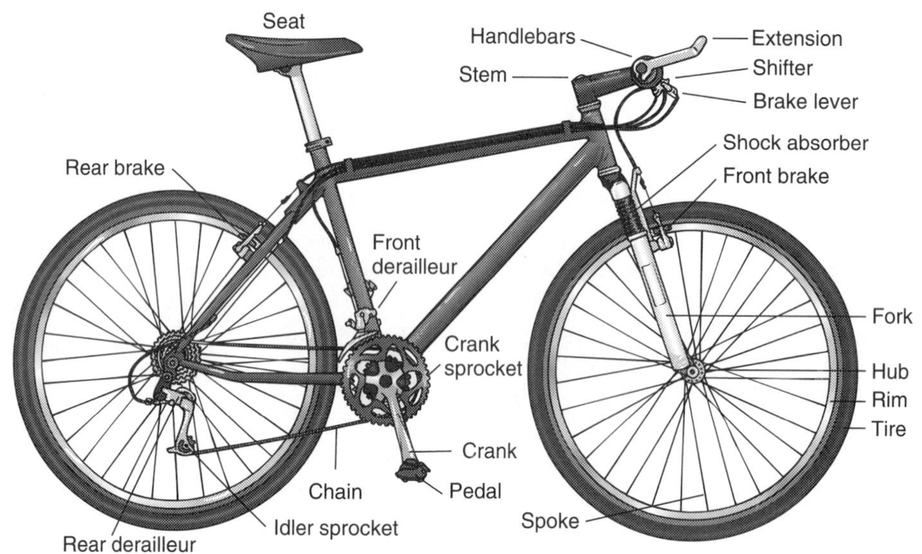

Section 5.2

Bicycles

A bicycle is a wonderfully energy efficient human-powered vehicle. Its wheels allow its rider to make full use of inertia and gravitational potential energy, rolling forward easily on a level surface and accelerating down hills. Compare the easy motion of a bicycle to that of walking, which requires effort every step of the way. Bicycles are very simple machines and most of their moving parts are quite visible: the pedals, gears, brakes, and steering mechanisms, to name a few. Their simplicity and visibility make bicycles relatively easy to fix, even for a novice.

Questions to Think About: *Why is a two-wheeled bicycle so much preferred to the apparently more stable three-wheeled tricycle? Why do you lean into a turn on a bicycle? How is it possible to ride a bicycle without hands on the handle bar? How does pedaling a bicycle make it move forward? Why does a bicycle have gears? Why can you pedal backward on a multispeed bicycle and what is the clicking sound you hear when you do? How do the brakes stop a bicycle? Why are most bicycle tires hollow and filled with air?*

Experiments to Do: *If you know how to ride a bicycle, pay attention to its stability as you ride it. Notice how you lean the bicycle into turns and how the bicycle naturally steers into the turn as you begin to lean. This automatic steering is part of the bicycle's stabilizing behavior. While you pedal, observe which part of the chain becomes taut and the relative rates at which the pedals and the rear wheel turn. How does this relative rate depend on your choice of gears? How does the ease of pedaling depend on this choice? Notice also that when you stop pedaling, the gear on the rear wheel stops turning but the wheel itself keeps going. Something inside the gear or wheel hub is only connecting the gear and wheel some of the time. How does that device determine when to connect the gear to the wheel?*

A Bicycle's Stability

A bicycle has only two wheels to support it. No matter how hard you try, you will be unable to get a riderless, stationary bicycle to stay up on its two wheels for more than a few seconds. While it's up, it's in an unstable equilibrium and the slightest wind or tremor in the ground will start it falling. Why then do we use two wheeled bicycles for transportation?

We will begin to answer that question by looking at **static stability**, the stability of an object that is not in motion. An object that is **statically stable** can be in a stable equilibrium when at rest. Static stability is one reason why so many stools are built with three legs. A three-legged stool is very stable and doesn't rock, even when the floor is not flat. If the stool had fewer than three legs, it would almost certainly fall over because it has no base of support. If the stool had more than three legs, it would rock unless the legs were all of just the right length to touch the floor simultaneously. You have probably sat at a table with four legs that rocked back and forth when you leaned on it. Only a three-legged stool or table is free of the rocking problem, yet has enough legs to avoid falling over. Three legs for our three dimensional world.

Fig. 5.2.1 - A tricycle is very stable when it's standing still. However, it tips over easily during a high-speed turn because the rider can't lean into the turn. The rider must also pedal furiously to move at a reasonable speed.

The same rule applies to wheeled vehicles. Three wheels are uniquely stable. So why don't all vehicles have three wheels? An automobile avoids the rocking problem by suspending its wheels on springs. The lengths of the automobile's "legs" adjust automatically so that they all touch the ground together. For people-powered vehicles, spring suspension is impractical and so four wheels are rarely used. If stationary stability were the only issue for people-powered vehicles, then we really would all ride tricycles. Bicycles are inherently unstable and fall over easily. Unicycles are out of the question.

However, static stability is not the only criterion for choosing the number of wheels on a vehicle. It's the most important criterion only for people who have difficulty balancing, such as children. When it's not moving, a tricycle is hard to tip over, which is why children ride tricycles (Fig. 5.2.1). But when a child rolls down a steep hill and then suddenly makes a sharp turn, he or she will frequently flip over. What has gone wrong?

A tricycle is stable only as long as the rider makes no sudden accelerations. A turn at high speed is an example of a sudden acceleration. When a child makes such a turn, he needs a force to produce the necessary sideways acceleration and that force comes from friction between the wheels and the pavement. As the child turns, the pavement pushes horizontally on the tricycle wheels. If the turn is slow enough, the wheels convey this force to the child and the tricycle turns safely. But if the turn is too abrupt, the child's momentum keeps his body going straight while the wheels turn. Crash. The tricycle has good static stability but its **dynamic stability** isn't so good. An object that's **dynamically stable** can be in a stable equilibrium when in motion. A tricycle isn't always dynamically stable.

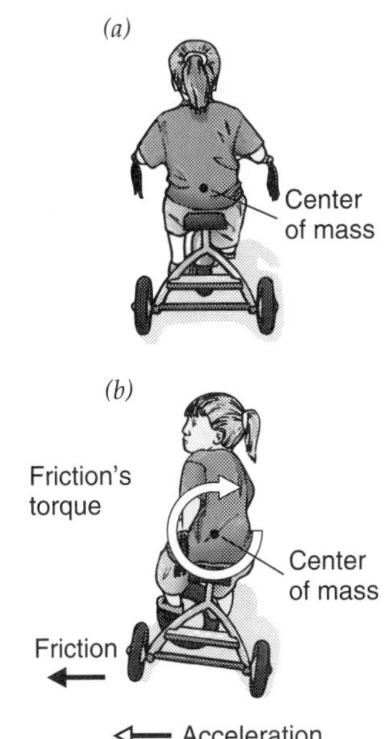

Fig. 5.2.2 - (*a*) A tricycle that is heading straight is stable because any tip causes its center of gravity to move upward. (*b*) However, during a fast left turn, the tricycle accelerates left and friction exerts a large leftward force on the wheels. This frictional force exerts a torque about the tricycle and rider's combined center of mass and can cause the tricycle to tip over.

The tricycle's problem is that it can't handle the torque that friction exerts on its wheels during a turn. The horizontal frictional force that causes the tricycle to accelerate during a turn is exerted well below the tricycle and rider's center of mass (Fig. 5.2.2). Friction exerts a torque about that center of mass, tending to tip the tricycle over. The tricycle's static stability resists this tipping, up to a point. If the turn is too fast, the horizontal force on the wheels just flips the tricycle over. During high speed turns, the tricycle is dynamically unstable.

So four or more wheels are out because of the rocking problem and three wheels are statically stable but dynamically unstable. There aren't many possibilities left. Since the goal of a wheeled vehicle is to go somewhere, dynamic stability is much more important than static stability. So what about two wheels? A stationary two wheeled bicycle has no base of support and falls over as soon as it

Fig. 5.2.3 - In 1970, British physicist David Jones investigated the origins of a bicycle's dynamic stability. He built a series of "unridable" bicycles, including one that had a front wheel so small that it became almost red hot when the bicycle was ridden. Jones found that gyroscopic effects alone did not account for a bicycle's stability. He discovered that this stability also comes from the shape of the front fork.

starts to lean to one side. The closest a stationary bicycle comes to stability is an unstable equilibrium that occurs when its center of gravity lies perfectly above the line between where the front and rear wheels touch the ground. But despite its lack of static stability, a moving bicycle is remarkably stable. Its dynamic stability is so great that it's almost hard to tip over. It can even be ridden without any hands on the handlebars. This feat is a popular daredevil stunt among children who haven't yet realized how easy it is.

As British physicist David Jones discovered (Fig. 5.2.3), the bicycle's incredible dynamic stability comes from *two* sources. First, the wheels behave as gyroscopes. Because of their rotation, they have angular momentum and tend to continue spinning at a constant rate about a fixed axis in space. A wheel's angular momentum can be changed only by a torque so it naturally tends to keep its upright orientation.

But angular momentum alone doesn't keep the bicycle from tipping over. Instead, the front wheel's gyroscopic action automatically steers the bicycle in whatever direction the bicycle is leaning. If the bicycle's frame begins to lean to the left, the front wheel will automatically steer toward the left so as to return the bicycle to an upright position. While a stationary bicycle falls over when it's disturbed from its unstable equilibrium, a moving bicycle steers itself automatically and returns almost magically to equilibrium.

The front wheel steers into leans because of gyroscopic **precession**. When you exert a torque on a spinning gyroscope, you change the amount and/or direction of its angular momentum. In most cases, the gyroscope begins to turn about a new axis of rotation. When the bicycle begins to lean to the left, the pavement exerts a torque on the front wheel and the wheel undergoes precession. It turns left!

While gyroscopic precession can keep a bicycle from tipping over all by itself, it's assisted by a second effect. The front wheel also steers into a lean because of the shape and angle of the fork that supports it. The bicycle is always acting to lower is center of gravity, which is why it falls over when stationary. Because of the shape and angle of the fork, the bicycle can lower its center of gravity by turning its front wheel away from straight ahead. The direction the wheel must turn to lower the bicycle's center of gravity depends on the bicycle's lean. When the bicycle is leaning toward the left, its front wheel steers toward the left. Once again the bicycle automatically steers in the direction of its lean and avoids falling over. These self-correcting effects explain why a rider-less bicycle stays up so long when you push it forward all by itself.

There is little a bicycle designer can do to change the gyroscopic stabilizing effect, but the fork shape and angle can be varied. To be stable, the front wheel must touch the ground behind the steering axis (Fig. 5.2.4). If the fork is bent or improperly made so that the wheel touches the ground ahead of the steering axis, the bicycle will steer away from leans and be unridable. Modern bicycles have front forks that arc forward so that the wheel touches the ground just behind the steering axis. This situation leaves the bicycle dynamically stable enough to ride easily yet very maneuverable. Some children's bicycles are built with straighter forks so that the wheel touches the ground far behind the steering axis. These bicycles are more stable than adult bicycles but not as easy to turn.

Fig. 5.2.4 - A bicycle is stable when moving forward in part because its front wheel touches the ground behind the steering axis. As a result, the front wheel naturally steers into a lean and returns the bicycle to an upright position.

So why does a bicycle rider lean during a turn? The lean ensures that the road doesn't exert a torque on the bicycle and rider about their overall center of mass. A torque would make them tip over. Because the centers of mass and gravity are coincident, gravity can't exert a torque on the bicycle and rider about their center of mass. But the road can exert a torque. When the bicycle is heading straight, the rider keeps the bicycle upright. That way, the contact force from the road points directly toward the center of mass and exerts no torque about it (Fig. 5.2.5*a*). But when the rider wants to turn left, she leans the bicycle to the left.

Since the bicycle is accelerating to the left, the road exerts both an upward contact force and a leftward frictional force on the wheel (Fig. 5.2.5*b*). Because of the lean, the road's overall force points directly toward the center of mass and exerts no torque about it. The bicycle doesn't tip over.

So the rider leans over during turns in order to keep the overall road force on the wheels pointing directly toward the overall center of mass. As a result, the bicycle remains in *rotational equilibrium*—having zero net torque on it—and doesn't tip over. After a while, this habit of leaning the bike becomes so automatic, you don't even think about it. As they say, you never forget how to ride a bicycle.

Leaning into turns is possible only with two wheeled vehicles. With the proper lean, you can always prevent the vehicle from tipping over. Even if you turn a bicycle or motorcycle too hard and begin to skid, you can still keep the vehicle upright. But vehicles that can't lean can and do tip over. Even a car can tip over on a high-speed turn and some cars or modified trucks are particularly prone to flipping. The higher a car's center of mass, the easier it is for frictional forces to tip it over it during a turn. There have been several recent production cars that have been found unsafe because their centers of mass are too high and their bases of support too narrow. Small trucks that have been lifted up to resemble tractor-trailer cabs may look cool, but they also tip over easily.

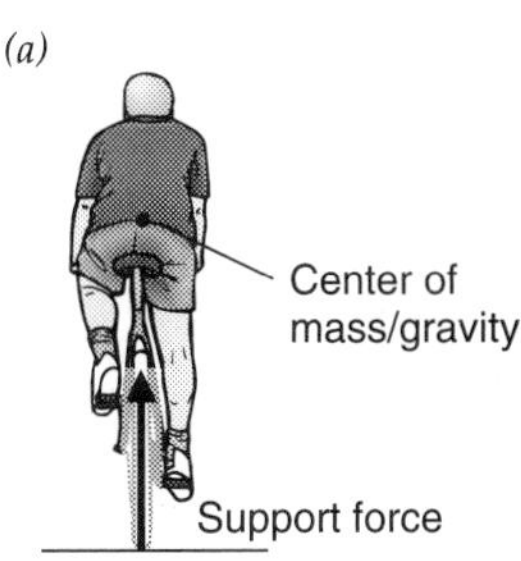

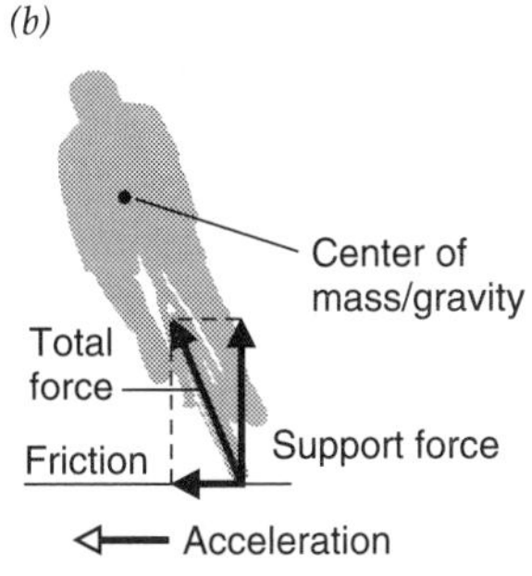

Fig. 5.2.5 - (*a*) A bicycle that is heading straight is in rotational equilibrium when it's perfectly upright. The contact force from the road exerts no torque about its center of mass. (*b*) A bicycle that is turning left is in rotational equilibrium when it's tilted to the left. Together, the contact and frictional forces from the road exert no torque about its center of mass.

CHECK YOUR UNDERSTANDING #1: Cutting Corners

Why does a skier lean into turns during a downhill run?

Pedaling Bicycles

So far, we have only discussed stability. But once we have settled on a bicycle as the most likely configuration for a useful person-power vehicle, we need to figure out how to person-power it. The rider could push his feet on the ground, but that would be pretty inconvenient and even dangerous at high speeds. Instead, we use foot pedals and exert a torque on one of the wheels. But which wheel and how to exert the torque?

The original answer was to power the front wheel by putting pedals and a *crank* directly on its axle. The axle was suspended with bearings so that exerting a torque on the pedals produced a torque on the front wheel and it began to turn. Friction between the turning wheel and the ground then pushed the bicycle forward. This method is still used in most modern children's tricycles but it has two drawbacks. First, you must keep pedaling as long as you are moving. If the front wheel is turning, so are the pedals. The second drawback is more serious: you can exert more than enough torque on the wheels of a tricycle but you will have trouble moving your legs quickly enough to get a tricycle going very fast. When you ride on level ground, you will find yourself pedaling furiously to move at any decent pace and will feel very little resistance from the pedals.

A solution to the too-much-torque problem on early bicycles was to use a gigantic front wheel. In such a configuration, one turn of the wheel would take you a considerable distance, so that you no longer had too much torque and too little speed. The pennyfarthing of the middle 19th Century was this sort of bicycle (Fig. 5.2.6). It still had the problem that you could not stop pedaling and it had a new problem: it was impossible to pedal up hill. You need much more torque to keep a bicycle wheel turning while going up a hill than you do on level ground.

Fig. 5.2.6 - The pennyfarthing used a large directly driven front wheel to permit the rider to travel at a reasonable speed without having to pedal very rapidly. Its names came from its resemblance to two coins, the large English penny and the smaller farthing.

The torque problem was solved long ago by removing the pedals from the wheel axle and using an indirect drive scheme for the rear wheel. This change

Fig. 5.2.7 - A modern bicycle. The rear wheel is driven by a chain that allows the rider to vary the mechanical advantage between the pedals and the rear wheel. A freewheel in the hub of the rear wheel lets that wheel turn freely in one direction. This free motion allows the bicycle to coast forward, even when the pedals are stationary.

makes it possible to use mechanical advantage, when necessary, to convert a large torque and low angular velocity at the pedals to a small torque and high angular velocity at the wheel. The problem of the perpetually turning pedals has been solved by incorporating a one-way drive or freewheel in the hub of the rear wheel. The freewheel allows the wheel to turn freely when you stop pedaling. A modern bicycle with these improvements appears in Fig. 5.2.7.

CHECK YOUR UNDERSTANDING #2: Bigger Isn't Always Better

Why is it so difficult to climb a hill on a Pennyfarthing?

Gears and Belts

Most bicycles use a chain drawn between two toothed wheels to transfer rotational work from the pedals and crank to the rear wheel. This indirect drive system is a combination of two simpler rotary drive mechanisms: gears and belts. To understand how the bicycle chain transfers work from the pedals to the rear wheel and how it obtains mechanical advantage in the process, let's first look at how gears and belts work by themselves.

Gears are toothed wheels that mesh together and exert torques on one another as they turn. Figure 5.2.8 shows a typical gear system, in which the first gear (gear 1) turns the second gear (gear 2). These two gears turn in opposite directions, one counter-clockwise and the other clockwise.

Suppose that you are exerting a counter-clockwise torque on gear 1 and that it's turning counter-clockwise. As we observed in section 4.2, you are doing work on gear 1 because you are exerting a torque on it and it's rotating in the direction of that torque. But gear 1 doesn't retain that work; instead it does work on gear 2. Gear 1 exerts a clockwise torque on gear 2, and gear 2 turns in the direction of that torque. But gear 2 doesn't retain the work either; it does work on whatever is attached to it, such as a wheel. Overall, you are doing work on gear 1, which is doing work on gear 2, which is doing work on the wheel. The gears are simply conveying work from you to the wheel.

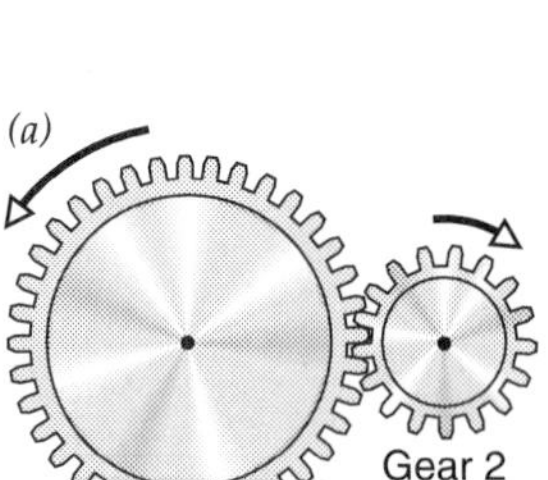

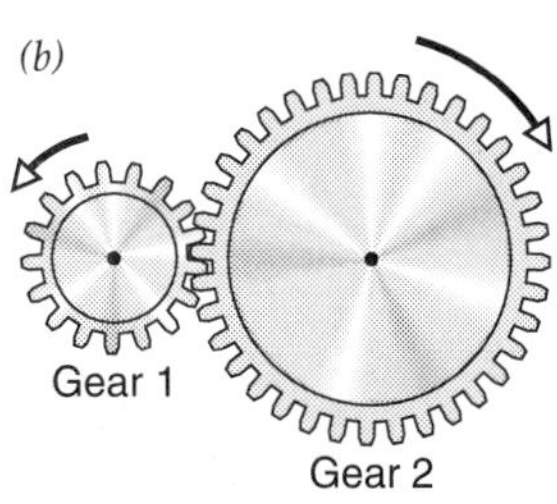

Fig. 5.2.8 - Two sets of gears. You exert a torque on gear 1, and gear 1 exerts a torque on gear 2. In (*a*), gear 2 turns more quickly than gear 1 but gear 2 exerts less torque on its load than you exert on gear 1. In (*b*), gear 2 turns less quickly than gear 1 but gear 2 exerts more torque on its load than you exert on gear 1. In either case, the two gears turn in opposite directions.

But the gears exhibit mechanical advantage. Although the work you do on gear 1 is transferred to gear 2, the gears alter the relationship between the angle of rotation and the amount of torque. If gear 1 has more teeth than gear 2 (Fig. 5.2.8*a*), then gear 2 turns more quickly than gear 1. However, the torque that gear 2 exerts on the wheel is less than you exert on gear 1. The work being done in each case, the product of torque times angle of rotation, is the same. Thus the work you do on gear 1 by turning it slowly with a large torque is equal to the work gear 2 does on the wheel by turning it rapidly with a small torque.

What happens if gear 1 has fewer teeth than gear 2? In that case, gear 2 turns more slowly than gear 1 (Fig. 5.2.8*b*). Again, all of your work is conveyed to gear 2 but now the work you do on gear 1 by turning it rapidly with a small torque is equal to the work gear 2 does on the wheel by turning it slowly with a large torque.

A *belt* is a continuous strap that runs between two or more disks and allows them to exert torques on one another. Figure 5.2.9 shows a typical belt system, in which the first disk (disk 1) moves the belt and causes the second disk (disk 2) to turn. Work is conveyed from disk 1 to disk 2 by tension in the belt, a subject we will discuss in detail in the next section. Unlike gears, the two disks turn the same direction.

Suppose that you are exerting a counter-clockwise torque on disk 1 and that it's turning counter-clockwise. You are doing work on disk 1. But disk 1 is using the belt to exert a counter-clockwise torque on disk 2 and disk 2 is turning in the direction of that torque. Thus disk 1 is doing work on disk 2. If there is a wheel attached to disk 2, then disk 2 is doing work on that wheel. Overall, you are doing work on disk 1, which is doing work on disk 2, which is doing work on the wheel.

The sizes of the disks in a belt drive system determine which disk turns fastest and which exerts the most torque. In Fig. 5.2.9*a*, the smaller disk 2 turns more rapidly than disk 1, but it exerts less torque on the wheel than you exert on disk 1. The work you do on disk 1 by turning it slowly with a large torque is equal to the work disk 2 does on the wheel by turning it rapidly with a small torque. In Fig. 5.2.9*b*, the larger disk 2 turns more slowly than disk 1, but it exerts more torque on its load than you exert on disk 1. The work you do on disk 1 by turning it rapidly with a small torque is equal to the work disk 2 does on the wheel by turning it slowly with a large torque.

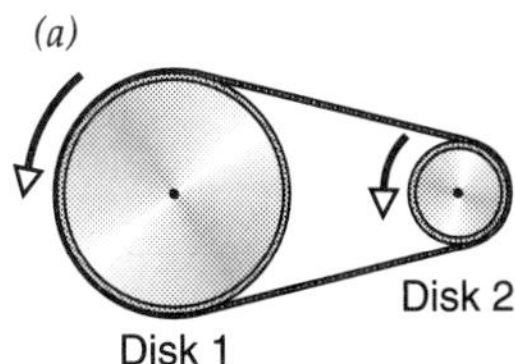

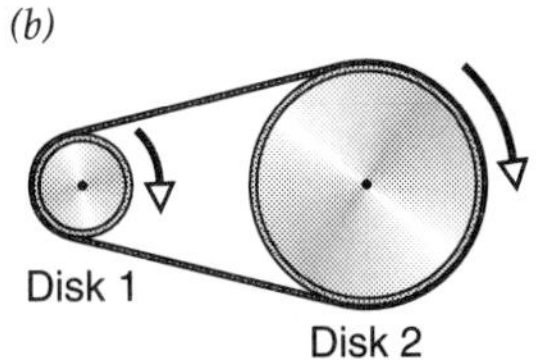

Fig. 5.2.9 - Two sets of belt driven disks. You exert a torque on disk 1 and disk 1 exerts a torque on disk 2. In (*a*), disk 2 turns more quickly than disk 1 but disk 2 exerts less torque on its load than you exert on disk 1. In (*b*), disk 2 turns less quickly than disk 1 but disk 2 exerts more torque on its load than you exert on disk 1. The two disks turn in the same directions.

A modern bicycle uses a chain drive to convey work from your legs to its rear wheel (Fig. 5.2.10). A chain drive makes use of ideas found in both gears and belts. It uses sprockets or toothed disks, connected by a chain, to convey torque and work from one rotating system to another. In a bicycle, the first sprocket (crank sprocket) is located on the crank so that your pedaling makes it rotate. This sprocket moves the chain and causes the second sprocket (freewheel sprocket) to turn. The freewheel sprocket is attached to the rear wheel and makes it turn. Thus as you pedal, you do work on the crank sprocket, which does work on the freewheel sprocket, which does work on the rear wheel.

The relative sizes of the two sprockets determine the mechanical advantage in a bicycle. The larger the crank sprocket, the faster the chain moves and the faster the rear wheel turns. Of course, you don't get something for nothing. Although the rear wheel may turn quickly, the chain will exert relatively little torque on it. You will not be able to climb a steep hill like this. The large crank sprockets are best for traveling downhill or on flat surfaces.

A small crank sprocket moves the chain slowly. While the rear wheel won't turn very fast, the chain will exert a relatively large torque on it. You will be able to climb hills easily. You may not go fast, but you will be able to make steady progress uphill without having to exert much torque on the pedals.

Of course, the size of the freewheel sprocket also matters. The overall mechanical advantage in the bicycle's drive system is determined by the ratio of the number of teeth on the crank sprocket to the number of teeth on the freewheel sprocket. The larger this ratio, the faster the rear wheel turns as you pedal but the smaller the torque you exert on it. The smaller this ratio, the more torque you can exert on the rear wheel but the slower you will go.

Many bicycles have several crank sprockets and as many as seven freewheel sprockets. A device called a derailleur moves the chain from one sprocket to another and thus varies the mechanical advantage. All of the important tension in the chain occurs along it top, between the crank sprocket and the freewheel sprocket. Beyond the freewheel sprocket, the tension diminishes and the chain passes over several spring-mounted idler sprockets before returning to the crank sprocket. The idler sprockets take up the extra length of the chain so that it doesn't drag on the ground or slip off the other sprockets.

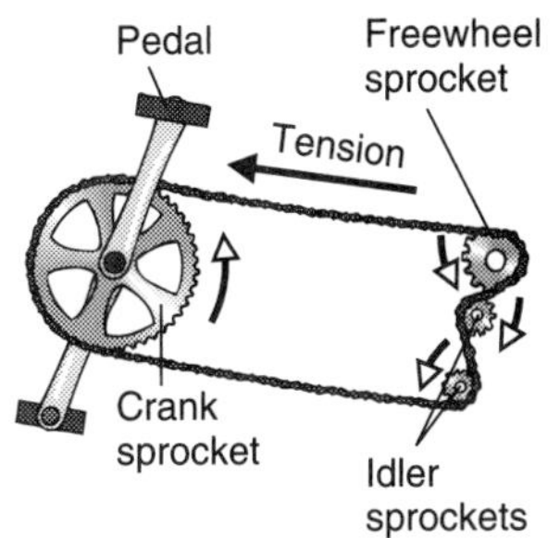

Fig. 5.2.10 - The drive system for a modern bicycle. The pedals supply torque that turns the crank sprocket. The crank sprocket produces tension in the chain which exerts a torque on the freewheel sprocket. The idler sprockets handle the extra chain.

CHECK YOUR UNDERSTANDING #3: Racing Down a Hill
If you wish to go particularly fast downhill, which sprockets should the chain be on?

The Freewheel

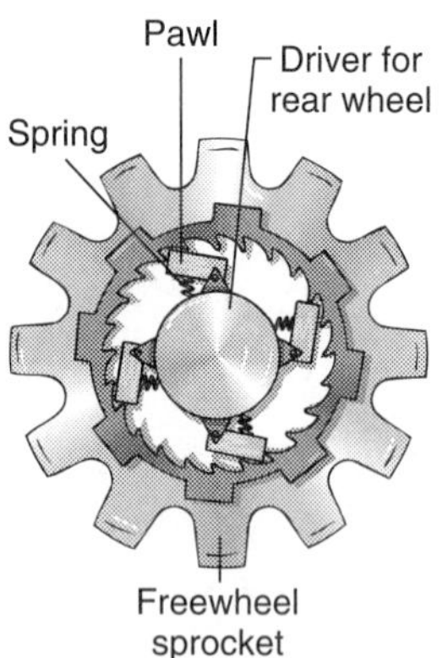

Fig. 5.2.11 - The ratchet in a bicycle freewheel. If the relative rotation of the inner and outer parts is in the correct direction, the pawls transmit torque from the outer part to the inner part. If the relative rotation direction is reversed, the pawls compress the springs and skip along the teeth on the inside of the outer part. No torque is transmitted.

Another feature of modern bicycles is the freewheel itself. The sprockets driving the rear wheel are not directly attached to the rear wheel. Instead, they are attached to a device that transmits torque to the rear wheel only in one direction. When you pedal forward, you exert torque on the rear wheel. When you stop pedaling, or pedal backward, the freewheel allows the rear wheel to turn freely. As a result, you don't have to keep pedaling whenever the bicycle is moving.

The actual freewheel mechanism is called a *ratchet* (Fig. 5.2.11). The outer part of the ratchet pushes on the inner part by way of several *pawls*. These pawls are mounted with springs so that they connect the inner and outer parts of the ratchet only when they try to rotate one direction relative to one another. When they try to rotate the other way, the pawls compress their springs and skip freely along the teeth of the outer part. You hear the sound of this skipping when you stop pedaling on a quiet road. Other devices that use ratchets include socket wrenches, pull-starters on gasoline engines, window shades, winding stems on wrist watches, and automobile seat belts.

CHECK YOUR UNDERSTANDING #4: Tickets Please

Turnstiles in sports stadiums, subways, stores, and amusement parks turn only one way, so that you can't return the way you came. What mechanism prevents them from turning backward?

Pneumatic Tires

Most bicycles have *pneumatic* tires: a hollow rubber inner tube inside a hollow tire. The inner tube is filled with air so that air pressure supports the weight of the bicycle and rider. But pneumatic tires can be punctured, so why not simply use solid rubber?

One problem with solid rubber tires is mass. Rubber is much denser than air so that a solid rubber tire is massive. Moreover, putting this extra mass so far from the axle gives the wheel a large moment of inertia. Since a spinning tire that's moving forward with the bicycle has kinetic energy both in its translational motion and in its rotational motion, adding to the tire's mass and moment of inertia will substantially increase its total kinetic energy. The rider would have to do extra work to bring this solid rubber tire up to speed. To minimize this energy requirement, bicycle tires are lightened by filling them with air.

A second reason for using pneumatic tires is to smooth out the ride. Solid rubber is nearly incompressible. When you hit a rock on a bicycle with solid rubber tires, the rock exerts a huge force on the tire. The tire conveys that force to the bicycle and the whole bicycle jumps violently. A pneumatic tire is more compressible and deforms around small rocks so that you barely even feel them.

But pneumatic tires aren't perfect either. If the tire deforms too easily, a large portion of its surface will touch the pavement. This flattening of the tire introduces sliding friction as the bicycle rolls forward. Energy is also wasted bending the rubber surface of the tire. To keep the tire from deforming very much, the air inside it must have a large pressure. The higher the air pressure, the less tire surface has to touch the ground to support the bicycle and rider's weight. Thus the most efficient bicycle tires are high pressure pneumatic tires. They provide a reasonably smooth ride, without wasting much energy as they roll and without adding much to the weight and mass of the bicycle.

CHECK YOUR UNDERSTANDING #5: Better Bouncing Bicycles

Which bounces better when dropped on the ground: a low-pressure tire or a high-pressure tire?

Brakes

Coasting downhill is certainly lots of fun, but sometimes you have to stop. Many bicycles have caliper brakes: rubber pads that rub against the wheel rims to slow their rotation. The more you squeeze the brakes, the more quickly sliding friction slows the wheels. Since the sliding friction force is approximately proportional to the force pressing the rubber pads against the wheel rims, you can control the bicycle's deceleration by how hard you squeeze the brakes.

As the wheels begin to turn more slowly, the pavement exerts a backward static frictional force on the tires. It's this frictional force that ultimately slows the bicycle's forward motion. If you try to stop too quickly, you will exceed the maximum force of static friction and the wheels will begin to skid. Sliding friction will continue to slow the skidding bicycle, but you will have no control over direction any more and will probably travel in a more or less straight line. This is a good way to have an accident.

Most bicycles have an independent brake on each wheel. The two wheels are not equivalent when braking. The rear wheel skids relatively easily during braking while the front wheel does not. As you brake, the pavement exerts backward forces on the bottoms of the tires. These forces exert torques on the bicycle about its center of mass, tending to lift the rear wheel off the ground and causing it to skid. Braking only with the rear wheel is a good way to experience "fishtailing." However, this torque presses the front wheel more tightly against the pavement and improves its traction. As long as you don't brake so hard that you throw yourself over the handlebars, the front brake is more effective at slowing the bicycle than the rear brake.

A typical child's bicycle doesn't have exposed brakes. Its coaster brake is located inside the hub of its rear wheel. As the rider pedals backward, the coaster brake activates and sliding friction inside the hub slows the rear wheel's rotation. Since the rear wheel does all of the braking, this type of bicycle tends to skid very easily. Some children enjoy slamming on their brakes at high speeds and skidding to a stop; leaving patches of black rubber on the pavement. To reduce this tendency to skid, some coaster brake bicycles also have a caliper brake for the front wheel.

When you brake, the rotating wheel is doing work against sliding friction in the brake pad. Where is that energy going? The answer is thermal energy. Whenever you do work against sliding friction, you are creating thermal energy. In the bicycle, this thermal energy appears in the metal of the wheel, in the rubber brake pad, and in the air inside the tire. They all become hotter. During a long downhill descent, you risk overheating the pads, the wheels, and the tires as your gravitational potential energy is converted into thermal energy by the brakes. Air that is heated in a confined space will experience a rise in pressure so that an overheated tire may explode. Instead of using the brakes to slow your descent, you do better to use air resistance. You will still be creating thermal energy, but in the air around you, not the air in your tires.

CHECK YOUR UNDERSTANDING #6: Hot Wheels

Some exercise bicycles are essentially bicycles with their brakes on all the time. To increase the resistance, you just turn the brakes on harder. Where does your energy go in such a bicycle?

Summary

How Bicycles Work: While a stationary bicycle tips over easily, a moving bicycle is remarkably stable. It remains upright with the help of two stabilizing effects of motion; one due to gyroscopic precession and the other due to the shape and angle of the front fork. Both effects work together to steer the bicycle in the direction that it's leaning. It drives under its center of mass and thus remains upright.

The rider propels the bicycle by exerting a torque on the pedals and crank. The crank's rotary motion is conveyed to the rear wheel by a chain and sprockets, often with mechanical advantage. As the rider pedals forward, the rear wheel turns and friction between the tire and the pavement pushes the bicycle forward. A freewheel at the hub of the rear wheel disconnects the rear sprocket from the wheel whenever the wheel is turning forward faster than the sprocket. This disconnection makes it possible to stop pedaling while the rear wheel is still turning forward.

Once in motion, the pneumatic tires absorb the impacts of small bumps in the pavement. Their low masses make the bicycle easy to accelerate and maneuver. To stop, rubber brake pads are pressed against the wheel rims so that sliding friction can convert the bicycle's kinetic energy into thermal energy.

The Physics of Bicycles

1. An object that's statically stable uses forces and torques that appear when it's at rest to create a stable equilibrium.

2. An object that's dynamically stable uses forces and torques that are present when it's moving to create a stable equilibrium.

3. An object's static stability or instability doesn't always predict its dynamic behavior. Some objects that are unstable when stationary become stable when moving and vice versa.

4. A torque exerted on a rotating object changes the amount or direction of its angular momentum. A gyroscope's axis of rotation often changes direction in response to a torque, an effect called precession.

5. Gears and belts are used to transfer work from one rotating system to another, often with mechanical advantage. While the torque and the angle of rotation may be different after passing through a system of gears or belts, the product of the two, the work, remains the same.

Check Your Understanding - Answers

1. To make sure that the force on her skis is directed toward her center of mass.

Why: There are many sports and situations where a person must lean into a turn. Since a turn always involves horizontal acceleration and horizontal force, the person must shift her center of mass toward the inside of the turn. When she leans just the right amount, her net force points directly toward her center of mass and exerts no torque on her about her center of mass. She doesn't begin to rotate and doesn't tip over.

2. Each turn of the huge wheel takes you a long distance up the hill and requires lots of work. To do this much work in a single turn, you must exert a very large torque on the wheel.

Why: As you pedal a bicycle, you are doing work on the bicycle. You exert a torque on the crank with your feet and the crank turns in the direction of your torque. Work is torque times angle of rotation. To do a very large amount of work on the bicycle in a single turn of the crank, you must exert a very large torque on the crank. That is exactly what you must do when pedaling a Pennyfarthing up a hill. Its huge wheel rapidly ascends the hill, doing lots of work against gravity, so you must exert a huge torque on the crank to provide that work.

3. The largest crank sprocket and the smallest rear wheel sprocket.

Why: On a downhill run, you want the rear wheel to turn as fast as possible. Torque is unlikely to be a problem. If you select the largest crank sprocket, the chain will be drawn forward quickly as you pedal. If you select the smallest rear sprocket, the movement of the chain will cause the rear wheel to turn as quickly as possible. With this arrangement, each turn of the crank will cause the rear wheel to turn about 5 times. This way you can pedal slowly even as the bicycle races down the hill.

4. A ratchet.

Why: As you walk through a turnstile in the permitted direction, pawls in the ratchet skip and it turns easily. You can often hear the pawls clicking. But when you try to go the other direction, the pawls engage and prevent the turnstile from rotating.

5. A high-pressure tire.

Why: A high-pressure tire is more efficient at returning energy when it pushes against the road. Like a properly inflated tennis ball, it bounces well. Low pressure or underinflated tires bend too much and waste energy in the rubber.

6. Your energy ends up as thermal energy in the bicycle and the air.

Why: As you pedal, you do work on the bicycle wheel and it does work against sliding friction. The energy flows out of you, through the bicycle as kinetic energy, and becomes thermal energy in the brakes. If you touch the brakes after a workout, you will find that they are hot.

Glossary

dynamic stability An object's stability when it's in motion.

dynamically stable An object that can be in a stable equilibrium when in motion.

precession The change in orientation of a spinning object's rotational axis that occurs when it's subject to an outside torque.

static stability An object's stability when it's not in motion.

statically stable An object that can be in a stable equilibrium when at rest.

Review Questions

1. Some objects that are quite stable at rest tip over when moving. Explain how torques that only appear when an object is moving can cause these disasters.

2. Some objects that are unstable at rest become stable when they're moving. How can a statically unstable object use torques that only appear when it's moving to become stable?

3. While you could balance a soda bottle upside down on a board, you would have trouble carrying it around the room without it falling over. Why is this an example of a statically stable but dynamically unstable object?

4. A dart has no stable equilibrium orientation when it's motionless above the ground, but it flies point forward when thrown. Explain why a dart is a dynamically stable object.

5. If you tried to pedal a normal tricycle down a hill at a moderate speed, you would have to move your legs up and down at a furious pace. Why?

6. If you tried to pedal a pennyfarthing bicycle up a hill at constant velocity, you would have to exert enormous forces on its pedals. Why?

7. How should you arrange the gears of a bicycle to travel downhill without having to pedal furiously?

8. How should you arrange the gears of a bicycle to travel uphill without having to exert enormous forces on its pedals?

9. If you were rolling downhill on a bicycle that directly connected the freewheel sprocket to the rear wheel, with no ratchet in between, what would happen when you tried to stop pedaling?

10. If a bicycle's tires are underinflated, the bicycle won't accelerate upward much as it rolls over small bumps. Why not?

11. When you slow yourself on a bicycle by applying the brakes, what becomes of your kinetic energy?

Exercises

1. Sprinters start their races from a crouched position with their bodies well forward of their feet. This position allows them to accelerate quickly without tipping over backward. Explain this effect in terms of torque and center of mass.

2. If the bottom of your bicycle's front wheel becomes caught in a storm drain, your bicycle may flip over so that you travel forward over the front wheel. Explain this effect in terms of rotation, torque, and center of mass.

3. Some bicycle races held on level tracks begin with the bicycles already moving at full speed. These bicycles often have extra mass added to the rims of their wheels. How does this added mass assist a rider during the race?

4. If a motorcycle accelerates too rapidly, its front wheel will rise up off the pavement. During this stunt the pavement is exerting a forward frictional force on the rear wheel. How does that frictional force cause the front wheel to rise?

5. When you turn while running, you must lean into that turn or risk falling over. If you lean left as you turn left, why don't you fall over to the left?

6. The riders of one amusement park ride stand against the inside rim of a cylindrical cage. The cage begins to spin rapidly about a vertical axis so that the riders find themselves pressed against the cage. The cage then tips on its side so that it's spinning about a horizontal axis. The ride's mechanism must exert a tremendous torque on the spinning cage to change its axis of rotation, even though the cage's angular speed remains constant. Why is this torque necessary?

7. When a toy top is spinning on a table but its axis of rotation isn't exactly vertical, gravity exerts a torque on it. The top's angular speed doesn't change, but its axis of rotation gradually changes directions, moving around in a circle. What sort of gyroscopic behavior is this top exhibiting?

8. Most racing cars are built very low to the ground. While this design reduces air resistance, it also gives the cars better dynamic stability on turns. Why are these low cars more stable than taller cars with similar wheel spacings?

9. As a skateboard rider performs stunts on the inside of a U-shaped surface, he often leans inward toward the middle of the U. Why does leaning keep him from falling over?

10. The starter motor in your car is attached to a small gear. This gear meshes with a large gear that's attached to the engine's crankshaft. The starter motor must turn many times to make the crankshaft turn just once. How does this gearing affect the amount of torque that the starter motor must exert while making the crankshaft rotate?

11. The crank of a hand-operated kitchen mixer connects to a large gear. This gear meshes with smaller gears attached to the mixing blades. Since each turn of the crank makes the blades spin several times, how is the torque you exert on the crank related to the torque that the mixing blades exert on the batter around them?

12. A bread making machine uses gears to reduce the rotational speed of its mixing blade. While its motor spins about 50 times each second, the blade spins only once each second. Why does this arrangement of gears allow the machine to exert enormous torques on the bread dough?

13. Your car contains several belts that provide power to the fan, water pump, and alternator, and to the power steering and air conditioner if you have them. The engine drives each belt by exerting a torque on a disk around which the belt passes. The belt then passes around a second disk that exerts a torque on the device attached to it. Show that the engine's disk does work on the belt and that the belt does work on the device's disk.

14. The chain of a motorcycle must be quite strong. Since the motorcycle has only one sprocket on its rear wheel, the top of the chain must pull forward hard to keep the rear wheel turning as the motorcycle climbs a hill. Why would replacing the motorcycle's rear sprocket for one with more teeth make it easier for the chain to keep the rear wheel turning?

15. A socket wrench is a device that grabs the head of a bolt and allows you to exert a torque on that bolt by pushing against a long handle that extends outward at right angles to the axis of rotation. The socket wrench contains a ratchet that only transmits torque to the bolt when you turn the wrench one way. When you twist the wrench the other way, it turns freely. Show that, while the wrench is turning at a steady angular velocity, you do work on it only when you turn it in the direction that transmits torque to the bolt.

16. A poorly inflated basketball bounces poorly (it has a low coefficient of restitution). Compare this behavior with that of a poorly inflated pneumatic tire on a bicycle or car.

17. If you inflated a bicycle's tires with water, you would enjoy a pretty soft ride but the bicycle would be more difficult to start, stop, or pedal up hills. Why?

18. If sliding friction causes wear and wear is undesirable, why can't your bicycle's brakes use static friction to slow you down? After all, static friction doesn't cause wear.

19. As you descend a long, steep hill in a car, you must be careful not to apply the car's brakes too much or too long, lest they overheat. You could lubricate the car's brakes so they experience less sliding friction and don't get as hot, but that wouldn't be a good idea. Why not?

20. Your car's brakes could stop your forward motion simply by preventing the wheels from turning. The wheels would then experience sliding friction with the ground and would convert the car's kinetic energy into thermal energy. Why isn't that a good technique for stopping?

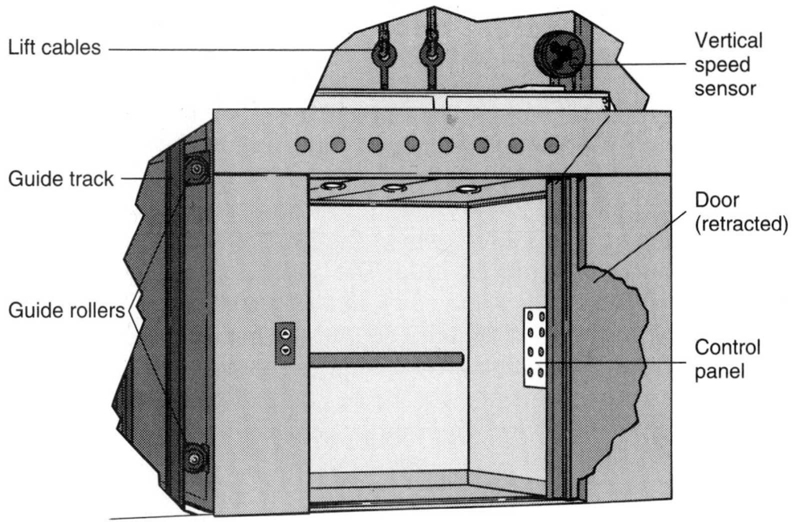

Section 5.3

Elevators

While the invention of steel made skyscrapers possible, the invention of elevators made them practical. Imagine life in a big city without elevators. Business at the top of World Trade Centers would be limited to a few world class athletes.

At the heart of an elevator is a very simple lifting machine. There are only a few different types of elevators and the techniques they use to raise or lower their cars have changed very little since Elisha Otis invented the safety elevator in 1853. What has changed is the source of power for operating the elevators and the sophistication of their control equipment. Electricity has long since replaced steam as the power source and elevator operators have been replaced by computers.

Questions to Think About: *Why do many elevators have counterweights that descend as the car moves upward? Why does an elevator have a weight limit? How fast do elevators actually move? Why do you feel particularly heavy as the elevator you are in starts to move upward and light as that elevator starts to move downward?*

Experiments to Do: *Glass elevators are a popular form of functional art, providing exciting views for the passengers and giving you opportunities to see the mechanisms that make these elevators work. If you look into the shaft of a glass elevator, you will see its cables or hydraulic piston, its counterweights and its control machinery. Find a glass elevator and watch it work. Look for a shiny metal piston pushing the car upward from below or for metal cables lifting the car upward from above. Even if you can't find an elevator with visible parts, take a ride on an elevator. You may be able hear its motors activating, feel the cable lifting the car, or sense a jerkiness in the piston pushing the car upward. Close your eyes and try to feel the elevator start or stop. Can you tell which way you are moving when the elevator is traveling steadily up or down? Why do you experi-*

ence the same feeling when the car stops moving downward as when the car starts moving upward?

Pushing Up from Below: Hydraulic Elevators

❐ The earliest reliable elevators were supported by jackscrews: screws used as lifting devices. A sturdy threaded shaft, the jackscrew, extended beneath the elevator platform and lifted the platform upward as it turned. The jackscrew provided mechanical advantage, so that only a modest torque was needed to turn the jackscrew and raise a heavy load on the platform. But sliding friction in the screw created heat and wear, limiting the speed at which the platform could rise and making jackscrew elevators impractical in modern skyscrapers.

The two main types of elevators are hydraulic elevators and cable-lifted elevators. A hydraulic elevator is lifted from below by a long metal shaft while a cable-lifted elevator is pulled up from above by a long metal cable. Let's begin by looking at hydraulic elevators. (For an earlier type of elevator, see ❐).

The car of a hydraulic elevator is lifted from below by a hydraulic ram (Fig. 5.3.1). A *hydraulic ram* is a long *piston* that is driven into or out of a hollow cylinder by pressure in a hydraulic fluid. The hydraulic fluid, usually oil or water, exerts a force on any surface it touches, including the base of the piston. If the pressure in the hydraulic fluid is high enough, the force it exerts on the base of the piston will exceed the weight of the piston and elevator car and they will accelerate upward.

But as the piston rises, the hydraulic fluid has more space to fill and its pressure drops. To keep the piston moving upward, something must continuously add high-pressure hydraulic fluid to the cylinder. That something is usually an electrically powered pump. This pump draws low pressure hydraulic fluid from a reservoir and pumps it into the cylinder. The pump does work on the fluid and this work is what lifts the elevator car.

When the elevator car has reached the proper height, the pump stops and the piston rests on the high-pressure hydraulic fluid beneath it. As long as the amount of fluid in the cylinder doesn't change, the piston and car will stay where they are as the passengers get on and off.

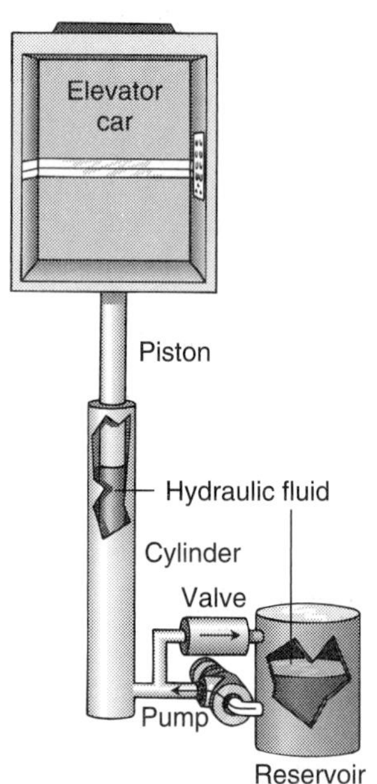

Fig. 5.3.1 - A hydraulic elevator supports the car with a hydraulic ram. The ram's piston rises as high-pressure hydraulic fluid is pushed into the hollow cylinder by a pump. The car is lowered by opening the valve and allowing the high-pressure hydraulic fluid to flow back into the storage reservoir.

To let the car descend, the elevator opens a valve and permits the high-pressure hydraulic fluid to return to the low pressure reservoir. The fluid naturally accelerates toward the lower pressure and the cylinder begins to empty. The car descends. However, the fluid in the cylinder has considerable pressure potential energy and that energy must go somewhere. As it flows through the valve, the fluid accelerates and it rushes into the reservoir at high speed. But its kinetic energy soon becomes thermal energy as the fluid swirls around randomly. When the swirling has stopped, the fluid in the reservoir will be warmer than it was before the elevator made its trip up and down.

Because it lifts the car from below like a jackscrew, the hydraulic elevator is naturally very safe. Even if the cylinder springs a leak, the hydraulic fluid will probably not flow out of the cylinder fast enough for the car to descend at a dangerous speed. But unlike a jackscrew, a hydraulic ram encounters very little friction and wear, so its piston can move in or out of the cylinder rapidly. As a result, the car of a hydraulic elevator can be lifted as fast as the pump can deliver high pressure hydraulic fluid. Of course, the pump has to do a great deal of work on that fluid in a short time, so it must be very powerful. Nonetheless, the speed of a hydraulic elevator is limited only by the power of the pump and the comfort of the passengers. Most passengers don't enjoy huge accelerations. While you could build a hydraulic elevator that would leap from one floor to another in the wink of an eye, it would require seat belts and airbags.

However, if speed is not important, even a very small pump can lift the elevator upward. With enough patience, you could actually lift a very heavy elevator with a hand-powered pump. That is just what you do when you lift an automobile with a hand-powered hydraulic jack or when you squeeze something together with a hand-powered hydraulic press. In these and many similar tools, hydraulic rams provide an interesting form of mechanical advantage.

To see how this mechanical advantage works, suppose that you have two hydraulic rams connected by a hose so that hydraulic fluid can flow freely from one cylinder to the other (Fig. 5.3.2). One hydraulic ram is much wider than the other. Since fluid accelerates toward lower pressure, the pressures in the two cylinders will tend to equalize. This pressure exerts an upward force on each piston equal to the pressure times the surface area of that piston. As a result, the upward force on the wide piston is enough to support the weight of an elevator car while the upward force on the narrow piston is only enough to support the weight of your hand. As things stand, neither the elevator nor your hand moves because each is supported by pressure in the hydraulic fluid.

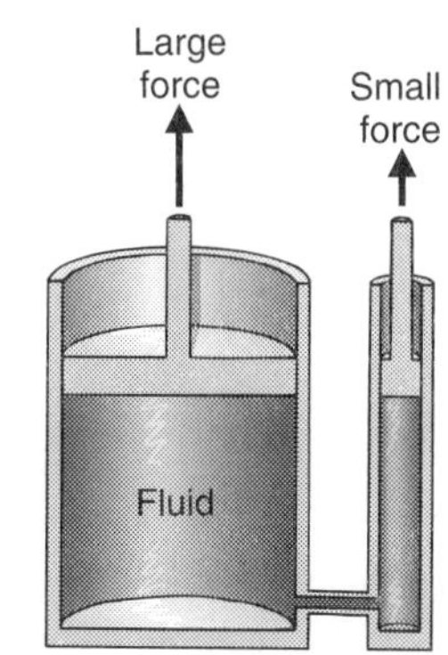

Fig. 5.3.2 - If fluid can flow freely between two hydraulic rams, then the pressures inside the two cylinders are equal. The force exerted on each piston is equal to that pressure times the surface area of the piston. The upward force on the wide piston is much greater than the upward force on the narrow piston.

Now imagine that you begin to push down a little harder on the narrow piston. The pressure inside that cylinder rises in order to exert an equal but oppositely directed force on your hand. Because of the pressure imbalance, fluid begins to flow out of the narrow cylinder and into the wide cylinder. With less fluid in the narrow cylinder, its piston descends and your hand moves downward. With more fluid in the wide cylinder, its piston rises and the elevator car move upward. You are raising a heavy elevator with a hand pump!

As usual, you didn't get something for nothing. Although you have lifted the elevator upward, it moves only a tiny distance. Pushing the narrow piston inward a long distance only squeezes a modest amount of fluid into the wide cylinder. The wide piston moves upward only a very short distance. You have produced a huge upward force on the elevator and lifted it a short distance by exerting a modest downward force on the narrow piston and moving it downward a very long distance. The work you do on the fluid is equal to the work the fluid does on the elevator car. Energy is conserved, as it must be.

The piston of the narrow cylinder will reach the bottom long before the elevator reaches the second floor. To make a more practical hand-powered elevator, you would need to add several one-way valves and a fluid reservoir to the narrow cylinder and convert it into a proper pump. That way you could slowly raise the elevator upward, with ever so many cycles of the pump: fill the narrow cylinder with fluid and then squeeze it into the wide cylinder, fill the narrow cylinder with fluid, and so on. To return the wide piston to its original position and lower the elevator, a bypass valve should allow the hydraulic fluid to flow back into the fluid reservoir.

While real hydraulic elevators don't use hand-pumps of this sort, many tools do. Small hydraulic rams allow a normal person to exert Herculean forces on objects. Hand-operated hydraulic rams are used in small cranes, presses, punches, shears, and jacks. When a motorized pump is added, hydraulic rams become even more useful. They are ubiquitous in construction and industrial machines. Nearly every motion of most digging, lifting, and pushing machines is powered by its own hydraulic ram.

Although hydraulic elevators are wonderful in many situations, they do have at least two draw backs. First, a hydraulic elevator is only as tall as its piston and cylinder. The piston has to reach all the way to the top floor and the equally tall cylinder must be hidden below the ground. Burying the cylinder is quite a procedure in a tall build. A deep hole must be drilled and the cylinder must be lowered into the hole with a crane. The difficulties involved in manufacturing the cylinder and piston and in assembling the completed hydraulic ram limit its height. However, some hydraulic elevators are over 30 stories tall.

The other deficiency of hydraulic elevators is that there is no mechanism for storing energy between trips. The energy expended in lifting people up 30 floors is not saved as those people descend. It becomes thermal energy in the hydraulic fluid as the hydraulic fluid returns to the reservoir. For a tall building with lots of up and down traffic, the elevator can turn a lot of electric energy into thermal energy in the hydraulic fluid.

Fig. 5.3.3 - The modern era of elevators began in 1853, when Elisha Otis first demonstrated his Safety Elevator. He stood in the elevator car, high above the ground, while an assistant cut the rope that supported it. A mechanism in the car immediately grabbed onto the side rails and prevented the car from falling.

CHECK YOUR UNDERSTANDING #1: Heavy Lifting

A typical hydraulic elevator is lifted by a piston 20 cm in diameter. Steel is very strong and a 10 cm steel rod could support the elevator. Why the thick piston?

Pulling Up from Above: Cable-Lifted Elevators

To eliminate the need for long hydraulic rams, most elevators are lifted from above by cables. Introducing cable-lifted elevators was not easy because people were wary of any system that would drop catastrophically if the rope broke. In 1853, the American inventor Elisha Graves Otis (1811–1861) demonstrated a "safety elevator" that would stop automatically if the rope broke (Fig. 5.3.3). In a further improvement, the ropes used to lift early elevators were replaced with metal cables, which were less prone to wear and aging and made cable failure a rare event. With safety no longer an issue, cable-lifted elevators soon became the dominant form of elevator. But before we look at how a cable-lifted elevator actually works, we'll need to know how a rope lifts an object and how pulleys redirect forces exerted on a rope. Let's take a moment to look at ropes and pulleys.

Suppose that the elevator in your building is broken. You decide to lift the empty elevator car by hand with a lightweight rope (Fig. 5.3.4). The elevator is on the ground floor and you are pulling the rope up from the fifth floor, where your apartment is located. The empty elevator weighs 500 N (112 pounds), which is about all you can lift. If your arms were long enough, you could pull the elevator up directly. The rope simply extends your reach so that you can exert an upward force on the elevator many meters below you.

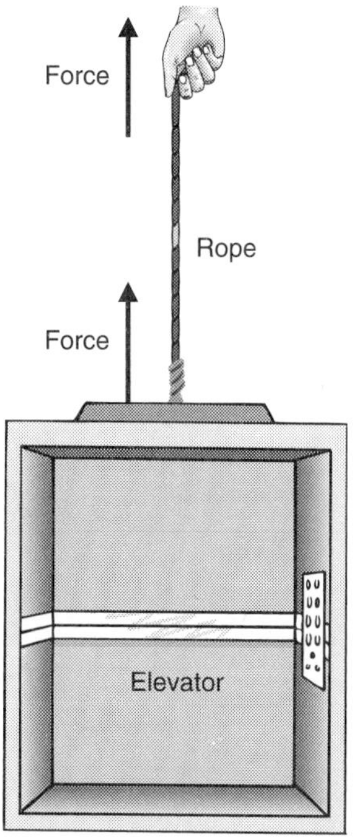

Fig. 5.3.4 - If you pull on a very light, stationary rope, you produce a uniform tension throughout the rope that is equal to the force you exert on it. Each portion of rope, for example the two end portions or the shaded portion near the middle, experiences an upward force from whatever is above and a downward force from whatever is below.

Pulling on a rope produces tension throughout the rope. **Tension** means that each portion of the rope pulls on the two adjacent portions with a certain amount of force. To keep the empty elevator hanging motionless from the bottom of the rope, you must pull upward on the rope with 500 N of force. Each portion of rope then exerts 500 N of upward force on whatever is below it and 500 N of downward force on whatever is above it. The bottom of the rope exerts 500 N of upward force on the elevator. Overall, your upward force of 500 N is conveyed meter by meter along the rope until it's exerted on the elevator far below. In effect, you are exerting an upward force of 500 N on the elevator and it's pulling back. As promised, the rope simply extends your reach.

Since the elevator weighs 500 N and you are exerting an upward force of 500 N on it, the net force on the elevator is zero and it doesn't accelerate. Because the elevator is initially stationary, it remains stationary. If you now exert a little more upward force on the rope, the elevator will experience a net upward force and will accelerate toward the fifth floor. Once the elevator has begun to move upward, you can reduce your force back to 500 N and the elevator will continue to move upward at constant velocity. You are now doing work on the elevator because you are pulling upward on it via the rope and it's moving upward.

Lifting the empty elevator to the fifth floor doesn't require an enormous amount of force, but that force must come from the middle of the elevator shaft. It would be nice to stand somewhere else as you pulled on the rope, so you suspend a *pulley* in the elevator shaft (Fig. 5.3.5). With the rope draped over the pulley, you can create tension in the rope from a different location. In fact, you can even pull downward on the rope. While each portion of rope continues to pull inward on its neighbors, the directions of these forces gradually change as the rope bends around the pulley. The pulley redirects the forces on the rope so that a downward force on one end of the rope can exert an upward force on the other end. This redirection makes it much easier to lift the elevator. A pulley even permits the weight of water to lift an elevator (see ❐ on p. 241).

CHECK YOUR UNDERSTANDING #2: Pulling on a Pulley

A rope is draped over a pulley near the ceiling and you and a friend are each pulling downward on opposite ends. If each of you exerts 100 N of force on an end, what is the tension in the rope?

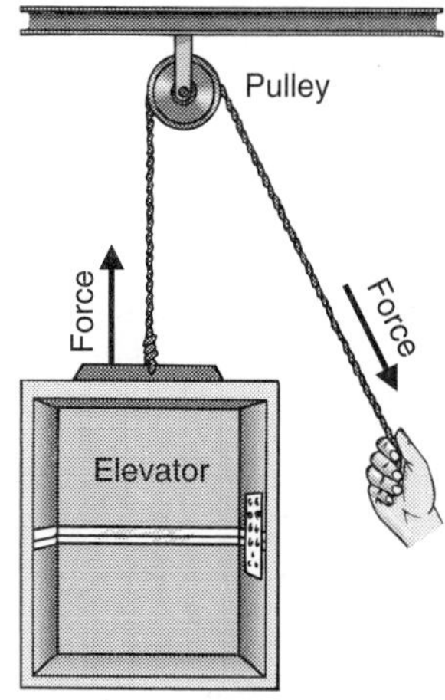

Fig. 5.3.5 - With the rope drawn over a pulley, you can lift the elevator by pulling from almost anywhere. You exert the same force and the tension in the rope is the same, but the pulley redirects the force to make the job more convenient.

Multiple Pulleys

However the elevator is not always empty. Last week the bathtub cracked and you and your friends pushed it off the fire escape. That was easy enough, although it ruined the flower garden next door. But the new bathtub weighs 1300 N (292 pounds), so how are you going to get it up to your apartment? It has to ride up in the elevator. You could rig up the same single pulley and get all of your friends to pull on the rope. But a better idea is to use a multiple-pulley, sometimes called a block-and-tackle. When you pull on a rope, you produce a tension all along that rope. If you could use that same tension several times, you could get mechanical advantage. Here is how a multiple-pulley works.

In a multiple-pulley, the cord goes back and forth between a fixed set of pulleys and a moving set of pulleys (Fig. 5.3.6). The far end is tied to one of the pulley sets. It's important that the cord pass easily over the pulleys. Now when you create tension in the cord, that same tension appears on every segment of cord between the two sets of pulleys. If you exert 500 N of force on the cord, each cord segment will have 500 N of tension. As a result, the two sets of pulleys will be pulled together with 500 N of force for each segment of cord connecting them. If there are 4 cord segments attached between the top of the elevator and the fifth floor, then the total lifting force on the elevator and bathtub will be 2000 N. Since the bathtub and elevator only weigh 1800 N, they will experience a net upward force and will accelerate upward.

While it takes less force on the cord to lift the bathtub and elevator with a multiple pulley than with a single pulley, you don't get something for nothing. To lift the elevator 1 m, you must shorten each segment of cord by 1 m. Since there are 4 segments, you will have to pull 4 m of cord through the system of pulleys. You are obtaining mechanical advantage, using a modest force exerted over a long distance to obtain a larger force exerted over a shorter distance. The amount of work required to lift the bathtub and elevator to your apartment is the same, whether you use a single or multiple pulley. The multiple pulley merely allows you to do this work more gradually, with a smaller force exerted over a longer distance.

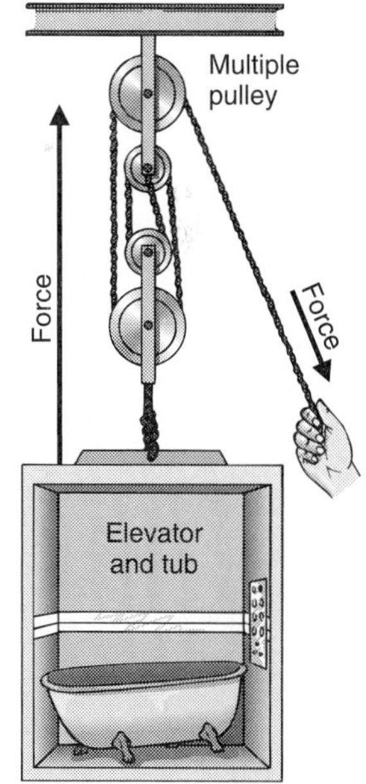

Fig. 5.3.6 - A multiple-pulley being used to lift an elevator and bathtub. The tension in the rope pulls upward on the elevator four times because there are four rope segments between the elevator and the support above it. The upward force on the elevator is thus four times the tension in the rope.

CHECK YOUR UNDERSTANDING #3: Pulling Yourself Up

If you stood inside the elevator shown in Fig. 5.3.6 and pulled on the free end of the rope, could you lift yourself up?

Cable-Lifted Elevators and Counterweights

True cable-lifted elevators (Fig. 5.3.7) resemble the hand-powered one we have just discussed, except that machines pull the cables. In early cable-lifted elevators, the cables were pulled by steam-powered hydraulic rams. Steam was used to pump fluid into or out of the ram and the ram's movement was used to pull the cables. Usually, the ram was used to separate the two halves of a multiple-pulley. The cable coming out of this multiple pulley ran over a pulley at the top of the elevator shaft and down to the elevator car itself. As the two halves of the

Fig. 5.3.7 - The car of this elevator is pulled upward from above by 4 cables and rides on rails to its left and right. The riders control the car through the electric cable hanging below the car.

multiple pulleys were drawn apart, they drew in more cable and lifted the elevator car. As fluid was released from the hydraulic ram, the multiple pulley released cable and the elevator car descended.

The first improvement that appeared in cable-lifted elevators was the counterweight (Fig. 5.3.8). Lifting the elevator car by itself requires a considerable amount of work because the car's gravitational potential energy increases as it rises. It would be nice to get back this stored energy when the car descends. Unfortunately, it's hard to turn gravitational potential energy back into high-pressure steam. However, it's possible to use that energy to lift a counterweight.

The counterweight in an elevator descends when the car rises and rises when the car descends. Because the two objects have similar masses, the total amount of mass that is rising or falling as the elevator moves is almost zero. The overall gravitational potential energy of the elevator is not changing very much; it's simply moving around between the various parts of the machine. The counterweight balances the car so that it takes very little power to move the system. The elevator and counterweight resemble a balanced seesaw, which requires only a tiny push to make it move.

The counterweight on most elevators hangs from its own cable attached to the elevator car. That cable travels from the car, over pulleys at the top of the elevator shaft, and down to the counterweight. The counterweight is usually equal to the mass of the empty elevator car plus about 40% of the elevator's rated load. Thus, when the elevator is 40% filled, the counterweight will exactly balance the car and very little work will be done in raising or lowering the car.

Most modern elevators are driven by electric motors. The advantages of electric motors are their variable speeds of rotation, high torque, and reliability. While we will save our discussion of electric motors for a later chapter, we will note here that electric motors can be made to operate efficiently at many rotational speeds, torques, and overall power-levels. The output power of an electric motor is frequently rated in horsepower and the motors used in elevators may be as large as several hundred horsepower.

Because early electric motors could not deliver so much mechanical power, the first electric elevators used *winches* to lift their elevator cars. The cable from the elevator car was actually wound up on a drum at the top of the elevator shaft. The counterweight was attached to a cable that was also wound on the drum. The two cables were arranged so that the counterweight cable unwound as the car cable wound up. An electric motor used gears to turn the drum.

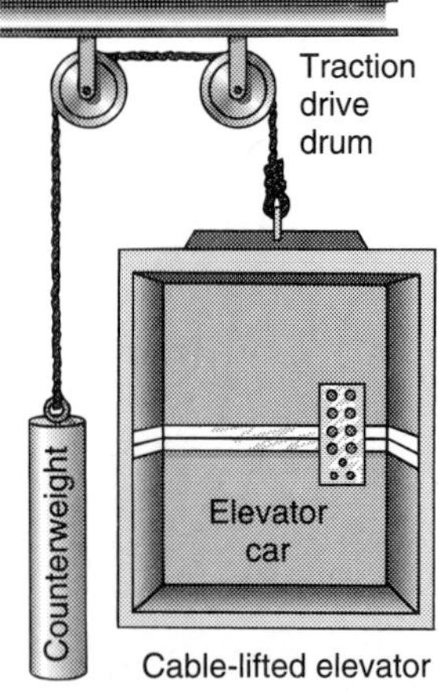

Fig. 5.3.8 - A cable-lifted elevator usually supports the car and a counterweight from opposite ends of its cable. A motor turns a traction drive that either raises or lowers the car. The counterweight moves in the other direction, assisting the motor in lifting the car or storing energy as the car descends.

This winch mechanism had a number of disadvantages. It raised or lowered the car relatively slowly because the gearing limited the rate at which the drum could be turned. The overall height of the elevator was limited because the drum had to be able to hold all of the cable when the elevator was at the top of its travel. The diameter of the drum was constrained by the need to keep torques low and only about 100 m of cable could be accommodated.

Instead of winding and unwinding cable from a drum, most modern elevators use traction to draw a cable over a drum. The cable rises from the elevator car, travels over the traction drive drum and then descends into the elevator shaft where it's attached to the counterweight. An electric motor turns the traction drive drum. When high speed is not important, the drum can be turned by a small motor through the use of gears. However, in tall buildings, the drum is usually turned directly by a large motor. Elevators of this type can run at speeds as high as 10 m/s (22 mph) in buildings of any height.

The mechanical power required from the drive motor depends on how well balanced the car and counterweight are. If the elevator car is loaded to 40% of capacity so that the two weights are balanced, the motor will have little difficulty in moving the car up or down. If the car is particularly empty or particularly full, the motor will have to provide considerable mechanical power when lifting the heavy side of the system and various brakes will have to absorb energy released by the elevator when the heavy side descends. The motor's maximum mechanical power, together with the strength of the cables, limits how much weight the elevator can lift.

In many freight elevators, the car is lifted by a multiple pulley so that a single segment of cable doesn't have to support the entire load. Even when a single pulley is used, several separate cables support the car, both for safety and to reduce cable stretching. Cable stretching is a serious problem in tall elevators. Tension always tends to pull things apart, so a cable becomes longer. Like most objects, a cable behaves as a spring when it's subject to tension. Its length increases by an amount proportional to the tension it experiences. As people enter the elevator car and its total weight increases, the tension on its support cable increases and that cable stretches slightly. Modern elevators are equipped with automatic leveling systems that turn the traction drum to make up for the stretching of the cables. The passengers are unaware of this careful adjustment taking place as they step on or off the elevator. Nonetheless, you may be able to feel the cable stretch if you bounce up and down on a cable-lifted elevator.

❒ The development of safe elevators had an enormous effect on people's interest in tall buildings. Suddenly the upper floors became more desirable than the lower floors. Speed became very important. A "water balance" elevator was tried in the New York Western Union Building in 1873. The elevator car was drawn upward by the weight of an enormous bucket of water. To descend, the bucket of water was emptied. Controlled only by braking and without any automatic safety system, this elevator was too scary to be popular.

CHECK YOUR UNDERSTANDING #4: A Light Load

If an elevator car is nearly empty and weighs much less than the counterweight, how much work must the motor do to lift that elevator car upward?

Balance

Elevator cars must remain level no matter where the passengers choose to stand. The only way to keep the car level is to make it run along a vertical track. To see why the track is necessary, consider the case of an empty car (Fig. 5.3.9). The lifting force on the car is exerted at the middle of the elevator car, at either its top or its bottom. The center of mass of the empty car is also at the middle of the elevator car so the lifting force exerts no torque on the car about its center of mass. The car remains level.

Now consider what happens when passengers enter the car and begin to walk around inside. The center of mass of the car moves with the people inside. Now the lifting force exerts a torque on the car about its new center of mass and

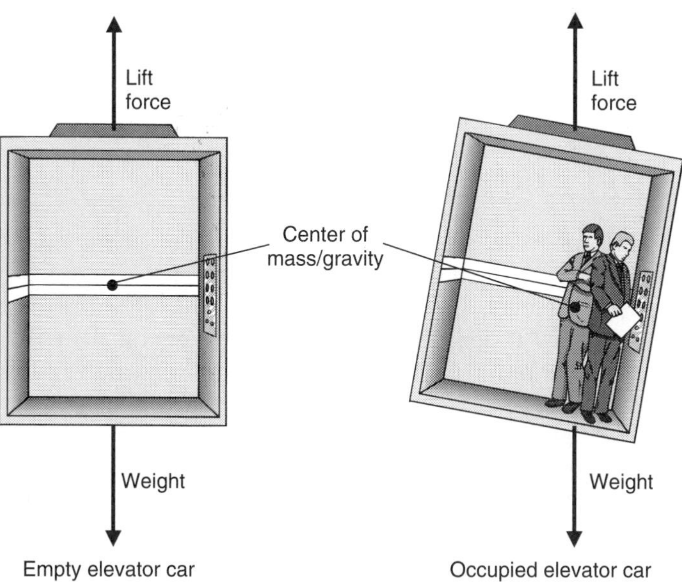

Fig. 5.3.9 - An empty elevator car (left) experiences no torque about its center of mass so it remains level. The occupied car (right) has a different center of mass. The lift force exerts a torque on this car about its center of mass and it begins rotating. For the car to remain level as the passengers move around, the car must run in a track that can exert leveling torques on the car.

❒ The only time a safety elevator plummeted to the bottom of its shaft was in 1945, when the Empire State Building was struck by a military airplane. The plane lodged in the elevator shaft near the 79th floor, cutting all of the cables to the elevator car on the 38th floor. The car dropped to the basement, but its descent was cushioned by the increasing air pressure beneath it and by a mountain of severed cables and an emergency bumper at the bottom of the shaft. The only occupant of the car, a 20 year old elevator operator, survived without serious injury.

it tends to rotate. The best way to prevent the car from tilting is to confine the car on a track. The rails of the track exert the torques needed to keep the car level.

CHECK YOUR UNDERSTANDING #5: Not on the Level

If the track supporting an elevator car were suddenly removed so that it could tilt, would its center of gravity move up or down?

Safety

All cable-lifted elevators have safety devices to keep them from falling if their cables break. Most modern elevators have more than one lifting cable, but they still require mechanisms to ensure that there are no accidents (see ❒).

The original safety device that Otis developed for his first elevators had jaws that would grab onto the rails of the elevator track if there was a loss of tension in the supporting cable. If the cable broke and its tension vanished, springs would force the jaws into the track.

Modern elevators use mechanisms that monitor the vertical speed of the elevator. If the speed exceeds a certain permissible value, brakes on the car grab the tracks. This speed control prevents a nearly empty elevator from moving upward too quickly just as it prevents a full elevator from falling. One such speed-sensing device is the *centrifugal governor*, a mechanism that senses how quickly a shaft is turning (Fig. 5.3.10). When it's used with an elevator, the shaft is turned by a pulley on a special cable attached to the elevator car. The faster the elevator moves, the faster the shaft turns. The centrifugal governor swings several masses around in a circle. Since the masses travel in uniform circular motion, they need some centripetal force to accelerate them toward the center of the circle. In the centrifugal governor, this centripetal force is exerted by several rods that are held apart by a spring.

As long as the shaft is turning slowly, the spring can keep the rods from moving together. But when the shaft is turning quickly, the centripetal force becomes very large and the rods compress the spring. As the rods move, they push on a lever. In the case of the elevator, this lever activates brakes that slow the elevator down.

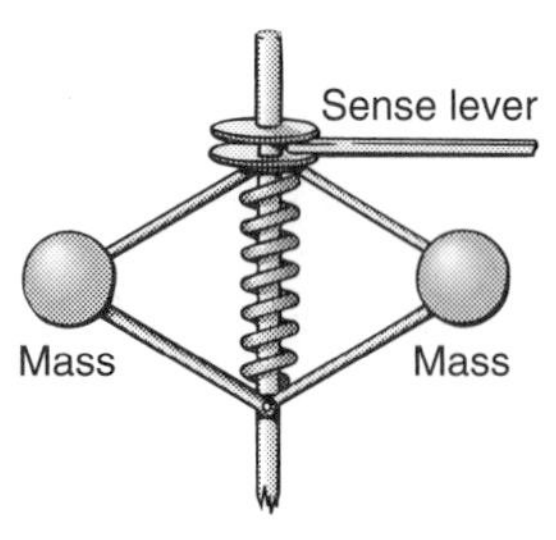

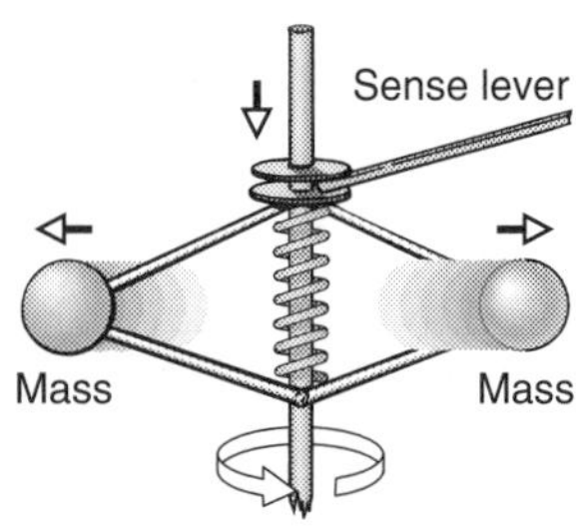

Fig. 5.3.10 - A centrifugal governor uses the principle that a central force is required to accelerate masses around in a circle. As long as the shaft is stopped or spinning slowly (top), the spring can keep the upper and lower rods apart. But once the shaft spins too quickly, the masses swing outward (bottom) and the sense lever is shifted.

CHECK YOUR UNDERSTANDING #6: Going for a Spin

How does the spinning centrifugal governor sense that the elevator is moving too fast in *either* direction?

Starting and Stopping

Simply moving the elevator car up or down is not enough. To be useful, an elevator must be able to stop at the proper level, exchange passengers or freight, and then start to move to a new level. To be pleasant to ride, the elevator must start and stop slowly enough that it doesn't knock the passengers off their feet. To meet these added requirements on the motion of the elevator car, variable speed electric motors are used.

Whether the elevator is handled by an operator or is run automatically, the torque exerted on the traction drive drum is carefully controlled in order to avoid sudden accelerations. Whenever the elevator you are in accelerates upward, as it does when it starts moving upward or stops moving downward, you feel particularly heavy. Your apparent weight increases because of the upward acceleration. If the upward acceleration is too great, you may be thrown to the floor of car. Whenever the elevator accelerates downward, as its does when it starts moving downward or stops moving upward, you feel particularly light. Your apparent weight decreases because of the downward acceleration. If the downward acceleration is too great, you may leave the floor of the elevator and bump against its ceiling. Only after the elevator reaches constant velocity, either up or down, does your apparent weight return to your true weight.

A well-designed elevator accelerates and decelerates smoothly and gradually. This need for smooth deceleration means that the operator or the automatic mechanism must anticipate stops and begin to decelerate before reaching the stopping point. Operating an antique elevator, with no machinery to help anticipate the stop, required great skill. In manually operated elevators, the operator's ability to stop at the correct height limited the maximum vertical speed that could be used effectively. Modern elevators anticipate stops automatically and gradually reduce the speed of travel so as to come to a stop at exactly the right height. These elevators can move up or down extremely quickly and still stop properly.

CHECK YOUR UNDERSTANDING #7: Fast Track to the Basement

What would it feel like to be in an elevator car if the cable broke on the top floor of a skyscraper and no safety mechanism turned on the brakes?

Summary

How Elevators Work: An elevator car is either lifted from above by a cable or pushed up from below by a hydraulic ram. In either case, an electric motor normally does the work of lifting the car against the force of gravity. The car travels along vertical rails that keep it from tilting as the passengers move about inside. In many elevators, particularly cable-lifted elevators, a counterweight descends as the car rises, providing some of the energy needed to lift the car. When the car later descends, the counterweight is lifted upward and stores some of the energy released by the car. The car's accelerations are carefully controlled so that the passengers don't experience abrupt changes in their apparent weights. The car accelerates smoothly up to speed, travels at constant velocity for much of its trip, and then accelerates smoothly to a stop. Although the elevator's speed is normally limited by the lifting and lowering mechanisms, a separate mechanism is attached to many elevators to monitor the rate at which the car rises or falls. If the car were to move too quickly in either direction, a safety mechanism in the car would activate brakes to slow it down.

The Physics of Elevators

1. The force exerted on a piston by a hydraulic fluid is equal to the fluid's pressure times the surface area of the piston's base. A wide piston experiences a much larger force than a narrow piston when the two are exposed to the same hydraulic fluid pressure.

2. When a rapidly moving fluid flows into a reservoir, its viscosity eventually slows it to a stop. The fluid's kinetic energy becomes thermal energy.

3. The tension in a massless rope is equal to the force exerted at each end of that rope. The rope essentially conveys the force from the object at one end to the object at the other end. The two forces at the rope's ends must be equal in amount but oppositely directed.

4. When a rope travels around a pulley, its tension is not changed but the forces at its ends no longer have to be oppositely directed. A rope and pulley can be used to redirect a force.

5. A multiple pulley can both redirect force and provide mechanical advantage. The forces drawing together the two

sets of pulleys are equal to the tension in the rope times the number of rope segments between the pulley sets.

Check Your Understanding - Answers

1. The width of the piston is set by the hydraulic fluid pressure, not the need for strength in supporting the elevator car.

Why: The wider the piston, the more upward force it experiences at a particular hydraulic fluid pressure. If the piston in an elevator were narrower, the hydraulic fluid pressure would have to be much higher. Furthermore, to keep the piston's weight down and its rigidity up, it's actually a hollow tube, except at its base.

2. 100 N.

Why: The rope conveys force from one of its ends to the other via tension. The pulley merely redirects that force so that it need not be in a straight line. You exert 100 N of force on the rope and it exerts 100 N on your friend. Because forces always occur in equal but oppositely directed pairs, she must pull back with 100 N of force on the rope and it must exert 100 N on you. The rope is just the "middleman" exerting 100 N of force on each of you. Thus, its tension is 100 N.

3. Yes.

Why: Pulling on the free end of the rope would create a tension in the entire rope. Each rope segment, including the free end in your hands, would exert an upward force on you and the elevator equal to the tension in the rope. When the overall upward force from the rope segments exceeds the total weight of the elevator and contents, it will begin to accelerate upward. This idea of lifting oneself with a rope and pulleys is commonly used by arborists working in trees or by window washers on the outsides of office buildings.

4. No work at all. In fact, energy is released as the car rises and brakes must be used to slow its ascent.

Why: If the counterweight is heavier than the car, it can lift the car unassisted and the system will actually accelerate. This dangerous situation must be controlled by brakes on the cable drum. Otis's original safety mechanism didn't address the possibility of an upward runaway elevator car.

5. The center of gravity would move downward.

Why: Objects are always trying to lower their centers of gravity. If the car's center of gravity is not directly below the lift cable, the car can lower its center of gravity by tilting. It will always tilt so that the heavy side moves downward.

6. No matter which way the governor rotates, its moving masses accelerate toward the center of the circle. The spring helps to accelerate the masses inward. If the rotation is too fast, the spring becomes compressed and the governor turns on the brakes.

Why: The governor's operation doesn't depend on which way it rotates. All that matters is that the masses undergo uniform circular motion and that they require a large centripetal force as they do.

7. You would feel weightless because you would be in free fall.

Why: With the elevator car accelerating downward in free fall, your apparent weight would drop all the way to zero. You would float freely around the car, along with everything else in it, until you reached the bottom of the shaft. Then the car would abruptly accelerate upward and you would have a sudden, unpleasant encounter with its floor.

Glossary

tension A stress that appears in a rope, cable, or other object when outside forces are trying to stretch it or pull it apart.

Review Questions

1. When the car of a hydraulic elevator isn't moving, how does the weight of the car and the diameter of the piston affect the pressure of the hydraulic fluid in the cylinder?

2. If two pistons are supported by hydraulic fluid at the same pressure, why will the wider piston experience a larger upward force than the narrower piston?

3. If two pistons and cylinders are connected to one another and share the same hydraulic fluid, why does a modest downward force exerted on the narrow piston produce a much larger upward force on the wider piston?

4. If two people are pulling on opposite ends of a rope, each with a force of 200 N, what is the tension in the rope?

5. Both ends of a rope are dangling from a pulley overhead. If you attach one end of the rope to your waist and pull downward on the other end with a force of 20 N, how much upward force will the rope exert on your waist?

6. Why is the force pulling the two ends of a multiple pulley together proportional to the number of rope segments connecting those two ends?

7. Explain how a multiple pulley provides mechanical advantage, letting you do the same work with a smaller force exerted over a longer distance.

8. Why is a counterweight an energy storage device?

9. How does a counterweight make it possible for a cable-lifted elevator to accelerate *upward* at a dangerous rate?

10. Why do you feel unusually light whenever an elevator begins its descent or a rising elevator stops its ascent?

Exercises

1. When you lie down on a water bed, the pressure inside the water bed rises by a very small amount. However, if you stand on the water bed, the pressure inside rises much more. Explain this effect.

2. An earth moving machine uses a piston and cylinder to lift an extremely heavy load 3 m upward. As hydraulic fluid is pumped into the cylinder, it pushes the piston out and the piston lifts the load. If the piston and cylinder are replaced and the new piston is twice the diameter of the old piston, lifting the load will require only a quarter of the original pressure in the hydraulic fluid. Explain this reduction.

3. While the wide piston in Exercise **2** needs less pressure in the hydraulic fluid to lift its load 3 m upward, it uses more hydraulic fluid than the narrow piston. How much more?

4. In Exercises **2** & **3**, two different pistons and cylinders were used to lift a heavy load 3 m upward. The narrow piston needed higher pressure hydraulic fluid but used relatively little of it, while the wide piston needed lower pressure hydraulic fluid but used much more of it. Why is the product of the fluid pressure times the fluid volume used the same in each case? How is that product related to energy?

5. If the pump of a hydraulic elevator does 800,000 J of work as it lifts the elevator 20 m upward, how much does the elevator weigh?

6. Some emergency road kits include a large plastic bag that you can inflate with the exhaust from your car. As the bag inflates, it lifts the corner of your car. How can a thin plastic bag with a pressure only slightly above atmospheric pressure lift something as heavy as an automobile?

7. The first breath you blow into an elastic rubber balloon is the hardest. Once the balloon begins to inflate, the pressure inside it drops and you don't have to blow as hard. However the tension in the rubber actually increases as the balloon inflates. Why then is the pressure inside the balloon largest when its diameter and surface area are the smallest?

8. One of the tires of your car is low on air. You have a small bicycle pump that's just like the water pump in Fig. 3.2.2, except that high pressure air emerges from it as you push the piston in and out of the cylinder with your hands. If you attach this bicycle pump to the air inlet of the tire and then push the piston in and out many times, you will slowly inflate the tire and lift the car. The inward force that you exert on the pump's piston is much smaller than the upward force needed to lift the car. Why are you able to lift the car?

9. If you know the air pressure in your car tires and the amount of tire surface area that is in contact with the pavement, you can make a reasonable estimate of how heavy your car is. Explain.

10. A flag pole has a loop of rope that goes over a pulley at the top of the pole. You attach a flag to one side of the loop and then pull down on the other side to lift the flag. If the flag weighs 100 N, how much downward force must you exert on the rope to raise the flag at a constant velocity?

11. As you raise the flag in Exercise **10**, what is the tension in the rope?

12. To lower an old radiator out of your third floor apartment, you've hung it from a multiple pulley that's attached to a roof beam. There are five rope segments between the radiator and the roof, and the end of the rope that you're holding in your hand descends from the top pulley. If the radiator weighs 1200 N, what is the tension in the rope as you lower the radiator steadily?

13. As you lower the radiator in Exercise **12** to the sidewalk, 10 m below your apartment, how much rope passes through your hands into the upper pulley?

14. How much work does the rope in Exercises **12** & **13** do on your hands as you lower the radiator to the pavement?

15. The laces of a shoe resemble a multiple pulley. If there are ten lace segments connecting one side of a shoe to the other and the laces can move without friction, how much tension would you have to exert on the laces to pull the two sides of the shoe together with a force of 400 N?

16. In Exercise **15**, how much lace must you pull out of one eyelet of the shoe to draw the two halves of the shoe 1 cm closer together?

17. A window sash in an older home slides up and down in its frame with the help of two sash weights. On each side of the window, a sash cord attached to the top of the sash passes over a pulley near the top of the window frame and supports a hanging weight. When you lift the sash, the weights descend. How do the descending weights contribute to the energy required to lift the sash?

18. If the sash in Exercise **17** weighs 100 N and each sash weight weighs 45 N, what force must you exert on the sash to lift it at a constant velocity (neglecting friction)?

19. When you jump down from a high place, it's important that you bend your knees when you land. Why does allowing your knees to flex during the landing reduce the chance of injury?

20. A mountain climber uses a rope for support while rappelling down the vertical face of a cliff. The climber starts and stops frequently by grabbing and releasing the rope. Why must the rope be able to withstand tensions that are much greater than the climber's weight?

Epilogue for Chapter 5

This chapter discussed the physical concepts behind three types of machines. In *rockets*, we studied the ways in which reaction forces propel rockets forward and control their orientations. We looked at how a rocket remains stable on the ground and how aerodynamic forces can help to stabilize it in flight. We found that a rocket's ultimate speed is not limited by its exhaust velocity so that it can lift itself into orbit around the earth. We also learned that gravity weakens as the distance between two gravitating objects increases, making it possible for a spaceship to break free from earth's gravity and travel to the stars.

In *bicycles,* we investigated the concept of dynamic stability, where an object that falls over when stationary becomes remarkably stable in motion. We saw how the need for mechanical advantage between the turning of the pedals and the rotation of the wheels led to the development of the modern chain-driven, multispeed bicycle. We looked at how pneumatic tires affect the bicycle's comfort and energy efficiency and at how the brakes dissipate excess energy to bring the bicycle to a stop.

In *elevators,* we looked at how a hydraulic system uses fluid to obtain the mechanical advantage needed to lift a heavy elevator car. We saw how tension in a cable can pull an elevator car up from above and how pulleys can redirect forces and provide mechanical advantage. We explored the usefulness of counterweights to store energy between trips and examined the need for speed controls and safety devices to keep the elevator car from accelerating too quickly or traveling too fast.

Explanation: High Flying Balls

The top ball doesn't bounce off the ground, it bounces off the bottom ball. The top ball actually completes its bounce as the bottom ball is heading upward. After colliding and storing energy in the air they contain, the two balls push off one another in just the same way that a rocket pushes off its exhaust. The balls exchange momentum and energy as they push against one another and the small top ball—which has less mass and accelerates much more easily than the bottom ball—ends up with a large upward momentum and more than its fair share of energy. It flies upward as though it were hit by a massive upward-moving *bat.* In fact, it has been hit by a massive upward-moving *ball* and it rebounds at high speed.

If the small ball's mass were really negligible in comparison to that of the large ball and the balls bounced without wasting any energy, then the small ball would rebound from the floor traveling 3 times as fast as when it arrived. Its kinetic energy would be 9 times as great as before and it would rebound to 9 times its original height. Of course, real balls aren't perfect so the tennis ball won't bounce quite so high. Still, the effect is pretty impressive.

Cases

1. You are consulting for a screenwriter who is working on a science-fiction movie about the crew of a large, intergalactic spaceship. She wants the script to be realistic from a scientific perspective so she has asked you to check her work so far. You read through the script, which includes descriptions of special effects, and quickly find several problems with it. Here are some of the scenes that have flaws.

a. In one scene, the spaceship must make an abrupt left turn to avoid hitting an asteroid. The crew members stand anxiously but motionlessly on the command bridge during this maneuver. In reality, what should happen to the crew members?

b. In another scene, a guard fires a projectile weapon at a massive alien creature. The guard remains stationary, but the creature is thrown backward by the weapon's impact, even though the weapon doesn't explode. In reality, what should happen to the guard as he fires the weapon?

c. The main spaceship carries several small fighter spacecraft that resemble high-tech modern military fighter aircraft. In the depths of space, these fighter spacecraft dodge and turn rapidly, even though the rocket exhaust is always sent straight out of the rear of each fighter. In reality, why couldn't such spacecraft turn in space while similar looking aircraft can turn near the ground?

d. One small spacecraft secretly shuttles supplies from one earth-like planet to another, completing 8 round trips without replenishing its chemical rocket fuel. You inform the writer that the spacecraft will barely be able to leave a planet even once using chemical rockets, unless that spacecraft is allowed to eject stages. Why are you right?

e. Near the end of the script, a main character falls from a cliff but is rescued only 2 m above the ground when she lands on the wing of a passing spacecraft. You inform the writer that the impact with the wing of the spacecraft would be just as fatal as one with the ground. Why is this the case?

2. Modern Olympic athletes are aided by a variety of scientific advances. Here are a few examples of how science enhances performance in athletics.

a. A ski-jumper is suspended in a wind-tunnel, where a device measures the horizontal and vertical forces on him as air flows past him horizontally. The jumper arranges his body to obtain the most favorable forces for a long flight. Describe the most favorable horizontal and vertical forces.

b. In cross-country skiing, the skier slides forward first on one ski and then on the other. The ski bottoms are coated with special waxes so that they exhibit large static frictional forces on snow but small sliding frictional forces. How does this difference in frictional forces allow the skier to use a stationary ski to push herself forward on a sliding ski?

c. Modern tracks are often constructed out of recycled rubber tires. The track surface is very resilient, deforming slightly when you push down on it and then bouncing back when you stop pushing. How does this surface improve runners' times as compared to the dirt, gravel, or cinder tracks of long ago?

d. Bicyclists now use aerodynamically designed bicycles and clothing to minimize air drag. Bicyclists also "draft" one another, a strategy in which one bicycle follows immediately behind another bicycle, easing the rear bicycle's passage through the air. If the front bicycle were so aerodynamic that it experienced no air drag at all, how would this affect the strategy of drafting and why?

e. During a downhill ski race, the racer tries to keep her skies in contact with the surface at all times. Even if she remains in a tight, aerodynamic tuck position, she slows down whenever she becomes airborne after hitting a bump. Why does contact with the ground keep her moving forward as she races downhill?

3. Hobby stores sell model rockets that are built of cardboard, plastic, and balsa wood. These devices range in length from about 15 cm to 2 m and have masses of up to 0.100 kg. They are driven skyward by disposable solid fuel rocket engines.

***a.** A typical model rocket engine exerts an average force of 5 N on its exhaust gas. How much average thrust does the gas exert on the engine and rocket?

***b.** If a rocket has a total mass of 0.030 kg at launch, including the engine from **a**, what is the net force on the rocket at liftoff?

***c.** How quickly does the rocket accelerate upward at liftoff?

***d.** If the rocket continued to accelerate at the same rate as in **c**, how fast would it be traveling when the engine stopped burning 2 s after liftoff? Compare the rocket's speed to the speed of a car or airplane.

e. Describe the rocket's motion after the engine stops burning.

f. The engine contains 0.006 kg of propellant which is gradually ejected out of the engine's nozzle. How does this loss of mass affect the rocket's acceleration?

g. How does air affect the rocket's speed and stability?

4. Pulleys are often used on sailboats to lift objects up their masts and to control their sails.

***a.** On a small boat, the top of the main sail is pulled up by a rope. This rope attaches to the top of the sail, passes over a pulley at the top of the mast and then drops downward to the deck. If the sail weighs 500 N, how much downward force must you exert on the rope to support the sail?

***b.** On a large boat, the top of the main sail is pulled up by a multiple pulley, with five segments of cord stretching between the sail and the top of the mast. If the sail weighs 2000 N, how much downward force must you exert on the rope passing through the multiple pulley to support the sail?

***c.** The mast of the large boat is 10 m tall. How much cord must you pull through the multiple pulley to raise the sail?

***d.** The rigid boom that supports the bottom of the large boat's sail is pulled inward by a multiple pulley. Three segments of cord extend between the boat and the boom. If the wind is pushing the sail outward with a force of 1000 N, how much work must you do to pull 3 m of cord through the multiple pulley?

5. When the propellant contained in a bullet cartridge burns, it fills the volume behind the bullet with extremely high pressure gas. This gas pushes on every surface it encounters, including the bullet in front of it and the base of the gun barrel behind it.

a. The bullet in front of the gas has the same diameter as the base of the gun barrel behind it. Which surface experiences the larger force, or are the forces equal in magnitude?

b. Which object receives the greater impulse, the gun or the bullet, or are those impulses equal in magnitude?

c. Compare the momentum of the gun and bullet before the bullet is fired to their momentum after the bullet is fired.

d. Why does the bullet accelerate more quickly than the gun?

e. Why does the bullet obtain a greater speed than the gun?

f. Which object receives more kinetic energy, the gun or the bullet? Why?

g. How does the mass of the gun influence its final speed?

6. You are standing in an elevator at the top of an extremely tall building when the lift cable suddenly breaks. The safety mechanisms fail and you and the elevator car begin to fall.

a. You are so surprised that you let go of the book you were holding in your hand. Instead of falling quickly to the floor of the car, what happens to the book?

b. You decide that you don't want to be on the floor of the elevator when it hits the bottom of the shaft, so you begin to climb up the wall to the ceiling of the car. Is it hard or easy to lift yourself to the ceiling?

c. Once you reach the ceiling, the elevator's safety mechanisms begin working and they slow the elevator's descent. What happens to you?

d. The elevator car comes to a reasonably graceful stop, 10 m below the lowest access door to the elevator shaft. You open a panel in the roof of the car and see that the rescuers have attached a multiple pulley to the car's top. However, they ask you to pull on the rope to lift yourself and the car upward. Is it possible to lift yourself with a multiple pulley, assuming that you can pull hard enough, or must someone else exert the downward force on the rope?

7. The brakes of an automobile are controlled by hydraulic fluid. When you step on the brake pedal, a brake lever attached to that pedal pushes a piston into the brake system's master cylinder. This cylinder is filled with brake fluid.

***a.** Because the piston of the master cylinder attaches to the brake lever close to its pivot, the piston moves only half as far as your foot does when you step on the brake pedal. If you exert 200 N of force on the brake pedal, how much force does the brake lever exert on the piston?

b. Assume for the moment that the brake fluid is trapped in the cylinder. What happens to that fluid when you push downward on the brake pedal? How far does the pedal move and what limits its motion?

c. In reality, brake fluid can flow out of the master cylinder through pipes that connect to brakes on the four wheels. In most cars there are two separate master cylinders, one for the front brakes and one for the rear brakes. That way, if one hydraulic system fails, you can still control half of the brakes. Each brake contains a slave cylinder with a piston inside it. As pressurized brake fluid flows into a slave cylinder, why does that slave cylinder's piston experience a force?

d. If the brake fluid from a particular master cylinder is supplying fluid to two slave cylinders and all of the pistons are the same diameter, what is the relationship between the distance you push the piston into the master cylinder and the distance a piston moves out of one of the slave cylinders?

e. Occasionally cars develop an air bubble in their brake lines (the pipes connecting the master cylinder to the slave cylinders). Air compresses easily as the pressure inside the brake lines increases. When you push the brake pedal down and squeeze brake fluid out of the master cylinder, the slave cylinders don't respond properly. Why do the pistons of the slave cylinders move out less than they should?

8. Fire extinguishers that use carbon dioxide gas can be used to fight all kinds of fires because carbon dioxide displaces air, doesn't support combustion, settles downward on surfaces, and is non-toxic.

a. Each carbon dioxide molecule is heavier than an average air molecule. Why does carbon dioxide gas sink in air?

b. The pressure inside the bottle of the extinguisher is so high—about 60 times atmospheric pressure—that most of the carbon dioxide inside it is liquid. This liquid turns to gas as you use the fire extinguisher. If you were to release the gas directly into the air through a narrow opening, why would the gas suddenly acquire a large forward velocity?

c. If you were to hold the extinguisher when the gas rushed out its nozzle, you would feel a strong backward force. Explain the origin of this force.

d. To reduce the force noted in **c**, a carbon dioxide fire extinguisher has a deflecting surface directly in the path of the carbon dioxide as it leaves the bottle. The carbon dioxide hits this surface and turns toward the right or left. Why does this surface allow you to use the fire extinguisher without being pushed backward?

e. The fire extinguisher collects the swirling gas after it has been deflected by the surface and directs that gas toward the fire. Because the gas travels slowly after hitting the deflecting surface, it carries much less momentum and kinetic energy. What has become of the energy that was stored in the compressed carbon dioxide before it left the extinguisher?

9. A one-wheeled cycle, or unicycle, is the ultimate in statically unstable vehicles.

a. Why does a unicycle always fall over when the rider doesn't try to keep it upright?

b. To keep the unicycle from falling over, the rider continually tries to position the wheel so that the force the ground exerts on the wheel points toward a particular point. What is that point?

c. A person riding a two-wheeled bicycle must lean left while making a turn toward the left. Does a unicycle rider also have to lean left during a left turn?

d. While a unicycle doesn't exhibit dynamic stability the way a two-wheeled bicycle does, there is a way to give it dynamic stability. If you spin the unicycle extremely rapidly about its vertical axis, it will act like a toy top and won't fall over for a very long time. (Unfortunately, it's hard to ride this way.) While gravity will exert a torque on this spinning unicycle if its axis isn't perfectly vertical, the unicycle doesn't simply fall over. Instead, its axis of rotation changes directions. What is this behavior an example of?

CHAPTER 6

HEAT

We can't see all of the motion that takes place around us. Some motion is hidden deep inside each object, where thermal energy keeps the individual atoms and molecules jiggling back and forth in an endless flurry of activity. In most situations, we are only aware of this thermal energy because it determines the object's temperature. You can feel with your hand when an object is hot or cold. The more thermal energy a particular object contains, the higher its temperature and the hotter it feels.

What you are feeling when you touch a hot object is its thermal energy flowing into your colder hand and raising the temperature of your skin. When thermal energy flows from a hotter object to a colder one, we call this moving thermal energy *heat*. In this chapter, we will examine temperature and heat in order to understand some of the objects in our everyday world.

Experiment: A Ruler Thermometer

One effect that a change in temperature has on a typical object is to change its size. You can use this size change to build a crude thermometer. While the size change is so tiny that it's normally hard to detect, you can use mechanical advantage to help you. Here is a way to make a thermometer using only a clear plastic ruler, a pin, a small weight, a piece of stiff paper, and some tape.

Lay the plastic ruler along the edge of a table and tape one end of it securely to the table. Cut a very thin strip of stiff paper, about 3 mm wide (0.1 inches) and 15 cm long (6 inches), and push the pin carefully through one end of the strip. Use a dot of tape to stick the pin's head to the paper. When you are done, the paper strip should be securely attached to the pin so that as the pin

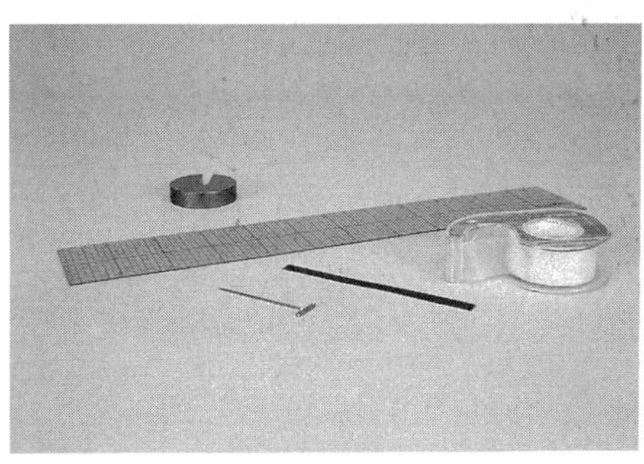

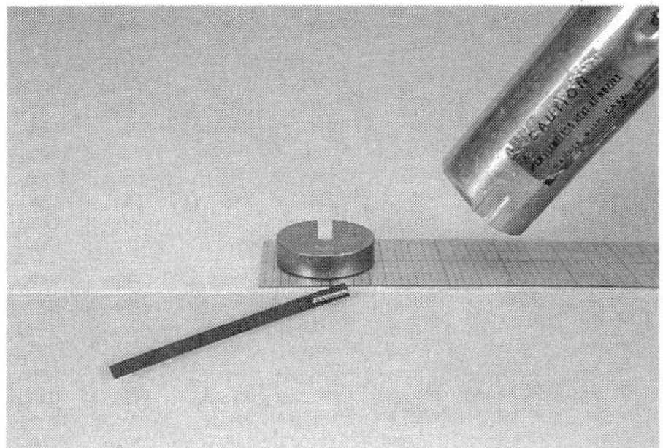

turns, the strip turns. This strip is your thermometer's pointer.

Now slide the pin under the free end of the ruler and place the small weight above it. The weight is there to push the ruler and pin together so that they experience plenty of static friction. That way, as the free end of the ruler moves left or right, the pin will rotate and turn the pointer.

Your thermometer is now complete. If you turn the pin and pointer carefully by hand so that the pointer is horizontal, you can "read" the thermometer by its position relative to the table top. If you heat the plastic ruler, by breathing on it, laying your hands on it, or warming it gently with a hair dryer, the ruler will become longer. Its free end will move away from the fixed end and will cause the pin to rotate. You will see a small change in the pointer's orientation as your thermometer reports its new temperature.

Chapter Itinerary

In this chapter, we'll examine thermal energy, temperature, and heat in the context of four common types of objects: (1) *wood stoves*, (2) *clothing and insulation*, (3) *incandescent light bulbs*, and (4) *thermometers and thermostats*. In *wood stoves*, we'll look at how thermal energy can be produced and at the three principal means by which thermal energy is transferred as heat from hotter objects to colder ones: conduction, convection, and radiation. In *clothing and insulation*, we'll consider the ways in which heat transfer can be reduced and how those methods are used in our clothing and our homes. In *incandescent light* bulbs, we'll study heat transfer by radiation to see that it can include the emission of visible light. And in *thermometers and thermostats*, we'll study changes in materials that accompany temperature changes, in order to understand some of the common methods of temperature measurement.

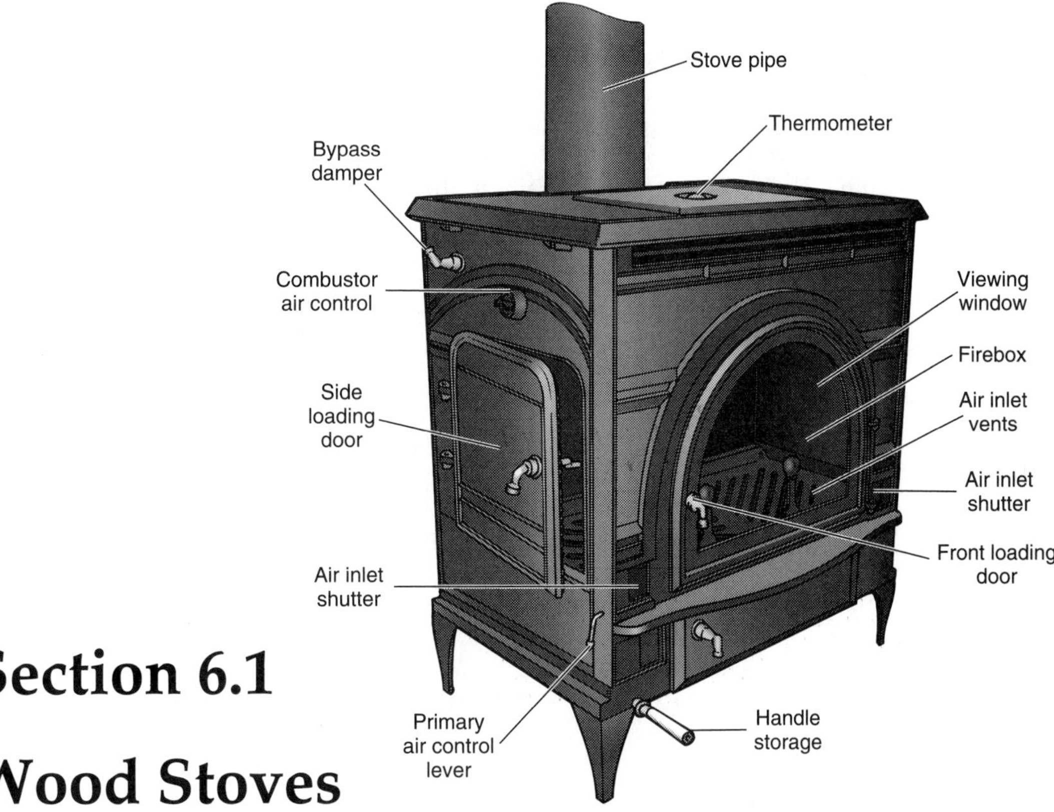

Section 6.1 Wood Stoves

Winter would be pretty unpleasant for most of us were it not for heating. Heating keeps our rooms warm even though the weather outside is cold. One of the most pleasant types of heating is a wood stove, which burns logs in its firebox and sends thermal energy out into the room. In this section we will look at how a wood stove produces thermal energy and how this thermal energy flows out of the wood stove to keep us warm.

Questions to Think About: *What happens to the chemical potential energy in a log when you burn it? Why do you get burned when you touch a hot object? Why does your skin feel warm when you face a campfire, even if the air around you is very cold? Why does your hand feel hotter when you hold it above a hot surface than next to that hot surface?*

Experiments to Do: *A burning candle produces thermal energy, providing both light and warmth to a small room. But were does this thermal energy come from and how does it flow into its surroundings?*

Light a short candle and look first at the way in which it releases thermal energy. The wax is slowly consumed by the flame. But the flame also needs air. Cover the candle with a tall glass, one that will not be touched or damaged by the flame. The glass should prevent room air from reaching the candle. How does the newly sealed environment affect the candle flame? Try to explain this result.

Relight the candle and consider the ways in which heat flows from the flame to you. Carefully pass your hand over the flame, keeping a safe distance above it to avoid being burned. Why does the flame warm your skin so quickly when your hand is directly above it? Now hold your hand beside the flame, again at a safe distance. You should again feel warmth from the flame. How is heat flowing from the flame to your hand now?

Take a wooden pencil and hold it a few centimeters above the flame for no more than 2 seconds. Then carefully touch the pencil's surface with your fingers. It should warm your fingers. How is heat flowing from the pencil to your hand in this case?

A Burning Log: Thermal Energy

A wood stove produces thermal energy and distributes it as heat to the surrounding room (Fig. 6.1.1). We've encountered thermal energy before, in the file cabinet sliding along the sidewalk, in the elevator's hydraulic fluid returning to its reservoir, and in the bicycle brakes slowing the turning wheels. In each case, ordered energy—energy that can easily be used to do work—became disordered thermal energy and the temperatures of the objects increased. But now that we are going to study heating machines themselves, let's reexamine thermal energy and temperature to see how thermal energy moves from one object to another.

When you burn a log in the fireplace or wood stove, you are turning the log's ordered chemical potential energy into thermal energy—a disordered form of energy, contained in the kinetic and potential energies of individual atoms and molecules. The presence of thermal energy in the log is what gives the log a temperature; the more thermal energy in the log, the higher its temperature.

Fig. 6.1.1 - This wood stove transfers heat to the room by conduction through its metal walls, convection of air past its surfaces, and radiation from its black exterior.

The nature of thermal energy depends somewhat on what it's in. In the hot, burning log, the thermal energy is mostly in the wood's atoms and molecules, which jitter back and forth rapidly relative to one another. When each of these particles moves, it has kinetic energy. When it pushes or pulls on its neighbors, it has potential energy.

But energies associated with the whole log are not included in thermal energy. Only disordered energy that is internal to the log contributes to its thermal energy. Thus if you stir the fire about with a poker, the moving log's total kinetic energy increases but its thermal energy does not. Since the moving log can do work on the external things it touches, the kinetic energy in its motion is neither disordered nor internal to the log. Only kinetic energy in the relative motions of individual atoms and molecules contributes to the log's thermal energy.

Similarly, if you lift the burning log with tongs, you increase its gravitational potential energy but not its thermal energy. Since the new potential energy is stored in the gravitational force between the log and the earth, it is ordered energy that can do work directly on external objects. The potential energy in thermal energy is stored between the atoms and molecules of the log, where it is disordered and can't do work directly.

The air near the burning log also contains thermal energy, but since the atoms and molecules in a gas are essentially free and independent, most of this thermal energy is kinetic energy. The air particles store potential energy only during the brief moments when they collide with one another.

CHECK YOUR UNDERSTANDING #1: A Warm-Up Pitch

If you drop a ball, will its thermal energy increase as it falls?

Forces Between Atoms: Chemical Bonds

If an atom couldn't exert forces on its neighbors, the log's thermal energy wouldn't include any potential energy. But that would hardly matter because, with nothing holding it together, the log wouldn't even exist. Fortunately, atoms do exert forces on one another. Let's take a moment to look at the nature of these forces and at how they can bind atoms together to form objects such as our log.

As you bring them close together, atoms exert attractive forces on one another (Fig. 6.1.2*a*). These forces are electromagnetic in origin and grow stronger as the two atoms approach. But the attraction diminishes when the atoms start to touch, and is eventually replaced by repulsion at very short distances (Fig. 6.1.2*b*). The separation between atoms at which the attraction ends and the repulsion begins is their *equilibrium separation;* that is, the separation at which the atoms exert no forces on one another (Fig. 6.1.2*c*). Since atoms are tiny, this equilibrium separation is also tiny, typically only about one ten-billionth of a meter.

Imagine holding two atoms in tweezers and slowly bringing them together. They pull toward one another as they approach, doing work on you and increasing your energy. They are giving up chemical potential energy and are beginning to bind together. Once the atoms reach their equilibrium separation, you can let go of them and they will not come apart. Like two balls attached by a spring, the atoms are in a stable equilibrium. They have lost some of their chemical potential energy and can only be separated by the return of that energy. Since the atoms did work on you as they pulled together, something must do work on them to pull them apart.

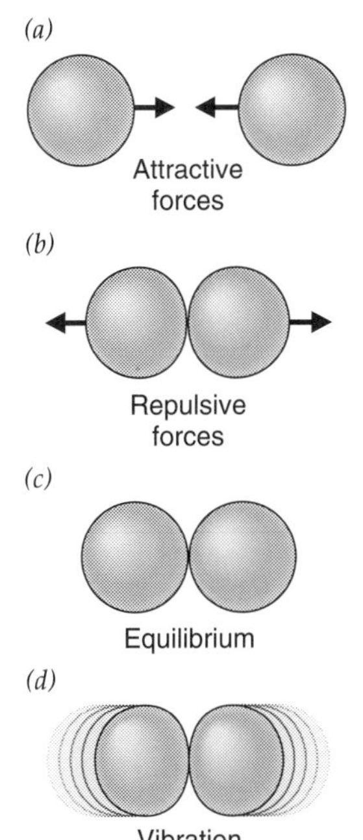

Fig. 6.1.2 - (*a*) Two atoms attract one another at moderate distances but (*b*) repel when they are too close. (*c*) In between is their equilibrium separation, at which they neither attract nor repel and are thus in equilibrium. (*d*) Pairs of atoms with excess energy tend to vibrate about their equilibrium separations.

The atoms are held together by a chemical bond. They have become a molecule. The strength of this bond is equal to the amount of work the atoms did when they drew together; the same amount of work required to pull the atoms apart. Bond strengths range from extremely strong in the case of two nitrogen atoms to extremely weak in the case of two neon atoms.

Now that the atoms have formed a molecule, they will store energy whenever something knocks them away from their equilibrium separation. As long as they don't have enough energy to separate completely, the two displaced atoms will vibrate back and forth about this equilibrium separation with a mixture of kinetic energy and chemical potential energy (Fig. 6.1.2*d*). At times when the atoms are moving quickly toward or away from one another, most of their energy will be kinetic. At times when the atoms are slowing to turn around, most of their energy will be chemical potential. Overall, the molecule's total energy will remain constant and it will vibrate back and forth until it transfers its extra energy elsewhere.

Chemical bonds and chemical potential energy are also important for molecules with more than two atoms. In a large molecule, each pair of adjacent atoms has a chemical bond and an equilibrium separation. If you give this molecule excess energy, it will vibrate in a complicated manner as the energy moves among the various atoms and chemical bonds. The atoms in the molecule will continue to jiggle about until something removes the excess energy from the molecule.

Chemical bonds are also present in the burning log. Like all liquids and solids, the log is just a huge assembly of atoms and molecules, held together by chemical bonds of various strengths. The atoms and molecules in the log push and pull on one another as they vibrate about their equilibrium separations. Their motion is *thermal motion* and the energy involved in this disorderly jiggling is thermal energy.

The form of an atom's thermal energy changes as it vibrates—from kinetic energy to potential energy and back again. The amount of its thermal energy also changes as the atoms do work on one another during their vibrations. Because this exchange of energy between atoms is unpredictable, the log's considerable thermal energy can't be used directly to do useful work. Thermal energy is fragmented among the atoms as countless tiny packets of energy, each of which is too small to do measurable work on a large object. Once energy has been fragmented, it can't be unfragmented. While it's easy to turn ordered energy into fragmented thermal energy, the reverse is much harder to do. A log that has been burned can't be "unburned."

CHECK YOUR UNDERSTANDING #2: When Atoms Collide...

As two independent nitrogen atoms collide with one another, what forces do they experience?

Heat and Temperature

Everything contains thermal energy, from the hot burning log to the cold metal poker you use to stir the fire. But what happens to thermal energy when two objects touch? When you push on the log with the poker, what happens to the thermal energy in the log or the poker?

When they touch, the poker and the log begin to exchange thermal energy. In effect, the two objects become one larger object and thermal energy that has been moving among the atoms in each individual object begins to flow across the junction between the two. Since each object starts with some amount of thermal energy, energy moves in both directions across this junction. Nonetheless, there will be some direction of average or net flow between the objects. To allow us to predict this direction of flow, we define a temperature for each of the objects.

Temperature is the quantity that indicates which way, if any, thermal energy will flow between two objects. If no thermal energy flows when two objects touch, then those objects are in **thermal equilibrium** and their temperatures are equal. But if thermal energy flows from the first object to the second, then the first object is hotter than the second.

A temperature scale classifies objects according to which way thermal energy will flow between any pair. An object with a hotter temperature will always transfer thermal energy to an object with a colder temperature and two objects with the same temperature will always be in thermal equilibrium. Thus the hot burning log will transfer thermal energy to the cold poker. We say that the burning log is *hot* because it tends to transfer thermal energy to most objects while the poker is *cold* because most objects tend to transfer thermal energy to it.

Energy that flows from one object to another because of a difference in their temperatures is called **heat**. Heat is thermal energy on the move. Strictly speaking, the burning log doesn't contain heat, it contains thermal energy. However when that log transfers energy to the cold poker because of their temperature difference, it is heat that flows from the log to the poker. (For a historical note about the understanding of heat, see ❐).

❐ Before the time of Benjamin Thompson, Count Rumford (American-born British physicist and statesman, 1752–1814), heat was believed to be a fluid called caloric that was contained within objects. Thompson disproved the caloric theory by showing that the boring of cannons produced an inexhaustible supply of heat. Among his scientific and technological contributions, Thompson improved cooking and heating methods. He reshaped fireplaces and developed the damper as ways to reduce smoking and improve heat transfer to the room. Thompson also had a life of sensational escapades and great rises and falls in fortune. He fled New Hampshire in 1775 because he was a British loyalist, he fled London in 1782 under suspicion of being a French spy, and was, at the time of his studies of heat, among the most powerful people in Baveria.

Our present definition of temperature can order the objects around us from hottest to coldest, but it doesn't quantify temperature in any unique way. While you could establish your own temperature scale by comparing every two objects in sight to see which way heat flows between them, that scheme wouldn't be very convenient. To make life simpler, we have established standard temperature scales such as Celsius, Fahrenheit, and Kelvin. When one object has a higher temperature than another, as measured on one of these scales, then heat will flow from the first to the second when they touch. If two objects have equal temperatures, then they will be in thermal equilibrium and no heat will flow.

These standard temperature scales are based on an object's average thermal kinetic energy per atom. The more kinetic energy each atom has, on average, the more vigorous the object's thermal motion and the more work the object's individual atoms will do on those of a second object. This microscopic work is what actually passes heat from one object to another. Thus an object with more average thermal kinetic energy per atom will pass heat to an object with less and it makes sense to assign temperatures according to average thermal kinetic energies per atom.

The Celsius, Fahrenheit, and Kelvin scales all measure temperature in this manner. In each scale, a 1 degree increase in temperature reflects a specific increase in average thermal kinetic energy per atom. The hot burning log has a large average thermal kinetic energy per atom and is assigned a high temperature. The cold poker has much less thermal kinetic energy per atom and is assigned a much lower temperature.

The relationship between average thermal kinetic energy per atom and assigned temperature is based on several standard conditions: absolute zero, water's freezing temperature, and water's boiling temperature. (Recall from Section 3.1 that absolute zero is the temperature at which all thermal energy has been removed from an object.) Once specific temperatures have been assigned to two of these standard conditions, the whole temperature scale is fixed. For example, the Celsius scale is built around 0° C being water's freezing temperature and 100° C being water's boiling temperature. Temperatures for the three standard conditions appear in Table 6.1.1.

Table 6.1.1 - Temperatures of several standard conditions, as measured in three temperature scales: Celsius, Kelvin, and Fahrenheit.

Standard Condition	*Celsius*	*Kelvin*	*Fahrenheit*
Absolute Zero	-273.15°	0°	-459.67°
Freezing Water	0°	273.15°	32°
Boiling Water	100°	373.15°	212°

CHECK YOUR UNDERSTANDING #3: Frozen Fingers

If you pick up an ice cube, your hand suddenly feels cold. Which way is heat flowing?

Open Fires

Suppose you needed an easy way to heat your room. The oldest and simplest method would be to start a campfire in the middle of the floor. You would arrange some wood and light it with a match. The wood would burn and the room would become warmer. The burning wood would produce thermal energy, which would flow as heat into the colder room. But how does burning wood produce thermal energy?

This thermal energy is released by a **chemical reaction** between molecules in the wood and oxygen in the air. Recall that atoms do work as they join together in a chemical bond and that the amount of work done depends on which atoms are being joined. While carbon and hydrogen atoms can bond together to form *hydrocarbon* molecules, these atoms form much stronger bonds with oxygen atoms. Thus while it may take work to disassemble a hydrocarbon molecule, the work done in joining its hydrogen and carbon atoms to oxygen atoms more than makes up for that work. As a hydrocarbon molecule burns in oxygen, new, more tightly bound molecules are formed and chemical potential energy is released as thermal energy. The *reaction products* produced by burning a hydrocarbon in air are primarily water and carbon dioxide.

Wood is composed mostly of cellulose, a long carbohydrate molecule that we'll examine in Section 17.3 on plastics. *Carbohydrates* contain carbon, hydrogen, and oxygen atoms. Despite the presence of a few oxygen atoms, carbohydrates still burn nicely to form water and carbon dioxide. When you light the wood with a match, you are supplying the energy needed to break the old chemical bonds so that the new bonds can form. This starting energy is called **activation**

energy, the energy needed to initiate the chemical reaction. Heat from the match flame gives the wood enough thermal energy to break chemical bonds between hydrogen and carbon atoms and start the reaction.

Unfortunately, wood is not pure cellulose. It also contains many complex resins that don't burn well and create smoke. If you plan to breathe the same air in which you burn fuel, wood is a poor choice. Better choices are kerosene and natural gas, both of which are nearly pure hydrocarbons and burn cleanly to produce water vapor and carbon dioxide. Actually, wood can be converted to a cleaner fuel by baking it in an airless oven to remove all of its volatile resins. This process converts the wood into charcoal, which burns cleanly to form water vapor, carbon dioxide, and ash.

However, there are a number of disadvantages to the direct fire-in-the-room heating concept:

1. The room's occupants must breathe any chemicals produced or released by the fire.
2. The fire consumes the room's oxygen, the component of the air that sustains life.
3. An open fire is a safety hazard.

Nonetheless, fires have been used for thousands of years to heat dwellings. Fireplaces that burn wood or peat have chimneys that carry away their noxious fumes. Unfortunately, a chimney reduces the value of a fire by carrying away much of the fire's thermal energy and the room's air. That is why a room that is being heated by a fireplace only feels warm near the fireplace itself. Elsewhere, the room is often cold because outside air is seeping in through cracks to replace air drawn up the chimney. Even when clean burning fuels are used without a chimney, there are no simple solutions to the oxygen or safety problems.

CHECK YOUR UNDERSTANDING #4: Strike Up the Bond
Why must you scrape a match on a matchbox in order to light it on fire?

Wood Stoves and Heat Exchangers

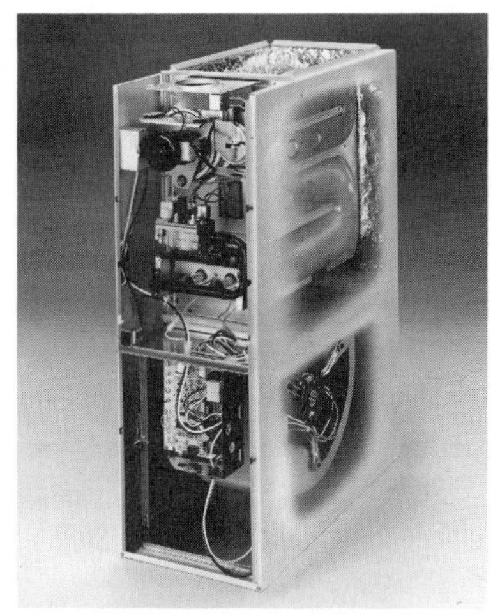

Fig. 6.1.3 - This modern furnace burns natural gas in an S-shaped firebox. A fan at the bottom of the furnace blows fresh air past the hot outer surfaces of the firebox and then circulates the heated air among the rooms.

It's nearly impossible to solve all the problems with simple fires simultaneously. However, a good first step is to separate the air used for the fire from the air that you breathe. If you burn the fuel in air obtained from outside and send the reaction products outside, the fire will not affect the room air. But it will not warm the room air very much, either.

You need a **heat exchanger**—a device that transfers heat from the fire to the air in your room without transferring the hot molecules themselves. A wood stove is an example of a heat exchanger. A well-designed wood stove burns wood in an enclosed fire box, drawing air in from outside and returning that air to the outside. What makes the wood stove different from a simple fireplace is that, before all of the thermal energy released by the fire can escape through the chimney, most of it is transferred into the room. The gas furnace in Fig. 6.1.3 also employs a heat exchanger.

The burning coals and hot gases inside the wood stove contain a great deal of thermal energy and are much hotter than the room air. Because of this temperature difference, heat tends to flow from the fire to the room. What is not so clear yet is how that heat is transferred from one to the other.

There are three principal mechanisms by which heat moves from the fire to the room: conduction, convection, and radiation. The wood stove makes wonderful use of all three so that most of the thermal energy released by the burning wood is transferred to the room. Let's examine these three mechanisms of heat transport, beginning with conduction.

CHECK YOUR UNDERSTANDING #5: Feeling the Heat

You can make a heat pack by wrapping hot, wet towels in a plastic sheet. This pack will warm an injured muscle but will not get it wet. Is thermal energy moving in this case?

Heat Moving through Metal: Conduction

Conduction occurs when heat flows through a stationary material. The heat moves from a hot region to a cold region but the atoms and molecules don't. For example, if you place one end of a metal poker in the fire, the poker's other end will gradually become warm as heat is conducted through the metal.

Some of this heat is conducted by interactions between adjacent atoms. Each time these vibrating atoms push on one another, one atom does work on the other and they exchange some kinetic energy. Such individual exchanges permits thermal energy to move about even though the atoms themselves remain fixed. The energy one atom receives from its neighbor is soon passed on to a third atom, a fourth atom, and so on. Since each exchange is random and unpredictable, thermal energy tends to flow about randomly within a material, ensuring that, on average, each atom has about the same amount of kinetic energy as all the other atoms.

But when one end of the poker is hotter than the other, the flow is no longer completely random. The atoms at the hot end of the poker have more kinetic energy to exchange with their neighbors than atoms at the cold end. The exchanges statistically favor the flow of thermal energy away from the hot end and toward the cold end. This flow of thermal energy from hot to cold through the poker is conduction (Fig. 6.1.4).

However, this atom by atom "bucket-brigade" is not the only way in which a material can conduct heat. In a metal, the primary carriers of heat are actually mobile **electrons**—the tiny negatively charged particles that make up the outer portions of atoms. When atoms join together to form a metal, some of the electrons stop belonging to particular atoms and travel almost freely throughout the metal. These mobile electrons can carry electricity (as we'll discuss in Chapter 11) and are also good at transporting heat.

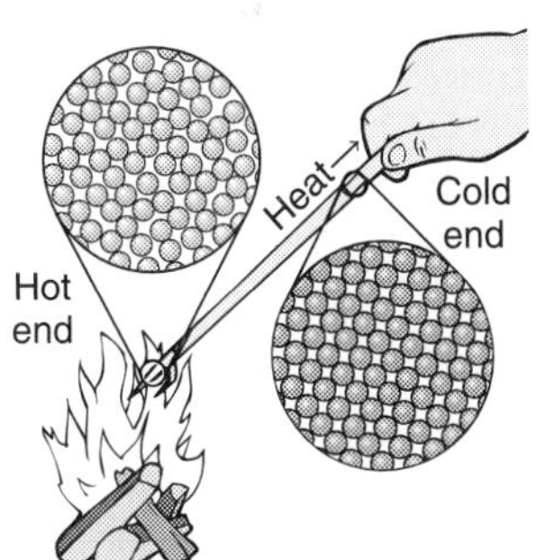

Fig. 6.1.4 - When one end of a metal poker is hotter than the other, the atoms at the hot end vibrate more vigorously than those at the cold end. The poker then conducts heat from the hot end to the cold end. Some of this heat is conducted by interactions between adjacent atoms. However, in the metal poker, most of the heat is conducted by mobile electrons, which carry thermal energy long distances from one atom to another.

When one of these mobile electrons collides with a metal atom, the two particles may exchange thermal energy. Thus the mobile electrons can participate in the bucket-brigade of heat conduction. But unlike atoms, which can only exchange energy with their neighbors, the mobile electrons can travel great distances between collisions and can move thermal energy quickly from one place to another.

The ease with which electrons move heat about a metal explains why metals are generally much better conductors of heat than non-metals. The best conductors of electricity—copper, silver, aluminum, and gold—are also the best conductors of heat. Poor conductors of electricity—stainless steel, plastic, and glass—are also poor conductors of heat. There are a few exceptions to this rule. Diamonds, for example, are terrible conductors of electricity but wonderful conductors of heat.

Conduction is what moves thermal energy from the wood stove's inside to its outside. No atoms move through the metal walls of the stove, just heat. So conduction serves as a filter, separating desirable thermal energy from the unwanted smoke and noxious gases that then go up the chimney.

Thus conduction makes the outside surface of the wood stove very hot so that heat should flow from it to the colder room. But what carries heat into the room? If you touch the stove, conduction will immediately transfer a huge amount of heat to your skin and you will be burned. But even without touching the stove, you are aware of its high temperature. It transfers heat into the room by convection and radiation.

CHECK YOUR UNDERSTANDING #6: Too Hot to Handle
Some pot handles remain cool during cooking while others become unpleasantly hot. What determines which handles remain cool and which become hot?

Heat Moving with Air: Convection

Convection occurs when a moving fluid transports heat from a hotter object to a colder object. The heat moves as thermal energy in the fluid so that the two travel together. The fluid usually follows a circular path between the two objects. It picks up heat when it comes in contact with the hotter object and gives up that heat when it touches the colder object. The fluid then returns to the hotter object to begin again.

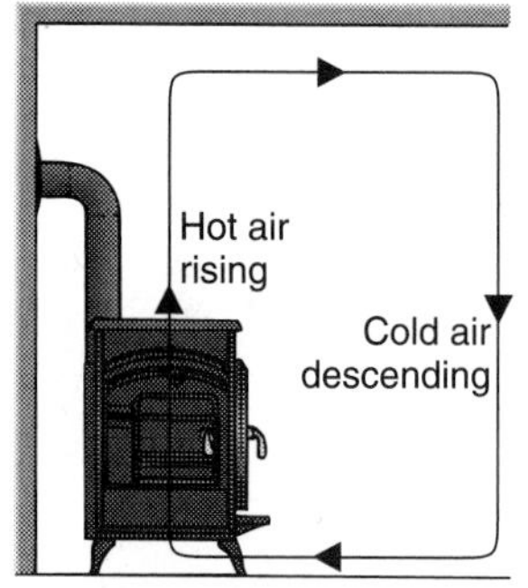

Fig. 6.1.5 - When the wood stove is hot, convection carries heat from its surfaces to the ceiling and walls of the room. Warm air rises upward, supported by the buoyant force, and is replaced by cooler air from the floor. The warmed air eventually cools and descends. It then returns toward the stove to repeat the cycle.

In most cases, this circulation develops naturally because of changes in the fluid's density with temperature. As the fluid warms near the hotter object, its density normally decreases and it floats upward, supported by the buoyant force. When the fluid cools near the colder object, its density normally increases and it sinks downward.

Thus air heated by contact with the wood stove rises toward the ceiling and is replaced by colder air from the floor (Fig. 6.1.5). Eventually, this heated air cools and descends. Once it reaches the floor, it's drawn back toward the hot wood stove to start the cycle over. This moving air is a *convection current* and the looping path that it follows is a *convection cell*. Within the room, convection currents carry heat up and out from the wood stove to the ceiling and walls. When you put your hand over the stove, you feel this convection current rising upward as it transfers heat to your hand.

Natural convection is good at heating the air above the wood stove, but most of that hot air ends up near the ceiling. While some of it will eventually drift downward to where you are standing, convection sometimes needs help. Adding a ceiling fan will help move the hot air around the room and make the wood stove more effective. This forced air circulation still transfers heat from the hot stove to the colder occupants of the room, but it doesn't rely on gravity and the buoyant force to keep the air circulating. The faster the air moves, the more heat it can transport from hot objects to cold objects.

CHECK YOUR UNDERSTANDING #7: Heat and Wind
When sunlight warms the land beside a cool body of water, a breeze begins to blow from the water toward the land. Explain.

Heat Moving as Light: Radiation

There is one more important mechanism of heat transfer: radiation. As the particles inside a material vibrate back and forth, they not only bump into one another and exchange energy, they also emit and absorb electromagnetic radiation. *Electromagnetic radiation* refers to **electromagnetic waves**, which travel through empty space at the speed of light. Radio waves, microwaves, gamma rays, and infrared, visible, and ultraviolet light are all examples of electromagnetic waves and are the subjects of Chapters 14 and 15.

For the moment, what is important is that electromagnetic waves (or equivalently electromagnetic radiation) can carry thermal energy. When heat flows from a hot object to a cold object as electromagnetic radiation, we say that heat is being transferred by **radiation**. Unlike conduction and convection, which need atoms, molecules, or electrons to assist in the heat transfer, radiation occurs directly through space and can transfer heat from one object to another even when there is nothing connecting them.

The types of electromagnetic waves that are present in an object's *thermal radiation* depend on its temperature. While a colder object emits only radio waves, microwaves, and infrared light, a hotter object can also emit visible or even ultraviolet light. The red glow of a hot coal in the wood stove is actually that hot object's thermal radiation.

Since our eyes are only sensitive to visible light, we can't see all of the thermal radiation emitted by an object. But whether we see it or not, electromagnetic radiation contains energy and will transfer heat to whatever object absorbs it. Thermal radiation is everywhere, because everything emits thermal radiation. However, the amount of that emission depends on temperature. The hotter an object gets, the more thermal radiation it emits. Since hot objects emit more thermal radiation than cold objects, any exchanges of thermal energy via radiation always transfers heat from the hotter object to the colder one.

Radiation transfers a great deal of heat from the wood stove's surface to the surrounding objects. The stove bathes the room in infrared light and this light warms the objects that it reaches. To encourage this radiative heat transfer, the wood stove and its chimney are usually painted black. Black is best because it not only absorbs light well, it's also particularly good at emitting light. If you heat a black poker red hot, it will glow much more brightly than a poker that is either white or silvery metal. White or silvery surfaces are poor absorbers of light when they are cold and don't emit it well when they are hot.

Even if the air in the room is cool, you can usually feel the invisible infrared light from a wood stove on your face. When you block this light with your hands, your face suddenly feels colder because less heat is reaching your skin. This thermal radiation effect is even more pronounced with a fireplace or campfire, where thermal radiation is the primary mechanism for heat transfer to the surroundings.

Overall, modern wood stoves are very good heat exchangers. Convection draws the hot smoke up a long black pipe and out of the building. This smoke heats the stove and the pipe, which conduct the heat to their outer surfaces. From there, the heat is distributed about the room by convection and radiation.

In principle, a wood stove should burn the wood in cold air drawn from outside the house. That way, heated room air doesn't flow into the stove and up the chimney. However, most wood stoves burn the wood in room air because it's easier to obtain. So long as a stove uses the room air carefully and doesn't send huge volumes of it up the chimney, it will still work well. The stove's damper controls the airflow so that it draws in only enough air to completely burn the wood. Because they are so effective at extracting thermal energy from the smoke before sending it outdoors, wood stoves are far better at heating rooms than are open fireplaces.

CHECK YOUR UNDERSTANDING #8: Keeping Warm

When you stand under a heat lamp, you feel warm even though the lamp emits very little light. How is heat reaching your skin?

Summary

How Wood Stoves Work: A wood stove burns a fuel in air to obtain hot gas. The burning process is actually a chemical reaction in which molecules in the fuel and the air are disassembled into fragments and then reassembled into new, more tightly bound molecules such as water and carbon dioxide. This reassembly process releases more energy than was required to disassemble the original fuel and oxygen molecules. This extra energy appears as thermal energy within the reaction products, so that they are hot.

Rather than distribute the hot burned gas directly to the rooms, a wood stove transfers heat from the burned gas to clean air or water. Heat is conducted through the walls of a wood stove and then flows into the room via convection and radiation. The stove's black surface improves its ability to radiate heat.

The Physics of Wood Stoves

1. Thermal energy consists of the individual kinetic energies and potential energies of the particles making up an object. Thermal energy doesn't include the overall kinetic energy of the object nor does it include potential energies due to outside forces such as gravity.

2. Atoms are attracted toward one another by chemical forces. These chemical forces become repulsive if the atoms come too close together. At their equilibrium separation, the atoms exert no forces on one another. Atoms near their equilibrium separation and with little kinetic energy may be held together by a chemical bond.

3. The work done by two atoms as they pull together to their equilibrium separation determines the strength of the chemical bond. This same amount of work is required to separate the two atoms again. Bond strength depends on the atoms involved. A chemical reaction releases energy if the bonds in the original molecules are weaker than the bonds in the reaction products.

4. Temperature is the quantity that indicates which way, if any, thermal energy will flow between two objects. Thermal energy naturally flows from a hotter object to a colder object. Two objects with identical temperatures are in thermal equilibrium and exchange no thermal energy.

5. Standard temperature scales are based on the average thermal kinetic energy per atom in a material. As temperature increases, the average thermal kinetic energy per atom also increases.

6. Heat is energy that flows from one object to another because of differences in their temperatures. There are three principal mechanisms of heat transfer: conduction, convection, and radiation.

7. In conductive heat transfer, thermal energy is transferred from one atom to the next as they vibrate against one another. Mobile electrons also help to conduct heat in metals.

8. In convective heat transfer, a moving fluid carries heat from a hotter object to a colder object. Circulation occurs naturally because of thermally induced density changes in the fluid and resulting buoyant effects.

9. In radiative heat transfer, heat moves directly through space as electromagnetic radiation. All objects emit thermal radiation but most of it is not visible to our eyes.

Check Your Understanding - Answers

1. No, its thermal energy will remain constant.

Why: As the ball falls, its gravitational potential energy is transformed into kinetic energy of the entire ball. However neither energy is internal to the ball so they aren't included in thermal energy. Thus the ball's thermal energy doesn't change.

2. At first, they experience attractive forces. But once they come too close together, the forces will become repulsive. As they separate, the forces will once again be attractive.

Why: As they approach one another, the two nitrogen atoms experience attractive forces and a chemical bond begins to form between them. They accelerate toward one another, converting chemical potential energy into kinetic energy. However when they are very close together, the forces become repulsive and they bounce off one another. They head apart and the forces again become attractive but their kinetic energy breaks the bond and they separate forever.

3. From your hotter hand to the colder ice cube.

Why: Heat naturally flows from a hotter object to a colder object. Since the ice cube is colder than your hand, heat flows out of your hand and into the ice cube. Since your hand is losing thermal energy, its temperature drops and you sense cold. While it's tempting to think of cold as something that flows out of an ice cube, the only thing that really moves about is heat. Ice cubes are wonderful absorbers of heat and cool our drinks by reducing their thermal energies.

4. The match requires activation energy before its chemical reactions can begin and it can ignite. Sliding friction supplies this activation energy.

Why: The chemicals in a match head have plenty of chemical potential energy but can't begin to react without activation energy. Something must weaken the original chemical bonds so that the reaction can begin. Friction heats the match head as you scrape it along the matchbox so that the chemical reactions can begin.

5. Yes, thermal energy is flowing from the hot towels, through the plastic, to the muscle.

Why: The plastic sheet is acting as a heat exchanger, allowing heat to flow from the hot towels to the cooler muscle but preventing any movement of the hot water itself.

6. The handle's ability to conduct heat.

Why: Some handles are made of plastics or stainless steel, which are poor conductors of both electricity and heat. These handles normally remain cool, unless they are directly heated by hot gases from the stove. Other handles are made from aluminum or copper, which are good conductors of electricity and heat. These handles often become unbearably hot.

7. Convection occurs, with warmed air rising over the land and being replaced by cooler air from above the water. The air moving from over the water to over the land creates the breeze.

Why: Winds are giant convection currents caused by solar heating. Air rises over warm spots on the earth's surface and surface winds blow toward those warm spots to replace the missing air.

8. Radiation transfers heat from the lamp's filament to your skin.

Why: A heat lamp emits large amounts of invisible infrared radiation. Although you can't see this radiation, you can feel it on your skin.

Glossary

activation energy The energy required to initiate a chemical reaction. This energy serves to break or weaken the bonds in the starting chemicals so that the reaction can proceed to form the reaction products.

chemical reaction An encounter between two or more atoms or molecules that results in a rearrangement of the atoms to form different atoms or molecules.

conduction The transmission of heat through a material by a transfer of energy from one atom or molecule to the next. The atoms themselves don't move with the heat.

convection The transmission of heat by the movement of a fluid. Convection normally entails the natural circulation of the fluid that accompanies differences in temperatures and densities.

electromagnetic waves Waves consisting of electric and magnetic fields that travel through empty space at the speed of light. These waves carry energy and momentum and are emitted and absorbed as particles called photons. Radio waves, microwaves, infrared, visible, and ultraviolet light, X-rays, and gamma rays are examples of electromagnetic waves.

electrons The tiny negatively charged particles that make up the outer portions of atoms and that are the main carriers of electricity in metals.

heat exchanger A device that allows heat to flow naturally from a hotter material to a colder material without any actual exchange of those materials.

heat The energy that flows from one object to another as a result of a difference in temperature between those two objects.

radiation The transmission of heat through the passage of electromagnetic radiation between objects.

thermal equilibrium A situation in which no heat flows in a system because all of the objects in the system are at the same temperature.

Review Questions

1. If you burn a stick in a sealed container, the total energy in that container won't change. However, some of the energy changes form. What happens to it?

2. The atoms in a molecule have formed chemical bonds with one another. In what sense are the atoms held together by these bonds?

3. When you touch the forehead of a person with a fever, you can feel heat flowing into your hand. Why does that heat flow indicate the person's temperature is higher than yours?

4. If two objects are in thermal equilibrium, do they have the same temperature?

5. If a piece of paper has chemical potential energy to release when it burns, why doesn't it begin burning on its own?

6. When you stir boiling soup with a metal spoon, the spoon's handle may become uncomfortably hot. A plastic spoon's handle remains cool. Explain the difference.

7. Does heat itself rise, or is it only heated fluids such as air or water that rise? What lifts a heated fluid?

8. The coals of a campfire emit a red glow. How is that glow related to the transfer of heat from the fire to you?

9. If thermal energy is contained in atoms and molecules, in what ways can a wood stove transfer heat to you without transferring atoms and molecules to you?

Exercises

1. If you pass your finger very quickly through a candle flame, it won't burn you. Why not?

2. Why do aluminum pans heat food much more evenly than stainless steel pans when you cook on a stove?

3. Why do meats and vegetables cook much more quickly when there are metal skewers sticking through them?

4. It's often a good idea to wrap food in aluminum foil before baking it near the red hot heating element of an electric oven. Why does this wrapped food cook more evenly?

5. When you bake a pie in a glass dish over a red hot electric element, why does the bottom of the pie cook relatively quickly?

6. You can use a blender to crush ice cubes, but if you leave it churning too long, the ice will melt. What supplies the energy needed to melt the ice?

7. Use the concept of convection to explain why firewood burns better when it's raised above the bottom of a fireplace on a grate.

8. Why are black steam radiators better at heating a room than radiators that have been painted white or silver?

9. Why does a blackened baking pan heat up more quickly in an oven than one that is shiny?

10. The space shuttle generates thermal energy during its operation in earth orbit. How is it able to get rid of that thermal energy as heat in an airless environment?

11. When you stand out in an open field on a dark, clear night, you feel colder than if you stand under a tree. The temperature of space is only a few degrees above absolute zero, which is clearly colder that the tree. But how does heat flow from the tree to you?

12. In a forced warm-air heating system, the ducts carrying warm air into a room are usually located near the floor. What's wrong with putting them near the ceiling?

13. Your body is presently converting chemical potential energy from food into thermal energy at a rate of about 100 J/s, or 100 W. If heat were flowing out of you at a rate of about 200 W, what would happen to your body temperature?

14. When you heat a pot of water on the stove, the water's temperature is almost uniform throughout the pot. Why?

15. Explain how convection contributes to the shape of a candle flame.

16. Astronauts orbiting the earth in a spaceship are in free fall. Why doesn't natural convection carry heat efficiently through the spaceship's cabin?

17. If an astronaut in an orbiting spaceship were to light a candle, why would the flame look peculiar?

18. A modern convection oven has a fan to circulate the air. Why does this forced circulation of air increase the oven's cooking speed?

19. When you turn on the jets in a whirlpool spa, the water suddenly feels much hotter than before. Explain.

20. The cold air returns of a properly designed hot air heating system are located near the floor of a room. Air entering these returns flows to the furnace to be reheated. Why locate the returns near the floor?

21. The cliffs west of San Francisco are ideal for hang gliding. They offer a strong land breeze that blows daily from the surface of the cold Pacific Ocean to the warm land nearby. Explain this breeze in terms of heat transfer.

22. Most refrigerators use the freezing compartment to cool the refrigerating compartment as well. Why do most refrigerators place the freezing compartment on top?

23. Ice boxes used melting ice to chill food. Why were the blocks of ice placed above the food compartment?

24. When you hold a lighted match so that its tip is lower than its stick, the flame travels quickly up the stick. Why?

Section 6.2

Clothing and Insulation

When you sit in front of a fire on a cold winter day, your skin is warmed by heat from the hot embers. But when you walk through the snow on your way to the store, the last thing you want is heat transfer. As the hottest object around, you will become colder, not warmer. Instead, you do your best to avoid heat transfer. So you bundle up tight in your new down coat. Its thermal insulation keeps you warm in your frigid environment. In this section, we will examine thermal insulation and see how it keeps heat from moving between objects.

Questions to Think About: *When you wear a thick coat on a cold day, what provides the heat that keeps you warm? How does hair keep your head warm? Why do people sweat on hot days or during exercise? Why is good cookware often made of several different materials? Why do Thermos bottles and Space Blankets have mirrored surfaces?*

Experiments to Do: *Examine the effects of thermal insulation by touching objects with and without insulation on your hand. As you pick up a piece of hot toast with your bare hand, why does your skin feel so hot? Now try again with a towel or napkin between your skin and the toast. What has changed? Perform the same experiment with an ice cube. The towel or napkin keeps your skin comfortable when you touch either hot or cold objects. How is that possible? What happens when you use aluminum foil instead of the towel or napkin?*

You can do similar experiments with a heavy coat. The coat will obviously keep you warm in a cold environment but it will also keep you cool (at least briefly) in a very hot environment. Sit in front of a fireplace with a heavy coat on. The parts of your body that are covered by the coat will barely notice the fire's presence. In fact, you should be very careful because it's possible to scorch or even ignite your coat without feeling the heat. Firefighters use heavy fireproof coats to keep cool as they combat building fires.

The Importance of Body Temperature

Thermal insulation slows the heat transfer between objects and keeps your home warm, your refrigerator cold, and your fingers comfortable when you pick up a cup of hot coffee. One of the most important examples of thermal insulation is your clothing. The principal non-aesthetic purpose of clothing is to control the rate at which heat flows into or out of your body. Clothing helps you maintain your proper body temperature.

The goal of maintaining body temperature is unique to mammals and birds. Cold-blooded animals such as reptiles, amphibians, and fish make no attempts to control their body temperatures. Instead, they exchange heat freely with their surroundings and are generally in thermal equilibrium with their environments.

Unfortunately, the chemical processes that are responsible for life are very temperature sensitive. Many chemical reactions proceed only when thermal energy provides the necessary activation energy. As a cold-blooded animal's temperature goes down, there is less thermal energy per molecule and these chemical reactions occur more and more slowly. The animal's whole metabolism slows down and it becomes sluggish, dimwitted, and vulnerable to predators.

In contrast, warm-blooded animals have temperature regulation systems that allow them to maintain constant, optimal body temperatures. Regardless of its environment, a mammal or bird keeps the core of its body at a specific temperature so that it functions the same way in winter as in summer. The advantages of uniform temperature are enormous. On a cold day, a warm-blooded predator can easily catch and devour its slower-moving cold-blooded prey.

But there is a cost to being warm-blooded. The thermal energy associated with an animal's temperature must come from somewhere and the animal must struggle against its environment to maintain its body temperature. Without realizing it, many of our behaviors are governed by our need to maintain body temperature. Our bodies are careful about how much thermal energy they create and we work hard to control the rate at which we exchange heat with our surroundings.

A resting person converts chemical potential energy into thermal energy at the rate of about 80 Calories-per-hour. Our bodies use that much ordered energy even when we are doing no work on the outside world. Our hearts keep pumping, we keep synthesizing useful chemicals and cells, and we keep thinking. Since the chemical energy is not doing outside work or creating much potential energy anywhere, most of it ends up as thermal energy.

80 Calories-per-hour is a measure of power, equal to about 100 W. A resting person is using about as much power as a 100 W light bulb and, as with the light bulb, most of that power ends up as thermal energy. If a person is more active, he or she will produce more thermal energy. This steady production of thermal energy is why a room filled with people can get pretty warm. 100 W may not seem like very much power, but when a hundred people are packed into a tight space, they act like a 10,000 W space heater and the whole room becomes unpleasantly hot.

If you had no way to get rid of this thermal energy of metabolism, you would become hotter and hotter. To maintain a constant temperature, you must transfer heat to your surroundings. Since heat flows naturally from a hotter object to a colder object, your body temperature must be hotter than your surroundings. This requirement is one reason why human body temperature is approximately 37° C (98.6° F). This temperature is higher than all but the hottest locations on earth so that heat flows naturally from your body to your surroundings. You produce thermal energy as a byproduct of your activities and transfer this thermal energy as heat to your colder surroundings.

Since the rate at which your resting body generates thermal energy is fairly constant, the principal way in which you maintain your temperature is by controlling heat loss. You and other warm-blooded animals have developed a number of physiological and behavioral techniques for controlling heat loss. Let's examine those techniques in terms of the three mechanisms of heat transfer: conduction, convection, and radiation.

CHECK YOUR UNDERSTANDING #1: Keeping the Room Toastie Warm

A toaster oven turns electric energy into thermal energy at a rate of 500 J/s or 500 W. If the appliance's temperature remains constant, how quickly is it transferring heat to its environment?

Retaining Body Heat: Thermal Conductivity

Overall, you must lose thermal energy at the same rate as you produce it; about 100 joules each second. This modest rate is relatively easy to achieve. Except on hot days or when you are exercising hard, your body must struggle to avoid losing heat too quickly. Since all three heat transfer mechanisms are involved in this heat loss, you must control them all in order to keep warm.

One way in which your body retains heat is by impeding conductive heat loss. Some materials are better conductors of heat than others; they have different thermal conductivities. **Thermal conductivity** measures of how rapidly heat flows through a material that is exposed to a difference in temperatures. Skin has a particularly low thermal conductivity, meaning that it conducts relatively little heat compared to materials such as glass or copper.

Because thermal conductivity is a characteristic of the material itself, not the object from which that material is made, it's defined for a small cube of material with a temperature difference of one degree across it. To determine how much heat will flow through your skin you must consider not only your skin's thermal conductivity, but also its size and shape and the temperature difference across it. The more skin surface you have and the greater the temperature difference across it, the more heat your skin will conduct. However, thickening your skin reduces the temperature difference across each cube of it and lessens the heat conduction through it.

Thus the amount of heat flowing through your skin depends on its thermal conductivity, its surface area, the temperature difference across it, and its thickness. Your body controls all of these factors in trying to minimize heat loss:

1. It uses materials with very low thermal conductivities in your skin.
2. It makes your skin as thick as possible.
3. It minimizes the surface area of your skin.
4. It minimizes the temperature difference across your skin.

Your skin and the layers immediately beneath it contain fats and other thermal insulators. Fat's thermal conductivity is about 20% that of water and only about 0.03% that of copper metal. Your body uses fat for energy storage anyway, but by locating the fat in and beneath your skin, your body improves its heat retention. Furthermore, the presence of a fatty layer beneath your skin effectively thickens your skin and reduces the temperature difference across each unit of thickness. "Thick-skinned" people retain body heat better than those who are "thin-skinned."

Minimizing surface area means that your body is relatively compact, shaped more like a ball than a sheet of paper. Many other adaptive pressures

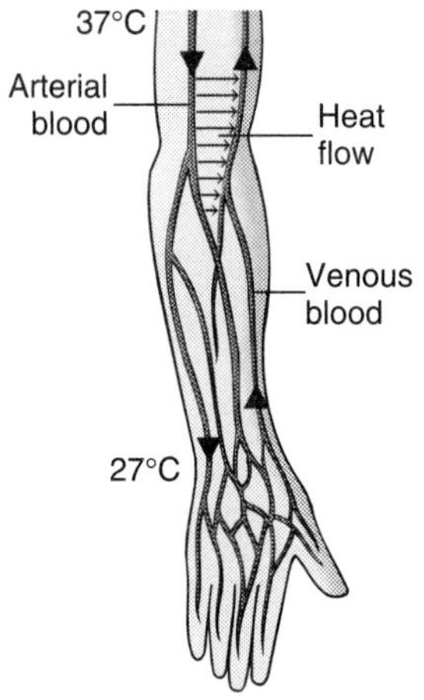

Fig. 6.2.1 - Blood flowing toward your hand through arteries exchanges heat with blood returning to your heart through veins. In this fashion, your blood is able to carry oxygen and food to your fingers without warming them all the way up to core body temperature. This adaptation reduces the rate at which you lose heat in cold weather.

have led to the evolution of arms, legs, and fingers that increase your total surface area. However, you have little superfluous surface through which to lose heat.

Finally, your body tries to lessen conductive heat loss by reducing the temperature difference between your skin and the surrounding air. It does this by letting your skin temperature drop well below your core body temperature. On a cold day, your hands and feet feel cold because they are cold. The colder they get, the less heat they lose to the cold air they touch.

Allowing your hands to become cold would be simple were it not for your circulating blood. Your blood must cool down from core body temperature as it approaches your cold fingers and must warm back up to core body temperature as it approaches your heart. This change in blood temperature occurs via a mechanism called *countercurrent exchange*. As the warm blood flows through arteries toward your cold fingers, it transfers heat to the blood returning to your heart through nearby veins (Fig. 6.2.1). The blood heading toward your fingers becomes colder while the blood returning to your heart becomes warmer.

CHECK YOUR UNDERSTANDING #2: Don't Get Burned

You have been cooking ears of corn in a pot of boiling water and it's time to fish them out. With which are you least likely to burn your hands: copper tongs or plastic tongs? (Copper has a much higher thermal conductivity than plastic.)

Retaining Body Heat: Convection

Heat leaving your skin warms the nearby air. How quickly the air's temperature increases depends on how much air there is and on that air's **specific heat capacity**; that is, the amount of heat it takes to warm one kilogram of air one degree Celsius (or Kelvin). The more air you are heating and the greater its specific heat capacity, the more heat it needs to warm one degree. The colder that air was to start with, the more heat you must give it to bring it to body temperature.

Different materials have different specific heat capacities. For example, it takes about 4 times as much heat to warm a kilogram of water one degree as it does to warm a kilogram of air one degree. This difference in specific heat capacities is part of the reason why you cool off faster swimming in cold water than standing in cold air. A material's specific heat capacity reflects the number of ways in which thermal energy can exist in that material. Since the molecules in 1 kg of water have about 4 times as many ways to hold thermal energy as the molecules in 1 kg of air, adding a certain amount of heat to each raises air's temperature about 4 times as much as it raises the water's temperature.

Since air is a very poor conductor of heat, your skin warms only a thin layer of it. As long as this layer of air doesn't move, its temperature will slowly approach that of your skin and the rate of heat flow from your body will decrease. Protected by this warm air, you will feel comfortable even on a cold day.

But air is rarely still. Convection gently removes the warmed air from your skin and replaces it with cooler air. With cooler air nearby, the temperature difference across your skin remains large and heat flows more quickly out through your skin. You feel cold. A wind worsens this heat loss because it blows away any warmed air near your skin. The enhanced heat loss caused by moving air is called *wind chill*—you feel even colder on a windy day.

To combat convective heat loss and wind chill, warm-blooded animals are covered with hair or feathers. Hair is itself a poor conductor of heat but its main purpose is to block airflow. Air passing through hair experiences large drag forces that slow its motion. In the dense tangle of a sheep's wool, air is trapped and can barely move at all. Since convection requires airflow, the sheep can only lose heat via conduction through the hair and the air. Since both are terrible conductors of heat, the sheep stays warm.

We humans have relatively little hair and are thus poorly adapted to living in cold, windy climates. Our lack of natural insulation is one of the reasons we wear clothing. Like hair and feathers, our clothing traps the air and reduces convection. Finely divided strands or filaments are particularly effective at stopping the flow of air. Not surprisingly, the best insulating clothing is made of hair (natural or synthetic) and feathers (also natural or synthetic). Since motionless air has a lower thermal conductivity than the hair or feathers that trap it, the ideal coat uses only enough material to keep a thick layer of air from moving.

This discussion also applies to water and swimming. If the water around you didn't move, you would soon be nice and warm. That's why some swimmers wear wet suits. The spongy material in a wet suit keeps the layer of water near the swimmer's skin from moving. As long it remains motionless, water is a respectable thermal insulator. This is evident in Fig. 6.2.2, where convection is inhibited by heating the *top* of a tube of water.

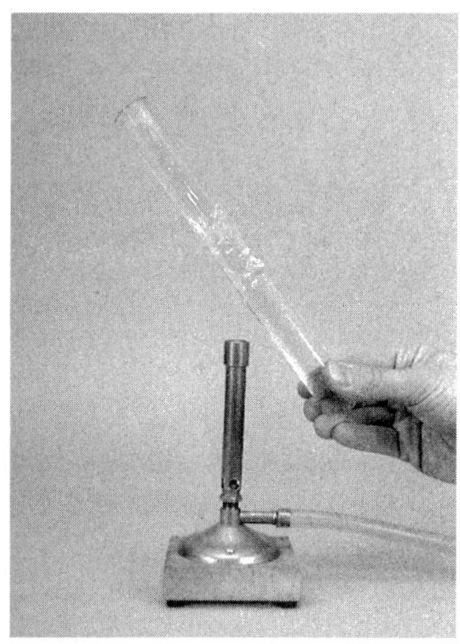

Fig. 6.2.2 - Convection only occurs if the hotter object is below or next to the colder object so that the heated fluid is able to rise. If you heat a tube of water near its top, the hot water stays near the top and the cold water remains at the bottom. Because water itself is a poor conductor of heat, the water at the top of the tube can boil while the bottom of the tube is cool enough to hold in your hand.

CHECK YOUR UNDERSTANDING #3: When a Cold Wind is Blowing

Why does wearing a thin, nylon wind breaker make such a difference in your ability to keep warm on a cool, windy day?

Retaining Body Heat: Radiation

You also exchange heat via radiation. Your skin emits electromagnetic waves toward your surroundings and they emit electromagnetic waves toward you. The amount of heat transferred by these waves depends on the temperature of each surface and on how well they absorb and emit light. The amount of heat radiated by a surface depends roughly on the fourth power of its temperature, measured in an absolute temperature scale, so that hotter objects radiate far more heat than colder objects.

As always, heat flows from the hotter object to the colder object. However, while conduction and convection transfer heat in proportion to the temperature difference between objects, radiation transfers heat in proportion to the difference between the *fourth powers* of their temperatures. That is why radiative heat transfer to or from your skin is most noticeable when you are exposed to an unusually hot or cold object.

The sun warms your skin quickly because it radiates more heat at you than the rest of your surroundings combined. Measured on an absolute temperature scale, the sun's surface temperature (6000° K) is about 20 times that of your skin (310° K). Though it's very distant and appears small to your eye, the sun radiates about 20^4 or 160,000 times as much heat toward you as you radiate toward it.

In contrast, the dark night sky cools you quickly because of its extremely low temperature. The mostly empty space beyond the earth's atmosphere is only a few degrees above absolute zero. When you stand in an open field at night, you radiate about a hundred watts of thermal power toward space but it radiates very little back toward you. Since you lose heat quickly, you feel cold. You can improve your situation by standing under a leafy tree. Even in cold weather, the tree is much hotter than space and emits far more thermal radiation. While the tree can't replace a crackling campfire, it will still help to keep you warm.

You might wonder why the air overhead doesn't radiate heat toward you to compensate for the heat you radiate toward the sky. The answer is that air is reasonably transparent to infrared light, absorbing and emitting relatively little of it. Only water vapor, carbon dioxide, and a few other gases in air interact with infrared light. Thus most of this exchange of energy by radiation is between you and empty space.

Not all surfaces absorb and emit thermal radiation well. A mirrored surface reflects thermal radiation while a white surface scatters it in all directions. Because they don't interact strongly with thermal radiation, they act as thermal insulators.

Fig. 6.2.3 - This Lunar Lander is wrapped in reflective foil to reduce its emissivity. As a result, it emits and absorbs relatively little thermal radiation.

A material's ability to absorb and emit thermal radiation is called its **emissivity**. A perfectly black object has an emissivity of 1, meaning that it absorbs all thermal radiation that hits it and emits thermal radiation of its own as efficiently as possible. A highly reflecting or purely white object will have an emissivity close to 0, meaning that it reflects or scatters almost all the thermal radiation that hits it and doesn't emit very much thermal radiation of its own (Fig. 6.2.3). Because most thermal radiation is infrared light, which we can't see, it's not always easy to guess an object's emissivity by looking at it. An object that is white or shiny to visible light may be nearly black to infrared light.

Because a larger surface has more opportunity to emit thermal radiation than a smaller surface, the heat an object radiates is also proportional to its surface area. We can combine that observation with our previous ones to obtain a single relationship between an object's temperature, emissivity, and surface area and the power it emits through thermal radiation. This relationship can be written as a word equation:

$$\text{radiated power} = \text{emissivity} \cdot \text{Stefan} - \text{Boltzmann constant} \cdot \text{temperature}^4 \cdot \text{surface area}, \qquad (6.2.1)$$

in symbols:

$$P = e \cdot \sigma \cdot T^4 \cdot A,$$

and in everyday language:

You don't have to expose much warm skin to radiate away lots of heat. You do better to expose only the cool, light-colored surfaces of your clothes.

This relationship is called the **Stefan–Boltzmann law** and the **Stefan–Boltzmann constant** that appears in it has a measured value of $5.67 \cdot 10^{-8}$ J/(s·m²·K⁴). Remember that the temperature must be measured in degrees Kelvin.

These issues of radiative heat transfer explain why we wear certain colors and why we are careful about exposing ourselves to the sun. On hot, sunny days, it makes sense to wear light colors and sit in the shade. Both actions reduce the amount of heat transferred to you by the sun. Light colored clothes have low emissivities, at least for visible light, so they don't absorb much sunlight. Sitting in the shade prevents the sun from exchanging heat with you directly.

Less obvious is the fact that white or reflective clothing also keeps you warmer in cold surroundings when there is no sun. Such clothing usually has a low emissivity in the infrared so that it doesn't radiate your body heat efficiently. Since you retain heat better in white or reflective clothing, you feel warmer. Highly reflective plastic blankets, found in emergency rescue kits, help to keep you warm in very cold surroundings by reducing the amount of heat you lose as thermal radiation.

CHECK YOUR UNDERSTANDING #4: Foil Wrapping

Wrapping a hot dish of food in shiny aluminum foil seems to keep it warm longer than wrapping it in clear plastic wrap. Aluminum is a good conductor of heat, so why does it impede the flow of heat so well?

CHECK YOUR FIGURES #1: The Cold of Deep Space

Suppose that an accident in the depths of space leaves you exposed to an environment near absolute zero. Since your surroundings radiate almost no heat toward you, you are losing heat fast. If your surface area is 2 m^2, your skin temperature is 310° K, and your emissivity is 0.5, how much power will you radiate?

Keeping Cool When It's Hot Outside

Slowing heat loss isn't always a good idea. If you retain heat too well, you will overheat. When exercising or on a very hot day, it may be necessary to encourage heat transfer to your surroundings by enhancing conduction, convection, or radiation.

You can increase conductive heat loss by moving into cold air or, even better, cold water. With a larger temperature difference across your skin, the rate of heat conduction through it will increase. You can increase convective heat loss by actively circulating the air or water with a fan or pump. The more cold air or water that directly touches your skin, the more heat you will lose. You can increase radiative heat loss by wearing black clothing while staying out of the sun. Actually, controlling radiative heat transfer is tricky, because even indirect sunlight can transfer heat to you. You may do better to avoid radiative heat transfer altogether by wearing white.

But what happens when you are put in an environment that is hotter than body temperature? If you are the coldest object around, you are going to get hotter and hotter. For a minute or two, insulating clothing can slow the rate at which your temperature rises so that you can pull a casserole from a hot oven or rescue a person from a fire. But even when you are perfectly insulated from your surroundings, your metabolism will cause your body temperature to rise. What does your body do to keep from overheating?

It sweats. By covering your skin with water, your body uses a new trick to eliminate heat. For water to evaporate, changing from a liquid to a gas, it needs energy. The molecules in liquid water are held together by chemical bonds that must be broken during evaporation. The energy that breaks these bonds is drawn from your body as heat. The faster the water evaporates, the more heat must flow out of your skin. Animals with hair can't sweat directly because there is little air circulation near their skin. Instead, these animals pant. Evaporation from their mouths and lungs draws heat from their bodies.

CHECK YOUR UNDERSTANDING #5: The Wind in Your Face

When you're traveling in a car on a warm day, opening the car windows cools you off. Explain.

Insulating Houses

The same techniques that keep people and animals warm are used to control heat flow in houses and household objects. However, because houses and their contents don't move much, they can make use of insulating methods that are heavy, bulky, rigid, or fragile. Let's take a look at some of the insulating schemes in the world around you.

The goal of housing insulation is to render a house's internal temperature effectively independent of the outside temperature. When it's cold outside, you want as little heat as possible to flow out of your warm house. When it's hot outside, you want as little heat as possible to flow into your cool house. So you or the builder fill its walls with insulating materials.

While there are many solid materials that are poor conductors of heat, including glass, plastic, hair, sand, and clay, the best insulator used in normal construction is air. Most modern buildings use air insulation. Unfortunately, air tends to undergo convection so it can't be used by itself. To prevent convection, air is trapped in porous or fibrous materials such as glass wool, saw dust, plastic foam, or narrow channels.

Glass wool or fiber glass is made by spinning glass into very long, thin fibers that are then matted together like cotton candy. Solid glass is already a poor conductor of heat but reducing it to fibers makes it even more insulating. The path that heat must take as it's conducted through the tangled fibers is very long and circuitous and very little heat gets through. Most of the volume in glass wool is taken up by trapped air. The glass fibers keep the air from undergoing convection so the air must carry heat by conduction.

Overall, glass wool and the air trapped in it are very good insulators. They also have the advantage of being nonflammable. In addition to its use in buildings, glass wool serves as insulation in ovens, hot water heaters, and many other machines that require nonflammable insulation. Most modern houses have about 10 cm to 20 cm of glass wool insulation built into their outside walls, along with a vapor barrier to keep the wind from blowing air directly through the insulation. (For a discussion of older insulating techniques, see Fig. 6.2.4.)

Fig. 6.2.4 - While stone is not a good conductor of heat, it's not nearly as good an insulator as air trapped in a fibrous mat. Medieval stone castles were notoriously cold in winter because heat flowed too easily out of them through their stone walls. The tapestry on the wall of this French Chateau slows the flow of heat to the outside air and helps to keep the room warm.

Because hot air rises and cold air sinks, the temperature difference between the hot air just below the ceiling and cold air just above the roof can become quite large. The ceiling/roof is thus a very important site of unwanted heat transfer and requires heavy insulation. Glass wool inserted between the ceiling and the roof of a new house may be more than 30 cm thick.

While glass wool is a very good insulator, other materials are used in certain situations. Urethane and polystyrene foam sheets are both waterproof and better insulators than glass wool. Unfortunately, they are also flammable and relatively difficult to work with. Nonetheless, they are used in construction and are particularly well suited for refrigerators and coffee cups, where rigidity and flammability are not problems. (For an even better insulator, see Fig. 6.2.5.)

Fig. 6.2.5 - The best thermal insulators currently under study are tenuous materials called aerogels. These porous glassy structures are almost entirely air and resemble frozen smoke. There's hope that aerogels and other high performance insulators will dramatically improve the energy efficiencies of refrigerators and other machines.

In older houses that were not insulated properly during construction, insulation can be blown into the walls or ceilings through holes drilled in the surfaces. As always, these insulators are porous or fibrous materials so that the main insulator is trapped air. Urea-formaldehyde foams are convenient for filling walls and ceilings because they can be pumped into cavities before they harden. However, concerns that they release toxic chemicals have reduced their appeal. Vermiculite and fireproofed cellulose chips are among the most common loose fill insulations.

CHECK YOUR UNDERSTANDING #6: Is More Always Better?

Glass wool insulation is easily compressed so that you can put two or three layers into the space that one layer will normally fill. To improve a building's insulation, why not pack as much glass wool insulation into the walls as possible?

Other Types of Insulation

While most household insulation revolves around air trapped in pores or around fibers, there are a few special circumstances in which finely divided materials just will not do. Windows have a special requirement that they must be transparent. They can't be filled with foam or fiberglass and solid glass is just not a good enough insulator.

The most common way to insulate windows is to use several panes of glass separated by narrow gaps of air or another gas. The air gaps prevent the easy conduction of heat from one side of the window to the other. While convection does occur in the air between the panes, their nearness creates tall, thin convection cells that are relatively ineffective at carrying heat from one side of the window to the other.

However, even a multiple-pane window transfers much more heat than a properly insulated wall. Glass conducts heat reasonably well and doesn't block radiative heat transfer completely. Shades and curtains not only block the view, they also reduce heat transfer through the window. Some energy efficient houses have special quilted shades that dramatically reduce this heat transfer .

A more sophisticated way to lower radiative heat transfer through a window is to use low-emissivity glass. This glass has a special coating to reduce the amount of infrared radiation it absorbs and emits. In effect, the glass acts like a mirror for infrared light. Coating the inner surfaces of a multipane window improves the window's insulating ability significantly because the panes exchange relatively little thermal radiation.

Food storage also depends on thermal insulations such as plastic foam and fiber mats. But if you try to keep food hot or cold for a very long time, you will find that even a fairly thick blanket of foam or fiber insulation will not be a sufficient barrier against heat transport. You do better with a glass or metal Thermos bottle, which makes use of a completely different technique of insulation: a vacuum.

A Thermos bottle is a consumer version of a Dewar flask, named after Sir James Dewar who invented it in the late 1800's. Instead of using air as insulation and inserting a tenuous material to prevent convection, a Thermos surrounds the food with a region that contains nothing at all, not even air (Fig. 6.2.6). In order to withstand atmospheric pressure, the Thermos has two strong walls. One wall surrounds the food and the other surrounds the first wall at a small distance. Since there is nothing between the two walls, there is no conduction and no convection. The two walls have mirror finishes so that they reflect thermal radiation and have very low emissivities. This mirroring dramatically reduces the radiative heat exchange between the walls. Since the only way heat can flow to or from the food is through its narrow mouth, a properly made Thermos bottle can keep food hot or cold for a remarkably long time.

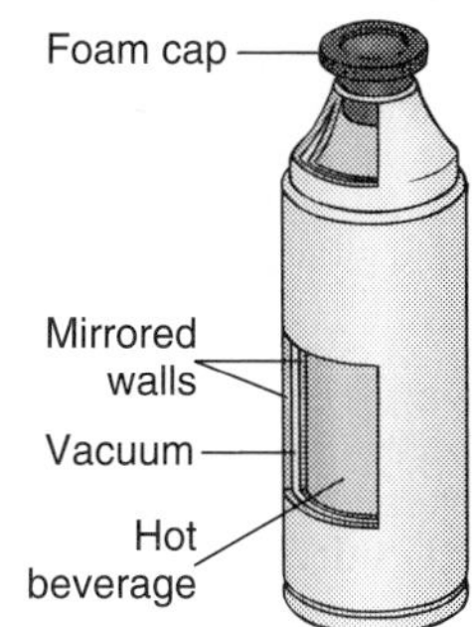

Fig. 6.2.6 - A Dewar flask or Thermos bottle uses a vacuum to insulate its inner volume. The vacuum can't conduct heat or undergo convection and mirrored walls reduce the role of radiative heat transport as well. The only significant heat transfer occurs through the narrow mouth of the vessel.

Cookware itself presents an interesting challenge to manufacturers. An ideal pot should cook food evenly by conducting heat from the burner outside to the food inside. Its surfaces should be non-toxic, non-stick, chemically inert, and resistant to discoloration and abrasion. And its handles should stay cool.

Naturally, one material can't meet all these requirements. At their core, pots are usually made of a good conductor of heat, such as aluminum or copper.

Pure stainless steel is a relatively poor conductor of heat and is not suitable for pots without help from aluminum or copper. To render them non-toxic, non-stick, inert, and resistant to discoloration and abrasion, pots are often coated with stainless-steel, anodized aluminum, or a non-stick plastic such as Teflon or Silverstone. Their handles are typically made out of a heat resistant thermal insulator such as a durable plastic.

CHECK YOUR UNDERSTANDING #7: Keeping a Satellite Cool

How does a satellite in earth orbit eliminate the waste heat produced by its electronic components?

Summary

How Clothing and Insulation Work: Clothing and thermal insulation help you control your temperature and that of your home by interfering with conductive, convective, and radiative heat transfer. Conductive heat transfer is reduced by using materials with low thermal conductivities, typically nonmetallic substances such as glass, plastic, ceramic, and wood. Since air has a very low thermal conductivity, trapped air is a particularly effective insulator. Finely divided insulators, such as hair, feathers, cloth, and foam, trap air to prevent convection and take advantage of its excellent insulating properties. Radiative heat transfer is reduced by using reflective or white materials. These materials are poor absorbers and emitters of thermal radiation and thus don't transfer much heat by radiation.

Important Laws and Equations

1. The Stefan–Boltzmann Law: The power an object radiates is proportional to the product of its emissivity times the fourth power of its temperature times its surface area, or

$$\text{radiated power} = \text{emissivity} \cdot \text{Stefan} - \text{Boltzmann constant} \cdot \text{temperature}^4 \cdot \text{surface area} \quad (6.2.1)$$

The Physics of Clothing and Insulation

1. For an object's temperature to remain constant, its thermal energy must not change. The rate at which it turns ordered energy into thermal energy must be equal to the rate at which it transfers heat elsewhere.

2. The amount of heat transferred through an object increases with that object's surface area and with the temperature difference between its ends. It decreases with the object's increasing thickness.

3. The amount of thermal radiation an object emits is roughly proportional to the fourth power of its temperature, measured on an absolute temperature scale.

4. The amount of heat transferred between two objects by radiation depends roughly on the fourth power of the temperature difference between the two when their temperatures are very different. The heat emitted by a very hot object as thermal radiation can be enormous.

5. An object's capacity to emit or absorb thermal radiation is measured by its emissivity. This emissivity ranges from 1 for a very black object to 0 for a highly reflective or white object.

Check Your Understanding - Answers

1. 500 W.

Why: Since electricity is adding 500 J of thermal energy to the toaster oven each second, without raising its temperature, heat must be flowing out of the appliance at a rate of 500 J/s or 500 W.

2. The plastic tongs are less likely to burn you than the copper tongs.

Why: When you immerse the tongs in hot boiling water, they begin to conduct heat toward your hands. With copper tongs, this heat flow would be so rapid that your skin would overheat and you'd receive a burn. But the plastic tongs conduct heat slowly enough that you can collect the corn without burning your fingers.

3. It allows you to maintain an insulating layer of warm air near your skin.

Why: Even though a thin nylon shell is not much of an insulator itself, it keeps the wind from blowing away the warm air near your skin or clothing and blocks wind chill. It also discourages convection so that there is a relatively motionless layer of air between you and the cold outdoor air.

4. The shiny aluminum is a poor emitter of thermal radiation.

Why: The plastic wrap and the aluminum foil are both too thin for their thermal conductivities to limit heat transfer and they are equally effective at reducing convection. But where aluminum foil really differs from plastic wrap is in its emissivity. Shiny aluminum is a very poor absorber and emitter of infrared light so it greatly reduces the radiative heat loss from the food.

5. The inrushing airstream blows the layer of warmed air off your skin and replaces it with cooler air. You lose heat more rapidly as a result.

Why: The rate at which you lose heat depends on the temperature of the air near your skin. When a breeze blows cool air across your skin, you begin to lose more heat and feel cooler.

6. The glass wool is not the insulator, air is. The more tightly you pack the glass wool, the worse insulator it becomes.

Why: The purpose of the glass wool is to prevent the air from moving. Glass itself is not such a great insulator. If you pack the glass wool too tightly, it will begin to conduct a fair amount of heat from one side to the other. You should only use just enough glass wool to fill the wall and stop convection. Similarly, an overstuffed down coat is not as warm as one that is fluffy and lightly filled. Air is the real insulator in the coat. If the down is crushed, matted, or wet, the coat is no longer very warm.

7. The satellite radiates the waste heat out into space.

Why: Because the satellite is in empty space, it can't eliminate heat through conduction or convection. It must rely on radiation. The waste heat from its electronics, together with heat absorbed from the sun, are emitted as thermal radiation. The satellite's temperature will naturally increase until it emits heat fast enough to maintain a constant temperature.

Check Your Figures - Answers

1. About 262 W.

Why: We can use Eq. 6.2.1 to obtain the radiated power:

$$\text{radiated power} = 0.5 \cdot 5.67 \cdot 10^{-8}\,\text{J}/(\text{s}\cdot\text{m}^2\cdot\text{K}^4)\cdot(310°\ \text{K})^4 \cdot 2\ \text{m}^2$$
$$= 262\ \text{J}/\text{s} = 262\ \text{W}.$$

This power is much more than the thermal power your body produces while you're standing still. You'll get cold quickly.

Glossary

emissivity A surface's capacity to emit or absorb thermal radiation, relative to that of a perfectly black object at the same temperature.

specific heat capacity The measure of how much heat it takes to raise the temperature of a standard amount of particular material by one degree.

Stefan–Boltzmann constant The constant of proportionality relating a surface's radiated power to its emissivity, temperature, and surface area. It has a measured value of $5.67 \cdot 10^{-8}\ \text{J}/(\text{s}\cdot\text{m}^2\cdot\text{K}^4)$

Stefan–Boltzmann law The equation relating a surface's radiated power to its emissivity, temperature, and surface area.

thermal conductivity The measure of a material's capacity to transport heat by conduction from its hotter end to its colder end.

Review Questions

1. Why does your body temperature affect how quickly you can move and think?

2. Even when a tile floor has the same temperature as a carpeted one, the tile floor feels colder to your bare feet. Why?

3. How does having a layer of fat beneath its skin help an animal keep warm?

4. You feel colder when it's windy, even though the temperature of the wind is the same as that of still air. Why?

5. If you put two pots on the stove, one containing 1 kg of water and one containing 1 kg of salad oil, the temperature of the oil will rise faster than that of the water, even though both pots receive similar amounts of heat each second. What characteristic of water and oil identifies this difference?

6. Why does trapped air make an excellent insulator?

7. How does the thermal radiation emitted by a space heater depend on the temperature of its heating element?

8. If you wrap cold food in aluminum foil, that food will stay cold longer. If you wrap hot food, it will stay hot longer. How is the aluminum foil affecting thermal radiation?

9. You feel cold when you stand in front of an open freezer, even if cold air from the freezer doesn't touch you. How does thermal radiation explain this effect?

10. Why is it difficult to keep cool when the environment around you is hotter than you are?

Exercises

1. Down sleeping bag manufacturers often brag about how much volume each gram of their goose down takes up. Why is this characteristic important in determining how warm a 0.5 kg sleeping bag keeps you?

2. Why does a thick, viscous soup retain its heat better than a thin, runny soup?

3. Why does a layer of whipped cream on top of a hot piece of pie help keep that piece of pie warm?

4. Why is it that you can put your hand into a hot oven briefly without getting hurt but will be burned almost immediately if you touch any metal inside the oven?

5. On a cold winter day, your car's metal door handle feels much colder than the glass window, even though both are at the same temperature. Explain.

6. When you step out of a hot shower, the ceramic tile floor of the bathroom feels much colder than the cloth mat nearby, even though they are at the same temperature. Why?

7. Milk is pasturized by raising its temperature briefly to kill harmful microorganisms. Draw a picture of a possible pasturization facility that uses countercurrent exchange to save energy.

8. Why does hot water feel warmer when you move your arm through it than when you keep your arm motionless?

9. Energy efficient homes are often so tightly sealed that the air inside becomes rather stale. How can countercurrent exchange be used to freshen a home's air without exchanging heat with the outside?

10. For dives into extremely cold water, divers replace their wet suits with dry suits. Instead of trapping droplets of water near a diver's skin, a dry suit traps bubbles of air. Why does a dry suit keep the diver warmer than a wet suit?

11. Some modern thermometers can measure your temperature almost instantly by observing thermal radiation emitted by the inside surface of your ear. Explain how this type of measurement can determine your temperature.

12. Polar bears are so enormous that they have no natural predators and no real need for camouflage. One explanation for their white coats is that their hairs slow heat loss via conduction and convection but are transparent enough to allow sunlight to reach their skins and help keep them warm. How does this arrangement resemble windows in a house?

13. When the space shuttle reenters the atmosphere, friction heats its underside to enormous temperatures. That surface is coated with a thick layer of porous ceramic tiles. How do these tiles keep the shuttle itself cool, and why are they porous rather than solid?

14. The outsides of many wooden homes have two layers: the exterior wall and the interior wall. These walls are nailed to the opposite sides of vertical boards (studs) that form the frame of the house. The air space between the studs is filled with glass wool insulation. Why do homes lose more heat through the studs than they do through the glass wool?

15. Why would a candle flame look very different if there were no gravity?

16. Donuts have a hole in the center so that they will cook more evenly. Why wouldn't the center of a complete disk of airy donut batter cook very well?

17. When you bake in an electric oven, heat comes from the red hot element at the bottom of the oven. Why does food in a black metal pan tend to brown more on the bottom than food in a shiny metal pan when you bake in an electric oven?

18. Aluminum heat sinks are attached to large computer chips to help those chips eliminate waste heat. They pass most of this waste heat to the air through rows of metal fins. Many of these heat sinks are painted black. Why?

Problems

1. If a burning log is a black object with a surface area of 0.25 m^2 and a temperature of 800° C, how much power does it emit as thermal radiation?

2. When you blow air on the log in Problem **1**, its temperature rises to 900° C. How much thermal radiation does it emit now? Why did the 100°C rise make so much difference?

3. If the sun has an emissivity of 1 and a surface temperature of 6000° K, how much power does each 1 m^2 of its surface emit as thermal radiation?

4. A dish of hot food has an emissivity of 0.4 and emits 20 W of thermal radiation. If you wrap it in aluminum foil, which has an emissivity of 0.08, how much power will it radiate?

5. A space vehicle uses a large black surface to radiate away waste heat. How does the amount of heat radiated away depend on the area of that radiating surface? If the surface area were doubled, how much more heat would it radiate, if any?

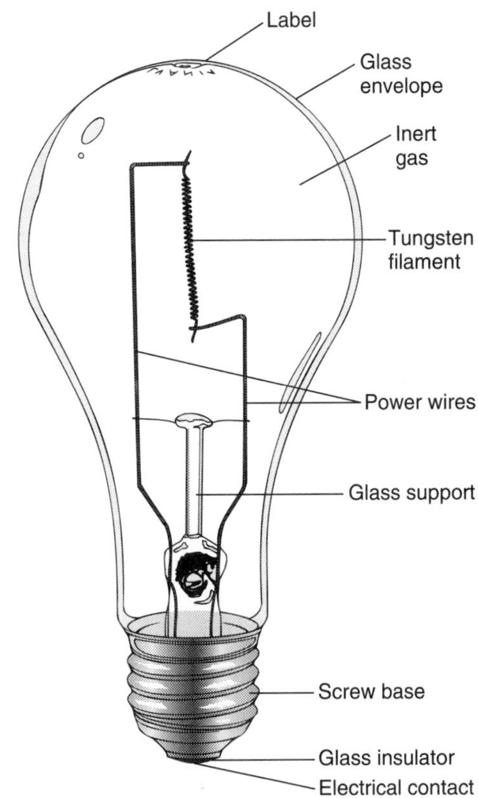

Section 6.3

Incandescent Light Bulbs

For more than a century, incandescent light bulbs have provided light at the flip of a switch. Their invention brought to a close the era of candles and gaslights and spurred the development of electric power. While the variety of incandescent bulbs has grown over the years to include everything from heat lamps to halogen headlights, all incandescent bulbs have at their hearts one simple object: an extremely hot wire filament.

Questions to Think About: *What part of a light bulb emits the light? How is a light bulb similar to a fire or a candle? How do they differ? What colors of light can a plain, unpainted light bulb emit? Why does the top of a light bulb darken with age? What happens when a light bulb burns out?*

Experiments to Do: *Take a look at a few incandescent light bulbs. Try turning one on and off. Are the transitions instantaneous? Stand in a darkened room with your eyes closed and open your eyes suddenly, just after you turn the bulb off. Can you see the bulb going dark? How do its brightness and color change with time?*

Compare the color of the light from a conventional bulb with that from an extended life bulb. Which one produces a better simulation of sunlight? Which bulb should you use in your desk lamp? in an inaccessible ceiling lamp?

Now compare both bulbs with a halogen bulb. How do their colors differ? Which bulb do you expect to live the longest in normal use? Is it surprising that halogen bulbs live longer than conventional bulbs?

Light, Temperature, and Color

The hot wire filament inside an incandescent bulb emits visible light as part of its thermal radiation. While most of the electromagnetic waves that transfer heat between objects are invisible, our eyes are sensitive to a narrow range of waves that we call *visible light*. Any object that is hotter than about 400° C emits enough visible light for us to see it in a dark room. At higher temperatures, that visible light brightens and shifts in color from red to orange to yellow to white. At 500° C, we see the object glowing a dull red. At 1700° C, it emits the orange light of a candle. By 5800° C, it gives off the same brilliant white light as the sun, because the sun's surface temperature is 5800° C.

Fig. 6.3.1 - As you increase the power to an incandescent light bulb, its filament becomes hotter and emits a brighter and whiter light. The cooler filament on the left is dim and red while the hotter one on the right is bright and yellow-white.

To reproduce pure white sunlight, the bulb's filament should also be heated to 5800° C. Unfortunately, nothing is solid at that high temperature. Even tungsten metal, the best filament material known, evaporates quickly at temperatures above 2500° C. Thus incandescent light bulbs operate at lower temperatures and can't really reproduce sunlight. Most give off the warm, orange-white light that is characteristic of tungsten metal at 2500° C.

As you can see, the filament's brightness and color depend on its temperature (Fig. 6.3.1). We can measure brightness as the number of watts of visible light the filament emits. But how do we characterize color and what distinguishes visible light from the invisible forms of electromagnetic radiation? The full answers to these questions will have to wait until Chapters 14 and 15, when we examine electromagnetic radiation and light in detail, but we can make a few essential observations about them now.

Visible light is part of a continuous spectrum of electromagnetic radiation that extends from radio waves at one extreme to gamma rays at the other (Fig. 6.3.2). Different types of electromagnetic radiation are distinguished by their **wavelengths**; that is, the distance between their wave crests. Wavelength is easy to see in the waves on a lake or sea, where the crests are visible and you can directly measure the distance from one to the other. Though electromagnetic waves are not so easy to see, they also have crests and it's possible to measure their spacings.

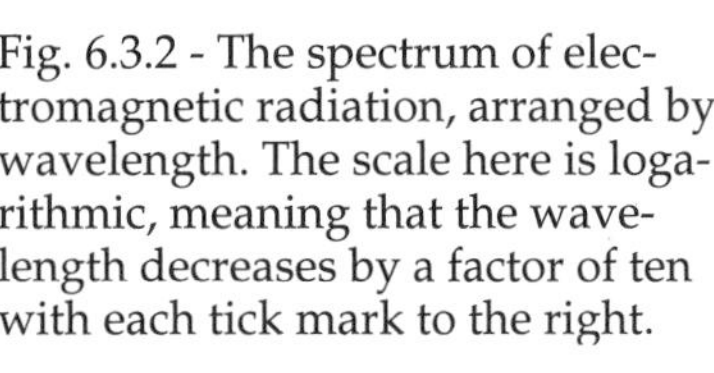

Fig. 6.3.2 - The spectrum of electromagnetic radiation, arranged by wavelength. The scale here is logarithmic, meaning that the wavelength decreases by a factor of ten with each tick mark to the right.

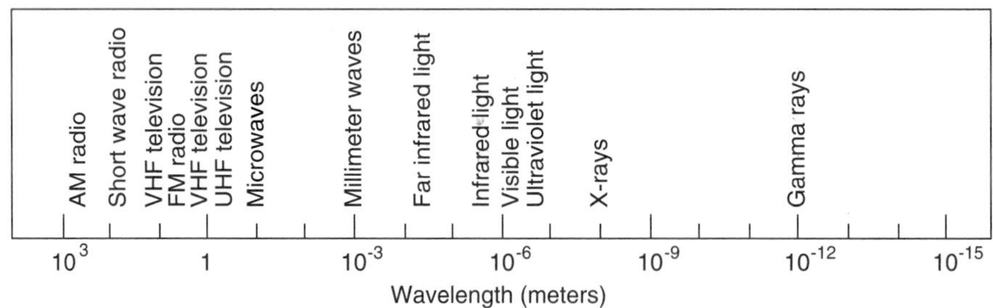

The electromagnetic radiation produced by a hot filament is mostly infrared, visible, and ultraviolet light. While this light is just a tiny portion of the overall electromagnetic spectrum, it's particularly important to our everyday world. Figure 6.3.3 gives an expanded view of the visible portion of the electromagnetic spectrum. Various colors that we see actually correspond to specific wavelength ranges. For example, light with a wavelength of 530 nanometers (billionths of a meter, abbreviated nm) appears green to our eyes.

But the thermal radiation emitted by a filament isn't a single electromagnetic wave with one specific wavelength. Instead, it is many individual waves that cover a broad range of wavelengths. Some of these waves are red light, some green, some blue, and some are invisible.

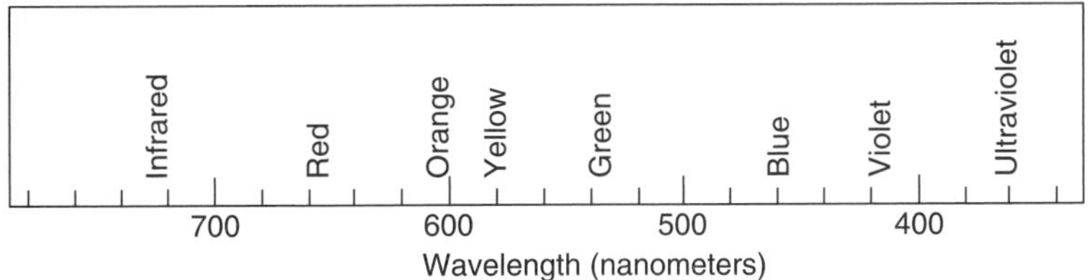

Fig. 6.3.3 - A portion of the electromagnetic radiation spectrum around visible light. Wavelengths are measured in nanometers (nm or billionths of a meter).

The distribution of wavelengths emitted by the filament depends on its temperature and surface properties, particularly its emissivity. Since the filament is essentially black, it emits and absorbs light efficiently and its emissivity is nearly 1. The distribution of wavelengths emitted by a black object is determined by its temperature alone and is called a **black body spectrum**. As you can see from the examples in Fig. 6.3.4, the spectrum of a black body brightens and shifts toward shorter wavelength as its temperature increases. An object that is not black emits somewhat less thermal radiation, but that radiation still brightens and shifts toward shorter wavelength as the object becomes hotter.

Our eyes make an average assessment of the distribution of wavelengths emitted by a black object and we observe reddish, orangish, yellowish, whitish, or bluish light, depending on the object's temperature (Table 6.3.1). The temperature associated with a particular distribution of wavelengths is the **color temperature** of that light.

We can already see two of the principal shortcomings of incandescent light bulbs: their poor efficiency at converting electric energy into visible light and their low color temperature. At 2500° C, only about 5% of the thermal radiation they emit is visible light. The rest is invisible infrared light. They would have to

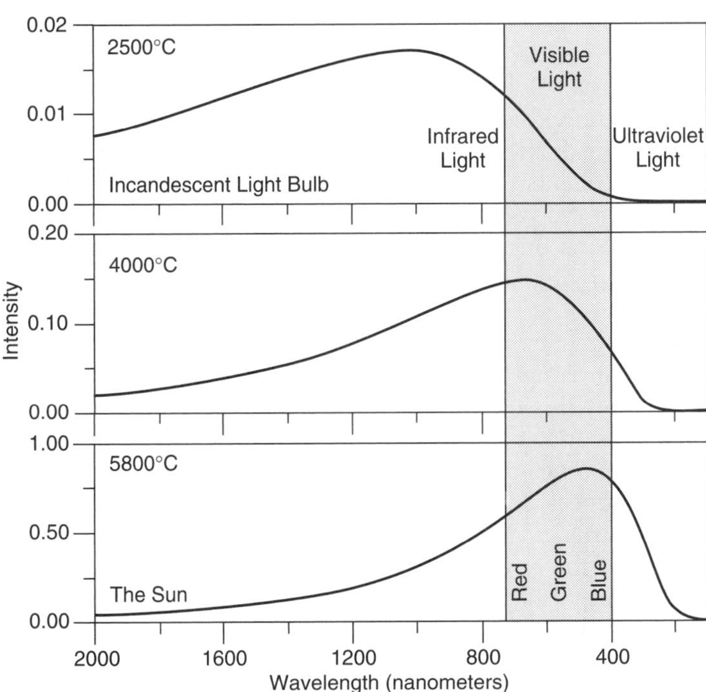

Fig. 6.3.4 - The distributions of light emitted by black objects at 2500° C (top), 4000° C (middle), and 5800° C (bottom). In addition to containing a larger fraction of visible light, the 5800° C object is much brighter than the 2500° C object (note the different intensity scales).

Object	*Temperature*	*Color*
Heat Lamp	500° C	Dull Red
Candle Flame	1700° C	Dim Orange
Bulb Filament	2500° C	Bright Yellow-White
Sun's Surface	5800° C	Brilliant White
Blue Star	>6000° C	Dazzling Blue-White

Table 6.3.1 - The temperatures and colors of light emitted by hot objects.

reach 5000° C before most of their thermal radiation would be visible. Moreover, their 2500° C color temperature makes them look yellowish when compared to sunlight because they don't emit enough blue light. Most of the developments in lighting over the past half century have focused on improving energy efficiency and color temperature.

CHECK YOUR UNDERSTANDING #1:

Satellites can measure the temperatures of agricultural regions from space, looking for signs of crop distress and disease. How do they make such measurements?

The Filament

The bulb's filament is heated by a current of electrically charged particles. These particles flow through the filament, where most of their electric energy is converted into thermal energy. The filament's temperature rises until it radiates away thermal energy as quickly as the electricity produces more. But while the concept of an incandescent bulb is simple, finding a material that can tolerate extremely high temperatures is not.

Early filaments were made of carbon and platinum. Of these materials, carbon showed the most promise. In 1879, Thomas Edison developed an incandescent lamp with a carbon filament that survived for several hundred hours. His was not the first incandescent bulb but rather the first practical incandescent bulb.

However, while carbon has the highest melting temperature of any element (3550° C), it evaporates atoms directly from the solid at much lower temperatures. This process by which a solid turns directly into a gas is called **sublimation** and occurs because individual atoms can occasionally gather together enough thermal energy to break free from the material. Because of carbon's tendency to sublime, a carbon filament that is heated close to its melting temperature quickly disappears as a gas. When a gap appears in the filament, it stops carrying electricity and "burns out." Furthermore, carbon is flammable, so it must be enclosed in an air-tight glass bulb that contains either inert gases or a vacuum.

A better choice for filaments, now used in virtually all modern incandescent bulbs, is tungsten metal. Tungsten melts at 3410° C and sublimes only very slowly at temperatures below this melting temperature. Tungsten filaments can be heated to higher temperatures than carbon filaments before the rate of sublimation becomes intolerable. Since a tungsten filament bulb runs hotter than a carbon filament bulb, it produces a richer, whiter light; more like that of the sun. Like carbon, hot tungsten burns in air and must be protected in a glass bulb.

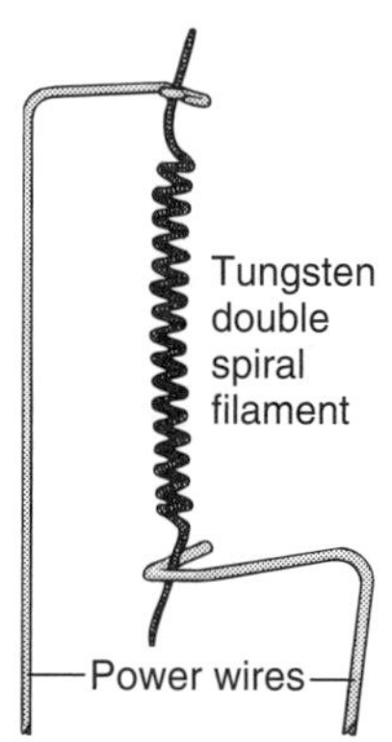

Fig. 6.3.5 - The tungsten filament of a modern incandescent light bulb is a double spiral; a coil wound from a smaller coil of extremely fine tungsten wire. The double spiral allows the manufacturers to put a long length of wire in a small space.

To ensure that most of the electric energy passing through the filament is converted into thermal energy, the filament must be long and thin. A typical 60 W light bulb has about 2 m of 25 microns (0.001 inches) tungsten wire, coiled up into a filament only about 2 cm long. To minimize the filament's length, it's wound into a double spiral. First it's wound into a thin spring-like coil about 0.25 mm wide. This coil is itself wound into a spring-like coil to form the actual filament (Fig. 6.3.5). Fabricating such a complicated tungsten filament is so difficult that it was not accomplished until 1937.

To keep the white hot filament from burning or subliming, it's surrounded by inert gas in a glass bulb. The inert gas, which is mostly argon and contains no oxygen, slows sublimation by bouncing some of the escaping tungsten atoms back onto the filament. While this inert gas extends the filament's life, it wastes energy by permitting conduction and convection to carry heat away from the filament. Tiny tungsten particles that form in this inert gas also rise with convection currents to produce a dark spot on the top of the bulb.

Krypton-filled bulbs transfer less heat from the filament to the envelope because krypton gas is a poor conductor of heat. Such bulbs are more energy efficient than conventional argon-filled bulbs, but they are also more expensive. Krypton is a rare inert gas that is present in the atmosphere at the level of about one part in a million. Krypton bulbs are used in flashlights where the energy efficiency of the lamp is critical to battery life.

CHECK YOUR UNDERSTANDING #2: Here Today, Gone Tomorrow

Snow often disappears from the ground of a period of weeks, even when the temperature remains below freezing. How does the snow disappear?

Extended Life, Halogen, and Three-Way Bulbs

Another way to increase the life of the filament is to make it longer than normal. This change spreads the thermal energy out more so that the filament doesn't get quite as hot and doesn't sublime as quickly. The result is an extended life bulb. Unfortunately, such a bulb is dimmer and redder than a conventional bulb and also less energy efficient. Because an extended life bulb produces less visible light than a conventional bulb, the extended life bulb must have a higher wattage to give equal lighting. Thus extended life bulbs aren't always a bargain. The money you save by not having to replace the bulb so often may well be spent on increased energy costs.

If you make the filament even longer, its temperature will be so low that the bulb will emit the dull, red glow of a heat lamp. Most of the light energy will be in the infrared range, too red for our eyes to see. Nonetheless, you can feel this radiation as heat. Because the filament in a heat lamp runs at a low temperature, it lasts indefinitely.

On the other hand, some bulbs use shorter filaments that operate very near the melting temperature of tungsten. Photoflood lamps used in photography attempt to mimic sunlight by using a very hot filament and a blue-tinted glass housing. While they live only a few hours, these bulbs are able to reach a color temperature as high as 4500° C. The filament itself is not actually that hot but the blue glass allows the lamp to imitate the light of a black object at 4500° C. This high color temperature is important for color photography because normal incandescent illumination gives pictures an orangish appearance.

A halogen bulb also produces whiter light than a normal incandescent bulb, but without the short life of a photoflood. The halogen bulb uses a chemical trick to rebuild its filament continuously during operation. The filament is enclosed in a small tube of quartz or aluminosilicate glass, which can tolerate high temperatures and reactive chemicals (Fig. 6.3.6). This tube contains molecules of the halogen element bromine, or sometimes iodine. During operation, the small tube becomes extremely hot and the halogen reacts with any tungsten atoms on the inside surface of the tube. They form a gas of tungsten-halogen molecules that drift around the tube. When they encounter the white hot filament, these molecules are torn apart and the tungsten atoms stick to the filament.

Fig. 6.3.6 - These halogen bulbs operate at higher temperatures than normal incandescent bulbs and produce whiter light. The large glass envelope of the upper bulb protects a smaller lamp inside.

The halogens act as scavengers, seeking out stray tungsten atoms and returning them to the filament. Although the tungsten filament continues to sublime, the halogens keep bringing the tungsten atoms back. Unfortunately, this rebuilding process slowly changes the structure of the filament. The returning tungsten atoms deposit unevenly on the filament, so that it gradually develops

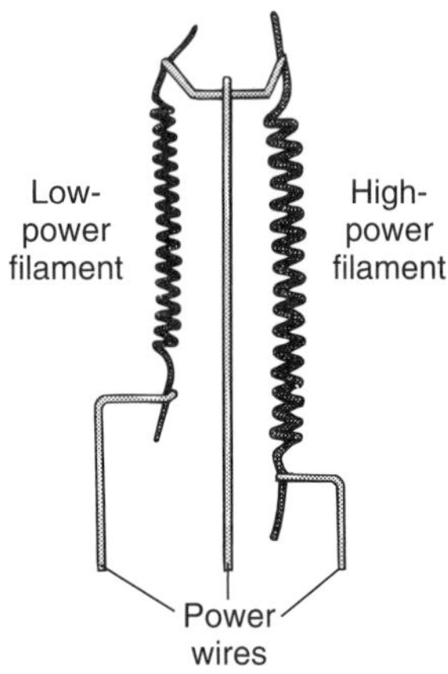

Fig. 6.3.7 - A three-way bulb has two independent filaments. The filament on the left is shorter and thinner than the one on the right and emits about half as much light. The three different light levels correspond to having the left filament on, the right filament on, and both filaments on.

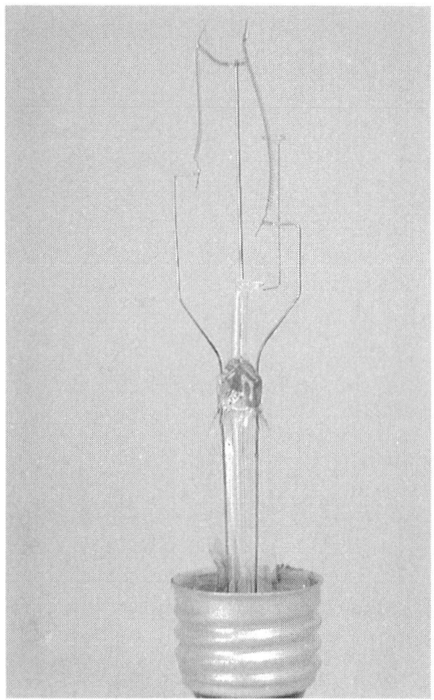

Fig. 6.3.8 - The glass envelope of this three-way light bulb has been removed to expose its two filaments. The shorter, thinner filament (left) uses 50 W, while the longer, thicker one (right) uses 100 W.

thin spots and eventually burns out. Nonetheless, the filament lasts so much longer than in a conventional bulb that it can run several hundred degrees hotter and still have a life of more than 2000 hours. This higher filament temperature allows halogen lamps to produce a whiter light than conventional bulbs with an increased energy efficiency.

Halogen bulbs do have some drawbacks. First, the quartz glass tube is small, extremely hot, and filled with toxic halogen gas. It's a fire and safety hazard, which is why it's often protected by an outer glass tube. Second, the quartz tube is sensitive to fingerprints, which discolor and damage it when it becomes hot. Third, the tungsten recycling system only functions at high temperatures. When a halogen lamp operates on a dimmer at less than full power, tungsten can accumulate on the quartz tube, darkening the bulb and reducing its life. The bulb needs to operate at full power periodically to clean off the quartz tube and return the tungsten to the filament.

Bulbs with different rated powers have different sized filaments. The filament of a 100 W bulb is four times as large as that of a 25 W bulb and emits four times as much light. Unlike the elongated filament of a cool-running extended life bulb, the filament of the 100 W bulb is both longer *and thicker* than that of the 25 W bulb. The 100 W filament draws four times as much electric power as the 25 W filament so that they both operate at the same temperature and emit light with the same color temperature.

One way to make an incandescent bulb with a variable light output is to divide the filament into several parts that can be turned on or off independently. A "three-way bulb" has two independent filaments (Figs. 6.3.7 and 6.3.8). In a typical three-way bulb, a 50-100-150 W bulb, one filament uses 50 W of electric power and the other filament uses 100 W. They are wired so that electric currents can flow through either filament separately or through both filaments simultaneously. If only the low power filament is on, the bulb appears to be a 50 W bulb. If only the high power filament is on, it appears to be a 100 W bulb. If both filaments are on, the bulb appears to be a 150 W bulb. Since the two filaments can burn out separately, the bulb fails by going from having three light levels to only one.

CHECK YOUR UNDERSTANDING #3: Double Wrapping

If you look inside a car's halogen headlight, you will see another small, clear bulb. Why does the manufacturer go to the trouble of putting two separate glass bulbs around the filament?

Summary

How Incandescent Light Bulbs Work: An incandescent light bulb produces light as thermal radiation from an extremely hot tungsten filament. The filament is heated by a steady stream of electrically charged particles passing through it. To prevent the tungsten filament from burning up, it's enclosed in a glass bulb filled with an inert gas.

The spectrum of light emitted by the filament depends on its temperature. The hotter the filament, the whiter its light. Atoms in the filament sublime during operation and the filament slowly disappears. Eventually it becomes so thin that it breaks. The tungsten filament of a normal bulb operates at 2500° C because it would burn out too quickly at higher temperatures. Reducing the operating temperature, as is done in an extended life bulb, prolongs the filament's life at the expense of color temperature and energy efficiency. In contrast, adding halogen gas actually increases the filament's life and improves both the color temperature and energy efficiency.

The Physics of Incandescent Light Bulbs

1. All objects emit thermal electromagnetic radiation. Electromagnetic radiation is characterized by its wavelength, which is what distinguishes between infrared, visible, and ultraviolet light.

2. The distribution of wavelengths in the thermal radiation emitted by a black object, an object with an emissivity of 1, is determined only by its temperature. As its temperature increases, its thermal radiation brightens and shifts toward shorter wavelengths.

3. An object that is not black has an emissivity of less than 1 and emits less thermal radiation than it would if it were black. However, that thermal radiation still brightens and shifts toward shorter wavelengths as the object heats up.

4. When the distribution of wavelengths emitted by a light source is equal to that emitted by a black object at a particular temperature, the black object's temperature is the source's color temperature.

5. The hotter the object, the more visible and ultraviolet light it emits. Below about 5000° C, a black object emits most of its thermal energy as infrared light.

6. Many solids can sublime directly into gases at temperatures near their melting points. This process occurs because individual atoms or molecules occasionally accumulate enough thermal energy to break free from their neighbors and leave the material.

Check Your Understanding - Answers

1. These satellites can measure the wavelength distributions of thermal radiation emitted by various patches of land and determine their temperatures.

Why: Even objects that are near room temperature emit thermal radiation, although this radiation is entirely in the infrared. While the land is not really black, the infrared light it emits is still an accurate indication of its temperature. A satellite can sense the exact form of the distribution and determine the temperature with great accuracy.

2. The snow sublimes to form water vapor in the air.

Why: Ice sublimes quickly at temperatures near its melting temperature, going from a solid to a gas without ever become liquid water.

3. The inner bulb aids the tungsten recycling process while the outer bulb directs the light and protects the inner bulb.

Why: The inner bulb must get hot enough for the halogen gas to react with and recycle the tungsten atoms on its surface. The outer bulb projects the light forward and ensures that nothing touches the hot inner bulb.

Glossary

black body spectrum The distribution of thermal electromagnetic radiation emitted by a black object. This distribution is the amount of radiation emitted at each wavelength and depends only on the temperature of the black object.

color temperature The temperature at which a black object will emit thermal electromagnetic radiation with a particular distribution of wavelengths.

sublimation The process by which atoms or molecules go directly from a solid to a gas.

wavelength A structural characteristic of a wave, corresponding to the distance separating adjacent peaks or troughs.

Review Questions

1. Which is hotter: a black body that is glowing red or one that is glowing blue? Why?

2. If an incandescent bulb only converts a few percent of the electric energy it receives into visible light, what becomes of the rest of this energy?

3. How can a hot filament gradually disappear without the filament ever melting?

4. Why do halogen car headlights produce brighter, whiter light than conventional headlights?

5. Why are extended life bulbs less energy efficient than conventional bulbs?

6. Why does a typical lamp bulb have to be replaced every year or two?

7. Why is there a glass bulb around the tungsten filament of an incandescent light bulb?

Exercises

1. Mothballs are made from a white solid (naphthalene) that has a strong odor. If you leave a mothball out, it slowly disappears. What happens to the mothball?

2. You have a table lamp with a dimmer switch. The dimmer allows you to adjust the temperature and brightness of the lamp's incandescent bulb from very dim red up to brilliant yellow-white. How does the bulb's energy efficiency, the amount of visible light it produces per unit of power consumed, depend on the dimmer's setting?

3. A biologist is studying nocturnal animals with a camera that records infrared light. What differences in this infrared light can the camera look for in order to distinguish the warmer animals from their cooler surroundings?

4. How would you estimate the temperature of a glowing coal in a fireplace?

5. The strongest evidence for the Big Bang theory of the origin of the universe is the thermal radiation emitted by that explosion. This radiation has cooled over the years to only 3° K and is now mostly microwaves. Why should 3° K thermal radiation be microwaves?

6. The heating element of a toaster glows red when it's on. About how hot is that element?

7. A metallurgist measures the temperature of hot metal by comparing its thermal radiation with that from a test object at a known temperature. Explain why this technique works.

8. A light bulb burns out when a small portion of its thinning filament overheats and vaporizes. Why is this event accompanied by a bright flash of blue-white light?

9. Frozen vegetables will "freeze dry" if they're left in cold, dry air. How can water molecules leave the frozen vegetables?

10. Astronomers can tell the surface temperature of a distant star without visiting it. How is this done?

11. When you operate a 50-100-150 W three-way bulb at its 50 W setting, it emits yellow-white light. When you use a dimmer to operate a regular 150 W bulb on only 50 W of electric power, it emits orangish light. Explain the difference.

12. A doctor can study a patient's circulation by imaging the infrared light emitted by the patient's skin. Tissue with poor blood flow is relatively cool. What changes in the infrared emissions would indicate such a cool spot?

13. The 100 W filament in a three-way bulb has twice the surface area of the 50 W filament. Both filaments operate at the same temperature. Why does the 100 W filament emit twice as much thermal radiation as the 50 W filament?

14. Which device delivers more heat per second into a windowless room: a 100 W space heater or a 100 W incandescent light bulb?

15. The chemical reactions that allow a halogen bulb to rebuild its filament only occur at high temperatures. Use the concept of activation energy to explain this behavior.

16. The filament of an incandescent bulb is quite small, yet it emits as much as 100 W of thermal radiation. That's almost as much as your whole body emits. What accounts for the strength of the filament's thermal radiation?

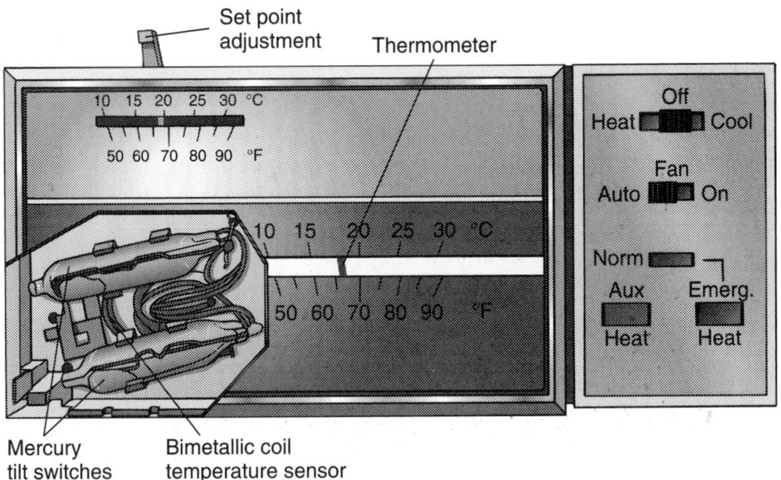

Section 6.4

Thermometers and Thermostats

Knowing temperature is important when you are going on a picnic, baking bread, or lying in bed with the flu. You use a thermometer to measure how hot it is outside, in the oven, or on your forehead. A thermometer is able to measure temperature because the characteristics of its components change with temperature. In this section, we will examine some of those changes to see how thermometers work.

There are also times when you want to control temperature. You don't just want to know how hot your house is, you want to maintain it at a certain level. In such cases, you need a thermostat, a thermometer that uses its temperature measurements to control other equipment.

Questions to Think About: *Where does the extra liquid come from when the red liquid in a glass thermometer rises with the temperature? What kind of mechanism inside a meat or candy thermometer can cause its needle to turn? On a plastic strip thermometer, the kind with the color-changing numbers printed on its surface, what happens to the numbers that are not visible? Why does it matter that the thermostat in your home is exactly level?*

Experiments to Do: *Find a thermometer and measure the temperatures of a few different objects. Watch how the thermometer changes as you move it from a hot object to a cold object. Can you see its parts moving?*

Observe the thermometer's rate of response. Does it read the temperature of a new object immediately or is there a delay? What would cause such a delay? Is the thermometer really measuring the temperature of the object or is it measuring its own temperature? Is there a difference?

Your home is probably full of "accidental" thermometers: windows or doors that stick in hot or cold weather, metal stripping that buckles in the heat, and floors that creak and pop at night as the temperature drops. Look around and see what you can find.

Liquid Thermometers: Thermal Expansion

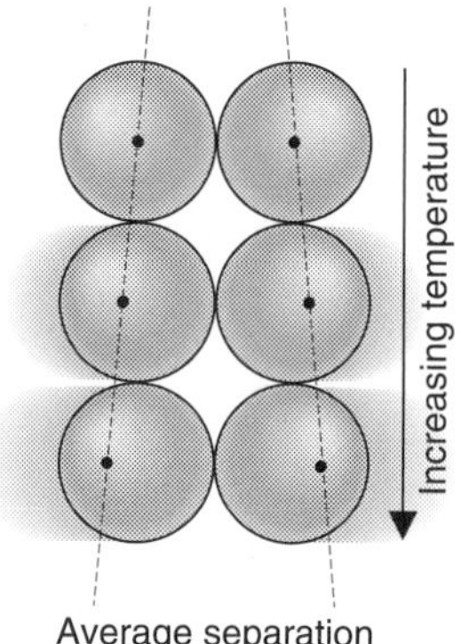

Fig. 6.4.1 - The internal kinetic energy of a solid increases with temperature, causing its atoms to bounce against one another more vigorously. As they vibrate, the atoms repel more strongly than they attract so their average separation increases slightly.

When the amount of thermal energy in an object changes, so does its temperature. But temperature isn't the only characteristic of the object that is sensitive to thermal energy. Materials change in many ways as you warm them or cool them and thermometers and thermostats are based on those changes.

One aspect of an object that changes with temperature is its volume. In most cases, the object expands uniformly as its temperature increases and contracts uniformly as its temperature decreases. Just how much an object's volume increases with temperature depends on what it's made of. There are a few special materials that contract when heated or that experience nonuniform expansions or contractions, but we won't consider those exceptional materials here.

These volume changes occur because thermal energy affects the average spacings of the object's atoms. In a solid or liquid, the atoms are touching and experience forces that push them toward their equilibrium separations. At absolute zero, the object's atoms would settle down at their equilibrium separations and form a neatly packed array that would give the object its minimal volume.

But absolute zero can't be reached because it's impossible to remove all of the thermal energy from an object. Since the object contains at least some thermal energy, its atoms vibrate back and forth about their equilibrium separations (Fig. 6.4.1). However, this vibrational motion is not symmetric. The repulsive force the atoms experience when they're too close together is stiffer than the attractive force they experience when they are too far apart. As a result, they push apart more quickly than they draw together and spend most of their time at more than their equilibrium separation. On average, their actual separation is thus more than their equilibrium separation and the object is bigger than it would be at absolute zero.

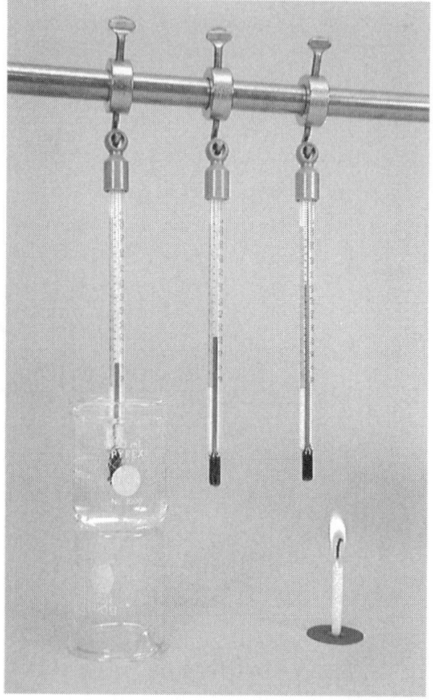

Fig. 6.4.2 - As their temperatures increase—from left to right—both the glass and the liquid in these thermometers expands. However the liquid expands more and flows up the tube inside the thermometer.

Increasing the object's temperature moves the atoms even farther apart and the object grows larger in all directions. But that is not the whole story. If it were, then a glass thermometer would simply become bigger as its temperature increased and the red or silver liquid would not move up the column. Fortunately, there is another complication. Different materials expand by different amounts as their temperatures increase. For example, the liquid inside the glass thermometer expands much more than the glass tube around it so the liquid flows up the column (Fig. 6.4.2).

The extent to which an object expands with increasing temperature is normally described by its **coefficient of volume expansion**: the fractional change in an object's volume caused by a temperature increase of 1° C. Fractional change in volume is the net change in volume divided by the total volume. Since most materials expand only a small amount when heated 1° C, coefficients of volume expansion are small, typically about 10^{-5} for metals, about 10^{-6} for special low-expansion glasses, and about 10^{-4} for liquids such as alcohol.

Now we can see how a glass thermometer works. Its hollow body contains just enough colored alcohol or mercury to completely fill the cavity in its bulb, plus a little extra (Fig. 6.4.3). The extra liquid projects part way up a fine, hollow capillary connected to the bulb. As the thermometer's temperature increases, both the liquid and the glass enclosure expand but the liquid expands more than the glass. Although the cavity in the bulb becomes slightly larger and can accommodate slightly more liquid, the liquid's expansion squeezes some of it out of the bulb and into the capillary. The column of liquid in the capillary becomes longer and you see a longer red or silver bar.

This mechanism for temperature measurement, with a liquid flowing out of a container as the temperature increases, is also found in thermostats. A typical oven thermostat uses a liquid-filled metal bulb to sense the temperature inside the oven. As the temperature increases, the liquid expands more than the metal and squeezes through a thin tube to the oven's control unit. There it pushes against a metal membrane, operating a switch that controls the flow of electricity or natural gas to the oven's burner. If the oven temperature rises above the desired value, the switch turns the burner off. The switch turns the burner back on when the oven temperature falls below the desired value. In this manner, the thermostat regulates the oven temperature.

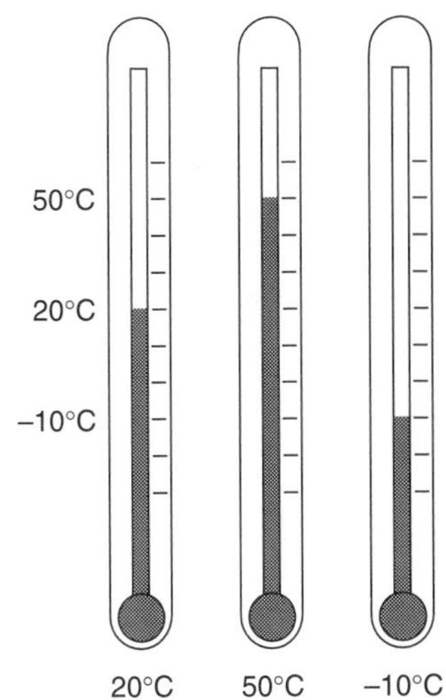

Fig. 6.4.3 - The liquid inside a glass thermometer has a larger coefficient of volume expansion than the glass. As the temperature increases, excess liquid is driven out of the cavity at the base of the thermometer and the thin column of liquid in the capillary rises. As the temperature decreases, liquid returns to the cavity and the column falls.

CHECK YOUR UNDERSTANDING #1: Overcooked

If you place a pot or saucepan on the stove and fill it to the brim with cold water, it will overflow as you heat it. Where does the extra water come from?

Metal Thermometers: Bimetallic Strips

Not all thermometers are based on expanding liquids. There are many other thermometers that use metal strips to measure temperature. However, because solids have much smaller coefficients of volume expansion than liquids, it's difficult to make sensitive metal thermometers. This distinction between liquids and solids is caused by fundamental differences in their microscopic structures. Let's take a moment to look at those structures.

The atoms and molecules of most solids are held together rigidly and form orderly structures called **crystals** (Fig. 6.4.4). Crystals are familiar to most of us as the beautiful, faceted minerals in geology exhibits and museum gift shops. The natural faceting that appears on these crystals reflects the extraordinary order present in *crystalline solids* at the atomic and molecular scale. The atoms and molecules in a crystal arrange themselves in nearly perfect **lattices**, repetitive and uniform arrangements that resemble stacks of oranges or food containers at the grocery store. *Crystalline order* is not unique to faceted minerals. Most solids, including metals, are actually *crystalline* because their particles are arranged in regular lattices.

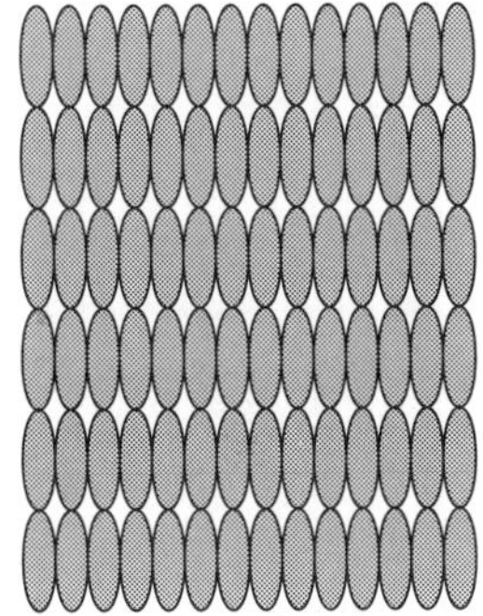
Fig. 6.4.4 - In a crystalline solid, the atoms or molecules are arranged in a highly ordered lattice.

In contrast, normal liquids are not crystalline (Fig. 6.4.5). Because the particles in a liquid don't form an orderly lattice, they have much more freedom in arranging themselves than they would have in a solid. This increased freedom is what gives liquids their large coefficients of volume expansion, as well as large specific heat capacities.

To understand how increased freedom affects specific heat capacity, we need to know what happens to a liquid when you add thermal energy to it. While some of the new thermal energy goes into making the particles vibrate harder, raising the liquid's temperature, a large fraction of it goes instead into breaking chemical bonds and separating the particles from one another. This alternative use of thermal energy increases the amount of heat you must add to the liquid to raise its temperature.

In a solid, added heat goes only into making its particles vibrate harder so that its temperature increases quickly and easily. Because it takes more heat to warm 1 kg of a typical liquid by one degree than to do the same for 1 kg of a typical solid, the liquid's specific heat capacity is larger than that of the solid.

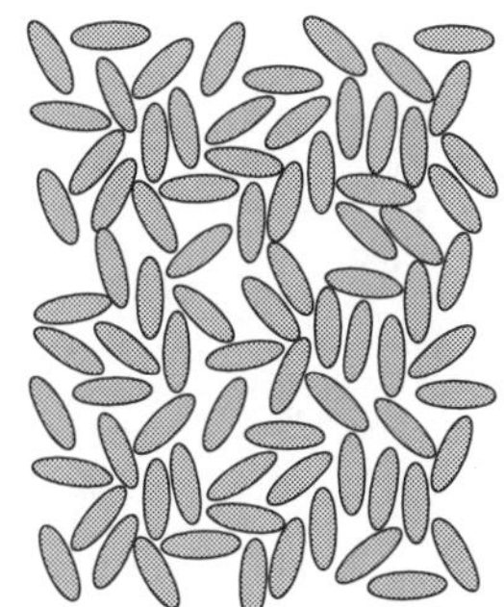
Fig. 6.4.5 - In a normal liquid, the atoms or molecules are disordered. They touch one another but don't form an ordered lattice.

There are two important consequences to the breaking of bonds in a liquid when you heat it. First, the liquid's viscosity decreases and it flows more easily.

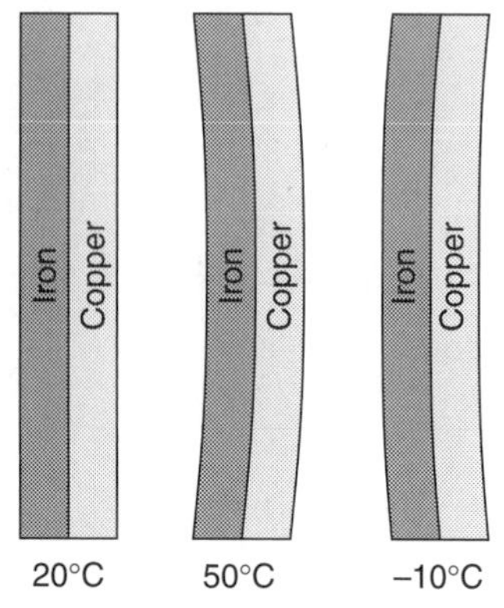

Fig. 6.4.6 - A copper and an iron strip are bonded together to form a bimetallic strip. Since copper has a larger coefficient of volume expansion than iron, the strip curls in response to changes in temperature. The strip's shape is an accurate indication of its temperature.

Fig. 6.4.7 - This refrigerator thermometer uses a coiled bimetallic strip to measure temperature. The coil unwinds as it cools.

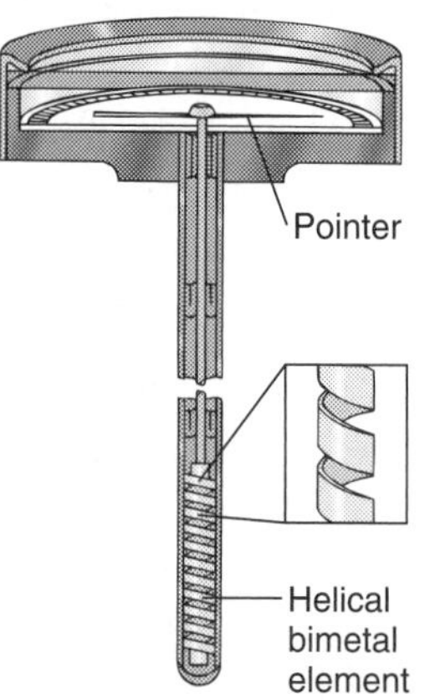

Fig. 6.4.8 - A kitchen thermometer is based on a bimetallic strip that has been wound into a helix. As the temperature changes, the helical element coils or uncoils and turns the pointer.

That's why pancake syrup is easier to pour if you heat it first. Second, the liquid's volume increases because the liquid's particles become even more loosely packed and take up additional space. This volume increase is larger than in a solid, where the particles remain in their lattice during heating, so the liquid expands more than the solid. Thus liquids generally have larger coefficients of volume expansion than solids.

Despite their small coefficients of volume expansion, metals are often used in thermometers. The most common type of metal thermometer is based on a *bimetallic strip*. In this design, narrow sheets of two different metals (often copper and iron) are permanently bonded together to make a thin metal sandwich (Fig. 6.4.6). Since these two metals have different coefficients of volume expansion, the bimetallic strip deforms as its temperature changes. Copper has the larger coefficient of volume expansion, so the strip's copper layer expands more when the strip is heated and shrinks more when the strip is cooled.

As the two layers of the bimetallic strip expand or shrink, the strip curls to one side or the other. There is only one temperature at which the strip is straight (Fig. 6.4.6). Above that temperature, the strip curls so that the longer copper layer is outside the iron and below that temperature, the strip curls so that the shorter copper layer is inside the iron. Since the strip's shape depends on its current temperature, it makes a fine thermometer.

Most dial thermometers, including meat and candy thermometers, are based on bimetallic strips. To increase the sensitivities of these thermometers, their bimetallic strips are wound into small coils (Fig. 6.4.7) or spirals (Fig. 6.4.8) that curl or uncurl with temperature. One end of the bimetallic coil is fixed to the thermometer's frame while the other end is attached to the pointer. As the thermometer's temperature changes, the curling bimetallic coil moves the pointer to indicate the current temperature.

Bimetallic coils are used in many home heating thermostats. Switches attached to the bimetallic coil in such a thermostat control the furnace. When the temperature becomes too high, the coil and switch turn the furnace off. When the temperature becomes too low, the coil and switch turn the furnace back on.

The furnace switch is usually a glass tube that is partially filled with liquid mercury metal (Fig. 6.4.9). This tube is attached to the movable end of the bimetallic coil; the end that normally turns the pointer of a dial thermometer. As the coil winds or unwinds, it tips the mercury from one end of the glass tube to the other. Two electric contacts are embedded in the wall of the tube. When the mercury is at one side of the tube, it connects these two contacts so that electricity can flow from one to the other. When the mercury tips to the other side of the tube, the contacts are not connected and no electricity flows. This mercury tilt switch allows the shape of the bimetallic coil to control the furnace.

When you set the temperature of the thermostat, you are actually changing the orientation of the bimetallic coil. This coil is attached to the temperature control knob so that turning that knob tilts both the coil and the mercury switch. When you turn up the temperature, you are tilting the coil so that it won't operate the mercury switch until the room becomes hotter. When you turn down the temperature, you are tilting the coil the other way so that it operates the switch at a relatively low temperature. Once the room reaches the desired temperature, the thermostat turns the furnace on and off as needed to maintain a nearly constant temperature.

Bimetallic strip thermostats are also used in clothes irons, toasters, and portable space heaters, where they directly control the flow of electricity through heating elements. As its temperature drops, the bimetallic strip in one of these simple thermostats bends until it touches a second piece of metal. Once contact is made, electricity flows from the strip to its contact and then through the heating element. That way, whenever the thermostat becomes too cold, it turns on the heating element. (For an interesting application of bimetallic strips, see ❒ at the top of the following page.)

Direct contact thermostats aren't as sensitive, precise, or durable as those that use mercury tilt switches, but they operate well in any orientation. Tilt switch thermostats are sensitive to orientation because they use gravity to move the mercury around. To keep them operating at the right temperatures, tilt switch thermostats must be permanently mounted so that they always remain level.

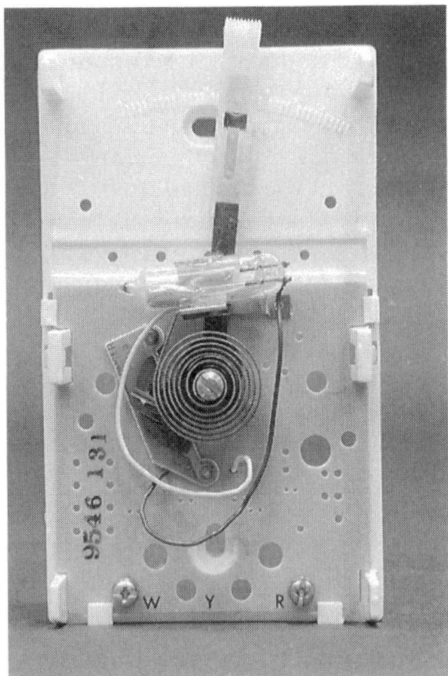

Fig. 6.4.9 - The coiled bimetallic strip in this home thermostat controls the orientation of a mercury tilt switch. As the ball of liquid mercury tips from one end of the glass switch to the other, it turns the furnace on or off.

CHECK YOUR UNDERSTANDING #2: Not So Flat Bottomed Pots

The bottoms of some stainless steel pots are copper coated so that they distribute heat more evenly to the food inside. When these pots are heated, they bow outward slightly and will rock when placed on a flat surface. What causes this bowing?

Plastic Strip Thermometers: Liquid Crystals

The plastic strip thermometer is a relatively recent development. This thermometer is a thin flexible strip that displays the current temperature as a brightly colored number. The strip has a whole range of temperatures printed on it, but only one number is easily visible at any given temperature. As the temperature changes, that number vanishes into the background and another number becomes visible. A plastic strip thermometer is particularly useful for measuring the temperature of a surface, such as the glass wall of a tropical fish aquarium or a child's fevered brow.

The strip is not really a solid piece of plastic. It's a multi-layer sandwich containing a remarkable material phase called a liquid crystal. A liquid crystal has properties intermediate between a solid and a liquid. Behind each number on the strip is its own drop of liquid crystal that reflects colored light only at the temperature associated with that number.

To understand liquid crystals, we must return again to the microscopic structures of solids and liquids. Crystalline solids are highly ordered materials. The spacings between particles in a crystal are so regular that, once you know exactly where a few of the particles are located, you also know exactly where millions of other nearby particles are located. This regularity is called **positional order**. This positional order is visible in Fig. 6.4.4. The particles in a crystal are also highly oriented, so that if you know the orientations of a few particles, you also know exactly how millions of other nearby particles are oriented. This second regularity is called **orientational order**. In contrast, normal liquids don't have positional or orientational order (see Fig. 6.4.5). Knowing the positions and orientations of a few particles in a liquid tells you little about the positions and orientations of nearby particles.

Liquid crystals lie in between solids and liquids. Like normal liquids, liquid crystals have little positional order. Knowing where some of the particles in the liquid crystal are won't help you locate other nearby particles. But liquid crystals do have substantial orientational order. Liquid crystals are composed of rod-like or disk-like molecules that align themselves with one another, even though their positions are free to change (Fig. 6.4.10). These molecules move about like those in a normal liquid but they remain highly oriented, like those in a crystalline solid. Hence the name "liquid crystal."

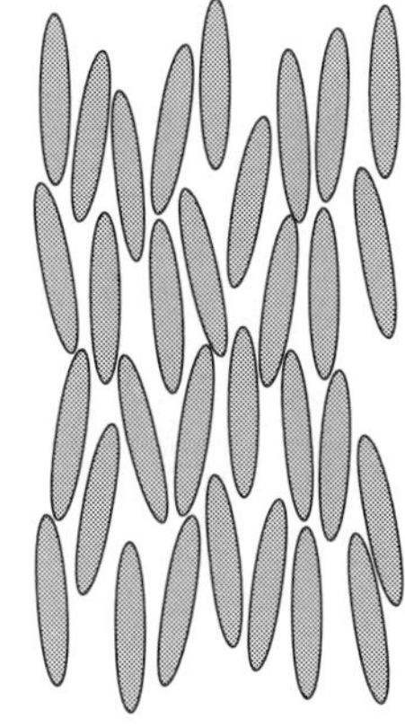
Fig. 6.4.10 - In a liquid crystal, the rod-like or disk-like molecules don't have positional order but they do have orientational order. Here all the rod-like molecules point in more or less the same direction, a behavior characteristic of a nematic liquid crystal.

Liquid crystals are actually quite common in biological systems. For example, cell membranes in animals are liquid crystals. Among the most familiar liquid crystals are the pearly, iridescent liquid hand soaps and shampoos. Their

❒ Light blinkers used in automobiles and holiday lights contain a bimetallic strip thermostat. Electricity flowing through the blinker heats its filament. When the temperature of the filament becomes high enough, the thermostat stops the current flow and turns off the filament. When it cools down, the thermostat again permits electricity to flow and the filament heats up. This process repeats over and over and causes any lights attached to the blinker to wink on and off endlessly.

striking optical properties come about because of their remarkable orientational order. Liquid crystals interact strangely with light, a feature that makes them useful in electronic watch displays and computer screens. It's just such a strange interaction between a liquid crystal and light that makes a plastic film thermometer work.

The liquid crystal used in a thermometer is not quite as simple as that depicted in Fig. 6.4.10. That figure shows a *nematic liquid crystal*, which has molecules that can be located anywhere, but that all point in roughly the same direction throughout the material. The liquid crystal used in a thermometer is a *chiral nematic liquid crystal*, which has a natural twist to it so that the preferred molecular orientation spirals around like a corkscrew as you look across the liquid in one particular direction (Fig. 6.4.11). A chiral nematic liquid crystal still has only orientation order, but that orientational order is a complicated spiral one.

The spiral orientation of its molecules gives the chiral nematic liquid crystal a twisting, wave-like appearance. It even has "crests," where the molecules all tend to point up and down, rather than to the side. The spacing between adjacent crests is called the *pitch*, the same word used to describe the distance between adjacent threads on a screw. The pitch of a chiral nematic liquid crystal is responsible for its remarkable optical properties.

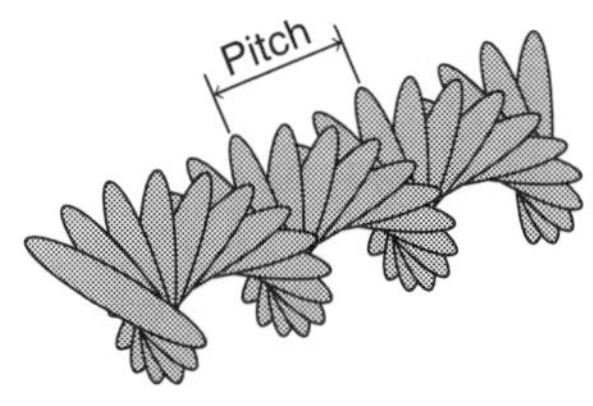

Fig. 6.4.11 - The molecules in a chiral nematic liquid crystal have an orientation that rotates in a smooth spiral along one direction through the liquid. The spacing between adjacent regions of upward pointing molecules is called the pitch.

This pitch may be only a few tens of nanometers or many microns, depending on the liquid crystal's chemical composition and *on its temperature*. Increasing the temperature shortens the pitch. This dependence of pitch on temperature is what makes liquid crystal thermometers possible. In a particular chiral nematic liquid crystal, there may be a narrow range of temperatures over which the pitch of the spiral is equal to the wavelengths of visible light in that liquid. When the liquid crystal's temperature is in this range, it suddenly begins to reflect colored light!

For example, if at 28° C the pitch of a particular liquid crystal is equal to the wavelength of blue light, then that liquid will appear brilliantly blue when illuminated by white light because it will reflect blue light back toward your eyes. If at 26° C, the pitch is longer and is equal to the wavelength of red light, then the liquid will appear red. If at 24° C, the pitch is longer than the wavelength of any visible light, then the liquid will reflect only infrared light and will appear transparent. This phenomenon is called **selective reflection** and is caused by *constructive interference*, a wave behavior that we will examine in Chapter 15. (This effect also appears in nature—see ❒.)

A plastic strip thermometer contains a series of different chiral nematic liquid crystals that are viewed through clear number-shaped openings in the otherwise opaque strip. Behind each temperature number is a liquid crystal that reflects visible light only at the temperature represented by that number. Because the liquid crystals have black plastic behind them, they appear black unless they are selectively reflecting light. For any given temperature, only one patch of liquid crystal is selectively reflecting light and it illuminates the number corresponding to the strip's temperature.

❒ Some insects obtain their striking colorations from chiral nematic liquid crystals. These liquid crystals contain oriented molecules that selectively reflect light of certain colors. In insects, these liquid crystal secretions harden to form solids that retain both the special spiral molecular order and the unusual optical effects.

CHECK YOUR UNDERSTANDING #3: Hot Spots

An engineer can measure the temperature distribution of a flat surface by laying a thin liquid crystal sheet on it. Different temperatures appear as different colors in the light reflected by that sheet. Explain.

Summary

How Thermometers and Thermostats Work: Most mechanical thermometers use differences in thermal expansion to measure temperature. In liquid thermometers, a liquid with a large coefficient of volume expansion moves up or down a glass capillary with a small coefficient of volume expansion. In metal thermometers, two metals with different coefficients of volume expansion are bonded together to form a bimetallic strip. This strip bends as its temperature changes and the two metals experience different changes in length.

Liquid crystal thermometers use selective reflection in chiral nematic liquid crystals to highlight numbers on a plastic strip. The liquid crystals are carefully made so that they reflect visible light and highlight their number only at the temperature corresponding to that number.

The Physics of Thermometers and Thermostats

1. Nearly all liquids and solids expand with increasing temperature because the stronger thermal vibrations increase the average separations between particles.

2. A material's tendency to expand with temperature is measured by its coefficient of volume expansion. The larger this coefficient, the more the material expands with each degree of increasing temperature.

3. Liquids exhibit larger coefficients of volume expansion than solids because, in a liquid, added thermal energy breaks the chemical bonds between molecules and increases their average separations substantially. In a solid, added thermal energy can only increase the vibrations of the crystalline lattice and cause small increases in the average separations.

4. Liquids exhibit larger specific heat capacities than solids because some of the heat entering a liquid goes into breaking chemical bonds rather than increasing vibrational strength and temperature.

5. Solids have both positional and orientation order while liquids have neither.

6. Liquid crystals have orientational but not positional order. This orientational order gives liquid crystals their unusual optical properties. Chiral nematic liquid crystals exhibit selective reflection and may appear brightly colored at certain temperatures.

Check Your Understanding - Answers

1. As you heat the pot of water, the water expands more than the pot. Since the water no longer fits in the pot, it overflows.

Why: The pot of water behaves like a glass (or metal) thermometer, but without the capillary to handle the expanding liquid. Both the pot and the water expand, but the water expands more. It flows out of the pot and onto the stovetop.

2. The copper-coated stainless steel bottom is a bimetallic strip. When heated, the copper expands more than the stainless steel so the bottom bows outward and is no longer flat.

Why: Any time two different materials are laminated together, they may curl as the result of a temperature change. This curling can be reduced by thickening the layers, but that can cause other problems. Instead of curling, one of the materials may crack or buckle, or the pair may delaminate and separate from one another. Manufacturers must consider these possibilities when designing cookware.

3. The chiral nematic liquid crystals in the sheet have a pitch that depends on temperature. At one temperature, the pitch is equal to the wavelength of red light and the sheet appears red. At a higher temperature, the pitch is equal to the wavelength of blue light and the sheet appears blue. The sheet's color at each point on the surface indicates the temperature of that point.

Why: The relationship between temperature and the liquid crystal's pitch can be tailored chemically so that the sheet responds to a particular range of temperatures. Near the bottom of this range, the sheet will appear red because the liquid crystal's pitch will be equal to the longest wavelengths of visible light. Near the top of this range, the sheet will appear blue because the pitch will be equal to the shortest wavelengths.

Glossary

coefficient of volume expansion The fractional change in an object's volume caused by a temperature increase of 1° C.

crystal An orderly solid arrangement of atoms or molecules that has both positional and orientational orders. The particles are organized in a simple, geometrical pattern that repeats endlessly in all directions.

lattice A regular geometrical pattern that repeats endlessly in all directions.

liquid crystal A liquid consisting of rod-like or disk-like molecules that exhibit orientational order.

orientational order The characteristic of particles that all point in the same direction.

positional order The characteristic of particles that are uniformly spaced and aligned so that they form a lattice.

selective reflection Reflection only of light with a particular wavelength or color.

Review Questions

1. How does the thermal motion of the atoms in a material cause it to expand as its temperature increases?

2. Why do virtually all materials expand as their temperatures increase? Why don't they shrink?

3. Why is it so important to divide a material's change in volume by its total volume when characterizing its thermal expansion?

4. Why does the liquid level in a glass thermometer rise rather than fall?

5. Which takes more heat to warm by 1° C: 1 kg of copper or 1 kg of water? Why?

6. If you made a bimetallic strip thermometer out of two metals with identical coefficients of volume expansion, would it still work? Why or why not?

7. Natural crystals such as quartz or salt often have beautiful facets on their surfaces. What microscopic aspect of a crystal gives rise to these facets?

8. What is the difference between positional order and orientational order?

9. In what ways is a liquid crystal like a normal liquid?

10. In what ways is a liquid crystal like a normal crystal?

Exercises

1. If you listen carefully after turning off a car that has been driven for a while, you can often hear the sounds of bending metal. What is causing the metal to bend?

2. A building that isn't heated in the winter often develops cracks in the walls. Why?

3. Hot water faucets are often made of several different materials with different coefficients of volume expansion. Explain how thermal effects in a hot water faucet might reopen the faucet slightly as it cools down after you close it.

4. Why do hot glass baking dishes often shatter when they're immersed in cold water?

5. Why do metal cookie sheets often distort when they're put in hot ovens?

6. Why do bridges have special gaps at their ends to allow them to change lengths?

7. Why are concrete sidewalks divided into individual squares rather than being left as continuous concrete strips?

8. If you were laying steel track for a railroad, what influence would thermal expansion have on your work?

9. A difficult-to-open jar may open easily after being run under hot water for a moment. Explain.

10. If you heat a window pane slowly and uniformly with a blowtorch, you can get it quite hot without breaking it. But if you heat one spot too quickly, it will crack. Explain.

11. Bottling companies never fill their bottles completely full of liquid. Why not?

12. Chemists use precision volumetric flasks to measure liquid volumes accurately. Why is the volume of a volumetric flask specified at a certain temperature?

13. How could temperature changes cause the creaks and pops that you hear in a house at night?

14. If you put a hot metal pot into a shallow basin of cold water, the pot won't warm the water up very much. Why does the metal cool off so much more easily than the water heats up?

15. If you put an empty pot on the stove, it will heat up quickly. But if you add only a modest amount of water to the pot first, it will heat up much more slowly. Why does the water slow the heating process so much?

16. When you add a tiny drop of detergent to a basin of water, it coats the water with a film that's only one molecule thick. Each detergent molecule is long and thin and they pack themselves tightly together like dry spaghetti noodles in a box. What kind of order do these molecules have?

17. The orientational order in a liquid crystal is temperature sensitive. Elevated temperatures tend to destroy that order. Compare this loss of order to what happens when you heat a normal crystal until it melts.

Epilogue for Chapter 6

This chapter examined the roles of thermal energy and heat in a variety of common objects. In *wood stoves*, we saw how combustion converts ordered chemical potential energy into disorder thermal energy. We also studied the principal ways in which this newly produced thermal energy flows out of the wood stove and into its surroundings: conduction, convection, and radiation. We learned about the importance of mobile electrons in conduction, about the need for a moving fluid in convection, and about the importance of emissivity in radiation.

We returned to conduction, convection, and radiation in *clothing and insulation*, but this time we considered ways of impeding these heat transfer mechanisms. We saw that porous materials that are mostly air are among the best insulators. We also saw that color choice is important for insulation because light colors emit and absorb less thermal radiation than dark colors.

In *incandescent light bulbs*, we found that a sufficiently hot object radiates some of its heat as visible light. The distribution of wavelengths in this light depends on the object's temperature, with white light coming from a hotter object.

In *thermometers and thermostats*, we examined some of the effects of temperature on materials. We investigated the reasons why materials normally expand when heated and why liquids expand more than solids. We also saw how the temperature-dependent orientational order of liquid crystals can be used to measure temperature.

Explanation: A Ruler Thermometer

Like most things, the clear plastic ruler expands when you heat it. Its length increases by an amount proportional to its increase in temperature. When you transfer heat to the ruler, by breathing on it, touching it, or exposing it to a hair dryer, its temperature rises, it expands, and its free end turns the needle and pointer. Since the pointer's movement is proportional to the ruler's length change, it's also proportional to the thermometer's change in temperature.

You can make the needle turn the other way by extracting heat from the ruler. Placing a few ice cubes on the ruler will cause the ruler to contract and the needle will turn in the opposite direction.

Cases

1. Glass transmits the visible light radiated by the white hot sun but it absorbs the infrared light radiated by objects near room temperature. This selective absorption and transmission gives rise to the "greenhouse effect."

a. A small patch of garden is bathed in sunlight. The sun is transferring heat to this garden by radiation but the garden's temperature remains essentially constant. Why doesn't the garden get hotter and hotter?

b. If you cover the garden with a glass dome, what happens to the garden's ability to get rid of heat via radiation?

c. How does the glass dome affect the garden's temperature?

d. Like glass, molecules such as water vapor, carbon dioxide, and methane transmit visible light while absorbing some infrared light. How does the presence of these "greenhouse gases" in the earth's atmosphere affect the earth's ability to get rid of heat via radiation?

e. How do these greenhouse gases affect the earth's temperature?

2. Your room is particularly cold at night. Here are several things that you decide to do to make your room more comfortable.

a. You tape sheets of clear plastic across the windows of your room. Why do these plastic sheets make the room warmer?

b. You install curtains on the windows and draw them at night. How do the drawn curtains make your room warmer?

c. Your room has a high ceiling. You install a ceiling fan to push the air downward from the ceiling. Why does the fan make the air warmer near the floor?

d. You put a thick, down comforter or quilt on your bed. Describe in terms of heat flow how the comforter keeps you warm in bed.

e. You put an electric blanket on the bed. It uses electricity to make heat inside the blanket itself. How does this heat flow from the electric blanket to you?

3. A child's toy oven uses an incandescent light bulb to bake cookies or cakes. It resembles a conventional oven, except that the heating element is simply a 60 W electric light-bulb.

a. The light bulb is completely enclosed inside the oven and is barely visible from the outside. Nonetheless energy enters the empty oven in one form and leaves in another. What are these two forms?
b. How much energy flows into the oven each second?
c. After the oven has been operating for a very long time, how much energy leaves it each second?
d. Inside the oven, the light bulb is located just below the food. What heat transfer mechanism carries most of the heat from the bulb to the food?
e. If the bulb were above the food rather than below it, what heat transfer mechanism would carry most of the heat from the bulb to the food?
f. The oven's door locks when the oven is hot. The lock uses a horizontal bimetallic strip. The lower piece of metal in the strip expands more when heated than the upper piece of metal. Which way does the strip curl when the oven is hot?

4. Baking a potato takes a long time, even in a hot oven, because the inside of a potato warms up very slowly.

a. What makes it hard for conduction to transfer heat to the center of the potato?
b. What makes it hard for convection to transfer heat to the center of the potato?
c. What makes it hard for radiation to transfer heat to the center of the potato?
d. Why does inserting a metal skewer through the potato help it to cook quickly?

5. An electric hot water heater takes cold water at its inlet pipe and delivers hot water at its outlet pipe. Between these two pipes is a large water container. Many electric hot water heaters have two separate heating elements: one near the top of the water container and one near the bottom. These elements use a lot of electric power so they don't operate simultaneously. Instead, the top element operates until the water near the top of the hot water heater reaches the desired temperature and then the bottom element operates until the water near the bottom is also at the desired temperature. The water heater then turns itself off and waits until more heating is required.

a. Both heating elements project into the same hot water container inside the water heater. When the top element is heating water at the top of the container, why doesn't convection occur and cause water at the bottom of the container to also become hot?
b. Convection dictates where the water pipes attach to the container. Where should hot water be extracted from the container and where should cold water be introduced into the container to replace that hot water?
c. Why does the water heater's bottom heating element operate more often and age more quickly than its top heating element?
d. Why is it important to thermally insulate the hot water heater's container?
e. Energy flows into the hot water heater as electricity. Since energy is conserved and can't accumulate forever inside the heater, it must leave somehow. At a time when the hot water is being used, how does most of the energy leave the hot water heater?

6. A typical electric oven has two separate heating elements: one on top and one on the bottom. The bottom element is used for baking while the top element is used to broil foods.

a. When only the bottom element is active and glowing red hot, what heat transfer mechanism carries most of the heat to the food in the oven?
b. When only the top element is on, what heat transfer mechanism carries most of the heat to the food?
c. How does wrapping the food in shiny aluminum foil affect the rate of heat transfer to the food in these two cases?
d. A convection oven has a fan that circulates air in the oven. How does the fan affect the rate of heat transfer to the food?

7. A duck has a thick layer of fat just beneath its skin. In addition to making the duck more buoyant, this fat helps it stay warm in cold water.

a. Why does the duck have trouble maintaining its body temperature in cold weather?
b. Why does the duck have more trouble staying warm in cold water than in cold air?
c. How does the layer of fat help the duck keep warm?
d. Fat is relatively rigid but is chemically almost indistinguishable from liquid oil. Why wouldn't liquid oil be as effective as semi-rigid fat at keeping the duck warm?
e. Muscle tissue is also relatively rigid, but having an extra layer of muscle wouldn't be as effective at keeping the duck warm as the layer of fat. Why not?
f. The duck's oily feathers also tend to trap air near its skin so that cold water doesn't actually touch the duck's skin. Why does this trapped air help keep the duck warm?

8. Sliding friction is useful for starting fires.

a. If you press a wooden peg against another piece of wood and spin the peg rapidly, it can become hot enough to ignite. The peg produces thermal energy no matter how fast it spins, so why is it important that it spin rapidly?
b. If you strike a piece of flint against a piece of steel, you can shave a red hot spark off the steel. However, moments later the surface of the steel itself barely feels warm. What happens to the thermal energy in that surface?
c. When you rub a match on a matchbook, it suddenly bursts into flames. The principal source of heat here is a chemical reaction in the match head. What purpose does the initial thermal energy serve in starting this reaction?

9. Changes in temperature can be quite useful in assembling or disassembling metal objects.

a. You can remove a tight-fitting iron ring from around a copper cylinder by carefully heating only the iron ring. Why does this technique work?
b. Even if you can't heat the iron ring without heating the copper cylinder, you can free the ring from the cylinder by putting them both in liquid nitrogen, which has a temperature of –195° C. Why should this technique work?
c. One scheme for making it impossible to remove an aluminum post from a hole in an aluminum plate involves temperature. If you start with a post that is slightly too wide to fit into the aluminum plate and then chill the post in liquid nitrogen, it will shrink enough to fit. When it warms up it will fit so tightly in the hole that nothing can remove it. Why won't heating or cooling the post and plate make it possible to extract the post?

CHAPTER 7

THERMODYNAMICS

Heat flows naturally from a hotter object to colder object, which is why the hot sun warms your skin as you sit on the beach and why a cold winter breeze cools your skin as you walk to the library. But not all objects in nature permit heat to flow passively. There are also some objects that control the flow of heat. For example, some objects actively transfer heat from colder objects to hotter objects or use the flow of heat in order to do useful work. In this chapter, we will examine the rules governing the movement of heat, a field of physics known as thermodynamics.

Experiment: A Paper Wind Turbine

One device that uses the flow of heat to do work is a paper wind turbine. You can make a wind turbine by cutting a spiral out of stiff paper or light cardboard. Take a paper disk about 15 cm in diameter and cut a spiral cut in it, about 3 cm wide. Punch a small hole in the center of the spiral and pass a piece of fine string or thread through it. Knot the string and use it to support the spiral.

This paper spiral is a wind turbine. With it hanging from the string, you can make it rotate by blowing down or up on it. The airflow does work on the turbine and the turbine begins to turn. But this chapter is about thermodynamics and our goal is to use heat to make the wind turbine rotate. So hold the wind turbine above a hot incandescent light bulb. The upward current of hot air leaving the light bulb will make the wind turbine rotate. You are using heat to do work, so the wind turbine is acting as a *heat engine*.

Chapter Itinerary

We'll examine thermodynamics by studying two everyday machines: (1) *air conditioners* and (2) *automobiles*. In *air conditioners*, we'll look at the rules governing the movement of thermal energy—the laws of thermodynamics—and see how those rules permit an air conditioner to use ordered electric energy to transfer of heat from the cold room air to the hot outdoor air. In *automobiles*, we'll investigate the ways in which thermal energy can be used to do work and see how the engine is able to convert thermal energy into work while heat flows from the hot burning gasoline to the cold outside air.

The principles illustrated by these two machines are also found in many other places. Refrigerators and home heat pumps both use ordered energy to transfer heat from cold objects to hot objects while steam engines and hot air balloons both use the natural flow of heat to make things move. Once you understand the concepts behind these *heat pumps* and *heat engines*, you'll see that they are quite common in the world around you.

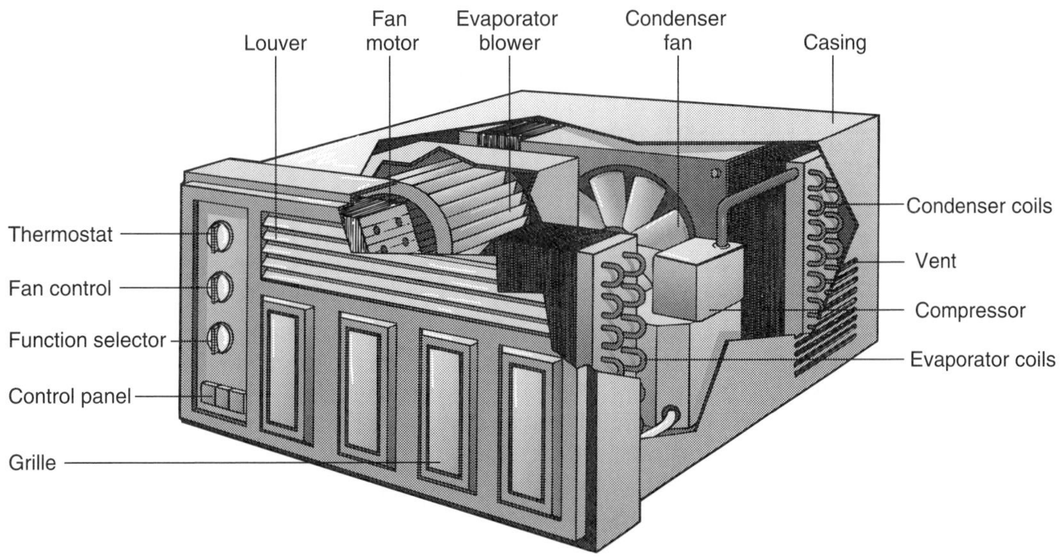

Section 7.1

Air Conditioners

On a summer day, your problem isn't staying warm; it's keeping cool. Instead of looking for something to burn in your wood stove, you turn on your air conditioner. An air conditioner is a device for removing thermal energy from room air to lower its temperature. But the air conditioner can't make thermal energy disappear. Instead, it transfers thermal energy from the cooler room air to the warmer air outside. Since the air conditioner transfers heat from a colder object to a hotter object, against its natural direction of flow, the air conditioner is a "heat pump." It's also a classic illustration of the laws of thermodynamics in action.

Questions to Think About: *Why doesn't heat naturally flow from a colder object to a hotter object? An air conditioner removes thermal energy from room air so why does the air conditioner require electric energy to operate? Where does this electric energy go? Why does an air conditioner always have an indoor component and an outdoor component? If you put a window air conditioner in the middle of a room and turn it on, what will happen to the temperature of the room?*

Experiments to Do: *Take a look at a window air conditioner. If you can't find one, examine a refrigerator instead because it's basically a food-storage closet with a powerful air conditioner built right in. As the air conditioner (or refrigerator) operates, feel the air leaving the indoor air vent (or inside the refrigerator) and compare its temperature to that of air leaving the outdoor air vent (or near the metal coils on the back of the refrigerator). Which way does the cooling mechanism move heat? If that mechanism were absent, which way would heat flow? Turn off the air conditioner or refrigerator and observe the resulting heat flow. Did you confirm your prediction?*

Moving Heat Around: Thermodynamics

On a sweltering summer day, the air in your home becomes unpleasantly hot. Heat enters your home from outdoors and doesn't stop flowing until it's as hot inside as it is outside. You can make your home more comfortable by getting rid of some of its thermal energy. But while we have already looked at several ways of *adding* thermal energy to an object, we haven't yet learned how to *remove* it. At present, the only cooling method we've discussed is contact with a colder object. Unless you have an ice house nearby, you need another scheme for eliminating thermal energy. You need an air conditioner.

An air conditioner transfers heat against its natural direction of flow. The heat moves from the colder air in your home to the hotter air outside, so that your home gets colder while the outdoor air gets hotter. There is a cost to transferring heat in this manner. The air conditioner requires ordered energy to operate and typically consumes large amounts of electric power. It's a type of **heat pump**; a device that uses ordered energy to transfer heat from a colder object to a hotter object, against its natural direction of flow.

Before learning how an air conditioner pumps heat, we should first show that pumping is necessary. There are a number of seemingly reasonable cooling alternatives that we should consider before turning to air conditioning. Three such alternatives are:

1. Letting heat flow from your home to your neighbors' home.
2. Destroying some of your home's thermal energy.
3. Converting some of your home's thermal energy into electric energy.

Unfortunately, these three alternatives can't be done. Still, it will be useful to us to examine them more closely because in doing so, we'll learn about the laws governing the movement of thermal energy, the **laws of thermodynamics**.

The first alternative raises an interesting issue. Your home is in thermal equilibrium with the outside air, meaning that no heat flows from one to the other and they are at the same temperature. Your neighbors' home is also in thermal equilibrium with the outside air. What will happen if you permit heat to flow between your home and your neighbors' home? Nothing. Since both homes are simultaneously in thermal equilibrium with the outside air, they are also in thermal equilibrium with one another. All three are at the same temperature.

This observation is an example of the **zeroth law of thermodynamics**, which says that two objects that are each in thermal equilibrium with a third object are also in thermal equilibrium with one another. This seemingly obvious law is the basis for a meaningful system of temperatures. If you had a roomful of objects at 20° C and some were in thermal equilibrium with one another while others were not, then "being at 20° C" wouldn't mean much. Fortunately, every object that has a temperature of 20° C is in thermal equilibrium with every other object at 20° C. The zeroth law is observed to be true in nature so that temperature does have meaning. And since your neighbors' home is just as hot as yours, they can relax because you're not going to be sending them any extra heat.

THE ZEROTH LAW OF THERMODYNAMICS:
Two objects that are each in thermal equilibrium with a third object are also in thermal equilibrium with one another.

The second alternative sounds unlikely from the outset. We have known since the first chapter that energy is special; that it's a conserved quantity. You can't cool your home by destroying thermal energy because energy can't be destroyed. To eliminate thermal energy, you must convert it to another form or transfer it elsewhere.

This concept of energy conservation is the basis for the **first law of thermodynamics**, which states that the change in a stationary object's internal energy is equal to the heat transferred into that object minus the work that object does on its surroundings. Internal energy includes both thermal energy and stored potential energy. This law says that heat added to the object increases its internal energy while work done by the object decreases its internal energy. Since energy is conserved, the only way the object's internal energy can change is by transferring it as heat or work.

THE FIRST LAW OF THERMODYNAMICS:
The change in a stationary object's internal energy is equal to the heat transferred into that object minus the work that object does on its surroundings.

CHECK YOUR UNDERSTANDING #1: Stirring Up Trouble

If you put cold water into a blender and mix it rapidly for several minutes, the water will become warm. Where does the additional thermal energy come from?

Disorder, Entropy, and the Second Law

The third alternative looks much more promising than the first two. It seems as though you should be able to convert thermal energy into electricity (or some other ordered form of energy). You could then sell it back to the electric company and get credit on your bill. Wouldn't that be great?

But there's a problem. Ordered energy and thermal energy are not equal. You can easily convert ordered energy into thermal energy but the reverse is much harder. For example, you can burn a log to convert its chemical potential energy into thermal energy but you'll have trouble converting that thermal energy back into chemical potential energy to recreate the log.

The basic laws of motion are silent on this issue. It isn't that the smoke doesn't have the energy to recreate the log. It's that the individual smoke particles must pool their thermal energies together to carry out the reassembly, a remarkably unlikely event. The particles would all have to move in just the right ways to turn the burned gases back into wood and oxygen; an incredible coincidence that simply never happens. Similarly, all of the air particles in your home would have to act together to convert their thermal energy into electricity. Since that coordinated behavior is ridiculously improbable, you are not going to be selling power to the electric company any time soon.

Once energy has been scattered randomly among the individual air particles, you can't collect that energy back together again. Creating disorder out of order is easy but recovering order from disorder is nearly impossible. As a result, systems that begin with some amount of order gradually become more and more disordered; never the other way around. The best they can do is to stay the same for a while so that their disorder doesn't change. From these observations, we can state the disorder of an isolated system *never decreases*.

This notion of never decreasing disorder is one of the central concepts of thermal physics. There is even a formal measure of the total disorder in an object: **entropy**. All disorder contributes to an object's entropy, including its thermal energy and its structural defects. Breaking a window or heating it both increase its entropy.

Because disorder never decreases, the third cooling alternative is impossible. Turning your home's thermal energy into electric energy would reduce its disorder and decrease its entropy. But our observations about entropy are not yet complete. There is one way to decrease your home's entropy: you can export that entropy somewhere else. In fact, you export entropy every time you take out the garbage, though that action also changes the contents of your home. But you can also export entropy without modifying your home's contents by transferring heat somewhere else. Heat carries disorder and entropy with it, so getting rid of heat also gets rid of entropy.

Our rule about entropy never decreasing is weakened by the possibility of exchanging heat and entropy between objects. Before asserting that an object or system of objects can't decrease its entropy, we must ensure that it is thermally isolated from its surroundings so that it can't export its entropy. With that in mind, the strongest statement that we can make concerning entropy is that the entropy of a thermally isolated system of objects never decreases. This observation is the **second law of thermodynamics**.

THE SECOND LAW OF THERMODYNAMICS:
The entropy of a thermally isolated system of objects never decreases.

Because of the second law, the only way to cool your home is to export its thermal energy and entropy elsewhere. Such a transfer would be easy if you had a cold object nearby to receive the heat. But lacking a cold object, you must use an air conditioner. Like all heat pumps, an air conditioner transfers heat and entropy in such a way that the second law of thermodynamics is never violated and the entropy of each thermally isolated system of objects never decreases. As we will see, the air conditioner lowers the entropy of your home but raises the entropy of the outside air so that overall, the entropy of the world actually increases.

There is a limit to how much entropy an air conditioner can remove from your home. As it exports thermal energy and entropy, the air conditioner lowers your home's temperature. In principle, your home will eventually approach absolute zero and, as it does, its entropy will approach zero. This relationship between the zero of temperature and the zero of entropy is the **third law of thermodynamics**, which states that as an object's temperature approaches absolute zero, its entropy approaches zero. The third law establishes absolute zero as a destination with no disorder left but the second law ultimately makes it impossible to extract all the disorder from an object. Because absolute zero is unattainable, the third law refers to *approaching it* rather than to *being there*.

THE THIRD LAW OF THERMODYNAMICS:
As an object's temperature approaches absolute zero, its entropy approaches zero.

CHECK YOUR UNDERSTANDING #2: Something for Nothing

People have tried for centuries to build a machine that provides an endless output of useful, ordered energy without any input of ordered energy. Unfortunately, such a perpetual motion machine violates the laws of thermodynamics. If the machine is thermally isolated, which law does it violate? What about if it's not thermally isolated?

Pumping Heat Against Its Natural Flow

While the second law of thermodynamics doesn't allow the entropy of a thermally isolated system to decrease, it does permit the objects in that system to redistribute their individual entropies. One object's entropy can decrease as long as the entropy of the rest of the system increases by at least as much. This shifting of entropy allows part of the system to become colder if the rest of the system becomes hotter.

For example, suppose that there's a pond of cold water behind your home. You could pump that water through your bathtub and let it draw heat out of the room air. Your home would become colder while the pond would become warmer. This transfer of heat from the hot air in your home to the cold water in the pond satisfies the second law of thermodynamics. The entropy of the combined system—your home and the pool of water—doesn't decrease. In fact, it actually increases!

This entropy increase occurs because heat is more disordering to a cold object than it is to a hot object. Each joule of heat that flows from your home to the pool creates more disorder in the pool than it creates order in your home.

A useful analog for this effect involves two parties taking place simultaneously: the garden society's annual tea party and a four year old's birthday party. The orderly tea party represents the cold pool while the disorderly birthday party represents your hot home. The analogy to letting heat flow from your hot home to the cold pool is to exchange one lively four year old from the disorderly birthday party for one quiet octogenarian from the orderly tea party. This transfer will reduce the birthday party's disorder only slightly, but it will dramatically increase the disorder of the tea party. The attendance at each party will be unchanged but their total disorder will increase.

When heat flows from your home to the pool, the overall entropy increases and the second law is more than satisfied. A similar increase in entropy occurs whenever heat flows from a hot object to a cold object, which is why heat normally flows in that direction.

But an air conditioner does the seemingly impossible: it transfers heat from a cold object, your home, to a hot object, the outside air. Heat flows the wrong direction and creates less disorder as it enters the hot outside air than it creates order by leaving the cold indoor air. The entropy of the combined system decreases, a violation of the second law of thermodynamics!

However, we've omitted an important feature of the air conditioner's operation: the electric energy it consumes. The air conditioner converts this ordered energy into thermal energy and delivers it as heat to the outside air. In doing so, the air conditioner creates enough extra entropy to ensure that the overall entropy of the combined system increases. The second law is satisfied after all.

An air conditioner always consumes ordered energy when it pumps heat from a colder object to a hotter object because it must create the extra entropy required by the second law of thermodynamics (Fig. 7.1.1). The amount of ordered energy it consumes depends on the temperatures of the two objects. If the two objects are close in temperature, the transfer of heat reduces the entropy only slightly so the air conditioner doesn't need to convert much ordered energy into thermal energy. But if the objects are far apart in temperature, the air conditioner must create lots of extra entropy to make up for the entropy lost in the transfer.

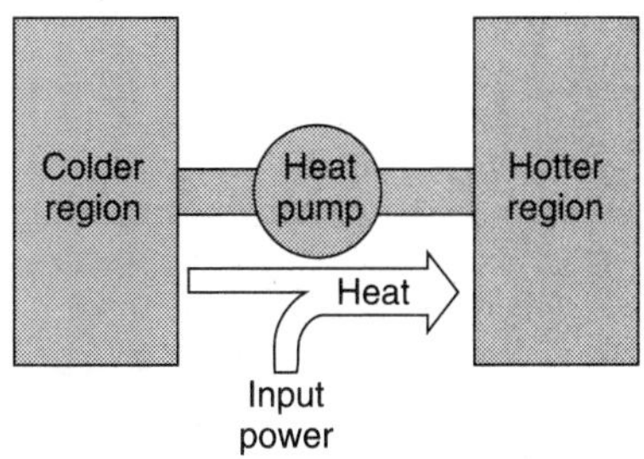

Fig. 7.1.1 - A heat pump transfers heat from a colder region to a hotter region. In doing so, it converts some amount of ordered energy into thermal energy in the hotter region. The larger the temperature difference between the two regions, the more ordered energy is required to transfer each joule of heat.

This need to keep entropy from decreasing explains why an air conditioner works best when it's cooling your home the least. The greater the temperature difference between the indoor air and the outdoor air, the more electric energy the air conditioner must consume in order to move each joule of heat. (This energy consumption would become infinite if you tried to cool your home to absolute zero, which is why absolute zero is unattainable!) Moreover, the heat flowing back into your home through its walls is roughly proportional to this temperature difference. Thus the colder you set the thermostat, the larger your electric bills will be.

CHECK YOUR UNDERSTANDING #3: Heat Pumps in Cold Weather

Homes located in mild climates are often heated by heat pumps during the winter. Home heat pumps are essentially air conditioners run backwards. They extract heat from the cold outdoor air and release it to the warm indoor air. Why are heat pumps most effective in mild weather, when the outdoor air isn't too cold?

How an Air Conditioner Cools the Indoor Air

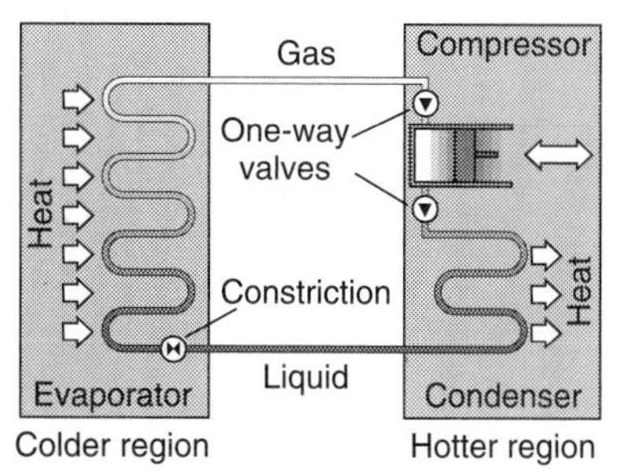

Fig. 7.1.2 - A heat pump transfers heat from a colder region to a hotter region. A typical air conditioner does this by condensing a gas to a liquid on the hot side and evaporating the liquid to a gas on the cold side. The necessary input of ordered energy is provided by a compressor.

Having determined the air conditioner's goals, it's time for us to look at how a real air conditioner meets them. In most cases, the air conditioner uses a fluid to transfer heat from the colder indoor air to the hotter outdoor air. This *working fluid* absorbs heat from the indoor air and releases that heat to the outdoor air.

The air conditioner has three main components: an evaporator, a condenser, and a compressor (Fig. 7.1.2). The evaporator is located indoors, where it transfers heat from the indoor air to the working fluid (Fig. 7.1.3). The condenser is located outdoors, where it transfers heat from the working fluid to the outdoor air. And the compressor is also located outdoors, where it does work on the working fluid and introduces the additional thermal energy needed to ensure that the total disorder of the system doesn't decrease.

To see how these three components pump heat out of your home, let's look at them individually. We'll begin with the evaporator, a long metal pipe that's decorated with thin metal fins. These fins assist heat flow between the working fluid inside the evaporator and the room air surrounding it so that whenever the working fluid is colder than the room air, heat will flow from the room air to the working fluid.

The working fluid flows toward the evaporator as a high-pressure liquid in a pipe. Its temperature is close to room temperature, so it can't cool the room air yet. But just before it enters the evaporator, the working fluid passes through a narrowing in the pipe. This constriction impedes the working fluid's flow and causes its pressure to drop dramatically. As a result, the working fluid enters the evaporator as a low-pressure liquid.

Fig. 7.1.3 - The evaporator of this central air conditioning unit extracts heat from the indoor air. This heat is transferred outside, where the compressor and condenser release it into the outdoor air.

The chemical bonds that hold the working fluid's particles together are weak and fragile. Near room temperature, the particles need a high pressure to keep them together as a liquid. Once the liquid working fluid passes through the constriction and loses its pressure, its particles begin to separate. One by one, these particles break away from each other so that the working fluid gradually **evaporates** from a liquid of attached particles to a gas of independent particles.

However, breaking the weak chemical bonds between particles takes energy; thermal energy. As the working fluid evaporates, some of its thermal energy becomes chemical potential energy in the newly separated gas particles. With its diminished thermal energy, the gaseous working fluid is quite cold. This cold gas chills the evaporator and its fins and heat begins to flow into the working fluid from the room air.

By the time the working fluid emerges from the evaporator, it has evaporated completely and has absorbed considerable thermal energy from the indoor air. It leaves the evaporator as a cool low-pressure gas and travels through a pipe toward the compressor.

Half the air conditioner's job is done: it has removed heat from the indoor air. But the remaining half of its job is more complicated: it must add heat to the outdoor air while ensuring that the total entropy of the combined system doesn't decrease. After all, there's no getting around the second law of thermodynamics.

CHECK YOUR UNDERSTANDING #4: Cooling at the Gas Grill

When liquid propane evaporates into a gas in the tank of a propane grill, the tank that contains that liquid propane becomes colder. Why?

How an Air Conditioner Warms the Outdoor Air

Satisfying the second law is the task of the compressor. The compressor receives low pressure gaseous working fluid from the evaporator, compresses it to much higher density, and delivers it as a gas to the condenser. The compressor may use a piston and one-way valves, like the water pump in Fig. 3.2.2, or it may use a rotary pump mechanism. In either case, the result is the same: the gaseous working fluid leaves the compressor at a much higher density and pressure than it had when it arrived.

Compressing a gas requires work because the compressor must push the gas inward while moving it inward; force times distance. Since work transfers energy, the compressor increases the energy of the gas. The air conditioner usually obtains this energy from the electric company and converts it into mechanical work with an electric motor.

But the only way that the gaseous working fluid can store its new energy is as thermal energy in its individual particles. These particles begin to move about more and more rapidly so that the gaseous working fluid leaves the compressor much hotter than when it arrived.

This hot working fluid then flows into the condenser. Like the evaporator, the condenser is a long metal pipe with fins attached to it. These fins assist heat flow through the walls of the condenser. Because the working fluid is so hot, heat flows from the working fluid to the outside air. The working fluid gradually becomes cooler as the outdoor air becomes hotter.

Near room temperature, the working fluid only remains gaseous if its pressure is low. Otherwise, its particles begin sticking to one another. Once the high pressure working fluid has cooled enough for this binding to begin, it gradually **condenses** from a gas of independent particles to a liquid of attached particles.

However, when the particles form chemical bonds with one another, they release their chemical potential energy as thermal energy and the working fluid becomes quite hot. This hot liquid warms the condenser and its fins and more heat flows from the working fluid to the outdoor air. By the time the working fluid emerges from the condenser, it has condensed from a hot high-pressure gas to a cool high-pressure liquid and has released considerable thermal energy to the outdoor air. This heat included both thermal energy extracted from the indoor air and electric energy converted into thermal energy by the compressor.

The second half of the air conditioner's job is now complete: it has converted ordered energy into thermal energy and it has released heat into the outdoor air. From here, working fluid returns to the evaporator to begin the cycle all over again. The working fluid passes endlessly around the cycle, extracting heat from the indoor air in the evaporator and releasing it to the outdoor air in the condenser. The compressor drives the whole process, thereby satisfying the second law of thermodynamics. The same technique is used to extract heat from the air inside a refrigerator and to release that heat to the room air (Fig. 7.1.4).

Before leaving air conditioners, we should take a moment to look at the

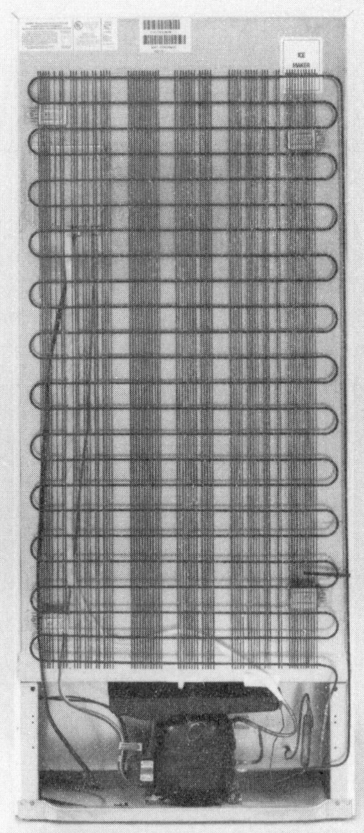

Fig. 7.1.4 - The compressor (bottom) and condenser coils (top) are visible on the back of this refrigerator. The compressor squeezes the working fluid into a hot, dense gas and delivers it to the condenser. There it gives up heat to the room air and condenses into a liquid. The working fluid evaporates inside the refrigerator, extracting heat from the food.

working fluid itself. This fluid must become a gas at low pressure and a liquid at high pressure, over most of the temperature range encountered by the air conditioner. For decades, the standard working fluids have been *chlorofluorocarbons* such as the various Freons. These compounds replaced ammonia, a toxic and corrosive gas used in early refrigeration.

Chlorofluorocarbons are ideally suited to air conditioners because they transform easily from gas to liquid and back again over a broad range of temperatures. They are also chemically inert and inexpensive. Unfortunately, chlorofluorocarbon molecules contain chlorine atoms and carry those chlorine atoms to the upper atmosphere. There they promote the destruction of ozone molecules, essential atmospheric constituents that absorb portions of the sun's ultraviolet radiation. Recently, *hydrofluorocarbons* have been developed as chlorine-free working fluids for use in air conditioners. Though not as energy efficient and chemically inert as the materials they replace, hydrofluorocarbons appear to be safe for the environment.

CHECK YOUR UNDERSTANDING #5: Air Conditioning or Space Heating?

What would happen to the average air temperature in your room if you placed a window air conditioning unit in the middle of the room and turned it on?

Summary

How Air Conditioners Work: An air conditioner moves heat against its natural direction of flow, using a working fluid that travels endlessly through an evaporator, compressor, and condenser. The working fluid is a material that, over the range of temperatures encountered in the air conditioner, tends to be a gas at low pressure and a liquid at high pressure.

The working fluid flows to the evaporator as a high-pressure liquid before passing through a constriction in the pipe. Its pressure drops at the constriction and the fluid enters the evaporator as a low-pressure liquid. In the evaporator, the working fluid absorbs heat from the indoor air and evaporates into a cold gas.

This cold low-pressure gas flows to the compressor, which compresses it into a hot, high-pressure gas and sends it to the condenser. As it passes through the condenser, the hot gaseous working fluid releases heat to the outdoor air and it condenses into a cool liquid.

The compressor uses ordered energy and delivers it as heat to the outdoor air. Without this input of ordered energy, the air conditioner could not move heat against its natural direction of flow.

Important Laws and Equations

1. The Zeroth Law of Thermodynamics: Two objects that are each in thermal equilibrium with a third object are also in thermal equilibrium with one another.

2. The First Law of Thermodynamics: The change in a stationary object's internal energy is equal to the heat transferred into that object minus the work that object does on its surroundings.

3. The Second Law of Thermodynamics: The entropy of a thermally isolated system of objects never decreases.

4. The Third Law of Thermodynamics: As an object's temperature approaches absolute zero, its entropy approaches zero.

The Physics of Air Conditioners

1. Heat naturally flows from a hotter region to a colder region. Heat can't spontaneously flow the other way because the overall energy disorder of the system, its entropy, would decrease.

2. A heat pump can transport heat against its natural direction of flow, but it must convert ordered energy into thermal energy in the process. This conversion ensures that the total entropy of the thermally isolated system does not decrease.

3. Thermal energy can't be directly converted into ordered energy because such a conversion would reduce the total entropy of the system. Thermal energy can only be removed from one object by transferring it to another object.

4. The amount of ordered energy needed to pump heat from a cold object to a hot object increases as the temperature difference between the two objects increases.

5. As the temperature of the cold object approaches absolute zero, the amount of ordered energy needed to pump heat out of it increases without bound. As a result, it is impossible to remove all of the disorder from an object and reduce its temperature to absolute zero.

Check Your Understanding - Answers

1. The blender's blade does work on the water and this work becomes thermal energy.

Why: The first law of thermodynamics states that the change in the water's internal energy is equal to the heat flowing into it minus the work it does on its surroundings. In this case, the water's surroundings are doing work on it by stirring it so its internal energy increases. Since the water can't store this new internal energy as potential energy, the energy becomes thermal energy and the water gets hotter.

2. A thermally isolated perpetual motion machine violates the first law while one that is not thermally isolated violates the second law.

Why: A thermally isolated perpetual motion machine clearly violates the conservation of energy aspect of the first law of thermodynamics. This isolated machine simply can't export energy forever because it will eventually run out. A perpetual motion machine that is not thermally isolated may not violate conservation of energy because it can absorb heat energy from its surroundings. Instead, it violates the second law of thermodynamics. This machine can't endlessly absorb heat energy and then export it as ordered energy. In doing so, the machine will eventually begin to reduce the entropy of the universe and violate the second law. Sad though it may be, perpetual motion machines can't exist.

3. A heat pump requires more ordered energy to pump heat from a cold object to a hot object when the temperature difference between them is large.

Why: A heat pump becomes less efficient at pumping heat when the temperature of the heat's source becomes much colder than the heat's destination. The colder it is outside, the more ordered energy it takes to move each joule of heat. On bitter cold days, heat pumps aren't able to move enough heat to keep their homes warm, which is why most home heat pumps have built in electric or gas furnaces to assist them during unusually cold weather.

4. The tank's liquid propane needs heat to evaporate into a gas and it extracts that heat from its surroundings.

Why: Just as in the evaporator of an air conditioner, the evaporating liquid propane absorbs heat.

5. The room air would become warmer, on the average.

Why: The air conditioner would begin to pump heat from its front to its back. The air right in front of the unit would become colder while the air behind the unit would become hotter. Since the unit would deliver more heat to the hotter air than it would absorb from the colder air, it would increase the total amount of thermal energy in the room. On the average, the room would become warmer.

Glossary

condense Transform from a gas to a liquid.

entropy The physical quantity measuring the amount of disorder in a system. The system's entropy would be zero at absolute zero.

evaporate Transform from a liquid to a gas.

first law of thermodynamics The change in a stationary object's internal energy is equal to the heat transferred into that object minus the work that object does on its surroundings. This law restates the conservation of energy.

heat pump A device that pumps heat against its natural direction of flow; transferring it from a cold object to a hot object. To satisfy the second law of thermodynamics, a heat pump must convert some ordered energy into thermal energy.

laws of thermodynamics The four laws that govern the movement of heat between objects.

second law of thermodynamics The entropy of a thermally isolated system of objects never decreases. This law recognizes that creating disorder is easy, restoring order is hard.

third law of thermodynamics As an object's temperature approaches absolute zero, its entropy approaches zero. This law points out that absolute zero is the unattainable state in which an object has no disorder.

zeroth law of thermodynamics Two objects that are each in thermal equilibrium with a third object are also in thermal equilibrium with one another. This law is the basis for a meaningful system of temperatures.

Review Questions

1. Suppose that you had two objects at the same temperature (as measured by a third object—a thermometer). If those two objects weren't in thermal equilibrium with one another, that would violate the zeroth law of thermodynamics. Why would temperature then be meaningless?

2. Compare work and heat to show how they are similar and how are they different.

3. Why is the first law of thermodynamics a restatement of energy conservation?

4. List five situations in which the entropy of a thermally isolated system increases and point out which aspects of each system become more disordered.

5. Why can't an object that's at room temperature have zero entropy?

6. How does being able to transfer entropy from one object to another make it possible to cool one object below the temperature of its surroundings?

7. Why does it take ordered energy to cool an object below the temperature of its surroundings?

8. Which experiences the larger rise in entropy when you transfer 1 J of heat to it: cold water or hot water?

9. As working fluid evaporates in an air conditioner's evaporator, what becomes of the thermal energy that the surrounding air transfers to it as heat?

10. As working fluid condenses in an air conditioner's condenser, where does the heat that flows into the surround air come from?

11. Why is it impossible to build an air conditioner that doesn't require an input of ordered energy?

Exercises

1. Drinking fountains that actively chill the water they serve can't work without ventilation. They usually have louvers on their sides so that air can flow through them. Why do they need this airflow?

2. If you open the door of your refrigerator with the hope of cooling your room, you will find that the room's temperature actually increases somewhat. Why doesn't the refrigerator remove heat from the room?

3. When the gas that now makes up the sun was compressed together by gravity, what happened to the temperature of that gas? Why?

4. There is interest in a commercial supersonic transport that would fly high in the atmosphere and carry passengers half way around the earth in just a few hours. This aerospace plane would move so quickly that it would temporarily compress the air through which it flies. As a result, the plane's surface would have to be built to withstand very high temperatures. High altitude air is normally very cold so why would the plane's surface become so hot?

5. The outdoor portion of a central air conditioning unit has a fan that blows air across the condenser coils. If this fan breaks, why won't the air conditioner cool the house properly?

6. If you block the outlet of a hand bicycle pump and push the handle inward to compress the air inside the pump, the pump will become warmer. Why?

7. Despite rubber's elastic character, the molecules inside rubber receive very little potential energy when you stretch it. Instead, these long, coiled molecules unwind and straighten out. When you stretch a rubber band, it becomes hot. Explain this effect in terms of work and thermal energy.

8. When you knead bread, it becomes warm. Where does this thermal energy come from?

9. A soda siphon carbonates water by injecting carbon dioxide gas into it. The gas comes compressed in a small steel container. As the gas leaves the container and pushes its way into the water, why does the container become cold?

10. You throw a ball into a box and close the lid. You hear the ball bouncing around inside as the ball's energy changes from gravitational potential energy, to kinetic energy, to elastic potential energy, and so on. If you wait a minute or two, what will have happened to the ball's energy?

11. When you switch the heat pump in a house to "cool", which way does it transfer heat between the house and the outdoors?

12. When you switch the heat pump in a house to "heat", which way does it transfer heat between the house and the outdoors?

13. If you drop a glass vase on the floor, it will become fragments. However, if you drop those fragments on the floor, they will not become a glass vase. Why not?

14. When you throw a hot rock into a cold puddle, what happens to the overall entropy of the system?

15. What prevents the bottom half of a glass of water from spontaneously freezing while the top half becomes boiling hot?

16. Suppose someone claimed to have a device that could convert heat from the room into electric power continuously. You would know that this device was a fraud because it would violate the second law of thermodynamics. Explain.

17. Why does snow blanket the ground almost uniformly rather than creating tall piles in certain areas and bare spots in others?

18. Why can't an object's entropy decrease as its temperature increases?

19. Suppose that an enormous spaceship has an internal flight deck—a huge sealed room with air inside it that serves as a boarding area for smaller ships. To let one of these small ships head into empty space, the big ship opens the doors at the end of its flight deck. The air farthest from the open doors then pushes the air nearest those doors out into space. What happens to the temperature of the air inside the flight deck?

20. To avoid the temperature problems that occurred in Exercise **19**, the designers of the spaceship install pumps to transfer the air from the flight deck to small storage containers before the doors of the flight deck are opened. What happens to the temperature of the air still inside the flight deck as the transfer takes place?

21. What happens to the temperature of the air as it's compressed into the storage containers in Exercise **20**?

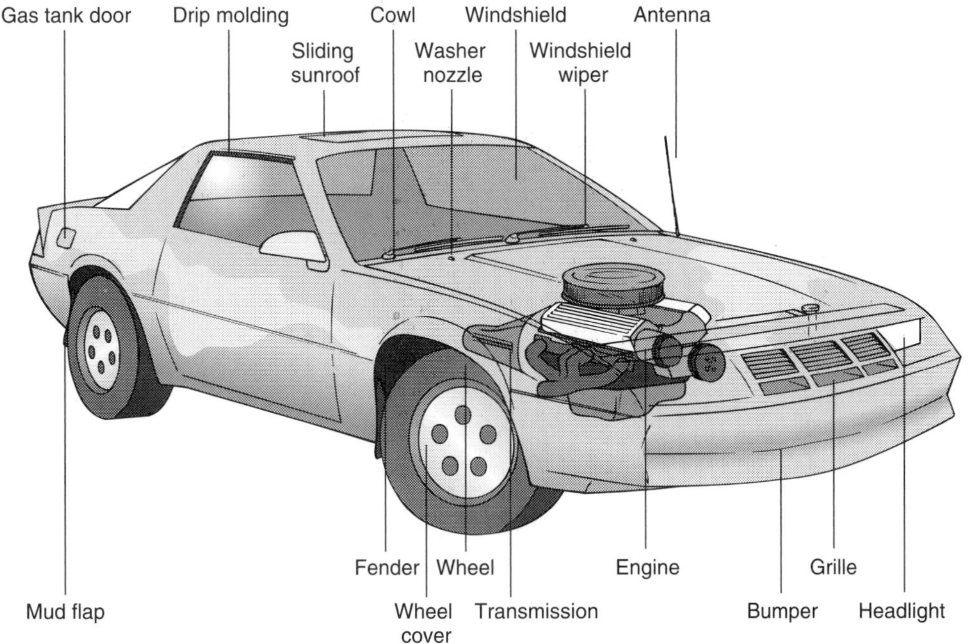

Section 7.2

Automobiles

Nothing is more symbolic of freedom and personal independence than an automobile. With car keys in your hand, you can go almost anywhere at a moment's notice. The mechanism that makes this instant transportation possible is the internal combustion engine. Though it has been refined over the years, this engine's basic design has changed little since it was invented more than a century ago. It uses thermal energy released by burning fuel to do the work needed to propel the car forward. That thermal energy can do mechanical work in certain situations is one of the marvels of the thermal physics and the primary focus of this section.

Questions to Think About: *What obstacles stand in the way of using the thermal energy in burning fuel to propel a car? Why are two objects, one hot and one cold, required in order to convert some thermal energy into useful work? What hot and cold objects does a car have? Why does a car have a cooling system to get rid of waste heat, rather than converting it into useful work? Why does a car have different gears?*

Experiments to Do: *The recent advances in automobile technology and the increasing demands for pollution control equipment have made automobiles exceedingly complicated. Nonetheless, take a moment to look under the hood of your car or that of a friend. You should be able to identify the engine and its electric support system. You should be able to count four or more spark plug wires heading toward the engine's cylinders. These cylinders convert thermal energy from burning fuel into work to propel the car. Why does the engine need so many cylinders, rather than relying on one larger cylinder?*

At the front of the engine compartment, you'll find the radiator. How does this device extract waste heat from the engine? Does heat flow naturally into the radiator and then into the outside air, or is there a heat pump involved?

Using Thermal Energy: Heat Engines

The light turns green and you step on the accelerator pedal. The engine of your car roars into action and, in a moment, you're cruising down the road a mile a minute. The engine noise gradually diminishes to a soft purr and vanishes beneath the sound of the radio and the passing wind.

The engine is the heart of the automobile, pushing the car forward at the light and keeping it moving against the forces of gravity, friction, and air resistance. It's not simply a miracle of engineering. It's also a wonder of thermal physics, because it performs the seemingly impossible task of converting thermal energy into ordered energy. But the second law of thermodynamics forbids the direct conversion of thermal energy into ordered energy, so how can a car engine use burning fuel to propel the car forward?

The car engine avoids conflict with the second law by being a **heat engine**, a device that converts thermal energy into ordered energy *as heat flows from a hot object to a cold object* (Fig. 7.2.1). While thermal energy in a single object can't be converted into work, that restriction doesn't apply to a system of two objects *at different temperatures*. Because heat flowing from the hot object to the cold object increases the overall entropy of the system, a small amount of thermal energy can be converted into work without decreasing the system's overall entropy and without violating the second law of thermodynamics.

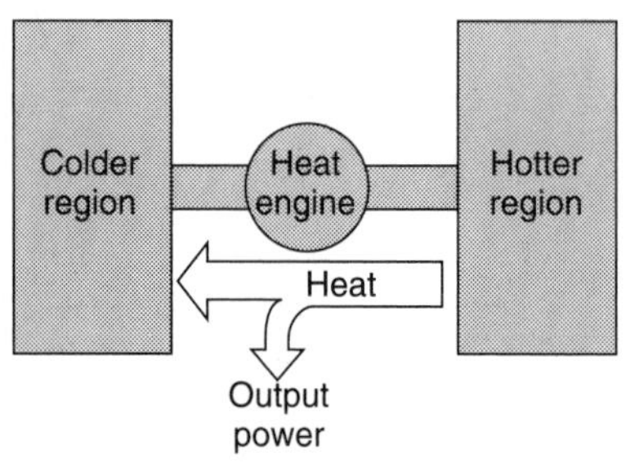

Fig. 7.2.1 - A heat engine converts thermal energy into ordered energy as heat flows from a hotter region to a colder region. The larger the temperature difference between the two regions, the larger the fraction of thermal energy that can be converted into ordered energy.

Another way to look at a heat engine is through the contributions of the two objects. The hot object provides the thermal energy that is converted into work. The cold object provides the order needed to carry out that conversion. As the heat engine operates, the hot object loses some of its thermal energy and the cold object loses some of its order. The heat pump has used them to create ordered energy. Since the heat engine needs both thermal energy and order, it can't operate if either the hot or the cold object is missing.

In a car engine, the hot object is burning fuel and the cold object is outside air. Some of the heat passing from the burning fuel to the outside air is diverted and becomes the ordered energy that propels the car forward. But what limits the amount of thermal energy the engine can convert into ordered energy?

To answer that question, let's examine a simplified car engine. We'll treat the burning fuel and outside air as a single, thermally isolated system and look at what happens to their total entropy as the engine operates. In accordance with the second law of thermodynamics, this total entropy can't decrease even if the engine produces ordered energy.

When the car is idling at a stop light, its engine is doing no work and heat is simply flowing from the hot burning fuel to the cold outside air. The system's total entropy increases because this heat is more disordering to the cold air it enters than to the hot burning fuel it leaves. In fact, the system's entropy increases dramatically because the burning fuel is extremely hot compared to the cold outside air.

This increase in the system's entropy is unnecessary and wasteful. The second law of thermodynamics only requires that the engine add as much entropy to the cold outside air as it removes from the hot burning fuel. Since a little heat is quite disordering to cold air, the car engine can deliver much less heat to the outside air than it removes from the burning fuel and still not cause the system's total entropy to decrease. As long as the engine delivers enough heat to the outside air to keep the total entropy from decreasing, there's nothing to prevent it from converting the remaining heat into ordered energy!

Conversion starts as soon as you remove your foot from the brake and begin to accelerate forward. Instead of transferring all of the thermal energy in the burning fuel to the outside air, your car then extracts some of it as ordered energy and uses it to power the wheels. The car engine really can convert thermal energy into ordered energy, as long as it passes along enough heat from the hot object to the cold object to satisfy the second law of thermodynamics.

Obeying the second law becomes easier as the temperature difference between the two objects increases. When the temperature difference is extremely large, as it is in an automobile engine, a large fraction of the thermal energy leaving the hot object can be converted into ordered energy—at least in theory. Unfortunately, theoretical limits are often hard to realize in real machines and the best automobile engines extract only about half the ordered energy that the second law allows. Still, obtaining even that amount is a remarkable feat and a tribute to generations of scientists and engineers who have labored to make automobile engines as energy efficient as possible.

CHECK YOUR UNDERSTANDING #1: Heat Pumps and Heat Engines

An air conditioner uses electric energy to make the air in your home colder than the outside air. Could you use this difference in temperatures to operate a heat engine and generate electric energy?

The Internal Combustion Engine

The internal combustion engines used in modern automobiles developed out of earlier external combustion engines during the latter half of the 19th century. An *external combustion engine* is a heat engine in which fuel is burned outside the engine itself. The classic external combustion engine is the steam engine, first made practical by Scottish inventor James Watt (1736–1819) in 1763.

In a steam engine, a fire boils water to produce high-pressure steam. This hot steam carries heat toward the cold outside air and does work on a moving surface along the way. By the time the steam is released from the engine, some of its thermal energy has been converted into ordered energy. In addition to powering early automobiles, steam engines were used to propel boats and locomotives and are still commonly used in electric power plants.

Of course, modern cars have long since abandoned steam engines in favor of internal combustion engines. Invented by the German engineer Nikolaus August Otto in 1867, an *internal combustion engine* burns fuel directly in the engine itself. Gasoline and air are mixed and ignited in an enclosed chamber. The resulting temperature rise increases the pressure of the gas and allows it to perform work on a movable surface.

To extract work from the fuel, the internal combustion engine must perform four tasks in sequence:

1. It must introduce a fuel-air mixture into an enclosed volume.
2. It must ignite that mixture.
3. It must allow the hot burned gas to do work on the car.
4. It must get rid of the exhaust gas.

In the standard, four-stroke fuel-injected engine found in modern gasoline automobiles, this sequence of events takes place inside a hollow cylinder (Fig. 7.2.2). It's called a "four-stroke" engine because it operates in four distinct steps or strokes: induction, compression, power, and exhaust. "Fuel-injected" refers to the technique used to mix the fuel and air as they are introduced into the cylinder.

Automobile engines have four or more of these cylinders. Each cylinder is a separate power source, closed at one end and equipped with a movable piston, several valves, a fuel injector, and a spark plug. The piston slides up and down in the cylinder, shrinking or enlarging the cavity inside. The valves, located at the closed end of the cylinder, open to introduce fuel and air into the cavity and to

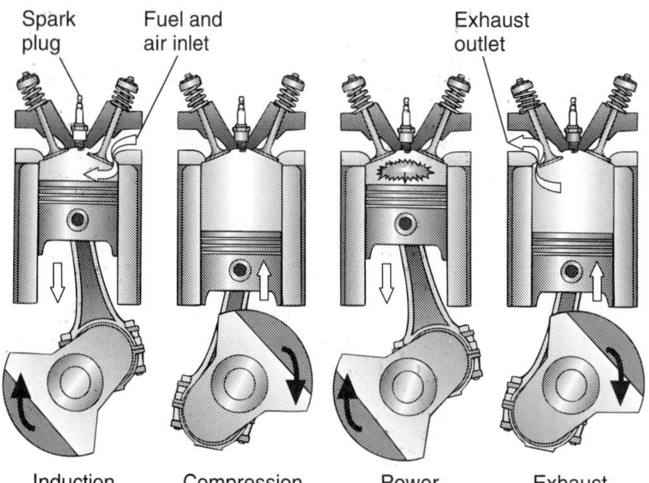

Fig. 7.2.2 - A four-stroke engine cylinder. During the induction stroke, fuel and air enter the cylinder. The compression stroke squeezes that mixture into a small volume. The spark plug ignites the mixture and the power stroke allows the hot gas to do work on the automobile. Finally, the exhaust stroke ejects the exhaust gas from the cylinder.

permit burned exhaust gas to escape from the cavity. The fuel injector adds fuel to the air as it enters the cylinder. And the spark plug, also located at the closed end of the cylinder, ignites the fuel-air mixture to release its chemical potential energy as thermal energy.

The fuel-air mixture is introduced into each cylinder during its *induction stroke*. In this stroke, the piston is pulled away from the cylinder's closed end so that its cavity expands to create a partial vacuum. At the same time, the cylinder's inlet valves open so that atmospheric pressure can push fresh air into the cylinder. The cylinder's fuel injector adds a mist of fuel droplets to this air so that the cylinder is soon filled with a flammable fuel-air mixture. Because it takes work to move air out of the way and create a partial vacuum, the engine does work on the cylinder during the induction stoke.

At the end of the induction stroke, the inlet valves close to prevent the fuel-air mixture from flowing back out of the cylinder. Now the *compression stroke* begins. The piston is pushed toward the cylinder's closed end so that its cavity shrinks and the fuel-air mixture becomes denser. Because it takes work to compress a gas, the engine does work on the mixture during the compression stroke. This work appears as thermal energy in the fuel-air mixture, which becomes hot. Since increases in a gas's density and temperature both increase its pressure, the pressure in the cylinder rises rapidly as the piston approaches the spark plug.

The fuel-air mixture is ignited at the end of the compression stroke, when a high voltage pulse is applied to the spark plug. This pulse produces a small electric spark that provides the activation energy needed to initiate combustion. The mixture burns quickly to produce very hot, very high pressure burned gas.

This hot burned gas does work on the car during the cylinder's *power stroke*. In this stroke, the piston moves away from the cylinder's closed end so that its cavity expands and the burned gas becomes less dense. The hot gas exerts a huge force on the piston as it moves outward, so the gas does work on the piston. It's this work that propels the car forward. As it does work, the burned gas gives up thermal energy and cools. Its density and pressure also decrease. At the end of the power stroke, the exhaust gas has cooled significantly and its pressure is only a few times atmospheric pressure. The cylinder has extracted much of the fuel's chemical energy as work.

The cylinder gets rid of the exhaust gas during its *exhaust stroke*. In this stroke, the piston is pushed toward the closed end of the cylinder while the cylinder's outlet valves are open. Because the burned gas trapped inside the cylinder at the end of the power stroke is well above atmospheric pressure, it accelerates out of the cylinder the moment the outlet valves open. These sudden bursts of gas leaving the cylinders create the "poof-poof-poof" sound of a running engine. Without a muffler on the exhaust pipes, the engine is loud and unpleasant.

Just opening the outlet valves releases most of the exhaust gas, but the rest is squeezed out as the piston moves toward the cylinder's closed end. The engine again does work on the cylinder as it squeezes out the exhaust gas. At the end of the exhaust stroke, the cylinder is empty and the outlet valves close. The cylinder is ready to begin a new induction stroke.

CHECK YOUR UNDERSTANDING #2: Getting Out More Than You Put In
How does the burned gas do more work on the piston during the power stroke than the piston does on the unburned fuel-air mixture during the compression stroke?

Multi-Cylinder Engines

Since the purpose of the engine is to extract work from the fuel-air mixture, it's important that each cylinder do more work than it requires to operate. Three of the strokes require the engine to do work on various gases and only one of the strokes extracts work from the burned gas. During the induction stroke, the engine does work to create a partial vacuum. During the compression stroke, the engine does work to compress the fuel-air mixture. During the exhaust stroke, the engine does work to squeeze the exhaust gas from the cylinder. Fortunately, the work done on the engine by the hot burned gas during the power stroke is much greater than the work the engine does on the various gases during the other three strokes.

Still, the engine has to invest a lot of energy into the cylinder before each power stroke. To provide this initial energy, most four-stroke engines have four or more cylinders (Fig. 7.2.3), timed so that there is always one cylinder going through the power stroke. The cylinder that is in the power stroke provides the work needed to carry the other cylinders through the three non-power strokes and there is plenty of work left over to move the car itself.

While the pistons move back and forth, the engine needs a rotary motion to turn the car's wheels. The engine converts each piston's reciprocating motion in rotary motion by coupling that piston to the *crankshaft* with a *connecting rod*. The crankshaft is a thick steel bar, suspended in bearings, that has a series of pedal-like extensions, one for each cylinder. As the piston moves back during the power stroke, it pushes on the connecting rod and the connecting rod pushes on its crankshaft pedal. The crankshaft rotates in its bearings and transmits torque out of the engine so that it can be used to power the car. So, while each cylinder initially produces a force, the crankshaft converts that force into a torque.

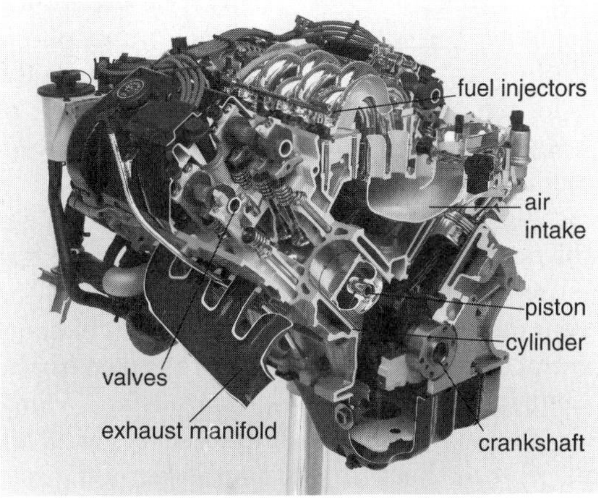

Fig. 7.2.3 - This cut-away view of a V-8 engine shows the eight cylinders arranged in two rows of four, one row in the foreground and the other behind.

CHECK YOUR UNDERSTANDING #3: Hard Starting

Modern cars use an electric motor to start the engine turning but early cars were started by a hand crank. Why was it so hard to turn the crank?

Engine Efficiency

The goal of an internal combustion engine is to extract as much work as possible from a given amount of fuel. In principle, all of the fuel's chemical potential energy can be converted into work because both are ordered energies. But it's difficult to convert chemical potential energy directly into work, so the engine burns the fuel instead.

This step is unfortunate, for when it burns the fuel, the engine converts the fuel's chemical potential energy directly into thermal energy. Burning creates a great deal of unnecessary entropy.

But all is not lost. Since the burned fuel is extremely hot, some of its thermal energy can be converted into ordered energy as heat flows from the burned fuel to the outside air. The hotter the burned fuel is and the colder the outside air, the more ordered energy the engine should be able to extract. To maximize its fuel efficiency, an internal combustion engine should burn its fuel to create the hottest possible gas, let that gas do as much work as it can on the pistons, and release the gas at the coldest possible temperature. That way, it will convert most of the thermal energy into work.

The best situation would be if, during the power stroke, the burned gas expanded and cooled until it reached the temperature of the outside air. The exhaust gas would then leave the engine with the same amount of thermal energy it had when it arrived and the engine would have extracted all of the fuel's chemical potential energy as work.

Unfortunately, that desirable result would violate the second law of thermodynamics by converting thermal energy completely into ordered energy. The engine simply can't extract all of the burned gas's thermal energy as work. Instead, the engine must release the burned gas before it cools to the temperature of the outside air. The engine's exhaust must be hot!

We can see the need for hot exhaust by comparing the normal power stroke with one in which the spark plug hasn't fired. In the latter case, the power stroke would simply reverse the compression stroke. The unburned fuel-air mixture would expand, do work on the piston and engine, and cool back down to the temperature of the outside air. It would leave the engine at the same temperature as when it entered and wouldn't carry excess thermal energy away with it.

But the spark plug does fire and the resulting hot burned gas has far more thermal energy than it can get rid of during the power stroke. Despite expanding and doing work on the piston and engine, this gas doesn't reach atmospheric pressure and temperature by the end of power stroke. It would have to push the piston much farther outward before its temperature would diminish to that of the outside air. Instead, the burned gas remains substantially hotter than the outside air and its pressure is several times atmospheric pressure.

The engine could actually extract a little more energy from the burned gas by allowing the piston to move still farther outward, but the expanding gas would soon reach atmospheric pressure and would still not have cooled to the outside temperature. Ultimately, the engine has no choice but to eject hot exhaust into the outside air because it can't extract any more thermal energy from it. The second law of thermodynamics can't be violated.

But a real internal combustion engine wastes energy and extracts less work than the second law allows. For example, some heat leaks from the burned gas to the cylinder walls and is removed by the car's cooling system. This wasted heat isn't available to produce work. Similarly, sliding friction in the engine wastes mechanical energy and necessitates an oil-filled lubricating system. Overall, a real internal combustion engine converts only about 20 to 30% of the fuel's chemical potential energy into work.

CHECK YOUR UNDERSTANDING #4: Avoiding the Burn

Fuel cells are essentially batteries that convert a fuel's chemical potential energy directly into electric energy, without burning the fuel first. Though harder to build and operate, fuel cells are potentially more energy efficient than combustion engines. Explain.

Improving Engine Efficiency

An engine is most efficient when the burned gas starts the power stroke at the highest possible temperature and pressure and is allowed to expand as much as possible before it's released. The burned gas can then convert most of its thermal energy into work during its expansion. Ideally, the burned gas should begin the power stroke in as small a volume as possible and finish the power stroke in a much larger volume.

The extent to which the cylinder's volume increases during the power stroke is measured by its *compression ratio*, its volume at the end of the power stroke divided by its volume at the start of the power stroke. The larger this compression ratio, the higher the initial temperature and pressure of the burned gas and the more energy efficient the engine. While normal compression ratios are between 8 to 1 and 12 to 1, those in high compression engines may be 15 to 1.

Unfortunately, the compression ratio can't be made arbitrarily large. The more tightly the engine packs the fuel-air mixture during the compression stroke, the greater its pressure, density, and temperature become. If the engine compresses the fuel-air mixture too much, it will ignite all by itself. This spontaneous ignition due to over compression is called *preignition* or *knocking*. When an automobile knocks, the gasoline burns during the compression stroke. The engine is not ready to extract work from the cylinder so much of the energy is wasted. Occasional knocking doesn't really hurt the engine, but it does waste energy.

There are two ways to reduce knocking. First, you can improve the uniformity of mixing between the fuel and air. Fuel injection provides excellent mixing, which is why it has replaced carburetion as the way of introducing fuel in the cylinders of all modern cars. Fuel injection also allows the car's computer to adjust the amount of fuel sprayed into a cylinder during its induction stroke. That control ensures complete combustion and helps to reduce knocking. Still, an occasional tune-up is important because a well-tuned engine is less likely to knock than one that is badly adjusted.

Second, you can use the right fuel. Fuels that ignite less easily are more resistant to knocking. A gasoline that contains larger or highly branched hydrocarbon molecules is harder to ignite than a gasoline that contains smaller, straighter molecules. Octane (actually isooctane) is one of the larger, branched molecules found in gasoline and it has become the standard measure of resistance to knocking. The higher a gasoline's "octane number," the more resistant it is to knocking. Regular gasoline has an octane number of about 87, premium has an octane number of about 93, and pure octane is 100.

Gasolines also contain anti-knock additives to impede ignition and minimize knocking. For decades, the main anti-knock additive was tetraethyl lead but this toxic substance has been largely replaced by less poisonous alternatives.

Choosing the right fuel is simply a matter of finding the lowest octane gasoline that your car can use without excessive knocking. A small amount of knocking during rapid acceleration or when climbing a steep hill is normal because each cylinder is then compressing large amounts of the fuel-air mixture. When less fuel-air mixture is introduced into the cylinder, it shouldn't knock. Most modern automobiles that are properly tuned work well on regular gasoline. Premium gasoline is only needed by a few high-performance cars with high compression engines. Premium gasoline in a normal car is a waste of money.

Since knocking sets the limit for compression ratio, it also sets the limit for efficiency in a gasoline engine. However, diesel engines avoid the knocking problem by separating the fuel and air during the compression stroke (Fig. 7.2.4). Invented by German engineer Rudolph Christian Karl Diesel (1858–1913) in 1896, the diesel engine has no spark plug to ignite the fuel. Instead, it compresses pure air with an extremely high compression ratio of perhaps 20 to 1 and then injects diesel fuel directly into the cylinder just as the power stroke begins. The fuel ignites spontaneously as it enters the hot, compressed air. Because of its high compression ratio, a diesel engine is more energy efficient than a standard engine. Unfortunately, it's also harder to start and requires carefully timed fuel injection.

In both gasoline and diesel engines, fuel injection is sometimes combined with a *turbocharger* or *supercharger*. These devices are essentially fans that pump outside air into the cylinders during their induction strokes. By squeezing more fuel-air mixture into the cylinders, a turbocharger or supercharger increases the engine's power output. The engine burns more fuel each power stroke and behaves like a larger engine. The fan of a turbocharger is powered by pressure in the engine's exhaust while the fan of a supercharger is driven by the engine itself.

The downside of turbochargers and superchargers, other than being expensive and wearing out rather quickly, is that they encourage knocking. As they pack air into the cylinders, they also do work on it and heat it up. Since the fuel-air mixture enters the engine hot, it ignites spontaneously during the compression stroke. To avoid knocking in a car equipped with one of these devices, you may need to use premium gasoline. Some cars are equipped with an *intercooler*, a device that removes heat from the air passing through the turbocharger. By providing cool, high-density air to the cylinders, the intercooler reduces the peak temperature reached during the compression stroke and avoids knocking.

CHECK YOUR UNDERSTANDING #5: Steam Heat

How can a steam engine be more energy efficient when it operates on 325°C steam than when it uses 300°C steam?

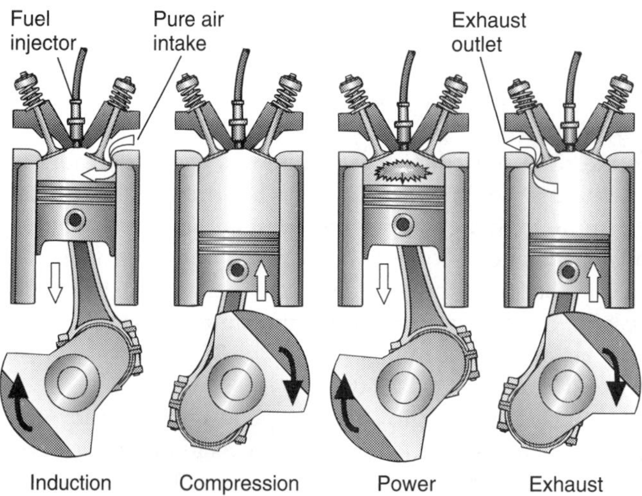

Fig. 7.2.4 - A diesel engine cylinder contains pure air during the compression stroke. As the piston does work on it, this air becomes extremely hot. At the start of the power stroke, diesel fuel is injected into the cylinder. The fuel ignites spontaneously and the hot burned gas does work on the piston and engine during the power stroke.

Manual Transmission: Friction and Gears

Just as a bicycle has gears and a freewheel between its pedals and rear wheel, an automobile has a *transmission* between its engine and wheels. This transmission receives torque from the engine and delivers it to the wheels, though it generally provides mechanical advantage along the way and can also decouple the engine from the wheels when necessary. Mechanical advantage allows the car to climb a hill and decoupling permits it to stop at a traffic light.

There are two types of transmissions: manual and automatic. In a car with *manual transmission*, the driver is responsible for decoupling the engine from the wheels and for adjusting the mechanical advantage. An *automatic transmission* accomplishes both tasks on its own (Fig. 7.2.5).

In a manual transmission, the decoupling is performed by the *clutch*. The clutch consists of two disks, one of which is turned by the engine and one of which ultimately turns the wheels. Normally, the two disks are pressed together hard, so that friction between them forces them to turn together. However, when you depress the clutch pedal inside the car, a mechanism separates the two disks so that they no longer rub together and the engine can turn independently of the wheels. As you release the clutch pedal, the two plates come back together and sliding friction begins to equalize their angular velocities. Once the plates have reached the same angular velocity, static friction keeps them turning together. Among other things, the clutch permits you to start a car moving gradually without stopping the engine from turning.

Because the two clutch plates experience sliding friction as you step on or off the clutch pedal, they convert work into thermal energy during those moments. Normally, you spend only a small fraction of the time turning the clutch on or off so that these periods of sliding friction are brief and infrequent. However, if you "ride the clutch" and allow the plates to slip against one another for a long time, sliding friction will overheat the clutch and it will burn out.

The clutch transmits torque from the engine to the *gearbox*. In the gearbox, incoming torque turns a set of gears. Each turning gear will transmit torque to any other gear that engages with it. The wheels are connected to another set of gears that are normally not engaged with the turning gears. However, a lever by the driver's seat allows him or her to engage a particular pair of gears so that the engine begins to turn the wheels.

In a typical car with five forward gears and one reverse gear, the driver can bring together five different combinations of gears for forward motion. The lowest gear is one in which the engine turns much more rapidly than the wheels. Although the wheels turn slowly, they receive a large torque from the gearbox. You can use first gear to start a car moving or to climb a very steep hill. In the highest gear, the wheels turn as fast or even faster than the engine. While the wheels turn rapidly, they receive only a small torque from the gearbox. You can use fifth gear to drive at high speed on a fairly level road or down a hill.

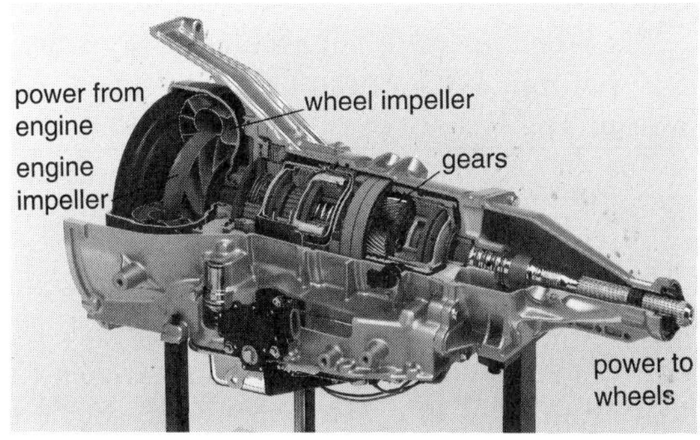

Fig. 7.2.5 - This cut-away view of an automatic transmission shows the two impellers on the left driving a series of hydraulically controlled gears. The transmission fluid is sealed inside the left portion of the transmission.

A mechanical gearbox also has a reverse gear. In reverse, the wheels turn the opposite direction from the forward gears. The gearbox makes this reversal simply by adding one extra gear between the engine and the wheels. Each time one gear turns a second gear, the direction of rotation reverses. So the reverse gear just involves one extra reversal.

CHECK YOUR UNDERSTANDING #6: Climbing in Low Gear

When you drive your car up a steep hill, it should be in first or second gear so that the engine turns much more rapidly than the wheels. Explain.

Automatic Transmission: Fluids and Gears

In a car with an automatic transmission, the clutch is replaced by one or more *fluid couplings* and the gearbox is designed so that the car can operate it automatically. In a fluid coupling, torque is transmitted from the engine to the wheels by a fluid that circulates between two fan-like *impellers*. The engine turns the first impeller, which pushes transmission fluid toward the second impeller. This moving fluid turns the second impeller, which ultimately turns the car's wheels. The fluid transfers power from the engine to the wheels because the first impeller does work on the fluid and the fluid does work on the second impeller.

What makes this funny, indirect coupling so useful is that it doesn't waste power, even if the two impellers are turning at very different speeds. When the second impeller is stopped while your car is at a red light, the transmission fluid can't do any work on it. The force on the impeller may be large, but it moves no distance. The transmission fluid stops circulating through the transmission and the first impeller spins easily so that it extracts very little power from the engine. A *torque converter* is useful variation on this fluid drive concept (see ❒).

❒ Some automatic transmissions combine the functions of the fluid coupling and part of the gearbox into one hydraulic mechanism called a *torque converter*. A torque converter is a fluid coupling that contains two differently shaped impellers. The first impeller turns rapidly, at the rotational speed of the engine. The second impeller turns more slowly but delivers more torque to the wheels.

When you permit the car to move, the second impeller begins turning and gradually picks up speed until it's turning at roughly the same angular velocity as the first impeller. Throughout this process, power flows efficiently from the first impeller to the second impeller and the car accelerates forward smoothly. An automatic transmission jerks and bumps less than a manual transmission.

The gearbox in an automatic transmission operates hydraulically. Hydraulic fluid pushes friction plates together so that they stop various parts of the transmission from turning. Complicated collections of gears are used so that stopping one gear from turning transfers torque to other gears. Mechanisms in the car sense its speed and choose which friction plates should be open and which should be closed. These plates determine the relative rate of rotation between the shaft that carries torque into the automatic transmission and the shaft that carries torque out of it.

CHECK YOUR UNDERSTANDING #7: No Need to Shift

When you stop at a light in a manual transmission car, it's best to put the car in neutral to reduce wear on the clutch. But when you stop an automatic transmission car, there is no need to put it in neutral. Why doesn't something in the automatic transmission wear out as a result?

Summary

How Automobiles Work: An automobile engine extracts work from its chemical fuel by burning that fuel inside its cylinders and making the resulting burned gas do work on the engine. Most engines have at least four cylinders, each of which requires four strokes to extract work from the fuel. During the induction stroke, a piston moves out of the cylinder and fuel and air enter it. During the compression stroke, the piston moves into the cylinder, compressing this fuel-air mixture to high density, pressure, and temperature. An electric spark then ignites the mixture and converts it into extremely hot burned gas. During the power stroke, the piston again moves out of the cylinder while the hot gas does work on it. This work is what powers the car. Finally, during the exhaust stroke, the piston moves into the cylinder and ejects the burned gas. The cylinder then begins again with fresh fuel and air.

The moving pistons turn a crankshaft so that power leaves the engine through a rotating shaft. This shaft powers the transmission, which in turn powers the car's wheels. The transmission provides adjustable mechanical advantage, so that the car can climb steep hills or drive at high speeds on a level roadway. It also allows the engine to be decoupled from the wheels so that the car can come to rest without stopping the engine from turning.

The Physics of Automobiles

1. A heat engine is a device that permits heat to flow from a hotter region to a colder region, extracting some of that heat as useful work in the process.

2. Moving heat from the hotter region to the colder region increases the total entropy of the system. The second law of thermodynamics then allows some heat to be converted to work.

3. A heat engine's maximum efficiency is limited by the temperatures of the hot and cold regions. The hotter the hot region and the colder the cold region, the more efficient the heat pump can be. Its efficiency is the fraction of the thermal energy extracted from the hot region that's converted into ordered energy.

4. Compressing a gas requires work and results in an increase in the temperature of that gas.

Check Your Understanding - Answers

1. Yes.

Why: A heat engine is essentially a heat pump operating backward. The air conditioner (a heat pump) uses ordered electric energy to pump heat from the cold air in your home to the hot outdoor air. The heat engine we are considering would use the flow of heat from the hot outdoor air to the cold air in your home to produce ordered electric energy.

2. The pressure is much higher in the burned gas than in the unburned fuel-air mixture.

Why: The amount of work done on the piston by the gas or done on the gas by the piston depends on the pressure inside the cylinder. The higher that pressure, the more outward force the piston experiences and the more work is done as it moves. The sudden rise in pressure that occurs when the fuel-air mixture burns explains why the burned gas can do so much work on the piston as it moves outward.

3. The person turning the crank had to do all the work needed to move the engine's pistons through the three non-power strokes.

Why: Before the engine started running on its own, it couldn't provide any of the energy the cylinders needed during the induction, compression, and exhaust strokes. The person turning the crank had to provide this energy. Once the fuel started burning, the power strokes could take over, but up to that point, turning the crank was hard work.

4. Because fuel cells don't turn the fuel's ordered energy into thermal energy, they don't have to operate as heat engines. In principle, they can convert all of the fuel's chemical potential energy into electric energy, unlike combustion engines.

Why: Fuel cells remain a promising alternative to combustion engines because they avoid the wasted energy that comes with burning fuel. However, making fuel cells that are efficient and robust is difficult and expensive. At present, fuel cells are only used in special situations, such as in spaceships, where energy efficiency is far more important than cost.

5. Like all heat engines, the steam engine can convert more thermal energy into work when the temperature of its hot object (the steam) increases.

Why: The steam engine converts thermal energy into work as heat flows from the hot steam to the outside air. The greater the temperature difference between those two objects, the more efficient the steam engine can be at turning thermal energy into work. That is why most steam engines use extremely hot steam.

6. Turning the wheels as your car climbs a hill requires an enormous torque. To keep them turning, the transmission must provide mechanical advantage, exchanging the engine's high angular velocity and modest torque for a modest angular velocity and high torque.

Why: On a hill, friction and the car's weight exert huge slowing torques on its wheels. The engine alone can't exert enough torque to keep the wheels turning. When you shift into a low gear, the transmission reduces the angular velocity while increasing the torque. The power leaving the transmission is equal to the power entering it, but that power arrives at the wheels as a high torque at a modest angular velocity.

7. The fluid coupling in the automatic transmission experiences no wear when you stop at the light.

Why: Stopping the car prevents the transmission's second impeller from turning so the first impeller simply turns idly, stirring the transmission fluid but doing little work. Since no surfaces rub against one another there is no wear in the automatic transmission. Shifting it out of gear at a light may satisfy the compulsive driver but is worse for the automatic transmission than simply leaving it alone.

Glossary

heat engine A device that converts thermal energy into work as heat flows from a hot object to a cold object.

Review Questions

1. When heat flows from a hot object to a cold object, what happens to the total entropy of the two objects?

2. When a heat engine does useful work on a machine, it transfers ordered energy to that machine. Where does the heat engine get this energy from?

3. Why can't a heat engine operate with only a hot object from which to obtain thermal energy? Why does it also need a cold object? What does it do with that cold object?

4. How do the relative temperatures of the hot and cold objects affect a heat engine's ability to convert heat into work?

5. How is the thermal energy in burning gasoline turned into ordered energy by the cylinder and piston of an engine?

6. Why must a car engine release heat into the air around it?

7. Why does the fuel and air mixture in a cylinder become hot during the compression stroke?

8. How is energy transferred from one cylinder to another as a multi-cylinder engine operates? When does a cylinder do work on another cylinder?

9. Why is a diesel engine more fuel efficient than a gasoline engine? What limits the efficiency of a gasoline or diesel engine?

10. A gasoline engine provides its maximum power when it's turning at a certain speed. How does the transmission make it possible for the engine to turn at approximately that speed, regardless of whether the car is moving slowly uphill or rapidly along level pavement?

11. How does the fluid in an automatic transmission transfer power from the engine to the wheels? Why is no power transferred when the wheels stop turning?

Exercises

1. Winds are driven by differences in temperature at the earth's surface. Air rises over hot spots and descends over cold spots, forming giant convection cells of circulating air. Near the ground, winds blow from the cold spots toward the hot spots. Explain how the atmosphere is acting as a heat engine.

2. If you transfer a glass baking dish from a hot oven to a cold basin of water, that dish will probably shatter. What produces the ordered mechanical energy needed to tear the glass apart?

3. Freezing and thawing cycles tend to damage road pavement during the winter, creating potholes. What provides the mechanical work that breaks up the pavement?

4. The air near a wood stove circulates throughout the room. What provides the energy needed to keep the air moving?

5. In passive solar heating, sunlight entering the windows of a house heats the air, which rises and circulates throughout the house. Since this circulation is a mechanical motion that's driven by heat, it's a heat engine. Describe the heat flow in this heat engine.

6. Hurricanes are giant heat engines powered by the thermal energy in warm ocean regions and the order in colder surrounding areas. Why are hurricanes most violent when they form over regions of unusually hot water at the end of summer?

7. On a clear sunny day, the ground is heated uniformly and there is very little wind. Use the second law of thermodynamics to explain this absence of wind.

8. A plant is a heat engine that operates on sunlight flowing from the hot sun to the cold earth. The plant is a highly ordered system with a relatively low entropy. Why doesn't the plant's growth violate the second law of thermodynamics?

9. A diesel engine burns its fuel at a higher temperature than a gasoline engine. Why does this difference allow the diesel engine to be more efficient at converting the fuel's energy into work?

10. A chemical rocket is a heat engine, propelled forward by its hot exhaust plume. The hotter the fire inside the chemical rocket, the more efficient the rocket can be. Explain this fact in terms of the second law of thermodynamics.

11. A lawnmower uses a one cylinder, two-stroke engine that combines the power stroke with the induction stroke and the compression stroke with the exhaust stroke. Explain why this engine won't run smoothly unless it's attached to the mower's massive cutting blade and can exchange energy and angular momentum with that blade.

12. Why is a car more likely to knock on a hot day than on a cold day?

13. Why does air become hot when you pump it into a bicycle tire with a hand pump?

14. An acquaintance claims to have built a gasoline-burning car that doesn't release any heat to its surroundings. Use the second law of thermodynamics to show that this claim is impossible.

15. A high-flying airplane must compress the cold, rarefied outside air before delivering it to the cabin. Why must this air be air conditioned after the compression?

16. If you arrange two fans face-to-face and turn on the first fan, the second fan will also begin to spin. How is power flowing from the first fan to the second fan?

Epilogue for Chapter 7

In this chapter we examined the ways in which heat flows between objects, and studied two mechanisms that are used to control the flow of heat. In *air conditioners*, we learned how heat pumps use ordered energy to pump heat against its natural flow. We explored the electrically powered mechanism used by an air conditioner to make the room air colder and the outdoor air hotter and found out why it becomes less energy efficient as the temperature difference increases. Finally, we recognized that there is no alternative way to cool room air; that its temperature can only be reduced by transferring its thermal energy elsewhere.

In *automobiles*, we saw that heat engines divert some of the heat flowing from a hot object to a cold object and convert it into work. We studied the internal combustion engine to see how it works as a heat engine and learned why it's most energy efficient when the burning gasoline is much hotter than the outside air. We examined the roles of the two objects in a heat engine, hot and cold, finding that the hot object provides the energy needed to do work while the cold object provides the order that makes the conversion of thermal energy into ordered energy possible.

Explanation: A Paper Wind Turbine

The paper wind turbine is an example of a heat engine. Heat flows from the hot bulb to the cold room via convection in the air. The wind turbine uses the air's convecting motion to convert thermal energy into work. Like any heat engine, the turbine is constrained by the laws of thermodynamics and works best when the bulb is much hotter than the room air. Though not a practical power source, this turbine is much easier to build than an internal combustion engine.

Cases

1. A solar panel is an electronic device that produces electric power when it's exposed to sunlight. It's also a type of heat engine.

a. What object supplies the heat that powers this heat engine?
b. What object receives the waste heat that passes through this heat engine?
c. If you try to run a solar panel by exposing it to light from an incandescent bulb, it will produce relatively little electric power. Explain this result in terms of the temperature of the bulb's filament.
d. Suppose that the solar panel could withstand the sun's temperature. If you were to immerse the panel in the sun, it wouldn't produce any electricity. Why not?
e. A thermoelectric cell produces electricity when you heat one side of it and cool the other. How does this device resemble a solar panel?

2. A refrigerator is a heat pump that cools the contents of a food locker. Like an air conditioner, it has an evaporator, a condenser, and a compressor.

a. Where is the evaporator located in a refrigerator?
b. Where is the condenser?
c. What will happen to the food's temperature if you wrap the refrigerator's condenser and compressor in thermal insulation, so that they can't exchange heat with the room?
d. What will happen to the temperature of the room if you open the door to the refrigerator and let it run for a few hours?
e. Why does it take more electricity to operate a freezer than a refrigerator of similar size and quality?

3. On a sunny morning at the seashore, the sun warms the land quickly while the sea remains cool.

a. The air above the land also warms and expands. What happens to its density? Why?
b. Because of its expansion, the column of air over the land becomes taller than the column of air over the nearby water. High altitude air begins to flow from over the land to over the water. What causes this flow?
c. With the weight of additional air pushing down on it, the air just above the water experiences a rise in pressure. It begins to flow toward the lower pressure air over the land and a land breeze forms. If the land is the hotter region and the water is the colder region, which way is heat flowing?
d. Explain why this mechanism for moving air back and forth near the seashore is a form of heat engine.

4. Desalinating salt water is one way arid regions are able to obtain drinking water. However it's not an easy process.

a. Which has more entropy: a container of salt water or a container of fresh water plus a box of salt?

b. Why does your answer to **a** indicate that the desalination process requires either the conversion of ordered energy into thermal energy or the movement of heat from a hotter region to a colder region?

c. One technique for desalination is reverse osmosis, in which salt water is squeezed through a special membrane that permits only pure water to pass. Show that squeezing water through this membrane takes work (ordered energy).

d. Another technique for desalination is distillation, in which salt water is boiled in a hotter region to form water vapor that is then condensed to pure water in a colder region. This desalination process is a heat engine that separates the water from the salt. Explain why it's a heat engine.

e. Still another desalination technique is freezing the salt water into pure ice and salt crystals. In this case the salt water has to be cooled to very low temperatures to freeze it, then warmed to a much higher temperature to melt it. Show that this technique is also using a heat engine to drive the desalination process.

5. Decompression sickness was first observed in workers preparing underwater foundations or *caissons* for bridges. The work had to be done in pressurized air to keep the water out. When the pressure was released, the workers would experience "caisson disease," a painful or even fatal disorder in which gas bubbles formed inside their tissues. Because they bent over in pain, their illness was called "the bends."

a. The workers would enter the caisson while it was at atmospheric pressure. The doors were then sealed and air was pumped in. The temperature in the caisson rose rapidly. What caused this temperature rise?

b. A few minutes after the caisson was brought up to full pressure, the air inside the caisson would return to normal temperature. What had happened to its thermal energy?

c. When the shift was over, the doors to the caisson were opened and the air was allowed to rush out. That outgoing air was ice cold. The air inside the caisson was also so cold that sometimes it would begin to snow. What caused this temperature drop?

d. When during this cycle of pressurizing and depressurizing was work done *on* the air inside the caisson

e. When was work done *by* the air inside the caisson?

f. What heat transfer took place?

6. It's your first day on the staff of the U.S. Patents Office and you're excited about having some new ideas come across your desk.

a. An inventor comes to you with a small box that is supposed to make batteries obsolete. The inventor claims that the box can produce electricity forever without having to be recharged. Why can you be sure that this claim is nonsense?

b. Another inventor comes in with a small motor-like device that's powered by a burning candle. The inventor claims that this device takes the heat from the candle and turns it entirely into work. The device thus doesn't warm the room at all. Your knowledge of the laws of thermodynamics assures you that this claim, too, is nonsense. Why?

c. As though the entire population of inept inventors was released on you in one day, another character comes in claiming to have a heat pump that can transfer heat out of a box of corn flakes for as long as you like. The cereal just gets colder and colder. Once again, you know that this is impossible. Explain.

d. Just when you thought you were out of the woods, a person comes in with a coffee mug that automatically reassembles itself if you accidentally break it. You're supposed to put the broken mug in a box and wait. In a minute or two, the mug will be as good as new. You agree with the inventor that this reconstruction trick doesn't violate any of the laws of motion. However, you can still be sure that it won't work. Why?

7. While it flies at high altitudes, a jet airplane must continually replace its cabin air with fresh air from outside. But because that outside air is very cold and its pressure and density are both quite low, the plane must compress the air.

a. The airplane already has an excellent compressor on board: its jet engines. It uses some of the air that has passed through the initial compressors of these engines, before fuel has been added. This compression process makes the air extremely hot. Why?

b. Has the air's total energy changed as a result of its passage through the compressor? If it has changed, what provided or absorbed some of the air's energy?

c. Rather than baking its occupants, the airplane refrigerates the compressed air. It uses an air conditioner to pump heat from the hot compressed air to the cold outside air. Why is it surprising that the airplane uses a heat pump to move heat in that direction?

d. As fresh air flows into the cabin, stale air is released from the plane. Overall, cold air enters the plane's engines and hot air leaves the plane through air outlets. Heat also flows into the outside air from the air conditioner. How is the entropy of the outside air affected by this process?

e. The only real change inside the airplane is that its fuel gradually diminishes. Since that fuel doesn't contribute to the plane's entropy, the plane's entropy isn't changing significantly. How can the plane continue to operate without an increase in its entropy? Doesn't this violate the second law of thermodynamics?

8. When air in the atmosphere rises or falls it often experiences a change in temperature. Interesting examples of this behavior are the chinook winds in the Rocky Mountains and the foehns in the Alps.

a. When a wind containing moist air blows up the side of a tall mountain, its altitude increases and its pressure drops. What happens to this air's density?

b. Why does this rising air experiences a drop in temperature?

c. The air's dropping temperature causes its moisture to condense into rain while the air itself continues over the mountain top without the moisture. As the moisture condenses from a gas into a liquid, it warms the air. Which portion of an air conditioner employs a similar process?

d. The warmed air then descends into the valley beyond the mountain and it warms up even more. The air reaching the valley is unusually hot and dry. Explain this additional warming during its descent.

CHAPTER 8

PHASE TRANSITIONS

Most of the materials around us take one of three simple forms or *phases*: solid, liquid, or gas. For example, salt is a solid, alcohol is a liquid, and air is a gas. But that list is only true near room temperature. At very high temperatures, all three substances are gases while at very low temperatures, all three are solids. Between these extremes, something happens to transform the materials from one form to another. Such transformations are called *phase transitions*. As we'll see in this chapter, phase transitions can be caused by changes in temperature, pressure, and chemical composition.

Experiment: Melting Ice with Pressure

Pressure can cause interesting phase transitions between ice and water. Here is an experiment that requires an ice cube, a piece of dental floss (or another strong, thin string), and a 2 to 5 kg weight (a dumbbell or hand weight is ideal). Tie the piece of dental floss into a loop about 30 cm (1 foot) in diameter and hang it from the midpoint of the ice cube. Prop the ice cube between two closely spaced chairs so that the dental floss loop dangles below. Hang the weight from the dental floss. Be prepared for everything to get a bit wet and for the weight to fall if something slips.

The ice cube will slowly melt but as it does, a strange thing will happen to the dental floss. It will slowly drift into the ice cube and will be covered by a transparent layer of new ice. After about ten or twenty minutes, the dental floss will have drifted about half way through the ice cube and will be trapped in the ice. If you wait long enough, the dental floss will emerge from the bottom of the ice cube, having passed all the way through it. The ice cube will remain in one

piece because new ice will have formed above the dental floss as it drifted downward.

Chapter Itinerary

This strange behavior of ice under pressure is called *regelation* and is unique to the phases of water. Because it exhibits both normal and unusual phase transitions, water is the main focus of this chapter. In *water, steam, and ice,* we'll examine the transformations between those three phases of water to see why and how they occur. We'll look at melting, freezing, evaporation, sublimation, condensation, and boiling to find the molecular reasons for these phase transitions. We'll also explore what happens when you dissolve chemicals in water and how the presence of these chemicals affects the temperatures at which water freezes and boils.

Despite this chapter's concentration on water, steam, and ice, the phase transitions that it describes can be found in a great many other materials as well. The wax melting in a candle and the solvent evaporating from a layer of acrylic spray paint are both undergoing phase transitions.

Section 8.1

Water, Steam, and Ice

Water is probably the single most important chemical in our daily lives. It plays such a central role in biology, climate, commerce, industry, and entertainment that it merits a whole section of its own. It also exhibits the three classic phases of matter, solid, liquid, and gas, so that much of what we learn about water is applicable to other materials as well. However, there are a few aspects of water that are almost unique in nature. Water really is a remarkable substance.

Questions to Think About: *Why does adding ice make a drink cold? Why is ice so slippery? How does snow disappear from the ground even when it's very cold outside? Why do icebergs float on water? How does perspiration cool you off? What is the difference between evaporating and boiling?*

Experiments to Do: *Experimenting with water is easy. Pick up an ice cube with wet fingers. What happens if the ice cube has just come out of the freezer? What if it has been melting on the table for a few minutes? Why is there a difference? Which ice cube feels the most slippery when you touch it with dry fingers? Put both ice cubes in water. Why do they float?*

Now heat tap water in a pot. Shortly before the water starts to boil, you will see mist begin to form above it. What is this mist and why does it form? Carefully feel the mist, but don't burn yourself! The mist feels damp because it contains water. How can water be leaving the pot before the water boils? Notice the small gas bubbles on the walls of the pot. It's not steam. Where did this gas come from?

Once the water boils, gaseous water or steam will appear in the mist. Don't touch this steam because it can burn you quickly. Why does steam release so much thermal energy when it touches your skin?

Solid, Liquid, and Gas: the Phases of Matter

Fig. 8.1.1 - The three phases of water: solid (ice), liquid (water), and gas (steam).

Like most substances, water exists in three distinct forms or **phases**: solid ice, liquid water, and gaseous steam (Fig. 8.1.1). These phases differ in how easily their shapes and volumes can be changed. Solid ice is rigid and incompressible, preventing you from altering an ice cube's shape or volume. Liquid water is an incompressible fluid, so that you can reshape a water drop but can't change its volume. Gaseous steam is a compressible fluid, so that you can vary both the shape and volume of the steam in a tea kettle.

These different characteristics reflect the different microscopic structures of steam, water, and ice. *Steam* is a gas, a collection of independent molecules that is also called *water vapor*. These water molecules bounce around their container, periodically colliding with one another or with the walls. The water molecules fill the container uniformly and easily accommodate any changes in its shape or size. Enlarging the container simply decreases the steam's density and lowers its pressure.

To remain independent of one another, the water molecules in steam need a certain amount of thermal energy. Without this energy, they stick together to form water. *Water* is a liquid, a disorderly, fluid collection of molecules that is held together by chemical bonds. These bonds pack the molecules tightly together and give water its fixed volume. However, water still has enough thermal energy for its individual molecules to separate briefly and form new bonds with different partners. Its evolving microscopic structure makes water fluid. While its volume is fixed, its shape is not.

When the water molecules have even less thermal energy, they are unable to rearrange and cling together stiffly as ice. *Ice* is a solid, a rigid collection of chemically bound molecules. Like most solids, ice is crystalline, with orderly arrangements of molecules that extend over large distances. This order produces the beautiful crystalline facets of snowflakes and frost.

Just as an orderly arrangement of cannon balls takes up less volume than a disorderly one, a crystalline solid usually occupies less volume than its corresponding liquid. That's why the solid phase of a typical substance sinks in the liquid phase of that same substance.

But there is an exception to this rule: water. Ice's crystalline structure is unusually open and its density is surprisingly low. Almost unique in nature, solid ice is slightly less dense than liquid water so that ice floats on water. That's why icebergs float on the open ocean and ice cubes float in your drink. In fact, water reaches its greatest density at about 5° C (40° F). Heated above that temperature, water behaves normally and expands. However water also expands as you cool it below that temperature, a very unusual effect.

CHECK YOUR UNDERSTANDING #1: Skating Anyone?
As a pond freezes, where does the ice begin to form?

Melting Ice and Freezing Water

When you heat ice, it remains solid until its rising temperature reaches 0° C. At that point, the ice stops getting warmer and begins to melt. **Melting** is a **phase transition**, a transformation from the ordered solid phase to the disordered liquid phase. This transition occurs when heat breaks some of the chemical bonds between water molecules and permits the molecules to move about. The ice transforms into water, losing its rigid shape and crystalline structure.

0° C is ice's **melting temperature**, the temperature at which heat goes into breaking bonds and converting ice into water, rather than making the ice hotter. The ice-water mixture remains at 0° C until all of the ice has melted. When only water remains, heating it can again cause its temperature to rise.

The heat used to transform a certain mass of solid into liquid, without changing its temperature, is called the **latent heat of melting**. The bonds between water molecules are relatively strong, so that water has an enormous latent heat of melting: it takes about 333,000 J of heat to convert 1 kg of ice at 0° C into 1 kg of water at 0° C. That same amount of heat would raise the temperature of 1 kg of liquid water by about 80° C so that it takes almost as much heat to melt an ice cube as it does to warm the resulting water all the way to boiling.

The latent heat of melting reappears when you cool the water back to its melting temperature and it begins to **solidify**. As you remove heat from water at 0° C, the water freezes into ice rather than becoming colder. Because the water molecules release energy as they bind together to form ice crystals, the water releases heat as it freezes. The heat released when transforming a certain mass of liquid into solid, without changing its temperature, is again the latent heat of melting. You must add a certain amount of heat to ice to melt it and you must remove that same amount of heat from water to solidify it.

Ice's huge latent heat of melting is what keeps a mixture of water and ice at 0° C. As long as both water and ice are present together in your glass, they are in the process of either melting or freezing. Any heat you add to the glass goes into melting more ice, not into raising its temperature. Any heat you remove from the glass comes from freezing more water, not from lowering its temperature. With ice floating in your drink, it will remain at 0° C, even in the hottest or coldest weather (Fig. 8.1.2).

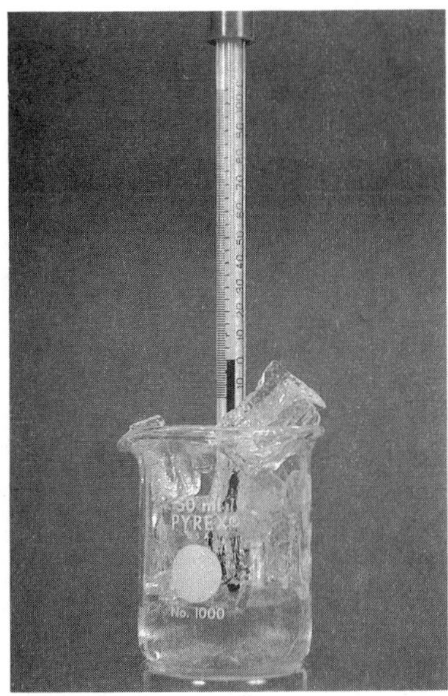

Fig. 8.1.2 - Ice and water can coexist only at 0° C, ice's melting temperature.

CHECK YOUR UNDERSTANDING #2: Keeping Crops Warm

Fruit growers often spray their crop with water to protect it from freezing in unusually cold weather. How does liquid water keep the fruit from freezing?

Sublimation, Evaporation, and Condensation

The surface of solid ice or liquid water is a busy place. Water molecules are constantly leaving it as water vapor and returning to it as ice or water. When molecules leave the surface, the water is *evaporating* or the ice is *subliming*. When molecules return to the surface, the water vapor is *condensing*. This simple picture of water molecules taking off and landing on the surface of ice or water explains many familiar phenomena. But to complete the picture, we must follow the flow of energy in this system.

Ice and water both contain thermal energy, which is exchanged between neighboring molecules and keeps them in motion. While the average water molecule is unable to break free from the surface, molecules occasionally obtain enough thermal energy from their neighbors to break their bonds and leave as water vapor. In doing so, these molecules carry away more than their fair share of the water or ice's thermal energy and it becomes colder. Your body uses this effect to keep cool on hot summer days, when perspiration that evaporates from your skin draws heat from you and lowers your temperature.

If you add enough heat to water or ice, it will evaporate or sublime without getting colder. The heat needed to transform a certain mass of solid or liquid into gas, without changing its temperature, is called the **latent heat of vaporization**. Water's latent heat of vaporization is truly enormous because water molecules are surprisingly hard to separate. About 2,300,000 J of heat are needed to convert 1 kg of water at 100° C into 1 kg of water vapor or steam at 100° C. That same amount of heat would raise the temperature of 1 kg of water by more than

500° C! Even more heat is required to convert ice directly into water vapor.

Although we are more familiar with the evaporation of water, we occasionally observe the sublimation of ice. For example, snow gradually disappears from the ground without ever melting and food that is left unprotected in a frostless freezer gradually dries out. While this "freezer burn" is a nuisance at home, sublimation from frozen food is used commercially to prepare freeze-dried food.

Evaporation and sublimation are most rapid when there isn't much water vapor above the water or ice. When the air above the water or ice is full of water molecules, they collide frequently with the surface and experience its attractive forces. Many molecules stick and the water vapor condenses into water or ice.

The latent heat of vaporization reappears when water vapor condenses. The gaseous water molecules bind together as water or ice and release their chemical potential energy as heat. The heat released when transforming a certain mass of gas into liquid or solid, without changing its temperature, is again the latent heat of vaporization. You must add a certain amount of heat to water or ice to vaporize it and you must remove that same amount of heat from water vapor to liquefy or solidify it.

The huge amount of heat released by condensing water vapor is often used to cook food or warm radiators in older buildings. When you steam vegetables, you are allowing water vapor to condense on the vegetables and transfer heat to them. A double-boiler uses condensing steam to transfer heat from a burner to a cooking container in a controlled manner.

CHECK YOUR UNDERSTANDING #3: Tea Time

A kettle of water heats up rapidly on the stove but takes quite a while to boil away. Why does the water take so long to turn into steam?

Relative Humidity

Something is missing from our description of evaporation, sublimation, and condensation. Neglecting ice for the moment, we have seen that sometimes water molecules leave the surface of water and sometimes they return. What determines whether water molecules are leaving or returning?

The answer is that both processes usually occur simultaneously. The airport picture for the surface of water, with water molecules constantly taking off and landing, is a pretty accurate one. There is a continual exchange of water molecules between the gaseous and liquid phases. What matters most is whether there is a net transfer from one phase to the other. If more molecules leave the surface than return, the water is evaporating. If more molecules return to the surface than leave, the water vapor is condensing.

The simplest indicator of whether water molecules will evaporate or condense is relative humidity. **Relative humidity** measures the returning rate as a percentage of the leaving rate. When the relative humidity is 100%, the two rates are equal and there is neither evaporation nor condensation. If the relative humidity is less than 100%, the returning rate is less than the leaving rate and the water evaporates. If the relative humidity is more than 100%, the returning rate is more than the leaving rate and the water vapor condenses.

Relative humidity depends on the temperature and on the density of water vapor in the air. The higher the temperature, the more frequently water molecules leave the water's surface as water vapor. Thus, increasing the temperature reduces the relative humidity. But the higher the density of water vapor, the more frequently water molecules collide with the surface and become liquid. Therefore, increasing the density of water vapor in the air increases the relative humidity.

Relative humidity plays an important role in countless experiences of everyday life. When the relative humidity is low, water evaporates quickly and the air feels dry. Perspiration cools you effectively. When the relative humidity is high (near 100%), water barely evaporates at all and the air feels damp. Perspiration clings to your skin and doesn't cool you much. The cooling of a wet surface can actually be used to measure relative humidity (Fig. 8.1.3).

When the relative humidity exceeds 100%, perhaps because of a sudden loss of temperature, water vapor begins to condense everywhere. Water droplets grow on surfaces as dew or form directly in the air as fog, mist, or clouds. When the humidity remains high, these droplets grow larger and eventually fall as rain. If the temperature is cold enough, water vapor can condense directly to ice, forming frost on surfaces or snowflakes or hail in the air.

CHECK YOUR UNDERSTANDING #4: Seeing Your Breath

When you breathe out on a cold day, you often see mist appearing from your mouth. Explain.

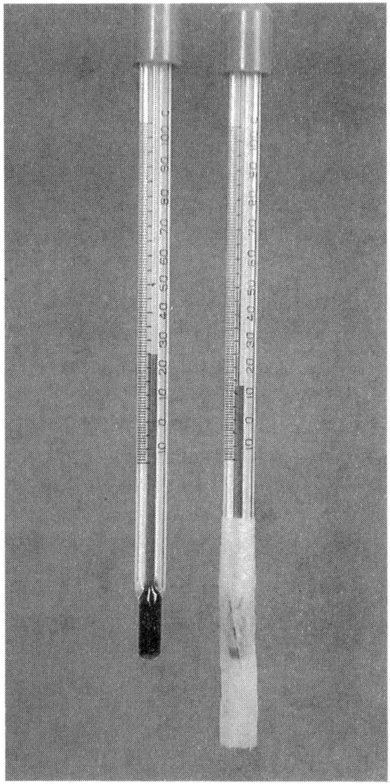

Fig. 8.1.3 - You can determine the air's relative humidity using two thermometers—one of which has a wet cloth wrapped around its bulb (right). Evaporation cools the wet bulb thermometer by an amount related to the air's relative humidity. The dryer the air, the colder the wet bulb thermometer becomes.

Boiling Water

If you seal water inside a container and keep its temperature constant, it will evaporate until the relative humidity inside the container reaches 100%. At that point, water molecules will return to the liquid as often as they leave it. The air will be **saturated** with water, containing just the right density of water molecules to ensure this balanced exchange.

Water molecules also generate part of the air pressure in the container. Water's contribution to the overall air pressure depends on its density in the air. When the air is saturated with water, water contributes its **saturated vapor pressure** to the air's overall pressure.

Water's saturated vapor pressure depends only on its temperature. The hotter the container, the more water vapor it will contain and the higher the water's saturated vapor pressure. Near room temperature, water's saturated vapor pressure is much less than atmospheric pressure. But as water approaches 100° C, its saturated vapor pressure approaches atmospheric pressure. Something special happens right when water's saturated vapor pressure reaches atmospheric pressure. The water begins to boil (Fig. 8.1.4).

Boiling occurs when bubbles of water vapor appear *inside* liquid water. A bubble forms when several water molecules simultaneously break free of their neighbors and create a tiny pocket of gaseous water inside the liquid water. The sudden appearance of this bubble allows evaporation to occur inside the water. More water molecules evaporate into the bubble and it grows in size and mass. Since gaseous water is much less dense than liquid water, the bubble floats upward to the water's surface.

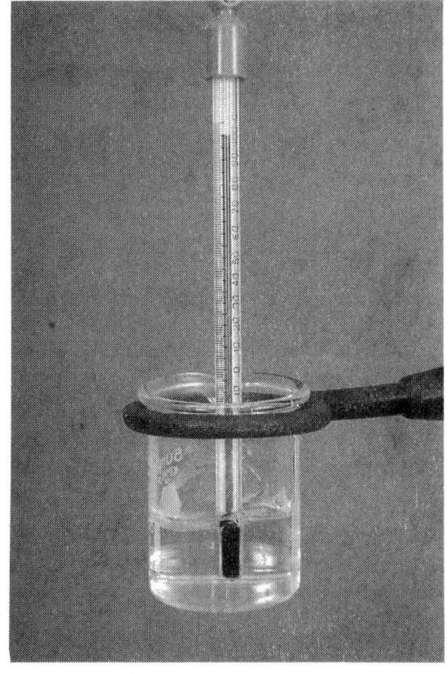

Fig. 8.1.4 - Water boils at 100º C, when atmospheric pressure can no longer smash the bubbles of water vapor.

The only gas present inside these bubbles is water vapor. If the water isn't hot enough, its saturated vapor pressure will be less than atmospheric pressure and any bubbles that begin to form will be crushed immediately by the surrounding water. Only when the saturated vapor pressure of water reaches atmospheric pressure can a steam bubble survive long enough to grow, float upward, and leave the surface of the water. When that occurs, the water is **boiling**. The temperature at which water's saturated vapor pressure reaches atmospheric pressure is water's **boiling temperature**.

Boiling converts water to steam much more rapidly than evaporation, so it takes an enormous input of heat to keep water boiling. Once water has reached its boiling temperature, the conversion of liquid water to gaseous steam consumes all of the heat you add to the water. In an open pot, the water's temperature stops increasing and remains constant at its boiling temperature until all of the liquid water is gone. Only when the water is entirely gaseous does its temperature begin to increase.

The constant, well-defined temperature of boiling water allows you to cook vegetables or an egg at a particular rate. When you place an egg in boiling water, it cooks in 3 minutes because it's in contact with water at its boiling temperature.

But water's boiling temperature depends on the ambient pressure above the water. For an open pot or pan, that pressure is atmospheric pressure. However atmospheric pressure decreases with altitude and depends somewhat on the weather. While water boils at 100° C near sea level, it boils at only 90° C at an altitude of 3,000 m. This reduction in the water's boiling temperature with altitude explains why many recipes must be adapted for use at higher elevations. At 3,000 m, an egg cooks slowly in boiling water because it's surrounded by 90° C water, not 100° C water. The same problem slows the cooking of rice, beans, and many other foods at high altitudes.

In fact, at sufficiently low pressures, water boils even at room temperature. With a vacuum chamber, you can actually get a glass of ice water to boil. At the other extreme, high pressures can make water boil at very high temperatures. In a boiler that produces high-pressure steam for a power plant, water may boil at a temperature exceeding 300° C.

One way to decrease cooking times is to use a pressure cooker. A pressure cooker seals in the steam so that the pressure inside can exceed atmospheric pressure. This increased pressure prevents boiling from occurring until the water's temperature is well above 100° C. If you subject water to twice sea-level atmospheric pressure, it doesn't boil until it reaches 121° C. An egg cooks very quickly at that temperature, as do vegetables and other foods.

CHECK YOUR UNDERSTANDING #5: Getting Into Not-So-Hot Water

If you put a thermometer in boiling water at the top of a 3,000 m mountain, what temperature will the thermometer read?

Why Ice Is Slippery

Pressure also affects the melting of ice. When you squeeze ice, you lower its melting temperature slightly. This behavior is extremely unusual. Pressure prevents virtually any other material from melting at its normal melting temperature because the material would expand during melting and would have to do work against the pressure surrounding it. But ice shrinks during melting because ice is less dense than water. Rather than keeping ice from melting, pressure actually liquefies ice, even below its normal melting temperature.

Melting is particularly easy at the surface of ice, where the crystalline structure is incomplete and disordered. Because they lack a full complement of neighbors, the outermost water molecules are relatively mobile and have a liquid-like character. This watery outer surface makes ice remarkably slippery. When you slide on a frozen walkway or ski down a mountain, you are riding on ice's liquid-like surface. The pressure your weight exerts on the ice thickens its liquid-like surface and makes it even more slippery. Ice skates concentrate your weight to increase this pressure so that they glide effortlessly.

Temperature also affects ice's liquid-like surface, making it thicker as the ice warms toward its melting temperature. That's why ice is most slippery when the weather is near freezing and the ice has a "wet" surface. When the weather is extremely cold, ice has a "dry" surface and is barely slippery at all.

Ice's liquid-like surface also permits ice crystals to fuse together easily, so that snowflakes pack together into snowballs or glaciers. Here, too, the thickness of the liquid-like surface is important. It's hard to make snowballs when the weather is cold and the snow is dry. Snow "packs" best when its temperature is just below freezing.

These behaviors are very unusual. Most solids aren't slippery and don't stick together well at all. Their surfaces are solid and experience friction when they slide across one another. They also don't fuse together on contact. Imagine "ice skating" on a slate sidewalk or trying to make a ball by packing dry sand together. Not likely.

CHECK YOUR UNDERSTANDING #6: Grooming the Ice

The ice of a speed skating rink is kept very close to ice's melting temperature, so that the skaters can travel as fast as possible. Explain.

Dissolving Chemicals in Water

Many chemicals dissolve in water. Those that dissolve best contain molecules that bind strongly to water molecules; more strongly than they bind to themselves. When you add a water soluble chemical to water, water molecules surround its molecules and carry them around individually. The chemical becomes separate molecules and **dissolves** in the water.

To understand this process better, let's look at the structure of water itself. A water molecule consists of two hydrogen atoms and an oxygen atom. For reasons that we'll discuss in Chapter 14, the water molecule has electrically charged ends. Its oxygen atom is negatively charged while its hydrogen atoms are positively charged. Since opposite electric charges attract (we'll discuss this concept in Chapter 11), the oxygen atom of one water molecule tends to stick to a hydrogen atom of another water molecule. This attachment is called a **hydrogen bond**. Hydrogen bonds are quite strong and the large energy needed to break them explains water's enormous latent heats of melting and vaporization.

Chemicals that dissolve well in water are those that have electric charges of their own. Salt dissolves easily in water because it consists of **ions**, electrically charged atoms, that attract the water molecules. When you put salt in water, it breaks up into its constituent ions and these ions are carried around by the water molecules. Sugar dissolves well in water because its oxygen and hydrogen atoms are electrically charged, just as they are in water molecules. Sugar molecules form hydrogen bonds with water molecules and are carried around in the water.

There are also a few gases that dissolve well in water because they bind with water molecules. Carbon dioxide gas is quite soluble in water, where it produces carbonated water or soda. But the second law of thermodynamics allows even those gases that aren't attracted to water molecules to dissolve. They dissolve in small amounts because they increase the overall disorder and entropy of the liquid. This increase in entropy is most important in cold water, so that it dissolves more gas than hot water. When you heat cold water, its dissolved gases reappear as bubbles on the walls of the pot.

The presence of dissolved chemicals in water lowers its melting temperature. This effect is a consequence of the second law of thermodynamics. Ice crystals are essentially pure water so the chemical-water mixture separates during freezing. Since this separation lowers the mixture's entropy, an additional source of order is needed to satisfy the second law. The extra order is provided by cooling the mixture below ice's normal melting temperature. That's why salt water freezes at a lower temperature than fresh water. And because a mixture of salt and water tends to be liquid at temperatures somewhat below water's normal melting temperature, you can melt ice on the sidewalk by sprinkling it with

❐ Concentrated sugar water boils at a much higher temperature than pure water. Boiling temperature is often used as a measure of sugar concentration when preparing candies. The higher the boiling temperature of the mixture, the less water there is in it. Eventually, the water evaporates completely and the temperature rises rapidly toward the melting and caramelizing temperatures of sugar itself.

salt. Actually, any chemical that dissolves easily in water will melt ice. Even sugar will do the trick.

Dissolved chemicals also raise water's boiling temperature. The chemicals interfere with evaporation, changing the balance between evaporation and condensation and lowering water's saturated vapor pressure. The mixture won't boil until water's saturated vapor pressure reaches the ambient pressure, so you must heat the mixture above water's normal boiling temperature (see ❐). Only if you add a chemical that boils more easily than water will the mixture boil at or below its normal boiling temperature.

CHECK YOUR UNDERSTANDING #7: Safe Sidewalks
Why does salt melt ice while sand does not?

Summary

How Water, Steam, and Ice Work: Water, steam, and ice are the liquid, gaseous, and solid phases of the chemical we call water. The water molecules in liquid water are bound together strongly enough to keep them tightly packed, but not strongly enough to give water a rigid shape. In steam, the water molecules are independent and the gas has neither a fixed shape nor volume. The molecules in ice are bound together rigidly in solid crystals, so that ice is rigid and its volume is constant.

Ice can transform into water or vice versa at its melting temperature. Heat added to ice at this temperature causes it to melt without becoming warmer. Heat removed from water at this temperature causes it to freeze without becoming colder. Similarly, ice and water can transform into steam or vice versa through sublimation, evaporation, and condensation. Turning ice or water into steam requires heat and this heat is released when the steam turns back into water or ice. Boiling is a special case of evaporation, in which evaporation occurs within the body of the water itself. For boiling to occur, the water's vapor pressure must equal the ambient pressure on the water, typically atmospheric pressure.

The Physics of Water, Steam, and Ice

1. The atoms or molecules in a gas are essentially independent of one another. In a liquid, the atoms or molecules are in contact with one another but continually change partners and are structurally disordered. In a solid, the atoms or molecules are in contact and don't change partners. The particles in most solids are arranged in orderly crystals.

2. Transitions between solid and liquid occur at the solid's melting temperature. Heat added to the solid at this temperature goes into melting the solid, rather than raising its temperature. Heat removed from the liquid at this temperature comes from freezing the liquid, rather than lowering its temperature.

3. Transitions between solid or liquid and gas occur over a broad range of temperatures as sublimation, evaporation, and condensation. Heat is required to convert the solid or liquid into gas and this heat is released when the gas reverts to a solid or liquid.

4. Boiling is when evaporation occurs in the body of a liquid. It requires that the material's vapor pressure equal the ambient pressure on the liquid.

5. Chemicals that are strongly attracted to water molecules can dissolve in water, separating into individual molecules that are carried around by the water molecules. The presence of dissolved chemicals normally lowers the mixture's melting temperature and raises its boiling temperature.

Check Your Understanding - Answers

1. At the surface of the pond.

Why: Because ice is less dense than water, it floats at the surface of the pond. The pond therefore freezes from its top down. The insulating layer of ice at the top of the pond impedes the freezing of the remainder of the water so that the pond is unlikely to freeze solid. This behavior allows animals to live in the pond during winter. If ice sank, the pond would freeze solid and the animals would die. Thus ice's tendency to float has profound implications for life on our planet.

2. Liquid water releases a large amount of heat as it solidifies and this heat helps to protect the fruit from freezing.

Why: Fruit freezes at a temperature somewhat below 0° C. As cold air removes heat from the water-coated fruit, the water begins to freeze. The liquid water gives up a great deal of heat as it solidifies and prevents the fruit's temperature from dropping below 0° C until the water has completely turned to ice. At that point, ice's poor thermal conductivity insulates the fruit and slows its loss of heat to the colder air.

3. The stove must operate for a long time to provide water's huge latent heat of vaporization.

Why: Converting a kettle of hot water into steam requires an enormous amount of heat. This heat separates the molecules without raising their temperature. Since the stove only provides a certain amount of heat each second, it may take many minutes to boil away the water.

4. As the warm, moist air from your mouth cools, its relative humidity increases above 100%. The water vapor condenses to form a mist of water droplets.

Why: Changes in temperature affect relative humidity. Although the warm air leaving your lungs can hold a large portion of water vapor, its relative humidity increases dramatically as it cools. At lower temperatures, condensation takes over and the invisible water vapor condenses into a mist of visible water droplets.

5. About 90° C.

Why: In the low pressure air found at high altitude, water boils at a relatively low temperature. The higher you go, the lower water's boiling temperature. In space, water boils at or below room temperature, so that you could hold it in your hand as it boiled.

6. Near its melting temperature, ice has a thick liquid-like surface that minimizes the friction experienced by the skaters.

Why: As ice warms toward its melting temperature, its liquid-like surface thickens and makes ice skating easier. Once it reaches its melting temperature, the ice begins to turn to water and becomes even more slippery. However, water exerts drag on skates, so the ice shouldn't melt too quickly.

7. Salt is soluble in water while sand isn't. Only chemicals that dissolve in water will melt ice.

Why: The trick of melting ice with chemicals requires that the chemicals dissolve well in water. Since sand is not soluble in water, it merely sits on the surface of the ice and provides increased traction. Salt actually dissolves in the water-like surface layer of the ice and causes the ice to melt. Unfortunately, at low enough temperatures the resulting salt water refreezes and becomes slippery once again.

Glossary

boil Undergo a phase transition from a liquid to a solid throughout the liquid.

boiling temperature The temperature at which the vapor pressure of a liquid reaches the ambient pressure and evaporation occurs within the body of the liquid.

dissolve Separate into individual particles and disperse throughout a fluid.

hydrogen bond A moderately strong chemical bond in which a positively charged hydrogen atom is attracted to a negatively charged oxygen atom.

ion An atom or molecule with a net electric charge.

latent heat of melting The heat required to disorder the atoms or molecules of a solid and to convert it from a solid to a liquid without raising its temperature.

latent heat of vaporization The heat required to separate the atoms or molecules of a solid or liquid and convert it from a solid or liquid to a gas without raising its temperature.

melt Undergo the phase transition from a solid to a liquid.

melting temperature The temperature at which a material undergoes a transformation between its solid and liquid phases.

phase (of matter) One of the common states of matter, particularly solid, liquid, and gas.

phase transition The process of converting from one phase to another.

relative humidity The rate at which water molecules return to the surface of water or ice as a percentage of the rate at which water molecules leave that surface.

saturated Containing as much of something as possible.

saturated vapor pressure The partial pressure of gas molecules in the vapor above a solid or liquid when there is no net evaporation or condensation occurring.

solidify Undergo a phase transition from a liquid to a solid.

Review Questions

1. In what ways are liquid water and gaseous steam similar? In what ways are they different?

2. In what ways are liquid water and solid ice similar? In what ways are they different?

3. If steam is a transparent gas, what is the mist emerging from a tea kettle?

4. What happens to the water molecules in ice when melting occurs? Why is energy needed to melt ice?

5. If ice and water can exist at the same temperature, why don't they turn into one another spontaneously without any exchanges of energy?

6. How can ice or water become gaseous steam without there being any boiling water involved?

7. When steam condenses on your skin, you feel the flow of heat into your skin. Where does this heat come from?

8. What role does the air's relative humidity play in the cooling you feel when your skin is wet?

9. Rain is usually caused by a sudden drop in the temperature of humid air. Explain.

10. Describe the motion of water molecules on the surface of a cool mirror in a hot, humid bathroom.

11. Why does water boil at a lower temperature on a mountain than in the valley beneath it?

12. Why is ice slipperiest when its temperature is very close to its melting temperature?

13. How does salt melt ice on a sidewalk?

14. What keeps carbon dioxide gas molecules in a glass of soda for such a long time?

Exercises

1. As molten iron solidifies during slow cooling, where in the container holding it does the solid portion form?

2. When you put a loaf of bread in a plastic bag, there is still air around the bread. Why doesn't the bread dry out as quickly in the bag as it does when there's no bag around it?

3. It's possible to run quickly through a bed of hot coals without burning your feet. The secret is to have water or sweat on your feet. How does this moisture reduce the amount of heat the coals transfer to your skin?

4. Why does it take longer to cook pasta properly in boiling water in Denver (the "mile high city") than it does in New York City?

5. A crock pot uses electricity and a thermostat to keep its contents at a constant elevated temperature. Why does a crock pot full of stew consume less electricity when it's covered than when it's opened to the surrounding air?

6. Why is it that you can put a metal pot filled with water on a red-hot electric stove burner without fear of damaging the pot?

7. The molecules of antifreeze dissolve easily in water. Why does adding antifreeze to the water in a car radiator keep the water from freezing in the winter and boiling in the summer?

8. Propane in the tank of a gas grill behaves much like water, except that propane evaporates and boils at a much lower temperature. As the grill consumes gaseous propane from the top of the tank, liquid propane in the bottom of the tank evaporates to replace it. The grill has a pressure regulator because the gas pressure in the tank increases with temperature. Why does the tank pressure increase with temperature?

9. Deluxe cappucino makers create foamed milk by bubbling "dry" steam through it. They first form "wet" steam by boiling water at atmospheric pressure and then convert this steam to "dry" steam by sending it through a second heater so that it's well above 100° C when it enters the milk. This "dry" steam doesn't condense into water until a significant amount of heat has been removed from it. Why not?

10. If you remove ice cubes from the freezer with wet hands, the cubes often freeze to your fingers. How can the ice freeze the water on your hands? Shouldn't they melt instead?

11. A propane tank contains a mixture of liquid and gaseous propane. Why does the tank become cooler when gas is released from the tank?

12. Some home ice cream machines have an insert that you put in the freezer and then use to chill the ice cream mixture so that it freezes. Why can't you simply use ice water to chill the sugary liquid so that it freezes?

13. Why does a pot of water heat up and begin boiling more quickly if you cover it?

14. A quality perfume is a mixture of *essential oils* and has a scent that changes with time and body temperature. Explain the gradual change of the perfume's scent with time in terms of evaporation.

15. If you make ice cubes directly from tap water, they will contain lots of tiny gas bubbles. If you boil the water first, the cubes will be clear. Explain.

16. Water pipes frequently burst when the water inside them freezes. How is this effect related to the fact that ice floats in water?

17. If you add a little hot tea to ice water at 0° C, the mixture will end up at 0° C so long as some ice remains. Where does the tea's extra thermal energy go?

18. If you try to cook vegetables in 100° C air, it takes a long time. But if you cook those same vegetables in 100° C steam, they cook quickly. Why does the steam transfer so much more heat to the vegetables?

19. When the outside temperature is –2° C (29° F), you can melt the ice on your front walk by sprinkling salt on it. But if you are out of salt, will baking soda do?

20. The frozen beverages sold in convenience stores are made by chilling juice to a few degrees below freezing. Rather than freezing solid, the juice becomes a mixture of tiny ice crystals and sweet syrupy liquid. Explain.

21. When you lift the lid of a hot tub or spa, its bottom surface is usually covered with water droplets. The tub's water doesn't reach the lid, so how did those droplets form?

22. Putting a clear plastic sheet over a swimming pool helps keep the water warm during dry weather. Explain.

23. You're floating outside the space station when your fellow astronaut tosses you a cold bottle of mineral water. It's outside your space suit but you open it anyway. It immediately begins to boil. Why?

24. Even on a very humid day the hot air from your blow dryer can extract moisture from your hair. Why is heated air able to dry your hair when the air around you can't?

25. You haven't been to the store lately so you're digging through the frozen foods in the back of your freezer. Most of these antique items are terribly dehydrated. What happened to the moisture in these foods?

Epilogue for Chapter 8

In this chapter, we examined phase transitions between solid, liquid, and gaseous substances. In *water, steam, and ice*, we saw that there are a variety of ways in which water molecules can transform between their solid form (ice), their liquid form (water), and their gaseous form (steam or water vapor). We explored the energy issues involved in converting one form to the other, explaining the need for prodigious amounts of heat in turning ice to water and water to steam. We also saw how dissolved materials cause water to remain liquid below its normal melting temperature and prevents it from boiling at its normal boiling temperature.

Explanation: Melting Ice with Pressure

The dental floss is able to drift downward through the ice because ice melts at a lower temperature when it's exposed to extreme pressure. The melting ice cube is at its normal melting temperature, so the pressure of dental floss causes the ice beneath the floss to melt fairly rapidly. As that ice melts, the dental floss descends. The water from beneath the dental floss flows upward and refreezes because its pressure drops and the melting ice below the floss draws heat out of it. Since ice melts below the floss and resolidifies above it, the dental floss moves downward. This effect is very unusual. Since virtually all other materials solidify under pressure, rather than melting, they don't permit a string to pass into them in this manner.

Cases

1. Puffed grain cereals such as puffed wheat or puffed rice are made by heating the moist grains in a tightly sealed container.

a. What happens to the pressure inside the container as the temperature of its moist contents increases to well above 100° C?
b. Why doesn't the water in the grains boil when the temperature reaches 100° C?
c. What happens to the water in the hot grains when the container is suddenly opened to the outside atmosphere?
d. Why do the grains suddenly puff up when the container is opened to the outside?

2. In hot, dry climates, many houses rely on evaporative coolers for "air conditioning." These coolers blow hot, dry air across wet surfaces to produce cool, moist air.

a. How does this evaporation process lower the temperature of the air?
b. What limits how much the evaporative cooler can reduce the air's temperature?
c. The second law of thermodynamics requires that entropy never decrease in a thermally isolated system. A sealed, insulated house is nearly thermally isolated, so its entropy can't go down during the evaporation process. In fact, its entropy goes up. What order is being lost as the evaporative cooler chills the air?
d. Why aren't evaporative coolers used in hot, wet climates where the relative humidity is 50 to 100%?

3. A dehumidifier removes moisture from the air. It resembles an air conditioner except that its condenser, evaporator, and compressor are all located indoors. The dehumidifier's fan blows room air first across the evaporator and then across the condenser. The evaporator cools the air so that its moisture condenses into water. The condenser then warms the air before sending it back into the room.

a. The evaporator removes heat from the air so that the air's temperature drops. What happens to the air's relative humidity and why does liquid water begin to form?
b. What is the chilled air's relative humidity as it leaves the evaporator?
c. The condenser adds heat to the air so that the air's temperature increases. What happens to the air's relative humidity?
d. How can the dehumidifier control the relative humidity of the air that leaves the condenser and reenters the room?
e. Although a dehumidifier first cools the room air and then reheats it, its overall effect is to warm the room air somewhat. Some of the additional heat that enters the room air comes from the electric company. But some of this heat comes from something else. What?
f. Why does the second law of thermodynamics make it impossible to operate a dehumidifier without a source of ordered energy, such as electricity or mechanical work?

4. In making freeze-dried coffee, a company encounters a number of interesting physical problems. Let's follow the production process from start to finish.

a. Coffee beans must be roasted before they're used for coffee. The coffee roasting machine stirs the pile of beans continuously during roasting. If it didn't, the beans on the inside of the pile wouldn't cook properly. Why do conduction, convection, and radiation all have trouble transferring heat to the inside beans?

b. Grinding the coffee beans requires work and thus a steady supply of electric energy. Using scissors as a simple example of a grinding machine, show that you must do work on a bean to cut it in half.

c. As hot water drips downward through the ground coffee beans, it extracts the flavors of coffee. The more finely ground the beans, the more slowly the water passes through them. What slows down the water's descent and why does a finer grind slow the water more?

d. The company then freezes the coffee in preparation for drying. On a warm summer day, what type of machine must the company use to remove heat from the coffee in order to freeze it, and what happens to that heat?

e. Even at temperatures below its freezing point, ice can sublime. The company places the frozen coffee in a vacuum chamber and the water in it vanishes. Only freeze-dried coffee crystals are left. Describe the way in which water leaves the coffee.

f. The company must add heat to the frozen coffee as its ice sublimes. Why does the ice need heat to sublime?

5. Espresso is a strong coffee made by pushing very hot water through finely ground coffee beans. An espresso maker heats water in a small chamber before passing it through the beans and into a cup. The heating chamber has two valves: one at its bottom which allows water to leave and one at its top which allows steam to leave. Initially, both valves are closed.

a. The espresso maker heats the water to more than 100° C in the sealed heating chamber, yet the water doesn't begin to boil. What keeps the water from boiling?

b. When the water valve is opened, the water begins to flow out of the chamber and through the beans. What pushes the hot water out of the heating chamber?

c. As the water drips out of the beans and into the cup, it emits a great deal of steam. It's as though the water begins to boil vigorously as it enters the air. How does a change in pressure explain this sudden onset of boiling?

d. Pushing the hot water through the finely ground coffee beans requires a substantial pressure imbalance across the beans. Why is it so hard for the water to flow through the fine particles?

e. The espresso maker can also make cappuccino. The classic topping for cappuccino is frothed milk, made by injecting steam and air into milk. When the steam valve is opened, 120° C steam flows out of the heating chamber and into the milk at a very high speed. What gives the steam its high speed?

f. The steam passes through a narrow orifice just before flowing into the milk. This orifice causes the steam's speed to rise even further. Holes surrounding the orifice allow air to accelerate toward the steam and that air becomes entrained in the jet that enters the milk. Explain why air accelerates toward the steam as the steam passes through the orifice?

g. The steam found just above an open pot of boiling water quickly condenses into water on any cooled surface. However, the "dry" steam emerging from the orifice doesn't condense so easily. Even after transferring some of its thermal energy to the milk and air, the steam can remain gaseous. Why?

6. You can make ice cream by putting a sealed container of ice cream mix in a bucket filled with melting ice and salt.

a. If you omit the salt and let the container sit in the bucket of melting ice, what temperature will the mix reach?

b. The ice cream mix can be thought of as water with a large amount of dissolved chemicals (although the fat doesn't actually dissolve). Does this mixture freeze at, above, or below the temperature at which water freezes?

c. By adding salt to the ice, you cause the ice to melt faster. This additional melting requires energy. Where does that energy come from?

d. What happens to the temperature of the ice cream mix after you add salt to the ice?

e. Use your answers to **a** through **d** to explain why salt is needed to make ice cream.

7. One of the most unusual but useful modern solvents is carbon dioxide. At atmospheric pressure, carbon dioxide is either a solid or a gas, but at pressures above 5 atmospheres, it can become a liquid. This liquid is used to extract caffeine from coffee, dissolve paints, and even clean up after oil spills.

a. If carbon dioxide is never a liquid at atmospheric pressure, how does it convert from solid to gas?

b. At atmospheric pressure, how does carbon dioxide convert from gas to solid?

c. Solid carbon dioxide is called *dry ice*. At atmospheric pressure a block of dry ice maintains a steady temperature of –78° C (–108° F). As you transfer heat to it, it just gets smaller, it doesn't get hotter. What happens to the heat?

d. To remove caffeine from coffee beans, the beans are immersed in the liquid carbon dioxide at a particular elevated pressure. The liquid dissolves the caffeine and washes it out of the beans. Before it entered the beans, the liquid carbon dioxide had a particular boiling temperature at the chosen pressure. What effect does the dissolved caffeine have on this boiling temperature?

e. Once most of the caffeine is gone, the beans are removed from the liquid and moved into the open air so that the carbon dioxide in them can escape. What happens to the temperature of the coffee beans as they enter the open air?

CHAPTER 9

RESONANCE

Many of the most fascinating motions in the world around us are repetitive ones. Our lives are filled with cycles of various durations, from the earth's slow orbit around the sun to a pond's rapid undulation when a rain drop strikes its surface. These cyclical motions are governed by physical laws and steadily mark the passage of time. Some of these cycles structure our lives out of necessity or tradition while others are simply there to be observed. And some of these cycles have become part of our everyday world because they are useful or enjoyable. This chapter is about two important uses for cyclical motions: in clocks and in musical instruments.

Experiment: A Singing Wine Glass

One simple experiment with cyclical motion involves a crystal wine glass, a little water, and a delicate touch. A crystal wine glass is hard and thin and easily supports a repetitive mechanical motion called a vibration. Its hardness allows the glass to retain energy in this vibrating motion for a long time so that it rings clearly when you tap it gently with a spoon and can acquire energy slowly in the manner described below.

The experiment itself is an activity discovered by many a bored youth, sitting too long in a fancy restaurant and having nothing to play with but the place settings. If you wet your finger slightly and run it gently around the lip of the wine glass at a slow, steady pace, you should be able to get the wine glass to sing loudly. The walls of the glass will vibrate back and forth, and emit a clear tone.

If you can't find a crystal wine glass, you're probably out of luck. A normal wine glass won't work well because it converts the energy from your finger into thermal energy rapidly and doesn't emit much sound. In any case, be sure that the wine glass has no sharp edges on its lip so that you don't cut your finger.

Chapter Itinerary

The crystal wine glass's vibration is an example of a natural resonance, a cyclic motion of the glass itself. Whether you tap the glass gently with a spoon or rub it with your finger, you're causing it to undergo this characteristic motion. In this chapter we'll examine several other objects that exhibit natural resonances as we examine: (1) *clocks* and (2) *violins and pipe organs*.

In *clocks*, we'll study pendulums, balance rings, and quartz crystals to see how their natural resonances are used to measure the passage of time. We'll also look at the structure of atoms to see how they're involved in even more accurate clocks. In *violins and pipe organs*, we'll look at the vibrations of strings and air columns to find out how these instruments produce their musical sounds.

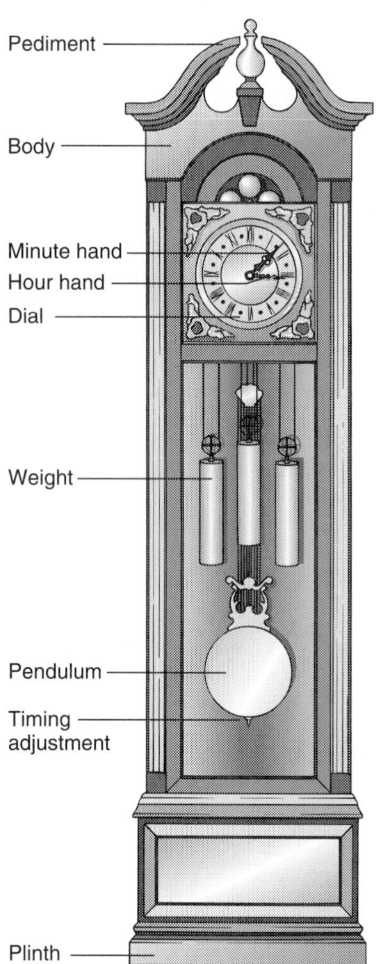

Section 9.1

Clocks

Long ago, people measured time by the motions of the sun, the moon, and the stars, and established days, months, and years to recognize their celestial passages. But day length varies with the seasons and the sky provides no easy way to measure short periods of time. So people invented clocks.

Early clocks were based on the time it took to complete simple processes, such as filling a container with sand or water or burning a candle. However, these clocks weren't very accurate and required constant attention. Better clocks measure time with repetitive motions, such as swinging or rocking. In this section, we'll examine the workings of modern clocks based on repetitive motions. As we do, we'll see that these repetitive motions are interesting in their own right and appear throughout nature in countless objects besides clocks.

Questions to Think About: *What exactly is time? Why do some objects swing or rock back and forth repetitively? How would you use a repetitive motion to measure time? How can you change the rate at which an object swings or rocks? Since repetitive motions don't normally continue forever, how can you keep them going without upsetting their time-keeping ability?*

Experiments to Do: *You can build the time-keeping portion of a pendulum clock by attaching a small weight to the end of a string and hanging that string from a table or doorway. Push the weight gently so that it swings back and forth. You will find that it completes this repetitive motion with great regularity. What limits its regularity?*

If the string is about 25 cm long, a complete swing (back and forth) will take almost exactly 1 s. Change the length of the string and observe its effect on the swing. Do you think that the weight affects the swing? Change the weight and see if you're right.

Now vary the extent of the swing to see if it affects the time each swing takes. How can it be that a small swing takes the same time as a large swing? Notice that you can get the weight swinging by pushing on it rhythmically, once each swing. Randomly timed pushes won't do; you must push the weight in synchrony with it motion. At what times should you push the weight to make it swing farther? To make it swing less far? Rhythmic pushes of this sort are what keep the time-keeper in a clock moving, hour after hour.

Time

Before examining clocks, we should take a brief look at time itself. Scientists treat time as a dimension, similar but not identical to the three spatial dimensions that we perceive in the world around us. In total, our universe has four dimensions: three spatial dimensions and one temporal dimension. Thus it takes four numbers to completely specify when and where an event occurs: three numbers identify the event's location and one identifies the time at which it occurs.

An obvious difference between space and time is that, while we can see space stretched out around us, we can only observe the passage of time. Despite the fact that we occupy only one location in space at a given moment, we are somehow more aware of the expanse of space around us. It's much harder to "see" the whole framework of time stretching off into the past and future. You must use your imagination.

Our whole perception of space is ultimately based on the need for forces, accelerations, and velocities to travel from one point to another. We know that a city is far away because we know that traveling there with reasonable forces, accelerations, and velocities would require a substantial amount of time. Our perception of time is based on the same mechanical principles. If two moments are separated by a large amount of time, then reasonable forces, accelerations, and velocities will permit you to travel large distances during the period between the two moments. In short, our perceptions of space and time are interrelated, and measurements of time and space are connected as well.

We measure space with rulers and we measure time with clocks. But how would you make a ruler? You could make a rather large ruler by driving a car at constant velocity and marking the pavement with paint every 10 s. Your ruler would not be very practical, but it would fit the definition of a ruler as having spatial markings at uniform distances. You would be using your movement through time to measure space.

How would you make a clock? You could make a rather strange clock by driving a car at constant velocity down your giant ruler and counting each time you saw one of your marks go by. You would then be using your movement through space to measure time. Most clocks really do use motion to measure time, but they normally use motions that are a bit more compact than a car ride.

CHECK YOUR UNDERSTANDING #1: The Ultimate Moon-Bounce

To measure the distance from the earth to the moon, scientists bounce light from reflectors placed on the moon by the Apollo astronauts. Light travels at a constant speed. How can a measurement of light's travel time to and from the moon be used to determine the distance from the earth to the moon?

Mechanical Clocks

The earliest clocks were based on the sun. Every day, the sun rises in the east and sets in the west. The mechanism responsible for this "clock" is the earth's steady rotation. Because the earth is an isolated object and virtually free of torques, its angular momentum doesn't change. Since the earth's moment of inertia is essentially constant, conservation of angular momentum gives it a constant angular velocity. Every hour, the earth turns about 15° on its axis. The sun serves merely as a reference mark in space and the earth's rotation places the sun nearly overhead once every 24 hours.

Other astronomical clocks known since antiquity include the annual motion of the earth around the sun and the 27.3 day orbit of the moon around the earth. These cycles still have enormous influence over life and culture. But dividing a single day into well-defined pieces proved difficult and measuring time at night was harder still. Astronomical clocks have peculiarities: a summer day is longer than a winter day so that when one tries to divide a day into equal pieces, the pieces change with the seasons. Eventually people began to develop clocks that didn't depend on the sky.

Early clocks used water, candles, and sand to measure the passage of time. They owed their limited accuracy to the laws of mechanical motion. In a water clock, the rate at which water flowed through a hole from one container to another was determined by the water's pressure and viscosity and by the dimensions of the hole. Similar laws governed the motion of sand in a sandglass. And the time it took for a candle to burn was determined by a complicated sequence of chemical reactions.

However, none of these early clocks could be left unattended. Each had to be started by hand and then ran only once for a specific amount of time. These clocks were better suited to timing individual intervals than they were to dividing the whole of time into well-defined segments.

Modern mechanical clocks are based on a particular type of repetitive motion called a natural resonance. In a **natural resonance**, the energy in an isolated object or system of objects causes it to perform a certain motion over and over again. Many objects in our world exhibit natural resonances, from tipping rocking chairs, to sloshing basins of water, to waving flag poles. Resonances are very regular motions and can measure the passage of time extremely accurately.

Clocks that used natural resonances as *time-keepers* were finally able to divide the entire day, including the night, into segments of equal length. At first clocks simply sounded the hours to provide a basic structure to life. Eventually clocks were given faces, minutes were added, and ultimately seconds. But until a few centuries ago, clocks were so rare and expensive that only a few cities or towns had even one. It wasn't until recently that mechanical clocks and watches became items that individuals could afford and own.

CHECK YOUR UNDERSTANDING #2: An Egg-Timer Clock

Although a sandglass can be made repetitive by turning it over every time the sand runs out, this manual restarting process introduces timing errors. If a 3 minute sandglass is always turned over within 10 s after the sand runs out, how accurately will it measure time over the course of a day?

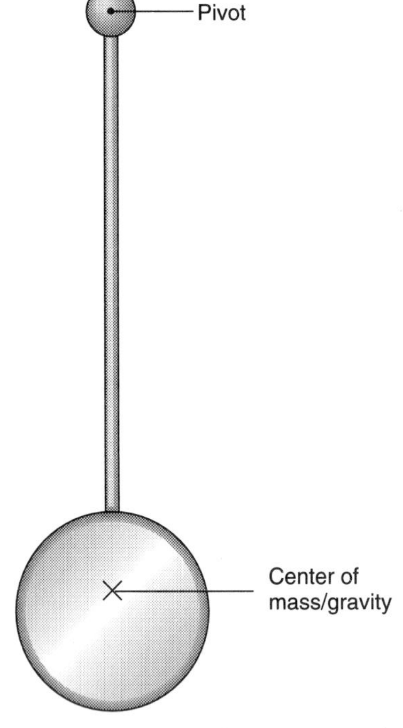

Fig. 9.1.1 - A pendulum consists of a weight hanging from a pivot. The pendulum is in a stable equilibrium when its center of mass/gravity is directly below the pivot.

Pendulums and Harmonic Oscillators

The two most common time-keeping mechanisms used in mechanical clocks are pendulums and balances. Let's begin by looking at a *pendulum*, a weight hanging from a pivot that swings back and forth in a natural resonance.

A pendulum swings back and forth because of its weight (Fig. 9.1.1). When the pendulum's center of mass (which coincides with its center of gravity) is directly below the pivot, it experiences zero net force and is in a stable equilibrium.

But when you tip the pendulum either way, its weight produces a restoring force that pushes it back toward its equilibrium position (Fig. 9.1.2). This restoring force is proportional to how far the pendulum is from equilibrium so that the more you tilt it, the stronger the restoring force. When you release the pendulum, it swings back and forth about its equilibrium position in a repetitive motion called an **oscillation**.

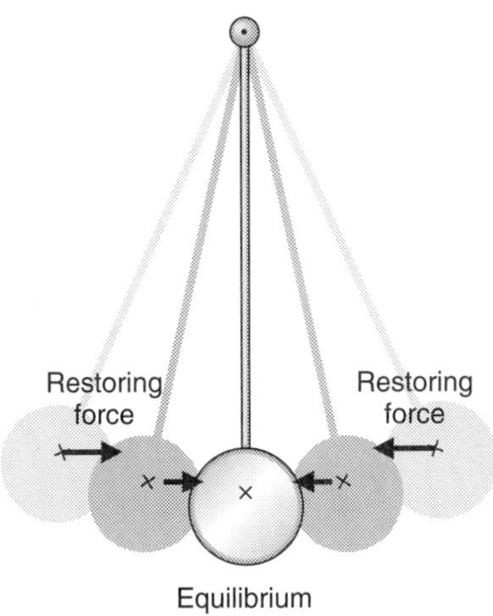

Fig. 9.1.2 - If you tilt a pendulum's center of mass away from its equilibrium position, it experiences a restoring force proportional to its distance from that equilibrium position.

As it swings, the pendulum's energy goes back and forth between potential and kinetic forms. When the pendulum comes to rest at the end of each swing, its energy is all potential, and when it swings rapidly through equilibrium in the middle of each swing, its energy is all kinetic. This repetitive transformation of energy from one form to another is part of any oscillation and keeps the *oscillator* moving back and forth until all of its energy is converted into thermal energy or is transferred elsewhere.

But the pendulum is not just any oscillator. Because its restoring force is proportional to its displacement from equilibrium, the pendulum is a **harmonic oscillator**, one of the simplest and best understood mechanical systems in nature. As a harmonic oscillator, the pendulum undergoes **simple harmonic motion**, a very regular and predictable oscillation that makes it a superb time-keeper.

The **period** of any harmonic oscillator, the time it takes for that oscillator to complete one full cycle of its motion, depends only on the stiffness of its restoring force and on the mass of its moving object. **Stiffness** measures how rapidly that restoring force increases as the object moves away from its equilibrium position. The stiffer the restoring force, the harder that force pushes on the displaced object and the faster it oscillates back and forth. But the object's mass opposes acceleration so that the larger its mass, the slower the object oscillates.

One of the most remarkable and important features of a harmonic oscillator is that its period doesn't depend on its **amplitude**, its furthest displacement from its equilibrium position. Whether that amplitude is large or small, the harmonic oscillator's period remains the same. This insensitivity to amplitude makes a harmonic oscillator ideal for time-keeping, since accidental changes in amplitude won't make the clock run fast or slow. Virtually all modern clocks are based on harmonic oscillators.

Actually, a pendulum is an unusual harmonic oscillator because its period *doesn't* depend on its mass. When you increase the pendulum's mass, you also increase its weight and stiffen the restoring force. These two changes exactly balance one another so that the pendulum's period remains the same.

However, a pendulum's period *does* depend on its length and on gravity. When you reduce the pendulum's length, the distance from its pivot to its center of mass, you stiffen the restoring force and shorten its period. Similarly, when you strengthen gravity (perhaps by traveling to Jupiter), you increase the pendulum's weight, stiffen the restoring force, and reduce the pendulum's period. Thus a short pendulum swings back and forth more rapidly than a long one and any pendulum swings faster on the earth than it would on the moon.

On the earth's surface, a 0.248 m pendulum has a period of 1 s (Fig. 9.1.3). Such a pendulum is well suited for a smaller mechanical clock, which must advance its second hand by 1 second for each complete cycle of the pendulum. A pendulum's period increases as the square root of its length, so that a 0.996 m pendulum (4 times as tall as a 0.248 m pendulum) takes 2 s to complete its cycle. Thus a tall clock with a 0.996 m pendulum must advance its second hand by 2 seconds per cycle.

Fig. 9.1.3 - The swinging pendulum controls the movement of the clock's hands. The pendulum is 0.248 m long, so each cycle takes 1 s to complete and advances the hands by one second.

CHECK YOUR UNDERSTANDING #3: Swing Time

A child swinging on a swing set travels back and forth at a steady pace. What determines the period of his motion?

Pendulum Clocks

While a pendulum can maintain a steady beat, it's not a complete clock. Something else must keep the pendulum swinging and display the current time. A pendulum clock does both. It sustains the pendulum's motion with gentle pushes and uses that motion to keep its hands advancing at a steady rate.

The top of the pendulum has a two-pointed *anchor* that controls the rotation of a *toothed wheel* (Fig. 9.1.4). This mechanism is called an *escapement*. A weighted cord wrapped around the toothed wheel's shaft exerts a torque on that wheel, so that the wheel would spin if the anchor weren't holding it in place. Each time the pendulum reaches the end of a swing, one point of the anchor releases the toothed wheel while the other point catches it. The wheel turns slowly as the pendulum rocks back and forth, advancing by one tooth for each full cycle of the pendulum. This wheel turns a series of gears, which slowly advance the clock's hands.

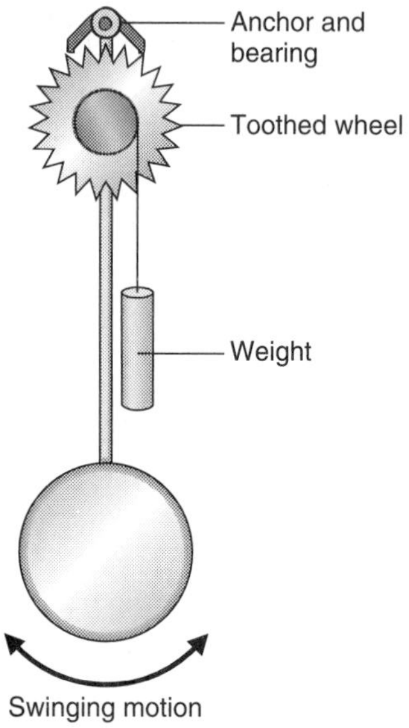

Anchor and bearing

Toothed wheel

Weight

Swinging motion

Fig. 9.1.4 - A pendulum clock uses a swinging pendulum to determine how quickly a toothed wheel turns and advances a series of gears that control the hands of the clock. The anchor permits the toothed wheel to advance by one tooth each time the pendulum completes a full cycle. The clock's energy comes from the descent of a weighted cord that is wrapped around the toothed wheel.

The toothed wheel also keeps the pendulum moving by giving the anchor a tiny forward push each time the pendulum completes a swing. Since the anchor moves in the direction of the push, the wheel does work on the anchor and pendulum and replaces energy lost to friction and air resistance. This energy comes from the weighted cord, which releases gravitational potential energy by descending a tiny distance. When you wind the clock, you lift this cord back around the shaft and replenish its potential energy.

Keeping the pendulum swinging steadily is extremely important. If the pendulum's amplitude varies significantly from cycle to cycle, the toothed wheel will turn erratically and the clock won't be accurate. A clock works best when its pendulum swings freely and easily, because any outside force—even a push from the toothed wheel—affects its period. The most accurate time-keepers are those that can oscillate on their own for thousands or millions of cycles. These precision time-keepers need only the slightest pushes to keep them moving and thus have extremely precise periods. That's why good clocks use aerodynamically designed pendulums and low-friction bearings.

But there is a second reason to keep the pendulum's amplitude steady: it's not quite a perfect harmonic oscillator. If you tip the pendulum too far, it becomes a decidedly **anharmonic oscillator**: its restoring force ceases to be proportional to its displacement from equilibrium and its period begins to depend on its amplitude. Think about the period of a pendulum that is tipping upside down! Since a change in period would spoil the clock's accuracy, the pendulum's amplitude must be kept small and steady. That way, the amplitude has a minimal effect on the pendulum's period.

Since the pendulum's period depends on its length and on gravity, both quantities must be carefully controlled. Temperature causes most materials to expand, so that precision pendulums must be thermally compensated. A compensated pendulum uses several different materials, with different coefficients of thermal expansion, to ensure that its center of mass remains at exactly the same distance from its pivot. Such a pendulum has several different shafts, some of which move it farther from its pivot as they expand and others of which move it closer to its pivot. These shafts work together as the temperature changes to keep the pendulum's center of mass from shifting.

While gravity doesn't change with time, it does vary slightly from place to place. A clock that worked well at the factory may run slightly fast or slow once it has been installed in your home. You can compensate for the change in gravity by modifying the length of the pendulum. Most pendulum clocks have a threaded adjustment that allows you to move the pendulum's center of mass slightly to adjust its period.

CHECK YOUR UNDERSTANDING #4: Swing High, Swing Low

When pushing a child on a playground swing, you normally push her forward as she moves away from you. What happens if you push her forward each time she moves toward you?

Balance Clocks

Because it relies on gravity for its restoring force, a pendulum clock isn't portable. Its mechanism can't tolerate being tipped or moved about. For a clock to be movable, its time-keeper must be completely independent of gravity. Since the best time-keepers are harmonic oscillators, a portable clock needs a gravity-independent restoring force that's proportional to displacement. It needs a spring!

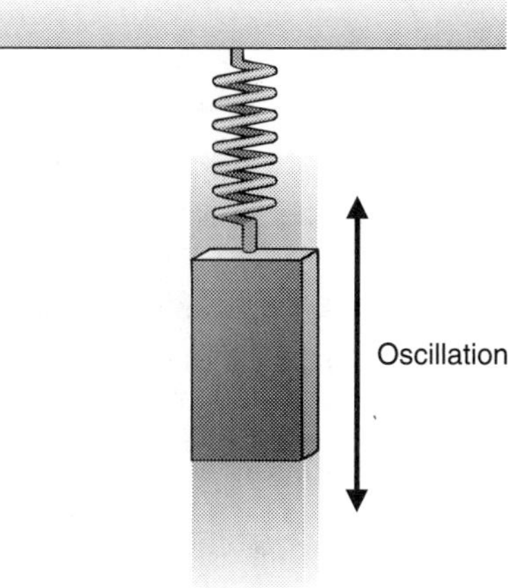

Fig. 9.1.5 - A block attached to a spring is a harmonic oscillator. The oscillator's period is determined only by the stiffness of the spring and the mass of the block.

As we saw in Section 2.1, the force exerted by a spring is proportional to its distortion. The more you stretch, compress, or bend a spring, the more it pushes back toward its equilibrium shape. If you attach a block to the free end of a spring, you will have a harmonic oscillator with a period determined only by the stiffness of the spring and the block's mass (Fig. 9.1.5). Since the period of a harmonic oscillator doesn't depend on the amplitude of its motion, the block will oscillate steadily about its equilibrium position and make an excellent time-keeper.

Unfortunately, this simple time-keeper is still affected by gravity. Although gravity doesn't alter the block's period, it does shift the block's equilibrium position downward. This shift is bad for a movable clock, which must work properly even when tipped upside down. However, there is another spring-based time-keeper that is completely unaffected by gravity and marks time accurately in any orientation or location. This ingenious device, used in most mechanical clocks and watches, is called a balance ring or simply a balance.

A *balance ring* is a symmetric metal ring, supported at its center of mass/gravity by an axle and mounted in nearly frictionless bearings. It rotates about this axle like a tiny bicycle wheel (Fig. 9.1.6). Since any friction the ring experiences in its bearings is too close to its axis of rotation to exert much torque on it, the ring turns very easily. Moreover, the ring pivots about its own center of gravity so that it experiences no gravitational torques.

The only thing exerting a torque on the balance ring is a tiny coil spring. One end of this spring is attached to the ring while the other is fixed to the body of the clock. When the spring is undistorted, it exerts no torque on the ring and the ring is in equilibrium. But if you rotate the ring either way, torque from the distorted spring will act to restore it to its equilibrium orientation. Because this restoring torque is proportional to the ring's rotation away from a stable equilibrium, the balance ring and coil spring form a harmonic oscillator!

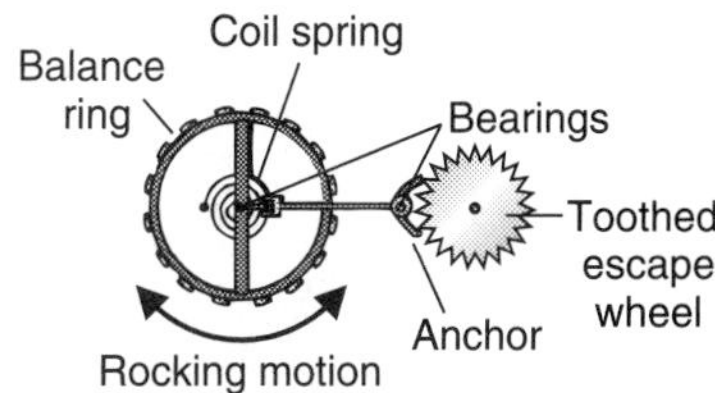

Fig. 9.1.6 - A balance clock uses a rocking balance ring to control the rotation of a toothed wheel and the gears that advance the clock's hands. The anchor permits the toothed wheel to advance by one tooth each time the ring completes a full cycle. The clock's energy comes from a main spring (not shown) that exerts a steady torque on the toothed wheel.

Because of the rotational character of this harmonic oscillator, its period depends on the torsional stiffness of the coil spring (how rapidly the spring's torque increases as you twist it) and on the balance ring's moment of inertia. Since the balance ring's period doesn't depend on the amplitude of its motion, it keeps excellent time. And because gravity exerts no torques on the balance ring, this time-keeper works anywhere and in any orientation.

The rest of a balance clock is similar to a pendulum clock (Fig. 9.1.7). As the balance ring rocks back and forth, it tips a lever that controls the rotation of a toothed wheel. An anchor attached to the lever allows the toothed wheel to advance by one tooth for each complete cycle of the balance ring's motion. The toothed wheel is connected by gears to the clock's hands and these hands slowly advance as the wheel turns.

Fig. 9.1.7 - The balance ring in this antique Swiss wrist watch is located at the left side the picture. Below it is the toothed escape wheel. Tiny jewel bearings minimize sliding friction and permit this watch to keep very accurate time.

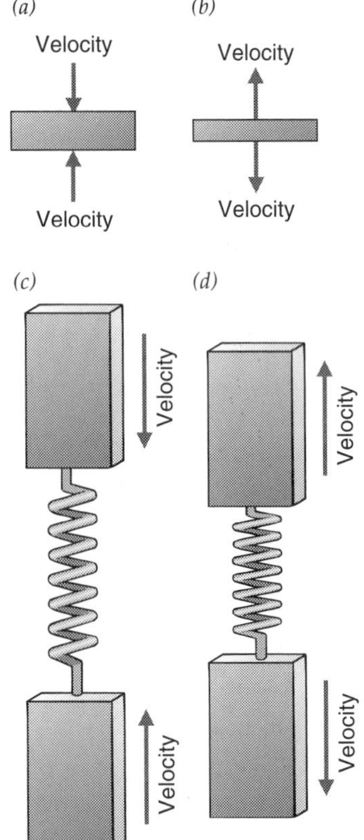

Fig. 9.1.8 - A quartz crystal acts like a spring with masses at each end. The vibrating crystal alternately (*a*) contracts and (*b*) expands, just as the masses alternately move (*c*) toward one another and (*d*) away from one another.

Because the balance clock is portable, it can't draw energy from a weighted cord. Instead, its *main spring* exerts a torque on the toothed wheel. This spring is a coil of elastic metal that stores energy when you wind the clock. Its energy keeps the balance ring rocking steadily back and forth and also turns the clock's hands. Since the main spring unwinds as the toothed wheel turns, the clock occasionally needs winding.

CHECK YOUR UNDERSTANDING #5: A Little Light Entertainment

When you accidentally strike a chandelier with a broom, this hanging lamp begins to twist back and forth with a regular period. What determines its period of oscillation?

Electronic Clocks

The best pendulum and balance clocks lose or gain about ten seconds per year. Their accuracy is limited by friction, air resistance, and thermal expansion. To do better, a clock needs a time-keeper that avoids these limiting effects. It needs a quartz oscillator.

A *quartz oscillator* is basically a crystal of quartz, the same mineral found in most white sand. Quartz is extremely hard and brittle, so that the crystal vibrates strongly after being struck. As it vibrates, the crystal acts as a harmonic oscillator with a period that doesn't depend on the amplitude of its motion. Since the vibrating crystal isn't sliding across anything or moving quickly through the air, it loses energy slowly and vibrates for a long, long time. And because quartz's coefficient of thermal expansion is extremely small, the crystal's period is nearly independent of its temperature. With its exceptionally steady period, a quartz oscillator can serve as the time-keeper for a highly accurate clock, one that loses or gains less than a tenth of a second per year.

A quartz crystal is a harmonic oscillator because it acts like a spring with objects at both ends (Fig. 9.1.8). The spring is the quartz crystal itself and the objects are its two halves. As shown in Fig. 9.1.8*c*,*d*, two objects on a spring oscillate in and out symmetrically about their combined center of mass, with a period determined only by the object's masses and the spring's stiffness. Since the forces on the objects are proportional to their displacements from equilibrium, they are harmonic oscillators.

A flat quartz crystal can undergo a similar distortion in which its upper and lower surfaces oscillate in and out symmetrically about its center of mass (Fig. 9.1.8*a*,*b*). Like almost any solid, a quartz crystal has an equilibrium shape and responds to a small distortion with restoring forces proportional to that distortion. When you strike the crystal and move its upper and lower portions closer together, these restoring forces act to return it to its equilibrium shape. But the struck crystal has excess energy and overshoots its equilibrium shape. Instead of coming to rest quickly, its two halves oscillate in and out for a long time while internal friction slowly converts its energy into thermal energy.

Because of its exceptional hardness, a quartz crystal's restoring force is extremely stiff. Even a tiny distortion produces a huge restoring force. Since the period of a harmonic oscillator decreases as its spring becomes stiffer, a typical quartz oscillator has an extremely short period. Its motion is usually called a *vibration* rather than an *oscillation*, because vibration implies a fast oscillation in a mechanical system. Oscillation itself is a more general term for any repetitive process and can even apply to such non-mechanical processes as electric oscillations or thermal oscillations.

Because of its rapid vibration, a quartz oscillator's period is a small fraction of a second. Rather than referring to its tiny period, we normally characterize a fast oscillator by its **frequency**, the number of cycles it completes in a certain amount of time. The SI unit of frequency is the **cycle-per-second,** also called the **hertz** (abbreviated Hz) after German physicist Heinrich Rudolph Hertz. Period and frequency are reciprocals of one another (period equals 1/frequency and vice versa) so that an oscillator with a period of 0.001 s has a frequency of 1,000 Hz.

Of course, a quartz crystal isn't a complete clock. Like the pendulum and balance, it needs something to keep it vibrating and to display the current time. While these tasks could conceivably be done mechanically, quartz clocks are normally electronic. There are two reasons for this choice. First, the crystal's vibrations are too fast and too small for most mechanical devices to follow. Second, a quartz crystal is intrinsically electronic itself. It responds mechanically to electric stress and responds electrically to mechanical stress. Because of this coupling between its mechanical and electric behaviors, crystalline quartz is a *piezoelectric material.*

This piezoelectric behavior makes an electronic quartz clock easy to build (Fig. 9.1.9). The clock's circuitry keeps the quartz crystal vibrating by exposing it to a series of electric stresses. Just as carefully timed pushes keep a child swinging endlessly on a playground swing, carefully timed electric stresses keep the quartz crystal vibrating endlessly in its holder. Because the crystal loses so little energy with each vibration, the clock only needs to do a tiny amount of work on it during each cycle to maintain its vibration.

Fig. 9.1.9 - This circuit board is part of a quartz wrist watch. The quartz crystal is located in the black cylinder at the bottom. Carefully polished to vibrate at a precise frequency, the crystal keeps the watch accurate to a few seconds a month.

The clock also detects the crystal's vibrations electrically. Each time its halves move in or out, the crystal experiences mechanical stress and emits an electric pulse. These pulses are synchronized with the vibration and have all the time-keeping accuracy of the quartz crystal itself. These pulses may operate an electric motor that advances clock hands or they may serve as input to a computer-like circuit that measures time by counting the pulses.

The quartz crystals used in clocks and watches are carefully cut and polished to vibrate exactly at specific frequencies. The thinner the crystal, the faster it vibrates. In effect, these crystals are tuned like musical instruments to match the requirements of their clocks. When a motorized clock uses a crystal that produces 10,000 pulses per second, it advances its second hand by 1 second for each 10,000 of pulses it receives. Similarly, a computer-like clock using the same crystal counts the pulses and increments the time it's displaying by 1 second for each 10,000 pulses it receives.

CHECK YOUR UNDERSTANDING #6: Heavy Metal Music

If you drop a metal rod on the floor, end first, you hear a high pitched tone. What's happening?

Atomic Clocks

Quartz clocks are accurate enough for all but the most demanding measurements of time. Still, there are occasions when an even greater accuracy is needed from a clock. The most accurate clocks in existence make use of atoms and atomic structure. Just as astronomical clocks are based on the motions of celestial objects orbiting one another in the heavens, atomic clocks are based on the motions of electrons and nuclei orbiting one another in atoms.

Because an atom is so small, the motions and behaviors of the particles inside it are dominated by the strange world of quantum physics. While the plan-

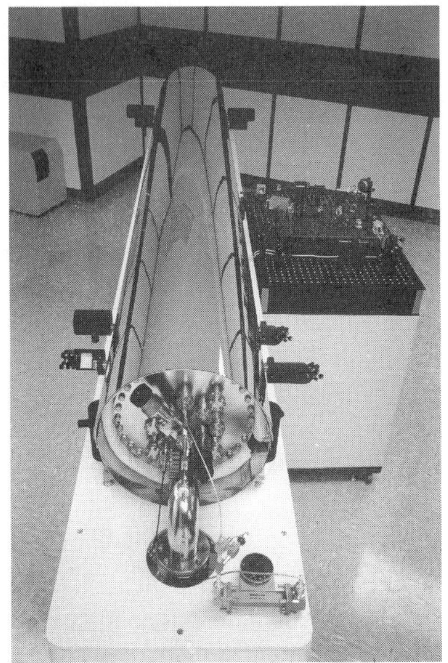

Fig. 9.1.10 - This National Institute for Standards and Technology (NIST) atomic clock uses cesium atoms as time-keepers, measuring time so accurately that it loses or gains less than 1 s every 10 million years. It's one of the world's standard clocks, used to set all other clocks.

ets in the solar system can travel in any orbit about the sun, the electrons in an atom can travel only in a few specific orbits about the atom's positively charged center, its **nucleus**. These allowed orbits depend only on the numbers of electrons and other particles in the atom so that two atoms with identical constituents also have identical allowed orbits. Thus one atom of cesium has the same allowed orbits as every other atom of cesium.

The quantum structure of an atom governs the spectrum of light wavelengths that it can absorb and emit. As we'll see in Chapter 15, an atom can only absorb or emit light when that light has the right wavelength to shift the atom from one arrangement of allowed orbits to another such arrangement. As a result, atoms absorb and emit only specific wavelengths of light.

Light is effectively a moving electromagnetic oscillation so that, in addition to its wavelength, it also has a frequency. This frequency is inversely related to its wavelength. Thus atoms are selective not only about which *wavelengths* they absorb and emit, but also about which *frequencies* they absorb and emit.

This selective absorption and emission of light makes atoms useful as time-keepers. In suitable conditions, atoms will emit one of their permitted frequencies of light and this emission can be used as the time-keeper for an electronic clock. Alternatively, atoms can be exposed to light from another source that is then carefully adjusted in frequency until the atoms begin to absorb the light. Both of these schemes are used in atomic clocks.

Because atoms are completely free of friction and most of the problems affecting other time-keepers, they produce exquisitely accurate clocks. The frequencies associated with some atomic rearrangements are so sharply defined that the atom can distinguish between lights differing in frequency by only 1 part in 10^{15} or more. This precise behavior has made it possible to define the second as 9,192,631,770 periods of a certain electromagnetic radiation absorbed by cesium atoms and the meter as the distance traveled by light during 1,650,763.73 periods of a certain light emitted by krypton atoms. In effect, the meter (a unit of length) is now officially defined in terms of time and the speed of light. Time and space really are closely related.

Modern atomic clocks use cesium, rubidium, or hydrogen as the time-keeper atom (Fig. 9.1.10). They are so accurate that they gain or lose as little as 1 s every 10 million years. But even this accuracy is inadequate for some purposes, so scientists are trying to develop even more accurate clocks. Current work with trapped and isolated atoms is intended to reach timing errors of perhaps 1 part in 10^{20}. A clock with this accuracy would have gained or lost no more than a few thousandths of a second since the creation of the universe.

CHECK YOUR UNDERSTANDING #7: Night Lights

Sodium vapor lamps used in street lights emit a yellowish or pinkish glow. Where does this peculiar coloring come from?

Summary

How Clocks Work: Clocks use various time-keepers to measure the passage of time. Most clocks are based on harmonic oscillators because of their extremely steady periods of oscillation. Most importantly, the period of a harmonic oscillator doesn't depend on the amplitude of its motion. Common harmonic oscillator time-keepers include pendulums, balance rings, and quartz crystals. Atomic clocks, on the other hand, use atoms and atomic structure to measure time with phenomenal accuracy.

In a pendulum clock, a swinging pendulum controls the rotation of a toothed wheel, which in turn controls the rotation of the clock's hands. Energy needed to keep the pendulum swinging and to advance the hands comes from the descent of a weighted cord. In a balance clock, a rocking balance ring controls the toothed wheel. Energy for this clock comes from a wound coil spring. A quartz crystal clock detects the vibration of its crystal electronically and uses that vibration to control a motor that advances its clock hands or an electronic circuit that measures time by counting the crystal's vibrations. Small, carefully timed electric pulses from the clock provide the energy that keeps the crystal vibrating.

The Physics of Clocks

1. Time is a dimension. Our universe has three spatial dimensions and one temporal dimension. Locating an event in space and time involves four quantities: three that place the event in space and one that places the event in time.

2. Once displaced from a stable equilibrium, an object will tend to oscillate about that equilibrium in a natural resonance. The displacement increases the object's energy and keeps it oscillating until that excess energy is transferred elsewhere or converted to thermal energy.

3. The object's period is determined by its restoring force and mass. For a rotational oscillation, the period is determined by the restoring torque and by the object's moment of inertia.

4. If the restoring force is proportional to the object's displacement from equilibrium, the object will be a harmonic oscillator and its period won't depend on the amplitude of its oscillation.

5. The restoring force of a spring is proportional to its distortion, so that springs are the basis for many harmonic oscillators. Since most solid objects behave as springs when distorted slightly, they vibrate as harmonic oscillators when struck.

6. If an oscillator's restoring force is not proportional to its displacement, that oscillator is anharmonic and its period depends on its amplitude. Any oscillator becomes anharmonic when its amplitude of motion is too large.

7. The excess energy needed to start an oscillator in motion or to sustain its oscillation can be supplied all at once or through many separate pushes that are carefully timed to do a little work on the oscillator during each cycle of its motion.

8. The orbits of electrons and nuclei in atoms are restricted by the laws of quantum physics. These laws structure the atoms and limit the wavelengths and frequencies of light that they can absorb or emit.

Check Your Understanding - Answers

1. Since light travels at a constant speed, the distance it travels is equal to the product of its speed times its travel time. If you know the travel time and the speed, you can determine that distance.

Why: Many distance measurements are made by measuring time. Surveyors routinely use light's travel time to measure distances. Some cameras use sound's travel time to measure distances. In general, the motion of an object at constant velocity can be used either to measure the distance traveled if you know the elapsed time or to measure the elapsed time if you know the distance traveled.

2. It may have lost as much as 80 minutes after 24 hours.

Why: If the operator of a sandglass waits 10 s before turning it over each time the sand runs out, the clock will lose 10 s every 3 minutes, 200 s every hour, and 4800 s every day. If you could be sure that the operator would always wait 10 s, you could include it in the clock's design. However, the operator might be faster one time than the next and this uncertainty makes the clock unreliable and inaccurate. For a repetitive clock to keep accurate time, it must repeat its motion with almost perfect regularity.

3. The strength of the earth's gravity and the length of the swing's chains.

Why: A child swinging on a swing set is a form of pendulum. As with any pendulum, the child's period of motion is determined only by the strength of gravity and the length of the pendulum. In this case, the length of the pendulum is approximately the length of the swing's supporting chains. Thus a tall swing has a longer period than a short swing.

4. The amplitude of her motion will gradually decrease so that she comes to a stop.

Why: To keep her swinging, you must make up for the energy she loses to friction and air resistance. By pushing her forward each time she moves away from you, you do work on her and increase her energy. But when you push her as she moves toward you, she does work on you and you extract some of her energy. You are then slowing her down rather than sustaining her motion.

5. The torsional stiffness of the chandelier's supporting cord and the chandelier's moment of inertia.

Why: The hanging chandelier is a harmonic oscillator. Its supporting cord opposes any twists by exerting a restoring force on the chandelier. Once you twist it away from its equilibrium orientation, the chandelier oscillates back and forth with a period determined only by the cord's torsional stiffness (its stiffness with respect to twists) and the chandelier's moment of inertia. As with any harmonic oscillator, the amplitude of the chandelier's motion doesn't affect its period.

6. The rod is vibrating as a harmonic oscillator, with its two halves first approaching one another and then moving apart.

Why: The metal rod vibrates in the same manner as a quartz crystal. The body of the rod exerts restoring forces on its two halves. After hitting the floor, these halves move toward and away from one another rapidly, emitting the tone that you hear. Because it's a harmonic oscillator the frequency (and pitch) of the tone doesn't change as the amplitude of motion decreases.

7. The structure of the sodium atom.

Why: Much of the light emitted by a sodium vapor lamp comes from structural changes within the sodium atoms themselves. These atoms first absorb electric energy and then emit light. When they emit light, they emit it at specific wavelengths, characteristic of the sodium atom's internal structure.

Glossary

amplitude The maximal displacement of a particular oscillator away from its equilibrium position.

anharmonic oscillator An oscillator in which the restoring force on an object is not proportional to its displacement from a stable equilibrium. The period of an anharmonic oscillator depends on the amplitude of its motion.

cycle-per-second The SI unit of frequency (identical to the hertz).

frequency The number of cycles completed by an oscillating system in a certain amount of time.

harmonic oscillator An oscillator in which the restoring force on an object is proportional to its displacement from a stable equilibrium The period of a harmonic oscillator doesn't depend on the amplitude of its motion.

hertz (Hz) The SI unit of frequency (identical to the cycle-per-second).

natural resonance A mechanical process in which an isolated object's energy causes it to perform a certain motion over and over again. The rate at which this motion occurs is determined by the physical characteristics of the object.

nucleus The positively charged central component of an atom, containing most of the atom's mass and about which the electrons orbit. Plural is nuclei.

oscillation A repetitive and rhythmic movement or process that usually takes place about an equilibrium situation.

period The time required to complete one full cycle of a repetitive motion.

simple harmonic motion The regular, repetitive motion of a harmonic oscillator. The period of simple harmonic motion doesn't depend on the amplitude of oscillation.

stiffness A measure of how rapidly a restoring force increases as the system exerting that force is distorted.

Review Questions

1. How do Newton's laws relate time and space?

2. Why are most modern clocks based on natural resonances?

3. List five common objects that exhibit natural resonances. Could these resonances be used to build clocks, at least in principle? How are these resonances similar or different? Can they be sustained by adding energy to them periodically?

4. Why does the period of a pendulum depend on the strength of gravity?

5. Why doesn't the period of a pendulum depend on the pendulum's mass?

6. What distinguishes a harmonic oscillator from an anharmonic oscillator?

7. Why is a harmonic oscillator so much better suited to time-keeping that an anharmonic oscillator?

8. Why does the balance ring of a balance clock maintain a steady period, independent of gravity and regardless of its orientation?

9. Why does the period of a balance ring depend on its moment of inertia rather than on its mass?

10. What is the relationship between an oscillator's period and its frequency?

11. A quartz crystal doesn't appear to have a spring. Why is it a harmonic oscillator?

12. What characteristics of an atom allow it to serve as the time-keeper of an atomic clock?

Exercises

1. The acceleration due to gravity at the moon's surface is only about $1/6^{th}$ that at the earth's surface. If you took a pendulum clock to the moon, would it run fast, slow, or on time?

2. If you pull a small tree to one side and suddenly let go, it will swing back and forth several times. The period of this motion won't depend on how far you bend the tree. How does the restoring force that returns the tree to its upright position depend on how far you bend the tree?

3. If you strike the top of a cube of gelatin, it will oscillate briefly up and down with a period that doesn't depend on the amplitude of its motion (within reason). Suppose that exerting a 1 N downward force on the gelatin with your spoon distorts its surface downward by 2 mm. Based on this information, you can predict that a 3 N downward force on the gelatin will distort its surface downward by 6 mm. Justify this prediction.

4. A clothing rack hangs from the ceiling of a store and swings back and forth. Why doesn't the period of this motion depend on how many dresses the rack is holding?

5. A ball bouncing on a table is poorly suited to use as a time-keeper because its period varies with amplitude. Explain why the period of this anharmonic oscillator varies with amplitude.

6. When you rap on a table, it vibrates and you hear a sound. While the amplitude (volume) of the vibration increases as you rap harder, the frequency (pitch) doesn't change. The table's mass is constant, so how is its surface able to complete a large cycle of its motion in the same amount of time it takes to complete a small cycle of its motion?

7. Depending on how the base of a rocking chair was cut, the period of its motion may or may not depend on how hard you're rocking it. What can you say about the restoring forces acting in these two cases?

8. If you had an extremely accurate timer, how could you use light, which travels at a speed of 299,792,458 m/s in empty space, to measure the distance between two locations?

9. If you had an extremely accurate distance measuring system, how could you use light (see Exercise **8**) to measure the time that elapses between two events?

10. Some cameras measure the distance to their subjects by bouncing sound off them. Sound travels at approximately 331 m/s. How can a camera use a sound emitter, a sound receiver, and a timer to measure how far away its subject is?

11. A spinning toy top has a motion that repeats over and over, but it wouldn't be suitable for a clock. Why wouldn't the top keep accurate time, even if you spun it with your fingers occasionally to keep it from stopping?

12. If there were no gravity, a pendulum wouldn't have gravitational potential energy. It also wouldn't oscillate. How are these two facts related?

13. A flag pole is a harmonic oscillator, flexing back and forth with a steady period. If you want to increase the amplitude of the pole's motion by pushing it near its base, when should you push—as it flexes toward you or away from you?

14. If a child stands up on the seat of a playground swing, how will the swing's period be affected?

15. Which of the following clocks would keep accurate time if you took them to the moon: a pendulum clock, a balance clock, and a quartz watch? Why?

16. The time-keepers in early electronic watches were tiny tuning forks rather than quartz crystals. A tuning fork has two metal tines that flex alternately toward one another and away from one another. A tuning fork is a harmonic oscillator, so how do the restoring forces on its tines depend on the distance between them?

17. If you strike a large tuning fork (see Exercise **16**) with a mallet, it will oscillate at a constant frequency, emitting a single pure tone even as the volume of that tone diminishes. But if you attach small weights to its tines, the fork will oscillate at a lower frequency than before. Why?

18. Why do thicker quartz crystals vibrate more slowly than thinner quartz crystals?

19. Most clocks have to be calibrated against a standard clock before they'll measure time accurately. One unusual aspect of an atomic clock is that it doesn't have to be calibrated. In fact, every atomic clock *is* a standard clock. Why?

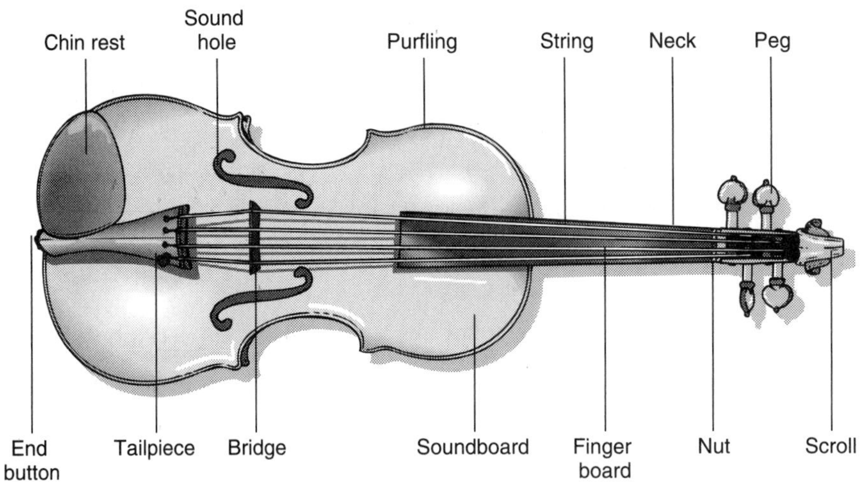

Section 9.2

Violins and Pipe Organs

Music has always been a part of human expression. While what qualifies as music is a matter of taste, it's always based on sound and often involves an instrument to create that sound. In this section, we'll examine sound, music, and two instruments: violins and pipe organs. As examples of the two most common types of instruments, strings and winds, these devices will help us to understand a great many other instruments as well.

Questions to Think About: *Why are the low pitch strings on a violin thicker than the high pitch strings? How does pressing a violin's string against the fingerboard change its pitch? Why does a violin sound different when it's plucked rather than bowed? What distinguishes a good violin from a mediocre one? What is vibrating inside a pipe organ? Why are some organ pipes longer than others?*

Experiments to Do: *Find a violin or guitar, or stretch a strong string between two rigid supports. Pluck the string with your finger and listen to the tone it makes. The string is vibrating back and forth at a particular frequency (pitch) even as its amplitude of motion decreases. What kind of oscillator has that behavior?*

Change the string's frequency of vibration by changing its tension or length. What happens when you shorten the string or prevent part of it from moving? What happens if you increase its tension by stretching it tighter? You can also increase the string's mass by wrapping it with tape. How does that affect its pitch?

You can imitate a pipe organ by blowing gently across the mouth of a bottle or soda straw. If done properly, you will get the air inside the bottle to vibrate up and down rhythmically and you will hear a tone. What happens to its pitch when you add water to the bottle or pinch off the straw at various points? Why does this tone sound different from that of the string, even when the two have the same pitch?

Sound and Music

To understand how violins and pipe organs work, we'll need to know a bit more about sound and music. As we noted in Section 4.4, **sound** in air consists of pressure waves, patterns of compressions and rarefactions that travel outward from their source at the speed of sound. When a sound passes by, the air pressure in your ear fluctuates up and down about normal atmospheric pressure so that it's sometimes less than atmospheric pressure and sometimes more. Even when these pressure fluctuations have amplitudes less than a millionth of atmospheric pressure, you hear them as sound.

When the fluctuation is rhythmic, you hear a *tone* with a *pitch* equal to its frequency. A bass singer's pitch range extends from 80 Hz to 300 Hz while that of a soprano singer extends from 300 Hz to 1100 Hz. Musical instruments can produce tones over a much wider range of pitches but we can only hear pitches between about 30 Hz and 20,000 Hz.

Most music is constructed around *intervals*, the frequency ratio between two different tones. This ratio is found by dividing one tone's frequency by that of the other. Our hearing is particularly sensitive to intervals, with pairs of tones at equal intervals sounding quite similar to one another. For example, a pair of tones at 440 Hz and 660 Hz sounds similar to a pair at 330 Hz and 495 Hz because they both have the interval 3/2.

The interval 3/2 is pleasing to most ears and is common in western music, where it's called a *fifth*. A fifth is the interval between the two "twinkles" at the beginning of *Twinkle, twinkle, little star*. If your ear is good, you can start with any tone for the first twinkle and will easily find the second tone, located at 3/2 the frequency of the first. Your ear hears that factor of 3/2 between the two frequencies.

The most important interval in virtually all music is 2/1 or an *octave*. Tones that differ by a factor of two in frequency sound so similar to our ears that we often think of them as being the same. When men and women sing together "in unison," they often sing an octave or two apart and the differences in the tones, always factors of 2 or 4 in frequency, are only barely noticeable.

The octave is so important that it structures the entire range of audible pitches. Most of the subtle interplay of tones in music occurs in intervals of less than an octave, less than a factor of 2 in frequency. Thus most traditions build their music around the intervals that lie within a single octave, such as 5/4 and 3/2. They pick a particular standard pitch and then assign *notes* at specific intervals from this standard pitch. This arrangement repeats at octaves above and below the standard pitch to create a complete *scale* of notes. (For a history of scales, see ❐.)

❐ In addition to his contributions to mathematics, geometry, and astronomy, the Greek mathematician Pythagoras (ca 580–500 BC) was perhaps the first person to use mathematics to relate intervals, pitches, and the lengths of vibrating strings. He and his followers laid the groundwork for the scale used in most Western music.

The scale used in western music is constructed around a note called A_4, which has a standard pitch of 440 Hz. At intervals of 9/8, 5/4, 4/3, 3/2, 5/3, and 15/8 above A_4 lie the 6 notes B_4, $C^{\#}_5$, D_5, E_5, $F^{\#}_5$, and $G^{\#}_5$. Similar collections of 6 notes are built above A_5 (880 Hz), which has a frequency twice that of A_4, and above A_3 (220 Hz), which has a frequency half that of A_4. In fact, this pattern repeats above A_1 (55 Hz) through A_8 (7040 Hz).

Actually, western music is built around 12 notes and 11 intervals that lie within a single octave. Five more intervals account for 5 additional notes, B^b_4, C_5, $D^{\#}_5$, F_5, and G_5. It's also not quite true that every note is based exclusively on its interval from A_4. While A_4 remains at 440 Hz, the pitches of the other 11 notes have been modified slightly so that they are at interesting and pleasing intervals from one another as well as from A_4. This adjustment of the pitches led to the well-tempered scale that has been the basis for western music for the last several centuries.

CHECK YOUR UNDERSTANDING #1: A Night at the Opera

A typical singing voice can cover a range of about 2 octaves, for example from C_4 to C_6. How broad is this range of frequencies?

How a Violin String Works

The tones produced by a violin begin as vibrations in its strings. But these strings are limp and shapeless on their own, and rely on the violin's rigid body and neck for structure. The violin creates tension in its strings and gives each of them a new equilibrium shape: a straight line.

To see that a straight violin string is in equilibrium, think of it as composed of many individual pieces that are connected together in a chain (Fig. 9.2.1). Each piece of the string pulls inward on the pieces adjacent to it and they pull outward on it. Since the string's tension is uniform, the two outward forces on a given piece balance one another perfectly. They have equal magnitudes and they point in exactly opposite directions, so that the piece experiences zero net force. With no net forces acting on its pieces, the string is in equilibrium.

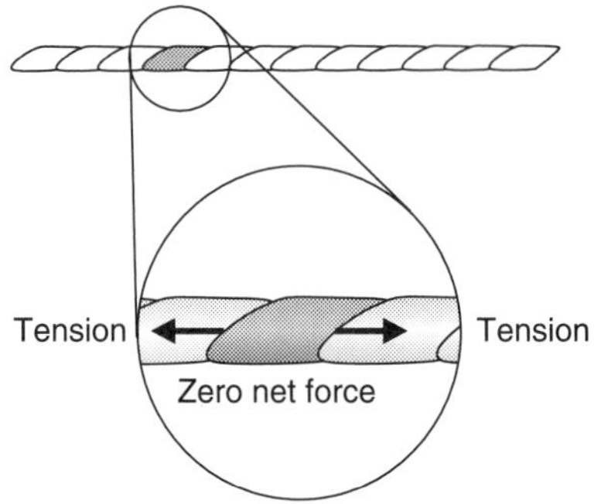

Fig. 9.2.1 - A taut violin string can be viewed as composed of many individual pieces. When the string is straight, the two forces exerted on a given piece by its neighbors cancel perfectly.

But when the string is curved, the outward forces on its pieces no longer balance one another (Fig. 9.2.2). While the two forces on a given piece still have equal magnitudes (the string's tension is still uniform), those two forces no longer point in exactly opposite directions. Instead, they leave that piece with a small net force.

The net forces on its pieces act to straighten the string. They are restoring forces. If you distort the string and release it, these restoring forces will cause the string to vibrate about its straight equilibrium shape in a natural resonance.

But the string's restoring forces are special. The more you curve the string, the stronger the restoring forces on its pieces become. In fact, the restoring forces increase in proportion to the string's distortion, so that the string is a form of harmonic oscillator!

Actually, the string is much more complicated than a pendulum or a balance ring. It can bend and vibrate in many different modes, each with its own period. Nonetheless, the string retains the most important feature of a harmonic oscillator: the period of each vibrational mode is independent of its amplitude. Thus a violin string's pitch doesn't depend on how hard it's vibrating. Think how hard it would be to play a violin if its pitch depended on its volume!

Like the baseball bat of Section 2.2, a violin string has a simplest vibration; its fundamental vibrational mode. In this mode, the entire string vibrates together in the simplest possible manner: its mid-point moves the farthest (the vibrational antinode) while its ends remain fixed (the vibrational nodes). The string's shape is a gradual curve based on the trigonometric sine function (Fig. 9.2.3).

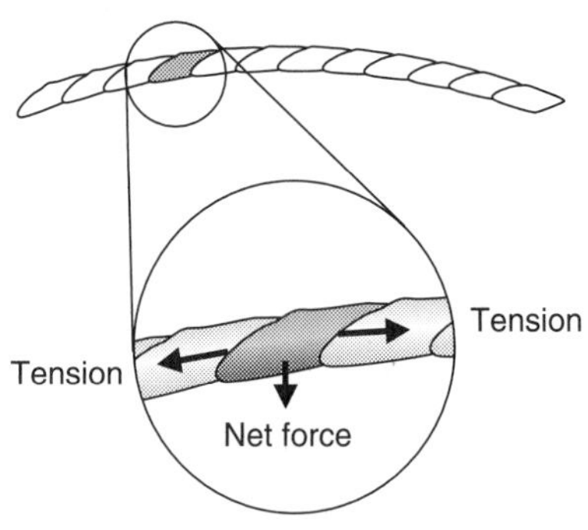

Fig. 9.2.2 - When a violin string is curved, the two forces exerted on a given piece by its neighbors don't point in exactly opposite directions and don't balance one another. The piece experiences a net force.

In this fundamental mode, the violin string behaves as a single harmonic oscillator. Its kinetic energy peaks as it rushes through its straight equilibrium shape and its potential energy (elastic potential energy in the string) peaks as it stops to turn around. As with any harmonic oscillator, the string's vibrational period depends only on its mass and on the stiffness of its restoring forces. Increasing the string's mass slows its vibration while stiffening the string speeds that vibration up. The frequency of this vibration establishes the violin string's *fundamental pitch.*

Each of a violin's four strings has a different mass and a different fundamental pitch. In a tuned violin, the notes produced by these strings are G_3 (196 Hz), D_4 (293 Hz), A_4 (440 Hz), and E_5 (660 Hz). The G_3 string, which vibrates quite slowly, is the most massive. It's usually made of gut, wrapped in a coil of heavy metal wire. The E_5 string, on the other hand, must vibrate quite rapidly and needs to have a low mass. It's usually a thin steel wire.

You can adjust a string's fundamental pitch by changing its stiffness, which itself depends on tension and length. Tightening a string stiffens it by increasing both the outward forces on its pieces and the net forces they experience during a distortion. You tune a violin by adjusting the tension in its strings, using *pegs* in its neck. However, the high pitched strings are under great tension and are difficult to tune accurately. To make tuning easier, many violins have additional fine tension adjusters at the other ends of the high pitched strings. And because a string's tension changes with temperature and time, you must always tune your violin just before a concert.

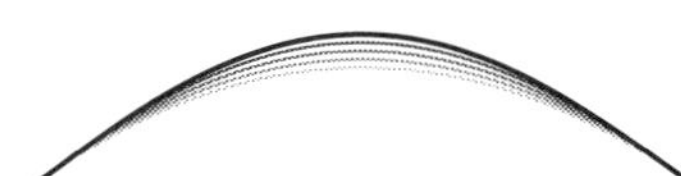

Fig. 9.2.3 - A string vibrating between two fixed points in its fundamental vibrational mode. The whole string moves together, traveling up and down as a single harmonic oscillator.

A string's length also affects its fundamental pitch. Shortening a string stiffens it by increasing its curvature during a distortion and subjecting its pieces to larger net forces. Shortening the string also reduces its mass. Since a stiffer string with a smaller mass vibrates faster, shortening a string increases its pitch. This effect allows you to raise the pitch of a string by pressing it against the fingerboard in the violin's neck. Part of a violinist's skill involves knowing exactly where on the string to press it against the fingerboard in order to produce a particular note.

CHECK YOUR UNDERSTANDING #2: Feeling Tense?

A common way to determine the tension in a cord is to pluck it and listen for how fast it vibrates. Why does this technique measure tension?

The Violin String's Harmonics

The fundamental vibrational mode is not the only way in which a violin string can vibrate. The string also has **higher-order vibrational modes**, in which the string vibrates as a chain of shorter strings that move in alternating directions (Fig. 9.2.4). For example, the string can vibrate as two half-strings that move in opposite directions and are separated by a motionless vibrational node.

In this mode, the violin string not only vibrates as two half-strings, it has the pitch of a half-string as well! Because the string must curve more as it bends both up and down at the same time, the restoring force on each portion of string becomes stiffer. This increased stiffness, together with the reduced mass of a half-string, causes the string to vibrate as though it were half its actual length.

Remarkably enough, halving the length of a violin string exactly doubles its vibrational frequency. Frequencies that are integer multiples of the fundamental are called **harmonics**. Because the two half-strings vibrate at twice the fundamental frequency, their frequency is the *second harmonic* and their mode of vibration is the string's *second harmonic mode.*

If the violin string can vibrate as two half-strings, then it ought to be able to vibrate as three third-strings. And so it can. The frequency of this vibrational mode is three times the fundamental, making it the *third harmonic*. The interval between the third harmonic and the fundamental is an octave and a fifth (2/1 times 3/2). Overall, the fundamental and its second and third harmonics sound very pleasant together.

While the violin string can vibrate in even higher harmonics, a more important fact is that the string often vibrates in more than one mode at the same time. For example, a violin string vibrating in its fundamental mode can also vibrate in its second harmonic and emit two tones at once. While this multiple vibration complicates the string's shape and motion, the tones it emits are still integer multiples of its fundamental pitch.

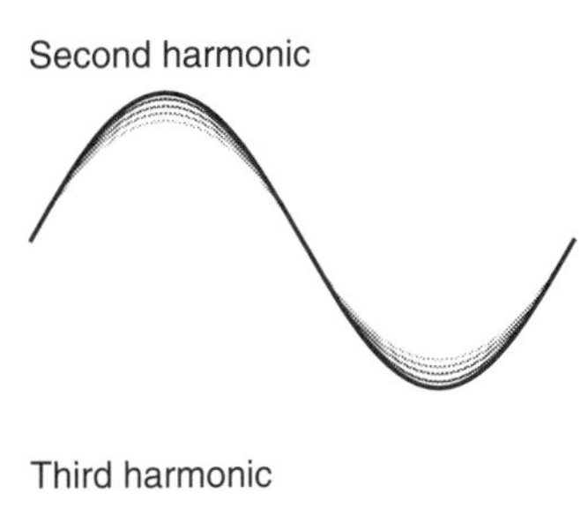

Fig. 9.2.4 - A string vibrating between two fixed points in its second and third harmonic modes. The string vibrates as two or three segments, completing cycles at two or three times the fundamental frequency, respectively.

Harmonics are important because bowing a violin excites many of its vibrational modes. The violin's sound is thus a rich mixture of the fundamental tone and its harmonics. Because this mixture of tones is characteristic of a violin, another instrument producing a different mixture won't sound like a violin.

CHECK YOUR UNDERSTANDING #3: Swinging High and Low Together

When two people swing a long jump rope, they can make it swing as a single arc or as two half-ropes arcing in opposite directions. To make the rope swing as two half-ropes, it must be turned faster or with a higher tension. Why?

Bowing and Plucking the Violin String

You play a violin by drawing a bow across its strings. The bow consists of horse hair, pulled taut by a wooden stick. Horse hair is rough and exerts frictional forces on the strings as it moves across them. Most importantly, horse hair exerts much larger static frictional forces than sliding ones.

When you draw the bow across a violin string, its hairs push on the string. At first, the string moves *with* the bow and static friction between the two allows the bow to push the string forward. But eventually the string's restoring force overpowers static friction and the string begins to slide *across* the bow in the opposite direction. Because the hairs exert little sliding friction, the string completes half a vibrational cycle before stopping to reverse direction. Soon the string is again moving *with* the bow so that the bow can push it forward with static friction. This process repeats over and over.

Fig. 9.2.5 - Resonant energy transfer makes it possible for sound to shatter a crystal wine glass. When the sound pushes on the glass rhythmically, doing work on it each cycle of its vibration, the sound slowly transfers energy to the glass. A crystal glass retains this energy for a long time, vibrating more and more vigorously until it finally shatters. Because the sound must be extremely loud and at exactly the resonant frequency of the glass, an unamplified singing voice probably can't break the glass.

Each time the bow pushes the violin string forward, it does work on the string and adds energy to the string's vibrational modes. This process is an example of **resonant energy transfer**, in which a modest force doing work in synchrony with a natural resonance can transfer a large amount of energy between objects. Just as gentle, carefully timed pushes can get a child swinging high on a playground swing, so modest, carefully timed pushes from a bow can get a string vibrating vigorously on a violin. Similar pushes can cause other objects to vibrate strongly, notably a crystal wine glass (Fig. 9.2.5) and the Tacoma Narrows Bridge near Seattle, Washington (Fig. 9.2.6).

How much energy the bow adds to each vibrational mode depends on where it crosses the violin string. When you bowing the string at the usual position, you produce a strong fundamental vibration and a moderate amount of each harmonic. Bowing the string nearer its middle reduces the string's curvature, weakening its harmonic vibrations and giving it a mellower sound. Bowing the string nearer its end increases the string's curvature, strengthening its harmonic vibrations and giving it a brighter sound.

The sound of a plucked violin string also depends on harmonic content and thus on where that string is plucked. But this sound is quite different from that of a bowed string. The difference lies in the sound's *envelope*, the way the sound evolves with time. This envelope can be viewed as having three time periods: an initial *attack*, a middle *sustain*, and a final *decay*. The envelope of a plucked string is an abrupt attack followed immediately by a gradual decay. In contrast, the envelope of a bowed string is a gradual attack, a steady sustain, and then a gradual decay. We learn to recognize individual instruments not only by their harmonic content but also by their sound envelopes.

Fig. 9.2.6 - The Tacoma Narrows Bridge collapsed in November, 1940, as the result of resonant energy transfer between the wind and the bridge surface. Shortly after construction, the automobile bridge began to exhibit an unusual natural resonance in which its surface twisted slowly back and forth so that one lane rose as the other fell. During a storm, the wind slowly added energy to this resonance until the bridge ripped itself apart. The entire collapse was beautifully recorded on film.

CHECK YOUR UNDERSTANDING #4: What's the Buzz?

Sometimes a tone from an instrument or an audio device will cause some object in the room to begin vibrating loudly. Why does this happen?

Turning the String's Vibration Into Sound

The violin's sound doesn't come directly from its strings because they're much too narrow to push effectively on the air. Instead, the violin creates sound with its top plate or *belly* (Fig. 9.2.7). The strings transfer their vibrational motions to the belly and the belly pushes on the air to create sound.

Most of this vibrational energy flows into the belly through the violin's *bridge*, which holds the strings away from the violin's body (Fig. 9.2.8). Beneath the G_3 string side of the bridge is the *bass bar*, a long wooden strip that stiffens the belly. Beneath the E_5 side of the bridge is the *sound post*, a shaft that extends from the violin's belly to its back.

As a bowed string vibrates across the violin's belly, it exerts a torque on the bridge about the sound post. The bridge rotates back and forth, causing the bass bar and belly to move in and out. The belly's motion produces most of violin's sound. Some of this sound comes directly from the belly's outer surface while some comes from its inner surface and must emerge through its f-shaped holes.

Any violin can convert string vibrations into sound, so what distinguishes a great violin from a mediocre one? The answer lies mostly in how the violin handles different tones. A great violin responds equally to all tones, so that a certain amplitude of string vibration always creates a certain volume of sound. In contrast, a mediocre violin responds unequally to different tones so that you must struggle to keep the volume uniform (see ❐ on following page).

A violin will misbehave whenever the music it's playing excites one of its own natural resonances. The violin body itself has vibrational modes, each with its own period and frequency of vibration. When a string is tuned to one of these

Fig. 9.2.7 - A violin's bridge transfers energy from its vibrating strings to its belly. The belly moves in and out, emitting sound. Some of this sound leaves the violin through the f-holes in its body.

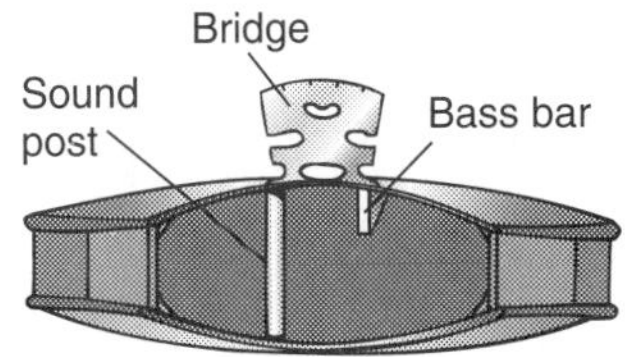

Fig. 9.2.8 - The bridge is supported by the bass bar on one side and the sound post on the other. As the strings vibrate back and forth, the bridge experiences a torque that causes the belly of the violin to move in and out and emit sound.

❐ Violins created 300 years ago by Antonio Stradivari are still among the finest in the world. After his death, violin makers attempted to duplicate his instruments but without success. The trouble lies at least partly in the wood itself, for no two pieces are exactly alike. Stradivari designed each violin around the individual pieces of wood he used. He tailored each piece of wood perfectly until the violin sounded exactly as he wanted it to. Without such individual attention, it's impossible to control natural resonances effectively.

frequencies, the string and violin body can exchange energy efficiently in a process called **sympathetic vibration**. When you start that string vibrating, the violin body will also vibrate and will emit an unusually loud tone. A great violin is carefully shaped and assembled to minimize its natural resonances and avoid trouble with sympathetic vibration. On the other hand, a mediocre violin has many natural resonances that make it hard to play.

A great violin is also very efficient at converting its strings' vibrational energy into sound and produces little thermal energy. Carefully chosen hardwoods and varnishes are important to this energy efficiency. Most great violins lose so little energy to anything but sound that they sing the moment the bow touches their strings.

CHECK YOUR UNDERSTANDING #5: Air Guitar
Why does an acoustic guitar have a sound box?

Sound From an Organ Pipe

Like a violin, a pipe organ uses vibrations to create sound and music. However, the vibrations in a pipe organ take place in the air itself. An organ pipe is essentially a hollow cylinder, open at each end and filled with air. Because that air is isolated, its pressure can fluctuate up and down about atmospheric pressure and it can exhibit natural resonances. Although this enclosed air can vibrate in many different ways, it has a fundamental vibrational mode resembling that of a quartz crystal, with air moving alternately toward and away from the pipe's center. (Fig. 9.2.9)

Let's follow the air's motion through one vibrational cycle. As air moves inward, toward the middle of the organ pipe, the density there rises and a pressure imbalance occurs. Since the pressure at the middle of the pipe is higher than at its ends, air accelerates *away from* the middle. It eventually stops moving inward and begins to move outward. As it does, the density in the middle of the pipe drops and a reversed pressure imbalance occurs. The pressure at the middle of the pipe is lower than at its ends so air accelerates *toward* the middle. It eventually stops moving outward and begins to move inward. This cycle repeats over and over.

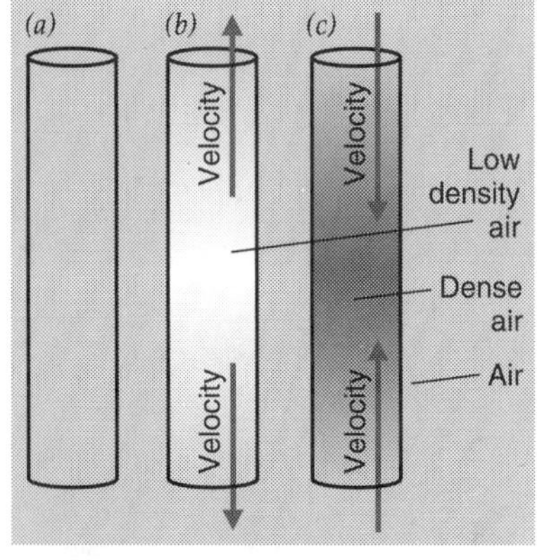

Fig. 9.2.9 - In a pipe that's open at both ends (*a*), the air vibrates in and out about the middle of the pipe. (*b*) For half a cycle, the air moves outward and creates a low pressure region in the middle and (*c*) for half a cycle, the air moves inward and creates a high pressure region there.

The air in an organ pipe behaves as a harmonic oscillator, vibrating about an equilibrium of uniform atmospheric density and pressure. The vibrating air's kinetic energy peaks as it rushes through this equilibrium arrangement and its potential energy (pressure potential energy) peaks as it stops to turn around. The period and frequency of this vibration depend only on the mass of the vibrating air and on the stiffness of the restoring force that air experiences.

Both the mass of the vibrating air and the stiffness of the restoring force depend on the length of the organ pipe. A shorter pipe not only holds less air than a longer pipe, it also offers stiffer opposition to any movements of air in and out of that pipe. Together, these effects make the air in a short pipe vibrate faster than that in a long pipe. In general, an organ pipe's vibrational frequency is inversely proportional to its length.

Unfortunately, the mass of vibrating air in a pipe also increases with the air's density, so that even a modest change in temperature or weather will alter the pipe's pitch. However, all of the pipes shift together so that an organ continues to sound in tune. Nonetheless, this shift may be noticeable when the organ is part of an orchestra.

The organ uses resonant energy transfer to make the air in a pipe vibrate. It starts this transfer by blowing air across the pipe's lower opening (Fig. 9.2.10*a*), although for practical reasons that lower opening is usually found on the pipe's side (Fig. 9.2.10*b*). As the air flows across the opening, it's easily deflected and tends to follow any air that's already moving into or out of the pipe. If the air inside the pipe is vibrating, the new air will follow it in perfect synchrony and strengthen the vibration.

This following effect is so effective at enhancing vibrations that it can even initiate a vibration from the random noise that is always present in a pipe. That's how the sound starts when the organ's compressor first blows air across the pipe. Once the vibration has started, it grows quickly in amplitude until energy leaves the pipe as sound and heat as quickly as it arrives as compressed air. The more air the organ blows across the pipe, the more power it delivers to the pipe and the louder the vibration.

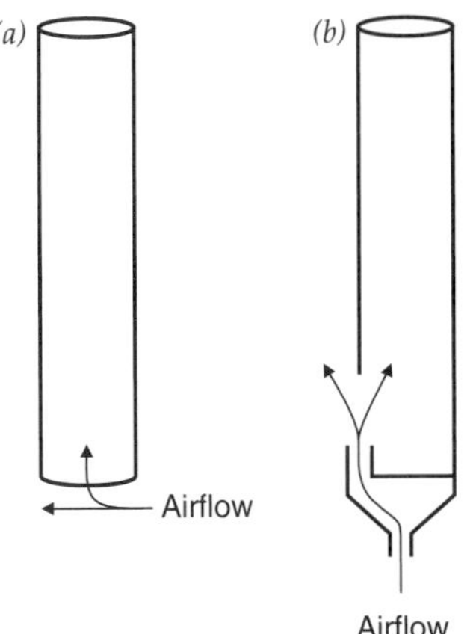

Fig. 9.2.10 - (*a*) Air blown across the bottom of an open pipe will follow any other air that's moving into the pipe. If the air in the pipe is vibrating, this effect will add energy to that vibration. (*b*) The lower opening in an organ pipe is cut in its side for practical reasons.

CHECK YOUR UNDERSTANDING #6: A Pop Organ

If you blow across a soda bottle, it emits a tone. Why does adding water to the bottle raise the pitch of that tone?

The Complete Pipe Organ

There is more to a pipe organ than a single pipe and a compressor. To begin with, the organ needs one pipe for every note of the scale. Each pipe has a different length, ranging from several meters for the lowest pitched pipe to a few centimeters for the highest pitched one.

But a complicated organ has several separate scales, each with its own sound. While the pipes of these different scales may play the same notes, those notes don't sound the same. One set of pipes may sound like a flute, another like a horn, and still another like a clarinet. What's going on?

The different pipes are producing different mixtures of harmonics. Like a violin string, an organ pipe can support many different modes of vibration. In its fundamental vibrational mode, the pipe's entire column of air vibrates together. In the higher-order vibrational modes, this air column vibrates as a chain of shorter air columns moving in alternating directions. In a cylindrical pipe, these vibrations occur at harmonics of the fundamental. When the air column vibrates as two half-columns, its pitch is exactly twice that of the fundamental mode. When it vibrates as three third-columns, its pitch is exactly three times that of the fundamental. And so on.

But the air column inside a pipe can vibrate in more than one mode simultaneously. By changing the shape of the organ pipe and where the air is blown across it, the pipe's harmonic content can be changed to imitate various instruments. To sound like a flute, the pipe should emit mostly the fundamental tone and keep the harmonics fairly quiet. To sound like a clarinet, its harmonics should be much louder. An organ pipe's volume always builds slowly during the attack, so it can't pretend to be a plucked string. However, a clever designer can make the organ imitate a surprising range of instruments.

You play the organ with a keyboard that controls the flow of air across a huge assortment of organ pipes. The control is fairly sophisticated because it carefully coordinates the simultaneous playing of several pipes at the press of a single key. While individual pipes already offer various mixtures of harmonics, the organ can produce even more variety by deliberately playing several pipes at the harmonics of the note chosen. For example, the organ can be set so that pressing the A_4 key causes not only the A_4 (440 Hz) pipe to play, but also the A_5

pipe (880 Hz) and the E_6 pipe (1320 Hz). The organ then sounds brighter and less flute-like than it would with only the A_4 pipe playing.

CHECK YOUR UNDERSTANDING #7: Shrill Harmonics

When you blow too hard into a whistle, recorder, or another wind instrument, its pitch often leaps upward by an octave. What is happening to the air inside it?

Summary

How Violins and Pipe Organs Work: A violin string exhibits natural resonances by vibrating back and forth about its equilibrium shape, a straight line. Bowing the string causes it to vibrate by pushing it away from its equilibrium shape and then allowing it to slip back. The bow continues to transfer energy to the vibrating string by pushing it whenever it moves in the bow's direction. The string's fundamental pitch is determined by its mass, tension, and length. After selecting a string of the correct mass, you tune that string by adjusting its tension. During play, you change the string's length and pitch by pressing it against the fingerboard. The belly of the violin serves to convert the string's vibration into sound by moving in and out as the string vibrates.

The air inside an organ pipe also exhibits natural resonances. As the organ blows air across the opening at the bottom of a pipe, the air inside that pipe begins to vibrate. The frequency of this vibration is determined principally by the pipe's length, so that a short pipe has a higher pitch than a long one. Pipes with different shapes produce different amounts of harmonics and yield different sounds. To produce an even wider variety of sounds, you can play several pipes together so that their sounds blend into one that is even richer in harmonics.

The Physics of Violins and Pipe Organs

1. Extended objects, such as strings and air columns, can exhibit natural resonances. In general, an object exhibits many different natural resonances with many different periods of oscillation.

2. Each mode of natural resonance in an extended object acts as its own oscillator, so that the object can exhibit several different modes of vibration simultaneously.

3. When an object's vibrational modes act as harmonic oscillators, their periods don't depend on their amplitudes of motion.

4. The vibrational frequency of a string increases with tension and decreases with length and mass.

5. The vibrational frequency of an air column decreases with length and density.

6. Strings and air columns (as well as most one dimensional extended objects) can also oscillate in halves, thirds, and so on at frequencies that are integer multiples (harmonics) of the fundamental vibrational frequency.

7. Energy can be transferred to a natural resonance by doing a little work on it once per cycle. Energy can be extracted from the resonance by having it do a little work on something else once per cycle.

8. Energy can flow quickly from one natural resonance to another via sympathetic vibration when the two share a common vibrational frequency.

Check Your Understanding - Answers

1. There is a factor of four in frequency between the lowest and the highest notes that voice can sing.

Why: Since notes separated by an octave are separated by a factor of two in frequency, notes separated by two octaves are separated by a factor of four.

2. The frequency of a string's fundamental vibrational mode depends on its tension.

Why: Any cord that is drawn taut from its ends will exhibit natural resonances like those in a violin string. The tauter the string, the higher the frequencies of those resonances.

3. A jump rope is essentially a vibrating string. The two half-rope pattern is the second harmonic mode, with a vibrational frequency twice that of the normal fundamental arc.

Why: Although it swings around in a circle, the jump rope is actually vibrating up and down at the same time it's vibrating forward and back. Together, these two vibrations create the circular motion. To make the rope vibrate in its second harmonic mode (as two half-ropes) without changing its tension, it must be swung twice as fast as normal.

4. The object has a natural resonance at the tone's frequency and sympathetic vibration is transferring energy to the object.

Why: Energy moves easily between two objects that vibrate at the same frequency. A note played on one instrument will cause the same note on another instrument to begin playing. Even everyday objects will exhibit sympathetic vibration when the right tone is present in the air.

5. To transfer the vibrational energy of the strings to the air.

Why: Guitar strings are too narrow to push effectively on the air and emit sound. They do better to transfer their energy to the body of the acoustic guitar so that its flat surfaces can push on the air. An electric guitar avoids the need for a sound box by converting the string's vibrations directly into electric currents and from there into movements of an audio speaker.

6. The water shortens the column of moving air inside the bottle and increases the frequency of its fundamental vibrational mode.

Why: A water bottle is essentially a pipe that is open at only one end. It has a fundamental vibrational mode with a frequency that is half that of an open pipe of equal length. As you add water to the bottle, you shorten the effective length of the pipe and raise its pitch.

7. That air column is vibrating in its second harmonic mode.

Why: The air in a wind instrument can vibrate in many different modes. If you are gentle with that air column, it tends to vibrate in its fundamental vibrational mode. However, it can also vibrate almost exclusively in its harmonic modes in certain situations. Blown too strongly, most wind instruments tend to vibrate in their harmonic modes.

Glossary

harmonic An integer multiple of the fundamental frequency of oscillation for a system. The second harmonic is twice the frequency of the fundamental and the third harmonic is three times the frequency of the fundamental. In principle, harmonics can continue forever.

higher-order vibrational mode A vibrational mode in which different parts of the system move in opposite directions.

resonant energy transfer The gradual transfer of energy to or from a natural resonance caused by small forces timed to coincide with a particular part of each oscillatory cycle.

sound In air, sound consists of pressure waves, patterns of compressions and rarefactions that travel outward from their source at the speed of sound.

sympathetic vibration The transfer of energy between two natural resonances that share a common frequency of oscillation.

Review Questions

1. If two tones differ from one another by two octaves, what is the relationship between their frequencies?

2. Why does a violin string's tension give it a straight equilibrium shape?

3. Any oscillating system has at least two forms for its energy. What are those two forms for a violin string? What are those two forms for the air in an organ pipe?

4. How do a violin string's length, tension, and mass influence the frequency of its fundamental vibrational mode?

5. When a violin string vibrates as two half-strings, its pitch is exactly twice that of its fundamental vibrational mode. Why?

6. Why does the volume a violin string's sound increase gradually when it's bowed and suddenly when it's plucked? How is energy transferred to the string in each case?

7. When you pluck a string on one guitar, the equivalent string on a nearby guitar may start to vibrate too. How is energy being transferred between the two strings?

8. Why can't a violin string create a loud sound by itself? Why does it need help from the body of the violin?

9. An organ pipe appears to be open only at its top, with a small whistle-like opening near its base. Why is it more accurately described as being open at both ends? Would it vibrate differently if it were truly closed at its base?

10. When the air in an organ pipe is vibrating, where is the air's pressure undergoing the largest fluctuations? Where is the air's velocity undergoing the largest fluctuations?

Exercises

1. When you press a piano key, a felt-covered hammer strikes a string and the string begins to vibrate. Describe the transfers of energy that take place as energy flows from your body, through the piano, and into the sound that the string emits.

2. If you stretch a rubber band between your fingers and pluck it, it will vibrate and emit a tone. If you now stretch that rubber band to twice its original length and pluck it again, it will emit almost the same tone. Why doesn't the rubber band's pitch change very much when you stretch it?

3. If you stretch the rubber band in Exercise **2** to its maximum length, so that it becomes stiff and taut, its pitch will increase significantly. Explain this increase in pitch.

4. Why do changes in temperature affect a violin string's pitch?

5. If you blow carefully across the top of a soda bottle, you can get it to whistle. If you fill the bottle half full of water, it will still whistle, but at a pitch roughly one octave higher than before. Why does its pitch increase?

6. Some wind chimes consist of sets of metal rods that emit tones when they're struck by wind-driven clappers. These rods vibrate like the baseball bat in Fig. 2.2.9. Why do the longer rods emit lower pitched tones than the shorter rods?

7. To modify the pitch of a guitar string you could change its mass, tension, or length. To raise its pitch, how should you change each of these three characteristics?

8. A violin string won't vibrate by itself. Why must you stretch the string across a violin before the string will vibrate?

9. If nothing exerted a tension on a violin string, it wouldn't be stretched and wouldn't have any elastic potential energy. It also wouldn't vibrate. How are these two facts related?

10. Why would replacing the air in an organ pipe with helium raise its pitch?

11. When a xylophone bar vibrates, its middle and ends move in opposite directions like the baseball bat in Fig. 2.2.9. The bar bends first ends down, middle up and then middle down, ends up. Why is the bar supported about a quarter of the way in from each end?

12. A bugle is effectively a pipe that's open at its flared "bell" and closed at its mouthpiece. As air vibrates in and out of the bell, the pressure at the mouthpiece fluctuates up and down. The bugler provides power to the vibrations by injecting pulses of air into the bugle through its mouthpiece. Even though there is no way to change the length of the bugle's pipe, a bugler can play several different tones, including those used in familiar pieces such as *reveille* and *taps*. What is happening to the air moving inside the bugle when it produces these different tones?

13. A trumpet resembles a bugle (see Exercise **12**), except that it has three valves. These valves let the trumpeter make small changes to the pipe's length. Why do these changes allow the trumpet to produce more tones than a bugle?

14. While a slide whistle resembles a small organ pipe that you blow with your mouth, it's not open at the far end. Instead, it has a plug in that end that you can slide toward or away from your mouth by moving a stick. What happens to the whistle's pitch as you slide the plug away from your mouth, so that the chamber inside the whistle gets longer, and what causes this change in pitch?

15. The most important difference between a trumpet and a tuba is in the lengths of their pipes. The tuba's pipe is much longer than that of the trumpet. How does this difference affect the relative pitches of these two instruments?

16. A flute and a piccolo are both effectively pipes that are open at both ends, with holes in their sides to allow them to produce more tones. The piccolo is very nearly a half-size version of the flute. How does this fact explain why the piccolo's tones are one octave above those of a flute?

17. The strings that play the lowest notes on a piano are made of thick steel wire wrapped with a spiral of heavy copper wire. The copper wire doesn't contribute to the tension in the string, so what is its purpose?

18. Why are the highest pitched strings on most instruments, including guitars, violins, and pianos, the most likely strings to break?

19. A recorder resembles an organ pipe that you blow with your mouth. It has holes in its side to change its pitch. These holes effectively shorten it by allowing air to flow in and out of the pipe before reaching the real end of the instrument. To play the lowest note, you cover all the holes. As you uncover the holes one at a time—starting with the hole farthest from your mouth—the recorder's pitch goes up. Explain this effect.

20. A clavichord is an ancestor of the piano. It has a number of taut strings that are played by pressing keys on a keyboard. When you press a key, it suddenly lifts a metal strip into contact with a string. The string's abrupt movement causes it to vibrate softly between this metal strip at one end and a fixed peg at the other end. Different keys lift different metal strips and allow different lengths of string to vibrate. If one key allows 25 cm of string to vibrate and another key allows 50 cm of that same string to vibrate, how will the pitches associated with those two keys be related?

21. A harpsichord is another ancestor of the piano. When you press a key on this instrument, it pulls a metal quill across a string and plucks it. This plucking action doesn't allow you to control the volume of the sound. How does plucking the string transfer energy to it and what determines how much energy the string receives?

22. When you press a key on a piano, a felt-covered hammer strikes one or more taut strings and causes them to vibrate. The hammer bounces off the string so quickly that the string is still moving away from its straight equilibrium shape when the hammer loses contact with it. Why is it important to the piano's volume that the hammer leave the string before the string begins to return toward its equilibrium shape?

23. Why doesn't the rebounding hammer in Exercise **22** take all the collision energy with it, leaving the string motionless?

24. While most woodwind instruments are silent while their players take breaths, bagpipes can play continuously. A piper blows air into the bag and then squeezes this air out of the bag with his arms. The air passes through several pipes that produce sound. Explain how the piper transfers energy to the bag and to the air in order to produce the sound.

Epilogue for Chapter 9

In this chapter, we looked at natural resonances in a variety of objects. In *clocks*, we examined these resonances in pendulums, balance rings, and quartz crystals and found that these objects are harmonic oscillators with periods that don't depend on the amplitudes of their motions. We learned how to control their periods in order to use them as time-keepers and we explored the role of atoms in even more accurate atomic clocks.

In *violins and pipe organs*, we explored the vibrations of strings and air columns to see that they too behave as harmonic oscillators. We learned how to adjust the pitches of their vibrations and examined the effects of higher-order harmonics on the sounds they produce. We also found that vibrations can be built up gradually through resonant energy transfer, in which energy is added a little at a time in synchrony with the vibrations.

Explanation: A Singing Wine Glass

Rubbing the lip of the glass is similar to bowing a violin string; both procedures cause resonant energy transfer. Your finger alternately pushes on the glass and then slips, helping the glass itself to vibrate back and forth beneath your finger in its fundamental vibrational mode. As the glass begins to vibrate, your finger does work on it a little bit at a time, whenever the glass is moving in the same direction as your finger. Once the glass is singing loudly, you can keep it singing by circling your finger steadily around the glass. Although the glass continues to emit its energy as sound, you keep giving it more energy by doing work on it with your finger.

Cases

1. Because sound carries energy, musical instruments need energy to operate. This energy is normally provided by the musician, who does mechanical work on the instrument. How does a musician transfer energy to

a. an accordion?
b. a piano?
c. a xylophone?
d. a violin?

2. When two people swing a jump rope between them, it forms a moving arc in the air. The rope's behavior is similar to that of a violin string, except that the jump rope swings around in a circle rather than vibrating directly back and forth.

a. What happens to the speed at which the jump rope turns if you increase the length of the rope but keep the tension on that rope the same?
b. What happens to the speed at which the jump rope turns if you use a thicker, more massive rope but keep the rope's length and tension the same?
c. What happens to the rope's motion if the people holding its ends pull harder on the rope and increase its tension?
d. How quickly must the people swing the rope to get it to form an S-shape, with a stationary node in the middle?

3. Imagine a stiff ruler that's lying at the edge of a table so that one end of it is on the table and the other end is free in the air. If you hold the ruler's supported end against the table with one hand and pluck its free end with the other hand, the ruler will vibrate up and down.

a. Why does the ruler's frequency of oscillation decrease when you extend more of it off the table?
b. Why doesn't the ruler's frequency of oscillation depend on how hard you pluck it?
c. What forms does the ruler's energy take as it vibrates up and down?
d. What eventually happens to the plucked ruler's energy?
e. If the ruler vibrates slowly enough, you can increase the energy of its vibration by pushing down on it rhythmically, once per cycle, just after it completes its upward motion and begins to head downward. Why must you push it downward as it heads downward, rather than pushing it downward as it heads upward?

4. A trumpet consists mainly of a long brass tube with a mouthpiece at one end and a flared opening or "bell" at the other end. This tube is coiled to make the trumpet more compact. The trumpet also has three valves that splice additional segments of tubing into the main tube, thus extending its overall length. When a musician seals his lips together and presses them against the mouthpiece of the trumpet, he creates a resonant system. The air can oscillate in and out of the open bell while the air pressure in the mouthpiece oscillates up and down.

a. To sustain the oscillation, the musician blows small bursts of high pressure air into the mouthpiece. Why are these bursts timed to coincide with the moments when the pressure inside the mouthpiece is already at its maximum?
b. As the musician presses the valves and lengthens the trumpet tube, the oscillation slows and the trumpet's pitch goes down. Explain.
c. Even without touching the valves, a good trumpet player can produce a variety of higher pitched notes using higher-order vibrational modes. What is the air inside the trumpet doing while it's oscillating in one of these higher-order modes?
d. The best way for a musical instrument to project sound is to move a large amount of air back and forth at low speeds rather than a small amount of air back and forth at high speeds. That's because room air can't respond quickly enough to the high-speed air near an instrument to extract its energy. Instead, that energy stays with the instrument. If the trumpet had no bell, the air at its open end would move back and forth very rapidly. Describe how the bell helps the trumpet project its sound.

5. In bungee jumping, a strong elastic cord or "bungee" is used to break the fall of a person who has leapt from a tall platform. This cord runs from the platform to the person's ankles and is short enough that the person doesn't hit the ground. After a few terrifying seconds of freefall, the cord pulls taut and the jumper bounces back upward. On this first bounce, the cord goes slack and the jumper once again experiences freefall. But in subsequent bounces, the cord remains taut and the jumper bounces up and down smoothly.

a. Why must the bungee cord stretch relatively easily to avoid injuring the jumper?
b. As a jumper bounces up and down on the cord, the time it takes to complete each bounce doesn't depend on how high that bounce is. Explain.
c. Why does a heavier person bounce up and down more slowly than a lighter person?
d. Who experiences the more extreme feelings of heaviness while coming to a stop at the end of the cord: a heavier person or a lighter person?
e. Why must the cords used in bungee jumping be able to withstand tensions that exceed the weights of the heaviest jumpers?
f. If the jumper used two bungee cords instead of one—perhaps one attached to each ankle—how would that affect the time it takes to complete each bounce up and down?

6. The body of a car is suspended above its wheels by four strong springs. These springs allow the wheels to respond quickly to bumps in the road without having the entire body respond, too. However, the wheels are also connected to the body by shock absorbers. Unlike the springs, which store energy when work is done on them, the shock absorbers use friction-like effects to convert work into thermal energy.

a. You remove the shock absorbers from your parked car and then jump into the driver's seat. The body bounces up and down for a long time. Explain this behavior.
b. If the car's first bounce takes 1 s, how long does its second bounce take? its tenth bounce?
c. You then drive the car off a high curb and onto the road. Describe the body's motion for the next ten seconds.
d. You reinstall the shock absorbers. Now every time the car body moves toward or away from the wheels, it does work on the shock absorbers and this work becomes thermal energy. You jump back into the driver's seat of the parked car. How does it respond?
e. You again drive the car off the high curb and onto the road. The shock absorbers make a difference. Describe the body's motion for the next ten seconds.

7. Diving off a spring board is much more complicated than diving off a platform because the spring board can bounce up and down. When you stand on the spring board without bouncing, it sags downward to a new equilibrium position. From that position you can begin experimenting.

a. If you begin to bounce up and down gently about the equilibrium position, what factors will determine the frequency of the bounce?
b. If you bounce more vigorously than in **a**, while still keeping your feet in contact with the board, how will the frequency of the bounce change?
c. If you begin to bounce so vigorously that your feet leave the surface of the board, the board's frequency of vibration changes abruptly. Why?
d. You're bouncing gently again. If you want to bounce more vigorously, you must add energy to the board. Your friend is floating under the board and can push on the board every time it bounces. When should your friend push on the board to add energy to it?
e. If you want to make the board bounce less vigorously, when should your friend push on the board?

8. A music box uses a row of small metal strips to play music. These strips are part of a metal plate that resembles a comb with many teeth. The free ends of these strips are plucked by pins projecting from the surface of a rotating metal cylinder. These pins are carefully placed so that, as the cylinder rotates, the music box plays a tune.

a. The metal strips are all of different lengths so that they play different notes. Why do the longer strips vibrate more slowly than the shorter ones?
b. The sound of each note starts abruptly and decays gradually. Why doesn't the sound build gradually?
c. How does the cylinder transfer energy to the strips?
d. Draw several sketches of a strip as it vibrates up and down. How is it similar to a violin string? How is it different?
e. When the strip is vibrating in its fundamental mode, where is its anti-node and where is its node?
f. A violin string can vibrate as two half-strings. Why can't a strip vibrate as though it were two half-strips?
g. Why does **f** imply that the strip can't produce the second harmonic of its fundamental pitch?

9. A toddler enjoys bouncing up and down in a seat that's suspended by a strong spring. The toddler pushes its feet on the floor and leaps up and down.

a. A toddler's bouncing motion in the seat is rhythmic. What determines the frequency of this bouncing?
b. As the toddler grows, what happens to the frequency of the bouncing or does it remain unchanged?
c. The toddler can change the amplitude of the bouncing motion by pushing on the floor. When should the toddler push on the floor to make the seat bounce higher?
d. When should the toddler push on the floor to make the seat bounce less high?

CHAPTER 10

MECHANICAL WAVES

Waves are fascinating objects. Most waves are peculiarly dependent on their surroundings because they depend on those surroundings for their very existences. Rather than being objects of their own, most waves are ripples or disturbances in other objects. For example, sound is a disturbance in air, an ocean swell is a disturbance in the water's surface, and light is a disturbance in the electric and magnetic structures of space itself. But despite their unusual relationships with their environments, waves can carry energy and other things around the room or around the universe. Our world is filled with waves that take many forms in many different systems. But the easiest waves to see and understand are mechanical ones, the focus of this chapter.

Experiment: Waves on a Jump Rope

You can make waves at the beach or in a pool of water. But you can also make waves on a rope. To do the latter, you'll need a heavy cord about 10 m long—a long jump rope is ideal. Fasten one end of the cord to a doorknob or another fixed object and then pull the cord reasonably taut from the other end. If you jiggle your end up and down, you'll create waves on the rope. But not all of these waves will be the same.

If you move your wrist up and down rhythmically at the right rate, it will vibrate like a violin string. By adjusting the rope's tension and the frequency of your motion, you should be able to make it vibrate in its fundamental mode (a single broad arc), its second harmonic (two arcs in opposite directions), and so on. These motions are actually waves of a type that don't go anywhere. Such waves are called *standing waves.*

But if you snap your wrist quickly up and down, something different will happen. You will launch an upward bump onto the cord and it will travel rapidly away from your hand. When it hits the doorknob, the bump will reflect and travel back toward your hand upside-down. You are creating a new type of wave that moves through space. Such waves are called *traveling waves.*

If you move your hand up and down rapidly several times, you will launch an even more complicated traveling wave and it, too, will reflect toward your hand. Why can't these waves simply disappear when they reach the fixed end of the rope? Can you think of another example of a traveling wave hitting a firm surface and reflecting back toward you?

To experiment further with these waves, try to **predict** what will happen when you change the tension in the rope. Now tighten or loosen the rope and **observe** the effect on the waves. Have a friend **measure** the time it takes for each wave to return to your hand as you vary the rope's tension. Did you **verify** your predictions?

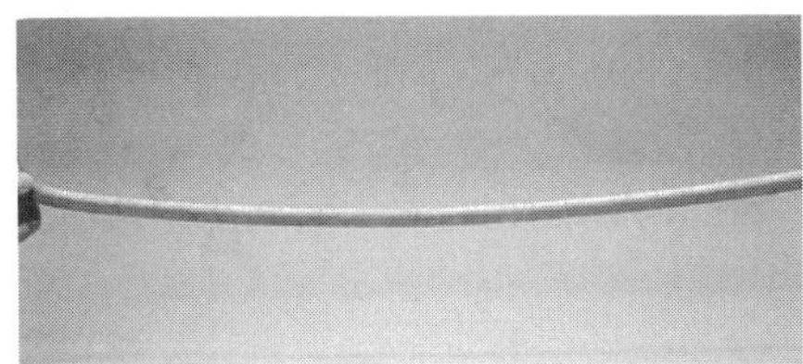
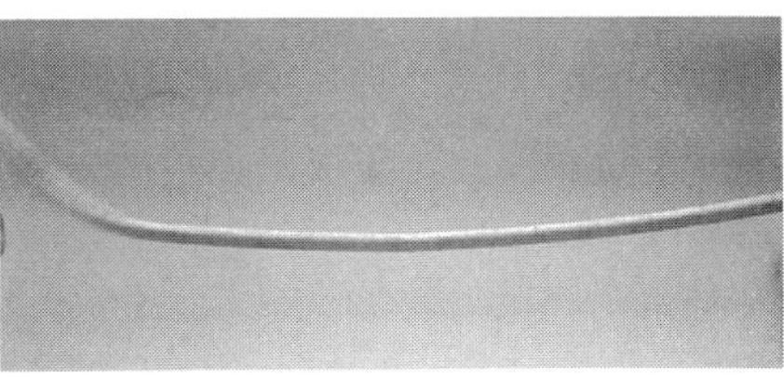
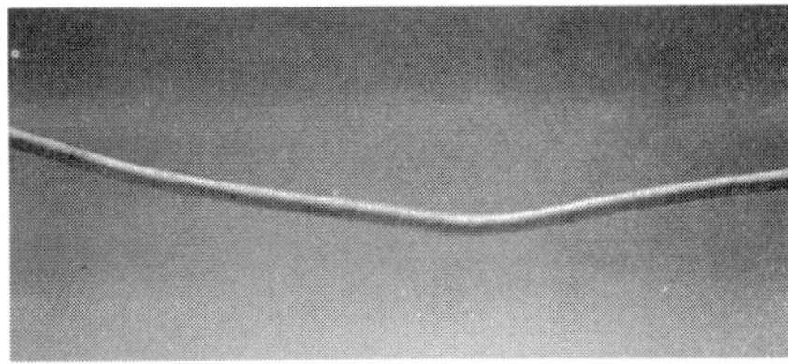

Chapter Itinerary

Mechanical waves appear in many objects and are particularly visible on the surfaces of fluids. In *the sea and surfing*, we'll look at how surface waves travel across the ocean and how they break on the beach. We'll also examine surfing to see what propels a surfer through the water and we'll look at the origins and properties of the tide.

But while the sea provides a beautiful example of mechanical waves, it's just one of many such examples. Waves are all around you, in the sounds you hear during a concert, the vibrations you feel on a roller coaster, and even the shaking you experience during an earthquake.

Section 10.1

The Sea and Surfing

While land normally remains stationary, the sea is always moving. This motion ranges from the global currents associated with climate to the tiny ripples caused by falling raindrops. When you visit the seashore, you'll probably notice two of the sea's most important motions: tides and surface waves. In this section, we'll examine the cycle of tides and look at how surface waves travel across water. These water waves can help us to understand other wave phenomena, including the electromagnetic waves that are responsible for light and the pressure waves that are the basis for sound. Water waves are also crucial for one of summer's ultimate sports: surfing.

Questions to Think About: *Why do the tides vary in height from place to place? Why are there no significant tides in a lake or a swimming pool? Why does high tide occur about every 12 hours? What moves in a water wave? Do all waves travel at the same speed? How deep is a wave? Why do waves break near shore? What propels a surfer forward? Why does a surfer travel faster when heading across a wave than when riding it directly toward shore?*

Experiments to Do: *While you can make water waves in a basin or tub, they move too quickly to see clearly. You do better to watch waves at the beach. On a calm day, the sea is in its flat equilibrium shape. But when the wind picks up, look out. How does wind affect the sea's surface? If water waves contain energy, where does that energy come from and what forms does it take in the water?*

Watch some floating debris move as a wave passes it. Does the debris move with the wave? If not, is the water itself moving with the wave? Can you think of other cases in which a disturbance moves steadily forward but the material carrying the disturbance remains behind?

Now watch as waves break near the beach. If you look around at different beaches, you should be able to find waves breaking in two different ways. In some cases, a wave will simply collapse into churning froth, while in other cases its top will dive forward over the water ahead of it. Can you see anything about the beaches or water that might account for this difference?

The Tide

If you visit the seashore and watch the water level for a few days, you will notice that it rises and falls with striking regularity. This cyclical change in water level is called the *tide* and it repeats every 12½ hours. Thus, if you arrive at the seashore at *high tide* (the moment at which the water level is at its maximum), the water level will drop for 6¼ hours until *low tide* (the moment at which the water is at its minimum) and will then rise until high tide occurs again, another 6¼ hours later. While the tide was once a wonderful mystery, we now know that it's caused by the moon's gravity and, to a lesser extent, that of the sun.

In Section 5.1, we observed that the earth's gravity becomes weaker as you move farther from its center of mass. Thus your weight diminishes with altitude. But to make a noticeable change in your weight, you must change your distance from the earth's center of mass by a significant fraction. Since the earth's center of mass is already almost 6,500 km beneath its surface, simply climbing around in the mountains won't do much. You need to move many, many kilometers upward to reduce your weight significantly.

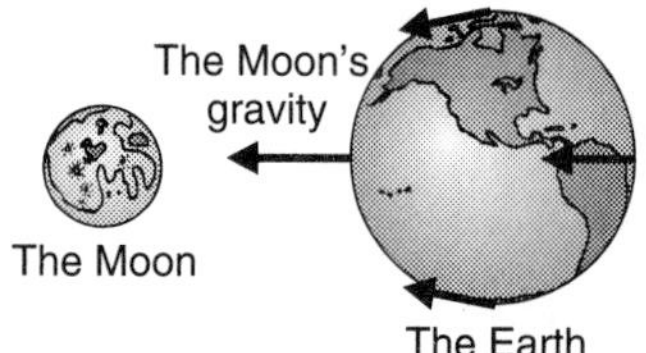

Fig. 10.1.1 - The moon's gravity varies over the surface of the earth. The closer a point is to the moon, the stronger the gravity it experiences. This variation in the moon's gravity produces the tide.

If you lived on the moon, you could make a similar observation about your moon weight. Hiking around on its surface wouldn't move you far enough toward or away from the moon's center of mass to have much effect on its gravity. But if you could move 13,000 km farther from the moon, your moon weight would change significantly, even if you started about 378,000 km from the moon's center of mass. Since gravity varies as the inverse-square of the distance, the moon's gravity would diminish by about 7% as you moved from 378,000 km away to 391,000 km away.

This apparently hypothetical situation isn't hypothetical at all. Although you don't live on the moon, you could do this experiment roughly twice a day. When the moon is overhead, you're about 378,000 km from its center of mass and 12½ hours later, when the earth has completed half a turn, you're about 391,000 km away. Thus the moon's gravitational force on you varies by about 7% as the earth rotates.

Because of the moon's small mass and great distance, it has only a tiny effect on your weight. The force it exerts on you is only a few thousandths of a newton, much too small for your bathroom scale to detect. But the moon's gravity is still strong enough to bend the earth's motion through space so that the earth and the moon orbit one another. And the variations in the moon's gravity over the earth's surface (Fig. 10.1.1) are enough to deform the oceans (Fig. 10.1.2) and produce the tide.

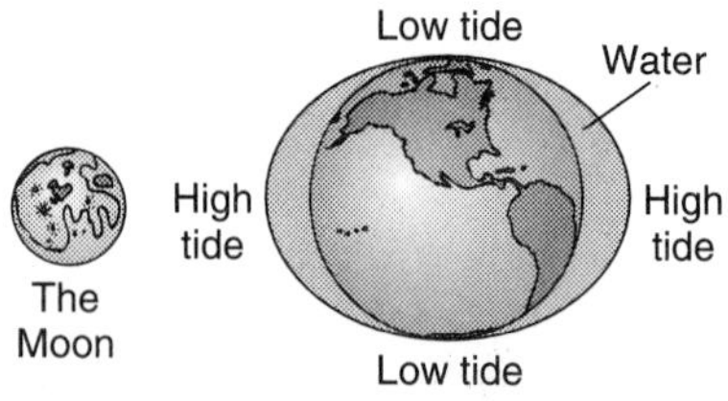

Fig. 10.1.2 - Variations in the moon's gravity produce tidal forces on the earth's surface and cause the earth's oceans to bulge outward in two places. These bulges, which are located nearest and farthest from the moon, move over the earth's surface as it rotates.

While the earth's path through space depends on the average gravity it experiences, variations in the moon's gravity with distance produce tidal forces on the earth's surface. These **tidal forces** are the differences between the moon's gravity at particular locations on the earth's surface and its average for the entire earth. The near side of the earth is pulled toward the moon more strongly than average so it experiences a tidal force toward the moon. The far side of the earth is pulled toward the moon less strongly than average so it experiences a tidal force away from the moon.

If the earth weren't so rigid, these tidal forces would stretch it into the shape of egg. The near side of the earth would bulge outward toward the moon, while the far side of the earth would bulge outward away from the moon. But while the earth itself is rigid, the oceans are not and they bulge outward in response to the tidal forces. Two *tidal bulges* appear: one closest to the moon and one farthest from the moon (Fig. 10.1.2). A beach located in one of these tidal bulges experiences high tide, while one in the ring of ocean between the bulges experiences low tide.

As the earth rotates, the locations of the two tidal bulges move westward around the equator. Since a particular beach experiences high tide whenever it's closest to or farthest from the moon, the full cycle from high to low to high tide occurs about once every 12 hours and 26 minutes. The extra 26 minutes is due to the moon's 27.3 day orbit around the earth. Because the moon isn't stationary, it passes overhead once every 24 hours and 52 minutes, rather than every 24 hours.

But the moon isn't the only source of tidal forces on the earth's oceans. Although the sun is much farther away than the moon, it's so massive that the tidal forces it exerts are almost half as large as those exerted by the moon. The sun's principal effect is to increase or decrease the strength of the tide caused by the moon (Fig. 10.1.3). When the moon and the sun are aligned with one another, their tidal forces add together and produce extra large tidal bulges. When the moon and the sun are at right-angles to one another, their tidal forces partially cancel and produce tidal bulges that are unusually small.

Twice each lunar month, the time it takes for the moon to orbit the earth, the tides are particularly strong. These *spring tides* occur whenever the moon and sun are aligned with one another (full moon and new moon). Twice each lunar month the tides are particularly weak. These *neap tides* occur whenever the moon and sun are at right-angles to one another (half moon).

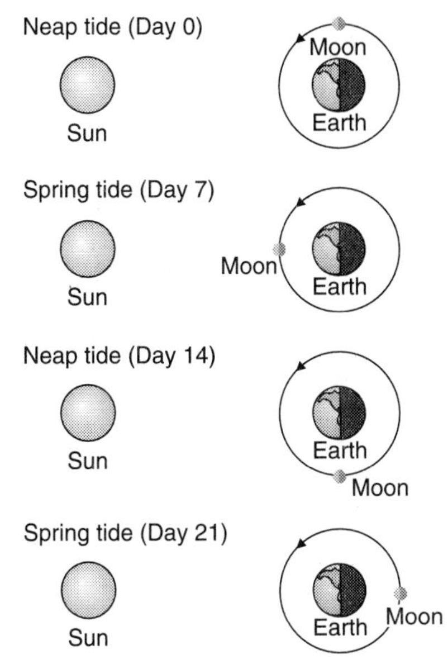

Fig. 10.1.3 - The tide varies over a lunar month. It's strongest when the sun and the moon are aligned (spring tides) and weakest when they are at 90° from one another (neap tides).

CHECK YOUR UNDERSTANDING #1: Seaside Vacations of the Tidal Rich
The Atlantic and Pacific Oceans come very close to one another in Central America. If the Atlantic side of this isthmus is experiencing high tide, what tide is the Pacific side experiencing?

Tidal Resonances

The size of the tide depends on where you are, but high tide is typically a few meters above low tide. Because the tidal bulges are located near the equator, the tides far to the north or south are smaller than that and tides in isolated lakes or seas are smaller still because water can't flow in to create the bulges. However, there are a few special places that have enormous tides. For example, the tide in the Bay of Fundy, an estuary between New Brunswick and Nova Scotia, can change the water level by as much as 15 m. How can a tide ever get this large?

The giant tides that occur in these special locations are the results of natural resonances in channels and estuaries. Just as a violin string can be made to vibrate strongly by a series of carefully timed pushes from a bow, the water in a channel or estuary can be made to oscillate strongly by a series of carefully timed pushes from the tide. Thus the giant tides are just another example of resonant energy transfer.

To see how these water oscillations work, consider the behavior of water in a wash basin. In equilibrium, the water's surface is smooth and horizontal. Since gravity acts to level the water, it exerts restoring forces on that surface. The water is thus in a stable equilibrium and will oscillate about that equilibrium if you disturb it. For example, if you shift water from one side of the basin to the other, the water will slosh back and forth until it turns its extra energy into thermal energy or transfers it elsewhere (Fig. 10.1.4).

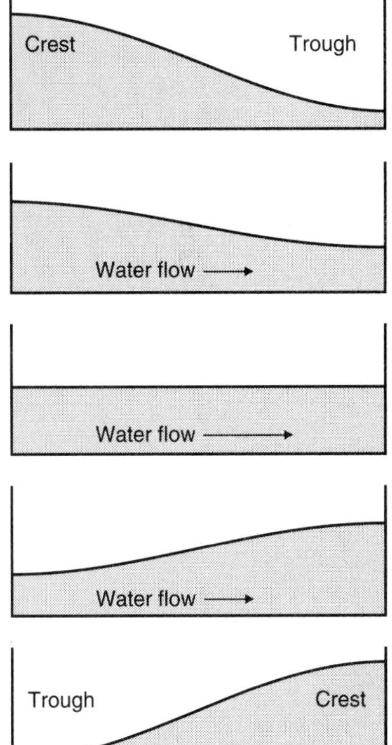

Fig. 10.1.4 - Water sloshing in a basin is an example of oscillation. The water oscillates back and forth, its crest becoming a trough and vice versa, over and over again.

Fig. 10.1.5 - The giant tides in the Bay of Fundy can cause its water level to change by as much as 15 m between high and low tide.

Because the restoring forces on water's surface increase in proportion to its displacement from equilibrium, water is yet another harmonic oscillator. While its modes of vibration are even more complicated than those of a string, each mode has a period that's independent of its amplitude. Thus water sloshing in the basin has a steady period and you can make that water slosh vigorously by pushing on it in synch with its period.

While the period of a wash basin's fundamental mode is only a second or two, that of a large channel may be hours or days. The Bay of Fundy has a period of roughly 12 hours and 26 minutes. Since this period matches the cycle of the tide, there is a resonant transfer of energy from the moon to the water flowing in the estuary. The tide drives water back and forth in this estuary until, after many cycles, that water is moving so strongly that its height varies dramatically with time (Fig. 10.1.5).

CHECK YOUR UNDERSTANDING #2: Potential Profits

People occasionally propose using giant tides to generate electric power. But there's a problem with this idea. If you extract all of the gravitational potential energy from the water at high tide in the Bay of Fundy, how long will it take for a giant high tide to appear again?

Waves on the Surface of Water

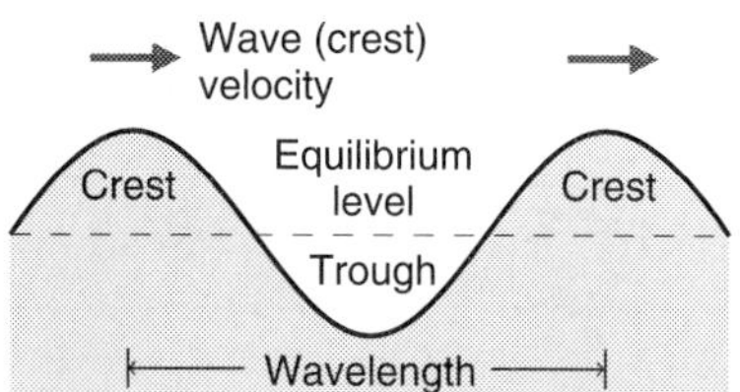

Fig. 10.1.6 - A surface wave distorts water's surface away from its equilibrium level. The highest points on the wave are its crests and the lowest are its troughs. The distance between crests is its wavelength. The entire pattern of crests and troughs moves across the water's surface at the wave velocity.

As you sit at the seashore, you can't help but notice waves crashing endlessly on the beach. The sea in front of you is covered with ridges that move steadily toward land. These ridges are **surface waves**, disturbances in what would normally be a smooth, level surface. Since distorting water's surface requires energy, it takes work to make waves. Most ocean waves are driven by the wind.

While it's customary to think of each breaking swell as a separate wave, we can also view the entire pattern of ridges as a single wave. From that perspective, a wave consists of many parallel crests and troughs (Fig. 10.1.6). In a **crest**, the water rises above its equilibrium level. In a **trough**, the water drops below its equilibrium level. As the wave passes across the surface of a deep stretch of open water, the pattern of crests and troughs is evenly spaced and the wave has a *wavelength* that is the distance between two adjacent crests.

Crests and troughs can't remain still. We learned in Section 1.4 that a system will accelerate in whatever direction reduces its potential energy most rapidly. Since the water in a crest can release gravitational potential energy by flowing into a trough, it will accelerate toward the trough. Thus only smooth, level water can remain stationary and a wave always involves motion.

However, there are two different types of waves: standing waves and traveling waves. In a **standing wave**, the water's surface oscillates up and down vertically, with its crests and troughs always going in opposite directions. This opposite motion turns each crest first into a trough and then back into a crest, at the same time it turns each trough first into a crest and then back into a trough. The standing wave's pattern of crests and troughs doesn't go anywhere, it simply flips up and down in place at the standing wave's frequency.

Water sloshing in a wash basin is usually a standing wave. The crest on one side of the basin gradually becomes a trough while the trough on the other side of the basin gradually becomes a crest (Fig. 10.1.4). This process reverses and repeats, over and over again. Like all standing waves, the sloshing water's energy goes back and forth between potential and kinetic forms. It's kinetic energy peaks as the water rushes through equilibrium and its potential energy (gravitational potential energy) peaks as the water stops to turn around.

In a **traveling wave**, the pattern of crests and troughs moves smoothly across the surface of the water (Fig. 10.1.7). Because this pattern moves in a particular direction with a certain speed, the traveling wave has a velocity; its **wave velocity**. It also has a momentum and that momentum points in the same direction as the wave velocity. The wave's crests and troughs move steady forward, so the water is never level or motionless. The energy in a traveling wave is always an even mixture of kinetic and potential energies.

Most waves on the open ocean are traveling waves. Despite their steady progress across the water, these traveling waves actually involve oscillation. You can see this oscillation by watching a fixed point on the water's surface. The point fluctuates up and down as the crests and troughs of a traveling wave pass through it, making the wave a form of oscillation and giving it a frequency.

This is not the first time we've encountered standing and traveling waves. The vibrational modes of a violin string are standing waves; the string bends to create crests and troughs, which interchange over and over again as the string vibrates. Sounds are traveling waves; compressions and rarefactions of the air that travel through space from their source to your ears.

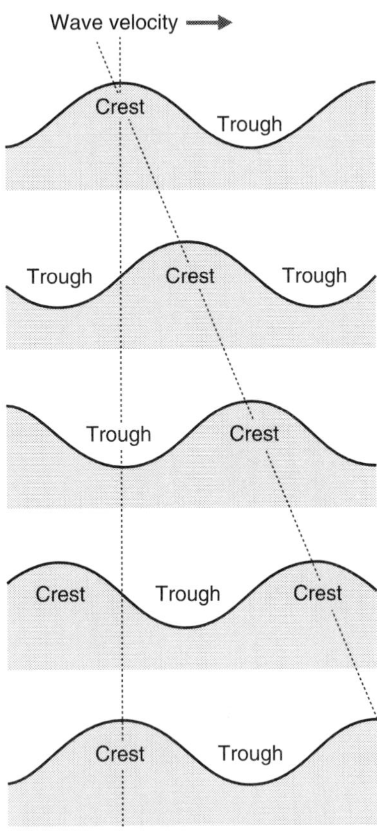

Fig. 10.1.7 - Waves on the open ocean are traveling waves. The crest on the left gradually moves to the right before passing out of view. If you watch the water's surface at one point, marked by the dotted line, you will see that it oscillates up and down.

CHECK YOUR UNDERSTANDING #3: The Frequency of Seasickness

You're sitting in a small boat on the open ocean, rising and falling with each passing wave crest. How could you measure the frequency of the wave passing under your boat?

The Structure of a Water Wave

We've seen that a traveling wave moves across the surface of the water with a certain velocity, wavelength, and frequency. But what is the water itself doing as the wave passes?

You can begin to answer that question by watching a bottle floating on the water's surface (Fig. 10.1.8). As a wave passes it, that bottle rises and falls with the crests and troughs and winds up traveling in circle. Overall, it doesn't go anywhere. Like the bottle, the water itself doesn't actually move with the wave. Instead, it simply bunches up or spreads out to create each crest or trough. The water moves only short distances as the wave passes and returns to its starting point once the wave has left.

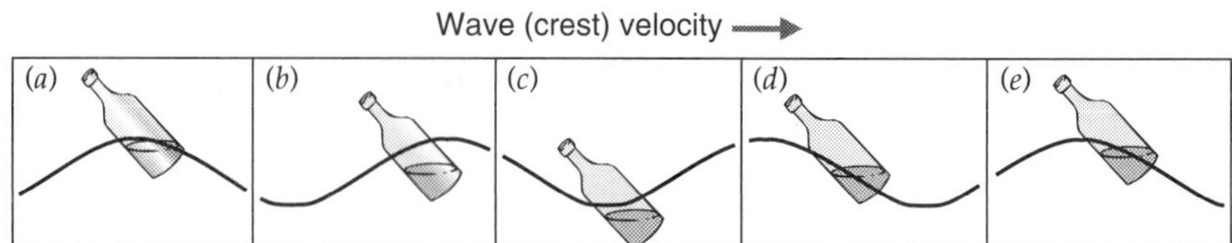

Fig. 10.1.8 - You can see that water doesn't move with a wave by watching a bottle floating on the water. The bottle moves in a circle as each crest passes by. Starting at (*a*) a crest, the bottle moves (*b*) down and right, (*c*) down and left, (*d*) up and left, and then (*e*) up and right. It returns to its original position just as the next crest arrives.

The wave's water moves in a circular pattern (Fig. 10.1.9). Water that starts out on top of a crest moves down and forward as the crest passes. It moves down and back as the trough passes, then up and back, and finally up and forward as the next crest arrives. Like the bottle in Fig. 10.1.8, the water itself just moves in a circle and returns to its original location each time a crest passes. The wave's direction of travel depends on which way the water is circling. It heads in the same direction as the water on top of a crest.

This circular motion occurs only for water that's fairly near the surface. As you look deeper into the water, the circular motion gradually diminishes until the water is barely moving at all. The depth at which the water stops moving with the surface wave is roughly equal to the wave's wavelength.

Wavelength also affects the wave's velocity. Since any material carrying a mechanical wave must accelerate in various directions as the wave moves through it, the wave moves fastest when the material is stiff and its density is low. That's because its stiffness gives rise to large forces when it's disturbed and its low density allows it to accelerate easily in response to those forces. But in some types of waves, particularly surface waves on water, the stiffness that the material exhibits actually depends the wave's wavelength. In those cases, waves of different wavelengths actually travel at different speeds!

The longer the wavelength of a surface wave on water, the faster that wave travels. Little ripples have short wavelengths and travel slowly while large ocean waves have long wavelengths and travel much more rapidly. Giant waves produced by earthquakes and volcanic eruptions—once called "tidal waves" and now called *tsunamis*—have extremely long wavelengths and can travel at hundreds of kilometers-per-hour. Because these giant waves travel so quickly and move water so deep in the ocean, they carry enormous amounts of energy.

CHECK YOUR UNDERSTANDING #4: Beneath the Waves

If you're swimming and want to dive beneath a wave, how deep must you go to avoid any significant motion in the water?

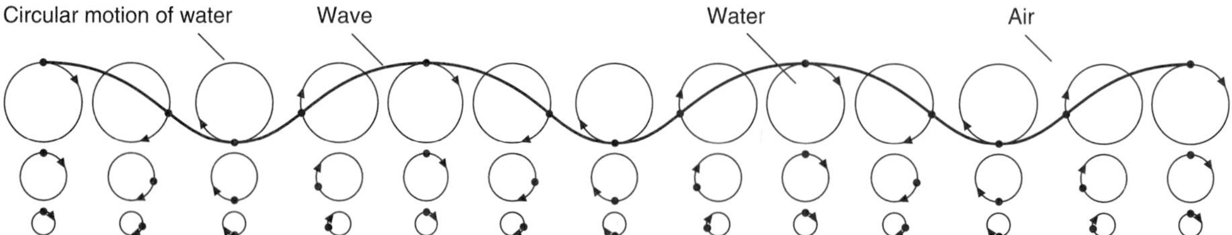

Fig. 10.1.9 - Surface water moves in a circular motion as a wave passes. The water currently located at each dark dot will follow the circular path outlined around it as time passes. The circles are largest at the surface and become relatively insignificant once you look more than a wavelength below the surface. The sense of the circular motion (clockwise or counter-clockwise) determines the direction in which the wave travels. This wave travels toward the right.

Waves at the Shore

As a wave approaches the shore, it begins to travel through shallow water. Since the water's circular motion extends below the surface, there comes a point at which the wave begins to encounter the bottom. Once the water is shallower than the wavelength of the wave, the bottom distorts the water's circular motion so that it becomes significantly elliptical. The velocity of the wave drops and the crests begin to bunch up. This explains why waves that looked broad and gradual on the open ocean look quite steep and dangerous as they approach the beach. Their crests and troughs have bunched together and the slopes between the two really have become steeper.

A wave forms each crest by bunching together water from in front of it and behind it. When the water becomes very shallow, there is not be enough water in front of the crest to completely build up that side of the crest. The crest becomes incomplete and begins to "break."

If the slope of the sea bottom is gradual, the wave breaks slowly to form a smooth, rolling, "boiling" surf (Fig. 10.1.10). If the slope of the sea bottom is steep, the wave breaks quickly by having the top of its crest plunge forward over the trough in front of it (Fig. 10.1.11). The steep slope essentially prevents half of the crest from forming. The half that does form continues through its normal circular motion and dives over the trough in front of it.

Fig. 10.1.10 - When the slope of the sea bottom is very gradual, the wave crests crumble gently into rolling surf.

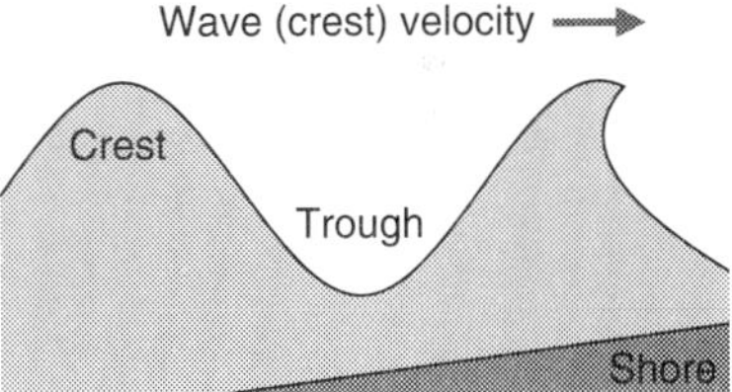

Fig. 10.1.11 - When the water becomes too shallow to form a complete crest for the wave, the wave "breaks." If the slope of the shore is steep enough, the crest will be quite incomplete on its shore side and will plunge forward over the trough in front of it.

CHECK YOUR UNDERSTANDING #5: Out Past the Surf

As you swim away from shore at the ocean, you go through a region where the wave crests are breaking and then reach a more distant region where they don't break. What distinguishes the two regions?

Surfing

Surfing is a lot like riding a skateboard down a hill. The forward edge of a wave crest is more or less a ramp and the surfer experiences a net downhill force. If the wave didn't move, a surfer starting at the top of the crest could slide down it toward shore. Because water exerts a drag force on moving objects, the downhill ramp force would soon be balanced by an uphill drag force and the surfer would descend at terminal velocity.

But the wave is moving (Fig. 10.1.12). A surfer located on the forward edge of the crest will find that the water below him is rising upward. You can see this effect by looking at the bottle at Fig. 10.1.8. The bottle in (*d*) is on the forward edge of the crest and it's in the process of moving upward toward its position in (*e*). So while the surfer is sliding down the front edge of the wave, that edge is rising up. As a result, the surfer neither rises nor falls. Instead, he maintains a constant height and moves forward with the wave crest behind him.

However, if the surfer rides the wave directly toward shore, he will find himself near the base of the crest. This happens because he doesn't move fast enough through the water for the uphill drag force to support him on the steep slope near the top of the crest. He slides forward almost into the trough and rides the wave leisurely to shore.

To obtain a better ride, the surfer can cut across the forward edge of the crest, so that he approaches the shore at an angle. By traversing the downhill slope of the wave, he reduces its apparent steepness and experiences a weaker downhill force. Now drag from the water can keep him from sliding far down

Fig. 10.1.12 - The surfer is riding this wave like a moving ramp. The wave rises beneath him as he slides forward, so he never reaches the trough. The wave is breaking because the water here is too shallow to form a complete crest.

the crest and he rides it much nearer its top. He also travels faster because he is moving across the wave as well as with it.

Just before the crest breaks, he turns the surfboard to travel directly across the wave. Suddenly he experiences no downhill force at all. Drag soon brings him to a stop and the crest passes him by. He can then paddle away from shore to catch a ride on another crest.

CHECK YOUR UNDERSTANDING #6: Water Works

Since drag extracts energy from a surfboard, the wave must constantly replace it. How does the wave transfer energy to the surfer?

Summary

How the Sea and Surfing Work: The sea exhibits two interesting motions: the tide and waves. The tide is caused by bulges in the earth's oceans, created by gravitational tidal forces from the moon and sun. The moon dominates the tide so that bulges appear on regions of the earth closest to and farthest from the moon. Since the earth is rotating, these bulges move with respect to land and give the tide its 12½ hour cycle.

Waves occur because water's surface has a stable equilibrium. When it's disturbed from its flat, level equilibrium, this surface will oscillate (or equivalently vibrate) in standing waves and/or traveling waves. The familiar ripples on the ocean's surface are traveling waves that head toward shore at speeds that increase with wavelength. As they approach the shore, these waves form incomplete crests which eventually break. A surfer riding one of these crests is effectively sliding down a hill. What makes surfing so interesting is that the "hill" moves. Its surface is always rising upward so that the surfer can continue to slide down the hill almost indefinitely.

The Physics of the Sea and Surfing

1. Variations in gravity with distance produce tidal forces on extremely large objects. These objects may deform if they aren't rigid.

2. A wave is a disturbance in a material that has a stable equilibrium. It forms a pattern of crests and troughs that is characterized by its wavelength, frequency, and velocity.

3. In a standing wave, the pattern of crests and troughs oscillates in place and the wave's energy goes back and forth between potential and kinetic forms.

4. In a traveling wave, the pattern of crests and troughs travels through space and the wave's energy remains equally distributed between potential and kinetic forms. The traveling wave carries momentum that normally points in the direction of its wave velocity. The material in which the wave exists doesn't travel with the wave.

5. The velocity of a mechanical wave depends on the stiffness of the material involved and on that material's density. The stiffer the material and the lower its density, the faster the wave will travel.

Check Your Understanding - Answers

1. High tide, too.

Why: When the Atlantic side of the Central American isthmus is experiencing high tide, there is a tidal bulge over all of Central America. Both sides of the isthmus experience the same (high) tide.

2. Many days.

Why: The sloshing water in the Bay of Fundy acquires its energy via resonant energy transfer. The tides do work on the water, slowly increasing its energy over many cycles of its periodic motion. Since each cycle takes half a day, many days are required to build up the energy needed for a giant tide. If you were to extract all this stored energy at once, the bay would have to start sloshing all over again.

3. Count the number of up and down cycles you complete in a certain amount of time.

Why: Your boat's up and down motion is caused by the oscillation of the ocean's surface. You are rising and falling at the frequency of the passing wave.

4. You must dive about one wavelength beneath the water's surface.

Why: A wave causes motion in the water up to a depth of roughly one wavelength. A typical wave has a wavelength of about 5 m, so you must dive 5 m under water before the water will remain essentially still.

5. The breaking region is too shallow for complete crests to form. The non-breaking region is deep enough to form complete crests.

Why: Wave crests break in shallow regions, where complete crests can't be formed. As long as the water is deep enough, the waves travel intact. But if there is a shallow region such as a sandbar, even quite far from shore, the crests may break as the wave passes through it.

6. The water pushes the surfer forward and upward and the surfer moves forward. Thus the wave does work on the surfer.

Why: Like any ramp, the side of the wave crest exerts a support force on the surfer that is perpendicular to the water's surface. That force points forward and upward. Since the surfer and wave head forward, partly in the direction of the force, the force does work on the surfer.

Glossary

crest The peak upward excursion of the system during the passage of a wave.

standing wave A wave that oscillates in place, with energy going back and forth between potential and kinetic forms.

surface waves Disturbances in the stable equilibrium shape of a surface.

tidal forces The differences between one celestial object's gravity at particular locations on the surface of a second object and the average of that gravity for the entire second object. Tidal forces tend to stretch the second object into an egg shape.

traveling wave A wave that moves through space, with energy always evenly distributed between potential and kinetic forms.

trough The peak downward excursion of the system during the passage of a wave.

wave velocity The speed and direction of the moving crests of a wave. In a typical water wave in the open ocean, the entire pattern of crests and troughs moves as a whole with the wave velocity.

Review Questions

1. Where are the two tidal bulges located on the earth's oceans and why are there two of them?

2. Why does your moon weight (your weight due to the moon's gravity) vary with time and your location?

3. Salt ponds often exchange water with the ocean through narrow channels. Sea water rushes in or out through these channels with each change of the tide. Why don't most of these salt ponds experience the tremendous tides of the Bay of Fundy? What's required for them to experience such tides?

4. List three examples of standing waves in common objects around you. What constitutes a crest or trough in each case?

5. List three examples of traveling waves in common objects around you. What constitutes a crest or trough in each case?

6. How does a mechanical wave travel through a material without actually transporting that material with it?

7. A diamond is much stiffer than a piece of plastic, although the two have similar densities. In which material will sound waves travel faster? Why?

8. How does the slope of the sea bottom affect the way waves break as they approach the shore?

9. How does a surfer extract energy from a passing wave?

10. If a surfer wants to stay on a wave, why must she ride across the wave, rather than heading straight down it?

Exercises

1. In the Mediterranean Sea, high tide is only 30 cm above low tide. Why?

2. The momentum of a water surface wave points in the direction of its wave velocity. Show that ocean waves always move in the same direction as the wind that created them.

3. A downhill skier goes much faster when she skis directly down a hill than when she traverses across the face of that same hill. Why?

4. We are fortunate that a sound wave's speed doesn't depend on its wavelength or frequency, the way a surface wave on water does. If the speed of sound did depend on its wavelength or frequency, what would you hear when you listened to an orchestra?

5. If you're carrying a full cup of coffee and take your steps at just the wrong frequency, the coffee will begin to slosh wildly in the cup. What is causing this energetic motion?

6. When you throw a stone into a pool of still water, small ring-shaped ripples begin to spread outward at a modest pace. Why do these ripples travel so much more slowly than waves on the ocean?

7. When one car stops suddenly in bumper-to-bumper traffic the cars behind it must also stop. Since each driver waits to brake until the brake lights of the preceding car turn on, the cars don't all stop at the same time. If you stand at the side of the road watching this phenomenon, the illuminating brake lights will appear to be a wave traveling backward along the road. The lights of each car will turn on slightly later than those of the preceding car. When the cars start up again, another wave of motion will travel along the road as each driver waits for the preceding car to begin moving before starting forward. Which direction does this new wave head?

8. While we're most familiar with sound in air, sound can also travel through solids. It still takes the form of compressions and rarefactions, but it usually travels much faster than in air. In lead, a dense and relatively soft metal, sound travels at 1,960 m/s. In beryllium, a stiff metal with a very low density, sound travels at 12,890 m/s. Why does a sound wave travel so much faster in beryllium than in lead?

9. The moon causes the earth's oceans to form two tidal bulges: one nearer the moon and one farther from the moon. These two bulges can be attributed to the different accelerations due to the moon's gravity at three locations: the surface of the earth nearest the moon, the earth's center of mass, and the surface of the earth farthest from the moon. Explain why this is so.

10. Why are the tides relatively weak near the north and south poles?

11. When you throw a rock into calm water, ripples head outward as several concentric circles. As the circumferences of these circles grow larger, their heights grow smaller. Explain this effect in terms of conservation of energy.

12. If you pull downward on the middle of a trampoline and let go, the surface will fluctuate up and down several times. Why is this motion an example of a standing wave?

13. Why is a violin string vibrating up and down in its fundamental vibrational mode an example of a standing wave rather than a traveling wave?

14. When you pluck the end of a kite string, a ripple will head up the string toward the kite. Why is this motion an example of a traveling wave rather than a standing wave?

15. Plucking the end of a kite string sends a ripple up the string toward the kite. The more tension there is in the string, the faster this traveling wave moves. In general terms, why does the increased tension increase the wave's speed?

16. If you replaced the kite string in Exercise **15** with a thinner, lighter string, how would that change affect the speed of traveling waves on the string?

17. When you make noise in one room, waves of sound flow through the air into the adjacent room. Why don't these air waves transfer air from the first room to the second room?

18. Why do the largest ocean waves also break farthest from shore?

19. As waves pass over a shallow sandbar, the largest ones break. What causes these waves to break and why only the largest ones?

20. Even when waves don't break as they pass over a sandbar (Exercise **19**), the sandbar is noticeable because you can see the wave crests move closer together. What is happening to cause this bunching?

21. When you place a glass of water on a vibrating counter, the water's surface often exhibits a pattern of standing waves. When bright light reflects from that surface, you can see rings or lines of motionless water and, between them, regions of water that are moving up and down rapidly. What portions of the wave structure in Fig. 10.1.4 correspond to the motionless water and to the moving water?

Epilogue for Chapter 10

In this chapter we examined waves and found that they are disturbances in a system with a stable equilibrium. In *the sea and surfing*, we found that there are two different types of waves: standing waves and traveling waves. We observed that both types of waves are found in water, with standing waves appearing in tidal resonances and traveling waves heading out over the open ocean. We explored the motion of surface waves on water, seeing that their speeds depend on their wavelengths and that they involve only the water near the sea's surface. We also examined the tide and the ways in which a surfer harnesses a wave's energy for an exciting ride.

Explanation: Waves on a Jump Rope

The cord supports two different types of waves: standing waves and traveling waves. The cord's standing waves are simply its fundamental and harmonic vibrations. Each of these has its own period and frequency, which depend on the cord's mass, length, and tension.

The cord's traveling waves are generally more complicated than its standing waves. You can give them almost any shape by moving your wrist properly. They travel at speeds that depend on the cord's mass and tension. You can make them travel faster by either tightening the cord or reducing its mass. Since these waves contain energy, they can't simply disappear. The cord can't do work on the fixed doorknob so a wave can't transfer its energy there. Instead, the wave reflects from the doorknob and returns its energy to your hand.

Cases

1. The simplest type of water bed is a large, flat bag that's filled with water. Such a bed is comfortable to sleep on but is susceptible to waves if something pushes on it.

a. In equilibrium, the bed's surface is almost perfectly flat. Why?

b. If you suddenly push one side of the bed downward and then let go, what will happen to that side during the next second or two?

c. As the result of your push, a traveling wave will move across the surface of the bed. Describe the motion of the water inside the bed as a wave crest passes by.

d. If instead of the sudden push, you push that side of the bed up and down gently and rhythmically at the right frequency, you will excite standing waves. The sloshing water will form crests and troughs that oscillate up and down in place. A particular crest will turn into a trough and then become a crest again, over and over. How can your gentle pushes create such large oscillations?

2. An earthquake generates waves that travel through the earth in all directions. While the earth's internal structure complicates the passages of these waves to distant sites, they follow simple paths to nearby places. However, there are at least two types of waves, and they travel at different speeds.

a. The waves that travel fastest in the earth's crust are those that resemble sound. They're called "P" or "primary" waves because they arrive first. As these waves travel forward, the ground moves alternately forward and backward, along the direction of the wave velocity. For example, if a "P" wave is heading north, the ground oscillates north and south as the wave passes. A *seismometer* can detect this motion by hanging a pendulum above a sheet of paper that's resting on the ground. When the wave passes, the pendulum marks the paper along the direction of the wave velocity. Why does the pendulum shift back and forth relative to the paper?

b. The ground along the path of a "P" wave doesn't all shift north or south at once. These shifts are actually the "P" wave's crests and troughs, and they move northward with the wave. When the ground underfoot shifts first northward and then southward, it's because a crest passed by, followed by a trough. At the moment the crest is under you, the trough is still a long way behind it. In fact, it's exactly half a wavelength behind it. How could you use several seismometers to determine the wavelength of a passing "P" wave?

c. How could you use one seismometer to determine the frequency of the "P" wave?

d. How could you use several seismometers to measure the wave velocity of the "P" wave?

e. Waves that involve sideways or vertical motion of the ground travel more slowly and arrive after the "P" waves. They're called "S" or "secondary" waves. For example, in one type of "S" wave that is heading north, the ground oscillates east and west as the wave passes. How will a hanging pendulum seismometer respond when this "S" wave passes?

f. In another northbound "S" wave, the ground moves up and down. How could you use a weight and a spring to detect this wave as it passes by?

g. The wave from an earthquake spreads out in all directions like a ripple on a pond. Use conservation of energy to explain why the wave's amplitude of motion must decrease as it gets farther from the earthquake's center and why the energy passing under a city near the earthquake must be much greater than the energy passing under a similar-sized city far away from the same earthquake. (This weakening with distance occurs even when no energy is lost as thermal energy. It protects us from all but relatively nearby earthquakes.)

3. When the paper cone in your stereo speaker moves forward and backward rhythmically, it produces a sound wave that travels through the air at about 331 m/s. The crests of this wave consist of slightly compressed air and the troughs consist of slightly rarefied air.

a. Suppose that you have an instrument that can monitor the pressure at one point in space. You place this instrument in front of the speaker and begin measuring. Draw a rough graph of the pressure that the instrument measures versus the time that has elapsed since the measurement started.

b. Suppose that instead of keeping the instrument in one place, you began to move the instrument so that it continues to detect a region of slightly compressed air (one of the crests in the wave). How fast must you move the instrument to keep up with the crest?

c. You spray a mist of water into the air so that you can see if the air itself moves as the sound travels through it. The mist remains still, so you know that the air isn't moving significantly, even though the sound wave is. Explain this surprising result.

d. How could you determine the wavelength of the sound wave emerging from the speaker?

e. What two types of energy does a sound wave contain?

f. The oscillation of air inside an organ pipe is actually a standing sound wave. How is it different from the traveling sound wave that's passing through your room?

4. You're near one end of a long rope that's stretched between two buildings. You reach out and pluck the rope, creating a bump that travels rapidly along the rope toward the other building. Although this bump is really a wave, it doesn't have a simple wavelength or frequency, so you begin to shake the rope back and forth quickly. Sure enough, a rippling wave heads off across the rope.

a. What determines the frequency of this wave?

b. How could you measure the wavelength of this wave, if you had the right equipment?

c. If you were to tighten the rope, why would the wave travel faster than before?

d. If you were to replace the rope with a more massive one, while keeping the tension unchanged, would the wave's velocity increase, decrease, or remain the same?

e. Show that you do work on the rope as you produce the wave.

f. If a friend at the far end of the rope sent a wave toward you, it could do work on you if you were touching the rope as it arrived. Show how the rope would do work on your hand.

5. If astronauts landing on Mars were to find liquid water on the surface, the waves that they would see would behave somewhat differently. That's because the acceleration due to gravity on Mars is only 3.71 m/s^2.

a. A wave always needs two forms for its energy. In a surface wave on water, one of those forms is gravitational potential energy. How would the gravitational potential energy of a wave on Mars compare with that of an identical looking wave on earth?

b. The "stiffness" of the water's surface depends on gravity. When gravity is weak, it takes less work to deform the water's surface. Use your answer to **a** to show that this is the case.

c. The surface of water on Mars is more easily deformed and thus less "stiff" than the surface of water on earth. On which planet would surface waves on water travel faster, Mars or earth?

d. Which wave carries more energy, a surface wave on Mars or an identical looking wave on earth?

e. On a spaceship that's far away from stars, planets, and other celestial objects, there is almost no gravity. If the astronauts on that spaceship put water in a small basin, would they be able to produce normal surface waves in that water? Why or why not?

CHAPTER 11

ELECTRIC AND MAGNETIC FORCES

Electric and magnetic forces are all around us. They play important roles in machines we use every day and figure prominently in industry and technology. While these forces are often useful—sticking food wrap to a bowl or a magnet to your refrigerator—they can also make your clothes cling together out of a hot dryer or erase the information on your credit card. Moreover, electric and magnetic forces are what hold atoms, molecules, and materials together.

Experiment: Moving Water Without Touching It

Unlike gravity, which always pulls objects toward one another, electric and magnetic forces can be either attractive or repulsive. You can experiment with electric forces using a thin stream of water and an electrically charged comb. First, open a water faucet slightly so that the flow of water forms a thin but continuous strand below the mouth of the faucet. Next, give your rubber or plastic comb an electric charge by passing it rapidly through your hair or rubbing it vigorously against a wool sweater. Finally, hold the comb near the stream of water, just below the faucet, and watch what happens to the stream. Is the electric force that you're observing attractive or repulsive? Why does this force change the path of the falling water?

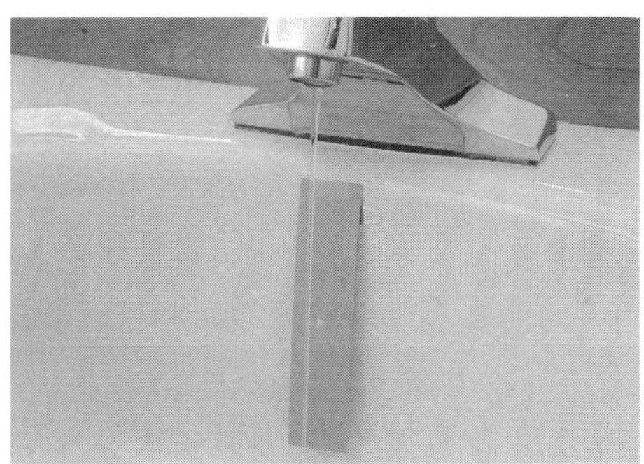

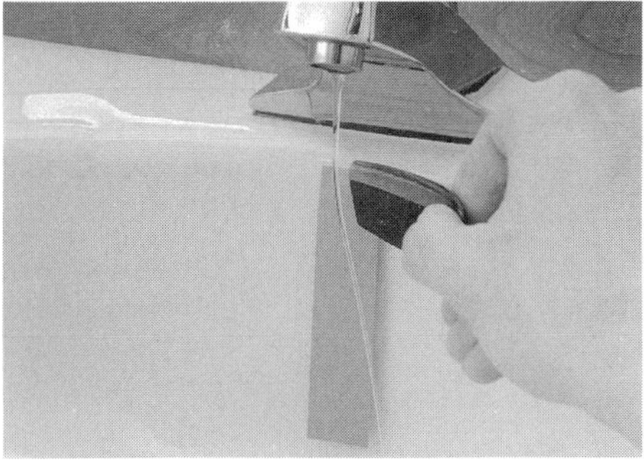

Rubbing the comb through your hair makes it electrically charged. Find other objects that can acquire and hold a charge when you rub them across hair or fabric. Try to **predict** which of these objects will work best. Now rub these objects to charge them and **observe** their behaviors. You can use the water stream to **measure** their charges. Did you **verify** your predictions? Which works better: a metal object or one that's an insulator? Why?

Chapter Itinerary

While many of us have experienced electric and magnetic forces as novelties or nuisances, there are also many important devices that depend on them. In this chapter, we'll examine three of these devices: (1) *air purification*, (2) *xerographic copiers*, and (3) *magnetically levitated trains*. In *air purification*, we'll look at the problems involved in removing tiny particles from polluted air and how the electric forces between charged particles makes such purification possible. In *xerographic copiers*, we'll see how these same electric forces work together with light to control the placement of colored powders, in order to reproduce an image on a sheet of paper. In *magnetically levitated trains*, we'll investigate the origins of magnetic forces and they ways in which these forces can be used to suspend high-speed trains above their tracks.

This chapter introduces the concepts of electric charges and magnetic poles and highlights the similarities and differences between the two. It also begins to describe the relationships between the two types of forces, electric and magnetic, for they are not truly independent of one another. This relationship will become even more apparent in subsequent chapters.

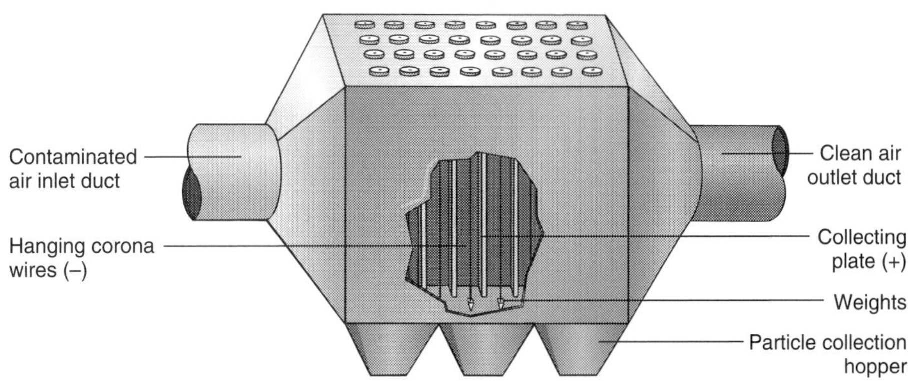

Section 11.1

Electronic Air Cleaners

When a ray of sunlight passes through an otherwise darkened room, you can see tiny pieces of dust, soot, and ash drifting in the air. While some of these particles occur naturally, many others are produced by our vehicles and factories. These particles are a cosmetic nuisance—darkening our air and dirtying our surroundings—and a health hazard. To control particle pollution, many industries filter particles from their smokestacks.

This filtering isn't easy because air resistance makes it hard to pull tiny particles out of the air. That's why many companies and households use electronic air cleaners. These cleaning devices pluck particles from the air with the forces of static electricity—the same forces that cause clothes to cling to one another as you remove them from the dryer. In this section, we'll examine static electricity to see how electronic air cleaners work.

Questions to Think About: *What keeps dust in the air and why doesn't gravity remove it? If gravity can't clear the air, how can any other forces succeed? Why don't companies use paper filters to clean dust from smoke? How can it be that some newly cleaned clothes push apart rather than clinging to one another?*

Experiments to Do: *You can study static electricity by rubbing a toy balloon vigorously through your hair or against a wool sweater. Though the balloon won't look any different, it'll begin to attract other things, particularly your hair. That's because sliding friction transfers electrons, tiny negatively charged particles, from your hair to the balloon, so that the balloon becomes negatively charged. Your hair, which gave up the electrons, becomes positively charged and is attracted toward the balloon because opposite electric charges attract one another. In contrast, like electric charges repel one another. Can you think of a way to observe this repulsion with balloons? How would you use these forces to remove dust from the air?*

Dust, Soot, and Ash

Let's begin our discussion of electronic air cleaners by looking at dust, soot, and ash particles in normal air. Dust is mostly rock, dirt, and organic matter that has been ground into tiny pieces. It also includes natural particles, such as pollens and spores. Soot is usually carbon particles formed when organic matter burns incompletely and often contains oils that give it a greasy or tarry character. Ash is the powdery, non-combustible residue of a fire.

Each dust, soot, or ash particle has a mass and a weight. After all, a particle of granite dust is just a tiny rock and a particle of carbon soot is just a tiny piece of charcoal. Since a granite boulder and a charcoal briquette both fall when you drop them, you have every reason to expect granite dust and carbon soot to fall, too. However, tiny particles stay aloft almost indefinitely. What holds them up?

The air holds them up. It exerts two types of forces on the particles: the buoyant force and the viscous drag force. The buoyant force, which we discussed in Section 3.1, is produced by the atmosphere to keep the air itself from falling. But the particles are more dense than air and each has far more mass and weight than the air it displaces. Thus the buoyant force alone can't support the particles.

Instead, small particles are supported mostly by the viscous drag force. This force, which we examined in Section 4.2, is a form of air resistance and acts like friction to slow a small particle's motion through the air. While the particle's weight pulls it downward, the viscous drag force keeps it from descending quickly. As a result, the particle goes down slowly at its *terminal velocity*, the velocity at which its downward weight is exactly balanced by the upward force of viscous drag. For small particles, this terminal velocity can be so slow, 1 mm/s or less, that they don't appear to descend at all.

Since drag forces oppose relative motion between dust and air, moving air tends to carry dust along with it. The slightest upward breeze can keep dust aloft. Its slow descent through air explains why dust is so difficult to remove with gravity alone. Gravity is too weak a force to pluck dust from the air. It needs help from stronger forces, the forces of static electricity.

CHECK YOUR UNDERSTANDING #1: Smoke Signals

The smoke from your campfire contains minute particles of soot and ash. What lifts these particles upward?

Dust and Electric Charge

Static electricity begins with electric charge, a concept that we've encountered before but have never discussed in detail. **Electric charge** or simply *charge* is an intrinsic property of matter. Many of the **subatomic particles** from which matter is constructed naturally have charge and their charge contributes to the overall charge of any object that contains them. Because electric charge is so intimately connected to the objects that possess it, we'll often refer to those objects as *charged objects* or just **charges**.

It has been known for millennia that there are two different types of electric charge, but it was Benjamin Franklin ❒ who finally gave them names. He called the charge that appears on glass when it's rubbed with silk "positive" and the charge that appears on hard rubber when it's rubbed with animal fur "negative."

❒ Although best remembered for his political activities, American statesman and philosopher Benjamin Franklin (1706–1790) was also the preeminent scientist in the American Colonies during the mid-1700s. His experiments, both at home and in Europe, contributed significantly to the understanding of electricity and electric charge. In addition to demonstrating that lightning is a form of electric discharge, Franklin invented a number of useful devices, including the Franklin stove, lightning rods, and bifocals.

Franklin's choice was unfortunate. It made electrons, the main carriers of electricity, negatively charged. Physicists have had to deal with the movement of negatively charged particles through circuits ever since. Imagine the awkwardness of having to carry out business using paper bills printed only in negative denominations.

The best evidence for the existence of electric charge is that charged objects exert forces on one another (Fig. 11.1.1). They either attract or repel. If both objects have like charges (both are positively charged or both are negatively charged), then each experiences a *repulsive* force that pushes it directly away from the other object. However, if the objects have opposite charges (one is positively charged and the other is negatively charged), then each object experiences an attractive force that pushes it directly toward the other object. These forces between stationary electrically charged objects are called **electrostatic forces.**

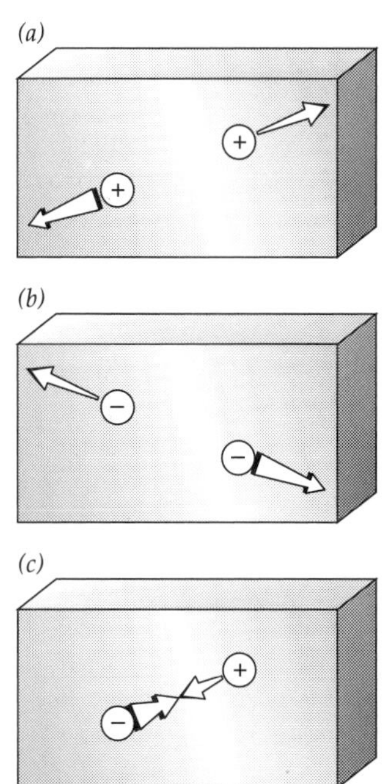

Fig. 11.1.1 - In (*a*), two positive charges experience equal but oppositely directed forces exactly away from one another. In (*b*), the same effect occurs for two negative charges. In (*c*), two opposite charges experience equal but oppositely directed forces exactly toward one another.

Electronic air cleaners use electrostatic forces to pull dust (and other particles) from the air. A typical air cleaner gives each dust grain a negative charge and then collects it on a positively charged surface. But how does the dust grain become negatively charged?

That question raises three important points about electric charge. First, the air cleaner can't simply create charge. Like momentum, angular momentum, and energy, charge is a conserved physical quantity and can't be created or destroyed. It can only be transferred from one object to another. So for the air cleaner to give the dust grain a negative charge, it must transfer that negative charge from somewhere else. Transfers of charge are all around us, from the flow of electricity through power lines to lightning bolts in the sky.

Second, by "becoming negatively charged," we don't mean that the dust grain can't contain any positive charge. Normal matter is always a mixture of positive and negative charges and the dust grain is no exception. In fact, the electrostatic forces between its positive and negative charges are what hold the dust grain together.

But the grain does have a **net electric charge**; the sum of all its individual charges. This net electric charge determines the overall electrostatic force that the grain experiences as it passes through the air cleaner. You can determine this net electric charge by adding up its positive charges and subtracting its negative charges. Therein lies the value of calling them "positive" and "negative."

Charge is measured in standard units and the SI unit of electric charge is the **coulomb** (abbreviated C). One coulomb is a lot of charge. On a dry winter day, the negative charge that you accumulate when you walk across a wool carpet in rubber-soled shoes is only about –0.000001 C.

Third, the dust grain probably enters the air cleaner with a net charge of zero. That's because a charged dust grain attracts opposite charges. As it picks up these opposite charges, the grain's net charge slowly diminishes toward zero. When it reaches zero, the grain is said to be electrically **neutral**.

But arriving exactly at zero net charge is possible only because of another remarkable feature of electric charge: it's **quantized**. Charge always appears in integer multiples of a specific quantity, the **elementary unit of electric charge**. No one has ever observed an isolated object with anything other than an integer multiple of this elementary unit of charge on it. Because of charge quantization, the dust grain will become neutral when it contains the same number of elementary units of positive charge as elementary units of negative charge.

The elementary unit of charge is extremely small, only $1.6 \cdot 10^{-19}$ C. This is the amount of charge found on most subatomic particles. An electron has one elementary unit of negative charge and a **proton**—one of the subatomic particles found in atomic nuclei—has one elementary unit of positive charge. Since the only charged subatomic particles in normal matter are electrons and protons, a neutral dust grain must contain the same number of electrons and protons.

Returning to the original question, we now know what an electronic air cleaner must do to a neutral dust grain in order to give the grain a negative charge. It must transfer negatively charged electrons to the grain and/or it must transfer positively charged protons from the grain. Either transfer will upset the charge balance of the neutral grain and give it a net negative charge.

CHECK YOUR UNDERSTANDING #2: A Bolt Out of the Blue

Just before a lightning strike, the clouds overhead have an enormous net charge. Where did this net charge come from?

Collecting Dust with Electrostatic Forces

We'll return to the problem of transferring negative charge to dust, but first let's see what becomes of that dust. It flows through the air cleaner, along with the air, until it passes a positively charged surface. It then begins to experience electrostatic forces. While gravity is too weak to remove dust from the air stream, electrostatic forces are strong enough to overwhelm viscous drag and pull dust quickly through the air. Though drag prevents the dust grains from accelerating forever, they move so swiftly that they leave the air stream and collect on the charged surface. The air continues on without the dust.

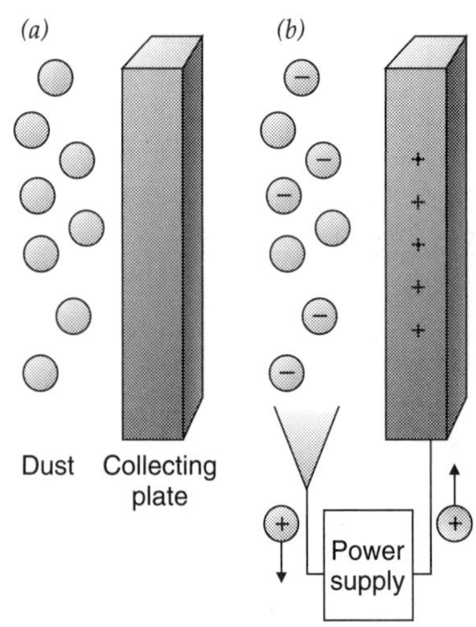

Fig. 11.1.2 - (*a*) By themselves, dust and a collecting plate soon become electrically neutral. (*b*) But the power supply in an electrostatic precipitator transfers positive charge from the dust to the collecting plate, giving each of them a net electric charge.

Because it precipitates clumps of dust on its collecting plates, this type of electronic air cleaner is called an *electrostatic precipitator*. Its chief advantage over a paper or fiber air filter is that huge amounts of dust can accumulate on its plates without blocking the airflow. It's also easy to clean.

But an electrostatic precipitator isn't a passive device. If it were, the negatively charged dust grains arriving on its positively charged collecting plates would quickly neutralize those plates and it would stop cleaning the air. To keep working, the precipitator must continuously transfer positive charge from the incoming dust to the collecting plates, using a device called a power supply.

A *power supply* pumps electric charge against its natural direction of flow. Because opposite charges attract, they tend to move together until everything is electrically neutral. But the precipitator's power supply reverses this flow by pulling positive charge away from the negatively charged dust and pushing it onto the positively charged plates (Fig. 11.1.2). Rather than becoming electrically neutral, the dust and plates acquire large but opposite net electric charges, an arrangement called **separated electric charge**. Motionless or slow moving accumulations of separated electric charge are known as **static electricity**.

It takes work to produce separated electric charge. The precipitator's power supply must pull positive charge away from the dust and push it onto the collecting plates. This work doesn't disappear. It's stored in the electrostatic forces between charges and is called **electrostatic potential energy**. Electrostatic potential energy is produced when electric charges are separated and is released when those separated charges flow back together.

The power supply transfers positive charge a little at a time from the dust to the collecting plate. Each transfer requires a small amount of work and this work is retained by the charge as electrostatic potential energy. Since the work the power supply does is proportional to the amount of charge it transfers, we can characterize its efforts by the amount of work it does per unit of charge transferred, a quantity called *voltage.*

Voltage measures the electrostatic potential energy of a unit of positive charge at a particular location. You can think of it as the work you'd have to do on a unit of positive charge to transfer it from a distant, electrically neutral object to that location. Equivalently, you can think of it as the work that a unit of positive charge at that location would do as it returned to the distant, electrically neutral object. Since the SI unit of energy is the joule and the SI unit of electric charge is the coulomb, the SI unit of voltage is the joule-per-coulomb, also called the **volt** (abbreviated V). A location's voltage can be negative if transferring a unit of positive charge to that location takes a negative amount of work!

In many situations, including the electrostatic precipitator, positive charge is transferred between electrically charged locations. In such cases, we're most interested in the *voltage difference* between those locations. This voltage difference is the work you'd have to do on a unit of positive charge to transfer it from its initial location to its final location. If transferring 1 C of charge takes 100 J of energy, then the voltage of the final location is 100 V higher than that of the initial location. A good example of this point is a household battery, a simple power supply that uses chemical reactions to pump positive charge from its negative terminal to its positive terminal. A typical battery does roughly 1.5 J of work on each coulomb of charge it pumps, so that the voltage of its positive terminal is about 1.5 V higher that of its negative terminal.

The power supply in an electrostatic precipitator is much stronger than a simple battery. It does so much work on the charge it transfers that the voltage difference between the dust and the collecting plates is more than 10,000 V. Because this powerful charge transfer leaves the dust grains negatively charged and the collecting plates positively charged, the dust grains move quickly through the air toward the collecting plates. Although drag turns most of a grain's electrostatic potential energy into thermal energy, the grain still moves fast enough to stick to the surface when it hits. There are a variety of techniques for producing high voltages, including the one shown in Fig. 11.1.3.

Fig. 11.1.3 - Static electricity can be produced by mechanical processes. In this van der Graaf generator, a moving rubber belt transfers negative charges from the base to the shiny metal sphere. This negative charge creates dramatic sparks as it returns through the air toward the positive charge it left behind.

CHECK YOUR UNDERSTANDING #3: Carpet Sparks

When you wipe your feet on the carpet, sliding friction transfers positive charge from your shoes to the carpet. What happens to your voltage and that of the carpet when this transfer occurs?

Charging the Dust

The most difficult part of an electrostatic precipitator is charging the dust. The precipitator uses an effect called a corona discharge to give the dust a negative charge. To understand a corona discharge, we must first examine the forces between charged particles in a little more detail.

These forces depend on three things. First, as we have already seen, they depend on the types of the charges. Like charges repel while opposite charges attract. Second, the forces are proportional to the amount of each charge. Doubling either charge doubles the forces. Third, the forces depend on the distance separating the two charges (Fig. 11.1.4). The forces are inversely proportional to the square of the separation, becoming weaker as the charges move away from one another.

These ideas can be combined to describe the forces acting on two charges, and written as a word equation:

$$\text{force} = \frac{\text{Coulomb constant} \cdot \text{charge}_1 \cdot \text{charge}_2}{(\text{distance between charges})^2}, \qquad (11.1.1)$$

in symbols:

$$F = \frac{k \cdot q_1 \cdot q_2}{r^2},$$

and in everyday language:

When two clouds with large opposite charges approach, expect lightning.

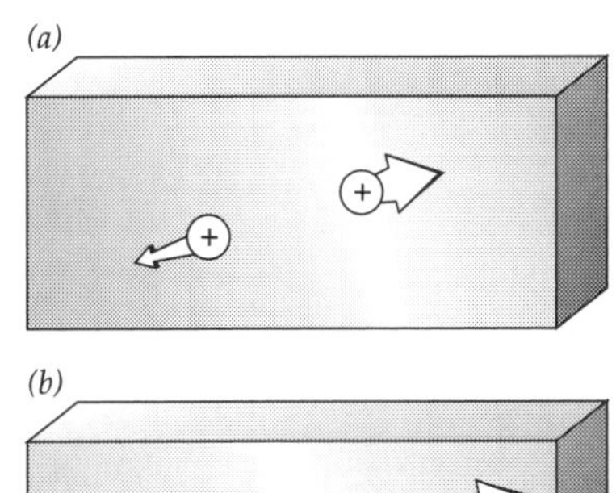

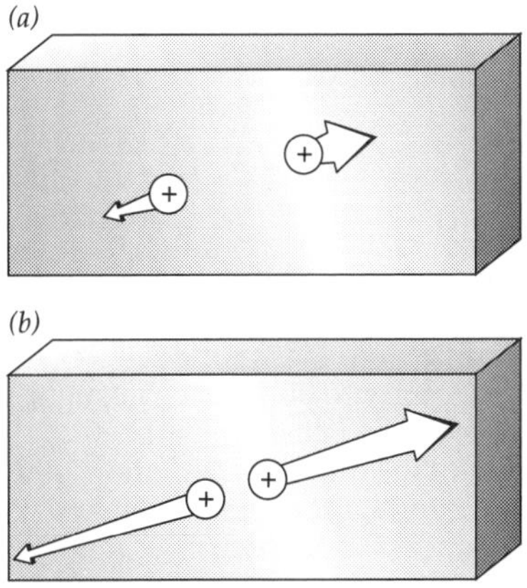

Fig. 11.1.4 - The electrostatic forces between two charges increase dramatically as they become closer. As the distance separating two positive charges decreases by a factor of two between (*a*) and (*b*), the forces those two charges experience increase by a factor of four.

❐ In 1781, after a career as a military engineer in the West Indies, French physicist Charles Augustin de Coulomb (1736–1806) returned to his native Paris in poor health. There he conducted scientific investigations into the nature of forces between electric charges and published a series of memoirs on the subject between 1785 and 1789. His research came to a close in 1789 when he was forced to leave Paris because of the French Revolution.

The force on $charge_1$ is directed toward or away from $charge_2$ and the force on $charge_2$ is directed toward or away from $charge_1$.

This relationship is called **Coulomb's law**, after its discoverer, French physicist Charles Augustin de Coulomb ❐. The **Coulomb constant** is about $8.988 \cdot 10^9 \text{ N} \cdot \text{m}^2/\text{C}^2$ and is one of the physical constants found in nature. Consistent with Newton's third law, the force exerted on $charge_1$ by $charge_2$ is equal in amount but oppositely directed from the force exerted on $charge_2$ by $charge_1$.

COULOMB'S LAW:
The magnitudes of the electrostatic forces between two objects are equal to the Coulomb constant times the product of their two electric charges divided by the square of the distance separating them. If the charges are like, then the forces are repulsive. If the charges are opposite, then the forces are attractive.

According to Coulomb's law, the repulsion between like charges increases dramatically as they approach one another. Thus when several like charges are bunched tightly together on a surface, their intense repulsions can actually push some of them out of the surface and onto passing gas molecules and dust grains. This flow of charge from a surface to nearby gas or dust is a **corona discharge**.

Corona discharges form easily around sharp metal points and thin metal wires. That's because like charges tend to bunch together on these structures and push one another into the air. This behavior is unusual because repulsive forces tend to spread like charges uniformly on the outside surface of a metal object. But a point or wire provides a place where a few like charges can get particularly far away from the rest of the like charges. Although these few charges repel one another ferociously, the other charges push on them strongly enough to keep them on the point or wire.

Placing positive charges on a metal point or wire takes energy, so a corona discharge requires a power supply. This power supply pushes positive charges onto the metal until they begin leaping into the air. The corona discharge won't start until the charges are packed extremely tightly, so it takes a great deal of work to push the last few charges onto the metal. When the corona begins, the positive charges on the metal surface have huge electrostatic potential energies and the voltage on that surface is roughly 10,000 V. A corona discharge involving negative charges precedes in exactly the same way except that, because voltage is defined as the electrostatic potential energy per unit of *positive* charge, the voltage on the negatively charged metal surface is –10,000 V.

Because electrostatic potential energy is released when a charge escapes from the surface, corona discharges are often accompanied by light. Corona discharges can be seen around high-tension wires or near the tips of lightning rods (see ❐). On ships, corona discharges are sometimes seen near masts, where they are known as St. Elmo's fire.

❐ Contrary to popular belief, lightning rods don't simply attract lightning strikes so as to protect the surrounding roof. Instead, they produce corona discharges that diminish any local build-ups of electric charge. By neutralizing the local electric charge, the lightning rod reduces the chances that lightning will strike the house. Similar devices, called static dissipaters, are found near the tips of airplane wings and protect planes from lightning strikes.

CHECK YOUR UNDERSTANDING #4: A Safety Pin

You can avoid the shocks of static electricity by carrying a very sharp needle in your hand and holding it out before touching a metal doorknob or wall. How does that needle protect you from static electricity?

CHECK YOUR FIGURES #1: Moving Out

You have two positively charged balls, each of which is experiencing a force of 1 N away from the other. If you separate them by twice as much, what force will each one exert on the other?

Real Electrostatic Precipitators

Inside an electrostatic precipitator, contaminated air is blown past many thin metal wires and undulating metal collecting plates (Figs. 11.1.5, 11.1.6, and 11.1.7). A power supply pumps positive charges from the wires to the plates, giving each wire a net negative charge and each plate a net positive charge. A corona discharge forms around each wire so that it transfers negative charges to the passing dust grains. This dust is then attracted to the positively charged collecting plates, where it builds into thick cakes. The plates' undulations keep the dust from reentering the precipitator's airstream. Because the plates bend in and out, the airstream stays near the middle and dust accumulates undisturbed in the calm pockets of the plates. (For the history of this device, see ❐.)

❐ In 1906, while an instructor at the University of California, Berkeley, Frederick Gardner Cottrell (American scientist and inventor, 1877–1948) became involved with local industries that were attempting to combat air pollution in the San Francisco Bay area. Within the year he had invented the electrostatic precipitator. Cottrell donated the patents for the precipitator to a nonprofit corporation he founded. For almost a century, Research Corporation has supported scientific work in the United States. Cottrell gave up a personal fortune to help build the careers of a great many young scientists.

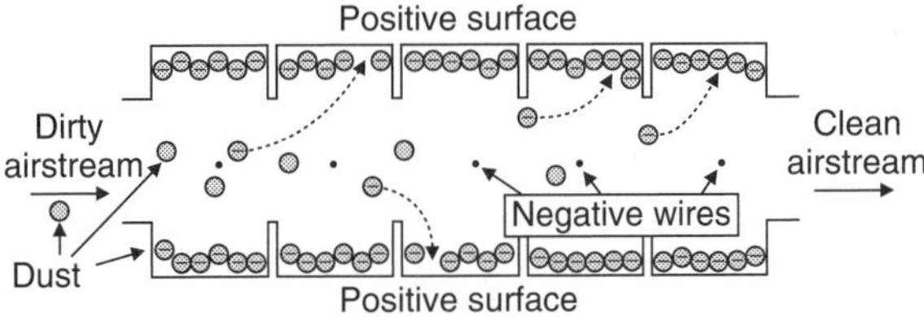

Fig. 11.1.5 - In an electrostatic precipitator, dust grains in dirty air are given negative charges so that they'll be attracted to positively charged surfaces.

When several centimeters of dust have accumulated, it's removed by rapping the plates with a stick. The sudden blow causes the plates to accelerate rapidly and they leave the dust behind. It falls in clumps to the bottom of the precipitator, where it's collected for recycling or disposal.

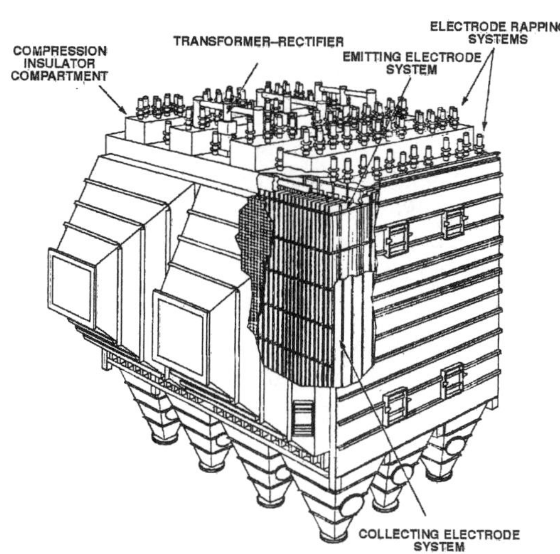

Fig. 11.1.6 - This commercial electrostatic precipitator is used to remove dust and ash from power plant emissions. Smoke enters through ducts on the front left and passes between the unit's electrodes. Particles accumulate on the collecting electrodes until the rapping system knocks them into hoppers at the bottom.

Not all electrostatic precipitators put negative charge on the dust. Some put positive charge on the dust and attract it to negatively charged surfaces. While negative corona discharges are less susceptible to accidental sparking and are easier to maintain, they also produce ozone, an irritating and toxic form of oxygen. Most industrial precipitators simply live with ozone production, but household units avoid it by using positive corona discharges.

CHECK YOUR UNDERSTANDING #5: Spring Cleaning

One way to remove dust from a rug is to hang it from a clothesline and smack it with a paddle. Why does this technique work?

Fig. 11.1.7 - This electrostatic precipitation facility is so effective at removing small particles from smoke that the chimney emissions become invisible when it's on.

Ion Generators

Household ion generators are also effective at removing dust and smoke from room air. These machines resemble electrostatic precipitators, except that they have no internal collecting plates. They use corona discharges to charge passing molecules and dust grains and then let those charged particles (or *ions*) drift about the room.

Some ion generators create both positive and negative ions while others produce only negative ions. In a positive and negative ion generator, the power supply pumps positive charge from one group of particles to another. In a negative ion generator, the power supply transfers positive charge from the particles to the earth itself.

A charged dust grain will stick to a surface even if that surface is electrically neutral. The neutral surface contains countless electric charges (Fig. 11.1.8) and a negatively charged dust grain will attract the surface's positive charges and repels its negative charges. As a result, the surface's positive and negative charges shift slightly in opposite directions and the surface becomes electrically **polarized**. While this polarized surface is still neutral overall, part of it has a slight positive charge and the rest a slight negative charge.

The dust grain is attracted to the polarized surface and they stick to one another. You can see this effect by charging a toy balloon in your hair and then touching it to a wall. The balloon's charge will polarize the wall and the balloon will stick to it. Thus an ion generator removes particles from the air by sticking them to the walls of the room. Although this method is cheap and effective, it slowly dirties the walls and furniture.

(a)

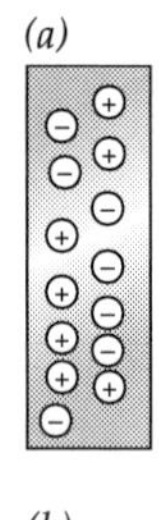

(b)

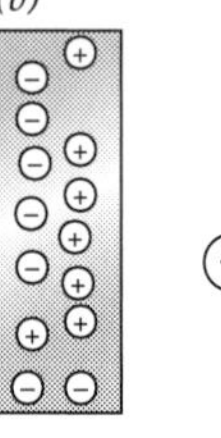

(c)

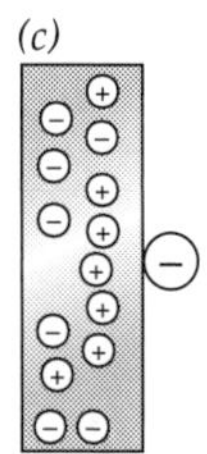

Fig. 11.1.8 - (*a*) A neutral wall contains countless positive and negative charges. (*b*) As a negatively charged dust grain approaches the wall, the positive charges move toward it and the negative charges move away. (*c*) The polarized wall continues to attract the grain and holds it in place.

CHECK YOUR UNDERSTANDING #6: Static Cling

When you remove socks from the dryer, they cling to everything. How can a sock cling to the wall or to a piece of lint?

Summary

How Electronic Air Cleaners Work: An electrostatic precipitator removes particles from the air by transferring electric charge to them and attracting them to oppositely charged collecting plates. Charging the particles is done with a corona discharge around a thin metal wire. A power supply pumps positive charge from the wire to the collecting plates, leaving the wire with a large negative net charge. As the negative charge accumulates, so does its electrostatic potential energy. When the voltage on the wire falls below about –10,000 V, negative charges begin to leave the wire and attach themselves to passing molecules and particles. The negatively charged particles are then attracted toward the positively charged collecting plates. Although air drag impedes their motion across the air stream, the charged particles experience such strong electrostatic forces that they collide with the collecting plates and remain there until they're removed.

Important Laws and Equations

1. Coulomb's Law: The magnitudes of the electrostatic forces between two objects are equal to the Coulomb constant times the product of their two electric charges divided by the square of the distance separating them, or

$$\text{force} = \frac{\text{Coulomb constant} \cdot \text{charge}_1 \cdot \text{charge}_2}{(\text{distance between charges})^2}. \quad (11.1.1)$$

If the charges are like, then the forces are repulsive. If the charges are opposite, then the forces are attractive.

The Physics of Air Purifiers

1. Electric charge is an intrinsic property of matter. There are two types of charge, called positive and negative.

2. Charged objects exert electrostatic forces on one another. Like charges repel one another while opposite charges attract.

3. The strengths of the electrostatic forces that two objects exert on one another depends on their electric charges and on the distance separating them, according to Coulomb's law.

4. Electric charge is a conserved physical quantity. It can't be created or destroyed and can only be transferred from one object to another.

5. The electrostatic force on an object depends on its net charge, the sum of all its individual positive and negative charges. Negative charges reduce this sum.

6. Electric charge is quantized, appearing in isolation only as integer multiples of the elementary unit of electric charge.

7. Separating positive and negative charge from one another requires work. This work is stored as electrostatic potential energy in the forces between the charges.

8. The voltage at a particular location is the work required to move a unit of positive charge from a distant, electrically neutral object to that location. It's also the electrostatic potential energy of each unit of positive charge at that location.

Check Your Understanding - Answers

1. The hot burned gas rises upward in the cooler air. This gas exerts upward drag forces on the smoke particles and carries them along with it.

Why: The smoke particles are so tiny that they would descend extremely slowly in still air. However, the gas surrounding the particles isn't still; it's rising quickly. While the particles descend relative to this gas, the gas is moving upward so quickly that the particles move upward. Only when the gas cools and stops rising can the smoke particles begin to settle back down toward the ground.

2. The ground.

Why: Since charge is a conserved physical quantity, the cloud can't simply manufacture it. Charge must migrate upward to the clouds from the ground to create static electricity. The precise mechanisms by which the clouds become charged are still not well understood.

3. The carpet's voltage increases relative to your voltage.

Why: As positive charge leaves you and enters the carpet, the carpet becomes positively charged and you become negatively charged. Since positive charge is attracted toward negative charge, a unit of positive charge on the carpet can release electrostatic potential energy by flowing to you. That means that the carpet's voltage is higher than your voltage. On a dry winter day, the carpet's voltage can exceed yours by as much as 50,000 V.

4. The needle emits charge into the air via a corona discharge.

Why: The needle acts as your personal lightning rod. When sliding friction gives you a net electric charge, some of that charge flows onto the needle. As soon as the repulsion between charges on the needle becomes strong enough, a corona discharge begins. This discharge limits your net electric charge and thus the size of any shock you may experience.

5. You cause the rug to accelerate so rapidly that it leaves the dust behind.

Why: Dust has inertia and it requires a force to accelerate. If you accelerate the rug rapidly enough, the dust will be unable to obtain the force it needs to follow the rug and it will be left behind.

6. The sock has a net electric charge and it polarizes nearby objects. Once polarized, these objects attract the sock.

Why: Your electrically charged sock exerts electrostatic forces on all nearby charges. The charges in the wall or the lint shift in response to these forces and these objects become electrically polarized. The end of each object that is closest to the sock has a charge that attracts the sock, so the two attract one another and stick.

Check Your Figures - Answers

1. 0.25 N.

Why: According to Coulomb's law, the force on each charge varies inversely with the square of the separation. By increasing that separation by a factor of two, you reduce the electrostatic force by a factor of four.

Glossary

charges Objects, particularly small particles, that carry electric charge.

corona discharge A faintly glowing discharge that surrounds a highly charged metal in the presence of a gas. In the discharge, electric charge is transferred from the metal to the gas molecules.

coulomb (C) The SI unit of electric charge. 1 coulomb is about 1 million times the charge you acquire by rubbing your feet across the carpet in winter.

Coulomb constant The fundamental constant of nature that determines the electrostatic forces two charges exert on one another. Its value is $8.988{\cdot}10^9\ \mathrm{N{\cdot}m^2/C^2}$.

Coulomb's law The magnitudes of the electrostatic forces between two objects are equal to the Coulomb constant times the product of their two electric charges divided by the square of the distance separating them. If the charges are like, then the forces are repulsive. If the charges are opposite, then the forces are attractive.

electric charge An intrinsic property of matter that gives rise to electrostatic forces between charged particles. Electric charge is a conserved physical quantity. There are two types of electric charge, called positive and negative.

electrostatic force The force experienced by a charged particle in the presence of other charged particles.

electrostatic potential energy Energy stored in the forces between electric charges.

elementary unit of electric charge The basic quanta of electric charge, equal to about $1.6{\cdot}10^{-19}$ C.

net electric charge The sum of all charges on an object, both positive and negative. Positive charges increase the net charge while negative charges decrease it. Net charge can be negative.

neutral Having zero net electric charge.

polarized A distribution of electric charge that is nonuniform so that the object has a negatively charged region and a positively charged region.

protons The positively charged subatomic particles found in atomic nuclei.

quantized Existing only in discrete units or quanta. Quantized physical quantities are only observed in nature in integer multiples of this elementary quanta.

separated electric charge Isolated quantities of positive and negative charge. Separated electric charge has electrostatic potential energy, stored in the forces attracting those charges back together.

static electricity Any stationary arrangement of electric charges, though often associated with the separated electric charge produced by sliding friction.

subatomic particles The fundamental building blocks of the universe, from among which atoms and matter are constructed.

volt (V) The SI unit of voltage (identical to the joule-per-coulomb). The voltage on the positive terminal of a common battery is about 1.5 volts above that on its negative terminal.

voltage The electrostatic potential energy of each unit of positive electric charge at a particular location.

Review Questions

1. If all objects are supposed to fall at the same rate near the earth's surface, why doesn't dust fall at that rate?

2. If you had two electrically charged balls, what could you do to determine if they had like charges or opposite charges?

3. What happens to the forces between two electrically charged balls as you move them closer together?

4. If you want to make a ball negatively charged, why must you transfer negative charge to it from somewhere else? Why can't you just make negative charge?

5. Compare the numbers of electrons and protons in a box that has a net negative electric charge.

6. Why do like charges arrange themselves uniformly on the outside surface of a metal ball? Why aren't any inside it?

7. Show that you do work as you transfer positive charge from a negatively charged ball to a positively charged ball.

8. You have three balls in front of you. The first has a net negative charge, the second is electrically neutral, and the third has a net positive charge. Which has the highest voltage? Which has the lowest voltage? Why?

9. If you put electric charge on a plastic spoon and hold it near small pieces of paper, the paper will be attracted to the spoon. What is happening inside the electrically neutral paper to make it attracted to the spoon?

Exercises

1. When you take clothes out of a hot dryer they often cling to one another. Two identical socks, however, usually repel. Explain their repulsion.

2. It may seem dangerous to be in a car during a thunderstorm, but it's actually relatively safe. Since the car is essentially a metal box, the inside of the car is electrically neutral. Why does any charge on the car move to its outside surface?

3. In industrial settings, neutral metal objects are often coated by spraying them with electrically charged paint or powder particles. How does placing charge on the particles help them to stick to an object's surface?

4. The paint or power particles discussed in Exercise **3** are all given the same electric charge. Why does this type of charging ensure that the coating will be highly uniform?

5. The moon has no atmosphere. Suppose that you were walking on its surface and threw a handful of moon dust and a moon rock into the sky together. Which would fall faster?

6. If two objects repel one another you know they have like charges on them. But how would you determine whether they were both positive or negative?

7. Pieces of lint often cling to clothes because the two are oppositely charged and attract one another. However, this attraction diminishes quickly when you remove the lint, so that when you drop it far from the clothes, it doesn't return to them. Why does the attractive force grow so weak?

8. Suppose that you had an electrically charged stick. If you divided the stick in half, each half would have half the original charge. If you split each of these halves, each piece would have a quarter of the original charge. Can you keep on dividing the charge in this manner forever? If not, why not?

9. A car battery is labeled as providing 12 V. Compare the electrostatic potential energy of positive charge on the battery's negative terminal with that on its positive terminal.

10. If the forces between electric charges didn't diminish with distance, an electrically charged balloon wouldn't cling to an electrically neutral wall. Why not?

11. If you hold a sharp needle in your hand as you run your shoes across a carpet, you won't accumulate much static charge. Why not?

12. If you observe high tension wires on a dark night, you may see a faint purple glow around them. This glow is particularly evident at joints where the wires connect to one another. What causes this glow and why is it strongest at the connections?

13. You're holding two oppositely charged balloons in your hands. Compare their voltages.

14. You begin to separate the two balloons in Exercise **13**. You do work on the balloons, so your energy decreases. Where does your energy go?

15. When you separate the balloons in Exercise **14**, do their voltages change? If so, how do those voltages change?

16. A bowling ball contains an enormous number of electrically charged particles. Why don't two bowling balls normally exert electrostatic forces on each other?

Problems

1. You remove two socks from a hot dryer and find that they repel with forces of 0.001 N when they're 1 cm apart. If they have equal charges, how much charge does each sock have?

2. If you separate the socks in Problem **1** until they're 5 cm apart, what force will each sock exert on the other?

3. If you were to separate all of the electrons and protons in 1 g (0.001 kg) of matter, you'd have about 96,000 C of positive charge and the same amount of negative charge. If you placed these charges 1 m apart, how strong would the attractive forces between them be?

4. If you place 1 C of positive charge on the earth and 1 C of negative charge 384,500 km away on the moon, how much force would the positive charge on the earth experience?

5. How close would you have to bring 1 C of positive charge and 1 C of negative charge for them to exert forces of 1 N on one another?

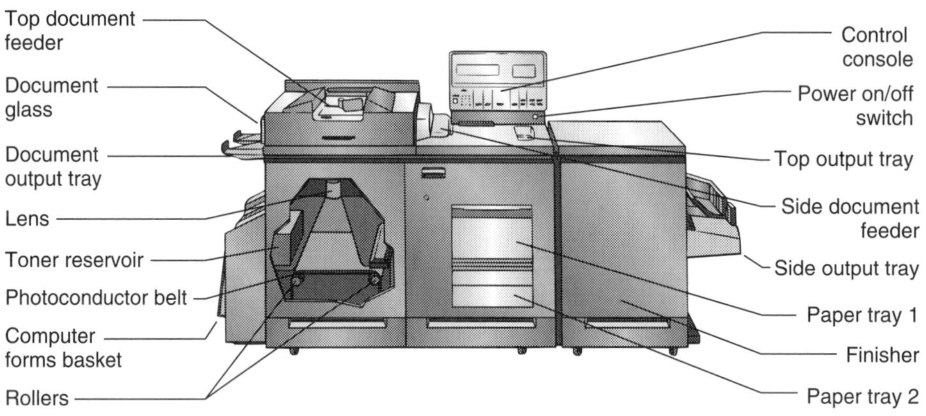

Section 11.2

Xerographic Copiers

What modern office could function without a xerographic copier? The days of carbon paper and mimeograph machines are long past. Our dependence on copiers has led to intense competition between copier companies. But while manufacturers often claim that their products are technologically superior to those of competitors and that they alone possess the inventive genius needed to create the finest copiers, these claims are mostly just salesmanship. That's because all xerographic copiers are based on the same experiments, done by Chester Carlson in 1938. In this section, we'll examine xerographic copiers to see the principles that make them possible.

Questions to Think About: *How could you use static electricity to position black powder on a sheet of paper? How would you put that static electricity on the paper? In order for characters to appear on the sheet, how should its static electricity be distributed? In a copier, what should light do to the static electricity to produce a copy of the original?*

Experiments to Do: *To get a feel for how a copier works, cut a small sheet of paper into tiny squares, about 1 mm on a side. Put the squares on a table and suspend a thin plate of clear plastic above them, a few millimeters away. The top of a clear plastic box will do. Now run a plastic comb through your hair or against a sweater several times and touch it to the top of the plastic plate. Squares of paper will leap off the table and stick to the plastic plate. What's supporting the squares?*

Controlling Static Electricity: Electrons in Solids

The image that a xerographic copier forms on a sheet of paper begins as a pattern of tiny black *toner* particles. The copier uses static electricity to arrange these particles on a drum or belt and then transfers them to the paper. Since the copier is duplicating an original document, it uses light from that document to control the placement of static electricity, so that the pattern of toner on the copy is identical to the printing on the original.

At the heart of the xerographic copier is a thin layer of photoconductor, which the copier uses to control the placement of static electricity. A **photoconductor** is a solid material through which electric charges can move only when it's exposed to light. In the dark, it's an **electric insulator**; preventing any net movement of electric charge. In the light, it's an **electric conductor**; allowing charges to move.

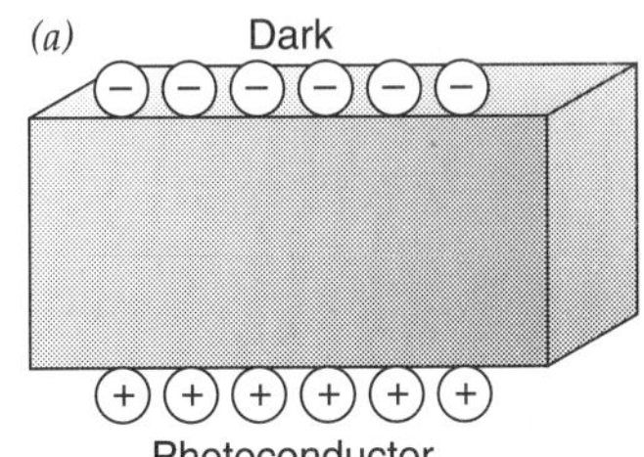

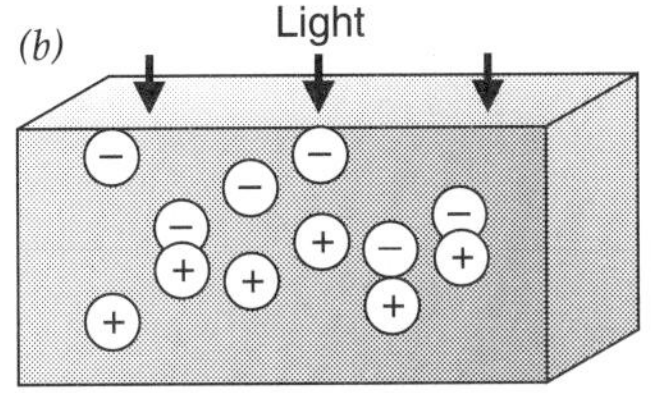

Fig. 11.2.1 - (*a*) In the dark, a photoconductor is an electric insulator so that separated electric charge on its surfaces remains there indefinitely. (*b*) But when the photoconductor is exposed to light, it becomes an electric conductor and the opposite electric charges soon join one another.

While a darkened photoconductor can keep positive and negative charges apart, these charges quickly attract one another when light hits the photoconductor (Fig. 11.2.1). This behavior allows light from the original document to determine the pattern of static electricity on the copying drum or belt and hence the placement of toner on a piece of paper. Photoconductors are so central to xerographic copying that we'll spend the next several pages examining photoconductivity and the behaviors of electrons in solids.

Photoconductivity is a consequence of quantum physics, which has enormous influence over the motions of small particles. In particular, quantum physics limits the possible paths that a small particle can take when it travels inside an object. This limitation is particularly important for electrons, which are the most mobile charged particles in matter and thus the main carriers of electricity. An electron in an atom can only orbit the atom's nucleus in one of the paths that quantum physics allows. These allowed paths are called **orbitals**. Similarly, quantum physics limits an electron's motion through a solid object to allowed paths or trajectories that are called **levels**.

One of the most remarkable observations of quantum physics is that every indistinguishable electron must have its own level. This law is called the **Pauli exclusion principle**, after is discoverer, Wolfgang Pauli ❐. The principle applies to a whole class of subatomic particles, the **fermi particles**, that includes all of the basic constituents of matter: electrons, protons, and **neutrons**, the electrically neutral subatomic particles that, together with protons, make up atomic nuclei. Two indistinguishable fermi particles can never travel in the same level.

THE PAULI EXCLUSION PRINCIPLE:
No two indistinguishable fermi particles ever occupy the same quantum level.

Because of the Pauli exclusion principle, no more than two electrons can be in one orbital or one level. The fact that as many as *two* electrons can share an orbital or level reflects a peculiar property of electrons: electrons have two possible internal states, usually called spin-up and spin-down. A spin-up electron is distinguishable from a spin-down electron, so that one spin-up electron and one spin-down electron can share a single orbital or level. However, two is the absolute maximum allowed by quantum physics.

An electron's level also determines its total energy, the sum of its kinetic and potential energies. Since electrons in different levels follow different paths at different speeds, each level corresponds to a particular energy. Electrons in the lowest energy levels generally travel slowly and stay close to the positively charged atomic nuclei while those in higher energy levels usually travel more quickly and avoid the nuclei.

Because of friction-like effects, an object's electrons lose energy until they're traveling in the lowest energy levels. However, since only two electrons can share each level, they can't all be in the lowest level. Instead, they fill the levels from lowest energy on up; two to each level. This filling continues up to the

❐ Austrian physicist Wolfgang Pauli (1900–1958) rose to fame at 21 by writing an article on relativity that impressed even Einstein. He went on to discover the exclusion principle, a fundamental part of quantum theory. He was renowned for his intensely critical attitude toward new ideas, considering them all "rubbish" until convinced otherwise. Pauli was also quite interested in psychology, corresponding with Carl Gustav Jung and writing a number of articles on the subject.

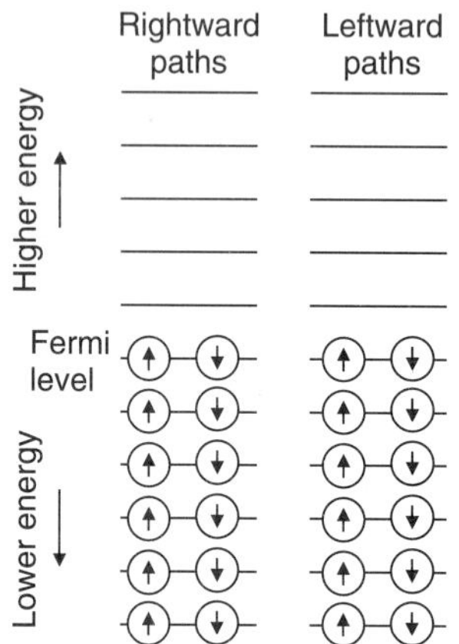

Fig. 11.2.2 - The levels in an object are normally filled from the lowest energy level up to the fermi level. Here, those levels are represented by horizontal bars, with spin-up and spin-down electrons drawn on the energy levels in which they travel. Rightward and leftward heading levels have the same energies and appear side by side.

fermi level; the highest energy level used by an electron. If we represent these levels with horizontal bars and arrange them vertically according to energy (Fig. 11.2.2), then the levels up to the fermi level each contain two electrons, while those above the fermi level are empty.

In an isolated solid, the levels traveled by electrons take them in every conceivable direction. In fact, electrons travel so uniformly in all directions that they have no average movement at all. For example, if we divide the levels below the fermi level into those that take their electrons toward the right and those toward the left, we find that the number of levels in each group is exactly equal. Each electron that heads right is balanced by another electron that heads left. The levels actually occur in left and right going pairs with the same energy, so that they appear side by side in Fig. 11.2.2. Thus while individual electrons are always moving through the material, they go nowhere on the average.

CHECK YOUR UNDERSTANDING #1: Going Nowhere Fast

A marble contains a great many electrons. If the marble is all by itself, what fraction of its electrons is heading left and what fraction is heading right?

Metals, Insulators, and Photoconductors

But Fig. 11.2.2 is incomplete. The levels in a solid actually occur in groups called **bands**. Each band corresponds to a particular type of path through the solid. Since the levels in a band involve similar trajectories, they also involve similar energies. Thus the levels in a solid are bunched together in energy bands. Between these bands of levels, there are occasional *energy gaps*, ranges of energy in which there are no levels at all.

Bands and energy gaps are what distinguish metals, insulators, and photoconductors. When an energy gap is located close to the fermi level, it can prevent the electrons in a solid from responding to outside forces. To see how that happens, let's examine a metal and an insulator.

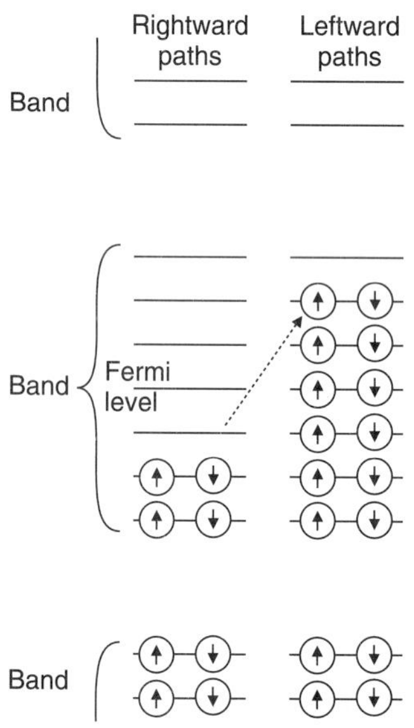

Fig. 11.2.3 - In a metal, the fermi level occurs in the middle of a band. When you put positive charges to the left of the metal, some of the electrons in levels just below the fermi level shift into the empty levels just above it. The new arrangement of electrons is unbalanced, with more electrons heading left than heading right. The metal is conducting electricity.

In a metal, there's no energy gap near the fermi level. When you put positive charge to the left of the metal, some of the metal's negatively charged electrons respond by shifting from the filled levels just below the fermi level to the empty levels just above it (Fig. 11.2.3). In effect, the metal's electrons accelerate toward the positive charge. Most importantly, some of the metal's electrons leave rightward heading levels for leftward ones, upsetting the balance. With more electrons traveling left than right, there's a net flow of electric charge through the metal. It's conducting electricity!

In a metal, shifting electrons from levels just below the fermi level to levels just above it takes only a tiny bit of energy. This energy is easily provided by the electrostatic potential energy that the positive charge releases when you move it near the metal. But in an insulator, there's an energy gap just above the fermi level (Fig. 11.2.4). When you put positive charge to the left of the insulator, its electrons can't respond because the nearest empty energy levels are way up above that energy gap. To shift into one of these empty levels, an electron near the fermi level would need more energy than it can get from the approaching positive charge. The motion of the electrons remains balanced so there's no net flow of electric charge through the insulator. It doesn't conduct electricity!

The levels below the fermi level are called **valence levels** and those above it are called **conduction levels**. In a metal, electrons can shift easily from the valence levels to the conduction levels because each shift takes only the tiniest amount of energy. In an insulator, the energy gap between the valence and conduction levels prevents such shifts.

But an electron can shift from a valence level to a conduction level if something provides the necessary energy. One energy source is light. When an insulator is exposed to light of the right frequency, that light can shift electrons from the valence levels to the conduction levels (Fig. 11.2.5). Such shifts can upset the balanced movement of electrons in the material and create a net flow of electric charge through it. Thus light can turn an insulator into a conductor, a *photoconductor*.

Light's frequency is important here because light is emitted and absorbed in energy packets called **photons**. The higher the frequency of light, the more energy each of its photons contains. To shift an electron across the large energy gap in a typical insulator, high energy, high frequency light is needed. The insulator must be exposed to violet or even ultraviolet light. But nature also provides materials with smaller energy gaps that can be crossed with the help of red or even infrared light. These materials are called **semiconductors**, because their properties lie somewhere in between those of conductors and insulators.

The photoconductors found in most xerographic copiers are semiconductors. They have energy gaps just above the fermi level, so they don't normally conduct electricity. But when they're exposed to visible light, electrons in these photoconductors shift from valence levels to conduction levels and begin to carry electric charge through the material. The photoconductor can thus use light to control static electricity so that the copier can reproduce the original.

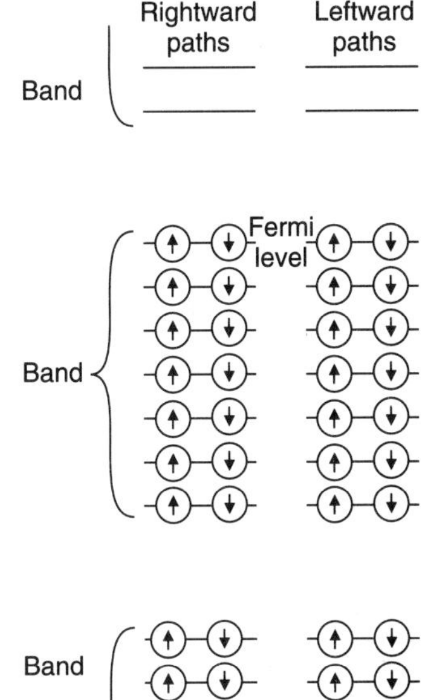

Fig. 11.2.4 - In an insulator, the fermi level occurs at the top of a band. When you put electric charges near the insulator, the electrons in the filled band can't rearrange to produce a net flow of charge through the insulator. The insulator can't conduct electricity.

CHECK YOUR UNDERSTANDING #2: Stop and Go Shopping

In many grocery check-out counters, a conveyor belt carries food to the register but stops when the food reaches the end and blocks a beam of light. How might a photoconductor be used to sense this blockage?

Moving Electric Charge with Electric Fields

The photoconductor permits light to control static electricity. The copier coats the photoconductor with electric charges and these charges move about when the photoconductor is exposed to light. But before we examine the actual xerographic copier, let's think again about what makes those charges move.

The copier puts charges on two sides of a thin sheet of photoconductor, negative charge on one side and positive charge on the other. When the photoconductor is illuminated, electrons flow from the negative side to the positive side and the separated charge disappears. Each electron that completes this journey does it by shifting among the valence and conduction levels so that it can accelerate toward the positive charges in front of it and away from the negative charges behind it.

But instead of thinking about the interaction between the traveling electron and all of the individual charged particles around it, we can view it as an interaction between the traveling electron and something local: an electric field. An **electric field** is an attribute of space that exerts forces on charged particles.

From this new perspective, the traveling electron accelerates forward because it's in an electric field, one that's created by all of the nearby charged particles. In this case, the electric field is simply an intermediary, allowing us to think separately about the traveling electron and the charges surrounding it; the surrounding charges create the field and the field causes the traveling electron to accelerate. But in later sections, we'll see that electric fields are sometimes created by things other than charges.

An electric field usually varies from place to place, so that the force it exerts on a charge depends on that charge's position. At each point in space, the field

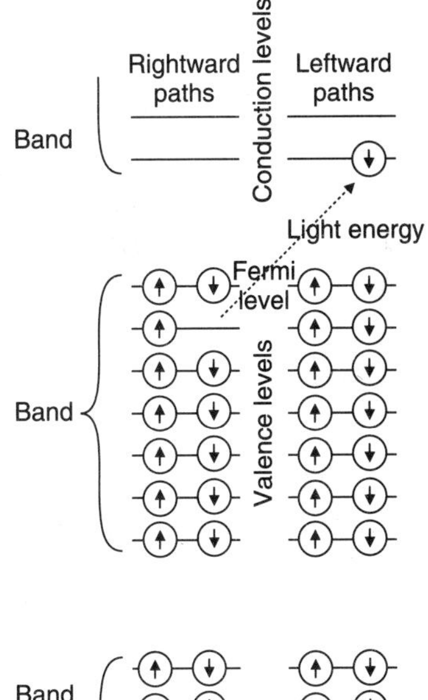

Fig. 11.2.5 - When light strikes an insulator, its energy can shift an electron from a valence level to an empty conduction level above the energy gap. Such shifts make it possible for there to be a net movement of electric charge through the insulator, turning it into an electric conductor (a *photoconductor*).

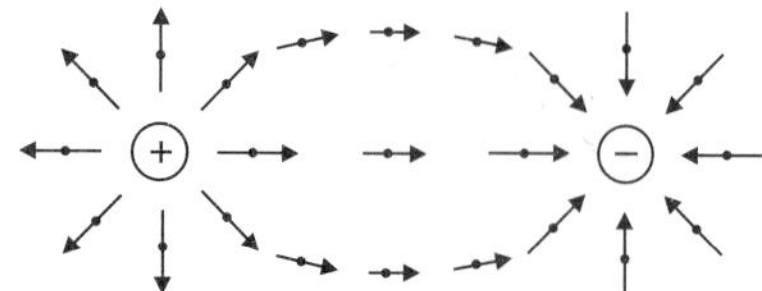

Fig. 11.2.6 - The electric field produced by two opposite electric charges. At each point in space marked by a black dot, the electric field's direction is indicated by an arrow and its strength by the arrow's length. A positive charge located at the dot would experience an electrostatic force in the direction of the arrow and of an amount proportional to the length of the arrow.

has a magnitude and a direction, where the magnitude specifies the amount of force it exerts on each unit of charge and the direction specifies which way it pushes on a positive charge (Fig. 11.2.6). Because an electron has a negative charge, it's pushed in the direction opposite the electric field.

The electric field inside the copier's photoconductor sheet always points from the positive side of the sheet to the negative side (the direction in which a positive charge would accelerate). When light strikes the photoconductor, electrons flow from the negative side to the positive side, pushed along by the electric field.

CHECK YOUR UNDERSTANDING #3: Medical Electrons

A medical linear accelerator uses a strong electric field to accelerate electrons forward and give them enormous kinetic energies. These high energy electrons enter the patient and kill cancer cells. In which direction does the accelerator's electric field point?

Xerography

The light sensitive component in a xerographic copier is a metal drum or belt that is covered with a thin layer of photoconductor. The copier coats this photoconductor with electrons, which remain in place as long as the photoconductor is in the dark. But wherever light strikes the photoconductor, it becomes conductive and allows the electrons to escape into the metal. Only the unilluminated portions of the photoconductor retain their static electric charge and eventually attract black toner particles. In that manner, the darkened parts of the photoconductor produce the dark parts of the final copy.

Modern xerographic copiers, such as the one shown schematically in Fig. 11.2.7, use the xerographic process outlined in Fig. 11.2.8. While this process was invented in 1938 by Chester Carlson ❐, the first commercially successful xerographic copiers weren't produced until 1960.

❐ Impoverished as a youth, American inventor Chester F. Carlson (1906–1968) supported his family by washing windows and cleaning offices after school. His work in a print shop as a teenager started him think about copying and he began to experiment with electrophotography. After attending Caltech, he worked for Bell Laboratories but was laid off in the Depression. While attending law school, he continued his experiments and invented the xerographic copying process in 1937–38. Development of commercial copiers was slow and it wasn't until 1960 that Haloid Xerox Corporation produced its first successful copier, Model 914. Carlson became extremely wealthy, but gave most of his money away anonymously.

The copier starts by applying a uniform charge to the surface of the photoconductor. While some copiers apply positive charge, for the present discussion, let's assume that it's negative. This charge is applied by a *corotron*, a fine wire centered in a half-cylinder of metal. A power supply pumps electrons onto the fine wire until they are emitted into the air as a corona discharge. When these electrons approach the photoconductor, they polarize it and stick to it, just as negative ions from a simple electrostatic air cleaner stick to the walls of your house. The photoconductor becomes uniformly charged (Fig. 11.2.8*a*), with about 10^{-7} C of negative charge per cm^2 of surface.

On the other side of the photoconductor is a grounded metal surface. It's grounded in the sense that it's electrically connected to the earth so that charges are free to flow between the two. The negative charge on the free surface of the photoconductor repels electrons so that they leave the metal for earth. As a result, the metal side of the photoconductor acquires a net positive charge.

After this charging, the copier exposes the photoconductor to light from the original document. It uses a lens to cast an image of the original onto the photoconductor's surface. We'll examine lenses and the formation of images when we look at photographic cameras in Chapter 16. For now, the important point is that light only hits the photoconductor in certain places, corresponding to white parts of the original document.

There are two standard techniques for exposing the photoconductor to light. Some copiers illuminate the original document with the brilliant light of a flash lamp and cast its entire image onto a flattened portion of a photoconductor

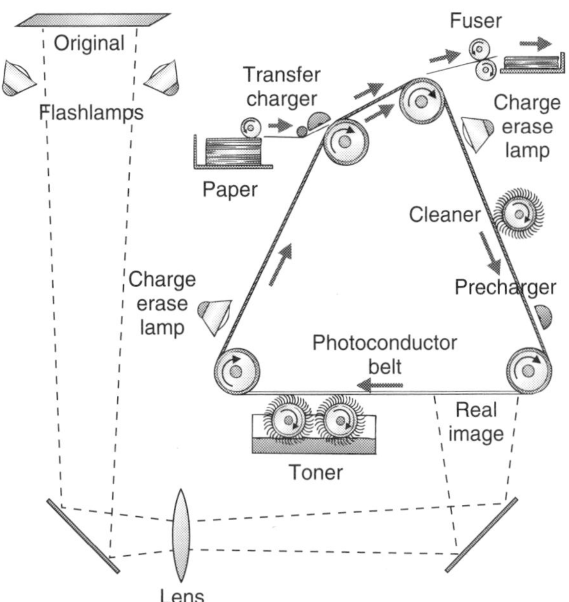

Fig. 11.2.7 - This xerographic copying machine uses a photoconductor belt to form black and white images of an original document. The copying process begins with the pre-charger, which coats the photoconductor with charge. The optical system then forms a real image on a flat region of the photoconductor belt, producing a charge image. After the charge image picks up toner particles, the first charge erase lamp eliminates the charge image and weakens the toner's attachment to the belt. The toner is then transferred and fused to the paper.

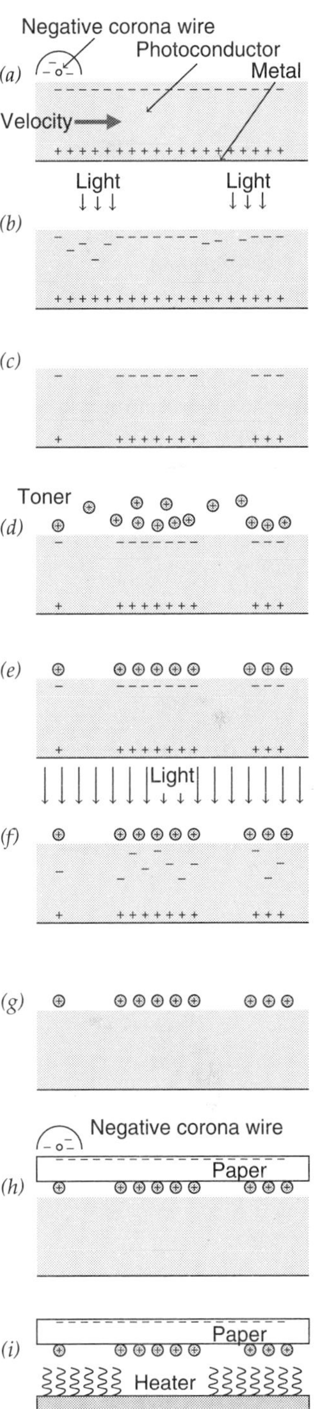

Fig. 11.2.8 - The photoconductor is first coated (*a*) with a uniform layer of negative charge. Exposure to light (*b*) erases some charge to form a charge image (*c*). The charge image attracts (*d*) positively charged toner particles (*e*). The charge image is erased (*f*) to release the toner particles (*g*). The toner is transferred to the negatively charged paper (*h*) and fused to the paper with heat (*i*).

belt. In an instant, charge flows through any regions of the photoconductor that were exposed to light, leaving these regions electrically neutral (Fig. 11.2.8*b*). In other copiers, a moving lamp or mirror illuminates the original a little at a time and the image is built piece by piece on a rotating photoconductor drum. The document's image moves with the turning drum, gradually exposing the photoconductor and allowing parts of it to become electrically neutral. When the exposure is over, the photoconductor carries a charge image of the original document (Fig. 11.2.8*c*).

To develop this charge image into a visible one, the xerographic copier exposes the photoconductor to charged toner particles (Fig. 11.2.8*d*). The toner is a fine insulating plastic powder that contains a colored pigment. Most often this pigment is black, but it can be any color. The toner is charged opposite to the photoconductor so that the two attract. In our case, the toner is positive.

Applying toner to the photoconductor must be done gently and it's often accomplished with the help of Teflon-coated iron balls. These tiny balls are held together in long filaments by a rotating magnetic shaft, so that the shaft resembles a spinning brush with extraordinarily soft bristles. These bristles wipe toner particles out of their storage tray and onto the photoconductor. During this transfer, the toner particles become positively charged so that they stick to the negatively charged portions of the photoconductor (Fig. 11.2.8*e*).

The photoconductor now carries a black image of the original document. But to create a copy, this black image must be transferred to paper. To begin this transfer, the copier illuminates the photoconductor with a charge erase lamp so that the photoconductor's negative charge escapes into the metal (Fig. 11.2.8*f*). The toner remains in place but it's very weakly attached (Fig. 11.2.8*g*).

The copier than transfers the toner to a nearby sheet of paper by applying negative charge to the paper's back (Fig. 11.2.8*h*). The positively charged toner is attracted to the negatively charged paper and the two leave the photoconductor

Fig. 11.2.9 - This xerographic copier places the photoconductor drum, toner supply, and a corona wire inside a disposable cartridge. After the paper passes through the cartridge from right to left, toner is fused to its surface and it leaves the copier.

together. The copier then heats and presses the copy, permanently fusing the toner into the paper (Fig. 11.2.8*i*). Sometimes, when a copier jams, you may remove a sheet before it has been fused. The image looks complete and normal but wipes off when you touch it because it's held in place only by electrostatic forces.

Once the image has been transferred to the paper, it's time to get the photoconductor surface ready for the next copy. The charge erasure and transfer processes leave a little charge and toner on the photoconductor so the copier performs an extra cleaning. A second charge erase lamp eliminates the charge and a brush or squeegee mops up the toner. Once this cleanup is done, the photoconductor is back to its original state and ready to be used again. In a personal copier, the photoconductor and toner supply are contained in a single disposable cartridge (Fig. 11.2.9).

CHECK YOUR UNDERSTANDING #4: Sticky Copies

When the copies emerge from a xerographic copier, they tend to stick to things and attract lint. What causes this effect?

Color Copiers and Laser Printers

Xerographic copying isn't limited to black and white. The toner itself can be any color, so it's easy to make red copies or blue copies. But to create full color copies, the copier needs to work with 3 or 4 different toners. Three of these toners are the *primary colors of pigment*: yellow, cyan, and magenta, while the fourth is black. As we'll discuss in the section on photographic cameras, yellow, magenta, and cyan pigments can be combined to make our eyes perceive any color imaginable. The black toner is optional but it improves the image contrast.

To make a full color copy, the color copier makes 3 or 4 different charge images and uses each of these images to transfer a particular toner to the paper. It forms these charge images through colored filters, so that the charge image that controls the yellow toner is formed by blue light (yellow toner absorbs blue light, just as black toner absorbs white light). The yellow, cyan, and magenta toner images are carefully assembled and transferred to the paper. Finally, a charge image is made with white light and used to control the placement of black toner. It, too, is transferred to the paper and the assembled image is fused in place. The result is a full color copy that looks just like the original document.

A laser printer is also a xerographic device, but it uses a laser beam to write a charge image directly onto its photoconductor drum. This rotating drum is charged by a corotron and then a spinning mirror scans laser light rapidly across its surface. Wherever laser light hits the drum, charge flows through the photoconductor. A computer in the printer turns the laser on and off as it systematically constructs the charge image, one dot at a time.

The printer then sticks toner to the charge image and transfers it to the paper. However, most laser printers use special electrostatic tricks to reverse the charge image, so that toner sticks only to the portions of the drum that were exposed to laser light. This reversal makes it easier for the printer to produce pure white backgrounds for its prints.

CHECK YOUR UNDERSTANDING #5: Hold Still

Why are color xerographic copiers very sensitive to movements of the original during copying?

Summary

How Xerographic Copiers Work: The photoconductor in a xerographic copier allows light to control the distribution of electric charge. An optical image of the original document is projected onto a charged region of the photoconductor. Wherever light hits, the photoconductor conducts electricity and the charge on that region escapes. The result is a charge image on the surface of the photoconductor. Tiny black toner particles, oppositely charged from the unilluminated portions of the photoconductor, are brought near the charge image. These toner particles stick to the charged portions of the photoconductor, forming a visible image of the document. The charge on the photoconductor is erased and the toner particles are transferred to an oppositely charged sheet of paper. The toner is fused into the paper by heat and pressure to create a finished copy.

Important Laws and Equations

1. The Pauli Exclusion Principle: No two indistinguishable fermi particles ever occupy the same quantum level.

The Physics of Xerographic Copiers

1. When small particles such as electrons travel inside objects, they must follow paths or trajectories that are allowed by quantum physics. In atoms, these allowed paths are called orbitals and in solids, they're called levels.

2. No two indistinguishable fermi particles can be in the same quantum level at the same time (the Pauli exclusion principle).

3. The electrons in a solid fill its levels from the lowest energy level to the fermi level. Two electrons travel in each level because one has spin-up and the other spin-down.

4. Because electrons fill left and right heading levels equally, there is no net movement of charge through an isolated solid object.

5. In a metal, the fermi level is located in the middle of a band, so that there are filled valence levels just below it and empty conduction levels just above it. Charge can move through the metal because its electrons can shift easily from valence to conduction levels. Such shifts make it possible for charge to move through the metal.

6. In an Insulator, there is an energy gap located just above the fermi level. Because of this gap, electrons can't shift from valence to conduction levels and charge can't move through the insulator.

7. When light strikes an insulator, its energy can shift electrons from valence levels to conduction levels and convert that insulator into an electric conductor. Semiconductors make the most practical photoconductors, because visible light carries enough energy to make them conduct.

8. An electric field exerts a force on any charged particle in its presence. Though usually created by surrounding electric charges, an electric field can have other origins as well. An electric field can vary from location to location, with a magnitude and direction proportional to the force it exerts on a unit of positive charge.

Check Your Understanding - Answers

1. 50% of the electrons are heading left and 50% are heading right.

Why: The electrons in the marble use every available level from the lowest energy level to the fermi level. Since half of these levels are right heading and half are left heading, the electrons are evenly divided between left and right. Because of this perfect balance, there is no net movement of electric charge in the isolated marble.

2. The beam of light shines on the photoconductor, allowing charge to move through it and operate the conveyer belt's motor. When food blocks the light beam, the photoconductor becomes an electric insulator and the belt stops moving.

Why: Photoconductors are commonly used in light sensors. When light hits the photoconductor, it behaves as an electric conductor and permits electricity to flow. This electricity can be used to operate machinery, trigger a burglar alarm, or turn lights on or off. In the present case, it runs the conveyer belt's motor.

3. It points backward, away from the patient.

Why: Since electrons are negatively charged, they accelerate in the direction opposite to the field. Since the accelerator must push the electrons forward, toward the patient, its field must point away from the patient.

4. The charge that was placed on the paper to attract the toner isn't always removed completely. Moreover, the toner itself is charged.

Why: The final transfer process, lifting the toner particles from the photoconductor to the paper is done by charging the paper, and some of this charge remains on the paper when it leaves the copier. Copier transparencies are particularly clingy because plastic retains charge so well.

5. The image is built up in four steps so a movement of the original will shift some of the toner layers with respect to the others.

Why: Printed images created with three or four color processes require perfect registration between the separate pigment layers. If one of the color layers is shifted with respect to the others, the image will look strange.

Glossary

band A group of levels that involve similar paths through a solid and thus involve similar energies.

conduction level A quantum level in a solid that involves more energy than the fermi level and that is normally unoccupied by electrons.

electric field An attribute of each point in space that exerts forces on electrically charged particles. An electric field has a magnitude and direction proportional to the force it would exert on a unit of positive charge at that location. While electric fields are often created by nearby charges, they can also be created by other electromagnetic phenomena.

electric conductor A material that permits electric charges to flow through it.

electric insulator A material that prevents any net movements of electric charge through it.

fermi level The highest energy level normally occupied by electrons in a solid.

fermi particles A class of fundamental particles that includes electrons, protons, and neutrons and that obeys the Pauli principle.

level One of the paths that an electron can take in a solid that is allowed by quantum physics.

neutrons The electrically neutral subatomic particles that, together with protons, make up atomic nuclei.

orbitals One of the orbits that an electron can take in an atom that is allowed by quantum physics.

Pauli exclusion principle An observed property of nature that no two indistinguishable fermi particles ever travel in the same orbital or level.

photoconductor A solid that is an electric insulator in the dark but that becomes an electric conductor when exposed to light of the correct wavelength.

photon A particle of light, having energy and momentum but no mass.

semiconductor A material in which there is a small energy gap above the fermi level, so that only a modest amount of energy is needed to shift an electron from a valence level to a conduction level.

valence level A quantum level in a solid that involves less energy than the fermi level and that is normally occupied by electrons.

Review Questions

1. What prevents two spin-up electrons from traveling in exactly the same path through a solid simultaneously?

2. When is it possible for two electrons to occupy the same quantum level?

3. The individual electrons in an object are always moving in their levels. Why doesn't this motion lead to a net flow of charge through the object?

4. The electrons in one material fill all the levels up to the middle of a band, while those in a second material fill all of the levels right to the top of a band. How does each material respond to nearby electric charge? Can either one conduct electricity?

5. What does light do to the electrons in a photoconductor that transforms it from an insulator to a conductor?

6. A positive charge produces an electric field around it. If you examine a point to the right of the charge, which way does the field point? What about if you examine a point to the left of the charge? Why does the direction of the electric field near a positive charge depend on which point you examine?

7. How does the magnitude of the electric field near a positive charge depend on the distance from that charge?

8. If you placed a small positive charge in an electric field, what would happen to that charge when you let go? What effect does the direction of the electric field have on the charge's motion? What effect does the magnitude of the electric field have on the charge's motion?

Exercises

1. If a very small piece of material contains only 10,000 electrons and those electrons have as little energy as possible, how many different levels do they occupy in that material?

2. If electrons had four different internal states that could be distinguished from one another, how many electrons could occupy the same level without violating the Pauli exclusion principle?

3. If electrons were not fermi particles, any number of them could occupy a particular level. How would these electrons tend to arrange themselves among the levels in an object?

4. Thermal energy can shift some of the electrons in a hot semiconductor from valence levels to conduction levels. What effect do these shifts have on the semiconductor's ability to conduct electricity?

5. Based on the result of Exercise **4**, would heating the photoconductor in a xerographic copier improve or diminish its ability to produce sharp, high-contrast images?

6. Many night lights and security lamps use a strip of photoconductor to tell when it's dark enough for them to turn on. Should the light turn on when the photoconductor is behaving as a conductor or an insulator?

7. Some photoconductors are insensitive to red light. Xerographic copiers based on such photoconductors turn red parts of an original document into black when preparing a copy because they can't detect red light. What prevents them from responding to red light?

8. Although Fig. 11.2.2 shows the levels in a material divided into rightward and leftward heading paths, the levels can actually be divided into opposite-going paths along any direction. If you suddenly brought a large positive charge near the *top* of a metal block, how would the electrons shift among the levels?

9. Photons of visible light carry enough energy to shift electrons from valence levels to conduction levels in semiconductors but not in normal insulators. How does this fact explain why semiconductors are generally opaque while normal insulators are generally clear or translucent?

10. Suppose that your room has an electric field in it and that this field is the same everywhere in the room, always pointing toward the right with the same magnitude. If you hold up a positively charged ball and move it slowly from left to right, where will the force exerted on that ball by the electric field be the strongest, or will it be the same everywhere?

11. If gravity weren't acting on the positively charged ball in Exercise **10** and you let go of it, what effect would the electric field have on its motion?

12. If the ball in Exercise **11** were *negatively* charged, what effect would the electric field have on its motion?

13. American physicist Robert Millikan (1868–1953) discovered the elementary unit of electric charge by balancing the downward weight of a tiny, electrically charged oil drop against the upward force from an electric field. Without any electric field, the drop would fall under its own weight. However, an electric field could exert an upward force on the charged drop to stop its descent. If the oil drop were negatively charged, which direction would the electric field have to point in order to stop the oil drop from falling?

14. Millikan (see Exercise **13**) observed that different oil drops would stop falling in different electric fields, an observation he attributed to their different electric charges. If the electric field needed to support one oil drop is twice as large as the field needed to support a second, similar oil drop, how do their electric charges compare with one another?

15. The electric field around an electrically charged hairbrush diminishes with distance from that hairbrush. Use Coulomb's law to explain this decrease in the magnitude of the field.

16. When a positively charged cloud passes overhead during a thunderstorm, which way does the electric field point?

17. The cloud in Exercise **16** attracts a large negative charge to the top of a tree in an open meadow. Why is the magnitude of the electric field larger on top of this tree than elsewhere in the meadow?

18. Corona discharges can occur wherever there is a very strong electric field. Why is there a strong electric field around a sharp point on an electrically charged metal object?

Section 11.3

Magnetically Levitated Trains

While jet airplanes still carry most travelers between distant cities, modern trains have eroded a jet's advantage for intermediate distances. Bullet trains are used in several countries, moving their passengers at about 200 km/h (125 mph). However, mechanical interactions between a train and its rails limit its maximum speed. The next generation of ultra-high-speed trains will probably fly above the rails on a cushion of magnetic force.

Questions to Think About: *What forces do a train's wheels exert on the rails and why should these forces become larger as the train travels faster? If two properly aligned magnets repel one another, how could they be arranged to support a train? How could you support a train with two magnets that attracted one another? How should the upward magnetic force on the train vary with height to keep the train hovering at a particular height? What would happen to the train if the force varied the wrong way?*

Experiments to Do: *Permanent magnets are wonderful toys. There is something fascinating about two objects that can attract or repel one another across centimeters of empty space. Play with two strong magnets to get a feel for how the forces between them work. Simple disk-shaped refrigerator magnets are best since some others, such as flexible strip magnets, have subtle complications. You should find that the two disks repel one another in one configuration but attract when you flip one disk around. The closer they are, the stronger the forces. Try to suspend one magnet above the other by repulsion. Did you succeed? Actually, it can't be done. Why not? Now try to suspend one magnet a centimeter or so below the other by attraction. Any luck this time? It can't be done, either. Evidently, magnetic suspension isn't so easy, after all.*

The Need for Magnetic Levitation

Trains operate more or less as they have since the early nineteenth century. Cars still roll on wheels, pulled forward by friction between locomotives and the rails beneath them. But while this scheme works well at speeds below about 160 km/h (100 mph), two serious problems appear at higher speeds.

First, a high-speed train has trouble following the rails because it has so little time to move up, down, left, or right. These sudden shifts in direction require rapid accelerations and involve large forces between the rails and the wheels. To minimize the forces and accelerations, the rails must be straight and exceedingly smooth, with only the most gradual curves. But even carefully constructed rails take a beating from the wheels and produce noise, vibration, and mechanical wear. Both the rails and the wheels must be replaced frequently so that, above about 320 km/h (200 mph), maintenance becomes prohibitively expensive.

Second, air resistance becomes so strong at high speeds that traction between the locomotive's wheels and the rails isn't enough to keep the train moving forward. The train can't maintain its cruising speed with friction alone and it can't stop quickly enough using friction, either. For safety reasons alone, conventional rail travel shouldn't exceed about 320 km/h (200 mph).

To overcome these limitations, a whole new technology for ultra-high-speed trains is being developed. Instead of running on metal wheels, these new trains are suspended above their tracks by magnetic forces. Because these magnetically levitated or *maglev* trains don't touch their tracks, those tracks don't have to be very smooth. The trains ride forward on magnetic cushions.

Conventional propulsion won't work for a maglev train because it doesn't have any wheels or friction. Instead, a maglev train uses a linear electric motor, in which magnets on the train push against magnets on the track. These magnets normally propel the train forward but they can also push the train backward to stop it during braking.

CHECK YOUR UNDERSTANDING #1: Bumper Cars

An otherwise straight track has a small bump in it. As it rolls over the bump, a train car experiences an upward force that exceeds its downward weight. How does this force affect the car's motion?

Supporting a Train with Magnetic Forces

Instead of rolling forward on wheels, a maglev train rides above its track on a magnetic suspension. Magnets in the train and track exert forces on one another and support the train without any direct contact. But what is magnetism and what are the forces between magnets?

Magnetism is a phenomenon that closely resembles electricity. Just as there are electric charges and electrostatic forces between them, so there are magnetic charges and **magnetostatic forces** between them. To distinguish magnetism from electricity, magnetic charges are called **magnetic poles**. There are two types of poles, north and south, and like poles repel while opposite poles attract. The forces between two poles grow weaker as they move apart and are inversely proportional to the square of the distance between them.

Since like poles repel one another, it might seem as though you could levitate a train simply by placing north poles in it and on the track. However this scheme will fail for two reasons. First, isolated magnetic poles don't seem to exist and second, even if they did, the train would soon fall off its magnetic cushion.

❐ With only a primary education, English chemist and physicist Michael Faraday (1791–1867) apprenticed with a bookbinder at 14. At 21, he became a laboratory assistant to Humphry Davy, a renown chemist. Faraday's experiments with electrochemistry and his knowledge of work by Oersted and Ampere led him to think that, if electricity can cause magnetism, then magnetism should be able to cause electricity. Through careful experimentation, he found just such an effect. Toward the end of his career, Faraday became a popular lecturer on science and made a particular effort to reach children. Faraday is discussed on p. 400.

❐ Self-educated before the French Revolution, during which his father was executed, French physicist André-Marie Ampère (1775–1836) became a science teacher in 1796. He served as a professor of physics or mathematics in several cities before settling at the University of Paris system in 1804. In 1820, only a week after learning of Oersted's experiment showing that an electric current causes a compass needle to deflect, Ampère published an extensive treatment of the subject. Evidently, he had been thinking about these ideas for a long time.

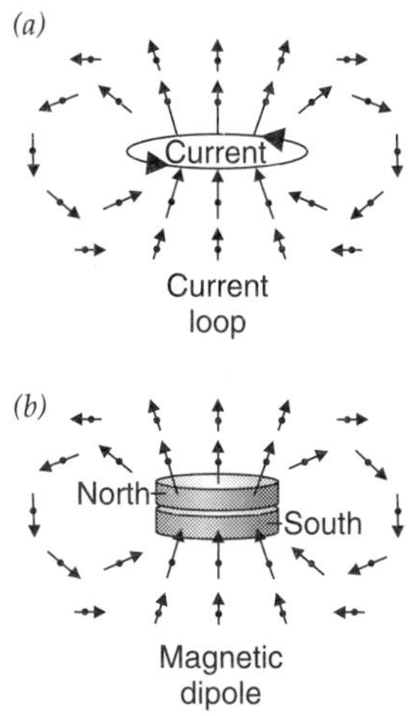

Fig. 11.3.1 - (*a*) The magnetic field around a loop of current-carrying wire points up through the loop and down around the outside of the loop. The arrow through each dot indicates the magnitude and direction of the force on a north pole at that location. (*b*) The field produced by two isolated magnetic poles, located one above the other.

Fig. 11.3.2 - These two disk magnets repel because their north poles are facing one another. However, the stick is needed to prevent the top disk from falling off the magnetic cushion that supports it.

The first problem points out an important difference between electricity and magnetism. While subatomic particles that carry a positive or negative electric charge are common, particles that carry a north or south magnetic pole have never been found. It's not even clear whether such **magnetic monopoles** exist in nature. But while isolated magnetic poles aren't available, pairs of magnetic poles are. These pairs consist of equal north and south poles, separated from one another in an arrangement called a **magnetic dipole**. Since the two opposite poles have equal magnitudes, they balance one another and the dipole has zero **net magnetic pole**.

These magnetic dipoles can't be made from isolated magnetic poles so they must be created by something else: electricity. Electricity and magnetism are intimately related, so that in addition to resembling one another, they can actually produce one another. Electric phenomena can produce magnetic phenomena and vice versa.

Electricity and magnetism are connected by their fields: electric fields and magnetic fields. Just as an electric field exerts a force on an electric charge, a **magnetic field** exerts a force on a magnetic pole. The magnetic field at a particular point in space indicates the magnitude and direction of the force that an isolated north pole would experience there (if there were an isolated north pole).

There are two electric phenomena that create magnetic fields: moving electric charge and an electric field that changes with time. The relationship between moving electric charge and magnetic fields was discovered by Danish physicist Hans Christian Oersted (1777–1851) in 1820 and studied by André-Marie Ampère ❐ over the next seven years. The connection between changing electric fields and magnetic fields was added in the 1865 by Scottish physicist James Clerk Maxwell (1831–1879), as part of his complete electromagnetic theory.

The most important source of magnetic fields for train levitation is moving electric charge, also called **electric current**. Electric currents are magnetic, meaning that whenever an electric current flows through a wire as "electricity" or leaps through the air as a spark, it produces a magnetic field. The structure of this magnetic field can be complicated, but there are a few special arrangements of current that produce simple magnetic fields.

One of these special arrangements is a loop of current. When current flows around a wire ring, it produces the symmetric magnetic field shown in Fig. 11.3.1*a*. This field is remarkably similar to one that would be produced by two opposite magnetic poles separated by a small distance (Fig. 11.3.1*b*). Unless you look at what's making the field, you can't tell them apart so that, even without isolated magnetic poles, it's possible to create dipole magnets.

A current loop dipole magnet behaves exactly like one made with real isolated magnetic poles. You can even imagine that the loop has a north pole on one side and a south pole on the other. If you take two of these magnets and arrange them so that their like poles are nearest, they repel. If their opposite poles are nearest, they attract. And if they are aligned at odd angles, they tend to twist one another around.

Even permanent magnets use currents to produce their magnetic fields. We noted in the previous section that electrons have an internal property called spin, and this spin is associated with angular momentum. In effect, electrons are spinning charged objects and behave like tiny loops of electric current. They're dipole magnets! In most materials, this magnetism is hidden because the electrons spin equally in all directions and their magnetic fields cancel one another. But there are a few magnetic materials that avoid this cancellation and can be used to make permanent magnets. A typical permanent magnet has a north pole at one end and a south pole at the other, both created by the spinning electrons inside.

The forces between dipole magnets can be surprisingly strong. If you cover the top of a track and the bottom of a train with dipole magnets, turned so that like poles face each other, the two will repel one another. This repulsion will become stronger as the train approaches the track and may even exceed the train's weight. When it does, the train will stop accelerating downward and can remain suspended over the track on a magnetic cushion. This is magnetic levitation.

Unfortunately, there are at least two problems with this simple levitation scheme. First, a track covered with magnets would attract magnetic materials such as iron and steel and make cleaning it a nightmare. But more importantly, a magnet suspended above another magnet is inherently unstable (Fig. 11.3.2). While it's possible to achieve dynamic stability (see Section 5.2) in a few special cases (Fig. 11.3.3), a train can't do this. Instead, the train will tend to slip sideways onto the ground, just as a marble sitting on top of a dome will tend to roll sideways onto the ground. The train may hover briefly on its magnetic cushion but it won't stay there for long without help.

Fig. 11.3.3 - This spinning magnetic top hovers above a repelling magnet, hidden in the wooden base. The top's dynamic stability involves gyroscopic effects that would be impractical for a maglev train.

CHECK YOUR UNDERSTANDING #2: Two Halves Make a Whole

You have a disk-shaped permanent magnet. The top surface is its north pole and the bottom surface its south pole. If you crack the magnet into two half-circles, the two halves will push apart. Why?

Stability and Feedback

A train supported by permanent magnets is in an unstable equilibrium. As we discussed in Section 5.1, an object that's in equilibrium experiences zero net force and doesn't accelerate. If the equilibrium is a stable one, the object will naturally return there after being disturbed because a restoring force will push it back (Fig. 11.3.4*a*). But when the equilibrium is unstable, the disturbed object won't return because it will begin to experience a force that pushes it away from the equilibrium (Fig. 11.3.4*b*).

There's a theorem in physics (Earnshaw's theorem) that states that no arrangement of electric charges can be in a stable equilibrium as the result of electrostatic forces alone. Similarly, no arrangement of magnetic poles can be in a stable equilibrium as the result of magnetostatic forces alone. Thus no matter how you arrange permanent magnets on the train and track, the train will be in an unstable equilibrium. While the repulsion between magnets can keep the train at a stable height (Fig. 11.3.5), the train's sideways motion is unstable. If the train isn't perfectly centered above the track, the repulsive forces will push it toward the side and it will fall off its magnetic cushion.

(*a*)

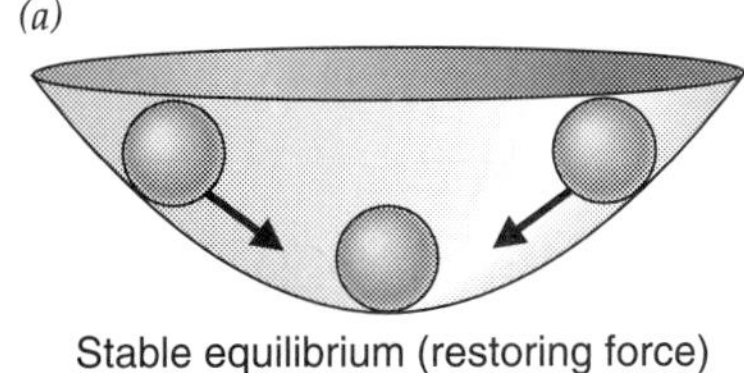

(*b*)

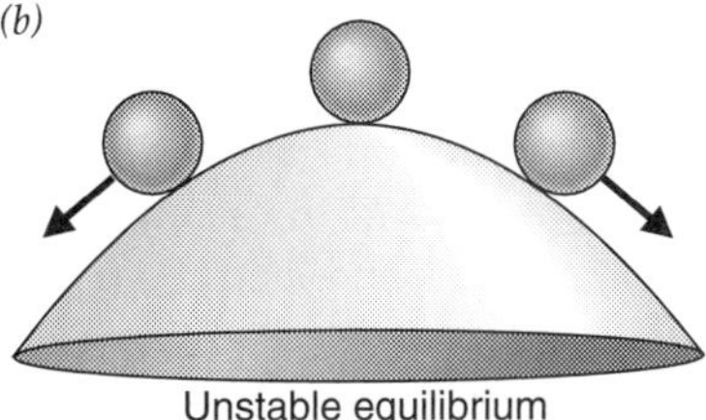

Fig. 11.3.4 - (*a*) A ball disturbed from a stable equilibrium position experiences a restoring force that pushes it back toward equilibrium. (*b*) A ball disturbed from an unstable equilibrium doesn't return.

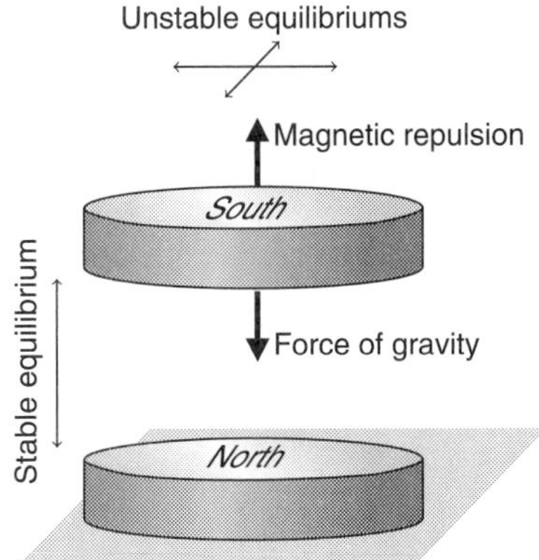

Fig. 11.3.5 - If the repulsion between permanent magnets is used to suspend a train above a track, the train's height will remain stable but its horizontal position will not. If the train isn't perfectly centered, the repulsive forces will push the train sideways until it falls.

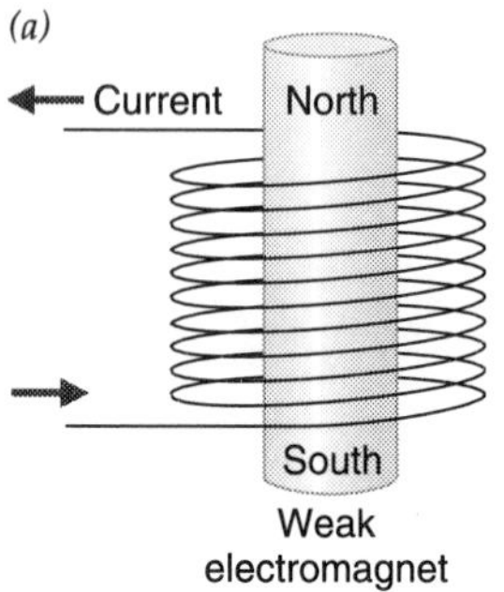

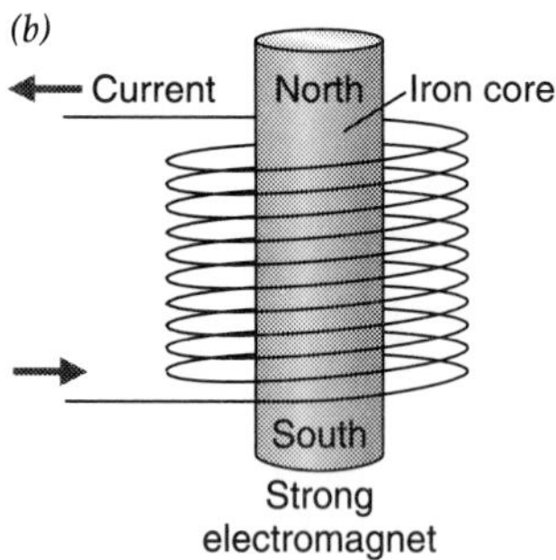

Fig. 11.3.6 - (*a*) When current flows through a wire coil, the coil becomes a weak dipole magnet. The direction of current flow through the coil determines which end is north and which end is south. (*b*) When iron is inserted into the coil, its spinning electrons align with the coil's magnetic field. The iron becomes a strong dipole magnet.

It might seem as if some clever arrangement of permanent magnets could make the train stable. However, there's no way around Earnshaw's theorem. To keep the train centered above the track, you have to push it there by turning some of the magnets on or off. Since magnetic fields are created by electric currents, you can turn a magnet off by stopping its current.

Because the currents inside a permanent magnet are hard to stop, you'll need electromagnets. An **electromagnet** is a coil of wire that's magnetic only when an electric current flows through it (Fig. 11.3.6). To make the electromagnet even stronger, its wire is usually wound around iron or another magnetic material. The iron's spinning electrons then align with and reinforce the coil's magnetic field. We'll discuss iron's magnetism more carefully in Section 12.5.

To keep the train stable, a control system must monitor the train's position and adjust the electromagnets accordingly. This technique of using information about the current situation to control the way in which the situation changes is called **feedback**. Feedback is used frequently in engineering to stabilize systems that are inherently unstable and to modify the ways in which systems respond to external stimuli (Fig. 11.3.7). Feedback can effectively create restoring forces where none exist naturally. Some maglev trains use feedback to stabilize their magnetic levitation (Fig. 11.3.8).

CHECK YOUR UNDERSTANDING #3: A Delicate Balance

Why is it much easier to balance a broom on your hand when you can look at the broom?

Fig. 11.3.7 - This ball and magnet are suspended in midair by the electromagnet above them. A control system uses light to measure the position of the ball. It then adjusts the strength of the electromagnet so as to keep the ball floating at a constant height.

Alternating Current Levitation

While feedback stabilization is used in some maglev trains, it requires very sophisticated control systems and nearly perfect trains and tracks. But there's another magnetic technique that can provide a restoring force in all directions and can levitate a train without any control system or feedback. Instead of putting permanent magnets or electromagnets on both the train and track, some trains use *electromagnetic induction* to make a conducting surface magnetic temporarily.

To understand induction, recall that there are two ways to produce a magnetic field: moving electric charge and an electric field that changes with time. We can make a similar observation about electric fields; there are two ways to produce an electric field: electric charge and a magnetic field that changes with time. We already knew that electric charge creates an electric field, but the changing magnetic field part is new. This effect, discovered in 1831 by Michael Faraday (see ❐ on p. 397), allows magnetism to create electricity. The sources of electric and magnetic fields are summarized in Table 11.3.1.

Table 11.3.1 - The sources of electric and magnetic fields.

Sources of Electric Fields	*Sources of Magnetic Fields*
Electric Charges	Moving Electric Charge
Changing Magnetic Fields	Changing Electric Fields

Knowing that changing magnetic fields create electric fields, think about what happens when a magnetic field changes inside an electric conductor. The changing magnetic field creates an electric field, which pushes on the conductor's mobile electric charges. These charges begin to move and form an electric current. This process, in which a changing magnetic field *induces* an electric current, is **electromagnetic induction**.

Since moving electric charge produces a magnetic field, this new electric current is itself magnetic. A changing magnetic field will turn an electric conductor into a magnet! Thus, if you hold an electromagnet near an electric conductor and turn it on, the magnetic field in the conductor will change and it will become magnetic. But how are the conductor's magnetic poles oriented?

The conductor becomes magnetic in such a way that it *repels* the electromagnet. For example, if the electromagnet turns on with its south pole nearest the conductor, then the conductor becomes magnetic with its south pole nearest the electromagnet (Fig. 11.3.9). Because the two magnetic fields point in opposite directions, they partially cancel one another. In effect, the induced current is opposing the increase in the electromagnet's magnetic field. In fact, currents produced by electromagnetic induction always flow in a direction that opposes the magnetic field change that produced them. This observation is called **Lenz's law**, after Estonian physicist Heinrich Emil Lenz (1804–1865) who discovered it.

Fig. 11.3.8 - This maglev train is supported by attractive forces between magnets located on the rail and magnets located in the train arms that reach beneath the rail. Feedback provides stability to this system.

LENZ'S LAW:
Current induced by a changing magnetic field always produces a magnetic field that opposes the change.

Because of Lenz's law, a light electromagnet can hover above a sheet of metal. When you turn the electromagnet on, it will repel the metal and rise above the surface. Remarkably, this form of magnetic levitation can be stable in all directions. For example, if the electromagnet is in a metal bowl, it will lift itself up and hover above the center of the bowl. If you bump it, it will experience restoring forces that return it to its equilibrium position.

Unfortunately, the metal's current and magnetic field gradually disappear as frictional effects slow the moving charged particles in the metal. To keep the electromagnet hovering above the metal, you must periodically reverse the electromagnet's poles. Each time you do, new currents will begin to flow in the metal's surface and these currents will repel the electromagnet.

You can reverse the poles of the electromagnet by reversing the direction in which current flows through its wires. An electric current that reverses directions periodically is called an **alternating current**, while a current that flows continuously in one direction is called a **direct current**. Sending an alternating current through the electromagnet, so that its poles reverse periodically, will keep it levitated indefinitely.

Scientists and engineers have tried to used *alternating current magnetic levitation* to suspend trains above U-shaped tracks. In some cases, they have put the electromagnets on the train and in others they have put the electromagnets on the track. But while both of these schemes work well, currents in the metal surfaces and the electromagnets encounter frictional effects that convert their electric energies into thermal energy. Because that electric energy must be replaced continuously, these maglev trains need too much power to be cost effective.

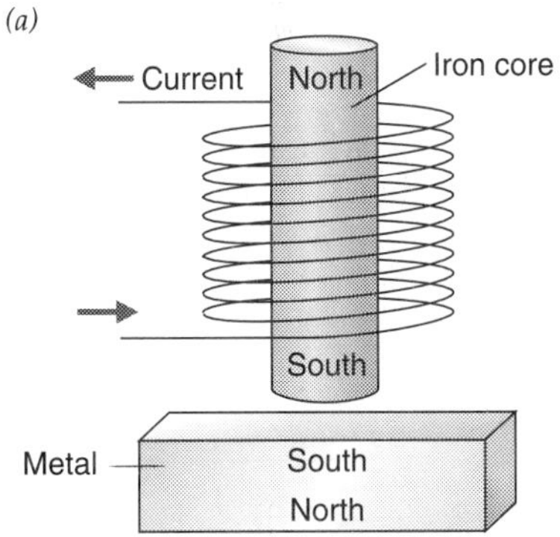

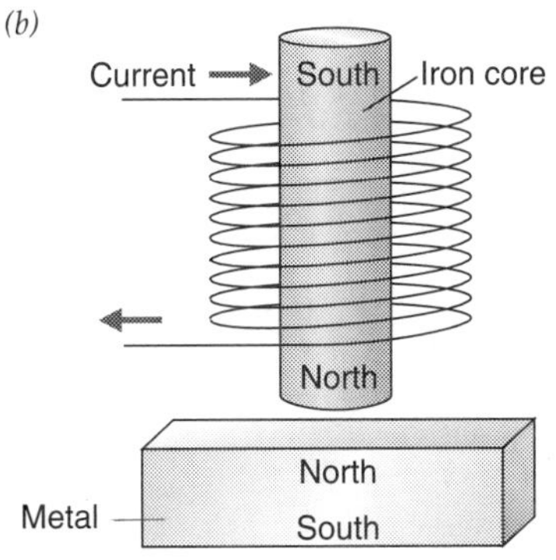

Fig. 11.3.9 - (*a*) When an electromagnet turns on with its south pole down, its increasing magnetic field induces a current in the metal below it. That metal becomes magnetic with its south pole up and repels the electromagnet. (*b*) If the current in the electromagnet is reversed, all the poles reverse and the metal still repels the electromagnet.

CHECK YOUR UNDERSTANDING #4: Field the Heat

Induction furnaces are common in industry. A piece of metal is placed near or inside of a strong electromagnet and the electric current through that electromagnet is reversed rapidly. The piece of metal becomes very hot. Why?

Electrodynamic Levitation

While the previous levitation schemes demand sophisticated control systems and excessive electric power, electrodynamic levitation is much more practical. In *electrodynamic levitation*, a train containing permanent magnets slides rapidly

(a)

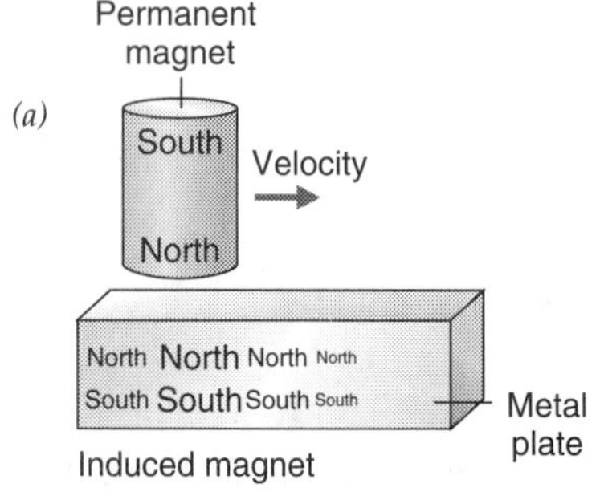

(b)

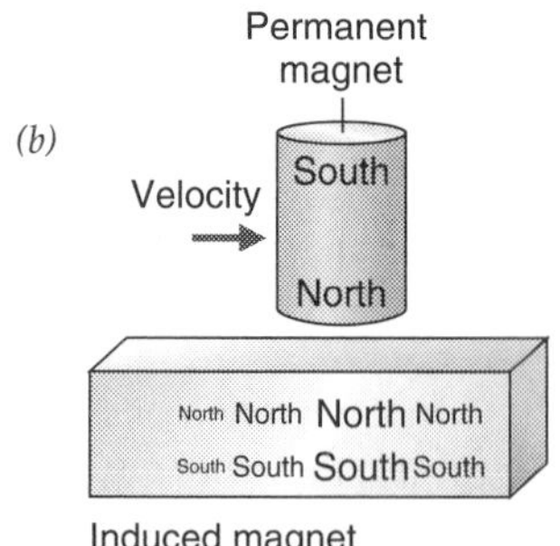

Fig. 11.3.10 - When a permanent magnet moves rapidly across a metal surface, it induces currents in that surface. The metal becomes magnetic and repels the moving magnet. The magnetized region of the metal surface moves with the permanent magnet.

Fig. 11.3.11 - This maglev train is supported by repulsive forces between superconducting magnets on the train and induced magnetic poles in the metal track.

across a metal track. As the magnets move by, the magnetic field in the track changes. Currents begin to flow through it and it becomes magnetic. The track's magnetic poles repel those of the train and the train rises on a cushion of magnetic force (Figs. 11.3.10 and 11.3.11).

Electrodynamic levitation works best when the train moves very fast. That way, the train's magnets pass quickly over each region of track. The magnetic poles that appear in that region don't just support the train; they also push it backward as it approaches and forward as it leaves. The forward push is particularly important, because without it the train would gradually slow down.

The strength of the forward push depends on the train's speed. If the train moves quickly, the track's magnetic poles will push the train forward when it leaves almost as strongly as they pushed it backward when it arrived. But if the train moves too slowly, the currents in the track will have plenty of time to decay and the track will lose much of its magnetism while the train is still nearby. Since the train will receive only a weak forward push as it leaves, it will slow down and some of its kinetic energy will become thermal energy in the track.

Even at high speeds, the train experiences a backward magnetic force. This force, called *magnetic drag*, is the sum of all the forward and backward magnetic forces on the train. Because the poles in the track ahead of the train's magnet are new and fresh, they are relatively strong. The poles in the track behind the train's magnet are weaker because they were created earlier and the currents that produce them have lost energy. With a stronger repulsion from the poles ahead of it than from those behind it, the train experiences a backward force.

This magnetic drag force is most severe at about 30 km/h (19 mph), the minimum speed at which the train can levitate, and diminishes with increasing speed. The faster the train moves, the less time there is for the poles in the track to lose energy and the more the forward and backward magnetic forces on the train balance one another. Above about 300 km/h (190 mph), magnetic drag is insignificant when compared to air resistance.

Of course, electrodynamic levitation can't support the train when it's stationary, so the train needs retractable "landing gear" to support it while it's starting or stopping. The train leaves the station on wheels, picks up speed, and then flies along on the magnetic cushion. It retracts its wheels during high-speed movement and lowers them again near the next station.

With suitable permanent magnets on board, an electrodynamically levitated train can hover more than 15 cm above an aluminum track. It easily clears small obstacles and provides an extremely smooth ride. Small imperfections in the railway can't be felt at all.

CHECK YOUR UNDERSTANDING #5: Magnetic Brakes

If you move a strong permanent magnet rapidly across a sheet of copper or aluminum, a few millimeters above the surface, you will feel a significant force opposing the magnet's motion. What sort of force is this?

Superconducting Magnets

While permanent magnets or normal electromagnets could be used for electrodynamic levitation, most maglev trains use *superconducting electromagnets*. These special electromagnets are light, relatively inexpensive, and create the intense magnetic fields needed to suspend a train well above its track. Superconducting electromagnets are based on exotic materials called **superconductors** that carry current without energy loss. While currents in normal metals experience friction-like effects and gradually slow down, currents in superconductors flow perfectly freely. With nothing to slow its motion, any current you start in a superconductor will continue to flow until you stop it. Since currents are magnetic, the superconductor will remain magnetic as long as you allow the current to flow in it.

Superconductivity is a manifestation of quantum physics and is based on the fact that charged particles in solids can only travel in allowed trajectories or levels. While the electrons in a normal metal can collide with the metal atoms and shift from one level to another, the electrons in a superconductor are prevented from shifting by a peculiar quantum effect. Confined to their original levels, the electrons in a superconductor can't collide with the atoms and must travel endlessly along fixed trajectories. If more of them are heading in one direction than the other, they are carrying current and will continue to carry that current forever.

Fig. 11.3.12 - A magnetic cube floats above the surface of a high temperature superconductor. Currents flowing in the superconductor make it magnetic and cause it to repel the magnetic cube.

Superconductivity appears in a fair number of metals, but only when they're cooled to extremely low temperatures. At only a few degrees above absolute zero, thermal energy upsets the order needed for superconductivity in most materials. But there are also some special ceramics that show superconductivity at temperatures of 125° K (–148° C) or more. Unfortunately, these recently discovered *high-temperature superconductors* have proven difficult to use in practical devices, so the old low-temperature ones are still used in most applications.

One such application is superconducting electromagnets. While a normal electromagnet needs a steady supply of electric power to keep current flowing through it, an electromagnet that uses superconducting wires can operate forever without electric power. All you have to do is get the current started in the superconducting electromagnet and it will flow all by itself for years.

In a real superconducting magnet, a coil of superconducting wire is initially connected to an external power supply that starts the current moving. When enough current is flowing through the wire, the two open ends of the coil are joined by a superconducting link to form an endless loop, superconducting throughout. The current flows around and around the coil forever and the power supply is no longer needed. Although you can't see this current directly, you can easily detect its magnetic field (Fig. 11.3.12).

Superconducting magnets are perfect for maglev trains that use electrodynamic levitation. Currents are introduced into the train's superconducting magnets while it's in the maintenance building. Once the superconducting links have been closed, the magnets need no additional electric power and remain magnetic as long as they're kept cold. The train has a small refrigeration unit to keep its magnets cold but requires no other on-board power for its suspension.

CHECK YOUR UNDERSTANDING #6: Batteries Not Included (or Needed)

Some superconducting magnets are shipped from the factory with currents flowing through their coils. Why doesn't the current stop by the time one of these magnets arrives?

(a)

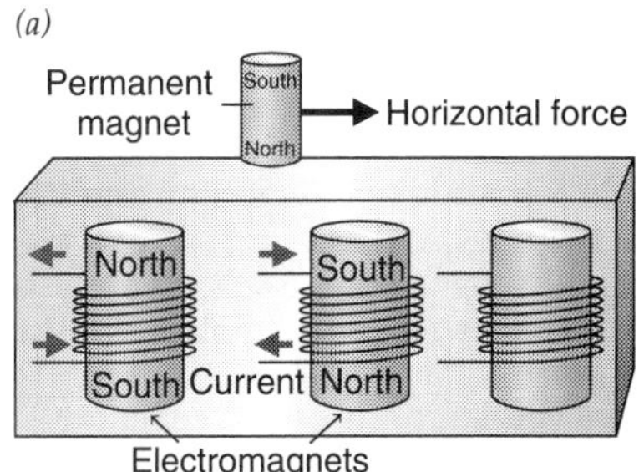

(b)

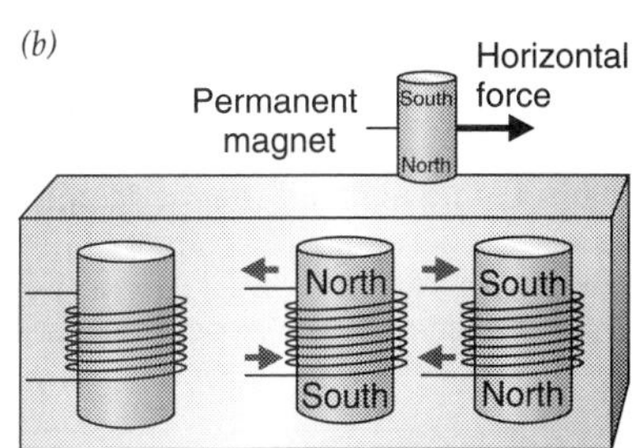

Fig. 11.3.13 - A linear motor works by turning electromagnets on and off at the proper times. The moving permanent magnet always experiences a horizontal force toward the right. In *(a)*, the left pair of electromagnets is active and in *(b)*, the right pair is active. The poles of the middle magnet are reversed by changing the direction of current through its coils.

Propulsion

However, the train still needs something to propel it forward. While a propeller or jet engine would do, most maglev trains use another novel magnetic concept: a linear motor. A linear motor is fairly complicated in practice, but the concept is simple. It uses attractive or repulsive forces between electromagnets on the track and train to push the train forward. These magnets are turned on and off at appropriate times so that the force on the train is always forward (Fig. 11.3.13).

In addition to pushing the train forward, the linear motor serves as a braking device. Changing the operation of the electromagnets can reverse the direction of the horizontal force on the train. The linear motor can stop the train much more effectively than sliding friction can. As we'll see when we examine electric motors, the linear motor converts electric energy into kinetic energy as the train accelerates forward and converts kinetic energy into electric energy as the train decelerates to a stop.

CHECK YOUR UNDERSTANDING #7: Heading Forward

Why can't a linear motor be built entirely out of permanent magnets?

Summary

How Magnetically Levitated Trains Work: The most promising levitation scheme is electrodynamic levitation. Strong, superconducting magnets on the train create large magnetic fields beneath it. As the train moves along its metal track, these magnets induce currents in the track and the track becomes magnetic. The poles of the track are opposed to those of the train's magnets and the train and track push against one another. This magnetic repulsion suspends the train above the track at a distance of 15 cm or more. To keep the train moving against air resistance and magnetic drag, the train employs a linear motor. This motor uses electromagnets on the train and track to pull and push the train forward. The linear motor can also slow the train.

Important Laws and Equations

1. Lenz's Law: Current induced by a changing magnetic field produces a magnetic field that opposes the change.

The Physics of Magnetically Levitated Trains

1. There are two types of magnetic poles, north and south.

2. Magnetic poles exert magnetostatic forces on one another. Like poles repel one another while opposite charges attract.

3. The strength of the magnetostatic force on each of two magnetic poles depends on the magnitude of each pole and the distance separating them.

4. Isolated magnetic poles don't exist in nature. However, magnetic dipoles (equal north and south poles separated by a distance) can be created by moving electric charges. Because there are no isolated magnetic poles, the net pole of an object is always zero.

5. A magnetic pole experiences a force in the presence of a magnetic field. The magnetic field points in the direction of the force that a north pole would experience at that location and the field's magnitude indicates the strength of that force.

6. Two things can produce a magnetic field: moving electric charge (a current) and a changing electric field.

7. Two things can produce an electric field: electric charge and a changing magnetic field.

8. A changing magnetic field induces an electric current in a conductor. This current produces a magnetic field that opposes the original change in field.

9. A superconductor is capable of sustaining an electric current without loss of energy as thermal energy.

Check Your Understanding - Answers

1. The car accelerates upward.

Why: By giving the car a net upward force, the track lifts the car over the bump. The time during which this lifting can be done depends on the car's speed. Doubling this speed halves the lifting time and actually involves four times the acceleration and four times the upward force from the track. That's because the car must move upward twice as fast as before, with only half the time to acquire the doubled upward speed. These rapidly increasing forces limit conventional trains to about 320 km/h (200 mph).

2. The top surfaces of the both halves are still north poles and the bottom surfaces are still south poles. The two tops repel, as do the two bottoms.

Why: This puzzling phenomenon, where a shattered permanent magnet opposes attempts to reassemble it, is an illustration of the potential energy contained in a permanent magnet. The magnet is a collection of many tinier magnets, all aligned with their north poles together and their south poles together. Like poles repel one another, so the tiny magnets are difficult to hold together. Given a chance, the magnet will push apart into fragments. Very strong permanent magnets release so much potential energy when they break and push apart that they can be dangerous.

3. A balanced broom is in an unstable equilibrium and you use visual feedback to keep it balanced.

Why: You need to know where the broom's center of gravity is located in order to keep it balanced. On the average, your hand has to be located beneath that center of gravity. Although you can feel the motion of the broom to some extent, it helps to be able to see it.

4. The changing magnetic field of the electromagnet induces currents in the piece of metal. As these currents decay, some of their energy is converted to thermal energy and the metal becomes very hot.

Why: In the presence of changing magnetic fields, a piece of metal has currents flowing through it. These currents can warm the metal just as they warm the filament of a light bulb.

5. Magnetic drag force.

Why: The moving magnet induces currents in the metal sheet and makes it magnetic. The poles of the sheet are opposite to those of the moving magnet, so there is a repulsion between the sheet and magnet. However, there is also a slowing force on the magnet. The strongest poles of the sheet are located ahead of the moving magnet and oppose its approach. The sheet pushes backward on the magnet and tries to slow it down.

6. Since the coil is superconducting, there is no way for the current to lose energy and come to a stop. It keeps on moving.

Why: While electric currents in normal conductors convert electric energy into thermal energy, those in superconductors do not. Once they are flowing around the coil, the electrons in a superconductor continue to flow forever. The magnetic field they produce travels along with the magnet, from factory to installation. The user doesn't need to own the equipment for initiating the current.

7. Each track magnet must reverse its poles as the train magnet passes over it in order to attract the train as it approaches and repel the train as it moves away.

Why: A permanent magnet can't reverse its poles. It will treat the passing train magnet equally as it approaches and leaves. Overall, it will do no work on the passing train. But an electromagnet can reverse its poles and do work on the passing train. The electromagnet pulls the train magnet forward until the magnets are right on top of one another. Then the electromagnet reverses its poles and pushes the train magnet forward.

Glossary

alternating current An electric current that periodically reverses its direction of flow.

direct current An electric current that always flows in one direction.

electric current The movement or flow of electric charge.

electromagnet A coil of wire, with or without an iron core, that becomes a magnet when an electric current flows around the coil.

electromagnetic induction The process whereby a changing magnetic field induces a current to flow in a conductor.

feedback The process of using information about a system's current situation to control changes you are making in that system.

Lenz's law The tendency of induced magnetic poles in a conductor to oppose the magnetic poles of the magnet that induced them.

magnetic dipole A pair of equal but opposite poles separated by a distance.

magnetic field An attribute of each point in space that exerts forces on magnetic poles. A magnetic field has a magnitude and direction proportional to the force it would exert on a unit of north magnetic pole at that location.

magnetic monopole An isolated magnetic pole, either north or south. Though there is speculation that they should exist, none has ever been observed.

magnetic pole A property of nature that gives rise to magnetostatic forces between magnetic poles. There are two types of poles, north and south.

magnetostatic force The force experienced by a magnetic pole in the presence of other magnetic poles.

net magnetic pole The sum of all poles on an object, both north and south. Since there are no isolated magnetic poles, an object's net magnetic pole is always zero.

superconductor An electric conductor that permits electrons to flow without losing any of their kinetic energy to thermal energy. Electrons will continue to flow in a superconductor indefinitely. Materials only become superconducting at extremely low temperatures.

Review Questions

1. How do we know there are two types of magnetic poles?

2. Why do magnets all have both north and south magnetic poles? Why don't some of them simply have north poles?

3. What's the difference between a magnetic pole and a magnetic dipole?

4. Which of the following items involve an electric current: a motionless ball with negative charge on its surface, a neutral ball that's falling, a negatively charged ball that's falling, a spinning household battery with positive charge at one end and negative charge at the other, and a loop of wire with electric charge moving through it.

5. If we think of an electron as a spinning, charge covered sphere, why should we expect it to be magnetic?

6. List five examples of objects that are in unstable equilibriums. What happens when each of these objects is disturbed?

7. What makes an electromagnet magnetic?

8. How can you use a permanent magnet to cause current to flow through a wire or a piece of metal?

9. An electromagnet can hover above a metal surface, supported by magnetic forces, but only if the current through the electromagnet changes direction periodically. Why must the current reverse?

10. Explain the origin of the upward force a magnet experiences as its travels rapidly across a horizontal metal surface.

11. Why does a magnet experience a backward force as it travels rapidly across a horizontal metal surface?

Exercises

1. Speed bumps are ridges that are put into streets to keep cars from driving too fast. Why does a speed bump affect a fast moving car so much more than a slow moving one?

2. Signs often caution you to slow your car down before you come to a sharp turn. That's because your car is more likely to skid during the turn when it's moving fast. Why is skidding more likely when you make a turn at high speed?

3. The faster a car moves through the air, the stronger the backward drag force it experiences. Friction between the tires and the ground must push the car forward to balance that wind resistance. Why will a car's tires begin to skid when its driver tries to make it exceed a certain speed?

4. To increase the maximum speed mentioned in Exercise **3**, some racing cars have wings that deflect the air upward as it passes over their powered rear wheels. As it deflects, the air pushes the rear wheels more tightly against the road. Why does this procedure help keep the rear wheels from skidding?

5. Trying to suspend one permanent magnet above another by facing their north poles toward one another is like trying to balance a pencil on its point. Use the concept of an unstable equilibrium to explain why this comparison makes sense.

6. You could open a mail order company that sold bottles of pure positive electric charges. You wouldn't get much business, but at least you'd be legal. If you tried to open a mail order company that sold bottles of pure north magnetic poles, you'd be subject to arrest for fraud. Why?

7. An inventor shows you a strange, rope-like object that supposedly has only a north magnetic pole. This pole is located at the object's end and your compass shows that this pole is there. However, you are certain that somewhere in the rope-like body of the object is a south pole of equal strength. Using your compass, you show that there are indeed south poles located at various places along its length and that these south poles together exactly balance the north pole. Why must those south poles be present?

8. Some decorative light bulbs have a loop-shaped filament that jitters back and forth near a small permanent magnet. When electric charge moves through the filament, it's attracted to or repelled by the magnet. The filament wire itself isn't magnetic, so why does the permanent magnet push or pull on the filament when the light is turned on?

9. A household battery has positive charge on its positive terminal and negative charge on its negative terminal. If you spin the battery very quickly about its center of mass, will it become magnetic?

10. You can probably balance a long pole on your hand, but only if you're allowed to watch it. If you close your eyes it's certain to fall. Why?

11. As you move the north pole of a permanent magnet toward an aluminum baking tray, the tray temporarily becomes magnetic. A pole appears at its surface, facing the approaching north pole. Is the tray's pole north or south?

12. The tray in Exercise **11** also develops a pole at its far surface, away from the approaching north pole. Is that second pole of the tray north or south?

13. Suppose you have a coil of wire that's connected to a source of alternating current. Current flows through the coil first one way and then the other, over and over again. Because the current reverses periodically, north and south poles appear alternately at one end of this coil. If you approach that end with a piece of aluminum, it will be repelled. Why?

14. One of the ways in which a coin operated vending machine checks to make sure that the coins you feed it are genuine is to roll them past a strong magnet. Why do coins made of good electric conductors such as copper slow down as they pass the magnet?

15. The needle of an automobile speedometer is connected to a spring and a small copper disk. The spring pulls the needle toward zero but a torque on the disk can pull the needle toward higher speeds. That torque is exerted on the disk by a spinning magnet near the disk's surface. The magnet spins with the car wheels so that the faster the car goes, the faster the magnet spins and the higher the speedometer reads. Why does the moving magnet exert a torque on the copper disk?

16. When you move a strong permanent magnet rapidly to the right just above the surface of a stationary aluminum table, what forces will that magnet experience?

17. Two superconducting magnets have recently been installed in a large hospital. They look identical, each consisting of a coil of superconducting wire that closes on itself to form a continuous loop. But at the moment, one magnet has a strong magnetic field around it while the other has no field at all. What is the difference between the two magnets?

18. If you move a permanent magnet across a piece of copper, you will induce electric currents in that copper. If you try the same with a piece of glass, no currents will flow. Why don't currents flow in the glass as the magnet moves by?

19. When you hold the north poles of two bar magnets together, the two magnets repel. But each magnet also has a south pole that is attracted toward the north pole of the other magnet. Why don't these attractive and repulsive forces balance one another so that each magnet experiences zero net force?

Epilogue for Chapter 11

This chapter dealt with three devices that depend on electric and magnetic forces for their operation. In *air purification,* we looked at how drag forces impede the removal of small particles from air. We introduced the concept of electric charge and discussed the attractive or repulsive forces that charged particles exert on one another. We learned that electric charge is quantized, coming in integer multiples of the elementary unit of electric charge, and found out how the distance separating two charges affects the forces they exert on one another.

In *xerographic copiers,* we saw that the Pauli exclusion principle makes it impossible for electrons to move through a photoconductor in the absence of light. We found that light energy can shift these electrons from filled valence levels to empty conduction levels so that charge can move through the photoconductor. We also examined electric fields as another way to look at the forces charged particles exert on one another: one charged particle creates an electric field and the second charged particle accelerates under its influence.

In *magnetically levitated trains,* we looked at the forces between magnetic poles. We discussed the apparent absence of isolated magnetic poles or monopoles from our universe and explained how magnetic fields and dipoles can be created by currents of moving electric charges. We found that electric currents slow down in normal metals but not in superconductors.

Explanation: Moving Water Without Touching It

The experiment uses sliding friction to give the comb an electric charge. Electrons rub off your hair and onto the rubber or plastic comb, leaving your hair positively charged and the comb negatively charged. Because the comb is an electric insulator, its negative charge remains trapped on its surface for a long time.

As you hold the negatively charged comb close to the water stream, the comb attracts positive charges present in the water and repels negative charges. Because the water conducts electricity somewhat, negative charges flow up the water stream and into the faucet. The positive charges are left behind in the water stream and give that water an overall positive charge. Thus the presence of the negatively charged comb makes the water stream positively charged and the water and comb attract one another. The water accelerates toward the comb and forms a graceful arc as it falls.

Cases

1. Lightning occurs when large amounts of electric charge flow between the clouds and the ground.

a. Lightning is most likely to strike when a cloud and the ground beneath it have accumulated large opposite charges. Why is lightning unlikely when they have large like charges?
b. The atmosphere does a great deal of work in creating the separated charge that produces lightning. Show that it takes work to move positive charge from the negatively charged ground to the positively charged cloud overhead.
c. When enough opposite charge has accumulated on the ground and in the cloud above it, lightning will strike. One indication that this dangerous charge accumulation has occurred is that your hair begins to stand up. Explain this effect.
d. A sharp lightning rod reduces any local buildup of electric charge and prevents nearby lightning strikes. How does the lightning rod get rid of local electric charge and allow it to flow gradually to the clouds overhead?

2. A Van der Graaf generator is an electrostatic device that uses a moving non-conductive belt to carry electric charge into a hollow metal sphere. This sphere is insulated from the ground and can accumulate charge until enormous voltages are reached. Small Van der Graaf generators are exciting novelties while large ones are used in research and industry.

a. A typical Van der Graaf generator uses a rubber belt to carry negatively charged electrons from its base to the sphere. As the belt passes through the sphere, it's touched by a metal brush. Electrons leave the belt and flow through a wire to the surrounding sphere. Why do they flow onto the outside of the sphere rather than staying near the belt?
b. As negative charge builds up on the sphere, the belt motor begins to strain. What makes it progressively more difficult for the motor to move the belt?
c. The amount of charge that can build up on the sphere depends on the sphere's surface. Why is it important that

there be no sharp points on the sphere?

d. As you move your hand close to the negatively charged sphere, your hand becomes positively charged and a spark may leap from the sphere to your hand. Why does your hand acquire a positive charge and where does it come from?

e. If you insulate yourself from the ground and put your hand on the sphere, negative charge will accumulate on you as well as on the sphere. Explain why your hair stands up.

3. As freshly laundered clothes tumble about in a hot dryer they rub against one another. Sliding friction transfers electric charges from one piece of clothing to the other so that some items become positively charged and others negatively charged. These charge accumulations produce static cling. Fabric softener is a soap-like chemical that binds to wet fabric fibers and lubricates them. This lubrication is what softens the clothes. But fabric softener also attracts moisture, which in turn reduces static cling.

a. As you unload the dryer you find several socks clinging to a shirt. How do the charges on the socks and shirt compare to one another?

b. As you pull the socks off the shirt, you do work on the socks. What form is your energy being transformed into?

c. Suppose that your socks are covered with positive charge. As you pull them away from the shirt, what happens to the voltage of the charges on the socks?

d. The farther each sock is from the shirt, the less that sock and shirt attract one another. Explain.

e. After removing them from the shirt, you find that the socks repel one another. Explain.

f. The next time you do your clothes you decide to add fabric softener. Your clothes emerge from the dryer with a thin layer of moisture on them and this moisture permits charge to move freely about the clothes. Why does this mobility prevent the build-up of separated charge?

4. The spark lighters used in gas stoves and grills are based on small piezoelectric crystals. When you compress a piezoelectric crystal, it produces a charge separation. One side becomes positively charged while the other becomes negatively charged. In a typical spark lighter, a spring-loaded mass strikes the crystal and produces a huge charge separation. Wires attached to the crystal allow these charges to approach one another across an air gap and create a spark.

a. Separating charge takes energy. Show that the spring-load mass does work on the crystal while compressing it.

b. Why does the amount of electric charge on each side of the crystal affect the likelihood that a spark will jump from one wire to the other?

c. To encourage a spark, the wire ends that make up the air gap are usually thin and pointed. Why do sharp points make it easier for charge to begin moving through the air?

d. If the two wires become separated by too great a distance, the spark lighter will stop working. Why does the distance separating the wires matter at all?

5. A nerve cell is an electrostatic device that operates very differently from a wire. It's a fluid-filled tube that's surrounded by another fluid. Both fluids contain ions (electrically charged atoms) of sodium, potassium, and chlorine. Each sodium or potassium ion has one elementary unit of positive charge while each chlorine ion has one elementary unit of negative charge. The wall of the nerve cell separates the two fluids and normally prevents these ions from passing through it.

a. When the nerve is in its resting state, the fluid outside it has slightly more sodium plus potassium ions than chlorine ions. The fluid inside it has slightly more chlorine ions than sodium plus potassium ions. What is the net electric charge of each of these two fluids?

b. How do you know that the resting nerve has electrostatic potential energy?

c. To produce its electrostatic potential energy, the nerve pumps sodium ions out of the nerve cell. Show that the nerve must do work on the sodium ions during this transfer.

d. Which part of the nerve has a positive voltage? Which part has a negative voltage?

e. When the nerve cell "fires," it abruptly allows sodium ions to pass through its walls. Which way do they move and what happens to the voltages of the various parts of the cell?

6. Your local market has an electric eye that rings a bell as you walk in the door and block a light beam. The light beam is emitted at one side of the door and normally strikes a small photoconductor on the other side of the door.

a. When the light beam is turned off or blocked, no light strikes the photoconductor. A battery pumps negative charges onto one side of the photoconductor and positive charges onto the other. These opposite charges exert attractive forces on one another so why don't they move together?

b. Because the photoconductor is in the dark, the battery soon stops sending charges to it. What makes it stop?

c. When the light beam reaches the photoconductor, charges can move through it. What has happened inside the photoconductor that allows it to conduct electricity?

d. When the photoconductor is exposed to light, the battery can continue to send charges to it. How has light made it possible for the battery to keep sending charges?

e. You bend down to play with the electric eye. You block the light beam with your back and shine your own flashlight onto the photoconductor. The bell turns off, as though you were out of the way. But when you shine red light from your bicycle taillight onto the photoconductor, it doesn't respond. Why doesn't red light (which has relatively low energy photons) affect the photoconductor?

7. A magnetic resonance imaging (MRI) machine uses an enormous and extremely strong magnet to study a patient's body. The magnet, which has its north pole at the patient's head and its south pole at the patient's feet, is actually a coil of superconducting wire through which electric charges flow.

a. This fancy electric system seems unnecessary; why can't the technicians simply put a large number of north magnetic poles near the patient's head and an equal number of south magnetic poles near the patient's feet?

b. The needle of your magnetic compass has its north magnetic pole painted red and its south pole painted white. If you stand a few meters from the MRI machine, at the end where the patient's head is, why does the white end of the compass turn toward the patient's head?

c. The compass is a magnetic dipole, with no net magnetic pole. So why do you feel it pulled toward the patient's head more and more strongly as you get closer to the magnet?

d. Aluminum isn't normally magnetic, but as you carry a large aluminum tray toward the magnet, you find that the magnet repels the aluminum. Explain.

e. You eventually manage to get the aluminum tray up to the magnet. As long as the tray doesn't move, it experiences no magnetic forces. But when you drop it, it falls past the magnet remarkably slowly. What slows down its fall?

CHAPTER 12

ELECTRODYNAMICS

Electricity and magnetism are related to one another in a great many ways, particularly when motion is involved. For example, moving electric charges give rise to magnetism and moving magnetic poles give rise to electricity. In this chapter, we'll examine several objects that make use of the relationships between electricity, magnetism, and motion to perform useful tasks. The general name for the effects arising from moving electric charges is electrodynamics. Since many of the magnetic poles we encounter are created by moving electric charges, electrodynamics also covers the effects arising from moving magnetic poles.

Experiment: A Nail and Wire Electromagnet

As an example of the relationship between electricity and magnetism, here is how to build a simple electromagnet. For this project, you'll need a large steel nail or bolt, a meter or so of insulated wire, a fresh 1.5 volt AA battery, and some small steel objects such as paper clips. The wire's metal conductor should be at least 0.4 mm in diameter so that it can carry the current you'll send through it without becoming too hot.

> **WARNING:**
> The electromagnet that you'll construct in this experiment **will become hot during use**. Be prepared to drop the electromagnet if it becomes uncomfortably hot. **Don't work near flammable materials.**

Wind the wire tightly around the bolt to form a coil. You should complete more than 50 turns of wire, all in the same direction. The exact number of turns isn't important and you can make several layers. Be sure that the two ends of the wire are still accessible and remove the insulation from each end so that you can connect them to the battery.

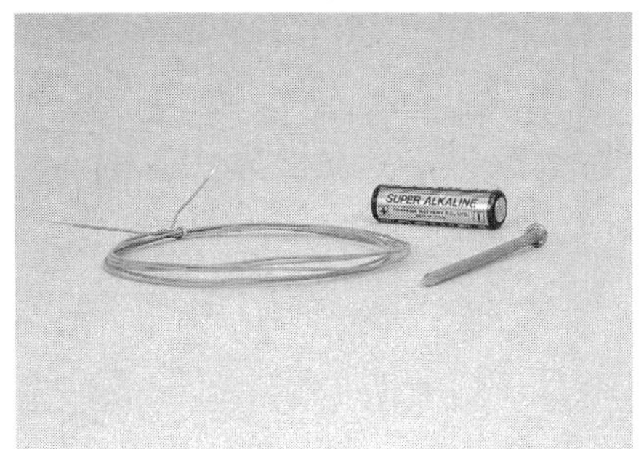

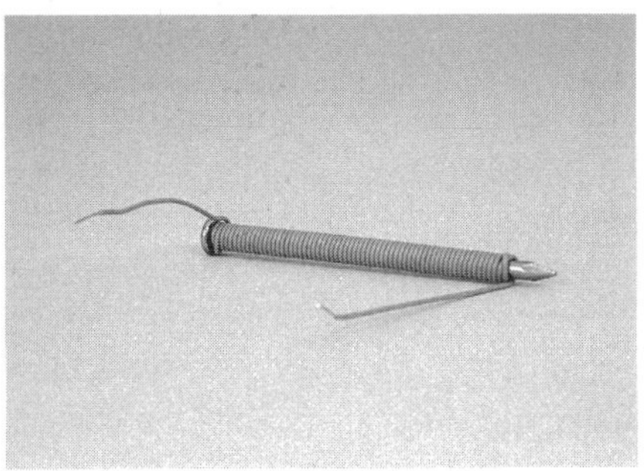

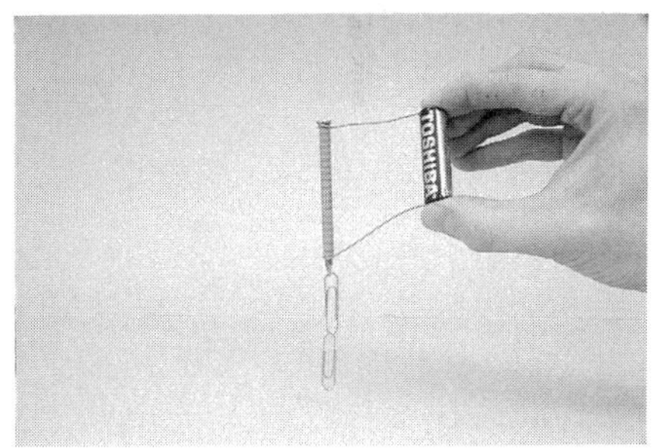

Now test your electromagnet. Connect one uninsulated end of the coil to each terminal of the battery. You can either hold the wires on the terminals with your fingers or use tape. A 1.5 volt battery can't give you a shock unless your skin is broken, but you should be prepared for the wire to get hot as current flows through it. If it gets too hot to hold, let go and make sure that the wire detaches from the battery so that it doesn't start a fire. Don't use a battery larger than AA or the wire may get dangerously hot.

While current is flowing through the wire, the nail will act as a strong magnet, an electromagnet. Try picking up steel objects with this electromagnet. As you touch each object with the nail, it should stick to the nail's surface. Your electromagnet will temporarily magnetize the steel object and attract it. Try to **predict** what will happen when you touch this magnetized steel object to a second object. **Observe** what happens and see if it **verifies** your prediction. Can you think of a way to **measure** how strong the magnet is? What will happen when you stop the flow of electric current through the coil of the electromagnet? Why does the coil get hot while current is flowing through it?

Chapter Itinerary

This chapter will examine (1) *flashlights*, (2) *electric power distribution*, (3) *electric power generation*, (4) *electric motors*, and (5) *tape recorders*. In *flashlights*, we'll look at how a current of electric charges carries power from several batteries to a light bulb. In *electric power distribution*, we'll see how electricity and magnetism are used to transport electric power from power plants to cities far away. In *electric power generation*, we'll study the mechanisms used to produce electric power, both from fossil fuels and from renewable resources such as sunlight and wind. In *electric motors*, we'll consider the ways in which these practical devices turn electric power into mechanical power. And in *tape recorders*, we'll explore the ways in which the magnetic structure of certain materials makes it possible to store sound and video information on strips of specially coated plastic. While there are many other electromagnetic objects that we encounter daily, from door bell buzzers to light switches to electric stoves, these five topics are representative of most of the basic electromagnetic phenomena.

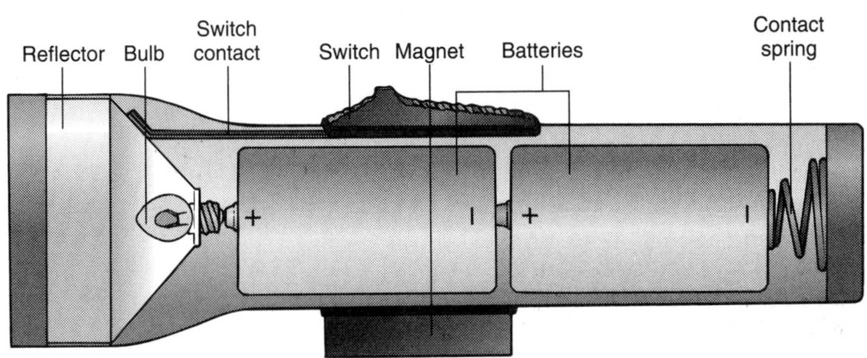

Section 12.1

Flashlights

There isn't much to a typical flashlight: you flip a switch and out comes a beam of light. The flashlight has only a few parts, all of which are visible when you open it to replace its batteries. But a flashlight is not a mechanical device, it's an electric one. The flashlight contains a complete electric circuit and most of its parts are involved in the flow of electricity. To understand how flashlights work, you must understand how an electric circuit works. That understanding in turn requires an understanding of electricity itself. Thus despite its simple appearance, a flashlight is a complicated device after all.

Questions to Think About: *Why are some flashlights brighter than others? Why is it important that all of the batteries point in the same direction? What is the difference between old batteries and new? What makes a flashlight suddenly become dim or bright when you shake it?*

Experiments to Do: *Find a flashlight that uses two or more removable batteries. Turn it on. What did the switch do to make the flashlight produce light? With the flashlight turned on, slowly open the battery compartment. The lamp will probably become dark. You should be able to turn the flashlight on and off by opening and closing the battery compartment. Why does this method work?*

Replace the flashlight's batteries with older or newer ones and compare its brightness. Turn one or more of the batteries around backward and see how that change affects the flashlight. What happens when you put a piece of paper or tape between two of the batteries? What happens when you carefully clean the metal surfaces of each battery with a pencil eraser before putting them in the flashlight?

Electricity and the Flashlight's Electric Circuit

A basic flashlight has only four components: a battery, a bulb, a switch, and a metal strip that connects them all together. The battery transfers energy to the bulb, which then turns it into light. The switch controls the energy transfer so that it only occurs when the flashlight is on. But what process carries the energy from battery to bulb and how does the switch start or stop that energy transfer? To answer these questions, we must first understand electricity and electric circuits, so that's where we'll begin.

In the previous chapter, we encountered one form of electricity: static electricity. That term refers to any distribution of stationary charged particles, such as the charge sitting on a rubber comb after you pass it through your hair. But there's a second form of electricity that involves moving charged particles: **dynamic electricity**. It's this form of electricity that powers flashlights, stereos, and microwave ovens. Since most of our discussions in this and subsequent chapters will focus only on the dynamic form of electricity, we'll leave out the word "dynamic" and just call it "electricity."

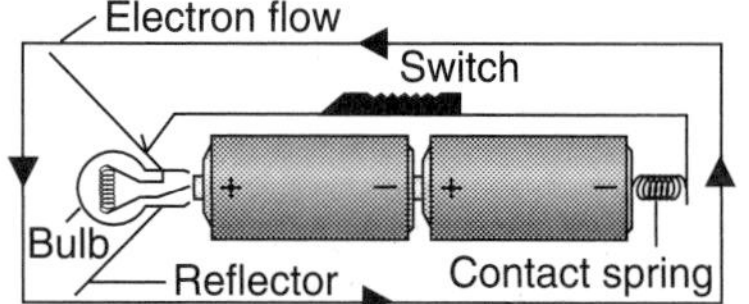

Fig. 12.1.1 - A flashlight contains one or more batteries, a light bulb, a switch, and several metal strips to connect them all together. When the switch is turned on (as shown), the components in the flashlight form a continuous loop of conducting materials. Electrons flow around this loop in the direction indicated.

When you turn a flashlight on, electricity transfers energy from the batteries to the bulb. An electric current flows through both parts, carrying the energy with it. The charged particles that make up this current are electrons and these electrons follow a circular path that takes them through the batteries and bulb many times each second (Fig. 12.1.1). As long as the flashlight is on, these electrons continue to travel around this circular path, receiving energy from the batteries and delivering it to the bulb.

The circular path taken by electrons in a flashlight is called an **electric circuit**. The flashlight needs this circuit so that its electrons don't accumulate somewhere along the path. Although a few electrons could flow from the batteries to the bulb without having to return, they would begin to repel any additional electrons that tried to join them. By giving the electrons a circular path, the flashlight allows them to carry energy from the batteries to the bulb and then return to the batteries to pick up more energy. This circular path or circuit concept is present in all electric devices and explains the need for at least two wires in the power cord of every home appliance: one wire carries electrons to the appliance to deliver energy and the other wire carries those electrons back to the power company to receive some more.

(a)

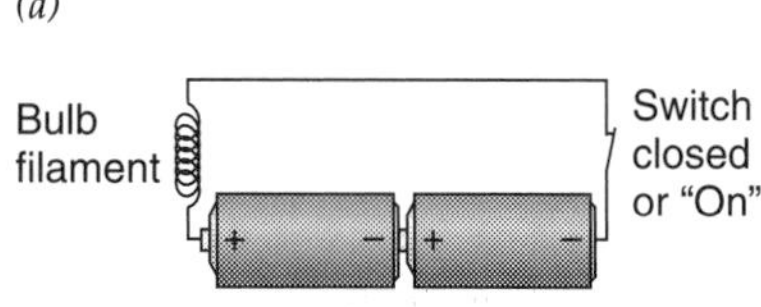

(b)

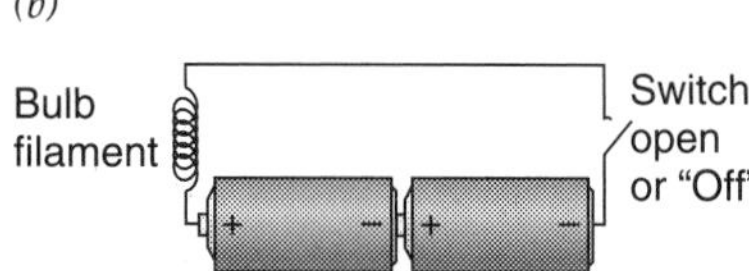

Fig. 12.1.2 -(*a*) When the flashlight's switch is "on", it closes the circuit so that current can flow continuously from the batteries, through the bulb's filament, and back through the batteries. It follows this circuit over and over again.
(*b*) When the flashlight's switch is "off", it opens the circuit so that current stops flowing.

But what role does the switch play in all of this? It works by making or breaking the flashlight's circuit (Fig. 12.1.2). The switch forms part of a conducting path that connects the batteries to the bulb. There are two of these paths, with electrons heading to the bulb through one and returning through the other. One of the paths is solid metal and always complete. The second can be interrupted by the switch. When the flashlight is on, the switch completes that conducting path so that electrons can flow continuously around the circuit (Fig. 12.1.2*a*). The circuit is **closed**. A closed circuit also appears on the left side of Fig. 12.1.3.

However, when you turn the flashlight off, the switch breaks that conducting path so that electrons can't flow around the circuit (Fig. 12.1.2*b*). The circuit is **open**. Although one path still connects the batteries and bulb, electrons can't travel all the way around the loop. With the two halves of the switch separated by a gap, charge accumulates on the metal strips and current stops flowing through the flashlight. Since no more energy reaches the bulb, it soon becomes dark. An open circuit also appears on the right side of Fig. 12.1.3.

There's one other type of circuit worth mentioning. A **short circuit** occurs when the two paths between the batteries and the bulb accidentally come in contact with one another (Fig. 12.1.4). When this occurs, the electrons have an alternative path through which to flow. Because the bulb must extract energy

Fig. 12.1.3 - When the switch is closed, current flows around this circuit and carries power from the battery to the light bulb. But when the switch is open, no current flows through the circuit.

from the electrons, it's designed to impede their flow and to convert their electrostatic and kinetic energies into thermal energy and light. We call this opposition to the flow of electricity **electric resistance**. Since the alternative path offers less resistance to their movement, most of the electrons flow through it and bypass the bulb. The bulb grows dim or goes out altogether.

Since the bulb is the only part of the flashlight that's meant to get hot, a short circuit will leave the electrons without a safe place to get rid of their energy. The electrons will deposit their energy in the batteries and the metal paths, which will become hot. Since short circuits can start fires, flashlights and other electric equipment are designed to avoid them.

Bulb filament — Path of short circuit — Switch closed or "On"

Fig. 12.1.4 - When an unwanted conducting path allows current to bypass the flashlight's filament, it creates a short circuit. Because it has no proper place for the current to deposit its energy, the short circuit becomes hot.

CHECK YOUR UNDERSTANDING #1: Cutting the Power

If you remove just one of the wires from an automobile's battery, the vehicle will not start at all. Why doesn't the other wire supply any power?

The Electric Current Through the Flashlight

The charged particles that flow through the flashlight's circuit are negatively charged electrons. They pass by in huge numbers, with each electron completing the journey many thousands of times per second. The rate at which these electrons flow through the circuit is important because each electron carries with it a certain amount of energy. If you want to determine the power reaching the light bulb, you must know how many electrons pass through the bulb each second and how much energy each of them delivers.

Rather than counting the electrons individually as they pass, we measure the current flowing through the metal strips. Current is the charge flowing past a particular point in the circuit per unit of time. The SI unit of current is the **ampere** (abbreviated A) and corresponds to 1 C (1 Coulomb) of charge passing by the designated point each second. 1 C is an enormous amount of charge so that when a current of 1 A flows through a wire, there are roughly $6.25 \cdot 10^{18}$ or 6,250,000,000,000,000,000 electrons passing through it each second.

Even though each electron only carries a tiny amount of energy, there are so many electrons flowing in a 1 A current that they transfer a substantial amount of power. Since each coulomb of 1 V electrons carries 1 J of energy, a 1 A current of these electrons carries 1 J per second or a power of 1 W. By itself, each 1 V electron carries only $1.602 \cdot 10^{-19}$ or 0.0000000000000000001602 J of energy, but these tiny amounts of energy add up when lots of electrons are involved.

However, current has a direction. When you turn on the flashlight in Fig. 12.1.1, electrons begin to flow around the circuit counter-clockwise. They flow from the battery chain's negative terminal, through the switch, through the bulb's filament, and into the battery chain's positive terminal. Given this counter-

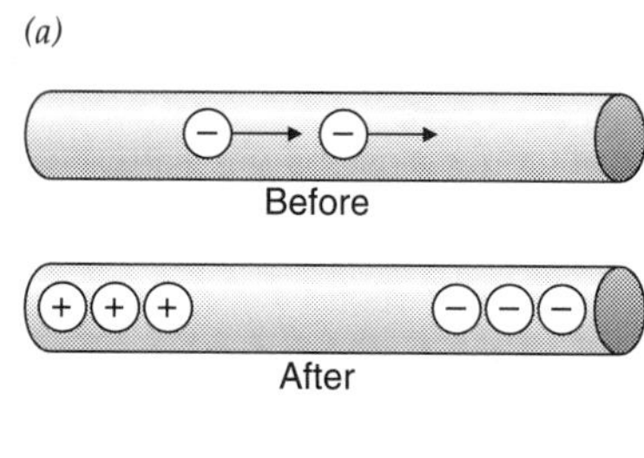

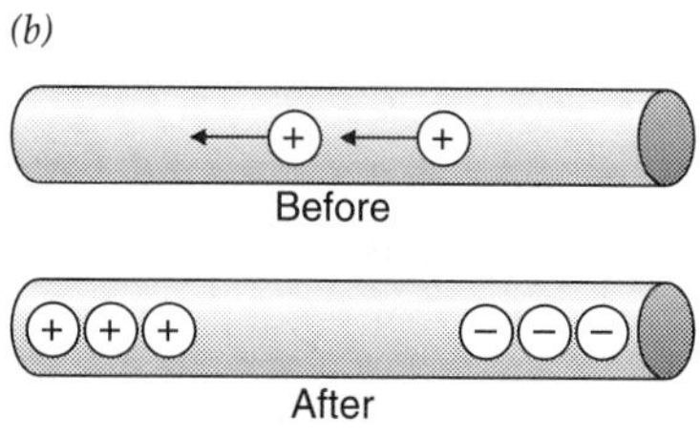

Fig. 12.1.5 - A current of negatively charged particles flowing to the right through a piece of wire (*a*) can't easily be distinguished from a current of positively charged particles flowing to the left (*b*). The end result of both processes is an accumulation of positive charge on the left end of the wire and negative charge on its right end.

clockwise flow of electrons, it would seem sensible to say that current flows counter-clockwise around the circuit.

Unfortunately, electrons are negatively charged. Rather than have current carry a negative charge with it, scientists and engineers decided long ago to have current point in the direction of positive charge flow. Because of this convention, current officially flows clockwise around the circuit in Fig. 12.1.1. In making this definition of the direction of current, the scientists and engineers chose to pretend that positive charges are carrying the current. These positive charges don't really exist; they are simply a useful fiction.

This fiction actually works extremely well, as illustrated by a simple example. If negatively charged electrons were to flow to the right through a piece of wire, its right end would become negatively charged and its left end would become positively charged (Fig. 12.1.5*a*). But exactly the same thing would happen if a current of fictitious positively charged particles were to flow to the left through that same piece of wire (Fig. 12.1.5*b*). Without very special equipment, you couldn't determine whether negative charges are flowing to the right or positive charges are flowing to the left, because the end results would be essentially indistinguishable.

We, too, will adopt this fiction and pretend that current is the flow of positively charged particles. In this and subsequent chapters, we'll stop thinking about electrons and imagine that electricity is carried by positive charges moving in the direction of the current. There are only a few special cases in which the electrons themselves are important and we'll consider those situations separately when they arise.

CHECK YOUR UNDERSTANDING #2: Don't Touch that Pipe

When you walk across wool carpeting in rubber-soled shoes, you accumulate a large number of negative charges on your body, particularly if the air is dry. These negative charges have considerable electrostatic potential energy. When you bring your hand near a large piece of metal, the negative charges will leap through the air as a spark in order to reach the metal. Which way is current flowing in this spark?

Batteries, Bulbs, and Metal Strips

To explain how a flashlight works, we'll have to look closely at how power is transferred through its circuit. But first we must understand more about the flashlight's individual components: its batteries, bulb, and metal strips. Let's start with the batteries.

A battery is a device that pumps positive charges from its negative terminal to its positive terminal, against the direction in which current naturally flows. Since the battery must pull a positive charge away from the negative terminal and push it onto the positive terminal, the battery does work on the charge. The battery uses its chemical potential energy to do this work, converting it into electrostatic potential energy. Positive charges accumulate on the battery's positive terminal and negative charges accumulate on its negative terminal. Overall, the battery produces separated electric charge.

Each time the battery transfers a positive charge, the work gets a little harder. As the amount of separated charge increases, the forces acting on the positive charges become stronger. The battery keeps on transferring positive charges until the work required to move the next charge is too much for the battery. At that point, the battery stops pumping charges and everything is in equilibrium. Positive charges on the positive terminal are attracted toward the negative charges on the negative terminal, but the battery's chemical forces balance the attraction and prevent the charges from flowing toward one another.

A typical alkaline battery pumps charges until the voltage of its positive terminal is 1.5 V above the voltage of its negative terminal. This voltage difference measures the amount of electrostatic potential energy that each unit of positive charge would release if it were permitted to flow from the positive terminal to the negative terminal. Since the voltage difference is 1.5 V (equivalent to 1.5 joules-per-coulomb), each coulomb of positive charge would release 1.5 J of energy in moving to the negative terminal.

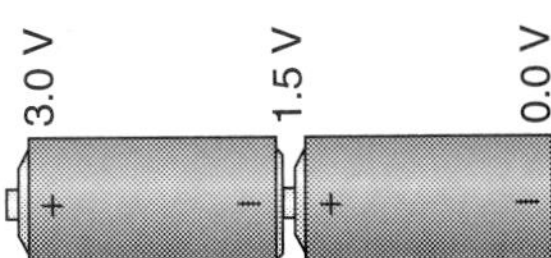

Fig. 12.1.6 - When two 1.5 V batteries are connected in a chain, their voltages add so that the chain's positive terminal has a voltage 3.0 V higher than the chain's negative terminal. If the chain's negative terminal is chosen to be at 0 V, then the chain's positive terminal is at 3.0 V.

An alkaline battery gets this energy through an electrochemical reaction, in which powdered zinc attached to its negative terminal forms a zinc oxide complex with a manganese dioxide paste attached its positive terminal. This reaction process resembles controlled combustion. In effect, the battery "burns" zinc to obtain the energy needed to pump charges from one of its terminals to the other.

As the battery consumes its store of chemical potential energy, its ability to pump charges diminishes. When its chemicals are nearly exhausted, the random effects of thermal energy in the battery begin to decrease its overall voltage. It also loses its ability to pump large amounts of current and wastes much of its energy as thermal energy. An aging battery can pump fewer charges per second than a fresh one and gives each charge it pumps less energy. Ultimately, less power reaches the flashlight's bulb and it grows dim.

When two batteries are connected together in a chain, so that the positive terminal of one battery touches the negative terminal of the other, the two batteries work together to pump positive charges from the chain's negative terminal to its positive terminal (Fig. 12.1.6). Each battery pumps charges until its positive terminal is 1.5 V above its negative terminal so that the chain's positive terminal is 3.0 V above its negative terminal. Since only *differences* in voltage matter in a flashlight's circuit, it will be convenient for us to define one part of that circuit as being at 0 V. If we define the battery chain's negative terminal as being at 0 V (Fig. 12.1.6), then its positive terminal must be at 3.0 V.

Fig. 12.1.7 - A 9 V battery actually contains 6 small 1.5 V cells, connected in a chain. Positive charge that enters the chain at the battery's negative terminal passes through all 6 cells before arriving at the battery's positive terminal.

The voltages of the two batteries add because each battery adds a certain amount of energy to the charge it pumps. Some flashlights use more than two batteries in the chain. The more batteries in the chain, the more energy the positive charges receive as they pass through it and the greater the voltage difference between the chain's positive terminal and its negative terminal. A flashlight that uses 6 batteries in its chain has a positive terminal that is 9 V above its negative terminal. A 9 V alkaline battery actually contains a chain of 6 miniature 1.5 V batteries, arranged so that their voltages add up to 9 V (Fig. 12.1.7).

If you reverse one of the batteries in the chain, the reversed battery will extract energy from the current passing through it (Fig. 12.1.8). Overall, the chain will still pump positive charges from its negative terminal to its positive terminal, but its overall voltage will be reduced. If the chain has 3 batteries, 2 of them will add energy to the current while 1 of them extracts energy. The overall voltage of the chain will be only 1.5 V.

Since current is flowing backward through the reversed battery, that battery extracts energy from the current and turns at least some of it into chemical potential energy. The reversed battery recharges. Battery chargers use a similar technique, pushing current backward through a battery, from its positive terminal to its negative terminal, in order to put chemical potential energy back into a rechargeable battery. However normal alkaline batteries are "non-rechargeable," meaning that they turn most of the recharging current's energy into thermal energy instead of chemical potential energy. Non-rechargeable batteries may overheat and explode during recharging.

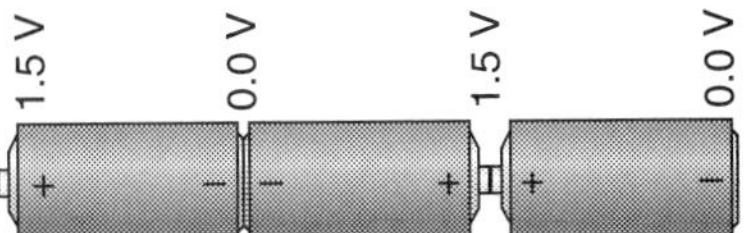

Fig. 12.1.8 - When one battery in a chain of 3 is reversed, the reversed battery's voltage is subtracted from the sum of the others. The chain's positive terminal has a voltage only 1.5 V higher than its negative terminal. The reversed battery recharges.

A bulb's behavior is roughly opposite that of a battery. Its filament is a tungsten wire that allows positive charges to move from its positively charged terminal to its negatively charged terminal, the direction in which current naturally flows. In traveling through the filament, each positive charge releases its electrostatic potential energy and that energy becomes thermal energy and light.

If the filament were a perfect conductor of electricity, a positive charge passing through it would speed up as its electrostatic potential energy became kinetic energy. But the filament is a coil of fine tungsten wire, which has a large electric resistance and significantly impedes the flow of electric current. As the positive charges flow through the filament, they collide with tungsten atoms and the filament becomes white hot. Overall, the current's ordered electrostatic energy becomes thermal energy in the atoms, and the filament emits light.

What about the metal strips connecting the flashlight's batteries and bulb? Because these thick pieces of metal exhibit very little electric resistance, they carry current easily and consume only a tiny fraction of the power flowing from the battery to the bulb. In some flashlights, the thick metal case itself helps to carry current. In general, the less electric resistance in the wires carrying current to and from the bulb, the more power reaches the bulb. That's why it's so important to use thick metal strips in the connections.

But a poor connection can spoil this efficient transfer of energy. If there is dirt or grease on a battery terminal or worn materials in the switch, the current will have to pass through a large electric resistance. When it does, it will waste a substantial fraction of its energy as thermal energy. The bulb will receive less power than it should and will emit relatively little light. Improving the connection, either by shaking the flashlight or by cleaning the metal surfaces, will increase the current flow through the circuit, reduce the wasted energy, and brighten the flashlight.

CHECK YOUR UNDERSTANDING #3: Car Batteries

A lead–acid automobile battery provides 12 V between its negative terminal and its positive terminal. It actually contains 6 individual batteries, connected in a chain. How much voltage does each individual battery produce?

Voltage, Current, and Power in Flashlights

To complete this discussion of flashlights, let's see how the electric current carries power from the batteries to the bulb. We'll consider a flashlight with two batteries, so that the voltage of the battery chain's positive terminal is 3.0 V above that of its negative terminal. When this flashlight is turned on, a current of 1.0 A flows through its circuit.

The bulb consumes electric power because the current passing through it experiences a drop in voltage. This **voltage drop**, the voltage before the bulb minus the voltage after it, measures the energy each unit of charge loses while passing through the bulb. Multiplying the voltage drop by the current passing through the bulb gives you the power consumed by the bulb. This observation can be written as a word equation:

$$\text{power consumed} = \text{voltage drop} \cdot \text{current}, \tag{12.1.1}$$

in symbols:

$$P = V \cdot I,$$

and in everyday language:

When a big current loses a lot of voltage, something will get hot.

Since the voltage drop across the flashlight bulb is 3.0 V and the current through it is 1.0 A, it's consuming 3.0 W of power.

The battery chain produces electric power because the current passing through it experiences a rise in voltage. This **voltage rise**, the voltage after the batteries minus the voltage before them, measures the energy each unit of charge gains while passing through the batteries. Multiplying the voltage gain by the current passing through the batteries gives you the power provided by the batteries. This observation can be written as a word equation:

$$\text{power provided} = \text{voltage rise} \cdot \text{current}, \qquad (12.1.2)$$

in symbols:

$$P = V \cdot I,$$

and in everyday language:

It takes a great deal of power to raise the voltage of a large electric current.

Since the voltage rise across the battery chain is 3.0 V and the current through it is 1.0 A, it's providing 3.0 W of power.

For this particular flashlight, the bulb has been designed to become white hot when it receives 3.0 W of power. It's also designed to allow 1.0 A of current to flow through it when the voltage drop across it is 3.0 V. The bulb is what controls the amount of current flowing in the flashlight's circuit and thus the amount of power transferred from the batteries to the bulb. If you put the wrong bulb in a flashlight, it will probably receive too much or too little power. If it receives too much, it will burn out quickly. If it receives too little, it will glow dimly. Powerful flashlights that use many batteries have bulbs that are designed to consume large powers and emit intense beams of light.

There are two more important components of the circuit: the strips connecting the battery chain to the bulb and the switch. In effect, these connecting parts are also devices that require power to operate. The power they consume keeps the current in them from coming to a stop and they convert this power into thermal energy. But because the strips and switch are short and thick, they have little electric resistance and consume little power. Most of the 3.0 W provided to the circuit by the battery chain reaches the bulb. Only when the connections are poor do they extract significant amounts of power from the circuit and reduce the current, voltage drop, and power delivered to the bulb.

CHECK YOUR UNDERSTANDING #4: Current Trends in Music

Your portable radio is powered by a large battery. Current enters the radio through one wire and leaves through another. Which wire has a higher voltage?

CHECK YOUR FIGURES #1: When You Turn the Ignition Key

When you start a car, a 200 A current flows through its starter motor. If that current experiences a voltage drop of 12 V, how much power is being consumed?

CHECK YOUR FIGURES #2: When You Turn the Ignition Key Again

When you restart the car, a smaller current of 150 A flows through the car's starter motor. The battery is supplying power to this current, providing it with a voltage rise of 12 V. How much power is the battery providing?

Summary

How Flashlights Work: A flashlight produces light when a current flows through its electric circuit. This circuit consists of a chain of batteries, a light bulb, a switch, and several metal strips, all connected in a continuous loop. The current obtains energy as it passes through the battery and it releases this energy as it passes through the filament of the light bulb. Collisions with tungsten atoms in the filament convert the current's electrostatic and kinetic energies into thermal energy and the filament becomes white hot. It radiates visible light.

The switch provides a way to control the flow of current in the circuit. When the switch is open (off), it breaks the circuit and prevents current from flowing completely around the circuit. Without a steady current to carry power from the batteries to the bulb, the flashlight is dark. When the switch is closed (on), it completes the circuit so that power can reach the bulb and the flashlight becomes bright. Any poor connection in the circuit or an aging battery will diminish the amount of power that the current delivers to the bulb and make the flashlight dim.

Important Laws and Equations

1. Power Consumption: The electric power consumed by a device is the voltage drop across it times the current flowing through it, or

$$\text{power consumed} = \text{voltage drop} \cdot \text{current}. \quad (12.1.1)$$

2. Power Production: The electric power provided by a device is the voltage rise across it times the current flowing through it, or

$$\text{power provided} = \text{voltage rise} \cdot \text{current}. \quad (12.1.2)$$

The Physics of Flashlights

1. Moving charged particles form an electric current. The current points in the direction of positive charge flow (or opposite the direction of negative charge flow).

2. A simple electric circuit consists of a power source (e.g. a battery) and a load (e.g. a light bulb) that are connected by electric conductors to form a continuous loop. An electric current flows through that loop, transferring power from the power source to the load.

3. Because the charged particles in an electric current repel one another, they can't accumulate anywhere. They must circulate endlessly through a circuit. Any break in that circuit creates an open circuit and stops the current flow.

4. A short circuit contains a shortcut through which current can bypass the intended load. Without an appropriate place to deposit its energy, the current flowing in a short circuit overheats the wiring itself.

5. Voltage measures the electrostatic potential energy of each unit of charge located at a particular point. Current measures the electric charge passing a particular point during each unit of time.

6. When the voltage decreases as current passes through a device, that device is consuming electric energy. When the voltage increases as current passes through a device, that device is providing electric energy.

Check Your Understanding - Answers

1. Removing a single wire from the battery breaks the circuit and prevents a steady flow of electric current through the car's electric system.

Why: Removing the negative wire from the battery stops electrons from leaving the battery. Removing the positive wire prevents electrons from returning to the battery. Either way, no steady current will flow, so the car will not operate.

2. The current flows from the metal toward your hand.

Why: The negative charges on your body flow toward the metal piece. Because current is defined as the flow of positive charges, it points in the direction opposite the flow of negative charges. Thus the current flows toward your hand. In this case, electricity doesn't involve a complete circuit. However the charges only move for a brief instant. For the charges to move continuously, they would have to be recycled and there would have to be a circuit.

3. Each individual battery produces 2 V.

Why: The voltages of the individual batteries add so that it takes 6 of the 2 V batteries to produce a 12 V chain. If one of the batteries becomes weak, through damage, loss of fluid, or consumption of its chemical potential energy, the voltage of the chain will drop below 12 V. Pushing current backward through a lead–acid battery recharges it.

4. The wire through which current enters the radio.

Why: The radio is consuming power, so the current passing through it is experiencing a voltage drop. The current has a higher voltage when it enters the radio than when it leaves.

Check Your Figures - Answers

1. 2400 W.

Why: The voltage drop across the starter motor is 12 V (each coulomb of charge loses 12 J of energy in passing through it) and the current through it is 200 A (200 C of charge pass through it each second). We can use Eq. 12.1.1 to determine the power the motor is consuming:

$$\begin{aligned}\text{power consumed} &= \text{voltage drop} \cdot \text{current}\\ &= 12\text{ V} \cdot 200\text{ A} = 2400\text{ W}.\end{aligned}$$

2. 1800 W.

Why: The voltage rise across the battery is 12 V (each coulomb of charge gains 12 J of energy in passing through it) and the current through it is 150 A (150 C of charge pass through it each second). We can use Eq. 12.1.2 to determine the power the battery is providing:

$$\begin{aligned}\text{power provided} &= \text{voltage rise} \cdot \text{current}\\ &= 12\text{ V} \cdot 150\text{ A} = 1800\text{ W}.\end{aligned}$$

Glossary

ampere (A) The SI unit of electric current (identical to the coulomb-per-second). 1 ampere is defined as the passage of $6.25 \cdot 10^{18}$ charged particles per second and is roughly the current flowing through a 100 W light bulb operating on household electric power.

closed circuit A complete electric circuit through which electric current can flow continuously in a loop.

dynamic electricity A process involving moving electrically charged particles.

electric circuit A complete loop of conductors, loads, and power sources through which an electric current can flow continuously.

electric resistance The measure of how much an object impedes the flow of electric current.

open circuit An incomplete electric circuit where a gap in the electric conductors stops electric current from flowing.

short circuit A defect in a circuit that allows current to bypass the load it's supposed to operate.

voltage drop The amount of electrostatic potential energy that each coulomb of positive charge loses in passing through a device. It's equal to the voltage of the charges entering the device minus the voltage of the charges leaving that device.

voltage rise The amount of electrostatic potential energy that each coulomb of positive charge receives in passing through a device. It's equal to the voltage of the charges leaving the device minus the voltage of the charges entering that device.

Review Questions

1. If electrons are flowing clockwise around an electric circuit, which way is current flowing around that circuit?

2. What is the difference between voltage and current?

3. When a battery is not installed in a flashlight, is there still a voltage rise from its negative terminal to its positive terminal? Is there a current flowing through the battery?

4. Why must the circuit of a flashlight be closed before power will flow from its batteries to its bulb?

5. Why does a short circuit that allows current in a flashlight to bypass the bulb keep that bulb from lighting?

6. When a flashlight is operating normally, which way are electrons flowing through the batteries? Which way is current flowing through the batteries?

7. The current leaving the positive terminal of one battery enters the negative terminal of a second battery. Will the voltage at the positive terminal of the second battery be more, less, or the same as the voltage at the positive terminal of the first battery? Will the current leaving the positive terminal of the second battery be more, less, or the same as the current leaving the positive terminal of the first battery?

8. The current leaving the positive terminal of one battery enters the positive terminal of a second battery. Will the voltage at the negative terminal of the second battery be more, less, or the same as the voltage at the positive terminal of the first battery?

9. Why do metals such as tungsten become hotter when electric currents flow through them?

10. When a current passes through a device that provides power to that current, does the voltage of that current increase or decrease?

11. When a current passes through a device that consumes power from that current, does the voltage of that current increase or decrease?

Exercises

1. The electric switch for the light in your room has two wires attached to it. One day the switch breaks so that the light won't turn off. What should the switch be doing with the electric circuit it's in and what is it now doing instead?

2. The cord running from the electric outlet to your table lamp has two wires inside it. Current flows to the lamp through one wire and returns from the lamp through the other. If a heavy metal file cabinet cuts into the cord and electrically connects those two wires to one another, will the lamp become brighter or dimmer? Explain your answer.

3. The rear defroster of your car is a pattern of thin metal strips across the window. When you turn the defroster on, current flows through those metal strips. Why are there wires attached to both ends of the metal strips?

4. When current is flowing through a car's rear defroster (Exercise **3**), the voltage at each end of the metal strips is different. Which end of each strip has the higher voltage, the one through which current enters the strip or the one through which current leaves, and what causes the voltage drop?

5. When you turn off the rear defroster (Exercises **3** and **4**), how are you changing the circuit connecting the battery to the defroster?

6. A bird can perch on a high voltage power line without getting a shock. Why doesn't current flow through the bird?

7. The power source for an electric fence pumps charge from the earth to the insulated fence wire. The earth can conduct electricity. When an animal walks into the wire, it receives a shock. Identify the circuit through which the current flows.

8. During the Cold War, a man escaped from East Germany by swinging along a high tension wire. Each time he came to a support pylon he had to drop onto the pylon and then leap to the wire at the other side. Why did he have to be so careful not to touch the wire and the pylon simultaneously?

9. The positive terminal of a battery has a net positive charge. You are normally electrically neutral. If you touch the positive terminal of the battery, does any current flow between the terminal and your finger? If so, for how long?

10. If you touch both the positive and negative terminals of a battery with your fingers, a small electric current will flow through your hand. Does your net electric charge change as this current flows through your hand? If so, do you become more negatively charged or more positively charged?

11. The two prongs of a power cord are meant to carry current to and from a lamp. If you were to plug only one of the prongs into an outlet, the lamp wouldn't light at all. Why wouldn't it at least glow at half its normal brightness?

12. One of the hazards of laying an extension cord across a busy walkway is that constant wear can cause its two internal wires to touch. One wire normally carries current to the appliances using the cord while the other wire carries that current back to the power company. What will happen to the current in the extension cord if the two wires inside it touch?

13. You're given a sealed box with two terminals on it. You use some wires and batteries to send an electric current into the box's left terminal and find that this current emerges from the right terminal. If the voltage of the right terminal is 6 V higher than that of the left terminal, is the box consuming power or providing it? Is it more likely to contain batteries or light bulbs? How can you tell?

14. Suppose that you have two 1.5 V batteries, some wires, and a light bulb. One of the batteries, however, is unlabeled and there is nothing on it to indicate which is its positive terminal and which is its negative terminal. The other battery is properly labeled. How can you determine which terminal of the unlabeled battery is positive and which is negative?

15. One time-honored but slightly unpleasant way in which a hobbyist determines how much energy is left in a 9 V battery is to touch both of its terminals to his tongue briefly. He experiences a pinching feeling that's mild when the battery is almost dead but one that's startlingly sharp when the battery is fresh. (Don't try this technique yourself.) What is the circuit involved in this taste test? How is energy being transferred?

Problems

1. An automobile has a 12 W reading lamp in the ceiling. This lamp operates with a voltage drop of 12 V across it. How much current flows through the lamp?

2. The rear defroster of your car operates on a current of 5 A. If the voltage drop across it is 10 V, how much electric power is it consuming as it melts the frost?

3. Your portable FM radio uses two 1.5 V batteries in a chain. If the batteries send a current of 0.05 A through the radio, how much power are they providing to the radio?

4. You have two flashlights that operate on 1.5 V "D" batteries. The first flashlight uses two batteries in a chain while the second uses five batteries in a chain. Each flashlight has a current of 1.5 A flowing through its circuit. What power is being transferred to the bulb in each flashlight?

5. You have two flashlights that have 2 A currents flowing through them. One flashlight has a single 1.5 V battery in its circuit while the second flashlight has three 1.5 V batteries connected in a chain that provides 4.5 V. How much power is the battery in the first flashlight providing? How much power is each battery in the second flashlight providing?

6. How much power is the bulb of the first flashlight in Problem **5** consuming? How much power is the bulb of the second flashlight consuming?

7. A 1.5 V alkaline "D" battery can provide about 40,000 J of electric energy. If a current of 2 A flows through two "D" batteries while they're in the circuit of a flashlight, how long will the batteries be able to provide power to the flashlight?

8. A radio-controlled car uses four "AA" batteries to provide 6 V to its motor. When the car is heading forward at full speed, a current of 2 A flows through the motor. How much power is the motor consuming at that time?

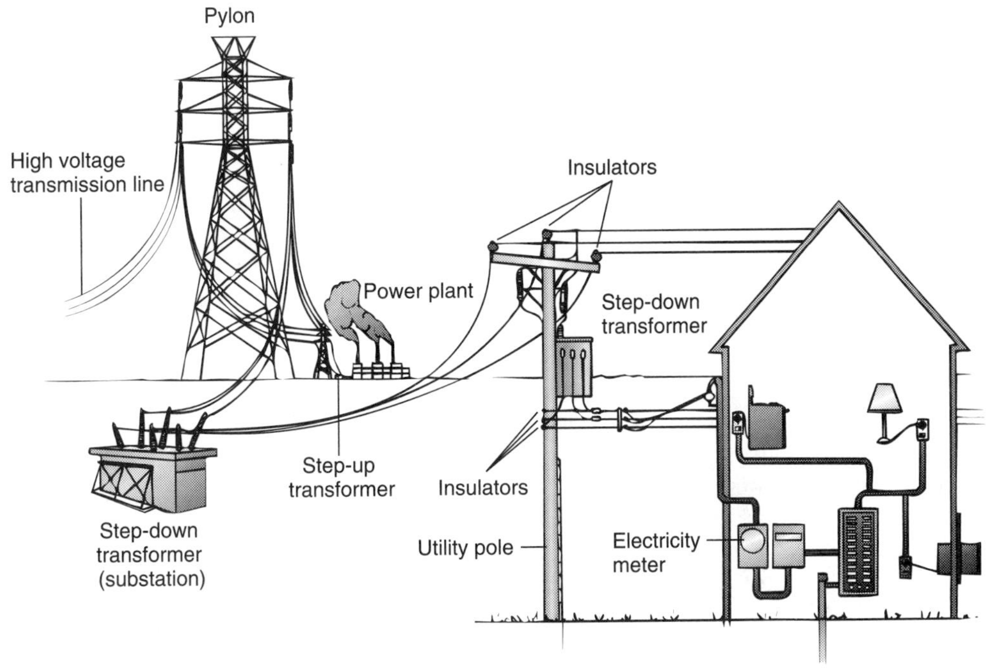

Section 12.2

Electric Power Distribution

Electricity is a particularly useful and convenient form of ordered energy. Because it's delivered to our homes and offices as a utility, we barely think about it except to pay the bills. The wires that bring it to us never plug-up or need cleaning, and work continuously except when there's a power failure, blown fuse, or tripped circuit breaker.

Just how does electricity get to our homes? In this section, we'll look at the problems associated with distributing electricity far from the power plant at which it's generated. To understand these problems, we'll examine the ways in which wires affect electricity and see how electric power is transferred and modified by devices called transformers.

Questions to Think About: *Why are power distribution systems always alternating current? What are high tension wires for? Why does the power company place large electric devices on the utility poles near homes or on the ground near neighborhoods? What are the advantages and disadvantages of 120 V versus 220 V electric power?*

Experiments to Do: *Experiments with electric power distribution are rather dangerous, but you can observe the way in which your local electric power is distributed. If your area is connected to a major electric power network, you'll be able to find an entire hierarchy of power conversion facilities. The power should travel from the power plant to your town on high tension wires, normally located overhead on tall pillars or pylons. These wires should end at a large power conversion facility, where enormous devices transfer power to the lower voltage power lines that fan out across your town. In some places, these wires are overhead but in others they are underground.*

But this power is still not ready for household use. It goes through at least one more level of conversion before it reaches individual homes. You may find the final transformers as boxes or cylinders on utility poles or on the ground outside houses. In cities, these

WARNING:

Electricity is dangerous, particularly when it involves high voltages. The principal hazard is that an electric current will pass through your body in the vicinity of your heart and disrupt its normal rhythm. While very little current is needed to cause trouble, your skin is such a poor conductor that it normally keeps harmful currents from passing through you. However large voltages can propel dangerous currents through your skin and put you at risk. While your body usually has to be part of a closed circuit for you to receive a shock, don't count on the absence of an identifiable circuit to protect you from injury—circuits have a tendency to form in surprising ways whenever you touch an electric wire. Be especially careful whenever you're near voltages of more than 50 V, even in batteries, or when you're near any voltages if you're wet or your skin is broken.

transformers are often located inside buildings, out of sight. Even though you may have trouble finding these transformers, they are almost certainly there. In this section, we'll see why they are necessary.

Direct Current Power Distribution

Batteries may be fine for powering flashlights, but they're not very practical for lighting a house. Early experiments that placed batteries in basements were disappointing because the batteries ran out of energy quickly and needed service and fresh chemicals much too frequently.

A more cost-effective source for electricity was coal- or oil-powered electric generators. Like batteries, generators do work on the electric currents flowing through them and can provide the electric power needed to illuminate homes. But while generators produce electricity more cheaply than batteries, early ones were large machines that required lots of fresh air and attention. These generators had to be built centrally, with people to tend them and chimneys to get rid of the smoke.

This was the approach taken by the American inventor Thomas Alva Edison (1847–1931) in 1882 when he began to electrify New York City. Each of the Edison Electric Light Company's generators acted like a mechanical battery, producing direct current that left the generator through one wire and returned to it through another. Edison placed his generators in central locations and conducted the current to and from the homes he served through copper wires. However, the farther a building was from the generator, the thicker the copper wires had to be. That's because wires impede the flow of current and making them thicker allows them to carry current more easily.

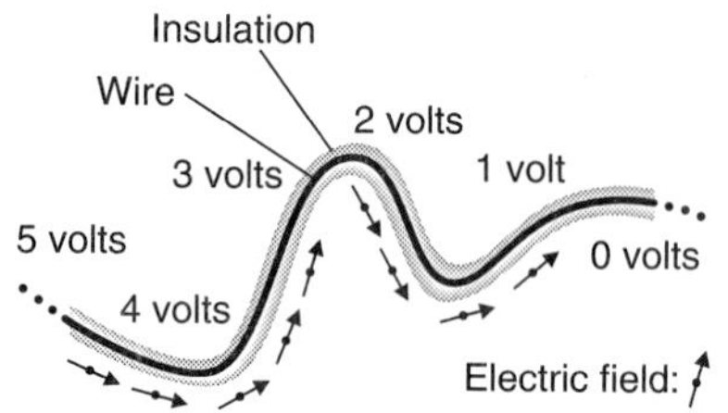

Fig. 12.2.1 - This wire is part of a current-carrying electric circuit. There is a voltage drop across the wire because the positive charges convert their electrostatic potential energies into kinetic energies in order to keep moving through the wire. Collisions with metal atoms continually turn their kinetic energies into thermal energies.

Wire thickness is important because the positively charged particles in a current lose kinetic energy whenever they collide with metal atoms. For these particles to keep moving forward, they must convert electrostatic potential energy into kinetic energy as they pass through the wire. Their electrostatic potential energy and voltage gradually diminish during their trip, so the wire has a voltage drop across it. There is an electric field associated with this voltage drop and that field is what pushes the positive charges through the wire (Fig. 12.2.1).

Collisions slow the charges down so that they move at a certain average speed. You can increase this average speed, and the current flowing through the wire, by increasing the electric field. When you increase the electric field, you also increase the voltage drop across the wire, so it's not surprising that the current in the wire is exactly proportional to the voltage drop across it.

Another way to increase the current in the wire is to make it thicker, so that more charges can move through it at the same time. With more charges moving, they don't have to move as quickly to maintain a large current and the wire exhibits a smaller electric resistance. Resistance incorporates not only thickness, but also length and composition. Overall, the current flowing through a wire is inversely proportional to its electric resistance.

We can combine these observations together as a word equation:

$$\text{voltage drop} = \text{current} \cdot \text{electric resistance}, \quad (12.2.1)$$

in symbols:

$$V = I \cdot R,$$

and in everyday language:

Long, skinny jumper cables lose so much voltage carrying current to your car that it probably won't start.

Equation 12.2.1 is called **Ohm's law**, after its discoverer Georg Simon Ohm ❐. The SI unit of electric resistance, the volt-per-ampere, is called the **ohm** (abbreviated Ω, the Greek letter omega) in his honor. Despite its simplicity, Ohm's law is extremely useful in physics and electrical engineering. It applies to an enormous number of systems so that nearly everything can be characterized by an electric resistance. Once an object's electric resistance is known, the voltage drop across it can be calculated from the current flowing through it or the current flowing through it can be calculated from the voltage drop across it. (For another useful form of Eq. 12.2.1, see ❐.)

❐ German physicist Georg Simon Ohm (1789–1854) served as a professor of mathematics, first at the Jesuits' college of Cologne and then at the polytechnic school of Nürnberg. His numerous publications were undistinguished, with the exception of one pamphlet on the relationship between current and voltage. This extraordinary document, written in 1827, was initially dismissed by other physicists, even though it was based on good experimental evidence and explained many previous observations by others. In despair, Ohm resigned his position at Cologne and it was not until the 1840's that his work was accepted. He was finally appointed professor of physics at Munich only two years before his death.

OHM'S LAW:
The voltage drop across a wire is equal to the current flowing through that wire times the wire's electric resistance.

In the case of a wire conducting current from a generating plant to a house, we care most about how much power the wire wastes as thermal energy. We can determine this wasted power by combining Ohm's law with Eq. 12.1.1:

$$\begin{aligned}\text{power consumed} &= \text{voltage drop} \cdot \text{current}\\ &= (\text{current} \cdot \text{electric resistance}) \cdot \text{current}\\ &= \text{current}^2 \cdot \text{electric resistance} \qquad (12.2.2)\end{aligned}$$

❐ Equation 12.2.1 can be rearranged algebraically as:

$$\text{current} = \frac{\text{voltage drop}}{\text{electric resistance}}.$$

This form makes it clear that the current flowing through a wire depends both on the voltage drop across it and on its electric resistance. Increasing the voltage drop increases the current while decreasing the wire's resistance increases the current.

The wire's wasted power is proportional to the square of the current passing through it! This relationship became all too clear to Edison when he tried to expand his power distribution systems. The more current he tried to deliver over a particular wire, the more power it lost as heat. Doubling the current in the wire quadrupled the power it wasted.

Edison tried to combat this loss by lowering the electric resistances of the wires. He used copper, because only silver is a better conductor of current. He used thick wires to increase the number of moving charges. And he kept the wires short so that they didn't have much chance to waste power. This length requirement forced Edison to build his generating plants within the cities he served. Even New York City contained many local power plants. (For an interesting tale about the early days of electric power, see ❐.)

Edison also tried to save power by delivering small currents at high voltages. Since the delivered power is proportional to the voltage drop times the current, Edison could reduce the current flowing through the wires by raising the voltages. Less current flowed through each house, but it underwent a larger voltage drop, so the power delivered was unchanged. However, high voltages are dangerous because they tend to create sparks as current jumps through the air. They also produce nasty shocks when current flows through your body. While high voltages could be handled safely outside a house, they could not be brought inside. Edison used the highest voltages that safety would allow.

❐ Love Canal is the United States' most famous toxic waste dump. The dump was created in the 1920's, at an abandoned section of canal constructed in 1892 by William T. Love. Love intended his canal to connect the upper and lower Niagara Rivers, so that the descending water could be used to generate DC electric power for the citizens of Niagara Falls, NY. The advent of AC power transmission systems in 1896 made the canal less useful and it was never finished.

CHECK YOUR UNDERSTANDING #1: The Trouble with DC Electric Power

If Edison doubled the length of his delivery wires, while keeping the currents through them the same, what would happen to the power they consumed?

CHECK YOUR FIGURES #1: Be Careful What You Touch

Anyone who has received a shock from a 110 V power outlet knows that it's unpleasant. But a shock from a 220 V outlet is far worse. How does doubling the voltage drop across your body change the current passing through you and the power you're "consuming"?

Alternating Current and Transformers

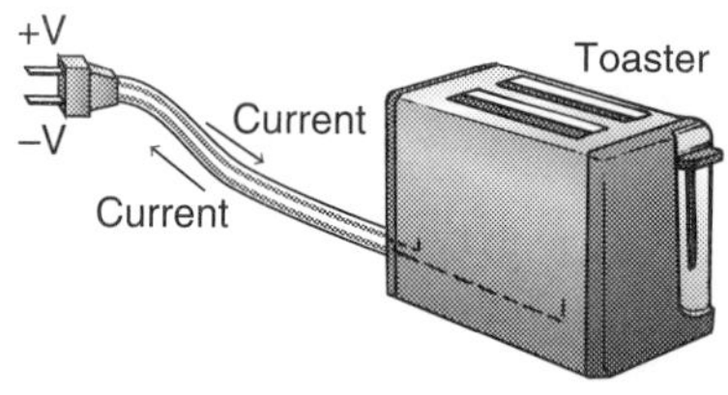

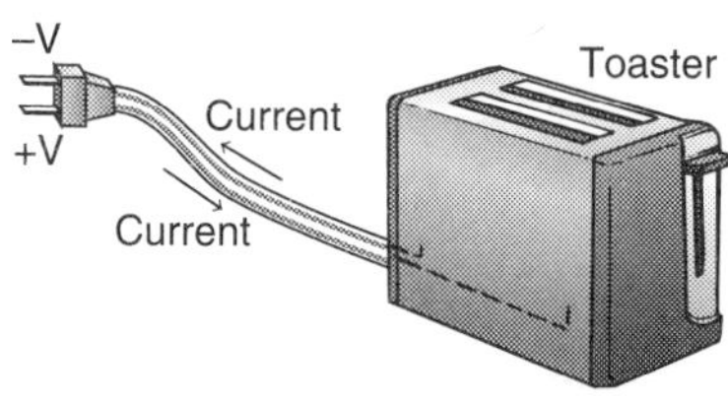

Fig. 12.2.2 - The current passing through this toaster reverses directions periodically as the outlet voltages reverse.

The real problem with distributing electric power with direct current (DC) is that there's no easy way to transfer power from one circuit to another. Because the generator and the light bulbs must be part of the same circuit, they can't use high voltages. Since delivering large amounts of power with low voltages requires large currents, direct current power distribution wastes much of its power in the wires connecting everything together.

However, alternating current (AC) makes it easy to transfer power from one circuit to another, so that different parts of the alternating current power distribution system operate at different voltages and with different currents. Since the wires that carry the power long distances operate at high voltages and low currents, they waste little power.

In alternating current, the direction of current flow through a circuit reverses periodically. At one moment, the current leaves the power plant through one wire and returns through a second, while a moment later it leaves through the second wire and returns through the first.

These reversals in direction are accomplished smoothly; the current slows to a stop and then gradually begins to flow backward. During each reversal there is actually a moment in which no current is flowing at all. In the United States, these reversals occur every 120th of a second so that the current completes 60 full cycles of reversal (back and forth) each second (60 Hz). In Europe, the reversals occur every 100th of a second and the current completes 50 full cycles of reversal (50 Hz) each second.

While these current reversals seem needlessly complicated, they have little effect on most household devices. Toasters (Fig. 12.2.2) and incandescent light bulbs still exhibit voltage drops when alternating currents pass through them and still consume electric power. In fact, power consumption is used to establish a voltage for alternating current. AC voltage is defined to be equal to the DC voltage that causes the same power consumption. Thus 120 V AC power delivers the same power to a toaster as 120 V DC power.

Edison was steadfastly opposed to alternating current, which he considered to be dangerous and exotic. The champions of alternating current were Nikola Tesla (1856–1943), a Croatian-American inventor, and George Westinghouse (1846–1914), an American inventor and industrialist. The advantage that Tesla and Westinghouse saw in alternating current was that its power can be *transformed*, or passed via electromagnetic action, from one circuit to another by a device called a **transformer**.

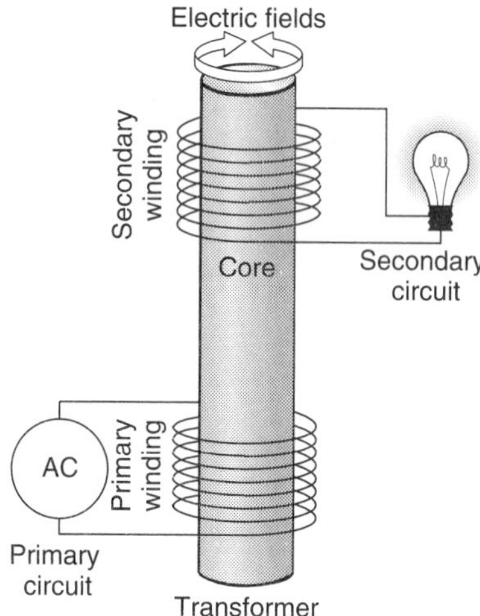

Fig. 12.2.3 - A transformer transfers power from its primary circuit to its secondary circuit. As the current in the primary circuit changes during the alternating current's cycle, the magnetic field in the transformer changes. This changing magnetic field delivers power to the current in the secondary circuit and extracts power from the current in the primary circuit. Overall, power is transferred from the power company in the primary circuit to the light bulb in the secondary circuit.

A transformer uses two of the electromagnetic principles we encountered in the previous chapter: an electric current creates a magnetic field and a changing magnetic field creates an electric field. Individually, these effects allow electricity to produce magnetism and magnetism to produce electricity. Together, they allow electricity to produce electricity.

Since a current creates a magnetic field, a changing current in one circuit creates a changing magnetic field around that circuit. But this changing magnetic field in turn creates an electric field. Such an electric field can produce a current in a second circuit, located near the first. Thus, through electric and magnetic processes, a changing current in one circuit can produce a current in a second circuit. If the changing current in the first circuit reverses directions smoothly and periodically, as it does in alternating current, the current in the second circuit will also reverse directions in just the same manner.

That's how a transformer works. Most transformers consist of two wire coils, the *primary coil* and the *secondary coil*, wrapped together around a central iron core (Fig. 12.2.3). When a current flows through the primary coil, it magnetizes the iron core and creates a magnetic field around it. If the current through the primary coil changes, the changing magnetic field around the core induces currents in the secondary coil.

One measure of the core's magnetization is the **magnetic flux** emerging from its ends. This flux takes the form of closed loops or "lines of force" that pass through the core and then return through the air outside it (Fig. 12.2.4). In a transformer, these flux loops run through both the primary and secondary coils before circling around the outsides of both coils. This flux is related to the magnetic field in an interesting way: the field points along the magnetic flux lines and its magnitude is proportional to how tightly those flux lines are packed together. Where the flux lines are closely spaced, as they are near the poles in Fig. 12.2.4, the magnetic field is relatively strong. Where the flux lines are farther apart, the magnetic field is relatively weak.

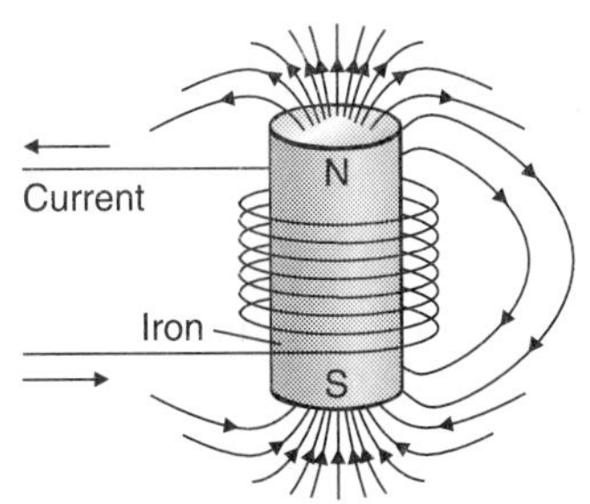

Fig. 12.2.4 - The magnetic flux around the iron core of a transformer while current flows through its primary coil. The flux forms complete loops.

When the current in the primary coil changes, the core's magnetization changes and so does the number of magnetic flux loops passing through it. A change in the magnetic flux through the iron core creates an electric field around it (Fig. 12.2.3) and can accelerate charges inside the two wire coils. In the secondary coil, this electric field pushes the charges forward and transfers power to the current. That current experiences a voltage rise as it flows through the secondary coil. In the primary coil, the field pushes the charges backward and extracts power from the current. The current experiences a voltage drop as it flows through the primary coil. Overall power is being transferred from the primary circuit to the secondary circuit (Fig. 12.2.5). And because the circuits aren't actually connected, no current can flow between them.

CHECK YOUR UNDERSTANDING #2: Use Only with AC Power

If you send direct current through the primary coil of a transformer, no power will be transferred to the secondary circuit. Explain.

Changing Voltages

The work that the electric field does on current in the secondary coil depends on how many times that coil circles the core. The more turns the coil has, the farther each charge travels in the direction of the force on it and the more work is done on it. Thus adding turns to the secondary coil increases the voltage rise across that coil.

When a transformer's primary and secondary coils have the same number of turns, the voltage drop across the primary coil is equal to the voltage rise across the secondary coil. As a result, the energy that one charge loses in going through the primary coil is picked up by another charge as it goes through the secondary coil. Since all of the power removed from the primary circuit is transferred to the secondary circuit, the currents passing through the two coils must also be equal.

Fig. 12.2.5 - When an alternating current flows through the primary coil on the left, it produces a changing magnetization in this transformer's ring-shaped iron core. An alternating current flows through the secondary coil on the right and lights the bulbs.

When a transformer's primary and secondary coils have different numbers of turns, the transfer of power is more complicated. If the secondary coil has fewer turns than the primary coil, then the voltage rise in the secondary coil is less than the voltage drop in the primary coil (Fig. 12.2.6). This transformer is called a "step-down" transformer, because the voltages in the secondary circuit are smaller than the voltages in the primary circuit. Since all of the power removed from the primary circuit must enter the secondary circuit, the secondary circuit must carry a larger current of less energetic charged particles.

Similarly, if the secondary coil has more turns than the primary coil, then the voltage rise in the secondary coil is more than the voltage drop in the pri-

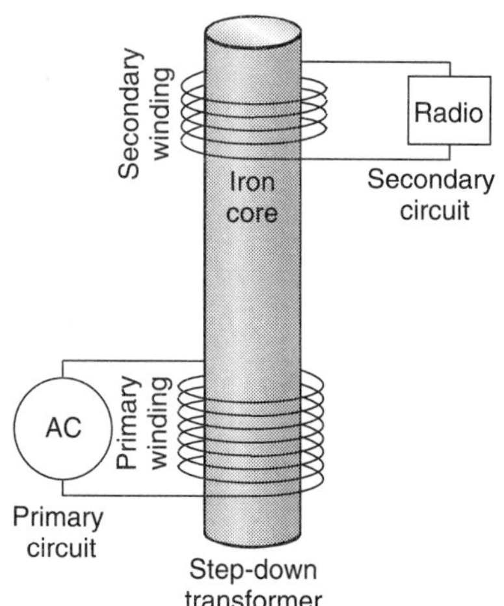

Fig. 12.2.6 - A step-down transformer has more turns in its primary coil than in its secondary coil. The voltage rise across the secondary coil is less than the voltage drop across the primary coil, hence the name "step-down." However the current flowing through the secondary circuit is larger than that flowing through the primary circuit.

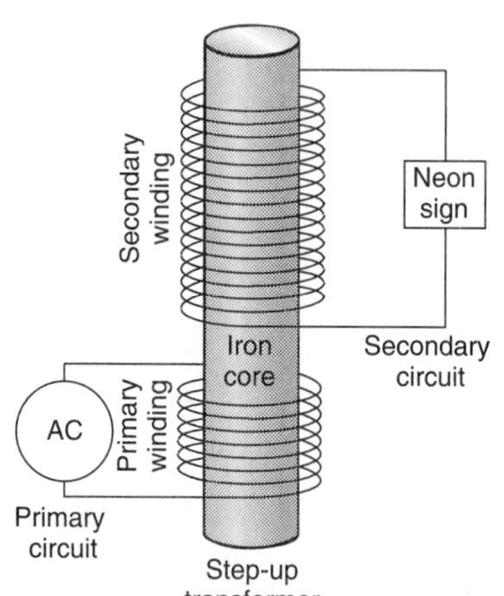

Fig. 12.2.7 - A step-up transformer has more turns in its secondary coil than in its primary coil. The voltage rise across the secondary coil is more than the voltage drop across the primary coil, hence the name "step-up." However the current flowing through the secondary circuit is smaller than that flowing through the primary circuit.

mary coil (Fig. 12.2.7). This transformer is called a "step-up" transformer, because the voltages in the secondary circuit are larger than the voltages in the primary circuit. The secondary circuit carries a smaller current of more energetic charged particles.

The voltage and current in the secondary circuit are related to those in the primary circuit by the ratio of secondary coil turns to primary coil turns. The specific relationships are

$$\text{secondary voltage} = \text{primary voltage} \cdot \frac{\text{secondary turns}}{\text{primary turns}}$$

and

$$\text{secondary current} = \text{primary current} \cdot \frac{\text{primary turns}}{\text{secondary turns}}.$$

The first relationship recognizes that the work done on a charge is proportional to the number of times it circles the transformer's core in one of the coils. The second relationship ensures that the power extracted from the primary circuit is the same as that transferred to the secondary circuit.

Transformers appear in many common electric devices. Battery eliminators for radios and tape recorders use transformers in which the primary coils have about 20 times as many turns as the secondary coils. These step-down transformers deliver low-voltage electricity to their secondary circuits (about 1/20th of line voltage at 20 times the current). Neon signs use transformers in which the secondary coils have about 100 times as many turns as the primary coils. These step-up transformers deliver high-voltage electricity to their secondary circuits (about 100 times line voltage at 1/100th the current).

Unfortunately, real transformers aren't 100% energy efficient and some of the power they extract from the primary circuit doesn't reach the secondary circuit. The wire coils waste power because they're not perfect conductors of electricity and the iron core of the transformer produces some thermal energy. Sometimes parts of the transformer move in response to the magnetic forces on them and make noise. Engineers designing transformers use the proper metal wires for the coils, carefully structure the metal cores so that they don't experience much current flow, and ensure that the devices are mechanically rigid. Large, well-designed transformers convert only about 0.5% of the power passing through them into heat.

CHECK YOUR UNDERSTANDING #3: High Intensity Lamps

In the United States, household voltage is typically 120 V. Some high-intensity lamp bulbs operate on 12 V power which they obtain from a step-down transformer. What is the ratio of turns in the coils of this transformer?

Alternating Current Power Distribution

We now have enough tools to formulate the solution to the power transmission problem. In order to transmit electric power over long distances, from the power plant where it's produced to the city where it's consumed, that power should travel as a small current of very high voltage charges. Since each charge then carries lots of electrostatic potential energy, a small current of them provides a large amount of power. Because the power loss in the wires is proportional to the square of the current it carries, a small current produces little loss. However, when that same electric power is delivered to a house, it must take the form of a larger current of lower voltage charges. Low voltage charges are safer and more practical in a house than very high voltage charges.

While it's impossible to meet these two requirements simultaneously with direct current, transformers make it possible to achieve both of them with alternating current. The power that is generated at a power station is converted to very high voltage by a step-up transformer at the power plant and travels cross-country in this form (Fig. 12.2.8). When it arrives at the destination city, the power is converted back to reasonable voltages by a step-down transformer.

Fig. 12.2.9 - This electric substation uses a large transformer to transfer power from the ultra-high voltage cross-country circuits above it to the medium-high voltage neighborhood circuits behind it. The small can-shaped transformer transfers some of this power to the low voltage circuits on the utility pole to the right.

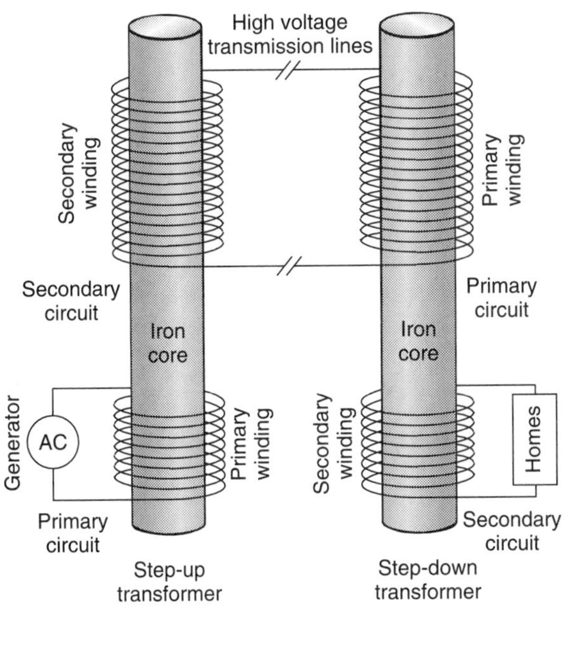

Fig. 12.2.8 - Power is transmitted cross-country by stepping it up to very high voltage near the power plant, transmitting it as a small current at very high voltage, and stepping it back down to low voltage near the houses that are to be served. The secondary circuit for the first (left) transformer is also the primary circuit for the second (right) transformer.

The generator drives a huge current through the primary circuit of the step-up transformer, producing a voltage drop of about 5000 V across its primary coil. The current flowing through the secondary circuit is only about 1/100th that in the primary circuit, but the voltage rise across the secondary coil is much higher, typically about 500,000 V. The secondary circuit is extremely long, extending all the way to the city where the power is to be used. Since the current is modest, the power wasted in heating the wires is within tolerable limits. The long stretches of wire from the power plant to the city are usually made of aluminum, which is cheaper and lighter than copper and almost as good a conductor of electricity.

Once it arrives in the city, the current passes through the primary coil of a step-down transformer, where it experiences a huge voltage drop (Fig. 12.2.9). The voltage rise across the secondary coil of this transformer is only about 1/100th as large, but the current flowing through the secondary circuit is about 100 times that in its primary circuit.

Now the voltage is reasonable for use in a city. Before entering houses, this voltage is reduced still further by other transformers. The final step-down transformers can frequently be seen as oil-drum sized metal cans hanging from utility poles (Fig. 12.2.10) or as green metal boxes on the ground (Fig. 12.2.11). Current enters the buildings at between 110 V and 240 V, depending on the local standards. While 240 V electricity wastes less power in the home wiring, it's more dangerous than 110 V power. The United States has adopted a 120 V standard while Europe has a 220 V standard.

Fig. 12.2.10 - The three metal cans on this utility pole are transformers. They transfer power from the medium-high voltage neighborhood circuits above them to the low voltage household circuits below them.

Fig. 12.2.11 - This transformer transfers power from a medium-voltage underground circuit to a low voltage underground circuit used by nearby houses. It handles 50KVA or 50,000 W of power.

But despite all these efforts, about 50% of the electric power generated in power plants is lost to thermal energy before it does useful work in our homes and offices. Efforts to reduce this waste continue as scientists and engineers look for more efficient transformers, for ways to use even higher voltage transmission systems, and for alternatives to copper and aluminum wires.

Superconducting wires, which don't waste electric power at all, would be ideal for power transmission. However, they remain impractical because they must be cooled to very low temperatures. The recently discovered high-temperature superconductors have revived interest in superconducting power transmission lines. However, these new superconductors have many practical limitations and will probably not appear in power transmission systems in the near future.

CHECK YOUR UNDERSTANDING #4: High Tension Wires

If a power utility were able to replace an existing 500,000 V transmission line with one operating at 1,000,000 V, how would that affect the power lost to heat in the wires?

Summary

How Electric Power Distribution Works: To minimize power losses in the transmission lines between power plants and cities, power distribution systems use alternating current and transformers. Near the power plant, relatively low voltage, high current electric power is converted into very high voltage, low current power for transmission on cross-country power lines. Because the power consumed by these high tension lines depends on the square of the currents they carry, the power losses are greatly reduced by this technique. When the power arrives at the city, it's converted into medium voltage, high current power for distribution to neighborhoods. Finally, in neighborhoods, step-down transformers convert this power to low voltage, very high current power for distribution to individual homes and offices.

Important Laws and Equations

1. Ohm's Law: The voltage drop across a wire is equal to the current flowing through that wire times the wire's electric resistance, or

$$\text{voltage drop} = \text{current} \cdot \text{electric resistance}. \qquad (12.2.1)$$

The Physics of Electric Power Distribution

1. The power consumed by a wire is proportional to the square of the current passing through it.

2. A transformer transfers electric power from its primary circuit to its secondary circuit.

3. The voltage rise across the secondary coil of a transformer is equal to the voltage drop across the primary coil times the ratio of the secondary coil turns to the primary coil turns.

4. The current in the secondary circuit of a transformer is equal to the current in the primary circuit times the ratio of the primary coil turns to the secondary coil turns.

5. Magnetic flux are lines of force that connect north and south magnetic poles. The magnetic field points along these flux lines and its magnitude is proportional to how tightly the flux lines are packed together.

Check Your Understanding - Answers

1. It would roughly double.

Why: Doubling the length of a wire is like placing two identical wires one after the next. If each wire uses 1 unit of power, then two wires should use roughly 2 units of power. Electric resistance is proportional to the length of a wire and inversely proportional to the cross-sectional area of that wire. Shortening and thickening a wire reduces its resistance.

2. When direct current flows through the transformer's primary coil, it creates a constant magnetic field around the iron core. Since that field doesn't change, it doesn't create any electric fields and doesn't induce current in the transformer's secondary coil.

Why: The current through the primary coil must change so that the magnetic field in the coils will change and current will be induced in the secondary coil. Transferring power from one circuit to another is so useful that there are many DC powered devices that switch their power on and off to mimic alternating current, so that they can use transformers.

3. About 10 primary coil turns for every secondary coil turn.

Why: To make the voltage rise across the secondary coil equal to 1/10th the voltage drop across the primary coil, the step-down transformer must have only 1 secondary coil turn for each 10 primary coil turns. The secondary circuit will have 10 times the current of the primary circuit.

4. It would reduce the amount of heat produced to only 25% of the previous value.

Why: A 1,000,000 V transmission line would be able to carry the same power as the 500,000 V transmission line with only half the current. Since the power wasted by the transmission lines themselves is proportional to the square of the current, halving the current reduces the power waste to 25%.

Check Your Figures - Answers

1. The current through you would double and the power you're consuming would increase by a factor of 4.

Why: Since the voltage drop across you would double, Eq. 12.2.1 indicates that the current through you should also double. That's because the average speed of the charges inside you would double. From Eq. 12.2.2, we see that doubling the current through you increases the power you're consuming by a factor of 4. It's not surprising that 220 V electric power delivers worse shocks than 110 V.

Glossary

magnetic flux The total amount of magnetic field passing through a surface or region of space.

ohm (Ω) The SI unit of electric resistance. A 1 ohm resistor exhibits a voltage drop of 1 volt when 1 ampere of current flows through it.

Ohm's law The law relating the voltage drop across an electric conductor to the electric current passing through it and to its electric resistance.

transformer A device that uses magnetic fields to transfer electric power from one circuit to another circuit. The two circuits are electrically isolated since no charges actually travel between the two circuits.

Review Questions

1. When a steady current flows through a wire, the voltage of that wire depends on where you look. Why is that voltage highest where the current enters the wire and progressively less as you look farther along the wire?

2. Why does a thick wire exhibit a smaller voltage drop when a particular current passes through it than a thin wire does when the same current passes through it?

3. Why does doubling the current passing through wires in your home cause them to waste four times as much power?

4. What is alternating about "alternating current"?

5. When an alternating current is used to power an electromagnet, how do the magnetic poles of that electromagnet behave as time passes?

6. Why does a changing magnetic field in a coil of wire exert a force on the charges inside that wire?

7. How does the changing magnetic field in the iron core of a transformer do work on the charged particles in the secondary coil of that transformer?

8. Why won't a transformer transfer power between two circuits carrying steady, constant currents?

9. Why is the work done on each charge passing through a transformer's secondary coil proportional to the number of turns that coil makes around the transformer's iron core?

10. How is a step-up transformer physically different from a step-down transformer? What is different about the ways they transfer power from the primary circuit to the secondary circuit?

11. If the power in a low-voltage transmission system is the same as that in a high-voltage transmission system, how do the electric currents in those two systems compare?

12. Why does sending electric power cross-country as a small current at high voltage waste less power than sending it cross-country as a large current at low voltage?

Exercises

1. Electric power is often sold on the basis of kilowatt-hours (kWh). A house that uses an average of 1000 W for an hour has consumed 1 kWh. Explain why a kilowatt-hour is a unit of energy.

2. Spot welding is used to fuse two sheets of metal together at one small spot. Two copper electrodes pinch the sheets together at a point and then run a huge electric current through that point. The two sheets melt and flow together to form a spot weld. Why does this technique only work with relatively poor conductors of electricity such as stainless steel and not with excellent conductors such as copper?

3. If you overload an extension cord so that it's carrying too much electric current, why does it become hot?

4. A fuse limits the current that can flow through an electric circuit to the value printed on its label. Why can putting a 50 A fuse in a circuit rated to carry only 30 A lead to a fire?

5. Your friends are installing a loft in their room and are using thin speaker wires to provide power to an extra outlet. If they only draw a small amount of current from the outlet, the voltage drop in each of the wires will remain small. Why?

6. When your friends from Exercise **5** plug a large home entertainment system into the outlet, it doesn't work properly because the voltage rise provided by the extra outlet is only 60 V. The power company provides a voltage rise of 120 V, so where is the missing voltage?

7. Many older buildings use a single electric circuit to provide power to a room's lights *and* its electric outlets. One wire delivers current to the lights and outlets and another wire returns that current to the power company. When you plug a hairdryer into an outlet in such a room and turn it on, the room lights dim. Why do the lights dim?

8. The electric company makes sure that the voltage drop across an incandescent light bulb is always 120 V. But as the bulb's filament warms up, its electric resistance increases. What happens to the current passing through the filament?

9. During a particular power surge, the voltage drop across an incandescent light bulb increases by 10%. If the bulb's electric resistance doesn't change, what effect will this surge have on the current passing through the bulb's filament?

10. Why will the power surge in Exercise **9** cause the bulb to consume approximately 20% more power than normal?

11. The recharger for a cordless electric toothbrush contains a coil of wire through which alternating current flows. This coil is magnetic, with poles at either end. How do these poles behave as time passes?

12. The cordless toothbrush in Exercise **11** contains an iron core with a coil of wire wrapped around it. When you put the toothbrush into the recharger, the coil in the recharger magnetizes the iron core and an alternating current begins to flow through the coil inside the toothbrush. How is power being transferred from the coil in the recharger to the coil in the toothbrush?

13. If the alternating current passing through a toaster actually reverses directions many times each second, are there moments when no current passes through the toaster and the toaster receives no electric power? Explain.

14. The primary coil of a transformer makes 240 turns around the iron core and the secondary coil of that transformer makes 80 turns. If the primary voltage is 120 V, why will the secondary voltage be only 40 V?

15. If a 3 A current is passing through the primary coil of the transformer in Exercise **14**, what current is passing through the secondary coil of that transformer?

16. Show that all of the power being consumed by the primary coil of the transformer in Exercises **14** and **15** is being provided to the secondary circuit by the transformer's secondary coil.

Problems

1. Your car battery is dead and your friends are helping you start your car with cheap jumper cables. One cable carries current from their car to your car and a second cable returns that current to their car. As you try to start your car, a current of 80 A flows through the cables to your car and back, and a voltage drop of 4 V appears across each cable. What is the electric resistance of each jumper cable?

2. If you replace the cheap cables in Problem **1** with cables having half their electric resistance, what voltage drop will appear across each new cable if the current doesn't change?

3. The wires in the wall of your home are probably 12 gauge copper, which exhibits 0.005 Ω of electric resistance for each meter of length. That means that if the wires leading to an outlet are 20 m long, each wire has an electric resistance of 0.1 Ω. If those wires are carrying their maximum rated current of 20 A, what voltage drop will each wire experience?

4. How much power is wasted in each wire of Problem **3**?

5. Each of the two wires in a particular 16 gauge extension cord has an electric resistance of 0.04 Ω. You're using this extension cord to operate a toaster oven, so a current of 15 A is flowing through it. What is the voltage drop across each wire in this extension cord?

6. How much power is wasted in each wire of the extension cord in Problem **5**?

7. The two wires of a high-voltage transmission line are carrying 600 A to and from a city. The voltage between those two wires is 400,000 V. How much power is the transmission line delivering to the city?

8. In bringing electricity to individual homes, the power in Problem **7** is transferred to low-voltage circuits so that the current passing through homes experiences a voltage drop of only 120 V. How much total current is passing through the homes of the city?

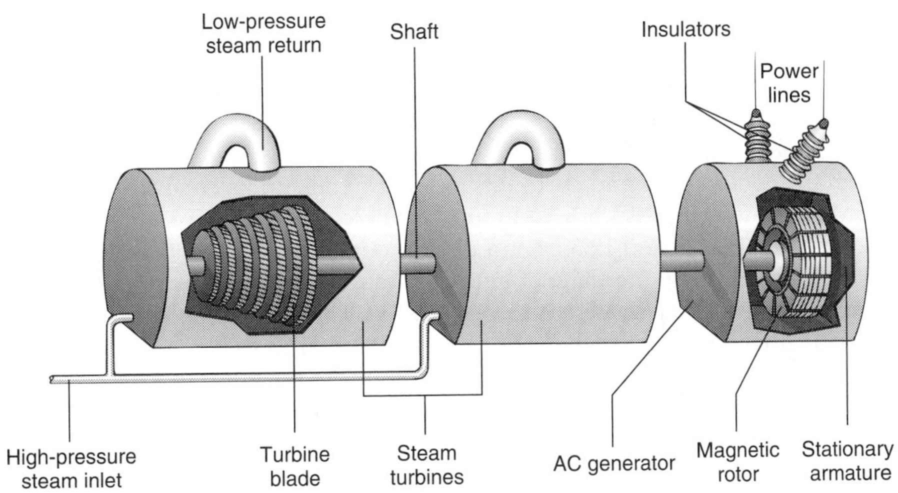

Section 12.3

Electric Power Generation

So far, we have looked at what electricity is and how it's distributed. In this section, we'll discuss how it's generated. We'll see how fossil fuels such as coal, oil, and natural gas are used to produce mechanical motion and how that mechanical motion is used to produce electric power. We'll also examine solar and wind power generation.

Questions to Think About: *How can a moving object push electric charges through a wire and produce electricity? Why are generating plants often built near bodies of water? What is the purpose of the giant cooling towers near some power plants? How does a power plant determine how much power it needs to generate and what happens to its generators when the demand for power suddenly increases or decreases?*

Experiments to Do: *There are a few household generators. One of these is a bicycle generator, a small device that uses the rotation of a bicycle's wheels to produce electricity for its lamp. Find a bicycle generator and turn its rotor (its central spindle) with your fingers. You will find that the rotor spins relatively easily when the generator is disconnected from the lamp but becomes much harder to turn when the generator is powering the lamp. The rotor contains a magnet, which moves near several coils of wire as the rotor spins. These coils can only carry current when they're connected in a circuit. How does that need for a circuit explain why the rotor is harder to turn when the generator is connected to the lamp? What is the relationship between the work you do on the generator and the power consumed by the lamp?*

Generating Electric Currents

In the last section, we saw that a change in the magnetic flux passing through a transformer's secondary coil causes current to flow in that coil. Since the magnetic flux through the secondary coil changes whenever the current through the primary coil changes, an alternating current in the transformer's primary coil induces an alternating current in its secondary coil.

But there is another way to change the magnetic flux passing through a coil of wire: move the magnetic flux. That's how a generator works. Whenever a magnet moves past a coil of wire or a coil of wire moves past a magnet, the flux through the coil changes and current flows in the coil and its circuit.

Most generators use rotary motion to produce electricity. The generator shown in Figs. 12.3.1*a* and 12.3.2 has a permanent magnet that spins between two fixed coils of wire. As the magnet spins, its magnetic flux lines sweep through the two coils and drive a current through them. This current experiences a voltage rise as it passes through the coils and a voltage drop as it passes through the light bulb, so it transfers power from the generator to the bulb.

The iron core inside each coil extends the magnet's flux lines so that they are sure to sweep through the coil each time a pole of the magnet passes by. These cores are temporarily magnetized by the nearby magnet and effectively increase its length (Fig. 12.3.1*a*). Without the iron cores, most of the rotating magnet's flux lines would bend around before passing through the entire coil (Fig. 12.3.1*b*) and the generator would be less effective at producing electricity.

A generator of this type produces an alternating current in the circuit it powers. This current flows in one direction as the magnet's north pole approaches a coil and in the opposite direction as the south pole approaches it. To generate the 60 Hz alternating current used in the United States, the generator must turn 60 times each second so that the current completes one full cycle of reversals every 1/60th of a second. In Europe, the generator must turn 50 times each second to supply 50 Hz alternating current. The generators throughout the continent-wide power distribution networks all turn together in perfect synchronization. That way, power can be redirected within each network so that any generator can provide the power consumed by any user.

Some devices require direct current electric power. A car is a good example. It generates DC electric power to charge its battery and to run its headlights, ignition system, and other electric components. While this power is actually produced by an AC generator or *alternator*, the car uses special electronic switches to send current from the alternator one way through its electric system. While the current in the alternator's coils reverses, the current through the car's electric system always travels in one direction.

Because large permanent magnets are extremely expensive, most industrial generators actually use iron-core electromagnets instead. These rotating electromagnets drive currents through generator coils just as effectively as permanent magnets would. Although these electromagnets consume some electric power, they are much more cost effective than real permanent magnets.

(*a*)

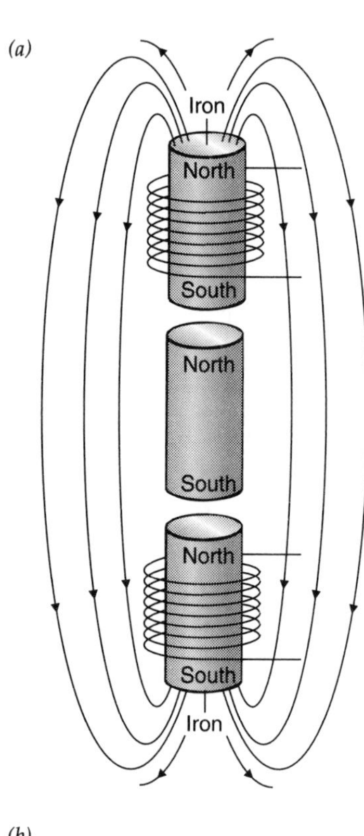

(*b*)

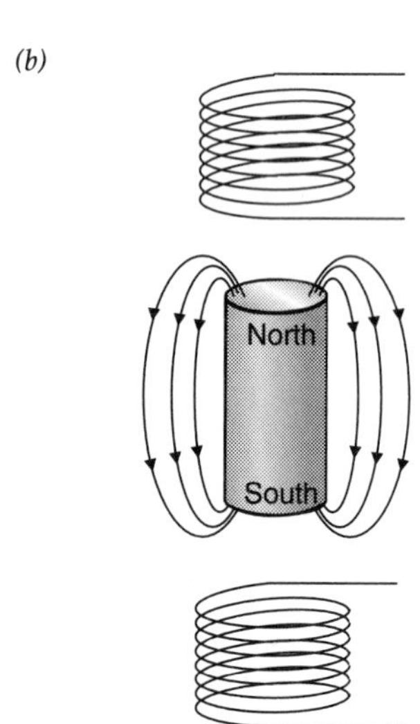

Fig. 12.3.1 - (*a*) As the rotating magnet of a generator turns, it periodically aligns with iron cores inside the generator's coils and temporarily magnetizes those cores. The changing flux through the cores and coils induces currents in the coil and generates electricity. (*b*) Without the iron cores, few of the flux lines from the rotating magnet would pass through the generator's coils.

CHECK YOUR UNDERSTANDING #1: Lights Out at the Stop Sign
When a bicycle light is powered by a wheel-driven generator, the light turns off when the bicycle comes to a stop. Explain.

Using Steam to Turn a Generator

Since the electric current extracts energy from the generator, something must do work on it to keep it turning. We can see why this work is needed by looking again at Fig. 12.3.2. As the permanent magnet turns its north pole toward the iron core above it, that core temporarily becomes magnetic with its south pole down. Opposite poles are then near one another and the permanent magnet is attracted toward the iron core. This attraction does (positive) work on the turning permanent magnet as the two poles approach but it also does negative work on the permanent magnet as the two poles separate. Overall, the iron core does zero net work on the permanent magnet.

But when the generator's coils are connected as part of a complete circuit, the currents induced in those coils make them magnetic, too. As required by Lenz's law, each magnetized coil repels the approaching permanent magnet. This effect is identical to the one we observed in electrodynamically levitated trains. An approaching magnet is always repelled by the currents it induces.

While a magnetized coil repels the turning permanent magnet, both as it approaches and as it leaves, the net work done by this repulsion isn't necessarily zero. If the current passing through the coils diminishes during the time between the permanent magnet's approach and its departure, the coil will repel the permanent magnet more strongly as it approaches than as it leaves and the net work done on the permanent magnet will be negative. Some of the permanent magnet's energy will be transferred to the electric current, which will deliver it to the devices in the circuit. The more power these devices extract from the circuit, the more work the permanent magnet must do to keep the current flowing.

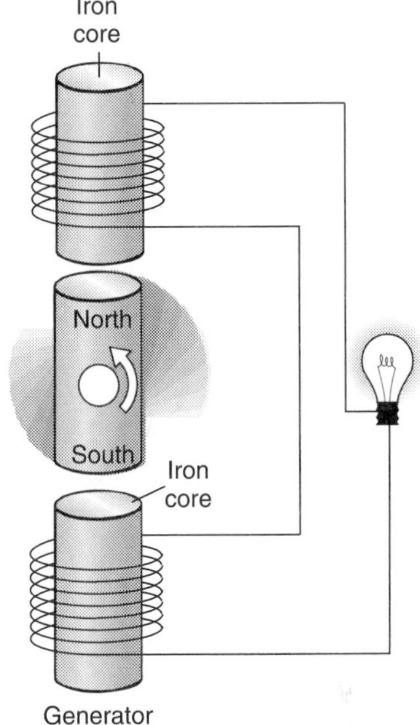

Fig. 12.3.2 - A generator works by sweeping magnetic flux through wire coils. Here a rotating magnet moves its flux past two coils and provides power to the electric circuit and the light bulb.

Since the permanent magnet loses energy to the electric current, something must exert a torque on it to keep it turning. While small generators are often turned by hand cranks, pedals, or internal combustion engines, most large power plant generators are driven by steam turbines. We encountered turbines in Section 4.4, where they appeared in jet engines. But instead of deriving their power from burning aviation fuel, power plant turbines operate on steam (Fig. 12.3.3).

A turbine resembles a fan run backward. While a fan uses rotary motion to push air from low pressure to high pressure, a turbine uses the flow of steam from high pressure to low pressure to propel a rotary motion. High-pressure steam exerts unbalanced pressures on the turbine blades and those blades rotate away from the steam. Since the steam's force and the direction of motion are both in approximately the same direction, the steam does work on the turbine and provides the ordered energy needed to spin the generator's permanent magnet.

This ordered energy comes from the steam's thermal energy. But as we saw in Section 7.1, thermal energy can't be converted directly into ordered energy. The steam turbine must and does operate as a heat engine, converting a limited amount of thermal energy into ordered energy as heat flows from a hotter object to a colder object. In the steam turbine, the hotter object is the high-pressure steam and the colder object is the outside air or water. As the steam's heat flows toward the colder world around, nature and thermodynamics allow us to extract some of that heat as ordered mechanical energy.

A steam turbine is ideally suited to electric power generation because both involve rotary motions. The shaft of the turbine, on which its blades are fastened, experiences a torque from the steam and transfers that torque to the generator. The turbine and the generator's permanent magnet rotate together, with power flowing from the high-pressure steam, to the turbine blades, to the shaft, to the permanent magnet, and finally to the electric current flowing through the generator (Fig. 12.3.4). The turbine spins at a steady, regulated rate to produce 60 Hz alternating current in the United States and 50 Hz in Europe.

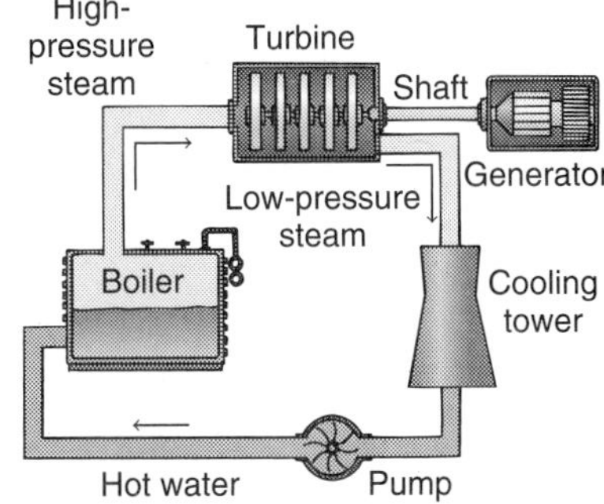

Fig. 12.3.3 - A steam turbine extracts work from steam as that steam flows from a high-pressure boiler to a low-pressure cooling tower. The cooling tower condenses the steam into hot water, which is then pumped back into the boiler.

Fig. 12.3.4 - A high-pressure steam turbine drives this electric generator, producing roughly 100 MW of electric power.

CHECK YOUR UNDERSTANDING #2: Light Work

When you turn a bicycle generator rapidly by hand, you can tell whether it's part of a completed circuit. If it isn't part of a circuit, the generator turns relatively easily. If it is, the generator is hard to turn. Why?

Heating and Cooling the Steam

Steam is produced in a boiler, where thermal energy is added to liquid water until its molecules separate into a dense, high-pressure gas. The hotter the steam, the higher its pressure and the more effective it is at spinning the turbine. This thermal energy can come from burned fossil fuels, from a nuclear reactor (see Section 18.2), or from sunlight.

As the steam flows through the turbine and does work on the rotating blades, its pressure drops. By the time the steam leaves the turbine, its pressure is only slightly above atmospheric and it's time to return it to the boiler for reuse. But the steam must first be converted back into water because the work required to pump steam into the boiler is proportional to its volume. Turning the steam into dense liquid water reduces that work enormously.

The low-pressure steam flows through a cooling tower, where it gives up heat to the surrounding air. Often that heat is used to evaporate additional water, which then condenses in the cooler air above the tower as a plume of white mist. Once the steam has given up enough heat, it condenses into water and can be returned to the boiler. Many power plants are built near large bodies of water, which also receive some of the steam's waste heat. And a few modern power plants use this waste heat for other industrial or commercial purposes such as heating buildings.

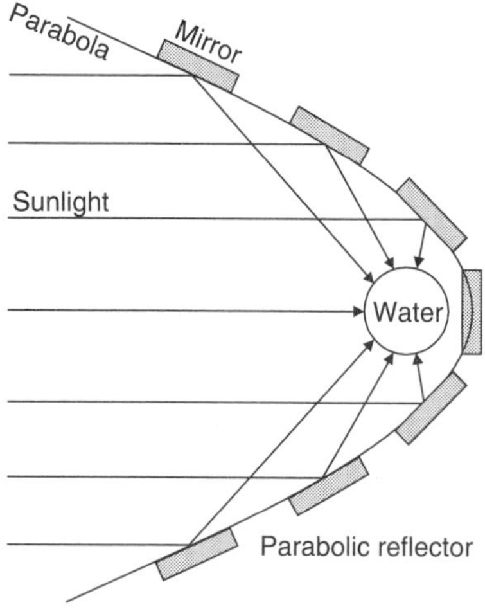

Fig. 12.3.5 - An array of mirrors can concentrate the sun's light onto a black tube containing water. The sunlight striking each mirror is redirected toward the water-filled tube. Here the mirrors form a parabolic arc and the tube rests at the focus of the parabola.

Virtually any source of heat can be used to generate electricity. Recently, several experimental power plants have been built that use sunlight to produce steam. The sun's thermal radiation delivers about 1,000 W of heat to each m^2 of the earth's surface. While that isn't enough heating for a normal boiler, concentrated sunlight can boil water fast enough to produce high-pressure steam. Most thermal solar power plants use mirrors to concentrate sunlight onto their boilers.

We will examine mirrors more carefully in later chapters. What is important here is that mirrors redirect the path of light and can be used to concentrate sunlight onto a boiler. One possible arrangement of mirrors is the parabolic reflector (Figs. 12.3.5 and 12.3.6). A parabolic mirror can take all of the light from

Fig. 12.3.6 - This solar electric power plant in the Mojave Desert uses curved mirrors to concentrate sunlight onto oil-filled pipes. The hot oil is then used to boil water and generate electricity.

one distant source (e.g. the sun) and focus that light together. If the mirror is parabolic all the away around, like a bowl, then it focuses light to a single spot. If it's parabolic only in one direction and extends as a trough in the other direction, then it focuses light to a line. In either case, it can concentrate sunlight tightly enough to boil water, produce steam, and operate a turbine.

One other interesting source of heat for a steam turbine is geothermal power. In many regions of the world, the earth's outer crust is relatively thin and high temperatures are found close to the earth's surface. Water sent far down into the ground returns to the surface as hot, high-pressure steam, which can be used to generate electricity. Natural geysers use a similar technique to produce beautiful jets of steam and water.

CHECK YOUR UNDERSTANDING #3: Following the Flow of Energy

How is energy transferred and transformed as it moves through a coal-fired electric power plant, from ancient history to your reading lamp?

Solar Cells

A second method of solar power generation is direct conversion of light into electricity by a *photoelectric cell* or photocell. A photoelectric cell uses light energy to pump charges from one of its terminals to the other. In effect, the photoelectric cell is a light-powered battery. But instead of using a chemical reaction to pump charges, as a normal battery does, the photoelectric cell uses light energy.

The photoelectric cell is built from semiconductors. When light strikes a semiconductor, that light can shift electrons from valence levels to conduction levels, so that the electrons can move through the material. As we saw in Section 11.2, this effect is the basis for xerography. Another consequence of light exposure is that electrons in the conduction levels can move from one part of the semiconductor to another, creating a temporary separation of charge. But such separations of charge are short lived because attractive forces between electrons and the positive charges they left behind quickly bring them back together.

To sustain these charge separations, the photocell must allow the electrons to flow in only one direction, so that they can't return to the positive charges. This one-way flow is achieved by turning the semiconductor into a diode. A **diode** is a one-way device for current, allowing it to flow only in one direction.

When light strikes a semiconductor diode and shifts some of its electrons to conduction levels, many of those electrons travel from the diode's electron emitting side, its **cathode,** to its electron-collecting side, its **anode**, and find it impossible to return. The anode becomes negatively charged while the cathode becomes positively charged and the diode begins to exhibit a voltage rise. It acts as a light-powered battery; a photoelectric cell.

A semiconductor diode is made by joining two different semiconductor materials. These two materials are not pure semiconductors and don't have perfectly filled valence levels and perfectly empty conduction levels. They have been **doped** with atomic impurities that either place a few electrons in the conduction levels (**n-type semiconductor**) or create a few empty valence levels (**p-type semiconductor**). These conduction level electrons or empty valence levels allow n-type and p-type semiconductors to conduct electricity, even in the dark.

But when a piece of n-type semiconductor touches a piece of p-type semiconductor, something remarkable happens (Fig. 12.3.7). Higher-energy conduction level electrons from the n-type semiconductor flow across the junction and fill in the empty lower-energy valence levels in the p-type semiconductor. Only a few electrons move because they create a charge imbalance when they do. The

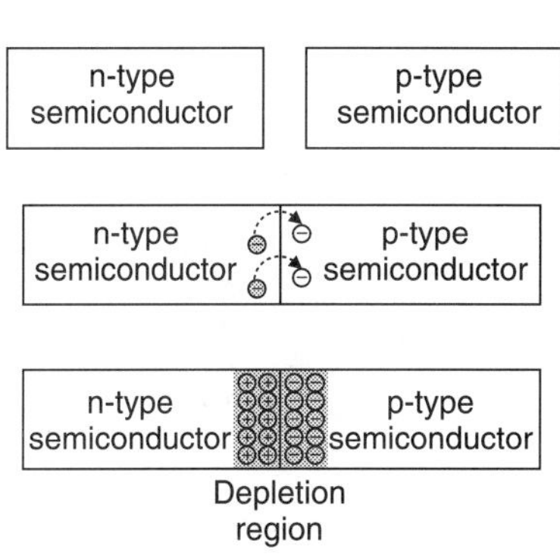

Fig. 12.3.7 - When p- and n-type semiconductors touch, conduction level electrons flow from the n-type material to the p-type material, creating a thin, electrically charged depletion region at the junction.

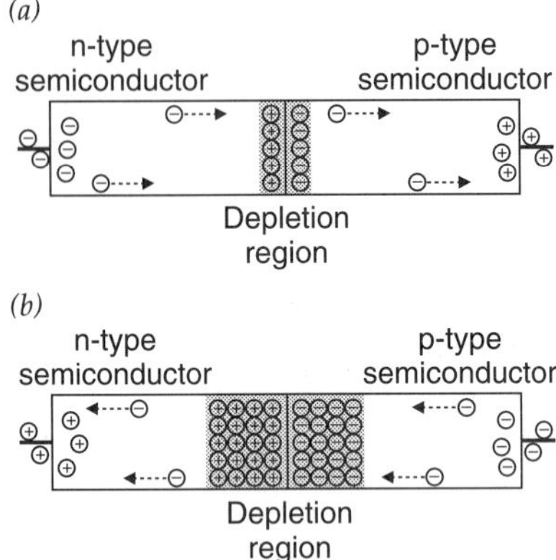

Fig. 12.3.8 - (*a*) When you add positive charges to the p-type side of a p-n junction and negative charges to the n-type side, the depletion region thins and current can flow across the junction. (*b*) Reversing the charges thickens the depletion region so that no current can flow across the junction.

n-type semiconductor then has too few electrons and a positive charge. The p-type semiconductor then has too many electrons and a negative charge.

A depletion region forms near the **p-n junction**, the place at which the p-type and n-type materials meet. Within this **depletion region**, the extra conduction level electrons from the n-type material fill the empty valence levels in the p-type material. With no conduction level electrons or empty valence levels left, the depletion region can't conduct electricity and charge can't move across the p-n junction.

However, when you attach wires to the diode and use them to add negative charge to the n-type material and positive charge to the p-type material, the depletion region becomes thinner (Fig. 12.3.8*a*). The added negative charge on the n-type material pushes conduction level electrons toward the depletion region and the added positive charge on the p-type material pulls valence level electrons away from the depletion region. As they thin the depletion region, these added charges also create a voltage drop across the p-n junction. When this voltage drop reaches about 0.6 V (for a silicon diode), the depletion region becomes so thin that conduction level electrons in the n-type material begin to flow across the junction into empty valence levels in the p-type material and the p-n junction conducts electric current.

The reverse happens when you add positive charge to the n-type material and negative charge on the p-type material (Fig. 12.3.8*b*). In that case, the depletion region becomes thicker as the added positive charge on the n-type material pulls conduction level electrons away from the depletion region and the added negative charge on the p-type material pushes valence level electrons toward the depletion region. The depletion region continues to prevent charge from moving and no current flows across the p-n junction.

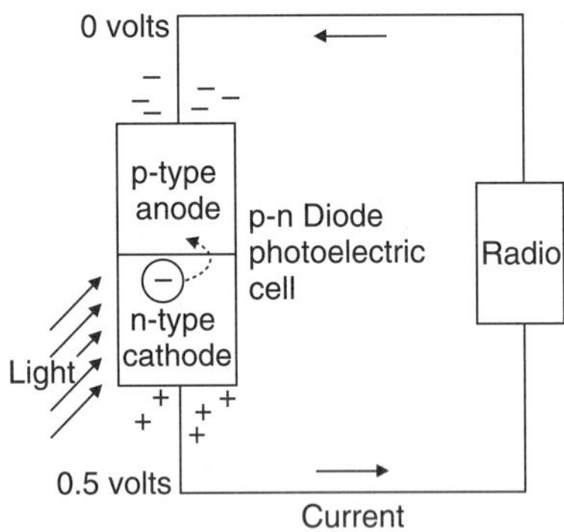

Fig. 12.3.9 - When light strikes a p-n junction, negative charge accumulates on its anode and positive charge accumulates on its cathode. A current flows in the circuit, obtaining power from the photocell and delivering it to the radio.

Since it allows current to flow in one direction but not the other, the p-n junction is a diode. It can also act as a photocell. Photons (light particles) with enough energy can propel conduction level electrons across the depletion region from the n-type material to the p-type material. Once they have made this trip, these electrons can't return. The p-type material becomes the diode's negatively charged anode and the n-type material becomes the positively charged cathode.

When exposed to light, a photocell can pump current through a circuit (Fig. 12.3.9). As this current passes from the anode to the cathode, it receives power from the photocell and experiences a voltage rise of about 0.5 V. It then delivers this power to the rest of the circuit. To obtain the larger voltage rises needed to power some devices, several photocells can be connected in a chain, just like batteries in a flashlight.

Fig. 12.3.10 - This solar cell is a diode that uses light energy to pump electrons from its n-type cathode (bottom surface) to its p-type anode (top surface). The thin metal strips on the anode collect those electrons while leaving most of the semiconductor surface exposed to light.

Photocells are usually made from thin sheets of highly purified silicon (Fig. 12.3.10). These sheets are chemically treated so that one surface is p-type and the other n-type, with a p-n junction between them. Unfortunately, fabricating such photocells is difficult and costly because it requires technology similar to that used in fabricating silicon computer chips.

A typical solar-powered house would require many m^2 of photocells, even in a sunny climate. If solar cells cost as much as computer chips, solar power would be prohibitively expensive. Fortunately, new techniques continue to simplify the fabrication of photocells and reduce their costs. Moreover, sunlight can be concentrated with mirrors or lenses so that smaller photocells can be used. Someday photocells will be cost effective for electric power generation on an industrial scale.

CHECK YOUR UNDERSTANDING #4: It's a One Way Street

What will happen if you include a p-n junction (a diode) in the AC circuit that connects the power company to your table lamp?

Wind and Water Power

Wind and water are also used to generate electricity. In a wind turbine, moving air exerts a torque on the turbine, which in turn spins a generator. The wind slows down while doing work on the turbine, as the generator transfers its energy to the electric current.

Wind power generating stations can be seen at various places in the United States, particularly in California. Sometimes whole hillsides are covered with wind turbines, all turning together. Their synchrony is more than an aesthetic issue. Each generator produces 60 Hz alternating current (50 Hz in Europe), so it must turn at just the right rate or it will be in conflict with the electric power network to which it's connected. A private wind generator can turn at whatever rate its owner chooses, but an industrial wind generator must turn at the rate specified by the power network.

Moving water, or water that is descending from an elevated river or lake, can also do work on a turbine and generate electricity. The water's total energy, the sum of its gravitational potential energy, kinetic energy, and pressure potential energy, diminishes as the generator transfers energy to the electric current.

Hydroelectric generating facilities are as old as the use of alternating current itself, since alternating current first made it feasible to transport power from generating sites to relatively distant cities. When AC hydroelectric power from Niagara Falls arrived at Buffalo, NY in 1896, it marked the beginning of the end for Edison's DC power distribution systems.

CHECK YOUR UNDERSTANDING #5: Worth the Weight

Why are hydroelectric power plants built at sites where water undergoes large changes in height?

Summary

How Electric Power Generation Works: Most electric power is generated electromagnetically by rotating magnets and coils. In a typical generator, a rotating permanent magnet or electromagnet moves past stationary wire coils. As magnetic flux from the moving magnet sweeps through each coil, it produces an electric field that pushes current through that coil. Power is transferred from the machine turning the generator to the rotating magnet, and then to the electric current. As always, energy is conserved.

In most industrial generators, the machine turning the generator is a steam turbine. Heat from burning fuel, a nuclear reactor, solar radiation, or a geothermal site boils water and produces high pressure steam. This steam does work on a steam turbine as it flows from high pressure to low pressure and the turbine spins the generator. The low pressure steam is condensed back into water and returned to the boiler to produce more high-pressure steam.

Photocells use light to separate electric charges and provide electric power. Light energy propels conduction level electrons across the junction of a semiconductor diode. Once they have crossed the junction from cathode to anode, the electrons can't go back and must flow through a circuit to return to the anode. On their way, they deliver the light's energy to that circuit.

The Physics of Electric Power Generation

1. Changes in the magnetic flux passing through a coil produce an electric field in that coil. This field exerts a force on current flowing through the coil and can do work on that current. A moving magnet can thus generate electric power.

2. While pure semiconductors have no empty valence levels and no filled conduction levels, adding impurity atoms can introduce either empty valence levels or conduction level electrons. Semiconductor doped so that it has conduction level electrons is called n-type while doped semiconductor with empty valence levels is called p-type.

3. When pieces of n-type and p-type semiconductor touch, conduction level electrons from the n-type material fill empty valence levels in the p-type material, producing a non-conducting depletion region. This p-n junction behaves as a diode, permitting current to flow only from its p-type anode to its n-type cathode.

4. Light can transfer conduction level electrons from the n-type cathode to the p-type anode of a p-n junction.

Check Your Understanding - Answers

1. When the bicycle stops, the generator's magnet no longer spins and the flux lines stop sweeping through the coils. No current is driven through the lamp circuit.

Why: Internally, a bicycle generator is just like Fig. 12.3.2. The poles of the permanent magnet must sweep through the coils in order to propel currents through the coils. When the bicycle stops moving, so do those currents.

2. When the generator is part of a circuit, current can flow through its coils and exert magnetic forces on the rotating permanent magnet. On the average, these forces extract energy from the rotating magnet.

Why: When current can flow through the generator's coils, they begin to exert repulsive forces on the rotating permanent magnet. Because current in the circuit diminishes with time, these repulsive forces do negative net work on the rotating magnet and reduce its energy. You must supply more energy with your hands, which is why the generator is hard to turn when it's electrically connected as part of circuit.

3. The energy arrived, probably as solar radiation, during the ancient period when the coal was formed and became chemical potential energy. In the power plant, coal and oxygen are combined and the energy becomes heat. The heat flows into the water and becomes thermal energy in the steam. The steam drives a turbine and the heat becomes mechanical work. The work turns the generator and becomes electrostatic potential energy (and some kinetic energy) in the current. The electrostatic potential energy heats the light bulb and becomes light.

Why: Since energy is conserved, you can follow its flow through any situation, from ancient times to the present. Although some energy is wasted along the route from coal formation to lighting your house, you can follow the path fairly easily.

4. Current will only flow through the circuit half the time and your lamp will be dim.

Why: If you include a diode in an AC circuit, it will prevent current from flowing in one direction. Current will only flow through the circuit during the half of each cycle when the diode's cathode is negatively charged and its anode positively charged. Since the lamp will receive only about half of its normal power, its filament won't reach normal operating temperature. The lamp will glow dimly but the bulb will last an extraordinarily long time. Diodes are often used in this manner to create dim light levels in lamps or appliances.

5. The greater its change in height, the more gravitational potential energy each kilogram of water has to release. This gravitational potential energy is what provides power to the electric current.

Why: Although there are power plants that use small changes in water's height to generate electricity (notably tidal generating plants), most use very large changes in height. After descending a long distance, each kilogram of water has lots of pressure potential energy or kinetic energy and can do considerable work on a turbine.

Glossary

anode The electron-collecting portion of a diode.

cathode The electron-emitting portion of a diode.

depletion region The non-conducting region near a p-n junction in which all of the valence levels are filled with electrons and there are no conduction level electrons.

diode A semiconductor device that permits electrons to flow through it only in one direction. Diodes are commonly formed by bringing n-type silicon into contact with p-type silicon.

doped Deliberately contaminated with an impurity that changes its physical properties.

n-type semiconductor A semiconductor such as silicon that contains impurity atoms such as phosphorus, arsenic, antimony, or bismuth that create conduction level electrons in the semiconductor.

p-n junction The interface between an n-type semiconductor and a p-type semiconductor that gives the diode its one-directional characteristic for electrons.

p-type semiconductor A semiconductor such as silicon that contains impurity atoms such as boron, aluminum, gallium, indium, or tellurium that create empty valence levels in the semiconductor.

Review Questions

1. As current flows through the coil of a generator, it experiences a voltage rise. What type of field does work on the current to make its voltage rise?

2. How does a generator use a magnet to produce an electric field in its coils?

3. As a generator turns its magnet past the coils of wire, it induces currents in those coils and produces electric power. Why do the forces that the coils exert on the turning magnet do negative work on the magnet and extract energy from it?

4. Why is a steam engine a form of heat engine?

5. Why are n-type and p-type semiconductors able to conduct electricity when pure semiconductors cannot?

6. Why is the depletion region around a diode's p-n junction unable to conduct electricity?

7. Why are electrons able to approach a diode's depletion region from the n-type cathode side but not the p-type anode side?

8. Why does adding positive charge to the n-type cathode side of a diode widen the diode's depletion region?

9. Suppose you installed a diode in a flashlight so that current from the battery had to flow through the diode before reaching the bulb. Would it matter which way the diode was installed? If so, why?

Exercises

1. If you turn the rotating portion of a bicycle generator slowly by hand, you'll feel it stick firmly at various orientations. What components in the generator's coils is the magnetic rotor grabbing onto?

2. If you move a magnet toward a piece of aluminum, you'll induce a current in that aluminum. The aluminum will become magnetic and will repel the approaching magnet. You will do work on the magnet as you push it toward the aluminum and this work will end up as energy in the aluminum. What forms will the energy in the aluminum take?

3. If the aluminum in Exercise **2** weren't solid but were instead a coil of aluminum wire, it would still become magnetic as you approached it with a magnet and you would still do work pushing the magnet toward the coil. If the coil included a light bulb as part of its circuit, the magnet's approach to the coil would cause the light bulb to light up. Use conservation of energy to argue that the light bulb will only light if you move the magnet forward (or backward) and that if you hold the magnet still, the bulb won't light.

4. If you push the north pole of a magnet toward an aluminum plate, a current will flow counter-clockwise in that plate. Which way will the current in the aluminum plate flow if you push the south pole of the magnet toward the plate?

5. If you hold the north pole of a magnet against an aluminum plate and then suddenly pull that magnet away, the changing magnetic field will induce a current in the aluminum. Since current contains energy, you must do work while removing the magnet. Which of the plate's induced magnetic poles points toward your magnet's north pole as you pull it away from the plate?

6. Suppose you have a coil of wire with a light bulb connecting the two ends of the coil. If you move the north pole of a magnet rapidly past this coil, you will induce a current in the coil and the bulb will light. The coil becomes magnetic with its north pole facing the north pole of your magnet. The coil will repel your magnet as it approaches and as it leaves. The repulsion is stronger as your magnet approaches the coil than as it leaves the coil. Why?

7. Is the total net work that you do while moving your magnet past the coil in Exercise **6** positive, negative, or zero? Why?

8. Draw a modified version of Fig. 11.2.4 to show roughly which levels of an n-type semiconductor are filled with electrons. Do the same for a p-type semiconductor.

9. Adding electrons to a pure semiconductor allows it to conduct electricity. Why does adding just the right number of electrons to a p-type semiconductor turn it from an electric conductor into an insulator?

10. Removing electrons from a pure semiconductor allows it to conduct electricity. Why does removing electrons from an n-type semiconductor turn it from an electric conductor into an insulator?

11. When you touch a piece of n-type semiconductor to a piece of p-type semiconductor, electrons move from the n-type semiconductor to the p-type semiconductor. Explain this electron movement using the principle that objects accelerate so as to reduce their potential energies as quickly as possible.

12. Draw a modified version of Fig. 11.2.4 to show which levels in the depletion region of a p-n junction are filled with electrons.

13. When you add electrons to the n-type semiconductor side of a p-n junction, they're attracted toward the junction. Why?

14. Light energy can send an electron across the p-n junction of a photocell, from the n-type cathode to the p-type anode. Why can't that electron return across the junction?

15. When a photocell operates from sunlight, it's a form of heat engine. Explain.

16. If a photocell isn't part of a circuit, it will soon stop transferring electrons from its n-type cathode to its p-type anode. What prevents light from transferring electrons indefinitely?

17. Some light fixtures have a low-light setting that uses a diode. Instead of flowing directly to and from the light bulb, alternating current from the power company must first pass through the diode. Why does the diode halve the average power reaching the light bulb?

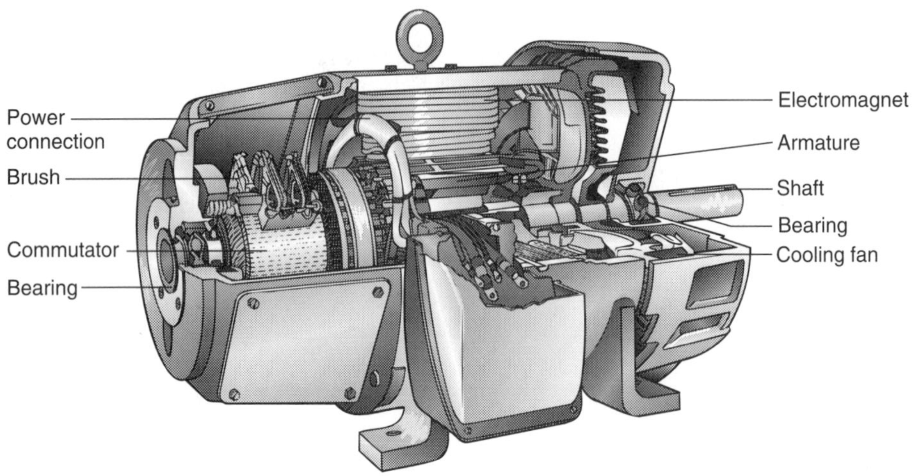

Section 12.4

Electric Motors

Electric motors spin the parts of many household machines. Sometimes this rotary motion is obvious, as in a fan or mixer, but often it's disguised, as in the agitator of a washing machine or the window opener of an automobile. Motors come in a variety of shapes and sizes, each appropriate to its task. No matter how much torque or power a motor must provide, you can probably find one that's suitable. Some motors operate from direct current and can be used with batteries while others require alternating current. There are even motors that work on either type of current.

In this section, we'll examine the basic mechanisms of electric motors. We'll look at what makes a motor turn and how motors differ from one another. As we do, we'll revisit many of the relationships between electricity and magnetism that we've been studying since Chapter 11.

Questions to Think About: *How can magnetic forces cause something to spin? If magnetic forces cause motors to spin, why can't a motor be built exclusively from permanent magnets? What determines which way a motor turns? Why are some motors safe for use with flammable chemicals while others are not?*

Experiments to Do: *Take a look at several different electric motors. You'll find motors in a cassette tape recorder, a room fan, a kitchen mixer, and in the starter of a car. Motors in the cassette recorder and the automobile starter both run on DC power, but one consumes far more electric power and provides far more mechanical power than the other. What aspects of a starter motor allow it to handle so much power?*

Motors in the fan and the mixer both run on AC power, but their internal structures are quite different. The fan motor starts slowly and gradually picks up speed while the mixer motor rotates at almost full speed only moments after you turn it on. How do the initial torques provided by those two motors compare? What do you think determines how fast these two motors turn?

What Spins an Electric Motor

The rotor of an electric motor needs a torque to start it spinning. This torque is normally produced by magnetic forces, exerted between magnetic poles on the rotor and those on the motor's stationary shell. Attractive or repulsive forces pull or push on the outside of the rotor, producing a torque that makes the rotor spin faster and faster until friction or the objects attached to it reduce its net torque to zero so that it reaches a steady angular velocity.

Both the rotor and the motor's fixed shell are magnetic. The forces between these magnets are what produce the torque. But while permanent magnets are often used in electric motors, at least some of a motor's magnets must be electromagnets. That's because the motor can only keep turning if some of its magnetic poles change or move as the rotor spins. That way, while the rotor turns to bring opposite magnetic poles as close together as possible, the poles keep changing or moving so that the rotor finds itself perpetually chasing the optimum arrangement of poles.

A simple motor is shown in Fig. 12.4.1. In this device, the rotor is a wire coil with current flowing through it. Since electric currents create magnetic fields, the coil is magnetic. Its magnetic field is similar to that of a permanent magnet, with its north and south poles shown in the figure.

The coil starts off in Fig. 12.4.1*a* with its magnetic poles arranged horizontally. Since opposite poles attract, the coil experiences a torque that acts to spin the coil counter-clockwise. The coil undergoes angular acceleration and begins to rotate counter-clockwise (Fig. 12.4.1*b*). The torque continues until the coil's poles reach the opposite poles of the stationary magnets. There is then no torque on the rotor because the magnetic forces act directly toward or away from the axis of rotation. The rotor is in a stable equilibrium.

But just as the opposite poles reach one another, the current in the coil reverses directions (Fig. 12.4.1*c*). Now the coil's magnetic poles are close to like

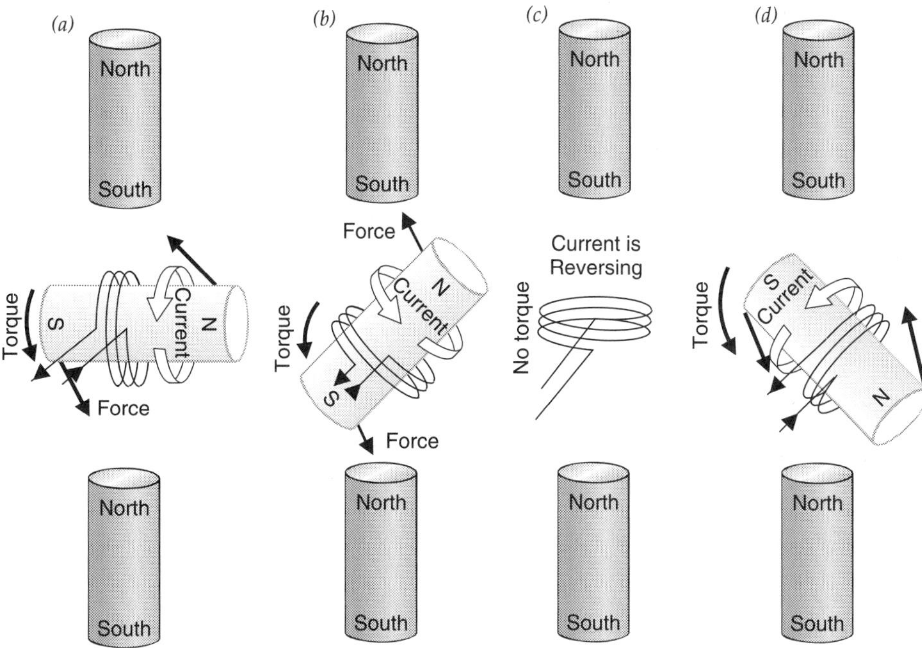

Fig. 12.4.1 - A simple motor consists of a coil rotating between two permanent magnets. (*a*) The coil's magnetic poles are attracted toward opposite poles of the stationary magnets. (*b*) The coil turns to bring the opposite poles as close to one another as possible, but just as it arrives (*c*) the current passing through it reverses directions. (*d*) The coil's magnetic poles also reverse and it continues turning.

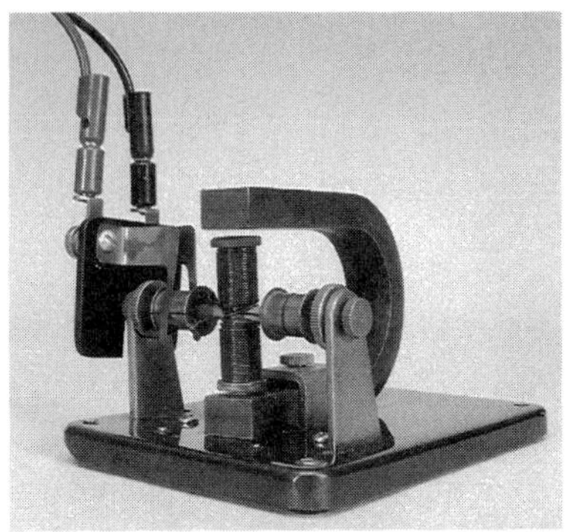

Fig. 12.4.2 - Current flows into and out of the rotor of this DC motor through the metal brushes on the left. These brushes touch the rotor's commutator so that current in the rotor's coil reverses every half turn.

poles of the stationary magnets and like poles repel one another. The rotor is suddenly in the worst possible orientation and must turn upside down to be optimally oriented again. Because it already has angular momentum, the coil continues its counter-clockwise rotation and once again experiences a counter-clockwise torque from the magnets.

But even when the coil has turned completely upside down, its travels aren't over. The current in the coil reverses again and the coil is obliged to continue its counter-clockwise rotation. Since the coil's current reverses every time it completes half a turn, the coil never stops. It keeps experiencing counter-clockwise torques and spins forever.

The rotor's endless pursuit of the optimal orientation is reminiscent of Sisyphus's efforts in Greek mythology. Sisyphus was condemned to spend eternity rolling a heavy stone up a hill, only to have it roll to the bottom every time he reached the top. This same endless pursuit appears in electric motors, although the way in which it's achieved varies from motor to motor. Let's examine how this effect is implemented in several different types of motors.

CHECK YOUR UNDERSTANDING #1: Like a Rolling Stone

Why can't you build an electric motor entirely out of permanent magnets?

DC Motors

Powering an electric motor with a battery isn't easy. The battery's direct current is fine for electromagnets, but because there's nothing that naturally changes with time, the poles of the electromagnets stay the same forever. Since a motor needs magnetic poles that change periodically, something must reverse the current flow through the electromagnets at the proper moments.

In most DC electric motors, the rotor is an electromagnet that turns within a shell of stationary permanent magnets (Fig. 12.4.2). To make the electromagnet stronger, the rotor's coil contains an iron core that's magnetized when current flows through the coil. The rotor will spin as long as this current reverses each time its magnetic poles reach the opposite poles of the stationary magnets.

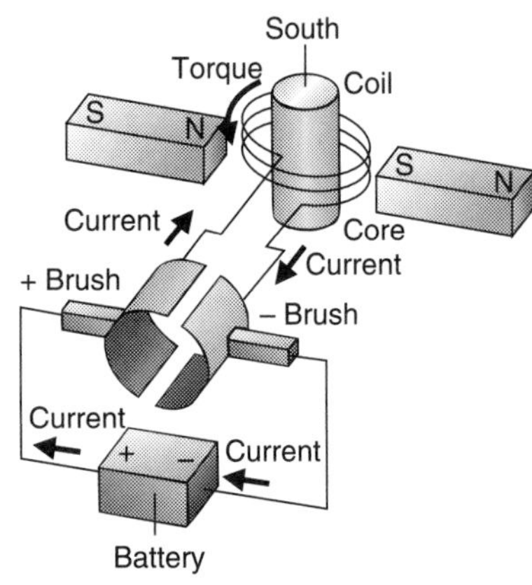

Fig. 12.4.3 - A commutator on the rotor of a DC motor connects its coil to the source of electric power. The commutator turns with the rotor and reverses the direction of current flow in the coil once every half turn of the rotor. Just as the rotor's north pole reaches the south pole to its left, the current in the coil reverses and so do the rotor's poles.

The most common way to produce these reversals is with a *commutator*. In its simplest form, a commutator has two curved plates that are fixed to the rotor and connected to opposite ends of the wire coil (Fig. 12.4.3). Electric current flows into the rotor through a conducting brush that touches one of these plates and leaves the rotor through a second brush that touches the other plate. As the rotor turns, each brush makes contact first with one plate and then with the other. Each time the rotor turns half of a turn, the brushes exchange plates. As a result, the direction of current flow around the coil reverses every half turn and the DC motor spins forever.

But the DC motor depicted in Fig. 12.4.3 has problems. First, there's nothing to determine which way the motor should turn when it starts, so that it will start randomly in either direction. Second, because there are times when a brush touches either both commutator plates at once or neither of them, the motor will sometimes not start at all.

To start reliably, the motor must make sure that its brushes always send current through the rotor and that there are no short circuits through the commutator itself. In most DC motors, this requirement is met by having several coils in the rotor, each with its own pair of commutator plates. As the rotor turns, the brushes supply current to one coil after another. The rotor is constructed so that each of its coils receives power only when it's in the proper orientation to experience a strong torque in the desired direction. Since the brushes are wide enough that they always supply current to at least one coil but not so wide that they directly connect the brushes to one another, the rotor will always start spinning when it's turned on.

The DC motor turns at a speed that depends on the amount of current flowing through it and on its load—the objects it's turning. The more current you send through the rotor's coils, the stronger the torque it experiences. It will spin faster and faster until something balances this torque. If nothing is attached to the motor, friction will eventually limit the rotor's rotational speed. But if you make the motor turn the wheels of a toy train, then torque from the railroad track will keep the motor's rotational speed from increasing forever. To make the train move faster, you must increase the current through the motor so that its rotor experiences more magnetic torque.

While the direction in which the rotor turns depends on the motor's asymmetry, it also depends on the direction in which current flows through the entire motor. If you reverse that current, the rotor will begin turning backward. Its electromagnets will still turn on and off as before, but now the magnetic forces between the rotor and the stationary permanent magnets will be reversed. Instead of being attractive, they will be repulsive or vice versa. The torque on the rotor will be reversed, too, and the motor will spin backward. So to make a toy train move backward, you simply reverse the direction of current flow through its DC motor.

DC electric motors of this sort are used in a wide variety of battery operated devices, from toy cars to electric screwdrivers. Unfortunately, their brushes experience mechanical wear as their rotors turn and eventually wear out and must be replaced. Furthermore, the brushes produce mechanical interruptions in the flow of current and these interruptions often create sparks. Motors with brushes are unsuitable for some environments because their sparking can ignite flammable gases. Sparking also produces radio waves so that an automobile's DC motors often interfere with its radio reception.

CHECK YOUR UNDERSTANDING #2: Energy Shortage
As the batteries in a toy car run out of energy, the car slows down. Explain.

Universal Motors

Before looking at true AC electric motors, let's look at an intermediate type of motor called a universal motor. This motor can run on either DC or AC electric power. A true DC motor can't tolerate AC power because its rotational direction will reverse with every half cycle of the power line and it will simply vibrate in place. A true AC motor can't tolerate DC power because, as we'll soon see, it depends on the power line's reversing current to keep the rotor moving.

However, if you replace the permanent magnets of a DC motor with electromagnets and connect these electromagnets in the same circuit as the commutator and rotor, you will have a universal motor (Fig. 12.4.4). This motor will spin properly when powered by either direct or alternating current.

If you connect DC power to a universal motor, the stationary electromagnets will behave as if they were permanent magnets and the universal motor will operate just like a DC motor. The only difference is that the universal motor will not reverse directions when you reverse the current passing through it. It will continue turning in the same direction because reversing the current through the rotor also reverses the current through the electromagnets. Since the universal motor contains no permanent magnets, every pole in the entire motor changes from north to south or from south to north. Because all the poles change, the

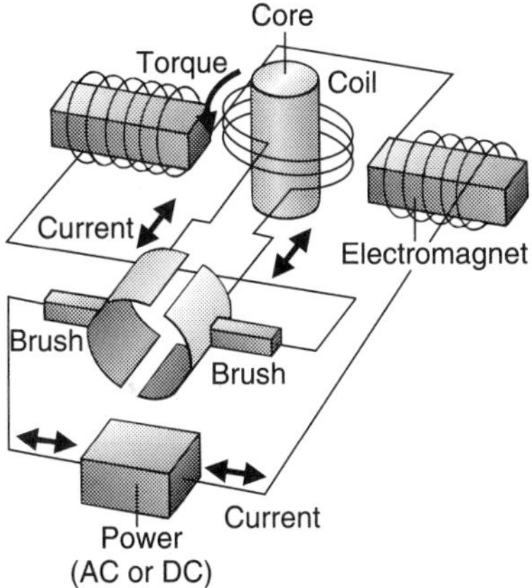

Fig. 12.4.4 - Because the stationary magnets of a universal motor are actually electromagnets, the motor doesn't notice changes in the direction of current flow from the power source—every pole in the entire motor reverses, leaving the forces between those poles unaffected. The universal motor works equally well on DC or AC electric power.

Fig. 12.4.5 - The top of this kitchen mixer contains a universal motor, which turns the mixing blade through a series of gears. The motor's graphite brushes can be serviced through black plastic ports, one of which is visible near the center of mixer. The speed switch controls the current to the motor.

motor's behavior does not. It keeps turning in the same direction. If you really want to reverse the motor's rotational direction, you must rewire the stationary electromagnets to reverse their poles.

Since the universal motor always turns in the same direction, regardless of which way current flows through it, it works just fine with AC electric power. There are moments during the current reversals when the rotor experiences no torque, but the average torque is still high and the rotor spins as though it were connected to DC electric power. Like a DC motor, its speed is governed by the amount of current flowing through it and by the torque exerted on it by whatever it's turning.

Universal motors are commonly used in kitchen mixers (Fig. 12.4.5), blenders, and vacuum cleaners. While these motors are cheap and reliable, their graphite brushes eventually wear out and must be replaced. To repair a motor with a worn brush, you simply remove what's left of the old brush and replace it with a new one from the hardware store. Some appliances even provide access ports through which you can replace the brushes without disassembling the motor.

CHECK YOUR UNDERSTANDING #3: Doing the Twist

A vacuum cleaner is advertised as having a "powerful, 8 amp motor." What are those 8 amps of current doing?

Synchronous AC Motors

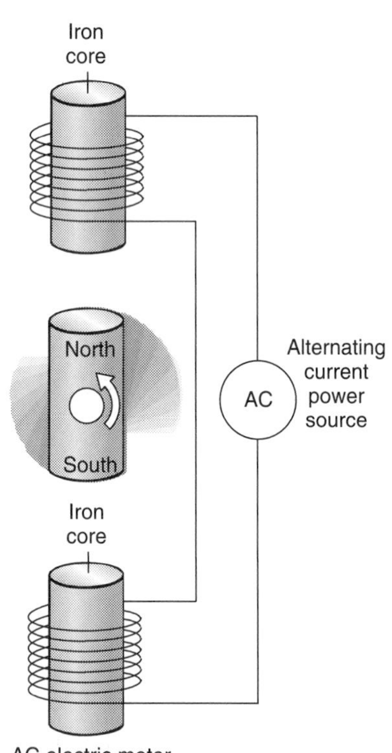

Fig. 12.4.6 - A synchronous AC motor uses electromagnets to turn a magnetic rotor. Each time the current reverses, so do the poles of the electromagnets. This motor turns once per cycle of the AC current.

Some motors are designed to operate exclusively on alternating current. One such motor is shown in Fig. 12.4.6. This motor is essentially identical to the generator from the previous section because generators and motors are closely related. A generator uses work to produce electric energy while a motor uses electric energy to produce work.

The rotor in Fig. 12.4.6 is a permanent magnet that spins between two stationary electromagnets. Because the electromagnets are powered by alternating current, their poles reverse with every current reversal. The rotor spins as its north pole is pulled first toward the upper electromagnet and then toward the lower electromagnet. Each time the rotor's north pole is about to reach the south pole of a stationary electromagnet, the current reverses and that south pole becomes a north pole. The rotor turns endlessly, completing one turn for each cycle of the AC current. Because its rotation is perfectly synchronized with the current reversals, this motor is called a synchronous AC electric motor.

Synchronous AC motors advance the hands in some electric clocks. These motors follow the cycles of the power line exactly and thus keep excellent time. However, synchronous motors are hard to start because their rotors must be spinning rapidly before they can follow the reversals of the electromagnets. Because they must be brought up to speed by other motor techniques, synchronous AC motors are only used when a steady rotational speed is essential.

However, synchronous motors illustrate an important point about motors and generators: they're essentially the same devices. If you connect a synchronous AC motor to the power line and let it turn, it will draw energy out of the electric circuit and provide work. But if you connect that same motor to a light bulb and turn its rotor by hand, it will generate electricity and light the bulb.

When the motor is acting as a generator, the turning magnetic rotor is inducing current in the coils of the electromagnets. But when the motor is acting as a motor, the turning magnetic rotor still affects the currents in the coils. In general, the currents in the electromagnets affect the motion of the rotor and the motion of the rotor affects the currents in the electromagnets. Whether the motor acts as a generator or a motor depends on which way energy is transferred.

The motor acts as a generator if you exert a torque on the rotor in the direction it's turning. You are then doing work on the rotor and this work ends up as electric energy. The motor acts as a motor if you exert a torque on the rotor opposite the direction it's turning. It's then doing work on you and this work is obtained from electric energy. Although some energy is wasted as thermal energy, it's always conserved. Thus the more work the motor does, the more electric energy it consumes. Similarly, the more work you do on the motor, the more electric energy it produces.

A motor's coils become hot when large currents flow through them. Whether a motor is consuming or producing electric power, it will overheat and burn out if it's required to handle too much current. Failures of this type occur in overloaded motors and in power plant generators during periods of exceptionally high electric power usage. Circuit breakers are often used to stop the current flow before it can cause damage.

AC synchronous motors aren't the only motors that can generate electricity. Most other motors will generate electricity when something does work on them. For example, a DC motor will generate direct current when you spin its rotor. In fact, a car's alternator is essentially a DC motor that's turned by the car's engine and powers the car's electric system. One exception to this rule is the universal motor, which contains no permanent magnets and thus can't begin to generate electricity.

CHECK YOUR UNDERSTANDING #4: Easy Come, Easy Go

A wind-turbine generator turns its magnetic rotor at a steady pace and pumps AC electric current through the power lines. When the wind dies, the turbine keeps turning anyway. Why?

Induction AC Motors

Some alternating current motors have rotors that are neither permanent magnets nor conventional electromagnets. These rotors are made of a non-magnetic metals such as aluminum and have no electric connections. But their electric isolation doesn't keep these rotors from becoming magnetic. When an aluminum rotor is exposed to changing magnetic fields, currents begin to flow through it and these induced currents make the rotor magnetic.

Motors that use induction to create magnetism in their rotors are called induction motors. Induction motors are probably the most common type of AC motor, appearing in everything from household fans to industrial pumps to cable-lift elevators. They provide lots of torque, start easily, and are inexpensive.

An induction motor works by moving a magnetic field around the rotor. The stationary shell surrounding the rotor contains a sophisticated electromagnet called a *stator*. While the stator doesn't move, the field it produces does. Through a clever use of various electromagnetic devices, the stator is able to create magnetic poles that move in a circle and travel around and around the rotor. In Fig. 12.4.7, the stator's north pole travels counter-clockwise around the rotor. Since the pole's motion is based on current reversals of the AC power line, the north pole usually travels once around the rotor for every full cycle of the current. A south pole also circles the rotor on the opposite side.

These moving magnetic poles pull the rotor along with them. They induce magnetic poles in the rotor and the rotor experiences magnetic torques. The rotor begins to spin, rotating faster and faster until it's turning almost as rapidly as the

(a)

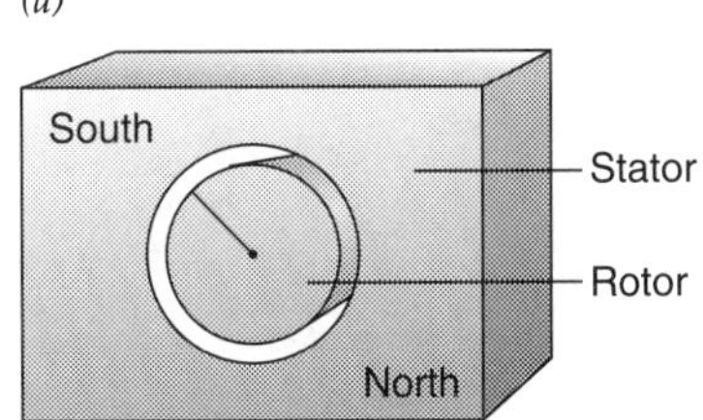

(b)

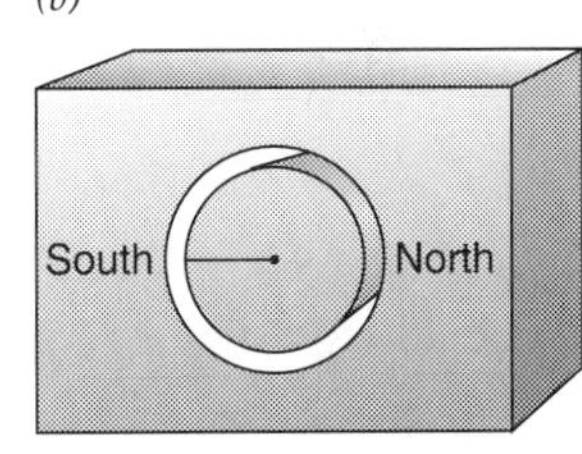

(c)

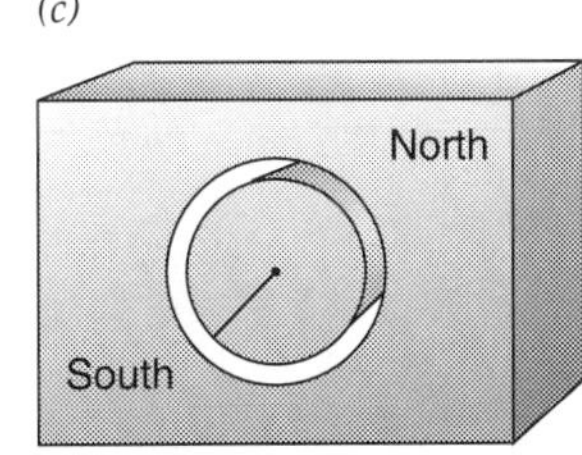

Fig. 12.4.7 - The magnetic poles produced by the stator of an induction motor circle around the rotor. Their moving magnetic fields induce currents in the metal rotor and it becomes magnetic, too. The rotor spins in order to keep up with the circling magnetic poles.

Fig. 12.4.8 - This 1 horsepower (750 W) induction motor is used in industrial and commercial equipment. The projection on top contains a capacitor that helps produce the rotating magnetic field the motor needs to spin its rotor.

moving magnetic poles of the stator.

This same effect appears in electrodynamically levitated trains. When you slide a strong magnet across a metal surface, magnetic drag forces tend to reduce the magnet's velocity relative to the metal surface. The metal surface acts to slow the magnet down while the magnet acts to speed the metal surface up.

In the induction motor shown in Fig. 12.4.8, the role of the moving magnet is played by the stator's circling magnetic poles. Even though no object is moving, the circling magnetic poles are perceived by the rotor as magnets that are traveling around it. The role of the metal surface is played by the rotor, which responds magnetically to the circling poles. The rotor acts to slow down the circling magnetic poles while the circling magnetic poles act to speed the rotor up. The stator is fixed and can't turn, but the rotor is free to spin. And so it does.

The rotor turns faster and faster until its surface moves right along with the circling magnetic poles. At that point, the magnetic poles no longer move relative to the rotor's surface, so they don't induce any currents in it. In actual use, the rotor never quite reaches this top speed. It lags a little behind so that it continues to experience a small magnetic torque. This torque keeps it rotating against the slowing effects of friction or whatever devices the motor is turning.

The rotor's direction of rotation is determined by the way in which the stator's magnetic poles circle. Most household induction motors make these poles circle with the help of a *capacitor*, a device we'll examine in the next chapter. To make the motor turn backward, the capacitor must be reconnected so that the magnetic poles circle backward. Since there are no electric connections to the rotor, it needs no brushes and requires little maintenance. Induction motors normally turn for years before failing. When they finally break, it's usually because their bearings wear out or because the stator coils overheat and burn up.

CHECK YOUR UNDERSTANDING #5: Low, Medium, and High

The induction motor in an electric fan often has three different rotational speeds. What changes to determine those different speeds?

(a)

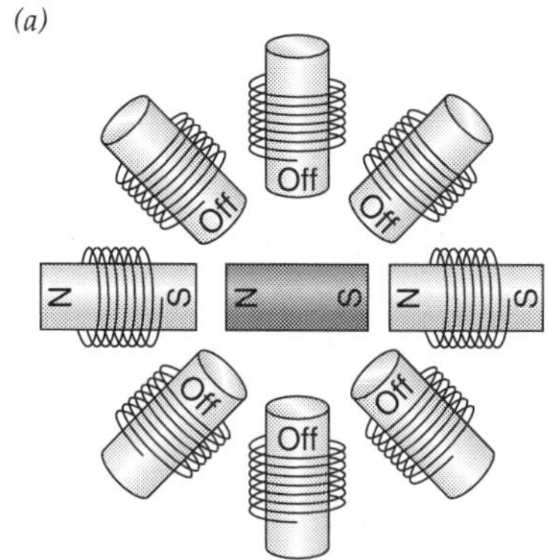

(b)

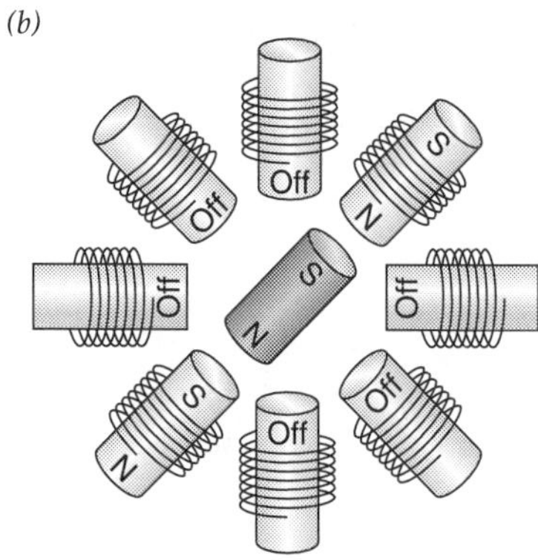

Fig. 12.4.9 - In a stepping motor, the permanent magnet rotor is attracted first to one pair of electromagnet poles and then to another. The rotor moves in discrete steps, pausing at each orientation until a computer activates a different set of electromagnet coils.

Stepping Motors

Many computerized devices use special motors that control the angles of their rotors. Instead of turning continuously, these rotors turn in discrete steps, so the motors are called stepping motors. The rotor of a stepping motor is a permanent magnet that's attracted sequentially to the poles of several stationary electromagnets (Fig. 12.4.9). These electromagnets are turned on and off in a carefully controlled pattern so that the rotor's magnetic poles move from one electromagnet to the next. The rotor is attracted so strongly by the nearby electromagnets that it snaps almost instantly from one pole to the next as the computer shifts the flow of currents through the electromagnets.

The speed with which the rotor can step from pole to pole is determined mostly by the rotor's moment-of-inertia. The smaller its moment-of-inertia, the more rapidly the rotor can respond to magnetic torques and the more steps it can make each second. Thus small stepping motors can respond more quickly than larger ones. Stepping motors are used in floppy disk drives, pen plotters, and numerically controlled machine tools.

CHECK YOUR UNDERSTANDING #6: Red Light, Green Light

After it makes a step from one pole to the next, the rotor oscillates back and forth briefly about the new pole. Why?

Summary

How Electric Motors Work: The rotor of an electric motor is spun by a magnetic torque. This torque is produced by the attractive or repulsive forces between magnetic poles on the rotor and poles on the motor's stationary shell. The rotor turns so as to bring opposite poles as close together as possible. But some of the motor's magnets are electromagnets that can exchange poles. Just as the rotor is about to reach its optimal orientation, some of the poles interchange and the rotor must continue turning in order to bring the new opposite magnetic poles as close together as possible. Thus the rotor spins forever, perpetually seeking an optimal orientation it can never reach.

Induction motors operate on a slightly different principle. These motors use rotating magnetic poles in a stationary stator to pull a metal rotor around in a circle. Although the rotor has no electric connections, induced currents make it magnetic and it experiences magnetic drag forces in the presence of the stator's moving poles. These drag forces produce a torque on the rotor, which spins so as to keep up with the rotating poles of the stator.

The Physics of Electric Motors

1. When a stationary electromagnet does work on a passing magnet, energy is removed from the current flowing through the electromagnet and transferred to the passing magnet. This effect is the basis for an electric motor.

2. When a passing magnet does work on a stationary electromagnet, energy is removed from the passing magnet and transferred to the current flowing through the electromagnet. This effect is the basis for an electric generator.

Check Your Understanding - Answers

1. In such a motor, the rotor would quickly orient itself so that its magnetic poles were as close as possible to opposite poles of the stationary magnets. After a brief period of settling down, the rotor would become motionless.

Why: There are two ways to look at this problem. First, the rotor needs a moving destination. Its magnetic poles must always seek the opposite poles of the stationary magnets but never quite reach them. If the destination stops moving, then so does the motor. Second, the motor needs a source of power to start it turning and to keep it turning when outside torques try to slow it down. Permanent magnets can't provide such power but electromagnets can.

2. The battery is no longer able to pump enough current through the rotor of the toy car's electric motor. The magnetic forces and torques in the motor decrease and it's not able to propel the car as rapidly.

Why: When batteries run out of chemical potential energy, they still give each charge that they pump the usual amount of energy but they can't pump those charges as quickly. The current the battery can send through the motor decreases and so does the motor's speed. Eventually the motor has so little torque that it can't even overcome its own internal friction and it stops turning altogether.

3. They are flowing through the stationary electromagnets, brushes, commutator, and rotor coils of a universal motor.

Why: The current flowing in and out of the vacuum cleaner through its cord is powering electromagnets in both the stationary and rotating portions of the vacuum cleaner's universal motor. The more current that flows through those electromagnets, the stronger the magnetic forces and the more mechanical power the motor can deliver to the vacuum cleaner's fan.

4. When the wind stops doing work on the generator, the generator becomes a motor and starts doing work on the wind.

Why: The wind turbine's generator is also a motor. If the wind is blowing hard enough to do work on the rotor, the generator will produce electric power in the power line. But if the wind is not blowing hard enough, electric power from the power line will begin turning the generator as motor and the rotor will begin doing work on the wind. The wind turbine will act as a fan. Instead of using wind to generate electricity, electricity will be creating wind.

5. The speed at which the magnetic poles circle in the stator.

Why: By varying the way electric current flows through the electromagnets in the stator, you can change the speed at which the magnetic poles circle the rotor. When the switch is set at high speed, the poles circle once every cycle of the alternating current. At medium speed, they circle once every two cycles. At low speed, they circle once every four cycles.

6. The rotor undergoes angular acceleration as it turns toward its new orientation. When it arrives, it has angular momentum and kinetic energy and must get rid of both of these conserved physical quantities. It oscillates back and forth as it eliminates excess energy.

Why: Each time it advances by one step, the stepping motor must start the rotor turning and then stop it. Getting rid of the kinetic energy and angular momentum that the rotor acquires as it starts turning is difficult. Fortunately magnetic torques are strong and can easily remove unwanted angular momentum. But getting rid of excess energy is harder. The moving rotor releases energy by generating currents in the stationary electromagnets. But it may oscillate back and forth about its equilibrium position during the process.

Review Questions

1. Why must an electric motor include at least one electromagnet?

2. In an electric motor, how do the magnetostatic forces between the poles of the rotor and those of the stationary magnets exert a torque on the rotor? (Draw a picture.)

3. At what point in its rotation should the poles of a DC motor's rotor reverse in order for it to continue to turn?

4. Why does a DC motor need brushes and a commutator?

5. Why do the brushes and commutator of a DC motor experience sliding friction rather than static friction?

6. Why does reversing the current through a DC motor cause it to spin in the opposite direction?

7. Why doesn't reversing the current through a universal motor cause it to spin backward?

8. Why does a synchronous AC motor turn exactly in synchrony with the current reversals of the AC power line?

9. The rotor of an induction motor has no electric connections and is made of a non-magnetic metal such as aluminum. How does it become magnetic?

10. List three factors that affect the time it takes for the rotor of a stepping motor to shift from one orientation to another.

Exercises

1. A motor's rotor always has north and south poles of equal strengths. Since each stationary pole near the rotor attracts the rotor's opposite poles and repels its like poles, one might expect the torques on the rotor to balance to zero. Why is there a net torque on the rotor? (Draw a picture.)

2. Show that the magnetostatic forces between a motor's stationary magnets and its magnetic rotor do work on the rotor as the rotor turns to bring opposite poles together.

3. Show that the magnetostatic forces between a motor's stationary magnets and its magnetic rotor do work on the rotor as it turns to separate like poles.

4. If a motor's poles remained unchanged after its rotor turned to bring opposite poles as close together as possible, the rotor's rotational kinetic energy would decrease as it continued turning. Why would it decrease?

5. Show that reversing the battery of the DC motor shown in Fig. 12.4.3 will reverse the torque on its rotor.

6. A universal motor relies on the fact that reversing the current through its electromagnets reverses all of their poles and leaves the forces they exert on one another unchanged. Draw two electromagnets and show that reversing their poles doesn't change the forces they exert on one another.

7. If you use your fingers to prevent the rotor of a small DC motor from turning, you feel it exerting a torque on your fingers. Why does this torque increase as you increase the current passing through the motor's electromagnets?

8. Why do the brushes of a DC or universal motor experience wear? Why is this wear unavoidable?

9. If you attach the wires from a DC motor to a light bulb and spin its rotor with your fingers, the light bulb will light. Since energy is conserved, you must be doing work on the rotor. Use the concept of rotational work (Section 4.2) to show that you are doing work on the rotor as you spin it.

10. Show that the synchronous AC motor spinning counterclockwise in Fig. 12.4.6 will operate equally well spinning clockwise.

11. Since synchronous AC motors have no natural direction of rotation (see Exercise **10**), what keeps a spinning synchronous AC motor from suddenly reversing directions?

12. The DC motor in your small fan operates on two batteries. If you reverse the fan's batteries, will the fan still turn? If so, will it turn the same way as before?

13. During each reversal of the AC current there is a moment when no current flows through an AC circuit. What keeps an AC motor turning during those periods of zero current?

14. In an induction motor, the magnetic poles of the stator circle the rotor in a particular direction. Why does the rotor always spin in the same direction as the circling poles?

15. If you float an aluminum pizza pan on water and circle that pan rapidly with a strong magnet, the pan will begin spinning with the moving magnet. How does the magnet exert a torque on the pan without touching it?

16. The pan in Exercise **15** always spins slightly slower than the circling magnet. Why can't it spin faster than the magnet?

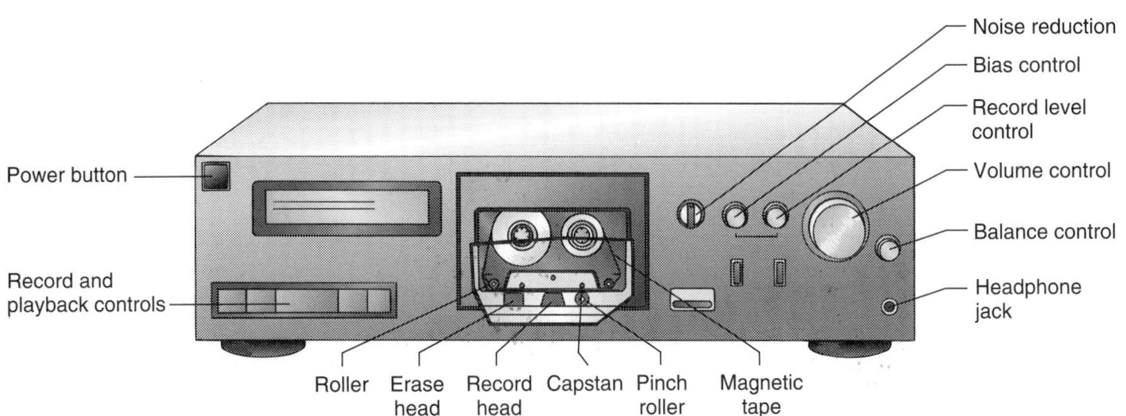

Section 12.5

Tape Recorders

Tape recorders save audio or video information on a thin plastic tape. It's already remarkable that audio or video information can be represented electronically, so it's even more surprising that something as simple as a plastic strip can store this information almost indefinitely. Magnetic recording tape and the devices that use it become better every year, yet the basic concepts behind magnetic recording have change relatively little since the early days of wire recorders and reel-to-reel tapes.

Questions to Think About: *What does it mean to represent sound electrically or as the magnetization of a magnetic tape? How can an electric current representing sound magnetize a tape? Why is it important that the magnetized tape move past the read head during playback? Does it matter how fast the tape moves during recording or playback? Why must you keep magnetic tapes and credit cards away from strong magnetic fields?*

Experiments to Do: *Take a look at a cassette tape recorder. While the actual recording structures are too small to see, the transport mechanisms are quite visible. The tape leaves the left spool of the cassette, travels past several components along the cassette's open end, and is wound back up on the right spool. Among the devices that you should be able to see along the tape's path are two or more shiny metal recording heads and a metal pin near a rubber roller. The two shiny metal heads are the erase head and the record/playback head. On a tape player, there is no erase head and all you'll see is one shiny metal playback head. The metal pin with its opposing rubber pinch roller is the capstan. The capstan is what pulls the tape along at a constant speed.*

While it's hard to record sound on tape without a tape recorder, you should have no trouble erasing it. Find an unimportant cassette and make sure that it has something recorded on it. Move a strong permanent magnet across its open end, close to but without touching the exposed magnetic tape. Now insert the cassette in a player, rewind it slightly, and play it back. You should find a two second quiet spot which you "erased" with the magnet. The tape recorder does a better job of erasing than you can do because it can actually touch the tape. But it too uses magnetic fields to erase and record information on the tape. In this section, we'll look at how the tape recorder magnetizes the tape and why the tape stays magnetized.

Representing Sound Electrically

Before looking at how a tape recorder records and retrieves sound information from magnetic tape, we must first study how sounds are represented electronically. As we saw in Section 9.2, sound consists of compressions and rarefactions of the air, which move outward from their source at about 300 m/s. In a region of compression, the air's density and pressure are somewhat higher than average and in a region of rarefaction, its density and pressure are somewhat lower than average.

A tape recorder measures these pressure fluctuations in order to recreate their sound. Its microphone produces an electric current that's proportional to how much the local air pressure differs from the average value. When the air pressure is higher than average, current flows in one direction through the recording circuit and when the air pressure is lower than average, current flows in the other direction. In this manner, the tape recorder represents anything you can hear as an electric current.

When it's time to recreate the sound, the tape recorder uses a speaker to compress and rarefy the air. The speaker produces a local air pressure that's proportional to the current in the tape recorder's playback circuit. Now the tape recorder is using an electric current to create something you can hear. Because the pressure fluctuations emerging from the speaker are identical to those detected previously by the microphone, the sound created by the tape recorder is indistinguishable from the sound it recorded earlier.

The tape recorder's task is to record the current produced by its microphone and then to recreate that current for its speaker. But while microphones and speakers are themselves quite interesting, this section will focus on recording and playback rather than on measuring and reproducing the sound.

CHECK YOUR UNDERSTANDING #1: Current Music

When the only sound reaching a tape recorder's microphone is a pure 500 Hz tone, the local air pressure is fluctuating up and down 500 times each second. Describe the current in the recording circuit.

❐ In 1898, while a research engineer for the Copenhagen Telephone Company, Danish engineer and inventor Valdemar Poulsen (1869–1942) developed a method for recording sound by magnetizing a steel wire. In his original recorder, the wire was stretched taut across his laboratory and the recording and playback devices were trolleys that traveled on that wire. The recording trolley would magnetize the wire to record sound and the playback trolley would measure the wire's magnetization to reproduce the sound. Poulsen would run along with the recording trolley and yell into the recording microphone. He would then listen to a reproduction of his voice as the playback trolley traveled along the wire.

Magnetic Tape

Magnetic tape is a thin strip of plastic, coated with a film of tiny permanent magnets. Each of these permanent magnets has a north magnetic pole at one end and a south magnetic pole at the other. During recording, a strong magnetic field can interchange the poles of these particles, altering the tape in a way that can be detected during playback. This controlled modification of the tape, together with its detection at playback, is the basis for magnetic recording. (Early recorders didn't use tape; see ❐.)

There are only a few materials that are noticeably magnetic and fewer still that can become permanent magnets. To be magnetic, a material must contain naturally occurring magnetic poles. For it to be a permanent magnet, those magnetic poles must be easily detectable from outside the material. Most magnetic materials contain tiny north and south poles that are mixed together in such a way that they tend to cancel one another. These materials generally appear non-magnetic when viewed from outside. But there are a few special materials in which the tiny north and south magnetic poles can be arranged to give the material large overall magnetic poles. When so arranged, these materials become permanent magnets.

Since isolated magnetic poles have never been observed, magnetic materials create their poles with moving electric charges. An electric charge traveling in a circular loop produces the same magnetic field as a pair of separated north and south magnetic poles (Fig. 12.5.1). This loop of current acts as a magnetic dipole.

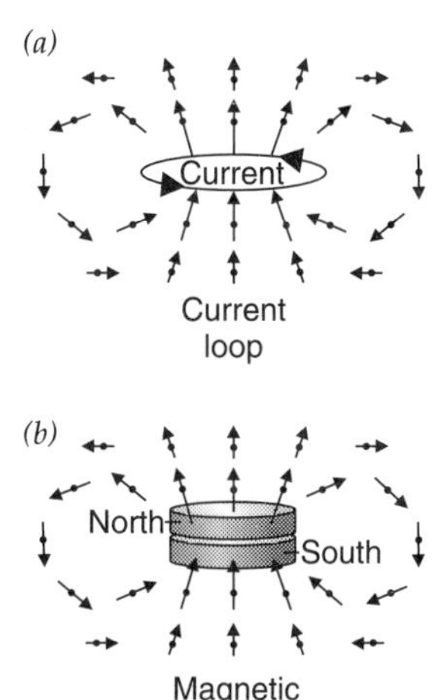

Fig. 12.5.1 - The magnetic field around a loop of electric current (*a*) is similar to that around a dipole magnet having equal north and south poles (*b*).

Even a single electron in an atom can create a magnetic field as it orbits the atom's nucleus. Furthermore, the electron itself has a spin (See Section 19.1) and behaves like a tiny, spinning ball that's covered with negative charge. This circulating electric charge makes even a stationary electron magnetic.

Thus each electron can create a magnetic field in two ways: through its orbital motion and its spin. But quantum physics severely limits the electron's allowed orbits and spins. As a result, the electron can only produce specific magnetic fields. In particular, its spin can only produce two different dipole magnetic fields. If the electron is spin-up, effectively rotating counter-clockwise when viewed from above, then it creates a field as though it had a south pole above a north pole. If the electron is spin-down, effectively rotating clockwise when viewed from above, then it creates a field as though it had a north pole above a south pole.

But when many electrons join together in an atom, they often form pairs in which one electron cancels the magnetic field of the other. An electron orbiting the nucleus clockwise tends to pair with an electron orbiting counter-clockwise and a spin-up electron tends to pair with a spin-down electron. The amount of magnetism in an atom depends on how many of its electrons remain unpaired. While most of the electrons do pair up, this pairing brings the electrons particularly close to one another. A few of the least tightly bound electrons often spread out in different orbits so that they can avoid the severe repulsive forces that come with being too close together. Adopting the same spin, either up or down, helps to keep these electrons in different orbits because the Pauli exclusion principle then forbids these indistinguishable particles from traveling in the same orbits. Most atoms have a few unpaired electrons and are magnetic.

If most individual atoms are magnetic, why are most solids non-magnetic? The answer is that when individual atoms join together to form a solid, their electrons usually pair up again and their magnetic fields cancel one another. In many cases, electron pairing is actually an important part of the bonding process between atoms. But there are a few materials in which the pairing is incomplete because the Pauli exclusion principle keeps electrons with identical spins away from one another. These magnetic materials are the basis for magnetic recording media.

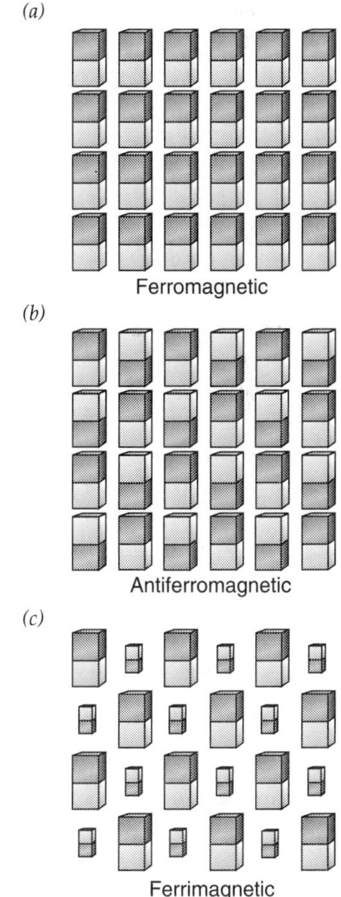

Fig. 12.5.2 - (*a*) The atomic magnets in ferromagnetic solids point in the same direction. (*b*) In antiferromagnetic solids, they cancel one another completely. (*c*) In ferrimagnetic solids, there is partial cancellation.

The three major types of magnetic solids are distinguished by the ordering of their individual magnetic atoms (Fig. 12.5.2). Iron, cobalt, nickel, and chromium dioxide are **ferromagnetic** solids, meaning that their individual magnetic atoms are all aligned in the same direction. Chromium is an **antiferromagnetic** solid, in which the individual magnetic atoms alternate back and forth so that their magnetic fields completely cancel one another. Ferrite (γ-iron oxide) is a **ferrimagnetic** solid, meaning that its individual magnetic atoms alternate back and forth but, because they have different magnetic strengths, they don't cancel one another completely. Ferromagnetic and ferrimagnetic materials can be used in magnetic recording media. Many other metals, including aluminum and copper, have no magnetic order at all.

Fig. 12.5.3 - (*a*) An ideal ferromagnet's individual magnetic atoms all point in the same direction, so that it behaves as though it had two giant magnetic poles. (*b*) However, a real ferromagnet's strong magnetic field twists the magnetic atoms out of alignment, so that they separate into magnetic domains with different orientations.

But despite the ferromagnetic ordering of its magnetic atoms, a piece of iron isn't obviously magnetic. That's because it contains many individual magnetic regions that all point in different directions. The magnetic fields produced by these isolated **magnetic domains** tend to cancel one another, so that the iron appears to be nonmagnetic.

To see why domains form, consider what would happen if they didn't. A piece of iron would then have a huge north magnetic pole at one end and an equally huge south magnetic pole at the other (Fig. 12.5.3*a*). These poles would exert torques on any nearby magnets, including the magnetic atoms *inside* the iron! This twisting action is so strong that it separates the magnetic atoms into magnetic domains (Fig. 12.5.3*b*) and turns the poles of each domain so that they all point in random directions.

But when a piece of iron is placed in a magnetic field, its magnetic character suddenly becomes obvious. Magnetic domains that are aligned with the field grow at the expense of other domains (Fig. 12.5.4*a*,*b*) and the iron develops large magnetic poles. The iron becomes *magnetized*. The electrons simply change the directions of their spins, leaving the iron atoms exactly where they were. This process explains why a permanent magnet attracts iron; the magnet magnetizes the iron (Fig. 12.5.4*b*) and then the two magnets attract one another.

But when you remove the permanent magnet, iron's domains change size again and it quickly *demagnetizes*, reverting to its nonmagnetic state. Iron is a **soft magnetic material**, meaning that it can't retain a magnetization on its own. Soft magnetic materials aren't useful for magnetic recording media, but they're essential in transformers and generators, and play an important role in a tape recorder's recording and playback heads.

Magnetic tapes use **hard magnetic materials**, materials that can remain magnetized on their own. There are several ways to obtain magnetic hardness. Some hard magnetic materials have microscopic structures and defects that impede the resizing of domains. These stiff magnetic domains make a material both hard to magnetize and hard to demagnetize. For example, the domains in common steel have some difficulty resizing, so that steel doesn't return completely to its non-magnetic state after exposure to a magnetic field. That's why steel nails, screwdrivers, and compass needles remain weakly magnetized after exposure to a strong magnet.

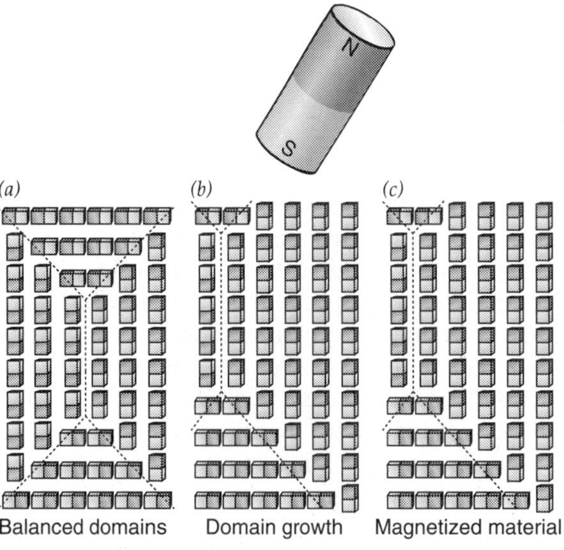

Fig. 12.5.4 - (*a*) Because its magnetic domains balance one another, a piece of iron is normally nonmagnetic. (*b*) But when the iron is exposed to a magnetic field, domains that are aligned with the field grow at the expense of other domains and the iron becomes a magnet. Since iron is a soft magnetic material, its domains return to normal (*a*) when the magnetic field is removed. (*c*) However the domains in a hard magnetic material can't change sizes easily so that the material remains permanently magnetized.

But magnetic recording media use a different type of hard magnetic material—tiny elongated particles of γ-iron oxide, chromium dioxide, cobalt-coated iron oxide, or pure iron. Each of these particles is so small that it contains only a single magnetic domain (Fig. 12.5.5). It has a north pole at one end and a south pole at the other, because that arrangement permits the opposite poles of adjacent magnetic atoms to be as close together as possible.

Each of these particles is a tiny permanent magnet that can't be demagnetized. It's always magnetized in one direction or the other. However the particle's magnetic poles can be interchanged by exposing it to a strong magnetic field. The particle itself won't flip, just its magnetic poles. This behavior makes such a particle ideal for magnetic recording. A tape recorder can set a particle's magnetic orientation during recording and expect it to remain that way indefinitely.

Magnetic tape is produced by coating a Mylar film with a mixture of these magnetic particles, a binder, and a solvent. Immediately after coating, the wet film passes through a strong magnetic field which rotates all of the tiny magnetic particles so that they're aligned along the direction in which the tape normally moves (Fig. 12.5.6). The tape is then dried and pressed so that its magnetic coating is dense, smooth, and shiny. The final magnetic coating is about 5 microns thick on audio cassette tape and video tape and 10 microns thick on audio reel-to-reel tape. Only the outer micron or so of these coatings is actually used for recording.

Magnetic particles are rated according to the magnetic fields needed to interchange their magnetic poles. Particles that are more resistant to interchange make better magnetic tapes because they are less affected by accidental magnetic fields. The conventional measure of this resistance to interchange is the *oersted*, with a higher number of oersteds meaning more resistance. Table 12.5.1 shows that iron particles make the best recording tapes. Although a large piece of iron is a soft magnetic material, with magnetic domains that resize easily, an elongated single-domain iron particle is an extremely hard magnetic material. Metal particle tapes are made from elongated iron particles.

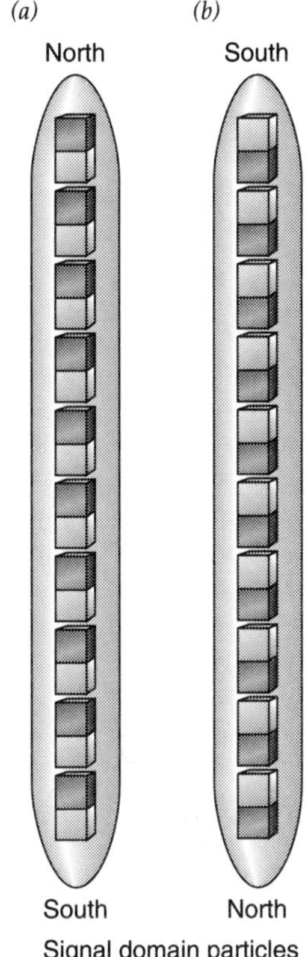

Fig. 12.5.5 - A tiny, elongated magnetic particle has only a single magnetic domain, which aligns with the long axis of the particle. This magnetic alignment is hard to change, so that only a strong magnetic field can convert (*a*) into (*b*) or (*b*) into (*a*) without actually rotating the particle itself.

Table 12.5.1 - Particles commonly used in magnetic tapes.

Type of Magnetic Particle	*Resistance to Interchange of Poles*
γ-iron oxide (γ-Fe_2O_3)	300 oersteds
chromium dioxide (CrO_2)	450 oersteds
cobalt-modified γ-iron oxide	600 oersteds
iron	1500 oersteds

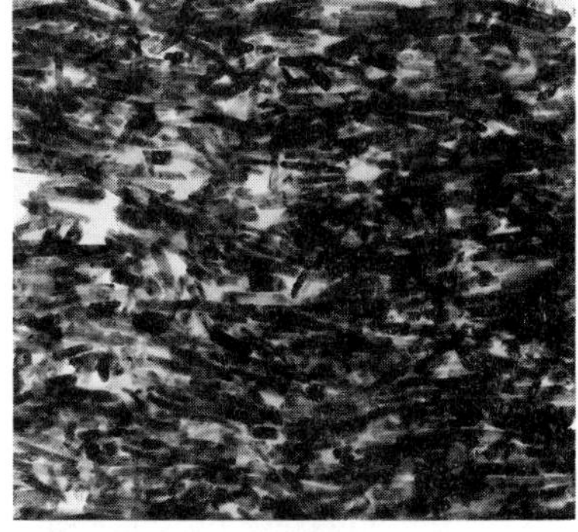

Fig. 12.5.6 - A metal particle tape records information in a layer of cigar-shaped iron particles. This electron micrograph of the recording layer shows that the particles were aligned with the tape during the manufacturing process.

CHECK YOUR UNDERSTANDING #2: Now for the Flip Side

If you bring the north pole of a strong magnet near the north pole of a tiny immobile magnetic particle, what will happen to that particle?

Recording Sound onto Magnetic Tape

We are now ready to look at how a tape recorder records sound information on a magnetic tape. This is done by a miniature electromagnet in the recording head, a tiny ring of soft magnetic material with a coil of wire wrapped around it (Figs. 12.5.7, 12.5.8, and 12.5.9). When an electric current flows through the coil, the coil's magnetic field temporarily magnetizes the ring.

The ring is actually incomplete. It has a tiny gap just at the point where it touches the magnetic tape (Fig. 12.5.7*a*). When current flows one way through the coil, the gap has a north pole on its left end and a south pole on its right end (Fig. 12.5.7*b*). When the current reverses, the poles also reverse (Fig 12.5.7*c*). During recording, sound is represented as a fluctuating current in the coil and therefore a fluctuating magnetic field in the ring's gap.

Before it encounters the magnetic recording head, the coating on the magnetic tape is demagnetized. Although the individual particles in the coating are always magnetic, they are arranged so that half of their north poles point forward and half point backward (Fig. 12.5.10*a*). The recording head's job is to reorient those tiny magnets so that their poles all point in the same direction (Fig. 12.5.10*b*). The particles don't actually flip over. Their magnetic poles simply interchange.

As the magnetic tape moves steadily past the recording head, the gap in the recording head magnetizes regions of the tape's magnetic coating in one direction or the other (Fig. 12.5.11). The depth to which the coating is magnetized depends on the strength of the magnetic field in the gap, which is proportional to the current in the coil. Since that current represents sound and air pressure, the magnetization depth is greatest during loud parts and least during quiet parts. The direction of magnetization, north poles forward or backward, depends on the direction of current flow through the coil, which depends on whether the recorder is trying to represent a compression or a rarefaction of the air.

There are two more important details about magnetic recording. First, the recording head expects to start with a blank tape, so the entire coating must be demagnetized by an erase head located just upstream from the recording head. The erase head is essentially a recording head with an extra large gap and a large, high-frequency alternating current in its coil. As the magnetic coating passes through the rapidly reversing magnetic field in the gap, the magnetic alignments of its particles flip back and forth many times. About half end up oriented in one direction and half in the other.

(*a*)

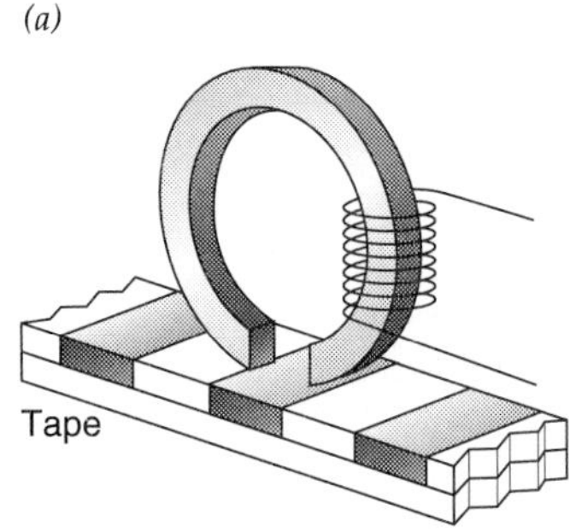

(*b*)

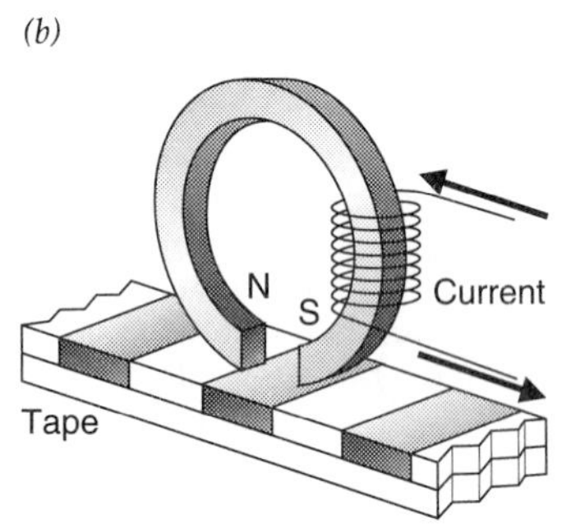

(*c*)

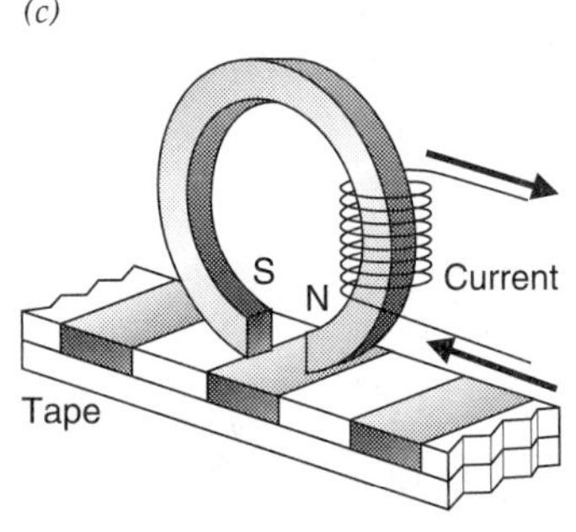

Fig. 12.5.7 - (*a*) A tape head is a ring of soft magnetic material with a small gap in it and a coil of wire wound around it. The ring is temporarily magnetized whenever a current flows through the coil (*b*,*c*) and it permanently magnetizes any hard magnetic material near its gap.

Fig. 12.5.8 - This stereo recording head has two soft iron rings, the shiny rectangles above and below the horizontal dividing line. The vertical gaps in these rings are so thin that they can only be seen with a microscope. The head records and plays back two separate channels.

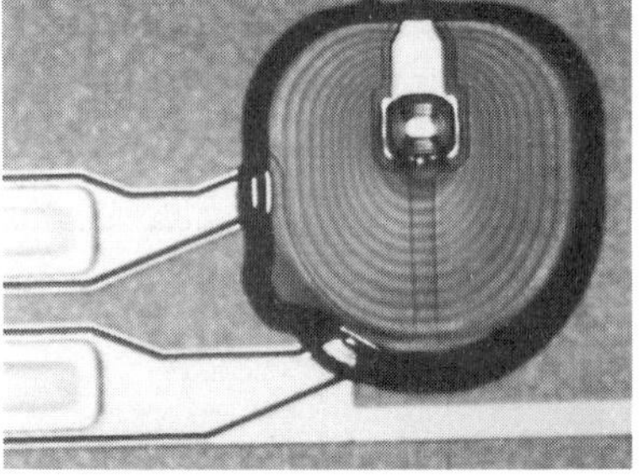

Fig. 12.5.9 - This thin film magnetic recording head is so small that it can only be viewed through a microscope. The head's iron ring is the shiny white object near the upper right, encircled by the coil of wire that connects to the circuit at the lower left.

A similar demagnetizing process is used in bulk tape erasers, which magnetically scramble all of the tape's particles simultaneously. Computer disks have no erase head at all, so the recording head itself must remove any prior magnetization of the magnetic coating. The coating used on a disk is extremely thin (less than 1 micron) so that the recording head can magnetize it completely.

Second, the recording head has trouble flipping the magnetic alignments of particles during quiet portions of the sound. To solve this problem, the tape recorder sends an additional high-frequency alternating current through the coil in the recording head. This *bias current* flips the particles' magnetic alignments back and forth during the recording process. While the bias current doesn't give the tape any particular overall magnetization, it allows the other current in the coil, the current representing sound, to magnetize parts of the coating even during quiet passages. Tape recorders adjust the bias current according to the magnetic hardness of the particles in the coating, either type I (normal bias), type II (high bias), or type IV (metal bias).

(*a*)

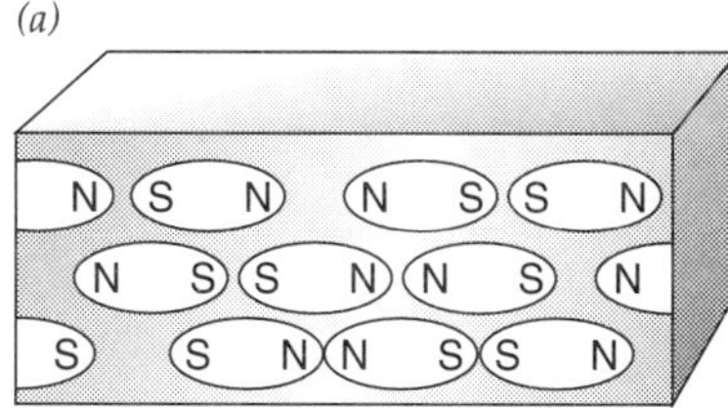

(*b*)

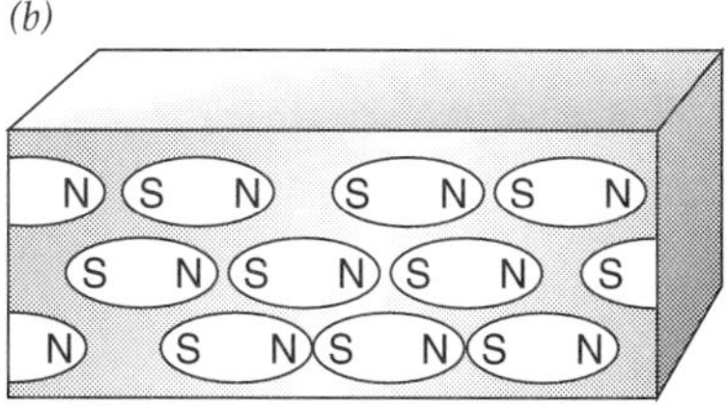

Fig. 12.5.10 - In an erased magnetic coating (*a*), roughly half the particles have north poles on the left and half on the right. In a magnetized portion of coating (*b*), all of the particles have their north poles aligned on the same side.

CHECK YOUR UNDERSTANDING #3: No More Rock and Roll

If you bring a strong magnet near a recorded magnetic tape, that tape may be erased. Why?

Playback, Noise Reduction, and Stereo

The tape recorder recreates sound by passing the magnetic tape across a playback head. The playback head is similar to the recording head and the two are often the same device. Fancy tape recorders have a separate playback head located downstream from the recording head, so that you can listen to the tape as you are recording it. Since the roles of the recording and playback heads are different, separate heads that are specialized to the two tasks outperform a single head that must perform both duties.

As the magnetized tape passes across the gap of the playback head, it temporarily magnetizes the ring (Fig. 12.5.12). The deeper the magnetization of the coating, the more the ring is magnetized and the more magnetic flux passes through the coil of wire wrapped around it. As the tape moves past the head, this flux changes and induces currents in the coil. The induced current is directly related to the original current that flowed through the recording head when the sound was recorded. This current is amplified and sent to a speaker, where it reproduces the compressions and rarefactions of the air that occurred in the original sound.

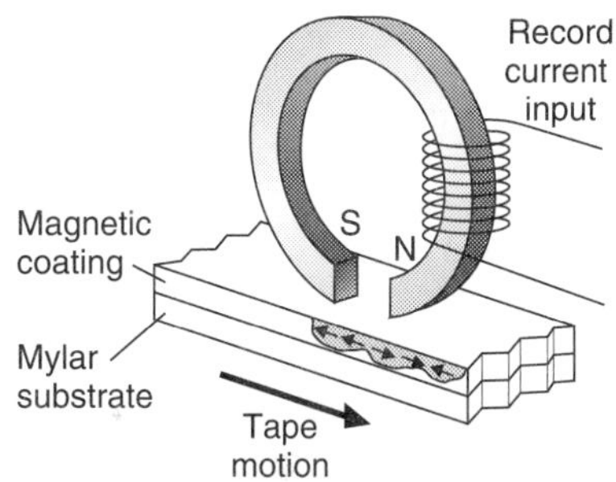

Fig. 12.5.11 - The recording head magnetizes regions of the tape's magnetic coating as the tape moves past the head. The depth of this magnetization is proportional to the amount of current in the coil.

Unfortunately, the tape recorder doesn't reproduce sound perfectly. It introduces some noise into the sound because of imperfections in the recording and playback processes. This noise is primarily due to:

(1) imperfect erasure of the tape.
(2) nonuniformities in the distribution of magnetic particles in the coating.
(3) imperfections in the bias current aiding low-level recording.
(4) surface irregularities in the coating.

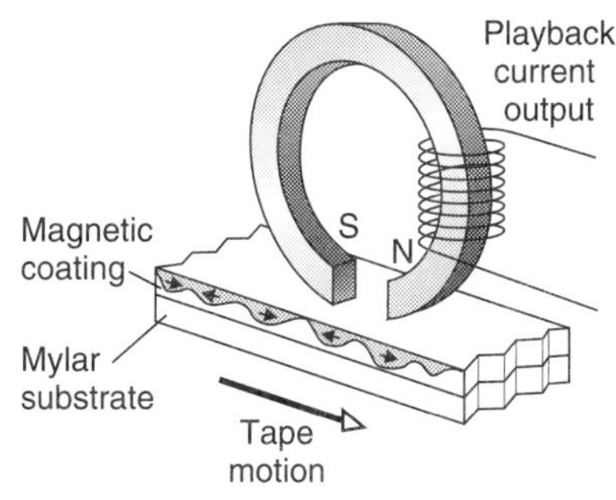

Fig. 12.5.12 - As the magnetized tape passes across the gap in the playback head's ring, the ring is magnetized and the magnetic flux passing through the coil changes. Current representing the recorded sound is induced in the coil.

Noise also appears when highly magnetized portions of a tape imprint their magnetizations onto adjacent layers of tape when the tape is wound tightly on a spool. Noise from all of these sources is particularly disturbing during quiet passages and is mostly high pitched hissing sounds.

To improve their fidelity, many tape recorders use noise reduction techniques such as Dolby A, B, and C. These techniques boost the volumes of high frequency sounds during recording and then reduce their volumes during playback. The desired sound is unchanged overall but any high-frequency noise that appeared between recording and playback is decreased in volume.

Many tape recorders work with more than one sound channel at a time. Stereo recorders write and read two or more audio tracks at once to represent sounds heard by microphones at different locations. These machines have at least two closely spaced recording heads and an equal number of playback heads. Each head writes or reads a narrow track of magnetization on the tape.

CHECK YOUR UNDERSTANDING #4: Feel Those Good Vibrations

Is the induced current in the coil of a playback head larger during a quiet passage or a loud passage?

The Tape Transport and Video Tape Recorders

Fig. 12.5.13 - This autoreverse cassette recorder mechanism moves the tape across the two magnetic recording heads at the center. During recording, one of these recording heads erases the tape while the second records. The tape speed is controlled by the capstans, the metal spindles to the left and right of the heads. These capstans are pressed against the tape by the rubber rollers below them.

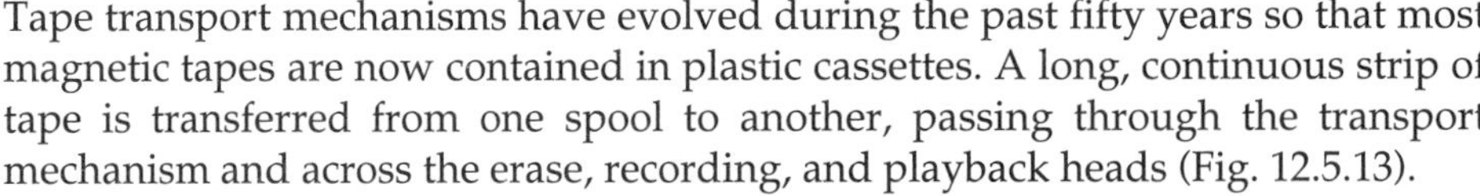

Tape transport mechanisms have evolved during the past fifty years so that most magnetic tapes are now contained in plastic cassettes. A long, continuous strip of tape is transferred from one spool to another, passing through the transport mechanism and across the erase, recording, and playback heads (Fig. 12.5.13).

The one transport component worth noting in an audio tape recorder is the capstan. The capstan is a cylindrical rod that rotates at a very steady rate and controls the tape's speed through the recorder. The tape is pressed against the capstan by the pinch roller and is drawn steadily forward as the capstan turns. The capstan has a flywheel attached to it to give it a large moment-of-inertia so that it always turns at exactly the same rate.

A video tape recorder is much more complicated because storing image information as well as sound takes lots of room on a tape. Since the smallest region of magnetization that a recording head can create is about 1 micron long, a video recorder uses about 2 m of track length on the tape every second. Instead of moving a long tape quickly past stationary heads, a video recorder sweeps the heads quickly across a wide tape. The heads are contained in a spinning drum and the tape is pressed against that drum at an angle. As the tape is pulled slowly across the drum, the spinning heads write or read a series of stripes across the tape's width. Because these stripes appear on the tape at an angle, the recorder stores information across the entire surface of the tape.

CHECK YOUR UNDERSTANDING #5: Audiophile Sound

High performance tape recorders move the tape more quickly past the heads than do common cassette recorders. Why?

Summary

How Tape Recorders Work: During recording, the air pressure fluctuations (sound) detected by a microphone become fluctuations in the current through a coil of wire. This coil is wrapped around a tiny ring of soft magnetic material, which is temporarily magnetized by the coil's current. The ring is broken by a narrow gap and the ring's magnetization creates poles on opposite sides of that gap. Demagnetized magnetic tape is drawn across the gap and is magnetized by its magnetic fields. Because the tape is coated with a layer of tiny permanently magnetic particles, it retains this magnetization indefinitely. The direction and depth of magnetization in the tape's coating are proportional to the difference between the air pressure at the microphone and the average air pressure.

During playback, the magnetized tape is drawn across the gap in the ring of a playback head (often the same as the recording head). As each magnetized region travels across the gap, the coating's magnetic field temporarily magnetizes the ring. This changing magnetization induces a current in the coil wrapped around the ring and this current is then amplified and sent to a speaker, where it reproduces the original sound.

The Physics of Tape Recorders

1. Electrons can create magnetic fields both by their actual motion and by their intrinsic spin.

2. Most atoms are magnetic because of the orbital motions and spins of their electrons. While most electrons form pairs with equal but opposite magnetic fields, a few unpaired electrons usually give the atom an overall magnetic field.

3. Most solids are non-magnetic because of complete pairing of their electrons. Only a few solids retain any magnetism in their atoms. The most common types of magnetic materials are ferromagnetic, antiferromagnetic, and ferrimagnetic.

4. Large pieces of ferromagnetic or ferrimagnetic solids form magnetic domains, each with its own magnetic alignment. The domains tend to cancel one another's magnetic fields. In the presence of an outside magnetic field, the domains resize and the material becomes magnetized.

5. Hard magnetic materials retain their magnetizations while soft magnetic materials quickly lose it.

6. Tiny particles have a single magnetic domain and are always magnetized. If they're sufficiently elongated, their magnetic poles will normally be located at their ends.

Check Your Understanding - Answers

1. It's an alternating current, reversing its direction 1000 times each second.

Why: The 500 Hz tone causes the air pressure to change up and down through 500 complete cycles each second. Each time the pressure goes above average, the current through the circuit flows in one direction. Each time the pressure goes below average, the current flows the other direction. That makes 1000 reversals in current each second.

2. The particle's magnetic poles will interchange.

Why: Although the particle itself can't move, its magnetic poles can. When the repulsion between the two north poles becomes strong enough, the poles of the magnetic particle will interchange and it will then present its south pole to the north pole of the other magnet.

3. The strong magnetic field magnetically realigns the particles in the tape's coating.

Why: The magnetic particles in the tape can respond to a strong magnetic field by interchanging their magnetic poles. When this happens, the recorded information is lost and is replaced by meaningless magnetization. That's why magnets must be kept away from magnetic recording media.

4. The current reaches its largest values during a loud passage.

Why: When loud passages are recorded, they produce deep regions of magnetization in the tape's coating. This extensive magnetization causes large changes in the magnetic flux through the ring of the playback head and induces large currents in the coil around that ring.

5. By using larger regions of tape coating to record each second of sound, the high performance recorders can save more information about that sound.

Why: The biggest limitation with modern cassette recorders is that the tape moves so slowly (only 4.75 cm/s) that too little coating passes by the heads to record all of the details clearly. Individual periods of compression or rarefaction in the air may be represented by only a micron or two of track length on the cassette tape's coating. That is not enough coating to capture all of the subtle detail.

Glossary

antiferromagnetic Composed of magnetic atoms that alternate back and forth in magnetic orientation so that their magnetic fields cancel.

ferrimagnetic Composed of magnetic atoms that alternate back and forth in magnetic orientation but have different magnetic strengths so that their magnetic fields don't cancel completely.

ferromagnetic Composed of magnetic atoms that all have the same magnetic orientation within a magnetic domain.

hard magnetic material A material that is relatively difficult to magnetize and that retains its magnetization once the magnetizing field is removed. Hard magnetic materials are suitable for permanent magnets.

magnetic domains Regions of uniform alignment within a magnetic material.

soft magnetic material A material that is relatively easy to magnetize and that loses its magnetization once the magnetizing field is removed. Soft magnetic materials are suitable for electromagnets.

Review Questions

1. What does it mean to "represent sound as an electric current"? Is the current actually sound?

2. If electrons are magnetic even at rest, why are so few materials magnetic?

3. If you were to examine the electrons that make a ferromagnetic material magnetic, would you find that nearby electrons are spinning in the same or opposite directions?

4. While iron is a ferromagnetic material, it normally has no detectable magnetic poles. How is it hiding its magnetism?

5. A large piece of soft magnetic material loses its magnetization when you remove it from all magnetic fields. What happens to it that makes its magnetization vanish?

6. A large piece of hard magnetic material retains its magnetization even after you remove it from all magnetic fields. What prevents it from losing its magnetization?

7. A magnetic recording particle is *always* magnetic in one direction or the other. Why can't it be made non-magnetic?

8. What does the magnetic recording head of a tape recorder do to the magnetic particles in the magnetic tape as it records music on that tape?

9. How does the bias current in the magnetic recording head make it easier for that head to magnetize the tape?

10. Why must a magnetic tape move past a playback head in order for that head to detect the tape's magnetization?

11. Why must the magnetic tape always move past a tape recorder's recording and playback heads at the same constant speed in order to reproduce the sound accurately?

Exercises

1. Since sound consists of pressure fluctuations above and below atmospheric pressure, the current that represents it fluctuates back and forth—it's an alternating current. A tape recorder records this alternating current on a magnetic tape. Which way is the tape magnetized? (Draw a picture.)

2. An aluminum atom is magnetic because it has one unpaired electron. How can it be that aluminum metal is nonmagnetic?

3. Aluminum-coated plastic films and tapes have a mirror-like surface and are frequently used for decorations. Why can't aluminum coated tape be used to record sound in a tape recorder?

4. Why does a magnet stick to the surface of a refrigerator?

5. A steel paper clip is normally not magnetic. However, when you touch one end of the paper clip to the north end of a permanent magnet, the paper clip becomes magnetic. Which end of the paper clip becomes a south pole?

6. A strong magnet can pick up a whole chain of steel paper clips. Explain how the first paper clip can pick up the second paper clip if the two aren't normally magnetic.

7. Junk yards use large direct current electromagnets to pick up scrap iron and steel. Why don't they use alternating current in those electromagnets?

8. Flexible or floppy disks are cut from coated plastic sheets that resemble magnetic recording tape. However, the magnetic particles on these sheets aren't aligned along any particular direction. Why wouldn't such alignment be helpful?

9. Suppose two magnetic tapes have pure tones recorded on them. One has a low frequency tone while the second has a high frequency tone. Explain how the magnetizations of the two tapes differ from one another.

10. When you have the magnetic strip on an ID or credit card read, it's always done by moving the card past a playback head. Why must the card be moving for the head to read it?

11. If you hold a permanent magnet the wrong way in an extremely strong magnetic field, its magnetization will be permanently reversed. What happens to the magnetic domains inside the permanent magnet during this process?

12. Heating a permanent magnet or striking it against a hard surface allows the magnet's magnetic domains to rearrange. The magnet will tend to reduce its potential energy by becoming somewhat less magnetic. Why does a hard magnetic material have more potential energy when it's magnetized than when it's not magnetized?

13. Why must the ring of a magnetic recording head be made from a soft magnetic material rather than a hard one?

14. Why must the recording particles in a magnetic tape be made from a hard magnetic material rather than a soft one?

15. What does exposing the magnetic strip of an ID or credit card to a strong magnetic field do to that strip that makes it unreadable?

16. Why must a tape recorder erase a tape before it records new sound on it, particularly when the new sound is soft?

17. A computer hard disk has no erase head. Why does using a very thin layer of magnetic particles allow the hard disk to do without an erase head in front of its recording head?

18. The recording particles on metal particle tapes are harder to magnetize or demagnetize than those on older types of magnetic tape. Why must a tape recorder use a larger bias current while recording on a metal particle tape than on the older types of tape?

19. If you sprinkle iron powder onto the magnetic strip of an ID or credit card, the powder will stick and form a pattern of stripes. Why does the powder stick to the magnetic strip?

20. Why does iron powder form a pattern of stripes on the magnetic strip of an ID or credit card (see Exercise **19**)?

Epilogue for Chapter 12

In this chapter we studied the relationships between electricity and magnetism, to see how one can be used to create the other in situations involving moving charge or moving objects. In *flashlights*, we examined circuits to see how a current of electric charges can transfer power from batteries to a light bulb. We found that both voltage and current contribute to this power transfer.

In *electric power distribution*, we saw how alternating electric current makes it possible to transfer power from one circuit to another by way of a transformer. We learned that transforming electric power to extremely high voltage and low current minimizes the power wasted between power plants and cities.

In *electric power generation*, we explored the machines used to produce electricity, both by electromagnetic induction in a generator and through the use of semiconductor electronics in a photocell. We found that moving magnets can drive currents in wires and that this effect is the most common way to produce electricity. We also learned about diodes and how their one-way handling of electric charge allows them to convert light energy into electric energy.

In *electric motors*, we investigated the relationships between generators and motors to see that they are essentially the same objects, distinguished only by whether electricity is doing work on the rotating system or the other way around. We saw that an alternating electric current can cause a rotor to turn endlessly and that even direct current can power a properly designed motor.

In *tape recorders*, we found that certain materials are intrinsically magnetic and this magnetism can be used to store the information needed to reconstruct either sound or image or both. We explored the ways of making permanent magnets and using those permanent magnets to construct recording tapes.

Explanation: A Nail and Wire Electromagnet

When you connect the wire from one terminal of the battery to the other, a current begins to flow from the positive terminal to the negative terminal through the wire. (In fact, negatively charged electrons move from the battery's negative terminal, through the wire, to its positive terminal.) This current produces a magnetic field around the wire. Because the wire is coiled around the nail, this magnetic field passes through the nail and causes its magnetic domains to resize until most of the nail's magnetic atoms are aligned with the field. Without any current in the wire, the magnetic domains in the steel point in many different directions, so the nail appears nonmagnetic. However with the current orienting the domains, they together produce a large magnetization. The nail becomes magnetic and exerts large magnetic forces on other nearby magnetic poles and dipoles.

Cases

1. Your neighborhood is experiencing an electric power outage. You have a huge box of 1.5 V flashlight batteries ("D" cells) in your garage and you decide to use them to operate some of your appliances until the power is restored.

a. You form a complete circuit that includes one "D" battery and a 120 V incandescent light bulb. Why doesn't the bulb light up?

b. You replace the single "D" battery in that circuit with 80 "D" batteries that are connected in a chain. You arrange these batteries alternately one way then the other, so that positive terminals touch positive terminals and negative terminals touch negative terminals. Why doesn't this arrangement light the bulb?

c. Now you rearrange the 80 "D" batteries in the chain so that they all point in the same direction, and the bulb finally lights. Why is this chain able to light the bulb?

d. You take another chain of 80 "D" batteries, like the one in **c**, and connect it to the AC induction motor in your refrigerator's compressor. Why doesn't the motor begin to turn?

e. After an hour, the 80 "D" batteries in **c** stop lighting the 120 V light bulb. Use energy considerations to describe what has happened to the batteries and the bulb.

2. A blender is a common kitchen appliance used to liquefy and combine various foods and beverages. It consists of a glass or plastic pitcher with a rotating blade at the bottom. The pitcher sits in a base containing an electric motor. When you push the "on" button, the motor spins very rapidly and turns the blade. The spinning blade cuts and mixes the contents of the pitcher.

a. If you put an ice cube into the pitcher and turn the blender on, the blade will chop the ice cube into tiny pieces. The pitcher is smooth so that it doesn't prevent the ice cube from moving to stay ahead of the blade. Nonetheless, the ice cube stays put and the blade slices through it. What holds the ice cube in place?
b. The blender has a universal motor and a two-wire electric cord. When you turn it on, an alternating current begins to flow through the cord and the motor. How do the currents through the two wires of the cord compare to one another?
c. The blender's universal motor can actually run on either AC or DC electric power. If you reverse the blender's plug so that its two prongs trade places in the outlet, how will that affect the direction in which the blade turns?
d. If you blend a frozen beverage for too long, the ice in it will melt. How does the spinning blade heat up the liquid?
e. The beverage tends to spin around with the blade and a deep well often forms near its center. This whirling shape is called a vortex. Why does the liquid in a vortex press so tightly against the walls of the pitcher?

3. A treadmill is an exercise machine that allows you to walk or jog without actually going anywhere. Your feet land on a wide fabric belt that moves between two rollers, one in front of you and the other behind you. An electric motor turns these rollers at a steady pace so that the top of the belt keeps moving toward you. You step forward on the belt, but because it's moving toward you, you stay in one place with respect to the room.

a. What is the net force on you as you walk on the treadmill?
***b.** The tilt of the belt is adjustable. Suppose for the moment that the belt is exactly horizontal. How much work are you and the treadmill doing on one another?
c. If you tilt the belt so that it rises in front of you, it becomes a much harder exercise. Why does it take so much more work to step forward on the tilted treadmill, even though you are not moving anywhere?
d. The motor provides the power necessary to keep the belt moving at a steady pace. When the belt is horizontal, the motor compensates for friction by pulling the belt along. But when the belt is tilted far enough, the motor begins to act as a brake, preventing the belt from turning too quickly. The motor is then extracting energy from the belt (and from you). What is the motor doing with the energy it receives?

4. An uninterruptable power supply (UPS) allows a computer system to operate during a brief power outage. Most personal computer UPS systems use an electronic device to convert 12 V DC power from a battery into 120 V AC power for the computer. The UPS circuitry first converts the 12 V DC into 12 V AC and then uses a transformer to increase the voltage to 120 V, as required by the computer. In Europe, the final voltage is 220 V.

a. Why can't the transformer directly convert 12 V DC into 120 V DC?
b. An electronic switching system converts the 12 V DC from the batteries into 12 V AC for the transformer. What is different about the current flowing through the two wires attached to the battery and the current flowing through the two wires attached to the transformer?
c. How does power move from the 12 V AC side of the transformer to the 120 V AC side of the transformer?
***d.** Energy is conserved. If there is an average current of about 20 A in the 12 V primary circuit of the transformer, what is the average current in the 120 V secondary circuit of that transformer?
e. The length of time that the UPS can deliver power to the computer is limited by the battery's chemical potential energy. Suppose you wish to extend that time by adding a second 12 V battery to the system. How should you connect the two batteries together so that the current arriving at the pair of batteries will still experience a 12 V rise in voltage and yet be able to extract power from both batteries?

5. Each electric outlet in a home is connected to the power company through a fuse or circuit breaker. When properly installed, current flows from the power company, through the fuse or circuit breaker, through the outlet and the appliances connected to it, and then back to the power company—a complete circuit. The fuse or circuit breaker limits the current that the circuit can carry.

a. If you plug too many appliances into a socket that has nothing to limit the current it can supply, the outlet and home wiring will become hot. Why?
b. A fuse contains a short wire with a relatively high electric resistance. Why will this wire melt first if too much current is drawn through the circuit?
c. Why will current stop flowing through the appliances when the fuse melts and opens the circuit?
d. Why is a short circuit that bypasses the appliances very likely to melt the fuse and open the circuit?
e. A circuit breaker contains an electromagnet. All of the current flowing through the circuit also flows through this electromagnet. If the electromagnet is to prevent too much current from passing through the circuit, should it open or close the circuit as it becomes more magnetic?

6. Many kitchen sinks are equipped with motorized garbage disposers which grind garbage into tiny pieces and send them down the drain. In a typical disposer, garbage washes down from the sink onto a spinning metal disk. This garbage moves to the disk's outer edge, where it's ground between the disk and the fixed body of the disposer. Water then carries the ground garbage into the sewer.

a. Why does garbage move toward the outer edge of the spinning disk?
b. The disk is driven by an AC induction motor. When you turn on the disposer, the disk quickly begins turning and soon reaches a fairly constant speed. What determines the disk's rotational speed?
c. Show that it takes work to grind something between two surfaces. You might show, for example, that it takes work to grind grain between a fixed stone and a movable one.
d. The motor that turns the spinning disk has a limited amount of mechanical power. The motor's angular velocity is constant so the power limit appears as a maximum torque that the motor can exert on the disk. Explain this maximum torque in terms of the magnetic forces present in the motor.

e. Grinding a hard piece of garbage may require more torque than the motor can provide. How does the disk's rotational inertia contribute to the grinding process?

f. Occasionally a disposer jams while grinding hard material. One way to unjam it is to reverse its direction of rotation. What must change for the induction motor to turn backward?

7. A recycling plant is trying to separate metals from other trash automatically. It grinds the trash up into small pieces and then sends these pieces past the poles of a strong electromagnet.

a. Iron and steel scraps in the trash are attracted to the poles of the electromagnet but aluminum scraps are not. Why do these metals behave so differently?

b. If they're moving fast enough, aluminum scraps experience a magnetic drag force that separates them from nonmetallic trash. Where does this force come from?

c. To look for metal hidden inside bundles of paper, the recycling plant exposes the bundles to an alternating magnetic field (one that reverses directions many times a second). If it finds that something inside a bundle creates another alternating magnetic field in response, it knows that the bundle contains metal. How can metal inside the trash create this second alternating magnetic field?

d. The company uses electric motors for its conveyor belts. It finds that the higher a belt lifts the trash before dumping it onto a heap, the more electric power that belt's motor consumes. Why should a motor's electric power depend on the height of the conveyor belt it drives?

e. To keep down the dust as people sort the trash, the company has a number of ion generators in its building. How do these ion generators get rid of the dust and where does that dust go?

8. Electric shavers come in two different types: reciprocating (back and forth motion) and rotary.

a. When you turn on a corded reciprocating shaver, AC current from the power company travels through a coil of wire inside the shaver. A permanent magnet attached to the cutting blades vibrates back and forth near this coil, taking the blades with it. What makes the permanent magnet vibrate back and forth?

b. How many complete cycles (back and forth) do the blades complete each second?

c. A cordless reciprocating shaver doesn't just send current from its battery through a coil of wire. Why wouldn't that arrangement work?

d. A corded rotary shaver uses a small universal motor to turn its circular blade. Why doesn't the shaver run backward if you plug it in backward?

e. A cordless rotary shaver uses a small DC motor. This shaver runs backward if you reverse its batteries. Why?

9. A home burglar alarm uses various sensors to detect an intruder. These sensors are connected to the alarm's control unit by wires. The control unit determines when a breakin has occurred, sounds an alarm, and notifies the authorities.

a. Each sensor is attached to the control unit via two wires rather than just one. It's very hard to signal a breakin electrically with only one wire. Why do two wires make it so much easier for a sensor to indicate that a breakin has occurred?

b. One of the simplest sensors is just a thin strip of metal foil that runs along the edge of a window. If a burglar breaks the window, the foil strip will be severed. How can the control unit determine if the strip has broken, using its two wires?

c. A more sophisticated sensor can tell when a door is opened. It uses a magnet attached to the door and a small device called a reed switch attached to the door frame. This switch contains two iron strips that are arranged head to tail in a line, but normally don't quite touch—they're bent slightly apart and must bend together to make contact. When the magnet's north pole is near the end of one of the iron strips, that strip becomes magnetic. Why?

d. The first iron strip, now magnetic, attracts the other strip and they pull together and touch. Why does the first strip attract the second strip?

e. When the door is closed, the magnet is near the reed switch. But as the door opens, the magnet moves away from the reed switch and the two iron strips spring apart. How can the control unit tell when the door opens?

10. A ground fault interrupter is a device that senses when some of the current flowing out one side of an electric outlet isn't returning through the other side of that outlet. The only way such an imbalance can occur is if some current is returning to the electric company through the ground. Because that accidental current path might include your body, the interrupter shuts off current as soon as it senses trouble.

a. Inside the interrupter, the two wires from the electric company pass together around the iron core of a single transformer, forming two identical primary coils. If the AC current passing through each coil is equal in magnitude but opposite in direction, how does the magnetization of the transformer's core change with time?

b. If the AC current passing through each coil isn't equal (perhaps because some current is escaping from a hair dryer plugged into the outlet), how does the magnetization of the transformer's core change with time?

c. The transformer's core has a secondary coil wrapped around it. If all of the current flowing to the hair dryer through one wire returns from it through the other wire, current passing through this secondary coil will experience no change in voltage. But if some of the current flowing to the hair dryer doesn't return, current in the secondary coil will experience a change in voltage. Explain.

d. The current from the secondary coil is used to trigger a switch that disconnects the outlet from the electric company. This switch has manual "test" and "reset" buttons. To test the ground fault interrupter, you press the "test" button. What can this button do to simulate a real current accident?

11. Your model train set operates from a small DC power unit that plugs into an electric outlet. This power unit has a knob that controls the train's speed and a switch that controls the train's direction of travel.

a. The train engine contains a DC motor that connects to the power unit through the two metal rails of the track. Why are both rails necessary rather than just one?

b. When the train is starting up, a large current flows through the rails. How does this current obtain power from the power unit and how does it deposit power in the engine?

c. The train is a good one that experiences little internal friction. When it's moving smoothly at high speed very little current flows through the rails. Why is the current so small?

d. If you turn down the power unit suddenly, the train will slow down rapidly and current will actually flow backward through the rails. What is pushing the current backward through the circuit?

e. To reverse the direction of the train, what does the power unit do to the current flowing through the engine?

f. The power unit has broken, so you replace it with a

small AC power unit. But the engine doesn't move. Instead, it just buzzes noisily. Why?

12. A typical audio speaker, such as that in a radio, consists of a permanent magnet, a coil of wire, and a moveable speaker cone. The coil of wire is connected electrically to the radio's amplifier so that the radio controls how much current flows through the coil. The coil is attached to the speaker cone so that if the coil moves, the speaker cone moves and creates sound waves in the air.

a. When the amplifier sends an electric current through the coil, the coil is either attracted or repelled by the permanent magnet. Its copper wire isn't normally magnetic, so why does the coil experience forces from a magnet?
b. If the radio reverses the direction of current through the coil 500 times each second (250 complete cycles each second, or 250 Hz), what will happen to the speaker cone?
c. Pushing the speaker cone forward quickly compresses the air in front of the cone and requires energy from the radio. Show that the speaker cone does work on the air while compressing it.
d. When you talk in front of a speaker, the sound of your voice moves the speaker cone and the coil. The coil moves back and forth near the permanent magnet. What happens to the mobile electric charges in the wire coil?
e. If you connect two speakers together, so that the two wires from one speaker's coil connect to the two wires from the other speaker's coil, you will form a complete circuit. What will happen to the first speaker when you talk into the second speaker?

13. The needle of a magnetic compass is a permanent magnet. When the compass is level, the needle rotates easily about a vertical axis so that its north magnetic pole points toward the earth's north geographic pole.

a. What type of magnetic pole is located near the earth's north geographic pole?
b. The compass needle is sensitive to any nearby steel or iron. Whenever a piece of iron is held near the compass, one end of its needle turns toward that iron. Why is either end of the needle attracted to pieces of iron?
c. The needle is also disturbed by nearby electric currents. Why?
d. Some compasses have mechanisms that can prevent their needles from turning. If you lock a compass's needle in place and pass its north pole near the north pole of a very strong permanent magnet, you will spoil the compass. When you release the needle, it will point toward the earth's south geographical pole. What will have happened to the needle?

14. In the United States, most electric outlets accept either two- or three-prong plugs. The two slot-shaped openings in the outlet permit electric current to flow to and from an appliance while the semicircular opening can electrically connect the appliance's frame to the earth itself.

a. Why must an appliance make connections to both slots? Why can't the appliance obtain electric power through just one of the slots?
b. The current provided by the two slots is alternating current. Describe how the current flowing through those two slots varies with time.
c. One of the slots in the outlet is usually wider than the other, just as one of the prongs on a power cord is usually wider than the other. The wider slot is called *neutral* because it has roughly zero net electric charge on it. If you were to touch *neutral* (which is attached to white wires to indicate relative safety), no current should flow into your hand even if you were also touching something connected to the earth. To make sure that neutral remains uncharged, it's connected to the ground as it enters your home. The outlet's narrow slot is called *power* because its net electric charge fluctuates above and below zero and it can propel current through an appliance. If you were to touch *power* (which is attached to black wires to indicate danger) and simultaneously to touch something connected to the earth, you would receive a serious and potentially life-threatening shock. Why must you also be connected to the earth to receive this shock?
d. Most appliances with plastic surfaces use two-prong plugs and keep *power* as far from your hands as possible. But appliances with metal surfaces often use three-prong plugs and connect their metal surfaces to the earth through the outlet's third opening. This third opening is called *ground* because it's connected to the ground (by a green wire to indicate its association with the earth). If the *power* wire in such a "grounded" appliance comes loose and touches the appliance's metal surface, current will begin to flow. Through what path will current go from *power* to *neutral* in this case?
e. If a metal appliance weren't "grounded" and *power* accidentally touched a metal surface, you would receive a shock if you touched the appliance and something connected to the earth at the same time. Why?

CHAPTER 13

ELECTRONICS

Electric currents and magnetized metals can do more than simply turn motors or power light bulbs. They can also represent many different things, from sounds, to video images, to computerized information. In this chapter, we'll see how electricity and magnetism give rise to the field of electronics.

Electronic devices are tools that manipulate electric currents in sophisticated ways. They first appeared in the early twentieth century with the development of the vacuum tube and have been maturing ever since. The invention of transistors has accelerated the electronic revolution so that it now permeates every facet of modern society.

Experiment: Listening to Yourself Talk

In this chapter, we'll discuss some of the components used in electronic devices and the ways in which these devices perform their tasks. One such task is the amplification of audio signals, electronic representations of sound. You can observe this amplification by playing with an audio system.

This activity is most instructive if the audio system has a microphone and a microphone input, so that you can hear your voice played through the audio system. Many tape recorders have microphones built into them or have microphone inputs that you can use. Even if you don't have a real microphone, you can use your headphones as microphones, because the speakers inside them can also work as microphones. Just as a motor becomes a generator when you do work on it yourself, a speaker becomes a microphone when you do work on it with your voice. If you don't have any way to introduce your voice into the audio system, you'll have to make do with prerecorded sound or radio.

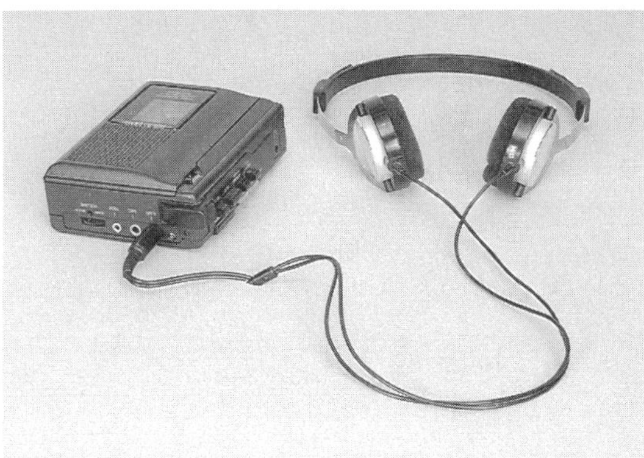

Turn on the audio system and connect the microphone (or headphone) to the microphone input. Talk into the microphone. If you have all the switches set properly—you must be in "record mode" if you're using a tape recorder—then your voice should be reproduced through the audio system's speakers or headphones.

Predict what will happen if you adjust each knob on the audio system and then **observe** the result. Were your predictions **verified**? Did the base and treble knobs affect its sound? How are these knobs and buttons controlling the behavior of this electronic device? What characteristics of the audio system would you **measure** to see how well it performs, if you had the right tools?

While this electronic reproduction of sound may seem trivial in these days of compact discs and digital stereo, it involves processes that would have astounded Thomas Edison or any other person in the 1800s.

Chapter Itinerary

To understand the remarkable properties of electronic devices, we'll examine two of them: (1) *audio amplifiers* and (2) *computers*. In *audio amplifiers*, we'll look at how small electric currents can be amplified into much larger currents and examine the components that make this amplification possible. In *computers*, we'll see how electricity and magnetism can represent both numbers and information and investigate the basic tools a computer uses to process them.

While this brief survey can't go beyond the most basic issues in electronics, it should provide a good basis on which to build. Once you understand how a few important electronic components work, it's not so hard to see how to combine them in ways that no one else has ever dreamed of. Like building blocks that control the movements of charged particles, electronic components can be put together to do almost anything.

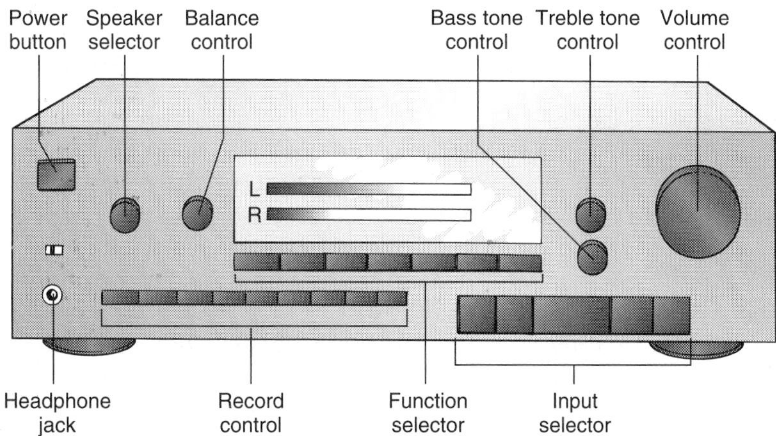

Section 13.1

Audio Amplifiers

Audio amplifiers are electronic devices that reproduce and enlarge small electric currents. They are commonly found in radios, stereos, and musical instruments, where they scale up small currents representing sound and provide the power needed by speakers.

Early audio amplifiers were built with vacuum tubes, relatively bulky devices that wasted power and aged quickly. Transistors have made audio amplifiers much more practical. They perform the task of amplification so efficiently so that most of the electric power used in the amplifier is delivered to the speakers. Moreover, transistors are tiny and extremely reliable. Transistor amplifiers can operate for decades or centuries without maintenance and are routinely built into other objects, such as games, toys, appliances, automobiles, and even greeting cards.

Questions to Think About: *Why do audio amplifiers need electric power to operate? What does an audio amplifier's power rating mean? Why does an audio amplifier become warm as it operates? How does the volume control work? What do the treble and bass knobs on an audio amplifier do?*

Experiments to Do: *You almost certainly own something with an audio amplifier in it. A component stereo system has a large one. Find an amplifier and look inside through the cooling holes, if there are any. You should see lots of small objects with wires protruding from their ends or sides. Are any of these familiar? Adjust the treble and base controls and listen to how they change the sound. Do the same with the volume control. Can you explain what these controls are doing to the currents delivered to the speakers?*

What an Audio Amplifier Does

There are many devices that produce audio signals, including microphones, musical instruments, tape recorders, and computers. A **signal** is a common term for any electric representation of information, and an **audio signal** is an electric current that represents sound.

The audio signal produced by a typical microphone or instrument is weak. While it's an accurate representation of the sound, it doesn't have the power that speakers need to produce that sound. Something must first enlarge the audio signal; it needs to be *amplified*.

Devices that enlarge various characteristics of signals are called **amplifiers**. An audio amplifier is an amplifier that's designed to boost signals in the frequency range that we hear or feel (20 Hz to 20,000 Hz). It has two separate circuits, an input circuit and an output circuit, and it uses the small current passing through its input circuit to control a much larger current passing through its output circuit. In this manner, the amplifier provides more power to its output circuit than it consumes from its input circuit.

There are many different kinds of audio amplifiers, but most have the same principal components: a preamplifier, a power amplifier, and a power supply (Figure 13.1.1). The preamplifier is designed to boost the small input signal from a tape deck, a CD player, or a microphone, to a level that the power amplifier is able to use. The preamplifier is part of two circuits. Its input signal arrives through one circuit and its output signal leaves through the other circuit. The preamplifier makes sure that the current reaching the power amplifier is within the proper range, protecting the power amplifier and improving its performance.

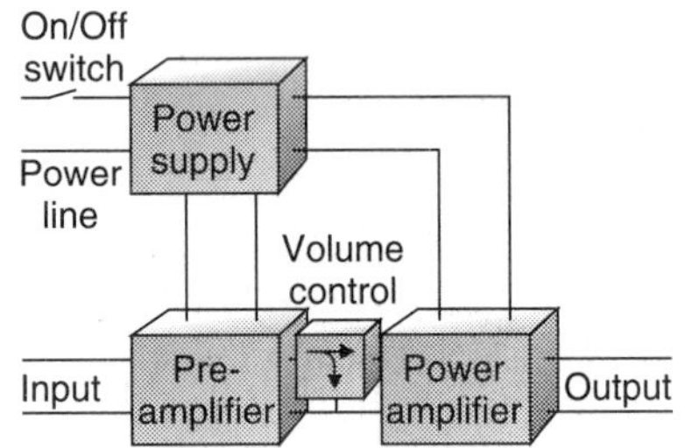

Fig. 13.1.1 - A block diagram of a typical audio amplifier.

Just how much the preamplifier must boost the input signal depends on the input device. Some input devices, such as tape decks or CD players, send relatively large currents to the preamplifier and their signals need little preamplification. Other devices, such as phonographs, produce such tiny currents that the preamplifier must boost their signals significantly. Because of the different preamplification requirements, a typical amplifier has several kinds of inputs, some for large signals and others for small signals.

The power amplifier resembles the preamplifier, except that the power amplifier starts with a relatively large input signal and creates an extremely powerful output signal, capable of delivering watts of power to a speaker system. A power amplifier doesn't have the subtlety of a preamplifier and is usually built of much larger electronic components.

Between the preamplifier and the amplifier is the volume control. The volume control is a device that splits the preamplifier's output current into two parts. One part is simply returned to the preamplifier directly while the other part passes through the power amplifier's input. Since the power amplifier only amplifies current that flows through its input, its output signal strength is proportional to the fraction of preamplifier current it receives. When you turn up the volume, you increase that fraction.

Larger currents carry more power, so both the preamplifier and the power amplifier provide more power to their output circuits than they consume from their input circuits. They must obtain this power from somewhere, which is why they're connected to a power supply. This power supply does work on the current passing through it and is either a battery or power-line operated device that imitates a battery. While we won't examine power-line operated supplies in detail, they are based on two components we've seen before: transformers and diodes. Transformers enable these supplies to exchange voltage for current to obtain the large currents and low voltages needed by audio amplifiers. Diodes prevent currents from flowing backward through these supplies so that, like batteries, they always pump currents from their negative terminals to their positive ones. In the end, a line-operated power supply performs just like a battery, except that its energy comes from a power plant far away.

While a typical audio amplifier contains hundreds of electronic components, often mounted together on *printed circuit cards* (Fig. 13.1.2), many of those components are relatively unimportant. They are there to ensure that the amplifier boosts all audio signals equally, regardless of strength or frequency. If we're willing to settle for less ideal performance, we can construct an audio amplifier from just a handful of parts. We'll do just that, after we first study those parts.

Fig. 13.1.2 - The electronic components on this "printed circuit" card are connected to one another by narrow copper strips. Because photographic techniques are used to form such strips on the surfaces of fiberglass cards, printed circuit cards are inexpensive to produce and extremely reliable.

CHECK YOUR UNDERSTANDING #1: Speak Up Please

Why can't you just connect the two wires of a microphone directly to the two wires of a speaker and have the speaker reproduce your voice loudly as you talk into the microphone?

Resistors

Our first electronic component is a **resistor**, which is simply two wires connected by a poor conductor of electricity (Fig. 13.1.3). Since current flows through a poor conductor only if there's an electric field pushing it forward, a current-carrying resistor must have a voltage drop across it. In accordance with Ohm's law (see Section 12.2), that voltage drop is proportional to the current in the resistor.

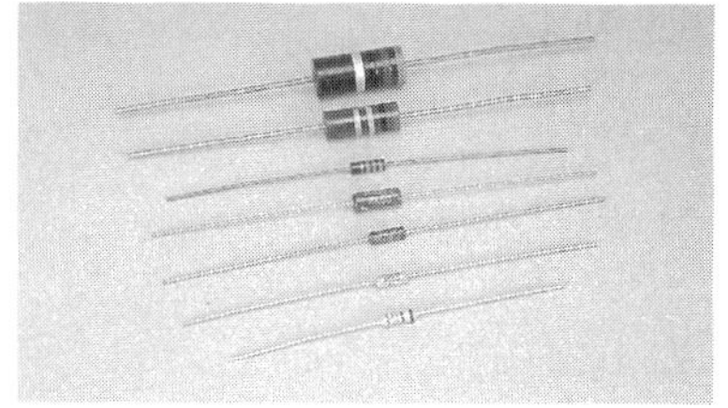

Fig. 13.1.3 - Resistors impede the flow of electric current, converting some of that current's electric energy into thermal energy. The larger resistors can handle more power without overheating.

But the voltage drop across a resistor is also related to the resistor's electric resistance—to how poor a conductor it is. The larger its resistance, the less current flows through the resistor. Some resistors are relatively good conductors ("low" resistance) while others are relatively poor conductors ("high" resistance).

As we saw in Section 12.2, a resistor's electric resistance is defined as the voltage drop across it divided by the current flowing through it and is measured in ohms. A resistor with a few ohms of resistance is nearly a good conductor while a resistor with a few million ohms of resistance is nearly an insulator. A k in front of the Ω means thousands (kΩ or kiloohms) and an M in front of the Ω means millions (MΩ or megaohms). Thus a 100 kΩ resistor has a resistance of 100,000 Ω and a 10 MΩ resistor has a resistance of 10,000,000 Ω.

The poor conductor in a resistor is usually graphite paste, metal film, or metal wire. It's encased in a cylindrical body, with a pair of wires that carry current to and from the poor conductor (Fig. 13.1.4). The body is labeled to indicate its electric resistance in ohms, often with colored bands. The first two or three bands indicate the digits (0 through 9), the next indicates the power of ten, and the final band indicates how precisely the resistance is controlled. For example, a resistor with bands yellow-violet-orange-gold is a 4, 7, 1000, 5% resistor; a 47 kΩ resistor that is guaranteed to be accurate to within 5%.

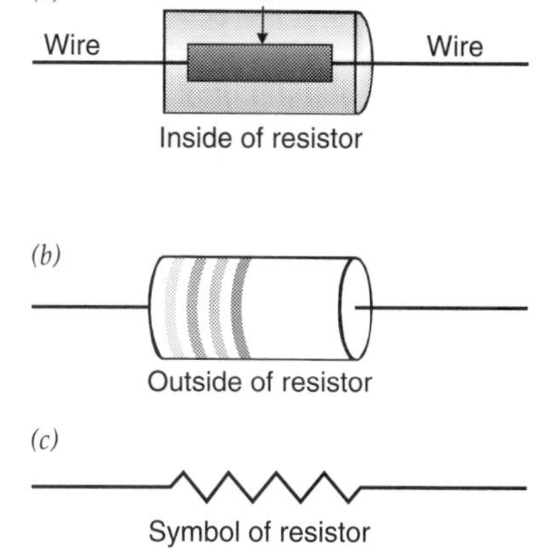

Fig. 13.1.4 - (*a*) A resistor is two wires with a poor conductor of electricity between them. (*b*) It's usually encased in a cylindrical shell, with colored stripes to indicate its resistance. (*c*) In a schematic diagram of an electronic device, the resistor is represented by a zigzag line.

Current loses energy as it flows through a resistor and the resistor becomes warm. Because it contains resistors and other parts that produce thermal energy, an amplifier must eliminate heat to the room air. Resistors are rated according to the maximum power they can safely convert into heat without burning up.

CHECK YOUR UNDERSTANDING #2: A Waste of Energy

Your radio produces sound when its amplifier allows current from its power supply to flow through its speaker. During a quiet moment, the amplifier sends only a small current through the speaker and this current retains most of its electric power. How can the radio use up the current's remaining power before returning it to the power supply for reuse?

Capacitors

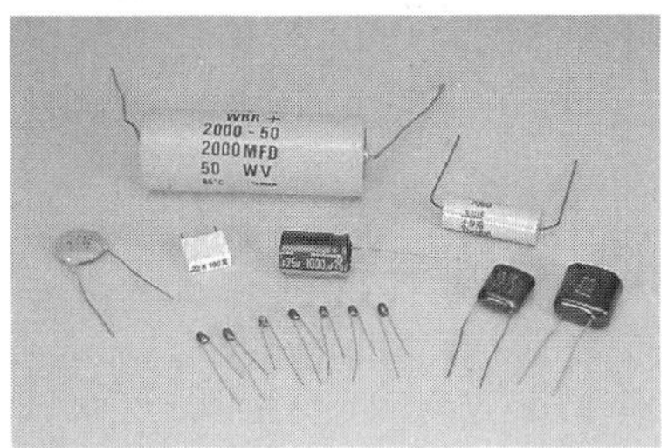

Fig. 13.1.5 - Capacitors store separated electric charge. Each of these capacitors contains two conducting surfaces separated by a thin insulating layer.

Our second electronic component is a **capacitor**, a device that stores electric charge (Fig. 13.1.5). It consists of two conducting plates separated by a thin insulating layer (Fig. 13.1.6). When one plate is positively charged and the other is negatively charged, the two opposite charges attract one another. This attraction allows the two plates to store large quantities of separated charge, while leaving the capacitor as a whole electrically neutral.

You can charge a capacitor's plates by transferring positive charge from its negative plate to its positive one. The work that you do during this transfer is stored in the capacitor as electrostatic potential energy and is released when you let the positive charge return to the negative plate. In fact, the charged capacitor resembles a battery because, when connected to a circuit, it pushes positive charge into the circuit from its positive plate and collects that charge from the circuit with its negative plate.

Since positive charge leaves the positive plate with more energy than it has when it returns to the negative plate, the voltage of the positive plate is higher than the voltage of the negative plate. The voltage difference between the plates is proportional to the separated charge on them. The more separated charge the capacitor is holding, the larger the voltage difference between its plates.

This voltage difference also depends on the structure of the capacitor. Enlarging the plates allows the like charges on each plate to spread out, so that they repel one another less strongly. Thinning the insulating layer between the plates allows the opposite charges on the two plates to move closer together, so that they attract one another more strongly. Both of these changes lower the separated charge's electrostatic potential energy and, consequently, the voltage difference between the plates.

Because these changes allow the capacitor to store separated charge more easily, they increase its **capacitance**, the separated charge it holds—the amount of positive charge that has been transferred from one plate to the other—divided by the voltage difference between its plates. The units are charge divided by voltage. The SI unit of capacitance is the coulomb-per-volt, also called the **farad** (abbreviated F). A capacitor with a farad of capacitance stores an incredible amount of separated charge while a capacitor with a billionth of a farad of capacitance is much more typical. The Greek letter μ in front of the F means millionths (μF or microfarads), the letter n in front of the F means billionths (nF or nanofarads), and the letter p in front of the F means trillionths (pF or picofarads). A capacitor's capacitance is marked on its wrapper, often in an abbreviated form.

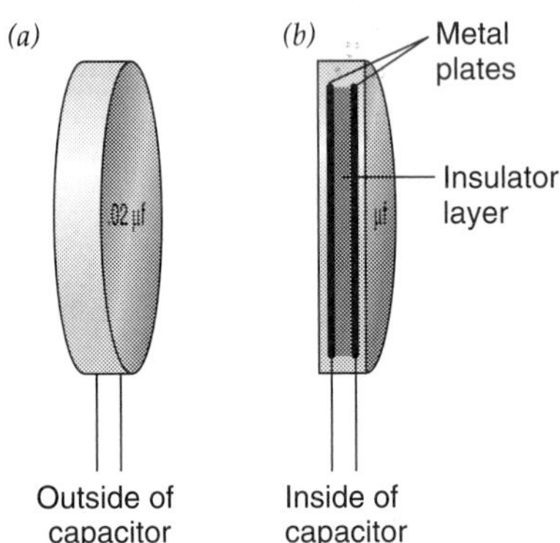

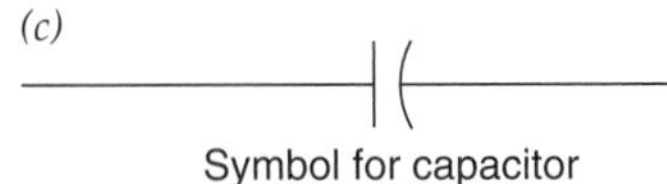

Fig. 13.1.6 - (*a*) A capacitor is usually a disk or cylinder with two protruding wires. Its capacitance is printed on its surface. Inside (*b*), the wires are connected to two conducting plates that are separated by a thin insulating layer.

While the simplest capacitors are just two flat plates separated by an insulator, other capacitors are made when a "sandwich" of metal and insulating sheets is rolled up like a jelly roll. Even a fairly small cylindrical capacitor formed in this manner may contain a square meter of sandwich inside. Some capacitors are created by a chemical process that forms enormous conducting surfaces separated by a fantastically thin chemical insulator. The two "plates" in this type of capacitor are not the same, since one of them is actually a conducting chemical called an *electrolyte*. Because of the differences in the plates, such *electrolytic capacitors* can only accommodate separated charge in one direction. One plate must always hold the positive charge while the other holds the negative charge.

CHECK YOUR UNDERSTANDING #3: Charging Up

Your stereo's amplifier obtains power from the power line. Since the power line delivers no power during the moments when its alternating current reverses directions, the stereo's power supply must store energy. How can it do this?

Transistors

Our third electronic component is a **transistor**, a device that allows a tiny amount of electric charge to control the flow of a large electric current (Fig. 13.1.7). Invented in 1948 by three American physicists, William Shockley (1910–1989), John Bardeen (1908–1991), and Walter Brattain (1902–), transistors are now the basis for nearly all electronic equipment. Like the diodes we studied in Section 12.3, transistors are built from doped semiconductors, a semiconductor such as silicon to which chemical impurities have been added. But unlike diodes, which simply prevent current from flowing backward through a single circuit, transistors allow the current in one circuit to control the current in a second circuit.

Fig. 13.1.7 - These MOSFET transistors make it possible for small electric charges to control large electric currents. The larger transistors can handle more electric power without overheating.

While there are many types of transistors, the simplest and most important type is the *field effect transistor*. Actually, even here there are several types, so we'll focus on one that is used widely in audio, video, and computer equipment: the *n-channel metal-oxide-semiconductor field effect transistor* or *n-channel MOSFET*. Despite its complicated name, the n-channel MOSFET is a relatively simple device, consisting principally of three semiconductor layers and a nearby metal surface (Fig. 13.1.8). The three layers are called the *source*, the *channel*, and the *drain*, while the metal surface is called the *gate*.

The three semiconductor layers are doped with chemical impurities so that they form two p-n junctions. One junction is between the n-type source layer and the p-type channel layer while the other junction is between the p-type channel layer and the n-type drain layer. Each junction behaves like a diode, permitting current to flow through it only in one direction. But because these two junctions are arranged back to back, they prevent any current from flowing between the source and the drain. If you were to include this transistor in a circuit and tried to send current through it from the drain to the source, no current would flow. But there is a way to turn on the transistor so that current *can* flow through it. To see how this works, we need to study electrons in the three layers.

The channel is p-type semiconductor, so it naturally has a few empty valence levels. The source and drain are n-type semiconductor, so they naturally have some conduction level electrons. Contact between the three layers allows conduction level electrons to move out of the source and drain and into the channel's empty valence levels. The channel is lightly doped so that it has relatively few empty valence levels and these are entirely filled by electrons from the source and drain. The channel is therefore one big non-conducting depletion region, which is why no current can flow through it.

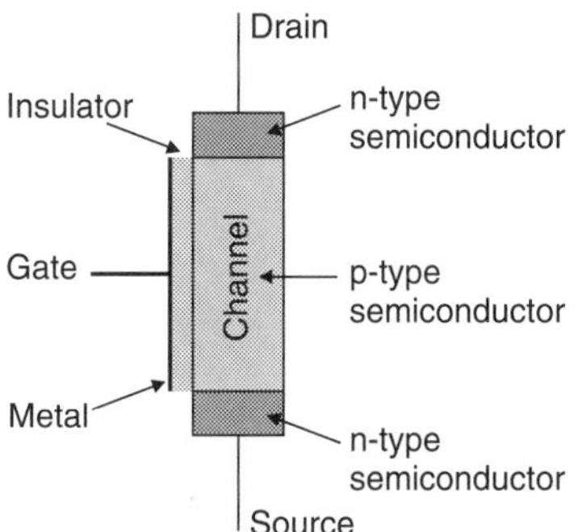

Inside an n-Channel MOSFET Transistor

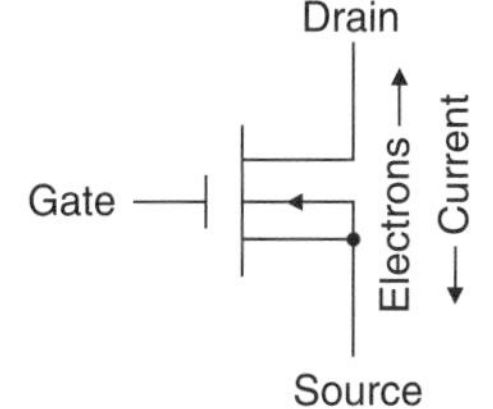

Symbol for n-Channel MOSFET Transistor

Fig. 13.1.8 - In an n-channel MOSFET transistor, current tries to flow through a narrow channel of weakly doped p-type semiconductor. Normally, the channel has no conduction level electrons or empty valence levels to carry current. But when positive charge is placed on the gate, it attracts conduction level electrons into the channel. The channel becomes n-type semiconductor and conducts current.

But if more electrons were somehow to enter the channel layer, they would have to go into its conduction levels. The channel layer would then have extra conduction level electrons, just like the n-type source and drain layers. In fact, the channel would then behave like n-type semiconductor and the p-n junctions would vanish. Current would then be able to flow through the transistor, from its drain to its source.

Drawing extra electrons into the channel is the task of the metal gate. Separated from the channel by an incredibly thin insulating layer, the gate controls the channel's ability to carry current. When a tiny positive charge is placed on the gate, it attracts conduction level electrons into the channel and the channel begins to conduct current. The more positive charge there is on the gate, the more conduction level electrons in the channel and the more current can flow. In effect, the transistor behaves like an adjustable resistor with a resistance that decreases as the positive charge on its gate increases.

We can now understand the n-channel MOSFET's name. The "n-channel" refers to the channel's n-type behavior when its gate is positively charged and it's carrying current. The "metal-oxide-semiconductor" indicates that the metal gate is separated from the semiconductor channel by a thin insulating layer. This insulator is easy to puncture, which is why computers and other electronic devices are so easily damaged by static electricity. The "field-effect transistor" indicates that the electric field from charge on the gate controls the current flow through the transistor.

CHECK YOUR UNDERSTANDING #4: Power and Control

Widening the channel of an n-channel MOSFET transistor allows it to handle more current between its source and drain. However the enlarged transistor needs more positive charge on its gate to control that current. Explain.

A Simple Audio Amplifier

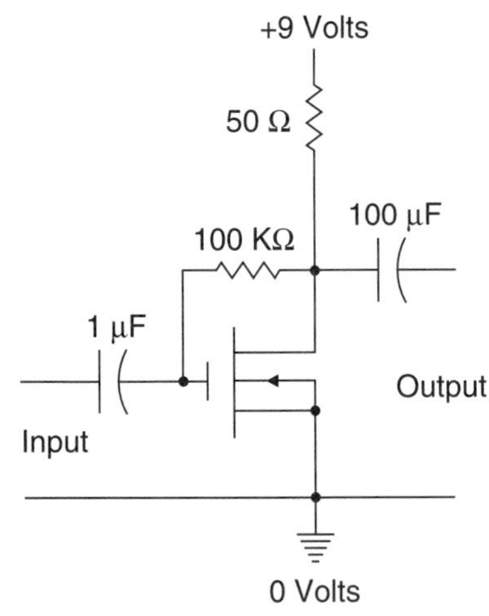

Fig. 13.1.9 - A simple audio amplifier can be built with one n-channel MOSFET, two resistors, and two capacitors. The device is powered by a 9 V battery.

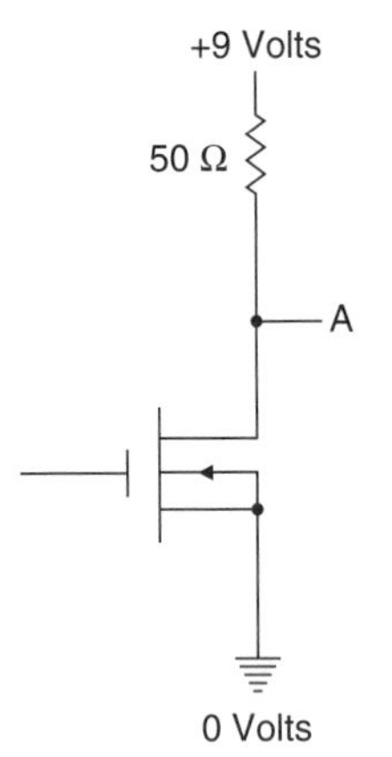

Fig. 13.1.10 - The voltage at **A** depends on the resistance of the MOSFET transistor. The lined triangle at the bottom signifies connection to ground (often the earth itself).

Figure 13.1.9 shows a simple audio amplifier, built from the components we've just studied. The figure is a **schematic diagram**, a conceptual drawing that uses symbols to represent each electronic component. The symbols used for the various components appeared in previous figures and the lines between them represent wires connecting them together.

The amplifier has only 5 components: an n-channel MOSFET transistor, two resistors, and two capacitors. It draws power from a 9 V battery (or an equivalent power supply) and amplifies a tiny alternating current in its input wires into a large alternating current in its output wires.

To understand how the amplifier works, let's first remove everything but the MOSFET transistor and the 50 Ω resistor (Fig. 13.1.10). Any current that passes through one of these components must also pass through the other. When the MOSFET transistor doesn't conduct current, no current flows through the 50 Ω resistor and it experiences no voltage drop. So the voltage at **A** is 9 V. But if the transistor does conduct current, a voltage drop will appear across the 50 Ω resistor and the voltage at **A** will decrease.

The transistor will conduct current only if positive charge is put on its gate. That can be done by connecting the gate to **A** with a 100 kΩ resistor (Fig. 13.1.11). Since **A** is at +9 V, it tends to push positive charge toward anything at lower voltage and is itself positively charged. Current flows slowly through the resistor from **A** to the gate. But as positive charge accumulates on the gate, the transistor begins to conduct current and the voltage at **A** drops. The positive charge at **A** gradually loses its electrostatic potential energy until it can no longer flow through the resistor to the gate.

At that point, the amplifier has reached a stable situation with the voltage at **A** near 5 V and a modest amount of charge on the transistor's gate. The 100 kΩ resistor has provided feedback to the transistor. If the transistor conducts too little current, positive charge flows onto its gate and makes it conduct more. If the transistor conducts too much current, positive charge flows off its gate and makes it conduct less.

The amplifier is now exquisitely sensitive to small changes in the charge on the transistor's gate. If you add just a tiny bit more positive charge, down goes the voltage at **A**. If you remove just a tiny bit of positive charge, up goes the voltage at **A**. The amplifier's input signal adds or subtracts positive charge from the gate and the amplifier's output signal emerges from **A**.

The amplifier has two input wires. Current from a microphone or other source flows into the amplifier through one wire and returns through the other. But the input signal is not connected directly to the gate. Instead, it's connected to the gate through a capacitor (Fig. 13.1.12). The capacitor allows the voltage on the transistor's gate to be different from the voltage on the top input wire. This flexibility is important in most audio amplifiers.

While the capacitor doesn't allow charge to flow from one plate to the other, it does respond to currents. As positive charge flows into the capacitor from the upper input wire, this charge attracts negative charge and draws it away from the gate. The capacitor remains electrically neutral but the gate becomes more positively charged. Even though the original positive charge can't reach the gate, the gate still receives a similar amount of positive charge.

When an alternating current flows back and forth between the two input wires, positive charge moves on and off the transistor's gate. As a result, the voltage at **A** varies up and down. Even a tiny alternating current on the input wires creates a large changing voltage at **A**.

All that remains is to use the changing voltage at **A** for some purpose. The amplifier has two output wires. Current from a speaker or other device flows into the amplifier through one wire and returns through the other. The amplifier provides power to this current. But the output signal isn't connected directly to **A**. Instead, it's connected to **A** through a capacitor (Fig. 13.1.13). Current flowing into one side of this capacitor causes a similar current to flow out the other side of the capacitor. The capacitor allows the voltage at **A** to be different from the voltage on the top output wire. Once again, this flexibility is useful in amplifiers.

Tiny alternating currents through the two input wires can produce large alternating currents through the two output wires. This amplifier works remarkably well, given its simplicity. If you connect a microphone to the input wires and a speaker to the output wires, the speaker will do a surprisingly good job of reproducing the sound in the microphone.

However our simple amplifier isn't perfect. It distorts the sound somewhat and it doesn't handle all frequencies or amplitudes of sound equally. It also wastes a large amount of electric power heating the 50 Ω resistor. Fancier amplifiers carefully correct for these problems. Many of them use feedback to make sure that the output signal is essentially a perfect replica of the input signal, only larger. They sense their own shortcomings and correct for them.

But perfect replication of the input signal is not always desirable. Sometime you want to boost the volume for part of the sound. The treble and bass controls on an amplifier allow you to selectively change the volumes for the high and low frequency portions of the sound, respectively. These controls often use resistors and capacitors to select and modify particular ranges of audio frequencies. Resistors slow the movement of charge and capacitors store charge. Together, these devices can respond differently to alternating currents of different frequencies. Such resistor-capacitor "filters" are common in audio equipment.

An amplifier's power rating indicates the peak power it can deliver to the speakers during a loud passage. The average power should never reach this value. But even during what seems like only a moderately loud passage, there will be instants during which the current must become extremely large. Different tones can work together to create a sudden high or low pressure at the microphone and the speaker must follow this pressure. Even when a 100 W stereo is only producing 20 W of sound on the average, there may be brief moments when it delivers 100 W of power to the speakers. If the amplifier is asked to deliver more power than it's capable of producing, the audio signal it sends to the speakers will be distorted and the sound will be unpleasant.

CHECK YOUR UNDERSTANDING #5: Sound Control

When you connect a microphone to the input of an MOSFET-based amplifier, the microphone sends current back and forth through the input wires. As a result, charge moves onto or off what critical control element in the amplifier?

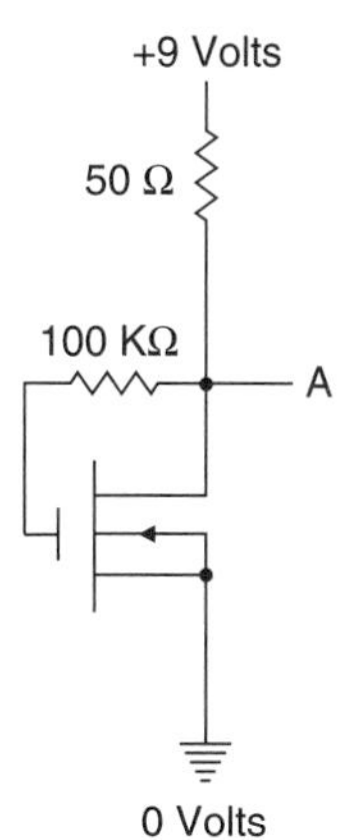

Fig. 13.1.11 - The 100 kΩ resistor transfers positive charge to the gate until the voltage at **A** drops to about 5 V.

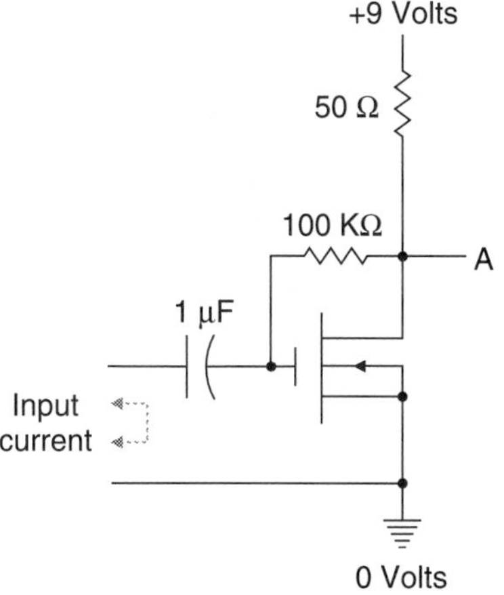

Fig. 13.1.12 - Because current flowing back and forth through the two input wires affects the charge on the transistor's gate, it also affects the voltage at **A**.

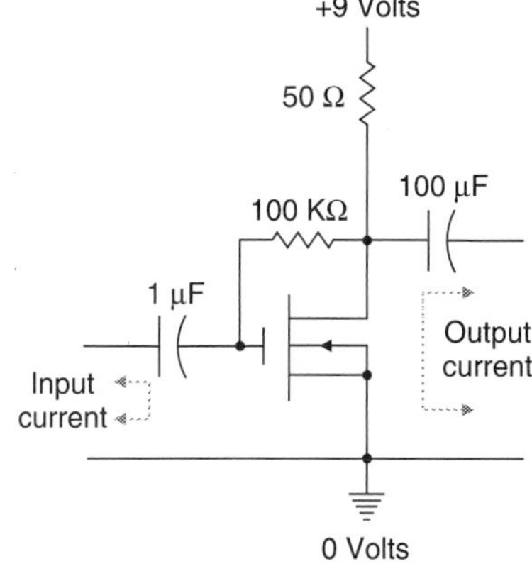

Fig. 13.1.13 - The amplifier causes currents to flow back and forth through the two output wires. The alternating current in the output wires is a good replica of the alternating current in the input wires, only larger.

Summary

How Audio Amplifiers Work: An audio amplifier permits a small current flowing through its input circuit to control a much larger current flowing through its output circuit. In a modern amplifier, this control is achieved with transistors, typically n-channel MOSFETs. The input current adds or subtracts charge from the gates of these transistors and changes their electric resistances. These transistors are connected to a power source, such as a battery or power supply, and determine how much power from the power source is transferred to the output wires. As the input signal varies, the transistors replicate its fluctuations in their output signal. However, the amplifier provides the output signal with more power than it consumes from the input signal. The audio amplifier thus provides power amplification.

The Physics of Audio Amplifiers

1. A resistor impedes the flow of current in a circuit. It experiences a voltage drop that is equal to the current flowing through it times its electric resistance.

2. A capacitor stores separated electric charge and electrostatic potential energy. It experiences a voltage difference between its plates that is equal to the amount of separated charge on its plates divided by its capacitance.

3. An n-channel MOSFET uses the electric field produced by a tiny amount of positive charge on its gate to control the electric resistance of its channel and thus the flow of current from its drain to its source.

Check Your Understanding - Answers

1. The tiny amount of power provided by the microphone is not sufficient to make the speaker emit loud, powerful sound.

Why: You can't get something for nothing. If you want loud sound, you must provide the power to create that sound. An amplifier allows a tiny signal from a microphone to control the flow of power from a power supply to a speaker.

2. The amplifier can send the current through a resistor.

Why: By passing the current through a resistor, the amplifier converts the current's electric power into thermal power. The current leaves the resistor with only enough electric power to carry it back to the power supply for reuse. Amplifiers often use resistors to get rid of unusable electric power. While such actions may seem wasteful, it's extremely hard to use all of the power supply's electric energy to make sound. Controlling current is most easily done by impeding its flow, and that often means resistors and wasted electric power.

3. It can store the energy as separated charge on a large capacitor.

Why: The stereo's power supply saves energy in large electrolytic capacitors. During the times when the power line is providing plenty of electric power, the power supply uses that power to increase the separated charge in its capacitors. When the power line isn't providing power, the power supply draws energy out of the capacitors. Unfortunately, electrolytic capacitors age quickly and often leak their liquid electrolyte. Once they leak, they can't store energy properly and your stereo will hum.

4. The larger transistor also has a larger gate. With more surface over which to spread its charge, the gate needs more positive charge in order to draw conduction level electrons into the channel.

Why: MOSFET transistors range in size from remarkably small (less than 1 square-micron) to relatively large (several square-millimeters). The smallest ones are used in computer chips, where millions of MOSFET transistors are created on a single wafer of silicon only a centimeter square. A tiny charge on the gate of one of these MOSFETs will allow it to conduct current. The largest MOSFET transistors are used in power control devices such as amplifiers and power supplies. These transistors have large gates and much more charge is needed to allow one of them to conduct current.

5. The gate of an MOSFET transistor.

Why: In all likelihood, the input current adds or removes charges from the gate of an MOSFET transistor in the amplifier's first stage of amplification.

Glossary

amplifier A device that replicates an input signal as a larger output signal.

audio signal An electric representation of sound in which current is usually proportional to the present air pressure relative to the average air pressure.

capacitance The amount of separated charge on the plates of a capacitor divided by the voltage difference across those plates. Capacitance is measured in farads.

capacitor An electronic component that stores separated electric charge on a pair of plates that are separated by an insulating layer.

farad (F) The SI unit of electric capacitance. A 1 farad capacitor will have a voltage difference between its plates of 1 volt when storing 1 coulomb of separated charge.

resistor An electronic component that impedes the flow of electric current, converting some of its energy into heat.

schematic diagram A symbolic picture of the conceptual structure of an electronic device.

signal An electric representation of information.

transistor An electronic component that allows a tiny amount of electric charge, either moving or stationary, to control the flow of a large electric current.

Review Questions

1. Why should the currents flowing in the input and output circuits of an ideal audio amplifier be proportional to one another?

2. Why should the voltage drop across the input wires of an audio amplifier be proportional to the voltage rise across the output wires of that amplifier?

3. What affect does an amplifier's volume control have on the current that amplifier sends through its output circuit?

4. How is the voltage drop across a resistor related to the current flowing through that resistor?

5. Why does a resistor become warm when current is passing through it?

6. What prevents the separated charge in a capacitor from flowing together?

7. In what form does a capacitor store energy?

8. Since the channel of an n-channel MOSFET is made from doped semiconductor, and doped semiconductor is an electric conductor, why can't current flow through that channel when the MOSFET's gate is uncharged?

9. What change must you make to an n-channel MOSFET to allow its channel to conduct current?

10. How can a small audio signal be used to change the charge on the gate of an n-channel MOSFET?

11. How can changing the amount of charge on the gate of an n-channel MOSFET control a large current flowing through a circuit?

Exercises

1. Suppose you connected a microphone directly to a large unamplified speaker. Why wouldn't the speaker reproduce your voice loudly when you talked into the microphone?

2. Why can't an audio amplifier operate without a power supply?

3. You like to listen to old phonograph records but your new stereo amplifier has no input for a phonograph. You connect the phonograph to the stereo's CD player input but find that the volume is extremely low. Why?

4. To correct the volume problem in Exercise **3**, you buy a small preamplifier and connect it between the phonograph and the CD player input of the stereo. The volume problem is gone. What is the preamplifier doing to fix the problem?

5. A screw came loose inside your clock radio and accidentally formed a good conducting path from one end of a resistor to the other. Is the electric resistance of this resistor-plus-screw more or less than that of the resistor itself?

6. The accident in Exercise **5** caused the current through a second resistor to double. Why did the voltage drop across that second resistor also double?

7. The resistor in Exercise **6** overheated and burned up. Why did doubling the voltage drop across this resistor cause it to consume four times as much electric power?

8. To repair the clock radio discussed in Exercises **5**, **6**, and **7**, you remove the screw and replace the burned resistor. That resistor was marked as being 47 Ω but all you have available is a 20 Ω resistor. If you were to install the 20 Ω resistor in the radio, either too much current would pass through it or the voltage drop across it would be too small (or both). Explain why each of these results could happen.

9. You have a small electronic game that runs on 6 V from four "AA" batteries. While you're on a trip, you want to operate it from the car's 12 V electric system. The game requires a current of 1 A, so you connect the game to the car's positive terminal through a 6 Ω resistor. Current from the car's positive terminal enters the resistor at 12 V but it's down to 6 V by the time it leaves the resistor and enters the game. The game then returns this current to the car's negative terminal and everything works beautifully. How does the resistor reduce the voltage and what happens to the power associated with the missing voltage?

10. Why is it important that the resistor in Exercise **9** be rated to handle at least 6 W of electric power?

11. When you turn off the game (see Exercises **9** and **10**), the voltage on both sides of the 6 Ω resistor becomes 12 V. Why did the voltage drop across the resistor change?

12. Suppose a battery is transferring positive charges from one plate of a capacitor to the other. Why does the work that the battery does in transferring a charge increase slightly with every transfer?

13. Two capacitors are identical except that one has a thinner insulating layer than the other. If the two capacitors are storing the same amount of separated electric charge, which one will have the larger voltage difference between its plates?

14. Even though capacitors can store large amounts of separated change, they normally remain neutral overall. When positive charge flows onto one plate of a capacitor in an electronic device, negative charge will flow onto the other plate to leave the capacitor neutral. If the plates were initially uncharged, what effect will this movement of charge have on the voltages of the two plates?

15. The large electrolytic capacitors used to store energy in an amplifier's power supply sometimes leak. Since its liquid electrolyte acts as one of the plates of a capacitor, a loss of electrolyte effectively reduces the surface area of that capacitor's plates. If the voltage difference between the plates isn't allowed to change, does the amount of separated electric charge stored in the capacitor increase or decrease? Why?

16. In an n-channel MOSFET, the source and drain are connected by a thin strip of *p-type* semiconductor. Why is this device labeled as having an *n-channel* rather than a *p-channel*?

17. The gate of an MOSFET is separated from the channel by a fantastically thin insulating layer. This layer is easily punctured by static electricity, yet the manufacturers continue to use thin layers. Why would thickening the insulating layer spoil the MOSFET's ability to respond to charge on its gate?

18. MOSFET transistors are often used as fast switches. They can open or close circuits in a few billionths of a second and never wear out. Suppose you have a battery, a light bulb, an n-channel MOSFET, and some wires. How could you arrange those parts so that adding or removing positive charge from the MOSFET's gate would turn the light bulb on or off, respectively? (Draw a schematic diagram of your circuit.)

19. When your calculator is on, current flows from the battery's positive terminal, through the calculator, and back to the battery. However, on its way back to the battery, the current passes through an n-channel MOSFET that acts as the calculator's real on-off switch. When you push the "on" button, you put positive charge on the MOSFET's gate. When you push the "off" button, you remove that charge. Explain how these movements of charge control the current flowing through the processor.

20. Your calculator (see Exercise **19**) turns itself off when you don't use it for a few minutes. How does it use the n-channel MOSFET transistor in its main circuit to do this?

21. An MOSFET transistor doesn't change instantly from a perfect insulator to a perfect conductor as you vary the charge on its gate. With intermediate amounts of charge on its gate, the MOSFET acts as a resistor with a moderate electric resistance. This flexibility allows the MOSFET to control the amount of current flowing in a circuit. Explain why an MOSFET transistor becomes warm as it controls that current.

22. Explain why the resistor in Fig. 13.1.10 becomes warm only when there is positive charge on the gate of the MOSFET.

23. Suppose there is initially no charge on the gate of the MOSFET in Fig. 13.1.11. Why does positive charge slowly flow onto the gate, rather than negative charge, and why does charge eventually stop flowing onto the gate?

24. When positive charge flows onto the left plate of the capacitor in Fig. 13.1.12, positive charge flows off the right plate of that capacitor and onto the gate of the MOSFET. Explain why this movement of charge allows the capacitor to remain electrically neutral.

25. As the MOSFET transistor in Fig. 13.1.13 responds to the input current, the voltage to the left of the 100 μF capacitor fluctuates up and down. Whenever that voltage drops, positive charge on the capacitor's left plate is attracted toward the electric components to its left and some of that charge flows leftward. Explain why a dropping voltage to its left causes positive charge to flow leftward from the capacitor's left plate.

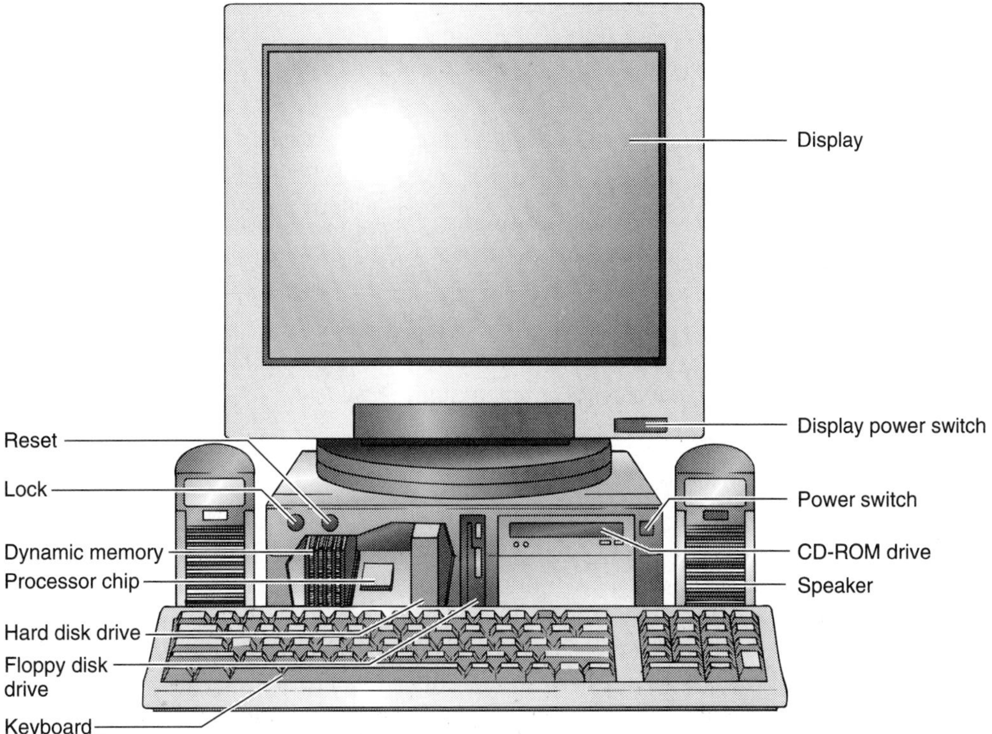

Section 13.2

Computers

Computers are devices that process numbers and other information according to a collection of programmed rules. They aren't really intelligent; they just do exactly what they're told and do it extremely rapidly. Whether computers make our lives easier or more frantic is an open question. But they certainly permit us to do things we couldn't otherwise do. Computers are found not only on desktops and in data processing centers, but also embedded inside everything from microwave ovens to automobiles. These embedded computers analyze information gathered by sensors and determine how the machines should respond. They have become so inexpensive that tasks once done by specialized electronic equipment are now better and more cheaply done by small, dedicated computers.

Questions to Think About: *Why are most computers electronic rather than mechanical? How can electricity represent numbers? How can numbers be used to represent letters, words, and symbols? If people use computers to write computer programs, how are the first programs for a new computer written?*

Experiments to Do: *Computers are so common that you should have little trouble finding one to study. Turn it on or start it up. The computer will begin to act on a series of rules that are stored in its memory until it eventually pauses for you to type or point. It will then act on what you type according to still more rules. Can you describe its process of starting up, waiting for you to type something, and then acting on what was typed in terms of a series of rules like "draw rectangular box" or "read typed character from keyboard"? How many of these simple rules do you think are contained in your favorite word processing package or computer game?*

Unlike a person who might respond differently each time you say hello, the computer responds exactly the same way each time you start it up. If it doesn't, then one of the rules it's following specifies the way in which it should vary its response. With no creativity of its own, the computer obeys its orders to the letter even when appearing unpredictable.

Some of the rules the computer follows affect what it displays on its screen. For example, it may display a character you typed. Does it have to do this or could it display a different character instead? Its rules may also involve storing the characters you type in its memory. In what situations should such storage occur?

Work with the computer for a while and try to think of it as a machine that's simply acting on your input—typing and pointing—according to specific rules and producing appropriate output—displayed material, printing, and sound. How does this view of a computer as a machine that responds to inputs with appropriate outputs apply to the computers that control automobiles or microwave ovens?

The History of Computers

Mechanical devices that can perform such simple numerical tasks as adding a column of numbers together or multiplying a pair of them to obtain a product have been around for a long time. The Chinese version of the abacus, with beads that slide back and forth on fixed shafts, is over 800 years old and allows a deft operator to perform basic numerical operations at a surprising rate. Numbers are represented as the positions of various beads and calculations are performed by moving these beads. However, all of the mechanical motions that carry out the computations on an abacus are done by the person using it. This aspect of its operation limits its ultimate speed and accuracy. People needed something that was faster, required less user expertise, and was less prone to errors.

By the nineteenth century, devices meeting these requirements had begun to appear. The new calculators and analytical engines were mechanical devices that represented numbers as the positions of various rotating dials and wheels, and used gears and cams to move these dials and wheels about during a calculation. Some even had push buttons with which to enter numbers into the calculation and electric motors to aid in the task of turning the gears and cams. But the calculations were still made entirely by mechanical devices at the speeds with which mechanical devices can move. These calculating devices were also inflexible. They performed only a few tasks, so that extensive calculations had to be coordinated by an operator.

In the past half-century, a whole new class of computing devices has developed. These machines incorporate at least two major innovations that quickly condemned mechanical calculators to the museums. First, modern computers are electronic devices and represent the numerical objects of their calculations as electric signals. Instead of representing a number as the position of a dial, a modern computer represents it as a current or charge.

Second, modern computers use stored programs to determine the sequence of calculations they'll perform on their numbers. Unlike a calculator, which requires the constant attention of an operator in order to perform a complex task, a stored program computer can be told once how to perform the complex task, and will then perform it repetitively without further help.

In addition to these fundamental changes in concept, modern computers make use of a number of changes in scale over their mechanical predecessors. Modern electronic computers operate much more rapidly than mechanical calculators, due to the ease and speed with which electrons move through conducting materials. Instead of requiring seconds to add two numbers together, a computer can do it in as little as a billionth of a second.

The other important change of scale is in the quantity of numbers that a modern computer can store and use in its calculations. While a mechanical calculator is limited to the two or three numbers on its dials, modern computers can work with millions, billions, even trillions of different numbers. With the advent of computer networks, there is essentially no limit to the quantity of numbers that a single computer can deal with. Since numbers can be used to represent letters, words, images, and sounds, computers can deal with enormous amounts of information of every type.

In this section, we'll look at some of the ways in which a computer represents and remembers numbers and how it uses electronic components to process those numbers.

CHECK YOUR UNDERSTANDING #1: Word and Numbers Get Together
List two ways to represent English words with numbers.

Representing Numbers

A computer represents numbers with physical quantities such as charge, current, voltage, and magnetization. There are two common techniques for this representation. The first technique relates the value of a physical quantity to the number itself. For example, a charge of 124 C could represent *124*, or a voltage of –2.313 V could represent *–2.313*. This representation technique is called **analog** and computers that use it are *analog computers*. But while analog computers are frequently used in electronic devices such as in the noise reduction part of your tape deck, they are rarely used for calculations.

Most of the devices we think of as computers use a second representation technique called **digital**, in which numbers are first decomposed into digits and then these digits are individually represented by specific values of a physical quantity. For example, *124* could be decomposed into the decimal digits 1, 2, and 4, and then these decimal digits could be represented as three separate charges, currents, or voltages. Because each charge, current, or voltage only has to represent the integers from 0 to 9, its value doesn't have to very accurate. If a voltage of 4 V is supposed to represent the digit 4, then 3.9 V or 4.1 V will still be understood to mean 4.

We usually break numbers into ones, tens, hundreds, thousands, etc.—the powers of 10—because we work in **decimal**. But we could also break numbers into ones, twos, fours, eights, sixteens, etc. Instead of using the powers of ten, as in decimal, we would be using the powers of two. This system for representing numbers is called **binary**.

In decimal, *124* is written as 124, meaning that *124* contains 1 hundred (10^2), 2 tens (10^1), and 4 ones (10^0). When these pieces are added together, 100+20+4, you obtain *124*. In binary, *124* is written as 1111100, meaning that *124* contains 1 sixty-four (2^6), 1 thirty-two (2^5), 1 sixteen (2^4), 1 eight (2^3), 1 four (2^2), 0 twos (2^1), and 0 ones (2^0). When these pieces are added together, 64+32+16+8+4, they again total *124*. This apparently complicated way to represent even a fairly small number is actually quite useful. The number has been broken into pieces that have only two possible values. There either *is* a thirty-two in the number being represented or there *isn't*. The only two symbols you need when representing a number in binary are 0 and 1.

Because *124* is 1111100 in binary, you could represent *124* by the charge, current, or voltage of seven separate objects. The first five objects would represent 1s while the last two would represent 0s. For example, if the seven objects were capacitors, the first five could hold separated charge while the last two held no charge. A device that needed to know what number the capacitors represented would measure their charges. Finding charge on the first five capacitors

(11111) and no charge in the last two (00), it would determine that the capacitors represent 1111100 binary or *124*.

Binary is useful because fast electronic devices that turn currents on and off are relatively easy to build. It's much harder to build fast devices that deliver the specific currents, charges, or voltages needed for analog representation. Analog representation is also susceptible to electronic imperfections and noise because an analog device that was trying to represent *124* with a voltage of 124 V might accidentally produce 123 V or 125 V. Imagine a bank computer that couldn't tell $124 from $123 or $125!

Although representing *124* in binary takes at least seven separate quantities, there is no confusion in the number being represented. Computers that decompose numbers into binary in this manner are called *digital computers*. Because they can reliably represent numbers of any size, digital computers are much more precise and accurate than analog computers.

CHECK YOUR UNDERSTANDING #2: New Math
What number does binary 10000001 represent?

Storing Numbers and Information

Inside a digital computer, numbers are represented in binary form. The number of binary digits used to represent a number determines how large that number can be. Each binary digit is called a **bit** and using more bits allows you to represent larger numbers. By associating these numbers with other things—words, sounds, colors, names, and so on—the computer can use bits to represent information. The more detailed that information, the more bits are needed to represent it.

Eight bits can be used to represent any number from *0* (which is 00000000) to *255* (which is 11111111). Since there are fewer than 256 of many common objects, these objects can be identified by groups of eight bits. For example, the keyboard symbols have all been assigned numbers between *0* and *255* and the letter A is *65*. Since the eight bits 01000001 represent *65*, they also specify an A. Groups of eight bits are so common and useful that they are called **bytes**.

There are several ways for a computer to store a bit. In the computer's main memory (often called random access memory or "RAM"), each bit is a tiny capacitor that uses the presence or absence of separated electric charge to denote a 1 or a 0. The computer stores a bit by producing or removing separated charge and recalls that bit by checking for that charge.

Each capacitor is built right at the end of its own tiny n-channel MOSFET. This MOSFET controls the flow of charge to or from the capacitor. To store or recall a bit, the computer places positive charge on the gate of the MOSFET so that the MOSFET becomes electrically conducting. The memory system can then transfer charge to or from the bit's capacitor.

Storing the bit is relatively easy; the computer simply sends the appropriate charge through the MOSFET to the capacitor. But recalling the bit is harder, because the charge on the capacitor is extremely small. Sensitive amplifiers in the memory system detect any charge flowing through the MOSFET from the capacitor and report what they find to the computer. Since this reading process removes any charge from the capacitor, the memory system must immediately store the bit again.

Unfortunately, these tiny capacitors can't hold separated charge forever because they're not surrounded by perfect insulators. Any separated charge they have eventually leaks away. Computer memory that uses charged capacitors to store bits is called *dynamic memory* and must be *refreshed* (read and restored) hundreds of times each second to ensure that a 1 doesn't accidentally switch to a 0 or vice versa.

The capacitors and MOSFET transistors used in dynamic memory and other computer components are formed on the surface of a thin wafer of silicon, creating an **integrated circuit** or "computer chip" (Figs. 13.2.1, 13.2.2, and 13.2.3). Careful processing, using photographic and chemical techniques, produces the tiny structures needed to make capacitors and MOSFETs that are only about 1 micron on a side. Since individual memory chips often store more than 16 million bits, they have more than 16 million capacitors and MOSFET transistors on their surfaces.

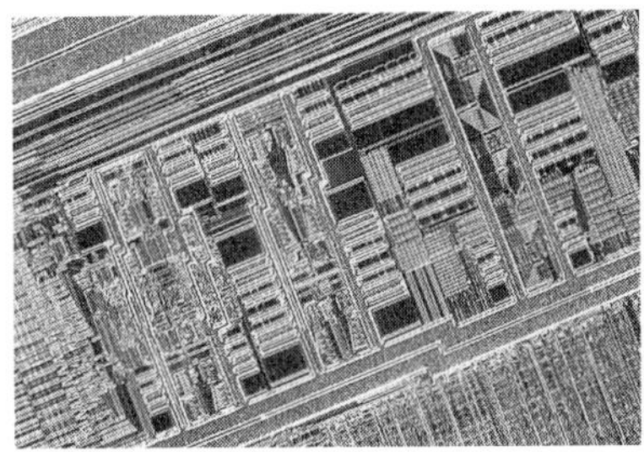
Fig. 13.2.1 - This microscope photograph shows a tiny portion of a Pentium integrated circuit microprocessor. The aluminum strips that you see connect millions of MOSFET transistors and other components that have been formed by photographic techniques on the surface of a thin wafer of silicon.

While dynamic memory is inexpensive, the time it takes to store and retrieve bits is relatively long and a computer with only dynamic memory will run rather slowly. That's why computers also use a faster but more costly type of memory, *static memory*, that employs a different technique for storing and retrieving bits. We'll discuss static memory later on, after we've studied the logic components from which it's built.

But using only static memory isn't cost effective either. Instead, a typical computer uses a large amount of cheap dynamic memory and a smaller amount of expensive static memory. The computer puts numbers it needs frequently in the faster static memory. When it needs numbers that are stored in the slower dynamic memory, it transfers them temporarily into the static memory, so that it can refer to them again quickly if necessary. This arrangement, in which numbers and information move from slow memory to fast memory on demand is called *caching* and the fast memory is called a *cache*.

Fig. 13.2.2 - Each of these integrated circuits contains vast numbers of transistors, resistors, and capacitors. Most of these devices are mounted in plastic or ceramic but the two tiny chips in the foreground are unmounted.

Numbers that are needed even less frequently than those in dynamic memory are usually stored magnetically. Hard disks, floppy disks, and computer tapes all use the magnetic recording concepts we discussed in Section 12.5. The recording surface consists of a thin plastic layer, filled with tiny elongated magnetic particles. In a hard disk, this layer rests on an extremely smooth aluminum disk (Fig. 13.2.4). In a floppy disk, it appears on both sides of a stiff plastic disk. And in a computer tape, this layer is coated on a thin strip of plastic.

In each case, the magnetic layer moves past a stationary recording head, which writes bits onto the magnetic layer by exposing it to a strong magnetic field. An electric current passing through a coil in the recording head temporarily magnetizes the head one way or the other and it imprints this magnetization into the magnetic layer. To read the bits back from the magnetic layer, the layer must pass by the recording head again. This time, the layer's magnetization temporarily magnetizes the recording head, inducing a current in its coil that indicates which way the layer is magnetized.

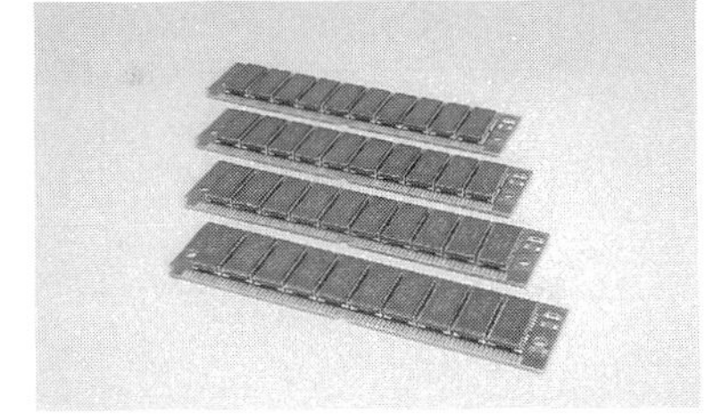
Fig. 13.2.3 - These dynamic memory modules use tiny capacitors to store millions of bits.

But there is a complication to this recording scheme. The computer must be careful about which bits it writes onto the magnetic layer because it needs some indication of where one bit ends and the next bit begins. When the magnetization suddenly changes directions, the computer knows that a new bit is starting. But if there are too many bits all magnetized in the same direction, the computer will have trouble distinguishing one bit from the next. To help it find the bits, the computer occasionally adds an extra reversal of the magnetization direction. When it retrieves bits from the magnetic layer, it recognizes and ignores these extra reversals.

Fig. 13.2.4 - Each of the 8 platters in this hard disk stores several billion bits. A motor hidden in the central column spins the platters while recording heads (not shown) read and write information on the platters' surfaces.

As a recording medium moves past a recording head, that head magnetizes only a thin *track* on the medium's surface. Shifting the head sideways allows hard disks, floppy disks, and computer tapes to record many parallel tracks on their recording media. A hard disk uses an electromagnet to push its recording head quickly from one position to another. The head's tiny mass and the strong force between the electromagnet and fixed permanent magnet allow the head to accelerate rapidly. It takes only a few thousandths of a second to leap from one track to another. Both floppy disks and computer tapes use stepping motors to move their heads, making them cheaper to build but slower in moving from one track to another.

The maximum number of bits that a computer can record on each square centimeter of disk or tape surface is determined by three things:

1. The size of the recording head.
2. The size of the magnetic recording particles.
3. The computer's ability to locate bits on the surface.

Each year, recording heads shrink further so that they're able to record bits on smaller patches of surface. Detecting these tiny bits is difficult, so a new head technology has developed that uses materials that change their electric resistances when exposed to magnetic fields. These *magnetoresistive* heads will allow bits to become significantly smaller than they presently are.

The size of recording particles will soon become a problem. These particles can't be made much smaller without becoming susceptible to the disordering effects of thermal energy. Random thermal agitation slowly weakens the magnetization of large permanent magnets and can flip the magnetizations of small magnetic particles. The smaller the particle, the more likely it is to flip.

Because smaller bits are harder to find, the positioning controls on recording heads are getting more and more sophisticated. Electronic and optical feedback techniques are now often used to find and follow the tracks as recording media move past recording heads.

Finally, computers often store bits using optical techniques. In some cases, the bits are still stored by magnetized particles but optical techniques are being used in the storage and retrieval processes. Although these *magnetooptic* devices are becoming common, we won't discuss them further. But there are also devices that store bits without magnetism. We'll explore fully optical storage techniques when we discuss compact disks in Section 16.3.

CHECK YOUR UNDERSTANDING #3: Like Sending a Letter to Yourself

Every few thousandths of a second, a computer stops briefly to read and then rewrite every bit in its dynamic memory. What is going on?

Processing Numbers and Information

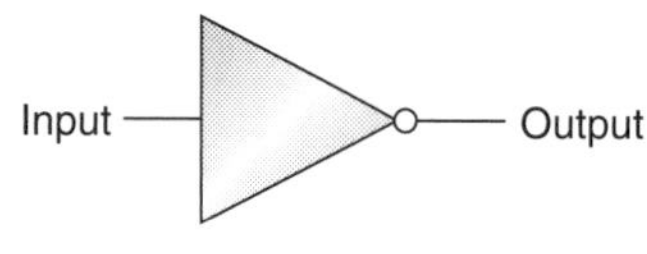

Input	Output
1	0
0	1

Fig. 13.2.5 - An inverter, shown here symbolically, produces one output bit that is the inverse of its one input bit.

While we have seen how numbers and information can be represented and stored as bits, we haven't yet looked at how a computer processes those bits. This processing is done by electronic devices that take groups of bits as their inputs and produce new groups of bits as their outputs. Since their output bits are related to their input bits by rigidly defined logic rules, these electronic devices are called *logic elements*.

The simplest logic element is the *inverter*, which has only one input bit and one output bit. Its output is the inverse of its input (Fig. 13.2.5). If an inverter's input bit is a 1, then its output bit is a 0 and vice versa. Inverters are used to reverse an action—turning a light on rather than off or starting a tape rather than stopping it. Inverters are also used to construct more complicated logic elements.

But inverters aren't just abstract logic elements; they are real electronic devices. They act on electric inputs and create electric outputs. In household computers, inverters and other logic elements represent input and output bits with electric charge. Positive charge represents a 1 and negative charge represents a 0.

Thus when positive charge arrives at the input of an inverter, it releases negative charge from its output.

Inverters and other logic elements are usually constructed from both n-channel and p-channel MOSFETs. We've already seen that n-channel MOSFETs conduct current only when their gates are positively charged. The p-channel MOSFETs are just the reverse, conducting current only when their gates are negatively charged. The drain and source of a p-channel MOSFET are made from p-type semiconductor while the channel is made from n-type semiconductor. Since n-channel and p-channel MOSFETs are exact complements to one another, logic elements built from them are called Complementary MOSFET or *CMOS* elements.

A CMOS inverter consists of one n-channel MOSFET and one p-channel MOSFET (Fig. 13.2.6). The n-channel MOSFET is connected to the negative terminal of the computer's power supply and controls the flow of negative charge to the inverter's output. The p-channel MOSFET is connected to the power supply's positive terminal and controls the flow of positive charge to the output. When negative charge arrives at the inverter's input and moves onto the gates of the MOSFETs, only the p-channel MOSFET conducts current and the output becomes positively charged. When positive charge arrives at the input, the n-channel MOSFET conducts current and the output becomes negatively charged.

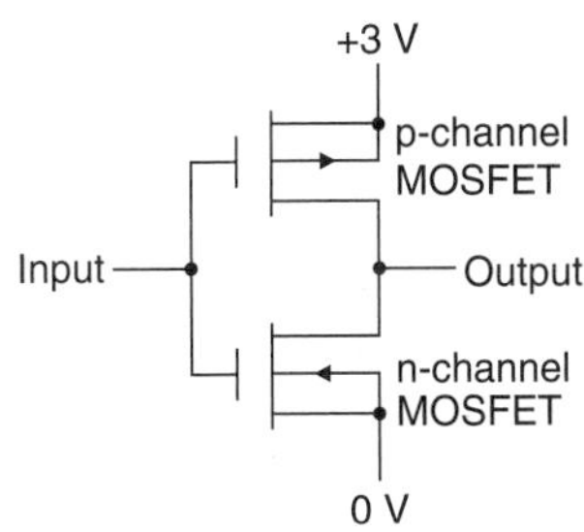

Fig. 13.2.6 - When negative charge arrives at the input of a CMOS inverter, its p-channel MOSFET (top) permits positive charge to flow to the output. When positive charge arrives at the input, the n-channel MOSFET (bottom) send negative charge to the output.

But a computer needs logic elements that are more complicated than inverters. One such element is the Not-AND or NAND gate. This logical element has two input bits and one output bit, and its output bit is 1 unless both input bits are 1s (Fig. 13.2.7). It's called a Not-AND gate because it's the inverse of an AND gate, which produces a 0 output unless both input bits are 1s. Simple memory-less logic elements are often called *gates*.

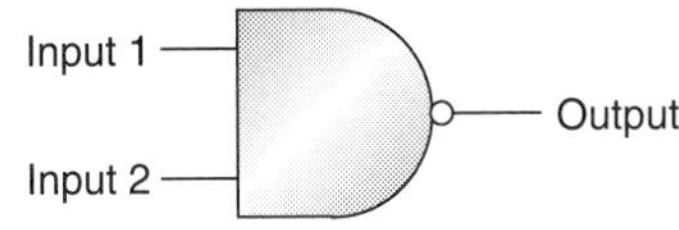

Input 1	Input 2	Output
1	1	0
1	0	1
0	1	1
0	0	1

Fig. 13.2.7 - The output bit of a Not-AND or NAND gate, shown here symbolically, is a 1 unless both input bits are 1s.

A CMOS NAND gate uses two n-channel MOSFETs and two p-channel MOSFETs (Fig. 13.2.8). The two n-channel MOSFETs are arranged in **series**—one after the next—so that current passing through one must also pass through the other. If either transistor has negative charge on its gate, no current can flow through the series. Components arranged in series must carry the same current, but they may experience different voltage drops.

The two p-channel MOSFETs are arranged in **parallel**—one beside the other—so that current can flow through either one of them to the output. If either transistor has negative charge on its gate, current can flow from one side of the pair to the other. Components arranged in parallel share the current reaching them through one wire and pass it on to the second wire. But while one may carry more of the current than the other, both experience the same voltage drop.

If either input of the CMOS NAND gate is negatively charged, the series of n-channel MOSFETs will be non-conducting and one of the p-channel MOSFETs will deliver positive charge to the output. But if both inputs are positively charged, both p-channel MOSFETs will be non-conducting and the series of n-channel MOSFETs will deliver negative charge to the output. Thus the CMOS NAND gate has the correct logic behavior.

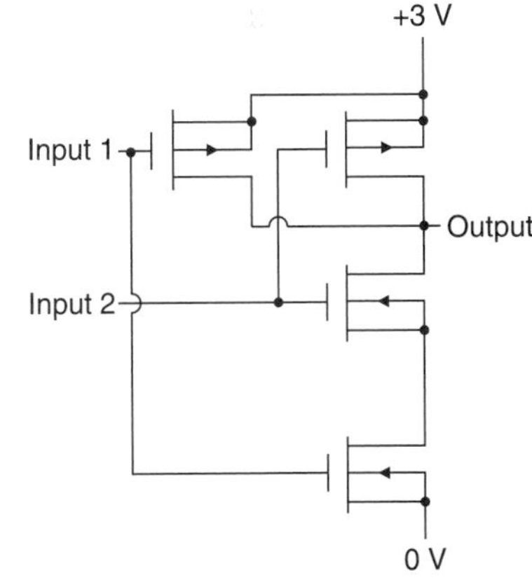

Fig. 13.2.8 - A CMOS NAND gate has two input bits. When negative charge arrives through either input, the chain of n-channel MOSFETs (bottom) stops conducting current and one of the two p-channel MOSFETs (top) permits positive charge to reach the output. Only if both inputs are positively charged will negative charge reach the output.

These two logic elements, inverters and NAND gates, can be combined to produce any conceivable logic element. For example, they can be used to build an adder, a device that sums the numbers represented by two groups of input bits and produces a group of output bits representing that sum. These adders can themselves be used to build multipliers and multipliers can be built into still more complicated devices. In this fashion, the simplest logic elements can be used to construct an entire computer.

In fact, two NAND gates are all that's required to make one bit of static memory (Fig. 13.2.9). That bit is a *flip-flop*, an element that's capable of adopting two different stable arrangements or *states*. The flip-flop has two inputs that normally receive positive charge and two outputs that normally emit opposite

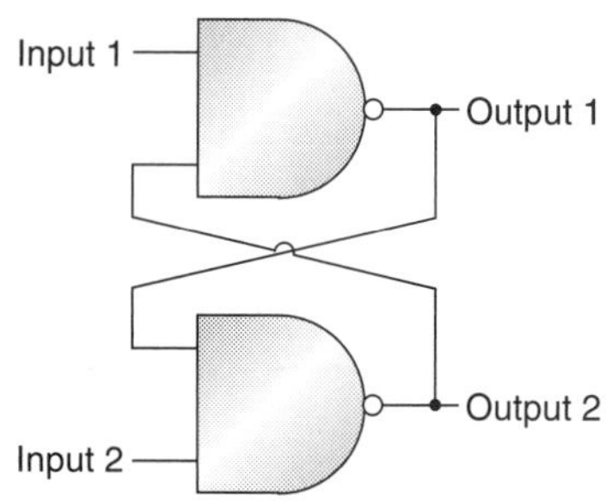

Fig. 13.2.9 - A flip-flop built from two NAND gates. Delivering negative charge to one of the two inputs sets the flip-flop's state. Static memory is based on flip-flops.

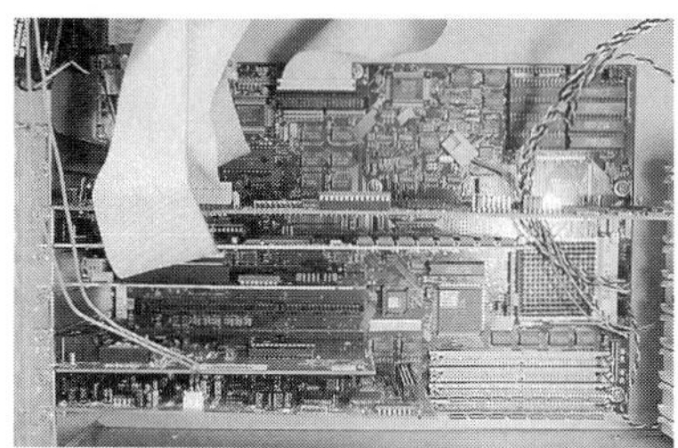

Fig. 13.2.10 - The processor chip of this personal computer appears at the right, covered by a heat sink—a metal grid that transfers heat from the processor chip to the air. Cards of dynamic memory fit into slots below the processor. Cables on the left connect the processor to its magnetic disk drives.

charges—one positive and the other negative. You can set the state of the flip-flop by temporarily delivering negative charge to one of its inputs. The associated output then becomes positively charged and the other output becomes negatively charged. The flip-flop remains like that until you temporarily deliver negative charge to its other input.

Each bit of static memory uses one flip-flop. Since a flip-flop can store its bit extremely quickly and retrieve it without then having to rewrite it, static memory is much faster than dynamic memory. But, since each flip-flop requires 8 transistors, static memory is much more expensive than dynamic memory.

Actually, a computer isn't built exclusively from NAND gates and inverters. To improve its speed and reduce its size, it uses a few other basic logic elements as well. Like the CMOS NAND gate and inverter, these elements are constructed directly from n-channel and p-channel MOSFETs.

All of these logic elements are wired together in an intricate pattern to create the final computer (Fig. 13.2.10). When you type a character at the keyboard, you send positive or negative charge to various logic elements and they act on that charge, storing a bit here, adding a bit there, and retrieving another bit from somewhere else. Moments later, the character appears on the display and becomes part of a document in your word processing package.

CHECK YOUR UNDERSTANDING #4: Getting It Together

How could you use inverters and NAND gates to create an AND gate, a logic element that has two inputs and one output, with the output only being 1 if both of the inputs are also 1s?

The Limits to a Computer's Speed

Computers perform calculations extremely quickly. The simple NAND gate we examined above acts in about one billionth of a second. That's the time it takes for the MOSFET transistors to start or stop conducting current and for the charge on the output wire to reach its proper value. Since scientists and engineers are continually improving the performances of MOSFETs, CMOS logic elements get faster every year.

But computer speed also depends on the time it takes for signals to travel between logic elements. While the signals from one element travel to other elements at almost the speed of light, that transfer speed limits how fast the computer can work. Each element must wait for its signals to arrive, and packing the elements closely together shortens the wait.

However, charged particles themselves travel much slower than light, so how can the signals move at nearly light speed? We can understand this by imagining a garden hose at the end of a closed water faucet. If you suddenly open the faucet, fresh water will begin to enter the hose. If the hose is already filled with water, water will begin to pour out of the hose long before fresh water from the faucet moves very far into the hose. In fact, water will begin to pour out of the hose almost as soon as you open the faucet. That's because the increase in water pressure that occurs when you open the faucet travels through the hose at the speed of sound.

Similarly, when charge enters a wire, its electric field pushes like charge ahead of it. Something resembling a pressure rise rushes through the wire at almost the speed of light and like charge begins to flow out of the far end of the wire long before the original charge arrives there. Since light takes about 3 billionths of a second to travel a meter or about 1 billionth of a second to travel a foot, putting the entire computer on a single integrated circuit is a good idea.

The logic elements are then so close together that they can exchange signals extremely quickly.

But there is another problem that slows a computer's calculations. Whenever a logic element changes its output from a 1 to a 0 or a 0 to a 1, it must change the charge on its output wire. If that wire is long and wide, it will have a fair amount of the old charge on it. Removing this old charge and replacing it with the new charge takes time because only so much current will flow through the tiny MOSFETs in the logic element.

Since the wires connecting logic elements have much more surface area than the gates of the MOSFETs they deliver charge to, the wires end up with most of the charge. The wires act as capacitors, storing and releasing charge. The best way to speed up a computer is to make these wires as short and as thin as possible. The wires in current computers are already well under half a micron in width and getting thinner all the time. Smaller is faster.

Moving charge onto and off the wires and gates isn't just a matter of speed, it's also a matter of energy. Each time positive charge flows onto a wire and is then replaced by negative charge, there's a net transfer of positive charge from the power supply's positive terminal to its negative terminal. The computer thus consumes a little electric energy and converts it into thermal energy. If the computer can't get rid of this thermal energy quickly enough, it will overheat.

The electric power consumed by the computer depends on the current flowing through it and on the voltage drop across the logic elements. One way to reduce this power is to run the computer more slowly, changing the charge on its wires less frequently and transferring that charge more slowly from one terminal of the power supply to the other. But this power can also be reduced by shrinking the wire size and using lower voltages to push charge onto and off those wires. Computers presently operate with voltage drops of about 3 V across their logic elements.

CHECK YOUR UNDERSTANDING #5: Thinner Wires are Faster Wires

Halving the width of the wires between logic elements roughly doubles the speed of the computer. Explain.

Synchronizing a Computer

Because each of the computer's logic elements takes a small but finite time to act, it needs something to keep faster logic elements from getting ahead of slower ones. For example, if the computer is trying to add two numbers that are themselves the results of other computations, its adder must be sure that those earlier computations are finished before producing a sum. To avoid any uncertainty, the computer needs a time-keeper to tell each logical element when it can be sure that all previous computations are finished.

This synchronizing device is the computer's clock. Just as a conductor keeps the instruments in an orchestra working together, the steady electric beat of the computer's clock keeps all of the logic elements in step. At one beat of the clock, new computations begin. At the next beat, all those computations are certain to be completed and a second group of computations begins. At the third beat of the clock, this second group of computations is completed and a third group of computations begins.

The computer's clock must beat slowly enough that all of the logic elements can complete their tasks between beats. Since a few tasks are far more complicated than others, they may be permitted two or more beats to complete their work. Division is an example of such a complex task. Rather than slowing the clock down so that division can finish in one beat, the computer allows the logic element handling division to work for several beats. Whenever the division element is activated, the computer's other logic elements generally wait for it to complete its work.

A computer that is specified as 100 MHz has a clock that beats 100 million times each second. This means that there are 10 billionths of a second between beats and that most of the logic elements in the computer can complete their tasks in that time. The clocks in modern computers beat about every 2 to 20 billionths of a second. As computers become smaller, these clock rates will continue to increase.

CHECK YOUR UNDERSTANDING #6: Short Order Chefs

If a computer's clock rate is 200 MHz, how long do most of the logic elements in that computer have to complete their tasks?

Summary

How Computers Work: A computer processes numbers and other information by operating on them according to preprogrammed logic rules. It assigns numbers to all information and then represents those numbers as groups of binary bits. These bits can only take values of 0 or 1. The computer processes bits using logic elements that follow specific rules relating their inputs to their outputs. These logic elements respond to their inputs without any creativity or intelligence, acting only as they were designed to act.

The electronic components that carry out these logic operations are formed on the surface of an integrated circuit. 1s and 0s are represented as positive and negative charge and this charge moves about on the computer chip under the control of n-channel and p-channel MOSFETs. The arrangements of these MOSFETs determine the logic operations that they perform and thus the relationships between their inputs and outputs. A clock synchronizes the processing of information so that the different logic elements don't get ahead of one another.

When not in use by the logic elements, bits are stored in several different ways. Static memory, itself a type of logic element, holds bits that are needed most frequently. Dynamic memory, which uses capacitors to store charge, holds bits that are needed less frequently. And bits that are rarely needed are stored as magnetization of a recording medium or on optical surfaces.

The Physics of Computers

1. In a parallel connection, current is shared between several devices. Since the devices have the same ingoing and outgoing wires, they experience identical voltage drops.

2. In a series connection, current passes sequentially through several devices so they all have the same current passing through them. Their voltage drops may be different.

Check Your Understanding - Answers

1. First, you could number the letters, A=1, B=2, etc., and use these assignments to turn each word into a short list of numbers (dog would be 4, 15, 7). Second, you could number the words in a dictionary, aardvark=1, aback=2, etc., and use these assignments to turn each of your words into a single number (dog might be 12255).

Why: Any piece of information—a word, sound, image, or whatever—can be represented as numbers with appropriate rules. We'll see how to represent sound with numbers in Section 16.3. These two ways of representing words as numbers are both used in computers. Raw text is usually represented character by character while words stored in giant databases (computerized encyclopedias or information retrieval programs) are often represented by single numbers. Single number representation makes word searches faster.

2. 129

Why: Binary 1000001 contains only 1 hundred-and-twenty-eight (2^7) and 1 one (2^0). All the other powers of two are not present. Since 128+1 is *129*, that is the number presented by this binary value.

3. The computer is making sure that the charge stored on each capacitor in the dynamic memory adequately represents that bit's contents.

Why: Since charge leaks quickly from the capacitors in dynamic memory, the contents of each bit must be refreshed many times a second. This refreshing process, reading each bit and storing it back into memory, slows the computer down slightly.

4. You could connect an inverter to a NAND gate so that the output signal of the NAND gate is the input signal of the inverter. When both of the NAND gate's input signals are 1s, it will produce an output of 0. This 0 will arrive at the inverter, which will invert it and yield an output of 1.

Why: Connecting logic elements together one after the next is the standard method for producing more complicated logic elements. In this case, two elements produce a third.

5. Halving the widths of the wires also halves the amount of charge they store. Since a logic element trying to change the charge on a wire now has to move only half as much charge, it can act twice as fast as before.

Why: The wires in the computer act as capacitors, storing separated charge. The wire itself stores one type of charge (positive or negative) while the other type of charge can be located elsewhere on the computer chip. This stored charge in its wires slows down the computer because the logic elements must provide that charge. Thinning a wire reduces its capacitance and makes it less of a burden on the logic elements.

6. 5 billionths of a second.

Why: A clock that beats 200 million times per second produces a beat every 5 billionths of a second. Since most of the logic elements in a computer must complete their tasks in the time between one beat and the next, those tasks must take no longer than 5 billionths of a second.

Glossary

analog The representation of numbers as values of physical quantities such as voltage, charge, or pressure.

binary The digital representation of numbers in terms of the powers of two. The number *6* is represented in binary as 110, meaning 1 four (2^2), 1 two (2^1), and 0 ones (2^0).

bit A single binary value, either a 0 or a 1.

byte Eight binary bits that collectively can represent a number from 0 to 255. Bytes are often used to represent letters and other characters, where a convention associates each character with a specific number.

decimal The digital representation of numbers in terms of the powers of ten. The number *124* is represented in decimal as 124, meaning 1 hundred (10^2), 2 tens (10^1), and 4 ones (10^0).

digital The representation of numbers by decomposition into digits that are then individually represented by specific values of physical quantities such as voltage, charge, or pressure.

integrated circuit A thin wafer of semiconductor that has been carefully structured, modified, and processed so that its surface contains the interconnected transistors, resistors, and/or capacitors needed to perform some electronic task.

parallel (wiring arrangement) An arrangement in which the current reaching two or more electric devices divides into separate parts to flow through those devices and then joins back together as it leaves them.

series (wiring arrangement) An arrangement in which the current reaching two or more electric devices flows sequentially through one device after the next before leaving them.

Review Questions

1. Why are digital and binary representations of numbers different from the numbers themselves?

2. How are the digits 0 through 9 used to represent numbers in decimal?

3. How are the digits 0 and 1 used to represent numbers in binary?

4. How does a computer use a capacitor to store a bit?

5. Why must a computer periodically read and restore the bits in its dynamic memory?

6. List five factors that limit the number of bits that can be stored on a particular magnetic disk.

7. Why does the time it takes for a computer to read bits stored on a magnetic disk depend on how rapidly that disk is rotating?

8. What is the relationship between the charge on the input of an inverter and the charge it places on its output?

9. If you connect two MOSFETs in series, why must they both be made conducting before electric current can flow through the pair?

10. If you connect two MOSFETs in parallel, why will the pair allow electric current to flow through them even if one of them is made insulating?

11. Why does it take time and energy for a computer logic element to put positive charge on its output wire and then remove that charge?

Exercises

1. We combine the three decimal digits 6, 3, and 1 to form 631 in order to represent the number *631*. What does the 6 in 631 mean? What are there 6 of?

2. We combine the three binary bits 1, 0, and 1 to form 101 in order to represent the number *5*. What does the left-most 1 in 101 mean? What is there 1 of?

3. What number does the binary byte 11011011 represent?

4. How is the number *165* represented in binary?

5. Do the decimal representations 453 and 0453 refer to the same number? Why?

6. Do the binary representations 11101111 and 011101111 refer to the same number? Why?

7. Why are there no 2's in the binary representation of a number? (In other words, why isn't 1101121 a valid binary representation?)

8. How could you use an n-channel MOSFET, a battery, a light bulb, and some wire to determine whether or not there is positive charge on one plate of a small capacitor?

9. Dynamic memory stores bits as the presence or absence of separated charge on tiny capacitors. It takes energy to produce separated charge and a computer that minimizes this energy will use less electric power. Why does making the insulating layers of the memory capacitors very thin reduce the energy it takes to store each bit in them?

10. The tiny MOSFETs that are used to move charge onto and off the capacitors in dynamic memory are so small that they're never very good conductors. Why do their modest electric resistances lengthen the time it takes to store or retrieve charge from the memory capacitors?

11. Why does the effect described in Exercise **10** limit the speed with which a computer can store or retrieve bits from its dynamic memory?

12. Computer memory is extremely sensitive to static electricity. If a spark were to introduce even a tiny amount of extra charge onto the capacitors in a dynamic memory chip, the bits stored in that chip would be spoiled. Why?

13. A floppy disk should never be exposed to strong magnetic fields. What effect would such a field have on the disk?

14. The recording head of a computer hard disk flies about a micron above the surface of the disk. That's so close that even a dust grain can't fit between the disk and the head. This narrow spacing is maintained by the airflow between the head and the disk. If they move too close, the air flowing between them becomes blocked and slows down. Why does this blocked air push the head and disk apart?

15. Because its recording heads should never touch its recording surfaces, a computer hard disk can't tolerate large accelerations. How can a sharp impact to the disk's case cause its recording heads to touch its recording surfaces?

16. If you connect the output of one inverter to the input of a second inverter, how will the output of the second inverter be related to the input of the first inverter?

17. When positive charge is being replaced by negative charge on the input of the CMOS inverter shown in Fig. 13.2.6, what is happening to the charge on its output?

18. Why does the absence of positive charge on either one of the n-channel MOSFETs at the bottom of Fig. 13.2.8 prevent current from flowing from the output to 0 V?

19. Why does the presence of negative charge on either one of the p-channel MOSFETs at the top of Fig. 13.2.8 allow current to flow from +3 V to the output?

20. If you supply negative charge to input 1 of the flip-flop in Fig. 13.2.9, what type of charge will soon be present on output 1? on output 2?

21. Each time a computer logic element places positive charge on a wire and then removes that positive charge, there is a net movement of charge from the computer power supply's positive terminal to its negative terminal. The more often the computer does this, the more energy it consumes. Why?

22. If two logic elements in a computer are connected by a wire that's twice as long as necessary, the second element will take twice as long to begin responding to the output of the first element. Why?

23. If any of a computer's main electronic components are separated by more than about 1 m (about 3 feet), the computer will have trouble operating with a clock ticking faster than about 150 MHz. Why?

24. A household extension cord allows you to plug several small appliances into the same electric outlet. Do the appliances end up wired in parallel or in series with one another?

25. Is the switch in a flashlight in series with the light bulb or in parallel with it?

26. If you connect two identical resistors in series, so that current must pass through one and then the other to get across the pair, will the electric resistance of this pair be larger or smaller than that of one of the individual resistors from which it's made?

27. If you connect two identical resistors in parallel, so that current can pass through either one of them to get across the pair, will the electric resistance of this pair be larger or smaller than that of one of the individual resistors from which it's made?

Epilogue for Chapter 13

In this chapter we looked at two electronic devices to see how they use electricity and magnetism to perform useful tasks. In *audio amplifiers*, we saw how resistors, capacitors, and transistors are combined to enlarge small electric currents representing sound. We learned that resistors impede the flow of current, that capacitors store separated electric charge, and that transistors allow the current in one circuit to control the flow of current in another circuit.

In *computers*, we encountered some of these same components again, but this time in a different context. We examined the ways in which numbers and information can be represented using binary bits and electric charge, and then learned how arrangements of transistors can perform complicated logic operations on that representation. We found out what limits a computer's speed and learned why further miniaturization will make computers even faster.

Explanation: Listening to Yourself Talk

The sound of your voice causes pressure fluctuations at the surface of the microphone and the microphone converts them into current fluctuations in an electric circuit. However, the fluctuating current flowing through the two wires of the microphone doesn't have the power needed to operate a speaker. That's where the amplifier comes in. When the small current from the microphone passes through the amplifier's input circuit, it produces a similar but larger current in its output circuit. This enlarged current flows through the two wires to the speaker and causes it to reproduce the sound of your voice. Adjustments to the audio system's controls, particularly the treble and base knobs, alter its amplification for specific frequency ranges and can change the volume and tone of the reproduced sound.

Cases

1. The bright red, green, and yellow lights that you find on many electronic devices are light emitting diodes or LEDs. Like any other diode, an LED carries current only in one direction. But unlike a normal diode, an LED emits a photon of light whenever an electron shifts from a conduction level in its n-type cathode to a valence level in its p-type anode.

a. Explain why an LED's brightness is proportional to the electric current flowing through it.
b. The current passing through an LED experiences a voltage drop. Use conservation of energy to explain why this voltage drop must occur.
c. A photon of green light has more energy than a photon of red light. Why must an LED that produces green photons have a larger voltage drop than an LED that produces red photons?
d. The power supply in a typical electronic device delivers current at too high a voltage for an LED. If that current were sent directly through the LED and then returned to the power supply, the LED would receive too much power, overheat, and burn out. To prevent such a disaster, the electronic device first sends the current through a resistor and then through the LED. How does the resistor protect the LED?
e. An LED's color is determined by the energy needed to shift an electron across the band gap in the semiconductor from which the LED is made. Because blue photons have more energy than green photons, development of blue LEDs has been slow. The materials needed to produce blue LEDs are almost true insulators rather than semiconductors. What distinguishes a semiconductor from an insulator?

2. A Leiden jar is an early form of capacitor, consisting of a glass jar with metal coatings on its inside and outside surfaces. Because the two coatings don't touch, the jar can store separated charge on them and they act as the two plates of a capacitor.

a. A metal post extends upward from the inner coating and rises above the top of the jar. Why does capping this post with a small metal sphere, rather than a sharp metal point, improve the jar's ability to store separated charge?
b. The jar is made of thick glass so that the two metal coatings are far apart. Why does transferring even a modest amount of charge from one coating to the other result in a huge voltage difference between the two coatings?
c. If the glass of the Leiden jar were made thinner, you would have to transfer far more charge from one plate to the other to produce the voltage difference in **b**. Why?
d. If a Leiden jar is storing separated charge and you connect the two coatings with a wire, a bright spark will appear. Where did the energy for this spark come from?

e. Some Leiden jars use removable metal cans in place of the metal coatings. If you disassemble such a jar while it's storing separated electric charge, the voltage difference between the two cans will rise. What provides the additional energy associated with this increased voltage difference?

3. The power supply in your stereo amplifier uses alternating current from the electric company to provide direct current to the amplifier. This supply converts electric power from one form to another through the use of transformers, diodes, capacitors, and transistors.

a. The power supply provides a relatively small voltage rise to the large current that it sends through the amplifier. The voltages provided by the electric company are too high for the amplifier and the currents are too small. How does a transformer make it possible for the power supply to provide a relatively small voltage rise to a relatively large current that flows between it and the amplifier?

b. The power supply's transformer provides AC electric power through two wires, but the amplifier needs DC electric power through two wires. To fix this mismatch, the stereo connects these two pairs of wires with 4 diodes so that even though the currents through the two wires of the transformer reverse, the currents through the two wires of the amplifier don't reverse. How are those 4 diodes connected between the transformer and amplifier? (Draw a picture and indicate which way current can flow through each diode.)

c. While the diodes (see **b**) ensure that current always flows in one direction through the amplifier, the transformer can't provide current when the current from the power company is reversing. To maintain a steady current through the amplifier, the power supply uses a large capacitor. When the transformer's current is large, some of that current is used to transfer charge between the capacitor's plates so that it stores separated charge. When the transformer's current is small, this separated charge is allowed to flow through the amplifier as current. Draw a graph of the voltage difference between the two plates of the capacitor versus time. Mark the times when the amount of separated charge is increasing and when it's decreasing.

d. The transformer, diodes, and capacitor do a pretty good job of sending direct current through the amplifier, but there still tend to be periodic fluctuations in that current. To keep the current stable, the amplifier uses a regulating device. Just before the current enters the amplifier, it passes through an n-channel MOSFET. An electronic sensor measures the current through the amplifier and determines if that current is too high or too low. It then adjusts the charge on the gate of the MOSFET to increase or decrease the MOSFET's electric resistance and lower or raise the current through the amplifier. If the sensor detects that the current is too high, should it increase or decrease the positive charge on the MOSFET's gate? Explain.

4. You enjoy listening to your little portable radio while jogging, but it hasn't been working properly since it got wet in the rain last week. The problem is that it keeps turning itself off. Because the radio makes no "click" sound when you press the "on" or "off" buttons and because it turns itself off automatically after an hour, you know that the switch that controls the radio's power is electronic. It's probably an n-channel MOSFET that's connected in series with the radio's electronics so that current from the battery must pass through both the electronics and the MOSFET before returning to the battery.

a. Why won't any power reach the electronics when the MOSFET isn't conducting current?

b. What must the "on" button do to make the n-channel MOSFET conduct current so that the radio will operate?

c. What must the "off" button or the automatic shutdown do to stop the n-channel MOSFET from conducting current, so that the radio will turn off?

d. Water is a poor conductor of electricity, but with patience you can send charge through it. If water is slowly turning off the radio, what is it probably doing?

e. You discover a small drop of water inside the "off" button, allowing positive charge to flow slowly from the gate of the n-channel MOSFET to the negative terminal of the battery. You remove the water and the radio works perfectly. Why did the drop cause trouble and why did removing the drop fix the radio?

5. Your new electric guitar and amplifier sound great, although all the neighbors seem to have gone on vacation since you bought it last week. After 6 straight hours of jamming you decide to take a break and figure out how it works.

a. You begin by examining the strings and pickups—the devices that sense the strings' motions and represent them as electric currents. The pickup near each string is a coil of wire wrapped around a small permanent magnet. The permanent magnet magnetizes the steel string so that it induces an alternating current in the pickup coil as it vibrates back and forth. What provides the power for this alternating current?

***b.** The electric power provided by the pickup is much too small to drive a speaker, so it must be amplified. The first step is to send it through a preamplifier. Current from the pickup flows through the preamplifier's input circuit and an amplified version of that current flows through the preamplifier's output circuit. The voltage rise at the preamplifier's output is 10 times as large as the voltage drop at its input, and the current passing through its output circuit is 10 times as large as the current passing through its input circuit. The power the preamplifier is providing is how many times as large as the power it's receiving from the pickup?

c. The output of the preamplifier is connected to the main power amplifier. You open a side panel of the amplifier and notice several large MOSFETs bolted to a heat sink. Since the amplifier was running recently (you wisely turned it off and unplugged it before opening it up), the heat sink is quite warm. These MOSFETs have been controlling the flow of current from the amplifier's power supply to your speaker, so their electric resistances have been fluctuating up and down. What has the amplifier been doing to the MOSFETs to make their electric resistances change?

d. Why have the MOSFETs been producing thermal energy?

e. You reinstall the amplifier's side panel and take a look at the speaker cabinet. This cabinet contains several speakers, each of which has a coil of wire that becomes magnetic when current flows through it. A nearby permanent magnet pushes on the magnetic coil and this force moves a paper cone to create sound. The currents needed for high volume are large and they flow to and from the speaker through the speaker wires. You notice that the speaker wires are warm—they have been wasting some of the power from your amplifier! You knew it was a mistake to buy cheap thin speaker wire, but it's too late now. You have extra wire which you can use to reduce the wasted power. Should you add a second wire in parallel to each of the present wires or should you add a second wire in series with each of the present wires? Explain your answer.

CHAPTER 14

ELECTROMAGNETIC WAVES

We have seen that changing electric fields produce magnetic fields and that changing magnetic fields produce electric fields. These relationships between electric and magnetic fields allow them to create one another even in empty space. In fact, they can form electromagnetic waves, in which the two fields re-create one another endlessly and head off across space at an enormous speed. These electromagnetic waves are all around us and are the basis for much of our communications technology, for radiative heat transfer, and for our ability to see.

Experiment: Boiling Water in an Ice Cup

You can experiment with electromagnetic waves using a microwave oven. As we'll see in Section 14.3, microwaves are a type of electromagnetic wave that falls between radio waves and light. For reasons that we'll discuss in that section, microwaves can transfer energy to water molecules in a liquid but not to water molecules in an ice crystal. That difference allows for some interesting cooking tricks.

Take a large ice cube from the freezer and melt a shallow bowl-shaped depression in its top surface. Then put the ice cube on a microwave-safe ceramic plate and return it to the freezer to cool. Once the ice cube and plate are cold and frozen, take them out of the freezer and put them quickly into the microwave. Place a few drops of water into the ice cube's bowl-shaped depression and immediately start the microwave.

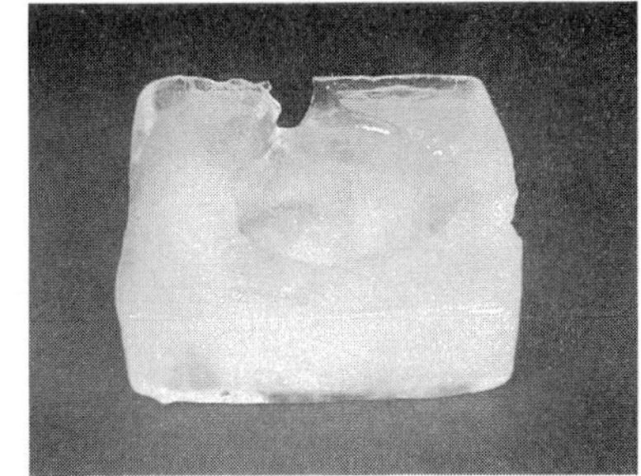

If you have been quick enough, the water drop on top of the ice cube will still be liquid when the oven starts producing microwaves. The liquid water will absorb power from the microwaves that fill the cooking chamber but the ice will not. The water drop will become extremely hot and will begin to melt the top of the ice cube. After a short while, the ice cube will have a hole "drilled" in its top surface by the heated water drop. Why was it important to chill the plate before putting the ice cube into the microwave oven?

If you repeat this experiment with a block of ice, you may be able to bring the water in the depression to a full boil before it melts its way through the block. Try to **predict** how the shape of the liquid portion will change as the microwaves heat it up. **Observe** what happens and try to **verify** your prediction. **Measure** how quickly the ice melts and think about how you could use that measurement to calculate how much power the microwave oven is delivering to the water.

Chapter Itinerary

In this chapter, we'll discuss how electromagnetic waves are formed and detected in three common situations: (1) *radio*, (2) *television*, and (3) *microwave ovens*. In *radio*, we'll examine the ways in which charge moving on an antenna can emit or respond to electromagnetic waves and how those waves can be controlled in order to send an audio signal from a radio transmitter to a radio receiver. In *television*, we'll look at how beams of electrons inside a picture tube form a picture on the screen and consider how electromagnetic waves carry the information needed to produce that picture to the television. In *microwave ovens*, we'll see how microwaves affect water molecules and how a sophisticated vacuum tube inside the oven creates those microwaves. While we can't see the electromagnetic waves that these three devices use, they clearly play important roles in our world.

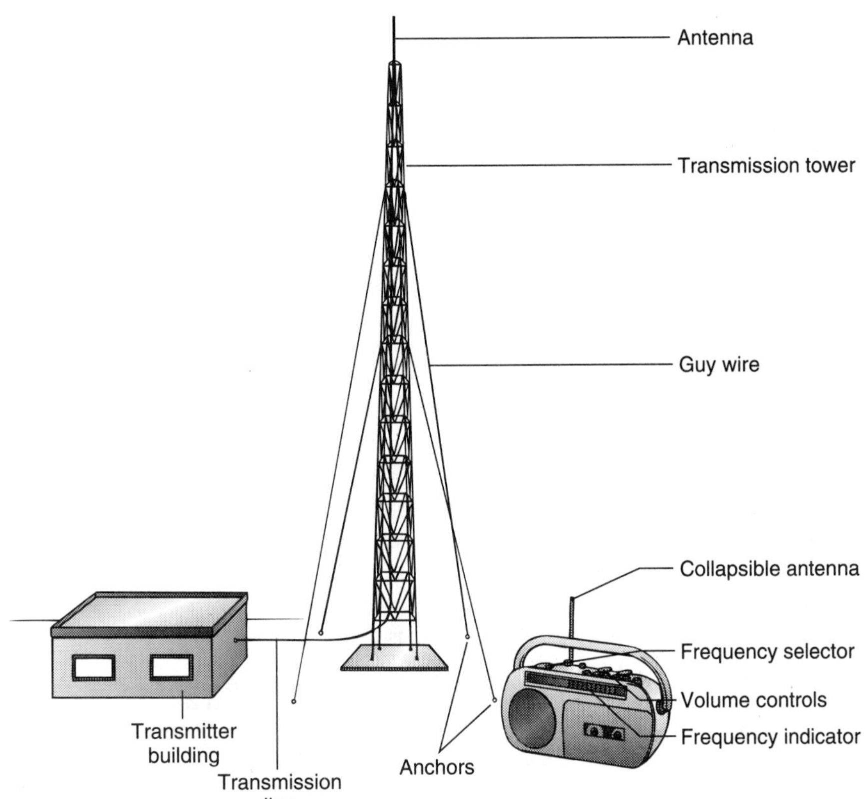

Section 14.1

Radio

We've seen that electric currents can represent sound and that these currents can carry speech and music anywhere wires will reach. But how would you send sounds to someone who is driving in a car or flying in a plane? You'd need a new way to represent sound that doesn't involve wires. You'd need radio.

This section describes how radio works. We'll look at how radio waves are transmitted and at how they're received. We'll also examine the common ways in which sound is represented by radio waves so that your voice can travel through space to a receiver somewhere far away.

Questions to Think About: *How might the movement of electric charge on one metal antenna affect electric charge on a second antenna located nearby? What about when the second antenna is far away from the first antenna? What does it mean when a radio station claims to transmit 50,000 watts? Why can you sometimes receive music from distant radio stations at dusk?*

Experiments to Do: *Listen to a small AM radio and notice that the volume of the sound depends on the radio's orientation or location. Radio waves are pushing electric charges back and forth along the radio's hidden internal antenna. You can often find an orientation where the radio is silent, because in that orientation the radio waves are unable to move charges along the antenna. If you put the radio inside a metal box, it will also become silent. Can you explain why?*

You can try similar experiments with a cordless telephone—actually a radio transmitter and receiver. See how far you can go with the handset before you lose contact with the base. Notice that the antenna's size and orientation affect its range. What happens to the reception if you stand near a metal object?

Antennas and Tank Circuits

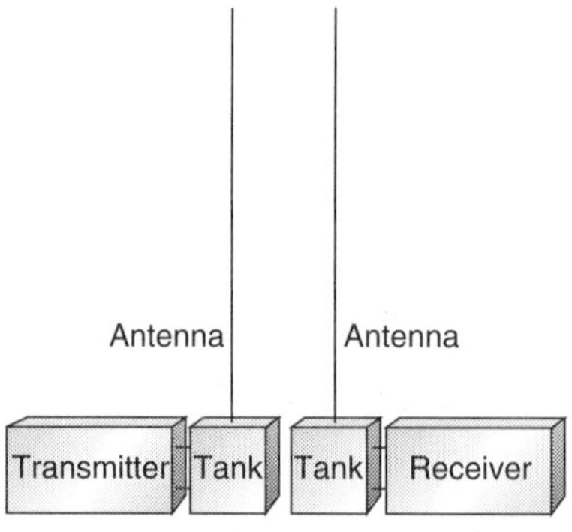

Fig. 14.1.1 - Electric charge rushing on and off the transmitting antenna causes a similar motion of electric charge in the receiving antenna.

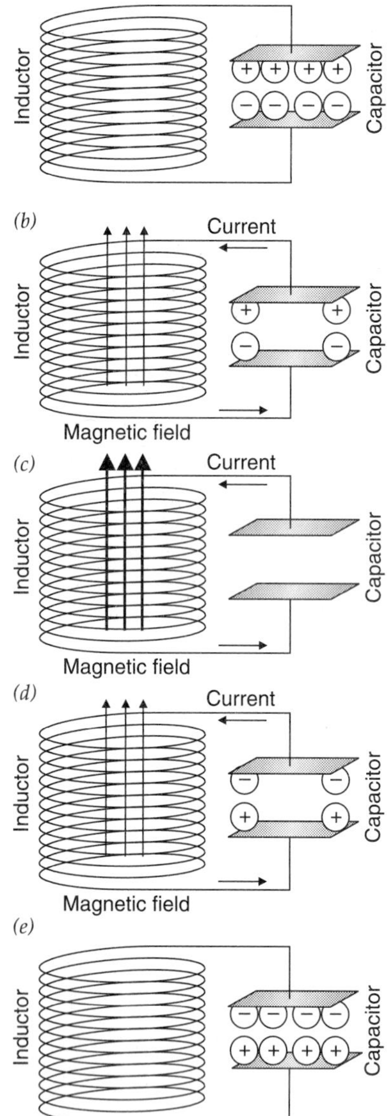

Fig. 14.1.2 - A tank circuit consists of a capacitor and an inductor. Energy sloshes rhythmically back and forth between the two components.

A radio transmitter communicates with a receiver via radio waves. These waves are produced by electric charge as it moves up and down the transmitter's metal antenna. We've already seen that electric charge produces electric fields and that moving charge produces magnetic fields. In the present situation, those two fields merge together as waves that can travel long distances through empty space. When these radio waves pass a radio receiver's antenna, electric charge in that antenna begins to move up and down, too. If this motion is vigorous enough, the receiver will be able to detect the transmission.

But before looking at how radio waves travel long distances through space, let's start with a much simpler situation. We'll look at how two nearby antennas affect one another. Fig. 14.1.1 shows a radio transmitter and a radio receiver, side by side. Because of their proximity, electric charge on the transmitter's antenna is sure to affect charge on the receiver's antenna.

To communicate with the nearby receiver, the transmitter sends charge up and down its antenna. The electric field from this charge extends through space to the receiver's antenna, where it pushes charge in the metal down and up. But the movement of charge on the receiver's antenna is so small that the receiver will have trouble detecting it. Therefore the transmitter adopts an important strategy—it first sends positive charge up its antenna, then negative charge, then positive charge again, over and over. And it does this rhythmically at a particular frequency. Since the resulting motion on the receiver's antenna is rhythmic at that same frequency, it's much easier for the receiver to detect.

Using this rhythmic motion has another advantage: it allows both the transmitter and the receiver to store energy in tank circuits. A **tank circuit** is a resonant electronic device consisting of a capacitor and an inductor (Fig. 14.1.2). We'll examine its details in a moment, but for now you can think of a tank circuit as a "playground swing" for electric charge. Charge sloshes back and forth through the tank circuit, from one end to the other, at a particular frequency. And just as you can make a child swing strongly at the playground by giving them a gentle push every swing, so the transmitter can make charge slosh strongly through the tank circuit by giving it a gentle push every cycle. By helping the transmitter move larger amounts of charge up and down the antenna, the tank circuit dramatically strengthens the transmission.

A second tank circuit attached to the receiving antenna helps the receiver detect this transmission. Gentle, rhythmic pushes by fields from the transmitting antenna cause more and more charge to move up and down the receiving antenna and to slosh back and forth in the receiving tank circuit. While the motion of charge on this antenna alone may be difficult to detect, the much larger charge sloshing in the tank circuit is relatively easy to measure.

We can understand how a tank circuit works by looking at how charge moves between its capacitor and its inductor. An inductor is essentially an electromagnet—a coil of wire that become magnetic when a current passes through it. Magnetic fields contain energy, so the inductor must extract energy from this current as its magnetic field grows stronger and return energy to the current as its magnetic field becomes weaker.

This exchange of energy between the inductor and the current gives rise to an interesting effect—an inductor opposes any change in the current passing through it. As this current increases, the inductor's rising magnetic field produces an electric field that pushes backward on the current and slows it down. As this current decreases, the inductor's collapsing magnetic field produces an electric field that pushes forward on the current and speeds it up. Thus while current in the inductor can change, that change always occurs relatively slowly.

Imagine that the tank circuit starts out with separated charge in its capacitor (Fig. 14.1.2*a*). Since the inductor conducts electricity, current begins to flow from the positively charged plate of the capacitor, through the inductor, to the negatively charged plate. But the current through the inductor must rise slowly and, as it does, it creates a magnetic field in the inductor (Fig. 14.1.2*b*).

Soon the capacitor's separated charge is gone and all of the tank circuit's energy is stored in the inductor's magnetic field (Fig. 14.1.2*c*). But the current keeps flowing, driven forward by the inductor's opposition to current changes. The inductor uses the energy in its magnetic field to keep the current flowing and separated charge reappears in the capacitor (Fig. 14.1.2*d*). Eventually, the inductor's magnetic field decreases to zero and everything is back to its original state—almost. While all of the tank circuit's energy has returned to the capacitor, the separated charge in that capacitor is now upside down (Fig. 14.1.2*e*).

This whole process repeats in reverse. The current flows backward through the inductor, magnetizing it upside-down, and soon the tank circuit is back the way it started. This cycle repeats over and over again, with charge sloshing from one side of the capacitor to the other and back again.

The time required for each cycle of the tank circuit, its period, depends on its capacitor and inductor. The larger the capacitor's capacitance, the more separated charge it can hold with a given amount of energy and the longer it takes that charge to move through the circuit as current. The larger the inductor's **inductance**—its opposition to current changes—the longer that current takes to start and stop. A tank circuit with a large capacitor and a large inductor may have a period of a thousandth of a second or more, while one with a small capacitor and a small inductor may have a period of a billionth of a second or less.

Inductance is defined as the voltage drop across the inductor divided by the rate at which current through the inductor changes with time. This division gives inductance the units of voltage divided by current per time. The SI unit of inductance is the volt-second-per-ampere, also called the **henry** (abbreviated H). While large electromagnets have inductances of hundreds of henries, a 1 μH (0.000001 H) inductor is more common in radio.

A tank circuit behaves as a harmonic oscillator, with a period that doesn't depend on the amount of charge sloshing in it. This resonant behavior makes the tank circuit useful in radio. Gentle, rhythmic pushes from the transmitter cause large amounts of charge to slosh back and forth in its tank circuit and travel up and down its antenna. Gentle, rhythmic pushes from that charge then make large amounts of charge travel down and up the receiving antenna and slosh back and forth in the receiver's tank circuit (Fig 14.1.3). The receiver can easily detect this large rhythmic motion in its tank circuit.

Power flows from the transmitter to the receiver via resonant energy transfer—from the transmitter, to the transmitting tank circuit and antenna, to the receiving antenna and tank circuit, to the receiver. This sequence of transfers can only work if all the parts have resonances at the same frequency. Tuning a radio to a particular station is largely a matter of adjusting the capacitor and inductor so that its tank circuit has the right resonant frequency.

The amount of charge sloshing in the receiver's tank circuit is proportional to the amount sloshing in the transmitter's tank circuit, so the transmitter can send information to the receiver by varying the amount of charge sloshing in its tank circuit. The receiver can obtain that information by measuring how much charge is sloshing in its tank circuit.

CHECK YOUR UNDERSTANDING #1: No Tanks

Why doesn't the radio transmitter simply push electric charge directly on and off the antenna, without using a tank circuit?

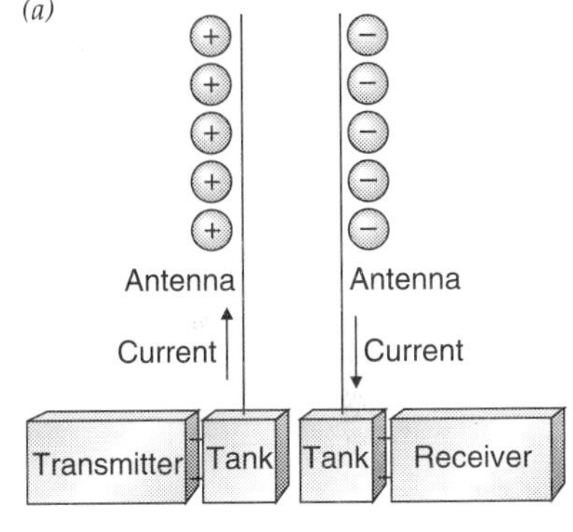

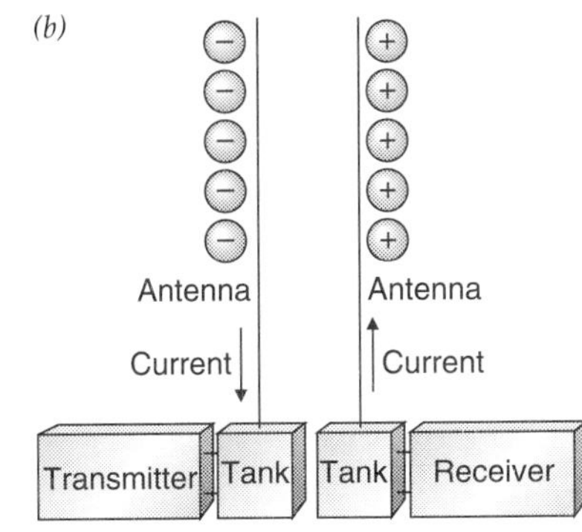

Fig. 14.1.3 - (*a*) As current flows *up* the transmitter's antenna, it causes current to flow *down* the receiver's antenna. (*b*) As current flows *down* the transmitter's antenna, it causes current to flow *up* the receiver's antenna.

Radio Waves

When the two antennas are close together, the charges in the transmitting antenna exert simple electrostatic forces on the charges in the receiving antenna. But when the antennas are far apart, the interactions between them become much more complicated. The transmitter's antenna must then emit a *radio wave*—an electromagnetic wave that travels through space on its own and pushes on charge in the receiver's antenna. Like a water wave, an electromagnetic wave is a disturbance that carries energy through space. But unlike water waves, electromagnetic waves don't need a fluid in which to travel. They can travel through empty space, from one side of the universe to the other.

An electromagnetic wave consists of a changing electric field and a changing magnetic field. It can exist all by itself because its changing electric field produces its magnetic field and its changing magnetic field produces its electric field. The wave's energy flows back and forth between these two fields as it travels through space at the **speed of light**—exactly 299,792,458 m/s (approximately 186,282 miles per second).

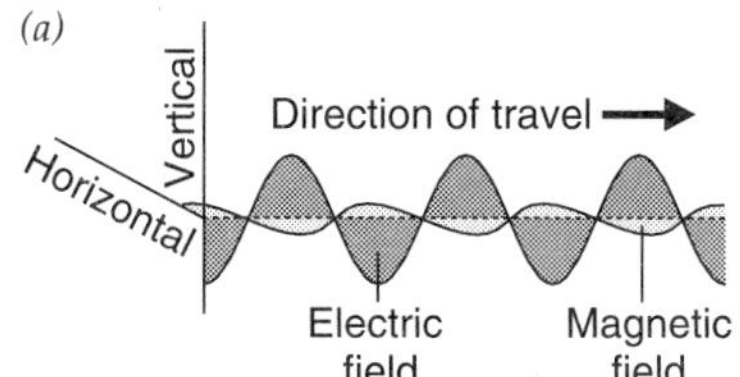

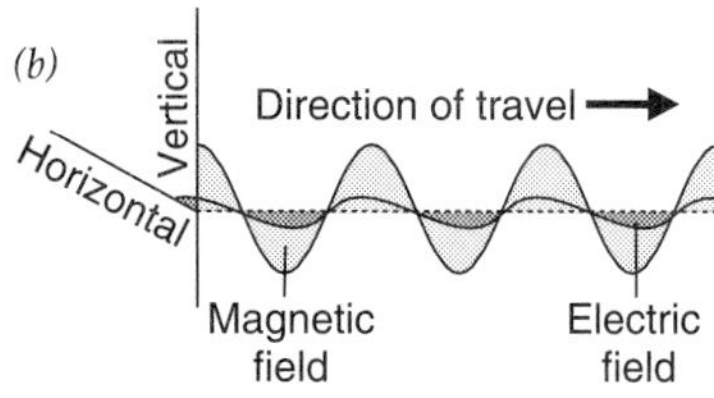

Fig. 14.1.4 - (*a*) In a vertically polarized electromagnetic wave, the electric field points alternately up and down while the magnetic field points alternately right and left. (*b*) In a horizontally polarized wave, the electric field points alternately left and right while the magnetic field points alternately up and down.

The wave is created when electric charge on the antenna accelerates. While stationary charge or a steady current produces constant electric and magnetic fields, accelerating charge produces fields that change with time. As charge flows up and down the antenna, its electric field points alternately up and down and its magnetic field points alternately left and right. These fields then recreate one another endlessly and sail off through space as an electromagnetic wave.

The wave emitted by the upright transmitting antenna is **vertically polarized**, meaning that its electric field points alternately up and down (Fig. 14.1.4). Had the transmitting antenna been tipped on its side, the wave's electric field would have pointed alternately left and right and it would have been **horizontally polarized**. Whatever the polarization, the electric and magnetic fields move forward together as a traveling wave, so the pattern of fields moves smoothly through space at the speed of light.

If you stood in one place and watched this wave pass, you'd notice its electric field fluctuating up and down at the same frequency as the charge that created it. When the wave passes a distant receiving antenna, it pushes charge up and down that antenna at the frequency at which charge moved up and down the transmitting antenna. If the receiver and transmitter both have the same resonant frequency, the amount of charge sloshing in the receiver's tank circuit should become large enough for the receiver to detect.

As the electromagnetic wave travels through space, it carries power away from the transmitter. When a radio station advertises that it "transmits 50,000 W of music," it's claiming that its antenna emits 50,000 W of power in its electromagnetic wave. The receiver's antenna must absorb enough of this power to detect the wave. But the farther the wave gets from transmitting antenna, the more spread out and weak it becomes. Trees and mountains also absorb or reflect some of the wave and interfere with reception.

A radio station can improve reception by using a vertical transmitting antenna of the proper length. When the transmitting antenna is exactly a quarter the length of the radio wave it transmits, a standing wave of electric charge forms on the antenna. Because current in the quarter-wavelength antenna moves at the speed of light, there is just enough time during one cycle of the radio wave for the antenna to (1) move positive charge to its tip, (2) return that positive charge to its base, (3) move negative charge to its tip, and (4) return that negative charge to its base. The quarter-wave antenna exhibits a natural resonance and there's a resonant energy transfer from the station's tank circuit to the antenna. As you might expect, these resonances help to produce a powerful radio wave.

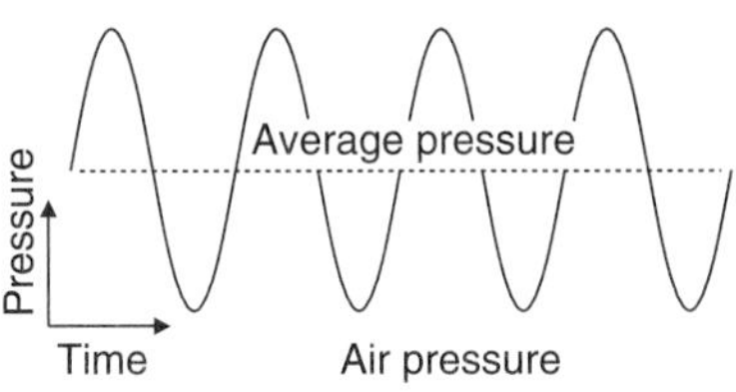

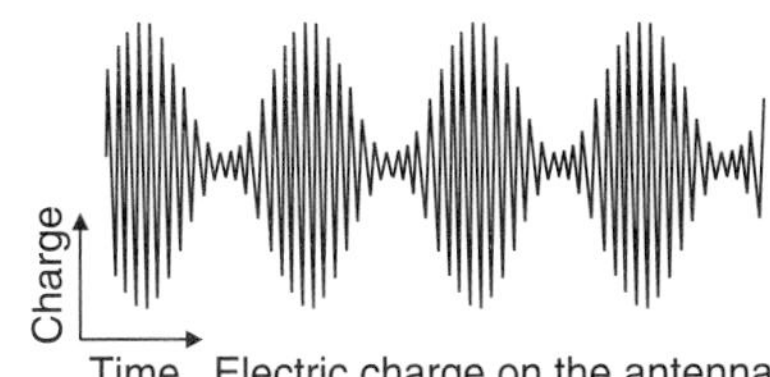

Fig. 14.1.5 - When sound is transmitted using amplitude modulation, air pressure is represented by the strength of the radio wave. A compression is represented by strengthening the radio wave and a rarefaction is represented by weakening it.

Because an antenna sends the strongest portion of its radio wave out at right angles to its length, orienting the antenna vertically sends most of the wave out horizontally, were people are most likely to receive it. Some stations use several separate antennas to direct the radio wave toward particular locations.

For the best reception, a listener should use a vertical receiving antenna that's a quarter wavelength long and make sure that there's a clear path from the transmitting antenna to the receiving antenna. Keeping the antenna vertical is important because the radio wave is vertically polarized and its vertical electric field should push charge *along* the antenna, not *across* it. That's why car radio antennas are vertical. Even the complicated antennas used by home radio receivers take the radio wave's vertical polarization into account—even though these antennas may not appear to be vertical themselves.

However, some receiving antennas respond to a radio wave's magnetic field rather than its electric field. This fluctuating magnetic field can induce currents in a loop or coil of wire. Coil antennas are particularly useful with very long wavelength radio waves because a normal quarter-wavelength antenna would be inconveniently long. Many AM radios have coil antennas hidden inside them. These antennas work best when the coil opens at right angles to the radio wave's path. That orientation sends the wave's magnetic flux back and forth through the coil and induces large currents in it.

CHECK YOUR UNDERSTANDING #2: There's No Place Like Home
Why do cordless telephones only work when they're close to their base units?

Representing Sound: AM and FM Radio

A radio transmitter does more than simply emit a radio wave. It uses that radio wave to *represent sound*. Because sound waves are fluctuations in air pressure and radio waves are fluctuations in electric and magnetic fields, a radio wave can't literally "carry" a sound wave. However, a radio wave can carry information to a radio receiver, instructing the receiver how to reproduce a particular sound. The radio wave is then *representing* the sound.

All the receiver needs to know in order to reproduce the sound is what the air pressure at its speakers should be. The radio wave provides this information by changing in one way to represent a compression of the air and in another way to represent a rarefaction of the air. There are two common techniques by which a radio wave can represent these pressure changes. One is called amplitude modulation and involves changing the overall strength of the radio wave. The other is called frequency modulation and involves small changes in the frequency of the radio wave.

In the **amplitude modulation** (AM) technique, air pressure is represented by the strength of the transmitted wave (Fig. 14.1.5). To represented a compression of the air, the transmitter is turned up so that more charge moves up and down the transmitting antenna. To represent a rarefaction, the transmitter is turned down so that less charge moves up and down the antenna.

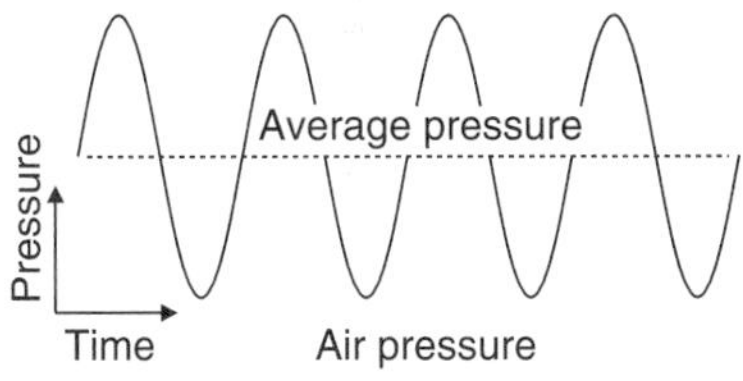

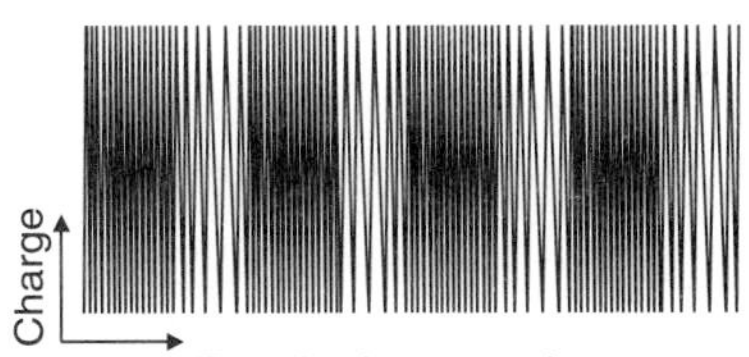

Fig. 14.1.6 - When sound is transmitted by frequency modulation, air pressure is represented by changing the frequency of the radio transmitter slightly. A compression is represented by increasing that frequency and a rarefaction by decreasing it.

The frequency at which charge moves up and down the antenna remains steady, so only the amplitude of the radio wave changes. The wave is strongest when the air should be compressed and weakest when the air should be rarefied. The receiver measures the strength of the radio wave and uses this measurement to recreate the sound.

In the **frequency modulation** (FM) technique, air pressure is represented by the frequency of the transmitted wave (Fig. 14.1.6). To represent a compression of the air, the transmitter's frequency is increased slightly so that charge moves up and down the transmitting antenna a little *more* often than normal. To represent a rarefaction, the transmitter's frequency is decreased slightly so that the charge moves up and down a little *less* often than normal. These changes in frequency are extremely small—so small that charge continues to slosh strongly in both tank circuits even though its frequency of motion changes slightly.

The amount of charge moving up and down the antenna remains constant and large, so only the frequency of the radio wave changes. The motion of charge on the antenna is most rapid when the air should be compressed and least rapid when the air should be rarefied. The receiver measures the radio wave's frequency and uses this measurement to recreate the sound.

Although the AM and FM techniques for representing sound can be used with a radio wave at any frequency, the most common commercial bands are the AM band between 550 kHz and 1,600 kHz (550,000 Hz and 1,600,000 Hz) and the FM band between 88 MHz and 108 MHz (88,000,000 Hz and 108,000,000 Hz). Elsewhere in the spectrum of radio frequencies are many other commercial, military, and public transmissions, including TV, short wave, amateur radio, telephone, police, and aircraft bands. These other transmissions use AM, FM, and a few other techniques to represent sound and information with radio waves.

CHECK YOUR UNDERSTANDING #3: Another Volume Control

When you are listening to AM radio in a car and drive through a tunnel, the volume becomes very low. Explain.

The Quality of Reception

You select radio stations by adjusting the resonant frequency of your receiver's tank circuit, either mechanically or electronically. Though many passing radio waves push charge up and down the antenna, only a radio wave with just the right frequency will cause large amounts of charge to slosh in the receiver's tank circuit. When you tune that tank circuit to match the frequency of the wave from your favorite station, your receiver will reproduce sound from that station.

However a receiver loses its ability to choose stations when it's extremely close to a powerful transmitter. The wave from the nearby transmitter is so strong that it pushes large amounts of charge up and down the receiver's antenna, even when that receiver's tank circuit is tuned to a different frequency. The receiver then plays only the nearby station, no matter where you tune it.

FM radio generally offers better fidelity than AM radio—it more faithfully reproduces the transmitted sound. To begin with, an AM transmitter operates below full power so that it can increase its power to represent a compression of the air. Its reduced power operation leaves the transmission more susceptible to electromagnetic noise. Our environment is filled with electromagnetic waves, both natural and artificial, that an AM receiver can't distinguish from real radio transmissions. When it responds to these unwanted waves, the AM receiver produces the sound we call "static."

In contrast, an FM radio transmitter is always operating at full power, so electromagnetic noise is less important. Moreover, a good FM receiver carefully tracks the transmitter's frequency and ignores other electromagnetic waves. This noise immunity makes FM radio better for music broadcasts. And because FM sound volume is determined by the radio wave's frequency rather than strength, you hear the music at full volume until you lose the station altogether.

Many FM stations transmit a wave representing stereo sound. This wave carries an audio signal that's meant to control two separate speakers. The audio signal has two parts, or rather two ranges of audio frequencies. Between 20 Hz and 19,000 Hz, is a part that contains the sum of the air pressures at the two speakers (called A+B). Exactly at 19,000 Hz is an inaudible tone that tells your radio that the transmission is stereo. And between 19,020 Hz and 38,000 Hz is a part that contains the different between the air pressures at the two speakers (called A–B). This latter signal has been shifted in frequency in a complicated way that can be undone with the proper electronics.

A stereo receiver undoes the frequency shift of the second part so that it can work with both parts of the audio signal: A+B and A–B. It electronically adds the two parts to obtain A, the air pressure that one speaker should produce. It electronically subtracts the two signals to obtain B, the air pressure that the other speaker should produce. In a non-stereo receiver, the A+B part is used by itself to reproduce sound, and stereo receivers automatically switch to this monaural mode if the FM signal is too weak to ensure clear decoding of the stereo signal.

Because high frequency radio waves travel in straight lines between antennas, it's hard to receive a commercial FM station from more than 100 km away. Even when the transmitting antenna sits on top of a tall tower, the earth's curvature and surface terrain severely limit the range of FM reception (see ❐).

But low frequency radio waves, such as those used by commercial AM stations, are reflected by charged particles in the earth's outer atmosphere so that portions of the radio wave that would otherwise be lost in space bounce back toward the ground. This returning power allows you to receive AM stations over a considerable distance, even when you have no direct line-of-sight to the transmitter's antenna. At sundown, these atmospheric layers become so effective at reflecting AM radio that you can hear a transmission from thousands of kilometers away as clearly as if it were a home town station.

❐ Like FM radio, the high frequency electromagnetic waves used by television broadcasts travel in straight lines and are blocked by the horizon less than 100 km from ground-based antennas. To reach broader audiences, engineers at Westinghouse experimented with aircraft-based antennas in a program called "Stratovision." Developed in the late 1940s, this program would have used 14 B-29 aircraft, carrying receivers and transmitters, to relay television broadcasts to more than ¾ of the population of the United States, including rural viewers. Despite successful test flights, the program foundered for political reasons and was never implemented.

CHECK YOUR UNDERSTANDING #4: Can You Hum a Few Bars?

When your AM radio is close to high tension wires, it hums loudly. Explain.

Summary

How Radio Works: A radio transmitter creates a radio wave by making electric charge accelerate up and down its antenna. To get as much charge moving as possible, the transmitter attaches a tank circuit to the antenna and slowly adds energy to that tank circuit until an enormous amount of charge is flowing up and down the antenna.

The radio receiver detects this radio wave by its effects on charge in the receiver's antenna. The passing wave causes that charge to accelerate up and down. If the receiver's tank circuit has the same resonant frequency as that of the transmitter and its tank circuit, large amounts of charge will slosh back and forth in the receiver's tank circuit and the receiver will detect the transmission.

This radio wave represents sound using either the AM or FM technique. In the AM technique, the strength of the wave is increased or decreased to represent compressions or rarefactions of the air, respectively. In the FM technique, the precise frequency of the transmission is increased or decreased to represent those compressions or rarefactions.

The Physics of Radio

1. An inductor opposes changes in the current passing through it. Its inductance is equal to the voltage drop across it divided by the rate at which the current through the inductor changes with time.

2. Accelerating electric charge emits electromagnetic waves.

3. Electromagnetic waves consist of changing electric and magnetic fields that continuously recreate one another. As it diminishes, the electric field recreates the magnetic field and as it diminishes, the magnetic field recreates the electric field. Together, these fields propagate through space at the speed of light.

4. Vertically polarized electromagnetic waves have vertical electric fields and horizontal magnetic fields.

5. Horizontally polarized electromagnetic waves have horizontal electric fields and vertical magnetic fields.

6. An antenna works best when it's about a quarter as long as the wavelength of its radio wave

Check Your Understanding - Answers

1. The amount of charge that the transmitter can move directly on and off the antenna is too small to create a strong radio signal.

Why: The tank circuit is useful because it allows the transmitter to move much more charge. Just as a tuning fork is inefficient at emitting sound waves by itself, so a radio antenna is inefficient at emitting radio waves by itself. You can make the tuning fork much louder by coupling it to an object that resonates at its frequency. Similarly, you can make the radio antenna emit a much stronger radio wave by coupling it to a tank circuit that resonates at its frequency.

2. When the cordless telephone is too far from its base unit, their electromagnetic waves become so spread out that they have trouble communicating.

Why: The powers emitted by the base unit and the handset are small, so that their waves are relatively difficult to detect. As long as the handset and base unit are nearby, they are able to detect each other's waves. But when the distance between them becomes too great, the waves become too spread out to detect and the handset and base unit lose contact with one another.

3. The tunnel blocks most of the radio wave. Since only a small fluctuating wave reaches your radio, the radio produces only small fluctuations in air pressure with its speaker.

Why: An AM radio has trouble distinguishing between a distant transmission representing loud music and a nearby transmission representing soft music. In both cases, the receiver detects only small variations in the current moving up and down its antenna. That's why you must turn up the volume of an AM radio as you move farther from the transmitting antenna or as you enter a tunnel.

4. The accelerating charge in the high tension wires emits a radio wave that's so strong it overwhelms the nearby radio.

Why: High tension wires act like giant antennas that transmit radio waves. Although these 60 Hz (50 Hz in Europe) radio waves are at the wrong frequencies for normal radio transmissions, they are so strong near the wires that your radio begins to hum anyway.

Glossary

amplitude modulation A technique for representing sound or data by changing the amplitude (strength) of a radio transmission.

frequency modulation A technique for representing sound or data by changing the exact frequency of a radio transmission.

henry (H) The SI unit of inductance. A 1 henry inductor will experience a 1 ampere change in the current flowing through it each second when subjected to a 1 volt voltage drop.

horizontal polarization An electromagnetic wave in which the electric field always points left or right (horizontally). The magnetic field always points vertically.

inductance The voltage drop across an inductor divided by the rate at which current through that inductor is changing with time. Inductance is measured in henries.

speed of light The speed with which an electromagnetic wave travels through space. In empty space, a vacuum, the speed of light is exactly 299,792,458 m/s.

tank circuit A simple resonant circuit consisting of a capacitor and an inductor. Energy is transferred back and forth between these two devices repetitively.

vertical polarization An electromagnetic wave in which the electric field always points up or down (vertically). The magnetic field always points horizontally.

Review Questions

1. In Fig. 14.1.2*d*, current flowing toward the lower capacitor plate is being repelled by the positive charge that's already there. What pushes this current forward?

2. What changes can you make to a tank circuit to lengthen its period and lower its frequency?

3. Why does accelerating charge emit electromagnetic waves while charge moving at constant velocity does not?

4 It's easy to see why charge moving up and down an antenna creates a changing electric field, but how does it create a magnetic field and why is that magnetic field changing?

5. Why must an electromagnetic wave have *both* electric and magnetic fields?

6. All waves have at least two forms for their energies. What are those two forms for an electromagnetic wave?

7. What is vertical about a vertically polarized electromagnetic wave?

8. Why does an AM radio station frequently change the amount of charge moving up and down its antenna?

9. How does an FM radio station represent the air pressure fluctuations in a sound?

Exercises

1. The transmitting antennas of FM radio stations are always oriented vertically. If you have a small portable FM receiver with a single antenna, why is it best for that antenna to be oriented vertically, too?

2. Cordless telephones often have a switch that allows you to change the radio frequencies they use to communicate with their base units. Why is this feature more important in cities than it is in rural areas?

3. The ignition system of an automobile produces sparks to ignite the fuel in the engine. During each spark process, charges suddenly accelerate through a spark plug wire and across a spark plug's narrow gap. Sometimes this process introduces noise into your radio reception. Why?

4. To diminish the radio noise in a car (see Exercise **3**), the ignition system uses wires that are poor conductors of electricity. These wires prevent charges from accelerating rapidly. Why does this change improve your radio reception?

5. The electronic components inside a computer transfer charge to and from wires, often in synchrony with the computer's internal clock. Without packaging to block electromagnetic waves, the computer will act as a radio transmitter. Why?

6. A transmitting radio antenna works best when it's a quarter wavelength long. If the antenna is too short, current flowing up it from the tank circuit will slow to a stop too early, before the time when it should begin flowing back toward the tank circuit. What forces stop the current flow?

7. If you connect the two wires of a large inductor to the terminals of a battery, current will gradually begin to flow through the inductor and it will develop a strong magnetic field. If you then suddenly disconnect the inductor from the battery, sparks will emerge from the wires of the inductor as the inductor pumps charge from one wire to the other. What provides the energy needed to make those sparks?

8. Suppose you include an inductor in an electric circuit that includes a battery, a switch, and a light bulb. Current leaving the battery's positive terminal must flow through the switch, the inductor, and the light bulb before returning to the battery's negative terminal. The current in this circuit increases slowly when you close the switch and it takes the light bulb a few seconds to become bright. Why?

9. When you open the switch of the circuit in Exercise **8**, a spark appears between its two terminals. As a result, the circuit itself doesn't open completely for about half a second, during which time the bulb gradually becomes dimmer. The bulb's behavior indicates that the current in the circuit diminishes slowly, rather than stopping abruptly when you open the switch. Why does the current diminish slowly?

10. An inductor stores energy in its magnetic field. Why does inserting an iron core into the coil of an inductor dramatically increase the energy it can store?

11. A junkyard crane uses a giant electromagnet to lift cars. The crane operator notices that it takes a second or two for the electromagnet to turn on or off, during which time the current through it increases or decreases slowly. Why are changes in the current through the electromagnet so slow?

12. While solar electric power is likely to become more important in the future, it's unclear how to store solar energy for nighttime use. People have proposed using giant superconducting inductors, hundreds of meters in diameter, to store solar energy in their magnetic fields. After allowing the current and magnetic field in an inductor to increase all day, the operators would let the inductor pump current through homes and businesses all night. How would the inductor do work on the current passing through it so as to transfer its energy to the homes and businesses as electric power?

13. A tank circuit consists of an inductor and a capacitor. Give a simple explanation for why the magnetic field in the inductor is strongest at the moment the separated charge in the capacitor reaches zero.

14. The metal wires from which most tank circuits are made have electric resistances. Why do these resistances prevent charge from sloshing back and forth forever in a tank circuit and what happens to the tank circuit's energy as time passes?

15. The best tank circuits in existence are made from superconducting metals (in contrast to the normal metals of Exercise **14**). Why does charge slosh so much longer in these superconducting tank circuits than in those made from normal metals?

16. The charge in a superconducting tank circuit (see Exercises **14** and **15**) gradually stops sloshing back and forth because its energy is radiated away as radio waves. How does that charge emit radio waves?

17. You have an extremely accurate electronic clock with you as you measure the electric field produced by a radio station's antenna. Since the charge on the antenna changes according to a regular schedule, you can use the clock to tell what charge is on the antenna at any moment. While you're at the base of the antenna, the electric field you measure seems sensible—for example, when positive charge is on the antenna above you, the electric field points downward. But when the distance between you and the antenna is about half the wavelength of its radio wave, everything is upside down! At a time when you know positive charge is on the antenna, the electric field you measure points upward! Explain why this happens.

18. The transmitting antennas used for radio links between buildings are sometimes oriented horizontally. What are the orientations of the electric and magnetic fields in the radio waves emitted by those antennas?

19. What is the polarization of the radio waves in Exercise **18**?

20. If the transmitting antenna used in a radio link is horizontal (see Exercise **18**), how should the receiving antenna be oriented? Explain.

21. While a particular AM radio station claims to transmit 50,000 W of music power, that's actually its *average* power. There are times when it transmits more power than this and times when it transmits less. Explain.

22. When your receiver is too far from an AM radio station, you can only hear the loud parts of the transmission. When it's too far from an FM station, you lose the whole sound all at once. Explain the reasons for this difference.

23. FM radio stations must operate at very different frequencies to avoid interfering with each other's broadcasts. Why?

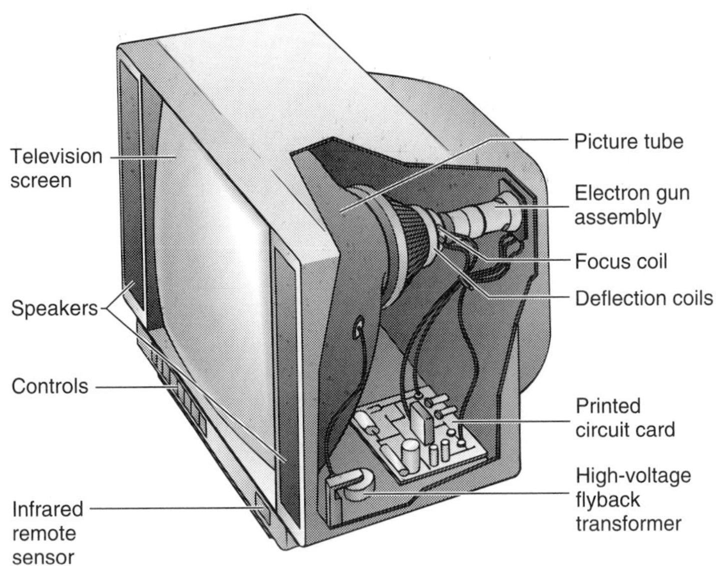

Section 14.2

Television

> **WARNING:**
> **Televisions contain dangerously high voltages, even after they have been unplugged.** Since a television uses beams of high energy electrons to create the images that you see, it needs high voltages. To avoid any risk of shock, you should never open a television while it's operating. But even turning the unit off doesn't necessarily make it safe. Because the television uses capacitors to store separated electric charge in its high voltage power supply, high voltages can persist inside it for minutes or more after you turn it off and unplug it. Don't open a television until you're sure that it has no more stored electric energy in its capacitors.

In the previous section, we learned how radio sends sound through space on an electromagnetic wave. But that process isn't nearly as complicated as sending pictures on an electromagnetic wave. Television performs this by breaking each picture up into tiny dots and handling these dots one by one at an incredible rate. In this section, we'll look at how a picture is represented by a video signal, how it's transmitted from one place to another, and how a television set turns that signal back into a picture.

Questions to Think About: *Why does it take a few seconds for a television set to warm up? Why are there tiny colored spots on the front surface of a picture tube? What happens when a picture tube "burns out"? Why is it so hard to make the front surface of a picture tube flat? Is it really possible to break a picture tube by hitting it with a shoe? If you take a photograph of the TV screen, why will you probably find that only part of the screen appears on the photograph?*

Experiments to Do: *One interesting but possibly costly experiment to do with a television set is to expose it to the field of a strong magnet. The magnet will deflect the stream of electrons flowing through the picture tube and create beautiful colors on the screen. However, just inside the surface of a color picture tube is a metal mask that is easily magnetized. If you hold a magnet to a color TV, this mask will probably have to be professionally demagnetized. But you can safely try this experiment with a black and white TV.*

A less risky experiment is to cut a narrow slot in a card and to watch a TV through this horizontal slot. Move the card up and down rapidly so that you only see the screen for a brief moment as the slot passes in front of your eye. You should see horizontal dark and light bands across the image. Can you explain why only part of the picture is illuminated? What happens when you hold the slot vertically and move it left and right?

Finally, use a magnifying glass to look at the screen of a color television set. You'll see a pattern of red, green, and blue dots or stripes. How can these simple colored dots create all of the colors that you see when you watch the television?

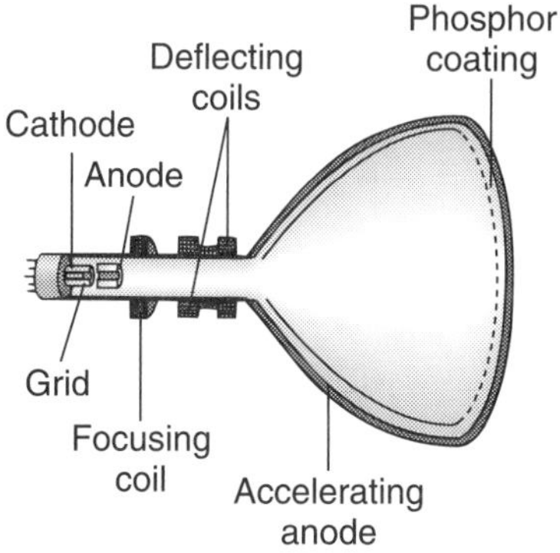

Fig. 14.2.1 - The main components of a black and white picture tube. Electrons travel from left to right and illuminate the screen.

Fig. 14.2.2 - A small picture tube.

Creating a Television Picture

A television builds its picture out of tiny colored dots, arranged in a rectangular array on the screen. The number of dots in this array depends on the television standard, which varies with country. For the following discussion, we'll consider only the color television standard used in the United States (*NTSC*), which specifies an array that is 525 dots high by about 700 dots wide. Other television standards differ somewhat in the details but not in the concepts.

While the television illuminates these dots one by one, it finishes with all of them so quickly that it appears as though they're all illuminated at once. Your eye responds slowly to changes in light and you see the whole pattern of dots on the screen as a single picture, a television picture. To create this picture, the television starts at the upper left hand corner of the screen and scans through the dots horizontally from left to right. Every $1/15{,}750^{\text{th}}$ of a second, it starts a new horizontal line. Since there are 525 horizontal lines, the television completes the entire picture every $1/30^{\text{th}}$ of a second.

Actually, if the television worked its way from the top of the screen to the bottom in $1/30^{\text{th}}$ of a second, our eyes would sense a slight flicker. To reduce this flicker, the television builds the image in two passes from top to bottom: it illuminates the odd numbered lines during the first pass and the even numbered lines during the second pass. That way, the television scans down the screen once every $1/60^{\text{th}}$ of a second and there is essentially no flicker at all.

CHECK YOUR UNDERSTANDING #1: Horizontal Hold

If you take a photograph of a television screen, using an exposure time of $1/250^{\text{th}}$ of a second, the finished photograph will contain only a band of the television image, about a quarter of the screen high. Explain.

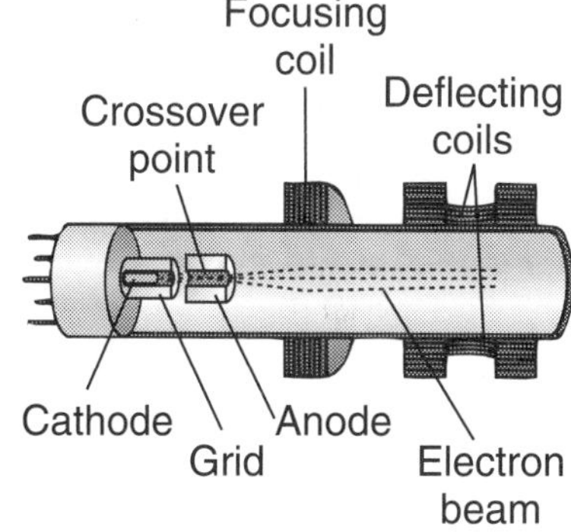

Fig. 14.2.3 - The neck of a picture tube, showing how the electron beam comes to a focus once inside the anode and again at the screen.

Black and White Picture Tubes

These dots of light are produced on the inside surface of a glass picture tube when electrons collide with a chemical **phosphor** (Figs. 14.2.1 and 14.2.2). These electrons are emitted by a hot surface in the neck of the picture tube and accelerate toward positive charge on the phosphor-coated screen. When the electrons strike the phosphor, they transfer energy to it and it glows brightly.

But the television exerts forces on the electrons as they fly through the empty space inside the picture tube. These forces focus the electrons into a narrow beam and then steer that beam to various points on the phosphor screen. This focusing and deflecting are done by components in the neck of the picture tube, a region that's shown in detail in Figs. 14.2.3 and 14.2.4. For the moment, let's consider only a black and white picture tube.

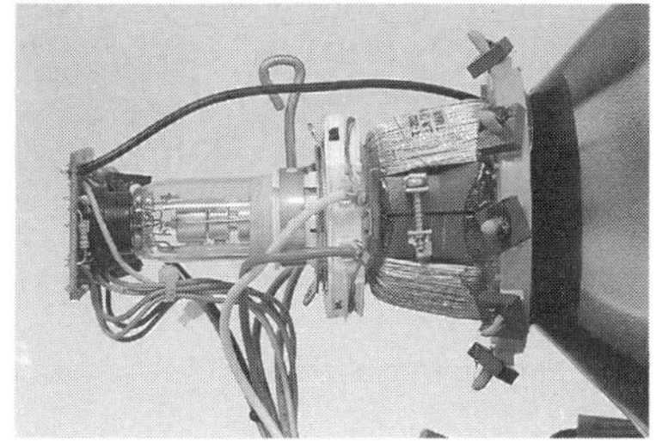

Fig. 14.2.4 - Electrons leaving the cathode, grid, and anode (in clear glass) pass through the focus coil (middle) and are steered by the wire deflecting coils (middle-right).

The electrons emerge from a negatively charged *cathode*, a device that emits electrons directly into space. In a picture tube, the cathode is hot, so that thermal energy tosses electrons from its surface. This cathode is heated by a nearby filament and takes a few seconds to warm up when you turn on the television. If the filament breaks or burns out, the picture tube is ruined.

Surrounding the cathode is a hollow *grid* with negative charge on it. Since this negative charge repels electrons, most of the electrons leaving the cathode

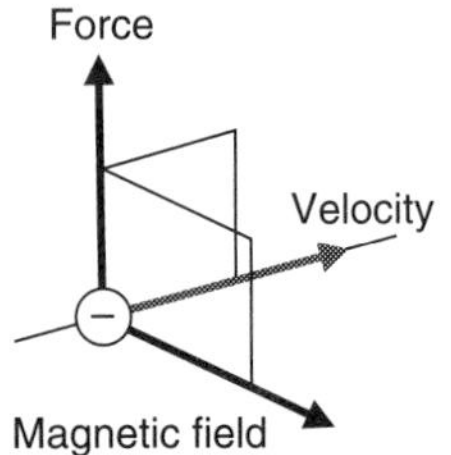

Fig. 14.2.5 - An electron moving through a magnetic field experiences a force that's at right angles to both its velocity and the magnetic field. A positively charged particle would experience a force in the opposite direction.

return to its surface. However, the grid has a small hole through which some of the electrons can escape. These escaped electrons are attracted by a positively charged *anode* and form a narrow beam.

The shape of the electric field between the cylindrical cathode, grid, and anode has an interesting effect on the electrons: it focuses them to an extremely narrow spot inside the anode. Regardless of which way they were heading when they left the cathode's surface, the electrons are all accelerated by the electric field and head toward the same point inside the anode, the *crossover point*.

But the electrons don't stop at the crossover point. Instead, they continue on through the anode and sail off toward the screen. After they have passed through the crossover point, the electrons are heading away from one another. They must be brought back together again so that they all strike the screen at exactly the same spot. This second focusing action is done by a magnetic field.

While it may seem surprising that a magnetic field can influence the path of an electron, we've already seen that moving charge is magnetic. A magnet pushes on charge flowing through the wires of an electric motor, so why shouldn't that magnet push on charges flying through an empty picture tube?

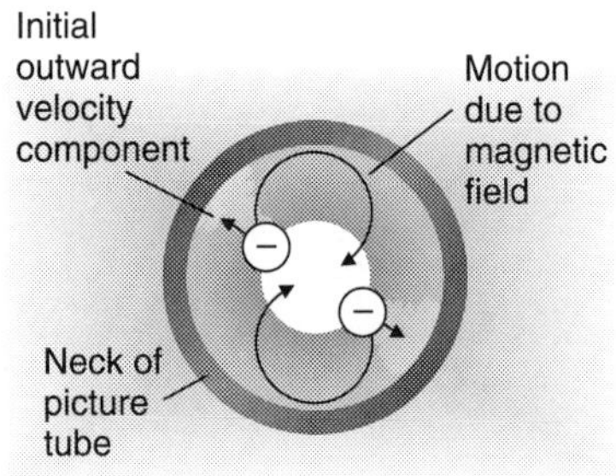

Fig. 14.2.6 - As diverging electrons move down the neck of the picture tube toward the screen, the focusing magnetic field makes them accelerate to their right. They travel in a spiral and return together just as they hit the screen.

As it passes through a magnetic field, an electron experiences a force that's proportional to its charge and velocity, and to the strength of the magnetic field. This force also depends on the angle between its velocity and the magnetic field. The force is strongest when the velocity and field are at right angles to one another and diminishes to zero when they both point in the same or in opposite directions. Oddly enough, this force acts at right angles to both the electron's velocity and the magnetic field (Fig. 14.2.5).

The magnetic field that focuses electrons to a spot on the screen is created by a wire coil that circles the neck of the picture tube. This field points directly toward the screen. Since the electrons' velocities also point toward the screen, the forces on them are small. But if we look down the neck of the tube at the screen (Fig. 14.2.6), we see that these electrons have small outward components of velocity that take them away from the center of the tube and the crossover point they just left behind. These outward components of velocity are at right angles to the magnetic field, so the electrons experience magnetic forces.

Each electron accelerates toward its right as it flies through the magnetic field so that it travels in a spiral. Viewed down the neck of the tube, this motion appears circular and is called **cyclotron motion**, after the particle accelerator based on this motion. Remarkably, the time it takes each electron to complete one full circle of cyclotron motion doesn't depend on how large that circle is. All of the electrons complete exactly one full circle on their flights from the crossover point to the screen and hit the screen at the same spot!

The television also uses magnetic fields to steer the electron beam to different parts of the screen. A pair of coils above and below the picture tube's neck produces a vertical magnetic field that deflects the electron beam horizontally (Fig. 14.2.7). By adjusting the amount and direction of current in these horizontal deflecting coils, the television can control the horizontal position of the beam spot on the screen.

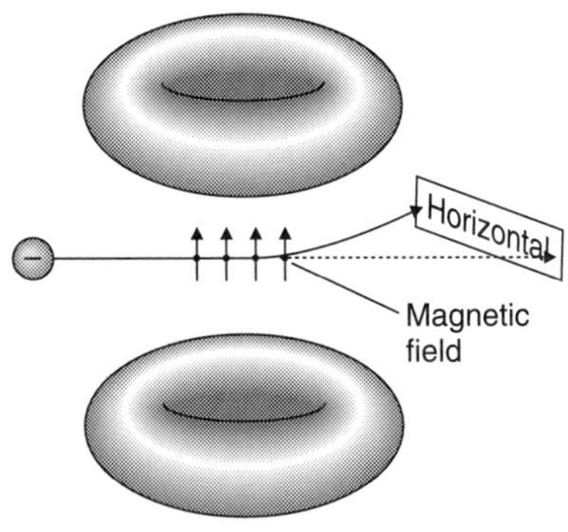

Fig. 14.2.7 - The horizontal deflecting coils are located above and below the neck of the picture tube. The vertical magnetic field they produce deflects the electron beam either left or right.

A second pair of coils mounted to the left and right of the tube's neck produces a horizontal magnetic field that deflects the electron beam vertically (Fig. 14.2.8). The amount and direction of current in these vertical deflecting coils determines the vertical position of the beam spot on the screen.

When the electron beam strikes the phosphor coating on the inside of the screen, it transfers energy to that phosphor, which then emits white light. Creating a bright image takes lots of energy, so the electron beam is accelerated on its way to the screen. A high voltage power supply (+15,000 V to +25,000 V) pumps positive charge onto the inside of the screen and the surrounding accelerating anode, and this charge attracts the electrons. By the time they hit, the electrons have enough kinetic energy to make the phosphor glow bright white.

But white isn't the only color in a television picture. To produce a gray or black spot, the television reduces the current of electrons in the beam. It controls this current by adjusting the charge on the picture tube's grid. The more negative charge on the grid, the harder it is for electrons to pass from the cathode to the anode and the fewer of them that strike the phosphor. The television carefully adjusts this grid charge as it sweeps the electron beam back and forth across the screen and thus creates a complete television picture, one dot at a time.

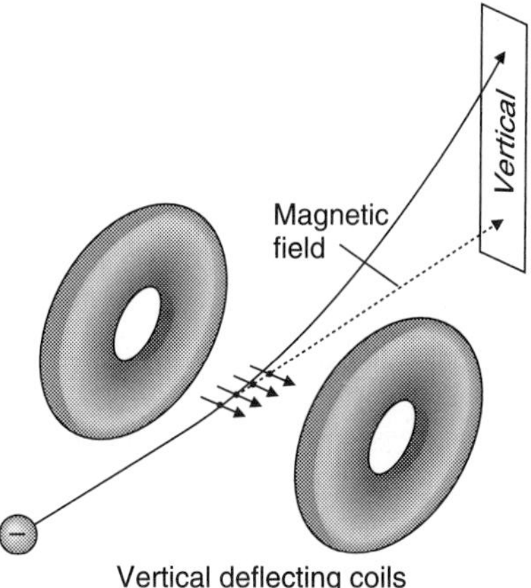

Fig. 14.2.8 - The vertical deflecting coils are located to the left and right of the picture tube's neck. The horizontal magnetic field they produce deflects the electron beam either up or down.

CHECK YOUR UNDERSTANDING #2: Currents in Time

The electron beam is swept from left to right by changing the current through the horizontal deflection coils. If each sweep takes $1/15{,}750^{th}$ of a second, how does the current through these deflection coils vary with time?

About the Picture Tube

Unfortunately, not all of the light from the phosphor coating travels directly out into the room where you can see it. Some is emitted backward into the tube itself or gets caught up in reflections in the glass. This lost light tends to reappear where it's not wanted and reduces the picture's contrast. To prevent light from bouncing around inside the picture tube, the phosphor coating is itself coated with a thin layer of aluminum. The electron beam can pass through this aluminum layer but light cannot. This light is reflected directly out of the tube, where it brightens and sharpens the image. The aluminum layer also provides a path for electrons to return from the phosphor coating so that they can be reused.

The flatness of the screen is limited by the picture tube's ability to focus and sweep the electron beam. Although a perfectly flat, rectangular screen is the ideal, it's virtually impossible to achieve in practice. The corners of a flat screen are farther from the focusing coil than is the center of the screen and the television can't get the beam to focus simultaneously in both places. Furthermore, the horizontal and vertical deflections interfere with one another to create images that aren't rectangular. While special electronics can correct for some of these focusing and deflection problems, truly flat picture tubes are still not practical.

Because the picture tube contains a vacuum, it experiences enormous inward forces from the surrounding air pressure. Any crack in the tube would cause it to implode violently. The neck and back of the tube are protected by the case, but the screen is exposed. It has a thick plate of glass glued to its surface, forming an armored surface that's almost indestructible.

CHECK YOUR UNDERSTANDING #3: There's Really Nothing Inside

If a picture tube will implode violently when cracked, it must contain potential energy. Where did it get this potential energy?

Color Picture Tubes

Color picture tubes work much like black and white tubes, except that they have three electron beams and phosphors that emit red, green, and blue light. As we'll discuss in the next chapter, mixtures of red, green, and blue light can make you perceive any color. A color television appears full color by carefully mixing these three colored lights.

The inside surface of a color television screen is coated with thousands of

(a)

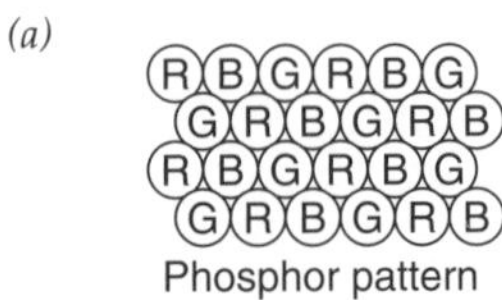

(b)

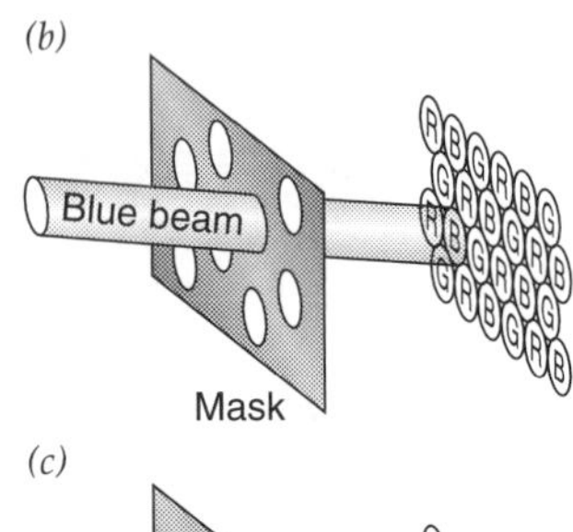

(c)

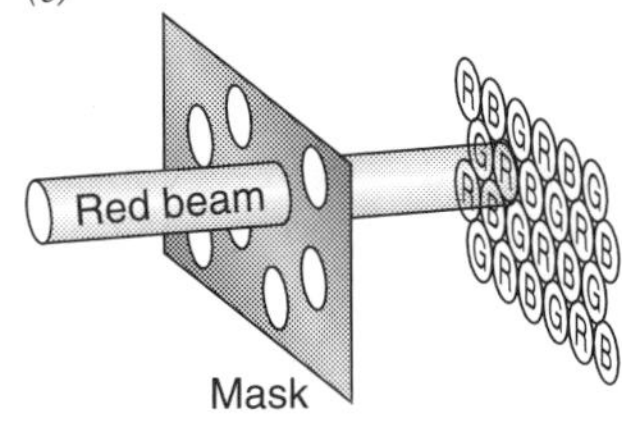

(d)

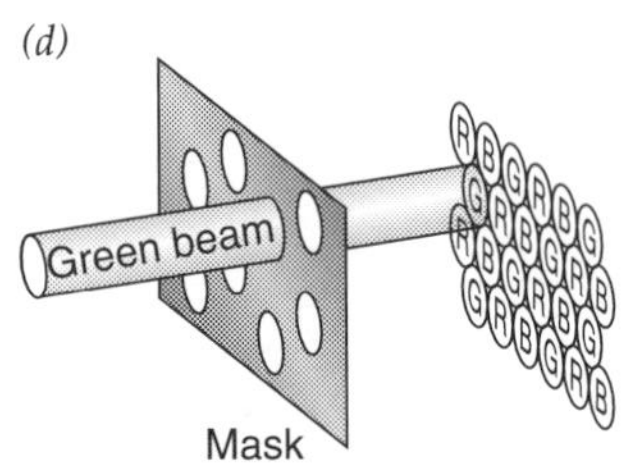

Fig. 14.2.9 - (*a*) The pattern of red, green, and blue phosphor dots coating the inside of a picture tube. The holes in the mask allow (*b*) the "blue" electron beam to illuminate only blue dots, (*c*) the "red" beam to illuminate only red dots, and (*d*) the "green" beam to illuminate only green dots.

tiny phosphor dots (Figs. 14.2.9*a* and 14.2.10). Some of these dots emit red light, some green light, and some blue light. The television directs electrons at these dots through holes in a metal mask. Three separate electron beams, coming from three slightly different angles, pass through the holes and strike the phosphors. Since each beam can only strike one color of phosphor dots (Fig. 14.2.9*b–d*), each beam controls the brightness of one of the three colors.

While some picture tubes use phosphor stripes rather than dots, and a grille rather than a mask, the basic idea is the same. Only one electron beam can hit a particular phosphor stripe. These masks and grilles are carefully aligned and mounted inside picture tubes and are made from special thermally compensated metals that stay aligned even when the temperature changes. These special metals are easily magnetized, which is why you shouldn't hold a strong magnet near a color television. The resulting magnetization would then deflect the electron beams and produce distorted images and colors.

CHECK YOUR UNDERSTANDING #4: The Blue and the Red

The three electron beams in a color picture tube always converge to a single spot on the screen, even when they're being deflected. How does the momentum transferred to electrons in the "blue" beam by this deflection compare to that transferred to electrons in the "red" beam?

Television Transmission

Representing a picture with a current or an electromagnetic wave is much more complicated than representing sound. This representation, which we'll call a *video signal*, must indicate the brightness and color of every dot in the picture 30 times a second. It must also indicate to the television which spot on the screen to illuminate at any given moment.

The basic structure of a black and white video signal is relatively simple. It just indicates how bright each dot on the screen should be as the television scans line by line through the picture (Fig. 14.2.11). To indicate when each line begins, the video signal briefly extends outside its normal range and takes a value beyond full blackness. The television responds to this *horizontal sync signal* by steering its electron beam back to the left edge of the screen and starting a new line. To indicate when a new picture begins, the video signal extends beyond blackness for a longer time (not shown in Fig. 14.2.11) and this *vertical sync signal* causes the television to steer its electron beam back to the top of the screen.

A color video signal contains the same brightness and synchronization signal, so that it's compatible with black and white televisions. But it also contains a separate color signal. To see how these two signals, called *luminance* and *chrominance* respectively, can be carried simultaneously by a single electromagnetic wave, let's look at how a television station produces that wave.

The luminance signal is transmitted in much the same way that sound is transmitted in AM radio—by allowing it to control the amount of charge sloshing in a tank circuit and moving up and down an antenna. When the luminance signal is off, the amount of moving charge doesn't change and the antenna emits a steady electromagnetic wave called the **carrier wave**. Figure 14.2.12*c* shows the

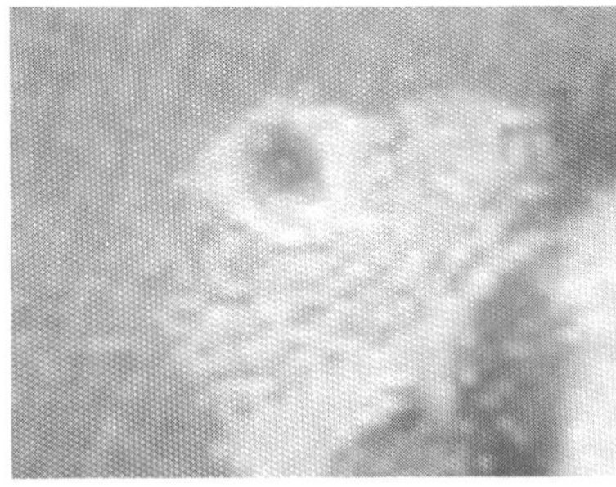

Fig. 14.2.10 - A color television creates its colored image with millions of tiny colored dots. This is a magnified photograph of those dots.

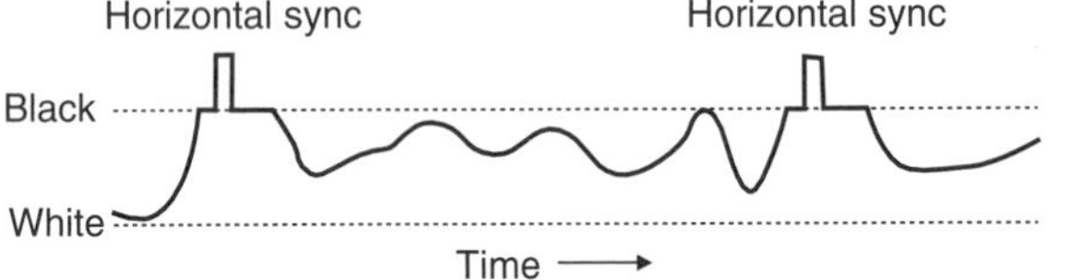

Fig. 14.2.11 - A black and white video signal represents the brightness of the picture line by line. A new horizontal line is indicated when the video signal briefly extends beyond black.

charge on the antenna as it emits this wave. When the luminance signal is on, it modulates the amount of charge moving up and down the antenna and the amplitude of the electromagnetic wave that the antenna emits. This **modulated carrier wave** (Fig. 14.2.12*b*) is just like the carrier wave itself, except shaped in amplitude by the luminance signal (Fig. 14.2.12*a*).

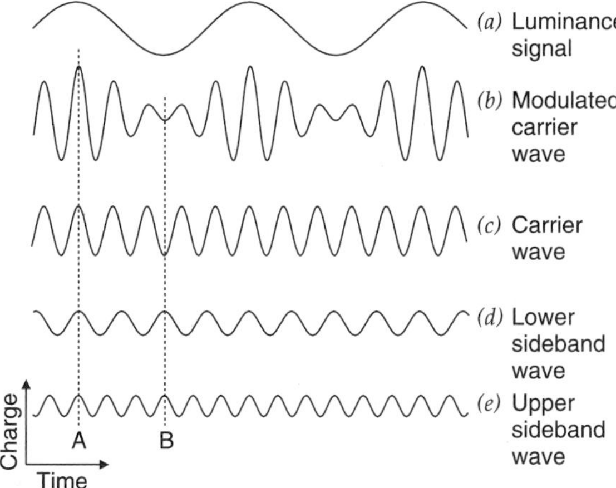

Fig. 14.2.12 - A television station represents this simple luminance signal by using it to amplitude modulate (AM) a carrier wave. This modulated carrier wave is actually the sum of the carrier wave and two sideband waves. Shown here are the amounts of charge on the transmitting antenna as it emits these various waves.

But we can also view the modulated carrier wave as the sum of three steady waves: the carrier wave and two **sideband waves** that have frequencies above and below the carrier wave (Fig. 14.2.12*d*,*e*). You can see this possibility in Fig. 14.2.12 by picking a specific time—such as *A* or *B*—and adding the antenna charge associated with the carrier, lower sideband, and upper sideband waves. They sum to the antenna charge associated with the modulated carrier wave! In effect, the modulated antenna charge can be thought of as three separate portions of charge that move up and down the antenna at slightly different frequencies.

This observation, that a modulated carrier wave actually contains several different waves at slightly different frequencies, has a profound impact on television and radio transmission. A television station that operates on Channel 2 and has a carrier wave frequency of 55.25 MHz also emits electromagnetic waves at other nearby frequencies! That's why competing television stations must be assigned widely different carrier wave frequencies—so that they don't accidentally emit sideband waves at the same frequency.

While the simple luminance signal in Fig. 14.2.12 has just one frequency and creates only one upper and one lower sideband, a more complicated luminance signal creates many upper and lower sideband waves. The range of frequencies present in upper or lower sideband waves is the same as the range of frequencies present in the luminance signal itself. Because it must specify all of the dots on the screen 30 times a second, the luminance signal includes frequencies as high as 5.5 MHz. Thus the sideband waves in a television signal extend as much as 5.5 MHz below and 5.5 MHz above the carrier wave frequency, giving them a total range of frequencies or a **bandwidth** of 11.0 MHz.

If television transmissions had 11.0 MHz bandwidths, competing stations would need carrier waves separated by at least 11.0 MHz. But that spacing would use up the precious spectrum of electromagnetic waves awfully quickly.

Consider radio stations for comparison. By international agreement, an AM radio station may use a bandwidth of 10 kHz, so its sideband waves can extend only 5 kHz above and below its carrier wave. Since the AM station is using an audio signal to amplitude modulate its carrier wave, that audio signal can't contain frequencies above 5 kHz. While this restricted frequency range is bad for music, it allows competing stations to function with carrier waves only 10 kHz apart, so that 106 different stations can operate between 550 kHz and 1600 kHz.

An FM radio station, which uses the frequency modulation technique (FM) to represent its audio signal, also produces sideband waves. In fact, the FM technique produces extra side bands and needs extra bandwidth. According to the international agreement, each FM station may use 200 kHz of bandwidth, 100 kHz on each side of the carrier wave frequency. This luxurious allocation permits FM radio to represent a very broad range of audio frequencies, in stereo, which is why an FM radio station can do a much better job of sending music to your radio than an AM station can.

The spectrum of electromagnetic waves is too valuable to allow a single television channel to occupy 11 MHz. The official allocation is 6 MHz, *including sound and color*. While this is like fitting an elephant into a shoe box, it turns out to be possible, although it does reduce the quality of the picture ever so slightly.

First, the upper and lower sideband waves carry redundant information. Since a particular frequency in the luminance signal creates both an upper *and* a lower sideband wave, only one of these sideband waves is really needed. So the television station uses electronic filters that stop it from producing most of the lower sideband waves. When this filtering is done, the lower sideband waves extend no more than 1.25 MHz below the carrier. Television receivers are able to compensate for the missing lower sideband waves.

To narrow the bandwidth of the television signal still further, the television station also filters out the highest frequency upper sideband waves. Instead of extending 5.5 MHz above the carrier wave, these upper sideband waves extend no more than 4.0 MHz. This filtering reduces the picture sharpness slightly.

The television signal now occupies 5.25 MHz of bandwidth about the carrier wave frequency—1.25 MHz below it and 4 MHz above it. The television station adds sound to its transmission by frequency modulating a separate carrier wave, just like in normal FM radio. This new audio carrier wave has a frequency that is 4.5 MHz above that of the video carrier wave.

With the inclusion of sound, the 6 MHz of allowed bandwidth is essentially full. Television channels are assigned to frequencies 6 MHz apart, although they are broken up into groups (VHF Low, VHF High, and UHF) in order to leave room in the electromagnetic spectrum for FM radio, cellular telephones, and other communication activities.

But what about color? The chrominance signal is inserted into the transmission by modulating an extra *chrominance subcarrier* wave located at a frequency 3.58 MHz above that of the video carrier wave. This new electromagnetic wave shares its bandwidth with the upper end of the luminance signal. For complicated reasons relating to the time structure of the video signal, these two electromagnetic waves don't interfere with one another and a black and white television doesn't even notice that the chrominance signal is there.

The chrominance signal controls the three electron beams in a color television. It contains two streams of color instructions which, together with the brightness instructions in the luminance signal, determine the coloring of the screen. Since your eyes are much more aware of brightness details than of color details, the chrominance signal is simplified so that it occupies only a modest bandwidth in the television transmission.

CHECK YOUR UNDERSTANDING #5: Beating the Bandwidth

Video tape recorders store video and audio signals on magnetic recording tape. The bandwidth available to the recording process is limited. Older VCR standards such as VHS, Beta, and 8 mm allow less than 6 MHz bandwidth for recording while some of the modern standards such as Super-VHS, Super-Beta, and Hi–8 mm approach or exceed 6 MHz. How does this recording bandwidth affect picture quality?

Cable Television

Cable television is similar to broadcast television except that it sends electromagnetic waves through a cable rather than empty space. Television cable consists of an insulated wire inside a tube of metal foil or woven metal mesh. This wire-inside-a-tube arrangement is called **coaxial cable** because the wire and the surrounding tube share the same center line. Electromagnetic waves can propagate easily through coaxial cable, following its twists and turns from the company that produces the waves to the television receiver that uses them.

Because the electromagnetic waves in a coaxial cable don't interact with those outside it, the whole electromagnetic spectrum inside the cable can be used for television channels. Since the cable can handle frequencies up to about 1000 MHz, it can carry about 170 channels.

To increase the number of channels, televisions are beginning to use compression techniques. Instead of transmitting the entire picture 30 times a second, a station can transmit only information about dots that have changed since the last picture. This compression narrows the bandwidth needed by each channel so that a coaxial cable can carry more than 500 of them simultaneously.

Compression is particularly important for high-definition television, which has about four times as many dots in each picture. To transmit this increased detail and remain within the allowed bandwidth for a television station, a high-definition television station must use compression techniques.

But compression becomes less important when coaxial cables are replaced by glass fiber cables (*fiber optic cable*) that guide light from one place to another. Light is also an electromagnetic wave and it can be amplitude modulated just like a radio wave. But light's frequency is extremely high; for visible light, it ranges from $4.5 \cdot 10^{14}$ Hz to $7.5 \cdot 10^{14}$ Hz. If we were to place television channels every 6 MHz throughout the visible spectrum, there would be about 50 million channels. Just imagine the newspaper television listings!

CHECK YOUR UNDERSTANDING #6: Networking

Computers often communicate with one another through networks of wire or glass fiber cables. How does this network resemble cable television?

Summary

How Television Works: A black and white television creates an image on its picture tube by scanning a beam of electrons rapidly across the inside of the screen. It starts at the top of the screen and scans left to right while gradually moving the beam down toward the bottom of the screen. It completes all 525 horizontal lines every $1/30^{th}$ of a second, but forms them in two passes down the screen. The brightness of any particular spot on the screen is determined by the current of electrons in the beam. The electron beam is focused to a tiny point on the screen by electric and magnetic fields and is scanned magnetically from left to right and top to bottom. A color television tube scans three separate electron beams across the inside surface of the screen, where they illuminate blue, red, or green phosphor patches through holes in a metal mask.

The electromagnetic waves that carry the television signal through space can be viewed either as three modulated carrier waves—luminance (brightness), chrominance (color), and audio—or as those carrier waves and their associated sideband waves. All of these waves fit within a 6 MHz range of frequencies. To operate in this narrow bandwidth, most of the lower sideband waves associated with luminance are filtered away, along with some of the upper sideband waves. This filtering reduces the detail of the television picture slightly but allows more television stations to operate simultaneously.

The Physics of Television

1. A moving charged particle experiences a force when it travels through a magnetic field. This force is at right angles to both its velocity and the field and is proportional to the particle's velocity and charge, and the strength of the field.

2. Moving electrons in magnetic fields tend to follow spiral trajectories.

3. Amplitude or frequency modulation of a wave introduces sideband waves at higher and lower frequencies.

4. The bandwidth of an amplitude modulated carrier wave is twice the bandwidth of the signal controlling the amplitude modulation. Half that bandwidth is upper sideband waves and half is lower sideband waves.

5. The bandwidth of a frequency modulated carrier wave is more than twice the bandwidth of the signal controlling the frequency modulation. Again, half the bandwidth is upper sideband waves and half is lower sideband waves.

Check Your Understanding - Answers

1. During the $1/250^{th}$ of a second exposure, the television will only complete about a quarter of its scan from the top of the screen to the bottom of the screen.

Why: While the television's scanning process easily fools your eye into seeing a complete image, a camera records exactly what it sees. If the exposure time is too short, the television will only be able to build part of the screen image during the exposure, and the photograph will contain only a horizontal band of the picture.

2. This current changes smoothly and steadily from flowing in one direction around the coils to flowing in the other direction around the coils. Every $1/15{,}750^{th}$ of a second, it restarts this process.

Why: The current flowing through the coils resembles a series of ramps when graphed as a function of time. It increases steadily up to a point and then suddenly returns to its starting value. As the current increases steadily, the electron beam is deflected less to the left and more to the right. Then the current is suddenly returned to its original value and the beam snaps back to the left. Because magnetic fields from this changing current can cause objects to move, some televisions emit an annoying tone at 15,750 Hz.

3. The machine that removed the air from the tube provided it.

Why: It takes mechanical work to remove air from the picture tube. The vacuum pump had to push the tube's air outward, into the surrounding atmosphere, and that air moved outward, in the direction of the pump's force. Thus rarefying the air in a container takes energy. When the tube cracks, the air rushes back into empty space and it can do work on the tube's walls. The tube is crushed violently.

4. They are exactly the same.

Why: The deflection process doesn't prevent the beams from coming together at a single point on the screen. Instead, it gives electrons in each beam exactly the same impulse, so that they all experience the same change in momentum and travel together to a specific point on the screen.

5. The higher the recording bandwidth, the finer the horizontal detail in the picture.

Why: A VCR that can record very high frequencies in the video signal will capture the finest details in each horizontal line of the picture. Recent high-bandwidth VCRs are able to record more detail than is currently present in broadcast television.

6. Both computer networks and cable television systems send information as electromagnetic waves in wires or glass fibers.

Why: The information that moves through a computer network travels as electromagnetic waves in cables. Sometimes those cables are coaxial cables, sometimes they are glass fibers, sometimes they consist of twisted pairs of wires (which can also carry and guide electromagnetic waves). A computer transmits information through the cable by creating and modulating an electromagnetic wave and another computer receives the information by detecting that modulated wave.

Glossary

bandwidth The range of frequencies involved in a group of electromagnetic waves.

carrier wave An electromagnetic wave with only a single frequency. This wave is modulated in order to represent an audio or video signal, or data.

coaxial cable A two-conductor electric cable in which an insulated central conductor is surrounded by a cylindrical outer conductor. Electromagnetic waves can travel inside a coaxial cable.

cyclotron motion The circular or spiral motion of a charged particle in a magnetic field. The charged particle tends to loop around the magnetic flux lines.

modulated carrier wave A wave that has been modulated so that it contains not only the carrier frequency, but also video, audio, or other information.

phosphor A solid that luminesces (emits light) when energy is transferred to it by light or by a collision with a particle.

sideband waves Waves at frequencies above and below the carrier wave that are produced when the carrier wave is modulated.

Review Questions

1. A television picture tube has a large surface, so why must the television focus its electron beam on a tiny spot?

2. Why do the electrons in a television's electron beam spread apart after they pass through the crossover point? Why can't they just stay together from then on?

3. Why is it important that all the electrons in the television's electron beam take the same amount of time to complete one full circle of cyclotron motion?

4. After the television's electron beam passes through the horizontal deflecting coils, its velocity points slightly to the left or right. Why can't the deflecting coils simply shift the beam to the left or right and leave its velocity pointing straight ahead?

5. Explain why less than a third of the electrons in the red beam of a color television strike the phosphors on the screen.

6. When a television station's carrier wave is amplitude modulated by the luminance signal from a camera, the station begins to transmit radio waves with frequencies that are slightly different from that of the carrier wave. Why does modulating the carrier wave introduce new frequencies?

7. To instruct a television to display alternating black and white dots, the video signal must shift from black to white to black 5.5 million times each second. If this video signal were used to amplitude modulate a carrier wave, why would the modulated carrier wave have a bandwidth of 11 MHz?

Exercises

1. A television's high-voltage power supply transfers electrons from the phosphor screen to the cathode for reuse. This supply does considerable work on the electrons it transfers. How is that work related to the light emitted by the screen?

2. The electrons in an electron beam experience forces as they pass through a magnetic field. How are those forces related to the forces electrons in a wire experience when they pass through a magnetic field?

3. Suppose you put a wire along the velocity arrow in Fig. 14.2.5 and send electrons through that wire in the direction of the arrow. Suppose also that there's a magnetic field pointing along the magnetic field arrow in that picture. Do the electrons in the wire experience a force and, if so, in which direction is that force?

4. A Penning ion trap suspends an ion (a charged atom or molecule) in empty space using electric and magnetic fields. In this trap, like charges on metal surfaces above and below the ion keep it from moving up or down, and a vertical magnetic field confines the ion horizontally by making it move in a circle. Why does the ion move in a circle?

5. A cyclotron is a particle accelerator invented in 1929 by American physicist Ernest O. Lawrence. It uses electric fields to do work on charged particles as they follow circular paths in a strong magnetic field. Lawrence's great insight was that all the particles take the same amount of time to complete one circle, regardless of their speed or energy. That fact allows the cyclotron to do work on all the particles at once as they circle together. Why is Lawrence's discovery also crucial to the sharpness of a television picture?

6. The force that a magnetic field exerts on a moving charged particle does no work on that particle. Use the definition of work to explain why there's no work done.

7. Computers and electronic games can tell where on a television screen you're holding a light sensor by studying when that sensor detects light from the screen. How is the time that the sensor observes light related to the sensor's position?

8. If the inside of the picture tube's screen didn't have a thin layer of aluminum on it, the electrons would accumulate there. Why would that be a problem?

9. X-rays are produced by a vacuum tube in which electrons from a hot, negatively charged cathode accelerate toward a positively charged metal anode. Collisions between the electrons and the metal atoms produce X-rays. Why is it important to shield the X-ray tube from magnetic fields?

10. The sun emits a stream of energetic electrons and protons called the *solar wind*. These particles frequently get caught up in the earth's magnetic field, traveling in spiral paths that take them toward the north or south magnetic poles. When they head northward and collide with atoms in the earth's upper atmosphere, those atoms emit light we know as the *aurora borealis*, or northern lights. These particles also interfere with radio reception. Why do they emit radio waves?

11. Some televisions whistle at a frequency of 15,750 Hz. Something inside the television must be moving back and forth 15,750 times each second. What television components exert forces on one another that fluctuate in strength at this frequency and thus might make the high-pitched sound?

12. When you put your hand near a picture tube that has just been turned on, negative charges flow from your hand to the glass screen. What attracts these charges to the screen?

13. If you hold a strong permanent magnet near the face of a color television picture tube, the image will be distorted in both shape and color. What causes this distortion? *(Warning: the distortion may not go away when you remove the magnet, so don't try this on a good television set!)*

14. Scrambled television stations (pay television) often remove or modify the horizontal synchronization signals. What happens to the picture if the television can't determine where each horizontal line starts?

15. When an AM radio station announces that it's transmitting at 950 kHz, that statement isn't quite accurate. Explain why it may also be transmitting at 948 kHz and 954 kHz.

16. The sidebands produced by AM radio broadcasts are legally prohibited from extending more than 5 kHz from the carrier wave's frequency. How does this limit affect the frequencies in the music that's transmitted by an AM station?

17. When a television signal travels through a coaxial cable, charge moves back and forth on both the central wire and the surrounding tube. Show that both electric and magnetic fields are present in the coaxial cable.

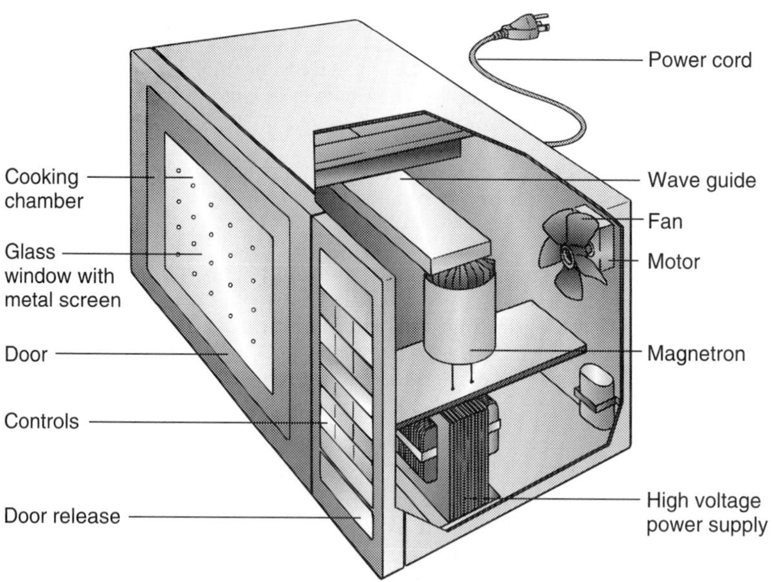

Section 14.3

Microwave Ovens

In addition to carrying sounds and pictures from one place to another, electromagnetic waves can carry power. One interesting example of such power transfer is a microwave oven. It uses relatively high frequency electromagnetic waves to transfer power directly to the water molecules in food, so that the food cooks from the inside out. This section discusses both how those waves are created and why they heat food.

Questions to Think About: *Why do microwave ovens tend to cook food unevenly if you don't move the food during cooking? How can part of a frozen meal become boiling hot while another part remains frozen? Why must you be careful with metal objects placed inside the oven? Why do some objects remain cool in the microwave oven while other objects become extremely hot? How does microwave popcorn work?*

Experiments to Do: *A microwave oven transfers power primarily to the water in food. You can see this effect by placing completely water-free food ingredients such as salt, baking power, sugar, or salad oil on a microwave-safe ceramic dish in a microwave oven. Cook the ingredients **briefly**. You will find that the ingredients and dish remain cool. Add just a little water to the collection. What happens when you cook them this time?*

Now try cooking a very cold ice cube. The cube should come directly from the freezer on an ice-cold plate, so that its surface is solid and dry. What happens? If ice contains water and water is what absorbs power in a microwave oven, why doesn't the ice absorb power and melt?

What Are Microwaves?

When we studied incandescent light bulbs in Section 6.3, we discussed the *wavelengths* of electromagnetic waves. While looking at radio and television, we considered the *frequencies* of electromagnetic waves. It's time to bring these two concepts together. A typical electromagnetic wave has both a wavelength *and* a frequency. Like Fig. 6.3.2, Fig. 14.3.1 shows the approximate wavelengths of many types of electromagnetic waves. But it also shows their frequencies.

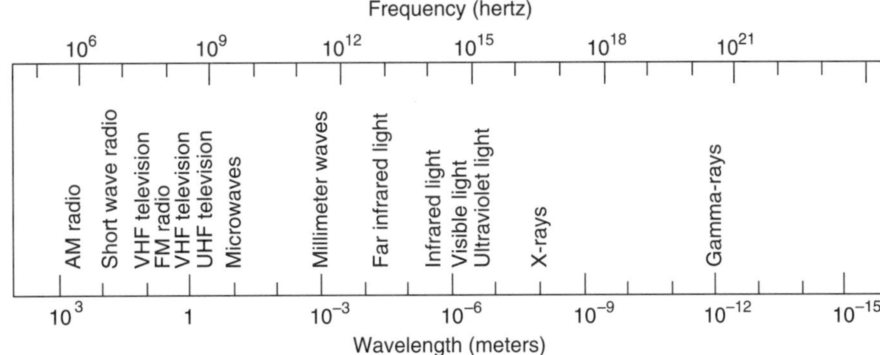

Fig. 14.3.1 - The electromagnetic spectrum. Microwaves have wavelengths between about 1 m and 1 mm, corresponding to frequencies from 300 MHz up to 300 GHz.

Although an electromagnetic wave is characterized by both its wavelength and its frequency, the product of these two values is always the speed of light (299,792,458 m/s in empty space). To see why this relationship is so, imagine a passing electromagnetic wave. If you measure how long each cycle of that wave is (its wavelength) and how often those cycles pass by (the wave's frequency), you can determine how fast the wave is moving (its speed) by multiplying those two values together. This relationship can be written as a word equation:

$$\text{speed of light} = \text{wavelength} \cdot \text{frequency}, \tag{14.3.1}$$

in symbols:

$$c = \lambda \cdot \nu,$$

and in everyday language:

The higher the frequency of an electromagnetic wave,
the shorter its wavelength.

Since wavelength and frequency aren't independent, you can calculate one from the other.

Radio and television broadcasts use the low frequency, long wavelength portion of the electromagnetic spectrum. Commercial AM radio is at frequencies of 550 kHz to 1600 kHz (wavelengths of 545 m to 187 m), commercial FM radio is at frequencies of 88 MHz to 108 MHz (wavelengths of 3.4 m to 2.8 m), and commercial television is at frequencies of 54 MHz to 806 MHz (wavelengths of 5.6 m to 0.4 m). Most of these waves have wavelengths longer than 1 m and are called radio waves. But some are shorter than 1 m and are called **microwaves**. Microwaves extend all the way down to wavelengths of 1 mm.

Microwaves are used extensively in communications. They travel easily through the air and through the upper atmosphere to and from satellites. Because of their short wavelengths, microwaves can be focused in much the same way that visible light is focused. Just as a curved mirror can focus sunlight onto the boiler of a solar power station, a curved satellite dish can focus microwaves from a satellite onto the antenna of a satellite receiver.

A satellite dish reflects microwaves because it's made out of metal. As an electromagnetic wave hits a metal surface, its electric field pushes on the mobile charges in that surface and they accelerate. These accelerating charges prevent the wave from entering the surface and reflect it instead.

However many satellite dishes are made out of wire mesh, which accommodates the wind and weather better than solid metal. The wire mesh still reflects microwaves because charges have enough time during each microwave cycle to flow all the way around each hole and compensate for its presence. As long as the wavelength of an electromagnetic wave is much larger than the holes in a metal surface, the wave can't get through that surface.

Because microwaves can be focused, nearby microwave communication links can operate independently. Each microwave transmitter uses a dish to focus its beam of microwaves toward the proper receiver and each microwave receiver uses a dish to focus that beam of microwaves onto its antenna. Similarly, a satellite dish can select between different satellites by adjusting its orientation to focus the correct beam of microwaves onto its antenna.

Microwaves are also used in radar, which beams microwaves into space to see if anything is out there. Microwaves reflect from objects, particularly metal ones, and the radar equipment detects these reflections. By scanning a focused microwave beam through space, the radar equipment can determine an object's direction. By measuring the time it takes for the wave to travel there and back, the equipment can determine the object's distance.

CHECK YOUR UNDERSTANDING #1: Staying in One Place

Most communications satellites are placed in geostationary orbits so that they remain above a particular location on the earth's surface. Why is it so important that the satellites maintain fixed positions overhead?

CHECK YOUR FIGURES #1: Shopping for Food

The red light used by many grocery store checkout stations to scan product codes is produced by a helium-neon laser. This light is an electromagnetic wave with a wavelength of approximately 633 nm. What is its frequency?

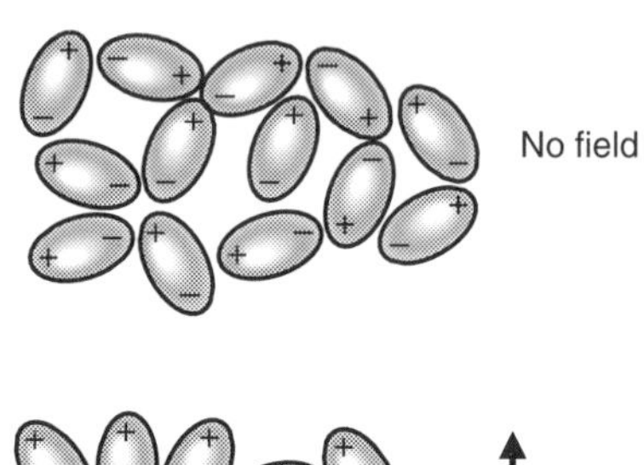

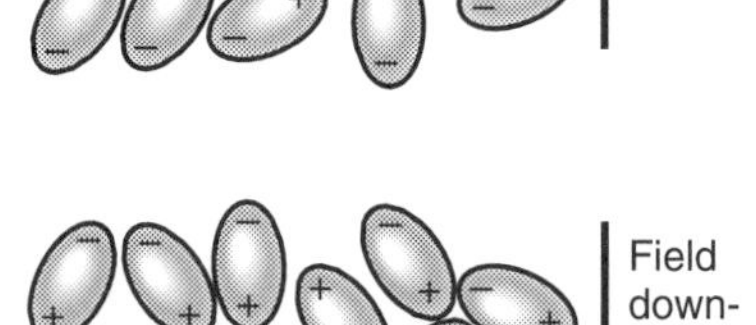

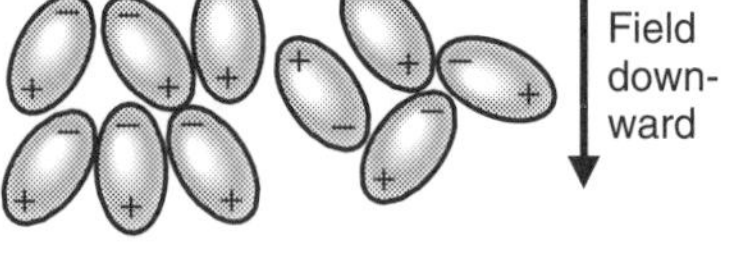

Fig. 14.3.2 - The water molecules in liquid water are randomly oriented when there's no electric field. But a field tends to orient them with their positive ends in the direction of the field.

Heating Food with Microwaves

To understand how a microwave oven heats food, recall that a water molecule is electrically polarized—it has a positively charged end and a negatively charged end. This polar character comes about because of the quantum mechanical structure of the water molecule and the tendency of an oxygen atom to pull electrons away from hydrogen atoms. The water molecule is bent, with its two hydrogen atoms sticking up from its oxygen atom like Mickey Mouse's ears. When the oxygen atom pulls the electrons partly away from the hydrogen atoms, its side of the molecule becomes negatively charged while the hydrogen atoms' side becomes positively charged. Water is thus a *polar molecule.*

In ice, these polar water molecules are arranged in an orderly fashion with fixed positions and orientations. But in liquid water, the molecules are randomly oriented (Fig. 14.3.2). Their arrangements are constrained only by their tendency to form hydrogen bonds with one another.

If you place water in a strong electric field, the water molecule will tend to rotate into alignment with the field. That's because a misaligned molecule has extra electrostatic potential energy and accelerates in the direction that reduces its potential energy as quickly as possible. In this case, the water molecule will experience a torque and will undergo an angular acceleration that makes it rotate into alignment. As it rotates, the molecule will bump into other molecules and convert some of its electrostatic potential energy into thermal energy.

A similar effect occurs at a very crowded party when everyone is suddenly told to face the stage. People brush against one another as they turn and sliding friction converts some of their energy into thermal energy. If the people were told to turn back and forth repeatedly, they would become quite warm. The same holds true for water. If the electric field reverses its direction many times, the water molecules will turn back and forth and become hotter and hotter.

The fluctuating electric field in a microwave is well suited to heating water. A microwave oven uses 2.45 GHz (2.45 gigahertz or 2,450,000,000 Hz) microwaves to flip the orientations of water molecules in food back and forth, billions of times per second. As the water molecules flip, they bump into one another and heat up. The water absorbs the microwaves and converts their energy into thermal energy. This particular microwave frequency was chosen because it was not in use for communications and because it gives the water molecules just about the right amount of time to complete a flip before the field reverses again.

This orientational effect explains why only foods or objects containing water—or other polar molecules—cook well in a microwave oven. Ceramic plates, glass cups, and plastic containers are water-free and usually remain cool. Even ice has trouble absorbing microwave power. The water molecules in ice are constrained by its crystal structure and can't change their orientations. But while ice melts slowly in a microwave, the liquid water it produces heats quickly. This peculiar heating behavior explains why a microwaved frozen burrito often has hot, liquid sections interspersed with cold, frozen sections.

The cooking chamber of a microwave oven has metal walls, which reflect the microwaves and keep them bouncing around inside. Even the oven's window is covered by a reflective metal grid. The reflection is so good that if there's nothing inside the oven to absorb the microwaves, they'll return to the magnetron and eventually cause it to overheat.

While metal surfaces help confine the microwaves inside the oven, cooking your food and not you, extra metal inside the microwave can cause trouble. If you wrap food in aluminum foil, the foil will reflect the microwaves and the food won't cook. However, food placed in a shallow metal dish cooks reasonably well because the microwaves pass through the food before reflecting from the dish.

Metals reflect microwaves because their mobile charges accelerate in electric fields. Sometimes those mobile charges do more than just reflect microwaves. If enough charges are pushed onto the sharp point of a metal twist-tie or scrap of aluminum foil, those charges will jump right into the air as a spark. This spark can start a fire, particularly when the twist-tie is attached to something flammable, such as a plastic or paper bag.

Metals can also become hot all by themselves in a microwave oven. When microwaves push charges back and forth in a metal, the metal experiences an alternating current. If the metal has a large electric resistance, this alternating current can heat it significantly. While thick oven walls and cookware have low resistances and remain cool, thin metal strips may overheat. Metallic decorations on porcelain dinnerware are particularly susceptible to damage in a microwave oven, so that warming up coffee in a metal-rimmed cup is a recipe for almost instant disaster. However, some foods come with special wrappers that conduct just enough current to become very hot in a microwave oven. These wrappers provide the high surface temperatures needed to brown the food.

Another peculiar feature of microwave ovens is that they don't cook evenly. That's because the amplitude of the microwave electric field isn't uniform throughout the oven. As the microwaves bounce around the cooking chamber, they pass through the same spot from several different directions at once. When they do, their electric fields combine with one another. At one location, the individual electric fields may point in the same direction and reinforce one another, so that food there heats up quickly. But at another location, those fields may point in opposite directions and cancel one another. Food at this second spot doesn't cook well at all.

If nothing is moving in the microwave oven, the pattern of microwaves inside it doesn't move either. There are regions in which the electric field has very large amplitudes and regions in which the amplitudes are very small. The larger the amplitude of the electric field, the faster it cooks food.

To avoid uneven heating, you can move the food around as it's cooking. Another solution to this problem is to stir the microwaves around the oven with a rotating metal paddle. The pattern of microwaves inside the chamber changes as the paddle turns and the food cooks more evenly. Still other microwave ovens use two separate microwave frequencies to cook the food. Because these two frequencies cook independently, it's less likely that a portion of the food will be missed altogether.

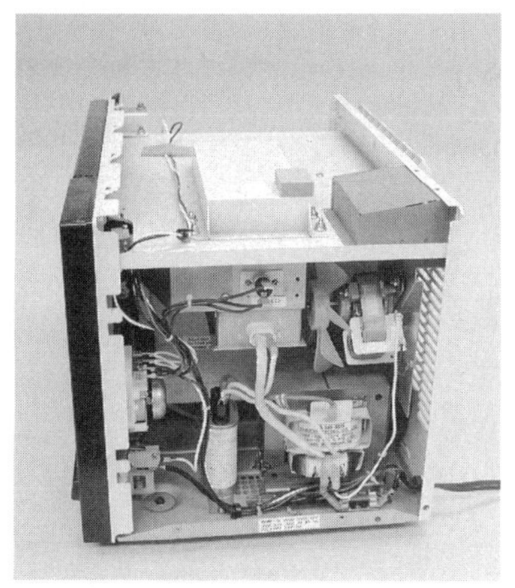

Fig. 14.3.3 - The magnetron of this microwave oven is located in the middle of this picture, just to the left of its cooling fan. Microwaves travel to the cooking chamber through the metal rectangle on top of the oven. The high voltage transformer at the bottom right provides power to the magnetron.

CHECK YOUR UNDERSTANDING #2: Microwave Popcorn

A popcorn kernel contains moist starch trapped inside a hard, dry hull. You can scorch this hull by cooking the corn in hot oil but not when you cook it in a microwave oven. How can the microwave oven pop the corn without risk of overheating the hull?

Creating Microwaves with a Magnetron

Clearly, changing electric fields cook the food as microwaves bounce around the inside of an oven. But how are these microwaves created? A reasonable guess would be that the oven creates an alternating current at 2.45 GHz and that this current causes charge to slosh in a tank circuit and move up and down an antenna. While this guess is not quite right, it highlights some of the problems involved in making microwaves.

First, creating an alternating current at 2.45 GHz is difficult. Transistor-based devices work easily below 250 MHz but struggle with alternating currents above 1 GHz. The wavelength of a 2.45 GHz microwave is only 12.2 cm, meaning that light itself only travels about 12.2 cm per cycle. Since electric currents can't go faster than light, a transistor device that creates a 2.45 GHz alternating current would have to be much smaller than 12.2 cm in diameter. Such small devices usually can't handle much power and a microwave oven needs about 700 W.

Second, a microwave antenna must be very short. If an antenna is longer than a quarter of the wavelength of the electromagnetic wave it emits, current doesn't have time to flow all the way to the end of the antenna before it has to turn around. The current ends up flowing in opposite directions at various points on the antenna and emits waves that partially cancel one another. To avoid such cancellations, radio and microwave transmitters use antennas that are a quarter as long as the wavelength of their electromagnetic waves. While AM radio antennas can be huge, microwave antennas must be very small.

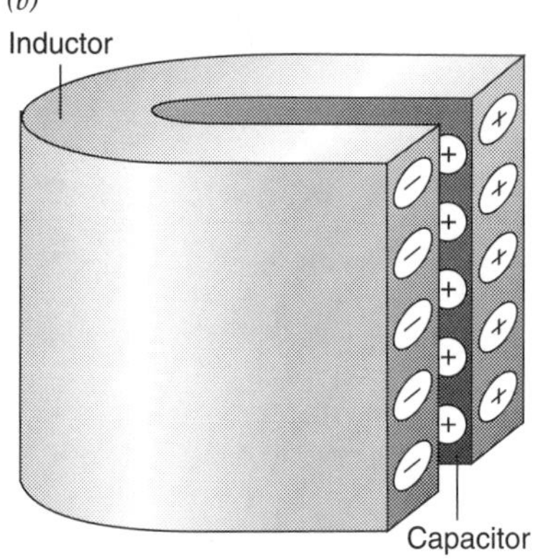

Fig. 14.3.4 - A tank circuit consists of an inductor and a capacitor. *(a)* At radio frequencies, the inductor is a coil of wire and the capacitor has large, closely spaced plates. *(b)* At microwave frequencies, the inductor is simply the curve in a C-shaped metal strip and the capacitor is just the two tips of that strip.

A microwave oven addresses these problems with a special vacuum tube called a **magnetron** (Fig. 14.3.3). A magnetron uses beams of electrons to make charge slosh in several microwave tank circuits. These tank circuits have resonant frequencies of 2.45 GHz, the operating frequency of the oven. With the help of a tiny antenna, the magnetron emits the microwaves that cook the food.

The microwave tank circuits are arranged in a circle and form the magnetron's outer edge. For one of these tank circuits to oscillate naturally at 2.45 GHz, its capacitor must have an extremely small capacitance and its inductor must have an extremely small inductance. The answer to this requirement is a C-shaped strip of metal (Fig. 14.3.4). Its curved side is the inductor and its opened side is the capacitor, making this strip a microwave tank circuit.

Electric charge sloshes back and forth on this C-shaped strip just as it does in a more conventional tank circuit. Let's follow its motion in Fig. 14.3.5, starting with separated charge at the capacitor side of the strip (Fig. 14.3.5*a*). Charge begins to flow around the strip, from the positively charged tip to the negatively charged tip (Fig. 14.3.5*b*). This current produces a magnetic field in the inductor side of the strip and this field grows in strength until the separated charge on the capacitor side is completely gone (Fig. 14.3.5*c*).

The inductor side of the strip stores energy in its magnetic field and opposes changes in the strip's current. The inductor side even propels current through the strip after the initial charge separation on the capacitor side has disappeared (Fig. 14.3.5*d*). The magnetic field gradually goes away, but not before separated charge is back on the capacitor side of the strip—upside-down (Fig. 14.3.5*e*). The whole process then begins again, in reverse.

The charge in the C-shaped metal strip sloshes back and forth in a natural resonance. It behaves as a harmonic oscillator, with a period that doesn't depend on the amount of charge that's sloshing. Because of its steady, rhythmic behavior, a microwave tank circuit is often called a **resonant cavity** or **resonator**.

The magnetron of a microwave oven typically contains eight of these resonators, each carefully adjusted in size and shape so that its natural resonance occurs exactly at 2.45 GHz. These resonators are arranged in a ring and each one shares its tips with the tips of the adjacent resonators (Fig. 14.3.6).

Because they share ends, these resonators tend to oscillate alternately. At the start of an oscillatory cycle, half the metal tips are positively charged and half are negatively charged (Fig. 14.3.6*a*). Currents begin to flow through the ring and produce magnetic fields in the resonant cavities (Fig. 14.3.6*b*). These magnetic fields propel the currents around the ring even after the charge separations have vanished. Soon the charge separations reappear but with the positive and negative tips interchanged (Fig. 14.3.6*c*).

These currents oscillate back and forth at 2.45 GHz, filling the magnetron with alternating electric and magnetic fields. But copper isn't a perfect conductor and the magnetron's copper resonators gradually turn their energy into heat. To make up for this energy loss and to replace energy that's extracted to cook the food, something must send power to the resonators. This power is provided by streams of electrons.

The resonators form the outer portion of a magnetron vacuum tube. At the center of this tube, surrounded only by empty space, is an electrically heated cathode (Fig. 14.3.7*a*). A high voltage power supply pumps negative charge onto this cathode so that a strong electric field points toward it from the positively charged resonator tips. If there were no other fields present in the magnetron, negatively charged electrons would emerge from the hot cathode and accelerate toward the positively charged tips as four beams of electrons (Fig. 14.3.7*b*).

However, the magnetron also contains a large permanent magnet. Why else would it be called a magnetron? This magnet creates a strong, steady magnetic field inside the ring. This field points out of the paper in Fig. 14.3.7*c*, meaning that a north magnetic pole would accelerate toward you as you looked at the figure. If there were no other fields inside the resonator, electrons would circle the magnetic field lines just as they do in the focusing coil of a television picture tube. This cyclotron motion would carry the electrons around the cathode in counter-clockwise loops and they would never go near the resonators.

But in a real magnetron, the electric field of Fig. 14.3.7*b* and the magnetic field of Fig. 14.3.7*c* are both present at once. Because both of these fields exert

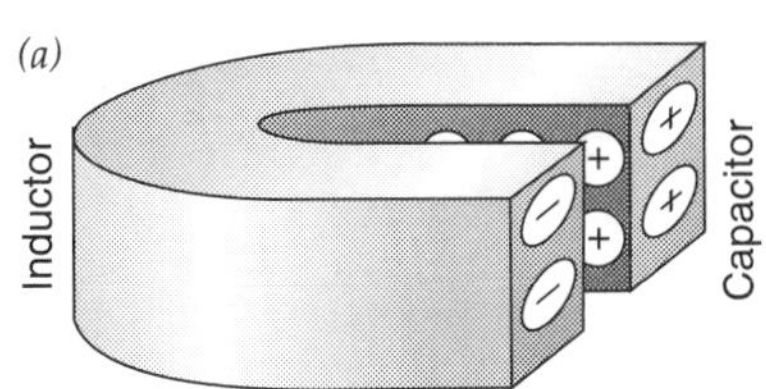

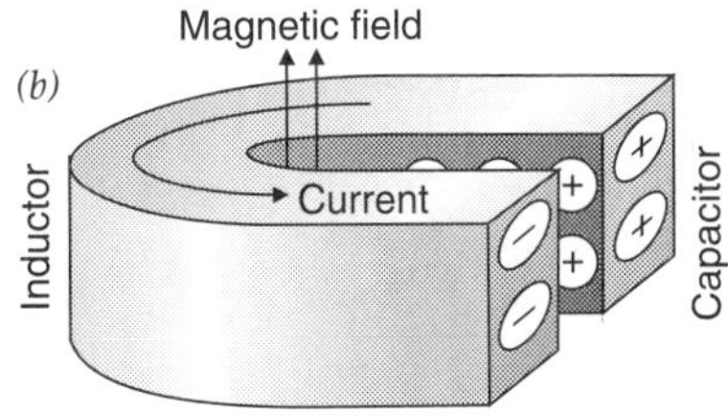

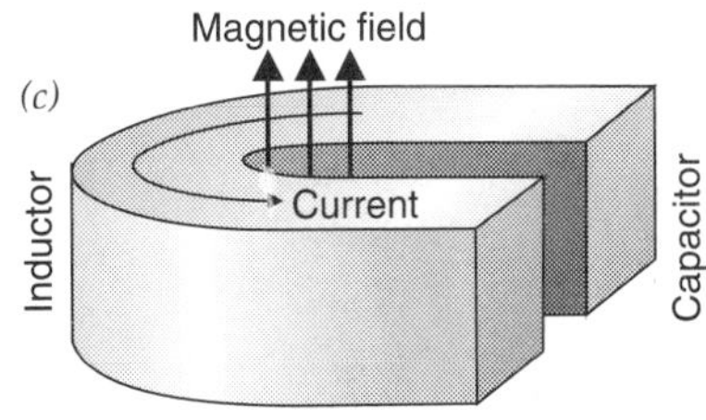

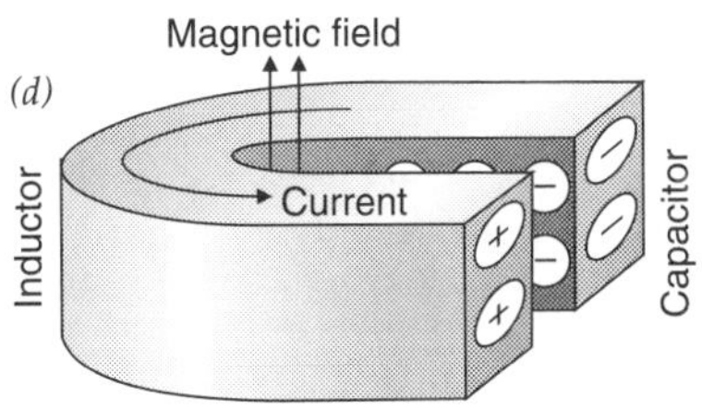

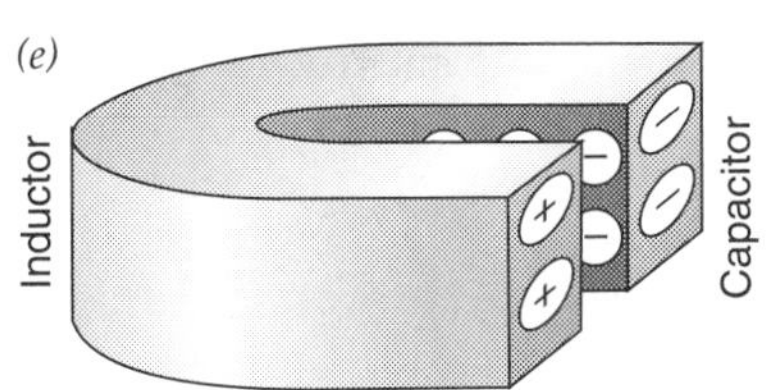

Fig. 14.3.5 - A curved metal strip is a microwave tank circuit. Energy moves back and forth between the capacitor (the ends of the strip) and the inductor (the curve of the strip) as current flows around the strip.

forces on moving electrons, the paths the electrons follow are extremely complicated (Fig. 14.3.7*d*). The outward directed and circulating motions merge together into four electron beams which arc outward and rotate counter-clockwise, like the spokes of a spinning bicycle wheel. An electron beam reaches each resonator not at its positively charge tip, as it would without the magnetic field, but at its negatively charged tip. The electron beams actually add to the charge separations in the resonators!

(a)

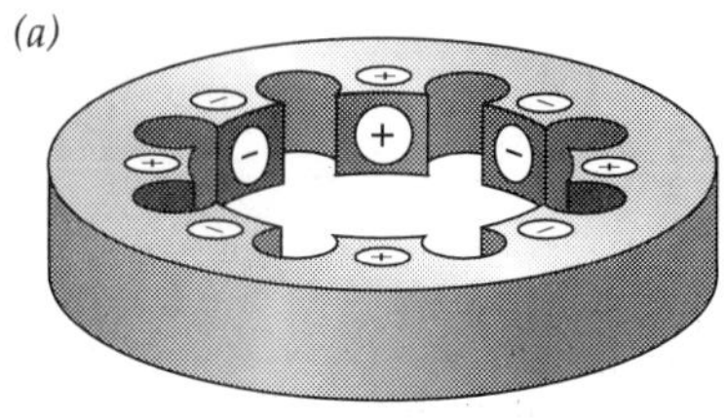

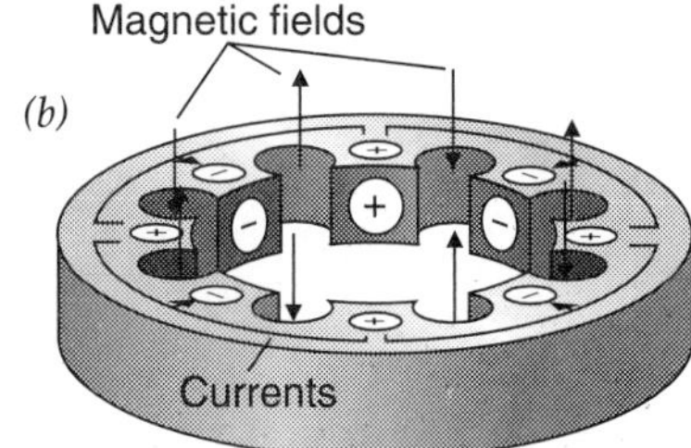

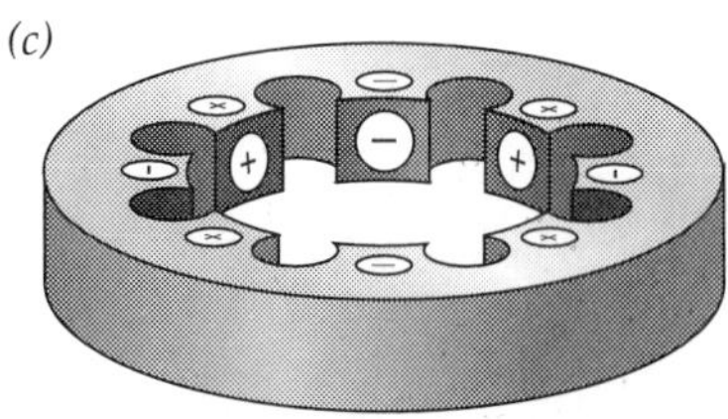

Fig. 14.3.6 - A typical magnetron has eight C-shaped resonators connected together and wrapped around in a ring. Separated charge on the tips (*a*) flows as currents through the ring (*b*) and becomes reversed (*c*). As the currents flow, magnetic fields appear in the eight cavities between the tips, pointing alternately up and down.

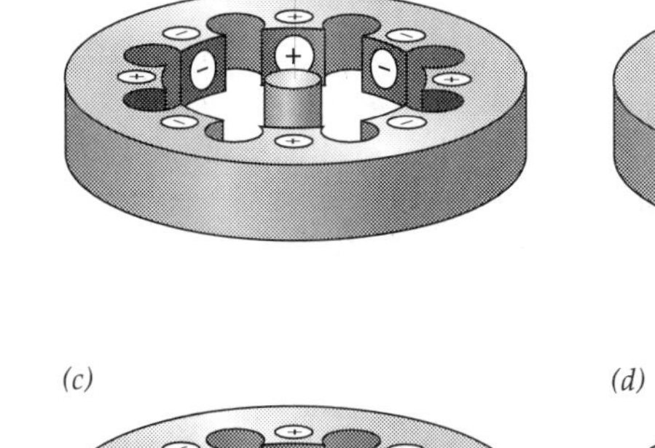

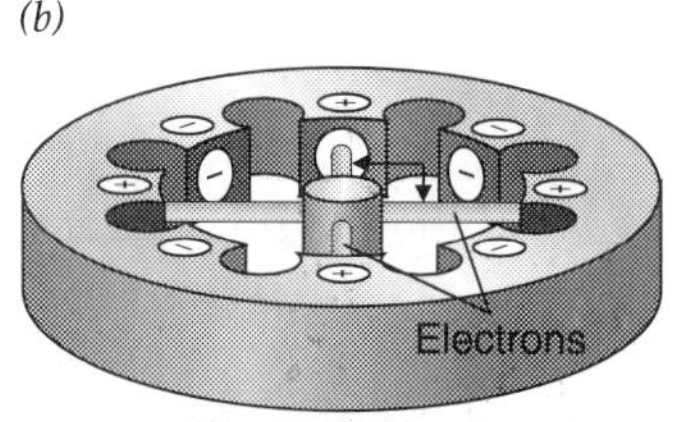

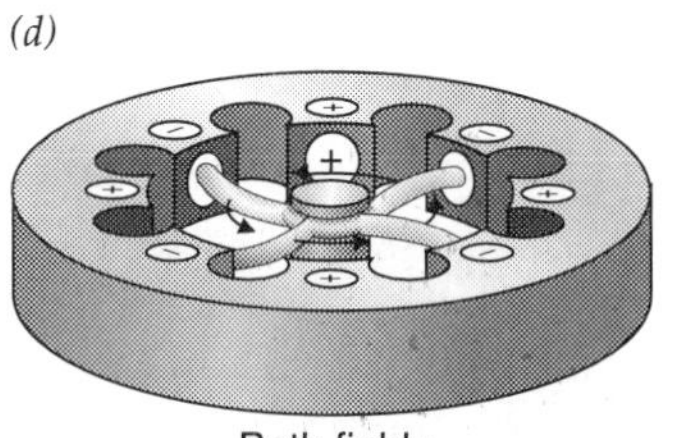

Fig. 14.3.7 - (*a*) Electrons are emitted by the hot cathode in the center of a magnetron's ring of resonators. (*b*) Electric fields alone would accelerate these electrons toward the positively charged resonator tips. (*c*) A magnetic field alone (pointing upward) would make the electrons orbit the cathode in counter-clockwise loops. (*d*) Together, these fields create spoke-like electron beams that circle the cathode counter-clockwise and always strike negatively charged tips of the resonators.

The electron beams turn in perfect synchronization with the oscillating charge on the resonators. The beams rotate from one tip to the next in the same amount of time it takes for the charge separation on the tips to reverse. As a result, the beams always arrive on the negatively charged tip. By adding to the charge separations, the electron beams provide power to the oscillations in the resonators, keeping them going and allowing them to transfer power to the food. The electron beams actually initiate the oscillation in the resonators by adding energy to tiny accidental oscillations that are always present in electric systems.

But how does the oscillating charge inside the magnetron create microwaves inside the oven's cooking chamber. There are many ways to extract microwaves from the ring of resonators. One extraction method is to insert a single-turn wire coil into one of the magnetron's cavities. As the magnetic field in that cavity changes, it induces 2.45 GHz alternating current in the coil. One end of this coil is attached to the ring but the other end passes out of the magnetron through an insulated, air-tight hole in the ring and connects to a small antenna. This antenna emits microwaves into a metal pipe attached to the cooking chamber. These microwaves reflect their way through the pipe and into the cooking chamber, where they cook the food.

CHECK YOUR UNDERSTANDING #3: Large Economy Size

If a manufacturer made a magnetron that was slightly larger than normal in every dimension, how would that magnetron behave?

Summary

How Microwave Ovens Work: A microwave oven uses a magnetron to create microwaves. These microwaves bounce around the cooking chamber, where they transfer energy to water molecules in the food. Because a water molecule is polar, having a positive end and a negative end, it tends to align with an electric field. The microwave's electric field causes the tightly packed water molecules to reverse their orientations rapidly and the ensuing collisions heat them up.

The magnetron creates microwaves when electrons flow from a hot cathode to the tips of a ring of resonators surrounding that cathode. A permanent magnet shapes the flow into four electron beams that circle the cathode like the spokes of a bicycle wheel. These electron beams provide separated charge to the resonators. This separated charge oscillates back and forth on the resonators at 2.45 GHz and is used to create the microwaves.

Important Laws and Equations

1. Relationship Between Wavelength and Frequency: The product of the frequency of an electromagnetic wave times its wavelength is the speed of light, or

$$\text{speed of light} = \text{wavelength} \cdot \text{frequency}\,. \tag{14.3.1}$$

The Physics of Microwave Ovens

1. The electric field in an electromagnetic wave exerts forces on charged particles. In conducting materials, these charged particles accelerate and tend to reflect the electromagnetic wave.

2. The electric field in an electromagnetic wave exerts torques on polar molecules. These molecules tend to align with the electric field.

3. An electromagnetic wave can't pass through holes in a conducting surface if those holes are significantly smaller than its wavelength.

4. An antenna is usually a quarter as long as the wavelength of the electromagnetic wave it emits. If it's longer than that, it will emit several waves that will partially cancel one another.

Check Your Understanding - Answers

1. So that satellite dishes don't have to move in order to receive the transmissions.

Why: To focus the microwaves from a communications satellite onto its antenna, a satellite dish must be pointed directly at that satellite. If the satellite were to move about overhead, the satellite dish would have to rotate to follow it.

2. The microwave oven transfers heat to water molecules in the starch so that the hull never becomes hotter than the material inside it.

Why: A corn kernel cooked in oil is heated by contact with the hot oil and pot. You can easily overheat the outer hull and burn it. But microwaves transfer heat to the water molecules inside the kernel. The hull can't overheat because the hottest thing it touches is the starchy insides of the kernel. When the pressure of steam inside the kernel becomes high enough, the hull breaks and the kernel "pops".

3. It would operate at a frequency below 2.45 GHz.

Why: The frequency of the microwaves emitted by a magnetron is determined exclusively by the natural resonances of its resonators. If those resonators are enlarged, both the inductances of their curved sides and the capacitances of their open sides will increase. Their resonant frequencies will decrease and the magnetron will emit lower frequency microwaves.

Check Your Figures - Answers

1. About $4.74{\cdot}10^{14}$ Hz.

Why: Since the product of frequency and wavelength for an electromagnetic wave is equal to the speed of light, its frequency is equal to the speed of light divided by its wavelength:

$$\frac{299{,}790{,}000\ \text{m/s}}{0.000000633\ \text{m}} = 4.74 \cdot 10^{14}\ \text{Hz}\,.$$

Glossary

magnetron A vacuum tube that produces microwaves at a single frequency established by the natural resonances of its resonant cavities.

microwaves Electromagnetic waves with wavelengths between about 1 meter and 1 millimeter.

resonant cavity A hollow electric structure in which the fields and currents exhibit natural resonances. There are also acoustic resonant cavities in which air pressure exhibits natural resonances.

resonator A structure that exhibits natural resonances.

Review Questions

1. What role does the speed of light play in relating the frequency of an electromagnetic wave to its wavelength?

2. What distinguishes a microwave from a radio wave?

3. What determines whether or not a radio wave can pass through a particular hole in a metal surface?

4. Why does water respond to and absorb microwaves while oil doesn't?

5. Why don't the water molecules in ice absorb microwaves?

6. Why does a metal spoon remain cool in a microwave oven while the metallic trim on a greeting card ignites the card?

7. What determines the frequency of the microwave that a magnetron produces?

8. Why does charge slosh back and forth so much more rapidly in the resonant cavity of a magnetron than it does in the tank circuit of a commercial AM radio transmitter?

Exercises

1. Why do many microwave ovens have turntables?

2. Why are most microwave TV dinners packaged in plastic rather than aluminum trays?

3. Porous, unglazed ceramics can absorb water and moisture. Why are they unsuitable for use in a microwave oven?

4. The Empire State Building has several television antennas on top, added in part to increase its overall height. These antennas aren't very tall. Why do short antennas, located high in the air, do such a good job of transmitting television?

5. To pop popcorn on a stovetop, you must add oil to the pot to help transfer heat to the kernels. But you can pop popcorn directly in a microwave oven simply by putting the corn in a paper bag. Why is there no need to add oil to the bag?

6. What won't a magnetron work without its magnet?

7. Why can't microwaves get through the metal screening on the window of a microwave oven?

8. If you tried to communicate with a friend using radios that operated at 2.45 GHz, you'd probably hear lots of noise instead of your friend's voice. Explain.

9. Why is it so important that a microwave oven turn off when you open the door?

10. Why can operating a microwave oven for a long time with only salad oil inside it damage the magnetron?

11. Why must a microwave oven heat the filament in its magnetron?

12. The high-voltage power supply that pumps electrons from the magnetron's resonant cavities to its filament provides a great deal of power. What becomes of all that power?

13. Why can a slight dent or deformation in the outer portion of a magnetron stop it from producing microwaves?

14. Dish-shaped reflectors are used to steer microwaves in order to establish communications links between nearby buildings. Those reflectors are often made from metal mesh. Why don't they have to be made from solid metal sheets?

15. When you're listening to FM radio near buildings, reflections of the radio wave can make the reception particularly bad in certain locations. Compare this effect to the problem of uneven cooking in a microwave oven.

16. Like any harmonic oscillator, a resonant cavity in a magnetron has two forms for its energy and that energy goes back and forth between the two forms. What are those forms?

17. Why do magnetrons always have even numbers of resonant cavities in the rings surrounding their filaments?

Problems

1. The frequency of the radio wave emitted by a cordless telephone is 900 MHz. What is the wavelength of that wave?

2. Citizens band radio (CB) uses radio waves with frequencies near 27 MHz. What are the wavelengths of these waves and how long should a quarter-wavelength CB antenna be?

3. The electromagnetic waves in blue light have frequencies near $6.5 \cdot 10^{14}$ Hz. What are their wavelengths?

4. Amateur radio operators often refer to their radio waves by wavelength. What are the approximate frequencies of the 160 m, 15 m, and 2 m wavelength amateur radio bands?

5. The radio waves used by cellular telephones have wavelengths of approximately 3.6 m. What are their frequencies?

Epilogue for Chapter 14

This chapter examined three common devices that are based on electromagnetic waves. In *radio*, we saw that electromagnetic waves can be created by accelerating electric charge and that these waves can be detected by looking for their effects on other electric charges. We also examined the techniques that are used to send audio signals through space by having them control either the amplitude or the frequency of electromagnetic waves.

In *television*, we looked at how electric and magnetic fields create, focus, and steer beams of electrons toward the screen of a picture tube. We studied the dot by dot way in which a picture is represented and examined the ways in which those dot by dot signals are broadcast via electromagnetic waves.

In microwave ovens, we explored the ways in which electromagnetic waves can interact directly with polar water molecules and can transfer energy to those molecules. We looked at the need for small antennas when dealing with short wavelength electromagnetic waves and how a magnetron uses microwave tank circuits and beams of electrons to create those waves.

Explanation: Boiling Water in an Ice Cup

The water molecules in ice are held in place rigidly by the ice's crystalline structure. Unable to move or turn, these water molecules can't follow the microwave radiation's rapidly changing electric field. As a result, ice is unable to absorb much energy from the microwave radiation.

In contrast, the water molecules in liquid water can move about relatively freely and turn back and forth with the changing electric field of the microwaves. These water molecules absorb energy from the microwave radiation and become hotter. The liquid water's temperature rises.

While heat tends to flow out of the hot water and into the ice, water is not a good conductor of heat. Convection stops working once the water temperature rises above 4° C and the hotter water begins to float on the colder water beneath it. In favorable circumstances, the water can begin to boil before it has melted the ice cup that contains it.

Cases

1. Microwave ovens allow for some interesting cooking and funny disasters.

a. Baked Alaska is a dessert in which a hot, baked meringue contains cold, frozen ice cream. The reverse is Frozen Florida, a dessert in which cold, frozen meringue contains boiling hot liqueur. Frozen Florida is prepared by taking a frozen meringue ball (cooked egg whites) containing liquid liqueur (a water-alcohol mixture that remains liquid at low temperature) out of the freezer and putting it in a microwave oven briefly. Why does the liqueur get hot while the meringue remains frozen?
b. If you try to cook an egg in a microwave oven, the egg may explode. Compare this result to the process of cooking popcorn in a microwave oven.
c. Some prepared foods come with browning sheets that contain very thin metallic layers. These sheets become hot in a microwave oven and help to brown the surface of the food. Why does a thin metallic film become so hot?
d. If you wrap a piece of food in sturdy metal screening with holes about 2 millimeters on a side, the food won't cook in a microwave oven. Why can't the microwaves get through the screen's holes the way light can?
e. A microwave oven has air vents and a fan to cool the magnetron tube. One reason that the tube gets hot is that heat flows out of the electrically heated cathode at the center of the tube. But there are at least two other important ways in which the tube's operation generates heat. What are they?

2. As a DJ at a very small local radio station, you often find yourself involved in technical issues for the station's two channels, one of which is AM at 1020 kHz and the other of which is FM at 89.5 MHz. The station is in the process of building a new transmitting system.

a. To save money, the director wants to use a single antenna for both channels. You warn him that the AM channel needs a taller antenna than the FM channel. Why is that true?

b. The director had planned to put the antennas next to the station, which is in a valley at the base of a small mountain. You suggest putting them at the top of the mountain, despite the extra cost of wires. Why is altitude important, particularly for the FM antenna?

c. A suggestion to orient the antennas horizontally is quickly dismissed as non-traditional. But it wouldn't work well, either. List two reasons why a horizontal antenna would produce a radio signal that most people would find hard to receive with their radios.

d. The AM channel must be careful not to "overmodulate" the radio wave during very loud passages because it distorts the sound people hear in their radios. You explain this effect as due to moments when the transmitter actually turns itself completely off. Why would the transmitter stop transmitting any wave at all?

e. The FM channel must also avoid overmodulation during loud passages because it will get in trouble with the FCC. Other FM stations in your area will also be angry with your station for spoiling the reception of their transmissions. How can your FM station affect those other FM stations when they operate at different carrier wave frequencies?

3. Your garage door opener is triggered by a radio transmitter. You press the button in your car to open the door.

a. When you press the button, the transmitter moves charge onto and off a small antenna. Why doesn't it just put charge on the antenna and leave the charge there?

b. If you're close enough to the garage door, it will open. Why must the transmitter and receiver be close together?

c. When you press the button of your transmitter, the radio in your car still sounds normal. Why doesn't the transmitter interfere with radio reception in your car?

d. In the early days of garage door openers you could drive around town opening other people's garages. To make things more secure, the transmitters now send codes to the receivers using AM modulation. Each transmitter turns itself on and off quickly in a pattern that's recognized only by the proper receiver. Graph the charge on a transmitter's antenna as a function of time during such an AM transmission.

4. When you talk on a cellular telephone, two radio waves connect you with the telephone network. Your telephone transmits one of these waves and the other is sent to you by a stationary base unit. Base units are located a few kilometers apart and each one provides service to the telephones in its vicinity—its "cell." A connection between a telephone and a base unit occupies one of 832 channels currently allocated to cellular communications. Each channel corresponds to radio waves of two frequencies: a lower frequency wave (in the range 825.03–849.96 MHz) that's emitted by the telephone and a higher frequency wave (in the range 870.03–894.96 MHz) that's emitted by the base unit. The waves of different channels are spaced 0.030 MHz (30 kHz) apart and those of your telephone and its base unit are 45 MHz apart.

***a.** When your telephone and its base unit are connected on Channel 1, your telephone's wave has a frequency of 825.03 MHz and its base unit's wave has a frequency of 870.03 MHz. What are the wavelengths of those two waves?

***b.** What is the proper length for the telephone's antenna if it's to have a natural resonance at the frequency of the radio wave it's transmitting?

c. Why will there be a problem if 900 people in a conference center all try to use their cellular telephones at once?

d. A base unit's antenna is oriented vertically because orienting it horizontally would leave people at certain angles from the base unit unable to receive its radio wave. Why?

e. Because the base unit's antenna is vertical, your telephone's antenna should also be vertical. Why?

f. If you're in a valley or a metal building, your telephone may lose contact with the base unit. Why?

g. Your telephone modulates its carrier wave in order to send your voice to the base unit. Why must it be careful not to produce sideband waves that have frequencies that are more than 15 kHz away from that of its carrier wave?

h. If you talk while traveling, a computer system makes sure that your telephone is always communicating with the nearest base unit. Why is it important that every telephone and base unit transmit the weakest radio waves that still permit reliable communication between them?

5. The Global Positioning System or GPS is a system of earth-orbiting satellites that provide position information to anyone with a GPS radio receiver. Each satellite transmits two microwaves, one at 1.57542 GHz and the other at 1.2276 GHz. These waves are modulated by carefully timed pulses so that by measuring exactly when it receives those pulses from four or more satellites, a GPS receiver can determine its position on earth to within 100 m. Recent changes in U.S. Government policy will allow GPS receivers to locate their positions even more accurately over the next few years.

a. Because each satellite carries a time-standard cesium atomic clock, its pulses are emitted at precisely known times. The GPS receiver knows exactly when it receives the pulses, so it can tell exactly how far it is from the satellite. How?

b. Because a GPS receiver only receives microwaves from satellites that are above the horizon, there must be at least 17 satellites in orbit at once. Why can't the receiver detect satellites that are below the horizon?

***c.** What are the wavelengths of the two microwaves?

***d.** One of the reasons for using microwaves in the GPS is that microwaves can be received by small antennas with well-defined locations. How long should an antenna be to receive the lower frequency GPS microwave?

e. Geologists are using the GPS to study motions of the earth's crust that are as small as a few millimeters. To do this, they actually study the electric and magnetic fields of the microwaves. Why would two GPS receivers located a few centimeters apart detect somewhat different electric and magnetic fields if they studied the microwave from one particular satellite at precisely the same moment?

6. A cordless microphone uses a 925 MHz carrier wave to carry sound information from a performer to the audio system of a theater.

a. The tank circuits in the microphone and in the theater's receivers have been carefully tuned to precisely the same resonant frequencies. Why is this adjustment important?

b. The frequency of the waves emitted by the microphone include waves with frequencies slightly above and below 925 MHz. Explain.

c. Because the performer moves and tips the microphone during the performance, the theater uses two receivers at different locations. This ensures that the radio wave from the microphone is always strong at one of the receivers. How is it possible for the wave to be weak at one of the receivers?

d. Each receiver has two antennas oriented at right angles to one another. Why does this ensure that the receiver can detect the radio wave regardless of how the microphone's antenna is tipped?

CHAPTER 15

LIGHT

While radio waves and microwaves are useful for communications and energy transfer, there is another portion of the electromagnetic spectrum that we find far more important: light. Light consists of very high frequency, very short wavelength electromagnetic waves. Light's frequencies are so high that it can't be handled by normal antennas. Instead, it's absorbed and emitted by the individual charged particles in atoms, molecules, or materials. Because of its special relationship with the charged particles in matter, light is important to physics, chemistry, and materials science. Moreover, it is one of the principal ways by which we interact with the world around us.

Experiment: Splitting the Colors of Sunlight

We see light because it stimulates cells in our eyes. This stimulation is an example of light's ability to influence chemistry. And because our eyes are able to distinguish between different wavelengths of light, we perceive colors. While sunlight normally appears uncolored, that's because it contains a broad mixture of wavelengths that our eyes interpret as whiteness. But there are situations in which sunlight becomes separated into its constituent colors.

You can observe this separation of colors by looking at sunlight passing through a cut crystal glass or bowl. Hold the cut crystal object in direct sunlight and observe the light that it redirects toward your eyes or projects onto a white sheet of paper nearby. While some of the light you see will still be white, you should see colors as well.

Turn the object slowly in your hand and observe how the colors change.

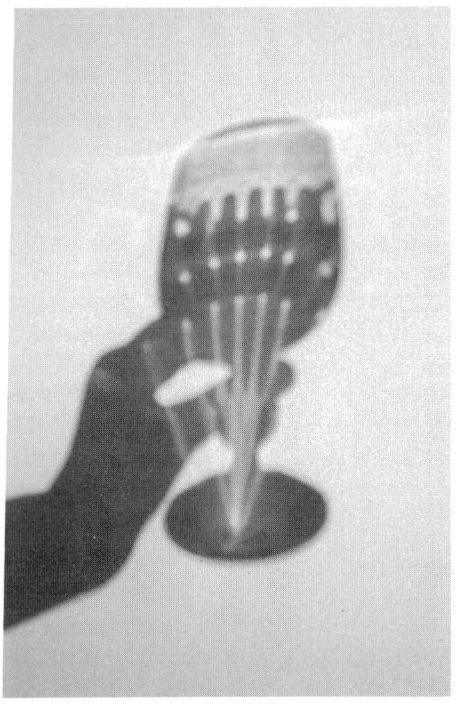

You will see gradual progressions from one color to the next. **Predict** the order of colors you will see and then **observe** the actual sequence. Were your predictions **verified**? Is the sequence of colors always the same? How does this sequence relate to the colors of the rainbow? What is the relationship between this sequence and the wavelengths of light?

Chapter Itinerary

In this chapter, we'll examine three sources of light: (1) *sunlight*, (2) *fluorescent lamps*, and (3) *lasers*. In *sunlight*, we'll see how sunlight travels to our eyes and how its passage through the atmosphere, raindrops, and soap bubbles can separate it into its constituent colors. In *fluorescent lamps*, we'll explore the ways in which atoms and molecules emit and absorb light, and how different atoms and molecules can be used to produce light of different colors. In *lasers*, we'll look at how atoms and molecules can duplicate or amplify the light passing through them and thus produce intense beams of highly ordered light.

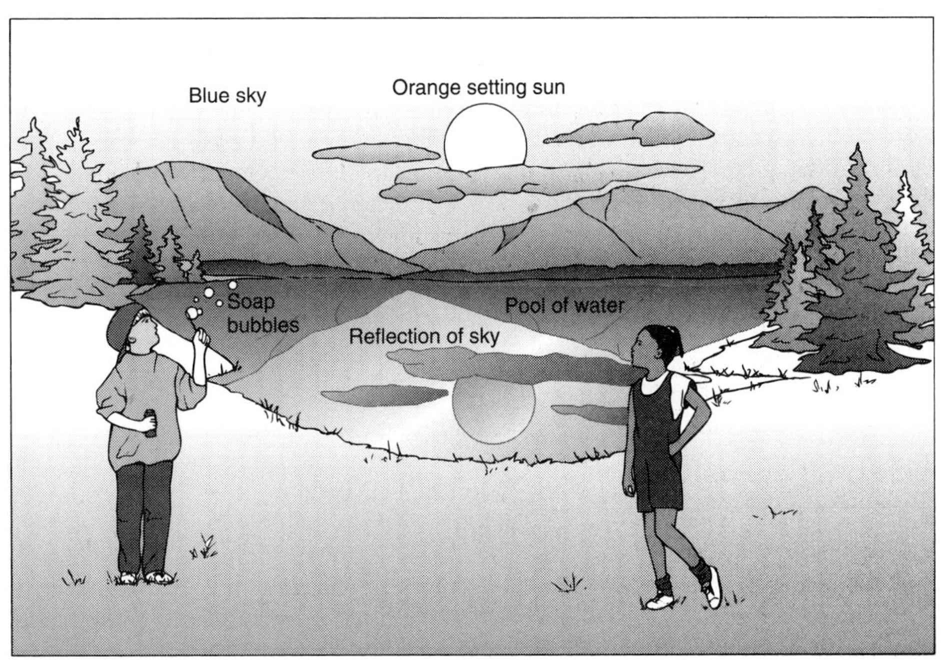

Section 15.1

Sunlight

For thousands of years, people have marked the passage of time by the rising and setting of the sun over the horizon. The sun first appears as a red disk in the east every morning, rises white in the blue sky, and then sets once again as a red disk in the west. The sunlight that we see takes about 8 minutes to travel the 150,000,000 km from the sun to our eyes and provides most of the energy and heat that make life on earth possible.

In this section, we'll examine sunlight to see how it interacts with the world around us. At a basic level, light is just another electromagnetic wave and could be considered part of the previous chapter. However, we can see light with our own eyes and it's so important to our activities that it deserves a chapter of its own.

Questions to Think About: *Why is the sun red at sunrise or sunset? Why is the sky blue during the day? Why can you occasionally see sunbeams when sunlight enters an otherwise dark room through a small window? Why do we see colors when sunlight passes through cut crystal or a soap bubble? Why do polarized sunglasses reduce glare and darken the sky?*

Experiments to Do: *Sunlight is actually composed of many different electromagnetic waves. These waves differ in frequency and wavelength, just as the radio waves from your two favorite stations differ in frequency and wavelength. But you don't need a machine to help you distinguish between various wavelengths of light, you can use your eyes. Take a look at a soap bubble on a bright, sunny day. You see the bubble because it reflects light. The bubble is colored, even though the sunlight hitting it is white and the soap bubble itself is clear. The bubble separates the sunlight according to wavelength and sends only certain wavelengths toward you.*

Sunlight and Electromagnetic Waves

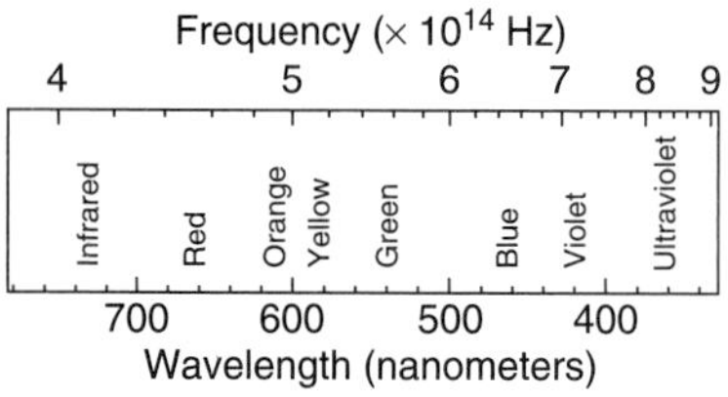

Fig. 15.1.1 - The visible portion of the spectrum of sunlight. Each wavelength of visible light has a particular frequency and is associated with a particular color. At the ends of the visible spectrum are invisible infrared and ultraviolet lights.

Electromagnetic waves can have any wavelength from thousands of kilometers to a tiny fraction of the size of an atomic nucleus. The radio waves and microwaves that we examined in the previous chapter have wavelengths longer than 1 mm. In this chapter, we'll turn our attention to shorter electromagnetic waves with wavelengths between 400 nm and 750 nm (recall that 1 nm, one nanometer, is a billionth of a meter). These are the electromagnetic waves that we perceive as **visible light** and the principal components of sunlight.

Because the electromagnetic waves in sunlight have such short wavelengths, their frequencies lie between 10^{14} Hz and 10^{15} Hz (Fig. 15.1.1). As one of these waves of sunlight passes by, its electric field fluctuates back and forth almost 1,000,000,000,000,000 times each second. Since producing microwaves, which have much longer wavelengths and much lower frequencies, already requires specialized components and tiny antennas, what can possibly emit or absorb light waves? The answer is the individual charged particles in atoms, molecules, and materials. These tiny particles can move extremely rapidly, often vibrating about at frequencies of 10^{14} Hz, 10^{15} Hz, or even more. As these charged particles accelerate back and forth, they emit light waves. Similarly, passing light waves cause individual charged particles in atoms, molecules, and materials to accelerate back and forth, thereby absorbing the light waves as well.

Sunlight originates at the outer surface of the sun, in a region called the *photosphere*. There atoms and other tiny charged systems (mostly atomic ions and electrons) jostle about at 5800° C. Since these charged particles accelerate as they bounce around, they emit electromagnetic waves.

Because the sun's surface emits light through the random, thermal motions of its charged particles, the distribution of wavelengths it emits is determined only by its temperature. It emits a black body spectrum, like the incandescent light bulbs we discussed in Section 6.3. Because the photosphere's temperature is 5800° C, the jostling motions are extremely rapid and most of the sunlight lies in the visible portion of the electromagnetic spectrum (Fig. 15.1.2).

However, not all sunlight is visible. On the long wavelength, low frequency side of visible light is **infrared light**. We can't see infrared light with our eyes but we feel it when we stand in front of a hot object. In sunlight, infrared light is produced by charges that are accelerating back and forth more slowly than average. On the short wavelength, high frequency side of visible light is **ultraviolet light**. We can't see ultraviolet light, either, but we are aware of its presence because it induces chemical damage in molecules. It causes sunburns and encourages skin to tan. In sunlight, ultraviolet light is produced by charges that are accelerating back and forth more rapidly than average.

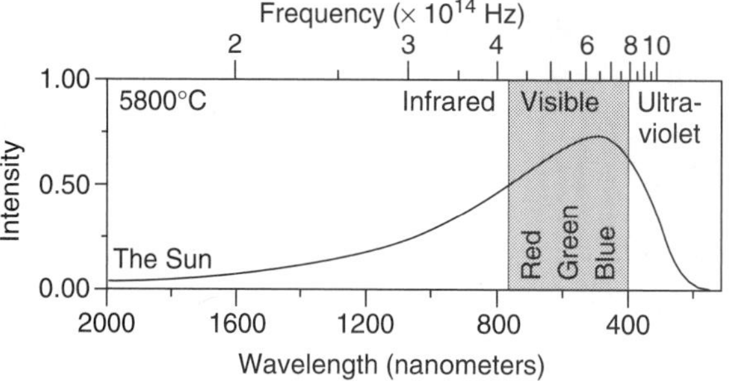

Fig. 15.1.2 - Sunlight comes from the sun's photosphere, where the temperature is 5800° C. This light has a black body distribution of wavelengths, with much of its intensity concentrated in the visible portion of the overall electromagnetic spectrum.

CHECK YOUR UNDERSTANDING #1: The Rosy Glow of Candlelight
Why does a burning candle emit reddish or yellowish light?

Sunlight's Passage to the Earth

Sunlight travels from the sun to the earth at the speed of light. But what sets the speed of light? Actually, it's one of the fundamental constants of nature, with a value of 299,792,458 m/s in empty space. While one could argue that the speed of light is set by the relationships between the electric and magnetic fields, that observation simply passes the buck. If you were then to ask what sets the relationships between the electric and magnetic fields, the answer would be the speed of light.

Rather than justifying why sunlight travels as fast as it does in empty space, let's look at what happens to it when it enters a region that's not empty. After all, sunlight eventually reaches the earth's atmosphere and, when it does, several interesting things happen.

First, the sunlight slows down as its electric and magnetic fields begin to interact with the electric charges and magnetic poles in the atmosphere. Light polarizes the molecules it encounters, a process that delays its passage and reduces its speed. Since most transparent materials respond much more strongly to light's electric field, we can safely neglect any magnetic effects. Thus as light passes through air or any other material, its electric field shifts the electric charges in that material back and forth and the wave slows down.

The factor by which light slows down in a material is known as the material's **index of refraction**. Light travels particularly slowly through materials that are easy to polarize and some of them have indices of refraction of 2 or 3. However, because air near sea level is only slightly polarizable, its index of refraction is just 1.0003. While this reduction in light's speed is too small to notice, we do notice the polarized air particles that cause it. These polarized air particles are what make the sky blue (Fig. 15.1.3).

The particles in air consist of individual atoms and molecules, small collections of atoms and molecules, water droplets, and dust. As a wave of sunlight passes through one of these particles, the particle becomes polarized. Its electric charges accelerate back and forth as the sunlight's electric field pushes them around and they reemit a new electromagnetic wave of their own.

This new wave draws its energy from the original wave. In effect, the particle acts as a tiny antenna, temporarily receiving part of the electromagnetic wave and immediately retransmitting it in a new direction. This process, where a tiny particle redirects the path of a passing light wave, is called **Rayleigh scattering**, after the English physicist Lord Rayleigh (John William Strutt, 1842–1919) who first understood it in some detail.

While most sunlight travels directly to our eyes, some of it undergoes

Fig. 15.1.3 - During the day, the sky above Monument Valley appears blue because the earth's atmosphere scatters mostly blue sunlight toward us. At night, the scattered sunlight is gone and the atmosphere is a clear window through which to observe the stars.

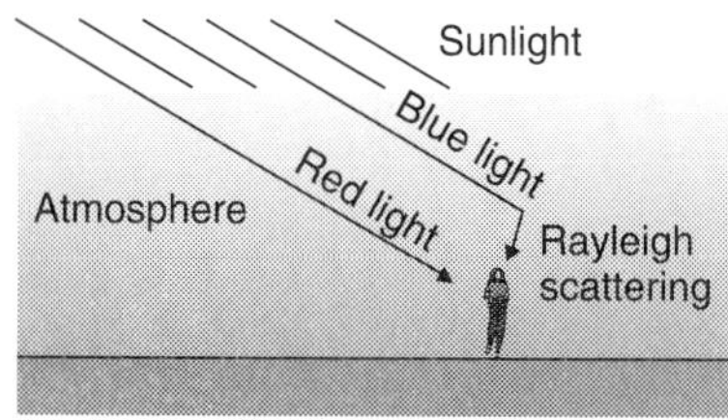

Fig. 15.1.4 - As sunlight passes through the atmosphere, some of its blue light Rayleigh scatters from molecules. We see this redirected blue light as the diffuse blue sky. The remaining light reaches our eyes directly from the sun and tends to be reddish, particularly at sunrise and sunset.

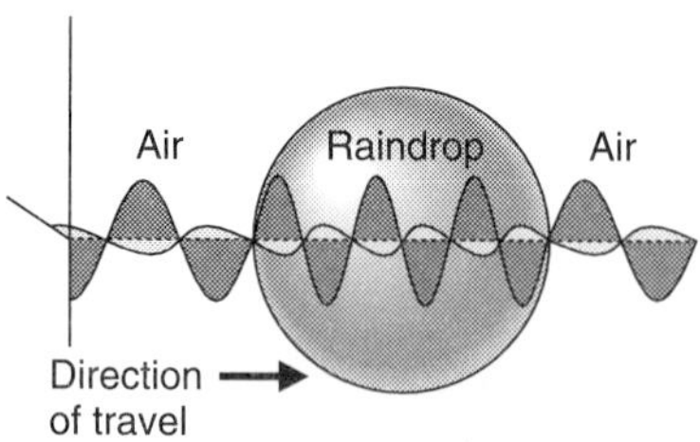

Fig. 15.1.5 - As an electromagnetic wave enters a material, its speed decreases and the waves bunch up together. Its wavelength decreases.

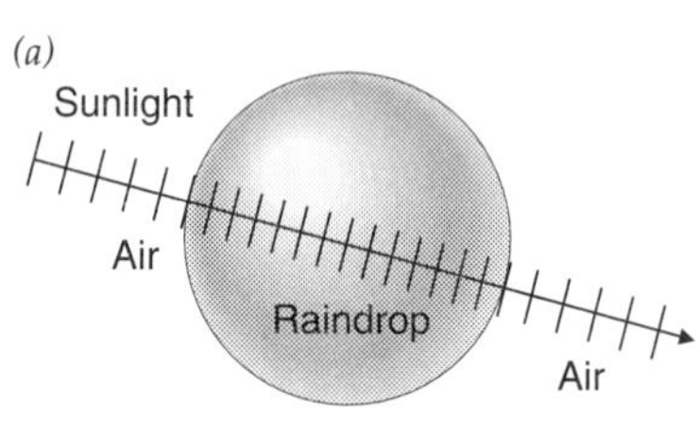

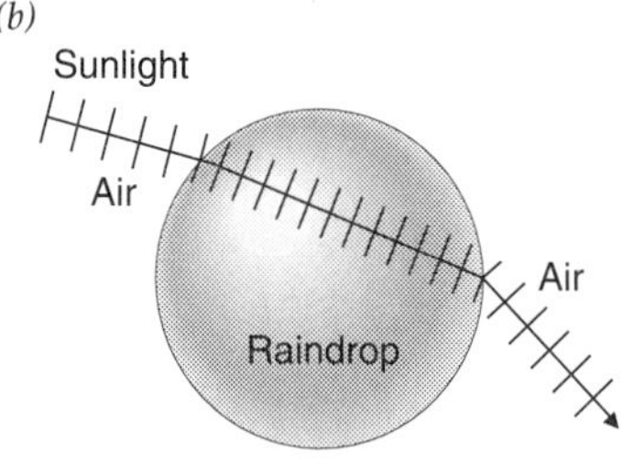

Fig. 15.1.6 - A side view of two narrow waves of sunlight entering and leaving raindrops. The lines drawn across each light wave indicate where the upward maximums in the electric field are located right now. These maximums are bunched more tightly in water than in air because the wave travels more slowly in water.

Rayleigh scattering and reaches us by more complicated paths. We see the direct light as coming from the brilliant disk of the sun, but the scattered light gives the entire sky a fairly uniform blue glow (Fig. 15.1.4). But why is this glow blue?

The sky's blue color comes about because the tiny air particles that Rayleigh scatter sunlight are too small to make good antennas for that light. We observed in Section 14.1 that an antenna works best when it's a quarter as long as the wavelength of its electromagnetic wave. The air particles make particularly bad antennas for longer-wavelength red light, so that very little red sunlight undergoes Rayleigh scattering on its way through the atmosphere. But the air particles are not such bad antennas for shorter-wavelength blue light. Some of the blue sunlight does Rayleigh scatter and reaches our eyes from all directions. We see this Rayleigh scattered light as the blue glow of the sky.

Rayleigh scattering not only makes the sky blue; it also makes the sunrises and sunsets red. As the sun rises or sets, its light must travel long distances through the earth's atmosphere in order to reach your eyes. Its path is so long that most of the blue light Rayleigh scatters away before it gets to you and all you see is the remaining red light. Sunrises and sunsets are particularly colorful when extra dust or ash is present in the atmosphere to enhance the Rayleigh scattering. Air pollution, forest fires, and volcanic eruptions tend to create unusually red sunrises and sunsets.

In contrast, clouds and fog appear white because they're composed of relatively large water droplets. These droplets are larger than the wavelengths of visible light and scatter all of the waves in sunlight equally well. This scattering is so effective that you can't see the sun through a cloud, but it doesn't give the cloud any color. The cloud simply looks white.

CHECK YOUR UNDERSTANDING #2: Seeing the Blues
The air in a dark, smoky room often looks bluish when illuminated by white light. What creates this bluish appearance?

Rainbows

Sometime water droplets do separate the colors of sunlight. When sunlight shines on clear, round raindrops as they fall during a storm, these raindrops can create a rainbow. To understand how clear spheres of water can bend sunlight's path and separate it according to wavelength, we must understand three important optical effects: refraction, reflection, and dispersion.

Let's begin by looking at what happens when a wave of sunlight passes directly through a raindrop. Because water is more polarizable than air, the wave slows down inside the raindrop and its cycles bunch together (Fig. 15.1.5). While this bunching effect reduces the light's wavelength inside the drop, the light's frequency remains the same. The cycles don't disappear as they go through the raindrop, they just move more slowly.

If a narrow wave of sunlight is aimed directly through the center of the raindrop, it will follow a straight path and emerge essentially unaffected from the other side (Fig. 15.1.6*a*). But if that wave strikes the raindrop near the top, it will bend as it enters the water (Fig. 15.1.6*b*). Because the lower edge of the wave will reach the water first and slow down, the upper edge will overtake it and the wave will bend downward. The wave will head more directly into the water.

As the wave in Fig. 15.1.6*b* leaves the raindrop, its upper edge emerges first and speeds up while the lower edge lags behind. The wave bends downward even further and heads less directly into the air and away from the water.

This bending of sunlight at the boundaries between materials is called

refraction. It occurs whenever sunlight changes speeds as it passes through a boundary at an angle. If sunlight slows down at a boundary, it bends to head more directly into the new material. If sunlight speeds up at a boundary, it bends to head less directly into the new material. The amount of the bend increases as the speed change increases.

However part of the sunlight striking a boundary doesn't pass through it at all. Instead, a small amount of the wave bounces back from that boundary. This **reflection** is a consequence of an *impedance mismatch*, an abrupt change in the way that light travels before and after it passes through the boundary. Impedance mismatches can occur for any kind of wave, including sound and water waves, and always cause reflections (see ❐).

In general, **impedance** is a measure of a system's opposition to the passage of a current or a wave. For an electromagnetic system, impedance measures how much voltage or electric field is needed to produce a particular current or magnetic field. In effect, impedance measures how hard is it for electric activity to produce magnetic activity.

The impedance of empty space is relatively high because the electric field there has nothing to help it produce the magnetic field. But inside most materials, the electric field has help. It polarizes the material and the material helps to create the magnetic field. Since it's easier for the wave's electric field to create its magnetic field, the impedance of most materials is much less than that of empty space. Because air behaves almost like empty space, the boundary between air and almost any material is an impedance mismatch for light.

Passing through an impedance mismatch upsets the balance between a light wave's electric and magnetic fields. To compensate for this imbalance, part of the incoming wave reflects off the boundary. Thus some sunlight reflects each time it moves from air to water or from water to air. While the amount of reflected sunlight increases with the severity of the impedance mismatch, the boundaries between air and most common transparent materials reflect about 4% of the sunlight. That's why you see a weak reflection of yourself when you look through a window and why the clear sand on a beach reflects enough light to appear white (see ❐). In contrast, metals polarize so easily that their impedances are essentially zero and they reflect light almost perfectly.

There is one more important point about sunlight's passage through water: Red light travels about 1% faster through water than violet light. That's because higher-frequency violet light polarizes the water molecules a little more easily than lower-frequency red light and that increased polarization slows down the violet light. This frequency dependence of the speed of light occurs in all materials and is called **dispersion**.

Because it changes the speed of light in a material, dispersion affects refraction. The more light slows as it enters a raindrop, the more it bends at the boundary. Since violet light slows more than red light, violet light also bends more and the different colors of sunlight follow somewhat different paths through the raindrop.

A rainbow is created when raindrops separate sunlight according to color (Fig. 15.1.7). To see the rainbow, you stand with the sun at your back and look up at the sky. When sunlight hits the raindrops, they deflect some of that light back toward you. Since the light from each raindrop deflects only in a narrow range of directions, you can't see the light from every raindrop. Only the raindrops in a narrow arc of the sky send their deflected light toward you. This arc appears brightly colored because raindrops at the inner edge of the arc send only violet light toward you while raindrops at the outer edge of the arc send only red light toward you. In between, you see all the colors of the rainbow.

To see how a raindrop can deflect different colors of light in different directions, look at Fig. 15.1.8. It shows the most important path that sunlight can

❐ When electromagnetic waves travel through wires, they reflect from impedance mismatches. These mismatches occur whenever the relationship between electric and magnetic fields changes and must be carefully avoided in television wiring. If you don't provide an impedance matching device when connecting an antenna wire (300 Ω) to a video cable input (75 Ω), you'll have severe reflections in the wires.

❐ Sand appears white because it redirects sunlight in all directions. One explanation for this effect is that the sand grains act as tiny antennas that respond to and reemit light's electromagnetic waves. A second explanation is that the sand grains present the sunlight with thousands of air–sand boundaries from which to reflect. However both explanations are descriptions of exactly the same physics—the charged particles in the sand grains are electrically polarized by the waves passing through them. These waves are randomly redirected without being absorbed and they give the sand its white appearance.

Fig. 15.1.7 - A rainbow forms when water droplets reflect sunlight back toward your eyes. Because the different wavelengths of light follow slightly different paths, we see the different colors coming from slightly different directions and observe bands of color.

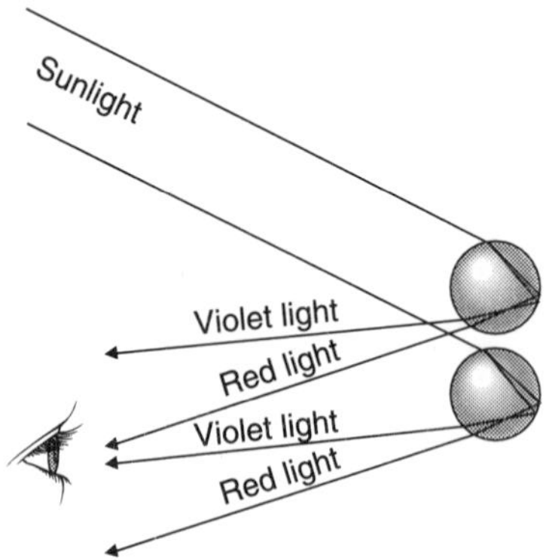

Fig. 15.1.8 - As sunlight passes through spherical raindrops, its colors separate. Violet light bends more at each air/water boundary than does red light, and the two emerge from the raindrops heading in different directions. You see red light coming toward you from the upper raindrops and violet light from the lower raindrops.

take through the raindrop. While there are many other possible paths, this is the path that's responsible for rainbows.

Sunlight enters near the top of the raindrop and is bent inward, to travel more directly into the water. Violet light bends more than red light, so the sunlight separates according to color. A small amount of the sunlight is also reflected from the raindrop, but this reflected portion doesn't contribute to rainbows.

When the light strikes the back surface of the raindrop, most of it leaves the drop and is lost. However a small fraction of the light reflects from that surface and continues to travel through the raindrop. This light strikes the raindrop's surface for a third time and most of it travels out of the drop. Violet light is again bent more strongly than red light as they move into the air, so the different colors of light leave the drop heading in different directions. Since the violet light is deflected more upward than the red light, you see violet light coming toward you from the lower raindrops. Red light is sent more downward so you see it coming toward you from the upper raindrops.

CHECK YOUR UNDERSTANDING #3: The Look of Diamonds

A diamond pendant sparkles with color when you look at it in sunlight. Where do the colors come from?

Soap Bubbles

Fig. 15.1.9 - Light reflected by the inner and outer surfaces of this soap bubble interferes with itself and gives the bubble its colorful appearance.

Soap bubbles also separate sunlight into its various colors (Fig. 15.1.9), using a wave phenomenon called interference. We encountered **interference** with microwave ovens, when we noted that microwaves passing through a particular location from several different directions sometimes help and sometime hinder the cooking at that location. When the electric fields there add together, they help one another and cooking proceeds quickly. This mutual assistance is called **constructive interference**. But when their electric fields cancel one another, cooking is slow and the effect is called **destructive interference**.

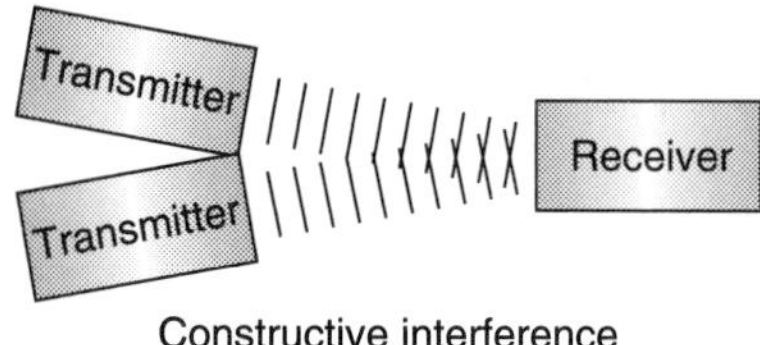

Constructive interference

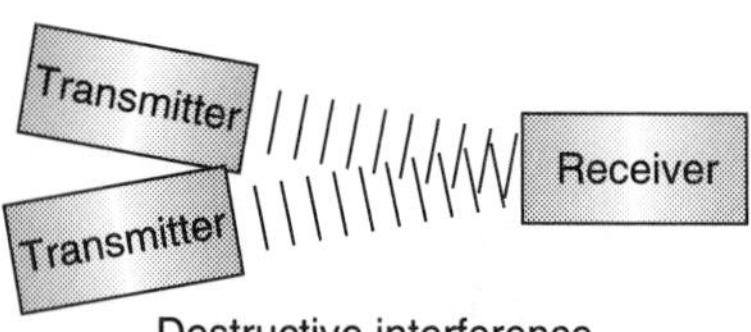

Destructive interference

Fig. 15.1.10 - When the waves from two separate transmitters arrive together in phase at the receiver, constructive interference produces a particularly large effect on the receiver. When the waves arrive together out of phase, destructive interference produces a particularly weak effect on the receiver.

Examples of these two types of interference for radio waves appear in Fig. 15.1.10. In each case, you are looking downward at two transmitters as they send their vertically polarized electromagnetic waves toward a receiver. As each wave travels to the right, it has regions where its electric field points most strongly upward. These regions are spaced one wavelength apart and are marked with a dark line.

In constructive interference, the two waves arrive together so that their electric fields both point in the same direction. They point up at the same time and down at the same time and are said to be **in phase** with one another. They work together to accelerate charge in the receiver's antenna and cause a particularly large response in the receiver.

In destructive interference, the two waves arrive **out of phase** with one another. One wave's electric field points up when the other wave's electric field points down. They oppose one another all the time and cause a particularly small response in the receiver.

Both of these forms of interference occur in sunlight as it passes through the upper surface of a soap bubble. The bubble is air, surrounded by a very thin soap film (Fig. 15.1.11). As a wave of sunlight passes through that film, about 4% of it is reflected from the film's top surface and another 4% is reflected from the film's bottom surface.

As you look at this reflected wave, it travels to your eyes via two different paths. Two waves reach your eyes—one from the top surface reflection and one from the bottom surface reflection. If these waves arrive in phase, so that their electric fields are synchronized with one another, you see a particularly bright reflection; constructive interference. If they are out of phase, you see a particularly dim reflection; destructive interference.

Whether you see constructive or destructive interference depends on the wavelength of the sunlight. The bottom surface reflection has to travel twice through the soap film, so it's delayed relative to the top surface reflection. If the delay is just long enough for the wave to complete an integral number of cycles, then the two reflected waves are in phase with one another as they head toward your eye and you see a bright reflection. If the delay allows the bottom surface reflection to complete an extra half a cycle, then the two reflected waves are out of phase with one another and you see a dim reflection.

Sunlight contains many different wavelengths of light and these wavelengths of light behave differently during the reflection process. You see a colored reflection, consisting mainly of those wavelengths of light that experience constructive interference. Because the delay experienced by the bottom surface reflection depends on the thickness of the soap film, you can actually determine the film's thickness by studying its colors.

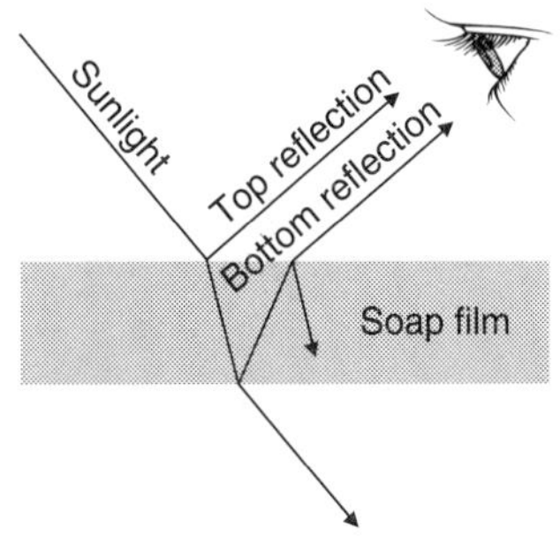

Fig. 15.1.11 - Sunlight reflects from both the top and bottom surfaces of a soap film. The bottom surface reflection is delayed relative to the top surface reflection because it must pass twice through the soap film itself. If the two reflected waves arrive in phase, you see a bright reflection. If they are out of phase, you see a dim reflection.

CHECK YOUR UNDERSTANDING #4: A Slick Oil Slick

A thin layer of oil or gasoline floating on water appears brightly colored in sunlight. Where do these colors come from?

Sunlight and Polarizing Sunglasses

Sunglasses absorb some of the sunlight passing through them. In effect, the sunglasses act as though they are full of tiny antennas that "receive" parts of the electromagnetic waves and turn some of their energy into thermal energy. But some sunglasses absorb horizontally polarized light much more strongly than vertically polarized light. These polarizing sunglasses dramatically reduce glare by absorbing most of the light reflected from horizontal surfaces.

When light strikes a transparent surface at right angles, about 4% of that light is reflected, regardless of its polarization. But when light strikes a horizontal surface at a shallow angle, horizontally polarized light reflects much more than vertically polarized light (Fig. 15.1.12).

That's because horizontally polarized light's horizontal electric field pushes electric charges back and forth along the surface. Charges move relatively easily

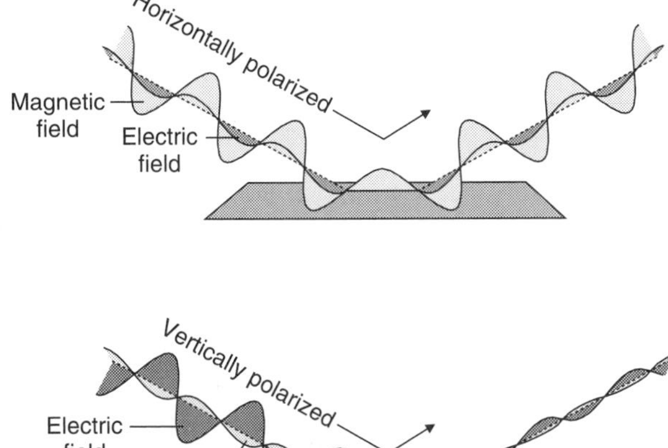

Fig. 15.1.12 - Horizontally polarized light reflects strongly from a horizontal surface, even at shallow angles. However, vertically polarized light doesn't reflect well from a horizontal surface at shallow angles because it has trouble polarizing the surface.

Fig. 15.1.13 - Polarizing sunglasses block horizontally polarized light, the main component of glare. While the remaining vertically polarized light will pass easily through a second pair of sunglasses, it will be blocked by sunglasses that are turned 90° so that they block vertically polarized light.

in that direction and the surface becomes even easier to polarize as the angle becomes shallower. As a result, the surface reflects more horizontally polarized light at shallow angles than it does at steeper angles.

But vertically polarized light's vertical electric field pushes electric charges up and down vertically. At shallow angles, this field acts to lift the charges in and out of a horizontal surface. Since the charges can't leave the surface, the surface becomes harder to polarize as the angle becomes shallower. A horizontal surface doesn't reflect much vertically polarized light and there is even a special angle, *Brewster's angle*, at which no vertically polarized light reflects at all.

Sunlight is an even mixture of vertically and horizontally polarized waves. Since horizontally polarized waves reflect most strongly from horizontal surfaces, sunglasses that absorb horizontally polarized light will prevent you from seeing most of this reflected light. That's why polarizing sunglasses are so effective at reducing glare.

To absorb this horizontally polarized light, the sunglasses contain the molecular equivalents of tiny horizontal antennas (Fig. 15.1.13). These tiny molecular antennas respond to the electric fields in horizontally polarized light, receiving that light and turning its energy into thermal energy. Vertically polarized light can't push charge up and down horizontal antennas, so it passes through the sunglasses without significant loss.

Polarizing sunglasses also darken the blue sky because most of the Rayleigh scattered light we see is horizontally polarized. This horizontal polarization occurs because, as sunlight is received and retransmitted by particles in the air, its new path is related to its polarization. This effect is similar to that in a vertical radio antenna, which tends to emit its vertically polarized waves in horizontal directions. The air particles tend to scatter horizontally polarized sunlight in vertical directions. It heads downward toward our eyes so that we can see it. These air particles tend to scatter vertically polarized sunlight in horizontal directions. It travels overhead where we can't see it.

CHECK YOUR UNDERSTANDING #5: Looking Into the Reflecting Pool

When you look into a pool of water, you see mostly a reflection of the sky. But when you wear polarizing sunglasses, you see into the water clearly. Explain.

Summary

How Sunlight Works: Sunlight originates at the outer surface of the sun, which has a temperature of about 5800° C. Thermal energy causes electrically charged particles there to accelerate back and forth rapidly and they emit the electromagnetic waves that are sunlight. This sunlight travels at the speed of light through empty space until its reaches the earth's atmosphere. There is slows slightly and some of it is Rayleigh scattered. This redirected sunlight causes the entire sky to glow. Short wavelength light is Rayleigh scattered more strongly than long wavelength light, so the sky appears blue.

As the sunlight passes through various objects, it slows down and its colors tend to separate. When sunlight passes through falling raindrops, its different wavelengths follow different paths and create rainbows. As sunlight reflects from thin films such as the outsides of soap bubbles, its waves are divided and may follow several different paths to the same destination. These light waves then interfere with one another so that some waves appear strong and bright while others are weak and dim. Interference depends on the wavelengths of light so that thin films appear brightly colored.

The Physics of Sunlight

1. The rapidly moving charged particles in atoms, molecules, and materials can emit and absorb light. Visible light has frequencies between about $4.0 \cdot 10^{14}$ Hz and $7.5 \cdot 10^{14}$ Hz and wavelengths (in empty space) between about 400 nm and 750 nm.

2. As a light wave enters matter, its electric field tends to polarize the matter's charged particles and the wave slows down. The easier it is to polarize the material, the slower light travels through it. The factor by which light slows down in a material is that material's index of refraction.

3. Tiny particles in the air can receive and retransmit light, changing the light's direction of travel. This Rayleigh scattering is most efficient for short wavelength light.

4. Because light travels slowly through materials, its wavelength is shortened in those materials. Its frequency remains the same.

5. When light strikes the boundary between two materials and changes speeds, it's refracted. If the light slows down, it bends to travel more directly into the new material. If the light speeds up, it bends to travel less directly into the new material.

6. When a wave encounters an impedance mismatch, part of the wave reflects. When a light wave moves from one material to another, the impedance usually changes and some of that wave reflects from the boundary between the materials.

7. The ease with which a material is polarized by a light wave's electric field depends on the wave's frequency. High-frequency violet light polarizes most materials more than low-frequency red light, so violet light travels more slowly than red light. This phenomenon is called dispersion.

8. Two or more electromagnetic waves can interfere with one another. When their electric fields point in the same direction, they experience constructive interference and produce a strong effect. When their electric fields oppose one another, they experience destructive interference and produce a weak effect.

9. When light strikes a surface at a shallow angle, the fraction that reflects depends on the light's polarization. When the light is polarized so that its electric field moves charge along the surface, the reflection is strong. When the light's electric field moves charge in and out of the surface, the reflection is weak.

Check Your Understanding - Answers

1. Charged particles in the hot flame accelerate back and forth and emit electromagnetic waves that include the low-frequency end of the visible spectrum.

Why: The accelerating charged particles in hot objects emit light. The hotter the object, the faster the charged particles move and accelerate and the higher the frequencies of light they emit. A candle isn't hot enough to emit the whitish light of the sun, so it emits mostly reddish or yellowish light.

2. Rayleigh scattering from the microscopic smoke particles scatters more blue light than red light and give the air a bluish glow.

Why: When smoky air in a dark room is illuminated by a bright spotlight or another white light source, the tiny particles in the air Rayleigh scatter some of the light. When you look at the air against a dark background you can see this glow. Since Rayleigh scattering affects blue light most strongly, the air appears bluish.

3. A diamond exhibits dispersion so that different frequencies of sunlight follow somewhat different paths through the diamond's polished facets. The different colors of sunlight emerge from the diamond traveling in slightly different directions, so that you can see them individually.

Why: One of the delightful aspects of a diamond is its strong dispersion. It bends violet light much more than red light so that sunlight is separated into its different colors as it passes through the stone. Cleverly cut facets help to isolate the individual colors.

4. From interferences between light reflected by the top and bottom surfaces of the floating layer of oil.

Why: A thin film of almost anything on water will appear colored because of interference. Part of each light wave striking the film reflects from the top surface and part reflects from the bottom surface. These two reflected waves interfere with one another in a wavelength-dependent manner and make the layer appear brightly colored.

5. Most of the light that reflects from the water is horizontally polarized. The sunglasses block horizontally polarized light so that you see mostly light from within the pool of water.

Why: Sunlight reflecting from a horizontally polarized surface is mostly horizontally polarized waves. By blocking horizontally polarized light, polarizing sunglasses virtually eliminated this reflected light and you see mostly light from below the surface. If you turn the sunglasses sideways, they will block the wrong polarization of light and you will see mostly the reflected light. Try it.

Glossary

constructive interference Interference in which two or more waves arrive at a location in phase with one another and produce a particularly strong effect.

destructive interference Interference in which two or more waves arrive at a location out of phase with one another and produce a particularly weak effect.

dispersion The dependence of light's speed through a material on the frequency of that light.

impedance A measure of a system's opposition to the passage of a current or a wave.

in phase The relationship between two waves in which they complete the same portions of their oscillatory cycles at the same time and place.

index of refraction The factor by which the speed of light in a material is reduced from its speed in empty space, equal to the speed of light in empty space divided by light's speed in the material.

infrared light Invisible light having wavelengths longer than about 750 nanometers.

interference A wave phenomenon in which waves passing through the same location from different directions reinforce or oppose one another.

out of phase The relationship between two waves in which they complete opposite portions of their oscillatory cycles at the same time and place.

Rayleigh scattering The redirection of a light wave due to its interaction with small particles of matter.

reflection The redirection of all or part of a light wave so that it returns from a boundary between materials.

refraction The bending of light's path that occurs when the light crosses a boundary between materials and experiences a change in speed.

ultraviolet light Invisible light having wavelengths shorter than about 400 nanometers.

visible light Light having wavelengths between about 400 nanometer (violet) and 750 nanometer (red). This small portion of the electromagnetic spectrum is all that we are able to detect with our eyes.

Review Questions

1. Why are the individual electrons in a hot object able to emit visible light while charge moving on a normal metal antenna is not?

2. Why is Rayleigh scattering stronger for blue light than it is for red light?

3. Dust particles are much more effective at causing Rayleigh scattering than are air molecules. Why?

4. Why is the wavelength of green light so much shorter in water than it is in air?

5. Why can you see your reflection in a basin of water?

6. When sunlight goes through a glass prism, its path changes and it separates into bands of color from red to violet. Explain the change in path and why the different colors appear from the white light.

7. Why is the arc of a rainbow violet inside and red outside?

8. If you hold a soap film up so that its surface is vertical, it will be thicker at the bottom than at the top. Why will this wedge-shaped film have horizontal strips of color to it?

9. How can polarizing sunglasses dramatically reduce reflections from the shiny surfaces in front of you without significantly darkening everything else around those reflections?

Exercises

1. How long should an antenna be to receive or transmit red light well?

2. How long should an antenna be to receive or transmit violet light well?

3. How rapidly should a charged particle on the sun's surface move back and forth to emit a wave of red light?

4. How rapidly should a charged particle on the sun's surface move back and forth to emit a wave of violet light?

5. Fig. 15.1.1 is accurate only for light waves traveling in empty space. To make this figure suitable for light passing through water, how would it have to change?

6. Although we can't see them, Rayleigh scattering also occurs for infrared and ultraviolet light. Which of these two types of light experiences the strongest Rayleigh scattering?

7. When astronauts aboard the space shuttle look down at the earth, its atmosphere appears blue. Why?

8. When astronauts walked on the surface of the moon, they could see the stars even though the sun was overhead. Why can't we see the stars while the sun is overhead?

9. A pencil lead is mostly graphite, a common form of carbon in which the atoms are arranged in sheets only one atom thick. Because these sheets can carry electric currents, graphite sometimes resembles a metal. When you rub a pencil back and forth across the same spot on a sheet of paper, you produce a layer of graphite in which the sheets are mostly lying flat on the surface. This layer looks shiny and metallic. How is this layer responding to the light that strikes it?

10. A laser light show uses extremely intense beams of light. When one of these beams remains steady, you can see the path it takes through the air. What makes it possible for you to see this beam even though it isn't directed toward you?

11. To make the beams at a laser light show (see Exercise **10**) even more visible, they're often directed into mist or smoke. Why do such particles make the beams particularly visible?

12. When you stand at the side of a calm lake and shine a flashlight into the water in front of you, the light beam hits the water at a shallow angle. Why does it bend as it enters the water so that it travels more directly downward?

13. If you shine a flashlight horizontally at a glass full of water, the glass will redirect the light beam. How?

14. Why can you see your reflection in a calm pool of water?

15. Mica is a mineral consisting of many thin transparent sheets. As long as these sheets are pressed tightly against one another, mica looks like a sheet of glass. But when the sheets are separated slightly and the spaces between them are filled with air, the mica looks almost like a mirror. Explain.

16. Use the concepts of refraction, reflection, and dispersion to explain why a diamond emits a spray of colored lights when sunlight passes through its cut facets.

17. When candlelight passes through a crystal candelabra, the light is bent with the various colors going in different directions. Why does the light change directions and separate according to color?

18. The light rays that you see in the candelabra of Exercise **17** are mostly red, orange, yellow, and green. Why is there very little blue light present?

19. If you hold a clear cylindrical glass of water in front of your eye and look at a bright light bulb through it, you should see a distorted image of the bulb. As you move the glass around, you may find that one edge of the image becomes reddish while the other edge of the image becomes bluish. What causes these colors?

20. Basic paper consists of many transparent fibers of cellulose, the main chemical in wood and cotton. Why does paper appear white and why does it become relatively clear when you get it wet?

21. Why is a pile of granulated sugar white while a single large piece of rock candy (solid sugar) is clear?

22. Diamond has an index of refraction of 2.42. If you put a diamond in water, you see reflections from its surfaces. But if you put it in a liquid with an index of refraction of 2.42, the diamond is invisible. Why is it invisible and how is this effect useful to a jeweler or gemologist?

23. Weather radar locates clouds by bouncing microwaves off them. Why do microwaves reflect when they move from clear air to air containing tiny water droplets?

24. When snow is packed into ice on the surface of a road, it appears white. But when the snow melts into water and refreezes, it can be nearly invisible "black ice." What makes packed snow ice so much more visible than black ice?

25. If you peel apart a large white onion, you will find it contains many nearly transparent layers. How do these layers work together to give the onion a white appearance?

26. The plastic used to make Styrofoam is completely clear, yet Styrofoam is white. Why is it white?

27. Before the era of paints, people would whiten their walls with whitewash, a mixture of lime particles and water. Lime is a mineral and a large crystal of lime is clear. So why does a wall coated with lime particles appear white?

28. Why is it easier to see into water when you look directly down into it than when you look into it at a shallow angle?

29. On a rainy day you can often see oil films on the surfaces of puddles. Why do these films appear brightly colored?

30. When two sheets of glass lie on top of one another, you can often see colored rings of reflected light. How do the nearby glass surfaces cause these colored rings?

31. The colored rings described in Exercise **30** change size when you squeeze the glass sheets closer together. Why do they change size?

32. You can see a strong reflection of the sky when you look at wet pavement far in front of you. But when you look at the wet pavement near your feet, you see only a weak reflection. Why is the reflection so strong when you look far ahead?

33. The light that reflects from the distant pavement in Exercise **32** tends to be polarized. Is it horizontally or vertically polarized? Why?

34. If you're wearing polarizing sunglasses and want to see who else is wearing polarizing sunglasses, you only have to turn your head sideways and look to see which people now have sunglasses that appear completely opaque. Why does this test work?

35. The test described in Exercise **34** assumes that everyone else's polarizing sunglasses block the same polarization as your sunglasses do. Why is that a safe assumption?

36. When you stand at the edge of a lake and try to look into the water, you mostly see a reflection of the sky. Why is it much easier to see into the lake when you're wearing polarizing sunglasses?

37. When you're wearing polarizing sunglasses and look into a lake (see Exercise **36**), there is a particular region of the surface in front of you that doesn't appear to be reflecting any light at all. Explain why you see no reflection at all from this region.

38. As you watch a soap bubble that's supported from the bottom, the film gradually dries out. The film becomes thinner and as it does, its colors changes. If you're watching a green reflection from the film as the film gets thinner, what color gradually replaces green?

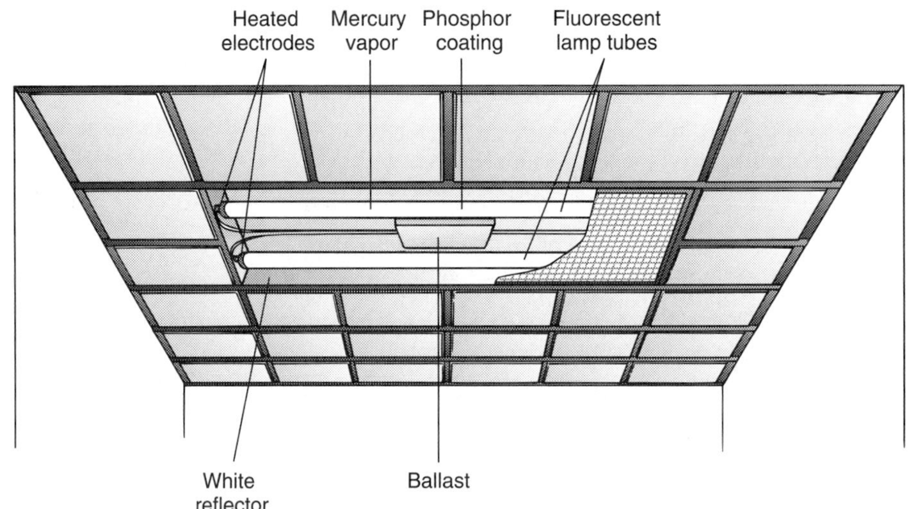

Section 15.2

Fluorescent Lamps

Energy efficiency is crucial to modern lighting. While incandescent lamps provide pleasant, warm illumination, most of the power they consume is wasted as infrared light. Gas discharge lamps produce far more visible light with the same amount of electric power and now dominate office, industrial, and street lighting. Even at home, small fluorescent lamps can replace incandescent light bulbs in table lamps and overhead light fixtures. These replacement lamps are somewhat expensive and can't be dimmed, but they pay for themselves in energy savings. While there are several types of discharge lamps, including fluorescent, mercury vapor, sodium vapor, and metal-halide lamps, they all share the same fundamental principles and are included in this section.

Questions to Think About: *Why do different fluorescent lamps often have slightly different colors? Why is the light from a neon sign red? Are neon atoms red? Why do some fluorescent lamps start instantly while others take a few seconds to turn on? Why are street lights so dim when they first turn on? Why are some highway lights orange?*

Experiments to Do: *You will find quite a variety of discharge lamps around you. Fluorescent lamps are normally frosted white tubes. Most of the visible light they emit originates in a white phosphor coating on the tube's inner surface. Look at several different fluorescent lamps and compare their colors. How do these colors differ? Now look at a red neon sign. Where does its light originate? Is there a phosphor coating on the inside of this tube or is the red light coming directly from the gas inside?*

Both fluorescent and neon lamps start almost immediately. But watch how long it takes a street light to start. While fluorescent and neon lamps remain cool during operation, the mercury, sodium, and metal-halide lamps used in street lighting get hot. Watch a street light warm up. Does its color change? If the street light loses power even for a moment, it must wait about 5 minutes before it can start again. Why must it wait?

How We See Light and Color

Before we examine fluorescent lamps, let's look at how our eyes recognize color. While it might seem that they actually measure the wavelengths of light, that's not the case. Instead, our retinas contain three groups of light-sensing *cone cells* that respond to three different ranges of wavelengths. One group of cone cells responds to light near 600 nm and makes us see red, another responds to light near 550 nm and makes us see green, and a third responds to light near 450 nm and makes us see blue (Fig. 15.2.1). These cone cells are most abundant at the center of our vision. While our retinas also contain *rod cells*, which are more light-sensitive than cone cells, rod cells can't distinguish color. They are most abundant in our peripheral vision and provide us with night vision.

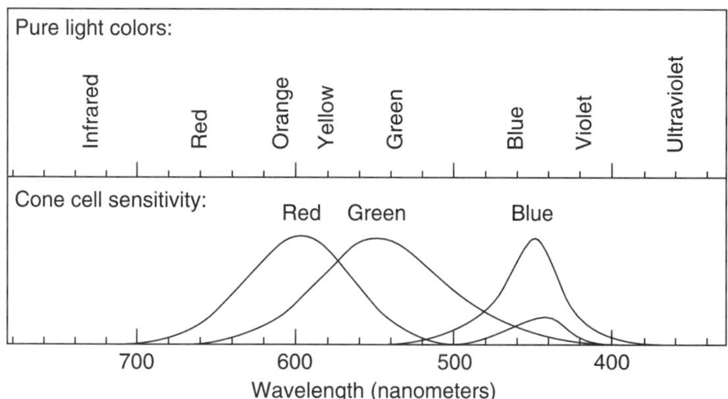

Fig. 15.2.1 - The red-sensitive cells in our retina detect light near 600 nm, the green-sensitive cells near 550 nm, and the blue-sensitive cells near 450 nm. The red-sensitive cells also respond near 440 nm, so that we see violet.

Having only three types of color sensing cells doesn't limit us to seeing just three colors. We perceive other colors whenever two or more types of cone cells are stimulated at once. Each type of cell reports the amount of light it detects and your brain interprets the overall response as a particular color.

While light of a certain wavelength will stimulate all three types of cone cells simultaneously, the cells don't respond equally. If the wavelength is 680 nm, the red cone cells will respond much more strongly than the green or blue cells. Because of this strong red response, you perceive the light as being red.

Other wavelengths of light stimulate the three types of cells somewhat more evenly. Light with a wavelength of 580 nm is in between red and green light. Both the red-sensitive and the green-sensitive cone cells respond about equally to this light and you see it as being yellow.

But you also see yellow when looking at an equal mixture of 640 nm light (red) and 525 nm light (green). The 640 nm light stimulates the red-sensitive cone cells and the 525 nm light stimulates the green-sensitive ones. Even though there is no 580 nm light entering your eyes, you see the same yellow color as before.

In fact, mixtures of red, green, and blue light can make you see virtually any color. For that reason, these three are called the **primary colors of light**. A color television uses red, green, and blue phosphors to produce full color pictures. While the television only produces mixtures of three different lights, it controls those mixtures carefully and you see all the possible colors.

CHECK YOUR UNDERSTANDING #1: Mixed Up Light

If you look at a mixture of 70% red light and 30% green light, what color will you see?

The Shortcomings of Incandescent Lamps

When all three of our color-sensing cells respond about equally, we see white light. That's because our vision evolved under a single incandescent light source: the sun. Sunlight stimulates the red-, green-, and blue-sensitive cells in our eyes about evenly, so any other source of "white light" will have to do same.

While an incandescent light bulb makes a good attempt at producing white light, it suffers from two serious drawbacks. First, because its filament can't reach the temperature of the sun's surface (5800° C), it emits a black body spectrum that is much redder than sunlight. Second, because most of its electromagnetic radiation is invisible infrared light, its energy efficiency is low.

The best alternatives to incandescent lamps are gas discharge lamps. Some discharge lamps do excellent jobs of producing white light and all of them are far more energy efficient than incandescent light bulbs. The most common and familiar of these lamps are fluorescent lamps. Instead of heating a filament until it glows white hot, a fluorescent lamp uses electrons to excite atoms, which then give off light. A fluorescent lamp remains cools as it operates, so it wastes little power heating the room or giving off infrared radiation.

CHECK YOUR UNDERSTANDING #2: Making Whiter Light

Some photographic lamps simulate sunlight by placing a blue filter in front of an incandescent bulb. This filter absorbs some of the red light so that the lamp appears whiter. Does this filter increase or decrease the lamp's energy efficiency?

Fluorescent Lamps: Atoms and Light

At the heart of a fluorescent lamp is a narrow glass tube that's sealed at both ends. This tube contains argon, neon, and/or krypton gases at a pressure and density only 0.3% that of the atmosphere outside the tube. The tube also contains a few drops of liquid mercury metal, a little of which evaporates to form mercury vapor. About one in every thousand gas atoms inside the tube is a mercury atom and it's these mercury atoms that create the light.

The tube lights up when an electric current passes through it. It has metal electrodes at each end so that current can enter the gas through one electrode and leave through the other. But as long as the gas contains no mobile electric charges, it can't conduct electricity. Something must inject electric charges into the gas so that electricity can move through it.

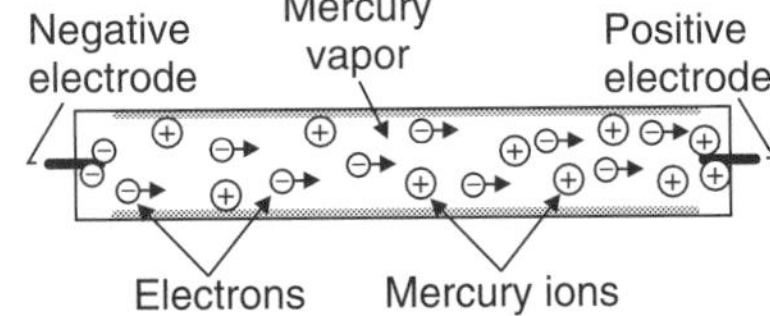

Fig. 15.2.2 - A fluorescent tube sends electrons through a low-pressure gas. These electrons collide with mercury atoms in the gas, transferring energy to those mercury atoms and causing them to emit ultraviolet light. Positively charged mercury ions, created by particularly energetic collisions, keep the electrons from repelling one another to the walls of the tube.

Fluorescent lamps use two techniques to introduce charges into the gas. Most heat their electrodes, so that thermal energy ejects electrons from their surfaces. Others use high voltages to squeeze like charges onto the electrodes until those charges push one another into the gas. But regardless of technique, the result is a gas that conducts electricity.

The lamp then starts a **discharge**—it sends current through the gas. This current consists of electrons, accelerating from the tube's negatively charged electrode to its positively charged electrode (Fig. 15.2.2). While these electrons collide frequently with various gas atoms, they have so little mass that they usually just bounce off the gas atoms without losing much energy. Like a Ping-Pong ball rebounding from an elephant, the electron does most of the bouncing and then continues on its way.

However, every so often an electron will collide with a mercury atom and something different will happen; the mercury atom will rearrange internally and absorb part of the electron's kinetic energy. The electron will rebound with less energy than it had before and the mercury atom will go on to emit a small amount of light.

To understand how the mercury atom manages to turn the electron's kinetic energy into light, we need to look at the structure of atoms. As we learned in the section on xerographic copiers, quantum physics permits the mercury atom's 80 electrons to orbit its nucleus only in certain allowed orbitals. Because of the Pauli exclusion principle, only one electron of each spin (up and down) can be in each orbital, so the atom's electrons must spread out into at least 40 different orbitals.

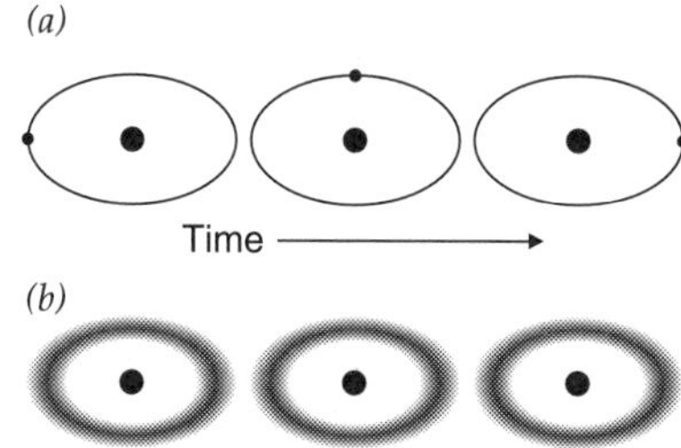

Fig. 15.2.3 - (*a*) In the classical picture of an electron in an atomic orbital, the electron is a tiny object that moves around the nucleus as time passes. (*b*) In the quantum picture, the electron is a diffuse wave that doesn't change in any measurable way as time passes.

Because there is a general tendency for an atom to minimize its overall energy, the mercury atom's 80 electrons normally fill the 40 orbitals that are closest to its nucleus and have the least energy. This arrangement is called the **ground state** of the atom. Like all isolated atoms near room temperature, the mercury atoms in a fluorescent lamp are almost always in their ground states.

But when a mercury atom is struck hard by an electron, its electrons may rearrange. One or more of its electrons may move into a formerly unoccupied orbital. This new arrangement of electrons is called an **excited state** of the atom, indicating that at least one of the electrons is not in its ground state orbital.

The mercury atom doesn't stay in this excited state long. It has too much energy and begins a series of rearrangements that quickly returns it to its ground state. In each rearrangement, an electron makes a transition from one orbital to another and the atom radiates light. Electronic transitions in which light is emitted or absorbed are called **radiative transitions**. We've seen that hot objects emit light, a process called **incandescence**, but here the mercury atom emits light without heat, a process called **luminescence**. While radiative transitions are too complicated for us to study in much detail, we can make some useful observations about atoms and light.

First, let's see why an electron that remains in its orbital *doesn't* emit light even though it's accelerating. The problem lies in our view of the orbiting electron. It isn't simply a negatively charged ball orbiting a positively charged nucleus (Fig. 15.2.3*a*). It's actually a negatively charged wave extending uniformly all the way around the nucleus (Fig. 15.2.3*b*). This wave doesn't change with time but remains steady. Since the wave's electric charge doesn't shift back and forth, it doesn't emit an electromagnetic wave.

The electron's wave nature is a consequence of quantum physics. Like all objects in our universe, electrons travel as waves when they move from place to place, and it's only when you go looking for them that you find them at particular locations (see ❐). When an electron orbits an atomic nucleus in a single orbital, it moves as a wave that fills the whole orbital evenly and doesn't change with time. Because of its unchanging form, an electron in an orbital doesn't emit any electromagnetic radiation at all.

❐ In 1927, American physicists Clinton Joseph Davisson (1881–1958) and Lester H. Germer (1896–1972) showed that electrons travel as waves by observing interference effects when electrons reflected from different atomic layers in a crystal of nickel metal. When the various electron waves arrived at a detector in phase, the detector found many electrons. When the waves arrived out of phase, the detector found few electrons. Their work was aided by a fortuitous accident, in which air entered the experiment's glass vacuum tube. While carefully eliminating oxygen from their nickel sample, they managed to perfect its crystalline structure, making it possible to observe the interferences.

But when that electron begins a transition to an empty, lower-energy orbital, its wave begins to change with time. The wave moves rhythmically back and forth during the transition so that by the time the transition is complete, the electron will have emitted a *particle* of light. Like an electron, light travels as a *wave* but behaves as a *particle* when you try to locate it. While it's being emitted or absorbed by an atom, light exhibits its particulate nature.

Each time the mercury atom undergoes a radiative transition, the atom emits a single particle of light, a photon, which then heads off through space as a wave. This photon carries with it the energy released by the mercury atom as its electron shifted to a lower-energy orbital. The photon's energy is equal to the frequency of its wave times a fundamental constant of nature known as the **Planck constant**, first used by German physicist Max Planck (1858–1947) in 1900 to explain the light spectrum of a hot object. This relationship between energy and frequency can be written as a word equation:

$$\text{energy} = \text{Planck constant} \cdot \text{frequency}, \tag{15.2.1}$$

in symbols:

$$E = h\nu,$$

and in everyday language:

Ultraviolet light and X-rays can injure your skin and tissues because each particle of high frequency light carries a great deal of energy.

The Planck constant is so small, only $6.626 \cdot 10^{-34}$ J·s, that light's particulate nature is difficult to observe. Because a photon of ultraviolet light with a frequency of 10^{15} Hz has an energy of only $6.626 \cdot 10^{-19}$ J, even a weak ultraviolet beam contains an incredible number of photons. With so many photons hitting your skin, you can't tell that they're arriving as particles. However, an ultraviolet photon is energetic enough to damage a molecule in your skin, contributing to a sunburn and inducing your skin to tan as a defensive response. X-rays have even higher frequencies and their energetic photons can cause more severe damage.

A mercury atom can only emit a photon corresponding to the energy difference between two of its **states**; that is its possible arrangements of electrons. This effect severely limits the spectrum of light that a mercury atom can emit. Since all mercury atoms have identical states, they all emit the same spectrum of light. Like mercury, every other atom emits its own unique spectrum of light.

Returning to the fluorescent lamp, we see that electrons moving through the tube can transfer energy to mercury atoms during collisions. Each time a collision leaves an atom in an excited state, that atom returns to the ground state through a series of radiative transitions. With each transition, a photon carries away some of the atom's excess energy until it returns to the ground state. The photons that the mercury atoms emit are characteristic of mercury and are determined by the energy differences between its orbitals.

CHECK YOUR UNDERSTANDING #3: Colorful Atoms
Many fireworks involve brilliantly colored lights. How do the atoms in burning chemicals produce particular colors of light?

CHECK YOUR FIGURES #1: The Particles in a Radio Wave
How much energy is carried by a photon from a 1000 kHz AM radio station?

The Phosphor Coating in a Fluorescent Tube

But the fluorescent tube has a complication. While its mercury atoms emit light, most of that light is ultraviolet. The final radiative transition that returns each mercury atom to its ground state releases a large amount of energy and produces a photon with a wavelength of 254 nm. This light can't go through the glass walls of the tube and you couldn't see it if it did. So the fluorescent lamp converts it into visible light, using phosphor powder on the inside of the glass tube.

Phosphors are solids that luminesce when something transfers energy to them. In the section on televisions, we saw that a beam of electrons can provide this energy. A phosphor emits light in a manner quite similar to that of an atom—an energy transfer shifts the phosphor from its ground state to an excited state, and it undergoes a series of transitions that return it to its ground state. Some of those transitions emit light.

In a fluorescent lamp, the phosphor receives energy from a photon of ultraviolet light. This transfer is actually a radiative transition, but one in which the photon is absorbed by the phosphor as one of its electrons makes a transition from a lower-energy level to a higher-energy level. (Recall that in solids the paths taken by electrons are called levels rather than orbitals.) The light's electric field pushes the electron's wave back and forth rhythmically until it shifts to the new level. The photon disappears and the phosphor receives its energy.

Once the phosphor is in an excited state, its electrons begin to make transitions back to their ground state levels. Most of these transitions radiate visible light, the light that you see when you look at the lamp. However some of these transitions radiate invisible infrared light or cause useless vibrations in the phosphor itself. Despite this wasted energy, phosphors are relatively efficient at turning ultraviolet light into visible light, a process called **fluorescence**.

The phosphors in a fluorescent lamp are carefully selected and blended to fluoresce over a broad range of visible wavelengths. While this light doesn't have the same spectrum as sunlight, it appears white because it stimulates the red-, green-, and blue-sensitive cells in your eyes about equally. The blending of phosphors is necessary because, like atoms, each phosphor fluoresces with a characteristic spectrum of light that's determined by the energy differences between its levels. Several different phosphors are needed to create the right balance between red, green, and blue lights. While some advertising and novelty lamps use brightly colored, unblended phosphors, fluorescent lighting tubes come in six standard color blends: cool white, deluxe cool white, warm white, deluxe warm white, white, and daylight.

When they were first introduced, fluorescent tubes used the daylight phosphor. This phosphor emits too much blue light and makes everything look cold and medicinal. Phosphors have improved over the years so that the most common phosphors today, cool white and warm white, look much more pleasant. Light from the "cool" phosphors resembles daylight while that from the "warm" phosphors resembles incandescent lighting. Household fluorescent lamps often use deluxe cool white and deluxe warm white, which are even more pleasant versions of white but which are slightly less energy efficient.

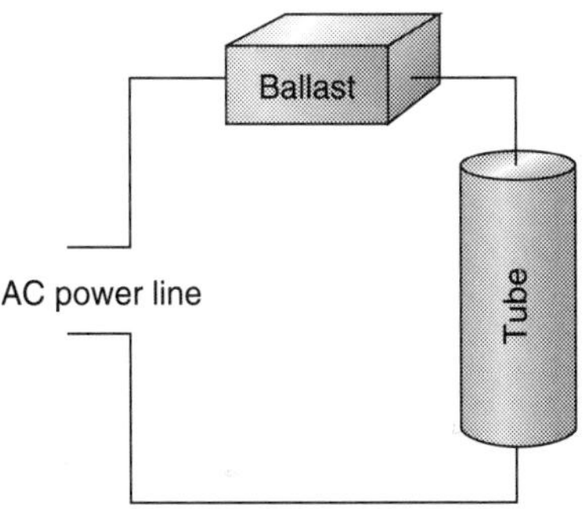

Fig. 15.2.4 - A fluorescent light fixture has a ballast in series with the fluorescent tube. The ballast is an inductor-like coil that regulates the current through the tube. The ballast is also partly responsible for starting the discharge when you turn on the light.

CHECK YOUR UNDERSTANDING #4: Sorry, We Forgot the Coating

If there were no phosphor coating on the inside of a fluorescent tube, what would you see when the lamp operated?

Fluorescent Lamps and Light Fixtures

Having seen what goes on inside the fluorescent tube, it's time to look at the rest of the lamp. Its electric system is more complicated than the coiled wire filament of an incandescent light bulb because gas discharges are inherently unstable. They're hard to start and difficult to control once started. Think about a spark of static electricity. At first nothing happens and then...zap!...an intense, uncontrolled discharge occurs. That spark isn't so different from the discharge inside a fluorescent tube. If nothing controls the current through a fluorescent tube, it will become so large that the tube will break.

One simple but effective current regulator is an inductor. Most fluorescent light fixtures connect their fluorescent tubes in a series circuit with an inductor-like device called a *ballast* (Figs. 15.2.4 and 15.2.5). The ballast opposes any changes in the current flowing through that circuit, thereby keeping the tube's current to a safe level. While the ballast can't limit the current forever, it's helped by the power line. With alternating current flowing through the circuit, the ballast only has to keep the current steady for a small fraction of a second before the current reverses and the discharge restarts in the opposite direction.

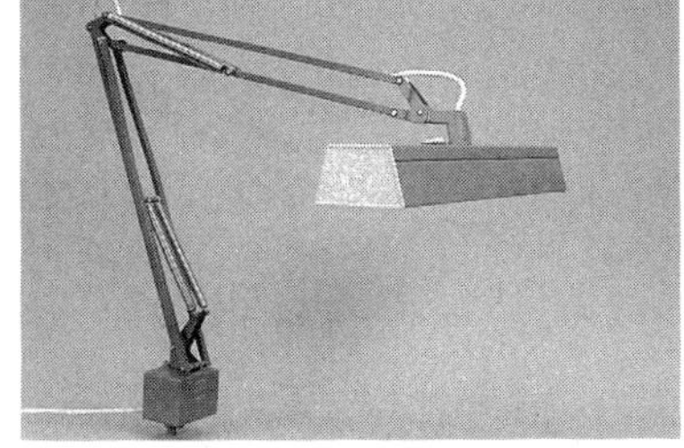

Fig. 15.2.5 - The ballast of this fluorescent desk lamp is built into its base. This ballast helps start the discharge in the lamp's mercury vapor-filled tubes and then limits the current in that discharge.

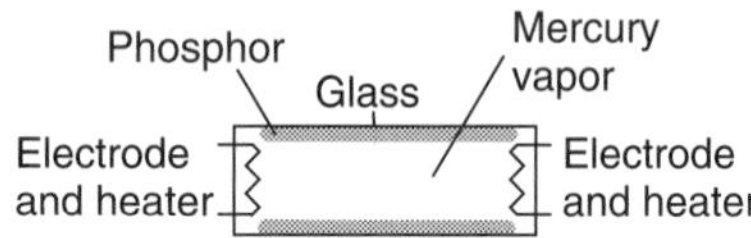

Fig. 15.2.6 - In a hot-electrode fluorescent tube, the electrodes are actually filaments and are heated by running currents through them.

The ballast also helps to start the discharge, either by heating the tube's electrodes until they emit electrons into the gas or by providing the high voltages needed to create corona discharges in the tube. Which of these starting techniques is used depends on the fluorescent light fixture. The hot-electrode fixtures heat the electrodes by running current through them (Fig. 15.2.6). In fact, the electrode at each end of the fluorescent tube is actually a tungsten filament. The tube has a pair of electric contacts on each end and the filament is heated by running current through it from one contact to the other.

There are two different kinds of hot-electrode light fixtures: preheat and rapid-start. *Preheat fixtures* send current through the filaments only while the lamp is starting. Once the discharge has begun, electrons colliding with the filaments keep them hot enough to sustain the discharge. In some preheat fixtures, you warm the filaments by holding down a start button and the discharge begins when the button is released. On automatic preheat fixtures, this starting process is handled electronically. In either case, the discharge won't start if the filaments haven't gotten hot enough. The automatic fixtures often blink when they try to start the discharge before the filaments are hot enough. Preheat fixtures can't be dimmed, since reducing their discharges would let the filaments cool off.

Rapid-start fixtures send current through the filaments continuously. When a rapid-start fixture is turned on, it immediately begins to heat the tube's filaments and the discharge begins as soon as there are enough electrons in the tube to initiate it. Since the discharge doesn't have to heat the filaments, rapid-start fixtures can be turned on or off easily. They can even be dimmed.

The third style of fluorescent light fixture is the *instant-start fixture*. Instead of heating the electrodes, the instant-start fixture's ballast applies a very high starting voltage to the tube. This voltage produces a corona discharge around each electrode, introducing electrons into the gas and initiating the discharge. All of this happens so quickly that the lamp begins producing light as soon as you turn it on. Since they have no filaments to heat, instant-start tubes have only a single electric contact at each end.

To keep electrons moving swiftly through a fluorescent tube, it must have a substantial electric field inside it. The longer the tube, the larger the voltage drop the tube needs to maintain this electric field. Power line voltages (110 V to 240 V) are appropriate for tubes up to 3 m in length. The tube must also have positively charged mercury ions in it to keep the electrons from repelling one another to the walls. These ions are produced in the discharge by particularly energetic collisions. The result, a gas-like mixture of positively charged ions and negatively charged electrons is called a **plasma**. All discharge lamps contain plasmas.

A neon sign resembles a fluorescent lamp except that its tube contains only neon gas and has no phosphor coating inside. A high voltage produces a corona discharge at each electrode and pushes current through the gas. The neon atoms emit red light after being struck by electrons because one of the neon atom's radiative transitions produces a 633 nm photon. While a true neon tube has no fluorescent phosphor coating to alter the wavelength of light it emits, many other colored advertising tubes use a mercury discharge and a colored phosphor coating inside the tube. If you look at them closely, you'll see that their light comes from the coating. In a true neon tube, the light comes from the neon gas itself.

Fluorescent tubes usually fail when their electrodes wear out. Most of this wear is caused by a process called **sputtering**, in which positive mercury ions from the plasma collide with the electrodes and chip away the tungsten atoms. Because sputtering is most severe when starting a preheat fixture, its fluorescent tubes fail after a few thousand starts. That's why you shouldn't turn a fluorescent lamp on and off more than once every few minutes.

Fluorescent tubes also have problems operating in extreme temperatures. Below 15° C, the density of mercury atoms becomes too low for the lamp to produce much light, and above 40° C, the density of mercury atoms becomes so high that the lamp wastes energy. The argon, neon, and/or krypton gases inside the tube help to start and sustain the discharge, particularly at lower temperatures.

CHECK YOUR UNDERSTANDING #5: Getting Red-y

When some fluorescent lamps are starting, you can see the ends glow deep red. Explain this observation.

Other Types of Discharge Lamps

There are four other important gas discharge lamps: high-pressure mercury lamps, metal-halide lamps, and low- and high-pressure sodium lamps. High-pressure mercury lamps differ from fluorescent lamps because they are able to extract visible light directly from the mercury atoms. Even in a low-pressure mercury discharge, the mercury atoms emit a little visible light. But when the density of mercury atoms in the discharge is very high, the mercury atoms begin to emit more visible light than ultraviolet light. This change occurs because ultraviolet light becomes trapped in the mercury atoms and only the visible light is able to escape from the discharge.

This **radiation trapping** is an interesting effect. Not only do mercury atoms often emit 254 nm photons, they also tend to absorb those same photons. That's because the same radiative transition that creates a 254 nm photon can also run backward to absorb that photon. In a dense gas of mercury, whenever one mercury atom tries to emit a 254 nm photon, another mercury atom snaps it up. So while the discharge keeps pouring energy into the mercury atoms, they are unable to get rid of this energy as 254 nm photons. Instead, they emit much of their energy through radiative transitions between other excited states. Light emitted by these other transitions is much less likely to be caught by other mercury atoms and emerges from the lamp. Many of these transitions produce visible light, particularly in the blue range of the spectrum.

When you first turn on a high-pressure mercury lamp, most of the mercury is liquid. The lamp starts like a small fluorescent tube without phosphor, so you see very little light. But the tube is designed so that it becomes hot during its operation and the liquid mercury evaporates to form a dense gas. As the tube heats up, the pressure of mercury atoms in the gas increases and the tube begins to emit a brilliant, blue-white light.

To make a lamp that is a little less bluish, some high-pressure mercury lamps contain additional metal atoms. These atoms are introduced into the lamps as metal-iodide compounds, making them metal-halide lamps. Sodium, thallium, indium, and scandium iodides all help to strengthen the red end of the spectrum and give metal-halide lamps a warmer light.

Pure sodium lamps resemble mercury lamps, except that they use different atoms. Sodium is a solid at room temperature, so both low- and high-pressure sodium lamps must heat up before they begin to operate properly. A low-pressure sodium lamp is extremely energy efficient because its 590 nm light comes directly from a sodium atom's strongest radiative transition. Many highways are illuminated by the yellow-orange glow of low-pressure sodium lamps.

But this monochromatic illumination is unpleasant and permits no color vision at all. While it may be acceptable on a highway, you wouldn't want it by your home. One way to make a sodium lamp more tolerable is to use high-pressure sodium rather than low-pressure sodium (Fig. 15.2.7). As the density of sodium atoms in the gas increases, less of the 590 nm light escapes from the discharge and more of the other wavelengths become visible.

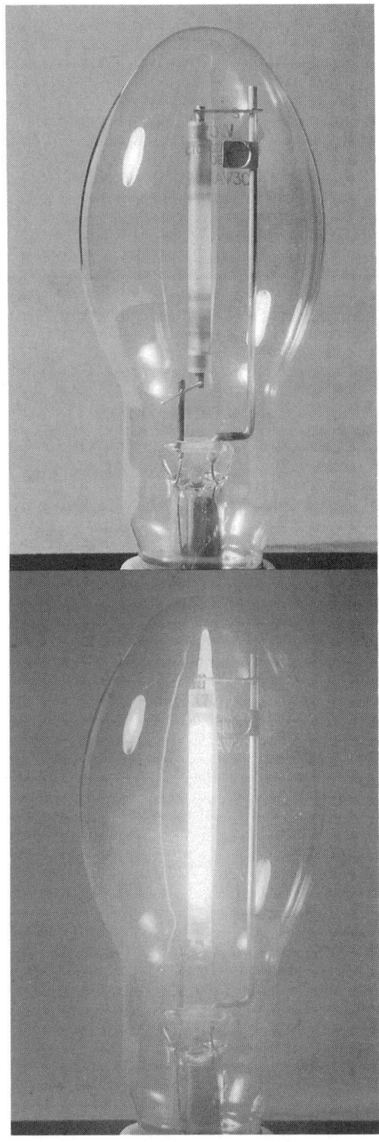

Fig. 15.2.7 - The active component of a high-pressure sodium vapor lamp is a small translucent tube. As the lamp warms up, sodium metal in the tube evaporates to form a brilliant yellow discharge.

The 590 nm emission in a high-pressure sodium lamp also smears out to cover a wide range of wavelengths, from yellow-green to orange-red. This smearing out occurs because of the many collisions suffered by the sodium atoms as they try to emit the 590 nm light. These frequent collisions in the high-pressure gas effectively distort the orbitals so that the atoms emit photons with somewhat different energies. Overall, a high-pressure sodium lamp emits remarkably little light exactly at 590 nm because that light is trapped by the atoms. There is actually a hole in the lamp's spectrum right at 590 nm.

The high-pressure discharge lamps suffer from a problem not found in low-pressure lamps: they are difficult to start when hot. It's much harder to initiate a discharge in a high-pressure gas than in a low-pressure gas. If the discharge in a high-pressure mercury, sodium vapor, or metal-halide lamp is interrupted, the lamp must cool down before it can be restarted.

CHECK YOUR UNDERSTANDING #6: Slow Glowing

At dusk, a mercury streetlight turns on. It glows dimly at first and gradually increases in brightness. What is happening during this warm-up period?

Summary

How Fluorescent Lamps Work: A fluorescent lamp emits visible light when the phosphor coating on the inside of its tube is exposed to ultraviolet light. This ultraviolet light is produced inside the tube by a mercury vapor discharge. Electrons are injected into the tube by heated electrodes at each end and these electrons carry current through the tube. When the electrons collide with mercury atoms, these atoms are excited and begin to make transitions that return them to their ground states. While making these transitions, the mercury atoms emit light. The strongest emission is ultraviolet light at 254 nm. This light excites the phosphor coating and causes it to emit visible light.

To keep the discharge stable, the fixture includes a ballast. This ballast behaves as an inductor and keeps a relatively steady current flowing through the tube. With each reversal of alternating current in the power line, the discharge stops and restarts in the reverse direction. The ballast prevents the current passing through the tube from becoming excessive until the next current reversal starts the discharge all over again.

Important Laws and Equations

1. Relationship Between Energy and Frequency: The energy in a photon of light is equal to the product of the Planck constant times the frequency of the light wave, or

$$\text{energy} = \text{Planck constant} \cdot \text{frequency}\,. \qquad (15.2.1)$$

The Physics of Fluorescent Lamps

1. Gases don't normally conduct electric currents. When charges are injected into a gas and then pushed through it by electric fields, the result is a gas discharge—a flow of current through the gas.

2. Because of the Pauli exclusion principle, at most two electrons can occupy each orbital of an atom. While the atom's electrons normally occupy its lowest energy orbitals, so that the atom is in its ground state, a transfer of energy to the atom can shift one or more of these electrons to a higher-energy, unoccupied orbital. The atom is then in an excited state.

3. An atom in an excited state can emit light in a series of radiative transitions.

4. In a radiative transition, an electron shifts from one orbital or level to another while emitting or absorbing a photon of

light. If the electron's energy decreases, it emits a photon. If its energy increases, it absorbs a photon.

5. Both light and electrons (as well as other objects) travel as waves but are detected as particles.

6. A particle of light, a photon, contains energy equal to the Planck constant times the frequency of its wave.

7. A photon emitted or absorbed by an atom must correspond to the energy difference between two of that atom's states.

8. Because atoms tend to absorb the same wavelengths of light they emit, such wavelengths of light can become trapped in high-pressure gases.

9. Collisions modify an atom's states and allow it to emit or absorb photons that it would otherwise not be able handle. Such collisions are common in a high-pressure gas.

Check Your Understanding - Answers

1. Orange

Why: Your red-, green-, and blue-sensitive cone cells have the same response to this light mixture as they would have to pure 600 nm light. Such light appears orange, so you see orange when viewing the mixture.

2. It decreases the lamp's energy efficiency.

Why: Although the bluish filter changes the spectrum of wavelengths so that the amount of blue light leaving the lamp increases relative to the red light, it does this by absorbing some of the red light. The filter gets hot and less electric power used by the bulb leaves as visible light.

3. When fire adds energy to an atom and shifts it to an excited state, it emits photons in order to return to its ground state. Each photon's energy, frequency, and color are determined by energy differences between the atom's orbitals.

Why: Each color in fireworks is produced by a particular type of atom. Some atoms, such as strontium and lithium, emit red light as they return to their ground states. Barium atoms emit green light, copper atoms emit blue light, and sodium atoms emit yellow-orange light.

4. You would see only a dim (blue-white) glow from the lamp because most of the light produced by the mercury discharge is invisible ultraviolet light.

Why: Without its phosphor coating, a fluorescent tube produces very little visible light. The phosphor coating is needed to convert the discharge's ultraviolet light into visible light. But even if you could see ultraviolet light, you would see very little of it leaving the tube. The tube is made of glass, which absorbs virtually all ultraviolet light with wavelengths shorter than 350 nm. Even in a normal fluorescent tube, any ultraviolet not converted to visible light by the phosphors is absorbed by the glass tube.

5. To start many lamps, their electrodes must be heated red hot. You can often see light emitted by these hot filament electrodes.

Why: In preheat fluorescent fixtures, the filaments are heated to high temperatures just before the discharge starts. Thermal energy then ejects electrons from the filament so that they can carry current through the gas.

6. The small high-pressure discharge tube is heating up, vaporizing more mercury.

Why: When the streetlight first turns on, the pressure of mercury atoms inside it is low and it emits very little visible light. As the discharge heats the tube, more mercury atoms evaporate until eventually all of the mercury inside the tube is gaseous.

Check Your Figures - Answers

1. $6.626{\cdot}10^{-28}$ J.

Why: Since the Planck constant is $6.626{\cdot}10^{-34}$ J-s and the frequency of the radio wave is 10^6 Hz or 10^6 cycles/second, the energy per photon is given by Eq. 15.2.1 as

$$\text{energy} = 6.626 \cdot 10^{-34}\,\text{J} \cdot \text{s} \cdot 10^{6}\,\text{cycles/s} = 6.626 \cdot 10^{-28}\,\text{J}.$$

This energy is so small that it's virtually impossible to observe the particulate character of radio waves.

Glossary

discharge A flow of electric current through a gas.

excited state A configuration of a system in which the electrons (or other particles) are not all in the lowest energy orbitals or levels.

fluorescence An emission of light that immediately follows an absorption of light from somewhere else.

ground state The lowest energy configuration of a system in which all of the electrons (or other particles) are in the lowest energy orbitals or levels.

incandescence The emission of thermal radiation from a hot object.

luminescence The emission of light by any means other than thermal radiation.

Planck constant The fundamental constant of quantum physics, equal to the energy of an object divided by the frequency of its wave. It is about $6.626{\cdot}10^{-34}$ J·s.

plasma A gas-like phase of matter consisting of electrically charged particles, such as ions and electrons.

primary colors of light The three colors of light—red, green, and blue—that are sensed by the three types of color-sensitive cone cells in our eyes. Mixtures of these three colors of light can make our eyes perceive any possible color.

radiation trapping The phenomenon in which a particular wavelength of light has trouble propagating through a material that eagerly absorbs and emits it. The light passes from one atom to the next and makes little headway.

radiative transition The movement of an atom or other system from one state to another through the emission or absorption of an electromagnetic wave.

sputtering Ejection of atoms from a surface caused by the impact of energetic ions, atoms, or other tiny projectiles.

state A possible arrangement of electrons (or other particles) in a system.

Review Questions

1. How does a color television set make you perceive yellow when all it has to work with are red, blue, and green dots?

2. Why does light from an incandescent light bulb appear redder than sunlight?

3. How do collisions between electrons and mercury atoms lead to the emission of ultraviolet light by those atoms?

4. What determines the wavelength of ultraviolet light that a mercury atom emits after being struck by an electron?

5. The light emitted by mercury atoms in a fluorescent lamp is not thermal radiation because the atoms are not hot. Identify two other cases in which a cold object emits light.

6. Detecting individual X-ray photons is relatively easy and is done routinely in airport security machines. In contrast, detecting individual radio wave photons is extremely difficult. Explain the difference.

7. When you walk into a room illuminated only by invisible ultraviolet light, some pieces of your clothing may begin to emit colored light. How are they producing this light?

8. Why can't a fluorescent tube operate directly from the AC power line; why does it require a ballast?

9. The 254 nm ultraviolet light emitted by excited mercury atoms has enormous difficulty traveling through a cell of high-pressure mercury vapor. Why?

Exercises

1. Light near 480 nm has a color called cyan. What mixture of the primary colors of light makes you perceive cyan?

2. Why can't you get a suntan through a glass window?

3. Incandescent light bulbs come with either frosted or clear glass bulbs. Why do fluorescent lamps always appear frosted?

4. Fancy makeup mirrors allow users to choose either fluorescent or incandescent illumination to match the lighting in which they'll be seen. Why does the type of illumination affect their appearances?

5. While many disposable products no longer contain mercury, a potential pollutant, fluorescent tubes still do. Why can't the manufacturers eliminate mercury from their tubes?

6. When white fabric ages it begins to absorb blue light. Why does this give the fabric a yellow appearance?

7. To hide yellowing (see Exercise **6**), fabric is often coated with fluorescent "brighteners" that absorbs ultraviolet light and emits blue light. In sunlight, this coated fabric appears white, despite absorbing some blue sunlight. Explain.

8. Increasing the power to an incandescent bulb makes its filament hotter and its light whiter. Why doesn't increasing the power to a neon sign change its color?

9. If the low-pressure neon vapor in a neon sign were replaced by low-pressure mercury vapor, the sign would emit almost no visible light. Why not?

10. A fluorescent tube is most efficient when the 254 nm photons emitted by its mercury atoms reach the phosphor coating without being absorbed and reemitted too many times by those atoms. Each absorption and reemission risks wasting energy as heat. Use this fact to explain why a fluorescent tube operating above 40° C is relatively inefficient.

11. Use the information in Exercise **10** to explain why fluorescent tubes are made relatively narrow.

Problems

1. The yellow light from a sodium vapor lamp has a frequency of $5.08 \cdot 10^{14}$ Hz. How much energy does each photon of that light carry?

2. If a low-pressure sodium vapor lamp emits 50 W of yellow light (see Problem **1**), how many photons does it emit each second?

3. A particular X-ray has a frequency of $1.2 \cdot 10^{19}$ Hz. How much energy does its photon carry?

4. A particular light photon carries an energy of $3.8 \cdot 10^{-19}$ J. What are the frequency, wavelength, and color of this light?

5. If an AM radio station is emitting 50,000 W in its 880 kHz radio wave, how many photons is it emitting each second?

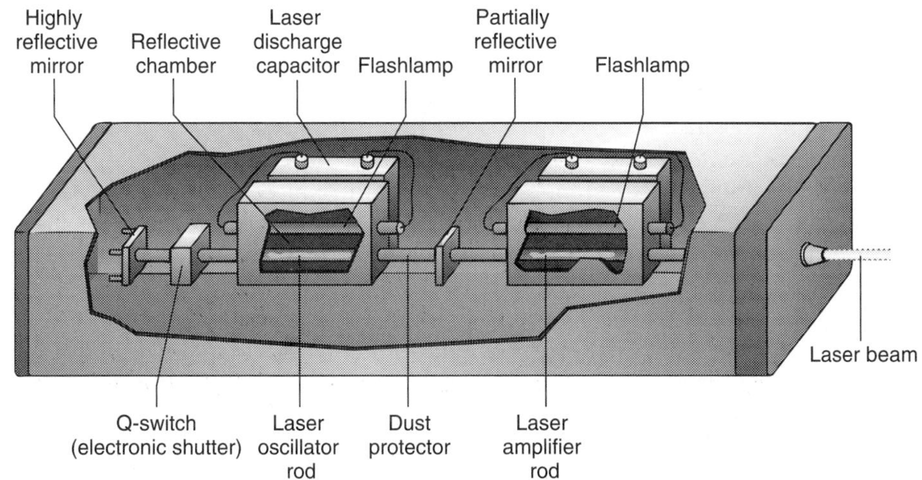

Section 15.3

Lasers

Few devices have inspired our imaginations more than lasers. Since their invention in the late 1950's, lasers have found countless uses, from cutting metal and clearing human arteries to surveying land and playing compact discs. Lasers are even being used to look for gravity waves from the collapse of distant stars. But lasers are more than just novel applications of old ideas. Instead, they bring together quantum and optical physics to produce a new type of light. This light is radically different from that produced by incandescent and fluorescent lamps and its properties make it particularly useful for many applications. In this section we will examine the nature of this new light and the ways in which lasers produce it.

Questions to Think About: *Why is laser light usually brightly colored? Why does laser light often appear as a narrow beam? In movies, "lasers" often emit bright streaks of light that can be dodged if one jumps quickly enough; is that view realistic?*

Experiments to Do: *While lasers are household objects, the ones in CD players and laser printers are hopelessly inaccessible. If you don't happen to own a laser pointer, look for a laser at a grocery store that scans bar-code labels at the register. The scanning system contains a gas or solid state laser that emits a very narrow beam of bright red light. This system also contains a rotating mirror or holographic disk that directs the beam as a pattern of thin stripes onto anything that passes over the scanner. A light sensor inside the window watches for this moving beam of light to travel across a braced label. If you look down at the laser light emerging from the scanner's window, you will see both the purity of its red color and its strange speckled character. The light appears to consist of tiny light and dark speckles. These speckles are caused by interference effects that are extremely pronounced in the ordered laser light. This light also appears unusually bright and, as when looking at the sun, you should keep you gaze brief to avoid eye injury.*

Lasers and Laser Light

To understand lasers, you must understand how laser light differs from the normal light emitted by hot objects or by individual atoms in an electric discharge. Each particle of normal light, each photon, is emitted willy-nilly without any relationship to the other light particles being emitted nearby. Because of this light's independent and unpredictable character, it's called *spontaneous light* and its creation is called **spontaneous emission of radiation**.

But theoretical work by Albert Einstein and others in the 1920s and 1930s predicted the existence of a second type of light, *stimulated light*, that can be created when an excited atom or atom-like system duplicates a passing photon. While this **stimulated emission of radiation** can only occur when the excited atom is capable of emitting the passing photon spontaneously, the copy that it produces is so perfect that the two photons are absolutely indistinguishable. Together, these two photons form a single electromagnetic wave.

To get a slightly better picture for how such stimulation occurs, think about an isolated atom in an excited state. That atom will eventually return to its ground state, but it must emit one or more photons in order for this to happen. That atom waits around in the excited state until it begins a spontaneous radiative transition. During the transition, one of the atom's electrons accelerates back and forth and the atom emits a photon.

But if a similar photon passes through the atom as it's waiting in the excited state, that photon's electric field can stimulate the radiative transition process through sympathetic vibration. The field pushes and pulls on the atom's electrons and makes them accelerate back and forth. While this effect is small, it may be enough to trigger the emission of light. If the atom does emit light, the photon it will produce will be a perfect copy of the stimulating photon.

When this stimulated emission process was first discovered, people immediately recognized that it made light amplification possible. If enough excited systems could be assembled together, a single passing photon could be duplicated exactly over and over again. Instead of a single particle of light, you would soon have thousands, or millions, or even trillions of identical light particles.

However, implementing this idea had to wait until the late 1950's, when the technical details for how to actually achieve light amplification were worked out. In 1960, the first laser oscillators were constructed. These were devices that emitted intense beams of light, in which each particle of light was identical to every other particle of light—a single particle of light had been duplicated by the stimulation process into countless copies.

When individual excited atoms or atom-like systems emit light through spontaneous emission, the particles of light head off separately as many independent electromagnetic waves (Fig. 15.3.1*a*). Light consisting of many independent electromagnetic waves is called **incoherent light**.

But when that same collection of excited atoms or atom-like systems emits light by stimulated emission, all of the light particles are *absolutely identical* and form a single electromagnetic wave (Fig. 15.3.1*b*). Light consisting of many identical photons and a single electromagnetic wave is called **coherent light**. Because of its single wave nature, coherent light exhibits remarkable interference effects. These effects are easily seen in the coherent light emitted by lasers.

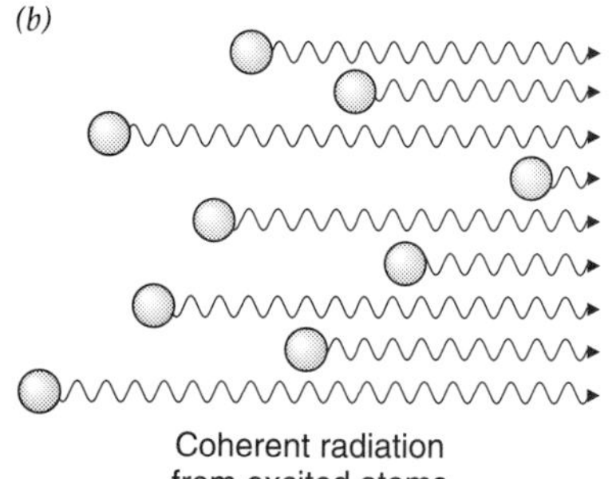

Fig. 15.3.1 - (*a*) Photons of incoherent light are created independently and have somewhat different wavelengths and directions of travel. (*b*) Photons of coherent light are produced by stimulated emission and are identical to one another in every way.

CHECK YOUR UNDERSTANDING #1: A Light Dusting of Photons
If you were to measure the electric field in the light from a flashlight, you would find that it fluctuates about randomly. Why is the light's electric field so disorderly?

Light Amplifiers and Oscillation

Producing coherent light requires amplification. You must start with only one particle of light and duplicate it many times. The basic tool for this duplication of light is a **laser amplifier** (Fig. 15.3.2). When weak light enters an appropriate collection of excited atoms or atom-like systems—the **laser medium**—that light is amplified and becomes brighter. The new light has exactly the same characteristics as the original light but it contains more photons.

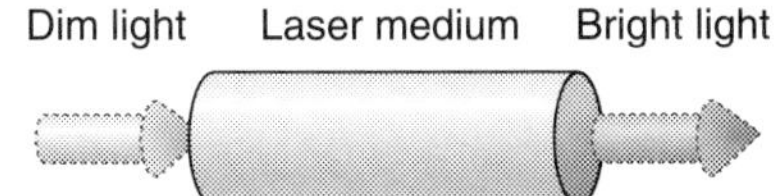

Fig. 15.3.2 - A laser amplifier uses excited atoms or atom-like systems to increase the number of light particles leaving the laser medium. The incoming light particles are duplicated by stimulated emission so that the outgoing light is a brighter copy of the incoming light.

But when we think of lasers, we rarely imagine a device that duplicates photons from somewhere else. We usually picture one that creates light entirely on its own. To do that, the laser must produce the initial particle of light that it then duplicates to produce others. A **laser oscillator** is a device that uses the laser medium itself to provide the seed photon, which it then duplicates many times (Fig. 15.3.3). If a laser medium is enclosed in a pair of carefully designed mirrors, it's possible for the stimulation process to become self-initiating and self-sustaining. However, the mirrors must be curved properly and must have the correct reflectivities. One mirror must normally be extremely reflective while the other must transmit a small fraction of the light that strikes its surface.

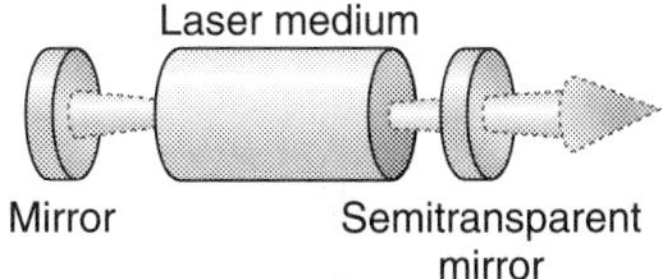

Fig. 15.3.3 - A laser oscillator is a laser amplifier enclosed in mirrors. Oscillation occurs when the laser medium spontaneously emits one photon in just the right direction. This photon bounces back and forth between the two mirrors and is duplicated many times. Some of the light is extracted from this laser by making one of the mirrors semitransparent.

When the laser medium is placed between these two mirrors, there is a chance that a photon, emitted spontaneously by one of the excited systems, will bounce off a mirror and return toward the laser medium. As that returning photon passes through the laser medium, it's amplified. Because the photon was emitted by one of the excited systems, it has the right wavelength to be amplified by other excited systems. (For a discussion of a photon's properties, see ❒.)

By the time the original photon leaves the laser medium, it has already been duplicated many times. This group of identical photons then bounces off the second mirror and returns for another pass through the laser medium. It continues to bounce back and forth between the mirrors until the number of identical photons in the collection is astronomical.

Eventually there are so many identical photons that the laser medium is no longer be able to amplify them. The laser medium has only so much stored energy and only so many excited systems in it. If the laser medium continues to receive additional energy, it may continue to amplify the light somewhat. But if it doesn't receive more energy, light amplification will eventually cease.

To let the light out of this laser oscillator, one of its mirrors is normally *semitransparent*—some of the photons that strike the surface of the mirror travel through it rather than reflecting. This transmission creates a beam of outgoing light; a *laser beam*. The laser beam continues to emerge from the mirror as long as the amplification process can support it. Because this laser beam consists of duplicates of one original photon, it's coherent light. For technical reasons, many lasers duplicate more than one original photon simultaneously, so that their laser beams are a little less coherent than they might be. However, with suitable fine-tuning, one original photon can usually be made to dominate the laser beam.

When you focus a flashlight's beam with a lens, its independent photons won't end up exactly together at the focus of the lens. That's because these photons leave the flashlight heading in somewhat different directions. But since virtually all of the photons in a laser beam are identical, they can all focus together to an extremely small spot. That's why a laser printer employs a laser; a laser beam can illuminate a very small spot on the photoconductor drum.

❒ While it might seem that a photon should have an exact wavelength and frequency, and travel in only one direction, that's not the case. Although they're emitted and absorbed as particles, photons travel as electromagnetic waves. Like all waves, photons spread while they move so that they can't follow a single direction of travel. And because each photon has a beginning and an end, its wave isn't perfectly steady and can't have only a single wavelength or a single frequency. Thus while lasers can produce some of the most perfect electromagnetic waves imaginable, consisting only of identical photons, those waves still spread outward slightly and still have a range of wavelengths and frequencies.

CHECK YOUR UNDERSTANDING #2: More of a Good Thing

If you take the laser beam from a particular laser oscillator and send it through a similar laser amplifier, what will happen to the laser beam?

How a Laser Medium Works

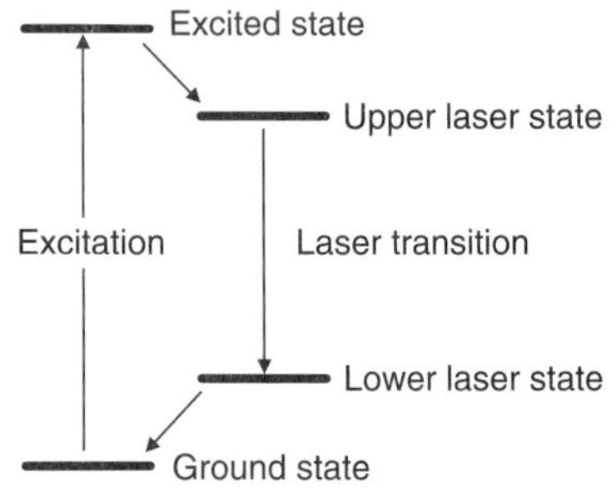

Fig. 15.3.4 - An ideal laser system passes through four different states during the laser's operation.

Obtaining the excited systems needed to amplify light is a critical issue for lasers. Ideally, a laser involves four different states of an atom or atom-like system: the ground state, an excited state, the upper laser state, and the lower laser state. The reason for having four separate states should become clear in a moment.

Let's consider an atom that acts as an ideal laser amplifier (Fig. 15.3.4). The atom starts in its ground state. A collision or the absorption of a photon shifts it to the excited state, giving it the energy it needs to amplify light. The atom then shifts to the upper laser state, either by emitting a photon or as the result of a collision. This preliminary shift is important because it prevents the excited atom from returning directly back to the ground state. Once it has shifted to the upper laser state, it's stuck there and is ready to amplify light.

When a suitable photon passes through the atom, that photon stimulates the emission of a duplicate photon and the atom undergoes a radiative transition to the lower laser state. However, if the atom remains in the lower laser state, it might absorb a photon of the laser light and return to the upper laser state. To prevent this reabsorption, the atom must quickly shift to the ground state, either by emitting a photon or as the result of another collision. The atom is then ready to begin the cycle all over again.

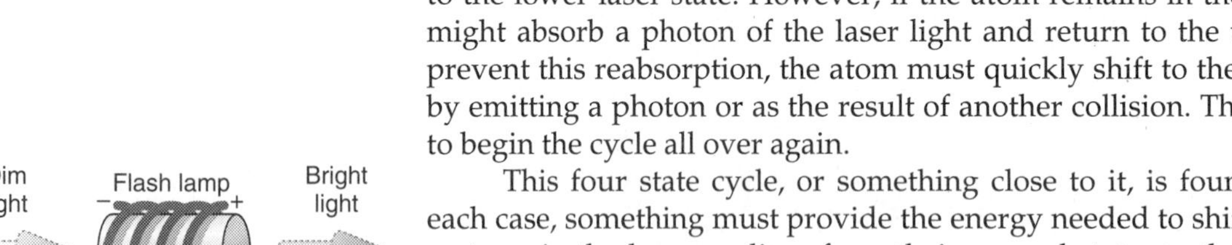

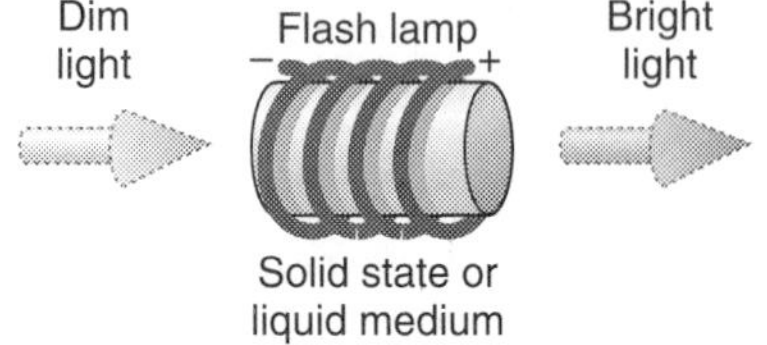

Fig. 15.3.5 - In an optically pumped laser, intense light from a flash lamp, an arc lamp, or even another laser transfers energy to the laser medium. Atoms or atom-like systems inside the medium store this energy and use it to amplify light.

This four state cycle, or something close to it, is found in most lasers. In each case, something must provide the energy needed to shift atoms or atom-like systems in the laser medium from their ground states to their exited states. This transfer of energy into the laser medium to prepare it for amplifying light is called *pumping*. How a particular laser medium is pumped depends on the laser.

In some lasers, the laser medium is pumped by intense light, such as that from a flash lamp, a high-pressure arc lamp, or another laser (Fig. 15.3.5). Two important examples of such *optically pumped lasers* are neodymium:YAG lasers and ruby lasers. YAG stands for yttrium aluminum garnet, a clear crystalline material that was once popular as an imitation diamond. The YAG crystal in a neodymium:YAG laser contains neodymium ions as impurities. These neodymium ions give the YAG crystal a purple hue and allow it to amplify light.

When it absorbs a photon of visible light, a neodymium ion shifts to one of several excited states. From there, it quickly loses energy and ends up in the upper laser state. It can remain in that state for as long as a thousandth of a second, waiting for light to come by at just the right wavelength. When the right light does come along, the ion duplicates it and shifts to the lower laser state. The ion then returns to its ground state.

Fig. 15.3.6 - The neodymium:YAG rod in this flashlamp-pumped laser amplifier is in the bottom half of the opened box, protected by a glass tube. Light from a long flashlamp in the top half of the box excites the neodymium ions so that they can amplify infrared light passing horizontally through the rod.

Neodymium:YAG lasers are common in industry, research, and medicine. They produce infrared light at 1064 nm and special optical techniques can combine the energy of several photons together to create visible and ultraviolet light. When they are pumped by flash lamps (Fig. 15.3.6), these lasers produce brief but extremely intense bursts of light and are called *pulsed lasers*. When they are pumped by arc lamps, they produce less intense but continuous beams of light and are called *continuous-wave lasers*.

A more well-known but less widely used laser is the ruby laser. A ruby laser uses a crystal of aluminum oxide (sapphire) containing chromium ions as impurities. These chromium ions give ruby its red color and are responsible for its activity as a laser medium. When the chromium ions in ruby are exposed to intense light, they shift to excited states and then to the upper laser state. A chromium ion can remain in the upper laser state for several thousandths of a second, waiting for a photon to duplicate.

However, a chromium ion's lower laser state is actually its ground state. Thus while chromium ions amplify red light at 694 nm, as they shift from the upper laser state to the lower laser state, they also tend to reabsorb that laser light. For this reason, ruby lasers only work when virtually all of the chromium atoms are shifted to the upper laser state at the same time. This requirement makes ruby lasers relatively impractical and they're rarely used now. Ruby lasers can only operate as pulsed lasers.

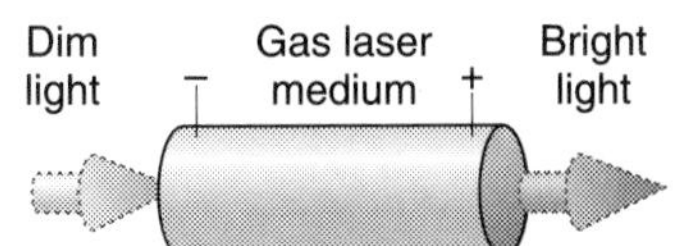

Fig. 15.3.7 - In a gas laser, an electric discharge excites atoms or molecules to the upper laser state. The gas atoms store energy until they are able to amplify light.

Some lasers use gas atoms or molecules as the laser medium, pumping them with an electric discharge (Fig. 15.3.7). In the discharge, collisions shift the atoms or molecules to their upper laser states and prepare them to amplify light. Two of the most common gas lasers are helium-neon lasers and carbon dioxide lasers. In a helium-neon laser (Fig. 15.3.8), the discharge actually excites the helium atoms, which transfer energy to the neon atoms during collisions. The collisions leave the neon atoms in their upper laser states.

Because neon atoms have several different lower laser states, they can amplify several different wavelengths of light. But the most widely used lower laser state produces red laser light. The 633 nm light from a red helium-neon laser can be seen in the product scanners of many grocery stores, although helium-neon lasers are gradually being replaced by solid state lasers in this application.

Carbon dioxide lasers use carbon dioxide gas as the laser medium. This gas is mixed with nitrogen molecules and these nitrogen molecules are excited by an electric discharge. Collisions transfer energy from the nitrogen molecules to the carbon dioxide molecules, leaving the carbon dioxide molecules in their upper laser states. In shifting to the lower laser states, the carbon dioxide molecules amplify 10,600 nm light. They then return to their ground states.

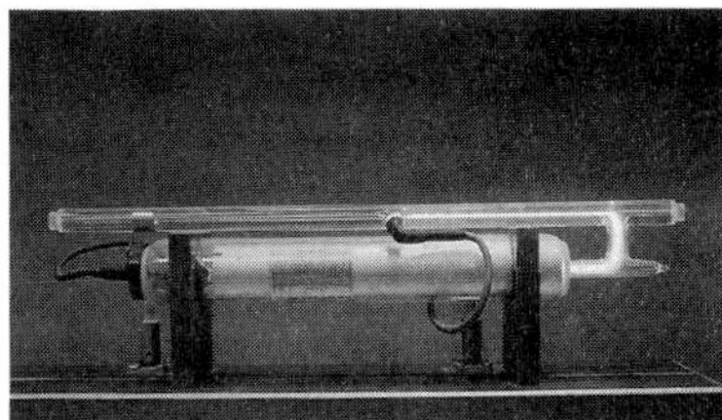

Fig. 15.3.8 - When current flows through a mixture of helium and neon gases, excited neon atoms are produced and form a laser medium. Mirrors at either end of the top tube turn this system into a laser oscillator that emits red light.

Although 10,600 nm (10.6 micron) light is in the infrared, carbon dioxide lasers are extremely useful for cutting and machining. They can convert as much as 30% of the electric power in the discharge into laser light. Industrial carbon dioxide lasers can produce beams of light containing over 10,000 W and can cut through heavy steel plates with ease.

Semiconductor lasers resemble diodes except that they emit light (Figs. 15.3.9 and 15.3.10). The excited systems are electrons in conduction levels that can emit light as they shift into empty valence levels. The lowest-energy conduction levels are the upper laser states and the highest-energy valence levels are the lower laser states. Sending an electric current through the laser transfers electrons out of the valence levels and into the conduction levels. As these electrons return to the valence levels, they amplify light. The wavelength and color of the amplified light depend on the exact characteristics of the semiconductors involved, particularly the energy separating its valence and conduction levels.

Semiconductor lasers or laser diodes are becoming extremely important in industrial and commercial applications. They are much cheaper and more reliable than any other kind of laser. They also have wonderful energy efficiencies, converting substantial fractions of the electric energy they consume into light. They are found most often in compact disc players, lightwave communication systems, and lecture pointers.

There are a few other ways in which to put energy into a laser medium, including violent chemical reactions, particle injection, shock waves, and even nuclear explosions. These methods aren't practical for household uses so you shouldn't expect them to show up in consumer products any time soon.

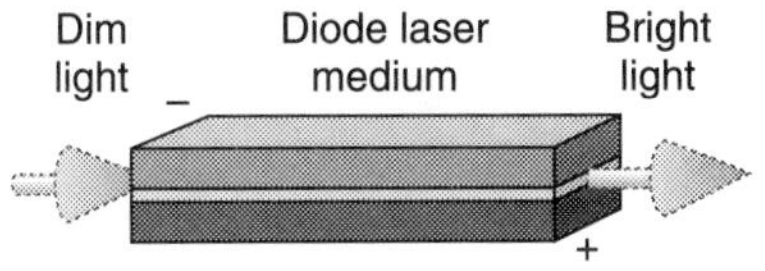

Fig. 15.3.9 - In a semiconductor laser, a current flowing through a diode moves electrons from valence levels to conduction levels. As the electrons return to the valence levels, they amplify light.

CHECK YOUR UNDERSTANDING #3: Not So Fast, 007

A secret agent in a move uses a tiny, hand-held laser to burn a hole through a thick metal plate. From the point of view of power, why is such a laser essentially impossible to make?

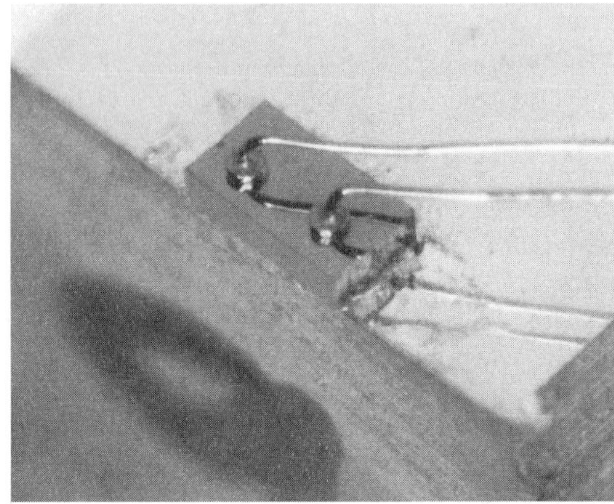

Fig. 15.3.10 - This tiny semiconductor chip is a diode laser that emits an intense beam of coherent light when a current flows through it. The light emerges from the lower left edge of the chip.

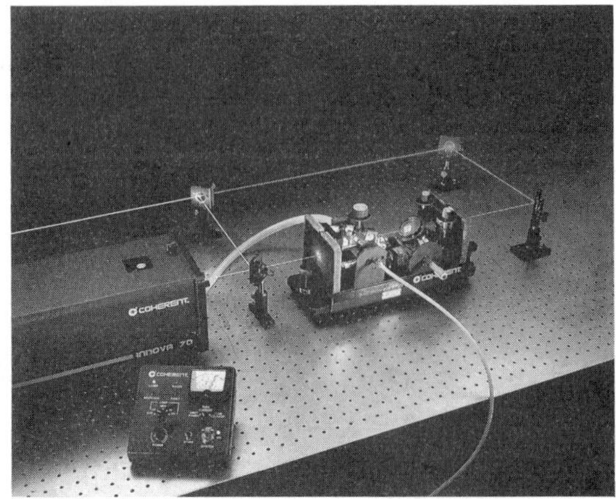

Fig. 15.3.11 - When the beam of an argon ion laser (left) illuminates the thin jet of liquid dye in the dye laser (right), the dye fluoresces brightly. Optical equipment in the dye laser causes the glowing dye to emit its own beam of intense coherent light. This new laser beam emerges toward the right.

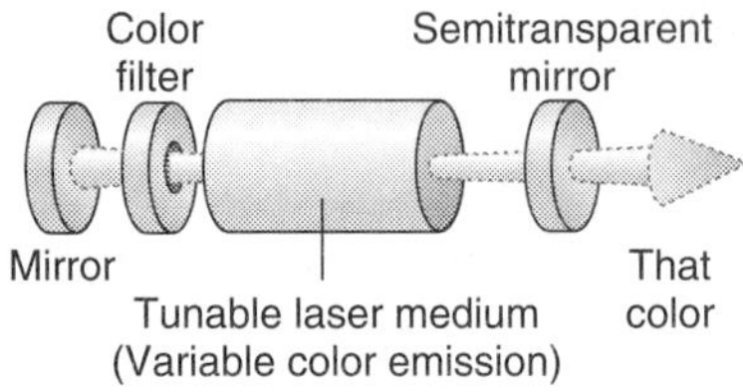

Fig. 15.3.12 - Some laser media can amplify many different wavelengths of light. Inserting a color filter into the region between the mirrors makes the laser oscillator emit only a particular wavelength of light.

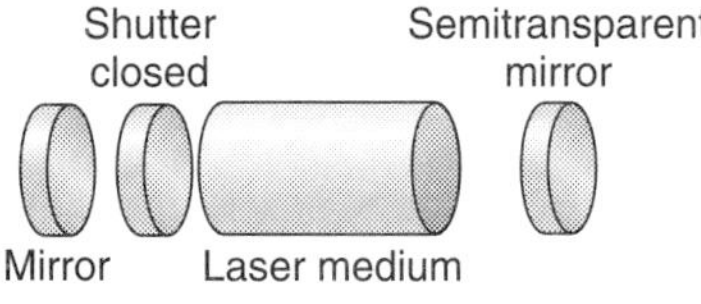

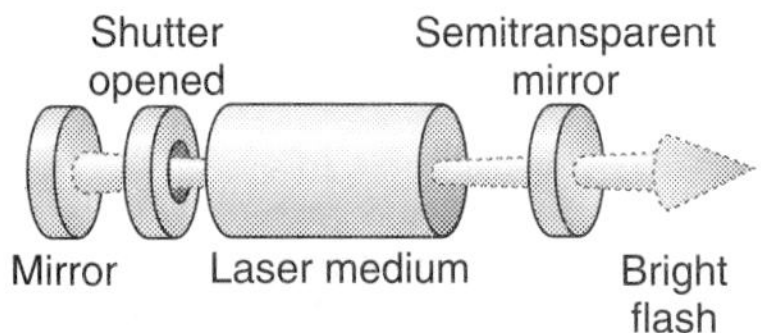

Fig. 15.3.13 - Inserting a shutter into the region between the mirrors allows a laser oscillator to store energy in the laser medium. When the shutter is opened, a giant laser pulse is created as all of the stored energy in the medium is quickly converted into light.

Variable Wavelength and High Intensity Lasers

A variable wavelength laser can be made by introducing a color filter into an appropriate laser (Fig. 15.3.11). Some fluorescent materials have many upper and lower laser states and can emit stimulated light over a range of wavelengths. The color filter selects which wavelength of light can reflect back and forth between the mirrors, so that the laser emits only light of that wavelength (Fig. 15.3.12).

If the color filter is adjustable, so that you can choose which wavelength of light it allows through, then the wavelength of the laser's light will change as you adjust the color filter. These adjustable-wavelength lasers are frequently used in research and medicine, where specific wavelengths of light can be used to control chemical or physical changes in atoms, molecules, or materials.

A high intensity laser is one that temporarily prevents any seed photons from entering the laser medium. With no seed photons to amplify, the atoms or atom-like systems in the laser medium will remain in their upper laser states for quite a while. This ability to store energy and wait makes it possible for a laser to emit extraordinarily intense bursts of light.

In addition to the two mirrors around the laser medium, these lasers include a shutter (Fig. 15.3.13). The shutter blocks light that is attempting to reflect from one of the two mirrors. Because light can't pass back and forth through the laser medium, the laser won't emit light while the shutter is closed and energy can be added to the laser medium gradually. When the shutter opens, light starts to bounce back and forth between both mirrors and is amplified over and over again. The laser emits a brilliant burst of light. The shutter is usually an *electrooptic device* that uses the electric properties of matter to control the passage of light.

In some electrically controlled lasers, the pulse of light that's produced when the shutter opens is only a few billionths of a second long. But the light emitted during that brief period is brighter than all the lights in a medium-sized city. Even more sophisticated short-pulse lasers can create bursts of light that are less than a trillionth of a second long (0.000000000001 s or 10^{-12} s) and that are brighter than all of the other lights on earth. The shortest pulses of laser light ever produced are less than 10^{-14} s long. These packets of light are only a few microns long as they move through space at the speed of light.

CHECK YOUR UNDERSTANDING #4: Short but Sweet

If a particular laser stores 1 J of energy in its excited atom-like systems over a period of 10^{-4} s and then emits all of that energy in a single burst only 10^{-8} s long, how much power is in the laser beam it produces?

Summary

How Lasers Work: Lasers amplify light through the process of stimulated emission. In this process, energy is transferred to atoms or atom-like systems contained in a laser medium. While these excited systems might emit light spontaneously, they can be stimulated into emitting duplicates of a passing photon. When a photon with just the right wavelength passes through an excited system, that system is likely to give up its stored energy by emitting an exact copy of the initial photon. In a laser oscillator, two mirrors cause light to bounce back and forth through the laser medium. An initial photon is duplicated endlessly to produce coherent light. One of the mirrors is semitransparent, so that part of the light emerges from the laser as a laser beam.

The Physics of Lasers

1. When a radiative transition happens spontaneously, it occurs at an unpredictable time and the photon it produces heads off in an unpredictable direction. Light from spontaneous transitions is incoherent.

2. When a radiative transition is stimulated by a passing photon, the emitted photon is an exact duplicate of the passing photon. The photons from stimulated transitions are absolutely identical and form coherent light.

Check Your Understanding - Answers

1. The flashlight produces incoherent light, with many independent light waves contributing their individual electric fields to the overall electric field. If you were to measure that overall field, you would find it very disorderly.

Why: Because the photons of incoherent light are independent, their individual electric fields don't fluctuate together. At any place and time, the individual electric fields will add in a complicated, random fashion. As time passes, this overall field will fluctuate randomly. But coherent light, in which each photon is the same as all the others, has a much more orderly electric field because all of the photons contribute equally.

2. It will become even brighter.

Why: A laser oscillator normally emits as intense a beam of light as it can, based on the amount of stored energy in the laser medium. This beam of light can be further amplified by sending it through a separate laser amplifier. Most high-powered lasers use a laser oscillator and one or more laser amplifiers to create particularly bright beams of light.

3. The light in such a laser beam would carry an enormous amount of power. Something must transfer that power to the laser medium, an impossible task in a tiny, hand-held unit.

Why: The power in a laser beam comes from the laser medium. The laser medium must have gotten it from somewhere else. Since batteries aren't up to delivering thousands of watts of electric power, it's unlikely that powerful hand-held lasers will ever be developed. Even if a suitable power source were available, these lasers would tend to overheat. The conversion of power to light is not perfectly efficient and most of the energy ends up as thermal energy in the laser's components. Lasers require cooling to remove this wasted energy.

4. About 10^8 W (the electric power used by a small city).

Why: In a typical high-power pulsed laser, the energy accumulates in the laser medium slowly. When all of that energy leaves the laser medium in a short burst of light, the light's power (its energy per second) is enormous. In this case, the laser beam's power is 100,000,000 W. More sophisticated and powerful lasers can produce light powers well in excess of 10^{12} W (the electric power used by the entire United States) for very short periods of time.

Glossary

coherent light Light consisting of identical photons that together form a single electromagnetic wave.

incoherent light Light consisting of individual photons, each its own independent electromagnetic wave.

laser amplifier A device that amplifies weak incoming light to produce brighter outgoing light. The outgoing light is a brighter copy of the incoming light.

laser medium An assembly of excited atoms or atom-like systems that is capable of amplifying passing light through stimulated emission.

laser oscillator A laser amplifier that is surrounded by mirrors so that it can amplify one or more spontaneously emitted photons to form an intense beam of coherent light.

spontaneous emission of radiation Light emission that occurs when an excited atom or atom-like system releases stored energy randomly through a radiative transition. The photon that results is independent and unique.

stimulated emission of radiation Light emission that occurs when an excited atom or atom-like system releases stored energy through a radiative transition by duplicating a photon passing through that system.

Review Questions

1. List three objects that emit light via spontaneous emission of radiation.

2. When one photon stimulates the emission of a second photon, why doesn't that new photon follow behind the original photon?

3. How does the electric field at one point in space change with time as a burst of incoherent light passes by?

4. How does the electric field at one point in space change with time as a burst of coherent light passes by?

5. Why can a laser amplifier only duplicate a photon if that photon has the right wavelength?

6. Why does a laser work best when the lower laser state of its atoms or atom-like systems isn't their ground state?

7. Why will a laser only work when more of its atoms or atom-like systems are in the upper laser state than in the lower laser state?

8. How does putting two mirrors, one of which is semi-transparent, around a laser amplifier cause it to begin emitting a laser beam?

Exercises

1. Explain why the electromagnetic wave emitted by a radio station is coherent—a low-frequency equivalent of coherent light.

2. If a radio transmitter and antenna can emit coherent electromagnetic radiation (see Exercise **1**), why aren't there similar transmitters and antennas for coherent light?

3. Does a radio transmitter form its coherent electromagnetic radiation (see Exercises **1** & **2**) by duplicating one photon over and over again, the way a laser does?

4. Is light emitted by a hot filament coherent or incoherent?

5. Is light emitted by a fluorescent lamp coherent or incoherent?

6. One of the most accurate atomic clocks is the hydrogen maser. This device uses excited hydrogen molecules to duplicate 1.420 GHz microwave photons. In the maser, the molecules have only two states: the upper maser state and the lower maser state (which is actually the ground state). To keep the maser operating, an electromagnetic system constantly adds excited state hydrogen molecules to the maser and a pump constantly removes ground state hydrogen molecules from the maser. Why does the maser require a steady supply of new excited state molecules?

7. Why must ground state molecules be pumped out of a hydrogen maser (see Exercise **6**) as quickly as possible to keep it operating properly?

8. While some laser media quickly lose energy via the spontaneous emission of light, others can store energy for a long time. Why is a long storage time essential in lasers that produce extremely intense pulses of light?

9. When an atom spontaneously emits a photon at a particular time, the photon's electromagnetic wave has a clear beginning and end. It resembles a radio wave that has been amplitude modulated so that there is no amplitude before a certain time and no amplitude after a certain time. Explain why having a beginning and end gives the photon a small range of frequencies and wavelengths, rather than a single frequency and a single wavelength.

10. A photon that's duplicated over and over again in a continuous-wave laser is smoothed by the laser until it has essentially no beginning or end (see Exercise **9**). Thus the wave emerging from the laser has essentially a constant amplitude. Why is this constant amplitude necessary for the laser light to have a single frequency and a single wavelength?

11. Even though light emerging from a short-pulse laser consists of many identical photons, it has a broad range of frequencies and wavelengths. Use your answers to Exercises **9** & **10** to explain this effect.

12. If you whistle the right tone near a violin string, the string will begin to vibrate. If you whistle the wrong tone, the violin string won't respond. Use these observations to explain why an excited atom will only duplicate a photon if the atom is capable of emitting that photon by spontaneous emission.

13. Astronomers occasionally detect stimulated emission in the electromagnetic waves coming from excited gases far out in space. There are obviously no mirrors out in space, so this radiation isn't coming from laser oscillators. What must be happening to produce this stimulated radiation?

14. A compact disc player uses a beam of laser light to read the disc, focusing that light to a spot less than 1 micron (10^{-6} m) in diameter. Why can't the player use a cheap incandescent light bulb for this task, rather than a more expensive laser?

15. When the shutter of the pulsed laser shown in Fig. 15.3.13 is suddenly opened, it isn't certain when the giant pulse of light will emerge from the laser. Why not?

16. To make sure that the pulsed laser in Fig. 15.3.13 emits its giant pulse of light immediately after the shutter opens (see Exercise **15**), the operators can insert a number of identical photons into the region between the two mirrors just before they open the shutter. Why will these photons ensure that the giant pulse begins immediately?

Epilogue for Chapter 15

In this chapter we explored the creation and movement of light. In *sunlight*, we looked at how light is scattered during its passage through the atmosphere and at how it bends and reflects when it moves from one material to another. We also examined the interference effects that occur when a light wave follows more than one path to a particular destination and learned how polarizing sunglasses can diminish glare.

In *fluorescent lamps*, we examined electric discharges in gases. We saw that atoms excited by collisions with charged particles can subsequently emit light through radiative transitions. We found that these radiative transitions emit light when an atom shifts from an excited state to one with a lower energy and absorb it when the atom shifts to a state of higher energy. We studied the primary colors of light to see how the phosphors on the inside surface of a fluorescent tube are able to produce a reasonable facsimile of white sunlight.

In *lasers*, we looked at the difference between common incoherent light and the unusual coherent light emitted by a laser. We saw that lasers use stimulated emission to duplicate photons, so that a small number of initial photons can be amplified into an enormous number. We learned how energy is stored in a laser medium, how a shutter can be used to release that energy as an extremely short burst of high intensity light, and how a wavelength filter can be used to control the wavelength of the laser light.

Experiment: Splitting the Colors of Sunlight

Light bends as it passes through the cut facets of the crystal glass or bowl. Because of dispersion, the angle of each bend depends slightly on the wavelength of the light involved, so that the different colors of sunlight follow slightly different paths through the crystal. When the light emerges from the crystal, its various wavelengths head in somewhat different directions and you see colors. The sequences of color progress from long wavelength to short wavelength, or vice versa, so that you see red, orange, yellow, green, blue, and violet, or the reverse. These are the colors of the rainbow.

Cases

1. The color of the sky depends on the weather and the air.

a. On a very humid day, when the air contains many water droplets that are about 1 micron (0.000001 m) in diameter, the air appears hazy and white. Why?
b. On a very clear, dry day when there is no dust or water in the air, the sky is very dark blue. Why?
c. Why is the sky black on the airless moon?
d. When you look at the surface of a calm lake, you see a reflection of the sky. The water is not a metal, yet you see a reflection. What change occurs as light passes from air to water that causes some of the light to be reflected?
e. Mirages appear on sunny days in the desert when very hot air from the earth's surface bends light from the horizon upward so that you see it as though it were coming from the ground in the distance. Hot air has fewer air molecules in it than colder air at the same pressure. Why does light bend slightly as it moves from colder air to hotter air?

2. The electronic flash in a typical camera is based on a xenon flashlamp, a tube filled with xenon gas at high pressure with an electrode at each end. This flashlamp is electrically connected to a small but powerful capacitor to form a circuit so that if the flashlamp were to conduct current, it would allow current to flow from one plate of the capacitor to the other.

a. Before you take a picture, the camera places separated electric charge on the two plates of the capacitor until a voltage drop of about 300 V appears across the xenon flashlamp. The flashlamp, however, conducts no current. Why not?
b. When you take a picture, the shutter opens and the camera causes a small high-voltage transformer to inject a few electrons into the gas in the flashlamp. The lamp suddenly allows current to flow from one plate of the capacitor to the other and the lamp "flashes." Why does this introduction of electrons into the flashlamp cause it to "flash"?
c. The flashlamp will only last for a certain number of flashes because each flash damages the electrodes. Why does the flash damage the electrodes?

d. The flashlamp uses high pressure xenon rather than low pressure xenon. Why does high pressure xenon give a more uniform spectrum of light than low pressure xenon?

e. The flashlamp uses xenon gas rather than sodium gas, in part because xenon emits light over a very broad range of wavelengths and does a good job of simulating sunlight. But why wouldn't a high-pressure sodium vapor flashlamp be practical, even if you didn't care that it was orange in color?

3. A paint is a plastic that contains dye molecules and small clear particles. While the dye molecules absorb certain wavelengths of light and give the paint its color, the particles are often completely clear. Nonetheless, these randomly shaped and oriented particles are very important to the paint's appearance.

a. The speed of light is different in the particles than it is in the plastic surrounding them. When light strikes the surface of one of these particles, why does some of it reflect?

b. A dyeless paint appears white because light bounces around among its particles and emerges heading in all directions. Sketch a few of the paths light could take in the paint and show how it could reach the surface beneath the paint.

c. Lead carbonate is a clear, toxic chemical that was put in paints before 1930 to make them white. The speed of light in a particle of lead carbonate isn't that much slower than in plastic. How does this fact help explain why "lead paints" weren't very good at covering dark surfaces, requiring lots of layers of paint before the surfaces really looked white?

d. Modern paints contain particles of clear, nontoxic titanium dioxide. Light travels very slowly in this material. Explain why titanium dioxide based paints look so white, even after only a single layer of paint.

e. Red paint usually contains a mixture of titanium dioxide particles and red dye molecules. The red dye molecules absorb any light that isn't red but they don't affect red light at all. How would the appearance of this paint change if it didn't contain any titanium dioxide particles?

4. Interior house paints contain dye molecules that absorb certain wavelengths of light and give the paints their colors.

a. Dye molecules only absorb light of certain wavelengths for the same reason as atoms. Explain that reason briefly.

b. The dyes in most house paints eventually convert the light energy they absorb into thermal energy. However, fluorescent or neon paints emit light of a new color. Why is the light emitted by fluorescent paints always longer in wavelength than the light they absorb?

c. A paint's appearance depends strongly on the light that illuminates it. Why are a white wall and a red wall indistinguishable when they're both illuminated by 650 nm light?

d. Two paints that contain different dyes can look indistinguishable when illuminated by sunlight, even though they absorb slightly different portions of the visible light range. In what way can their absorptions differ without your being able to see any difference between the paints?

e. Two paints that look indistinguishable in sunlight may look very different when illuminated by a fluorescent lamp. Why does the difference between the paints only appear for certain illuminations?

f. If you're trying to match new paint to old paint and want the match to work regardless of illumination, you'll have to find a new paint that uses exactly the same dye molecules as the old paint. Why?

5. While fluorescent lamps can be dimmed, it's not as easy as dimming an incandescent light bulb. Fluorescent lamps tend to turn off when they don't run at full power, so a dimmed fluorescent fixture must make sure that the discharge continues to operate, even at low power.

a. Why can a fluorescent fixture be dimmed only if it continuously heats the filaments of its fluorescent tubes?

b. While an incandescent light bulb can be dimmed by reducing the voltage drop across its filament, reducing the voltage drop across a fluorescent tube will spoil its discharge. With the voltage drop across the tube reduced, the average energy of its electrons would be substantially lower than normal and its mercury atoms would emit almost no ultraviolet light at all. Why is there a minimum energy that an electron must have in order to cause a mercury atom to emit ultraviolet light?

c. Reducing the voltage drop across the fluorescent tube would virtually stop the production of positive mercury ions in the vapor. That would make the tube's filaments last longer, but the discharge wouldn't operate. With no positively charged particles inside the tube, what would happen to the electrons as they flowed from one electrode to the other?

d. Rather than reducing the voltage drop across its fluorescent tubes, a dimmed fluorescent fixture shortens the time that the discharge is on following each reversal of current in the power line. An electronic switch allows current to flow through the tube and its ballast for only part of each half cycle of the alternating current. The switch senses the voltage reversal in the power line and then waits a certain amount of time before closing to allow current to flow. The switch remains closed until the current stops on its own at the next reversal of the power line, at which time the switch opens and begins waiting again. The longer into each half cycle the switch waits, the dimmer the fixture becomes. This arrangement allows the ballast to store energy as the current through it increases and use that stored energy to create light as the current through it decreases. Why shouldn't the switch try to open the circuit while there is a large current flowing through the fluorescent tube and the ballast?

e. Current passing through the electronic switching system of the dimmed fluorescent fixture experiences a small voltage drop. Why does the switching system get warm?

6. In 1965, American physicists Arno Penzias and Robert W. Wilson discovered that space is filled with thermal radiation, left over from the big bang that formed the universe. While this thermal radiation was once extremely hot, the universe's expansion has cooled it to only 3° K. It now consists mostly of microwaves that are extremely difficult to detect.

a. Satellite dish antennas are designed to detect coherent microwave radiation—a low-frequency equivalent of coherent light. Why are microwaves emitted by a normal microwave transmitter and antenna (e.g., a magnetron) coherent?

b. The microwaves reaching us from space are thermal radiation and effectively come from individual charged particles that moved back and forth to produce them. Why is this microwave radiation incoherent?

c. A normal satellite dish antenna will have trouble detecting microwaves from the big bang because they are incoherent. Why won't the charges on the antenna of a satellite dish respond strongly to these incoherent microwaves?

d. To detect the incoherent microwaves, Penzias and Wilson used a maser amplifier—a device that amplifies microwave photons via stimulated emission. How did introducing this amplifier make it possible to detect the microwave photons with a microwave antenna and receiver?

CHAPTER 16

OPTICS

Many of the devices around us perform useful tasks by manipulating light. Cameras record images of the objects in front of them while telescopes and microscopes allow us to look at things that would otherwise be too distant or small to see. Still other gadgets, such as compact disc players, use light in ways that have nothing to do with vision. But all of these objects manipulate light using similar techniques—the techniques of optics.

While optical tools such as lenses and prisms have been around for hundreds of years, the advances of modern technology have accelerated the development of optics. Just as the invention of transistors has sped the growth of the electronics industry, so the invention of lasers has sped the growth of the optics industry. The two fields are not far apart in many ways and there is hope that one day computers will be as much optical devices as they are electronic.

Experiment: Focusing Sunlight

There are many household devices that manipulate light and one of the most familiar is a magnifying glass. A magnifying glass bends light rays toward one another as they pass through it. In this chapter, we'll see how a simple converging lens of this sort can magnify an object or cast its image onto a sheet of film. For the moment, we'll simply use it to bring light rays together.

On a clear day with the sun roughly overhead, take a magnifying glass and a dark piece of paper outdoors to a place where there is nothing flammable around. You will use light from the sun to heat the paper until it smokes, so be prepared to extinguish a fire if necessary. Hold the magnifying glass just above the paper and slowly lift the lens upward toward the sun. Make sure that the lens is turned so that its surface faces the sun. You should see a bright circle appear on the dark paper and that circle should become smaller and smaller as you lift the lens. The sun's light rays are bending inward as they pass through the lens and are converging together as they head toward the paper. Be careful not to let this bright circle of light touch your skin because it will burn you. If the spot isn't round, try tilting the lens differently.

When the lens is a certain height above the paper, the circle of light will

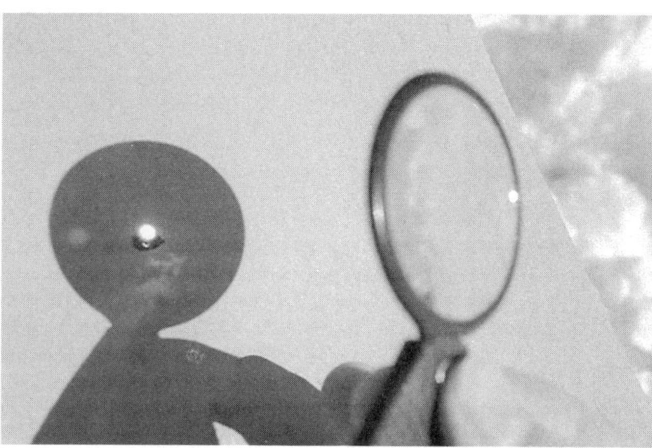

reach a minimum size. It will appear as a brilliant white disk that is actually an image of the sun itself. If the magnifying glass is large enough and the sun bright enough, the paper will begin to smoke. The image of the sun will be so bright that it will be able to heat the paper to its ignition temperature. If the lens is quite large, the paper may begin to burn with a flame.

Chapter Itinerary

This process of bringing light together to form an image of a distant object is a common theme in optics. In this chapter, we'll examine several devices that are based on this sort of manipulation of light: (1) *cameras*, (2) *telescopes and microscopes*, and (3) *compact disc players*. In *cameras*, we'll see how lenses bend light to form images and how those images are used to create photographs. In *telescopes and microscopes*, we'll explore the behavior of mirrors and see what happens when you look into a lens or mirror with your eyes. In *compact disc players*, we'll explore the roles of lasers in optics while investigating several novel optical effects.

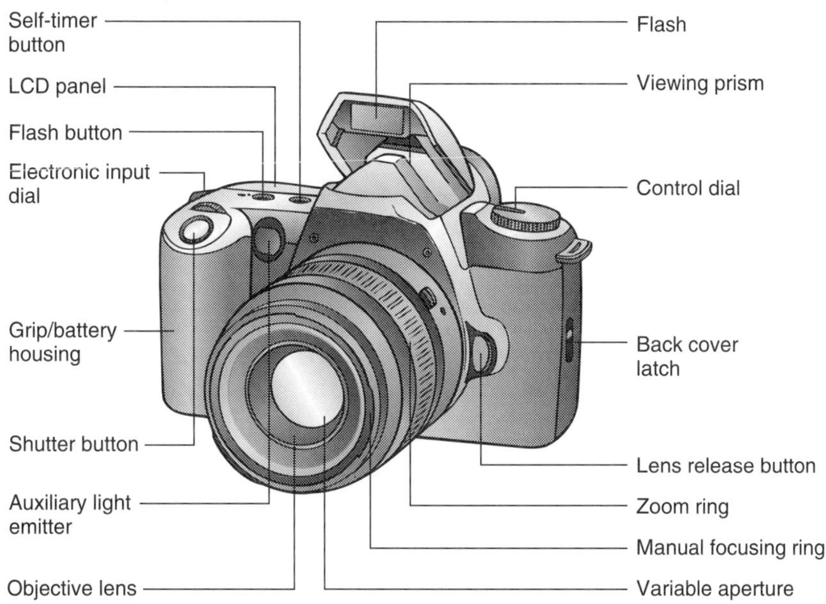

Section 16.1

Cameras

In the two centuries since photography was invented, cameras have become extremely easy to use. What started as a hobby for a few dedicated enthusiasts has evolved into an everyday activity. But despite all of the improvements in technology, photography still employs many of the same principles it did in the 1800's. Cameras still use lenses to record images on film and these images are still transferred to paper or projected on a screen. Electronic cameras are gradually emerging, but film cameras still dominate the market. In this section, we'll explore some of the principles that make these cameras work.

Questions to Think About: *Why are expensive camera lenses so complicated, with so many separate pieces of glass? Why does a longer lens seem to bring the objects nearer to you? What does a camera's aperture do? How does the camera's shutter speed affect the photograph? What is the difference between fresh and exposed film, before it's developed?*

Experiments to Do: *The basic activity of a camera—projecting an image of the scene in front of you onto the film—requires nothing more than a simple magnifying glass. In fact, you can use almost any simple lens that is bowed outward in the middle, including the eyeglasses of a farsighted person with no astigmatism. Stand in a darkened room, opposite a window. Hold a sheet of white paper so that it's parallel to the window and move the lens back and forth in front of the paper. Make sure that the lens itself is parallel to the window.*

You should find a distance at which the lens casts a clear image of the window onto the sheet of paper. You'll find that the scene appears upside down and backwards and that the image becomes fuzzy if you move the lens toward or away from the paper. You will also find that, when the window's image is sharp, the outside scene's image is fuzzy and vice versa. Which way must you move the lens to bring more distant objects into focus?

Cut a hole in a piece of cardboard and use it to cover all but the center portion of the lens. The image will be substantially darker but it will also be sharper over a wider range of distances from the paper. If the hole is narrow enough, virtually everything focuses at once. Evidently, the diameter of the lens and its ability to focus on several things at once are closely related. Why?

Lenses and Real Images

When you take a picture of the scene in front of you, the lens of your camera bends light from that scene into a real image on the camera's film. A **real image** is a pattern of light, projected in space or on a surface, that exactly reproduces the pattern of light in the original scene. Since the real image that's projected onto the film looks just like the scene you're photographing, recording the light in that image is equivalent to recording the appearance of the scene itself.

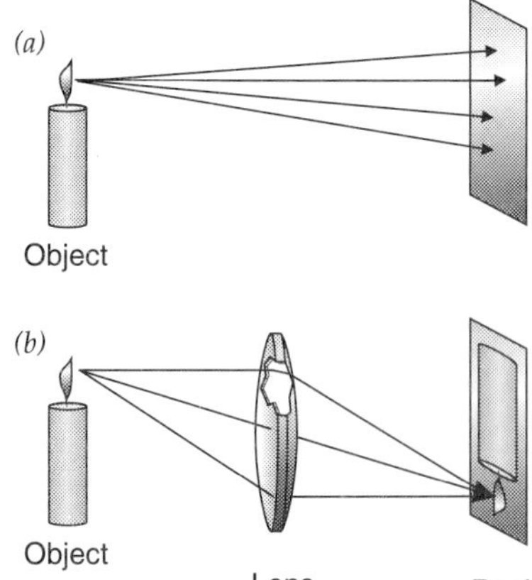

Fig. 16.1.1 - (*a*) Without a lens, light from a candle uniformly illuminates a sheet of photographic film. (*b*) When a lens is introduced between the candle and the film, it brings light from each point on the candle back together on the film's surface, forming an upside down and backward real image of the candle. The distance between the lens and the film must be chosen correctly or the image will be blurry.

Real images don't occur without help. When the light from a candle falls directly on a sheet of photographic film, it produces only diffuse illumination (Figs. 16.1.1*a* and 16.1.2). You couldn't tell by looking at the film what the candle looks like because the light that leaves the candle travels in all directions and is as likely to hit the top of the film as it is to hit the bottom.

That's why a camera needs a **lens**, a transparent object that uses refraction to form images. The light passing through a lens bends twice, once as it enters the glass or plastic and again as it leaves. In a camera lens, this bending process brings much of the light from one point on the candle back together at one point on the film. As you can see in Figs. 16.1.1*b* and 16.1.2, the real image that forms is upside and backward. This inversion of the real image relative to the object always happens when a single lens creates a real image.

The curved shape of the camera lens allows it to form a real image. Light passing through the upper half of the lens is bent downward while light passing through the lower half is bent upward. Because the camera lens bends light rays toward one another, it's a **converging lens**. You can see how it forms an image in Fig. 16.1.1*b* by following the rays of light leaving one point on the candle.

The upper ray from the candle flame travels horizontally toward the top of the lens. As it enters the lens and slows down, this ray of light bends downward. It bends downward again as it leaves the lens and travels downward toward the bottom of the film.

The lower ray from the candle flame travels downward toward the bottom of the lens and bends upward as it enters the lens. It bends upward again as it leaves the lens and travels horizontally toward the bottom of the film.

These two rays of light reach the film at the same point. They are joined there by many other rays from the same part of the candle flame, so that a bright spot forms on the film. Overall, each part of the candle illuminates a particular

Fig. 16.1.2 - When candlelight falls directly on a sheet of paper, it produces no image. But a lens inserted between the candle and paper forms an inverted real image of the flame on the paper.

spot on the film, so the lens creates a complete image of the candle on the film.

However, the lens can only bring the light back together to form a sharp image on the film if the lens and film are separated by just the right distance (Fig. 16.1.3). If the film is too close to the lens, then the light doesn't have room to come together. If the film is too far from the lens, then the light begins to come apart again before reaching the film. In either case, the image on the film is blurry. The candle's real image is only *in focus* at one distance from the lens.

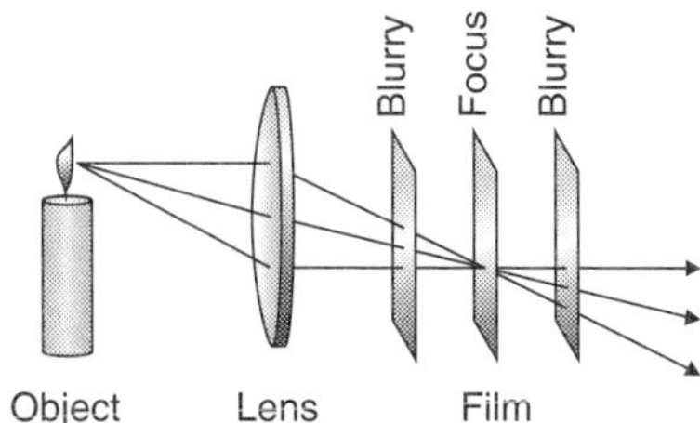

Fig. 16.1.3 - The real image is only in focus when the film is just the right distance from the lens. If the film is too near or far from the lens, the image is blurry.

If the candle moves toward or away from the camera lens, the distance between the lens and the film must also change (Fig. 16.1.4). When the candle is far away, all of the light rays reaching the lens arrive traveling almost parallel to one another and the lens has no trouble bending them toward one another. The rays come into focus relatively near the lens and that's where the film must be (Fig. 16.1.4*a*). The candle's image on the film is much smaller than the candle itself, because the light rays have only a short distance over which to move up or down after leaving the lens.

When the candle is nearby, its light rays diverge terribly as they approach the lens and the lens has trouble bending them toward one another. As a result, the rays come into focus relatively far from the lens (Fig. 16.1.4*b*). The candle's image on the film is quite large because the light rays have considerable distance over which to move up or down after leaving the lens. This image size behavior agrees with our intuition about photography—distant objects should be rendered as small images while nearby objects should be rendered as large images.

Because distant and nearby objects form real images at different distances from the camera lens, they can't both be in focus on the same sheet of film. When you take a picture of a person standing in front of a mountain, only one of them can be in sharp focus. However, if you're willing to compromise a little bit on sharpness, a lens can sometimes form acceptable images of both objects.

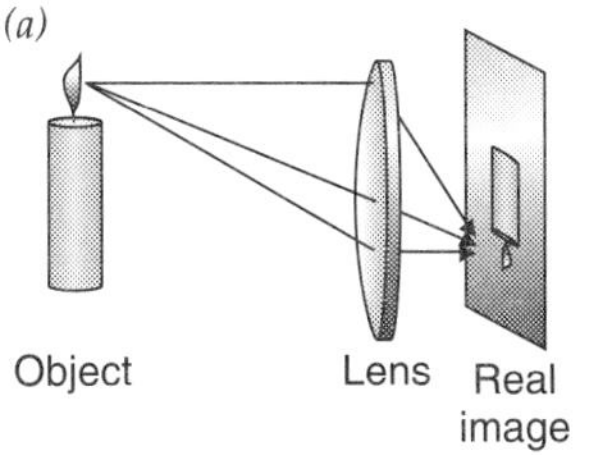

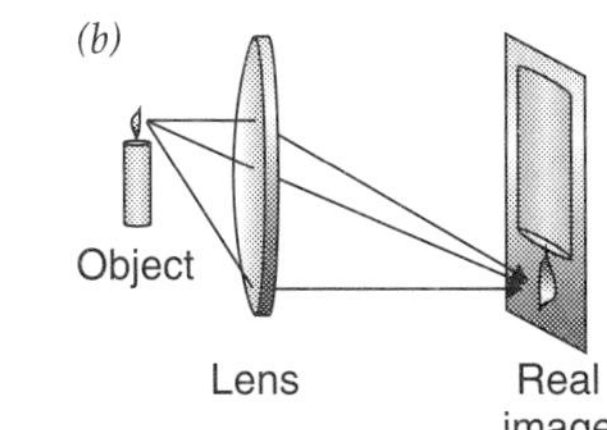

Fig. 16.1.4 - (*a*) The light from a distant candle is traveling in almost the same direction and the lens focuses it easily. The real image forms close to the lens. (*b*) The light from a nearby candle diverges and the lens has more difficulty bending it back together. The real image forms far from the lens. If the candle is too close to the lens, no real image forms at all.

CHECK YOUR UNDERSTANDING #1: Seeing the Lights

If you hold a magnifying glass at the proper distance above a sheet of white paper, you will see an image of the overhead room lights on the paper. Explain.

Focusing and Lens Size

A disposable camera is little more than a box with a lens. The lens projects a real image of the scene in front of the camera onto the film. Light hitting the film exposes it and, once developed, the film displays the image permanently. While it also has a shutter that starts and stops the exposure and a mechanism for advancing the film, there isn't much to this simple camera.

However, there are limitations to the disposable camera design. One of the most severe limitations is that you can't focus it—the camera has a fixed distance between the lens and film. Nonetheless, it manages to form relatively sharp real images on the film, even when there are objects at various distances from the camera. How can this scheme possibly work?

These simple cameras work because they use small lenses. A small lens gathers less light than a large lens but doesn't require *focusing*—adjusting the distance between the lens and the film. Because a large lens brings rays together from many different directions, it needs careful focusing (Fig. 16.1.5); if the film is even slightly too near or too far from the lens, the film's image will be blurry.

But a small lens forms a reasonably clear image even without focusing. Because rays from one part of the scene start nearby after passing through the small lens, they come together gradually and illuminate only a small part of the film even when the film isn't exactly the right distance from the lens. Since the film

(a)

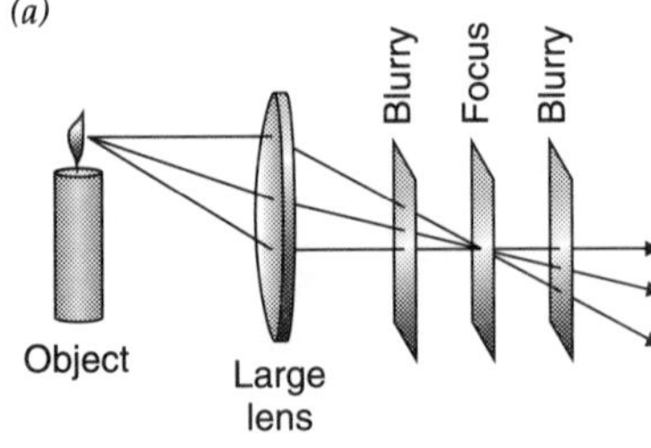

(b)

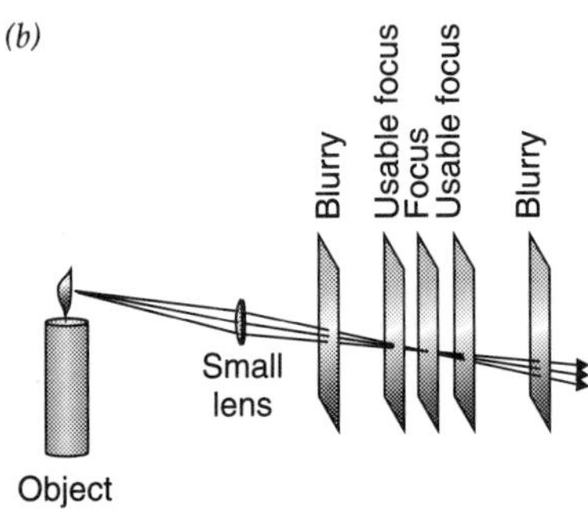

Fig. 16.1.5 - *(a)* A large lens collects lots of light but its focus is critical. The image is blurry only a small distance in front of or behind the actual focus. *(b)* A small lens collects little light but its focus is not critical. The image is relatively sharp over a range of distances around the focus.

can't record every minute detail anyway, the image that forms on it doesn't have to be in absolutely perfect focus. As a result, a camera with a small lens and no focus adjustment manages to take pretty good pictures.

However, sophisticated cameras use large lenses because they gather more light and expose the film much more rapidly. Simple cameras collect very little light and need extremely light-sensitive film. This high-speed film doesn't produce as sharp photographs as low-speed film. Furthermore, the pictures produced by simple cameras aren't very sharp; while everything is almost in focus, most things are just a little bit fuzzy if you look carefully or make enlargements. The pictures lack fine detail.

Sophisticated cameras automatically adjust the distance between the lens and the film. They identify the object you are photographing and position the camera lens so that it projects a sharp, real image on the film. As the camera focuses, you can usually see components in the lens move backward or forward to arrive at the correct distance from the film.

Even a camera with a large lens can take advantage of the small lens trick for focusing. Its lens contains an internal diaphragm that reduces its **aperture** or effective diameter. The *diaphragm* is a ring of metal strips with a central opening. These strips can swing in or out, changing the diameter of the diaphragm's opening and with it the aperture of the lens (Fig. 16.1.6).

When lens aperture is small, the sophisticated camera imitates a simple camera. Nearly everything is essentially in focus simultaneously. In such a situation, the camera has a large *depth of focus*. Actually, the sophisticated camera can bring the most important object into perfect focus, so it produces pictures that are superior to those from simple cameras. However, reducing the aperture of the lens also reduces the amount of light reaching the film. The scene in front of the camera must either be very bright or the exposure of the film must be very long. You don't get something for nothing.

While opening the aperture of a large lens makes full use of its light gathering capacity, focusing then becomes crucial. Even a small error in the lens to film distance produces a blurry picture, so the depth of focus is very small. This trade-off, between light gathering and depth of focus, is a continual struggle for photographers. However photographers sometimes take advantage of the small depth of focus in large lenses to blur the background or foreground of a photograph deliberately. At other times photographers choose very long exposures at small apertures in order to bring an entire scene into sharp focus.

Fig. 16.1.6 - The aperture of this lens can be adjusted by opening or closing its internal diaphragm. Closing the diaphragm reduces the lens's light gathering ability but increases its depth of focus.

CHECK YOUR UNDERSTANDING #2: Portrait Photos

While taking a photograph of your friend, with the diaphragm of your large camera lens wide open, you notice that the background is blurry. Explain.

Focal Lengths and F-numbers

Lenses are characterized by two quantities: focal length and f-number. The **focal length** of a lens is the distance between the lens and the real image it forms *of a very distant object*. For example, if a real image of the moon forms 100 mm behind a particular lens, then that lens has a focal length of 100 mm. The focal lengths of camera lenses range from about 1 cm in disc cameras to about 2 m in cameras used for nature photography.

When light from a scene passes through a short focal length lens, it comes to a focus near that lens and produces a relatively small image on the film. Because a long focal length lens permits the light passing through it to spread out more before coming to a focus, it produces a larger real image on the film.

The "normal" lens for a particular camera has a focal length that allows all of the objects in your central field of vision to fit onto the piece of film. When you hold the finished photograph about 30 cm from your eyes, the objects in the picture appear about the same size they did when the photograph was taken. The focal length of a camera's normal lens is about 1.5 times the horizontal width of its film (Table 16.1.1).

Fig. 16.1.7 - The 24 mm lens (top) forms a small image, close to the lens. The 300 mm lens (bottom) forms a large image, far from the lens.

Table 16.1.1 - Several cameras, the widths of the films they use, and their normal lenses.

Type of Camera	*Width of Film*	*Normal Lens*
Disc Camera	8 mm	10 mm
35 mm Camera	36 mm	50 mm
2¼" Medium Format Camera	2¼"	80 mm
5" Portrait Camera	5"	180 mm

A wide-angle lens has a shorter focal length than the normal lens (Fig. 16.1.7). The image it projects onto the film is smaller and most of the objects in your entire field of vision appear in the photograph. A telephoto lens has a longer focal length than the normal lens. The image it projects onto the film is larger, with only objects at the center of the scene appearing in the photograph.

The focal length of the camera lens relates two distances: the **object distance** from the object you're photographing to the lens and the **image distance** from the lens to the real image (Fig. 16.1.8). This relationship is called the **lens equation** and can be written in a word equation:

$$\frac{1}{\text{focal length}} = \frac{1}{\text{object distance}} + \frac{1}{\text{image distance}}, \qquad (16.1.1)$$

in symbols:

$$\frac{1}{f} = \frac{1}{o} + \frac{1}{i},$$

and in everyday language:

The farther away an object is, the closer to the lens its image forms.

THE LENS EQUATION:
One divided by the focal length of a lens is equal to the sum of one divided by the object distance and one divided by the image distance.

According to the lens equation, the image distance for a distant object is equal to the focal length of the lens. When the object is nearby, the image distance must be larger than the focal length. And when the object distance becomes less than the focal length, the image distance becomes negative and no real image forms at all. That's why you can't focus on an object that's too close to the lens.

A lens's **f-number** characterizes the brightness of the real image it forms on the film, with smaller f-numbers indicating brighter images. The f-number is calculated by dividing the lens's focal length by its diameter. Since long focal length lenses naturally produce larger and dimmer images on the film, the f-number takes into account both the light gathering capacity of the lens and its focal length. Increasing a lens's diameter increases its light gathering capacity and decreases its f-number. Increasing a lens's focal length decreases the brightness of its real image and increases the f-number.

Most sophisticated cameras use large diameter lenses so that their f-numbers are generally less than 4. Since it's difficult to fabricate a lens that is larger in diameter than its focal length, the smallest practical f-number is about 1.

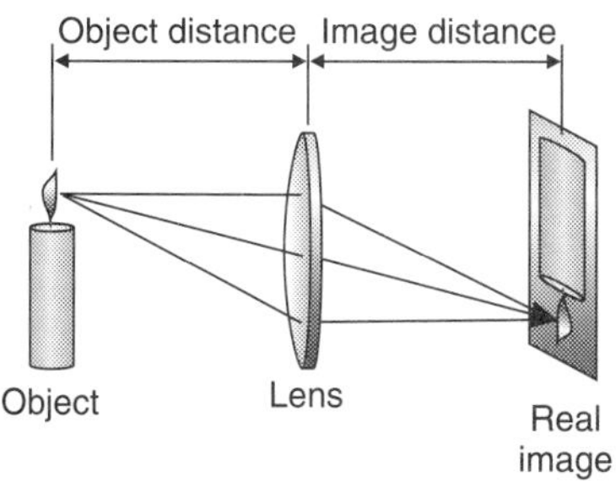

Fig. 16.1.8 - The relationship between the object distance, the image distance, and the focal length of the lens is given by the lens equation.

Since long focal length lenses need large apertures to keep their f-numbers small, some telephoto lenses are huge.

The diaphragm inside a lens allows you to decrease the lens's aperture and thus increase its f-number. A factor of 2 increase in f-number corresponds to a factor of 2 decrease in the lens's effective diameter and a factor of 4 decrease in lens's light gathering area. Thus when you double the f-number of the lens, you must compensate by quadrupling the film's exposure time. Although closing the aperture increases the lens's depth of focus, it requires a longer exposure.

CHECK YOUR UNDERSTANDING #3: Bright and Sharp Photographs

On bright, sunny days, your automatic camera takes photographs with large depths of focus, while on gray, overcast days its pictures have much smaller depths of focus. What causes this difference?

CHECK YOUR FIGURES #1: The Image of an Apple

If the distance from an apple to a converging lens is twice the focal length of that lens, where will the real image of the apple form?

Improving the Quality of a Camera Lens

A high quality camera lens isn't a single piece of glass or plastic. Instead, it's composed of many separate elements that function together as a single lens. This complexity is needed to improve the quality of the real image. To begin with, because dispersion causes different colors of light to bend differently as they pass through a single element lens, they focus at different distances behind that lens. This problem is called *chromatic aberration*. If you tried to take a photograph with a large single element lens, you would find that you couldn't get all of the colors to focus on the film simultaneously.

To fix this dispersion problem, most camera lenses use elements made of different types of glass, with different amounts of dispersion. These elements compensate for one another so that the overall lens has no dispersion. It's then called an *achromat* because of its freedom from color focusing problems.

Good lenses also correct for a number of other image imperfections, always trying to project all of the light from one part of the scene onto one part of the film. These image imperfections include spherical aberration, coma, and astigmatism. *Spherical aberration* occurs because lenses are normally ground with their two curved surfaces shaped like parts of large spheres. While spherical shapes are easy to form mechanically, they don't make ideal lenses. *Coma* occurs when trying to focus light from an object that's far from the center of the picture. Lenses have trouble focusing light passing through them at shallow angles. *Astigmatism* occurs because the corners of the flat film are actually farther from the lens than is the film's center.

A highly corrected lens may contain more than ten individually ground and polished glass elements. For the purposes of the lens equation, this complicated lens has an effective center from which to calculate object and image distances. You can think of it as a single element, located at that center.

However some light reflects each time it passes from air to glass or vice versa. If a lens's elements were simply assembled together, enough light would bounce around between them to produce an overall haze on the photographs. Instead, the glass elements are *anti-reflection coated* with thin layers of transparent materials. The best coatings use interference effects to cancel out the reflected light waves. All you can see from a well-coated lens is a weak violet reflection.

A zoom lens is a complicated lens that can change the size of the real image it projects onto the film (Fig. 16.1.9). By carefully rearranging the individual lens elements within the overall lens, the zoom lens can adjust its effective focal length as well as the distance between the effective lens and the film. As the zoom lens changes from short focal length to long focal length, the image it projects on the film becomes larger. This effect allows you to compose the picture so that the scene fills the photograph completely. A lens that can change its focal length while retaining the same f-number and still keep the real image in focus on the film is a truly remarkable achievement.

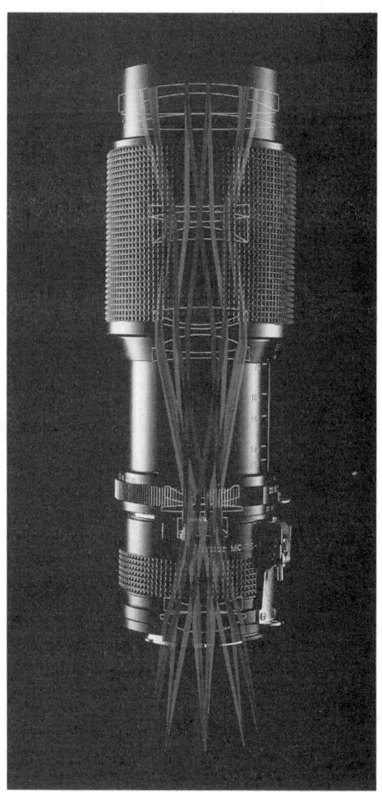

Fig. 16.1.9 - A zoom lens contains many optical elements. This complicated structure allows the photographer to change the lens's focal length and f-number, and corrects for color and focus imperfections that are present in simpler lens systems.

CHECK YOUR UNDERSTANDING #4: Not Such a Good Picture

If you use a large magnifying glass to form a real image of an overhead fluorescent light fixture, you will find that the corners of the light fixture tend to be blurry when the center is in focus. Why is the image quality so poor?

Photographic Film

The film is a collection of light sensitive grains, suspended in an emulsion on a plastic substrate. When the lens projects a real image onto the film, the light energy causes a subtle chemical change in the grains and forms a latent image. This latent image only becomes visible when the film is developed.

The light sensitive grains used in black and white photography are silver halides (silver chloride, silver bromide, and silver iodide). These are salts in which each silver atom gives up an electron to a halogen atom (chlorine, bromine, or iodine). Because of this transfer of electrons, the salt grain is composed entirely of electrically charged ions. When light strikes a silver halide particle, its energy transfers electrons back to the silver ions so that a tiny cluster of neutral silver atoms forms on the surface of the grain. While this cluster contains only three silver atoms, that's enough to mark the grain for later development. These tiny silver clusters are what form the latent image.

When the film is exposed to a *developer*, one silver cluster on the outside of a silver halide grain triggers the conversion of that entire grain into pure silver. The particle that forms is a tangled mass of metallic fibers, so it appears black and opaque. Silver halide grains that have no silver clusters on their surfaces don't develop and are later washed away in a chemical bath called a *fixer*.

Various sensitizing chemicals are added to the silver halide particles to make them more light-sensitive. Without this sensitization, clusters of four silver atoms would be needed in the latent image. With it, three silver atoms are enough. The sensitization also helps to place the silver clusters on the outsides of the silver halide grains, where the developer can find them. Early photographic emulsions were much less light sensitive and required extremely long exposures with bluish light. Modern emulsions are so sensitive to all wavelengths of light that a short exposure is all that's needed, even in fairly dim light.

The amount of light needed to create a latent image depends on how big the silver halide grains are. Since any grain that absorbs a few photons turns black during development, a big grain that can collect lots of light is more likely to turn black than a small grain. Therefore an emulsion of large silver halide grains is more light sensitive than one of small grains. The range of grain sizes present in an emulsion contributes to that emulsion's contrast. The narrower the size range, the more abruptly all of the particles go from unexposed to exposed and the greater the contrast in the developed image.

Following development, light-exposed areas of the film are black or gray while unexposed areas are clear. The film is a "negative," appearing dark where the original scene was light and light where the original scene was dark. When this negative image is printed on photographic paper, the same brightness reversal is repeated and the final print looks like the original scene.

Color photographic film contains three separate color sensitive layers—one for each of the three primary colors of light (red, green, and blue). This color sensitivity is aided by the fact that silver halide particles aren't normally sensitive to red or green light and must be chemically treated before they will respond to either of those two colors. The film's first layer responds to blue light, the second to green light, and the third to red light. Because the green and red layers are also sensitive to blue light, blue light that makes it through the blue sensitive layer is blocked before it can reach the other two layers.

During development, each layer becomes colored rather than opaque. The developer first converts exposed silver halide particles into metallic silver. This chemical reaction changes the developer molecules themselves and they then convert colorless dye couplers contained in the layer into brightly colored dye molecules. The silver halide particles and silver grains are no longer needed, so they are removed. The remaining film is transparent except for the dye molecules. Wherever the blue layer saw blue light, a yellow dye forms that absorbs blue light. Exposed portions of the red sensitive layer form a cyan dye and absorb red light. Exposed portions of the green sensitive layer form a magenta dye and absorb green light.

These three colors, yellow, cyan, and magenta, are called the **primary colors of pigment** because each one absorbs one of the primary colors of light. When white light passes through a particular mixture of these pigments, the amounts of red, green, and blue light that emerge depend on the respective amounts of cyan, magenta, and yellow pigments in the mixture. The primary colors of pigment create full color images in photography and color printing.

Following development, a negative color image is present on the film. When this negative is printed on photographic paper, a second color reversal occurs. The final color image looks almost exactly like the original scene. But the film's color sensitivity isn't exactly the same as that of our eyes. For example, the film's red layer may be a little too sensitive to a particular wavelength of yellow light so that what should be rendered as yellow may appear slightly orange in the photograph. The dyes used to form the final paper print may also be slightly different from perfect yellow, cyan, and magenta. As a result, the print may not have exactly the same colors as the original scene.

Different color films, printing chemicals, and papers can affect the final color characteristics of the photographs. Some processes produce bright, strong colors while others produce softer, more pastel colors. It's also possible to enhance or suppress individual primary colors by changing the light used during the printmaking process. The skill of the processing laboratory can have a substantial impact on the appearance of the final prints so that there is considerable artistry to the apparently mechanical process of printing photographs.

CHECK YOUR UNDERSTANDING #5: Hot and Hazy Photographs
Why is photographic film susceptible to damage from heat and X-rays?

The Shutter

The lens and film produce the photograph but the camera's *shutter* controls the exposure. There are three types of shutters: moving disks, opening diaphragms, and traveling slits. The moving disk shutter is an opaque plate that blocks the light leaving the lens of a simple camera. When you take a picture, a spring moves the disk aside briefly and then returns it to its original position. The disk's motion is usually driven by the force of pressing the shutter button. This simple but effective shutter works well for cameras with a small lens and only one exposure time or "shutter speed." The exposure time is determined by the stiffness of the spring and the masses of the shutter parts.

A *diaphragm shutter* works as part of the lens. While it resembles the diaphragm used to increase the lens's f-number, it's designed to close completely. It stays closed except during the exposure itself, when it's driven open and then closed by two springs wound when the film is advanced. Diaphragm shutters are used in sophisticated cameras with non-removable lenses. Since opening and closing the diaphragm are separate actions, the exposure time can be adjusted.

The traveling slit shutter is used on cameras with interchangeable lenses. The film is covered by a thin sheet of metal or fabric that is pulled away by a spring to initiate the exposure. To end the exposure, a second sheet is pulled over the film from the opposite side so that the two sheets chase one another across the film. The sheets are moved by two springs that are wound when the film is advanced. The sheets lie just in front of the film, in the focus of the lens, and form a *focal plane shutter*.

During very short exposures, the second sheet begins to cover one side of the film before the first sheet has finished uncovering the other side. The result is a moving slit-shaped opening that races across the film at high speed. Only during relatively long exposures, typically 1/60th of a second or longer, is the whole sheet of film exposed at once. In flash photography, the brilliant light from a xenon flash lamp lasts for at most 1/1000th of a second, so the flash must not be triggered until the whole sheet of film is uncovered. If the camera's exposure is made too short, then the whole sheet of film is not uncovered during the flash and part of the film remains unexposed.

CHECK YOUR UNDERSTANDING #6: Photography and Newton's Laws
Why are focal plane shutters made out of very light materials?

Focusing and Exposure Control

All but the simplest cameras have some control over the focusing of the lens, its aperture, and the length of the exposure. These adjustments are handled in a coordinated way, often in a completely automated fashion. But how does the camera know how to make these adjustments?

The camera studies light coming from the scene, either directly through the photographic lens or through a separate optical system. In cameras that use a separate optical system, the focus is often adjusted by a *range-finder*, a device that determines how far away objects are by observing them through two separate windows on the front of the camera (Fig. 16.1.10). Mirrors superimpose the views of the objects observed through these two windows.

Just as your two eyes help you determine the distance to an object, so the two windows of the range-finder help the camera. The camera tilts one of the mirrors until the two images overlap. When they do, the mirror's tilt indicates the distance to the object. A range-finder camera moves the photographic lens as it tilts the mirror so that when the range finder sees only a single image of the object, that object will be in focus on the film. But because the optical systems of a range finder are shifted from the lens taking the picture, nearby objects appear differently in the photograph than they do in the viewfinder.

Cameras with interchangeable lenses use the actual photographic lens for focusing. Prior to the exposure, a mirror between the lens and the film projects the lens's real image onto a translucent glass sheet (Fig. 16.1.11). While you look

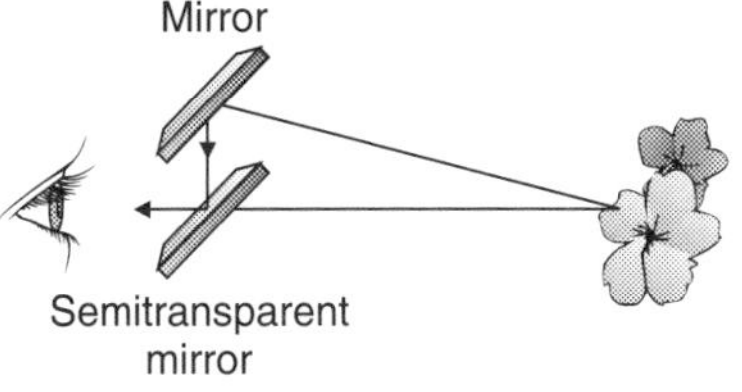

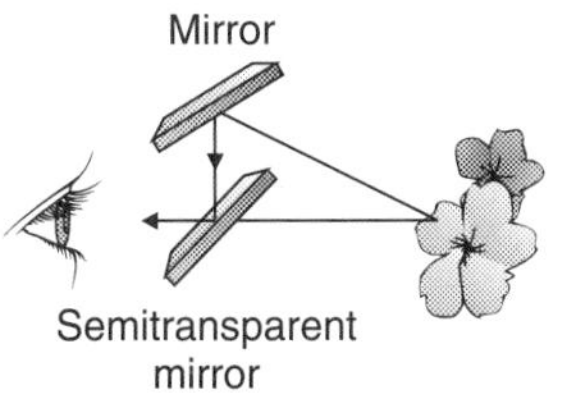

Fig. 16.1.10 - A range-finder overlaps the views through two separate optical systems to determine the distance to an object. The nearer the object is, the more the mirror must be tilted in order to overlap the two views.

Fig. 16.1.11 - The mirror in the center of this reflex camera directs light from the lens (removed for this photograph) onto the focusing screen above it. During the exposure, the mirror swings upward to permit light from the lens to strike the film at the back of the camera.

at that real image through the camera's viewfinder, the camera's electronics analyze it for contrast. A blurry, out-of-focus image has little contrast while a well-focused image is usually full of light and dark regions. The camera focuses the lens until it achieves the highest possible contrast in the image.

When the exposure is taken, the camera quickly moves the mirror out of the way and lets the lens project its real image onto the film. After the exposure, the mirror returns to its original position. Cameras with moving mirrors of this type are called reflex cameras and offer a true view of the final image, including exactly which objects will be recorded on the film.

In both of these camera systems, semiconductor light sensors measure the light reaching the camera from the scene in front of it. These sensors resemble photoelectric cells and produce a current that's proportional to the intensity of the light. The camera analyzes that current and adjusts the lens's aperture and the shutter speed (exposure time) so that the film receives just the right amount of light to create a latent image. In a range finder camera, the light sensor is outside the photographic lens. In a reflex camera, the light sensor is located on the reflex mirror or the viewing screen and monitors light passing through the photographic lens.

CHECK YOUR UNDERSTANDING #7: Out of Focus

Most automatic focus cameras can't focus on a snowy hillside with little contrast. Why not?

Summary

How Photographic Cameras and Projectors Work: A camera lens projects a real image of the scene in front of the camera onto the film inside the camera. This real image is formed when the lens bends all of the light reaching it from one part of the scene onto one part of the film. For this imaging process to work well, the distance between the lens and the film must be adjusted so that the light converges together (focuses) just as it reaches the film. If the film is too close to or far from the lens, the image is blurry. The depth of focus depends on the effective diameter of the lens; its aperture. The smaller the lens's aperture, the less critical the focus but the less light the lens gathers. A long focal length lens brings the light to a focus far behind it, forming a relatively large but dim real image on the film. To brighten this image, the long focal length lens must have a large aperture so that it gathers lots of light. The f-number, the ratio of a lens's focal length to its aperture, characterizes the brightness of the lens's image.

Important Laws and Equations

1. The Lens Equation: One divided by the focal length of a lens is equal to the sum of one divided by the object distance and one divided by the image distance, or

$$\frac{1}{\text{focal length}} = \frac{1}{\text{object distance}} + \frac{1}{\text{image distance}}. \qquad (16.1.1)$$

The Physics of Photographic Cameras and Projectors

1. A lens forms an image of an object at a location given by the lens equation. If the image distance from that equation is positive, the image is real and can be projected onto surfaces.

2. A lens's focal length is a measure of how strongly it bends light, with small focal length lenses bending light more than large focal length lenses. Larger focal length lenses generally produce larger real images.

3. A lens's f-number determines the brightness of the image it forms. The smaller the f-number, the brighter the image. Longer focal length lenses generally have larger f-numbers.

Check Your Understanding - Answers

1. It is a real image of the room lights, created by the converging magnifying glass.

Why: A magnifying glass is a converging lens and can form a real image. It brings light spreading outward from the room lights together onto the sheet of paper as a real image.

2. Light from the more distant background focuses closer to the lens than light from your friend. Since the camera is focusing on your friend, the background appears out of focus.

Why: When the aperture of your camera lens is wide open, focusing becomes critical. Objects in front of or behind your friend are out of focus on the film and appear blurry.

3. On a bright day, your camera needs only a small aperture to gather enough light for an exposure, so the depth of focus is large. On a dark day, it uses the largest aperture available to gather light and has a small depth of focus.

Why: Light gathering and depth of focus go hand in hand. If you must open the aperture of your camera's lens to gather enough light for an exposure, focusing will become critical and your photographs will have small depths of focus.

4. The magnifying glass has only one glass element and suffers from many image imperfections.

Why: A single converging lens can't form a high quality image on a flat surface because it has chromatic aberration, spherical aberration, coma, and astigmatism. If the lens is small, these failings are often invisible. But in a large magnifying glass, they are all readily apparent and you can't form a sharp image of the entire light fixture.

5. Anything that causes silver halide to break down into silver atoms will fog the film.

Why: In fresh film, the silver halide particles have essentially no silver atoms on their surfaces. But exposure to heat, X-rays, and other forms of penetrating radiation can deposit energy in the silver halide particles and cause them to release silver atoms. Since a few silver atoms on a particle produce a latent image, large-grained, high-speed films are particularly susceptible to damage.

6. They must accelerate quickly to extremely high speeds.

Why: A focal plane shutter moves with blinding speed. Its opening and closing surfaces must move across the film in only a few thousandths of a second. These surfaces must also move at almost constant velocity, so their accelerations must occur very early on in their motions. To reach such high speeds in such short times, the elements of the focal plane shutter use strong springs and very low mass components.

7. The image of the snow has little contrast and any camera that uses contrast to find the focus will have trouble.

Why: Many automatic focus cameras are baffled by low contrast scenes. They move their lenses back and forth, expecting that bright and dark regions will suddenly appear in the image. If there are no bright and dark regions to be had, this search is hopeless.

Check Your Figures - Answers

1. The image will form at a distance twice the focal length behind the lens.

Why: Since the object distance is twice the focal length, we can use Eq. 16.1.1 to find the image distance:

$$\frac{1}{\text{image distance}} = \frac{1}{\text{focal length}} - \frac{1}{\text{object distance}}$$

$$= \frac{1}{\text{focal length}} - \frac{1}{2 \cdot \text{focal length}}$$

$$= \frac{1}{2 \cdot \text{focal length}}.$$

The image distance is twice the focal length of the lens.

Glossary

aperture The diameter or effective diameter of a lens or opening.

converging lens A lens that bends the individual light rays passing through it toward one another so that they either converge more rapidly than before or at least diverge less rapidly from one another. A converging lens has a positive focal length.

f-number The ratio of a lens's focal length to its effective aperture.

focal length The distance after a converging lens at which a sharp real image of a distant object forms.

image distance The distance between the lens and the real or virtual image that the lens creates.

lens A transparent optical device that uses refraction to bend light, often to form images.

lens equation The equation relating a lens's focal length to the object and image distances.

object distance The distance between the lens and the object that it is imaging.

primary colors of pigment Three pigments—yellow, magenta, and cyan—that each absorb only one of the primary colors of light—blue, green, and red, respectively. Mixtures of these three pigments can filter white light to make our eyes perceive any possible colors.

real image A pattern of light, projected in space, that exactly reproduces the pattern of light at the surface of the original object. A real image forms after the lens that creates it and can be projected onto a surface.

Review Questions

1. Why is the real image formed by a converging lens upside down relative to the object?

2. If an object moves farther from a converging lens, does its real image move toward or away from the lens? Why?

3. Why does increasing the diameter of a converging lens increase the brightness of the real image it forms?

4. Why is the real image formed by a long focal length lens larger and dimmer than one formed by a short focal length lens?

5. How far must an object be from a converging lens for the distance between the lens and the real image to equal the focal length of the lens?

6. Why must a long focal length lens use large diameter optical components in order to have a low f-number?

7. If you look carefully at the real image formed by a simple magnifying glass on a sheet of paper, you should see colored bands resembling little rainbows in the image. What causes this color separation and what can be done to get rid of it?

8. Why is it surprisingly hard to see your reflection in the optical surfaces of a high-quality camera lens?

9. If you look at a piece of exposed photographic film before it's developed, you won't see any image at all. What invisible change has exposure to light caused in the film? How does developing make that change visible?

10. In what sense is the image on developed color print film a "negative" of the original scene recorded in the image?

11. Camera shutters used for short exposures are made from stiff springs and parts with small masses. Why?

Exercises

1. Your new 35 mm camera comes with two lenses, a 50 mm focal length "normal" lens and a 200 mm focal length "telephoto" lens. The minimum f-number for the 50 mm lens is 1.8. Although the 200 mm lens has much larger glass elements in it, its lowest f-number is 4. Why is the telephoto lens's f-number so much larger?

2. Sports photographers often use very long lenses on their cameras. How do the lengths of these lenses affect the photographs they produce?

3. If you hold your camera up to a small hole in a fence, you will be able to take a picture of the scene on the other side. However, the exposure will have to be quite long and the depth of focus will be surprisingly large. Explain.

4. If you look into a large glass marble, you will find that decorations in the glass on its far side will appear greatly enlarged. Why?

5. If you are taking a photographic portrait of a friend and want objects in the foreground and background to appear blurry, how should you adjust the camera's aperture and shutter speed?

6. On a bright, sunny day you can use a magnifying glass to burn wood by focusing sunlight onto it. The focused sunlight forms a small circular spot of light that heats the wood until it burns. Why is the spot of light circular?

7. Your eye is similar to a camera. Why does squinting your eye increase your depth of focus and make it easier to focus?

8. Why is it easiest for the lens of your eye to form sharp images on your retina when the scene in front of you is bright and the iris of your eye is very small?

9. People who need glasses find it hardest to see clearly without them in dimly lit situations when the irises of their eyes are wide open. Why?

10. As you zoom the lens of your video camera inward to take a close-up of one person, what characteristic of the lens changes and does it increase or decrease?

11. As you zoom a video camera in for a close-up of a person (see Exercise **10**), how does the distance between the effective center of the lens and the real image of that person change?

12. One part of a fiber optic communications system uses a lens to focus light from a semiconductor laser onto the end of an optical fiber. The tiny light source and fiber are on opposite sides of the lens and each is 1.0 cm away from the lens. Why must the lens have a focal length of 0.5 cm?

13. Why can't a black and white photographic negative be enlarged beyond a certain size without appearing grainy?

Problems

1. Your camera has a 35 mm focal length lens. When you take a photograph of a distant mountain, how far from that lens is the real image of the mountain?

2. If you use a 35 mm focal length lens to take a photograph of flowers 2 m from the lens, how far from that lens does the real image of the flowers form?

3. You're trying to take a photograph of two small statues with a 200 mm telephoto lens. One statue is 4 m from the lens and the other statue is 5 m from the lens. The real image of which statue forms closer to the lens? How much closer?

4. When you place a salt shaker 30 cm from your magnifying glass, a real image forms 30 cm from the lens on the opposite side. You can see this image dimly on a sheet of paper. What is the focal length of the magnifying glass?

5. When light from your desk lamp passes through a 50.0 mm focal length lens, it forms a sharp real image on a sheet of paper located 50.5 mm from the lens. How far is the desk lamp from the lens?

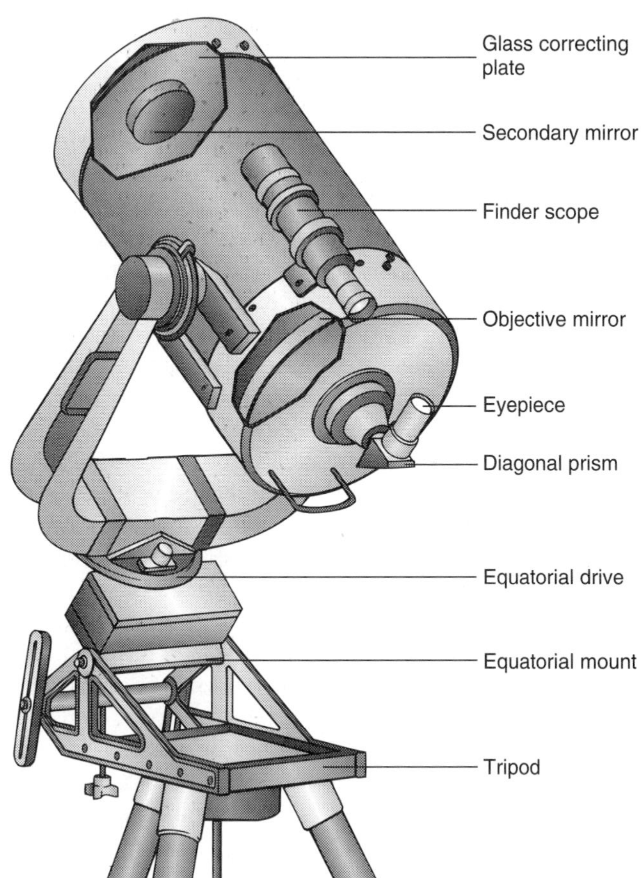

Section 16.2

Telescopes and Microscopes

Not everything that we wish to see is visible to the naked eye. We have trouble making out details in distant objects and can't resolve the tiny structures in nearby objects, either. We need help. To enlarge and brighten distant objects, we use telescopes, and to do the same with nearby objects, we use microscopes. Because these devices are almost identical in structure, though specialized to apparently different tasks, it's natural to describe them together in this section.

Questions to Think About: *Where is the image that you see when you look into a telescope or microscope? Why does that image often appear upside-down and backward? How do the eyepieces in telescopes and microscopes contribute to their magnifications? Why does the image become dimmer as the magnification increases? Is there a limit to a telescope or microscope's usable magnification? Why do major telescopes have such large diameters? What is the advantage to a telescope in space?*

Experiments to Do: *Building a simple telescope is easy if you have two different magnifying glasses. You need one with a long focal length (low magnification) and one with a short focal length (high magnification). Use the long focal length lens to form a real image on a sheet of paper. Once you know how far from the lens the real image forms, you can begin looking at that image with the more powerful magnifying glass. But instead of looking at the real image on the sheet of paper, hold the two lenses up in line with one another and look right through both. If you look into the high powered lens and adjust the separation between the two, you should find a special distance at which you see a clear image of the scene in front of you. This image will be enlarged, but upside-down and backward. Can you explain this inversion?*

(a)

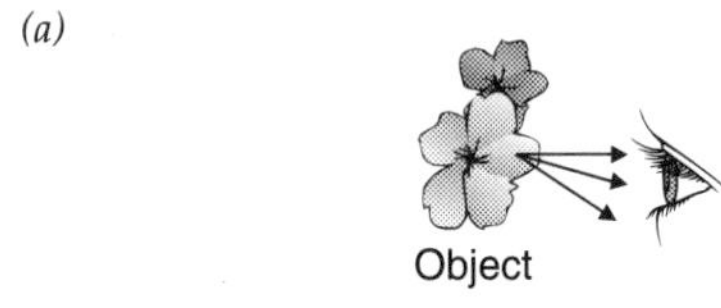

(b)

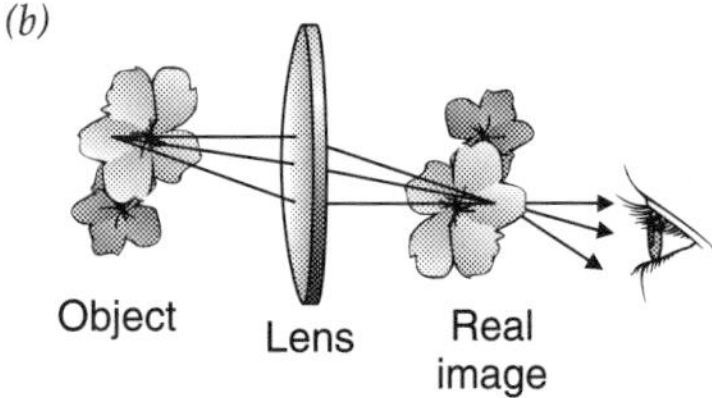

Fig. 16.2.1 - (*a*) The lens of your eye takes light diverging from an object and focuses it onto your retina. (*b*) When you look at a real image, you again see light diverging from a region of space and the lens of your eye focuses it onto your retina.

(a)

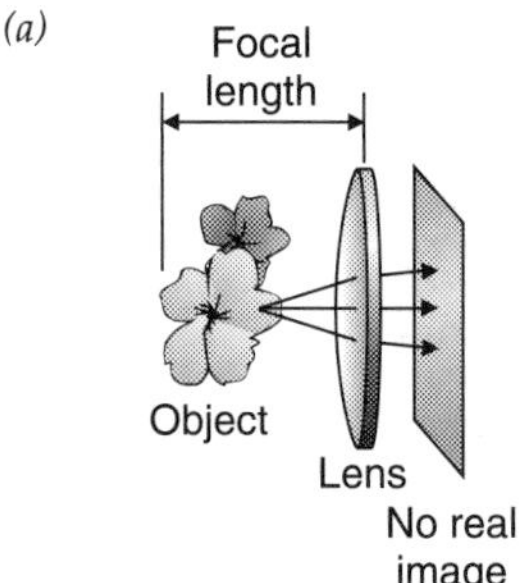

(b)

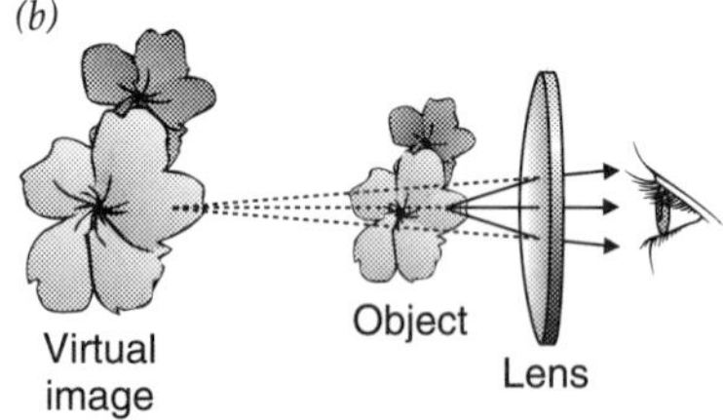

Fig. 16.2.2 - (*a*) Light from an object very near a converging lens diverges after passing through the lens. (*b*) Your eye sees a virtual image that is large and far away.

Fig. 16.2.3 - This magnifying glass creates an enlarged virtual image located far behind the printed text.

Photographic and Visual Telescopes

Let's begin by looking at light telescopes, which form images of distant objects either on film or in front of your eyes. Most telescopes used by professional astronomers are photographic—essentially gigantic telephoto lenses that form real images of stars on pieces of film or electronic light sensors. However visual telescopes are still popular among amateur astronomers and for terrestrial work. Binoculars are essentially a pair of visual telescopes, carefully matched to one another and equipped with devices that produce upright images.

A photographic telescope produces a real image, in which light from one part of the sky is brought together on one part of the film or light sensor. But a visual telescope doesn't create a real image behind its final lens. Instead, it creates a **virtual image** in front of its final lens.

To understand a visual telescope, imagine removing the film or light sensor from a photographic telescope. With nothing to stop it, light continues past the real image and forms a stream of diverging light. If you look into this stream of light, you'll see the real image floating there in space (Fig. 16.2.1). Because light from the real image diverges in the same way it would from a solid object, the real image looks like a solid object. In fact, it's sometimes hard to tell whether you're looking at an object or at a real image of that object.

But in a visual telescope, you don't look directly at the real image; you look at it through a magnifying glass. A magnifying glass is just a converging lens that's so close to the object you're looking at that a real image doesn't form *after* the lens (Fig. 16.2.2*a*). The object distance is less than the lens's focal length, so the image distance is negative and a virtual image forms *before* the lens (Fig. 16.2.2*b*). This means that light rays from the object continue to diverge after passing through the lens and appear to come from an enlarged virtual image, located farther away than the object itself. While you can't touch this virtual image or put a piece of paper into it, it's quite visible to your eye (Fig. 16.2.3).

In a visual telescope, the object you look at through the magnifying glass is actually the real image from the first part of the telescope. You use the magnifying glass—actually called the **eyepiece** or *ocular lens*—to enlarge the real image so that you can see its detail more clearly. You can focus the eyepiece, like any magnifying glass, by moving it back and forth. The farther the eyepiece is from the real image you're looking at, the less the light rays diverge after passing through the lens and the farther away the virtual image becomes. When the object distance is exactly equal to the eyepiece's focal length, the virtual image appears infinitely distant. You can then view the enlarged image with your eye relaxed, as though you were looking at something extremely far away.

The eyepiece provides magnification because when you look at an object through it, that object covers a wider portion of your field of vision. This magnification increases as the eyepiece's focal length decreases. That's because a short focal length eyepiece must be quite close to the object, where it can bend light rays coming from a small region so that they fill your field of vision. A long focal length eyepiece must be farther away from the object, where it can bend light rays coming from a large area to give you a broad but relatively low-magnification view of the object.

CHECK YOUR UNDERSTANDING #1: Real and Virtual Images

As you move a magnifying glass slowly toward the photograph in front of you, you see an inverted image that grows larger and nearer to your eye. This image eventually becomes blurry and then a new upright and enlarged image appears on the photograph's side of the lens. What is happening?

Refracting Telescopes

A common type of visual telescope uses two converging lenses: one to form a real image of the distant object and the second to magnify that real image so that you can see its fine detail (Fig. 16.2.4). The first lens is called the **objective** and the second lens is the eyepiece. Because this telescope uses refraction in its objective to collect light from the distant object, it's called a *refracting telescope*. More specifically, it's called a *Keplerian telescope*, after the German astronomer Johannes Kepler (1571–1630). An alternative design that uses one converging lens and one **diverging lens**—a lens that bends light rays away from one another—is called a *Galilean telescope*, after the Italian scientist Galileo Galilei (1564–1642).

(a)

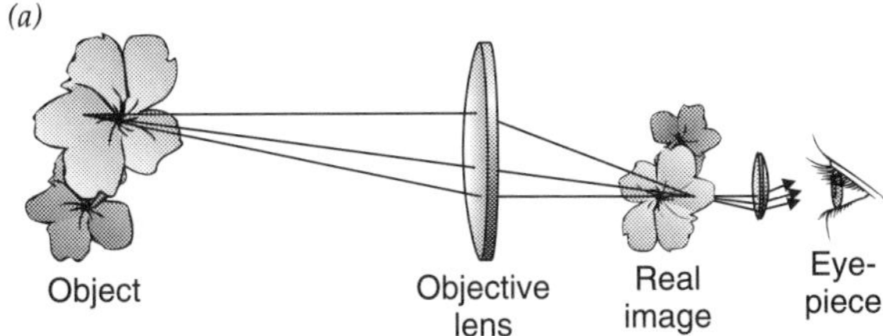

(b)

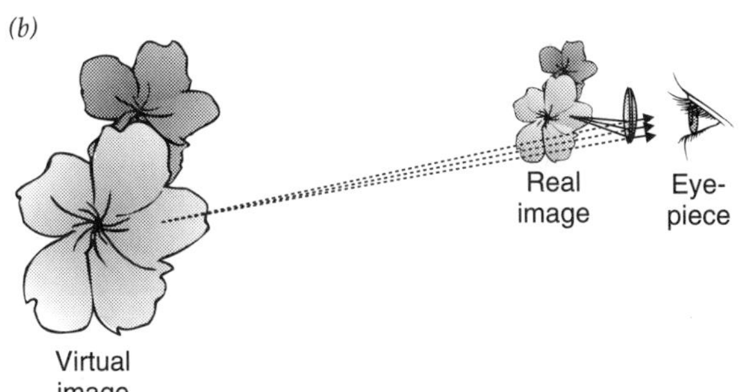

Fig. 16.2.4 - (*a*) A Keplerian telescope consists of two converging lenses. The objective lens forms a real image of the distant object. The eyepiece acts as a magnifying glass, allowing you to inspect that real image in great detail. (*b*) As you look through the eyepiece, you see a very large, far away, and inverted virtual image of the object.

Because a Keplerian telescope's objective forms an inverted real image, the virtual image you see through the eyepiece is also inverted. This inversion is tolerable for astronomical use but is a problem for terrestrial use. It's hard to watch birds when they appear upside down and backwards. That's why binoculars, which are basically Keplerian telescopes, include erecting systems that make the final virtual image upright. These erecting systems fold the optical paths of the binoculars and give them their peculiar shape.

When you look at an object through a visual telescope, you see a magnified virtual image. The magnification of a Keplerian telescope depends on the focal lengths of the two lenses. As the focal length of the objective increases, its real image becomes larger and the telescope's magnification increases. As the focal length of the eyepiece becomes shorter, you look more closely at the real image so the telescope's magnification again increases. Overall, a Keplerian telescope's magnification is its objective's focal length divided by its eyepiece's focal length.

However, as we discussed in the section on cameras, simple lenses create poor quality images. Most telescope objectives and eyepieces are actually constructed from several optical elements. To avoid chromatic aberration (color focusing problems), these lenses are built from *achromatic doublets*, pairs of matched optical elements that are joined by clear cement. The two elements in a doublet are made from different types of glass so that their dispersions cancel and they act as a single lens that's almost free of chromatic aberration.

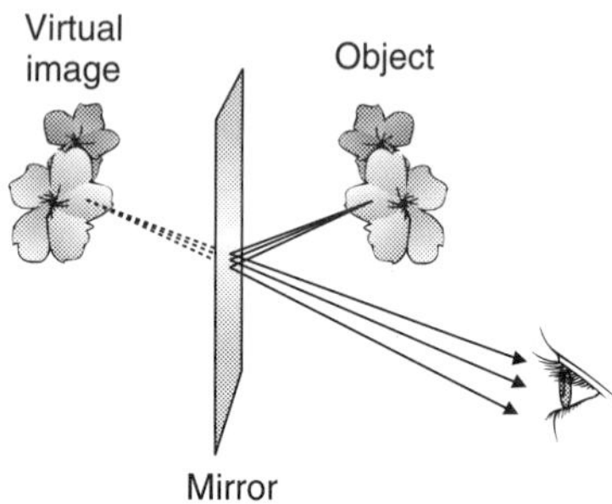

Fig. 16.2.5 - A flat mirror forms a virtual image of any object in front of it. The virtual image is located exactly opposite the original object and is flipped side-for-side.

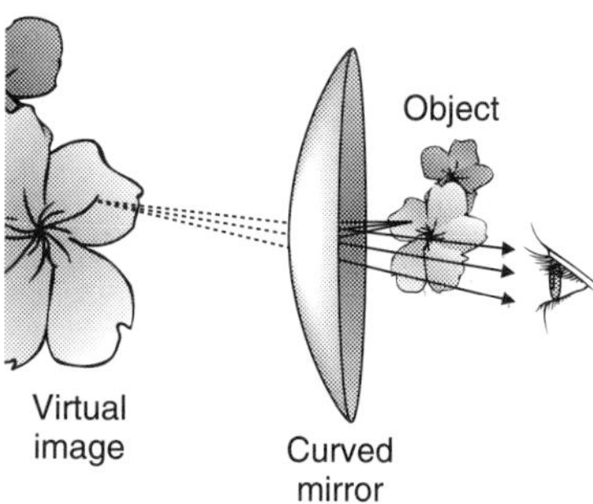

Fig. 16.2.6 - A concave mirror forms a virtual image of any object in front of it and close to its surface. The virtual image appears larger than the original object and is more distant from the mirror than the object itself. Like the flat mirror, the virtual image is flipped side-for-side.

Fig. 16.2.7 - A curved mirror produces an upright virtual image of a light bulb located just in front of it.

But even achromatic doublets don't create perfect images. That's why telescope eyepieces frequently contain four or five individual optical elements. These precision eyepieces produce images that are sharp from edge to edge. As in camera lenses, these individual elements are anti-reflection coated so that you don't see any reflections when you look into the telescope.

CHECK YOUR UNDERSTANDING #2: A Long Look at the Stars
Why are refracting telescopes and binoculars with high magnifications so long?

Reflecting Telescopes

But not all telescopes use lenses as their objectives. A *reflecting telescope* uses an objective mirror to create the initial real image. For reasons we'll discuss shortly, reflecting telescopes have almost completely replaced refracting telescopes for astronomical observation.

A flat mirror is already an interesting optical device. It forms a virtual image of any object (Fig. 16.2.5). Light diverging from the object bounces off the mirror and continues to diverge until it reaches your eye. This light appears to come from a virtual image located on the other side of the mirror, at just the same distance from the mirror as the object itself. This virtual image is reversed side-for-side from the object, so you see a "mirror image." When you view this mirror image, it looks the same size as the object itself.

If the mirror is curved rather than flat, it still creates virtual images. But the virtual images are no longer located exactly opposite the original objects and they don't appear the same size. A mirror that is bowed inward like the inside of a bowl is **concave**. When an object is located near a concave mirror, light diverging from the object bounces off the mirror and continues to diverge, but not as quickly (Fig. 16.2.6). It appears to come from a large virtual image that's located far behind the mirror (Fig. 16.2.7). This effect explains why makeup mirrors give you an enlarged view of your face. As long as your face is close to the bowl-shaped mirror, you see an enlarged virtual image of your face located far behind the mirror. (In contrast, a mirror that is bowed outward in the middle, a **convex** mirror, creates a reduced virtual image located just behind the mirror.)

However, when an object is located far from a concave mirror, light diverging from the object bounces off the mirror and converges to form a real image (Figs. 16.2.8 and 16.2.9). This real image is just like one formed by a lens except that it occurs *in front* of the mirror. You can project this real image onto a sheet of paper or onto a piece of film, which is how a photographic reflecting telescope works. As usual, the real image is inverted and flipped side-for-side.

A visual reflecting telescope simply adds an eyepiece, through which you

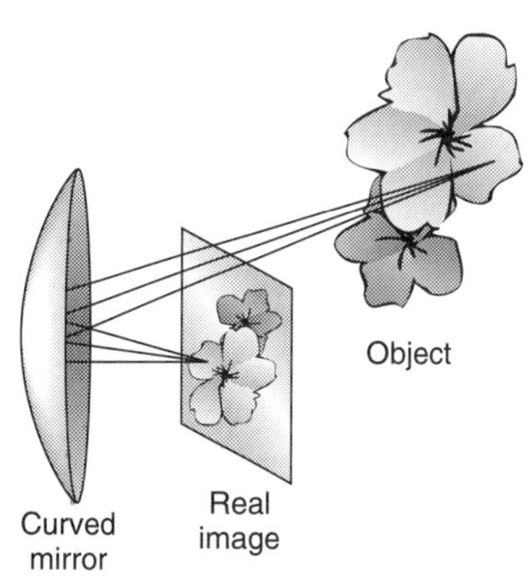

Fig. 16.2.8 - When an object is far from a concave mirror, the mirror forms a real image of the object. In such a reflecting telescope, a photographic film or other light-sensitive surface is usually placed where the real image forms.

Fig. 16.2.9 - A curved mirror produces an inverted real image of a light bulb located on a table far in front of it.

can inspect the real image in great detail (Fig. 16.2.10). Through that eyepiece, you see an enlarged virtual image of the original object.

Like the objective lens of a refracting telescope, the mirror of a reflecting telescope must be carefully shaped to create a sharp real image. Since it uses reflection rather than refraction, it has no chromatic aberration and handles colors well. But a single mirror suffers from a variety of image imperfections which, though usually negligible near the center of the real image, become progressively worse toward its periphery. The mirrors used for observing astronomical objects generally have parabolic shapes and create real images that are nearly perfect at their centers but relatively poor at their edges. Such telescopes normally only let you see the central portions of their real images, where those images are sharp and clear.

The visual reflecting telescope of Fig. 16.2.10 places your head in front of the curved mirror as you look into the eyepiece. While some giant telescopes allow you to sit in front of their mirrors to inspect their real images, most smaller telescopes try to get you out of the mirror's way. They redirect light from the objective mirror so that you aren't in front of it. The two most common ways of doing this are the *Newtonian reflecting telescope* (Fig. 16.2.11*a*) and the *Cassegrainian reflecting telescope* (Fig. 16.2.11*b*).

In a Newtonian telescope, a flat secondary mirror deflects light from the concave objective mirror by 90°, so that the real image forms at the side of the telescope. You can then look at this real image through an eyepiece without blocking light on its way to the primary mirror (Fig. 16.2.12).

In a Cassegrainian telescope, a convex secondary mirror reflects light from the concave objective mirror back through a hole in the center of the primary mirror. Since the secondary mirror bends the light outward slightly, it delays the light's convergence so that a real image forms behind the telescope. You look at this real image through an eyepiece at the end of the telescope (Fig. 16.2.12).

In both the Newtonian and Cassegrainian telescope designs, something

(a)

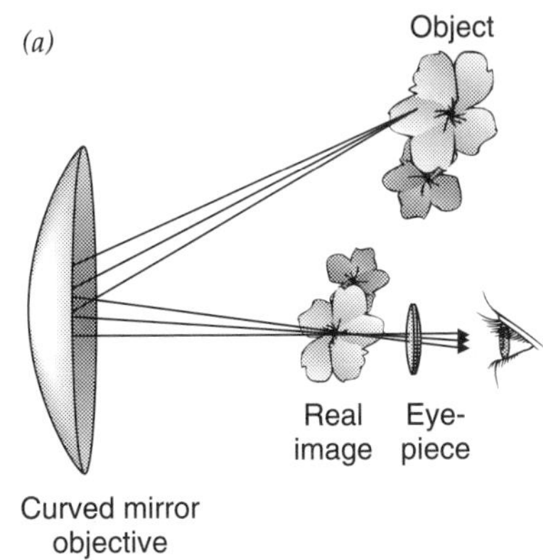

(b)

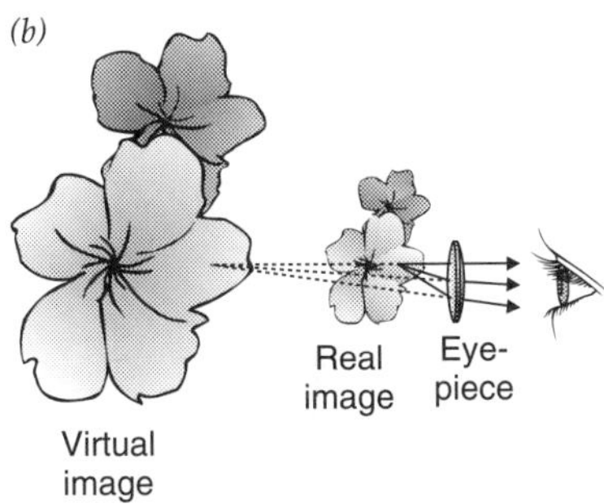

Fig. 16.2.10 - (*a*) A visual reflecting telescope consists of a curved mirror objective and a converging lens eyepiece. The concave mirror forms a real image of a distant object while the eyepiece allows you to inspect that real image in great detail. (*b*) As you look through the eyepiece, you see an enlarged and inverted virtual image of the object out in the distance.

(a)

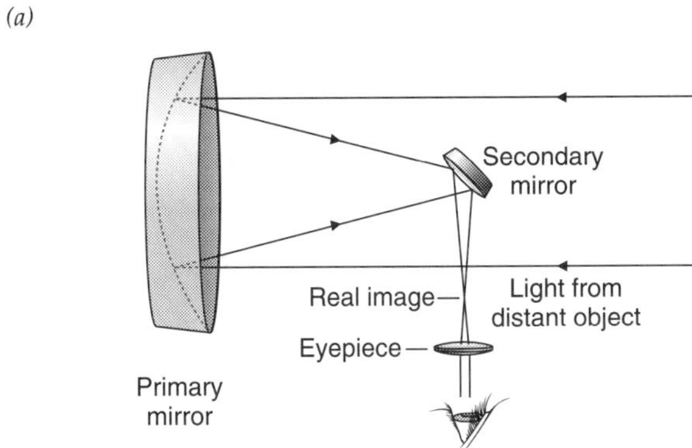

(b)

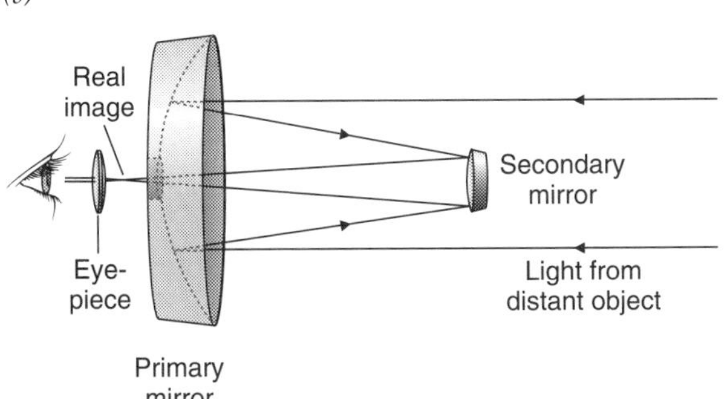

Fig. 16.2.11 - (*a*) A Newtonian telescope bends light from the primary mirror by 90° so that you can view it from the side of the telescope. (*b*) A Cassegrainian telescope sends light from the primary mirror back through a hole in the center of the primary mirror so that you can view it from the end of the telescope.

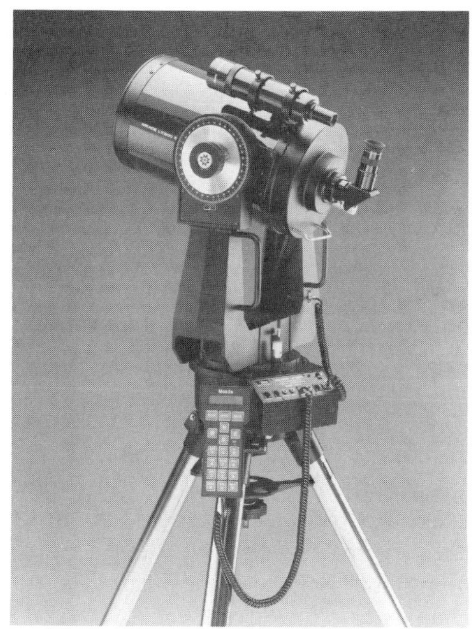

Fig. 16.2.12 - Light entering the right end of a Newtonian telescope (left) reflects from a large curved primary mirror at the far end. A small secondary mirror then turns the reflected light toward an eyepiece located on the side of the tube. In a Cassegrainian telescope (right), a small secondary mirror sends light reflected by the large primary mirror toward an eyepiece located beyond a hole in the primary mirror. An added prism just before the eyepiece bends the light 90° to make viewing easier.

must support a small mirror in the center of the telescope's field of view. These rigid supporting structures or "spiders" often appear in the photographs of stars because of interference effects and give each star an artificial cross shape.

CHECK YOUR UNDERSTANDING #3: Your Smile, Upside Down

If you hold a shiny spoon at arm's length in front of you and look into its bowl, you will see an inverted image of yourself hovering a few centimeters in front of the spoon. Explain.

The Telescope's Aperture

The size of the telescope's lens or primary mirror determines two important features of the telescope: its ability to gather light and its ability to resolve two nearby objects from one another. The issue of light gathering is straightforward—the more light that the telescope can collect and focus into a real or virtual image, the brighter that image will be. Since objects in the night sky are dim and the telescope is spreading the light it collects over large images, the telescope should gather as much light as possible. High magnification is useless if all you see is enlarged darkness.

The other reason for building large mirrors is resolving power. Light travels as a wave. When a wave passes through a narrow opening, it tends to spread out as it travels into the region beyond the opening, an effect called **diffraction**. When the light waves coming from a distant star pass through the lens or mirror of a telescope, they diffract and begin to spread out somewhat. The larger the lens or mirror, the less diffraction occurs and the smaller the spread. This spreading may seem like a trivial detail. After all, the photographs that you take of a meadow or a schoolhouse appear extremely sharp and clear even if you use a small lens. But what if you were trying to count a person's eyelashes from a kilometer away? That's equivalent to what some telescopes try to do.

When the light from two nearby stars passes through a telescope, that light diffracts. The spreading waves of light from one star may overlap with the spreading waves of light from the other star. In this case, the two stars will be unresolved. They will appear as a single bright spot on the film or in the virtual image that you are viewing. To resolve the two stars, you need a lens or mirror that is large enough that light waves from the stars don't spread into one another. To reach the level of resolution needed by modern astronomers, telescopes have to be several meters in diameter.

Unfortunately, lenses more than about 0.5 m in diameter are extremely hard to fabricate because the glass becomes so thick and heavy that it deforms under its own weight. The glass also acquires stresses and strains as it cools from its molten state and these degrade its optical quality. To make matters worse, two separate elements are needed to make an achromatic doublet. While there are a few refracting telescopes with lenses more than 0.5 m in diameter, each of these lenses represents the lifetime's work of a master lens maker.

Because enormous lenses can't be made, astronomers have turned to mirrors to photograph or observe the dim, distant objects of our universe. Mirrors are easier to fabricate than lenses, although producing a mirror larger than about 3 m in diameter is still a heroic effort. Because mirrors don't have to be transparent, their internal structures can be optimized for strength and stability. Mirrors are often made of translucent glass-ceramics that tolerate temperature changes much better than clear glasses. In recent years, it's become possible to build composite mirrors, in which several mirror segments are designed and crafted by computer-controlled machines and positioned by computers and laser beams. These giant segmented mirrors can gather light from enormous areas.

Unfortunately, the earth's atmosphere limits the resolution of large mirror telescopes. As the light from a star passes through the atmosphere, density variations in the air bend it just enough that it can't be brought to a single point in a real or virtual image. To reach the full resolving potential of a giant mirror, it must be removed from the atmosphere and put into space. The Hubble Space Telescope achieves this potential with spectacular results.

CHECK YOUR UNDERSTANDING #4: A Good View of the Sky
Why are major observatories usually located on mountain tops?

Microscopes

Microscopes are almost identical to Keplerian telescopes, except for the focal lengths of their objective lenses and the locations of their objects. A Keplerian telescope uses a long focal length objective lens to project a large real image of a distant object while a microscope uses a short focal length objective lens to project a large real image of a nearby object (Fig. 16.2.13). The eyepieces in telescopes and microscopes are virtually identical and serve as high quality magnifying glasses through which to inspect the real images.

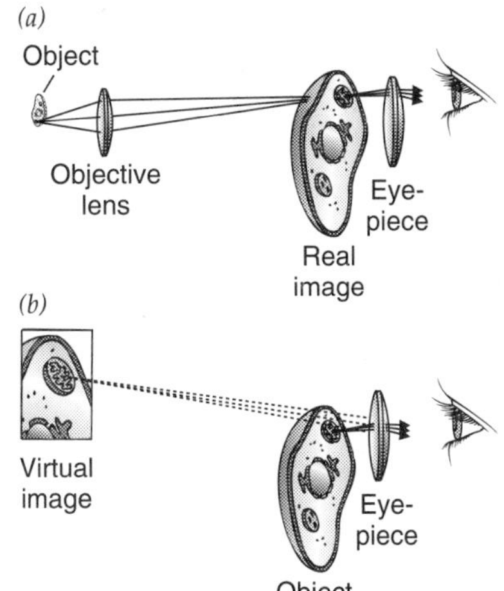

Fig. 16.2.13 - (*a*) A visual microscope consists of a two converging lenses. The objective lens forms a real image of a tiny, nearby object. The eyepiece acts as a magnifying glass, allowing you to inspect that real image in great detail. (*b*) As you look through the eyepiece, you see a very large, far away, and inverted virtual image of the object.

Although microscopes are technically similar to telescopes, they have different constraints on their construction. While a telescope objective must be very large, a microscope objective must be very small. The microscope objective must move in close to the object, catch its rapidly diverging light rays, and make those rays converge together as a real image far from the lens. Because the objective's focal length must be even shorter than the object distance, the objective must be built from small, highly curved optical elements. Several elements are needed to form a high quality real image, one that will still appear sharp and crisp when you look at it through the magnifying eyepiece.

A microscope's magnification is difficult to calculate exactly, mostly because it depends on what you choose as the unmagnified view—an object held close to your eye looks bigger than one held farther away. Nonetheless, shortening the focal length of the objective lens always increases a microscope's magnification because the objective must then move closer to the object and will form a real image of a smaller region of the object's surface. Shortening the focal length of the eyepiece also increases the microscope's magnification because it then allows you to inspect the real image more closely.

Illumination is important for a microscope. In a telescope you must use the light that's available naturally, but in a microscope you can add whatever light you like. Most microscopes use a bright lamp to send light upward through the object or downward onto the object. What you observe in these two cases is different. If you send light through the object, you see how much light it absorbs in different places. If you bounce light off its surface, you see how much light it reflects in different places.

Many modern microscopes have two separate eyepieces so that you can use both eyes to look at the object. These microscopes have only one objective lens but light from this lens is separated into two parts with a partially reflecting mirror. Two identical real images form and each eyepiece magnifies its own real image. In fact, there are also microscopes that separate additional portions of the light for other viewers or for photography. In the latter case, some of the light from the objective lens is diverted and forms a real image right on a piece of film or on the optical sensor of a video camera.

CHECK YOUR UNDERSTANDING #5: When the Objective is Downsizing...

A typical microscope has a rotating turret housing 3 or 4 different objective lenses. If you turn this turret to replace the 10 mm focal length objective lens that you have been using with a 5 mm focal length lens, what will happen to the magnification of the microscope?

Non-Light Telescopes and Microscopes

We observe objects in the universe almost exclusively through the electromagnetic radiation they send toward us. Electromagnetic waves travel easily through empty space and can cross the vast distances from one side of the universe to the other. But light isn't the only kind of electromagnetic wave. The earth is also exposed to radio waves, microwaves, X-rays, and gamma rays from sources elsewhere in the universe.

To study these electromagnetic waves, astronomers generally use reflecting telescopes. But the characteristics of those telescopes are somewhat different from light telescopes. A *radio telescope* uses a huge concave metal dish to form a real image of the electromagnetic waves from a distant object. In most cases, these waves are actually microwaves. Instead of using film or a magnifying glass to look at that real image, a radio telescope uses a small microwave receiver. The telescope studies the real image one spot at a time either by moving the microwave receiver around the image or by turning the dish and the receiver to point toward different regions of space.

A home satellite dish antenna is just a small radio telescope. It forms a real image of the microwaves coming toward it from the sky and examines the center of that real image with a microwave receiver. When the antenna is pointed at a satellite, that satellite's waves come to a focus exactly on the receiver's antenna.

Radio telescopes can be made of metal mesh rather than solid metal because electromagnetic waves are insensitive to holes that are substantially smaller than their wavelengths. This same insensitivity to details smaller than a wavelength works against optical microscopes. Because a light microscope can't observe details much finer than the wavelength of visible light, you'll need another kind of microscope to see structures smaller than about 1 micron on a side.

Two important types of non-optical microscopes are electron microscopes and scanning probe microscopes. *Electron microscopes* use electrons to form images of small objects. *Scanning probe microscopes* use extremely sharp needles to study the structures of surfaces, one point at a time.

The electrons in an electron microscope travel as waves and, like light, are unable to resolve structures substantially smaller than their wavelengths. However an electron's wavelength decreases as its kinetic energy increases so that the energetic electrons used in electron microscopes have wavelengths much less than atomic diameters. Some electron microscopes can resolve individual atoms.

Like optical microscopes, electron microscopes use lenses to form images with beams of electrons. However electron lenses aren't simply pieces of glass. Instead, they are electromagnetic devices that use electric and magnetic fields to bend and focus the moving electrons. In a *transmission electron microscope*, electrons pass through an object, through various electromagnetic lenses, and finally form a real image on a phosphor screen or a piece of film. The focal lengths of the lenses are chosen so that this real image is thousands or even millions of times larger than the object. A *scanning electron microscope* uses an electromagnetic lens to focus a beam of electrons to a tiny spot on the object and then detects reflected electrons. The microscope sweeps this spot back and forth across the object's surface and gradually creates an image of that surface.

Scanning probe microscopes originated in the early 1980's with the *scanning tunneling microscope*. The idea behind a scanning probe microscope is simple: move a sharp probe back and forth across a surface and use its interactions with that surface to study the surface. But few people imagined that a point could be made sharp enough to touch or interact with individual surface atoms.

Naturally, people were surprised when German physicist Gerd Binnig (1947–) and Swiss physicist Heinrich Rohrer (1933–) found that their metal probes could exchange electric currents with individual atoms. By scanning a probe across a surface, they could map out the locations of atoms and obtain images of the surface at the atomic scale. For their invention of the scanning tunneling microscope, this pair received the 1986 Nobel Prize in Physics.

Since then, a wide variety of scanning probe microscopes have been developed. In each case, an ultra-sharp probe scans back and forth across a surface, interacting with a few atoms on the surface as it moves along. A computer measures these interactions and builds a map of the surface. These interactions include the exchange of electrons, electric and magnetic forces, and even friction.

CHECK YOUR UNDERSTANDING #6: Aiming High

Why are many home satellite dishes so large?

Summary

How Telescopes and Microscopes Work: Telescopes provide enlarged views of distant objects. An objective lens or mirror bends light from the object so that it converges together to form a real image. In a photographic telescope, that real image is projected directly onto a piece of film or an electronic light sensor. In a visual telescope, you look at the real image through an eyepiece. The eyepiece acts as a magnifying glass, allowing you to see fine details in the real image by spreading it across more of your field of vision.

Microscopes provide enlarged views of small objects. An objective lens bends light from the object to form a real image. As with the telescope, that real image can be projected onto a piece of film or an electronic light sensor, or you can view its details through a magnifying eyepiece.

The Physics of Telescopes and Microscopes

1. If the lens equation predicts that the image of an object should appear at a negative image distance, then it's a virtual image. A virtual image is located before the lens that creates it and can be seen by looking into the lens. Because the virtual image forms before the lens, it can't be projected onto a surface.

2. If the object distance is greater than that of a converging lens's focal length, a real image forms after the lens. If the object distance is less than that lens's focal length, a virtual image forms before the lens.

3. A concave mirror forms an enlarged virtual image of a nearby object and an inverted real image of a distant object. A convex mirror always forms a reduced virtual image of an object.

4. The wavelengths of electromagnetic waves and electrons limit the resolution of imaging systems. Waves can't be used to resolve structural details that are substantially smaller than the wavelengths of those waves.

Check Your Understanding - Answers

1. The lens is initially creating a real image near your eye. This real image moves past you as the lens approaches the photograph. Finally, the lens is close enough to create an enlarged virtual image of the photograph.

Why: A magnifying glass forms either a real or virtual image, depending on object distance and the lens's focal length. If the lens and photograph are separated by more than the focal length, the image is real and you see it near your eye. You can touch this inverted image or project it on a piece of paper. If the lens and picture are separated by less than the focal length, the upright image is virtual and you see it on the far side of the lens.

2. They use long focal length lenses to create large real images. These real images form far behind their objective lenses.

Why: To achieve high magnifications, these telescopic devices use long focal length objectives and short focal length eyepieces. Since the image distance to a real image can't be shorter than the focal length of the lens, these devices are longer that the focal lengths of their objectives. Binoculars hide some of this length by folding their optical paths.

3. The concave spoon forms a real image of you at that location.

Why: Many common objects act as concave or convex mirrors. A spoon causes light from a distant object, such as you, to converge together as an inverted real image. You can see this real image hovering in front of the spoon or you can project it onto a small piece of paper.

4. To get above as much of the earth's atmosphere as possible.

Why: Because bending of light by the earth's atmosphere limits the resolution of all earth based telescopes, observatories are located at places with relatively little atmosphere overhead, no nearby light sources, and clear, calm weather. A few remote mountain tops have proven to be the best terrestrial locations for observatories.

5. It will double.

Why: Since a microscope focuses when the object distance is a little more than the focal length of the objective lens, the new lens will be twice as close to the object and will form a real image that's twice as wide and twice as high as before.

6. Large satellite dishes can gather more microwaves and can resolve between the transmissions of nearby satellites.

Why: A satellite dish is a small radio telescope that forms a real image of the microwaves reaching it from the sky. Each satellite overhead appears as a bright region in the microwave image created by the dish. If the dish is small, diffraction makes these bright regions broad and blurry, and they overlap with one another.

Glossary

concave mirror A mirror that is bowed inward at the center, like the inside of a bowl.

convex mirror A mirror that is bowed outward at the center, like the outside of a bowl.

diffraction Wave effects that limit the focusability of light and alter the way in which it travels after passing through an opening.

diverging lens A lens that bends the individual light rays passing through it away from one another so that they either diverge from one another more rapidly than before or at least converge less rapidly. A diverging lens has a negative focal length.

eyepiece The last lens of a telescope or microscope; the lens that prepares the light for your eye.

objective The first lens or mirror of a telescope or microscope; the first significant optical component that light encounters after leaving the object.

virtual image A pattern of light that appears to come from a particular region of space and reproduces the pattern of light at the surface of the original object. A virtual image forms before the lens that creates it and can't be projected onto a surface.

Review Questions

1. Why can't you put a piece of paper at the site of a virtual image and expect to see that image as a sharp pattern of light on the paper?

2. What kind of image does a magnifying glass form if the distance between the object you're studying and the magnifying glass is less than its focal length? Where is that image?

3. When you hold a magnifying glass close to a newspaper and look at the print, you see an enlarged image of the print. Is this image in front of or behind the print itself? Why?

4. Why is the image that you observe through a Keplerian telescope upside down?

5. Why does using a longer focal length objective on a Keplerian telescope increase its magnification?

6. A cheap toy telescope uses only one optical element for the objective lens. If you look at a bright light through this telescope, you'll see bands of color around it. Explain.

7. Why must an object be quite far from a concave mirror before the mirror can form a real image of it?

8. Since you can see the real image produced by a telescope's objective floating in space in front of your eye, what is the value of putting an eyepiece on that telescope?

9. Many small and inexpensive telescopes claim to have enormous magnifications. How does diffraction make these high magnifications virtually worthless?

10. How does using large diameter optics in a telescope reduce the image blurring effects of diffraction?

11. What limits the smallest detail that a light microscope can resolve? Explain.

12. The receiving antenna of a satellite dish is placed at a particular distance from that dish. Why?

13. What limits the resolution of a scanning tunneling microscope?

Exercises

1. Your eye contains a converging lens and a light-sensitive surface; the retina. Explain how your eye resembles a photographic telescope.

2. A makeup mirror is a concave mirror. If you sit near one and look into it, you'll see an enlarged, upright image of your face behind the mirror. Is that image real or virtual? How can you tell?

3. As you move farther from the makeup mirror in Exercise **2**, the image you see grows larger and more distant. Explain why the image moves away from you.

4. When you are quite far from the makeup mirror in Exercise **2**, you again see an image of your face, but now it appears upside down and located between you and the mirror. Is that image real or virtual? How can you tell?

5. You have accidentally misadjusted the focus of your telescope and are trying to figure out which way to move the eyepiece—forward or backward. You remove the eyepiece and use a small piece of paper to find the real image formed by the objective. Where should you place the eyepiece relative to the real image? Explain.

6. While the eyepiece is out of the telescope in Exercise **5**, you hold it close to a newspaper and find that it magnifies the newspaper dramatically. How does it achieve this effect?

7. Binoculars are a pair of Keplerian telescopes with prisms added after their objectives so that they form upright real images. Approximately where are those real images located in the binoculars, relative to the eyepieces?

8. Why do binoculars have eyepieces (see Exercise **7**)?

9. While focusing your binoculars (see Exercises **7** & **8**), you move the eyepieces away from the objectives. Although the virtual images you see through the eyepieces always occupy roughly the same fraction of your visual field, you find that the images move away from you. Draw pictures of the light rays to show why moving the eyepieces away from the objectives shifts the virtual images away from you.

10. You can use a concave makeup mirror as a primitive reflecting telescope. If you hold a small sheet of paper in front of the mirror, you will be able to find real images of objects located far behind you. Why won't you be able to find real images of nearby objects?

11. If you want to observe the details of a real image formed by a makeup mirror (see Exercise **11**), you can use a magnifying glass. How will the focal length of the magnifying glass affect the magnification you obtain?

12. The reflector of a good flashlight is a concave mirror and the bright spot of light the flashlight projects is actually a real image of its filament. Shifting the filament away from the mirror, as some flashlights let you do, shifts the real image of that filament toward the flashlight. Explain this shift.

13. If you turn a Keplerian telescope around backward and look through its objective, you'll see a reduced view of the world. In this case, the eyepiece is forming a real image. Roughly where is that real image? (Draw a picture.)

14. In a backward Keplerian telescope (see Exercise **13**), the objective forms a virtual image. Why is this image so small?

15. Two similar looking magnifying glasses have different magnifications. One is labeled 2 X (two times) and the other 4 X (four times). Which lens has the longer focal length?

16. Which magnifying glass from Exercise **15** must you hold closer to the object you're looking at in order to see a virtual image of that object far in the distance?

17. The high power objectives of a microscope must come precariously close to the objects you're studying. Why?

18. When you use the highest powers of an optical microscope, the image can appear dark even when the object is well lit. Why is the enlarged image so dim?

19. Why must scientists use extremely high energy particles to resolve the finest details in subatomic matter?

20. Images in cheap microscopes exhibit bands of color. Why?

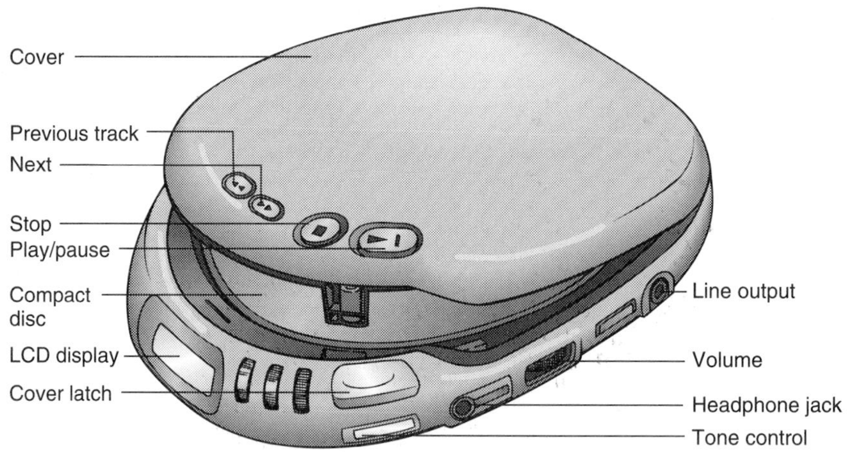

Section 16.3

Compact Disc Players

Compact disc players are remarkably sophisticated devices. They combine state of the art physics, computer science, and engineering to a degree that's almost unique among household objects. The introduction of compact disc players in the early 1980's revolutionized the music industry virtually overnight and had far reaching effects on computers, publishing, and communications. Compact disc players are revolutionary devices, unlike anything that existed before them. While compact discs and the phonograph records they replaced have comparable shapes, this similarity is only superficial. The methods use to represent sound in the two cases are entirely different, as are the techniques for reading that information.

Questions to Think About: *Why are compact discs so colorful? Why don't compact discs have a flip side, the way phonograph records do? Why can you see a bright light through a compact disc? Why is a laser involved in recording and playing back a compact disc? What limits the duration of a compact disc recording? Why are compact discs so free of noise? How can compact disc players ignore fingerprints, dust, and other surface contamination during playback?*

Experiments to Do: *Hold a compact disc by its edges with your fingers and look at its unlabeled surface. Beneath the clear plastic face is a smooth metal layer that reflects a rainbow of colors. This coloration is caused by tiny ridges in the metal layer. These ridges form a spiral around the center of the disc and are so closely spaced that light waves reflected from neighboring ridges interfere, as from a soap film. Although the colors are visible from both sides of the disc, it contains only one metal layer. The ridges on the unlabeled side of the metal layer appear as troughs on the labeled side. The metal layer is so thin that you can see through it if you hold it in front of a bright light. The metal layer is* *located on the labeled side of the disc, covered only by a thin protective lacquer and by the label itself. This metal layer is well below the surface of the unlabeled side, a feature that's extremely important during playback. Can you see the reflections of dust particles on the unlabeled surface? How deep below the plastic surface is the shiny metal?*

Digital Recording

A compact disc sounds better than a phonograph record because a compact disc represents sound in digital form—as a series of numbers. Because a compact disc player can read those numbers even from a dirty or damaged disc, its sound is virtually free of noise. In contrast, a phonograph record represents sound in analog form, as the shape of a thin spiral groove cut in its surface. Any imperfections in that groove appear directly in the reproduced sound.

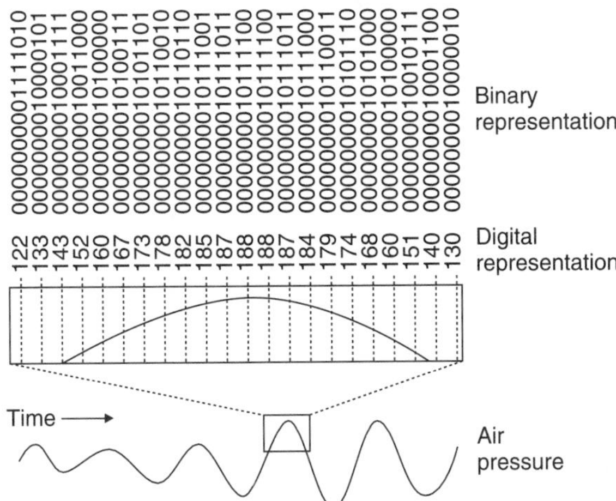

Fig. 16.3.1 - Sound can be represented as a series of numbers. Each number corresponds to the air pressure at a particular moment in time.

On a compact disc, compressions and rarefactions of the air are represented by numbers (Fig. 16.3.1). These numbers are essentially air pressure measurements taken over and over again during the recording process, and the compact disc player uses them to reproduce the recorded sound. While even a brief sound involves thousands of numbers, this digital representation allows for precise control of that sound at all times.

The air pressure measurements are made 44,100 times each second for two independent audio channels—to permit stereo recording. These measurements are recorded on the disc in binary form, using 16 bits for each measurement. (See Section 13.2 to review binary.) Since the air pressure can go down as well as up, these bits represent the positive and negative integers from –32768 to 32,767, which in turn represent how much the air pressure is above or below the average pressure. Pressure measurements with 16 bits of precision are sufficient to reproduce both loud and soft music with almost perfect fidelity.

But instead of measuring the air pressure directly as numbers, the microphone first represents it as an electric current. A device called an *analog-to-digital converter* then measures the microphone's current 44,100 times each second and obtains the series of numbers that's recorded on the compact disc. For stereo sound, the currents from two separate microphones or groups of microphones are converted to numbers and recorded simultaneously.

During playback, the compact disc player must convert these numbers into electric currents so that the amplifier and speakers can reproduce the sound. To make this conversion, a device called a *digital-to-analog converter* takes numbers 44,100 times a second and produces a current that's proportional to each number.

As the numbers change, so does the current. For stereo sound, there are either two separate digital-to-analog converters or a system that shares one converter between the two separate audio channels.

Overall, the recording studio converts the microphone current into numbers and saves those numbers sequentially on a compact disc. The compact disc player later recovers them from the compact disc and converts them back into a nearly perfect copy of the microphone current. But while digital recording is extremely accurate, it's not without its complications.

First, very sophisticated electronic filters are used in both the recording and playback processes. These electronic filters eliminate some of the problems that occur when a continuously varying electric current is represented as an evenly spaced series of integer numbers. One such problem is called *aliasing* and happens when high frequency sounds are included in the numbers that are recorded on a disc. Although you can't hear these sounds, the microphone responds to them and the digitization process doesn't measure them often enough to reproduce them properly. Instead, they reappear as audible noise during playback. To avoid aliasing, compact disc recorders and players carefully eliminate the inaudible frequencies above 20 kHz.

Second, digital recording suffers from *quantization error*. The recording equipment turns each value of the current into an integer and integers can't represent smoothly varying quantities such as current perfectly. A current of 10000.49 units would round down to 10000 if you converted it into an integer, while a current of 10000.51 units would round up to 10001. This rounding up and down distorts the sound slightly. In compact disc recording and playback, this distortion is reduced by the electronic filters.

Finally, the air pressure or current measurements aren't simply recorded one after the next on the disc's surface. Instead, these numbers are extensively reorganized before they're stored. This reorganization allows the compact disc player to reproduce the sound perfectly even if the disc can't be read completely. As we'll see shortly, the disc reading process is a technological *tour de force* and is subject to various imperfections. To be sure that the sound can be reproduced completely and without interruption, the numbers are recorded in an encoded manner. They appear redundantly so that even if one copy of the number is illegible, there is still enough legible information to completely recreate that missing number. This duplication of information reduces the playing time of the compact disc but is essential for reliability.

The complete encoding scheme, the Cross Interleave Reed-Solomon Code (CIRC), leaves the compact disc almost completely immune to all but the most severe playback problems. In principle, you can damage or obscure a 2 mm wide swath of the disc, from its center to its edge, and the compact disc player will still be able to reproduce the sound perfectly.

CHECK YOUR UNDERSTANDING #1: Cordless Clarity Away From Home

Some cordless telephone systems transmit your voice in digital form. In general terms, how does this digital transmission work?

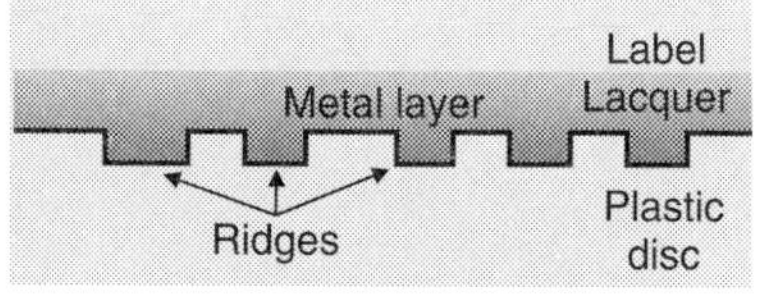

Fig. 16.3.2 - A compact disc has several layers on one side. Ridges in the metal layer store information needed to reproduce the music. This metal layer is protected by lacquer and the disc's label. The metal layer is read by light passing through the clear plastic disc from its unlabeled side.

The Structure of a Compact Disc

A standard compact disc is 120 mm in diameter and 1.2 mm thick. One side of the disc is clear and smooth but the other side contains a sandwich of layers: a thin metal layer, a protective lacquer, and a printed label (Fig. 16.3.2).

The metal layer is the recording surface. It's made from aluminum so thin that it actually transmits a small amount of light. While aluminum's electrons accelerate in response to the light's electric field and normally reflect that light completely, there aren't enough electrons in this 50 nm to 100 nm thick layer to do the job and some light gets through.

The metal layer is also not really flat. When viewed through the disc's clear plastic side, it has a series of narrow ridges on it that form a tight spiral pattern around the center of the disc (Fig. 16.3.3). The lengths of the individual ridges and the flat valleys between them represent numbers. The compact disc player examines these ridges and valleys as the disc turns, measuring their lengths and converting those lengths into numbers and sound.

Even though the individual ridges are only 110 nm high, 500 nm wide, and from 833 nm to 3,560 nm long, a compact disc player is still able to measure their sizes. These sizes aren't arbitrary—they're chosen to match the 780 nm wavelength of the laser beam that reads the disc. The compact disc player sends this laser beam through the disc's plastic surface so that it reflects from the metal layer inside. But the light's wavelength is only 780 nm when it's traveling in vacuum or air. Inside the disc, its wavelength is reduced to 503 nm—short enough to detect the ridges. This wavelength reduction occurs because polycarbonate plastic has an index of refraction of 1.55, meaning that the light's speed in that plastic is reduced from its vacuum speed by a factor of 1.55. Its wavelength is reduced by the same factor, from 780 nm in vacuum to 503 nm in plastic.

Fig. 16.3.3 - A compact disc (top) contains a thin layer of aluminum, sandwiched between two layers of plastic. The aluminum layer contains tiny ridges (bottom) that are detected by a laser beam in a CD player.

The compact disc player detects a ridge by bouncing light from the disc and determining how much of it reflects (Fig. 16.3.4). As the focused laser beam passes over a ridge, the reflection becomes weak. This weakening occurs in part because the curved ridge scatters light in all directions and in part because of interference effects. Light that's reflected back from the ridge travels about 220 nm less distance than light that's reflected back from the flat valley around the ridge. Since the wavelength of the laser light in the plastic is 503 nm, the two reflected waves are almost perfectly out of phase. They experience destructive interference and the player's light sensors detect relatively little light.

The 780 nm light is produced by a laser diode. This infrared wavelength standard was adopted in 1980, when 780 nm laser diodes were relatively inexpensive and reliable. But technology has advanced since then and new standards are likely to emerge that use shorter wavelength lasers. Blue light lasers will significantly increase the amount of music that can be stored on a compact disc because blue light can resolve much finer details on the metal layer. In the meantime, multi-layer compact disc standards are also being developed.

CHECK YOUR UNDERSTANDING #2: The Blue Light Special

When compact disc players begin to use blue lasers, compact discs will have to be redesigned. If the new laser light has a wavelength of 400 nm in air or vacuum, how should they change the height of the ridges in the metal layer?

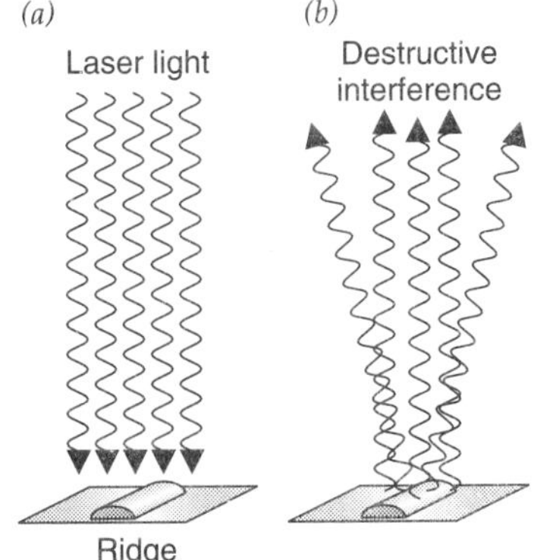

Fig. 16.3.4 - (*a*) A compact disc player directs coherent laser light onto the metal layer in a compact disc. (*b*) When that light encounters a ridge, some of it scatters in all directions. But even light that reflects directly back from the ridge interferes destructively with light reflected from the surrounding valleys.

The Optical System of a Compact Disc Player

It's remarkable that a compact disc player can locate the proper ridges within a disc and read them quickly and reliably simply by bouncing light off them. The ridges actually form a single spiral around the center of the compact disc, much like the spiral groove in a phonograph record, and the player follows this spiral as the disc turns. It slowly works its way from the center of the disc to the periphery in one long, continuous process. The disc turns at a rate that keeps the ridges moving passed the reading unit at a constant speed of either 1.2 m/s or 1.4 m/s. To maintain this constant speed, the compact disc player slowly decreases the disc's angular velocity as the reading unit moves farther away from the center of the disc. In all, the series of ridges can be up to 5.378 km long and represent as much as 74 minutes and 33 seconds of music.

But this spiral places the ridges very near one another as they circle the center of the disc. Adjacent turns or *tracks* of this spiral are separated from one another by only 1600 nm. They are so close together that about 40 tracks would fit side by side on the edge of this sheet of paper. The compact disc player must follow the correct track as the disc turns, always keeping the laser beam tightly focused on the metal layer.

Since a compact disc is neither perfectly flat nor perfectly round, the compact disc player must continuously adjust its reading unit during playback. The reading unit must keep its laser beam focused on the metal layer (auto-focusing) and must follow the ridges as they pass by (auto-tracking). These two automatic processes are beautiful examples of the use of feedback.

There are two standard approaches to auto-focusing and auto-tracking: the three-beam approach and the one-beam approach. The basic structure of a three-beam compact disc player is shown in Fig. 16.3.5. Apart from auto-focusing and auto-tracking issues, the one-beam compact disc player is quite similar.

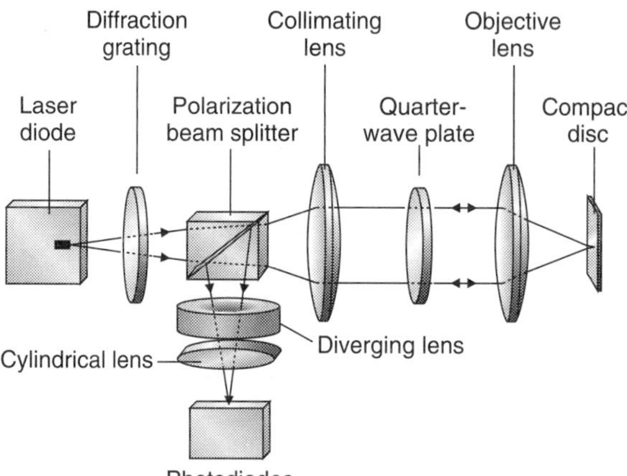

Fig. 16.3.5 - In a three-beam compact disc player, laser diode light passes through a diffraction grating, a polarization beam splitter, a collimating lens, a quarter-wave plate, and an objective lens before focusing on the metal layer inside the compact disc. Reflected light turns 90° at the polarization beam splitter and then passes through diverging and cylindrical lenses before focusing on an array of photodiodes.

In the three-beam system, light from a laser diode passes through several optical elements on its way to the compact disc's metal layer. It comes to a tight focus on the metal layer, where it illuminates only a single track. Some light reflects from the metal layer and returns through the optical elements. Finally, the reflected light turns 90° at a special mirror called a polarization beam splitter and focuses on an array of light detectors. The compact disc player measures the electric currents flowing through the detectors and uses those measurements both to reproduce the music and to control the focusing and tracking systems.

Shortly after it's emitted by the laser diode, the laser light passes through a *diffraction grating*. This optical element has a fine grillwork cut into its surface so that it breaks the light wave up into many separate components. Because the light is coherent, these individual wave pieces interfere with one another. In most directions the wave pieces interfere destructively so that little light travels in those directions. But in a few special directions the wave pieces interfere constructively and the light heading in those directions is bright. One of these bright beams travels in the same direction as the light did before it encountered the diffraction grating. However, this main beam is joined by two additional waves that travel at slight angles with respect to it. These two extra beams are used for auto tracking, as we'll see in a moment.

Next, the light passes through a *polarization beam splitter*. This device analyzes the light's polarization. It allows light of one polarization to pass directly through it but turns light of the other polarization by 90°. The laser diode used in the compact disc player produces polarized light and its polarization is such that its light passes directly through the beam splitter. We'll examine this beam splitter more carefully later on.

Light from the laser diverges rapidly as it passes through the beam splitter. As we discussed in the previous section, light waves always diffract and spread outward after they pass through openings—the smaller the opening, the worse the spreading. Because the emitting surface of the laser diode is essentially a very small opening, the laser beam becomes wider as it heads away from the diode. The compact disc player stops this light from diverging by placing a converging lens in its path. This lens is designed to *collimate* the laser beam, meaning that the beam maintains a nearly constant diameter after passing through the lens.

The laser light then passes through a *quarter-wave plate*. This remarkable device performs half the task of converting horizontally polarized light into vertically polarized light or vice versa. The light it produces is something new. Horizontally and vertically polarized lights are said to be **plane polarized** because their electric fields always oscillate back and forth in one plane as they move through space. But the light leaving the quarter-wave plate isn't plane polarized. Instead it's **circularly polarized**, with an electric field (and a magnetic field) that actually rotates about the direction in which the light is traveling.

If you look at vertically polarized light that's coming directly toward you, the electric field at each point in space will fluctuate up and down—oscillating in the vertical plane. But if you look at circularly polarized light that's coming directly toward you, the electric field at each point in space will spin around in a circle. For example, it might point up, then left, then down, then right. It can spin either clockwise or counter-clockwise, so there are two different circular polarizations—*left-handed* and *right-handed*. While circularly polarized light is hard to visualize, it's quite important in optical science. The important point at present is that it's neither vertically nor horizontally polarized, but half way in between.

Now the light passes through an objective lens that focuses it onto the metal layer of the compact disc. This lens actually contains several carefully designed optical elements and brings the light to an extremely tight focus. The lens resembles a good microscope objective, directing its light to the tiniest of spots on the metal layer. The lens also collects light that reflects from that spot.

On its way to the metal layer, the light enters the plastic surface of the compact disc. At its entry point, the beam is still about 0.8 mm in diameter. This large size explains why dust or fingerprints on the disc's surface don't cause much trouble. While contamination may block some of the laser light, most of it continues onward to the metal layer.

The light comes to a tight focus just as it arrives at the metal layer. Although it might seem that all of the light should be able to converge together to a single point on that surface, it actually forms a spot roughly 1000 nm in diameter. This spot size is limited by the wave nature of light. No matter how perfectly you try to focus light from a laser (or any other source), you can't make a spot that is much smaller than one wavelength of that light.

When you focus a laser beam with a converging lens, the light first comes together to a focus and then it spreads apart (Fig. 16.3.6). At its narrowest point, the beam forms a *waist*. This beam waist is about one wavelength of the light in diameter and a few wavelengths long, depending on the f-number of the converging lens—its focal length divided by its useful aperture. The compact disc player positions the metal layer of the disc exactly in this beam waist. Because the player's beam waist is only about 2000 nm long, its auto-focusing system must keep the objective lens just the right distance from the metal layer.

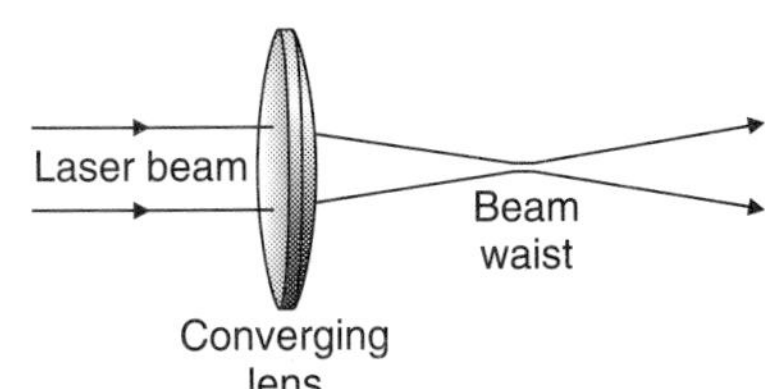

Fig. 16.3.6 - When a laser beam is focused by a converging lens, the light doesn't meet at a single point in space. Instead, it reaches a narrow waist with a diameter roughly equal to the wavelength of the light.

This fundamental limitation on how tightly a beam of light can be focused is another example of diffraction; the focusing lens acts as a narrow opening for the light wave. But even reaching this ideal focusing limit requires careful design and fabrication of the optical elements. While most other optical systems fall short of their ideal limits, a compact disc player's optical system does as well as can be done within the constraints of diffraction itself. Its optics are essentially perfect and are said to be "diffraction limited."

The amount of light that reflects from the metal layer depends on whether the laser spot hits a ridge. The light that is reflected then follows the optical path in reverse. It's collimated by the objective lens and travels back toward the polarization beam splitter. As this light passes through the quarter-wave plate for the second time, the plate finishes the job it started earlier on. The light ends up plane polarized, but with the opposite polarization from what it was when it left the laser. Horizontally polarized light is now vertically polarized or vice versa.

The reflected light then passes through the collimating lens, which causes it to begin converging to a focus, and then enters the polarization beam splitter. Because the light's polarization has changed, the polarization beam splitter no longer allows the beam to pass directly through. Instead, it turns this reflected beam 90° and directs it toward the detector array. This clever redirection scheme is important for two reasons. First, it conserves the laser light by allowing most of it to travel from the laser diode to the detector. Second, it prevents reflected light from returning to the laser diode, where that light would be amplified and would cause the laser diode to misbehave.

After turning 90°, the reflected light passes through a diverging lens and a cylindrical lens on its way to the detector array. The diverging lens delays the focus of the reflected beam while the cylindrical lens hastens that focus in one direction only. A **cylindrical lens** bends light in one direction, leaving the other direction unaffected. A glass of water is a cylindrical lens, focusing light horizontally but not vertically as you look at objects through the glass. In the compact disc player, the cylindrical lens affects the shapes of the light spots on the detector array and may make them elliptical.

The three-beam compact disc player uses the shapes of these spots to control the auto-focusing system. Each spot's shape actually depends critically on how well the light focuses on the metal layer inside the compact disc. If the focus is perfect, the spots are circular. If the focus is imperfect, the spots are elliptical. The player uses four individual light detectors to determine the shape of the main light spot. These detectors are arranged like slices of a pie that has been cut into quarters. When the spot is circular, each detector receives the same amount of light but when the spot is elliptical, the light intensities are unbalanced.

The compact disc player carefully adjusts the spacing between the objective lens and the compact disc so as to keep that light spot circular and the intensities balanced. The result of this feedback process is that the laser beam remains perfectly focused on the metal layer of the disc, even if the disc wobbles about.

Since the player's auto-focusing system must be able to move the objective lens quickly, the lens is attached to a coil of wire that's suspended near a permanent magnet. By varying the electric current flowing through the coil of wire, the player uses magnetic forces to move the objective lens back and forth rapidly.

But these light detectors are also responsible for reading music from the disc. As the disc turns, ridges in the disc's metal layer move through the focused laser beam and affect the amount of reflected light striking all four detectors. The detectors identify these overall changes in reflectivity with ridges on the disc and the compact disc player uses them to reproduce the music.

The light detectors are **photodiodes**, light-sensitive devices that resemble photoelectric cells. But while a photoelectric cell uses light to separate electric charges, a photodiode uses light to control the flow of an electric current. In the dark, a photodiode is like any other diode, carrying current only in one direction and preventing charge from traveling backward across the depletion region of the p-n junction. However, when it's exposed to light, a photodiode carries a backward current that's proportional to the light intensity. The photons provide the energy needed to send charges backward across the depletion region.

The two extra light beams in a three-beam compact disc player also reflect from the metal layer and travel toward the light detectors. But these reflected beams land on two additional photodiodes. When the main beam is perfectly aligned on a ridge, these extra beams are aligned symmetrically on either side of that ridge and their reflections have equal intensities. If the intensities of the two reflections are unequal, those beams are evidently misaligned and the player must move the beams sideways to find the ridge.

The mechanism the compact disc player uses to move the laser beam sideways varies from player to player. Some players use a screw to push the entire optical assembly from the center of the disc to its periphery as the music plays (Fig. 16.3.7). Other players swing the optical assembly on a pivoting arm like that on phonograph player. Because the laser beam must move quickly enough to follow the wobbling track of a poorly manufactured disc, players usually use one mechanism for slow movements across the disc and another for rapid minor adjustments. Often, the same coil and magnet that moves the objective lens up and down to control focusing also moves the lens sideways to control tracking.

One-beam compact disc players have no diffraction grating or cylindrical lens and don't create extra beams for the auto-tracking system to use. Instead, they use four photodiodes to study the main beam itself and are able to control both auto-focusing and auto-tracking. Although the one-beam system seems less sophisticated than the three-beam system, they both work about equally well.

Fig. 16.3.7 - This CD player's optical system is mounted on a small metal assembly (center) that moves downward during play. The lens appears in the middle of the assembly, supported by tiny electromagnetic coils and springs.

CHECK YOUR UNDERSTANDING #3: That's Why It's Called a *Laser* Disc

Why must the compact disc player use a laser diode rather than a more conventional light source such as an incandescent bulb?

The Polarization Beam Splitter

The polarization beam splitter is made out of two pieces of a clear crystalline mineral called calcite (calcium carbonate) (Fig. 16.3.8). These two pieces are carefully cut and polished to form two prisms that almost touch one another in the completed beam splitter. While light of one plane polarization manages to pass from one prism to the other without difficulty, light of the other polarization reflects from the gap between the prisms and turns 90°.

The two physical effects that make this beam splitter work are birefringence and total internal reflection. **Birefringence** occurs when a material has different indices of refraction for the two plane polarizations of light. In a crystal such as calcite, the molecules are arranged in an orderly fashion, like stacks of apples in a grocery store. This orderly arrangement of molecules may respond differently to vertical and horizontal electric fields. If one of these fields produces more electric polarization in the crystal than the other, the crystal will respond differently to vertically and horizontally polarized lights and will have two different indices of refraction. Light of the plane polarization that affects the crystal most strongly will travel slower than light of the other polarization.

Calcite exhibits birefringence. The index of refraction for one plane polarization of light is 1.70. This light travels relatively slowly through the crystal and is said to be polarized along the crystal's *slow axis* (Fig. 16.3.9). The index of refraction for the other plane polarization is 1.52. This light travels relatively quickly and is said to be polarized along the crystal's *fast axis*. The calcite prisms of the polarization beam splitter are carefully cut so that horizontally and vertically polarized lights travel through them at different speeds.

(a)

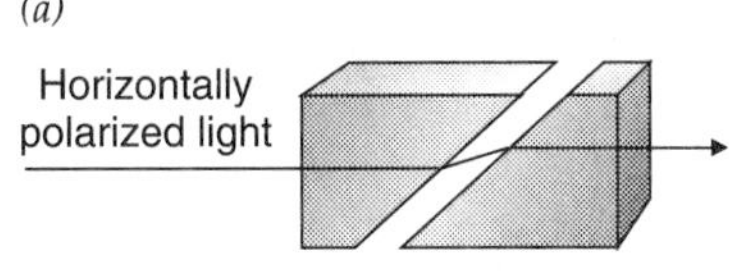

(b)

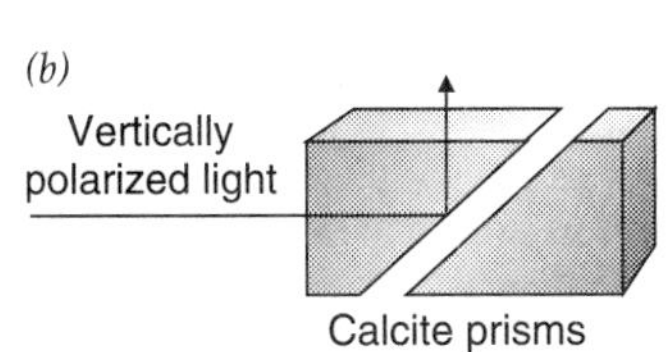

Fig. 16.3.8 - A polarization beam splitter separates horizontally and vertically polarized light. *(a)* Light polarized along the calcite prisms' fast axis passes from one prism to the next. *(b)* Light polarized along the prisms' slow axis experiences total internal reflection at the gap between prisms and turns 90°. The beam splitter above turns vertically polarized light 90° while letting horizontally polarized light pass straight through.

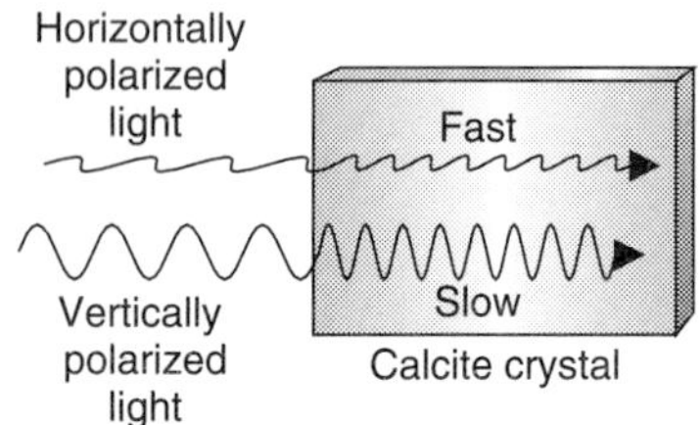

Fig. 16.3.9 - The two plane polarizations of light travel at different speeds in calcite. The crystal shown above has been cut and oriented so that horizontally polarized light travels faster than vertically polarized light.

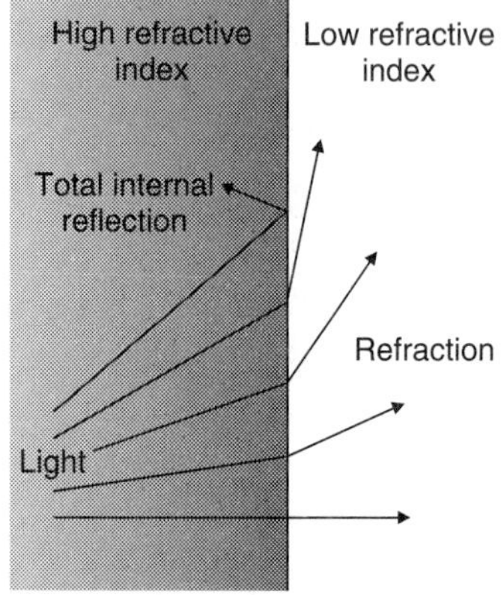

Fig. 16.3.10 - As light passes through a surface and enters a material with a lower refractive index, it bends toward that surface. If its approach angle is too shallow, the light will bend so much that it will simply reflect from the surface.

The two polarizations of light also refract differently as they enter the air gap between the two prisms. Light polarized along the crystal's fast axis bends considerably as it leaves the first prism but still manages to cross the air gap and enter the second prism. However, light polarized along the crystal's slow axis bends so severely as it tries to leave the first prism that it never leaves at all. This light is perfectly reflected from the surface of the prism in a process called total internal reflection.

Total internal reflection is an extreme case of refraction. When light passes through the surface between two materials, refraction causes the light to bend (Fig. 16.3.10). If the material it enters has a smaller refractive index than the one it leaves, the light bends away from a line perpendicular to the surface. The amount of the bend depends on the two indices of refraction and on the angle at which the light approaches the surface. As long as the approach angle is steep enough, light will succeed in leaving the surface. But if it's too shallow, the light won't enter the second material at all. Instead, the light will reflect from the surface and remain in the first material—as though it bounced off a perfect mirror. In fact, total internal reflection is much more efficient at reflecting light than a conventional metal mirror.

The polarization beam splitter is designed so that one plane polarization of light is able to negotiate the air gap between the calcite prisms while the other polarization of light experiences total internal reflection. Light polarized along the calcite's fast axis travels straight through the beam splitter with minimal disturbance. But light polarized along the calcite's slow axis is turned 90° and leaves the first calcite prism through a separate surface. In the compact disc player, light from the laser diode travels directly to the surface of the compact disc while light returning from the compact disc with the other polarization is turned toward the photodiodes.

CHECK YOUR UNDERSTANDING #4: If It Looks Like a Mirror...

When you look into an aquarium from the front, the sides of the aquarium often look like mirrors. Those sides are actually clear glass, so why do they reflect light so well?

Summary

How Compact Disc Players Work: A compact disc player uses a beam of laser light to read the ridges on a thin metal layer inside the compact disc. As the disc rotates, the ridges pass through the focused laser beam and the amount of reflected light fluctuates up and down. The player monitors the reflected light and uses it both to recreate the music and to keep its optical system properly aligned. This continual realignment allows the player to follow the spiral pattern of ridges as the disc turns and to keep the laser beam tightly focused on the metal layer. The optical system, which includes a laser diode, several photodiodes, and a variety of lenses and other optical elements, is so well designed and executed that the spot it creates on the metal layer is limited in smallness only by diffraction effects due to the wave nature of light.

The player starts near the center of the disc and works its way outward with time. The disc's angular velocity gradually decreases so that the ridges pass by the optical system at a constant speed of either 1.2 m/s or 1.4 m/s. The entire spiral pattern of ridges can be more than 5 km long and take over 74 minutes to play. Careful encoding of the digital information on the disc makes the playback almost immune to errors in the optical reading process and the digital information can be used to recreate nearly perfect and noiseless audio. Because the laser spot is large as it enters the disc, dust and fingerprints have no effect on the playback.

The Physics of Compact Disc Players

1. Because of diffraction effects, light can't detect structures that are significantly smaller than its wavelength or be focused to spots that are significantly smaller than its wavelength.

2. Because of diffraction, a laser beam that passes through a converging lens doesn't focus to a single point in space. Instead, it reaches a narrow waist and then begins to diverge again. This waist is typically about one wavelength of the light in diameter.

3. In addition to the two plane polarizations of light, horizontal and vertical, there are two circular polarizations, left-handed and right-handed. In circularly polarized light, the electric and magnetic fields rotate around the light's direction of travel.

4. A quarter-wave plate converts plane polarized light into circularly polarized light or vice versa.

5. Birefringent materials exhibit different refractive indices for different plane polarizations of light. Light of one polarization travels faster through a birefringent material than does light of the other polarization. This birefringence also affects refraction at that material's surfaces.

6. Light that approaches the surface between two materials at a shallow angle is refracted at that surface. If it tries to leave a material with a high index of refraction for one with a low index of refraction, the light may experience total internal reflection. When this effect occurs, the light is completely reflected from the surface as though the surface were a perfect mirror.

Check Your Understanding - Answers

1. A microphone converts the sound of your voice into a fluctuating current. This current is measured periodically by an analog-to-digital converter and represented as a sequence of numbers. These numbers are then represented with radio waves and transmitted through space. At the receiver, the radio waves are converted back into numbers and the numbers are converted back into a fluctuating current by a digital-to-analog converter. Finally, the current is amplified and sent through a speaker.

Why: Many modern communication devices use a digital representation of sound. In radio wave communication, digital representations are particularly useful because they are immune to the noise that plagues analog radio transmissions and also permit more efficient use of the radio bandwidth. Digital audio tape recorders also benefit from the noise immunity of digital techniques.

2. The ridges should become about half as high as they are now.

Why: The ridges in the metal layer should be about a quarter of a wavelength tall so that light reflected from the top of a ridge interferes destructively with light reflected from an adjacent valley. If the light wavelength changes to 400 nm, half the present value, then the ridges should change to half their present height.

3. The light in a compact disc player must be coherent so that it can be focused to a single diffraction limited spot on the metal layer and experience destructive interference when it encounters a ridge.

Why: You can't focus light from an incandescent bulb to a diffraction limited spot. Each photon leaving the bulb is independent and will focus at a somewhat different point in space. The bulb's incoherent light doesn't experience the strong interference effects of laser light.

4. The light you see is experiencing total internal reflection as it tries to leave the glass at a shallow angle. You see perfect reflections of the fish inside the aquarium.

Why: Light has trouble passing from water or glass into the air at a shallow angle. When it tries, it experiences total internal reflection and bounces off the surface without losing any intensity at all. Fiber optics use this effect to carry light through thin filaments of glass or plastic. The light experiences total internal reflection over and over again and remains trapped in the fiber. It follows the fiber from beginning to end with little loss of intensity.

Glossary

birefringence Refractive effects that occur when a material has more than one index of refraction.

circularly polarized light A light wave in which the electric and magnetic fields rotate about the direction in which that wave is heading.

cylindrical lens A lens that is shaped like part of a cylinder and focuses light only in one direction. It brings light together to a line rather than a point.

photodiode A diode that permits current to flow backward across the p-n junction when exposed to light. Light provides the energy needed to move charges across the junction's depletion region in the wrong direction. The current flowing in the reverse direction through a photodiode is proportional to the light intensity.

plane polarized light A light wave in which the electric field (and the magnetic field) fluctuates back and forth in a plane as the light travels through space.

total internal reflection Complete reflection of a light wave that occurs when that wave tries unsuccessfully to leave a material with a large refractive index for a material with a small refractive index at too shallow an angle.

Review Questions

1. When a phonograph record represents sound, what does it use to represent the air pressure at each moment in time?

2. When a compact disc represents sound, what does it use to represent the air pressure at each moment in time?

3. If a compact disc were made from diamond, which has an index of refraction of 2.42, would the ridges in its aluminum layer have to be taller or shorter than they are now?

4. Why does a compact disc player use a polarization beam splitter, rather than a polarizing filter like that in sunglasses?

5. Once a laser beam emerges from an opening, it may narrow temporarily but it will eventually spread out. Why?

6. Why can't a laser beam be focused to a point in space?

7. Why is the current flowing in a photodiode proportional to the intensity of the light striking its surface?

8. The number of wavelengths of 780 nm light that can fit in a calcite crystal depends on that light's polarization. Why?

9. If you shine a flashlight beam at the surface of a swimming pool from beneath the surface, the beam will sometimes reflect perfectly. How does this reflection depend on the angle at which the beam strikes the water's surface?

Exercises

1. Like a phonograph record, a normal audio tape represents sound in analog form. Explain why the tape's representation is analog.

2. Like a compact disc, a digital audio tape represents sound in digital form. The air pressures associated with sound are recorded as a series of numbers on the magnetic tape. Why is a digital audio tape essentially free of noise?

3. A compact disc could store much more music if it recorded fewer air pressure measurements each second. Why must so many measurements be made each second?

4. Why is a laser beam's focus delayed by its entry into the plastic of a compact disc? (Draw a picture.)

5. To prepare a sample for an electron microscope, the operator embeds the sample in clear plastic and then shaves off ultra-thin sheets with a diamond knife. The operator can tell how thick those sheets are by observing their colors through a microscope. Why do the sheets appear colored?

6. When using color to determine the thickness of an electron microscope sample (see Exercise **5**), the operator must consider the plastic's index of refraction. Why?

7. Some of the laser light striking the compact disc's aluminum layer hits the valley around a ridge and reflects back toward the photodiode. How does this reflected wave actually *reduce* the amount of light reaching the photodiode?

8. Scientists measure the distance to the moon by bouncing laser light from reflectors left on the moon by the Apollo astronauts. This light is sent backward through a telescope so that it begins its trip to the moon from an enormous opening. This procedure reduces the size of the laser beam when it reaches the moon. Explain.

9. As it passes through the lens that focuses it onto a compact disc's aluminum layer, the compact disc player's laser beam is more than 1 mm in diameter. Why does its large size as it leaves the lens allow the beam to focus to a smaller spot?

10. Why can light from a blue laser form a narrower beam waist than light from an infrared laser?

11. In a three-beam compact disc player, the three beams of light approach the objective lens from slightly different angles. Each beam is effectively a group of light rays traveling parallel to one another. Draw a picture of the light rays showing that the three beams focus at slightly different locations on the aluminum layer inside the compact disc.

12. Light from a distant object approaches the objective lens of a telescope as a group of parallel light rays. Draw a picture of the light rays from three stars passing through a converging lens to show that they focus at three separate locations.

13. The objective lens in a compact disc player must move quickly to keep the laser beam focused on the disc's aluminum layer. Why must this lens have a very small mass?

14. A glass of water is effectively a cylindrical lens. When you send a narrow laser beam directly through the center of the glass, the beam will form a horizontal stripe on a sheet of paper behind the glass. Why does its passage through the glass spread the light out horizontally but not vertically?

15. A compact disc player uses photodiodes because they respond extremely rapidly to light. The player puts positive charge on each photodiode's n-type cathode and negative charge on its p-type anode. A photon absorbed near the diode's p-n junction can send an electron across the depletion region from anode to cathode. The electron then accelerates rapidly, giving the photodiode its speedy response to light. What force causes the electron to accelerate so rapidly?

16. Why don't portable compact disc players play properly if you shake them back and forth too quickly?

17. Sometimes the surfaces of a glass of water look mirrored when you observe them through the water. Explain.

18. When you look into the front of a square glass vase filled with water, its sides appear to be mirrored. Why do the sides appear so shiny?

19. Squeezing clear plastic or glass distorts its atoms so that these atoms respond differently to horizontal and vertical electric fields. The material becomes birefringent. How could you use interference between reflections from the front and back surfaces of a material to show that squeezing it changes its index of refraction for vertically polarized light by a different amount than that for horizontally polarized light?

20. A quarter-wave plate is made from a birefringent material. Like a quarter-wave plate, a piece of squeezed glass or plastic (see Exercise **19**) can partly or completely convert horizontally polarized light into vertically polarized light, or vice versa. If you put a piece of plastic between two polarizing sheets, one that blocks horizontally polarized light and another that blocks vertically polarized light, you will only see light that passes through portions of the plastic that are squeezed (or stretched). Why is only that light visible?

Epilogue for Chapter 16

In this chapter we looked at four common objects and studied the ways in which these devices use optics. In *cameras,* we saw how a converging lens can form a real image of an object and how that real image can be used to record a scene on a piece of photographic film. We learned about the roles of focal length and f-number in determining the size and brightness of the real image and explored some of the complications that must be considered when designing lenses.

In *telescopes and microscopes,* we considered the ways in which lenses and curved mirrors can be used to form both real and virtual images. We saw how objective mirrors or lenses can be used to make real images of distant or small objects and looked at how eyepieces help us to observe the fine details in those images. We also studied the limitations that the diffraction of light imposes on these instruments and some of the ways for dealing with those limitations.

In *compact disc players,* we looked at how laser light is affected by various optical devices and at what happens to it when it's focused to an extremely small spot. We learned about birefringence and total internal reflection and at how these effects make it possible to separate the two plane polarizations of light. We also looked briefly at the nature of circularly polarized light.

Explanation: Focusing Sunlight

The magnifying glass forms a real image of the sun on the sheet of paper. Because the sun is so far away, the image distance between the lens and the paper is equal to the focal length of the lens. The real image is relatively small, yet it contains all of the sunlight that strikes the entire surface of the lens. With so much light focused onto a small region of the paper, the radiative heat transfer from the sun to the paper becomes quite large and the paper's temperature rises until it begins to smoke or burn. The lens's image also contains light from other parts of the sky and this light can be seen surrounding the sun's disk. Because of the way light is bent during the formation of the real image, the clouds in the sky appear upside down and backward on the sheet of paper.

Cases

1. A fiber optic light guide consists of a narrow glass fiber core coated by a second layer of glass with a lower refractive index. The core is made of an extremely pure and homogeneous glass so that it absorbs very little light.

a. Light in the fiber core hits the core's surface at a very shallow angle and is reflected. Why can't the light propagate into the surrounding layer of glass?

b. The glass is free of bubbles. What would happen to the light if it had to pass through a series of air bubbles on its way through the fiber?

***c.** Light passing through an optical fiber must travel through an enormous amount of glass. If impurities in the glass halved the light intensity each time it traveled 1 km, how much light would remain after 10 km of fiber?

d. Information is usually sent through a fiber as brief pulses of light. Those pulses are formed by effectively blocking and unblocking the light from a laser. Since that technique is equivalent to amplitude modulating a carrier light wave, the light pulses contain sideband waves at slightly different frequencies and wavelengths from the carrier wave. Because each pulse of light is formed by the combination of many waves with different frequencies and wavelengths, the fiber must exhibit very little dispersion. What would happen to a single pulse of light if it passed through a long fiber with strong dispersion?

e. Since even the best glass fibers absorb some light, light passing through a long fiber must be amplified by short segments of laser amplifier fiber that are spliced into the main fiber. As light passes through a segment of laser amplifier fiber, each photon is duplicated hundreds of times. Explain why this laser amplifier fiber needs power to operate.

2. The clear plastic safety reflector on a bicycle has a flat front surface but its back interior surface is covered with the corners of cubes. These cube corners or "corner cubes" stick up from the surface into the air like hundreds of small triangular pyramids. What makes the reflector so useful is that light striking it is sent directly back toward its source.

a. When light hits the reflector at a right angle to its front surface, most of that light propagates to the back of the reflector where it encounters one of the corner cubes. It hits the surface of the cube at a shallow angle and instead of leaving the plastic, the light reflects perfectly. In fact, it reflects perfectly from all three faces of the corner cube and ends up returning directly toward the source of the light. But the back surface of the reflector is clear plastic followed by air. Why does light reflect perfectly off each surface of the corner cube?

b. While the reflector works best for light that hits it at a right angle to its surface, it also works for light that arrives at other angles. But when that angle becomes too shallow, the reflector stops working and light passes through its back surface. Explain why the light is able to get through in this case.

c. How does refraction at the front surface of the reflector help to widen the range of angles over which the reflector is able to send light directly back toward its source?

d. Why won't a flat mirror work well as a safety reflector?

3. The lens of your eye resembles that of a camera—light from the scene in front of you focuses to a real image on your retina. But unlike a camera lens, the lens of your eye can actually change its focal length by changing its curvature. The more curved the lens's surfaces are, the more strongly it bends light together and the shorter its focal length. The lens's variable focal length allows you to focus the real image of a particular object onto your retina without having to change the distance between the lens and the retina.

a. Explain why focusing on a distant object requires less lens curvature than focusing on a nearby object.

b. The lens of a farsighted person has trouble becoming curved enough to form a real image of a nearby object on the retina. Explain why wearing glasses containing converging lenses helps a farsighted person see nearby objects.

c. The lens of a nearsighted person has trouble becoming flat enough to form a real image of a distant object on the retina. Explain why wearing glasses containing diverging lenses helps a nearsighted person see distant objects.

d. The iris of your eye changes diameter to control how much light reaches your retina. The brighter the light, the smaller the iris's opening. Why is your eye's depth of focus greater in bright light than in dim light?

4. The inexpensive disposable cameras that are sold in grocery stores produce remarkably good pictures considering the financial constraints on their design. Despite their apparent simplicity, these cameras are sophisticated devices.

a. A disposable camera has no focus adjustment yet it produces reasonably sharp images. How does its small lens contribute to this result?

b. The camera lens is normally made from a single piece of clear plastic with a refractive index of 1.55. If they accidentally used a plastic with a refractive index of 1.45 and left the rest of the camera unchanged, the photographs would be blurry. Why?

c. The plastic used in the camera lens exhibits very little dispersion. How would dispersion affect the photographs?

d. The shutter in a disposable camera is a thin metal plate with a hole in it. When you take a picture, a spring pulls this plate past the lens and the hole allows light to reach the film. How do the spring's tension and the plate's mass affect the exposure time?

5. A video camera uses a converging lens to form a real image of a scene on an electronic imaging device. This device forms a video signal out of the pattern of light on its surface.

a. Why must the lens of the video camera be a converging lens, rather than a diverging lens?

b. The camera focuses automatically by analyzing the contrast in the video image. As you shift the camera from a distant object to one that's closer, the lens moves. Does it move toward the imaging device or away from it? Explain.

c. The camera has an automatic iris—a diaphragm that controls the lens's aperture. When you turn the camera toward a bright scene, the iris reduces the lens's aperture. What effect does this change have on the camera's depth of focus?

d. The camera has a zoom lens that changes its focal length at the touch of a button. When you zoom in on a person for a close-up, does the focal length increase or decrease?

e. As the lens changes its focal length, it also moves toward or away from the electronic image device. As you zoom out to view more of the scene, does the lens move toward or away from the imaging device?

6. A slide projector is essentially the reverse of a camera. Light from an illuminated slide passes through a converging lens and forms a real image on the screen.

a. Since the slide is the object, the object distance is the separation between the slide and the lens. Compare this object distance to the lens's focal length when the projector casts a real image on a screen far at the other side of a room.

b. You focus the projector by moving the lens toward or away from the slide. If you want to focus the real image on a closer screen, which way should you move the lens?

c. The projector has a zoom lens that changes its focal length when you turn a dial. This lens makes it possible to change the size of the image on the screen. Zooming the lens also moves it toward or away from the slide. As the lens's focal length increases, should it move toward or away from the slide to keep the real image focused on the screen?

d. As you change the focal length of the zoom lens and move it toward the slide, the real image on the screen grows larger. Draw pictures of the light rays to show why moving the lens toward the slide causes the real image to grow.

e. Why must the slide be upside down in the projector in order to produce an upright real image on the screen?

7. Doctors use endoscopes to look inside patients through narrow openings. An endoscope contains a one- or two-meter-long bundle of optical fibers that are carefully arranged and bonded together at their ends but that are independent and flexible in between. Each optical fiber is a pipe for light—light entering one end emerges from the other. When one end of the endoscope's fiber bundle is illuminated with a particular pattern of light, that same pattern of light appears at the bundle's other end.

a. The core of each optical fiber is made from a glass with a large refractive index. This core is wrapped in a glass with a lower refractive index. Each time light tries to leave the core at a shallow angle, it is perfectly reflected. Why?

b. The endoscope uses an objective lens to form a real image of nearby tissue on one end of the fiber bundle. Why must the objective lens have a very short focal length in order to form a real image of nearby tissue?

c. The doctor uses an eyepiece to view light emerging from the other end of the bundle. How does the focal length of that eyepiece affect the magnification of the endoscope?

d. To obtain a large depth of focus, should the aperture of the endoscope's objective lens be large or small?

***e.** To illuminate the patient's tissue, the endoscope has a second bundle of fibers that carries light from a bulb to the tissue. It uses a converging lens to form a real image of the bulb's filament on the entry surface of the fibers. If the bulb's filament and the fibers are 4 cm apart and the lens is halfway between them, what should the lens's focal length be?

CHAPTER 17

MATERIALS SCIENCE

Between natural resources such as water and air, and finished products such as bulbs and bicycles, lies a world of modified and synthesized materials. The study and development of these basic ingredients of technology have enormous influences on our daily environment and are part of a field called materials science.

Experiment: Glue Putty

To remind yourself of how wonderful and varied materials can be, you can make an interesting material we'll call "glue putty." Glue putty resembles two commercial products: Silly Putty and Gaak. It's neither a simple solid nor a simple liquid, but something in between. It actually contains long polymer molecules of a type that we'll study in the section on plastics.

To make glue putty, start with two clean plastic containers. In one container, mix together about 125 milliliters (4 ounces) of white glue and about 125 milliliters (4 ounces) of water. In the other container, mix together about 125 milliliters (4 ounces) of water and about 5 milliliters (1 teaspoon) of borax powder (from the laundry aisle of the grocery store). Now combine the two liquids and stir. A lumpy, gooey substance should begin to form and cling to your spoon. Collect this substance and knead it with your hands on a hard surface. Be careful not to let it touch paper or clothing, to which it will stick. After a little kneading, you should have a smooth, elastic blob of glue putty.

When you pull slowly on glue putty, it flows like a liquid. It contains long chain-like molecules and patient stretching allows those chains to work their way past one another so that the putty can flow. It will even flow slowly under its own weight so that a round ball gradually flattens when you place it on a table.

But if you pull glue putty suddenly, it will break like a brittle solid. Since you don't give it time to flow, you create huge stresses within the material. A defect in the glue putty's surface soon becomes a crack, the crack propagates into the putty, and the putty suddenly breaks into two pieces.

Chapter Itinerary

In this chapter, we'll examine three areas of materials science: (1) *knives and steel*, (2) *windows and glass*, and (3) *plastics*. In *knives and steel*, we'll look at how steel is made and at how the quality of a knife depends on the microscopic structure of its steel. In *windows and glass*, we'll explore the nature of non-crystalline solids to see why glass is different from most other materials. In *plastics*, we'll study the structures of various plastics to see what distinguishes them from one another. While these systems represent only a fraction of the materials we frequently use, they should provide a flavor for the problems and solutions that are possible when working with the basic building blocks of technology.

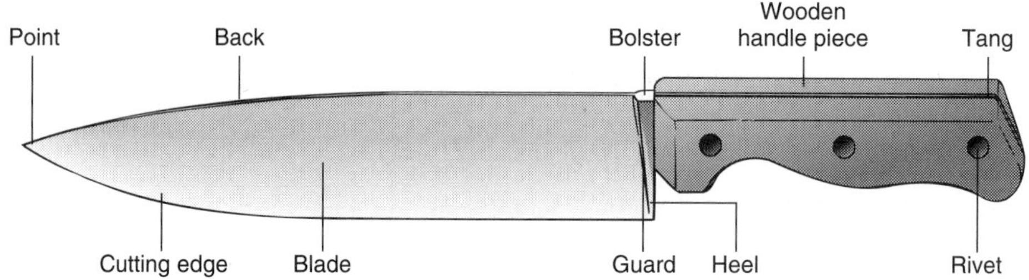

Section 17.1

Knives and Steel

A dull knife is useless in the kitchen—you might as well tear a tomato or a loaf of crusty bread apart by hand. But sharpening a knife isn't always easy because some knives take edges better than others. While there are many knives that simply won't stay sharp, a few wonderful knives never seem to dull. Though they are all made of steel, there is clearly something different about their blades.

Knives have always been one of the greatest tests of a steel maker's skill, from the days when swordmaking was an art to the present era of science and technology. A knife's blade must be tough and flexible, while its cutting edge must be hard but not brittle. Giving steel these properties requires great control over its composition and processing. To understand a knife blade, you must understand its steel.

Questions to Think About: *Why are some knife blades softer and more flexible than others? Why are some knives more corrosion resistant than others? Why does a knife blade spring back when you bend it a little but deform permanently when you bend it further? Why do some knife blades bend or stretch before they break while others simply shatter?*

Experiments to Do: *Take a new steel paper clip and carefully unbend one of its ends, making sure that you notice just how stiff the wire is. Now rebend that end and straighten it repeatedly. Do you notice any change in the stiffness of the metal? Eventually a crack will develop at the edge of the wire and the wire will break into two pieces. Cracking is typical for hard objects such as rocks. Has the steel in the paper clip become harder than it was originally? If so, what has happened inside the steel to cause this change?*

Stresses and Strains, Bends and Breaks

Knives are simple tools that we use to cut things into pieces. They are essentially wedges that use mechanical advantage to convert small forward forces into large separating forces. When you push down on the blade of a knife as you cut a carrot (Fig. 17.1.1), its cutting edge penetrates the carrot while the two inclined surfaces of the blade exert huge horizontal forces on the two halves of the carrot. One half accelerates to the left while the other half accelerates to the right and the carrot divides neatly in half.

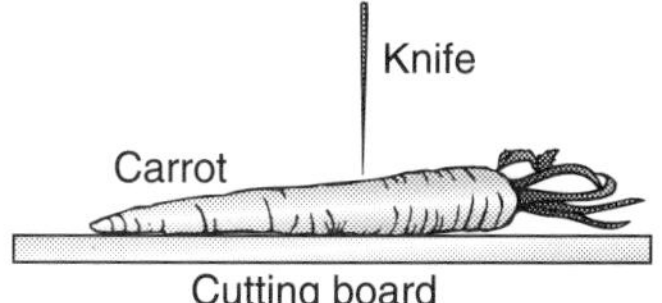

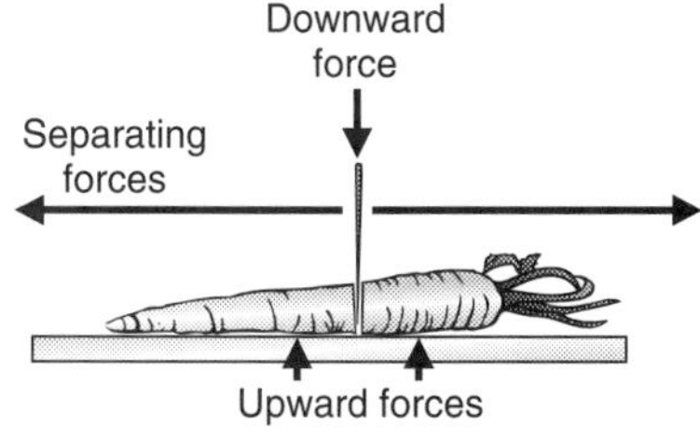

Fig. 17.1.1 - When you push down on a knife to cut a carrot, the cutting board pushes up. Overall, the vertical forces on the carrot, including its weight, cancel completely. But the ramp-like surfaces of the knife blade create large horizontal forces on the two halves of the carrot and these forces divide the carrot.

But while the mechanical action of cutting is simple, the physical structure of the knife is not. The secret to its ability to cut through the carrot lies in the properties of its blade. This blade is almost certainly made of steel, so the story of how a knife works is really the story of how steel works.

Steel is not a specific material but rather a whole range of iron-based metals. These metals differ from one another in their specific chemical compositions and in how they have been processed. The variety of steels is so broad that it's difficult to encompass them all in a single definition. However *steels* are generally mixtures of iron and other elements that contain no more than 2.06% carbon by weight. A mixture with more carbon than this is usually called *cast iron*. It's an unfortunate historical accident that cast irons actually contain more carbon and less elemental iron than many steels.

Carbon content is important in distinguishing steel from cast iron because it affects the hardness, brittleness, and other characteristics of these metals (Fig. 17.1.2). **Hardness** is a measure of a material's resistance to penetration, deformation, abrasion, and wear. **Brittleness** is the tendency for a material to fracture while it's being deformed. A good knife blade should be hard but not brittle; distorting very little as you push it through the carrot but resisting fracture even if you use the knife to open a metal can.

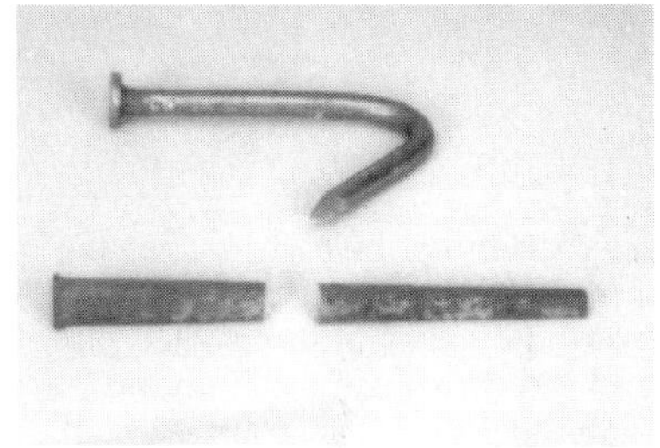

Fig. 17.1.2 - Soft steel containing little carbon bends easily while hard, carbon-rich steel breaks instead of bending.

Controlling the hardness, brittleness, and other characteristics of steel is a complicated task that involves all aspects of steel production. These characteristics can even vary within a single steel object—the cutting edge of a good knife blade is actually harder than the rest of the blade. To understand how these characteristics are controlled, we'll first look at how materials respond to outside forces and then examine the microscopic basis for steel's properties. What we learn in the process will apply not only to steel, but to many other materials.

When you push gently on a solid object, that object distorts by an amount proportional to the force you exert on it. This relationship is simply Hooke's law, which we first encountered in the section on spring scales. If the object is a steel block resting on the floor and you're stepping on it, then its tiny distortion is proportional to your weight. By measuring how much your weight distorts this block, you can learn something about the steel from which it's made.

But the distortion is also related to the block's dimensions. The broader its surface, the more your weight is spread out and the less the block distorts. Since we are trying to learn about the steel rather than the block, we divide your weight by the surface area of the block to obtain the stress on the steel (Fig. 17.1.3). **Stress** is the amount of force exerted on each unit of a block's surface area and is the measure of how much the steel is being squeezed.

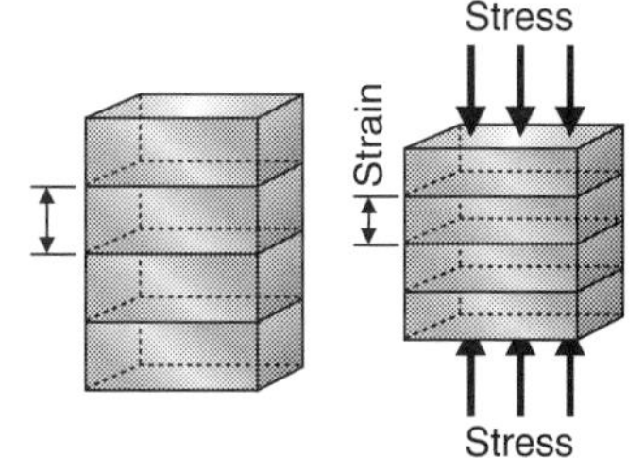

Fig. 17.1.3 - Steel is compressed when it experiences an inward stress. Stress is equal to the inward force exerted on each unit of surface area. This stress causes a strain in the steel. This strain is equal to the fractional change in the steel's length—its change in length divided by its original length.

However, the distortion also depends on the block's height. Each centimeter of metal shrinks a little bit so that a tall block shrinks more than a short block. Again, we are more interested in the steel than the block, so we divide the change in height by the original height of the block to obtain the strain in the steel (Fig. 17.1.3). **Strain** is the change in length per unit of a block's length and is a measure of how much the steel responds to being squeezed.

Hooke's law relates the stress on the steel to the strain that results:

$$\text{strain} = \frac{\text{stress}}{\text{Young's modulus}}, \tag{17.1.1}$$

where *Young's modulus* is a measure of how difficult it is to compress the steel. Strain is just a number and has no dimensions, while stress and Young's modulus are both pressures and are measured in pascals (Pa).

Young's modulus is related to the interatomic forces that hold a material together. The atoms in a solid exert attractive and repulsive forces on one another and these forces only balance when the atoms have the proper separations. If you push the atoms closer together or pull them farther apart, the forces no longer balance and they oppose your action. The stiffer the interatomic forces, the larger Young's modulus and the less the material shrinks as you squeeze it.

Because the interatomic forces in steel are primarily between iron atoms, it has roughly the same Young's modulus as iron: about 195 GPa (195 gigapascals or 195,000,000,000 Pa). This large value makes steel extremely difficult to compress. A 1 m cube of steel would lose somewhat less than a micron of height while supporting a city bus. A similar cube of lead would shrink 14 times more while a cube of tungsten would shrink only half as much.

Although we arrived at Eq. 17.1.1 while thinking about squeezing, this equation also applies to situations where the steel is exposed to gentle tension. In that case, the stress is negative (pulling apart rather than squeezing together) and the strain that results is also negative (stretching rather than compressing). Whether you squeeze or stretch steel, you are still measuring the stiffness of the interatomic forces and arrive at the same value for Young's modulus. Young's modulus is usually measured with **tensile stress** (stretching) rather then with **compressive stress** (squeezing) because compression can cause a thin piece of material to buckle while tension pulls it straight and true.

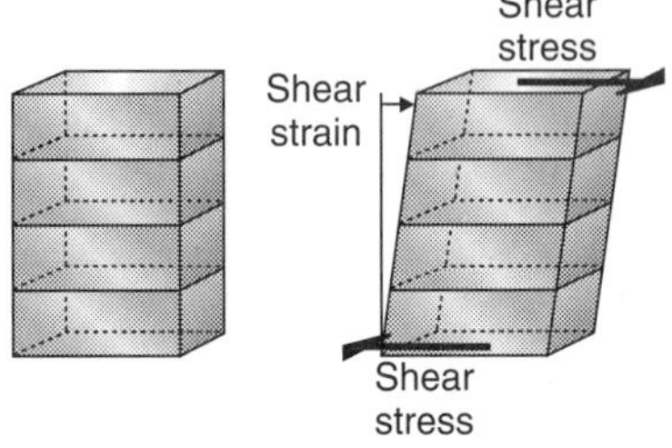

Fig. 17.1.4 - A material bends when it experiences shear stress. Shear stress is equal to the sideways force exerted on each unit of surface area and causes a shear strain in the material. The shear strain is equal to angle of the bend.

But compression and tension aren't the only stresses a steel block can experience. If you push the bottom of the block to the left and the top of the block to the right, the metal experiences **shear stress** (Fig. 17.1.4). This stress bends the block and causes shear strain in the metal. **Shear strain** is the angle of the bend caused by a shear stress. As long as the forces involved are relatively small, the shear strain is proportional to the shear stress,

$$\text{shear strain} = \frac{\text{shear stress}}{\text{shear modulus}}, \tag{17.1.2}$$

where the *shear modulus* is a measure of how difficult it is to bend the metal.

Stress and strain help to characterize a particular steel. The crucial test for a knife is how it responds to stress. Because all steels have similar Young's and shear moduluses, it's hard to tell them apart with gentle stresses. The real differences between steels only appear when the stresses become large and Eqs. 17.1.1 and 17.1.2 stop being valid. That's when the steels begin to bend and break and when the good knives begin to distinguish themselves from the poor ones.

CHECK YOUR UNDERSTANDING #1: Holding Up Under Stress

Which is experiencing the greater stress: a single brick supporting a 100 kg post or two bricks, side by side, supporting a 200 kg beam?

Plastic Deformation and Crystalline Materials

A spring has an elastic limit beyond which it stops obeying Hooke's law and bends permanently. A gentle force distorts the spring elastically and it bounces back while a strong force deforms it forever. The same holds true for a piece of steel. As long as the stress is small, steel undergoes **elastic deformation**—it distorts temporarily while stressed, according to Eqs. 17.1.1 and 17.1.2, and then returns to its original shape. However when the stress is large, steel undergoes **plastic deformation**—its atoms rearrange and it deforms permanently.

Steels differ considerably when it comes to plastic deformation. They all start off responding elastically to small stresses but gradually shift over to plastic deformation as the stresses increase. Just how much stress a particular steel can tolerate before it begins plastic deformation is its **yield strength** and is an important measure of that steel's load-bearing capacity. A weak steel yields easily to a modest stress while a strong steel responds elastically to all but enormous stresses. The cutting edge of a knife must tolerate severe stresses without yielding so that it doesn't become dull.

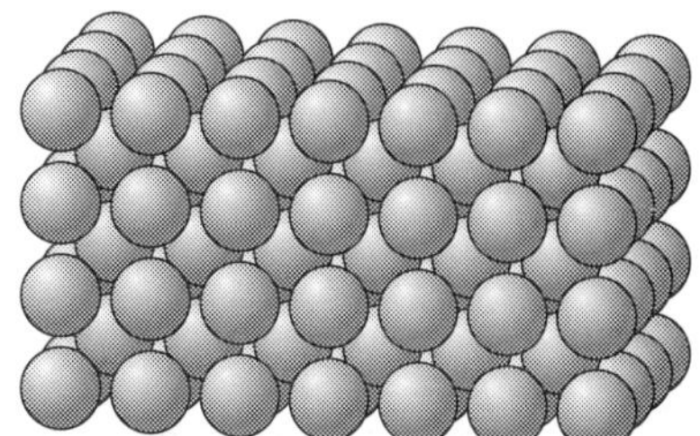

Fig. 17.1.5 - In ferrite, iron atoms are arranged in a body-centered cubic lattice. This lattice is built out of eight-atom cubes (2 by 2 by 2 atoms) with an additional atom located in the center of each cube.

To see how plastic deformation occurs in a knife blade, we must examine the microscopic structure of steel. But since steel has a rather complicated structure, let's start with pure iron instead. Like most metals, iron is a crystalline solid. It may not have the pretty natural facets typical of some minerals but it does have an orderly arrangement of atoms. At room temperature, iron forms ferrite crystals. *Ferrite* is a ferromagnetic material in which the iron atoms are arranged in a body-centered cubic lattice (Fig. 17.1.5).

As you can see by looking at this lattice, it has smooth surfaces between sheets of atoms. When it experiences enough shear stress, this lattice can undergo a phenomenon called **slip**, in which sheets of atoms slide across one another. Slip is the most common mechanism for plastic deformation. When you squeeze, stretch, or bend a piece of pure iron so that it yields, you are probably causing slip. As the sheets slide across one another, the crystals change shape just like a stack of cards does when you push the top card to one side. The atoms lose track of their original positions and the iron doesn't return to its original shape when you stop pushing on it.

Slip occurs fairly easily in iron because the bonds that hold it together are relatively non-directional. Its **metallic bonds** are formed by a general sharing of electrons between countless atoms. This sharing lowers both the kinetic and potential energies of the electrons and causes the atoms to cling to one another. Because these metallic bonds are rather insensitive to the relative positions of its atoms, iron crystals are quite susceptible to slip.

But there are other factors that affect when and where slip occurs. These factors appear because real metals are never perfect crystals. Nature tends to introduce randomness wherever possible so that even the purest crystals contain a few random mistakes. Typical iron crystals are filled with imperfections.

One common defect in iron is a **dislocation**, a sheet of atoms that ends abruptly in the middle of the crystal (Fig. 17.1.6). In a perfect crystal, a whole sheet of atoms must slip at once. But a dislocation breaks up the uniformity of the crystal and allows the sheet to slip gradually, one row of atoms at a time. A crystal containing a dislocation has an unusually low yield strength along the *slip plane*, the sheet of atoms perpendicular to the end of the dislocation.

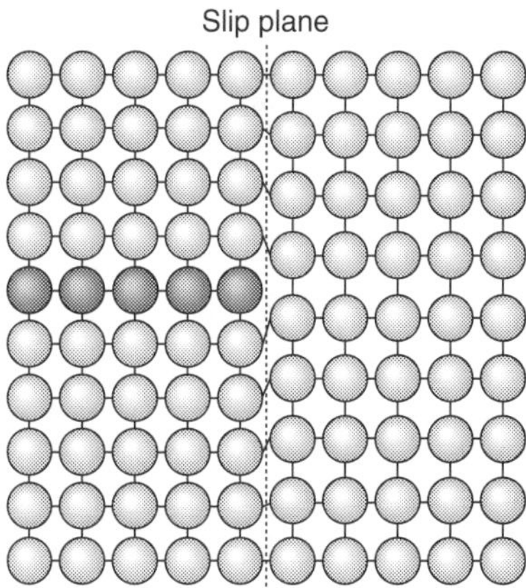

Fig. 17.1.6 - Dislocations occur when a sheet of atoms ends abruptly, part way through a crystal. In this top view of an imperfect crystal, the darkened sheet of atoms on the left ends suddenly and the sheets on the right are stretched to compensate. Slip occurs relatively easily along the slip plane. Thermal energy allows dislocations to move about in many materials, further softening those materials.

While dislocations weaken iron by easing slip, there are other imperfections that strengthen iron by preventing slip. One such imperfection is iron's **polycrystalline** structure—rather than being a single crystal, iron is normally composed of many individual crystallites that meet one another at random angles. These individual crystallites are called **grains** and the boundary layers of atoms between grains are called **grain boundaries**.

Grains and the grain boundaries strengthen iron. Within each grain, slip only occurs between particular sheets of atoms and along specific directions. Since the grains are randomly oriented, they aren't all able to slip along the same direction. They must coordinate their slips along many directions to allow the piece of iron to yield along one direction, a process that can only occur when the

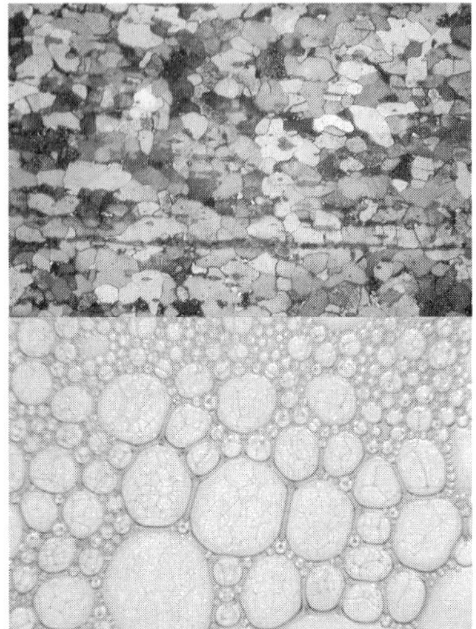

Fig. 17.1.7 - A piece of steel is composed of many tiny crystals (top). These crystals are jumbled together in an arrangement that resembles soap bubble froth (bottom). Surface tension acts to minimize the surface areas of both the metal grains and the bubble froth, so that both have similar features.

❒ When the Titanic struck an iceberg and sunk on April 14, 1912, it was thought that the iceberg had torn a long hole in its steel hull. However recent explorations of the wreck have found that the real culprit was probably brittle fracture. Instead of denting when it collided with the ice at 41 km/h, the ship's cold steel hull appears to have shattered.

❒ When the Purity Distilling Company's 30 m tall molasses tank burst in Boston's North End on January 15, 1919, it released 2.2 million gallons of molasses in a wave that destroyed several city blocks and killed 21 people and dozens of horses. The noontime disaster was initially attributed to an explosion but an official investigation found that the tank's brittle steel had fractured in the 4° C (40° F) winter weather. Purity, a subsidiary of United States Industrial Alcohol Company, paid $1,000,000 in damages. Ironically, the accident occurred one day before ratification of the 18th Amendment to the U.S. Constitution, prohibiting the sale of alcoholic beverages.

iron is experiencing enormous stress. As a result, iron with lots of tiny grains has a higher yield strength than iron with only a few large grains.

The sizes and shapes of the grains depend on how the iron has been handled and processed. Because the atoms in a grain boundary don't fit into the crystallites they connect, they have relatively high potential energies and create a surface tension around each grain. **Surface tension** appears whenever a surface has extra potential energy and it acts to make that surface as small as possible. Like a soap froth that tries to minimize the surface area of its soap bubbles, the iron tries to minimize the surface area of its grain boundaries (Fig. 17.1.7). But the iron atoms can only rearrange while the iron is hot. **Annealing** iron—heating it to high temperature and then cooling it slowly—allows its larger grains to grow by consuming the smaller grains. Annealing is the principal method for softening iron, steel, and most other metals.

In contrast, deforming iron at low temperatures breaks up its grains and makes it harder. Like most metals, iron can be **work hardened** by pounding, rolling, folding, and twisting it. Wrought iron is a good example of work hardening, a technique that has been used for millennia to strengthen metals.

But annealing and work hardening have consequences beyond their effects on yield strength. Instead of yielding to stress, work hardened iron may fracture into pieces. This catastrophe is called **brittle fracture** because the metal doesn't yield before it breaks. Brittle fracture occurs in metals that are so hard that sheets of atoms separate completely rather than sliding across one another. Overworked iron may also suffer from **metal fatigue**, in which the metal tears as cracks work their way inward from surface defects.

Annealed iron yields before it breaks. In fact, it's **ductile**, meaning that it can be stretched quite a bit during plastic deformation. The main reason that the annealed iron breaks at all is that it gets thinner as it stretches and usually develops a narrow neck. Stretching work hardens the neck and the high stress there eventually pulls the atoms apart. Because the iron yields before it breaks, this type of breakage is called **ductile fracture**. The peak stress that the iron can withstand before such fracture occurs is its **tensile strength**.

Plastic deformation and ductility are actually quite useful in a knife. When the stress on the knife is too great for the blade to handle elastically, you would rather have it bend than break. A metal's ductility increases with temperature because thermal energy allows dislocations and other defects to move through crystals so that slip can occur between many different sheets of atoms. In a cold metal, those defects are immobile and slip is reduced. Because of its low ductility, cold iron is much more susceptible to brittle fracture than hot iron. (For some interesting examples of brittle fracture in cold metal, see ❒s.)

Moreover, plastic deformation requires energy. When you push on an iron knife and it bends in the direction of your push, you are doing work on the blade and the blade absorbs energy. A metal's ability to absorb energy during deformation is called **toughness**. Toughness is extremely important in car bodies, swords, and armor, all of which dent during collisions to absorb energy. If they shattered instead of yielding, they wouldn't be nearly as useful. Glass car windows and eyeglass lenses are dangerous precisely because they don't exhibit plastic deformation. In contrast, plastic windows and eyeglass lenses are much safer because they're able to dent and absorb energy without shattering.

CHECK YOUR UNDERSTANDING #2: Softness and Strength

Pure annealed copper is so soft that you can bend thick rods of it with your hands. Nonetheless, ancient coppersmiths managed to hammer pure copper into useful tools and weapons. What gave these tools and weapons their strength?

Steel

Adding carbon to iron produces steel, an *alloy* or metallic mixture that is far tougher and harder than pure iron. But different amounts of carbon produce different types of steel. Moreover, the ways in which this material is processed mechanically and thermally can dramatically change its character.

The first small amount of carbon added to iron dissolves in the iron to form a solid solution. It may seem strange for a solid to have something dissolved in it, but there is no rule that says a solvent must be liquid. Solution occurs whenever energy and entropy (randomness) make it favorable for one material to divide into atoms, molecules, or ions and become dispersed throughout a second material. Whether it's liquid or solid, iron can dissolve a small amount of carbon.

The iron remains in its ferrite form and the dissolved carbon atoms arrange themselves randomly in the *interstitial spaces* between iron atoms. Ferrite can dissolve up to 0.01% carbon by weight at room temperature and is somewhat harder than pure iron. The carbon atoms introduce local distortions in the ferrite crystals so that sheets of atoms can't slide as easily across one another. Since the dissolved carbon reduces slip, it increases the yield strength of the steel.

Only hydrogen, nitrogen, and carbon atoms are small enough to fit in between the iron atoms and cause *interstitial solution hardening* of the steel. However, there are a number of other atoms that harden iron and steel by substituting for iron atoms in ferrite crystals. These atoms also distort the crystals, impeding slip and increasing the yield strength of the steel. Phosphorus, silicon, manganese, chromium, and nickel atoms are often added to steel to cause this *substitutional solution hardening*.

When the carbon fraction in ferrite exceeds 0.01%, it can't all remain dissolved in the ferrite at room temperature. A new material appears in the steel: iron carbide (Fe_3C). Iron carbide is an extremely hard and brittle crystalline material, also called *cementite*. The presence of tiny cementite crystals dispersed throughout the ferrite impedes slip and increases the steel's yield strength. Strengthening of this sort is called *dispersion hardening*.

The arrangement of ferrite and cementite particles in the steel depends on the amount of carbon in the mixture and on the thermal and mechanical histories of the steel. When it's cooled slowly from high temperatures, steel consists of a material called *pearlite*, which is alternating layers of ferrite and cementite and has a carbon content of about 0.8% by weight. If the steel has less than 0.8% carbon, it forms some extra ferrite and consists of pearlite interspersed with ferrite. If it has more than 0.8% carbon, it forms some extra cementite and consists of pearlite interspersed with cementite.

However good knives use characteristics of steel that don't appear until steel is subject to more sophisticated heat treatments. The microscopic structure of steel changes remarkably as it's heated and cooled. Above 723° C, iron forms austenite crystals. *Austenite* is a nonmagnetic material in which the iron atoms are arranged in a face-centered cubic lattice (Fig. 17.1.8).

Austenite can dissolve more carbon than ferrite; up to 0.8% by weight at 723° C and up to 2.06% at 1148° C. The latter figure sets the upper limit for what is considered steel. As you slowly heat steel above 723° C, its ferrite, pearlite, and cementite all convert into austenite—a structural change called a *solid to solid phase transition*. When you cool the steel back down slowly, the reverse phase transition occurs and the austenite turns into ferrite, pearlite, and cementite.

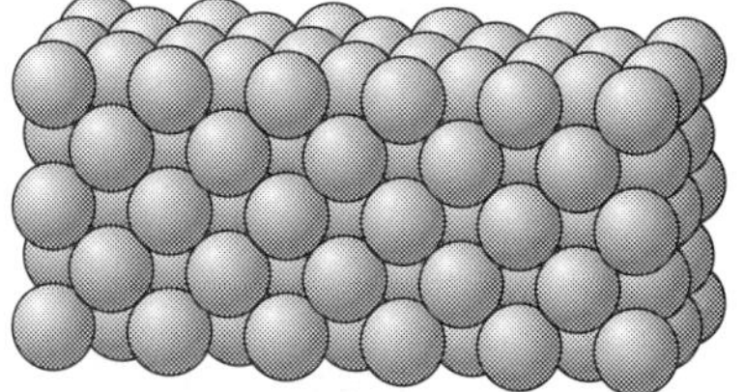

Fig. 17.1.8 - Above 723° C, the atoms in solid iron are arranged in a face-centered cubic lattice. This lattice is built out of eight-atom cubes (2 by 2 by 2 atoms) with an additional atom located in the center of every face of that cube.

But even more remarkable effects occur during rapid cooling. If the austenite is suddenly cooled to about 600–650° C and kept there, it transforms into *fine pearlite*—pearlite with extremely thin layers. To form pearlite, carbon atoms must diffuse through the iron. At lower temperatures, they can't travel the long distances needed to form thickly layered *coarse pearlite*, so they form fine pearlite instead. Fine pearlite doesn't undergo slip as easily as coarse pearlite so it has a higher yield strength.

Austenite that is suddenly cooled to about 260–400° C and kept there doesn't form pearlite at all. The carbon atoms diffuse such short distances that they form tiny nodules of cementite. These nodules are arranged between sheets of ferrite in a layered material called *bainite*. Bainitic steel has a higher yield strength than fine pearlite.

Austenite that is suddenly cooled below about 200° C forms an entirely new material. At 200° C, there is so little thermal energy around that carbon atoms don't diffuse through the iron at all, so cementite, pearlite, and bainite can't form. Instead, the austenite tries to turn into ferrite without first getting rid of all its dissolved carbon. What forms is *martensite*, a distorted ferrite that is stretched in one direction.

Since ferrite can't have more than about 0.01% carbon in it by weight, martensite is essentially ferrite with way too much dissolved carbon. A solution containing more dissolved material than is stable at the current temperature is said to be **supersaturated**. The dissolved material will eventually come out of solution but it may take quite a long time. Martensite is a supersaturated solution of carbon in ferrite that lasts almost forever at room temperature. Because martensite is very resistant to slip, steel containing it has an extremely high yield strength and hardness. However martensite also makes the steel brittle by preventing it from undergoing plastic deformation. Steel containing martensite isn't as tough as steel containing bainite or fine pearlite.

Clearly sudden cooling or **quenching** of austenitic steel produces a harder material than slow cooling. The faster you cool the austenite and the colder you take it in that first step, the harder the steel becomes. Unfortunately, red hot carbon steel must be plunged into water to cool it quickly enough to form martensite. The steel contracts during this harsh treatment and traps stresses that weaken the metal. To relieve these internal stresses, quenched steel is often **tempered** by reheating it just enough to let some of the stresses resolve themselves. The hotter and longer this tempering process, the more stress is relieved but the softer the steel becomes.

Quenching carbon steel in water is dramatic, with steam billowing up from the hot metal, but only a thin surface layer cools quickly enough to harden properly. To ease the formation of martensite and reduce the cooling rate needed to harden steel, it's often alloyed with other elements. Such alloy steels harden easily and deeply when quenched unspectacularly in oils or air and are important in knives and other tools. Some alloy steels also undergo *precipitation hardening*, in which small crystals of various compounds precipitate out of solution in the steel as it cools and strengthen the metal. Titanium carbide, niobium carbide, and vanadium nitride frequently appear in precipitation hardened steel.

Part of a knife maker's skill comes in choosing just the right steel and just the right thermal processing to bring out that steel's best characteristics in the blade and its cutting edge. Knife blades require particularly careful heat treatment. A blade that retains too much stress will be hard but brittle while one that has been tempered at too high a temperature will dull easily.

In a fine knife, the cutting edge may be hardened more than the body of the blade by adjusting both the chemical composition of the edge and its heat treatment. Flames and laser beams are often used to reheat the edge and to add more carbon to the steel there. The result is a knife with the toughness of fine pearlite or bainite in its body and the hardness of carefully tempered martensite in its cutting edge. You can occasionally see evidence for this special treatment in the color and appearance of the blade. (Joining different materials to perform a task that neither could do alone is common in modern construction, see ❐.)

❐ Many buildings are supported by a mixture of steel and concrete. These two materials enjoy a symbiotic relationship. Steel has enormous tensile strength but bends easily when compressed. Concrete can withstand immense compressive stress but fractures easily when stretched. However they can be combining into a composite material that has the tensile strength of steel and the compressive strength of concrete. Prestressing the pair by stretching the steel while the concrete is drying further improves the performance of the composite.

CHECK YOUR UNDERSTANDING #3: Hot Tips

Quality drill bits are usually made from a high carbon steel called tool steel (0.8% to 1.1% carbon). Each bit is shaped and hardened so that its tip can bore holes in most metals, including softer steels. When drilling in steel, you should lubricate the tip with a cutting oil so that sliding friction doesn't overheat it. What will happen to the tip if it becomes too hot?

Stainless Steel Blades

One last feature of steel that's important in knife blades is corrosion resistance. Because carbon steel is susceptible to rust, knife blades are usually made of stainless steel. This corrosion resistant material is formed by replacing at least 4% of steel's iron atoms with chromium atoms. Most stainless steels contain more than 11% chromium by weight and some contain nickel as well. Nickel enhances the steel's corrosion resistance and also makes it more ductile.

But not all forms of stainless steel are appropriate for knives. Alloying the steel with chromium and nickel affects more than just its chemical properties. The crystalline structures of these alloys depend on the precise mixtures of the various elements and on their thermal histories. Perhaps the most remarkable effect of adding chromium and nickel to the steel is that these elements can make austenite stable at room temperature.

The most common stainless steel is called *18–8 stainless* and contains roughly 18% chromium and 8% nickel. It consists exclusively of nonmagnetic austenite grains. Since 18–8 stainless can't form cementite or martensite, it can't be hardened by thermal processing. In fact, overheating it can spoil its corrosion resistance (see ❐). The only way to harden 18–8 stainless steel is to work harden it. Because it's difficult to make a good knife edge by work hardening, 18–8 stainless is a poor choice for knives. However 18–8 stainless is inexpensive, ductile, and easy to work with so it's often used in cafeteria grade cutlery. The next time a dull knife bends while you're cutting your food, you'll know that it's probably made from soft, austenitic stainless steel.

❐ Austenitic stainless steels like 18–8 are susceptible to subtle damage when they're overheated. At temperatures between about 500° C (932° F) and 800° C (1472° F), the chromium and carbon in the steel can precipitate out at the grain boundaries as chromium carbide. This process depletes the grain edges of chromium and leaves them extremely vulnerable to corrosion. Corrosion at the grain edges cuts the grains right out of the metal so that the metal falls apart. That's why a badly overheated stainless pot is never the same again.

There are two ways to harden stainless steels: martensite formation and precipitation hardening. *Martensitic stainless steels* contain little or no nickel and moderate amounts of carbon. With too little nickel to stabilize austenite, martensitic stainless steels crystallize like normal steels and are magnetic. Quenching hot martensitic stainless steel produces tiny martensite crystals and hardens the metal.

Precipitation hardening stainless steels contain alloying elements that form hard precipitate crystals within the steel during quenching. Titanium, niobium, aluminum, copper, and molybdenum dissolve in the steel when it's hot but form tiny precipitate crystals within the steel as it cools. These crystals make the stainless steel hard.

High quality knives and cutlery are generally made of martensitic or precipitation hardening stainless steels. These metals are tough, keep their edges well, and are hard to bend. But knives and utensils made from these steels require careful heat treatment during manufacture, so they are considerably more expensive than common cafeteria dinnerware.

CHECK YOUR UNDERSTANDING #4: Steel vs. Steel

With some care, you can cut a cheap stainless steel knife with a good stainless steel knife. How is that possible?

How Steel Is Made

Steel is produced on enormous scales using techniques that have become almost as high-tech as those involved in the semiconductor industry. The days of dirty, grimy steel mills are over because modern high quality steels require exceptional chemical purity.

Steel is generally made in two steps: iron ore is converted into pig iron and pig iron is converted into steel. These two steps were once quite separate and followed by many further independent processing steps. However modern steel mills convert iron ore into finished steel in an uninterrupted manner.

Iron ores are essentially iron oxides (Fe_2O_3, Fe_3O_4, FeO) so something must remove the oxygen atoms to convert iron ore into iron. That something is carbon. When iron oxide and carbon are heated together, the carbon reacts with the oxygen to form carbon monoxide and carbon dioxide. These two gases escape and leave behind *pig iron*, a mixture of iron and carbon.

The carbon needed for this process is obtained by heating coal to high temperatures. The coal cracks chemically and releases compounds such as acetylene, light oil, coal tar, and ammonia, all of which are collected for use outside the steel mill. What remains is nearly pure carbon, a material called *coke*.

But both the iron ore and the coke contain rocks and other undesirable materials that would contaminate the pig iron. To remove these contaminants, the iron ore and coke are mixed with *lime* (CaO) obtained from limestone ($CaCO_3$). The lime acts as a liquid *flux* at high temperature, floating on top of the molten pig iron and dissolving many of the undesirable materials.

The mixture of iron ore, coke, and lime is converted into pig iron in a blast furnace. The coke burns in a stream of air, providing the intense heat needed to initiate and sustain the chemical reactions between carbon and iron oxide. The lime flux carries away most of the unwanted materials as slag, which can then be used to make concrete.

Until about 1856, this was all the processing that iron received. In fact, most iron ore was converted to iron at relatively low temperatures so that it never melted at all. Carbon and carbon monoxide diffused through the iron ore and reacted with the oxygen atoms, thereby converting solid iron oxide into solid iron. This iron was then hammered into shape as wrought iron (Fig. 17.1.9). It contained various amounts of carbon and lots of slag.

Good techniques for melting pig iron and removing carbon and slag from it to make steel appeared in the last century and a half. The Bessemer converter, the

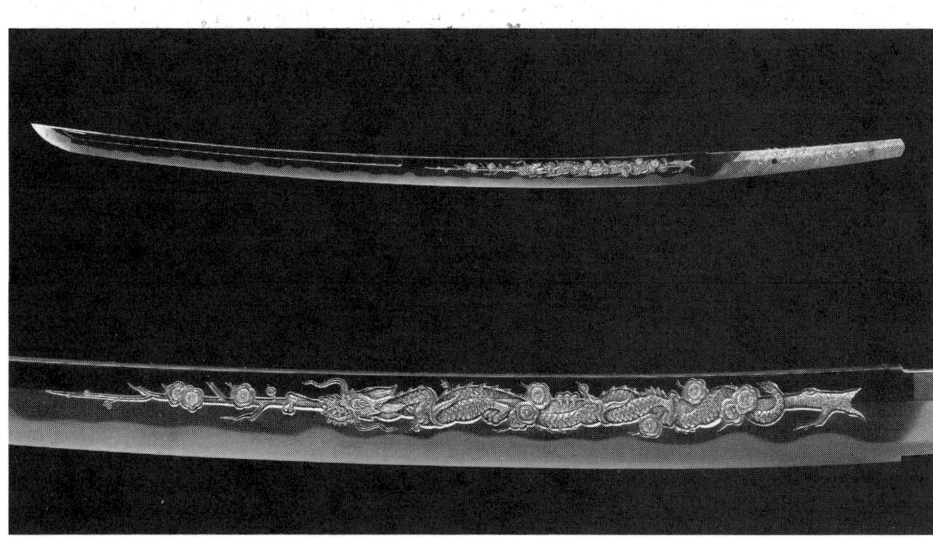

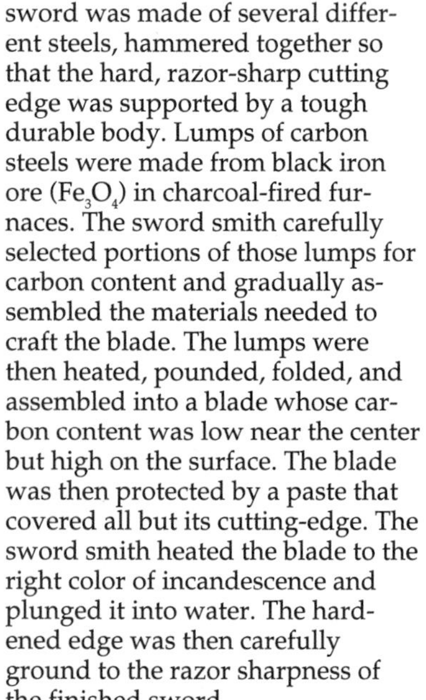

Fig. 17.1.9 - The blade of a Samurai sword was made of several different steels, hammered together so that the hard, razor-sharp cutting edge was supported by a tough durable body. Lumps of carbon steels were made from black iron ore (Fe_3O_4) in charcoal-fired furnaces. The sword smith carefully selected portions of those lumps for carbon content and gradually assembled the materials needed to craft the blade. The lumps were then heated, pounded, folded, and assembled into a blade whose carbon content was low near the center but high on the surface. The blade was then protected by a paste that covered all but its cutting-edge. The sword smith heated the blade to the right color of incandescence and plunged it into water. The hardened edge was then carefully ground to the razor sharpness of the finished sword.

open-hearth process, and the basic oxygen process all remove excess carbon and other impurities by burning them out of the pig iron. The basic oxygen process now dominates the steel industry. It uses a water-cooled pipe or "lance" to inject pure oxygen gas into a vessel of liquid pig iron. The impurities burn away, leaving behind almost pure iron. Even oxygen and other gases are often removed from the iron by vacuum pumps. The conversion from pig iron to steel takes roughly 30 minutes and is monitored by disposable sensors that are plunged into the liquid metal to examine its composition.

To make the appropriate steel, alloying elements are added to the liquid iron and the resulting material descends out of the bottom of the converter through a series of containers. In a continuous casting machine, the steel is formed into a red-hot solid bar that descends from a mold. This bar is bent until it's horizontal and it then enters a series of rolling machines. These machines shape the steel into anything from beams to sheet metal. The whole procedure is so smoothly orchestrated that the steel making never stops. The mill just keeps on making more steel and it keeps flowing out to the finishing equipment. Even changes in the steel's composition are handled without interrupting the flow.

The finishing equipment rolls the steel into various shapes, gradually changing its dimensions with enormous pressures and high temperatures. While steel is occasionally worked with as a liquid, it's generally much easier to handle as a hot solid.

While the work hardened steel that emerges from these rolling machines is fine for construction, it must be softened before it can be molded into car body panels or cooking pots. Softening is done by annealing the steel in huge ovens. At the opposite extreme is spring steel, which is reheated and quench hardened so that it snaps back to its original shape after being deformed.

Many tools, including locks and wrenches, need a hard surface layer on a softer core. The hard surface makes the tool impenetrable while the softer core allows it to withstand energetic blows. This layered arrangement can be made by *case hardening* the tool—exposing its heated surface to extra carbon. The carbon atoms diffuse into the surface and increase its hardenability. Once it has been heat treated, a case hardened tool is extremely difficult to break.

CHECK YOUR UNDERSTANDING #5: Going Up in Smoke

Pig iron contains roughly 4% carbon by weight. If a basic oxygen converter is initially charged with 250 metric tons (250,000 kg) of pig iron, how much carbon must be burned away to make steel?

Summary

How Knives and Steel Work: The steel in a knife blade enters the steel mill as iron ore. It's heated with coke and lime in a blast furnace and converted to pig iron. The pig iron is poured into a basic oxygen converter, where oxygen burns away all of the carbon and other contaminants from the iron. Carbon and other alloying elements are then added to the iron in carefully measured amounts to make the appropriate steel. This steel is then rolled into thin strips.

Knife blades are cut from these strips, ground to the proper shapes, and sent to ovens to be heat treated. The heat treatment begins at high temperature, where the steel is austenite, with all of the alloying elements in solution. As the blade cools, the austenite may become pearlite, bainite, or martensite and some of the alloying elements may come out of solution as precipitate crystals. By carefully controlling the blade's temperature and the chemical composition of its cutting edge, the manufacturer makes the cutting edge extremely hard and the body of the blade extremely tough. Once the hard edge has been ground sharp, it remains sharp indefinitely, supported by the tough steel to which it's attached.

In use, the blade's hard edge penetrates most materials, shearing apart rows of atoms and pushing them aside. It uses mechanical advantage to convert gentle downward forces exerted over large distances into huge horizontal forces exerted over small distances. The edge responds elastically to the enormous stresses it experiences during cutting because it has a high yield strength. The body of the blade has a lower yield strength and bends permanently during severe stresses, but its high tensile strength prevents the blade from breaking apart. Chromium and nickel in the stainless steel keep the blade from reacting with chemicals in the food and from rusting.

The Physics of Knives and Steel

1. When you exert a gentle stress on a solid material, it undergoes a strain that is proportional to the stress. This effect is equivalent to Hooke's law in springs.

2. When the stress exceeds a certain limit, the material either undergoes brittle fracture or yields to plastic deformation. Plastic deformation involves a permanent rearrangement of the atoms within the material so that it doesn't return to its original shape when the stress is removed.

3. Plastic deformation is caused by slip in the crystals that make up the material. Defects such as dislocations that assist slip reduce the yield strength of a material. Defects such as grain boundaries, substituted atoms, and precipitate crystals that impede slip increase the yield strength of a material.

4. Materials that can withstand severe stresses without undergoing plastic deformation have high yield strengths and are very hard. Materials that can withstand severe stresses without breaking have very high tensile strengths. Materials that can absorb lots of energy without breaking are very tough.

5. Metal atoms are bound together by a general sharing of electrons. This sharing lowers both the kinetic and potential energies of the electrons.

6. Materials can experience phase transitions where one crystalline form converts into another as the temperature or chemical composition of the material changes.

Check Your Understanding - Answers

1. All of the bricks experience the same stress.

Why: The two bricks may be supporting the heavier load, but they are doing it together. By dividing the force on them by the amount of surface area involved, stress measures just how much support each brick, or rather each square meter of brick, is providing.

2. The copper was work hardened during the process of hammering it into shape.

Why: Carefully annealed copper has such large grains that you can easily see them by eye. These huge crystallites slip easily, making annealed copper very soft. In contrast, hammered copper has much smaller grains that inhibit slip. Work hardened copper isn't as strong as iron or steel, but it's adequate for many purposes.

3. It will lose its hardness and soon become dull.

Why: Tool steel relies on martensite to give it the hardness it needs to cut through softer steels. To form martensite, a drill bit is heated and quenched. It's then minimally tempered to reduce its brittleness while retaining most of its hardness. If you accidentally overheat the tool steel, its martensite will gradually convert into grains of pearlite and cementite and the steel will become relatively soft.

4. While the good knife has a hardened cutting edge of martensitic or precipitation hardening stainless steel, the cheap knife is made of an unhardenable austenitic stainless. The hardened knife can shave away metal from the unhardened knife.

Why: All stainless steels are not the same. The cutting edge of the good knife has a much higher yield strength than the metal in the cheap knife. When you push the two together, they each experience the same stress, but the cheap knife yields first. The harder knife causes slip to occur in the softer knife and the softer knife comes apart.

5. About 10 metric tons.

Why: 4% of 250 metric tons are 10 metric tons. Burning away that much carbon in the basic oxygen process requires about 25 metric tons of oxygen. This oxygen is produced by fractional distillation of liquefied air.

Glossary

annealing Heating a material to allow recrystallization and grain growth and then allowing the material to cool gradually to room temperature.

brittle fracture Material fracture that occurs without plastic deformation when layers of atoms separate completely during stress.

brittleness A material's tendency toward brittle fracture during stress.

compressive stress A stress involving compression.

dislocation An incomplete sheet of atoms that appears as a defect in a crystal.

ductile Capable of undergoing extensive plastic deformation before fracturing.

ductile fracture Material fracture that occurs during plastic deformation when layers of atoms separate completely during stress.

elastic deformation A stress-induced change in shape that doesn't involve permanent rearrangement of the atoms in a material. The material returns to its original shape when the stress is removed.

grain A tiny crystallite that is part of a polycrystalline material.

grain boundary The thin layer of atoms between two separate crystals or grains of a polycrystalline material.

hardness A material's ability to resist penetration.

metal fatigue Material fracture that occurs when manipulation allows surface blemishes to propagate into a metal as tears.

metallic bonds The form of bonding that occurs in metals. Electrons are shared generally among many atoms and this sharing binds the atoms together.

plastic deformation A stress-induced change in shape that involves permanent rearrangement of the atoms in a material. The material doesn't return to its original shape when the stress is removed.

polycrystalline Composed of many individual crystallites.

quench Cool rapidly from high temperature so that the material doesn't have time to adopt its normal low-temperature state.

shear strain The angle by which a material bends when experiencing a shear stress.

shear stress A stress in which the top of a material is pushed to the left while the bottom is pushed to the right.

slip Plastic deformation of a crystal that occurs when one sheet of atoms slides across another sheet.

strain A change in a material's length caused by a stress divided by its original length.

stress A compressive or tensile force exerted on a material divided by the area over which that force is exerted.

supersaturated Containing more of something than is normally allowed.

surface tension An inward tension that appears whenever a surface has extra potential energy. It acts to make that surface as small as possible.

temper (metals) Bake a material at moderate temperatures to relieve trapped stresses.

tensile strength The maximum tensile stress that a material can withstand before fracturing.

tensile stress A stress involving tension.

toughness A measure of how much mechanical energy a material can absorb without fracturing.

work hardening Hardening that occurs when extensive manipulation reduces the size of grains in a polycrystalline material.

yield strength The stress at which a material first begins to undergo plastic deformation.

Review Questions

1. While a diamond is much harder than a steel knife, the knife is tougher than the diamond. Explain the difference between hardness and toughness, and why the diamond is more likely to break than the knife when the two collide.

2. Suppose you have two wooden blocks that are identical except that one is 5 times as tall as the other. If you stack the two blocks and then stand on the pair, why will the taller block lose 5 times as much height as the shorter block?

3. If you stack a rubber block on top of a steel block of the same size and then stand on the pair, the rubber block will lose much more height than the steel block. Explain this result in terms of the Young's moduli of the two materials.

4. A flagpole bends in the wind. If the wind pushes 5 times as hard on the flagpole, it will bend 5 times as far. Explain this effect in terms of shear stress and shear strain.

5. As we'll see in the next section, glass is unable to undergo slip. How does the absence of slip in glass contribute to its brittleness?

6. How does annealing a piece of iron soften it and prevent it from experiencing brittle fracture?

7. How does heating and rapidly quenching carbon-rich steel make it hard and brittle?

8. Hardening the steel of a car would make it wonderfully resistant to minor dents but would make it dangerous in a collision. Why is softer steel safer than harder steel in the body and frame of a car?

9. Why does forming tiny metal carbide crystals in steel make that steel harder?

10. Adding chromium and nickel to steel turns it into soft, corrosion resistant stainless steel. What change occurs in the metal's crystal structure that leaves it soft and unsuitable for heat treatment hardening?

11. Why are cheap 18-8 stainless steel utensils so soft and flexible, and why can't the manufacturer use heat treatments to make them stiffer?

Exercises

1. An inexpensive screwdriver is made from a piece of thick low-carbon steel wire. After this wire is stamped into shape at room temperature by a huge press, a plastic handle is attached and the finished screwdriver is packaged for sale. How does the structure of its steel allow this screwdriver to deform so easily when you try to tighten a screw?

2. A good screwdriver is made from a piece of medium-carbon steel rod. This rod is too hard to stamp into shape at room temperature (see Exercise **1**), so it's either shaped at high temperature or cut with an abrasive grinder at room temperature. How does the structure of its steel allow this screwdriver to resist deformation when you tighten a screw?

3. A premium quality screwdriver is heat treated (see Exercises **1** & **2**). While its tip is given an extremely hard surface that won't deform as it twists a screw, its shaft is made tough so that it won't break under stress. How must the screwdriver be heat treated to give it these properties? Explain.

4. A can opener and the can it opens are both made of steel. Explain why the opener cuts the can rather than the reverse.

5. Some fireplaces have soft, low-carbon steel linings. The instructions for one of these fireplaces warn you not to let hot coals and ashes build up against its lining because that will make the lining hard and brittle. How does trapping heat and carbon-rich coals near the lining make it brittle?

6. A tooth has a hard layer of enamel surrounding a softer core of dentin. Why is this composite tooth more durable than one that's solid enamel or solid dentin?

7. Thieves have been known to use liquid nitrogen to break locks. A lock will dent without breaking if you strike it with a hammer at room temperature, but it may shatter if you first cool it to –196° C (–320° F) with liquid nitrogen. Why does cooling the lock change its behavior so dramatically?

8. At room temperature, a bell made of lead is soft. If you strike it with a mallet it just dents. But if you first cool it with liquid nitrogen, it will ring beautifully when you hit it. It also won't dent. Why does cooling the metal make it harder?

9. Some dump trucks have extra wheels that are lowered onto the pavement when the truck is carrying a very heavy load. How do these extra wheels reduce the chance that the truck will cause plastic deformation of the roadway?

10. A theft protection bar bolts to a car's steering wheel and makes the wheel difficult to turn. How does case hardening make the bar both difficult to cut and difficult to break?

11. Steel furniture often breaks at welds, where the metal was heated briefly to its melting temperature in order to join separate pieces together. Why is the welded area of the steel harder and more prone to fracture than the rest of the steel?

12. The body of an aluminum can is formed out of a flat disk of aluminum by working it at room temperature. How does this shaping process strengthen the finished can?

13. Pure gold (24 carat) is generally too soft for jewelry. Alloys such as 14 carat and 18 carat are much harder. Why?

14. Steel drums are traditionally made from the tops of 55 gallon oil drums. The steel of an oil drum is soft and is easily worked into the proper shape for a steel drum. Before the metal will emit tones, however, it must be treated with heat and carbon. How does this treatment alter the metal?

15. The surface of a hardwood floor bends slightly when you step on it but immediately returns to its original shape. When you drop a pointed object on the floor, however, it dents permanently. Explain these two responses to stress.

16. The Young's modulus of diamond is approximately four times that of iron. If you stack a small cube of iron on top of a similar cube of diamond and press down on the stack, which cube will compress more? How much more?

17. Soft low-carbon steel and hard high-carbon tool steel have virtually identical Young's moduli. If you stack a small cube of low-carbon steel on top of a similar cube of tool steel and step on them, how will the strains they exhibit compare?

18. If soft low-carbon steel and hard high-carbon tool steel have equal Young's moduli (see Exercise **17**), why can the tool steel cut through the low-carbon steel so easily?

19. Pure aluminum is too weak for structural use. Most aluminum building materials have a few percent copper, magnesium, or silicon added to the aluminum. Why?

20. Oxygen-free high-conductivity copper is nearly pure copper and an excellent conductor of electricity. When this metal has been annealed carefully, its crystals are enormous and it's so ductile that it can be squeezed through tiny holes to make electric wires. But while a thick bar of this metal is easy to bend the first time, if you straighten it and bend it over and over again it gradually becomes stiffer. Explain this effect.

21. After bending a bar of soft copper until it becomes stiff (see Exercise **20**), what can you do to make it soft again?

22. Scientists have recently been able to fabricate metals with grains containing only a few thousand atoms each. These "nanocrystalline materials" are much harder than the normal metals from which they're prepared. Why?

23. The twist ties used to seal plastic food bags can only be used a few dozen times before they break. How does the repeated bending of the metal in the tie cause it to break?

24. Before drawing a steel bar through a narrow opening to make steel wire, the bar is heated to a high temperature and then cooled extremely slowly. Why does this heat treatment make the wire drawing procedure easier?

25. Magnetic wall strips can hold sharp stainless steel kitchen knives but not cheap 18–8 stainless steel knives. Why won't the 18–8 stainless steel knives stick to the strip?

26. Machine tools such as drill bits are made from high-carbon and alloy steels. Their edges are heat treated to give them enormous yield strengths and no capacity for plastic deformation. How does the microscopic structure of the edge make it resist deformation so well and shatter rather than bend when it's over-stressed?

27. The steel used to make springs is hardened to stop plastic deformation. Why is plastic deformation bad for a spring?

28. The solid–solid phase transition that occurs in iron when heated ferrite converts to austenite isn't unique. Many crystals rearrange as they are heated or cooled through certain temperatures. If the density of the new crystal structure is very different from that of the old crystal structure, the crystal may shatter as it rearranges. Why?

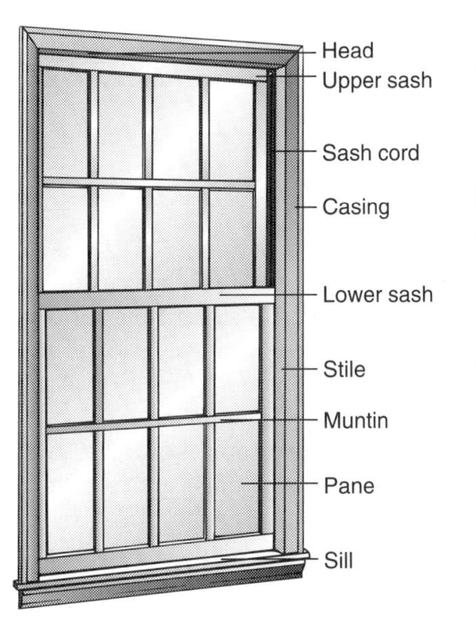

Section 17.2

Windows and Glass

When you look through a clean glass window on a sunny day, you hardly know the glass is there. Light passes through each pane with so little loss and distortion that it's difficult to see the window at all. But you have no trouble feeling glass—it's a tough, rigid material that keeps out the wind and weather while admitting light. Because glass is also a good thermal insulator, it helps to keep buildings cool in the summer and hot in the winter. Produced from cheap and plentiful raw materials with minimal effort and equipment, glass is truly a remarkable substance.

Questions to Think About: *How do they make glass sheets that are so flat and smooth? Why does some glass shatter when you heat it or cool it too rapidly? Since most solids melt suddenly at specific temperatures, why does glass becomes soft and pliable instead? Since most solids abruptly form narrow necks and snap when you try to stretch them, why is it possible to stretch softened glass into so many shapes? Why does tempered glass shatter completely into tiny pieces when it breaks? What colors the glass used in stained glass windows?*

Experiments to Do: *Prove to yourself that glass is a hard, brittle, elastic material. To show that glass is hard, see how difficult it is to scratch a glass bottle. Can you find anything that scratches it? To show that glass is brittle, try to dent a glass bottle without breaking it. Wrap the bottle in heavy cloth or paper and then tap it carefully with a rock or hammer. Were you able to dent it? To show that glass is elastic, push gently on a large glass mirror. Be careful not to break it! Look closely at the image to see if the glass is distorting. What happens to the glass when you stop pushing?*

Examine a piece of broken glass. You will find that the fractured surfaces are basically smooth and featureless. That's because glass has no orderly crystalline structure. Instead, its atoms are arranged almost randomly, like those in a liquid. In fact, glass is essentially a super-thick liquid. To allow glass to flow visibly, you must heat it dangerously hot. However, you can observe similar flow in cold honey. Honey stretches and flows and sticks to itself in much the same way as softened glass. And like glass, honey doesn't crystallize as you cool it—it just gets stiffer and stiffer.

What Is Glass?

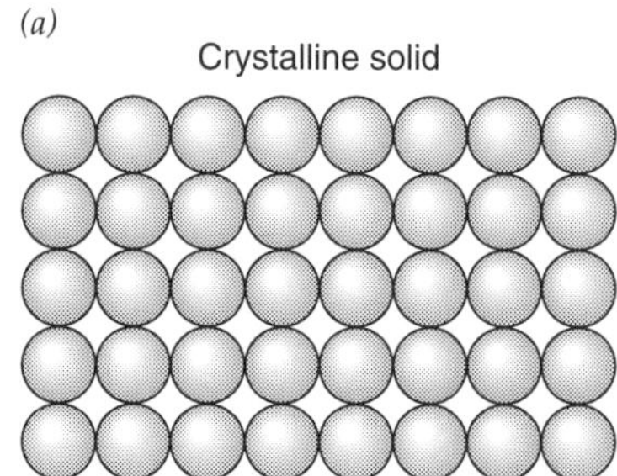

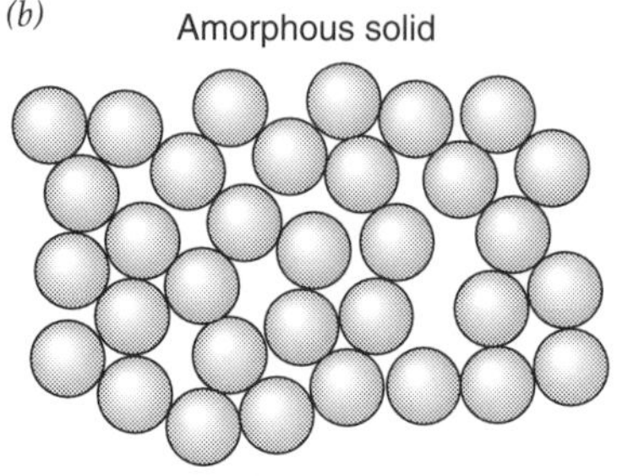

Fig. 17.2.1 - (*a*) A crystalline solid has long-range order, meaning that knowing where atoms are at one end of the crystal tells you where atoms are at the other end of the crystal. (*b*) An amorphous solid has no long-range order.

Let's begin with what glass is not—it is not a crystal. The atoms in a crystal are organized in a regular, repetitive lattice so you need only locate a few atoms in order to predict where all of their neighbors are (Fig. 17.2.1*a*). The atoms are so neatly arranged that, except for occasional crystal defects, you can predict positions for thousands or even millions of atoms in every direction. This spatial regularity is called **long-range order**.

Glass is an **amorphous solid**, a material without long-range order (Fig. 17.2.1*b*). Locating a few glass atoms tells you next to nothing about where to find any other atoms. The atoms in glass are arranged in the random manner of a liquid because glass is essentially a super-stiff liquid. Its atoms are jumbled together in a sloppy fashion but they can't move about to form a more orderly arrangement. Glass arrives at this peculiar amorphous state when hot liquid glass is cooled too rapidly for it to crystallize.

If molten glass were an ordinary liquid, it would begin to solidify abruptly during cooling once it reached its freezing temperature. At that point, its atoms would begin to arrange themselves in crystals that would grow in size until there was no liquid left. That's what's normally meant by freezing.

However some liquids are slow to crystallize when you cool them slightly below their freezing temperatures. While they may be cold enough to grow crystals, they must get those crystals started somehow. If crystallization doesn't start, a material's atoms and molecules will continue to move about and it will behave as a liquid. When that happens, the liquid is said to be **supercooled**.

Supercooling is common in liquids that have difficulties forming the initial **seed crystals** on which the rest of the liquid can crystallize. Because almost all of the atoms in a seed crystal are on its surface, it has a relatively large surface tension and surface energy. Below a certain critical size, a crystal is unstable and tends to fall apart rather than grow. However once the first seed crystals manage to form, a process called **nucleation**, the rest of the supercooled liquid may crystallize with startling rapidity.

Just below its freezing temperature, the atoms in glass don't bind to one another long enough to form complete seed crystals and nucleation takes almost forever. The glass is a supercooled liquid. At somewhat lower temperatures, seed crystals begin to nucleate, but glass's large viscosity (thickness) prevents these crystals from growing quickly. The glass remains a supercooled liquid for an unusually long time. At even lower temperatures, glass becomes so viscous that crystal growth stops altogether. The glass is then a stable supercooled liquid, one that will remain in that form indefinitely. At this temperature range, glass still pours fairly easily and can be stretched or molded into almost any shape, including window panes.

However, when you cool the glass still further, it becomes a ***glass***. Here the word *glass* refers to a physical state of the material—a type of amorphous solid. To distinguish this use of the word *glass* from the common building material, it is italicized here and elsewhere in this section. Glass, the material, becomes a *glass*, the state, at the ***glass*** **transition temperature** (T_g). Below T_g, the atoms in the glass rarely move past one another; they continue to jiggle about with thermal energy but they don't travel about the material.

It's hard to tell by looking at the glass just when this *glass* transition takes place. It generally occurs when the glass's viscosity exceeds about 10^{12} Pa·s, where the Pa·s is the SI unit of viscosity. At that point, the great difficulty the fluid has flowing past itself reflects the microscopic difficulty the atoms have in rearranging. The atoms are practically frozen in place and the fluid flows so slowly that the glass is almost indistinguishable from a solid.

To see that glass behaves like a solid, let's look at how viscous fluids and solids respond to shear stress. If you exert shear stress on a viscous fluid, by pushing its top right and its bottom left, the top portion will flow right and the bottom portion will flow left (Fig. 17.2.2). But if you exert shear stress on a solid, it will experience shear strain. It won't flow anywhere. To get the top half of a steel bar to "flow" right while the bottom half "flows" left, you'd have to break it. In contrast, even a small shear stress will eventually cause a liquid to flow. Thus there seems to be a fundamental difference between solids and liquids in their responses to shear stress: solids undergo shear strain while liquids flow.

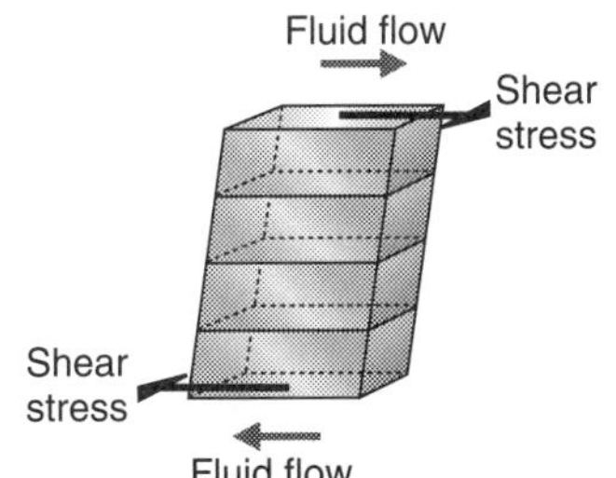

Fig. 17.2.2 - Viscosity measures the difficulty a fluid has flowing past itself. The higher this fluid's viscosity, the more shear stress it must experience to keep its bottom layer flowing quickly to the left and its top layer flowing quickly to the right.

But what happens when you suddenly expose an extraordinarily viscous liquid to shear stress? The liquid has so much trouble flowing that it undergoes shear strain instead. It bends elastically, just like a crystalline solid! If you release the stress quickly enough, the liquid will spring back almost to its original shape. However if you wait a while before releasing the stress, the liquid will have time to flow and the stress will go away on its own. You can see this effect by bending a stick of taffy. If you bend it briefly, the taffy is almost as springy as a solid. But if you bend it and wait, the taffy flows to relieve the shear stress.

The time a liquid takes to relieve shear stress by flowing increases with that liquid's viscosity. Thin liquids such as water relieve stresses quickly, but with a viscosity of 10^{12} Pa·s, it takes glass minutes or hours to relieve stresses and you would have to be very patient to detect anything non-solid about that glass. It would feel just as hard and elastic as an ordinary crystalline solid unless you were willing to wait for hours to see it flow (Fig. 17.2.3).

There is one more change that occurs as the *glass* transition takes place: the glass stops shrinking like a liquid with decreasing temperature and starts shrinking like a solid. Liquids shrink rapidly as you cool them because their mobile atoms pack together more tightly as their thermal energies decrease. Solids shrink less rapidly as they're cooled because their atoms can't rearrange. Above the glass transition temperature, glass shrinks quickly as it's cooled, like a liquid. Below the glass transition temperature, glass shrinks slowly as it's cooled, like a solid.

CHECK YOUR UNDERSTANDING #1: Flash Frozen

If you sprinkle some small ice crystals into supercooled liquid water, what will happen?

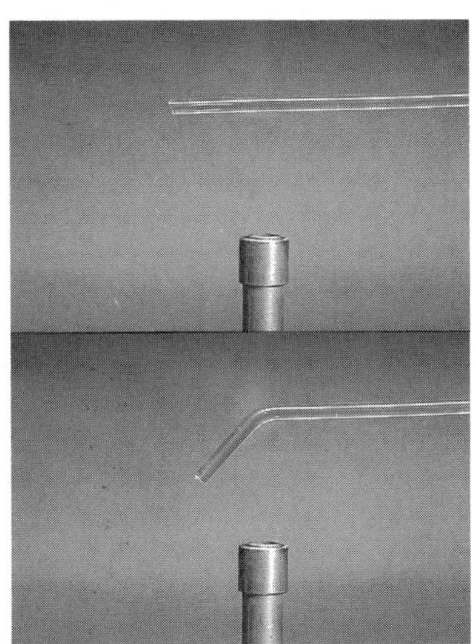

Fig. 17.2.3 - Below its *glass* transition temperature (top), this rod behaves as a solid and responds elastically to forces. But above its *glass* transition temperature (bottom), the rod behaves as a viscous liquid and bends slowly under its own weight.

What is in Glass?

With that introduction to the behavior of *glasses*, it's time to look at what's inside glass. Window glass is mostly silicon dioxide (SiO_2), the chemical found in quartz and quartz sand and commonly called *silica*. Silica is extremely common in nature, making up much of the earth's crust. It's hard and clear and resistant to chemical attack. It's also the best *glass* forming material in existence.

To be a good *glass former*, a material must have trouble nucleating seed crystals and must prevent those seed crystals from growing quickly. Because silica is extraordinarily viscous at its freezing temperature, it's an excellent *glass* former and is easily converted into quartz *glass* or *vitreous quartz*.

(a)

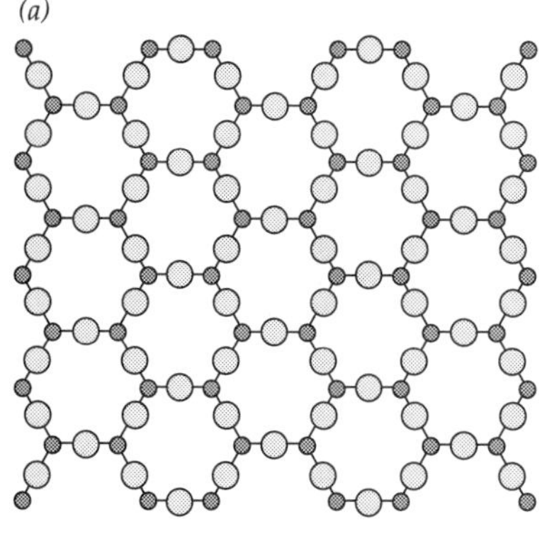

(b)

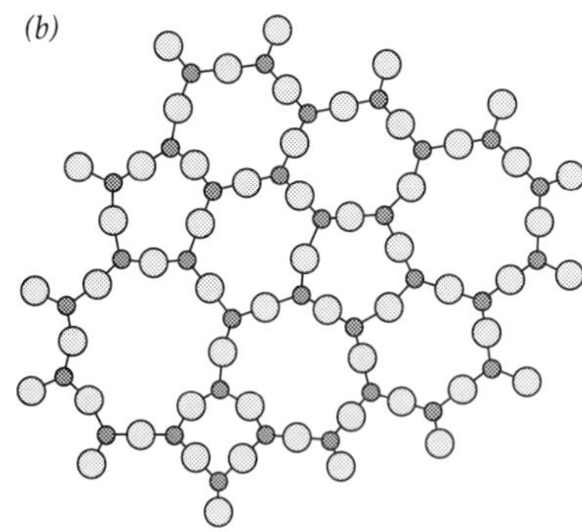

(c)

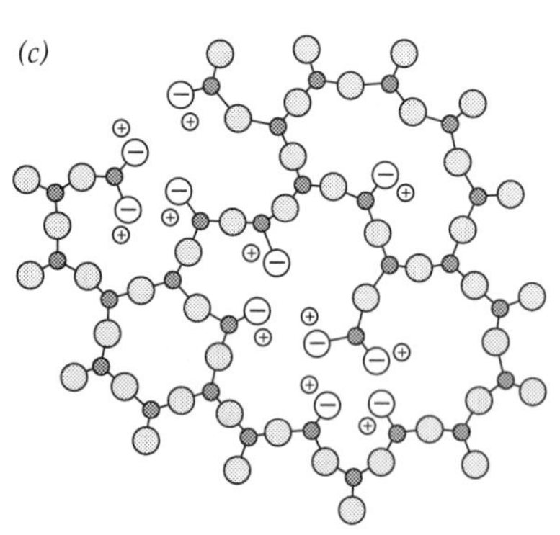

● "Silicon" Atom
○ "Oxygen" Atom
⊕ "Sodium" Ion
⊖ "Oxygen" Ion

Fig. 17.2.4 - Two-dimensional illustrations of the structures of (*a*) crystalline silica, (*b*) liquid or glassy silica, and (*c*) glassy or vitreous silica containing some sodium oxide.

Silica is held together by covalent bonds, even as a liquid. A **covalent bond** forms between two atoms when they share a pair of electrons. This sharing reduces the electrostatic potential energy of the atoms by placing extra negative charge in between their positive nuclei. It also reduces the kinetic energies of the shared electrons by letting them spread out as waves onto both atoms. Their wavelengths increase as they spread and lengthening an electron's wavelength, like lengthening a photon's wavelength, reduces its energy.

Covalent bonds are highly directional. Each silicon atom in the silica forms covalent bonds with four oxygen atoms and orients those atoms at roughly the corners of a tetrahedron. Each oxygen atom in the silica forms covalent bonds with two silicon atoms and orients them at roughly two corners of a tetrahedron. The result is an intricate network of silicon and oxygen atoms in which each oxygen atom acts as a bridge between two adjacent silicon atoms. Because of its interlinking structure, silicon dioxide is often called a *network former*.

In crystalline silica, this networking process creates an orderly three-dimensional lattice that's hard to visualize. To help us discuss the basic character of this lattice, let's examine a two-dimensional lattice with similar features (Fig. 17.2.4*a*). The "silicon" atoms in this simplified lattice bind to only three "oxygen" atoms, at the three corners of an equilateral triangle. The "oxygen" atoms bind to two "silicon" atoms, at opposite ends. While this lattice is oversimplified, it illustrates how intricate and orderly a quartz crystal is. If you know the locations of two or three atoms in this lattice, you can predict where all of the other atoms are.

When you melt silica, its atoms remain covalently bound, but they change partners frequently. The orderly crystalline lattice vanishes and is replaced by a tangled network of interconnecting rings (Fig. 17.2.4*b*). While the silicon and oxygen atoms still have the right numbers of neighbors, they often form rings with the wrong numbers of atoms in them.

Once the crystalline order of silica has been destroyed, it's hard to recover. The tangled networks of the liquid are relatively stable and the atoms can't tell which rings have too many or two few members. Moreover liquid silica is still held together by covalent bonds, which constrain the motions of its atoms and make it an extremely viscous fluid. Silica is thus an ideal *glass* former and supercools all the way to a *glass* when you cool it rapidly.

CHECK YOUR UNDERSTANDING #2: Instant Glass
The first test of the atomic bomb near Alamogordo, New Mexico on July 16, 1945, turned the desert sand into glass. Explain what happened.

Soda–Lime–Silica Glass

But pure silica has a serious drawback. Its covalent bonds hold it together so tightly that it doesn't melt until its temperature reaches 1713° C. That's far above the melting temperatures of most metals, including iron and steel. While pure quartz glass can be made, it's a specialty item and quite expensive. Instead, most common glass contains other chemicals that lower the mixture's melting temperature so that it's easier to work with.

The principal addition to window glass is sodium oxide or *soda* (Na_2O). Although soda contains oxygen atoms, it's held together by ionic rather than covalent bonds. **Ionic bonds** form when the atoms in a material become oppositely charged ions and attract one another. In this case, each oxygen atom in soda removes an electron from two nearby sodium atoms, producing a mixture of negatively charged oxygen ions and positively charge sodium ions.

When soda is added to silica and the two are heated, the mixture melts at a temperature that is below the melting temperature of either material. It's a **eutectic**, a mixture that melts more easily than the materials from which it's made. A mixture containing 25% soda by weight melts at a temperature of only 793° C. Equally remarkable is the fact that a fine mixture of soda and silica powders can be melted together at temperatures near this value. The soda acts as a flux to melt the silica and the glass manufacturer never has to heat the mixture to silica's normal melting temperature.

The reason for the low melting temperature of this *soda–silica glass* is illustrated in Fig. 17.2.4*c*. The sodium atoms in the mixture donate electrons to the oxygen atoms, leaving the substance full of positively charged sodium ions and negatively charged oxygen ions. In this case, the oxygen atoms have just a single charge. But an oxygen atom with an extra electron only needs to share one electron to complete is electronic shell. As a result, it binds to only one silicon atom and doesn't form a bridge between a pair of silicon atoms.

Soda–silica glass is full of *non-bridging oxygen atoms* that end the network rather than extending it. Their presence weakens the glass and lowers its melting temperature. While this ease of melting is important for glass manufacturing, soda–silica glass is less robust in almost every way than pure silica glass. It's softer and weaker than silica glass so it scratches and breaks more easily. It has more internal friction so it wastes energy in bending and emits a dull tone when struck. It's also much less chemically resistant than quartz *glass*. In fact, it's water soluble.

Sodium ions are so water soluble that they allow water to enter the glass's molecular network and chop it to pieces. As a result, soda–silica glass can be dissolved in water and painted onto surfaces. It's called *water glass* and is used to seal the outsides of eggs so that they don't dry out. It's also included in laundry detergent, where it protects the washer from corrosion.

Soda–silica glass can be made much less water soluble by adding calcium oxide or *lime* (CaO) to it. Calcium oxide is an ionic solid, but it's not soluble in water and makes the glass much more durable. *Soda–lime–silica glass* is almost insoluble in water and is the principal commercial glass. Windows, bottles, and jars are all made of soda–lime–silica glass.

In each of these glasses, silica is the *glass* or network former and soda and lime are modifiers. The *glass* former's job is to create the tangled network that gives the liquefied material its high viscosity and allows it to supercool all the way to the *glass* transition temperature. A modifier's job is to ease the path to the *glass* transition and to alter the properties of the glass that's produced. Most importantly, modifiers help to ensure that the *glass* former doesn't crystallize during cooling, a serious problem called *devitrification*.

CHECK YOUR UNDERSTANDING #3: Low Temperature Liquids

A mixture of soda and silica melts at a temperature below the melting point of either material. How does this resemble a mixture of ice and salt?

Other Glasses

There are many other compounds and mixtures that can form *glasses*. Most of these systems involve oxygen atoms, which are excellent bridging atoms and good at producing covalent networks. But there are some *glasses* that don't contain oxygen and there are even *glasses* that contain only metal atoms (see ❐). Because these exotic *glasses* are rarely used in windows, we'll focus instead on the more common oxygen-containing glasses (Table 17.2.1).

❐ Amorphous or *glassy* metals are made by ultrafast cooling of liquid metals. In "splat" cooling, a thin stream of molten metal pours onto a spinning refrigerated wheel. The liquid cools and solidifies in millionths of a second to produce a paper thin ribbon of metallic *glass*. However, even when cooled incredibly quickly, only special alloys form metallic *glasses*. Such *glasses* are much harder than normal metals because they have no crystalline structure and can't undergo slip.

Table 17.2.1 - The approximate compositions and uses of various common forms of glass.

Type of Glass	*Composition (by weight)*	*Uses*
Soda–Lime–Silica	73% silica, 14% soda, 9% lime, 3.7% magnesia, and 0.3% alumina	Glass Windows, Bottles, and Jars
Borosilicate	81% silica, 12% boron oxide, 4% soda, and 3% alumina	Pyrex Cookware and Laboratory Glassware
Lead (Crystal)	57% silica, 31% lead oxide, and 12% potassium oxide.	Lead Crystal Tableware
Aluminosilicate	64.5% silica, 24.5% alumina, 10.5% magnesia, 0.5% soda.	Fiberglass Insulation and Halogen Bulbs

Silica isn't the only *glass* forming chemical. Other materials that create tangled networks and form *glasses* during rapid cooling include phosphorus pentoxide (P_2O_5), germanium oxide (GeO_2), and boron oxide (B_2O_3). Of these, only boron oxide is commercially important and then only when mixed with silica.

The problem with pure boron oxide glass is that it's not very durable and dissolves easily in hot water. However, boron oxide-silica glasses are quite stable and play important roles in laboratories and kitchens under trade names such as Pyrex and Kimax. These *borosilicate glasses* tolerate temperature changes much better than soda–lime–silica glasses.

Soda and lime aren't the only modifiers used in commercial glasses. All of the alkali oxides and alkaline earth oxides act as modifiers when mixed with silica or other *glass* formers. Changing the modifiers in soda–lime–silica glasses produces subtle alterations in the mechanical, chemical, and optical properties of the glass. The most common alternative modifiers are potassium oxide (K_2O), which is substituted for soda, and magnesium oxide or *magnesia* (MgO), which is substituted for lime. Barium oxide (BaO) and strontium oxide (SrO) are frequently substituted for lime in glass that must block X-rays, such as that on the front of a television picture tube.

There are also some chemical compounds that are intermediate between *glass* formers and modifiers. While these compounds don't form *glasses* on their own, they participate in the tangled networks initiated by other compounds such as silica. The most important intermediate compounds are aluminum oxide or *alumina* (Al_2O_3) and lead oxide (PbO). *Aluminosilicate glasses* tolerate high temperatures better than soda–lime–silica glasses and are used in halogen lamps, furnaces, and fiberglass insulation. A small amount of alumina is included in most glasses to help them tolerate weather better.

Lead oxide is included in special optical glasses to increase their indices of refraction. It also makes them dense, hard, and X-ray absorbing. The *lead crystal* used in decorative glassware and windows isn't crystalline at all. Instead, it's a *glass* consisting mostly of soda, lead oxide, and silica. This soda–lead oxide-silica glass refracts light strongly and disperses its colors. Because its networks are more complete than those in soda–lime–silica glass, lead crystal is stronger and it has less internal friction. Lead crystal rings beautifully when you strike it gently with a hard object.

One other important variation in window glass is color. The pure glasses that we've considered up until now are colorless because they have no way of absorbing photons of visible light. But impurities and imperfections can give these glasses colors. Many of the colored glasses that appear in stained glass windows are that way because they contain metal ions such iron, cobalt, or copper. The electrons in these ions have many empty orbitals to which they can move and they absorb photons of visible light. Just which colors these ions absorb depend on their structures and on how many electrons they're missing.

Because iron ions that are missing three electrons appear green, iron impurities in common window glass make it appear slightly green. Copper and cobalt ions that are missing two electrons appear blue (blue bottle glass). Manganese ions that are missing three electrons appear purple. Chromium ions that are missing three electrons appear green (green bottle glass). Vanadium ions that are missing four electrons appear red. And so it goes. There is also special ruby glass that contains tiny particles of gold metal and appears red because these metal particles absorb green and blue light.

Most well mixed glasses are completely uniform throughout and have no internal variations in refractive index to scatter and reflect light. But there are also *opal* or *milk glasses* that are not uniform and scatter light like a cloud. These translucent glasses are used in privacy windows and are produced when particles of one type of glass precipitate out of solution in another type of glass during cooling. Materials in which crystalline particles precipitate out of solution in a glass during cooling are called *glass-ceramics* and are generally translucent rather than transparent.

CHECK YOUR UNDERSTANDING #4: Play It, Sam

How can you distinguish a lead crystal wine glass from one made of soda–lime–silica glass?

Making Glass Windows

Making glass is relatively easy compared to making steel. In principle, all you have to do is mix the raw materials, melt them together into a liquid, and form the liquid into a finished product as it cools. As long as you cool it quickly through the temperature range at which seed crystals can nucleate and grow, it will turn into glass. Nonetheless, there are some interesting procedures involved in creating glass and its products, particularly windows.

The raw materials in soda–lime–silica glass are commonly occurring minerals. Silica (SiO_2) is obtained from quartz sand and sandstone. Soda ash (Na_2CO_3), which decomposes into soda (Na_2O) and carbon dioxide (CO_2) when heated, is a naturally occurring mineral called trona and can also be produced by reactions between limestone and salt. Limestone ($CaCO_3$), which decomposes into lime (CaO) and carbon dioxide (CO_2) when heated, is a ubiquitous sedimentary rock.

These three minerals are heated together in a large ceramic chamber to a temperature of roughly 1500° C. At this temperature, the soda acts as a flux and helps the other two minerals to melt. Soon the entire mixture is liquid. Because it contains several different compounds, the liquid glass is carefully stirred to ensure that it's uniform throughout.

Gases released from the minerals, particularly carbon dioxide, water vapor, and trapped air, cause the liquid to froth. One of the hardest tasks in glass making is eliminating the bubbles. This process, called *fining*, involves waiting for the bubbles to float to the surface and adding small quantities of additional chemicals—often arsenic and antimony compounds—to help the bubbles escape.

Finally, the glass is cooled to a working temperature of about 800° C, passing quickly through the temperature range in which it can crystallize. As it cools, the liquid's viscosity increases dramatically. In the melting region of the chamber, its viscosity is about 10 Pa·s or 1000 times that of water. At 800° C, its viscosity has risen to 1,000 Pa·s or 100,000 times that of water. It flows like a thick syrup and is ready for use in making bottles or windows.

Glass bottles are formed by injecting gobs of glass into molds and then blowing air into them to create their hollow interiors. The injection step creates the neck of the bottle, attached to a carefully shaped but uninflated glass bubble.

Fig. 17.2.5 - Float glass is made by pouring liquid glass onto the surface of liquid tin. As the ribbon of liquid glass flows away from the melting chamber, it cools and hardens. Finally, it's annealed in a long tunnel-shaped lehr before being cut into sheets.

The blowing step inflates that bubble to create the bottle's storage region. The glass is then cooled slowly through its glass transition temperature to produce a finished bottle.

Window panes are produced by pouring liquid glass onto a pool of molten tin (Fig. 17.2.5). This "float" method was developed in 1959 and revolutionized window making. The biggest problem in window making has always been producing perfectly flat surfaces on both sides (Fig. 17.2.6). The top surface of liquid glass is naturally flat but the bottom surface conforms to whatever it's lying on. When the glass is poured onto a liquid metal, its bottom surface is supported by another perfectly flat surface. The finished plate of float glass is thus perfect on both sides.

Fig. 17.2.6 - Early glass windows were made by blowing a large glass sphere on the end of an iron pipe. This sphere was cut off the pipe and attached to a metal rod called a punty. Spinning the punty caused the hot sphere to open up into a large flat disk, like a pizza. When this disk cooled, it was cut into a number of small panes. Because each pane was precious, even the central one was used. The bullseye mark left by the punty is plainly visible in these old windows.

That this process works is a wonderful confluence of behaviors. First, tin melts at the relatively low temperature of 232° C but doesn't boil until 2260° C. This range allows the liquid glass to harden while the tin remains liquid. Second, liquid tin is far denser than liquid glass so the glass floats easily on top, supported by the buoyant force. Third, tin and glass are immiscible—tin is held together by metallic bonds while glass is bound by covalent and ionic bonds. The two liquids don't bind strongly to one another so they keep to themselves. The glass remains chemically unaffected by the liquid tin below it.

As the hot glass flows out onto the tin, it naturally spreads until it's about 6 mm thick. A glass maker can produce thicker window panes by corralling the spreading liquid so that it remains relatively thick. To produce thinner sheets, the glass maker stretches the liquid to spread it more thinly. The float technique can produce extremely flat glass in thicknesses ranging from less than a millimeter to several centimeters. (As shown in Fig. 17.2.7, glass's tolerance of stretching also makes it possible to fabricate incredibly thin glass fibers and tubes).

The soda–lime–silica mixture used for float glass has a *glass* transition temperature of about 540° C. When the glass leaves the liquid tin, it's already slightly below that temperature. However, its rapid change in temperature on the tin surface has caused it to shrink and this shrinkage creates stresses in the glass. If the glass were simply cooled to room temperature at this point, the stresses would remain trapped forever and would weaken the glass. Stresses also make the glass birefringent, which can distort the light passing through it.

To reduce the stresses trapped in the glass, it's carefully annealed in a long tunnel called a *lehr*. It's kept near the glass transition temperature for a long time so that the atoms in the glass can rearrange just enough to relieve the stress. When most of the stress has been eliminated, the finished glass is finally allowed to cool to room temperature. In the optical glass used in lenses and telescopes, the annealing is done much more slowly. In many cases, the glass is cooled through the *glass* transition temperature at a rate of less than 1° C per hour. The glass disk of the 200 inch telescope mirror at Mt. Palomar was cooled from 500° C to 300° C at the phenomenally slow rate of only 1° C per day.

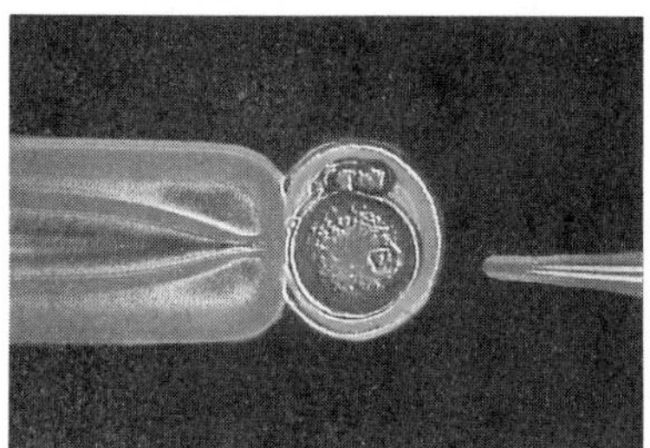

Fig. 17.2.7 - Glass tubing can be heated and stretched to form tubes with microscopic dimensions; miniature copies of the original tubes. Here one glass tube holds a mouse embryo while a much smaller glass tube prepares to inject DNA into it.

CHECK YOUR UNDERSTANDING #5: Glassy Sweets

Many hard sheet candies are smooth and glossy on one side but dull on the other. What causes the difference?

Strengthening Windows

Once it leaves the annealing lehr, the float glass is ready for use. But first it must be cut to size. Cutting glass is a tricky business because it lacks a crystalline structure and its amorphous nature gives it some peculiar properties. To begin with, glass can't undergo any plastic deformation because it can't undergo slip. If you bend glass, it deforms elastically up until the moment it breaks via brittle fracture. When that break finally occurs, a crack propagates uncontrollably through the glass because the glass has no crystalline grains to stop it or change its path.

The crack always begins at the surface of the glass, where a defect weakens the molecular structure. Even the most perfect glass surface has a few defects in it and any one of these can initiate a crack when it's under tensile stress. (For an interesting way to minimize these defects, see ❐.) When you bend a piece of glass, you stretch one surface while compressing the other and create a tensile stress on the stretched surface. If this tensile stress is large enough, a defect there will become a crack and will propagate through the glass so that it shatters.

❐ Fiberglass is made by pulling hot aluminosilicate glass into thin strands. Hot glass is a simple liquid and will stretch without breaking to form fantastically thin fibers. These fibers are so narrow that they have almost no surface area and thus very few surface defects. As a result, glass fibers are difficult to break and are very strong. They are used as structural materials and as thermal insulation.

Simply bending glass and hoping that it will break along a straight line isn't a practical scheme for cutting glass. Instead, you use a diamond scribe to scratch the glass along the intended break. The scratch introduces defects right where you want the cracks to occur. Wetting the scratch helps because water creates defects in glass and weakens it. If you then stress the glass carefully, either mechanically or by a sudden change in temperature, you can usually get it to crack along the scratch. Still, the glass often breaks in an undesired direction and you are left with useless fragments.

Another way to cut glass is with an abrasive saw. Here a rapidly turning disk containing extremely hard crystals chips out tiny fragments from the glass. The chipping exerts only a modest stress on the glass so that it doesn't crack. Although slow, this method is quite reliable. Similar abrasive techniques are used to cut decorative glass.

Once it's been cut to size, the window is finished. It's ready for installation in a house or building. But it's not ready for use in an oven or an automobile. This window would not be able to tolerate the sudden changes in temperature present in an oven and would pose a serious hazard during a collision in an automobile. So the glass must be heat stabilized and strengthened.

To make an oven window more tolerant of thermal shocks, its chemical composition must be changed. Because soda–lime–silica glass has a large coefficient of volume expansion, it expands considerably as its temperature rises and this expansion can produce huge stresses in the glass. If one half the oven window is suddenly heated, it will expand and try to stretch the other half. If the resulting tensile stress is large enough, that window will crack and break. However, the oven window is made out of a borosilicate glass such as Pyrex. Borosilicate glasses have coefficients of volume expansion that are about a third those of soda–lime–silica glasses. Thus it takes a much larger thermal shock to break a Pyrex window or container. For that reason, most cookware and laboratory glassware is made out of borosilicate glass.

To strengthen a car window, its mechanical structure must be changed. Glass breaks when its surface begins to tear apart. If you modify the glass so that its surface is normally under compressive stress, it will be much harder to get a tear started and the glass will be stronger.

The best way to put the glass's surface under compressive stress is to **temper** the glass. Tempering is done by heating the glass until it softens and then suddenly cooling its surfaces with air blasts. When the glass's surface cools through the *glass* transition temperature, it becomes solid-like and no longer

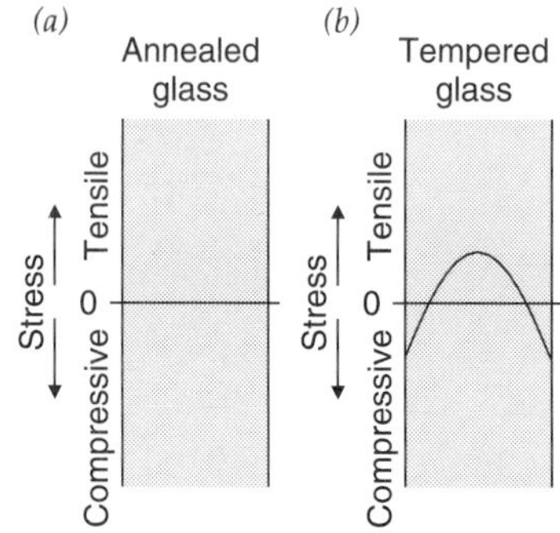

Fig. 17.2.8 - (*a*) Properly annealed glass has no internal stresses. (*b*) Tempered glass is specially heat treated so that its surfaces are under enormous compressive stresses while its body is under substantial tensile stress. Tempered glass is very hard to break.

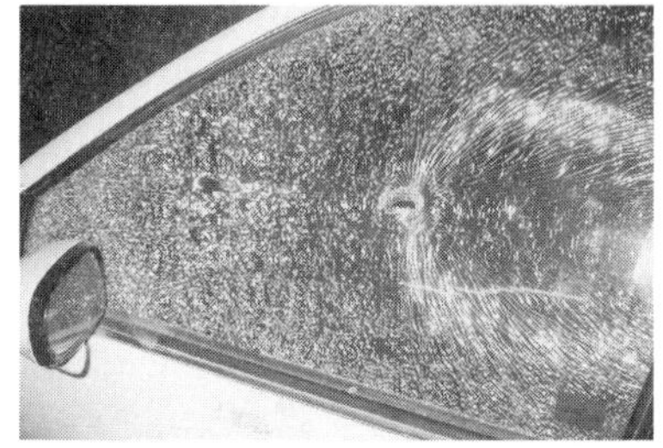

Fig. 17.2.9 - When a small rock cut through the compressed outer skin of this tempered glass car window, it exposed the tense inner layer. That layer immediately began to tear and the entire window broke into thousands of tiny fragments.

shrinks rapidly with temperature. However the inside of the glass is still liquid-like and continues to shrink rapidly as it cools. As the glass inside shrinks, the surface layers are placed under enormous compressive stresses (Fig. 17.2.8).

This compressive stress makes it difficult to initiate a crack in the outer surface of tempered glass. You must first stretch that outer surface so much that the compressive stress disappears and becomes tensile stress. Breaking tempered glass takes about three times as much force as breaking ordinary glass.

But tempered glass has an interesting complication. When it breaks, it undergoes *dicing fracture*—the glass crumbles completely into tiny pieces less than a centimeter on a side (Fig. 17.2.9). This catastrophic failure is what makes tempered glass safe in automobile windows. Tempered glass windows are hard to break but once they do break, they crumble into little cubes.

Dicing fracture occurs whenever the body of tempered glass becomes the surface. The body is under severe tensile stress so it will crack and tear whenever possible. As long as it's protected by a shell of compressed glass, the body won't crack. But any penetration into the body will cause the whole sheet to self-destruct. For that reason, tempered glass can't be cut in any way once it has been tempered. Car windows are cut to size before they're tempered because they would crumble if you tried to cut them.

Oven windows and refrigerator shelves are also tempered to give them additional resistance to thermal shocks. By tempering the glass, you make it harder for thermal shocks to put the glass surface under tension and cause it to break.

However, a car's front windshield is not tempered because it would crumble when struck by road debris. Instead, it's made by laminating a plastic sheet between two sheets of annealed glass. This three layered *safety glass* sandwich can tolerate minor breaks without falling apart. The plastic keeps the glass together even if it does break and provides a barrier so that a crack in one sheet of glass can't propagate into the other sheet. *Bullet proof glass* is a natural extension of this idea, with many alternating layers of glass and plastic.

CHECK YOUR UNDERSTANDING #6: Bigger Can Be Better

When soda–lime–silica glass is placed in a liquid containing lots of potassium ions, the large potassium ions can replace sodium ions and cause the glass to swell. Since only the surface layer swells, does this effect strengthen or weaken the glass?

Summary

How Windows and Glass Work: Window glass starts as a mixture of quartz sand, soda ash, and limestone. This mixture is heated to about 1500° C in a melting chamber, where the three compounds merge together into a uniform liquid. Once all the bubbles of gas have risen out of the molten glass, it's cooled quickly to its working temperature of about 850° C and poured onto a pool of liquid tin. There it spreads out into a thin, smooth layer and flows forward as a river of glass. It moves to a cooler portion of the tin pool and its viscosity increases until it's stiff enough to handle without injury to its surface. It passes into an annealing oven where any residual stresses in the glass are relieved and is finally cut into sheets when it reaches room temperature.

To prepare an individual window, a large sheet is usually scratched with a hard tool and then exposed to a carefully placed stress. A crack appears in the glass and propagates along the scratch from one end of the sheet to the other. Once the glass has been cut to the proper size, it can be installed in its frame. However, if the window needs extra strength, then it may be reheated and tempered. To temper the glass, it's heated until it softens and then its surfaces are cooled quickly by blasts of air. This processing puts the window's surfaces under compressive stress so that starting a tear in the glass becomes very difficult. If the tempered window does break, it crumbles into tiny fragments.

The Physics of Windows and Glass

1. *Glasses* are amorphous solids. They don't have crystalline structures and have no long-range structural order. They can't undergo slip so they don't exhibit plastic deformation. They deform elastically in response to stress up until the moment they break.

2. Liquids that have trouble nucleating seed crystals when cooled through their freezing temperatures can become supercooled. These liquids often crystallize suddenly when nucleation finally does occur.

3. Glass or network formers have difficulties both in forming seed crystals and in growing on existing crystals. They supercool as liquids until their increasing viscosities prevent crystal growth altogether. At still lower temperatures, their atoms become essentially immobile and they become *glasses*.

4. The *glass* transition usually occurs when a supercooled liquid's viscosity reaches about 10^{12} Pa·s. At this enormous viscosity, the material is so stiff that it takes minutes or hours to flow in response to stress. At that point, it's virtually indistinguishable from a solid.

5. *Glasses* break when tensile stress separates the atoms and causes a crack to propagate through the material. Bending a piece of glass creates tensile stress on one surface and can cause a defect there to introduce a crack into the glass.

Check Your Understanding - Answers

1. The water will crystallize onto those ice crystals and become solid.

Why: Adding seed crystals to a supercooled liquid will cause the liquid to crystallize. By adding ice to supercooled water, you are by-passing the nucleation stage and going directly to the crystal growth stage. Water crystals grow easily and quickly so the water turns to ice, although it must get rid of heat as it does. But some liquids grow crystals so slowly that adding seed crystals doesn't do much. These liquids form *glasses* when cooled.

2. The blast melted the quartz sand into a liquid, which then cooled quickly enough to form *glassy* or vitreous quartz.

Why: The rapidly cooled liquid quartz formed quartz *glass* because the network of covalent bonds between atoms didn't have time to create the complicated and orderly arrangement of crystalline quartz. Volcanic processes can also form silica-based glasses such as obsidian.

3. A mixture of ice and salt also melts at a temperature below the melting point of either material.

Why: Ice and salt form a eutectic mixture that melts below 0° C (32° F). They produce salt water with a freezing temperature that's several degrees below zero. That's why salt sprinkled on an icy sidewalk causes the ice to melt.

4. The lead crystal wine glass is harder, so it will emit a clear tone when you tap it with your fingernail. It has a higher index of refraction so it will reflect and refract light more strongly than the soda–lime–silica one. And the lead crystal wine glass is more dense.

Why: Lead oxide participates in the molecular network, making lead crystal harder than soda–lime–silica glass. It responds more to electric fields, increasing the glass's index of refraction. And its massive lead atoms make lead crystal more dense.

5. Like window glass, sheet candy is made by solidifying a layer of liquid. While the upper surface of the candy is exposed only to air and remains smooth and glossy, the lower surface forms an impression of the surface on which it rests.

Why: Many candies form *glasses* when they cool and take on the surface structure of whatever they touch when they pass through their *glass* transition temperatures.

6. It strengthens the glass in the same manner as tempering does.

Why: Replacing the sodium ions at the surface of glass with larger potassium ions makes the surface of glass expand. Since the body of the glass remains unchanged, the surface finds itself under compressive stress. Since glass breaks by tearing as the result of tensile stress, it's much harder to break such chemically strengthened glass. This technique is quite important in the glass industry.

Glossary

amorphous solid A solid that is non-crystalline and without long-range order.

covalent bond A chemical bond formed between two atoms when they share a pair of electrons.

eutectic An alloy or solution of two or more materials that has a lower melting temperature than any of the individual materials.

glass A non-crystalline material having no long-range order.

glass transition temperature (T_g) The temperature below which the atoms in a liquid are no longer able to rearrange and the material responds elastically to shear stress.

ionic bond A chemical bond formed when the atoms in a material become oppositely charged ions and attract one another.

long-range order Uniform spatial structure that extends for many nanometers in a solid.

nucleation The formation of initial seed crystals through the random fluctuations in a liquid.

seed crystal An initial crystal on which a liquid can crystallize.

supercooled liquid Existing as a liquid at a temperature below the normal freezing temperature of the material.

temper (glass) To soften glass at high temperature and then cool its surface quickly so as to put that surface under compressive stress.

Review Questions

1. Compare the arrangement of atoms in a salt crystal with the arrangement of atoms in molten salt.

2. Compare the arrangement of atoms in a glass window with the arrangement of atoms in molten glass.

3. Very pure water can be cooled a few degrees below its normal freezing temperature, 0° C (32° F), without solidifying. Explain this effect.

4. As glass cools below its freezing temperature, there's a range of temperatures at which the glass can devitrify—form large crystals. Above or below this temperature range, there's no danger of devitrification. What occurs in this temperature range that makes devitrification possible?

5. When moisture from the air forms frost on a cold window, you can see that it's composed of large crystals that meet at random angles. Explain this structure in terms of crystal nucleation and growth.

6. How do stiff, directional covalent bonds make it easier for materials to form glasses?

7. Why is silicon dioxide so good at forming *glasses*?

8. Why does adding a chemical that forms ionic bonds soften glass?

9. Why is glass cookware made from glasses with small coefficients of volume expansion?

10. Why is tempered glass so hard to break?

11. Why can't glass be cut to size after being tempered?

Exercises

1. Many candies are made by boiling a mixture of sugar and water. When only a little water remains, the mixture is cooled rapidly and forms a clear, uniform material that resembles window glass. If the cooling is done too slowly, the mixture turns into damp sugar granules. How do the sugar molecules arrange themselves in these two cases?

2. When candy is cooled too slowly and turns into damp sugar granules (see Exercise **1**), it becomes translucent rather than clear. Why?

3. When transparent glass is cooled too slowly through its freezing temperature, it becomes translucent or white. What happens inside the glass to cause this change in appearance?

4. Glass-ceramics are inhomogeneous materials consisting of tiny ceramic crystals in glass. Even if none of the materials in a glass-ceramic absorb light, the glass-ceramic appears translucent or white rather than clear. Why?

5. Scientists continue to search for ways of producing amorphous or glassy aluminum metal. This metal can be as light as normal aluminum but as strong as steel. Why is glassy aluminum harder than normal crystalline aluminum?

6. The asphalt used to bind gravel together in roadway paving is an amorphous solid at room temperature. What part does asphalt's glass transition temperature play when a road is being built out of hot asphalt and gravel?

7. Ice cream is stirred during freezing to prevent large ice crystals from forming. Another way to form smooth ice cream is to freeze it very quickly by immersing it in liquid nitrogen. Why does this rapid cooling technique prevent large crystals from forming?

8. While honey can remain a liquid for long periods of time, it eventually crystallizes. You can return it to liquid form by heating it. But you must make sure that all the crystals are gone or it will quickly crystallize again. Why?

9. Streams of bubbles rise to the surface of a mug of soda. These bubbles form when carbon dioxide gas comes out of solution in the soda. But carbon dioxide molecules don't join together often enough on their own to start the bubbles. Instead, bubbles usually form at defects on the inside of the mug and rise upward from those defects. How is the difficulty of bubble formation in soda related to the effects that allow *glasses* to form?

10. People preparing electron microscope samples shave them into incredibly thin sheets with knives made of diamond or glass. While knives made of crystalline diamond remain sharp indefinitely, glass knives must be freshly prepared since they gradually lose their atomic-scale sharpness even without being used. Explain this spontaneous loss of sharpness in a glass knife.

11. There are conflicting opinions about why the glass windows in some ancient European cathedrals are thicker at their bottoms than at their tops. While some people think that this merely reflects the difficulty early glass makers had in producing glass of uniform thickness, others suggest that the glass has slowly shifted downward over the years. Explain why it's possible for glass to shift downward in certain circumstances and why such a downward shift is almost certainly *not* responsible for the thickening observed in the cathedral windows.

12. Taffy doesn't freeze as you cool it—it just gets stiffer and stiffer until it's rigid. How can the taffy become rigid without there being a particular temperature at which it freezes from a liquid to a solid?

13. Because a crystal tends to contain only a single chemical, chemists frequently purify chemicals by crystallizing them. To avoid trapping impurities in a rapidly growing crystal, they use conditions where crystal growth is slow. Under these conditions the crystals often won't start growing on their own and the chemists must add a few impure crystals to start the crystallization. Why won't the crystals start growing on their own?

14. Some high quality stainless steel sinks are coated underneath with a type of rubber that exhibits lots of internal friction. How does this rubber coating affect the sound the sink emits when you drop a spoon in it?

15. Why does an ice cube usually crack when you take it directly from the freezer and drop it into a glass of water?

16. A board makes a simple bridge across a stream. If you overload the board, it will begin to crack. Why will the crack first appear on the bottom surface of the board?

17. Why isn't tempered glass used in the windows of the armored cars used to deliver money to banks?

18. Why do recyclers ask you not to mix glass cookware with glass bottles and jars?

19. A diamond can be reshaped by cleaving it—exposing it to severe shear stress along a particular plane so that it undergoes brittle fracture. If done properly, the fracture occurs along a plane of atoms and is almost perfectly smooth. If you try this technique with a piece of glass, you won't end up with a smooth fracture. Why not?

20. A tempered glass cup is remarkably hard to break. If you drop it onto a tile floor, it may bounce rather than break. But if its bouncing causes it to scrape hard against the sharp edge of a tile, or even a grain of sand, the cup will crumble into tiny pieces. Explain this behavior.

21. A Bologna bottle is a peculiar version of tempered glass. It's a thick glass bulb that was heated until soft and then cooled rapidly from its outside in. The last portion of the bottle to cool through the *glass* transition temperature was its inner surface. While the bottle's outer surface is strong enough to hammer in nails, the bottle's inner surface is extraordinarily fragile. The slightest injury to that surface will tear the bottle apart. Why is the inner surface so fragile?

22. The sprinkler systems that protect public buildings from fire release water when pieces of metal inside them melt. This metal melts at about 74° C (165° F), well below the melting temperature of most metals. How is it possible to make a metal that melts at such a low temperature?

23. The glass used to make lenses for cameras and eyeglasses is cooled extremely slowly after it solidifies. Why would more rapid cooling reduce the glass's optical quality?

24. Glass that's under stress is birefringent, responding differently to horizontally and vertically polarized light. Because of its bifringence, stressed glass can partially or completely convert light of one polarization into the other. Use this information to explain why the tempered rear windows of some cars exhibit colored blotches when you look through them at the blue sky while wearing polarizing sunglasses.

25. A glassblower can shape heated glass in endless ways to create works of art. Why isn't this craft possible with normal solids such as salt?

26. When a glassblower is finished forming a piece of glass artwork, the piece should be allowed to cool slowly in an oven. Why?

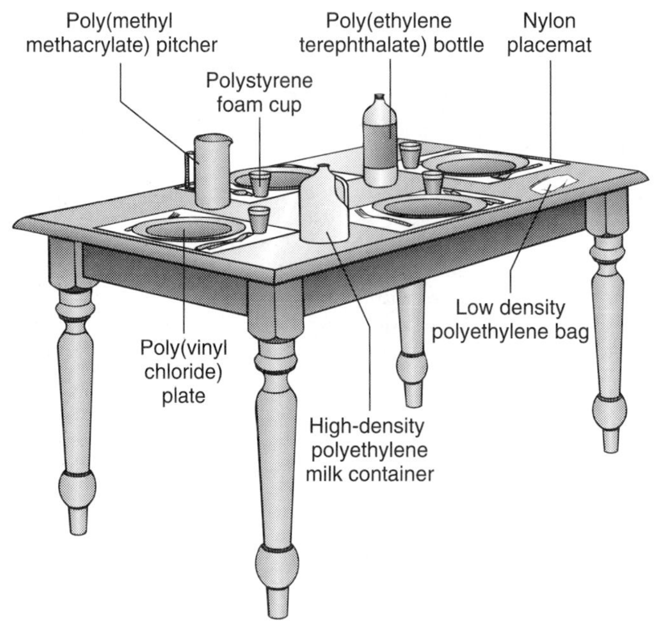

Section 17.3

Plastics

Until the middle of the nineteenth century, virtually all objects of everyday life were made from naturally occurring or naturally derived materials such as wood, glass, metal, paper, wool, and rubber. While this short list of available materials was sufficient for most purposes, there were situations that demanded something new. In 1863, a billiard ball manufacturing company, Phelan and Collander, offered a prize of $10,000 to anyone who could find a substitute for natural ivory in billiard balls. In response to this offer, American printer John W. Hyatt and his brother Isaiah figured out how to form billiard balls out of a recently discovered synthetic chemical called nitrocellulose. By 1871, Hyatt had established two companies to work with this new material, under the name celluloid, *and the plastics industry was born.*

The word plastic *has a wide usage and a broad definition. Most often it refers to materials that, while solid in their final forms, take on liquid or shapable states during earlier phases. Modern plastics can assume many shapes and forms and exhibit a rich variety of physical and chemical properties. Most importantly, the development of plastics has made it possible to design materials that exactly suit their uses.*

Questions to Think About: *Why do some plastics melt when heated while others simply burn? Why are some plastics flexible while others are hard and brittle? Why do clear plastics often turn hazy and white when you bend them too far? What makes Kevlar, the plastic fiber used in bulletproof vests, so strong? Are all plastics recyclable?*

Experiments to Do: *Plastic eating utensils are made of relatively hard, brittle plastics that soften considerably when you put them in boiling water. To see this softening in action, heat a plastic spoon until it melts. Find a safe location and hold the middle of the spoon above a candle flame. Keep the spoon and your fingers out of the flame itself. As the* *spoon heats up, it will initially soften to a leathery consistency. At still higher temperatures, it will be elastic, like a rubber band. And at still higher temperatures the plastic will flow, first like chewing gum and then like syrup.*

These five behaviors are characteristic of plastics, although not all plastics exhibit them clearly. Just which behavior a particular plastic exhibits depends on its composition and temperature. The plastic spoon is hard and brittle at room temperature. Can you find plastic items that exhibit one of the four other behaviors at room temperature?

❐ Polymers were discovered long before anyone understood what they were. It wasn't until 1920 that German chemist Hermann Staudinger (1881–1965) made his macromolecular hypothesis, suggesting that polymers are actually giant molecules formed by the permanent attachment of countless smaller molecules. Through careful experiments, he proved his hypothesis to be correct and was rewarded with the 1953 Nobel Prize in Chemistry.

Polymers

Plastics are based on **polymers**, enormous chain-like molecules containing thousands or even millions of atoms. Like all organic molecules, and the glasses of the previous section, the atoms in a polymer are held together by covalent bonds. But while propane, diesel fuel, and paraffin wax consist of chain-like molecules roughly 3, 16, and 30 carbon atoms long respectively, the chain-like molecules of high density polyethylene (HDPE) are between 1000 and 3000 carbon atoms long (Fig. 17.3.1). Once a molecule is more than about 1000 atoms long, it's considered a polymer. (For a history of the understanding of polymers, see ❐.)

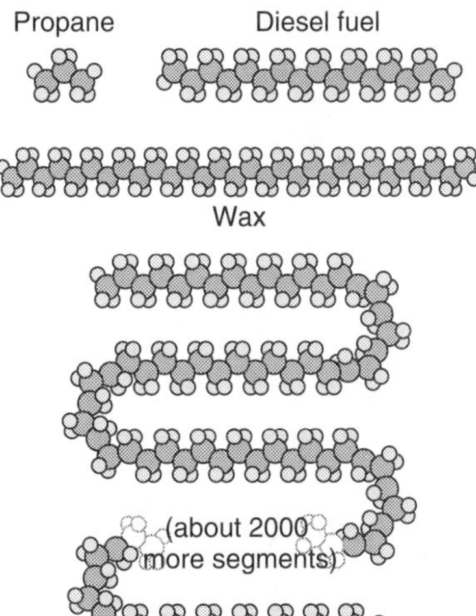

Fig. 17.3.1 - Paraffin chains are found in many materials. Each paraffin chain consists of a backbone of carbon atoms (the larger balls), decorated by pairs of hydrogen atoms (the smaller balls). A chain is terminated at each end by an extra hydrogen atom. Propane, diesel fuel, and wax contain relatively short paraffin chains while high density polyethylene, the plastic used in milk containers, contains paraffin chains of between 1000 and 3000 carbon atoms.

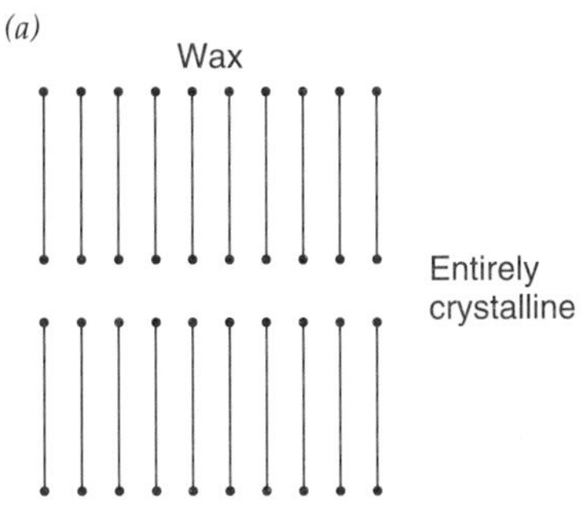

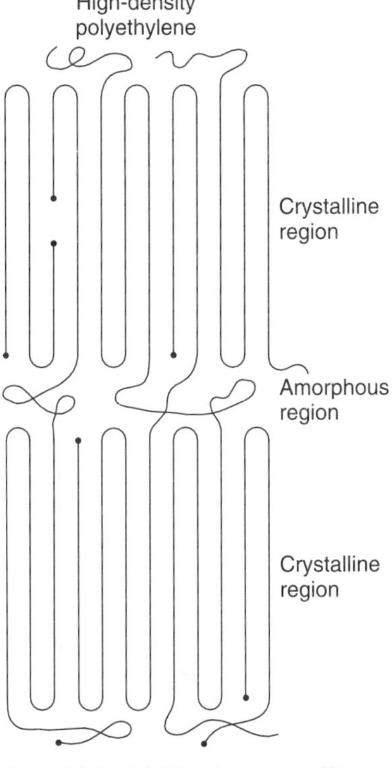

Fig. 17.3.2 - (*a*) Wax is a crystalline material, held together only by van der Waals forces. (*b*) High density polyethylene is about 80% crystalline and 20% amorphous, with entanglements helping to hold it together.

Its simple structure makes HDPE a good starting point for a discussion of polymer structure. It's also a commercially important polymer because it's sturdy, nontoxic, and quite resistant to chemical attack. HDPE is used extensively in bottles and containers and is familiar as the cloudy white plastic of milk jugs.

HDPE looks and feels like wax because the two are closely related chemically. In both materials, the molecules consist of hydrogen atoms attached to chains of carbon atoms. But the great lengths of these chains in HDPE distinguish it from wax in several important ways. First, HDPE has a substantially higher melting temperature than wax, which isn't surprising because its molecules are so much larger. In fact, what *is* surprising is that HDPE melts at only about 140° C. The way in which it melts is complicated and is one of the reasons that polymers are so interesting.

Second, while wax is completely crystalline, HDPE is partly crystalline and partly amorphous (Fig. 17.3.2). In about 80% of the plastic, the molecules are neatly oriented like dry spaghetti in a box. In the other 20%, the molecules are coiled randomly and entangled with one another like cooked spaghetti. These amorphous regions are as unavoidable as the knots that appear in shoulder-length hair. HDPE's mixed structure gives it a non-uniform index of refraction—its denser crystalline regions slow light more than its amorphous regions. This non-uniformity gives solid HDPE a milky haziness that disappears when the plastic melts to become a uniform liquid.

Third, HDPE doesn't crumble like wax when you bend it. The amorphous regions of tangled polymer chains provide the plastic with considerable flexibility. When you stress HDPE, its coiled chains unwind. They can't cross one another where they're entangled but they still allow the plastic to stretch by as much as 50% in length. The material responds elastically to small stresses but undergoes plastic deformation when the stresses are severe. If you push gently on a milk container, it will bounce back, but if you crush it, it will stay crushed.

Finally, HDPE has a much greater tensile strength than wax. Wax molecules are relatively short chains, clinging to one another only with van der Waals forces. **Van der Waals forces** are weak, non-directional intermolecular forces that are created by temporary fluctuations in the positions of electric charges in molecules. All molecules exert van der Waals forces on nearby molecules. Van der Waals forces are so weak that the wax molecules slide across one another easily and the whole crystal can be pulled apart. Although HDPE chains are also bound together only by van der Waals forces, their great lengths and frequent entanglements make the plastic hard to pull apart. You must actually break some of the molecules in order to divide the plastic at room temperature.

While low density polyethylene (LDPE) is also built from long wax-like molecules, these molecules have short branches or side chains. Side chains get in the way of crystallization because they make it hard for the molecules to line up neatly. As a result, LDPE is about half amorphous—heaps of coils and tangles.

With its mechanical properties dominated by the amorphous regions, LDPE is limp and flexible at room temperature. However, it has considerable tensile strength because its entangled chains can't be separated easily. The major uses of LDPE are in trash bags, shopping bags, food and dry goods wrappers, and electric insulation.

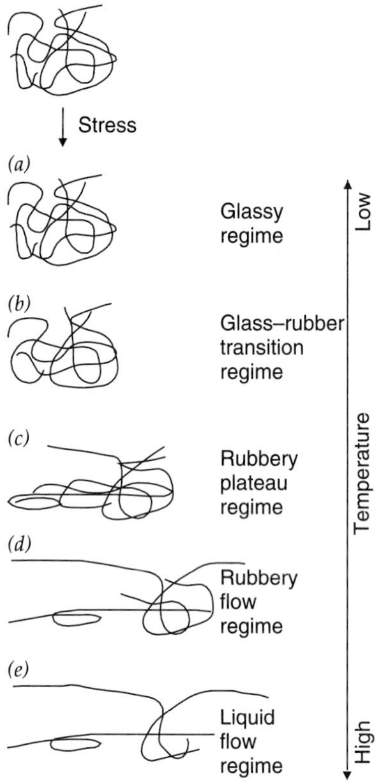

Fig. 17.3.3 - As you heat a polymer, it goes through the five stages shown here from top to bottom. (*a*) When cold, its atoms don't shift during stress so it's glassy. (*b*) When cool, its atoms shift slowly but the chains remain fixed so it's between glassy and rubbery. (*c*) At intermediate temperatures, its atoms and chains move but entanglements prevent them from getting past one another so it's rubbery. (*d*) When warm, the chains begin to untangle so it exhibits rubbery flow. (*e*) When hot, the chains easily untangle and it exhibits liquid flow.

CHECK YOUR UNDERSTANDING #1: All For One or One For All?
How does a polymer chain differ from a chain of many individual diesel fuel molecules?

Polymers and Temperature

At room temperature, low density polyethylene is neither a solid nor a liquid. It bends almost effortlessly so it isn't a normal solid. It doesn't flow so it isn't a normal liquid. Instead, LDPE lies in between solid and liquid in what is called the *rubbery plateau regime*. There are five temperature regimes for most polymers and the rubbery plateau is the middle one. At lower temperatures, there are the *glassy regime* and the *glass–rubber transition regime* and at higher temperatures, there are the *rubbery flow regime* and the *liquid flow regime*.

These five regimes exist because the long polymer chains can't crystallize completely at low temperatures and can't avoid entanglements at high temperatures. With trouble forming either a normal solid or a normal liquid, a polymer's characteristics change rather gradually with temperature.

Within this evolving behavior, polymer scientists have identified the five separate regimes (Fig. 17.3.3). There are no sharp boundaries between these regimes, so a particular polymer moves smoothly from one to the next as it warms up. While the presence of crystalline regions in a polymer complicates this picture, those regions can be safely ignored in polymers that are substantially amorphous. We'll examine the five regimes in LDPE, which is amorphous enough that we won't have to worry about its crystalline regions.

At extremely low temperatures, LDPE is a hard, brittle solid. Because the atoms of LDPE can't move relative to one another when it's below its glass transition temperature (about –128° C), it's a glass (Fig. 17.3.3*a*). Your shopping bag never gets cold enough to reach this **glassy regime**, but there are many common polymers that are glassy at room temperature. One of the most important of these is poly(methyl methacrylate) or Plexiglas.

At about –80° C, the atoms in LDPE have enough thermal energy to move relative to one another. However, the overall chains still can't move (Fig. 17.3.3*b*). LDPE is in the **glass–rubber transition regime** and has a leathery character—stiff but pliable. A familiar polymer that is in this regime at room temperature is poly(vinyl acetate), the principal polymer in most latex interior house paints.

Near room temperature, LDPE is in the **rubbery plateau regime**, with its atoms moving freely and even its chains moving slightly. Because the chains are still too entangled with one another to permit flow (Fig. 17.3.3*c*), the plastic has a rubbery feel—flexible and elastic. The chains pull taut when you stretch it and snap back to normal when you let go. But if you pull too hard on the plastic, the chains break at the entanglements and the plastic stretches irreversibly or tears. These behaviors explain the familiar stretchiness of LDPE trash and shopping bags and the tearing that occurs when you overload them with sharp objects.

At about 100° C, the chains in LDPE are mobile enough to disentangle themselves and the material is in the **rubbery flow regime** (Fig. 17.3.3*d*). The chains disentangle themselves through a process called **reptation**, in which thermal energy causes the chains to slide back and forth along their lengths until they leave entanglements behind them. This motion was first postulated by French physicist Pierre-Gilles de Gennes (1932–) and is named for its similarity to the motions of snakes (Figs. 17.3.4 and 17.3.5). A common polymer that is in this rubbery flow regime just above room temperature is chicle, the main constituent of chewing gum. Silly Putty, a silicon polymer, is also in this regime.

At higher temperatures, the chains in LDPE slide across one another freely and the polymer is in the **liquid flow regime** (Fig. 17.3.3*e*). Reptation occurs so rapidly in this regime that it's hard to distinguish a liquid polymer from any other viscous liquid. One polymer that exhibits liquid flow at room temperatures is poly(dimethyl siloxane), a silicon-based compound. Because its long molecular chains slide across one another so easily, this liquid makes an excellent lubricant.

Even at high temperatures, polymers exhibit unusual flow characteristics as a consequence of reptation (Fig. 17.3.6). Because the chains have to slither large fractions of their lengths to get past each entanglement, polymers with longer chains are more viscous than polymers with shorter chains. Reptation theory accurately predicts that a polymer's viscosity should be proportional to the third power of the average chain length.

It's sometimes possible to shift a polymer's behavior from one regime to another by adding chemicals to it. **Plasticizers** are chemicals that dissolve in a polymer and soften it. They lower its glass transition temperature (T_g) and usually decrease the sizes of any crystalline regions. Plasticizers added to poly(vinyl chloride) or "vinyl" convert it from a glassy solid at room temperature to the leathery material used in some upholstery. In hot weather, you can smell these plasticizers evaporating. When enough of them leave, the aging vinyl returns to its glassy state and becomes susceptible to cracking.

CHECK YOUR UNDERSTANDING #2: On An Unusually Cold Day...

How would the plastics you use behave if the temperature were –100° C?

(a)

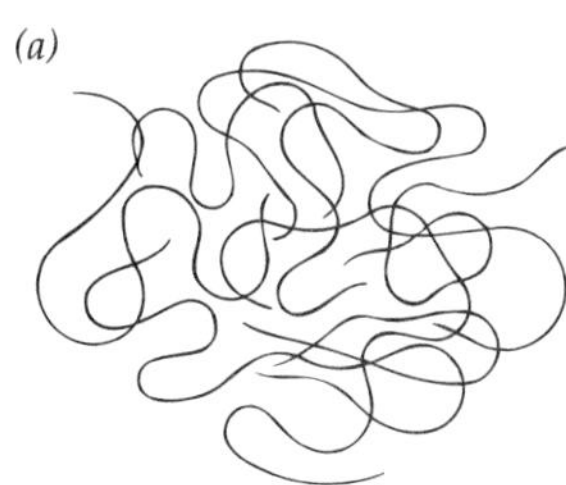

(b)

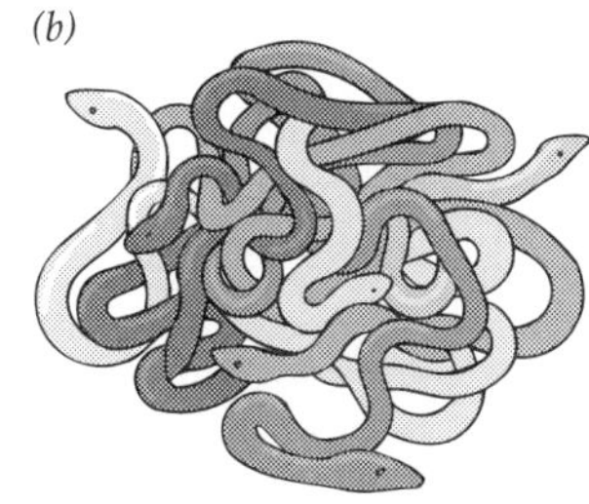

Fig. 17.3.4 - (*a*) Tangled polymer chains can only untangle themselves by moving forward or backward along their lengths. (*b*) Because this motion is reminiscent of snakes, it's called reptation.

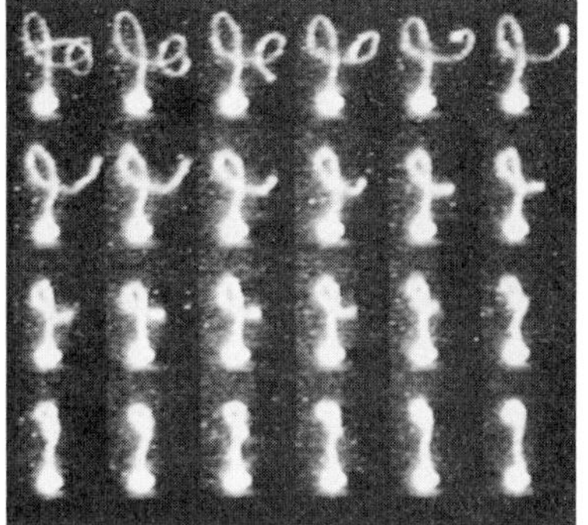

Fig. 17.3.5 - While the theory of reptation correctly predicted many features of polymer behavior, reptation itself wasn't directly observed until 1993, when American physicist Steven Chu first watched it through a microscope. Chu and his colleagues coated a single polymer molecule with fluorescent dye and used a laser beam to drag that molecule through another polymer. The molecular chain snaked its way through entanglements, just as predicted by reptation theory.

Fig. 17.3.6 - When you place a spinning rod into normal liquids, the molecules of liquid are thrown outward and move away from the rod. But in a liquid polymer, entanglements draw the polymer chains toward the rod, creating the Weissenberg effect.

Thermoplastics and Thermosets

Because the molecules in HDPE and LDPE can move independently at high temperatures, these polymers behave as liquids when heated. They are both **thermoplastics**, a class of plastics that can be reshaped at high temperatures. While some thermoplastics burn or char before they melt, even those materials can usually be dissolved in solvents, reshaped, and allowed to dry.

But not all polymers can be reshaped. Some polymers have chemical cross-links between their chains that prevent them from reptating. These **thermosets** can't flow. They won't melt when you heat them and they won't dissolve in solvents unless those solvents destroy their molecular structures.

Because they can't be reshaped, thermosets must be produced in their final forms. Many thermosets begin as thermoplastics that are then cross-linked, a process generally called **vulcanization**. In other contexts, this cross-linking is known as tanning (leather), curing (resins), and drying (oil paints). Many important polymers are cross-linked, as we'll soon see.

Most polymer molecules are built by tacking together much smaller molecules. These small molecules are referred to as **monomers** and the tacking together process is called **polymerization**. The final chains are then named after their monomers (e.g. polyethylene from an ethylene monomer, and poly(methyl methacrylate), from a methyl methacrylate monomer).

In many cases, a single monomer is used over and over again to form a *homopolymer*. Representing that monomer as the letter "A", the finished homopolymer looks like "AAAAAAAAAA...". However, there are also cases in which several different monomers are incorporated in the same chain. Such chains are called *copolymers*. Representing two monomers by the letters "A" and "B", the finished copolymer might be "ABABABABAB...".

Distinguishing between homopolymers and copolymers is particularly important when the various monomers in a copolymer come together in random order. Then the finished copolymer might be "ABAABBBABAABBAAA...". The resulting molecular chains are different from one another and are unlikely to form crystals. Most of these *statistical copolymers* are amorphous.

Even homopolymers can be disordered if their monomers randomly adopt different orientations during polymerization. If a monomer can enter the chain as either "A" or "∀" then the homopolymer might be "∀AA∀A∀AA∀∀...". These *atactic homopolymers* are amorphous.

❐ Most animals can't digest cellulose because they're unable to break its molecular chains into individual glucose molecules. Fortunately, bacteria and protozoans produce a catalytic enzyme that breaks up cellulose so that dead trees and plants decompose quickly. Cows and other ruminants carry bacteria and protozoans in their stomachs and let those tiny animals convert cellulose into small digestible molecules.

CHECK YOUR UNDERSTANDING #3: A Really Bad Hair Day
Hair is actually a thermoset composed of long protein chains that are cross-linked by sulfur atoms. At what temperature does hair melt?

Natural Polymers: Cellulose and Natural Rubber

Several of the most important polymers occur naturally, most notably cellulose and natural rubber. Cellulose is the principal structural fiber in wood and plants and by far the most abundant polymer on earth. We use it as a building material, in making paper, clothing, rope, and chemicals, and as a fuel. Cellulose molecules are made of polymerized glucose sugar molecules (Fig. 17.3.7). These chains are unbranched and orderly so cellulose crystallizes easily. Because natural cellulose is about 70% crystalline and 30% amorphous, it appears translucent or white. But while it's built from sugar molecules, we can't digest it (see ❐).

The oxygen and hydrogen atoms in adjacent cellulose molecules form strong hydrogen bonds and bind the chains together so tightly that overheated cellulose decomposes instead of melting. That's why wood burns rather than turning into liquid in your fireplace. Cellulose's glass transition temperature is also quite high so wood is rigid. Cellulose is only flexible as extremely thin fibers, such as those in cotton. Cotton is nearly pure cellulose.

Since cellulose chains also form hydrogen bonds with water molecules, cellulose eagerly soaks up water. The water dissolves in cellulose, forming a solid solution. The fibers swell and soften as water enters them because water acts as a plasticizer. When the water dries up, the swollen fibers become hard and stiff again. These changes lead to the shrinkage that occurs when you launder cotton clothes in water. This shrinkage can be reduced by lubricating the fibers with fabric softeners. (For more on solid solutions in plastics, see ❐.)

Natural rubber forms when a chemical called isoprene polymerizes into long chains of *cis*-polyisoprene (Fig. 17.3.8). The prefix "*cis*-" indicates that the isoprene monomers alternate up and down, as shown in the figure. Natural rubber is extracted from the sap of several tropical trees, where it forms a *latex* of tiny polymer particles suspended in a watery liquid. As the sap dries, its re-

❐ Oxygen permeable contact lenses swell when wet, allowing oxygen molecules to dissolve in the plastic and diffuse to the surface of the eye. The plastic bottles used for carbonated beverages similarly dissolve water, oxygen, and carbon dioxide molecules. Because carbon dioxide and water slowly diffuse out of a bottle while oxygen diffuses in, the beverage it contains eventually goes flat.

Cellulose

Fig. 17.3.7 - Part of a cellulose molecule, showing five repetitions of the glucose monomeric units from which it's built. Each carbon atom (C) is bound to four other atoms, including oxygen atoms (O) and hydrogen atoms (H). The covalent bonds between the atoms are indicated by lines. Cellulose is a stiff, mostly crystalline solid at room temperature and is the main structural material in trees and plants.

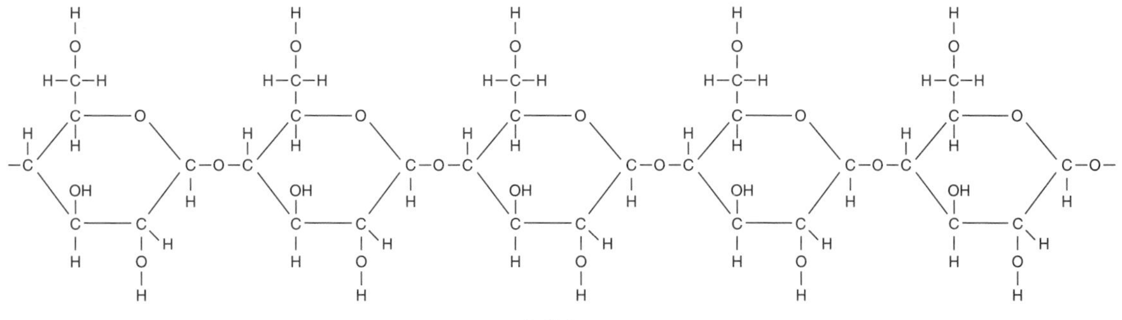

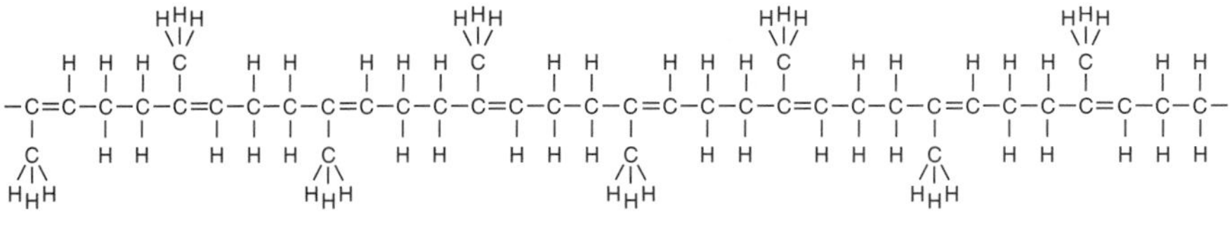

Fig. 17.3.8 - Part of a natural rubber molecule showing eight repetitions of the basic isoprene monomeric units from which it's built. The orientations of the isoprene units alternate up and down. The chains coil up randomly on their own but straighten when you stretch the rubber. The rubber reaches its elastic limit when the chains are straight.

Natural rubber

maining water pulls the polymer particles together and the rubber coagulates into an elastic solid.

Without oxygen atoms, the molecules in rubber can't form hydrogen bonds. They stick together so weakly that rubber's melting temperature is only 28° C and its glass transition temperature is a frigid –70° C. However, rubber rarely crystallizes at any temperature because its molecules naturally wind themselves up into random coils. These coils are what give rubber its wonderful elasticity. As you stretch a piece of rubber, its polymer molecules straighten out. When you let go, those molecules return to their randomly coiled shapes.

This coiling is an example of nature's tendency to maximize randomness and entropy. A straight rubber molecule is orderly. Once you stop pulling on it, thermal motions in its atoms give it a bend here and a twist there, and soon the molecule is wound into a jumble of random coils. While thermal motions could one day straighten that molecule back out, such an event is quite unlikely.

Rubber first came into use in 1820, when Englishman Thomas Hancock constructed the first rubber factory. Three years later, Scottish chemist Charles Macintosh began using rubber to waterproof fabrics. But natural rubber is sensitive to temperature and therefore hard to use. Moderate or cold weather puts rubber in the rubbery plateau regime, where it's a firm, elastic material. But hot weather takes it to the rubbery flow regime, where it's as gooey as glue.

The famous accident of the American inventor Charles Goodyear (1800–1860) changed all that. In 1939, after ten years of trying to prevent rubber from softening at high temperatures, Goodyear accidentally dropped a mixture of rubber and sulfur onto a hot stove. The material that formed was exactly what he had been looking for. It was stiff but elastic and remained that way regardless of temperature. He called this process *vulcanization*, after the Roman god of fire.

What Goodyear had done was to form cross-links between individual molecular chains in the rubber (Fig. 17.3.9). He had turned natural rubber, which is a thermoplastic, into vulcanized rubber, a thermoset. Since the chains in the vulcanized rubber were interconnected, they couldn't flow and the rubber couldn't melt—it was a single giant molecule.

Vulcanized rubber exhibits the elastic behavior of the rubbery plateau regime over a broad range of temperatures. As long as there aren't too many cross-

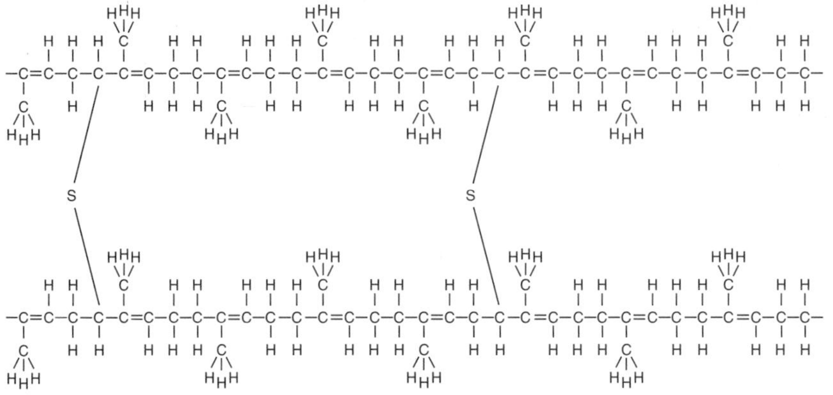

Fig. 17.3.9 - Vulcanized rubber is formed by heating a mixture of natural rubber and sulfur. The sulfur links individual rubber chains to form a thermoset. Vulcanized rubber no longer melts the way natural rubber does. The more sulfur that's added to the natural rubber, the harder the vulcanized rubber becomes.

Vulcanized natural rubber

links between chains, the random coils can still wind and unwind, and the vulcanized rubber still stretches when you pull on it. But the more sulfur you add to the mixture, the more cross-links are formed and the stiffer the rubber becomes. Sulfur content thus controls the stiffness of vulcanized rubber.

Because natural rubber is in limited supply, most modern rubber is synthesized from petroleum. While many of these synthetic rubbers or *elastomers* are chemically different from vulcanized natural rubber, they all have random coils in their polymer chains and cross-links between those chains. Their various molecular structures give them new and useful properties such as resistance to chemical attack and better stability at high temperatures. For a discussion of silicon rubber, see ❐, and for a note about another natural polymer, DNA, see ❐.

❐ Poly(dimethyl siloxane) is a particularly important silicon polymer, with an extremely low glass transition temperature and thus a low viscosity near room temperature. Silicone rubber sealants are based on this polymer, but with an added twist—the ends of each polymer chain are treated so that they cross-link in the presence of moisture. After you apply the liquid sealant, it undergoes room temperature vulcanization (RTV) in air to form a soft silicon rubber. This process releases acetic acid, the main ingredient in vinegar, so you can smell that the rubber is vulcanizing.

CHECK YOUR UNDERSTANDING #4: Wagon Wheels and Tires Don't Melt

Neither cellulose nor vulcanized rubber can melt but for different reasons. What are those reasons?

Synthetic Polymers: Celluloid, Plexiglas, Nylon, and Teflon

❐ Genetic information is encoded in a naturally occurring copolymer called deoxyribonucleic acid or DNA, where it's contained in the exact sequence of monomers. Medical, biological, and forensic scientists can now produce this copolymer, using the polymerase chain reaction (PCR). This technique selectively polymerizes DNA monomers (nucleic acids) to copy portions of a particular DNA chain. PCR can locate particular patterns in the DNA and is coming into frequent use in trials as a genetic equivalent of fingerprinting.

When Phelan and Collander offered $10,000 to anyone who could find a replacement for ivory in billiard balls, neither cellulose nor rubber was up to the task. Instead, the Hyatts used a chemically modified cellulose called nitrocellulose (Fig. 17.3.10) to produce the first practical synthetic plastic: celluloid.

In contrast to cellulose, which can't be reshaped, chemically softened or plasticized celluloid can be molded fairly easily. The objects that were produced with celluloid resemble those we use today, including transparent plastic films, combs, toys, and synthetic silk-like cloth. However celluloid darkens with long exposure to light and is extremely flammable. Nitrocellulose itself is a high-explosive and the main constituent of smokeless gunpowder.

Instead of nitrating cellulose, chemists found that they could attach other useful chemical groups to cellulose to form useful and less dangerous plastics. Cellulose acetate was the first important alternative to nitrocellulose. Referred to as "acetate" or "triacetate", cellulose acetate continues to be used in many forms. Chemists also found that they could take cellulose through a series of chemical transformations that finally ended where it began, with regenerated cellulose. This chemical process allowed cellulose to be reshaped into fibers or films. Rayon and cellophane are both examples of regenerated cellulose.

Plexiglas and Lucite are made from the thermoplastic poly(methyl meth-

Fig. 17.3.10 - Nitrocellulose is formed in a chemical reaction between cellulose and an acid mixture. Unlike cellulose, nitrocellulose can be shaped and was the first synthetic plastic. However, nitrocellulose is both flammable and explosive.

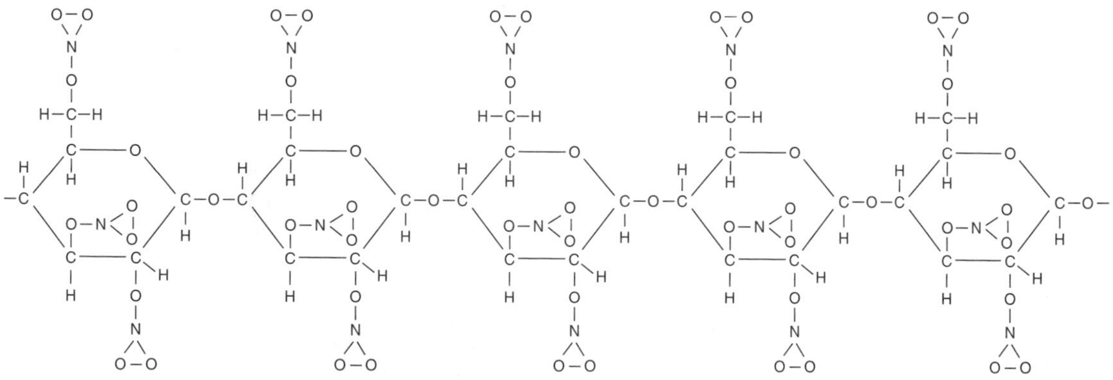

Nitrocellulose or celluloid

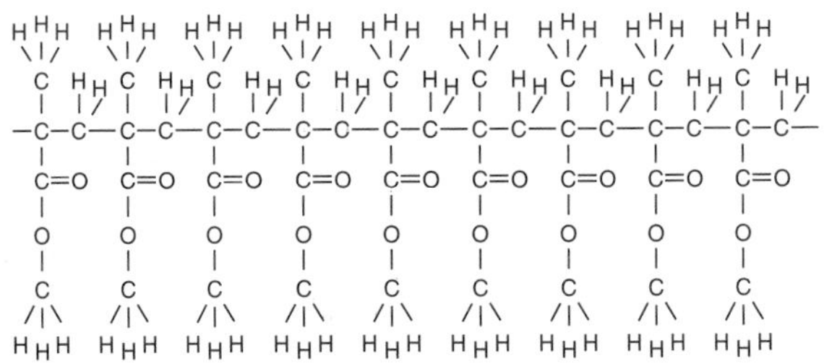

Fig. 17.3.11 - Plexiglas is long chains of poly(methyl methacrylate). These chains bind relatively strongly to one another to form a glassy, transparent material at room temperature.

acrylate) (Fig. 17.3.11). Like cellulose, poly(methyl methacrylate) contains oxygen atoms and forms hydrogen bonds between chains. These bonds are so strong that Plexiglas is a glassy solid below about 105° C. It's hard, clear, and durable.

Instead of cracking like glass when you bend it too far, Plexiglas *crazes*. Voids appear on the stretched plastic surface but they're prevented from separating completely by taut polymer chains that span the open fissures. Because light's speed changes as it passes in and out of the voids, it reflects randomly from the crazed plastic and gives the plastic a hazy, whitish look. However, the Plexiglas doesn't break. This impact resistance makes Plexiglas an important alternative to glass.

Plexiglas has a number of close relatives that are also useful. Poly(methyl acrylate) is more flexible and elastic than Plexiglas and is the principal polymer in acrylic paints. Methyl cyanoacrylate is a monomer that polymerizes rapidly in the presence of moisture to form a hard, glassy plastic. It's the active ingredient in Superglue and Crazy Glue.

Nylon was developed at Du Pont in 1931 by Wallace Carothers ❐ and Julian Hill. Their work was based on the chemical reaction that occurs when an organic acid group (of atoms) on one molecule encounters an organic base group on another molecule. This reaction binds the two molecules together while releasing a water molecule. Carothers and Hill realized that hydrocarbon molecules with these reactive groups at both ends would tend to polymerize into long chains. To test their idea, they began producing double-ended organic acids and double-ended organic bases. Their work was a success. The acids and bases bound together to form chains of alternating monomers (Fig. 17.3.12). This strong, tough, elastic, and chemically inert copolymer was named nylon.

❐ At 31, American chemist Wallace Hume Carothers (1896–1937) left an instructor's position at Harvard University to head an extraordinarily successful research group at Du Pont. Not only did his group invent nylon, but they also prepared neoprene, one of the most important synthetic rubbers. His pioneering work even included polyesters, although his early fabrics melted during ironing. But despite his scientific successes, he was plagued by terrible bouts of depression. Following the sudden death of his sister and three weeks after applying for a patent on nylon, Carothers took his own life.

Because the acid and base groups in the chains allow them to hydrogen bond to one another, nylon's properties depend on the lengths of its monomer molecules. The shorter the monomer molecules, the more acid and base groups are incorporated into the nylon and the stiffer and harder the nylon becomes. Nylons are named according to the lengths of their base and acid monomers. Nylon–6,6 is made from a six carbon-atom long base and a six carbon-atom long acid and is stiffer than nylon–6,12, which uses a twelve carbon-atom long acid.

Another plastic that's extremely resistant to chemical attack is polytetrafluoroethylene, known as Teflon or PTFE (Fig. 17.3.13). Teflon's polymer chains resemble those in polyethylene except that its carbon chains are surrounded by fluorine atoms rather than hydrogen atoms. Because fluorine atoms are slightly larger than hydrogen atoms, they fully enclose the carbon backbones of the Teflon polymer chains. With its chains sheathed in fluorine atoms, Teflon is one of the most chemically inert materials in existence. It's also slippery and makes an

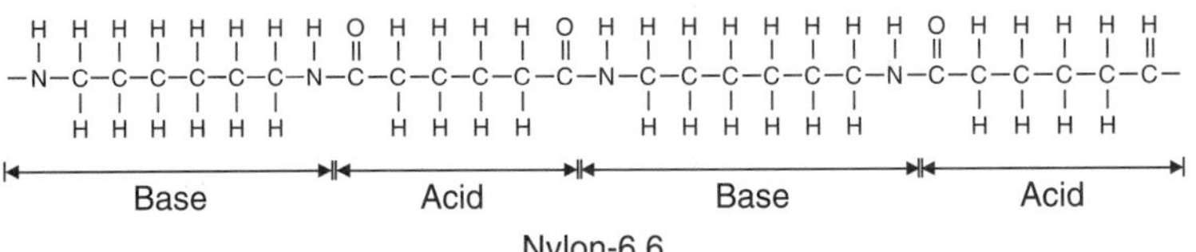

Fig. 17.3.12 - Nylon forms when a double-ended acid reacts with a double-ended base. The reaction permanently binds the acid to the base and they become alternating segments in long copolymer chains. This particular nylon is called 6,6 because the double-ended base monomer contributes 6 carbon atoms to the copolymer chain (the first 6) and so does the double-ended acid (the second 6).

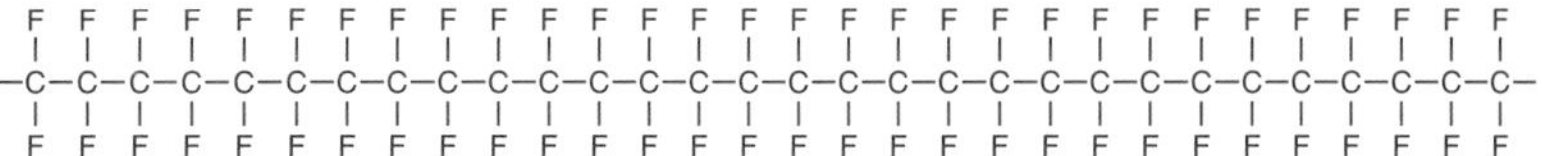
Polytetrafluoroethylene or Teflon

Fig. 17.3.13 - Polytetrafluoroethylene or Teflon is composed of enormous carbon chains, completely surrounded by fluorine atoms (F). The fluorine atoms make Teflon harder and even more resistant to chemical attack than its hydrogen-based relative, polyethylene.

excellent nonstick surface for cookware and laboratory equipment.

Unfortunately, Teflon pulls apart easily unless its molecules are at least 100,000 carbon atoms long. This length makes reptation so slow that Teflon is almost incapable of flowing. Even at its melting temperature of 330° C, Teflon is so viscous that it appears solid. That's why Teflon objects are formed by filling a mold with powdered Teflon and heating it until it fuses together.

CHECK YOUR UNDERSTANDING #5: Hard to Get Hold Of

Why is Teflon more resistant to chemical attack than nylon?

Liquid Crystal Polymers: Kevlar

Not all polymers form tangles of random coils, even at high temperatures. Some polymers are stiff, rod-like molecules that line up with one another, even as they reptate forward and backward. As liquids, these polymers are liquid crystals and as solids, they are remarkably strong fibrous materials.

One of these liquid crystal polymers is poly(*p*-phenylene terephthalamide), also called PPD-T or Kevlar. This copolymer resembles nylon except that nylon's hydrocarbon chains are replaced by aromatic rings (Fig. 17.3.14). Because these rings don't bend, the copolymer chains are rigid and straight.

At room temperature, the molecules of Kevlar form long, uniform crystals that give it an enormous tensile strength. Because the molecules are already straight, they don't have to uncoil or disentangle themselves when the Kevlar is subjected to tensile stress along the direction of the chains. Instead, they all oppose the stress together. To break a Kevlar fiber, you must break all of its molecules at once. Because Kevlar and other liquid crystal polymers stretch very little and are extremely hard to break, they are used in bulletproof vests, sails, parachutes, and ropes, and as reinforment in ultrahigh-strength composites.

Even ordinary polymers can be strengthened by straightening their polymer chains, a step that often occurs when a fiber is spun. Spinning is done by squirting a liquid out of a fine nozzle and allowing it to solidify. Some polymers, such as nylon, are spun from hot liquids and solidify by cooling. Others, such as acrylics, are spun from solutions and solidify as the solvent dries. Still others, such as rayon, are spun from a solution that reacts with a chemical outside the spinneret to produce the final polymer fiber. In each case, the synthetic fiber is drawn out of the nozzle under tension. This tension stretches the fiber along its length and unwinds many of its polymer chains. When the fiber solidifies, it retains this modified structure and is stronger as a result.

CHECK YOUR UNDERSTANDING #6: You Can't Be Good at Everything

While a thin Kevlar fiber has an enormous tensile strength, a large block of solid Kevlar exhibits a similar tensile strength only along one direction. Explain.

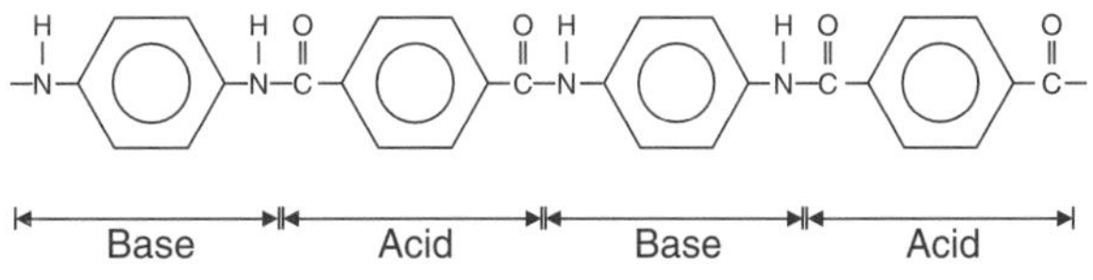

Base Acid Base Acid
Kevlar

Fig. 17.3.14 - Kevlar resembles nylon, except that the flexible hydrocarbon chains of nylon are replaced by rigid aromatic rings. Each chain is stiff and straight, so the polymer forms a liquid crystal at high temperatures. When liquid Kevlar cools and solidifies, it becomes a fibrous material with extraordinary tensile strength.

Summary

How Plastics Work: Plastics are composed of polymers, long chain-like molecules that are thousands or even millions of atoms long. Each polymer is built out of small units or monomers that are connected to one another with covalent bonds to form a strong, continuous structure. Because of their long molecules, plastics don't behave like ordinary solids or liquids. At lower temperatures, they form amorphous or partly amorphous solids and at higher temperatures, they form extremely viscous liquids. In between, they are usually elastic and rubbery.

The character of a particular plastic depends on the nature of its polymer molecules. Chains that are orderly and repetitive often form crystalline solids while those that have some randomness usually remain amorphous. The more strongly the chains cling to one another, the stiffer the plastic is at room temperature. Some plastics are rigid solids at room temperature while others are liquids. Thermosets, in which the polymer chains are cross-linked, can't flow and remain elastic solids even at fairly high temperatures. They are effectively giant molecules.

The Physics of Plastics

1. Polymers are such long molecules that they generally become entangled with one another and produce at least partly amorphous solids. Disorder in a polymer's chains further reduces its crystallinity. These amorphous regions give plastics considerable elasticity at or above their glass transition temperatures.

2. At higher temperatures, the polymer molecules try to flow but are constrained by their entanglements. They free themselves from these entanglements by slithering backward and forward with thermal energy, a process called reptation. The shorter the chains and the higher the temperature, the more effective reptation becomes and the lower a polymer's viscosity.

3. Entanglements produce five principal regimes of behavior in the amorphous regions of a thermoplastic polymer: (1) glassy, (2) leathery, (3) rubbery, (4) rubbery flow, and (5) liquid flow.

4. Thermosets are prevented from flowing by cross-links between polymer chains. These materials remain in the glassy, leathery, or rubbery plateau regimes, regardless of temperature.

5. Liquid crystalline polymers form orientationally ordered liquids. These polymers solidify as highly ordered materials that have extremely high tensile strengths.

Check Your Understanding - Answers

1. The atoms in a polymer chain are bound to one another by strong covalent bonds while the individual oil molecules are bound to one another by much weaker van der Waals forces.

Why: A polymer chain is strong because it's held together exclusively by strong covalent bonds. To pull a polymer chain apart, you must break one of these covalent bonds, a difficult task. In contrast, a chain of oil molecules can be pulled apart easily because there are only weak bonds between the molecules.

2. Hard plastics such as Plexiglas would be unaffected, but soft ones such as rubber would become a whole lot stiffer. Rubber bands, among other things, would break rather than stretch.

Why: Many of the behaviors we expect from plastics depend on the ambient temperature. Changing that temperature can shift a plastic from one behavioral regime to another. For example, 200° C weather would convert many synthetic clothes into liquids.

3. It can't melt because cross-links prevent it from flowing.

Why: Like all thermosets, hairs are giant molecules and can't melt or flow. The only way to permanently change a hair's shape is to break the cross-links. That's what a "permanent" does. Chemicals applied to the hair sever the cross-links, converting the hair to a thermoplastic so that it can be reshaped. Once the new shape is reached, more chemicals reestablish the cross-links and the hair is again a thermoset.

4. The polymer chains in cellulose are held together so tightly that, although it's a thermoplastic, cellulose decomposes before it reaches its melting temperature. Vulcanized rubber is a thermoset and can't melt.

Why: Different thermoplastics go through the various behavioral regimes at different temperatures. Cellulose remains a partly crystalline solid to very high temperatures because of the strong hydrogen bonds between its chains. Sugar is hard to melt without scorching so a polymerized sugar such as cellulose has even more trouble. In contrast, rubber melts easily but vulcanized rubber, a thermoset, can't melt.

5. Teflon polymer chains are armored with fluorine atoms so that no chemicals can attack the carbon atoms. The carbon atoms in nylon's polymer chains are more open to chemical attack.

Why: While Teflon chains are sheathed in a protective layer of fluorine atoms, nylon, like most other polymers, isn't protected in that manner. Moreover nylon can be attacked by chemicals that react with the acid and base groups along its chains. Nylon is particularly vulnerable to attack by acids and even boiling water slowly breaks up the polymer chains.

6. The chains in Kevlar align in one direction when it's a liquid and retain this orientation when that liquid solidifies. If you pull on the solid at right angles to this special direction, it will be no stronger than an ordinary plastic.

Why: Kevlar and the other liquid crystalline polymers are strong along the direction of their aligned chains. When you pull on the material in that direction, you stress the chains themselves. But if you pull on the plastic across the chains, they separate relatively easily. To give it strength in many directions, Kevlar is often made into woven fabrics.

Glossary

glass–rubber transition regime A regime in which the atoms of an amorphous polymer can move enough to make it pliable and leathery.

glassy regime A regime in which the atoms of an amorphous polymer can't move about, making it rigid and hard.

liquid flow regime A regime in which the chains of an amorphous polymer reptate quickly enough to support liquid flow. The polymer is a viscous liquid.

monomer A small molecule from which a polymer forms.

plasticizer A chemical additive that lowers a polymer's glass transition temperature and reduces its crystallinity.

polymer A giant molecule that forms when many monomer molecules bind together in a permanent chain. To be called a polymer, the chain must be about a thousand or more atoms in length.

polymerization The process in which individual monomer molecules are chemically bound together to form a polymer chain.

reptation A thermally induced slithering motion of polymer chains, backward and forward, that allows them to get past entanglements in order to flow. It's named for its similarity to the motion of snakes.

rubbery flow regime A regime in which the chains of an amorphous polymer can reptate enough to flow very slowly past one another. The polymer is elastic and flows very slowly.

rubbery plateau regime A regime in which the individual atoms of an amorphous polymer can move freely but the chains can't flow past one another because of entanglements. The polymer is soft and elastic.

thermoplastic A polymer composed only of individual chains. A thermoplastic can melt and flow.

thermoset A polymer in which the individual chains have been cross-linked with covalent bonds so that it's one giant molecule. A thermoset can't melt or flow.

van der Waals forces Weak, non-directional intermolecular forces that are created by temporary fluctuations in the positions of electric charges in molecules.

vulcanization A process in which the individual chains of a thermoplastic are cross-linked to produce a thermoset.

Review Questions

1. Except at very low temperatures, polymers are less brittle than crystalline solids. Explain their resistance to fracture.

2. What role do entanglements play in establishing the rubbery plateau regime of a thermoplastic polymer?

3. Polymers in the rubbery flow regime can become longer and longer if you stretch them slowly, but they respond elastically to sudden intense stress. Why?

4. List three cases in which water acts as a plasticizer for a natural polymer, making that polymer softer.

5. Why is a bicycle tire essentially one giant molecule?

6. Natural rubber is firm and elastic in cold weather. Why does it become sticky in warm weather?

7. Why don't logs melt when you put them in a fire?

8. Why is Teflon so chemically inert?

9. Why does rubber stretch so much more than Kevlar?

Exercises

1. Styrofoam coffee cups are made from a frothy mixture of polystyrene and gas. Which of the five regimes is polystyrene in at room temperature? at the temperature of hot coffee?

2. Polystyrene is a clear plastic. Why is Styrofoam white?

3. Starch is a digestible natural polymer that's very similar to cellulose. If you mix starch with a little water you'll find that it behaves very strangely. If you stir it slowly it will behave as a liquid, but if you poke it suddenly with your finger it will feel very hard. Explain this behavior in terms of entanglements and reptation.

4. Dental floss is a strong polymer thread that's used to clean between teeth. Is this polymer above or below its glass transition temperature (T_g) at room temperature and how would the floss behave if it were on the other side of T_g?

5. When you inflate a balloon and then let the air out of it, it returns to its original shape. When you inflate a piece of bubble gum and then let the air out of it, it just sags. Explain these different behaviors.

6. Some cheap plastic measuring cups are made out of amorphous polystyrene, which has a glass transition temperature of about 100° C (212° F). What would happen to such a cup if you put 120° C (248° F) cooking oil into it?

7. A rubber cooking spatula will scorch if you touch it to a very hot pan, but it will not become soft or sticky. Why not?

8. A glue gun heats a rod of leathery thermoplastic and squeezes the softened plastic onto the pieces being glued together. Once the glue cools, it bonds the pieces together tightly. Describe how this process works in terms of the five regimes of polymer behavior.

9. A good way to keep the end of a nylon rope from unraveling is to melt it together. Why doesn't that technique work for an elastic cord made from strands of vulcanized rubber?

10. A Splat ball is made from an extremely soft silicon polymer. If you throw it against a wall it spreads out into a pancake as though it were liquid, but it gradually pulls itself back together into a sphere. What regime is the polymer in and is it a thermoplastic or a thermoset?

11. Why does vinyl upholstery tend to crack in very cold weather?

12. If you apply latex interior house paint to a damp basement wall, the moisture in that wall will still be able to escape into the air. How does it get through the paint?

13. Some drugs are administered in sealed plastic patches that are applied to the skin. How do the drug molecules get through one of these plastic patch?

14. The more mobile the atoms of a polymer are, the more easily gas molecules can diffuse through that polymer. Use this information to explain why balloons made of rubber (an elastic thermoset) lose helium faster than those made of Mylar (an amorphous thermoplastic with a glass transition temperature well above room temperature).

15. The plastic cup covers provided by fast food restaurants often have buttons you can push to indicate a cup's contents. When you push the button, you bend the glassy plastic and it turns white. Why?

16. Your fingernails are made of a glassy semitransparent polymer called keratin. If you bend your nail too far, it turns white. What causes this sudden color change?

17. The adhesive on the back of a postage stamp is a polymer that softens when you lick it and hardens when it dries out. What is the role of water in changing the behavior of this polymer?

18. Hair is made from a protein polymer that dissolves a moderate amount of water. Why is it so much easier to shave wet hair than dry hair?

19. Wet hair is much limper than dry hair. Hair is made from a protein polymer that dissolves a moderate amount of water. Why is wet hair relatively limp?

20. Woods such as ash can be softened and bent by exposing them to hot, wet steam. Why is the cellulose in ash wood normally rigid and why does heat and moisture make it relatively flexible?

21. It's important to wet a paper tissue before cleaning plastic eyeglass lenses since wet paper is softer than dry paper. Paper is made from cellulose. Why does water soften paper?

22. Poly(vinyl chloride) is a glassy plastic at room temperature, so plasticizers are added to it to produce the soft, leathery vinyl used in notebook covers. If you leave one of these notebooks pressed against a xerographic copy, lettering from the copy will stick to the notebook. Explain.

23. Pure concrete is brittle. But when Kevlar fibers are added to the concrete, it becomes much stronger. How do the Kevlar fibers strengthen the concrete?

24. Pure vulcanized rubber is both flexible and elastic. To keep some rubber tires from stretching when they are inflated, belts of Kevlar fibers are built into them. Why don't the Kevlar belts simply stretch along with the rubber?

25. Silicon rubber flying disks are great for dogs because they're nearly indestructible. They're made from a silicon thermoset that has a glass transition temperature of about –100° C (–148° F). A normal Frisbee is a thermoplastic with a glass transition temperature above room temperature. Why is the silicon rubber disk so much less brittle than a normal Frisbee?

26. Rayon is a form of cellulose fiber. Why do clothes made of rayon shrink when you launder them in water?

27. Spectra is a polyethylene fiber in which all of the polyethylene molecules have been drawn out so that they are almost perfectly straight. Why is cord made from this material so much stronger than normal polyethylene cord?

Epilogue for Chapter 17

In this chapter, we examined three areas of materials science. In *knives and steel*, we saw how solids respond to stresses with strains and how those strains can involve both elastic and plastic deformations. We examined slip in crystals and learned that controlling slip is critical to producing the hard metals needed for knives. We also explored the value of producing composite systems—an extremely hard cutting edge bonded to a tough, elastic knife blade.

In *windows and glass*, we learned that *glasses* are amorphous solids resembling frozen liquids. We saw how glasses are formed and why their amorphous natures prevent them from undergoing plastic deformations. We investigated the processes used to make glass windows and studied the differences that come with different chemical compositions and heat treatments.

In *plastics*, we saw why their extremely long molecules give plastics unusual properties. We examined the five behavioral regimes of thermoplastics and explained how temperature and chemical composition can move a plastic from one regime to another. We also looked at thermosets to see why they can't flow and why they remain in the three colder behavioral regimes.

Explanation: Glue Putty

White glue is a thermoplastic containing many independent polymer molecules. These molecules are plasticized by a liquid (mostly water) to form a thick fluid that's in the liquid flow regime. As the plasticizing liquid evaporates, the glue becomes thicker and thicker, passing through the various regimes on its way to the glassy state. However, when you add borax to white glue, you create cross-links between the independent polymer molecules. In effect, you vulcanize the glue so that it becomes a thermoset. Unable to slide freely past one another, the polymer molecules turn the fluid into a thick, elastic material in the rubbery plateau regime.

However the cross-links formed by the borax are relatively weak and thermal energy causes them to break and reattach frequently. When you pull slowly on the glue putty, you give the cross-links time to open and close so that the polymer molecules can slowly move past one another. When you pull quickly on the glue putty, the cross-links have no time to adjust and you break the polymer molecules directly. The glue putty then cracks like a solid.

Cases

1. Although giants are popular in fairy tales, they aren't very practical from a structural point of view.

***a.** The bone in a human ankle is roughly 3 cm in diameter and has a cross sectional area of about 7 cm^2. What is the compressional stress on each ankle bone of a standing person weighing 700 N?

***b.** If a giant is identical to the standing person, except for being 5 times larger in every respect, what is the cross sectional area of the giant's ankle bone?

***c.** How many times more volume does the giant's body occupy than does the body of the normal person?

***d.** What is the giant's weight?

***e.** How does the compressional stress on the giant's ankle bone compare to that of the person?

2. Many people reshape their hair. Hair is a thermoset consisting of protein polymer chains that are cross-linked by sulfur atoms.

a. How does the number of cross-links between the polymer chains affect the stiffness of a person's hair?

b. Why is it difficult to reshape hair with heat alone?

c. When a hairdresser performs a permanent on someone's hair, the first step is to chemically break the cross-links. Why does removing the cross-links make it possible to change the shape of the hair?

d. After the hair has been set into its new shape, chemicals are added to create new cross-links. Why does the hair retain the new shape after the cross-links have formed?

3. Anodized aluminum cookware has many attractive features. "Anodizing" refers to an electrochemical process that coats the pots with a thick layer of aluminum oxide, or alumina, the ultrahard mineral in rubies and sapphires.

a. Aluminum atoms are bound together by metallic bonds while atoms in alumina are bound by covalent bonds. Why is aluminum so much easier to bend and deform than alumina?
b. Pure aluminum is too soft for cookware. Why does adding a small amount of other metals to the aluminum produce a much stronger aluminum alloy?
c. How does having an ultrahard coating on this strong aluminum alloy create a particularly sturdy pot?
d. Aluminum is a better conductor of heat than stainless steel, so an aluminum pot cooks more evenly than a stainless steel pot. Aluminum is also a better electric conductor. Why is aluminum's excellence in both areas related?
e. The alumina coating makes it difficult for food to stick to the pot. The atoms in alumina share electrons with one another and bind strongly together. However, they don't form strong chemical bonds with nearby food. What weak forces between molecules hold the food to the alumina?

4. Ancient people had to make do with simple metals that either occurred naturally in metallic form or were easy to extract from their ores. Among these were copper and tin.

a. While early metal smiths could melt copper and cast it into finished shapes, they found that this technique produced objects that were too soft to be useful. Why were these objects so easy to bend or dent?
b. To make copper objects that were significantly stiffer than those produced by casting, the metal smiths would pound and hammer the copper into shape. Why was beaten copper so much harder than cast copper?
c. Tin is a very soft metal that melts at only 232° C (450° F). But when a small amount of tin is added to copper, the two metals form bronze, an alloy that is much harder than either copper or tin. How might the presence of two different atoms in the metal crystals make this metal so hard?
d. One important use of bronze was in bells. Bell metal is a copper–tin alloy containing about 25% tin. It melts quite easily, a useful feature for casting, and is extremely hard. Why is hardness important in a bell?
e. Unfortunately, bronze bells were also susceptible to cracking. Why did they crack rather than dent when they were struck too hard?

5. Some metal alloys become harder with age. Though relatively soft and easy to shape when new, they become hard and strong by the time they're ready for use. The aluminum alloys used in aircraft construction are good examples of this behavior. They contain roughly 4% copper in aluminum. At temperatures above about 500° C (932° F), copper dissolves in aluminum. If the metal is then cooled rapidly to room temperature, the copper will remain trapped in the aluminum crystals.

a. The presence of copper atoms in the aluminum crystals hardens the aluminum somewhat. Why?
b. With time, tiny crystals of a new material, $CuAl_2$, begin to grow inside the aluminum crystals at room temperature. This growth occurs as copper atoms diffuse through the aluminum. The crystals grow more rapidly at elevated temperatures and more slowly at reduced temperatures. Why does temperature affect the rate at which these crystals grow?
c. As the $CuAl_2$ crystals grow inside the aluminum crystals, they make it difficult for dislocations to move through the aluminum crystals. How does preventing the dislocations from moving make the metal harder?
d. Exposing the metal to elevated temperatures allows the largest $CuAl_2$ crystals to grow at the expense of the smaller crystals. The largest crystals soon become quite large while the smaller crystals dissolve and disappear. Why does this over-aged aluminum alloy become weaker?
e. How can you soften a piece of this age-hardened aluminum to make it easier to cut and shape?

6. Latex house paint is a latex—a water-based emulsion of tiny polymer particles. While the polymer particles aren't actually water soluble, they can't settle out of the water-based solvent because they're coated with a layer of surfactant—a soap or detergent. This coating provides a stable interface between the polymer molecules in the particles and the water molecules around them, allowing the particles to remain suspended in the solvent indefinitely.

a. When you apply a relatively thin layer of the paint to a wall, it basically remains in place. Why doesn't this fluid flow rapidly down the wall?
b. As the water in the paint evaporates, pairs of polymer particles find themselves connected by tiny drops of water. The surface of each water drop has a large potential energy. Explain why surface tension in an evaporating water drop pulls the polymer particles together until they touch.
c. On a warm day, the polymers used in house paints are between the glass–rubber transition regime and the rubbery plateau regime. Explain how, with time, the polymer particles merge together to form a continuous plastic sheet.
d. Why is it a bad idea to apply latex house paints in very cold weather?
e. While most latex house paints contain dyes and pigment particles, there are latex coatings that contain nothing more than the polymer particles and the solvent. Why do these coatings look milky white when you apply them but become transparent when the solvent has evaporated?

7. The glass in a large rectangular fish tank must endure severe stresses when it's full of water.

a. Glass near the bottom of the tank experiences enormous outward forces. Why do these forces depend on the depth of water inside the tank?
b. Is the inside surface of the glass near the bottom of the tank under compressive stress or tensile stress?
c. Is the outer surface of the glass near the bottom of the tank under compressive stress or tensile stress?
d. If the glass were to break, would the crack begin at the inner or outer surface of the glass?
e. Water weakens glass by creating defects in its surface. Is that fact important for the aquarium tank? Why or why not?
f. How would tempering the glass affect that tank's ability to handle the stresses?

8. Plastic lenses have replaced glass lenses in most eyeglasses because they weigh only about half as much. But the plastics used in eyeglasses must be carefully selected to avoid a variety of problems.

a. Plastic lenses are made from plastics that are entirely amorphous and contain no crystalline regions. Why would having crystalline regions be a problem in lenses?
b. The plastics used in lenses have glass transition temperatures well above room temperature. Why?
c. The plastics used in lenses must have large indexes of refraction and small amounts of dispersion. Why?
d. The lenses are molded from molten plastic. Why can't thermosets be used in lenses?
e. Plastic lenses are much harder to shatter than glass lenses. Why do they turn white rather than breaking?

CHAPTER 18

NUCLEAR PHYSICS

It has been almost a century since Ernest Rutherford discovered the atomic nucleus. One day in 1911, Rutherford thought about his experimental results and realized that most of an atom's mass must be located in an unimaginably small object at the atom's center. He called this object the atom's nucleus and went on to create a new field of science: nuclear physics. Nuclear physics soon led to an understanding of the differences between atoms, to an explanation of radioactivity, and to a recognition that enormous energies lie trapped within nuclei. Only a few decades after its inception, nuclear physics served as the basis for two developments that have had profound effects on our society: nuclear weapons and nuclear power.

Experiment: Thoughts About the Sun's Fire

It's difficult to detect atoms and more difficult still to show that they have nuclei. There are thus no simple experiments to be done with nuclei. Instead, we'll make a thought experiment; an experiment that we'll do only with our imaginations. Scientists often use thought experiments to test the feasibility of an idea they might have. Let's try to figure out how long the sun would last if it were made of burning wood. We won't worry about where it gets the oxygen needed to burn—we'll just see how many years it would take to burn up completely.

The sun has a radius of about 700,000 km. Since our wooden version is burning inward from all sides, it will disappear completely once the flames have advanced 700,000 km. Since it takes about a day for flames to burn a quarter meter into a good hardwood log, we can expect the flames to advance inward on our wooden sun at about the same rate. Each day, our wooden sun will lose about a quarter meter of its radius. After a full year, the radius will have lost about 90 m and after about 11 years, it will have lost 1 km. Finally, after about 8 million years, the whole 700,000 km of radius will have been consumed and the wooden sun will cease to exist.

Even if the sun had started with a radius so large that the earth was actu-

ally rolling on its surface, it would still only last a total of 1.6 billion years. Since the solar system is at least 4 billion years old, the sun can't be a piece of burning wood. Instead, its energy must come from something other than the chemical reactions of fire. Can you think of other potential sources of energy that might power the sun? Can you show whether or not they're possible?

Chapter Itinerary

As you might suspect, the sun's energy source is nuclear energy. In this chapter, we'll examine the two principal uses of nuclear energy here on earth: (1) *nuclear weapons* and (2) *nuclear reactors*. In *nuclear weapons*, we'll explore the structures of atomic nuclei and see how taking them apart or joining them together can release enormous amounts of energy. In *nuclear reactors*, we'll see how these nuclear reactions can be controlled and harnessed to produce useful energy such as electricity.

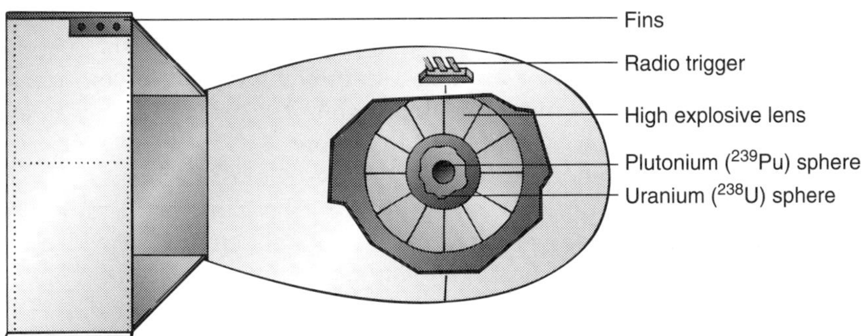

Section 18.1

Nuclear Weapons

The atomic bomb is one of the most remarkable and infamous inventions of the twentieth century. It followed close on the heels of various developments in the understanding of nature, developments which in many ways made the invention of nuclear weapon inevitable. By the late 1930's, scientists had discovered most of the principles behind nuclear energy and were well aware of how those principles might be applied. The onset of World War II prompted concern that Germany would choose to follow the military path of nuclear energy. Propelled by fear, curiosity, and temptation, the scientists, engineers, and politicians of that time brought nuclear weapons into existence.

Questions to Think About: *Where is nuclear energy stored in the atoms? Where did this nuclear energy come from? How do nuclear weapons release nuclear energy? Why are nuclear weapons so difficult to build? Why do we associate uranium and plutonium with nuclear weapons? How much uranium or plutonium does it take to build a bomb?*

Experiments to Do: *Since uranium and plutonium aren't sold in hardware stores, you won't be able to build your own bomb. However, you can get a feel for the way a chain reaction works by playing with a box of dominos. If you stand the dominoes on end on a flat table, each of them will have extra gravitational potential energy that it can release by tipping over. If you spread the dominos widely about the table and then give the table a gentle shake, they will tip over one by one.*

However if you pack the dominos tightly together, so that one falling domino can knock over others, they will no longer be independent. As you jiggle the table, nothing will happen until the first domino falls but then many or even all of the dominos will tip over in quick succession. You will have created a chain reaction, where a single event triggers an ever increasing number of subsequent events. What characteristics of the dominos and their arrangement determine whether or not such a chain reaction occurs? Can you envision a scenario in which a single tipping domino could trigger the release of an enormous amount of stored energy? Another chain reaction, this time in the decay of atomic nuclei, is what makes nuclear weapons possible.

Background

At the end of the 19th Century, "Classical Physics" reigned supreme. Here classical physics means the rules of motion and gravitation identified by such people as Galileo, Newton, and Kepler, and the rules of electricity and magnetism developed by such people as Ampere, Coulomb, Faraday, and Maxwell. It was generally felt that most of physics was well understood: physicists knew all of the laws governing the behavior of objects in our universe and what remained was to apply those laws to more and more complex examples. It was a time when physicists didn't know what they didn't know.

However, there remained a few nagging problems; particular failures of classical physics; specific difficulties that couldn't be explained by the known rules. Among these problems were the spectrum of light emitted by a black body, the photoelectric effect in which electrons are ejected from metals by light, and the apparent absence of an aether or medium in which light traveled. At the beginning of the twentieth century, the whole of classical physics collapsed under the weight of these seemingly trivial difficulties and a largely new understanding of the universe emerged. The major advances took 25 years, from 1901 to 1926, and the years since then have largely been spent applying those new laws to more and more complicated examples.

The two main developments, both essential to the making of the atomic bomb, were the discoveries of quantum physics and relativity. Often these are called *quantum theory* and the *theory of relativity*. But while the word *theory* might imply that they're somehow on shaky ground, they're not theories in the sense of hypotheses waiting to be tested. In fact, they have been confirmed countless times since they were developed and have been shown to have enormous predictive power. Rather, they're theories in the sense of being carefully constructed and codified sets of rules that model the behavior of the physical universe in which we live. Between them, these two theories made the discovery of nuclear forces and nuclear energy inevitable. Finally, given peoples' love for gadgets and power, they also made the development of nuclear weapons inevitable.

CHECK YOUR UNDERSTANDING #1: In Theory, It Means That...
If something is referred to as a "theory," how likely is it to be true?

The Nucleus

Though the name "atomic bomb" has stuck for more than half a century, the more correct name would be "nuclear bomb." The items that are responsible for the energy released by nuclear weapons are not atoms, but tiny pieces of atoms—their nuclei (plural for nucleus). But before we can discuss nuclei, let's put them into context. Let's start by looking at atoms.

To get an idea of just how tiny atoms are, imagine magnifying a grain of table salt, 1 mm on a side, until it was the size of the State of Colorado. That grain would then appear as an orderly arrangement of spherical particles, each about the size of a grapefruit (Fig. 18.1.1). These spherical particles would be single atoms and there would be about 7.2 million of them along each edge of the grain.

Like most solids, table salt is a crystal and its atoms are bound to one another by their outermost components: their electrons. Electrons dominate the chemistry of atoms and molecules. Sodium is a reactive metal because of its electrons and chlorine is a reactive gas because of its electrons. When mixed, these two chemicals react violently to form table salt and release a considerable amount of light and heat. This, then, is a true "atomic bomb."

Obviously, something is missing here. If a crazy person could buy a kilogram or two of sodium and a tank of chlorine from the chemistry stockroom and destroy an entire city, rural living would be a whole lot more popular. Fortunately the energy released by chemical reactions is fairly limited. A kilogram of chemical explosives just can't do that much damage. But atomic bombs tap an entirely different store of energy deep within the atoms.

While all of nuclear weaponry is often attributed to Einstein's famous equation, $E=mc^2$, that notion is vastly oversimplified. Nonetheless, this equation is quite significant. One of Einstein's discoveries at the beginning of the twentieth century was that matter and energy are in some respects equivalent. In certain circumstances mass can become energy or energy can become mass. This equivalence is part of the theory of relativity and has some interesting consequences. It implies that an object can reduce its mass by transferring energy to its surroundings. Thus, if you weigh an object before and after it undergoes some internal transformation, you can use any weight loss to determine how much energy was released from the object by that transformation.

Because of this equivalence, mass and changes in mass can be used to locate energy that is hidden within normal matter. This technique is important in nuclear physics but it also applies to chemistry. When sodium and chlorine react to form table salt, their combined mass decreases by a tiny amount. What is missing is some chemical potential energy, which became light and heat and escaped from the mixture. In leaving, this chemical potential energy reduced the mass of the sodium and chlorine mixture by about one part in 10 billion. This tiny change in mass is too small to detect with present measuring devices but scientists are working on techniques that will soon make it possible to measure mass changes due to chemical bonds.

However, electrons are by far the lightest part of an atom and thus have relatively little mass to release as energy. Most of an atom's mass is located deep within it in its nucleus. The nucleus is fantastically small; only a little more than 10^{-15} m in diameter. If you were to peer into one of the grapefruit-sized sodium ions of our giant salt crystal, you would see a tiny particle at its center. There, just at the threshold of visibility, would be the ion's nucleus, only 1 micron in diameter. The remaining 99.9999999999999% of the ion is occupied only by its 10 electrons, moving about in their orbitals.

The sodium nucleus is constructed of 11 protons and 12 neutrons (Fig. 18.1.2). Each of these nuclear particles or **nucleons** has about 2000 times as much mass as an electron, so that 99.975% of the sodium ion's mass is in this nucleus. Thus, while the electrons are certainly important to chemistry and matter as we know it, their contribution to the ion's mass is insignificant. The ion is mostly empty space, lightly filled with fluffy electrons and having a tiny nuclear lump at its center.

The nucleons that make up this nucleus experience two competing forces. The first of these forces is the familiar electrostatic repulsion between like electric charges. Because each proton in the nucleus has a single positive charge, they're constantly trying to push one another out of the nucleus. However, the second of these forces is attractive and holds the nucleus together. This new force is called the **nuclear force** and it overwhelms the weaker electrostatic repulsion. However, the nuclear force only attracts the nucleons toward one another when they're touching. As soon as they're separated, they're on their own.

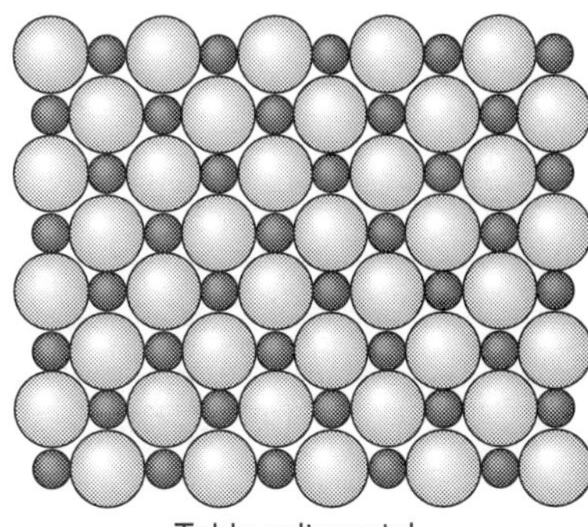

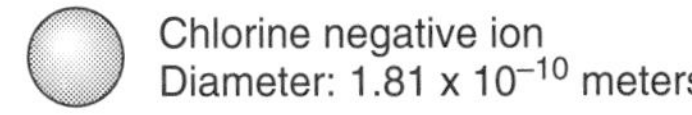

Fig. 18.1.1 - A salt crystal is an orderly array of sodium and chlorine ions. These ions are held together by the attractive forces between oppositely charged ions. The ions are so small that there are about 7.2 million ions on the edge of a 1 mm wide salt crystal.

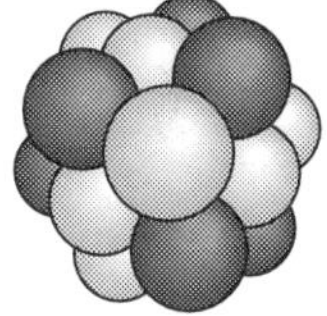

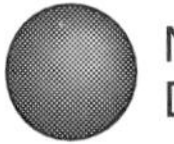

Fig. 18.1.2 - At the center of a sodium ion is a tiny nucleus containing about 99.975% of the ion's mass. It consists of 11 positively charged protons and 12 uncharged neutrons. The protons repel one another at any distance but the protons and neutrons are bound together by the highly attractive nuclear force as long as they touch one another.

Fig. 18.1.3 - These three hopping toys store energy as their springs are compressed and retain that energy while their suction cups grip their bases. When the suction cups let go, the toys leap into the air.

The competition between these two forces, the repulsive electrostatic force between like charges and the attractive nuclear force between nucleons, is analogous to a common toy (Figs. 18.1.3 and 18.1.4). This hopping toy has a suction cup attached to a spring, so that the spring tries to separate the toy's top from its base while the suction cup tries to keep the two parts together. When the two parts are well separated, the spring is the only source of forces and it tries to separate the parts further. But when the two parts touch, the suction cup begins to act and it can hold the two parts together by exerting an attractive force that is stronger than the spring's repulsive force.

What makes the hopping toy exciting is that its suction cup leaks. Eventually, the suction cup lets go and allows the spring to toss the toy into the air. But suppose that the suction cup didn't leak. Once pushed together, the pieces would never separate and the spring would retain its stored energy indefinitely. To get the suction cup to let go, you would have to pull it away from the base. Only then could the spring release its stored energy.

In effect, there would then be an energy barrier that prevented the leak-free toy from hopping. Until you did a little work on it by pulling the suction cup off the base, it would not be able to release its stored energy. The same effect occurs in a bottle of champagne, where you must push on the cork to help it out of the neck. After that initial investment of energy, a great deal of energy is released as gas blasts the cork across the room.

The nucleus is in a similar situation. The attractive nuclear force prevents the nucleus from coming apart, despite the enormous amount of electrostatic potential energy it contains. The nuclear force creates an energy barrier that prevents the nucleons from separating. Unless something adds energy to the nucleus to help the nucleons break free of the nuclear force, the nucleus will remain together forever. At least that's the prediction of classical physics.

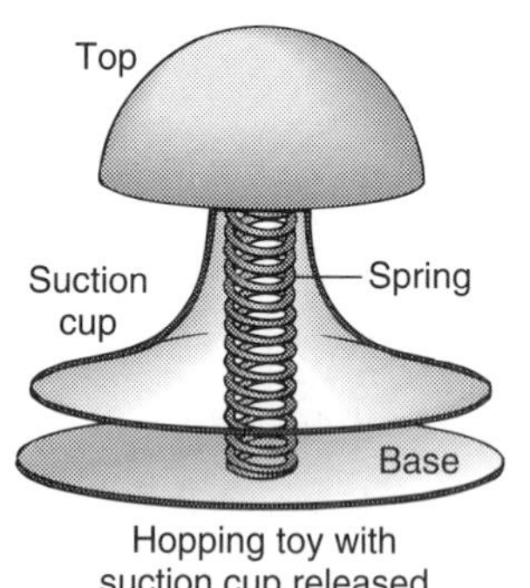

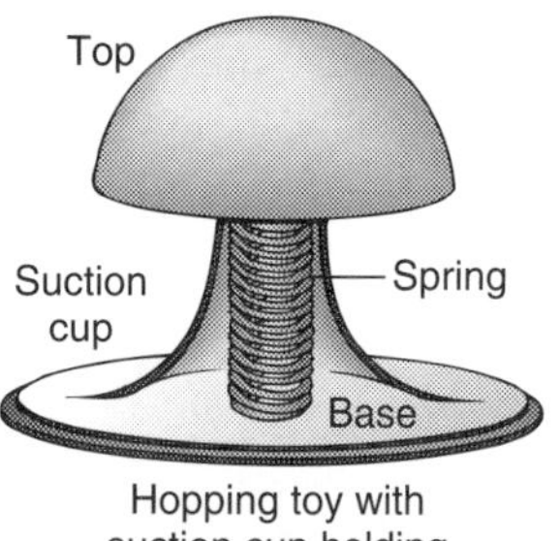

Fig. 18.1.4 - A spring and a suction cup combine to form a toy that hops suddenly after a long wait. The spring tries to separate the top from the base while the suction cup tries to hold the two parts together. Energy you store in the spring is released when the leaking suction cup eventually allows the spring to expand.

However quantum physics has an important influence on the behavior of the nucleus. One of the many peculiar effects of the quantum physics is that you can never really tell exactly where an object is located. That fuzziness is a manifestation of the **Heisenberg uncertainty principle**, which is itself a result of the partly wave and partly particle nature of objects in our universe. Since waves are normally broad things that occupy a region of space rather than a single point, objects in our universe normally don't have exact locations.

The smaller an object's mass, the fuzzier it is and the more uncertain its location. While the fuzzy nucleons in a nucleus will normally stay in contact with one another for an extremely long time, there is always a tiny chance that they will find themselves separated by a distance that is beyond the reach of the nuclear force. The nucleons will then suddenly be free of one another and electrostatic repulsion will push them apart in a process called **radioactive decay**. The quantum process that allows the nucleons to escape from the nuclear force without first obtaining the energy needed to surmount the energy barrier is called **tunneling** because the nucleons effectively tunnel through the barrier.

The more protons there are in a nucleus, the more they repel one another and the more likely they are to cause radioactive decay. Adding additional neutrons to the nucleus reduces this proton–proton repulsion by increasing the size of the nucleus without adding to its positive charge. However adding too many neutrons also destabilizes the nucleus for reasons that we'll discuss in the next section. So constructing a stable nucleus is a delicate balancing act.

In nuclei with only a few protons, the attractive nuclear force wins big over the repulsive electrostatic force and the nucleons stick like crazy. These nuclei resemble hopping toys with weak springs and big suction cups—once brought together, the pieces never come apart. In fact, the average binding energy of the nucleons would increase if these nuclei had even more protons and neutrons.

In nuclei with many protons, the electrostatic repulsion is so severe that the nuclear force can't hold the nucleons together for long. These nuclei decay rapidly. They resemble hopping toys with strong springs and small suction cups. The average binding energy of the nucleons would increase if these nuclei had fewer protons and neutrons.

In nuclei with roughly 26 protons, in between the two extremes we've just considered, the attractive nuclear force and repulsive electrostatic force are nicely balanced. These nuclei are extremely stable and you can't increase the average binding energy of their nucleons by adding or subtracting nucleons. Smaller nuclei tend to grow to reach this intermediate size while larger nuclei tend to shrink toward the same goal.

A small nucleus can grow if something pushes more nucleons toward it. Electrostatic repulsion may initially oppose this growth but once everything touches, the nuclear force will bind everything together and release a large amount of potential energy. This coalescence process is called **nuclear fusion**.

A large nucleus can shrink if quantum effects separate its pieces beyond the reach of the nuclear force. Electrostatic repulsion will then push the fragments apart and release a large amount of potential energy. This fragmentation process is called **nuclear fission**.

The energies released when large nuclei undergo fission or when small nuclei undergo fusion is enormous when compared to chemical energies. Uranium, a large nucleus, converts about 0.1% of its mass into energy when it breaks apart. Hydrogen, a tiny nucleus, converts about 0.3% of its mass into energy when it fuses with other hydrogen nuclei. Kilogram for kilogram, nuclear reactions release about 10 million times as much energy as chemical reactions. Fortunately, they're much harder to start.

CHECK YOUR UNDERSTANDING #2: A Sticky Nuclear Problem
If you take two intermediate-sized nuclei and combine them to make a single uranium nucleus, will the process release or consume energy?

Radioactive Decay and the Fission Bomb

Natural radioactive decay was discovered accidentally by French physicist Antoine-Henri Becquerel (1852–1906) in 1896. Intrigued by the recent discovery of X-rays, he began looking for materials that might emit X-rays after exposure to light. To his surprise, he found that uranium fogged photographic plates, even through an opaque shield and even without exposure to light.

In 1911, British physicist Ernest Rutherford (1871–1937) discovered that atoms have nuclei. He subsequently found that these nuclei could be shattered by energetic helium nuclei. And in 1932, British physicist James Chadwick (1891–1974) discovered a fragment of the nucleus, the neutron, that had no electric charge and could thus approach a nucleus without any electrostatic repulsion. It was soon discovered that neutrons can stick to the nuclei of many atoms.

But the crucial discovery that made the atomic bomb possible, was neutron-induced fission of nuclei. In 1934, Italian physicist Enrico Fermi (1901–1954) and his colleagues were trying to solve a particular riddle about the nucleus, a radioactive decay process called beta decay. They were adding neutrons to the nuclei of every atom they could get hold of. When they added neutrons to uranium, with its huge nucleus, they observed the production of some very short-lived radioactive systems. They thought that they had formed ultraheavy nuclei and even went so far as to give these new elements tentative names.

However four years later, Otto Hahn, Fritz Strassmann, Otto Frisch, and Lise Meitner collectively showed that what Fermi's group had actually done was

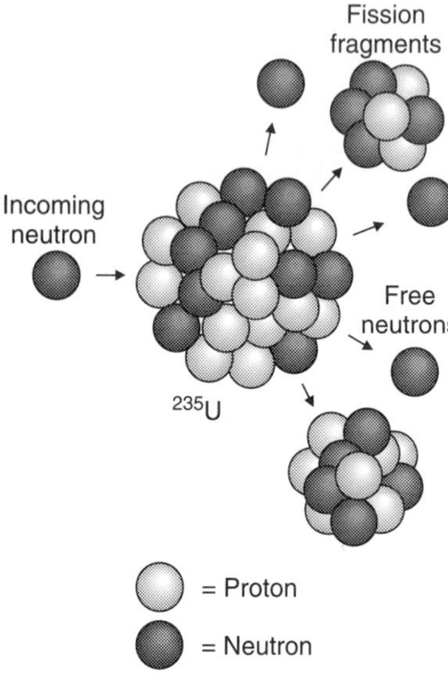

Fig. 18.1.5 - When a uranium nucleus is struck by a neutron, there's a good chance that it will fall apart into fragments. This process is called induced fission. Among the fragments of induced fission are other neutrons.

to fragment uranium into lighter nuclei (Fig. 18.1.5). Many of the fragments created by this **induced fission** were neutrons which could themselves cause the destruction of other uranium nuclei.

It had suddenly become possible to create a **chain reaction** in which the fission of one uranium nucleus would induce fission in two nearby uranium nuclei, which would in turn induce fission in four other uranium nuclei, and so on (Fig. 18.1.6). The result would be a catastrophic nuclear process in which many or even most of the nuclei in a piece of uranium would shatter and release fantastic amounts of energy.

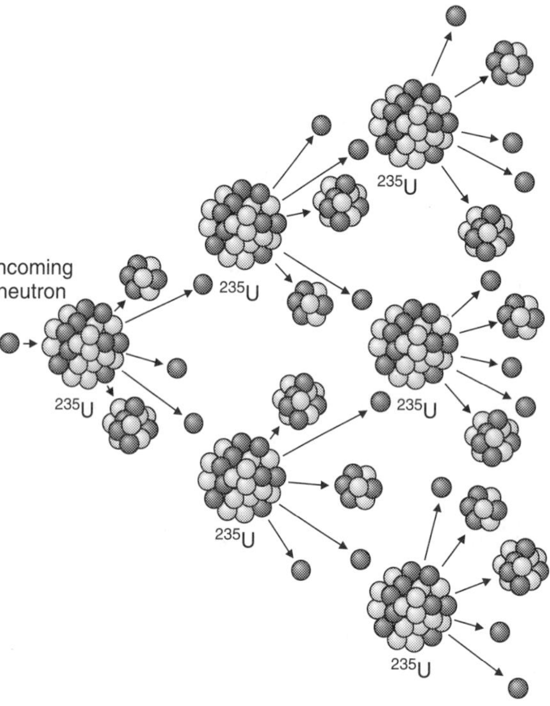

Fig. 18.1.6 - A chain reaction occurs when the fragments of one fissioning nucleus induce fission in at least one additional nucleus. Such a chain reaction is particularly easy in ^{235}U, the light isotope of uranium, where each fissioning nucleus releases an average of 2.5 neutrons.

In a sense, the remaining work toward both the atomic bomb and the hydrogen bomb was a matter of technical details. Only four conditions had to be satisfied in order for an atomic or *fission bomb* to be possible:

1. A source of neutrons had to exist in the bomb to trigger the explosion.
2. The nuclei making up the bomb had to be **fissionable**, that is they had to fission when hit by a neutron.
3. Each induced fission had to produce more neutrons than it consumed.
4. The bomb had to use the released neutrons efficiently so that each fission induced an average of more than one subsequent fission.

Meeting the first condition was easy. Many radioactive elements emit neutrons. But meeting the second and third conditions was more difficult. Here is where uranium fit into the picture. It was known to be fissionable and it was known to release more neutrons than it consumed.

But not all uranium nuclei are the same. While a uranium nucleus must contain 92 protons, so that it forms a neutral atom with 92 electrons and has all of the chemical characteristics of uranium, the number of neutrons in that nucleus is somewhat flexible. Nuclei that differ only in the numbers of neutrons they contain are called **isotopes** and natural uranium nuclei come in two isotopes: ^{235}U and ^{238}U, where the number specifies how many nucleons are in each nucleus. The ^{235}U nucleus contains 92 protons and 143 neutrons, for a total of 235 nucleons. In contrast, the ^{238}U nucleus contains 238 nucleons—92 protons and 146 neutrons.

It turns out that only ^{235}U is suitable for a bomb. It's marginally stable, with too many protons for the nuclear force to keep together, even with the diluting effects of 143 neutrons. Because this nucleus has a 50% chance of falling apart spontaneously over a period of 710 million years, it's said to have a **half life** of 710 million years. But when struck by a neutron, ^{235}U fissions easily and releases about 2.5 neutrons in the process.

^{238}U is slightly more stable than the other isotope and its half life is 4.51 billion years. But this nucleus absorbs most neutrons without undergoing fission. Instead, it undergoes a series of complicated nuclear changes that eventually convert it into plutonium, an element not found in nature. It becomes ^{239}Pu, a nucleus with 94 protons and 145 neutrons. As we'll see later on, plutonium itself is useful for making nuclear weapons.

So ^{238}U actually slows a chain reaction rather than encouraging it. Since only ^{235}U can support a chain reaction, natural uranium had to be separated before it could be used in a bomb. But ^{235}U is quite rare. The earth's store of uranium nuclei was created long ago, in the explosion of a dying star. That *supernova* heated the nuclei of smaller atoms so hot that they collided together and stuck. Uranium nuclei were formed, with the supernova's energy trapped inside them. They were incorporated in the earth during its formation about 4 or 5 billion years ago and have been decaying ever since. The only uranium isotopes that remain in any quantity are ^{235}U and ^{238}U. Since ^{235}U is less stable, it has dwindled to only 0.72% of the uranium nuclei. The remaining 99+% of the uranium is ^{238}U.

Separating ^{235}U from ^{238}U is extremely difficult. Since atoms containing these two nuclei differ only in mass, not in chemistry, they can only be separated by methods that compare their masses. Because the mass difference is relatively small, heroic measures are needed to extract ^{235}U from natural uranium. During World War II and the Cold War era, the United States government developed enormous facilities for separating the two uranium isotopes. The need for such installations is one of the major obstacles to the proliferation of nuclear weapons.

The last condition for sustaining a chain reaction is that the bomb use neutrons efficiently, so that each fission induces an average of more than one subsequent fission. That means that the bomb's contents can't absorb neutrons wastefully and that it can't let too many of them escape without causing fission. A lump of relatively pure ^{235}U wouldn't absorb neutrons wastefully, but it might allow too many of them to escape through its surface. For a chain reaction to occur, the lump must be large enough that each neutron has a good chance of hitting another nucleus before it can escape from the lump. The lump also should have a minimal amount of surface. It should be a sphere.

But how large must that sphere be? Since atoms are mostly empty space, a neutron can travel several centimeters through a lump of uranium without hitting a nucleus. Thus a golf ball-sized sphere of uranium would allow too many neutrons to escape. For a bare sphere of ^{235}U, the **critical mass** required to initiate a chain reaction is about 52 kg, a ball about 17 cm in diameter. At that point, each fission will induce an average of one subsequent fission. But for an **explosive chain reaction**, in which each fission induces an average of much more than one fission, additional ^{235}U is needed; a **supercritical mass**. About 60 kg will do it.

By 1945, the scientists and engineers of the Manhattan Project had found ways to meet these four conditions and were prepared to initiate an explosive chain reaction. They had accumulated enough ^{235}U to construct a supercritical mass. Carefully machined pieces of ^{235}U would be put into the bomb so that they would join together at the moment the bomb was to explode. When the critical mass was reached, a few initial neutrons would start the chain reaction. When the uranium became supercritical, catastrophic fission would quickly turn it into a tremendous fireball.

However, assembly was tricky. The supercritical mass had to be completely

Fig. 18.1.7 - The Little Boy (top) used a cannon to fire a wedge of uranium into an incomplete sphere of uranium. The Fat Man (bottom) used high explosives to crush a sphere of plutonium to extreme density. Little Boy destroyed Hiroshima while Fat Man destroyed Nagasaki.

assembled before the chain reaction was too far along, otherwise the bomb would begin to overheat and explode before enough of its nuclei had time to fission. In pure ^{235}U, the time it takes for one fission to induce the next fission is only about 10 ns (10 nanoseconds or 10 billionths of a second). In a supercritical mass, each generation of fissions is much larger than the previous generation so it takes only a few dozen generations to shatter most of the uranium nuclei. The whole explosive chain reaction is over in less than a millionth of second, with most of the energy released in the last few generations (about 30 ns).

To make sure that the assembly was complete before the bomb exploded, it had to be done extraordinarily quickly. In the ^{235}U bomb called "Little Boy" (Fig. 18.1.7), exploded over Hiroshima on August 6, 1945 at 8:15AM, the supercritical mass was assembled when a cannon fired a wedge of ^{235}U into a second piece of ^{235}U (Fig. 18.1.8). When the wedge arrived, it completed a 60 kg sphere of uranium, housed in a tungsten carbide and steel container. This container confined the uranium, holding it together with its inertia as the explosion began. An explosive chain reaction started immediately and by the time the uranium blew itself apart, 1.3% of the ^{235}U nuclei had fissioned. The energy released in that event was equal to the explosion of about 15,000 tons of TNT.

But "Little Boy" was actually the second atomic explosion. Its concept was so foolproof and its ^{235}U so precious that "Little Boy" was dropped without ever being tested. However, the Manhattan Project had also developed a plutonium-based bomb that involved a much more sophisticated concept. This bomb was much less certain to work, so it was tested once before it was used.

"The Gadget", as the first atomic bomb was called, didn't use ^{235}U. Instead, it used plutonium that had been synthesized from ^{238}U in nuclear reactors. As we'll see in the next section, a nuclear reactor carries out a controlled chain reaction in uranium, and neutrons from this chain reaction can convert ^{238}U into ^{239}Pu.

Like ^{235}U, ^{239}Pu meets the conditions for a bomb. The ^{239}Pu nucleus is relatively unstable, with a half life of only 24,400 years. It fissions easily when struck by a neutron and releases an average of 3 neutrons when it does. Thus ^{239}Pu can be used in a chain reaction. For a bare sphere of ^{239}Pu, the critical mass is about 10 kg; a ball about 10 cm in diameter.

But ^{239}Pu has a problem. It's so radioactive and releases so many neutrons when it fissions that a chain reaction develops almost instantly. There is much less time to assemble a supercritical mass of plutonium than there is with uranium. The cannon assembly method won't work because the plutonium will overheat and blow itself apart before the wedge ever touches the sphere.

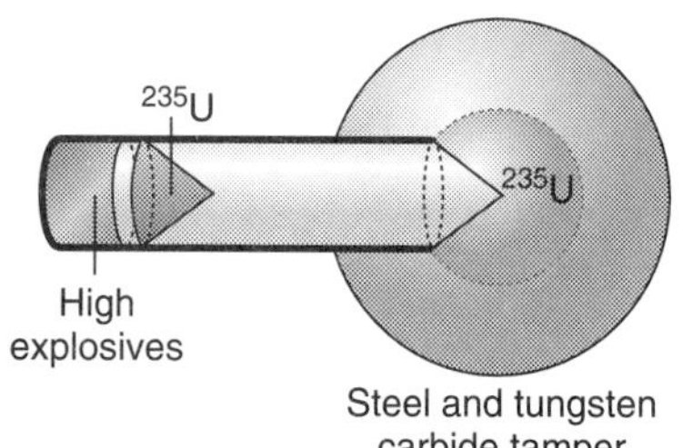

Fig. 18.1.8 - The concept behind "Little Boy" was simple: a sphere of ^{235}U was divided into two parts so that neither was a critical mass on its own. One part was an incomplete sphere and the other part was a wedge that would complete that sphere. When the bomb was detonated, a cannon fired the wedge into the incomplete sphere, creating a supercritical mass and initiating an explosive chain reaction.

Thus a much more sophisticated assembly scheme was employed. When "The Gadget" (Fig. 18.1.9) was detonated at Alamogordo on June 16, 1945 at 5:29AM, over 2000 kg of carefully designed high explosives crushed or *imploded* a sphere of plutonium (Fig. 18.1.10). By itself, the 6.1 kg sphere wasn't large enough to be a critical mass; it was **subcritical**. But it was surrounded by a *tamper* of ^{238}U, whose massive nuclei reflected many neutrons back into the plutonium like marbles bouncing off bowling balls. And the implosion process compressed the plutonium well beyond its normal density. With the plutonium nuclei packed more tightly together, they were more likely to be struck by neutrons and undergo fission.

The scheme worked. The chain reaction that followed caused 17% of the plutonium nuclei to fission and released energy equivalent to the explosion of about 22,000 tons of TNT. The tower and equipment at the Trinity test site disappeared into vapor and the desert sands turned to glass for hundreds of meters in all directions. A nearly identical device name "Fat Man" (Fig. 18.1.7) was dropped over Nagasaki on August 9, 1945 at 11:02AM.

In the years following the first fission bombs, development focused on how best to bring the fissionable material together. The longer a supercritical mass

Fig. 18.1.9 - The first atomic bomb, nicknamed "The Gadget," was detonated on a tower at a remote desert site near Alamogordo, New Mexico on June 16, 1945. Here employees of the top secret Manhattan Project are seen hoisting parts of the plutonium bomb onto the tower before the explosion.

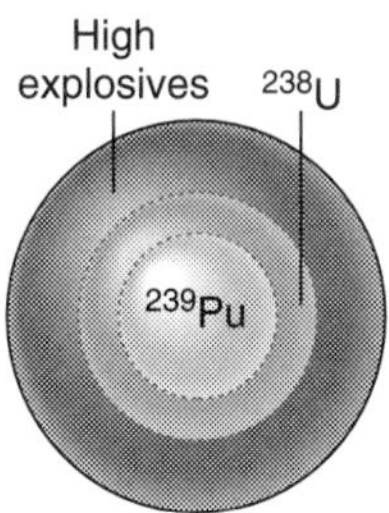

Fig. 18.1.10 - The concept used in "The Gadget" and "Fat Man" was relatively sophisticated. Carefully shaped high explosives crushed a baseball-sized sphere of ^{239}Pu inside a neutron-reflecting shell of ^{238}U. The plutonium was compressed to high density and quickly reached supercritical mass, initiating an explosive chain reaction.

could be held together before it overheated and exploded, the larger the fraction of its nuclei that would fission and the greater the explosive yield. The crushing technique of "The Gadget" and "Fat Man" became the standard and bombs grew smaller and more efficient at using their nuclear fuel. The implosion process reduced the amount of plutonium needed to reach a supercritical mass so that very small fission bombs were possible. The smallest atomic bomb, the "Davy Crockett," weighed only about 220 N (50 pounds).

CHECK YOUR UNDERSTANDING #3: Tickling the Dragon's Tail

What would happen if you slowly moved two 30 kg hemispheres of ^{235}U together to form a single sphere?

The Fusion or Hydrogen Bomb

Since fissionable material begins to explode as soon as it exceeds critical mass, this critical mass limits the size and potential explosive yield of a fission bomb. In search of a way around this limit, bomb scientists took a look back at small nuclei and figured out how to extract energy by sticking them together.

The fission bomb brought to the earth, for the first time, temperatures that had previously only been observed in stars. Stars obtain most of their energy by fusing hydrogen nuclei together to form helium nuclei, a process that releases a great deal of energy. Because hydrogen nuclei are protons and repel one another with tremendous forces, hydrogen doesn't normally undergo fusion here on earth. To cause fusion, something must bring those protons close enough for the nuclear force to stick them together. The only practical way we know of to bring the nuclei together is to heat them so hot that they crash into each other. That is what happens in a *fusion bomb*, also called a *thermonuclear* or *hydrogen bomb*.

In a fusion bomb, an exploding fission bomb heats a quantity of hydrogen to about 100 million degrees Celsius (Fig. 18.1.11). At that temperature, hydrogen nuclei begin to collide with one another. To ease the nuclear fusion processes, heavy isotopes of hydrogen are used: *deuterium* (^{2}H) and *tritium* (^{3}H). While the normal hydrogen nucleus (^{1}H) contains only a proton, the deuterium nucleus contains a proton and a neutron. The tritium nucleus contains a proton and two neutrons. When a deuterium nucleus collides with a tritium nucleus, they stick to form a helium nucleus (^{4}He) and a free neutron. Because this process converts about 0.3% of the original mass into energy, the helium nucleus and the neutron fly away from one another at enormous speeds.

Since hydrogen won't explode spontaneously, even in huge quantities, a hydrogen bomb can be extremely large. A fission bomb is used to set it off, but

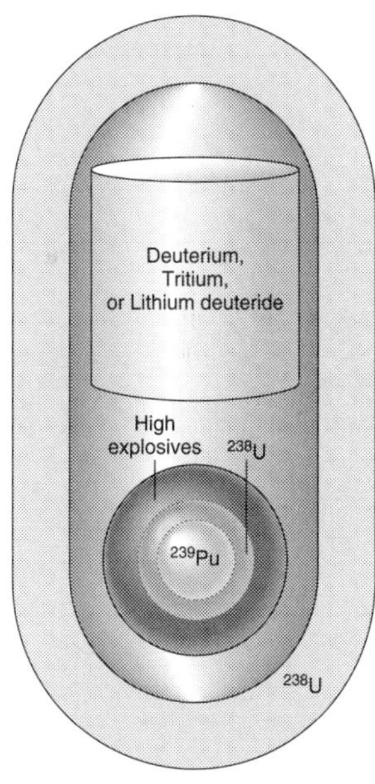

Fig. 18.1.11 - A fusion bomb releases energy by fusing deuterium (^{2}H) and tritium (^{3}H) nuclei together to form helium (^{4}He) and neutrons. This fusion is initiated by heating the hydrogen to more than 100 million degrees Celsius with a fission bomb. High energy neutrons released by the fusion process then induce fission in the ^{238}U tamper, releasing still more energy. Lithium (^{6}Li) produces tritium when exposed to neutrons.

after that, the sky's the limit. Some enormous fusion bombs were constructed and tested during the early days of the Cold War. These bombs usually consisted of a fission starter and a hydrogen follower, all wrapped up in a tamper of ^{238}U. The tamper confined the hydrogen as the fusion began. Once fusion was underway, converting deuterium and tritium into helium and neutrons, the neutrons collided with ^{238}U nuclei in the tamper. These neutrons were so energetic that they induced fissions in the ^{238}U nuclei, releasing still more energy. Overall, this structure is sometimes called a fission-fusion-fission bomb.

A variation on this bomb is the so called "neutron" or "enhanced radiation" bomb. This bomb has no ^{238}U tamper so that the energetic neutrons from the fusion process traveled out of the explosion and irradiated everything in the vicinity. This bomb is lethal to humans but not particularly destructive to property.

Tritium is a radioactive isotope created in nuclear reactors. It has too many neutrons to be a stable nucleus and slowly decays into a light isotope of helium (^{3}He). Because tritium has a half life of 12.3 years, fusion bombs containing tritium require periodic maintenance to replenish their tritium.

Many fusion bombs use solid lithium deuteride instead of deuterium and tritium gases. Lithium deuteride is a salt containing lithium (^{6}Li) and deuterium (^{2}H). When a neutron from the fission starter collides with a ^{6}Li nucleus, that nucleus fragments into a helium nucleus (^{4}He) and a tritium nucleus (^{3}H). In the bomb, lithium deuteride is quickly converted into a mixture of deuterium, tritium, and helium, which then undergoes fusion.

CHECK YOUR UNDERSTANDING #4: It's Hard to Get Together

Why must hydrogen be heated to extremely high temperature to initiate fusion?

Heat, Radiation, and Fallout

Once a nuclear weapon has exploded—after its fissionable material has fissioned and its fusible material has fused—what then? First, a vast number of nuclei and subatomic particles emerge from the explosion at enormous speeds, many at nearly the speed of light. These particles crash into nearby atoms and molecules, heating them to fantastic temperatures and producing a local fireball around the bomb itself. They also cause extensive radiation damage in the surrounding area.

Second, a flash of light emerges from the explosion, caused partly by the fission and fusion processes themselves and partly by the ultra-hot fireball that follows. This light is not only visible light, but every portion of the electromagnetic spectrum from infrared to visible to ultraviolet to X-rays to gamma rays. It burns things nearby, inside and out.

Third, the explosion creates a huge pressure surge in the air around the fireball. A shock wave propagates outward from the fireball at the speed of sound, knocking over everything in its path for a considerable distance. Fourth, the rarefied and superheated air rushes upward, lifted by buoyant forces, to create a towering mushroom cloud.

But the most insidious after-effect of a nuclear explosion is *fallout*, the creation and release of radioactive nuclei. Fission converts uranium and plutonium nuclei into smaller nuclei. Each new nucleus has several dozen protons and its share of neutrons from the nucleus that fissioned. These new nuclei attract electrons and become seemingly normal atoms like iodine or cobalt. But while large nuclei like those of uranium need extra neutrons to dilute their protons and reduce their electrostatic repulsions, intermediate and small nuclei like those of iodine and cobalt don't need as many neutrons. The new nuclei wind up with too many neutrons and are radioactive.

These new nuclei live for anywhere from millionths of a second to millions of years before they spontaneously decay into fragments. Until they decay, the atoms that contain them are almost indistinguishable from normal atoms. These atoms are radioactive isotopes of common atoms and our bodies naively incorporate them into our tissues. There they sit, performing whatever chemical tasks our bodies requires of them. But eventually these radioactive atoms fall apart and release nuclear energy. Because each radioactive decay that occurs near us or inside us releases perhaps a million times more energy than is present in a chemical bond, these decays cause chemical changes in our cells. They can kill cells or damage the cells' genetic information.

CHECK YOUR UNDERSTANDING #5: Not So Good to Eat

Normal iodine ^{127}I is stable forever. But the fission product ^{131}I is radioactive and has a half life of about 8 days. What are the consequences of eating ^{131}I?

Summary

How Nuclear Weapons Work: A fission bomb releases nuclear energy through a chain reaction in the fissionable isotopes of uranium or plutonium. When a neutron strikes one of these fissionable nuclei, that nucleus shatters into fragments that include several neutrons. These new neutrons can then induce subsequent fissions and a chain reaction ensues. However, for the chain reaction to be self sustaining, these neutrons must not be allowed to escape from the material without causing fissions. Thus the bomb must assemble a large enough portion of material to ensure that each fission induces an average of much more than one subsequent fission. Assembling such a supercritical mass must be done rapidly so that the bomb can't blow itself apart before most of the nuclei have undergone fission. This assembly is accomplished either by firing the materials together with a cannon or by crushing those materials together with high explosives.

A fusion bomb uses the heat from a fission bomb to initiate fusion in hydrogen nuclei. The heavy isotopes of hydrogen, deuterium and tritium, are placed near the fission bomb so that they experience extremely high temperatures and undergo fusion. The two components are surrounded by a tamper of ^{238}U, which uses its inertia to confine the explosion so that fission and fusion proceed as far as possible before the bomb blows itself apart. Fusion releases energetic neutrons, which are able to induce fission in the ^{238}U, adding to the energy released by the explosion. Because tritium has a short half life, it must be replaced frequently. In some fusion bombs, the tritium is created during the explosion by neutron collisions with lithium.

The Physics of Nuclear Weapons

1. An atom consists of a dense nucleus, surrounded by a diffuse orbiting cloud of electrons. The nucleus is composed of positively charged protons and uncharged neutrons. Because each of these nucleons is about 2000 times as massive as an electron, most of the atom's mass is in its nucleus.

2. Atoms are roughly 10^{-10} m in diameter while their nuclei are roughly 10^{-15} m in diameter. Thus an atom is mostly empty space, occupied only by the electrons.

3. The nucleons in a nucleus are held together by the nuclear force, an attractive force that only acts when the nucleons are touching one another. The protons in the nucleus also experience electrostatic repulsion, but the nuclear force is able to keep the nucleus together indefinitely in most cases.

4. Because of their quantum natures and the Heisenberg uncertainty principle, the nucleons in some radioactive nuclei can suddenly find themselves beyond the reach of the nuclear force. These nuclei then undergo spontaneous fission—electrostatic repulsion pushes the fragments apart. Fission may release neutrons.

5. Collisions can induce fission in some nuclei. Neutrons can cause fissions easily because they're not repelled by nuclei.

6. A chain reaction occurs when a material reaches critical mass. Then each fission induces an average of one subsequent fission. If the mass becomes supercritical, each fission induces an average of more than one subsequent fission and an explosive chain reaction can occur.

7. Compressing a fissionable material to ultrahigh density can initiate a chain reaction because the closely spaced nuclei are then more likely to absorb neutrons and fission. Although the material's mass doesn't change, it exceeds "critical mass."

Check Your Understanding - Answers

1. That depends completely on the theory and the extent to which it has been compared to the real world.

Why: A great many theories have been formulated to explain the world around us. Each of these theories is an attempt at describing some particular behavior of our universe in terms of various rules or mechanisms. But until a theory has been tested by comparing it carefully to the system it's trying to explain, you can't tell if it's true or not. Some theories are eventually proven true, others false, and many remain uncertain. The theories of relativity and quantum physics have long since been proven true, although there is always the possibility that they may only be parts of a more complete theory.

2. It will consume energy.

Why: To merge two smaller nuclei and form a uranium nucleus, you will have to push the two together with considerable force because they contain many protons. You will have to do considerable work to bring these nuclei close enough for the nuclear force to bind them. The energy you invested in this new nucleus is the same energy that is released when it undergoes fission.

3. Before they actually touched, a chain reaction would occur.

Why: As soon as the hemispheres became close enough that each fission began to induce an average of one subsequent fission, a chain reaction would occur. You would have assembled a critical mass. The hemispheres would begin to get hotter and hotter and would eventually melt or explode. Because they would blow apart before most of the nuclei could fission, any explosion would be rather small. But you would suffer severe radiation injury. Just such an accident killed Louis Slotin during the Manhattan Project.

4. The protons in hydrogen nuclei repel one another quite strongly at short distances. They must be moving very quickly with lots of thermal energy for them to touch.

Why: In a gas, thermal energy takes the form of kinetic energy. The hotter the gas, the faster the particles are moving. At 100 million degrees Celsius, the hydrogen nuclei are moving so fast that they can overcome their electrostatic repulsion and touch. When they do, the nuclear force pulls them together and they fuse.

5. The ^{131}I is incorporated into your body and exposes you to radiation, particular during the first few weeks before most of it has decayed away.

Why: Your body can't distinguish ^{127}I from ^{131}I because they're chemically identical. Since your body uses iodine in its functions, any ^{131}I that you ingest is likely to become part of your body's iodine supply. Over the next 8 days, half of this iodine will decay and expose you to radiation. The remaining ^{131}I is also radioactive and half of this amount will decay in the next 8 days. Thus after 16 days, only 1/4 of the original amount will remain. After 24 days, only 1/8 will remain. And so on.

Glossary

chain reaction A process in which one event triggers an average of one or more similar events so that the process becomes self-sustaining.

critical mass A portion of a fissionable material that is able to sustain a fission chain reaction. The amount of material required depends on its mass, shape, and density.

explosive chain reaction A chain reaction in which each fission induces an average of much more than one subsequent fission and the fission rate skyrockets.

fissionable Able to undergo induced fission.

half life The time needed for half the nuclei of a particular radioactive isotope to undergo radioactive decay.

Heisenberg uncertainty principle A quantum physical law that states that an object's position and momentum can't be sharply defined at the same time. This principle gives objects with small masses a fuzzy character.

induced fission A fission event that's caused by a collision, usually with a neutron.

isotopes Chemically indistinguishable atoms containing nuclei that differ only in their numbers of neutrons.

nuclear fission The shattering of a heavy nucleus into smaller fragments. During fission, the positively charged fragments repel one another and release energy.

nuclear force An attractive force that binds nucleons together once they touch one another.

nuclear fusion The merging of two small nuclei to form a larger nucleus. During fusion, the nuclear force binds the nucleons together and releases energy.

nucleon A general name given to the particles that make up atomic nuclei: protons and neutrons.

radioactive decay The spontaneous decay of a nucleus into fragments.

subcritical mass A portion of fissionable material that is too small to sustain a chain reaction.

supercritical mass A portion of fissionable material that is well in excess of a critical mass so that it undergoes an explosive chain reaction.

tunneling Because of the Heisenberg uncertainty principle, small objects have somewhat ill-defined positions and occasionally move through energy barriers to places they can't reach classically. That process is tunneling.

Review Questions

1. The nucleus of a uranium atom contains more than 99.95% of the atom's mass. Why does this fact suggest that the nucleus may contain most of the atom's energy?

2. A teaspoonful of matter consisting of neutrons touching one another, as they do in the nucleus of an atom, would have a mass of about $8 \cdot 10^{12}$ kg. If neutrons weigh so much, why does a teaspoonful of normal matter weigh only a tiny fraction of a kilogram?

3. What force makes it difficult to bring nucleons together to form an atomic nucleus?

4. What force binds the nucleons in an atomic nucleus?

5. When a ^{235}U nucleus undergoes fission, what force does work on the fragments as the nucleus releases its potential energy?

6. When a ^{2}H nucleus and a ^{3}H nucleus undergo fusion, what force does work on the two nuclei and releases potential energy as they fuse together to become a larger nucleus?

7. A proton can induce fission when it strikes a fissionable nucleus. However, it's much easier to induce fission with a neutron than with a proton. Why?

8. The Heisenberg uncertainty principle recognizes that you can't determine exactly where an object is located. Explain how this principle is related to the fact that all objects travel as waves.

9. Small nuclei tend to release energy when they fuse together while large nuclei tend to release energy when they fall apart into fragments. Why?

10. A fission chain reaction is only possible if each fissioning nucleus emits an average of at least one neutron. Why?

11. Why can't ^{235}U be extracted from natural uranium through the use of chemical reactions?

12. Why does ^{239}Pu's severe radioactivity make assembling a critical mass so difficult?

Exercises

1. Alchemists tried for centuries to turn base metals into precious metals through chemical reactions. Why were they doomed to fail?

2. Why is it difficult or impossible to make very small atomic bombs?

3. Naturally occurring copper has two isotopes, ^{63}Cu and ^{65}Cu. What is different between atoms of these two isotopes?

4. Why is it extremely difficult to separate the two isotopes of copper, ^{63}Cu and ^{65}Cu?

5. If a nuclear reaction adds an extra neutron to the nucleus of ^{57}Fe (a stable isotope of iron), it produces ^{58}Fe (another stable isotope of iron). Will this change in the nucleus affect the number and arrangement of the electrons in the atom that's built around this nucleus? Why or why not?

6. If a nuclear reaction adds an extra proton to the nucleus of ^{58}Fe (a stable isotope of iron), it produces ^{59}Co (a stable isotope of cobalt). Will this change in the nucleus affect the number and arrangement of the electrons in the atom that's built around this nucleus? Why or why not?

7. When two medium-size nuclei are stuck together during an experiment at a nuclear physics lab, the result is usually a large nucleus with too few neutrons to be stable. The nucleus soon falls apart. Why could more neutrons make it stable?

8. To stick the two nuclei together in Exercise **7**, the nuclear physics laboratory had to accelerate them to enormous speeds and let them collide head on. Why was it so hard to get these nuclei together?

9. When a large nucleus is split in half during an experiment at a nuclear physics lab, the result is usually two medium-sized nuclei with too many neutrons to be stable. These nuclei eventually fall apart. Why don't these smaller nuclei need as many neutrons as they received from the original nucleus?

10. The uranium sold by chemical supply companies is "spent uranium"—natural uranium from which the ^{235}U has been removed. There's no concern that this spent uranium will be used to make an atomic bomb. Why not?

11. Suppose that there were a non-radioactive isotope that was fissionable and released several neutrons as it underwent fission. Building a nuclear weapon from this isotope would be extremely easy because it wouldn't require any explosives to detonate. Why wouldn't explosives be needed to detonate this weapon and how could it be detonated?

12. A hanging flower basket that's taped to the ceiling has a great deal of stored energy that won't be released until the tape gives way. Where is that energy stored?

13. Compare the system described in Exercise **12** to a radioactive nucleus—what features of the flower basket system play the roles of the repulsive electrostatic force, the attractive nuclear force, and the process of radioactive decay?

14. How could you arrange many hanging flower baskets (see Exercises **12** & **13**) so that a chain reaction could occur? What plays the role of critical mass in this arrangement?

15. When high explosives crush a sphere of plutonium so that it undergoes a chain reaction, the mass of the plutonium doesn't change, yet it exceeds "critical mass." Critical mass is clearly a misnomer. What *does* change about the plutonium as it's crushed that allows the chain reaction to proceed?

16. Once fallout from a nuclear blast distributes radioactive isotopes over a region of land, why is it virtually impossible to separate many of those radioactive isotopes from the soil?

17. Explain the strategy of putting highly radioactive materials in a storage place for many years as a way to make them less hazardous. That wouldn't work for chemical poisons, so why does it work for radioactive materials?

18. Burning chemical poisons in a gas flame often renders those poisons harmless. Why won't that strategy make radioactive materials less dangerous?

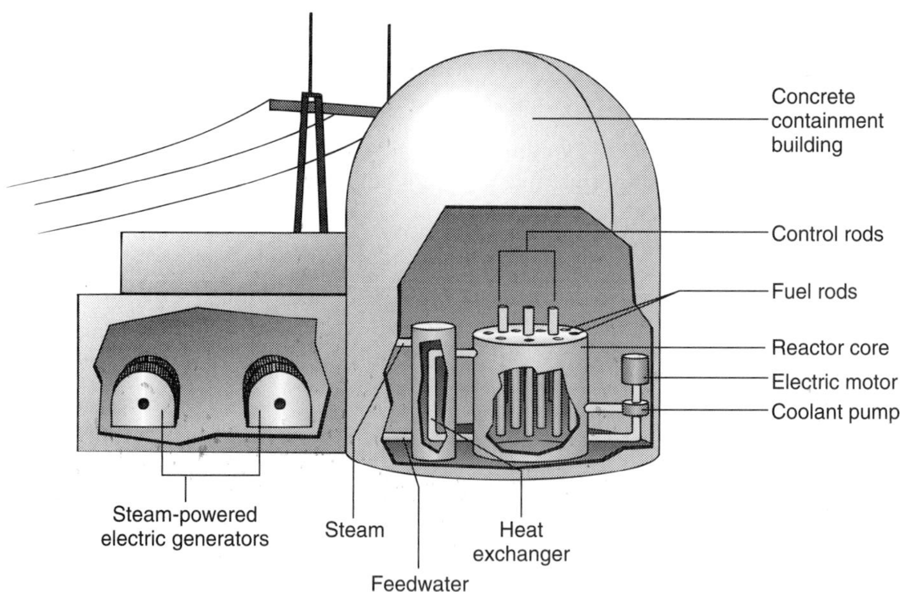

Section 18.2

Nuclear Reactors

Nuclear weapons were only one of two paths that lay before scientists and engineers at the end of the 1930's. While nuclear fission chain reactions and thermonuclear fusion were clearly ways to unleash phenomenal destructive energy, they could also provide virtually limitless sources of useful energy. By controlling the same nuclear reactions that occur in nuclear weapons, people have managed to extract nuclear energy for constructive uses. In the half century since their conception, nuclear fission reactors have developed into a fairly mature technology and have become one of our major sources of energy. Nuclear fusion power remains an elusive goal, but efforts continue to harness this form of nuclear energy as well.

Questions to Think About: *Why doesn't a nuclear reactor explode once it reaches a critical mass? How is the energy released in a nuclear reactor turned into electricity? What caused the famous accidents at Three Mile Island and Chernobyl? How can hydrogen fusion be initiated in a container? How is hydrogen held together at the temperatures needed for fusion?*

Experiments to Do: *A nuclear reactor operates very close to critical mass. Below critical mass, each spontaneous fission induces a limited number of subsequent fissions. Above critical mass, the number of subsequent fissions becomes unlimited. A similar effect occurs in a sand pile as it becomes steeper. If you pour sand slowly onto the pile, its shape will change and its steepness will increase. Initially, the grains will stay where they land, but once the pile becomes quite steep, they will begin to roll down the pile's sides. If the pile becomes too steep, a single grain of sand can trigger an avalanche. That same behavior occurs in a reactor that's near critical mass—a single spontaneous fission can trigger an enormous number of subsequent fissions.*

Nuclear Fission Reactors

Assembling a critical mass of uranium doesn't always cause a nuclear explosion. In fact, it's rather hard to cause a big explosion. The designers of the atomic bomb had to assemble not just a critical mass but a supercritical mass and they had to do it in much less than a millionth of a second. That's not something that happens easily or by accident. It's much easier to reach a critical mass slowly, in which case the uranium will simply become very hot. It may ultimately explode from overheating, but it will not vaporize everything in sight.

This slow assembly of a critical mass is the basis for nuclear fission reactors. Their principal product is heat, which is often used to generate electricity. Fission reactors are much simpler to build and operate than fission bombs because they don't require such purified fissionable materials. In fact, with the help of some clever tricks, nuclear reactors can even be made to operate with natural uranium.

Let's begin by showing that a fission chain reaction doesn't always lead to an explosion. What's important is just how fast the fission rate increases. In an atomic bomb, it increases breathtakingly quickly. At detonation, the fissionable material is far above the critical mass so the average fission induces not just one, but perhaps two, subsequent fissions. With only about 10 ns (10 nanoseconds) between one fission and the two it induces, the fission rate may double every 10 ns. In less than a millionth of a second, most of the nuclei in the material undergo fission, releasing their energy before the material has time to explode.

But things aren't so dramatic right at critical mass, where the average fission induces just one subsequent fission. Since each generation of fissions simply reproduces itself, the fission rate remains essentially constant. Only spontaneous fissions cause it to rise at all. The fissionable material steadily releases thermal energy and that energy can be used to power an electric generator.

A nuclear reactor contains a core of fissionable material. Because of the way in which this core is assembled, it's very close to a critical mass. Whether it's above or below critical mass is determined by several neutron-absorbing rods, called *control rods,* that are inserted into the reactor's core. Pulling the control rods out of the core increases the chance that each neutron will induce a fission and moves the core toward supercriticality. Dropping the control rods into the core increases the chance that each neutron will be absorbed before it can induce a fission and moves the core toward subcriticality.

A nuclear reactor uses feedback to maintain the fission rate at the desired level. If the fission rate becomes too low, the control system slowly pulls the control rods out of the core to increase the fission rate. If the fission rate becomes too high, the control system drops the control rods into the core to decrease the fission rate. It's like driving a car. If you're going too fast, you ease off the gas pedal. If you are going too slow, you push down on the gas pedal.

The car driving analogy illustrates another important point about reactors. Both cars and reactors respond relatively slowly to movements of their controls. It would be hard to drive a car that instantly stopped when you lifted your foot off the gas pedal and leaped to supersonic speed when you pushed your foot down. Similarly, it would be impossible to operate a reactor that immediately shut down when you dropped the control rods in and instantly exploded when you pulled the control rods out.

But reactors, like cars, don't respond quickly to movements of the control rods. That's because the final release of neutrons following a fission is slow. When a ^{235}U nucleus fissions, it promptly releases an average of 2.47 neutrons which induce other fissions within a thousandth of a second. But some of the fission fragments are unstable nuclei that decay and release neutrons long after the original fission. On average, each ^{235}U fission eventually produces 0.0064 of these *delayed neutrons*, which then go on to induce other fissions. It takes seconds or minutes for these delayed neutrons to appear and they slow the response of the reactor. The reactor's fission rate can't increase quickly because it takes a long time for the delayed neutrons to build up. The fission rate can't decrease quickly because it takes a long time for the delayed neutrons to go away.

To further ease the operation of modern nuclear reactors, they are designed to be stable and self-regulating. This self-regulation ensures that the core automatically becomes subcritical if it overheats. As we'll see later on, this self-regulation was absent in the design of Chernobyl Reactor Number 4.

CHECK YOUR UNDERSTANDING #1: No Delayed Decays

If the fission of ^{235}U produced only stable fission fragments, would a nuclear reactor be easier or harder to operate?

Thermal Fission Reactors

The basic concept of a nuclear reactor is simple: assemble a critical mass of fissionable material and adjust its criticality to maintain a steady fission rate. But what should the fissionable material be? In a fission bomb, it must be relatively pure ^{235}U or ^{239}Pu. But in a fission reactor, it can be a mixture of ^{235}U and ^{238}U. It can even be natural uranium. The trick is to use **thermal neutrons**—slow moving neutrons that have only the kinetic energy associated with the local temperature.

In a fission bomb, ^{238}U is a serious problem because it captures the fast moving neutrons emitted by fissioning ^{235}U nuclei. Natural uranium can't sustain a chain reaction because its many ^{238}U nuclei gobble up most of the fast moving neutrons before they can induce fissions in the rare ^{235}U nuclei. The uranium must be *enriched*, so that it contains more than the natural abundance of ^{235}U.

But slow moving neutrons have a different experience as they travel through natural uranium. For complicated reasons, the ^{235}U nuclei seek out slow moving neutrons and capture them with unusual efficiency. ^{235}U nuclei are so good at catching slow moving neutrons that they easily win out over the more abundant ^{238}U nuclei. Even in natural uranium, a slow moving neutron is more likely to be caught by a ^{235}U nucleus than it is by a ^{238}U nucleus. As a result, it's possible to sustain a nuclear fission chain reaction in natural uranium if all of the neutrons are slow moving.

❒ The first nuclear reactor was CP–1 (Chicago Pile–1), a thermal fission reactor constructed in a squash court at the University of Chicago. Each of the graphite bricks used in this pile contained two large pellets of natural uranium. By December 2, 1942, the pile was complete and would reach critical mass once the control rods were removed. As Enrico Fermi, the project leader, directed the slow removal of the last control rod, the pile approached criticality and the neutron emissions began to mount. It was noon, so Fermi called a famous lunch break. When everyone returned, they picked up where they had left off. At 3:25PM, the pile reached critical mass and the neutron emissions increased exponentially. The reactor ran for 28 minutes before Fermi ordered the control rods to be dropped back in.

But the previous paragraph seems to be hypothetical because ^{235}U nuclei emit fast moving neutrons when they fission. That's why pure natural uranium can't be used in a fission bomb. However, most nuclear reactors don't use pure natural uranium. They use natural uranium plus another material that's called a **moderator**. The moderator's job is to slow the neutrons down so that ^{235}U nuclei can grab them. A fast moving neutron from a fissioning ^{235}U nucleus enters the moderator, rattles around for about a thousandth of a second, and emerges as a slow moving neutron, one with only thermal energy left. It then induces fission in another ^{235}U nucleus. Once the moderator is present, even natural uranium can sustain a chain reaction! Reactors that carry out their chain reactions with slow moving or thermal neutrons are called *thermal fission reactors*.

To be a good moderator, a material must simply remove energy and momentum from the neutrons without absorbing them. When a fission neutron leaves a good moderator, it has only thermal energy left. The best moderators are nuclei that rarely or never absorb neutrons and don't fall apart during collisions with them. Hydrogen (^{1}H), deuterium (^{2}H), helium (^{4}He), and carbon (^{12}C) are all good moderators. When a fast moving neutron hits the nucleus of one of these atoms, the collision resembles that between two billiard balls. Because the fast

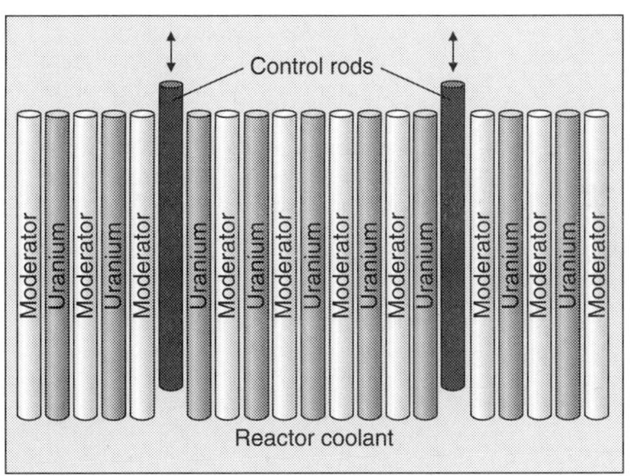

Fig. 18.2.1 - The core of a thermal fission reactor consists of uranium pellets, interspersed with a moderator that slows the fission neutrons to thermal energies. Neutron-absorbing control rods are inserted into the core to control the fission rate. A cooling fluid such as water flows through the core to extract heat.

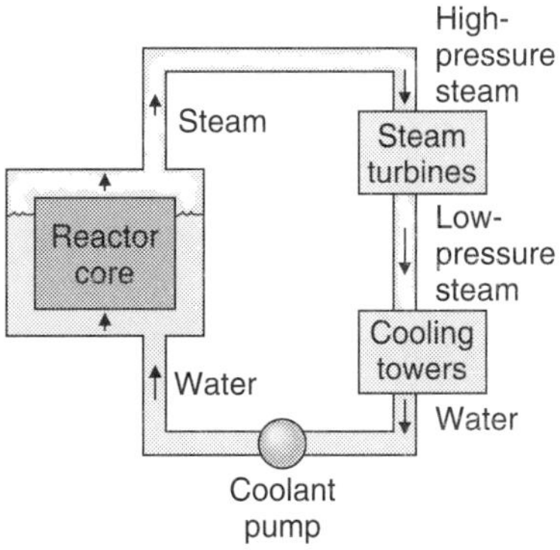

moving neutron transfers some of its energy and momentum to the nucleus, the neutron slows down while the nucleus speeds up.

Water, *heavy water* (water containing the heavy isotope of hydrogen: deuterium or ^{2}H), and graphite (carbon) are the best moderators for nuclear reactors. They absorb few neutrons while slowing those neutrons down to thermal speeds. Of these moderators, heavy water is the best because it slows the neutrons quickly without absorbing them at all. However, heavy water is expensive because only 0.015% of hydrogen atoms are deuterium and separating that deuterium from ordinary hydrogen is difficult.

Graphite moderators were used in many early reactors because graphite is cheap and easy to work with (see ❐). However, graphite is a less efficient moderator than heavy water, so graphite reactors had to be big. Furthermore, graphite can burn and was partly responsible for two of the world's three major reactor accidents. Normal or "light" water is cheap, safe, and an efficient moderator, but it absorbs enough neutrons that it can't be used with natural uranium. For use in a light water reactor, uranium must be enriched slightly, to about 2–3% ^{235}U.

The core of a typical thermal fission reactor consists of small uranium oxide (UO_2) fuel pellets, separated by layers of moderator (Fig. 18.2.1). A neutron released by a fissioning ^{235}U nucleus usually escapes from its fuel pellet, slows down in the moderator, and then induces fission in a ^{235}U nucleus in another fuel pellet. By absorbing some of these neutrons, the control rods determine whether the whole core is subcritical, critical, or supercritical. The ^{238}U nuclei are basically spectators in the reactor since most of the fissioning occurs in the ^{235}U nuclei.

In a practical thermal fission reactor, something must extract the heat released by nuclear fission. In many reactors, cooling water passes through the core at high speeds. Heat flows into this water and increases its temperature. In a *boiling water reactor*, the water boils directly in the reactor core, creating high pressure steam that drives the turbines of an electric generator (Fig. 18.2.2). In a *pressurized water reactor*, the water is under enormous pressure so it can't boil (Fig. 18.2.3). Instead, it's pumped to a heat exchanger outside the reactor. This heat exchanger transfers heat to water in another pipe, which boils to create the high pressure steam that drives a generator (Fig. 18.2.4).

When properly designed, a water-cooled thermal fission reactor is inherently stable. The cooling water is actually part of the moderator. If the reactor overheats and the water escapes, there will no longer be enough moderator around to slow the fission neutrons down. The fast moving neutrons will be absorbed by ^{238}U nuclei and the chain reaction will slow or stop.

Fig. 18.2.2 - In a boiling water reactor, cooling water boils inside the reactor core. It creates high-pressure steam that drives steam turbines and an electric generator. The spent steam condenses in a cooling tower and is then pumped back into the reactor.

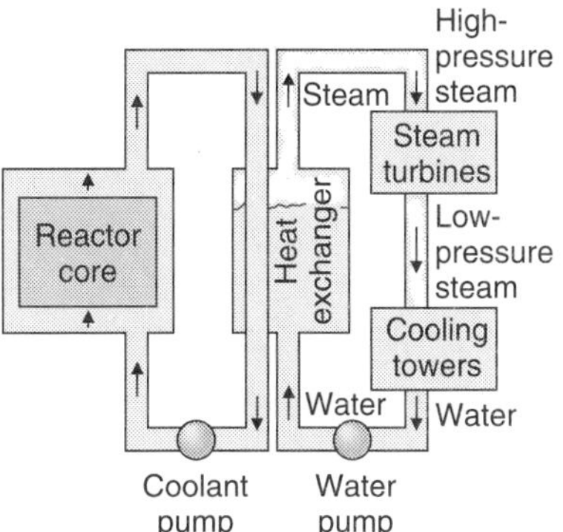

Fig. 18.2.3 - In a pressurized water reactor, liquid water under great pressure extracts heat from the reactor core. A heat exchanger allows this cooling water to transfer heat to the water used to generate electricity. Water in the generating loop boils to form high pressure steam, which then powers the steam turbines connected to the electric generators. The steam condenses back into liquid water and returns to the heat exchanger to obtain more heat.

CHECK YOUR UNDERSTANDING #2: Nuclear Candles

Would wax made from hydrogen and carbon atoms be a good moderator?

Fig. 18.2.4 - The North Anna Nuclear Power Station in Mineral, Virginia uses two Westinghouse pressurized light water reactors to generate a total of 1900 MW of electric power. At capacity, each reactor produces more than 2,900 MW of thermal power and can generate 950 MW of electric power. Waste heat from the plant is deposited in nearby Lake Anna.

Fast Fission Reactors

Thermal reactors require simple fuel and are relatively straightforward to construct. But they consume only the ^{235}U nuclei and leave the ^{238}U nuclei almost unaffected. Anticipating the day when ^{235}U will become scarce, several countries have built a different kind of reactor that contains no moderator. Such reactors carry out chain reactions with fast moving neutrons and are thus called *fast fission reactors*.

A fast fission reactor operates much like a controlled fission bomb and it requires highly enriched uranium fuel as well. While a thermal fission reactor can get by with natural uranium or uranium that has been enriched to about 2–3% ^{235}U, a fast fission reactor needs 25–50% ^{235}U fuel. That way, there are enough ^{235}U nuclei around to maintain the chain reaction.

But there is a side effect to operating a fast fission reactor. Many of the fast fission neutrons are captured by ^{238}U nuclei, which then transform into ^{239}Pu nuclei. Thus the reactor produces plutonium as well as heat. For that reason, fast fission reactors are often called *breeder reactors*—they create new fissionable fuel. The ^{239}Pu can eventually replace the ^{235}U as the principal fuel used in the reactor.

A thermal fission reactor makes some plutonium, which is usually allowed to fission in place, but a fast fission reactor makes a lot of it. Because plutonium can be used to make nuclear weapons, fast fission reactors are very controversial. However, because they convert otherwise useless ^{238}U into a fissionable material, they use natural uranium far more efficiently than thermal fission reactors.

One interesting complication of the unmoderated design is that fast fission reactors can't be water cooled. If they were, the water would act as a moderator and slow the neutrons down. Instead, they are usually cooled by liquid sodium metal. The sodium nucleus ^{23}Na rarely interacts with fast moving neutrons so it doesn't slow them down.

CHECK YOUR UNDERSTANDING #3: A Mix-up at the Uranium Plant
What would happen if they fueled a fast fission reactor with natural uranium?

Fission Reactor Safety and Accidents

One of the greatest concerns with nuclear fission reactors is the control of radioactive waste. Anything that comes in contact with the reactor core or the neutrons it emits becomes somewhat radioactive. The fuel pellets themselves are quickly contaminated with all sorts of fission fragments that include radioactive isotopes of many familiar elements. Some of these radioactive isotopes dissolve in water or are gases and all of them must be handled carefully.

The first line of defense against the escape of radioactivity is the large and sturdy containment vessel around the reactor. Because most of the radioactive materials remain in the reactor core itself or in the cooling fluid, they are trapped in the containment vessel. Whenever the nuclear fuel is removed for reprocessing, care is taken not to allow radioactive materials to escape.

The other great concern is the safe operation of the reactors themselves. Like any equipment, reactors experience failures of one type or another and a safe reactor must not respond catastrophically to such failures. Toward that end, reactors have emergency cooling systems, pressure relief valves, and many ways to shut down the reactor. For example, injecting a solution of sodium borate into the core cools it and stops any chain reactions. The boron nuclei in sodium borate absorb neutrons extremely well and are the main contents of most control rods. But the best way to keep reactors safe is to design them so that they naturally stop their chain reactions when they overheat.

There have been three major reactor accidents in the past half century. The first of these accidents occurred in 1957 at Windscale Pile 1, one of Britain's two original plutonium production reactors. This thermal fission reactor was cooled by air rather than water and had a graphite moderator. During a routine shutdown, the reactor overheated. The graphite's crystalline structure had been modified by the reactor's intense radiation and had built up a large amount of chemical potential energy. When that energy was suddenly released during the shutdown, the graphite caught fire and distributed radioactive debris across the British countryside.

The second serious accident occurred at Three Mile Island in 1979. This pressurized water thermal fission reactor shut down appropriately when the pump that circulated water in the power generating loop failed. Although this water loop wasn't directly connected to the reactor, it was important for removing heat from the reactor core. Even though the reactor was shut down with control rods and had no chain reaction in it, the radioactive nuclei created by recent fissions were still decaying and releasing energy. The core continued to release heat and it eventually boiled the water in the cooling loop. This water escaped from the loop through a pressure relief valve and the top of the reactor core became exposed. With nothing to cool it, the core became so hot that it suffered permanent damage. Some of the water from the cooling loop found its way into an unsealed room and the radioactive gases it contained were released into the atmosphere.

The third and most serious accident occurred at Chernobyl Reactor Number 4 on April 26, 1986. This water-cooled, graphite-moderated thermal fission reactor was a cross between a pressurized water reactor and a boiling water reactor. Cooling water flowed through the reactor at high pressure but didn't boil until it was ready to enter the steam generating turbines.

The accident began during a test of the emergency core cooling system. To begin the test, the operators tried to reduce the reactor's fission rate. However, the core had accumulated many neutron-absorbing fission fragments which made it incapable of sustaining a chain reaction at a reduced fission rate. The chain reaction virtually stopped. To get the chain reaction running again, the operators had to withdraw a large number of control rods. These control rods were motor-driven, so it would take about 20 seconds to put them back in again.

The operators now initiated the test by shutting off the cooling water. That should have immediately shut down the reactor by inserting the control rods, but the operators had overridden the automatic controls because they didn't want to have to restart the reactor again. With nothing to cool it, the reactor core quickly overheated and the water inside it boiled. The water had been acting as a

Fig. 18.2.5 - Several hours after a misguided experiment caused it to exceed the critical mass and vaporize parts of its core, Chernobyl Reactor Number 4 continues to emit radioactive smoke and steam.

moderator along with the graphite. But the reactor was overmoderated, meaning it had more moderator than it needed. Getting rid of the water actually helped the chain reaction because the water had been absorbing some of the neutrons. The fission rate began to increase.

The operators realized they were in trouble and began to shut down the reactor manually. However, the control rods moved into the core too slowly to make a difference. As the water left the core, the core went "prompt critical." The chain reaction no longer had to wait for neutrons from the decaying fission fragments because prompt neutrons from the ^{235}U fissions were enough to sustain the chain reaction on their own. The reactor's fission rate skyrocketed, doubling many times each second. The fuel became white hot and melted its containers. Various chemical explosions blew open the containment vessel and the graphite moderator caught fire.

The fire burned for 10 days before firefighters and pilots encased the wreckage (Fig. 18.2.5) in concrete. Many of these heroic people suffered fatal exposures to radiation. The burning core releasing all of its gaseous radioactive isotopes and many others into the atmosphere, forcing the evacuation of more than 100,000 people.

CHECK YOUR UNDERSTANDING #4: A Breath of Not-So-Fresh Air

One of the common fission fragments of ^{235}U is ^{131}I, a radioactive isotope of iodine. Iodine sublimes easily and becomes a gas. What will happen if you sit in a room full of used uranium fuel pellets?

Nuclear Fusion Reactors

Nuclear fission reactors use a relatively rare fuel: uranium. While the earth's supply of uranium is vast, most of that uranium is distributed broadly throughout the earth's crust. There are only so many deposits of high-grade uranium ores that are easily turned into pure uranium or uranium compounds. Fission reactors also produce all sorts of radioactive fission fragments that must be disposed of safely. There is still no comprehensive plan for safe keeping of spent reactor fuels. These must be kept away from any contact with people or animals virtually forever. No one really knows how to store such dangerous materials for hundreds of thousands of years.

An alternative to nuclear fission is nuclear fusion. By joining hydrogen nuclei together, heavier nuclei can be constructed. The amount of energy released in such processes is enormous. However fusion is much harder to initiate than fission because it requires that at least two nuclei be brought extremely close together. These nuclei are both positively charged and they repel one another fiercely. To make them approach one another closely enough to stick, the nuclei must be heated to temperatures of more than 100 million degrees Celsius.

The sun combines four hydrogen nuclei (^{1}H) to form one nucleus of helium (^{4}He), a very complicated and difficult nuclear fusion reaction. For fusion to occur on earth, it must be done between the heavy isotopes of hydrogen: deuterium and tritium. These are the isotopes used in thermonuclear weapons. If a mixture of deuterium and tritium are mixed together and heated to about 100 million degrees, their nuclei will begin to fuse and release energy. The deuterium and tritium become helium and neutrons.

In contrast to fission reactions, there are no radioactive fragments produced. Tritium itself is radioactive, but can easily be reprocessed into fuel and retained within the reactor system. The dangerous neutrons can be caught in a blanket of lithium metal, which then breaks into helium and tritium. It's convenient that new tritium is created because tritium isn't naturally occurring and must be made by nuclear reactions. Thus, fusion holds up the promise of producing relatively little radioactive waste. If the neutrons that are released by fusion events are trapped by nuclei that don't become radioactive, then there will be no radioactive contamination of the fusion reactor either. This is easier said than done, but it's better than in a fission reactor.

Unfortunately, heating deuterium and tritium and holding them together long enough for fusion to occur isn't easy. There are two main techniques that are being tried: inertial confinement fusion and magnetic confinement fusion.

Inertial confinement fusion uses intense pulses of laser light to heat and compress a tiny sphere containing deuterium and tritium (Fig. 18.2.6). The pulses of light last only a few trillionths of a second, but in that brief moment they vaporize and superheat the surface of the sphere. The surface explodes outward, pushing off the inner portions of the sphere. The sphere's core experiences huge inward forces as a result and it implodes. As it's compressed, the temperature of this core rises to that needed to initiate fusion. In effect, it becomes a tiny thermonuclear bomb with the laser pulses providing the starting heat.

To date, inertial confinement fusion experiments have observed fusion in a small fraction of the deuterium and tritium nuclei. The technique is called inertial confinement fusion because there is nothing holding or confining the ball of fuel. The laser beams crush it while it's in free fall and its own inertia keeps it in place while fusion takes place. Unfortunately, the lasers and other technologies needed to carry out inertial confinement fusion are sufficiently complex and troublesome that it may never be viable as a source of energy. Nonetheless, these experiments provide important information on the behaviors of fusion materials at high temperatures and pressures.

The other technique being developed to control fusion is *magnetic confinement*. When you heat hydrogen atoms hot enough, they move so quickly and hit one another so hard that their electrons are knocked off. Instead of a gas of atoms you have a gas of free positively charged nuclei and negatively charged electrons, a plasma. Along with being hot, a plasma differs from a normal gas because it's affected by magnetic fields.

As we saw in the section on television, moving charged particles tend to circle around magnetic field lines, a behavior called cyclotron motion (Fig. 18.2.7). If the magnetic field surrounding a charged particle is carefully shaped in just the right way, the charged particle will find itself trapped by the magnetic field. No matter what direction it heads, the charged particle will spiral around the magnetic field lines and will be unable to escape.

Magnetic confinement makes it possible to heat a plasma of deuterium and tritium to fantastic temperatures with electromagnetic waves. Since the heating is done relatively slowly, it's important to keep heat from leaving the plasma. Magnetic confinement prevents the plasma from touching the walls of the container, where it would quickly cool off.

One of the most promising magnetic confinement schemes is the *tokamak*. The main magnetic field of the tokamak runs around in a circle to form a magnetic doughnut or toroid, the geometrical name for a doughnut shaped object (Fig. 18.2.8). The magnetic field is formed inside a doughnut-shaped chamber by running an electric current through coils that are wrapped around the chamber. Plasma nuclei inside the chamber travel in spirals around the magnetic field lines and don't touch the chamber walls. They are confined inside the chamber and race around the doughnut endlessly. The nuclei can then be heated to the extremely high temperatures they need in order to collide and fuse.

Magnetic confinement fusion reactors have observed considerable amounts of fusion (Fig. 18.2.9). They are now close to scientific break-even, the point at which fusion is releasing enough energy to keep the plasma hot all by itself.

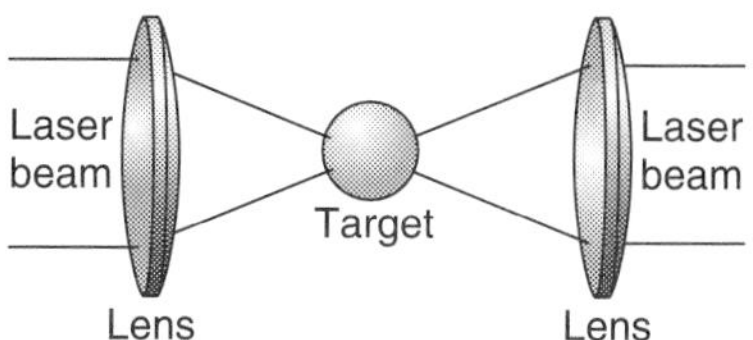

Fig. 18.2.6 - In inertial confinement fusion experiments, several laser beams are focused onto a tiny sphere containing deuterium and tritium. These ultra-intense pulsed beams compress and heat the sphere so that fusion occurs.

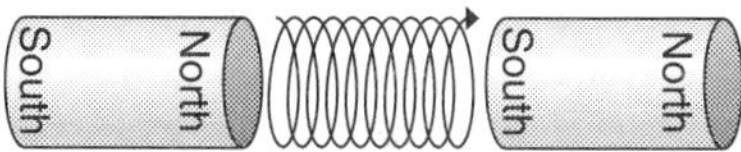

Fig. 18.2.7 - When a charged particle moves in the magnetic field between two magnetic poles, it travels in a spiral path around the magnetic field lines connecting them. The particle is confined to a particular region of space.

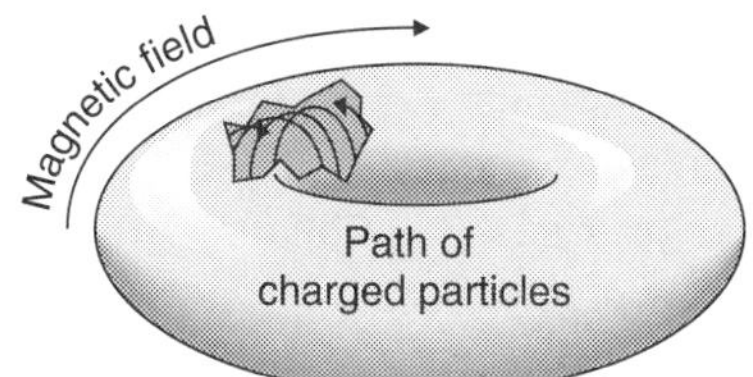

Fig. 18.2.8 - A tokamak magnetic confinement fusion reactor consists of a doughnut shaped chamber with a similarly shaped magnetic field inside it. Plasma particles moving inside the tokamak's chamber travel in spirals and circle around the field lines inside the chamber. Because the plasma doesn't touch the walls of the tokamak, it retains heat well enough to reach temperatures at which fusion can occur.

Fig. 18.2.9 - The Tokamak Fusion Test Reactor at Princeton University is an experimental facility used to learn how to induce fusion between hydrogen nuclei.

More development is needed to meet and exceed practical break-even, where the entire machine produces more energy than it needs to operate.

CHECK YOUR UNDERSTANDING #5: Trying for Fusion in Your Basement

Why can't you initiate fusion by running an electric discharge through deuterium and tritium in a glass tube?

Summary

How Nuclear Reactors Work: A nuclear fission reactor releases energy by controlling a fission chain reaction. The reactor's core contains a fissionable material such as ^{235}U or ^{239}Pu. By carefully adjusting the positions of control rods in that core, the reactor is able to go from subcritical to critical to supercritical and back again. The reactor initially allows chain reactions to start and then moves the control rods in and out so as to keep the fission rate essentially constant at the desired value. The fissions release energy into the material where it can be extracted as heat by a cooling fluid such as water or a gas. This heated fluid can then be used to generate electricity.

In a thermal fission reactor, fast moving fission neutrons are slowed to thermal energies in a moderator such as water, heavy water, or graphite. These slow neutrons are preferentially absorbed by ^{235}U, which then undergoes fission. Because ^{235}U is so effective at absorbing slow neutrons, a thermal fission reactor can operate with natural or slightly enriched uranium. In a fast fission reactor, fast moving fission neutrons must directly induce subsequent fissions. Because ^{235}U is not particularly good at absorbing fast neutrons, a fast reactor must use enriched uranium. However the fast neutrons convert many of the ^{238}U nuclei in the fuel to ^{239}Pu, which can then be induced to fission.

Nuclear fusion reactors remain in the development stage. They heat the heavy isotopes of hydrogen, deuterium and tritium, to extremely high temperatures and confine them together until their nuclei undergo fusion. In inertial confinement fusion, pulses of laser light crush a small sphere of these materials and initiate a thermonuclear explosion. In magnetic confinement fusion, a hot plasma of the heavy hydrogen isotopes is confined in a magnetic field and heated until the nuclei begin to fuse.

The Physics of Nuclear Reactors

1. At critical mass, the fission rate in a reactor core is essentially constant because each fission induces an average of one subsequent fission. Above critical mass, the fission rate increases rapidly while below critical mass, the fission rate decreases rapidly.

2. While most fission neutrons are released promptly, radioactive fission fragments sometimes emit neutrons seconds or minutes after the original fission. These delayed neutrons slow down any increases or decreases of the fission rates in a reactor's core, making that reactor easier to control.

3. Fast fission neutrons that collide with moderator nuclei give up much of their energies and momenta. After several collisions with the moderator, the neutrons emerge with only thermal energies. These thermal neutrons are particularly effective at inducing fission in ^{235}U and are able to sustain a chain reaction, even in natural uranium.

Check Your Understanding - Answers

1. It would be much harder to operate.

Why: Without any delayed neutrons, one generation of fissions would finish inducing the next generation of fissions in about a thousandth of a second and the response of the reactor core to changes in the criticality would become extremely rapid. When the reactor core became supercritical, its fission rate would skyrocket and when it became subcritical, its fission rate would plummet. Trying to keep the fission rate steady would be a wild roller coaster ride and the reactor would be impossible or impractical to control. Fortunately the delayed neutrons slow the reactor's response to changes in criticality so that it can be controlled fairly easily.

2. Yes.

Why: Paraffin wax is made up of carbon and hydrogen atoms, both of which are good moderators. The fast fission neutrons don't care what molecules the carbon or hydrogen nuclei are inside. They just collide with those nuclei and lose energy and momentum. The problem with using wax as a reactor moderator is that it decomposes chemically in the intense radiation. However, wax can safely slow small numbers of neutrons in a laboratory situation.

3. It would not sustain a chain reaction.

Why: Fast neutrons aren't good at inducing fission in ^{235}U. If the reactor doesn't slow the neutrons, most of them will zip right by the ^{235}U nuclei and be captured by the many ^{238}U nuclei of the natural uranium.

4. The ^{131}I will enter the room air and you will breathe it in. It will be incorporated in your body's iodine supply.

Why: The fission products of ^{235}U include radioactive isotopes of almost every element. Some of these isotopes are gaseous and can be absorbed into your lungs. Others can be carried away in water. To safely contain the fission fragments from used nuclear fuel, these fragments must be carefully separated and stored. Because the radioactive isotopes may continue to change from one element to another, such storage is particularly tricky.

5. The gases will lose heat to the glass tube and will never reach the temperatures needed to start fusion.

Why: Very hot gases lose heat extremely easily. The hotter the gas, the faster heat flows out of it. Without something to keep the gas from touching the walls of the glass tube, it will never get close to the 100 million degrees Celsius needed for fusion.

Glossary

moderator A material with which fast neutrons can collide and give up energy without being absorbed. Neutrons leave a moderator with only thermal energies.

thermal neutrons Neutrons that move with only thermal energies so that their kinetic energies are consistent with the local temperature.

Review Questions

1. Enrico Fermi's original nuclear reactor, CP–1, was built from pellets of natural uranium in graphite blocks. What was the purpose of the graphite blocks?

2. How does inserting control rods into a nuclear reactor slow or stop its chain reaction?

3. If there were no delayed neutrons in a nuclear reactor—if every fission released a few neutrons promptly and left only stable nuclei as fragments—the nuclear reactor would be almost impossible to control. Why?

4. Why does slowing down the neutrons in a reactor make it possible to operate that reactor using natural or slightly enriched uranium?

5. A nuclear reactor that uses ordinary water as a moderator must be fueled with slightly enriched uranium, while a reactor that uses heavy water can use natural uranium. Explain.

6. Why must a fast fission reactor operate with highly enriched uranium?

7. Explain how a fast fission reactor is able to "produce" more nuclear fuel than it consumes.

8. Fast fission reactors are often cooled by liquid sodium metal. Why not simply use water as a coolant?

9. The coolant in a modern thermal fission reactor acts as part of that reactor's moderator. Why does this design ensure that the reactor's chain reaction will stop if there's a loss of coolant in the reactor core?

10. Why would one of the overmoderated Chernobyl-style reactors in the former Soviet Union experience an increase in its fission rate if it were to lose cooling water?

11. Why does initiating fusion between hydrogen nuclei require enormous temperatures?

12. Normal hydrogen atoms experience no forces as they move through magnetic fields. Why is the ultra-hot hydrogen in a tokamak confined by the tokamak's magnetic fields?

Exercises

1. The uranium fuel in a pressurized light water reactor must be replaced every so often. When that fuel is removed, it's still mostly uranium. Why can't this uranium be cleansed of impurities and simply put right back into the reactor over and over again until the uranium is completely used up? What changes are occurring in this uranium that make it less able to support a chain reaction?

2. While a fission bomb can be used to initiate fusion in hydrogen as part of a hydrogen bomb, a power plant can't use a fission reactor to initiate fusion in hydrogen as a way to generate electricity. Why not?

3. Control rods are usually installed on top of the reactor where their weights tend to pull them into the core. Why is this arrangement much safer than putting the control rods at the bottom of the reactor?

4. One popular scenario for a reactor accident has the reactor core melting and then drilling deep into the ground as the chain reaction in the molten core continues. But that won't actually happen for a thermal fission reactor. While the loss of cooling water could lead to the core melting, the chain reaction would stop and the core wouldn't be able to drill a deep hole. Explain why the chain reaction would stop.

5. Some satellites have been launched with small nuclear reactors on board to provide electric power. Why must these small reactors use highly enriched uranium or plutonium rather than natural or slightly enriched uranium?

6. The nucleus of a normal hydrogen atom is a proton which has almost exactly the same mass as a neutron. When a fast moving neutron hits the nucleus of a stationary hydrogen atom head on, the neutron and proton exchange their momentums. The proton rebounds at high speed while the neutron practically stops moving—a behavior similar to that of the billiard balls discussed in Section 2.2. Use this information to explain why water makes an excellent moderator for thermal fission reactors.

7. Suppose that oxygen nuclei frequently absorbed neutrons. How would that behavior affect water's usefulness as a moderator?

8. The nucleus of a carbon atom has about 12 times the mass of a neutron. A collision between a neutron and a carbon nucleus is roughly analogous to the collision between a softball and a junior-weight bowling ball. Explain why a fast moving neutron must collide with many carbon nuclei before it slows to thermal speeds.

9. Graphite—a crystalline form of carbon—is used as a moderator in some thermal fission reactors. After use as a moderator, however, the graphite is not entirely crystalline; much of it is amorphous. Why does acting as a moderator gradually convert the carbon from a crystalline material to an amorphous material?

10. When water is used as a moderator in a thermal fission reactor, some of it is converted into hydrogen and oxygen gases. Explain why some of the water molecules decompose while the reactor is operating.

11. Among the fission fragments that gradually accumulate in the fuel rods of a nuclear reactor are nuclei that are strong absorbers of neutrons. The presence of these nuclei in used fuel rods is one reason that those rods must be removed from the reactor periodically for reprocessing. As the neutron absorbing nuclei accumulate in the fuel rods, what must the reactor operators do to keep the chain reaction going?

12. Inertial confinement fusion initiates fusion by using the temperature rise that occurs when a pellet of hydrogen gas is compressed. Why does compressing the hydrogen gas cause its temperature to rise?

13. A gas of hydrogen atoms or molecules won't respond to microwaves. However, a plasma formed from hydrogen atoms will become extremely hot when intense microwaves pass through it. Why does hydrogen plasma respond to microwaves while hydrogen gas does not?

14. The magnetic field of a tokamak prevents the hydrogen plasma from touching the walls of its container and losing heat by conduction or convection. But the plasma does radiate away some of its thermal energy whenever its particles collide. Why must the moving charged particles collide with one another in order to emit electromagnetic waves?

Epilogue for Chapter 18

In this chapter we examined the nature of atomic nuclei and the ways in which they can be altered through nuclear reactions. In *nuclear weapons*, we examined nuclear fission to see how nuclear chain reactions can be used to release electrostatic potential energy stored in the large nuclei of uranium and plutonium atoms. We also studied nuclear fusion and found that when small nuclei of hydrogen bind together, they release a potential energy associated with the nuclear force. We saw how a fission explosion can be used to reach the enormous temperatures needed to initiate fusion here on earth. We learned about radioactive isotopes and explored issues related to nuclear proliferation and fallout.

In nuclear reactors, we saw how both fission and fusion can be controlled in order to obtain useful energy. We learned how moderators facilitate a chain reaction in natural or only slightly enriched uranium and how unmoderated reactors running on highly enriched uranium can be used to breed plutonium, another reactor fuel. We examined the major reactor accidents to see what went wrong and looked at many of the safety issues associated with the use of nuclear reactors for the production of electric power.

Explanation: Thoughts About the Sun's Fire

The sun's source of energy is nuclear fusion. Deep within the sun, hydrogen nuclei (^{1}H) are joining together to form helium nuclei (^{4}He) in a process that releases prodigious amounts of energy. These nuclear reactions have provided the enormous power needed to keep the sun glowing with incandescent heat for billions and billions of years. This recognition that nuclear energy heats the sun and stars is a relatively new idea, dating back only to the 1920's and 30's. Before that time, no one really knew what made the sun shine.

Cases

1. Nuclear power plants inevitably convert some of the uranium in their fuel rods into plutonium. One of the great concerns with the presence of nuclear reactors and power plants in so many countries is that those countries might try to extract plutonium from the used reactor fuel and use that plutonium in nuclear weapons.

a. Instead of worrying about plutonium, perhaps we should be worried about the uranium fuel rods themselves. Fortunately, the fuel rods themselves aren't a danger. Why can't a typical nuclear reactor's uranium fuel be used to make nuclear weapons?

b. Why is it much easier to extract ^{239}Pu from the used fuel rods than it is to extract ^{235}U from those same fuel rods?

c. In most nuclear power reactors, the plutonium that's produced isn't pure ^{239}Pu. About 25% of the plutonium is actually ^{240}Pu which is about 4 times as radioactive as ^{239}Pu (^{240}Pu has a half life of only 6,580 years, compared to ^{239}Pu's half life of 24,400 years). Some people argue that plutonium from nuclear power reactors is "safer" than weapons grade plutonium because the large fraction of ^{240}Pu in the reactor grade plutonium makes assembling a supercritical mass of reactor grade plutonium very difficult. Why would it be hard to assemble a supercritical mass of reactor grade plutonium?

d. Reactor grade plutonium also contains a percentage or two of ^{238}Pu. This highly radioactive isotope (with a half life of 88 years) contributes to the problem discussed in **c** and adds a new problem: heat. Even without a chain reaction, the core of a nuclear weapon built from reactor grade plutonium would need a cooling system like that in a nuclear reactor. This cooling system would have to run all the time while the weapon was "sitting on the shelf." Why?

e. Why is reactor grade plutonium, which has these relatively large proportions of ^{238}Pu and ^{240}Pu, a more serious health risk to the people working with it than weapons grade plutonium, which is relatively pure ^{239}Pu?

2. Carbon dating is a technique used to determine the ages of carbon-containing objects that obtain their carbon from contemporary plants. It's based on a radioactive isotope of carbon, ^{14}C (carbon–14), that's present in our atmosphere in small quantities.

a. The ^{14}C in our atmosphere is produced indirectly by cosmic rays—high energy particles that enter our atmosphere from elsewhere in the universe. When these cosmic rays collide with nuclei in the atmosphere, they often produce neutrons, and some of these neutrons interact with the nuclei of nitrogen atoms. Why is it easy for a neutron to approach a ^{14}N nitrogen nucleus in order to stick to it?

b. When it sticks to a ^{14}N nitrogen nucleus, a neutron causes that nucleus to rearrange and release a proton. The result is a ^{14}C nucleus. Adding a neutron and subtracting a proton leaves the number of nucleons in the nucleus unchanged, but it converts one chemical element into another. Why is ^{14}C chemically different and distinguishable from ^{14}N?

c. ^{14}C is radioactive with a half life of 5,730 years. That means there's a 50% chance of each ^{14}C nucleus spontaneous falling apart during a period of 5,730 years. But because radioactive decay is a true chance phenomenon and doesn't depend on what happened before, a ^{14}C nucleus that survived the first 5,730 years intact has a 50% chance of surviving the next 5,730 years. If it survives the second 5,730 years, then it has a 50% chance of surviving the third 5,730 years, and so on. After 28,650 years, what fraction of the original ^{14}C nuclei in an object will remain?

d. From the 5,730 year half life of ^{14}C, it's possible to calculate the fraction of ^{14}C nuclei that will survive 1 year without decaying: 99.9879%. The fraction of nuclei that survive 2 years without decaying is 99.9879% of 99.9879%, or 99.9758%. And so on. If you know how many ^{14}C an object contained when it was new, how could you determine how old the object is?

e. Most carbon nuclei are not radioactive. The carbon in rocks and petroleum is about 99% ^{12}C and about 1% ^{13}C. But because ^{14}C is introduced into the atmosphere at a steady rate and plants obtain their carbon from the atmosphere, ^{14}C nuclei make up a small but important fraction of the carbon nuclei in a living plant. This fraction can be measured in a modern plant to determine its value in an ancient plant while that plant was alive. After the plant died, the fraction of carbon nuclei that were ^{14}C decreased as the ^{14}C nuclei decayed. How does measuring the fraction of ^{14}C nuclei in a plant's carbon allow scientists to determine how long ago it lived?

f. Scientists can also determine how long ago an animal lived if that animal ate plants to obtain carbon. But if the animal obtained a substantial fraction of its carbon from minerals, that dating process wouldn't work. Why not?

3. Impressed by your mastery of video games, the local nuclear power plant has offered you a part-time job controlling their reactor. The graphics are lousy and the controls aren't very responsive, but the pay is good so you take the job.

a. You are supposed to move the control rods up and down to maintain a steady rate of nuclear decays in the reactor core. You find that when you insert the control rods too far into the core, the rate of nuclear decays diminishes. Since natural radioactive decay is completely random and can't be controlled by outside influences, how do the control rods affect the rate at which ^{235}U nuclei decay in a reactor?

b. Water flows through the reactor core at an enormous rate. Why must the water keep moving?

c. The power plant produces electricity for weeks without needing any fuel shipments and without releasing anything but heat into the environment. What is being used up to produce this electric energy?

d. One day you decide to see what will happen when you pull the control rods completely out of the reactor core. It's a moment reminiscent of your video game days, with a spectacular sound and light show, but sadly no free game. What has probably happened to the reactor core?

4. Radon is a colorless, odorless gas that's extremely radioactive. It has three naturally occurring isotopes—^{219}Rn, ^{220}Rn, and ^{222}Rn—that are made when radioactive elements in the earth's crust decay. While the nuclei of these isotopes each contain 86 protons, they contain different numbers of neutrons. Radon is thought to be the single most important cause of lung cancer in nonsmokers. It's particularly abundant in indoor air above geological formations that include radon-producing minerals.

a. ^{219}Rn, ^{220}Rn, and ^{222}Rn nuclei are found in the atoms of chemically inert gases. In fact, radon is one of the noble gas elements which include helium, neon, argon, krypton, and xenon. It's not surprising that the atoms of all three isotopes are gases. It would be much more surprising if one isotope were normally a solid, one a liquid, and the third a gas. Why?

b. The first isotope, ^{219}Rn, is produced when the radioactive element actinium (^{227}Ac) decays. The half life of ^{219}Rn is extremely short, only 3.92 s. If there were 800 of these ^{219}Rn nuclei in a box, how long would it be before only about 25 of them were left?

c. The second isotope of radon, ^{220}Rn, is produced when the radioactive element thorium (^{232}Th) decays. This isotope has a half life of 54.5 s and decays into yet another radioactive nucleus. The ensuing series of radioactive decays usually produces a stable isotope of lead (^{208}Pb). In the long series of unstable nuclei that appear briefly as thorium turns into lead, the only nucleus that forms a gaseous atom is ^{220}Rn. From that information, why can you safely assume that very little ^{220}Rn is found in building air, even if it's built above rocks that are rich in thorium?

d. The third radon isotope, ^{222}Rn, is produced when uranium (^{238}U) decays. This isotope has a half life of 3.8 days. Why is this isotope much more likely than the others to be found in building air, particularly if that building is built above uranium-containing rocks?

5. There are two types of smoke detectors in common use: photoelectric and ionization. A photoelectric smoke detector uses a beam of light to "see" the smoke particles. Unfortunately, it isn't sensitive to the smallest smoke particles and requires considerable electric power to operate. An ionization smoke detector is more sensitive and consumes less power. It uses a radioactive material to ionize the air so that a small amount of electric current can pass through that air. When smoke is present, the ionized air molecules stick to the smoke particles, less current flows through the air, and the detector's alarm goes off. The radioactive material in an ionization smoke detector is an isotope of americium, a radioactive element that's not found in nature. The specific isotope of americium is ^{241}Am. This isotope is made by adding neutrons to ^{238}U in a nuclear reactor. In fact, ^{241}Am is made in exactly the same way that ^{239}Pu is made, except that ^{241}Am has been exposed to a few more neutrons.

a. ^{241}Am has a half life of 458 years. What would happen if an ionization smoke detector used a radioactive isotope with a half life of 458 days instead?

b. Many ^{241}Am nuclei were made during the supernova explosion that produced the earth's uranium. Why aren't there any naturally occurring ^{241}Am nuclei around today?

c. When uranium nuclei undergo spontaneous fission in the earth's crust, their fragments are often radioactive nuclei. Why aren't any of these fragments ^{241}Am nuclei?

d. An ^{241}Am nucleus decays into a ^{239}Np nucleus, an isotope of neptunium with a half life of 2 million years. Both americium and neptunium are radioactive solids. Why is it important that the radioactive isotope used in the smoke detector not decay into a radioactive gas such as radon?

CHAPTER 19

MODERN PHYSICS

In recent years, scientists have been looking deeper into the atom to see how it's made, farther into space to see how the universe works, and more carefully into objects to see how complicated things can be understood in terms of simple laws. Among the most important tools that these scientists have to work with are quantum theory and the theory of relativity. Many results of modern physics have already appeared in this book, particularly in the sections on audio amplifiers, lasers, nuclear weapons, and nuclear reactors. But there are still other ways in which modern physics contributes to our lives. This chapter will take a look at some of those contributions.

Experiment: Radiation Damage to Colored Paper

One path of modern physics involves the control and use of high energy radiation. While most sources of high energy radiation are carefully controlled so that you can't experiment with them, there's one source of somewhat high energy radiation that anyone can use: the sun. Because the sun's ultraviolet light is energetic enough to damage chemical bonds and rearrange molecules, it can provide a glimpse of the effects that occur with X-rays and beyond.

To see sun damage for yourself, expose a sheet of colored construction paper or a cheaply printed product wrapper to direct sunlight for a few days. Cover the paper with a few opaque objects such as coins and place it outdoors. Don't cover the paper with glass because glass absorbs enough of the ultraviolet light to slow the damage process. After a few days, you should find that the exposed portions of the paper have lightened. The sun's ultraviolet radiation has destroyed some of the dye molecules in the paper. If nothing happens, the dye is evidently robust enough to tolerate ultraviolet light for a while. Try the experiment again with a different page and a different color. Some dyes fade much faster than others and reds tend to be particularly vulnerable to damage.

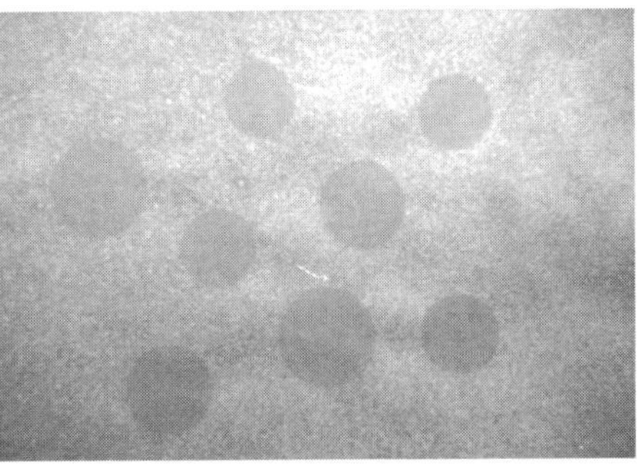

This same optical bleaching appears on items that are displayed in shop windows and on outdoor furniture. It was once the only method people had for whitening fabrics. The sun's ultraviolet light also damages your skin when you sit in the sun—a sunburn isn't thermal damage, it's radiation damage.

Chapter Itinerary

Fortunately ultraviolet light can't penetrate far into your body. In *medical imaging and radiation,* we'll examine some much more penetrating forms of high energy radiation and we'll see how those forms of radiation are used to help rather than hurt. We'll study the ways in which X-rays are produced and how they interact with the atoms and molecules in a patient. We'll also look at the particle accelerators that produce high energy particles for radiation therapy. Finally, we'll discuss the bases for CT and MRI imaging, which make it possible to prepare detailed maps of patients' insides without ever touching their bodies.

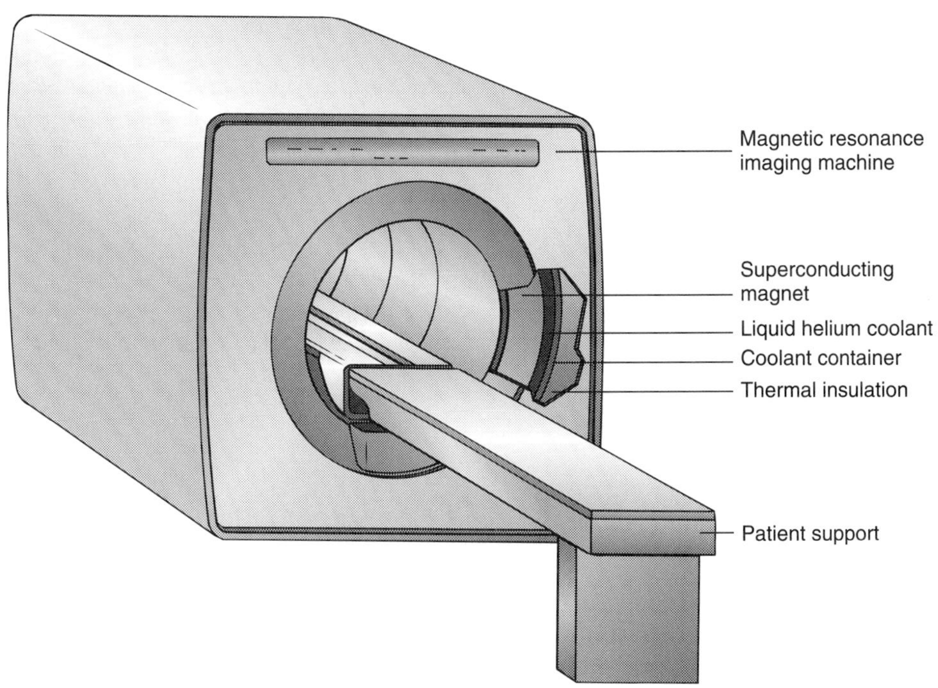

Section 19.1

Medical Imaging and Radiation

Some of the most important recent advances in health care have occurred at the border between medicine and physics. As scientists have refined their understanding of atomic and molecular structure and learned to control various forms of radiation, they have invented tools that are enormously valuable for diagnosing and treating illness and injury. The developments continue, with new applications of physics appearing in clinical settings almost every time you turn around. In this section, we'll examine two of the most significant examples of medical physics: the imaging techniques that are used to detect problems and the radiation therapies that are used to treat them.

Questions to Think About: *What is different about bones and tissues that bones appear white in an X-ray image while tissues appear dark? How can a CT scan or an MRI image show a cross-sectional view of a living person without touching them? If X-rays are a form of electromagnetic radiation, why aren't they absorbed by opaque materials? Why does an MRI image show primarily tissue while a CT scan shows primarily bone? The subatomic particles used in radiation therapy often have enormous energies; where do these energies come from?*

Experiments to Do: *One of triumphs of medical imaging is its ability to locate objects inside a person without actually entering their body. This is often done by using X-rays to look through the person from several different angles. From each vantage point, the imaging machine determines which objects are to the left or right of one another, but it can't tell how far those objects are from the observing equipment. Nonetheless, by piecing together information from many different observations, the imaging machine is able to locate each object exactly.*

You can experiment with this process by sprinkling a number of small objects onto the surface of a small table, without looking to see where they come to rest. Close your eyes and bring your head to the height of the table's surface. Open only one eye and take a brief look at the objects. If the lighting is bright and uniform and you don't move your head, you will have trouble telling how far the objects are from your eye. While you will know something about where each object is, you won't know enough to locate it exactly on the table's surface. Now move your head to a new position around the table and take a second brief look. Again, you won't be able to tell how far away the objects are, but you will learn more about their relative positions. How many such views will it take for you to learn exactly where the objects are? How does the presence of many opaque objects complicate the problem? Why does opening two eyes at once make it much easier for you to locate the objects?

X-rays

Since their discovery in 1895, X-rays have played an important role in medical treatment. Their usefulness was obvious from the very evening they were discovered. It was November 8 and German physicist Wilhelm Conrad Röntgen (1845–1923) was experimenting with an electric discharge in a vacuum tube. He had covered the entire tube in black cardboard and was working in a darkened room. Some distance from the tube a phosphored screen began to glow. Some kind of radiation was being released by the tube, passing through the cardboard and the air, and causing the screen to fluoresce. Röntgen put various objects in the way of the radiation but they didn't block the flow. Finally, he put his hand in front of the screen and saw a shadowed image of his bones. He had discovered X-rays and their most famous application at the same time.

The first clinical use of X-rays was on January 13, 1896, when two British doctors used them to find a needle in a woman's hand. In no time, X-ray systems became common in hospitals as a marvelous new technique for diagnosis. But this imaging capability was not without its side effects. Although the exposure itself was painless, overexposure to X-rays caused deep burns and wounds that took some time to appear. Evidently the X-rays were doing something much more subtle to the tissue than simply heating it.

X-rays are a form of electromagnetic radiation, as are radio waves, microwaves, and light. These different forms of electromagnetic radiation are distinguished from one another by their frequencies and wavelengths—while radio waves have low frequencies and long wavelengths, X-rays have extremely high frequencies and short wavelengths. But they are also distinguished by the energies carried in their photons. Because of its low frequency, a radio wave photon carries little energy. A medium frequency photon of blue or ultraviolet light carries enough energy to rearrange one covalent bond in a molecule. But a high frequency X-ray photon carries so much energy that it can break many covalent bonds and rip molecules apart.

In a microwave oven, the microwave photons work together to heat and cook food. The amount of energy in each microwave photon is unimportant because they don't act alone. But in radiation therapy, the X-ray photons are independent. Each one carries enough energy to damage any molecule that absorbs it. That's why X-ray burns involve little heat and appear long after the exposure—the molecular damage caused by X-rays takes time to kill cells.

CHECK YOUR UNDERSTANDING #1: Forms of Radiation

Which is more closely related to X-rays: the beam of electrons traveling through a television picture tube or the infrared light from the hot filament of a toaster?

Making X-Rays

Medical X-ray sources work by crashing fast-moving electrons into heavy atoms. These collisions create X-rays via two different physical mechanisms: bremsstrahlung and X-ray fluorescence.

Bremsstrahlung occurs whenever a charged particle accelerates. This process is nothing really new, since we know that radio waves are emitted when a charged particle accelerates on an antenna. But in a radio antenna, the electrons accelerate slowly and emit low energy photons. Bremsstrahlung usually refers to cases in which a charged particle accelerates extremely rapidly and emits a very high energy photon. In X-ray tube bremsstrahlung, a fast-moving electron arcs around a massive nucleus and accelerates so abruptly that it emits an X-ray photon (Fig. 19.1.1). This photon carries away a substantial fraction of the electron's kinetic energy. The closer the electron comes to the nucleus, the more it accelerates and the more energy it gives to the X-ray photon. However the electron is more likely to miss the nucleus by a large distance than to almost hit it, so bremsstrahlung is more likely to produce a lower energy X-ray photon than a higher energy one.

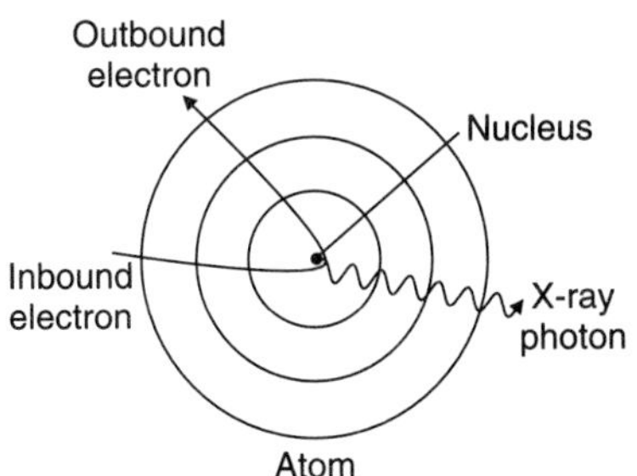

Fig. 19.1.1 - When a fast-moving electron arcs around a massive nucleus, it accelerates rapidly. This sudden acceleration creates a bremsstrahlung X-ray photon, which carries off some of the electron's energy.

In **X-ray fluorescence**, the fast-moving electron collides with an inner electron in a heavy atom and knocks that electron completely out of the atom (Fig. 19.1.2). This collision leaves the atom as a positive ion, with a vacant orbital close to its nucleus. An electron in that ion then undergoes a radiative transition, shifting from an outer orbital to this empty inner one and releasing an enormous amount of electrostatic potential energy in the process. This energy emerges from the atom as an X-ray photon. Because this photon has an energy that's determined by the ion's orbital structure, it's called a **characteristic X-ray**.

To discuss the energies carried by X-ray photons, we need an appropriate energy unit. While the joule is a useful unit of energy for describing collisions between bats and balls, it's too big to be practical for X-rays and collisions of subatomic particles. Instead, we'll use **electron-volts** (abbreviated eV). 1 eV (1 electron-volt) is about $1.602 \cdot 10^{-19}$ J and is the amount of kinetic energy an electron acquires as it accelerates through a voltage difference of 1 V.

Photons of visible light carry energies of between 1.6 eV (red light) and 3.0 eV (violet light). Because the ultraviolet photons in sunlight have energies of up to 7 eV, they are able to break chemical bonds and cause sunburns. But X-ray photons have much larger energies than even ultraviolet photons.

In a typical medical X-ray tube, electrons are emitted by a hot cathode and accelerate through vacuum toward a positively charged metal anode (Fig. 19.1.3). The anode is a tungsten or molybdenum disk that is spun rapidly by an induc-

(a)

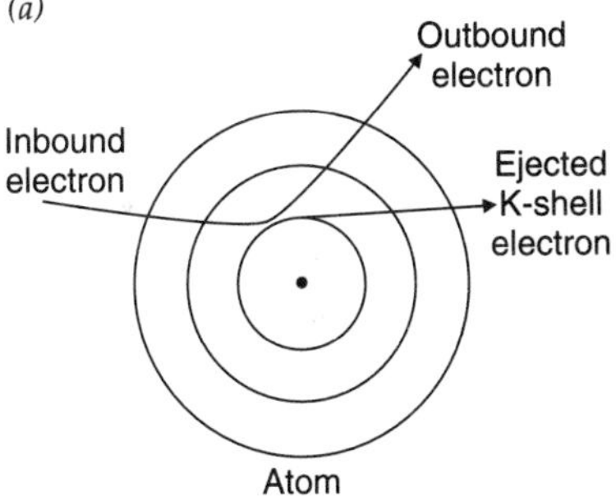

(b)

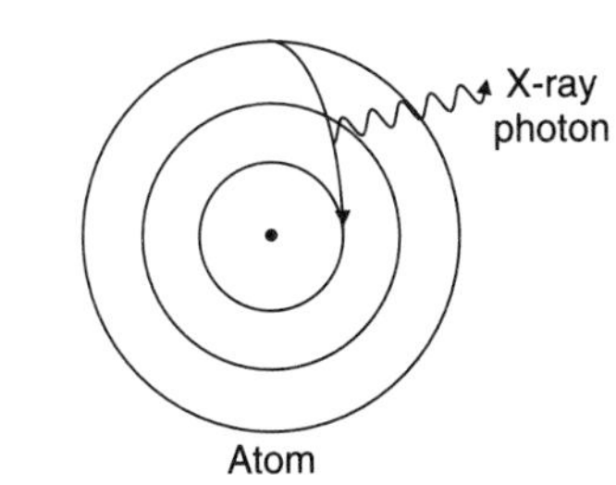

Fig. 19.1.2 - (*a*) When a fast-moving electron collides with an electron in one of the inner orbitals of a heavy atom, it can knock that electron out of the atom. (*b*) An electron from one of the atom's outer orbitals soon drops into the empty orbital in a radiative transition that creates a characteristic X-ray.

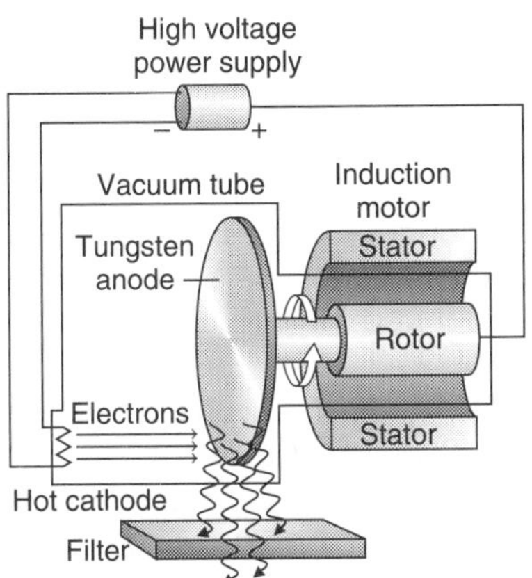

Fig. 19.1.3 - In a medical X-ray machine, electrons from a hot filament accelerate toward a positively charged metal disk. They emit X-rays when they collide with the disk's atoms. To keep the disk from melting, it's spun by an induction motor. The filter absorbs useless low energy X-rays.

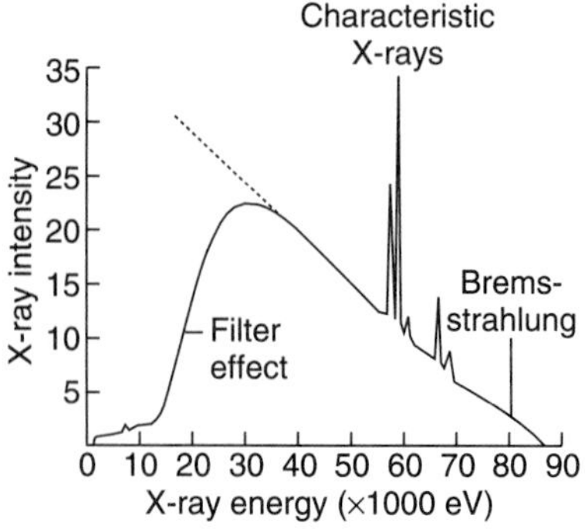

Fig. 19.1.4 - When electrons with 87,000 eV of energy collide with tungsten metal, they emit X-rays via bremsstrahlung and X-ray fluorescence. While the bremsstrahlung X-rays have a broad range of energies, an absorbing filter blocks the low energy ones. X-ray fluorescence produces characteristic X-rays with specific energies.

tion motor to keep it from melting. The energy of the electrons as they hit the anode is determined by the voltage difference across the tube. In a medical X-ray machine, that voltage difference is typically about 87,000 V, so each electron has about 87,000 eV of energy. Since an electron gives a good fraction of its energy to the X-ray photon it produces, the photons leaving the tube can carry up to 87,000 eV of energy. No wonder X-rays can damage tissue!

When the electrons collide with a target of heavy atoms, they emit both bremsstrahlung and characteristic X-rays (Fig. 19.1.4). The characteristic X-rays have specific energies so they appear as peaks in the overall X-ray spectrum. The bremsstrahlung X-rays have different energies but are most intense at lower energies. Because lower energy X-ray photons aren't useful for imaging or radiation therapy and injure skin, medical X-ray machines use absorbing materials, such as aluminum, to filter them out.

CHECK YOUR UNDERSTANDING #2: The Origins of Synchrotron Radiation

The giant particle accelerators used in high energy physics are often built as rings so that they can use the same electrically charged particles over and over again. As these particles travel in circles around the rings, they emit X-rays. Explain.

Using X-Rays

X-rays have two important uses in medicine: imaging and radiation therapy. In X-ray imaging, X-rays are sent through a patient's body to a sheet of film or an X-ray detector. While some of the X-rays manage to pass through tissue, most of them are blocked by bone. The patient's bones form a shadow image on the film behind them. In X-ray radiation therapy, the X-rays are again sent through a patient's body, but now their interaction with diseased tissue is what's important. The X-rays deposit some of their energy in this tissue and kill it.

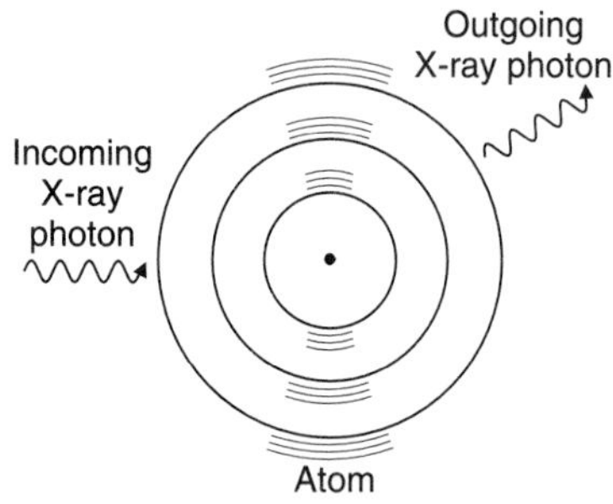

Fig. 19.1.5 - When an X-ray photon scatters elastically from an atom, the whole atom acts as an antenna. The passing photon jiggles all of the charges in the atom and these charges absorb the photon and reemit it in a new direction.

X-ray photons interact with tissue and bone through four major processes: *elastic scattering*, the *photoelectric effect*, *Compton scattering*, and *electron–positron pair production*. **Elastic scattering** is already familiar to us as the cause of the blue sky—an atom acts as an antenna for the passing electromagnetic wave, absorbing and reemitting it without extracting any of its energy (Fig. 19.1.5). Because this process has almost no effect on the atom, elastic scattering isn't important in radiation therapy. However, it's a nuisance in X-ray imaging because it produces a hazy background—some of the X-rays passing through a patient bounce around like pinballs and arrive at the film from odd angles. To eliminate these bouncing X-ray photons, X-ray machines use filters that block X-rays that don't approach the film from the direction of the X-ray source.

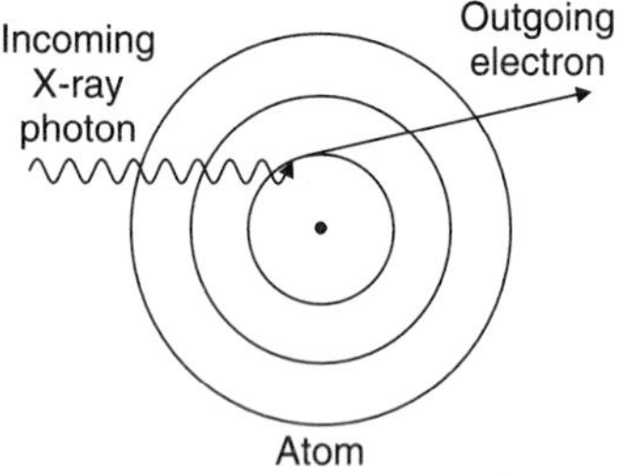

Fig. 19.1.6 - In the photoelectric effect, an absorbed photon ejects an electron from an atom. Part of the photon's energy is used to remove the electron from the atom and the rest becomes kinetic energy in the electron.

The **photoelectric effect** is what makes X-ray imaging possible. In this effect, a passing photon induces a radiative transition in an atom—one of the atom's electrons absorbs the photon and is tossed completely out of the atom (Fig. 19.1.6). If the atom were using the X-ray photon to shift an electron from one orbital to another, that photon would have to have just the right amount of energy. But because a free electron can have any amount of energy, the atom can absorb any X-ray photon that has enough energy to eject one of its electrons. Part of the photon's energy is used to remove the electron from the atom and the rest is given to the emitted electron as kinetic energy.

However the likelihood of such a *photoemission* event decreases as the ejected electron's energy increases. This decreasing likelihood makes it difficult for a small atom to absorb an X-ray photon. All of its electrons are relatively weakly bound and the X-ray photon would give the ejected electron a large kinetic energy. Rather than emit a high energy electron, a small atom usually just ignores the passing X-ray.

In contrast, some of the electrons in a large atom are quite tightly bound and require most of the X-ray photon's energy to remove. These electrons would depart with relatively little kinetic energy. Because the photoemission process is most likely when low energy electrons are produced, a large atom is likely to absorb a passing X-ray. Thus the small atoms found in tissue (carbon, hydrogen, oxygen, and nitrogen) rarely absorb medical X-rays, while the large atoms found in bone (calcium) absorb X-rays frequently. That's why bones cast clear shadows onto X-ray film. Tissue shadows are also visible, but they're much less obvious.

While one shadow image of a patient's insides may help to diagnose a broken bone, more subtle problems may not be visible in a single X-ray image. To determine exactly where things are located inside a patient, the radiologist needs to see shadows from several different angles. Better yet, the radiologist can turn to a *computed tomography* or *CT scanner*. This device automatically forms X-ray shadow images from hundreds of different angles and positions and uses a computer to analyze these images. The computer determines where objects are located inside the patient.

The CT scanner works one "slice" of the patient's body at a time. It sends X-rays through this narrow slice from every possible angle, including the two shown in Fig. 19.1.7, and determines where the bones and tissues are in that slice (Fig. 19.1.8). The scanner then shifts the patient's body to work on the next slice.

Radiation therapy also uses X-rays, but not the ones used for medical imaging. Even though tissue absorbs fewer imaging photons than bone, most imaging photons are absorbed before they can pass through thick tissue. For example, only about 10% of the imaging photons make it through a patient's leg even when they miss the bone. That percentage is good enough for making an image, but it won't do for radiation therapy because most imaging X-rays would be absorbed long before they reached a deep seated tumor. Instead of killing the tumor, intense exposure to these X-rays would kill tissue near the patient's skin.

To attack malignant tissue deep beneath the skin, radiation therapy uses extremely high energy photons. At photon energies near 1,000,000 eV, the photoelectric effect becomes rare and the photons are much more likely to reach the tumor. Photons still deposit lethal energy in the tissue and tumor, but they do this through a new effect: Compton scattering.

Compton scattering occurs when an X-ray photon collides with a single electron so that the two particles bounce off one another (Fig. 19.1.9). The X-ray

(a)

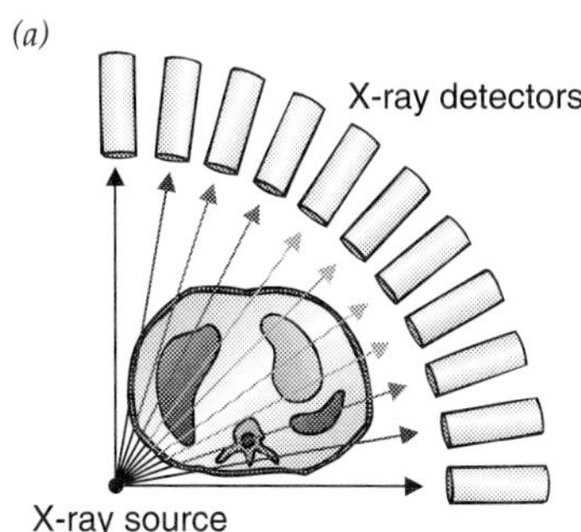

(b)

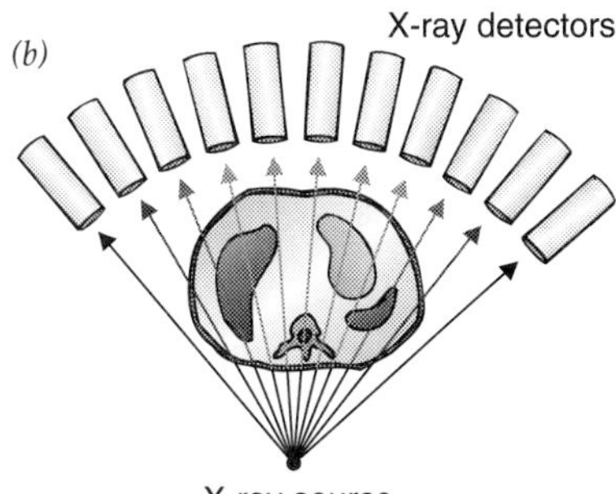

Fig. 19.1.7 - A computed tomography or CT scan image is formed by analyzing X-ray shadow images taken from many different angles and positions. An X-ray source and an array of electronic X-ray detectors form a ring that rotates around the patient as the patient slowly moves through the ring.

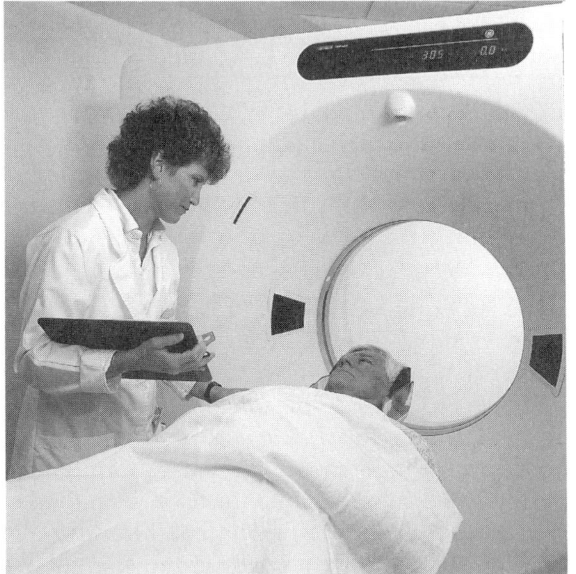

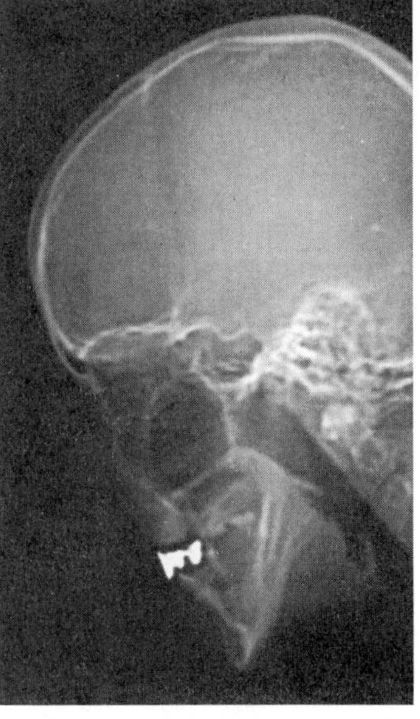

Fig. 19.1.8 - The computed tomography (CT) scanner on the left uses X-rays to image layer after layer of the patient's body. With the help of a computer, it produces a three-dimensional map of heavy elements in the patient. Part of that map, showing the patient's head, appears in the image on the right.

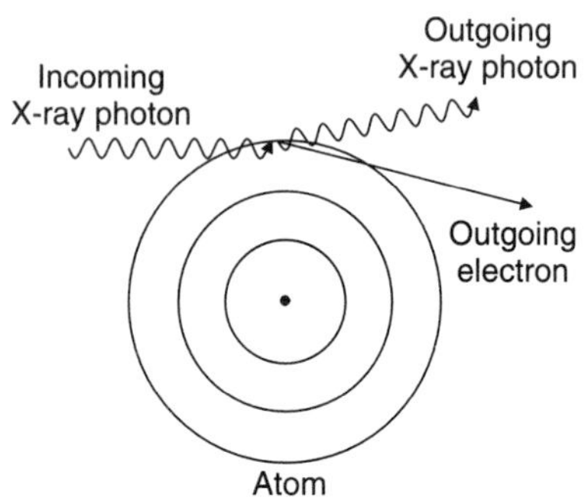

Fig. 19.1.9 - In Compton scattering, an X-ray photon collides with a single electron and the two bounce off one another. The electron is knocked out of the atom.

photon knocks the electron right out of the atom. This process is different from the photoelectric effect because Compton scattering doesn't involve the atom as a whole and the photon is scattered (bounced) rather than absorbed. The physics behind this effect resembles that of two billiard balls colliding, although it's complicated by the theory of relativity. The fact that it occurs at all is proof that a photon carries both energy and momentum and that these quantities are conserved when a particle of light collides with a particle of matter.

Compton scattering is crucial to radiation therapy. When a patient is exposed to 1,000,000 eV photons, most of the photons pass right through them but a small fraction undergo Compton scattering and leave some of their energy behind. This energy kills tissue and can be used to destroy a tumor. By approaching a tumor from many different angles through the patient's body, the treatment can minimize the injury to healthy tissue around the tumor while giving the tumor itself a fatal dose of radiation.

But Compton scattering isn't the only effect that occurs when high energy photons encounter matter. X-rays with slightly more than 1,022,000 eV can do something remarkable when they pass through an atom—they can cause **electron–positron pair production**. A **positron** is the **antimatter** equivalent of an electron. Our universe is symmetrical in many ways and one of its nearly perfect symmetries is the existence of antimatter. Almost every particle in nature has an antiparticle with the same mass but opposite characteristics. A positron or antielectron has the same mass as an electron, but it's positively charged. There are also antiprotons and antineutrons.

Antimatter doesn't occur naturally on earth, but it can be created in high energy collisions. When an energetic photon collides with the electric field of an atom, the photon can become an electron and positron. This event is an example of energy becoming matter. (In Section 18.1, we discussed matter becoming energy.) It takes about 511,000 eV of energy to form an electron or a positron, so the photon must have at least 1,022,000 eV to create one of each. Any extra energy goes into kinetic energy in the two particles.

The positron doesn't last long in a patient. It soon collides with an electron and the two annihilate one another—the electron and positron disappear and their mass becomes energy. They turn into photons with a total of at least 1,022,000 eV. So energy became matter briefly and then turned back into energy. This exotic process is present in high energy radiation therapy and becomes quite significant at photon energies above about 10,000,000 eV. Not surprisingly, it also helps to kill tumors.

CHECK YOUR UNDERSTANDING #3: Aluminum X-ray Windows

Aluminum atoms are much smaller than calcium atoms. While aluminum metal blocks visible light, it's relatively transparent to high energy X-rays. Explain.

Gamma Rays

Producing very high energy photons isn't quite as easy as producing those used in X-ray imaging. In principle, a power supply could create a huge voltage difference across an X-ray tube so that very high energy electrons would crash into metal atoms and produce very high energy photons. But million volt power supplies are complicated and dangerous, so other schemes are used instead.

One of the easiest ways to obtain very high energy photons is through the decay of radioactive isotopes. The isotope most commonly used in radiation therapy is cobalt 60 (^{60}Co). The nucleus of ^{60}Co has too many neutrons and that makes it unstable. It eventually decays, undergoing a series of transformations that produce two high energy photons: one with 1,170,000 eV and one with 1,330,000 eV. These photons penetrate tissue well and are quite effective at killing tumors. So ^{60}Co sources, carefully shrouded in lead containers, are often used in radiation therapy.

The process by which ^{60}Co produces those two high energy photons is complicated but worth describing briefly. It shows that protons, electrons, and neutrons are not immutable and that there are other subatomic particles in our universe. The decay begins when one of the neutrons in a ^{60}Co nucleus abruptly turns into three particles: a proton, an electron, and a *neutrino*. This process is called *beta decay* and can occur because neutrons that are by themselves or in nuclei with too many neutrons are unstable and radioactive. When one of the neutrons in a neutron-rich ^{60}Co nucleus decays, it leaves behind a positively charged proton and the nucleus becomes nickel 60 (^{60}Ni). The negatively charged electron and the neutral neutrino escape from the nucleus and travel outward.

The **neutrino** is a subatomic particle with no charge and little or no mass. Neutrinos aren't found in normal atoms. Though important in nuclear and particle physics, neutrinos are difficult to observe directly because they travel at or near the speed of light and hardly ever collide with anything. Without charge, they don't participate in electromagnetic forces and, unlike the electrically neutral neutron, they don't experience the nuclear force. They experience only gravity and the **weak force**, the last of the four **fundamental forces** known to exist in our universe. (The other three fundamental forces are the gravitational force, the electromagnetic force, and the **strong force**—that's a more complete version of the nuclear force that we discussed in the previous chapter.) Because it's weak and occurs only between particles that are very close together, the weak force rarely makes itself apparent. One of the few occasions where it plays an important role is in beta decay.

With almost no way to push or pull on an another particle, a neutrino can easily pass right through the entire earth. Neutrinos can be detected once in a while, but only with the help of enormous detectors. That's why physicists first showed that neutrinos are emitted from decaying neutrons by measuring energy and momentum before and after the decay. The proton and electron produced by the decay don't have the same total energy and momentum as the neutron had before the decay. Something must have carried away the missing energy and momentum and that something is the neutrino.

Once ^{60}Co has turned into ^{60}Ni, the decay isn't quite over. The ^{60}Ni nucleus that forms still has extra energy in it. Nuclei are complicated quantum physical systems just as atoms are and they have excited states, too. The ^{60}Ni nucleus is in an excited state and it must undergo two radiative transitions before it reaches the ground state. These radiative transitions produce very high energy photons or **gamma rays** that are characteristic of the ^{60}Ni nucleus—one with 1,170,000 eV of energy and the other with 1,330,000 eV. These gamma rays are what make ^{60}Co radiation therapy possible.

CHECK YOUR UNDERSTANDING #4: A Visit from the Snake Oil Salesman

If someone offered to sell you a bottle of neutrinos, you'd be foolish to buy it. What's wrong with the idea of a bottle of neutrinos?

Particle Accelerators

Electromagnetic radiation isn't the only form of radiation used to treat patients. Energetic particles such as electrons and protons are also used. Like tiny billiard balls, these fast moving particles collide with the atoms inside tumors and knock

(a)

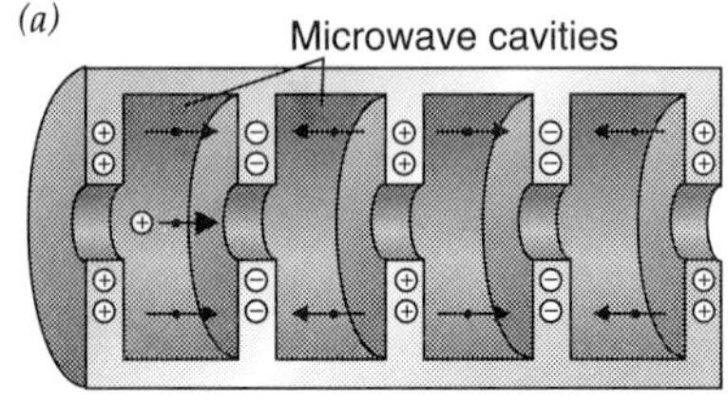

(b)

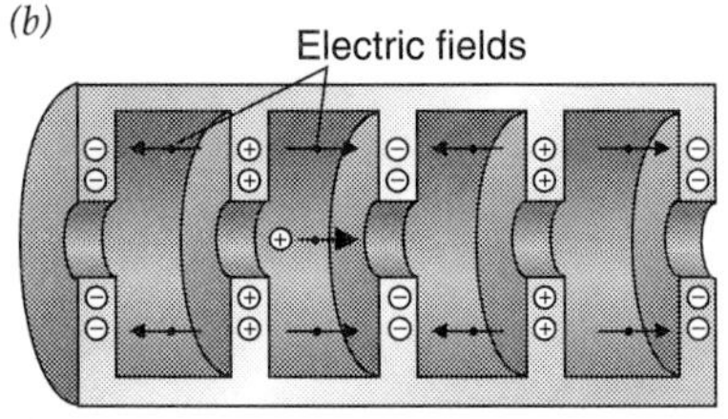

Fig. 19.1.10 - In a linear accelerator, moving charged particles are pushed forward by electric fields that change with time. (*a*) While the moving positive charge is passing through the first of a series of microwave cavities, the field there pushes it forward. (*b*) By the time the moving charge has entered the second cavity, the fields have reversed and the field there pushes it forward again.

them apart. As usual, this atomic and molecular damage tends to kill cells and destroy tumors.

However, obtaining extremely energetic subatomic particles isn't easy. High voltage power supplies can be used to accelerate an electron or proton to about 500,000 eV, but that isn't enough. When a charged particle enters tissue, it experiences strong electric forces and is easily deflected from its path. To make sure that it travels straight and true, all the way to a tumor, the particle must have an enormous energy. Giving each charged particle the millions or even billions of eV it needs for radiation therapy takes a particle accelerator.

Particle accelerators use many of the concepts we encounter in Section 14.3, when we discussed magnetrons. Charge that's sloshing back and forth in a magnetron not only creates microwaves, it also creates huge electric fields that change with time. In a particle accelerator, similar electric fields are used to push charged particles through space until they reach incredible energies.

One important type of particle accelerator is the *linear accelerator*. In this device, charged particles are pushed forward in a straight line by the electric fields in a series of resonant cavities (Fig. 19.1.10). Each of these cavities has charge sloshing back and forth rhythmically on its wall, just as charge sloshes back and forth in the cavities of a magnetron. When a small packet of charged particles enters the first cavity through a hole, it's suddenly pushed forward by the strong electric field inside that cavity (Fig. 19.1.10*a*). The packet accelerates forward and leaves the first resonant cavity with more kinetic energy than it had when it arrived—the electric field in that cavity has done work on the packet.

If the fields in the cavities were constant, the electric field in the second cavity would slow the packet down. In Fig. 19.1.10*a*, you can see that the electric field in the second cavity points in the wrong direction. But by the time the packet reaches the second cavity, the charge sloshing in its walls has reversed and so has the electric field (Fig. 19.1.10*b*). The packet is again pushed forward and it emerges from the second cavity with still more kinetic energy.

Each resonant cavity in this series adds energy to the packet, so that a long string of cavities can give each of the packet's charged particles millions or even billions of eV. This energy comes from magnetron-like microwave generators that are attached to the accelerator. Microwaves from these generators are what cause charge to slosh in the accelerator's microwave cavities. The linear accelerator then only has to inject charged particles into the first cavity, using equipment resembling the insides of a television picture tube, and those charged particles will come flying out of the last cavity with incredible energies (Fig. 19.1.11).

However there are a few complications to this acceleration technique. Most importantly, the cavities have to be built and operated so that their electric fields reverse at just the right moments to keep the packet accelerating forward. For simplicity of operation, the cavities have the same resonant frequency and all

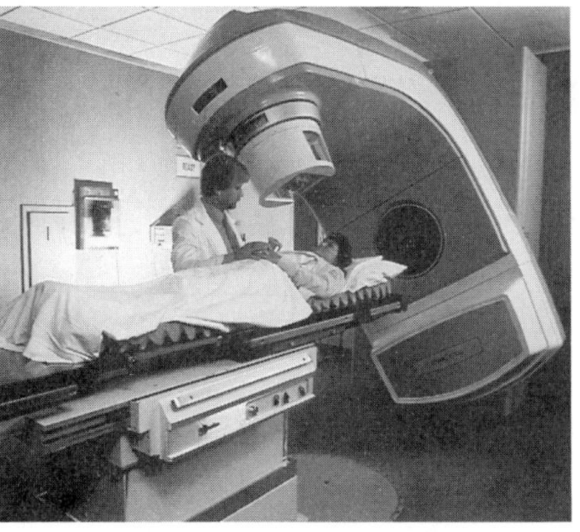

Fig. 19.1.11 - This radiation therapy unit uses a linear accelerator to produce extremely high energy subatomic particles. These particles penetrate deep inside the patient to destroy a cancerous tumor. The linear accelerator itself is hidden from view in the room behind this one. Its beam is steered by magnets in the rotateable arm toward a particular spot in the patient.

reverse their electric fields simultaneously. Since the packet spends the same amount of time in each cavity and it travels faster and faster as it goes from one cavity to the next, each cavity must be longer than the previous one.

But as the packet approaches the speed of light, something strange happens. The packet's energy continues to increase as it goes through the cavities, but its speed stops increasing very much. What this means is that the simple relationship between kinetic energy and speed given in Eq. 1.4.7 isn't valid for objects moving at almost the speed of light! This breakdown is part of Einstein's **relativity theory**, the rules governing motion at speeds comparable to the speed of light. As a further consequence of relativity, the packet can approach the speed of light but can't actually reach it. Though each charged particle's kinetic energy can become extraordinarily large, its speed is limited by the speed of light.

Because the packet's speed stops increasing significantly after it has gone through the first few cavities of the linear accelerator, the lengths of the remaining cavities can be essentially constant. Only the first few cavities have to be specially designed to account for the packet's increasing speed inside them. The charged particles emerge from the accelerator traveling at almost the speed of light. They pass through a thin metal window that keeps air out of the accelerator and enter the patient's body. They have so much energy that they can penetrate deep into tissue before coming to a stop.

CHECK YOUR UNDERSTANDING #5: Particle Recycling

Many research accelerators send each packet of electrons through the same series of resonant cavities several times. After the packet leaves the last cavity, magnets steer it around in a circle and sent back through the cavities again. With each pass through the cavities, the packet acquires more energy, so how can it possibly stay synchronized with the reversing electric fields in the cavities?

Magnetic Resonance Imaging

While X-rays do an excellent job of imaging bones, they aren't as good for imaging tissue. A better technique for studying tissue is *magnetic resonance imaging* or *MRI*. This technique locates hydrogen atoms by interacting with their magnetic nuclei. Since hydrogen atoms are common in both water and organic molecules, finding hydrogen atoms is a good way to study biological tissue.

The nucleus of a hydrogen atom is a proton. Like the electrons we studied in Section 12.5, a proton is effectively a spinning object with charge on its surface and this moving charge gives it a magnetic field. The proton behaves like a tiny bar magnet, with a north pole at one end and a south pole at the other. If you put the proton in a magnetic field, it will tend to align itself with that field (Fig. 19.1.12*a*). But while protons would align perfectly with the field at absolute zero, thermal energy flips the protons about so that they're only aligned with the field on the average. Some of them are upside down.

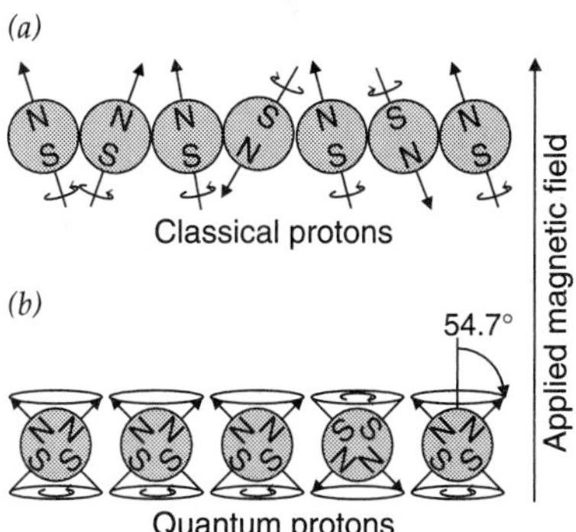

Fig. 19.1.12 - (*a*) Classical protons are spinning magnetic objects that tend to align roughly with an applied magnetic field. (*b*) Quantum protons are also magnetic, but their axes of rotation are not so well defined. Their angular momenta can only partly align with the applied field and they appear either spin-up (partly aligned with the field) or, less frequently, spin-down (partly anti-aligned with the field).

However this classical view of the protons as little spinning magnetic balls isn't quite right. They're so small that their behavior is complicated by quantum physics. Because of the wave character of matter, the proton's axis of rotation isn't well defined. Just as the Heisenberg uncertainty principle prevents us from knowing exactly where an object is, we are prevented from knowing exactly what its axis of rotation is. For large objects like a bowling ball, this axis uncertainty is too small to matter. But for a proton, it's significant. The end result here is that the protons adopt one of only two possible alignments with an applied magnetic field: they either partly align with that field or they partly anti-align with it (Fig. 19.1.12*b*). In either case, each proton's axis of rotation is tilted 54.7°

away from alignment or anti-alignment, but you can't tell whether it's tilted toward the left, right, back, or front! The direction of its tilt is hidden by the Heisenberg uncertainty principle.

We'll call the protons that partly align with the applied magnetic field "spin-up" and the ones that partly anti-align "spin-down." As before, thermal energy keeps these quantum protons from all being spin-up. In fact, even in a strong magnetic field, the spin-up protons only slightly outnumber the spin-down protons. To make the following discussion simpler, we'll only consider the extra spin-up protons. The other protons are balanced and don't affect MRI.

When a patient enters the strong magnetic field of an MRI machine, the protons in their body respond to the field and an excess of spin-up protons appears. Because aligning with a magnetic field reduces a bar magnet's potential energy, each of these spin-up protons has less energy than it would have if it were spin-down. While it's possible to convert one of the extra spin-up protons into a spin-down proton, that process takes energy.

An MRI machine uses radio wave photons to flip the extra spin-up protons. When a radio wave photon with just the right amount of energy passes through a spin-up proton, the proton may absorb it in a radiative transition and become a spin-down proton. How much energy the radio wave photon must have depends on how much energy it takes to flip the spin. That energy in turn depends on how strong the magnetic field is around the proton. Since the energy needed to flip the proton's spin increases as the magnetic field gets stronger, the energy of the radio wave photon must also increase with the field.

If the protons in the patient's body were all experiencing exactly the same magnetic field, they would all require the same radio wave photon energy to flip. But the protons don't all experience the same field. The MRI machine introduces a slight spatial variation to its magnetic field (Fig. 19.1.13). Because the magnetic field is different for different protons, only some of them can absorb radio waves of a particular energy. This selective absorption is how the MRI imager locates protons within a patient.

In its simplest form, an MRI machine applies a spatially varying magnetic field to the patient's body. It then sends various radio waves through the patient and looks for those radio waves to be absorbed by protons. Since only a proton that is experiencing the right magnetic field can absorb a particular radio wave photon, the MRI machine can determine where each proton is by which radio waves it absorbs. By changing the spatial variations in the magnetic field and adjusting the energies of the radio waves, the MRI machine gradually locates the protons in the patient's body. It builds a detailed three-dimensional map of the hydrogen atoms. A computer manages this map and can display cross-sectional images of the patient from any angle or position.

Since its invention in the mid–1970s, MRI has become increasingly sophisticated. A modern MRI machine (Fig. 19.1.14) doesn't fully invert the spin-up protons in the manner we've just discussed. Instead, it flips each proton only half way upside down. The result isn't a "spin-sideways" proton—as it might be in a classical universe. Instead it's a quantum proton that's 50% spin-up and 50% spin-down. That means that if you were to measure its current spin, you'd have a 50% chance of finding it spin-up and a 50% chance of finding it spin-down. Before that measurement, the proton is simultaneously in both states.

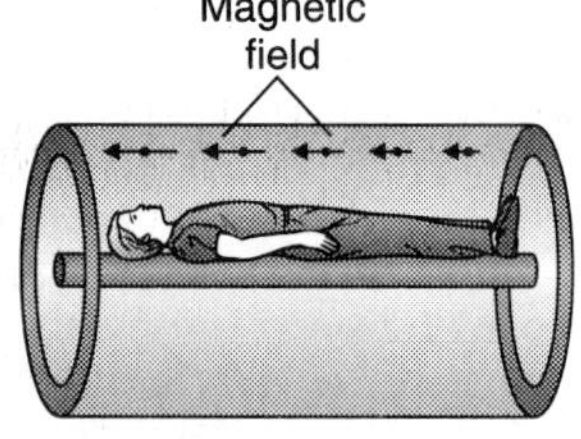

Fig. 19.1.13 - An MRI machine places the patient in a strong magnetic field. This field varies spatially, so that protons at different places in the patient's body experience different fields and absorb different radio wave photons.

These two possible states, spin-up and spin-down, correspond to quantum waves and, like the light waves we studied in Section 15.1, these matter waves can exhibit interferences. The interferences of a half-flipped proton are extremely sensitive to the magnetic field near the proton and to the radio waves that pass through it. Using these interferences, the MRI machine can not only locate each proton, it can even tell what molecule the proton is in and what the chemical environment of that molecule is like. In this way, an MRI machine can generate

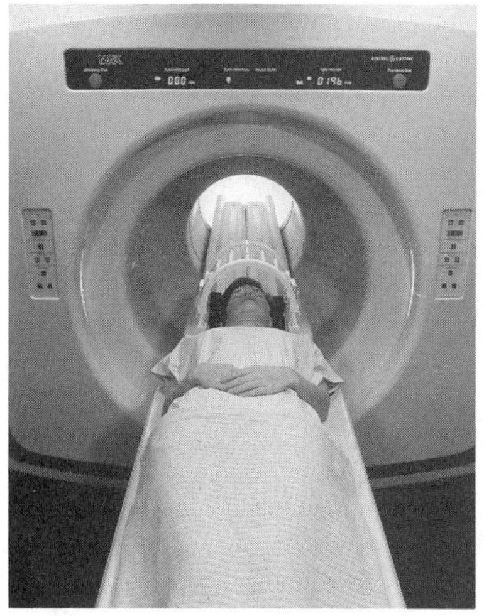

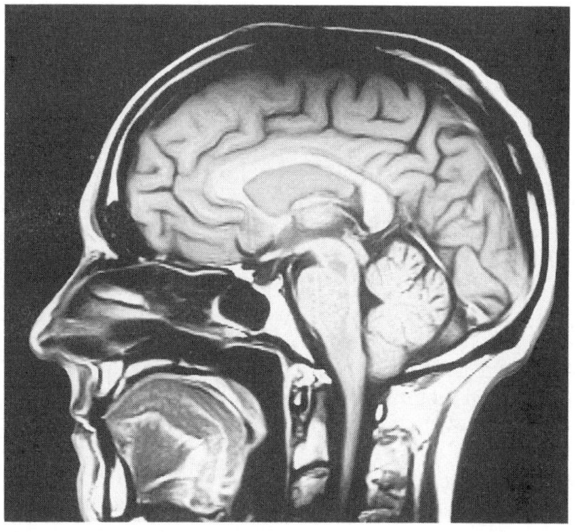

Fig. 19.1.14 - The patient on the left is entering the intense magnetic field of a magnetic resonance imaging (MRI) system. Using the interactions between electromagnetic waves and protons that take place in magnetic fields, it produces a three-dimensional map of hydrogen atoms in the patient's body. A portion of this map, showing the patient's head, is displayed in the image on the right.

detailed images of where water molecules are in the patient, where fat molecules are, and even where the patient's blood is flowing (Fig. 19.1.14).

CHECK YOUR UNDERSTANDING #6: Major Magnets

Some of the most modern MRI machines use extremely strong magnetic fields. As a machine's magnetic field gets stronger, what must happen to the radio waves that are used to flip the spins of protons in a patient?

Summary

How Medical Imaging and Radiation Work: Because X-rays pass more easily through tissue than they do through bone, X-rays form shadow images of a patient's bones. While the heavy atoms in bone can absorb X-rays and emit electrons, the lighter atoms in tissue are less able to interact with X-rays.

A basic X-ray machine crashes a beam of high energy electrons into a metal target. X-rays are produced both when the electrons accelerate near the metal nuclei and when the moving electrons knock other electrons out of the metal atoms. A filter near the X-ray source blocks low energy X-rays so that only high energy X-rays pass through the patient toward a sheet of film. A second filter in front of the film prevents X-rays that scattered elastically on their way through the patient from fogging the image.

Radiation therapy is done with very high energy X-rays and gamma rays because they pass more easily through tissue. These electromagnetic waves kill tumor cells by depositing energy in the atoms and molecules of those cells. As they collide with the electrons and electric fields of an atom, these waves can knock the electrons out of the atom or create electron–positron pairs. Gamma rays are usually obtained from radioactive nuclei. Some radiation therapy is done with energetic particles that are given enormous energies by particle accelerators. These accelerators use the changing electric fields in resonant cavities to push particles forward until they travel at almost the speed of light.

Magnetic resonance imaging uses the magnetic nature of hydrogen nuclei (protons) to locate hydrogen atoms in a patient's body. The patient is put in a strong magnetic field and the protons tend to align with this field. The imaging machine then uses radio waves to flip the alignment of those protons. By making the magnetic field vary slightly from place to place, the machine is able to find the protons it's flipping. A computer records and analyzes the results so that it can present cross-sectional images of the hydrogen atoms in a patient's body.

The Physics of Medical Imaging and Radiation

1. X-rays and gamma rays are high frequency, short wavelength forms of electromagnetic radiation. They are emitted and absorbed as photons that carry relatively large amounts of energy.

2. When a charged particle undergoes rapid acceleration, it may emit an electromagnetic wave in a process called bremsstrahlung. If the acceleration is rapid enough, this wave may be an X-ray or gamma ray.

3. When an electron is ejected from one of the inner orbitals of a large atom, an electron from one of the outer orbitals will usually shift into the empty inner orbital in a radiative transition that emits an X-ray photon. This process is X-ray fluorescence.

4. An electron passing through an atom can emit an X-ray either via bremsstrahlung as it bends around the atom's nucleus or via X-ray fluorescence after knocking an electron out of the atom.

5. An atom can briefly absorb and reemit an X-ray photon in an elastic scattering event.

6. An atom can absorb an X-ray photon by using its energy to eject an electron. This is the photoelectric effect. Part of the photon's energy is used to remove the electron from the atom and the rest goes into the electron's kinetic energy. The higher the kinetic energy of the outgoing electron, the less likely this process is to occur.

7. A very high energy photon can bounce directly from a single electron in a process called Compton scattering. This effect is proof that photons carry both energy and momentum and can interact with matter as a particle.

8. When a photon with more than 1,022,000 eV of energy interacts with an electric field, it can become an electron–positron pair. A positron is the anti particle of an electron—the two particles have identical masses but many of their characteristics are reversed. For example, a positron has a positive charge. When an electron and positron collide, they annihilate and become energetic photons.

9. Radioactive decays are one source of the subatomic particles that aren't found in ordinary matter. One such particle is the massless and chargeless neutrino. When isolated or in a neutron-rich nucleus, a neutron can decay into a proton, an electron, and a neutrino (actually an anti-neutrino, to be accurate).

10. When a particle such as an electron or proton is traveling at almost the speed of light, you can increase its kinetic energy substantially without increasing its speed much. This effect is a consequence of relativity. The particle may approach the speed of light but can never reach it.

11. Protons behave as spinning, charged objects and thus have magnetic fields. Placed in a magnetic field, the protons tend to align with that field. However their quantum natures leave their axes of rotations partially undefined so that they adopt one of only two states: spin-up and spin-down. In each case, the proton's axis of rotation is tilted with respect to the applied magnetic field, but the direction of that tilt is uncertain. The protons can be flipped between the spin-up and spin-down states by a radio wave photon with the proper energy.

12. Careful treatment with radio wave photons can put a proton into a state in which it's partly spin-up and partly spin-down at the same time. If you then measure its state, there is a certain likelihood of finding it in the spin-up state and a likelihood of finding it in the spin-down state. This mixed state is subject to interesting interference effects.

Check Your Understanding - Answers

1. The infrared light from the hot filament.

Why: Infrared light and X-rays are both forms of electromagnetic radiation. All that distinguishes them is their frequencies and wavelengths, and the energy of their photons. The beam of electrons in a television tube is also a form of radiation, but it involves particles of matter, not electromagnetic waves.

2. Because a particle traveling in a circle is accelerating, it emits electromagnetic waves. In this case, those waves are X-rays.

Why: A rapidly accelerating charged particle will emit an X-ray, whether it's accelerating around the nucleus of a heavy atom or around the ring of a particle accelerator. In an accelerator, these X-rays are called synchrotron radiation. Synchrotron radiation is useful in research and industry and is often deliberately enhanced by adding special magnets to the ring of the accelerator.

3. The electrons in an aluminum atom are so weakly bound that they are unlikely to absorb high energy X-ray photons through the photoelectric effect.

Why: Like the small atoms in biological tissue, aluminum atoms rarely use the photoelectric effect to absorb high energy X-rays. This result makes it possible to use thin films of aluminum as windows and filters for X-ray sources.

4. Neutrinos travel at the speed of light and can't stay in a bottle. Moreover the bottle couldn't confine them even if they didn't move so fast.

Why: Like all massless particles, neutrinos travel at the speed of light. Because they interact so weakly with matter, they don't reflect from anything and a bottle couldn't trap them for long. Actually, you're being bathed in neutrinos from the sun all the time without even noticing it. They're as common as dirt and you can't do anything with them anyway.

5. The packet is traveling at almost the speed of light, so its speed barely changes as its energy increases.

Why: If the packet were speeding up with each trip through the cavities, it wouldn't stay synchronized with the reversing electric fields inside them. But the packet's speed is so nearly constant once it nears the speed of light that it can travel through the cavities over and over again without any problem.

6. Each radio wave photon must carry more energy (the radio waves must have higher frequencies).

Why: The stronger the magnetic field it's in, the more energy a proton needs to flip its spin. The MRI machine must use higher frequency, more energetic radio waves to cause such spin flips. There are many advantages to using extremely strong magnetic fields, including lower noise and better spatial resolution in the image. However the magnetic fields of these advanced MRI machines are so strong that they can erase credit cards from across the room and rip steel objects right out of your pockets.

Glossary

antimatter Matter resembling normal matter, but with many of its characteristics such as electric charge reversed.

bremsstrahlung The process in which a rapidly accelerating charge emits electromagnetic radiation, usually an X-ray or gamma ray photon.

characteristic X-ray An X-ray emitted by X-ray fluorescence from an atom. The energy of the characteristic X-ray is determined by the atom's orbital structure.

Compton scattering The process in which a photon bounces off a charged particle, usually an electron. The photon and charged particle exchange energy and momentum during the collision.

elastic scattering The process in which two particles bounce off one another without losing any of their kinetic energies. When a photon scatters elastically, its overall energy remains unchanged.

electron–positron pair production The formation of an electron and a positron during an energetic collision.

electron-volt (eV) A unit of energy equal to the energy obtained by an electron as it moves through a voltage difference of 1 V. 1 eV is equal to about $1.602 \cdot 10^{-19}$ J.

fundamental forces The four basic forces that act between objects in the universe—the gravitational force, the electromagnetic force, the strong force, and the weak force.

gamma ray Extremely high energy photons of electromagnetic radiation, often produced during radioactive decays.

neutrino A chargeless particle created during radioactive decays and other nuclear events. The neutrino appears to have zero mass and to travel at the speed of light. It rarely interacts with matter.

photoelectric effect The process in which an atom absorbs a photon in a radiative transition that ejects one of the electrons out of the atom.

positron The antimatter counterpart of the electron. The positron is positively charged.

relativity theory The physical rules governing motion at speeds comparable to the speed of light. The general theory of relativity considers motion near massive objects.

strong force The fundamental force that gives structure to nuclei and nucleons and is the basis for the nuclear force.

weak force The fundamental force that allows electrons and neutrinos to interact and that's responsible for beta decay.

X-ray fluorescence The process in which an electron in one of the outer orbitals of an atom undergoes a radiative transition to an empty inner orbital, emitting an X-ray photon.

Review Questions

1. If you want to cook food, you should expose it to microwaves. If you want to sterilize the food by killing bacteria, you should expose it to X-rays. Explain why microwaves and X-rays have such different effects on food.

2. A moving electron emits electromagnetic waves whenever it accelerates, as it does when it passes near another charged particle. However, to emit an X-ray, the electron must pass near the nucleus of a massive atom. Why?

3. When an electron is removed from an orbital near the nucleus of a heavy atom, the atom soon emits an X-ray. Why?

4. The spectrum of X-rays emitted from an X-ray tube is complicated. While the tube produces X-ray photons with a broad range of energies, a substantial fraction of the X-rays have only a few specific energies. What mechanism produces X-rays with the broad range of energies and what mechanism produces X-rays with a few specific energies?

5. Why are the calcium atoms in your bones much more likely to absorb diagnostic X-ray photons than the carbon or hydrogen atoms in your tissues?

6. Why aren't diagnostic X-rays used to treat deep-seated tumors? Why are much higher energy therapeutic X-rays or gamma rays used instead?

7. Scientists have recently learned how to make anti-hydrogen, the antimatter equivalent of hydrogen. Its mass is expected to be exactly equal to the mass of normal hydrogen. If a normal hydrogen atom consists of an electron and a proton, what does anti-hydrogen consist of?

8. When it undergoes radioactive decay, a ^{60}Co nucleus emits two gamma rays with specific energies. Why is that behavior an indication that nuclei are quantum systems with ground states and excited states, like those of atoms?

9. As its energy increases in a particle accelerator, an electron's speed also increases. However, while the electron's energy increases without limit, its speed can't reach or exceed the speed of light. As it continues through the accelerator, the electron's energy keeps increasing steadily but its speed increases by smaller and smaller amounts. Why is this behavior inconsistent with the classical laws of motion discussed in Chapter 1 and thus an indication that there must be a more complete theory—the theory of relativity?

10. How does magnetic resonance imaging determine where hydrogen atoms are located in a patient's body?

Exercises

1. Before the dangers associated with X-rays were well understood, X-ray machines were often used as marketing gimmicks. For example, shoe stores had X-ray machines to show customers how their feet fit in their shoes. The customers had no idea that they were being injured by the strong exposure to X-rays. Had they been exposed to intense microwaves, they'd have felt their feet grow warmer. Why is the injury caused by X-rays so much more subtle than that caused by microwaves?

2. Sunlight bleaches the colors out of clothing while indoor lighting generally does not. Explain this difference.

3. Why is it important to keep foods and drugs out of direct sunlight, even when they're not in danger of overheating?

4. Why are many drugs packaged in amber-colored containers that block ultraviolet light?

5. Museums often display priceless antique manuscripts under dim yellow light. Why not use white light?

6. Sunscreen absorbs ultraviolet light while permitting visible light to pass. Why does this coating reduce the risk of chemical and genetic damage to the cells of your skin?

7. An X-ray technician can adjust the energy of the X-ray photons produced by a machine by changing the voltage drop between the X-ray tube's cathode and anode. Explain.

8. The anode of an X-ray tube is made from atoms with relatively large nuclei and many electrons. Why would the X-ray tube have trouble creating X-rays through the bremsstrahlung process if its anode were made of beryllium, a metallic element with an extremely small and light nucleus?

9. An X-ray tube with a beryllium anode (see Exercise **8**) would also be unable to produce characteristic X-rays via X-ray fluorescence. Since a beryllium atom has only four protons in its nucleus, it would emit only a visible or ultraviolet light photon when one of its outer electrons underwent a radiative transition to an empty inner orbital. Why must an X-ray machine's anode be made of atoms with many protons in their nuclei before the anode can emit X-ray photons through radiative transitions in those atoms?

10. Visible light undergoes very little Rayleigh scattering as it passes through individual atoms or small molecules. That's why it takes many kilometers of atmosphere to scatter enough light to make the sky appear blue. But X-rays undergo Rayleigh-like elastic scattering much more frequently in atoms so that it's a serious problem for X-ray imaging. Why are atoms so much more effective at scattering X-ray photons than they are at scattering photons of visible light?

11. A thick steel pipe can be made by rolling a steel plate into a cylinder and then welding the seam between the two edges. But even a weld that looks perfect may have gaps that weaken it and could cause disaster. The manufacturer makes sure that the weld is flawless by studying it with X-rays. An X-ray source is placed outside the pipe and the shadow of the weld is recorded by an X-ray detector inside the pipe. This technique works wonderfully, but it can't be done with the same X-rays that are used for medical imaging. The iron atoms in the steel would absorb the medical imaging X-rays so strongly that virtually nothing would arrive inside the pipe. How must the X-ray source differ from that of a diagnostic medical X-ray source in order to penetrate into the steel and cast a shadow of the weld on the detector inside the pipe?

12. Helium gas, which has atoms with only 2 electrons each, is almost completely transparent to X-rays. In contrast, xenon gas, which consists of atoms with 54 electrons each, absorbs X-rays reasonably well. Explain this difference.

13. The idea that electromagnetic waves behaved as particles during collisions was one of the surprising realizations of early quantum physics. It was strange enough that light appeared as photons or quanta of energy; it was even stranger that one of these photons had zero mass and yet carried momentum with it. However, that's what the 1922 experiments of American physicist Arthur Holly Compton (1892–1967) indicated. Why does the Compton effect, in which an X-ray photon scatters off a free electron and deflects the path of that electron, prove that the X-ray photon has momentum?

14. Electric charge is strictly conserved in our universe, meaning that the net charge of an isolated system can't change. Why doesn't the production of an electron–positron pair in a patient cause a change in the patient's net charge?

15. Which has more mass: a positron or an antiproton?

16. Scientists have always been curious about the nuclear reactions that occur deep inside the core of the sun. Unfortunately, the sun is so thick and opaque that no electromagnetic waves reach us directly from the core. However, we can observe neutrinos produced in the sun's core. Why are those neutrinos able to reach us when light can't?

17. Which subatomic particle isn't contained in an iron atom: a proton, a positron, an electron, or a neutron?

18. What is the electric charge of an antiproton?

19. Modern high energy physics facilities study extremely high energy subatomic particles. In designing the equipment for handling these particles, it's frequently safe to assume that the particles are traveling at almost exactly the speed of light. Given that the kinetic energies of these particles may change significantly during different experiments, why is this assumption about speed still reasonable?

20. Magnetic resonance imaging (MRI) differs from computed tomography imaging in that it involves no "ionizing radiation." What electromagnetic radiation is used in MRI and why aren't the photons of this radiation able to remove electrons from atoms and convert those atoms into ions?

21. Magnetic resonance imaging isn't good at detecting bone. Why not?

22. No magnetic metals such as iron or steel are permitted near a magnetic resonance imaging machine. In part, this rule is a safety precaution since those magnetic metals would be attracted toward the machine. But the magnetic fields from these magnetic metals would also spoil the imaging process. Why would having additional magnetic fields inside the imaging machine spoil its ability to locate specific protons inside a patient's body?

Epilogue for Chapter 19

In this chapter we examined some of the applications of modern physics to the field of medicine. In *medical imaging and radiation,* we learned how X-rays are produced when energetic electrons collide with a metal target and why those X-rays pass more easily through tissue than through bone. We saw how X-rays can be used to make simple shadow images and how, with the aid of a computer, they can form complete three-dimensional maps of a person's body. We explored the uses of higher energy X-rays and gamma rays for radiation therapy and saw how gamma rays are produced by radioactive decays. We continued with a discussion of high energy particle radiation produced by particle accelerators. We concluded by examining magnetic resonance imaging and saw how it's based on the interactions between protons, magnetic fields, and microwave radiation.

Explanation: Radiation Damage to Colored Paper

The paper fades because the ultraviolet light photons in sunlight have enough energy to shift electrons out of the orbitals that bond the dye molecules together. These dye molecules then fall apart and leave the paper without color. In some cases, the photons of light completely remove the electrons from the dye molecules. That process is the same photoelectric effect that makes it possible to distinguish tissue from bone in X-ray imaging.

Cases

1. Physicists use accelerators to study the smallest objects in the universe: subatomic particles. They need these accelerators because enormous energies are required to resolve extremely fine details. Since a particle's wavelength decreases as its energy increases, higher energy particles are able to resolve tinier features.

a. In an electron accelerator, electrons pass through a long chain of resonant cavities, powered by strong microwave generators. The electric fields in these cavities push the electrons forward. Why must the electrons be bunched together in packets rather than sent into the cavities continuously?
b. A packet of electrons takes the same amount of time to pass through each cavity of the chain. The first few cavities of the chain are shorter than the rest. Why?
c. The remaining cavities of the chain are essentially equal in length. Why don't the cavities continue to get longer?
d. Some accelerators also work with positrons. They inject packets of positrons into the chain of cavities so that each positron packet is exactly one cavity behind an electron packet. Explain why this arrangement leads to the acceleration of both the electrons and the positrons.
e. When the packets of electrons and positrons emerge from the chain of cavities, they pass through a magnetic field. The electrons turn one way while the positrons turn the other way. Explain why these two particles turn in different directions.
f. The electrons and positrons are steered with magnets until they collide with one another head on. When they hit, an electron and positron annihilate each other in a burst of energy that makes new particles from empty space. Why is it that the more energy the electron and positron have, the more mass the particles they create can have?

2. Police forensic laboratories often use the neutron activation technique to look for trace elements on objects left at a crime scene or belonging to a suspect. In this technique, the objects are made temporarily radioactive by exposing them to neutrons. Instruments then study the radioactive decays of the activated atoms to determine what atoms they were.

a. Neutron activation is most easily done at a nuclear reactor facility. Why can't a lamp or an X-ray tube be used to add neutrons to an object's nuclei?
b. A neutron can attach itself to almost any nucleus in an object, even if the neutron is only moving with thermal energy. Why doesn't the nucleus push the neutron away before the neutron makes contact with it?
c. Once the neutron has attached itself to a nucleus, the nucleus may become radioactive. Detectors studying an object that has been exposed to neutrons detect electrons and gamma rays emerging from that object. Describe one likely mechanism by which neutron-activated nuclei may produce these electrons and gamma rays.
d. By measuring the energies of the gamma rays emitted by an object following neutron activation, the detectors can determine what elements are present in that object, even if those elements are there in extremely small quantities. For example, they can determine whether trace amounts of barium, residue of the primer in a bullet cartridge, are present on a suspect's glove. Why are the gamma rays emitted by a neutron-activated barium nucleus different from those emitted by a neutron-activated iron nucleus?
e. While neutron activation studies can tell what elements are present in an object, they can't tell what chemicals those elements are part of. Why can't neutron activation tell whether the barium atoms it detects are present as barium

metal or as a barium compound such as barium chloride?

f. How can neutron activation techniques be used to detect an art forgery made with pigments that were not known to the painting's alleged artist?

g. An object is radioactive following neutron activation, but that radioactivity fades quickly. The object is normally safe in a few weeks or months; you just have to wait. Why won't heating or chemically treating the object speed the decrease in radioactivity?

3. Scientists often use electromagnetic waves to study the structure of atoms and nuclei. But while there are many ways of producing electromagnetic waves at the long wavelength end of the spectrum, it's relatively difficult to produce gamma rays. One technique that's used to prepare gamma rays is Compton scattering.

a. When an X-ray photon collides with a stationary electron, energy and momentum are transferred between the two particles. Why is the energy carried by the photon after the collision always less than the energy it had before the collision?

b. Suppose an X-ray photon collides with a moving electron. What must happen to the electron in order for the photon to leave the collision with more energy than it had before the collision?

c. Suppose that an X-ray machine emits an X-ray photon directly behind an extremely high energy electron that's just emerged from a particle accelerator. The electron has far more energy than the X-ray photon, yet the X-ray photon is able to overtake the electron. Why can the X-ray photon always catch up to the electron?

d. As it overtakes the electron, the X-ray photon can undergo Compton scattering from the electron. The photon can emerge from this collision with considerably more energy than it had before. Instead of being an X-ray photon, it's now a gamma ray photon. However, because momentum is also conserved, the electron must retain a substantial fraction of its original energy. In fact, the electron continues onward at almost the same speed it had before the collision. How can it give up a substantial fraction of its energy to the photon without slowing down dramatically?

4. When a positron enters matter, it quickly finds an electron and the two annihilate one another. In the brief moments before that annihilation, however, they often form a strange atom consisting only of an electron and a positron. This short-lived atom is called *positronium*.

a. A positronium atom resembles a hydrogen atom because both have one electron and one positively charged particle. In a hydrogen atom that positively charged particle is a proton, but in a positronium atom it's a positron. While it's sensible to think of the electron in a hydrogen atom as orbiting the massive proton, that's not the case for positronium. In positronium, the electron and positron orbit one another as equals. Why doesn't the positron remain essentially stationary while the electron orbits it?

b. The electron and positron in a positronium atom can only annihilate one another when they actually touch. If they orbited one another as particles, they would never touch. How does quantum physics make it possible for these two orbiting objects to touch one another?

c. When the electron and positron in a stationary positronium atom annihilate one another, they turn into two or more gamma ray photons and nothing else. Noting that a photon carries with it both energy and momentum, explain why a stationary positronium atom can't turn into a single photon when its electron and positron annihilate.

d. When a stationary positronium atom turns into two gamma ray photons, the momentums of those two photons are exactly equal in magnitude but opposite in direction. Why must these two photons have equal but oppositely directed momentums?

5. Magnetic resonance imaging (MRI) is an outgrowth of a technique used by scientists to identify the chemical and magnetic environments of nuclei—usually hydrogen nuclei, which are individual protons. The original technique is called nuclear magnetic resonance, or NMR. When nuclear magnetic resonance was used in imaging, the word "nuclear" was dropped from its title because that word made the imaging technique sound far more dangerous than it actually is.

a. When a sample of a chemical containing hydrogen atoms is placed in a magnetic field, it will absorb only specific electromagnetic waves. How are these electromagnetic waves affecting the protons in the sample when the sample is absorbing them?

b. If the magnetic field in which the sample resides is increased, the frequency of the electromagnetic waves must also be increased or the sample will stop absorbing them. Why?

c. The electrons in the sample itself affect the magnetic field that the protons experience. In and around each proton are a number of electrons which can move somewhat in response to electric fields. When the sample is placed in a magnetic field, the field induces a response that is analogous to a current in the sample, even if the sample isn't actually metallic. This current partially cancels the magnetic field that the proton experiences so that it absorbs electromagnetic waves at a lower frequency than it would absorb if it had no nearby electrons. Why does the "induced" current tend to cancel the magnetic field being applied to the sample rather than strengthen it?

d. The frequency of electromagnetic waves that a particular proton will absorb also depends on any nearby protons and whether they are spin-up or spin-down. How are these nearby protons able to change this frequency?

APPENDIX A

VECTORS

Many of the quantities of physics are vectors, meaning that they have both magnitudes (amounts) and directions. Among these vector quantities are position, velocity, acceleration, force, torque, momentum, angular momentum, and electric and magnetic fields. Of these vector quantities, position is probably the easiest to visualize—you specify an object's position by giving its position vector: its distance and direction from a reference point. For example, you can specify the library's position with respect to your home by giving both its distance from your home (say 3.162 km or 1.965 miles) and its direction from your home (18.43° east of due north). That position vector is all the information someone would need to travel from your home to the library.

In illustrations, such as Fig. A.1, vector quantities are drawn as arrows. The length of each arrow indicates the vector's magnitude, while the direction of the arrow indicates which way the vector points. Suppose that you live in a city with major east–west and north–south streets spaced 1 km apart. Figure A.1 is four aerial views of your city. The vector **A** in Fig. A.1*a* shows the position of the library with respect to your home. It begins at your home and ends at the library, thus indicating both the magnitude and direction of the library's position.

Let's look at another position vector. The vector **B** in Fig. A.1*b* begins at the library and ends at your friend's home. This position vector shows the position of your friend's home with respect to the library and is 2.828 km long and points 45°east of south. If you happen to be at the library, you could use this vector to find your friend's home.

But how can you go from your home to your friend's home? To make this trip, you must add two vectors: you first follow vector **A** from your home to the library and then follow vector **B** from the library to your friend's home. This combined trip is shown as the upper path in Fig. A.1*c*. But you could also go directly from your home to your friend's home by following a new vector in Fig. A.1*c*—vector **C**. This vector from your home to your friend's home is the sum of vectors **A** and **B**, and is 3.162 km long and points 18.43° north of east. Using bold letters to indicate that **A**, **B**, and **C** are vectors, we can write **A+B=C**, meaning that vector **C** is the sum of vectors **A** and **B**.

Another interesting path from your home to your friend's home is shown in Fig. A.1*d*: you first travel along vector **B** and then along vector **A**. On this path, you would arrive at your friend's home without visiting the library. The first leg of this journey would take you into new territory but the second leg would leave you at your friend's home. The sum of vectors **B** and **A** is still vector **C**, or **B+A=C**. Thus vectors added in any order yield the same sum.

While you can estimate the sum of two vectors by drawing arrows on a sheet of paper, obtaining an accurate sum requires some thought. Adding their magnitudes is unlikely to give you the magnitude of the sum vector and adding their directions doesn't even make sense. To add two vectors, it helps to specify them in another form: as their *components* along two or three directions that are at right angles with respect to one

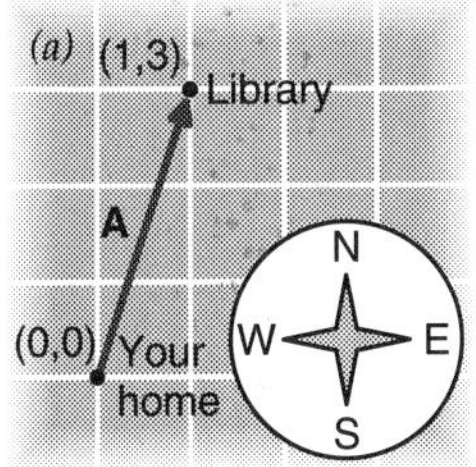

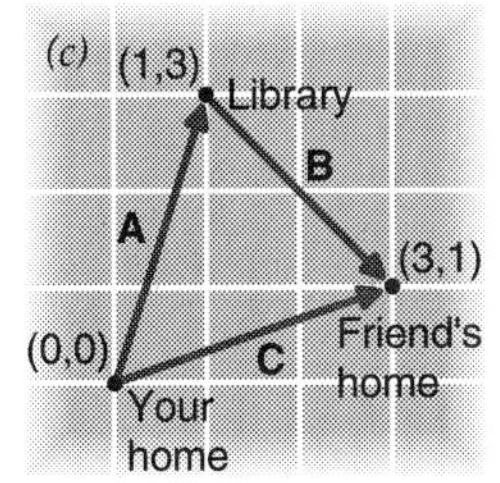

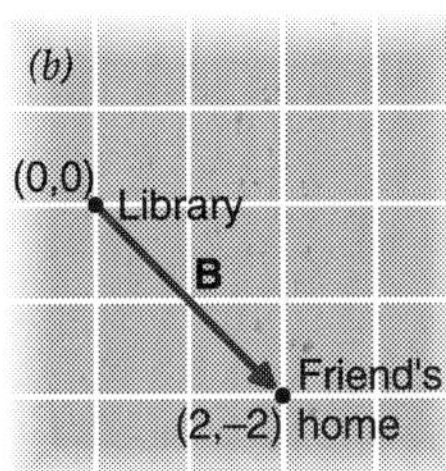

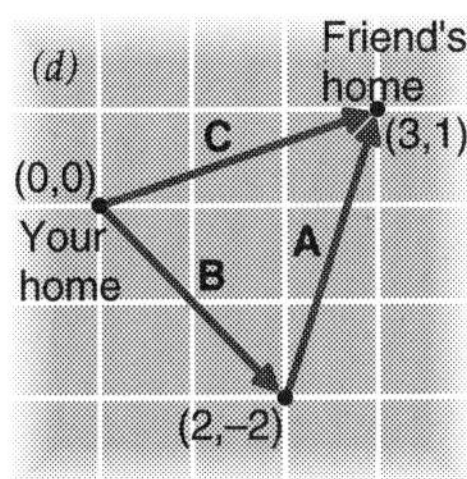

Fig. A.1 - Four aerial views of your city, showing the major north–south and east–west streets that are spaced 1 km apart. (a) To go from your home to the library, you must travel 3.162 km at a direction 18.43° east of north—vector **A**. (b) To go from the library to your friend's home, you must travel 2.828 km at a direction 45° east of south—vector **B**. (c) You can go from your home to your friend's home either by going first to the library along vector **A** and then going to your friend's home along vector **B**, or you can travel directly along vector **C**, which is the sum of vectors **A** and **B**. (d) You can also reach your friend's home by going first along vector **B** and then along vector **A**. The sum of these two vectors is still vector **C**. However, you will not visit the library on that trip.

another. In the present example of travel in a city, two right angle directions are all we need. If height were also important, we'd need three right angle directions.

For the two right angle directions we need in the city, let's choose east and north. We can then specify vector **A** as its component toward the east and its component toward the north. Its east component is 1 km and its north component is 3 km, so vector **A**, the position of the library with respect to your home, is 1 km to the east and 3 km to the north.

This new form for vector **A**, a pair of distances, is often more convenient than the old form, a distance and a direction. If you go to the library by walking 3.162 km in the direction 18.43° east of north, you'll have to pass through a lot of other buildings and backyards. Walking to the library by heading 1 km east and 3 km north allows you to travel on the sidewalks.

If we designate the position of your home as 0 km east and 0 km north, then the position of the library is 1 km east and 3 km north. These positions are labeled in Fig. A.1*a* as (0,0) and (1,3) respectively. The first number in the parentheses is the distance east, measured in kilometers, and the second number is the distance north, also in kilometers.

To go from the library to your friend's home, you must go 2 km east and 2 km south. That's the new form of vector **B**. Because a southward position has a negative component along the northward direction, the position of your friend's home with respect to the library is 2 km east and –2 km north. In Fig. A.1*b*, the library is at (0,0) and your friend's home is at (2,–2).

Now adding these vectors **A** and **B** is relatively easy. To go from your home to your friend's home by way of the library, you must move east first 1 km and then 2 km for a total of 3 km, and you must move north 3 km and then –2 km for a total of 1 km. Thus the position of your friend's home with respect to your home, vector **C**, is 3 km east and 1 km north. In Fig. A.1*c*, your home is at (0,0) and your friend's home is at (3,1).

Similarly, you could go from your home to your friend's home by following first vector **B** and then vector **A**, as shown in Fig. A.1*d*. This trip would take you through unknown regions, but you'd still arrive at the right place. You would move east first 2 km and then 1 km and you'd move north first –2 km and then 3 km. In the end, you'd be 3 km east and 1 km north of your home, the position of your friend's home. Thus the sum of vectors **B** and **A** is still vector **C**.

As you can see, vectors are useful for specifying physical quantities in our three-dimensional world. When you encounter vectors, remember that their directions are just as important as their magnitudes and that these directions must be taken into account when you add two vectors together.

APPENDIX B

UNITS, CONVERSION OF UNITS

When you return from a camping trip and begin to describe it to your friends, there are a number of physical quantities that you may find useful in your conversation. Distance, weight, temperature, and time are as important in everyday life as they are in a laboratory. And when you explain to your friends how far you hiked, how much your backpack weighed, how cold the weather was, and how long the trip took, you'll have to relate those quantities to standard units or your friends won't appreciate just how difficult the trip was.

Most physical quantities aren't simple numbers like 7 or 2.9. Instead, they have units like length or time, and are specified in multiples of widely accepted standard units such as meters or seconds. When you say a door is 3.0 meters tall, you're comparing the door's height to the meter, a widely accepted standard unit of length. With this comparison, most people can determine just how tall the door is, even though they've never seen it.

But the meter isn't familiar to everyone; many people prefer to measure length in multiples of another standard unit: the foot. These people might be more comfortable hearing that the door is 9.8 feet tall. These quantities, 3.0 meters and 9.8 feet, are the same length.

Determining the door's height in feet doesn't require a new measurement because we can convert the one in meters to one that's in feet. To perform this conversion, we need to know how to express one particular length in both units. Any length will do. For example, Table B.1 states that 1 foot is the same length as 0.30480 meters. Because of that equality, we know that the following equation is true:

$$\frac{1\text{ foot}}{0.30480\text{ meters}} = 1.$$

We can multiply 3.0 meters, the height of the door, by this version of 1 and obtain the door's height in feet:

$$3.0\text{ meters} \cdot \frac{1\text{ foot}}{0.30480\text{ meters}} = 9.8425\text{ feet}.$$

Notice that the original units, meters, cancel and are replaced by the new units, feet. Since we only knew the door's height to 2 digits of precision in meters, we can't report the door's height to any more than 2 digits of precision in feet. So we round the result to 9.8 feet.

You can change the units of almost any physical quantity by multiplying that quantity by a version of 1. You should form this 1 by dividing new units by old units, where the number of new units in the numerator is equivalent to the number of old units in the denominator. You can obtain these pairs of equivalent quantities from Table B.1 below. When you do this multiplication, the old units will cancel and you'll be left with the physical quantity expressed in the new units.

One last note about units: when you use physical quantities in a calculation, make sure that you keep the units throughout the process. They're as important in the calculation as they are anywhere else. Some of the units may cancel, but in all likelihood the result of the calculation will have some units left and these units must be appropriate to the type of result you expect. If you're expecting a length, the units of your result should be meters or feet or another standard unit of length. If the units you obtain are seconds or kilograms, you've made a mistake in the calculation.

Table B.1 - Conversion of Units. This table lists pairs of equivalent quantities, one in SI units and one in other units. Each pair can be used to convert measurements expressed as multiples of one unit into measurements expressed as multiples of the other unit. These pairs are grouped according to physical quantity and are given to a precision of 5 digits.

1. Acceleration: (SI unit: 1 meter/second^2 or 1 m/s^2)

1 foot/second^2	= 0.30480 m/s^2

2. Angle: (SI unit: 1 radian)

1 degree (1°)	= 0.017453 radians

3. Area: (SI unit: 1 meter^2 or 1 m^2)

1 foot^2	= 0.092903 m^2
1 inch^2	= $6.4516 \cdot 10^{-4}$ m^2

4. Density: (SI unit: 1 kilogram/meter^3 or 1 kg/m^3)

1 pound/foot^3	= 16.018 kg/m^3

Table B.1 (continued)

5. Energy: (SI unit: 1 joule or 1 J)

1 Btu	= 1054.7 J
1 calorie, thermochemical	= 4.1840 J
1 electron-volt (1 eV)	= $1.6022 \cdot 10^{-19}$ J
1 foot-pound	= 1.3558 J
1 kilocalorie (Food calorie)	= 4,184.0 J
1 kilowatt-hour	= 3,600,000 J

6. Force: (SI unit: 1 newton or 1 N)

1 pound	= 4.4482 N

7. Length: (SI unit: 1 meter or 1 m)

1 angstrom (1 Å)	= 10^{-10} m
1 centimeter (1 cm)	= 0.01 m
1 fermi (1 fm)	= 10^{-15} m
1 foot	= 0.30480 m
1 inch	= 0.02540 m
1 kilometer (1 km)	= 1,000 m
1 light year	= $9.4606 \cdot 10^{15}$ m
1 micron (1 μm)	= 10^{-6} m
1 mil	= $2.5400 \cdot 10^{-5}$ m
1 mile	= 1,609.3 m
1 millimeter (1 mm)	= 0.001 m
1 nanometer (1 nm)	= 10^{-9} m
1 picometer (1 pm)	= 10^{-12} m

8. Mass: (SI unit: 1 kilogram or 1 kg)

1 gram (1 g)	= 0.001 kg
1 metric ton	= 1000 kg
1 pound (at the earth's surface)	= 0.45359 kg
1 slug	= 14.594 kg

9. Power: (SI unit: 1 watt or 1 W)

1 Btu/hour	= 0.29307 W
1 horsepower	= 745.70 W

10. Pressure: (SI unit: 1 pascal or 1 Pa)

1 atmosphere	= 101,325 Pa
1 millimeter of mercury (1 torr)	= 133.32 Pa
1 pound/inch^2 (1 psi)	= 6,894.8 Pa

11. Temperature: (SI units: Celsius or C; Kelvin or K)

Because temperature in the three common units, C, K, and F, aren't multiples of one another, you must use special formulas to convert between them:

Temperature in C = 5/9·(Temperature in F – 32)
Temperature in C = Temperature in K – 273.15
Temperature in K = Temperature in C + 273.15

12. Time: (SI unit: 1 second or 1 s)

1 day	= 86,400 s
1 femtosecond (1 fs)	= 10^{-15} s
1 hour	= 3,600 s
1 microsecond (1 μs)	= 10^{-6} s
1 millisecond (1 ms)	= 0.001 s
1 minute	= 60 s
1 nanosecond (1 ns)	= 10^{-9} s
1 picosecond (1 ps)	= 10^{-12} s

13. Torque: (SI unit: 1 newton-meter or 1 N·m)

1 foot-inch	= 0.11298 N·m
1 foot-pound	= 1.3558 N·m

14. Velocity: (SI unit: 1 meter/second or 1 m/s)

1 foot/second	= 0.30480 m/s
1 kilometer/hour (1 km/h)	= 0.27778 m/s
1 knot	= 0.51444 m/s
1 mile/hour (1 mph)	= 0.44704 m/s

15. Volume: (SI unit: 1 meter^3 or 1 m^3)

1 cup	= $2.3659 \cdot 10^{-4}$ m^3
1 fluid ounce	= $2.9574 \cdot 10^{-5}$ m^3
1 foot^3	= 0.028317 m^3
1 gallon	= 0.0037854 m^3
1 liter (1 l)	= 0.001 m^3
1 milliliter (1 ml)	= 10^{-6} m^3
1 quart	= 0.00094635 m^3

APPENDIX C

FORMULAS OF PHYSICS

1. Newton's Second Law of Motion (p. 7): An object's acceleration is equal to the force exerted on that object divided by the object's mass, or

$$\text{force} = \text{mass} \cdot \text{acceleration}. \qquad (1.1.1)$$

2. Relationship Between Mass and Weight (p. 10): An object's weight is equal to its mass times the acceleration due to gravity, or

$$\text{weight} = \text{mass} \cdot \text{acceleration due to gravity}. \qquad (1.1.2)$$

3. The Velocity of an Object Experiencing Constant Acceleration (p. 12): The object's present velocity differs from its initial velocity by the product of its acceleration times the time since it was at that initial velocity, or

$$\text{present velocity} = \text{initial velocity} + \text{acceleration} \cdot \text{time}. \qquad (1.1.3)$$

4. The Position of an Object Experiencing Constant Acceleration (p. 13): The object's present position differs from its initial position by the product of its average velocity since it was at that initial position times the time since it was at that initial position, or

$$\text{present position} = \text{initial position} + \text{initial velocity} \cdot \text{time} + \tfrac{1}{2} \cdot \text{acceleration} \cdot \text{time}^2. \qquad (1.1.4)$$

5. The Definition of Work (p. 26): The work done on an object is equal to the product of the force exerted on that object times the distance that object travels in the direction of the force, or

$$\text{work} = \text{force} \cdot \text{distance}. \qquad (1.2.1)$$

6. Gravitational Potential Energy (p. 28): An object's gravitational potential energy is its mass times the acceleration due to gravity times its height above a zero level, or

$$\text{gravitational potential energy} = \text{mass} \cdot \text{acceleration due to gravity} \cdot \text{height}. \qquad (1.2.2)$$

7. Newton's Second Law of Rotational Motion (p. 41): The torque exerted on an object is equal to the product of that object's moment of inertia times its angular acceleration, or

$$\text{torque} = \text{moment of inertia} \cdot \text{angular acceleration}. \qquad (1.3.1)$$

The angular acceleration points in the same direction as the torque. This law doesn't apply to objects that are wobbling.

8. Relationship Between Force and Torque (p. 43): The torque produced by a force is equal to the product of the lever arm times that force, or

$$\text{torque} = \text{lever arm} \cdot \text{force}, \qquad (1.3.2)$$

where we include only the component of the force that's at right angles to the lever arm.

9. Linear Momentum (p. 61): An object's linear momentum is its mass times its velocity, or

$$\text{linear momentum} = \text{mass} \cdot \text{velocity}. \qquad (1.4.1)$$

10. The Definition of Impulse (p. 62): The impulse given to an object is equal to the product of the force exerted on that object times the length of time that force is exerted, or

$$\text{impulse} = \text{force} \cdot \text{time}. \qquad (1.4.2)$$

11. Angular Momentum (p. 63): An object's angular momentum is its moment of inertia times its angular velocity, or

$$\text{angular momentum} = \text{moment of inertia} \cdot \text{angular velocity}. \qquad (1.4.4)$$

12. The Definition of Angular Impulse (p. 64): The angular impulse given to an object is equal to the product of the torque exerted on that object times the length of time that torque is exerted, or

$$\text{angular impulse} = \text{torque} \cdot \text{time}. \qquad (1.4.5)$$

13. Kinetic Energy (p. 66): An object's translational kinetic energy is ½ of its mass times the square of its speed, or

$$\text{kinetic energy} = \tfrac{1}{2} \cdot \text{mass} \cdot \text{speed}^2. \qquad (1.4.7)$$

An object's rotational kinetic energy is ½ of its moment of inertia times the square of its angular speed, or

$$\text{kinetic energy} = \tfrac{1}{2} \cdot \text{moment of inertia} \cdot \text{angular speed}^2. \qquad (1.4.8)$$

14. Hooke's Law (p. 82): The restoring force exerted by an elastic object is proportional to how far it is from its equilibrium shape, or

$$\text{restoring force} = -\text{spring constant} \cdot \text{distortion}. \qquad (2.1.1)$$

15. Acceleration of an Object in Uniform Circular Motion (p. 109): An object in uniform circular motion has an acceleration equal to the square of its speed divided by the radius of its circular trajectory, or

$$\text{acceleration} = \frac{\text{speed}^2}{\text{radius of circular trajectory}}. \quad (2.3.1)$$

16. The Ideal Gas Law (p. 132): The pressure of a gas is equal to the product of the Boltzmann constant times the particle density times the absolute temperature, or

$$\text{pressure} = \text{Boltzmann constant} \cdot \text{particle density} \cdot \text{absolute temperature}. \quad (3.1.3)$$

17. Bernoulli's Equation (p. 146): For an incompressible fluid in steady-state flow, the sum of its pressure potential energy, its kinetic energy, and its gravitational potential energy is constant along a streamline, or

$$\begin{aligned}\frac{\text{energy}}{\text{volume}} &= \frac{\text{pressure potential energy}}{\text{volume}} + \frac{\text{kinetic energy}}{\text{volume}} + \frac{\text{gravitational potential energy}}{\text{volume}} \\ &= \text{pressure} + \tfrac{1}{2} \cdot \text{density} \cdot \text{speed}^2 + \text{density} \cdot \text{acceleration due to gravity} \cdot \text{height} \\ &= \text{constant } (\textit{along a streamline}). \quad (3.2.5)\end{aligned}$$

18. Poiseuille's law (p. 162): The volume of fluid flowing through a pipe each second is equal to $(\pi/128)$ times the pressure difference across that pipe times the pipe's diameter to the fourth power, divided by the pipe's length times the fluid's viscosity, or

$$\text{volume flow rate} = \frac{\pi \cdot \text{pressure difference} \cdot \text{pipe diameter}^4}{128 \cdot \text{pipe length} \cdot \text{fluid viscosity}}. \quad (4.1.1)$$

19. The Definition of Work in Rotational Motion (p. 176): Work is the torque exerted on an object times the angle through which the object rotates in the direction of that torque, or

$$\text{work} = \text{torque} \cdot \text{angle of rotation}. \quad (4.2.1)$$

The angle is measured in radians.

20. The Law of Universal Gravitation (p. 219): Every object in the universe attracts every other object in the universe with a force equal to the gravitational constant times the product of the two masses, divided by the square of the distance separating the two objects, or

$$\text{force} = \frac{\text{gravitational constant} \cdot \text{mass}_1 \cdot \text{mass}_2}{(\text{distance between masses})^2}. \quad (5.1.1)$$

21. The Stefan–Boltzmann Law (p. 268): The power an object radiates is proportional to the product of its emissivity times the fourth power of its temperature times its surface area, or

$$\text{radiated power} = \text{emissivity} \cdot \text{Stefan – Boltzmann constant} \cdot \text{temperature}^4 \cdot \text{surface area}. \quad (6.2.1)$$

22. Coulomb's Law (p. 379): The magnitudes of the electrostatic forces between two objects are equal to the Coulomb constant times the product of their two electric charges divided by the square of the distance separating them, or

$$\text{force} = \frac{\text{Coulomb constant} \cdot \text{charge}_1 \cdot \text{charge}_2}{(\text{distance between charges})^2}. \quad (11.1.1)$$

If the charges are like, then the forces are repulsive. If the charges are opposite, then the forces are attractive.

23. Power Consumption (p. 416): The electric power consumed by a device is the voltage drop across it times the current flowing through it, or

$$\text{power consumed} = \text{voltage drop} \cdot \text{current}. \quad (12.1.1)$$

24. Power Production (p. 417): The electric power provided by a device is the voltage rise across it times the current flowing through it, or

$$\text{power provided} = \text{voltage rise} \cdot \text{current}. \quad (12.1.2)$$

25. Ohm's Law (p. 422): The voltage drop across a wire is equal to the current flowing through that wire times the wire's electric resistance, or

$$\text{voltage drop} = \text{current} \cdot \text{electric resistance}. \quad (12.2.1)$$

26. Relationship Between Wavelength and Frequency (p. 511): The product of the frequency of an electromagnetic wave times its wavelength is the speed of light, or

$$\text{speed of light} = \text{wavelength} \cdot \text{frequency}. \quad (14.3.1)$$

27. Relationship Between Energy and Frequency (p. 538): The energy in a photon of light is equal to the product of the Planck constant times the frequency of the light wave, or

$$\text{energy} = \text{Planck constant} \cdot \text{frequency}. \quad (15.2.1)$$

28. The Lens Equation (p. 561): One divided by the focal length of a lens is equal to the sum of one divided by the object distance and one divided by the image distance, or

$$\frac{1}{\text{focal length}} = \frac{1}{\text{object distance}} + \frac{1}{\text{image distance}}. \quad (16.1.1)$$

SOLUTIONS TO SELECTED PROBLEMS

Section 1.1

E.1 The dolphin's inertia carries it upward, even though its weight makes it accelerate downward and gradually stop rising.

E.3 Any collision in which the car accelerates forward, such as when the car is hit from behind by a faster moving car.

E.5 The anvil's large mass slows its acceleration, so the hot metal is squeezed between the moving hammer and the stationary anvil.

E.7 The pad's inertia tends to keep it in place. If you pull the paper away too quickly, the pad won't be able to accelerate with the paper.

E.9 No (their directions of travel are different).

E.11 Toward the south.

E.13 The forward component of his velocity remains constant as he falls and he follows an arc that carries him forward over the rocks.

E.15 In the absence of air effects, a ball hit at 45° above horizontal will travel farthest. A ball hit higher or lower won't travel as far.

E.17 The child who simply runs forward reaches the water first.

E.19 Backward, in the direction opposite your forward velocity.

E.21 Regardless of their horizontal components of velocity, all objects fall at the same rate. The ball and bullet descend together.

E.23 As it turns left, the car accelerates left. The loose objects remain behind and end up on the right side of the dashboard.

P.1 3,200 N.
P.3 11.13 m/s.

P.5 Your Mars weight would be about 38% of your earth weight.

P.7 About 0.64 s (0.32 s on the way up and 0.32 s on the way down).

P.9 48 N.
P.11 About 0.20 s.

P.13 4,800 N.
P.15 19.6 N.

P.17 2 s.
P.19 About 102 kg and 1,020 kg.

P.21 About 0.49 m/s^2.

Section 1.2

E.1 While they're touching, the floor exerts a large upward force on the barbell, accelerating the barbell upward, and the barbell exerts an equal but oppositely directed force on the floor. Both have weights.

E.3 The egg and car would push on one another so hard that the car would dent. The egg would do (positive) work on the car by pushing its surface downward as that surface moves downward.

E.5 A person does more work. Movements that appear large compared to the ant's height still involve small distances and little work.

E.7 The marble's speed is greatest at the bottom of the slide.

E.9 Its magnitude must equal the suitcase's weight.

E.11 Each car experiences zero net force.

E.13 The downhill force on it is less than its weight so it accelerates slowly down the ramp and also doesn't head directly downward.

E.15 Both forces have exactly the same magnitude.

E.17 As it rolls on the surface, it always accelerates downhill.

E.19 The astronaut exerts an upward force of 850 N on the earth.

E.21 Yes, you pushed it inward and it dented inward.

E.23 He does work on his hand by pushing it forward as it moves forward. His hand does work on his opponent in a similar manner. Energy flows from him, to his hand, to the opponent.

E.25 Zero net force.
E.27 Zero net force.

E.29 Your kinetic energy becomes gravitational potential energy.

E.31 Your feet accelerate upward rapidly when they hit the ground and the snow continues downward, leaving your feet behind.

P.1 9,800 N.
P.3 14,700,000 J.

P.5 60 J.
P.7 12.5 J.

P.9 8,000 N.
P.11 36,480 J.

Section 1.3

E.1 A force exerted at the hinges produces no torque about them.

E.3 Your force far from the hinges produces a large torque. To oppose this torque, the nut must exert a huge force near the hinges.

E.5 It would rotate about its center of mass, not its geometric center. It would appear to wobble as it turned.

E.7 A long pole has a large moment of inertia. The tightrope walker can exert large torques on it without making it rotate rapidly.

E.9 It reduces the car's moment of inertia so that the car can undergo rapid angular accelerations and change directions quickly.

E.11 To prevent the umbrella from rotating about its tip when the wind blows, the sand must exert its forces far from the tip.

E.13 Weight on the shelf produces a torque about its attachment to the wall. The brace produces an opposing torque to support it.

E.15 When they're near the pivot, these metal objects can't produce enough torque to stop the blades from rotating and they're cut.

E.17 By pushing far from the pivot, you exert more torque on the lid.

E.19 Your small effort exerted on the crowbar far from its pivot produces a large force on the box, located near the crowbar's pivot.

E.21 The weights of your chest and your feet exert torques in opposite directions about your knees. They partially balance one another.

E.23 The farther the water is from the water wheel's pivot, the more torque its weight produces on the wheel.

E.25 The strip exerts its force far from the wheel's pivot and thus produces a substantial torque on the wheel to slow its rotation.

E.27 When the nipple is on one side, the wheel is unbalanced. There is a net torque on the wheel and it undergoes angular acceleration.

E.29 The force was exerted far from the crank's pivot, producing a torque that caused the crank and the engine to rotate.

P.1 About 122 N·m to the left.
P.3 About 122 N·m to the right.

P.5 3 times as much torque
P.7 12.5 N·m, slowing the blade.

P.9 100 N.

Section 1.4

E.1 Its angular momentum keeps it turning, even without torques.

E.3 You're exerting a torque on it for a time—an angular impulse.

E.5 As you descend, you land hard and your knees and legs must convert your kinetic energy into thermal energy. Injuries can occur.

E.7 Skidding sideways does work against sliding friction, converting some of the skier's kinetic energy into thermal energy.

E.9 The pin's surface turns with the crust and doesn't slide across it.

E.11 The nearer the frictional force is to the pivot, the less torque it produces to slow the yo-yo's rotation. Slipperiness reduces friction.

E.13 To avoid accelerating when pushed on with a force, the villain would have to have infinite mass. That's impossible.

E.15 Angular momentum.

E.17 Sliding friction opposes its motion and it decelerates.

E.19 A static frictional force from the pavement pushes you forward.

E.21 Pressing the wheels more tightly against the pavement increases the maximum force that static friction can exert on the wheels.

E.23 The collapsing star's angular momentum can't change. Since its moment of inertia decreases, its angular velocity must increase.

E.25 When you land on the merry-go-round, it transfers some of its angular momentum to you, so you turn with it.

E.27 Angular momentum.

E.29 Like static friction, the binding opposes relative motion between the boot and ski. But once a limit is exceeded, relative motion occurs.

P.1 Its energy would be only 0.2 times as large as before.

P.3 If your mass is 100 kg, your velocity changes by 10^{-6} m/s.
P.5 2400 kg·m/s forward. **P.7** 3,600 J.
P.9 450 kg·m/s to the right. **P.11** 120 N.
P.13 3 times as much as usual. **P.15** 9 times as much as usual.

Section 2.1

E.1 As you draw the string away from its equilibrium shape, it experiences a restoring force proportional to its displacement.
E.3 The top of the curl supports more weight than the bottom.
E.5 The bump can compress the truck's springs rather than having to accelerate the massive truck out of the way.
E.7 Place the pony's front feet on one scale and back feet on the other scale. The sum of the two measurements is the pony's weight.
E.9 15 N.
E.11 A scale must be calibrated regularly with standard weights to remain accurate. Typical scales are out of calibration and inaccurate.
E.13 Use your friend's scale to calibrate some standard weights (you and your friends) and then calibrate your scale with those weights.
E.15 Correct.
E.17 You support your clothes and the scale supports you.
E.19 The scale reports the force it exerts on you. However, you accelerate with each bump, so this force isn't exactly your weight.
P.1 15,000 N/m. **P.3** 10 mm.

Section 2.2

E.1 It has a coefficient of restitution near 0 (lots of inertial friction).
E.3 Your relative velocity is small—in your frame of reference the car in front of you is barely moving, so the impact is very gentle.
E.5 During a bounce, the work done on a RIF ball—to store energy in it—involves a smaller force exerted for a longer distance.
E.7 At high pressure, the shoes are stiffer and exert larger forces when distorted. They accelerate more rapidly and bounce faster.
E.9 The bat accelerates more than the ball during their collision.
E.11 The bat transfers some of its energy and momentum to the ball.
E.13 Because the coins are lively and have identical masses, each coin passes its energy and momentum to the next coin during the event.
E.15 The ball is stiffer than the floor, so the floor distorts more and receives and stores most of the collision energy.
E.17 You do work on the sand as you step on it, but the sand doesn't return this energy to you as you lift your foot back up again.
E.19 The force that the hammer exerts on the nail during their collision would diminish if the hammer's surface distorted easily.
E.21 An increase in the balls' coefficients of restitution.
E.23 A 30% loss of rebound height is a 30% loss of gravitational potential energy—equal to the energy that became thermal energy.
E.25 Because the trains' relative velocity is zero, a person jumping between them views them both as essentially stationary.

Section 2.3

E.1 While you can feel accelerations, you can't feel velocity.
E.3 The ball rolls up the side of the bowl during an acceleration until the horizontal component of the support force exerted on it by the bowl is enough to make it accelerate with you and the bowl.
E.5 The sharper the curve, the more centripetal force the train needs to accelerate around the curve. If the track can't supply it...disaster.
E.7 The nail exerts an enormous force on the hammer to slow it down. The hammer pushes back, driving the nail into the wood.
E.9 Yes. **E.11** At the bottom of each swing.
E.13 When the rattle accelerates, the beads inside it continue on and hit the walls of the rattle. The rattle then makes noise.
E.17 As the salad undergoes rapid centripetal acceleration, the water travels in straight lines and runs off the salad.
E.19 Static friction between the tires and the road accelerates the car inward around the curve. At excessive accelerations, the car skids.
E.21 The bucket makes the water accelerate downward faster than the acceleration due to gravity and the water remains inside it.
E.23 As the car accelerates upward, you must pull upward on the briefcase extra hard to make it accelerate upward too.
E.25 After going over a bump, the car begins to accelerate downward and your apparent weight is briefly less than your real weight.
E.27 The pilot makes the plane follow the arc that a freely falling object would follow. **E.29** It will travel toward the left.
E.31 The water will travel in the direction it was moving when it left the tire's surface—toward the rear of the truck, not downward.
E.33 As your hands accelerate, the water is left behind.
P.1 It must turn at about 31 m/s. **P.3** 20 m/s^2.

Section 3.1

E.1 An upward force equal in magnitude to the fish's weight.
E.3 It will sink. Its average density is larger than that of helium.
E.5 As water enters the car, the car's average density will increase.
E.7 The road exerts an upward pressure on the tire to balance its internal pressure. The higher this pressure, the less tire surface the road must exert its pressure on to support the car's weight.
E.9 The air in the whale would compress during the dive. The whale's average density would increase and it would sink.
E.11 Cooling the air in the container reduced its pressure. The resulting pressure imbalance across the lid pushes the lid inward.
E.13 Warm air floats on top of the cold, dense air in the bins.
E.15 The upper liquid is less dense than the lower liquid. It floats.
E.17 1 kg of gasoline takes up more space than 1 kg of water.
E.19 Oil is less dense than water and floats on water.
E.21 Heated air expands, its density decreases, and it floats upward.
E.23 The crust has a lower density than the mantle. The crust floats.
E.25 Helium atoms will enter the balloon and it will inflate further.
E.27 A dead fish couldn't maintain the same density as water.
E.29 Because it contains lots of air, the boat's average density is less than that of water. The solid metal anchor has a much higher density.
P.1 101,400 Pa. **P.3** 3 times as much as before.
P.5 0.94 times as much as before. **P.7** 122.4 kg or 0.1224 m^3.
P.9 28.4 N.

Section 3.2

E.1 Water will flow downhill through the siphon, *into* the boat!
E.3 The water loses too much pressure flowing uphill.
E.5 There is no limit, as long as the straw is always 0.5 m tall.
E.7 It won't travel as far upward because some of its kinetic energy becomes gravitational potential energy on the way up.
E.9 As wind slows in the bag, its pressure rises and inflates the bag.
E.11 The pressure imbalance on the base of the dam is larger.
E.13 It's stored as elastic potential energy in the skin of the balloon.
E.15 Your foot (which does work on the foil packet's surface).
E.17 From pressure potential energy provided by your lungs.
E.19 As the air warmed, its pressure increased. The resulting pressure imbalance pushed the ink out of the pen.
E.21 The more force, the greater the pressure and pressure potential energy in the bellows and the faster the air moves as it enters the fire.
E.23 The pressure outside your lungs is above atmospheric pressure.
E.25 Pressure potential energy provided by the atmosphere.
P.1 About 250,000 Pa above atmospheric pressure.
P.3 About 50 m. **P.5** About 50 m.
P.7 141 m/s or 510 km/h **P.9** About 3,100,000 Pa.

Section 4.1

E.1 The increased flow in the cold water pipe requires a larger pressure difference in it, leaving too little pressure to operate the shower.

E.3 Squeezing highly viscous frosting through a narrow "pipe" requires an enormous pressure difference across that pipe.

E.5 Flow through an artery scales with the fourth power of radius.

E.7 Drawing more water through the pipes requires a larger pressure difference in those pipes. Less pressure and pressure potential energy are left to operate the showers and provide a hard spray.

E.9 Viscous ketchup requires a large pressure difference to flow rapidly. Gravity alone can't provide enough pressure difference.

E.11 The pressure at the can bottom rises suddenly due to water hammer. The pressure at the can top falls or remains the same.

E.13 In turbulent flow, adjacent regions of water become separated.

E.15 With no restoring force pushing it into place, it stopped sealing.

E.17 As the water stops, its pressure rises suddenly and produces a pressure imbalance across the window that can break the window.

E.19 The bucket stops the water, so its pressure rises. The higher pressure above the bucket pushes it downward hard.

E.21 Elastic rubber lids would experience restoring forces that would push them tightly into place. They would thus form leak-tight seals.

E.23 In warm weather, the molecules in a viscous liquid have more thermal energy and are able to move past one another more easily.

E.25 As the walls of a cavity collide, they experience water hammer and their pressures rise abruptly.

P.1 For a fish with an obstacle length (width) of about 2 cm, turbulence will appear if it moves faster than about 0.1 m/s.

P.3 About 84 times as long. **P.5** About 400,000 Pa.

P.7 About 160,000 times as fast. **P.9** About 18.3 m/s.

Section 4.2

E.1 Viscous drag in the air prevents them from falling rapidly.

E.3 A large raindrop falls faster. **E.5** They experience more drag.

E.7 The pearl experiences more viscous drag in the viscous honey.

E.9 The air's speed decreases and its pressure increases.

E.11 As the air and water enter the wide bucket, they slow down and the water has time to fall out of the air before they enter the fan.

E.13 The wind in a tornado exerts tremendous drag forces on objects.

E.15 The air speeds up in the narrow channel between vehicles and its pressure drops. Surrounding air pushes the vehicles together.

E.17 Air speeds up as it squeezes together to flow around the curved side of the car. As the airflow speeds up, its pressure drops.

E.19 High-speed air exerts enough drag force on dust to remove it.

E.21 The imbalance of pressures above and below the lid dents it in.

E.23 Blood slows down in the wide aneurysm and its pressure rises.

E.25 The drag forces on you would decrease much less than your weight would and they would be able to keep you from falling fast.

E.27 The drag forces on the meteor are so huge that they slow it.

E.29 The water would speed up and its pressure would drop each time it entered a narrow region. The opposite for a wide region.

E.31 As the filter becomes dirty and blocked, the drop in pressure that the air experiences by passing through it will become larger.

E.33 Air rushing through the narrow neck has a low pressure, so surrounding air pinches the neck closed. This stops the air in the neck and its pressure rises, pushing the neck opened. This repeats.

P.1 6π or about 18.8 radians. **P.3** About 0.64 N·m.

P.5 Low speed: 2 W, medium speed: 16 W, high speed: 128 W.

Section 4.3

E.1 The water slows down just upstream and downstream of the posts and speeds up on the sides of the posts.

E.3 Lift supports the skier and drag pulls the skier backward.

E.5 The massive head shifts the arrow's center of mass forward so that its center of pressure is well behind its center of mass.

E.7 The front runner drags the air forward so that you experience less drag while running through forward-moving air.

E.9 The dimpled ball will hit first.

E.11 It balances the backward force due to air drag.

E.13 Both objects are slowed by similar pressure drags. But the spear's larger mass and momentum keep it from slowing as quickly.

E.15 Without backspin, it can't obtain lift and won't go as far.

E.17 It increases pressure drag and reduces your terminal velocity.

E.19 Dimples produce a turbulent boundary layer and delay the flow separation behind the hull. This decreases pressure drag on the boat.

E.21 Air experiences a rise in pressure when it slows down at the front of a car. This high pressure air presses the paper against the car.

E.23 Small disturbances in the airflow around the sides of the nonspinning volleyball produce lift forces that push the ball to the side.

Section 4.4

E.1 As water speeds up to flow around the sides of the sphere, its pressure drops and air accelerates toward it.

E.3 Adding energy to the burned gas makes it expand, so it flows out of the engine even faster than it would without afterburners.

E.5 When the air becomes too thin, the wings can't obtain enough of a pressure imbalance and lift to support the plane's weight.

E.7 The water speeds up around the curve and its pressure drops. The resulting pressure imbalance pushes the spoon into the stream.

E.9 An airfoil, round in front and tapered behind, would be better.

E.11 By pushing exhaust downward, they obtain upward thrusts.

E.13 You experience a "sonic boom" whenever a plane passes at supersonic speeds, not just when it "breaks the sound barrier."

E.17 Turbulent air pockets form behind its moving blades, so swirling vortices and eddies flow out of the fan.

E.19 Gravity steadily transfers downward momentum to the plane.

E.21 Air travels faster over your hand than under it, so the pressure above your hand is less than the pressure under your hand.

E.23 The lift is vertical but the plane's motion is horizontal. No work.

E.25 The drag forces push the air forward over time—an impulse.

E.27 As the plane's speed decreases, the pressure imbalance between the top and bottom of its wing decreases and it loses its lift.

E.29 If the wings' angle of attack in the upward rushing air becomes too great, the airflow will separate from their tops and they will stall.

Section 5.1

E.1 The fan pushes air backward and air pushes the boat forward.

E.3 The round-bottomed bottle's center of gravity descends as it tips while the flat-bottomed bottle's center of gravity rises as it tips.

E.5 A rocket pushes backward on its exhaust, not on the launcher.

E.7 The cup's low center of gravity rises as the cup tips.

E.9 The boat's center of gravity must rise upward a considerable distance before the boat will tip over. This takes lots of energy.

E.11 If the boat's center of gravity is above its center of buoyancy, its center of gravity will descend as it tips and it will be unstable.

E.13 Standing raises the center of gravity so the canoe is unstable.

E.15 200 kg·m/s southward. **E.17** They are equivalent.

E.19 It tips when its center of gravity isn't above its base of support.

E.21 Yes. **E.23** Directly toward the earth.

E.25 A rocket must do much less work against the moon's gravity.

Section 5.2

E.1 As a sprinter accelerates, the ground must push toward the sprinter's center of mass to avoid exerting a torque on the sprinter.

E.3 The spinning, massive wheels store energy for the rider.

E.5 The leftward frictional force on your feet keeps you balanced.

E.7 Precession.

E.9 As long as the inward force from the U-shaped surface points toward the skater's center of mass, the skater doesn't begin rotating.

E.11 The torque you exert on the crank is much larger than the torque the blades exert on the batter.

E.13 The engine's disk exerts a static frictional force on the belt, which moves in the direction of that force. The belt exerts a similar force on the device's disk, which moves in the direction of that force.

E.15 To do work, you must exert a torque on the wrench and it must turn in the direction of that torque. With no torque, there's no work.

E.17 The water's mass would store energy, making the bicycle hard to stop and start. Lifting the water uphill would also be hard work.

E.19 The main purpose of the brakes is to convert the car's kinetic energy into thermal energy, to slow it down. That wouldn't happen.

Section 5.3

E.1 As you lie on the bed, the bed's pressure acts over a large surface to support you. Standing reduces that supporting surface.

E.3 4 times as much. **E.5** 40,000 N.

E.7 When the balloon is uninflated, the stretching force is produced by air pressure acting over a small area and a large pressure is needed to obtain enough force to begin expanding the balloon.

E.9 Where the tires touch the pavement, the pressures are about equal above and below the rubber. The force the pavement is exerting on the car is the tire contact area times the tire air pressure.

E.11 100 N. **E.13** 50 m.

E.15 40 N.

E.17 The weights provide some of the force needed to lift the sash and they transfer their gravitational potential energies to it.

E.19 Flexing your knees allows them to absorb your kinetic energy gradually as you land, with small forces exerted over long distance.

Section 6.1

E.1 Although heat flows rapidly from the hot flame to your colder finger, there isn't time for your finger's temperature to rise much.

E.3 They are poor conductors of heat. The metal carries heat better.

E.5 The glass permits thermal radiation to heat the pie directly.

E.7 The grate permits convection to carry air upward to the wood.

E.9 The blackened pan absorbs thermal radiation from the oven while the shiny pan reflects that thermal radiation.

E.11 The tree sends thermal radiation toward you.

E.13 Your body temperature would decrease.

E.15 The rising hot air of convection draws the flame upward.

E.17 There would be no convection to draw the flame "upward."

E.19 Still water cools off near your skin and insulates you. The forced circulation of hot water in the spa removes this insulating layer.

E.21 Heat from the hot land flows to the cold ocean by traveling up and over in a convection cell. Cold air returns to the land as a breeze.

E.23 Chilled air near the ice was denser and sank downward.

Section 6.2

E.1 The real insulator is air. The more air the down traps, the better.

E.3 It traps bubbles of air, which insulate the pie from the cooler air.

E.5 The metal handle conducts heat better, so your hand can't warm the metal's surface the way it can warm the glass's surface.

E.9 If air entering the home passes near air leaving the home, heat can flow from one to the other. Entering air arrives at the indoor temperature while leaving air exits at the outdoor temperature.

E.11 Your ear emits thermal radiation characteristic of its temperature. The thermometer measures your ear's spectrum and brightness.

E.13 Porous tiles are terrible conductors of heat, impede convection, and block radiation, and thus prevent heat from reaching the shuttle.

E.15 Without gravity, there'd be no convection to shape the flame.

E.17 The black pan absorbs thermal radiation from the heating element and becomes extremely hot, browning the food touching it.

P.1 18,800 W. **P.3** 73,500,000 W.

Section 6.3

E.1 The naphthalene sublimes and the mothball becomes gaseous.

E.3 The animals emit brighter infrared light at shorter wavelengths.

E.5 The cooler the object, the longer the wavelengths of its black body spectrum. The spectrum of a very cold object peaks in the microwave portion of the electromagnetic spectrum.

E.7 An object's thermal radiation is characteristic of its temperature. When you compare thermal radiation, you compare temperature.

E.9 The water sublimes to form a gas, leaving only dry vegetables.

E.11 In a three-way bulb, any filament that's on is at full temperature. A dimmed bulb's filament operates below full temperature.

E.13 The amount of thermal radiation emitted by a hot object is proportional to its surface area, as stated in the Stefan-Boltzmann law.

E.15 Activation energies are needed to break chemical bonds in the bulb's molecules, so they can rearrange. They need thermal energy.

Section 6.4

E.1 As car parts cool, they contract. Parts that contract more or faster than others may slip past them and emit sounds as they do.

E.3 If the faucet stem contracts more than the rest of the faucet, it will lift the washer away from the seat as it cools and let water out.

E.5 A cookie sheet that's made of different metals, or that heats unevenly, will bend to accommodate uneven expansions of its parts.

E.7 The sidewalk has a different coefficient of volume expansion than the ground and would break when the temperature changed.

E.9 The metal lid has a larger coefficient of volume expansion than the glass jar and it pulls away from the jar when you heat them both.

E.11 If a full bottle were heated, the liquid inside would expand more than the glass and it would push the top off the bottle.

E.13 As the house's different materials warm or cool, they expand or contract differently and slide past one another, causing noises.

E.15 Water has an enormous specific heat capacity, so the water-filled pot requires lots of heat before its temperature rises much.

E.17 When a crystal melts, the positional and orientational orders of its molecules are lost in the random structure of the liquid.

Section 7.1

E.1 They are heat pumps and transfer heat to the surrounding air.

E.3 Its temperature increased. Gravity did work on the gas and this work appeared in the gas as thermal energy. The gas became hot.

E.5 Without the fan, the air conditioner won't be able to transfer its waste heat to the surrounding air. It will stop pumping heat.

E.7 As you stretch the rubber band, you do work on it and increase its internal *kinetic* energy so that its temperature rises.

E.9 The gas still in the container does work pushing the other gas into the siphon and uses up some of its thermal energy. It cools.

E.11 It transfers heat from the house to the outdoors.

E.13 Though not forbidden by the laws of motion, it's extraordinarily unlikely for the vase fragments to reassemble themselves.

E.15 This uneven distribution of thermal energies is extraordinarily unlikely. It would violate the second law of thermodynamics.

E.17 Though not forbidden by the laws of motion, such uneven distributions of snow are remarkably unlikely.

E.19 It decreases. **E.21** It rises.

Section 7.1

E.1 As heat flows from the hot spots to the cold spots, by way of huge convection cells, some heat is becoming ordered kinetic energy.

E.3 Heat flowing into or out of the pavement during weather changes allows the pavement to do work as it tears itself apart.

E.5 Heat flows from the hotter sun to the colder house.

E.7 Without temperature differences, heat won't naturally flow. Such heat flow is what powers a heat engine such as the winds.

E.9 The hotter the burned gas, the larger the fraction of heat that can be converted to work as it flows to the outdoor air.

E.11 The blade does work on the cylinder during the compression stroke. Without the blade, the cylinder couldn't compress the air.

E.13 You do work on the air and the air's thermal energy increases.

E.15 The plane does work on the air by compressing it and this raises the thermal energy and temperature of the compressed air.

Section 8.1

E.1 At the container's bottom, because the solid sinks in the liquid.

E.3 The moisture evaporates rapidly as its temperature rises, a phase transition that requires enormous amounts of heat.

E.5 The cover prevents evaporation from taking heat from the pot.

E.7 Dissolved chemicals such as antifreeze decrease water's melting temperature and raise its boiling temperature.

E.9 Hot, "dry" steam remains a gas even when its temperature drops. It must give up considerable heat before it begins to condense.

E.11 As gas is released from the tank, additional propane evaporates from the liquid to replace it. The liquid requires heat to evaporate.

E.13 Covering the pot prevents evaporation from taking heat away.

E.15 Dissolved gases come out of solution when water freezes or boils. Boiling the water before making ice allows the gases to escape.

E.17 It goes into melting ice at 0° C to form water at 0° C.

E.19 Yes. Anything that dissolves well in water will melt ice.

E.21 They formed when water vapor in the hot, moist air above the spa's water condensed on the cooler lid.

E.23 There is no atmospheric pressure pushing inward on the water so bubbles that form inside the liquid can grow and expand easily.

E.25 The moisture sublimed, leaving your food freeze-dried.

Section 9.1

E.1 It would run slow.

E.3 Since its period is independent of its amplitude, it is a harmonic oscillator and its restoring force is proportional to its distortion.

E.5 Above the ground, the ball experiences a *constant* restoring force, its weight, not one that's *proportional* to its displacement.

E.7 If the chair's period doesn't depend on how far it rocks, its restoring force must be proportional to its tilt. If the extent of the rock affects its period, the restoring force isn't proportional to the tilt.

E.9 Release a pulse of light at the first event and measure how far the light travels by the second event. Divide by the speed of light to see how long the light was in flight and the time between the events.

E.11 The top isn't an oscillator and its motion isn't a natural resonance, so the time it takes to complete each turn can vary.

E.13 Push on it as it flexes away from you (so you do work on it).

E.15 The time-keepers in a balance clock and a quartz watch don't involve gravity and would keep accurate time. A pendulum clock's time-keeper, a pendulum, has a period that depends on gravity.

E.17 The restoring forces won't change but the moving masses will. The fork's tines will accelerate less and its frequency will decrease.

E.19 Atomic clocks use atoms as time-keepers. Since properly selected atoms are perfectly identical, they don't need to be calibrated.

Section 9.2

E.1 You do work on the key, the key does work on the hammer, the hammer does work on the string, and the string does work on the air.

E.3 Once the rubber band can't stretch any more, increasing its tension doesn't make it longer. It behaves like a violin string, with a pitch the rises as its tension increases.

E.5 The shorter air column is stiffer and its mass is smaller.

E.7 Less mass, more tension, or less length.

E.9 To oscillate, the string's energy must shift back and forth between two forms. Without tension, it has no elastic potential energy.

E.11 The bar's vibrational nodes are at those support locations.

E.13 A small change in the pipe's length alters the stiffness and mass of the air column and changes its vibrational frequency slightly.

E.15 The tuba's longer air column is less stiff and more massive than that of the trumpet and takes longer to complete each cycle.

E.17 The copper wrap adds mass to the string to lower its pitch.

E.19 Opening holes effectively shortens the recorder. Its air column becomes stiffer and less massive, so it vibrates more rapidly.

E.21 By pulling the string away from equilibrium, the quill does work on the string. The farther it pulls, the more work it does.

E.23 Because the hammer rebounds from the string as the string moves away from it, the hammer bounces weakly and receives only a small fraction of the rebound energy.

Section 10.1

E.1 Ocean water can't enter or leave the Mediterranean Sea quickly enough to allow its high tide to be very different from its low tide. The tide simply rearranges the water levels within the sea itself.

E.3 The downhill force is stronger when the skier heads directly down the hill so that the skier reaches a larger terminal velocity.

E.5 Resonant energy transfer occurs when your steps are synchronized with the rhythmic motion of the coffee sloshing in the cup.

E.7 The new wave heads backward, too.

E.9 The earth's center of mass falls faster than water farthest from the moon, leaving that water behind as a bulge. Water nearest the moon falls faster than the earth, pulling that water out into a bulge.

E.11 The wave's energy is proportional to its circumference and its height. As its circumference grows, its height must diminish.

E.13 Crests and troughs don't travel along the string. Instead, the center of the string becomes alternately a crest and a trough.

E.15 With more tension, the restoring forces in the string are stiffer and the accelerations are more rapid. Everything happens faster.

E.17 Sound is a disturbance in the air, not the air itself. As waves of compressions and rarefactions travel through it, the air moves little.

E.19 There isn't enough water in front of the largest waves to complete their crests as they pass over the sandbar.

E.21 The moving water corresponds to the crests and troughs, while the motionless water is in the middle of the figure.

Section 11.1

E.1 Because they are identical, the socks acquire like charges as the result of sliding friction with clothes in the dryer. Like charges repel.

E.3 The charged paint particles electrically polarize the surface being coated, so that the paint particles are attracted to it and stick.

E.5 They would fall at the same rate.

E.7 Electrostatic forces diminish as one over the square of distance.

E.9 A coulomb of positive charge has 12 J more electrostatic potential energy at the positive terminal than at the negative terminal.

E.11 A corona discharge will occur at the needle's tip, emitting much of your charge into the air. Less charge will remain on you.

E.13 The positive balloon has a higher voltage than the negative one.

E.15 Their voltages change. The positive balloon's voltage increases, the negative balloon's voltage decreases.

P.1 $3.3 \cdot 10^{-9}$ C. **P.3** $8.3 \cdot 10^{19}$ N.

Section 11.2

E.1 5,000 levels. **E.3** All in lowest energy level.

E.5 Heating would diminish the contrast of the images.

E.7 The red light photons don't have enough energy to shift electrons from filled valence levels to empty conduction levels.

E.9 Light entering a semiconductor is absorbed as it shifts electrons between levels. This same light passes through an insulator.

E.11 It would experience a constant acceleration toward the right.

E.13 The electric field should point directly downward.

E.15 The forces between a positive charge and the hairbrush decrease with their separation. Thus the electric field that acts on that positive charge also decreases with distance from the hairbrush.

E.17 The increased charge on the treetop exerts particularly strong forces on nearby charge. In other words, the charge on the treetop produces a strong local electric field.

Section 11.3

E.1 A fast moving car must accelerate upward much faster to acquire the upward speed it needs to move over the bump quickly.

E.3 The force of static friction is limited and the tires will skid when the forward frictional force can't balance backward wind resistance.

E.5 While a pencil balanced on its point is in equilibrium, any disturbance will cause it to tip over. Similarly, one magnet balanced above another is in equilibrium, but can't tolerate any disturbances.

E.7 There are no isolated magnetic poles, only magnetic dipoles.

E.9 Yes, it will be magnetic. **E.11** The tray's pole is north.

E.13 The changing magnetic field induces currents in the aluminum, making the aluminum magnetic in such a way that it repels the coil.

E.15 Magnetic drag pulls the copper along with the moving magnet.

E.17 The magnet with the field has a current in it. The other does not.

E.19 The forces between magnetic poles decrease rapidly with distance. The two closest poles experience the strongest forces.

Section 12.1

E.1 When on, the switch should close the circuit by connecting the two wires electrically. When off, it should open the circuit. The broken switch is unable to disconnect the wires and open the circuit.

E.3 Current arrives through one wire and leaves through the other.

E.5 You are opening that circuit so current stops flowing through it.

E.7 Current flows from the power source, through the fence wire, through the animal, through the earth, and back to the power source.

E.9 Current flows very briefly until you're positively charged, too.

E.11 The circuit would be open so current wouldn't flow through it.

E.13 Batteries, because something is supplying power to the current.

E.15 Current flows from the battery's positive terminal, through the hobbyist's tongue, to the battery's negative terminal. Energy is being transferred from the battery to the hobbyist's tongue by the current.

P.1 1 A. **P.3** 0.15 W.

P.5 Each battery supplies 3 W. **P.7** About 13,333 s, or 3.7 hours.

Section 12.2

E.1 A kilowatt is a unit of power, or energy per time. Multiplying power by the time over which it's used yields the total energy used.

E.3 Each wire of the overloaded cord has a large voltage drop and consumes considerable electric power. It produces thermal energy.

E.5 The voltage drop in each wire is proportional to the current.

E.7 As you draw more current through the building's wires, the voltage drops in those wires increase. Less power reaches the lights.

E.9 It will rise by 10%. **E.11** They reverse rapidly.

E.13 Yes, for a moment each time the alternating current reverses.

E.15 9 A.

P.1 0.05 Ω. **P.3** 2 V.

P.5 0.6 V. **P.7** 240,000,000 W.

Section 12.3

E.1 The rotor is attracted to the iron cores in the stationary coils.

E.3 You must move the magnet so that you are doing work in order to transfer energy to the light bulb so that it will light.

E.5 The plate's south pole points toward your magnet's north pole.

E.7 You do positive net work. **E.9** They fill the valence levels.

E.11 Electrons in the conduction levels of the n-type semiconductor can reduce their potential energies by accelerating toward the p-type semiconductor, with its empty, lower-energy valence levels.

E.13 The junction's n-type semiconductor side is positively charged.

E.15 A photocell converts some heat (thermal radiation) into ordered energy (electric energy) as it flows from the hot sun to the cool earth.

E.17 Current flows through the bulb only half the time; whenever the electric company is trying send current in one of the two directions. The bulb thus receives about half its normal operating power.

Section 12.4

E.3 As like poles move apart, their repulsive magnetostatic forces and their motions are in the same direction, so the forces do work.

E.5 The rotor's poles will reverse, as will the forces between those poles and the poles of the permanent magnets, and thus the torque.

E.7 The poles of the electromagnets grow larger as the current through them increases and the forces and torques also increase.

E.9 The rotor spins in the direction of your torque, so you do work.

E.11 Its rotational inertia. **E.13** Its rotational inertia.

E.15 It exerts a magnetic drag force on the aluminum beneath it.

Section 12.5

E.3 Aluminum isn't intrinsically magnetic and can't be magnetized.

E.5 The end touching the north pole becomes a south pole.

E.7 An AC electromagnet would induce currents in the scrap iron and steel, leading to repulsive rather than attractive forces.

E.9 The distances between magnetization reversals are wider on the lower frequency tape than on the higher frequency tape.

E.11 The domains aligned with the new applied field grow while those that are anti-aligned with that field shrink.

E.13 The ring must be easily magnetized by small currents in its coil.

E.15 It magnetizes the recording particles in a new direction.

E.17 Its recording head can magnetize the whole layer all by itself.

E.19 It's attracted to the poles of the magnetic recording particles.

Section 13.1

E.1 The microphone can't provide enough power.

E.3 The phonograph produces a much smaller voltage rise than a CD player would provide. The amplifier expects a larger voltage.

E.5 Less than the resistor itself.

E.7 The current through the resistor also doubles. With twice the current experiencing twice the voltage drop, the power quadruples.

E.9 The resistor extracts power from the current passing through it, turning some of that current's energy into thermal energy.

E.11 When it carries no current, the resistor has no voltage drop.

E.13 The one with the thicker insulating layer.

E.15 As its plate area grows smaller, the capacitor has less room to store separated charge and thus stores less of it at a given voltage.

E.17 The gate's charge must be very near the channel so that it can attract opposite charge and draw that charge into the channel.

E.19 With positive charge on its gate, the MOSFET allows current to flow through the calculator. With no charge, the current flow stops.

E.21 When current experiences a voltage drop as it flows through an MOSFET, some of its energy is converted into thermal energy.

E.23 Resistors initially connect the gate only to +9 V but when the MOSFET conducts current, the gate also becomes connected to 0 V.

E.25 The dropping voltage is associated with negative charge, which attracts positive charge toward it from the capacitor's left plate.

Section 13.2

E.1 There are 6 hundreds. **E.3** *219.*

E.5 Yes. There are 0 thousands in both decimal representations.

E.7 Instead of reporting 2 twos, the binary representation should report it as 1 four, the next higher power of two.

E.9 Form a circuit in which the MOSFET controls current flowing between the battery and the bulb. Touch the MOSFET's gate to the capacitor plate. If the bulb lights, the plate has positive charge on it.

E.11 It takes time for an MOSFET to change the charge on a wire in order to store or retrieve a bit.

E.13 It would change the disk's magnetization and erase its memory.

E.15 If the heads and disk don't experience identical accelerations during the impact, their separation will change and they may touch.

E.17 Negative charge is being replaced by positive charge.

E.19 Current can flow from +3 V to output through either MOSFET.

E.21 Each charge change consumes a certain amount of energy. The more such changes occur each second, the more power is consumed.

E.23 Even at the speed of light, currents won't have time to go back and forth between these components during one clock tick.

E.25 In series. **E.27** Smaller.

Section 14.1

E.1 The charges in a vertical antenna can respond to the vertically polarized electromagnetic wave emitted by the transmitting antenna.

E.3 As charges accelerate in the wires, they emit radio waves.

E.5 As charges accelerate in the computer, they emit radio waves.

E.7 The inductor's magnetic field provides this energy.

E.9 The inductor opposes a more rapid decrease in current.

E.11 The electromagnet is an inductor, opposing current changes.

E.13 The inductor's magnetic field contains energy and both reach their largest values when the capacitor's energy is zero.

E.15 It produces no thermal energy as it flows in the superconductor.

E.17 The field you are observing half a wavelength from the antenna was emitted half a cycle ago, when the charges were reversed.

E.19 Horizontally polarized.

E.21 The AM station changes the power of its transmission in order to represent air pressure fluctuations with the radio wave.

E.23 Their sideband waves include a range of frequencies near the carrier wave frequency. These sideband waves mustn't overlap.

Section 14.2

E.1 The work done by the power supply becomes the light's energy.

E.3 Yes, along the force arrow shown in the figure.

E.5 Electrons in the picture tube complete one full circle of cyclotron motion, regardless of outward speed, and focus on the screen.

E.7 The television must be illuminating a dot near the sensor's position at the moment the sensor detects light.

E.9 Electrons would steer in the magnetic field and miss the anode.

E.11 The horizontal deflection coils often produce such sounds.

E.13 The magnetic field changes the electron beams' trajectories.

E.15 Amplitude modulation introduces sideband waves that extend as much as 5 kHz above and below the carrier wave's frequency.

E.17 The charges on the wire and tube produce electric fields while the currents in the wire and tube produce magnetic fields.

Section 14.3

E.1 Because the pattern of microwaves in the cooking chamber isn't uniform, food cooks more evenly when it moves about the chamber.

E.3 Water trapped in the ceramic would absorb microwaves and the ceramic would become extremely hot. It might even shatter.

E.5 The microwaves heat the water in the kernels directly.

E.7 The holes in the screen are much smaller than the wavelength of the microwaves, so the screening acts like a solid metal surface.

E.9 Releasing the microwaves into the room wouldn't be healthy.

E.11 Heating the filament releases electrons into the tube's vacuum.

E.13 A dent would change the resonant frequency of one of the resonant cavities. It wouldn't produce microwaves as efficiently.

E.15 Both involve destructive interference in electromagnetic waves.

E.17 The charges of the resonator tips must alternate, so there must be an even number of tips and thus an even number of resonators.

P.1 0.333 m. **P.3** $0.461 \cdot 10^{-7}$ m (461 nm).

Section 15.1

E.1 About 160 nm. **E.3** About $4.7 \cdot 10^{14}$ Hz.

E.5 The wavelength scale should shift to the left.

E.7 Rayleigh scattering deflects blue light in *all* directions.

E.9 Light polarizes the graphite so strongly that the light reflects.

E.11 The large particles scatter light better than air molecules.

E.13 The light refracts as it enters and leaves the glass.

E.15 Some light reflects from each air/mica boundary in the separated sheets of mica. In the solid mica, there are no such boundaries.

E.17 The candlelight is bent by refraction. Dispersion causes the various colors of light to bend differently and follow separate paths.

E.19 Dispersion in the materials causes the colors to bend differently.

E.21 Each surface in the granulated sugar reflects some of the light passing through it. These random reflections make the sugar white.

E.23 The microwaves polarize the cloud's water droplets and the microwaves slow down. Their speed change causes a reflection.

E.25 Light changes speed each time it enters or leaves a layer of the onion, so part of the light reflects from each layer.

E.27 Some light reflects from each surface of a lime particle.

E.29 Light reflects from the top *and* bottom surfaces of an oil film and the two reflections interfere with one another. The type of interference depends on the film's thickness and the light's wavelength.

Section 15.2

E.1 Green and Blue **E.3** That's the phosphor coating.

E.5 Mercury atoms themselves produce the tubes' ultraviolet light.

E.7 Fluorescence from brighteners replaces the missing blue light.

E.9 Excited mercury atoms emit primarily invisible ultraviolet light.

E.11 Each mercury atom in a narrow tube is near the phosphors.

P.1 $3.37 \cdot 10^{-19}$ J. **P.3** $7.95 \cdot 10^{-15}$ J.

Section 15.3

E.1 The radio station's photons are identical—part of a single wave.

E.3 No, it produces all the photons at once with a single antenna.

E.5 Incoherent.

E.7 They'll absorb the microwaves the maser is trying to produce.

E.9 Starting and stopping the light wave produces sideband waves.

E.11 In a very short light pulse, the sideband waves extend over a broad range of frequencies and wavelengths.

E.13 Initial photons that were emitted at random are evidently passing through enough excited atoms to be duplicate many times.

E.15 The initial photon is created in a random, unpredictable event.

Section 16.1

E.1 To have an f-number of 1.8, the elements in the 200 mm lens must be four times the diameter of the elements in the 50 mm lens.

E.3 The hole acts as an aperture, reducing the diameter of the lens.

E.5 You should use a large aperture and a short exposure time.

E.7 Reducing the size of your eye's lens increases its depth of focus.

E.9 The depth of focus is smallest when the whole lens is used.

E.11 The distance increases.

E.13 The film's image itself consists of black grains of silver.

P.1 35 mm.

P.3 The real image of the more distant statue forms 2.2 mm closer.

Section 16.2

E.1 Both form real images of objects on light-sensitive surfaces.

E.3 As you move away from the mirror, the rays hitting its surface diverge less and it bends the reflected rays more nearly parallel to one another. They appear to come from a distant virtual image.

E.5 A few centimeters from the real image, on your side of that image, so that it can act as a magnifying glass for the real image.

E.7 They're located a short distance in front of the eyepieces.

E.11 The shorter its focal length, the greater the magnification.

E.15 The lower magnification 2 X glass has the longer focal length.

E.17 Their extremely short focal lengths let them form highly magnified real images. The object distance is just over the focal length.

E.19 The higher the energy of a particle used to study a small object, the shorter the particle's wavelength and the greater its resolution.

Section 16.3

E.1 The tape's magnetization is proportional to the change in air pressure that the tape is representing.

E.3 The disc wouldn't be able to represent high pitched sound well.

E.5 Waves reflected from a sheet's front and back surfaces interfere.

E.7 Waves reflected by the valley and ridge interfere destructively.

E.9 The lens's large opening reduces diffraction effects.

E.13 The low mass lens can be accelerated rapidly by modest forces.

E.15 The separated charge in the photodiode creates a strong electric field that pushes on the electron and makes it accelerate rapidly.

E.17 Light is experiencing total internal reflection inside the glass.

Section 17.1

E.1 Its soft steel contains little carbon and doesn't oppose slip.

E.3 The tip should be quenched rapidly from high temperature to room temperature to form hard martensitic steel. The shaft should be quenched less fully to form tough bainitic or pearlitic steel.

E.5 Dissolving extra carbon in the hot steel lining will harden it.

E.7 At low temperatures, the dislocations can't move to assist slip.

E.9 Spreading out the load reduces the local stress on the roadway.

E.11 Heating followed by rapid cooling hardens carbon steel.

E.13 By disrupting gold's crystal structure, alloying elements make it less able to undergo slip. The gold alloys are harder than pure gold.

E.15 Wood deforms elastically in response to mild stresses, but deforms plastically when exposed to a severe stress.

E.17 They'll exhibit almost identical strains in elastic deformation.

E.19 These other elements disrupt the aluminum crystals and impede slip. This change gives the aluminum a higher yield strength.

E.21 Heat the bar and cool it slowly to let its crystals grow larger.

E.23 Repeated bending hardens the metal and it eventually cracks.

E.25 A spring should respond elastically and not bend permanently.

Section 17.2

E.1 Candy sugar is amorphous, while sugar granules are crystalline.

E.3 Some of the glass crystallizes, so the material is inhomogeneous.

E.5 Glassy aluminum can't undergo slip or plastic deformation.

E.7 Growing large crystals takes mobile molecules and time. Rapid cooling stops the molecules from moving before large crystals form.

E.9 Nucleation of bubbles in soda is a relatively rare, random process that sets the limit on how quickly soda loses its gas. The slow nucleation of seed crystals similarly sets the limit on how quickly a supercooled liquid crystallizes, allowing it to become a glass instead.

E.11 Above its glass transition temperature, glass can flow slowly as a liquid. But at much colder temperatures, there is virtually no flow.

E.13 Nucleation of a seed crystal is a rare, random phenomenon.

E.15 Its non-uniform thermal expansion produces huge stresses.

E.17 Any deep injury to the glass's skin would cause it to shatter.

E.19 There is no crystalline structure to guide the fracture.

E.21 It's under tremendous tension and an injury causes it to tear.

E.23 Internal stresses would leave the glass inhomogeneous.

E.25 Crystalline solids can't flow like hot glass can.

Section 17.3

E.1 Polystyrene is in the glassy regime at both temperatures.

E.3 It takes time for starch molecules to reptate past entanglements. If you're patient, it will flow. If you poke at it, it doesn't have time to.

E.5 The balloon is a thermoset and can't flow. Bubble gum is a thermoplastic and can flow if you're patient.

E.7 Rubber is a thermoset and can't melt. It can only burn.

E.9 The rubber in the strands is a thermoset and can't melt or flow.

E.11 Cold shifts it to the glassy regime and it becomes brittle.

E.13 The drugs dissolve in the plastic and then diffuse through it.

E.15 The plastic crazes when it's bent, forming gaps that scatter light.

E.17 Water plasticizes the adhesive, temporarily softening it.

E.19 Water plasticizes hair, temporarily softening it.

E.21 Water plasticizes paper, temporarily softening it.

E.23 The Kevlar fibers are extremely difficult to tear and give the concrete substantial tensile strength.

E.25 The silicon rubber is a thermoset in the elastic regime. It isn't brittle at all, although it can be torn if overstressed.

E.27 Like a liquid crystal polymer, the molecules in Spectra work together to oppose tensile stress and undergo remarkably little strain.

Section 18.1

E.1 Changing one element into another requires *nuclear* reactions.

E.3 Nuclei of ^{65}Cu atoms contain 2 more neutrons than those of ^{63}Cu.

E.5 No, the number of electrons is set by the number of protons.

E.7 Neutrons would experience the attractive nuclear force while diluting the repulsion between protons.

E.9 With fewer protons, the diluting effect of neutrons matters less.

E.11 A nuclear chain reaction can't occur without the first neutron, so a supercritical mass would remain inert until that neutron arrived.

E.13 Gravity acts to pull the basket away from the ceiling while the tape acts to hold them together. When the tape releases...decay!

E.15 Its nuclei move closer together and block neutrons better. The chance decreases that a neutron will escape from the sphere.

E.17 With time, radioactive nuclei decay spontaneously and the materials become less and less hazardous.

Section 18.2

E.1 Fissionable ^{235}U is used up while non-fissionable ^{238}U remains.

E.3 Gravity can assist in inserting the control rods into the reactor only when they're mounted above the reactor.

E.5 The only reactors that can use natural or slightly enriched uranium are large thermal fission reactors that have moderators.

E.7 Water would be a poor moderator if the oxygen in it absorbed many neutrons.

E.9 Collisions with neutrons knock the carbon atoms out of position and disrupt graphite's crystalline structure.

E.11 They must pull the control rods farther out of the reactor.

E.13 Hydrogen atoms are electrically neutral while the individual charged particles in a plasma are not.

Section 19.1

E.1 X-rays cause chemical damage that kills cells slowly.

E.3 Photons in sunlight can cause chemical damage.

E.5 Photons in blue or ultraviolet light can cause chemical damage to the molecules in the manuscripts. Yellow light generally can't.

E.7 The greater the voltage drop from the anode to the cathode, the more kinetic energy the electrons will acquire as they fly through the tube and the more energy they'll release when they hit the anode.

E.9 To emit an X-ray photon, an electron must shift into an orbital that's extremely tightly bound near a nucleus with a large positive charge. That can't occur in an atom with a small nucleus.

E.11 The X-ray source used to image welds must produce X-ray photons so energetic that many of them pass right through steel.

E.13 Since the velocity of the electron changes during the collision, its momentum changes. Because momentum is conserved, the photon must have a momentum and that momentum must change, too.

E.15 An antiproton has more mass than a positron.

E.17 A positron isn't found in an iron atom.

E.19 Near the speed of light, changes in a particle's kinetic energy are accompanied by only modest changes in the particle's speed.

E.21 MRI imaging detects hydrogen. Bone contains little hydrogen.

PHOTO CREDITS

Chapter 1 *Opener*: Courtesy Louis Bloomfield. Figure 1.1.2: Focus on Sports. Figure 1.1.8: ©Estate of Harold Edgerton, courtesy of Palm Press, Inc. Figure 1.1.10: Richard Megna/Fundamental Photographs. Figure 1.2.1: Courtesy Jerry Ohlinger's Movie Material Store. Figure 1.2.7: Larry Mulvehill/Photo Researchers. Figure 1.3.8: Dugald Bremmer/Tony Stone Images/ New York, Inc. Figure 1.4.2: From *Friction and Lubrication of Solids*, by F.P. Bowden and S. Tabor, Clarendon Press, 1950. Figure 1.4.4: Courtesy D. Kulmann-Wilsdorf, Department of Material Science, University of Virginia. Figure 1.4.6: Bettmann Archive. Figure 1.4.11: Steven E. Sutton/Duomo Photography, Inc.

Chapter 2 *Opener* and figure 2.1.4: Courtesy Louis Bloomfield. Figure 2.1.6: Courtesy NASA. Figure 2.2.2: ©Estate of Harold Edgerton, courtesy of Palm Press, Inc. Figure 2.2.3: Courtesy Education Development Center. Figure 2.2.4: Ben Rose/The Image Bank. Figure 2.2.7: ©Estate of Harold Edgerton, courtesy of Palm Press, Inc. Figure 2.3.4: Luis Castañeda/The Image Bank. Figure 2.3.5: Jon Reis/The Stock Market. Figure 2.3.6: Bob Torrez/Tony Stone Images/ New York, Inc. Figure 2.3.8: Chad Slattery/Tony Stone Images/ New York, Inc.

Chapter 3 *Opener*: Courtesy Louis Bloomfield. Figure 3.1.5: Coco McCoy/Rainbow. Figure 3.1.8: UPI/Bettmann Archive. Figure 3.1.9: Thomas Braise/The Stock Market. Figure 3.1.10: Adrian Davies/Bruce Coleman, Inc. Figure 3.2.3: Antje Gunnar/Bruce Coleman, Inc. Figure 3.2.6: Courtesy Department of Water and Power of the City of Los Angeles. Figure 3.2.9: David W. Hamilon/The Image Bank.

Chapter 4 *Opener*: Courtesy Louis Bloomfield. Figure 4.1.2 (left): David Stoecklein/The Stock Market. Figure 4.1.2 (right): Jeff Gnass/The Stock Market. Figure 4.1.3: Courtesy Louis Bloomfield. Figure 4.1.4: Courtesy Peter Bradshaw and P.W. Beatman, Imperial College, London. Figure 4.3.3: Bettmann Archive. Figure 4.3.7: Courtesy Rebus, Inc. Figure 4.3.13: Scott Halleran/Allsport. Figure 4.4.2: Courtesy Thomas Miller. Figure 4.4.4: Courtesy Thomas Miller. Figures 4.4.8 and 4.4.9: Courtesy Smithsonian Institution.

Chapter 5 *Opener*: Courtesy Louis Bloomfield. Figure 5.1.3: Courtesy NASA. Figure 5.1.6: Photo by Esther C. Goddard, courtesy AIP Neils Bohr Library. Figure 5.1.8: Courtesy New York Public Library. Figure 5.2.3: Courtesy Dr. David Jones. Figure 5.3.3: Courtesy Otis Elevator Co. Figure 5.3.7: Courtesy Louis Bloomfield.

Chapter 6 *Opener*: Courtesy Louis Bloomfield. Figure 6.1.1: Wardene Weisser/Bruce Coleman, Inc. Figure 6.1.3: Courtesy Bryant Heating and Cooling Co. Figure 6.2.2: Courtesy Louis Bloomfield. Figure 6.2.3: Courtesy NASA. Figure 6.2.4: The Granger Collection. Figure 6.2.5: Courtesy Lawrence Livermore Laboratory. Figures 6.3.1, 6.3.6, 6.3.8, 6.4.2, 6.4.7 and 6.4.9: Courtesy Louis Bloomfield.

Chapter 7 *Opener*: Courtesy Louis Bloomfield. Figure 7.1.3: Courtesy Courrier Corporation. Figure 7.1.4: Courtesy Frigidaire Co. Figure 7.2.3: Courtesy Ford Motor Co. Figure 7.2.5: Courtesy Ford Motor Co.

Chapter 8 *Opener* and figures 8.1.1-8.1.4: Courtesy Louis Bloomfield.

Chapter 9 *Opener*: Courtesy Louis Bloomfield. Figure 9.1.3: Robert Mathena/Fundamental Photographs. Figure 9.1.7: Gregory G. Dimijian, M.D./Photo Researchers. Figure 9.1.9: Tibor Bognar/The Stock Market. Figure 9.1.10: Courtesy National Institute of Standards and Measures. Figure 9.2.5: Ben Rose/The Image Bank. Figure 9.2.6: AP/Wide World Photos.

Chapter 10 *Opener*: Courtesy Louis Bloomfield. Figure 10.1.5: Courtesy National Film Board, Canada. Figure 10.1.10: Fritz Herle/Photo Researchers. Figure 10.1.12: Don & Liysa King/The Image Bank.

Chapter 11 *Opener*: Courtesy Louis Bloomfield. Figure 11.1.3: Adam Hart-Davis/Photo Researchers. Figure 11.1.6: Courtesy GE Environmental Systems. Figure 11.1.7: Courtesy Research-Cottrell. Figures 11.2.9, 11.3.2, 11.3.3 and 11.3.7: Courtesy Louis Bloomfield. Figure 11.3.8: Courtesy Deutsche Bundesbahn/Transrapid. Figure 11.3.11: Takeshi Takahara/Photo Researchers. Figure 11.3.12: Richard Megna/Fundamental Photographs.

Chapter 12 *Opener* and figures 12.1.3, 12.1.7 and 12.2.5: Courtesy Louis Bloomfield. Figure 12.2.9: Coco McCoy/Rainbow. Figure 12.2.10: Roy Morsch/The Stock Market. Figure 12.2.11: Courtesy Louis Bloomfield. Figure 12.3.4: Jeff Zaruba/The Stock Market. Figure 12.3.6: Hank Morgan/Photo Researchers. Figures 12.3.10, 12.4.2, 12.4.5 and 12.4.8: Courtesy Louis Bloomfield. Figure 12.5.6: Courtesy BASF Magnetics Corporation. Figure 12.5.8: Courtesy Louis Bloomfield. Figure 12.5.9: Courtesy of Read-Rite Corporation, Milpitas, California. Figure 12.5.13: Courtesy Louis Bloomfield.

Chapter 13 *Opener* and figures 13.1.2, 13.1.3, 13.1.5, 13.1.7, 13.2.2, 13.2.3, 13.2.4 and 13.2.10: Courtesy Louis Bloomfield. Figure 13.2.1: Michael W. Davidson/Photo Researchers.

Chapter 14 *Opener* and figures 14.2.2, 14.2.4, 14.2.10 and 14.3.3: Courtesy Louis Bloomfield.

Chapter 15 *Opener*: Courtesy Louis Bloomfield. Figure 15.1.3 (left): S. Achernar/The Image Bank. Figure 15.1.3 (right): Grant Faint/The Image Bank. Figure 15.1.7: John Edwards/Tony Stone Images/ New York, Inc. Figure 15.1.9: Pierre Kopp/West Light. Figure 15.1.13: Diane Schiumo/Fundamental Photographs. Figures 15.2.5, 15.2.7, 15.3.6 and 15.3.8: Courtesy Louis Bloomfield. Figure 15.3.10: Courtesy SDL, Inc. Figure 15.3.11: Courtesy Coherent, Inc.

Chapter 16 *Opener* and figures 16.1.2 and 16.1.6: Courtesy Louis Bloomfield. Figure 16.1.7: Courtesy Canon USA, Inc. Figure 16.1.9: Courtesy Vivitar Corporation. Figures 16.1.11, 16.2.3 and 16.2.7: Courtesy Louis Bloomfield. Figure 16.2.12: Courtesy Meade Instruments Corporation. Figure 16.3.3 (top): Dale E. Boyer/Photo Researchers. Figure 16.3.3 (center): Andrew Syred/Photo Researchers. Figure 16.3.7: Courtesy Louis Bloomfield.

Chapter 17 *Opener* and figure 17.1.2: Courtesy Louis Bloomfield. Figure 17.1.7 (top): Mike McNamee/Photo Researchers. Figure 17.1.7 (bottom): Courtesy Louis Bloomfield. Figure 17.1.9: Courtesy Museum of Fine Arts, Boston. Figure 17.2.3: Courtesy Louis Bloomfield. Figure 17.2.6: Vu/T. Havill/Visuals Unlimited. Figure 17.2.7: Jon Gordon/Phototake. Figure 17.2.9: Courtesy Viracon. Figure 17.3.5: Courtesy Stephen Chu. Figure 17.3.6: From L.H. Sperling's *Introduction to Physical Polymer Science*, page 491, figure 10.19, John Wiley & Sons, Inc.

Chapter 18 *Opener*: Richard D. Estes/Photo Researchers. Figure 18.1.3: Courtesy Louis Bloomfield. Figure 18.1.7 (top): Courtesy U.S. Department of Energy. Figure 18.1.7 (bottom): Courtesy Defense Nuclear Agency. Figure 18.1.9: Courtesy Los Alamos Scientific Laboratory, University of California. Figure 18.2.4: Richard Ustinich/The Image Bank. Figure 18.2.5: Igor Kostin/Imago/Sygma. Figure 18.2.9: Courtesy Princeton University Plasma Physics Laboratory.

Chapter 19 *Opener*: Courtesy Louis Bloomfield. Figure 19.1.8 (left): Pete Saloutos/The Stock Market. Figure 19.1.8 (right): Richard Megna/Fundamental Photographs. Figure 19.1.11: Larry Mulvehill/Photo Researchers. Figure. 19.1.14 (left): Charles Thatcher/Tony Stone Images/ New York, Inc. Figure 19.1.14 (right): SBHA/Tony Stone Images/ New York, Inc.

INDEX

page: 123 - in-text reference
page: **123** - in-text definition
page: ***123*** - glossary entry
page: *123* - exercise, problem or case